Fourth-Generation Wireless Networks:
Applications and Innovations

Sasan Adibi
Research in Motion (RIM), Canada

Amin Mobasher
Research in Motion (RIM), Canada & Stanford University, USA

Tom Tofigh
AT&T Labs, USA

INFORMATION SCIENCE REFERENCE

Hershey · New York

Director of Editorial Content:	Kristin Klinger
Senior Managing Editor:	Jamie Snavely
Assistant Managing Editor:	Michael Brehm
Publishing Assistant:	Sean Woznicki
Typesetter:	Kurt Smith, Sean Woznicki
Cover Design:	Lisa Tosheff
Printed at:	Yurchak Printing Inc.

Published in the United States of America by
   Information Science Reference (an imprint of IGI Global)
   701 E. Chocolate Avenue
   Hershey PA 17033
   Tel: 717-533-8845
   Fax: 717-533-8661
   E-mail: cust@igi-global.com
   Web site: http://www.igi-global.com/reference

Copyright © 2010 by IGI Global. All rights reserved. No part of this publication may be reproduced, stored or distributed in any form or by any means, electronic or mechanical, including photocopying, without written permission from the publisher.
   Product or company names used in this set are for identification purposes only. Inclusion of the names of the products or companies does not indicate a claim of ownership by IGI Global of the trademark or registered trademark.

Library of Congress Cataloging-in-Publication Data

Fourth-generation wireless networks : applications and innovations / Sasan
Adibi, Amin Mobasher, and Tom Tofigh, editors.
   p. cm.
 Includes bibliographical references and index.
 Summary: "This book presents a comprehensive collection of recent findings
in access technologies useful in the architecture of wireless networks"--
Provided by publisher.
  ISBN 978-1-61520-674-2 (hardcover) -- ISBN 978-1-61520-675-9 (ebook) 1.
Wireless communication systems--Technological innovations. 2. Cellular
telephone systems--Technological innovations. 3. Mobile communication
systems--Technological innovations. I. Adibi, Sasan, 1970- II. Mobasher,
Amin, 1978- III. Tofigh, Tom, 1955-
 TK5103.2.F683 2010
 621.384--dc22
                            2009040025

British Cataloguing in Publication Data
A Cataloguing in Publication record for this book is available from the British Library.

All work contributed to this book is new, previously-unpublished material. The views expressed in this book are those of the authors, but not necessarily of the publisher.

Dedication

To our wives
Negar, Mona, & Mojgan

List of Reviewers

Nada Golmie, *National Institute of Standards and Technology (NIST), USA*
Pouya Taaghol, *Intel Mobility Group, USA*
Subhas Mondal, *Wipro Inc., USA*
Behrouz Maham, *University of Oslo (UiO), Norway*
Masoud Ebrahimi, *Research in Motion (RIM), Canada*
Yi Yu, *Research In Motion (RIM), Canada*
Shirook Ali, *Research In Motion (RIM), Canada*
Farzaneh Kohandani, *Research In Motion (RIM), Canada*
Yongkang Jia, *Research In Motion (RIM), Canada*
Hadi Baligh, *Huawai Technologies, Canada*
Alireza Bayesteh, *Research In Motion (RIM), Canada*
Hamid Farmanbar, *Nortel Networks, Canada*
Xinhua Ling, *Research In Motion (RIM), Canada*
Heunchul Lee, *Stanford University, USA*
Cagatay Buyukkoc, *AT&T Labs, USA*
Wei Wu, *Research In Motion (RIM), Canada*
Biplab Sikdar, *Rensselaer Polytechnic Institute, USA*

Table of Contents

Foreword xxiv

Preface xlvii

Acknowledgment xlix

Section 1
Network Architectures

Roadmaps and Architectural Models

Chapter 1
Evolution of Personal Wireless Broadband Services from 3G to 4G 1
    Sudhir K. Routray, Krupajal Engineering College, India

Chapter 2
A New Global Ubiquitous Consumer Environment for 4G Wireless Communications 20
    Ivan Ganchev, University of Limerick, Ireland
    Máirtín S. O'Droma, University of Limerick, Ireland
    Jený István Jakab, Tecnomen Ltd, Ireland
    Zhanlin Ji, University of Limerick, Ireland
    Dmitry Tairov, University of Limerick, Ireland

Chapter 3
4G Access Network Architecture 46
    Young-June Choi, NEC Laboratories America, USA

Chapter 4
Architecture for IP-Based Next Generation Radio Access Network 61
    Ram Dantu, University of North Texas, USA
    Parthasarathy Guturu, University of North Texas, USA

Chapter 5
Long Term Evolution (LTE): An IPv6 Perspective 77
    Nayef Mendahawi, Research In Motion (RIM), Ltd., Canada
    Sasan Adibi, Research In Motion (RIM), Ltd., Canada

Chapter 6
HWN* Framework Towards 4G Mobile Communication Networks 100
   Chong Shen, Tyndall National Institute, Ireland
   Dirk Pesch, Cork Institute of Technology, Ireland
   Robert Atkinson, University of Strathclyde, Scotland
   Wencai Du, Hainan University, China

Forthcoming 4G Challenges: Problems and Solutions

Chapter 7
User Experience in 4G Networks 125
   Pablo Vidales, Deutsche Telekom Laboratories, Germany
   Marcel Wältermann, Deutsche Telekom Laboratories, Germany
   Blazej Lewcio, Deutsche Telekom Laboratories, Germany
   Sebastian Möller, Deutsche Telekom Laboratories, Germany

Chapter 8
Fourth Generation Networks: Adoption and Dangers 146
   Jivesh Govil, Cisco Systems Inc., USA
   Jivika Govil, Carnegie Mellon University, USA

Chapter 9
Potential Scenarios and Drivers of the 4G Evolution 181
   Elias Aravantinos, Stevens Institute of Technology, USA
   M. Hosein Fallah, Stevens Institute of Technology, USA

Chapter 10
Knowledge Sharing to Improve Routing and Future 4G Networks 193
   Djamel F. H. Sadok, Federal University of Pernambuco, Brazil
   Joseilson Albuquerque de França, Federal University of Pernambuco, Brazil
   Luciana Pereira Oliveira, Federal University of Pernambuco, Brazil
   Renato Ricardo de Abreu, Federal University of Pernambuco, Brazil

Next Generation Technologies

Chapter 11
Personal Environments: Towards Cooperative 4G Services... 228
   Tinku Rasheed, Create-Net, Italy
   Usman Javaid, Vodafone Group, UK

Chapter 12
Next Generation Broadband Services from High Altitude Platforms... 249
   Abbas Mohammed, Blekinge Institute of Technology, Sweden
   Zhe Yang, Blekinge Institute of Technology, Sweden

Chapter 13
Radio-over-Fibre Networks for 4G 268
    Roberto Llorente, Universidad Politécnica de Valencia, Spain
    Maria Morant, Universidad Politécnica de Valencia, Spain
    Javier Martí, Universidad Politécnica de Valencia, Spain

Section 2
Radio Access Protocols

Scheduling and Quality of Service

Chapter 14
MAC Protocol of WiMAX Mesh Network 292
    Ming-Tuo Zhou, National Institute of Information and Communications Technology, Singapore
    Peng-Yong Kong, Institute for Infocomm Research, Singapore

Chapter 15
Advanced Scheduling Schemes in 4G Systems 313
    Arijit Ukil, Tata Consultancy Services Ltd., India

Chapter 16
End-to-End Quality of Service in Evolved Packet Systems 361
    Wei Wu, Research In Motion, Limited, USA
    Noun Choi, Research In Motion, Limited, USA

Mobility and Handover

Chapter 17
An End-to-End QoS Framework for Vehicular Mobile Networks 377
    Hamada Alshaer, University of Leeds, UK
    Jaafar Elmirghani, University of Leeds, UK

Chapter 18
LTE Mobility Solutions at Network Level for Global Convergence 405
    Titus-Constantin Bălan, Siemens SIS PSE, Romania
    Florin Sandu, "Transilvania" University of Brasov, Romania

Chapter 19
Handover Optimization for 4G Wireless Networks 424
    Dongwook Kim, Korea Advanced Institute of Science and Technology, South Korea
    Hanjin Lee, Korea Advanced Institute of Science and Technology, South Korea
    Hyunsoo Yoon, Korea Advanced Institute of Science and Technology, South Korea
    Namgi Kim, Kyonggi University, South Korea

Cross-Layer Designs

Chapter 20
Survey of Cross-Layer Optimization Techniques for Wireless Networks 453
Han-Chieh Chao, National Ilan University, Taiwan
Chi-Yuan Chang, National Dong Hwa University, Taiwan
Chi-Yuan Chen, National Dong Hwa University, Taiwan
Kai-Di Chang, National Dong Hwa University, Taiwan

Chapter 21
Cross-Layer Joint Optimization of Multimedia Transmissions over IP Based
Wireless Networks 469
Catherine Lamy-Bergot, THALES Communications S.A., France
Gianmarco Panza, CEFRIEL, Italy

Chapter 22
Video Streaming Based Services over 4G Networks: Challenges and Solutions 494
Elsa Mª Macías, Universidad de Las Palmas de Gran Canaria, Spain
Alvaro Suarez, Universidad de Las Palmas de Gran Canaria, Spain

Section 3
Physical Layer Advances

Advanced Multiple Access Transmission Schemes

Chapter 23
Aspects of OFDM-Based 3G LTE Terminal Implementation 526
Wen Xu, Infineon Technologies AG, Germany
Jens Berkmann, Infineon Technologies AG, Germany
Cecilia Carbonelli, Infineon Technologies AG, Germany
Christian Drewes, Infineon Technologies AG, Germany
Axel Huebner, Infineon Technologies AG, Germany

Chapter 24
The Use of Orthogonal Frequency Code Division (OFCD) Multiplexing in Wireless
Mesh Network (WMN) 565
Syed S. Rizvi, University of Bridgeport, USA
Khaled M. Elleithy, University of Bridgeport, USA
Aasia Riasat, Institute of Business Management, Pakistan

Chapter 25
A New Approach to BSOFDM: Parallel Concatenated Spreading Matrices OFDM 582
Ibrahim Raad, University of Wollongong, Australia
Xiaojing Huang, University of Wollongong, Australia

Chapter 26
The Next Generation CDMA Technology for Futuristic Wireless Communications:
Why Complementary Codes? 596
    Hsiao-Hwa Chen, National Cheng Kung University, Taiwan

Enhanced Decoding Techniques

Chapter 27
Configurable and Scalable Turbo Decoder for 4G Wireless Receivers 622
    Yang Sun, Rice University, USA
    Joseph R. Cavallaro, Rice University, USA
    Yuming Zhu, Texas Instruments, USA
    Manish Goel, Texas Instruments, USA

Chapter 28
Parallel Soft Spherical Detection for Coded MIMO Systems 644
    Hosein Nikopour, Huawei Technologies Co., Ltd., Canada
    Amin Mobasher, Stanford University, USA
    Amir K. Khandani, University of Waterloo, Canada
    Aladdin Saleh, Bell Canada, Canada

Collaboration and Capacity

Chapter 29
Capacity Estimation of OFDMA-Based Wireless Cellular Networks 666
    André Carlos Guedes de Carvalho Reis, Universidade de Brasília, Brazil
    Paulo Roberto de Lira Gondim, Universidade de Brasília, Brazil

Chapter 30
Wireless Collaboration: Maximizing Diversity through Relaying 684
    Patrick Tooher, Concordia University, Canada
    M. Reza Soleymani, Concordia University, Canada

Compilation of References 710

About the Contributors 759

Index 777

Detailed Table of Contents

Foreword xxiv

Preface xlvii

Acknowledgment xlix

Section 1
Network Architectures

Roadmaps and Architectural Models

Chapter 1
Evolution of Personal Wireless Broadband Services from 3G to 4G 1
    Sudhir K. Routray, Krupajal Engineering College, India

The chapter covers the basic conceptual model of the 4G system and the operations of its physical systems. It starts from the very basics of the wireless communication services, and then the author goes though the different standards of the systems which provide wireless broadband services, like the 3G, and other wireless broadband systems, like WiMAX etc. The author then looks through the 3GP project and its visions, then goes through the 3GPP2. The vision and the achievements of the 3GPP LTE are then discussed including the 4G and its successor systems. After that the authors turn towards the technical ideas behind the wireless broadband services like the 3G and 4G. 4G system architecture and its features are looked into, the differences between the 3G and 4G are discussed, and then the whole chapter is concluded with the impact of the 4G system on the present mobile communication scenario.

Chapter 2
A New Global Ubiquitous Consumer Environment for 4G Wireless Communications 20
    Ivan Ganchev, University of Limerick, Ireland
    Máirtín S. O'Droma, University of Limerick, Ireland
    Jený István Jakab, Tecnomen Ltd, Ireland
    Zhanlin Ji, University of Limerick, Ireland
    Dmitry Tairov, University of Limerick, Ireland

A changed wireless environment for 4G and future generations of wireless communications is addressed in this chapter. This change is primarily focused on making the end user of wireless services more central and more a consumer in the global wireless environment than heretofore. In the 'Ubiquitous Consumer Wireless World (UCWW)' –the descriptive name for this new wireless environment paradigm– the global supply of wireless communications services is founded on the new Consumer-centric Business Model (CBM). This is a radical change and departure from present globally pervasive business model for service delivery based on the user being a subscriber, called appropriately the Subscriber-centric Business Model (SBM). The reasons and background for the drive to bring about this changed wireless environment are reviewed, with the main body of the chapter focusing on descriptions of the technological composition of two of the new core infrastructural enabling elements. These are the third-party authentication, authorization, and accounting (3P-AAA) service, and the service advertisement, discovery, and association (ADA) through newly defined wireless billboard channels (WBCs). The former, 3P-AAA, arises from the need to bring about a separation of the supply of the AAA service from the supply of the communications service, which is necessary to ensure the consumer character of the user and to promote and safeguard all the new benefits that will flow for users, new consumer-oriented wireless access network providers, and other stakeholders through this new wireless environment. As there will be restructuring implications for the operation and location of charging and billing functions, treatment of this aspect is also included. The latter, ADA & WBCs, arises in tandem with this separation of services, the consequent metric of business success changing from 'number of subscribers' to 'number of consumer transactions and service purchases' and the need, therefore, for a new direct 'push' advertisement means for service providers to attract consumers and for consumers to be continually up-to-date on new service offerings. The proposals for protocol interfaces and architectures for both these elements are explored and discussed, with those aspects needing to be addressed in global standardization activities highlighted.

Chapter 3
4G Access Network Architecture 46
Young-June Choi, NEC Laboratories America, USA

Although all-IP networking is the ultimate goal of 4G wireless networks, 3G LTE and WiMAX systems have designed semi all-IP network architectures for efficient radio resource and mobility management. These semi all-IP networks separate layer 2 and layer 3 handoff operations by grouping many base stations (BSs) as a subnet, thus alleviating handoff, while the pure all-IP networks provide a simple network platform at the cost of high handoff overhead. The authors compare the semi all-IP networks to the pure all-IP networks, and provide an overview to WiMAX access service networks and 3G LTE backhaul networks. They then present advanced architectures that support efficient radio resource and mobility management. First, they present a semi hierarchical cellular system with a super BS that behaves like a normal BS as well as a supervisor over other BSs within the group. They further extend this model to a system that combines multiple access techniques of OFDMA and FH-OFDMA with microcells and macrocells. Also, to alleviate the handoff latency, a dual-linked BS model is presented in order to support seamless handoff. Finally, as an integrated approach to supporting diverse QoS requirements, the authors consider an IP-triggered resource allocation strategy (ITRAS) that exploits IntServ and DiffServ of the network layer to interwork with channel allocation and multiple access of MAC and PHY layers, respectively. These cross layer approaches shed light on designing a QoS support model in a 4G network that cannot be handled properly by a single layer based approach.

Chapter 4
Architecture for IP-Based Next Generation Radio Access Network 61
Ram Dantu, University of North Texas, USA
Parthasarathy Guturu, University of North Texas, USA

High call volumes due to novel mobile data applications necessitate development of next generation wireless networks centered on high performing and highly available radio access networks (RANs). In this chapter, the authors present an innovative IP-based wireless routing architecture (for a RAN) with mechanisms for seamless handoff operations and high Quality of Service (QOS). Algorithms for dynamic configuration of the RAN, and efficacious network bandwidth management through traffic control are also presented. The authors establish the superiority of their system with real-life data indicating significant cost and availability improvements with our system over the traditional networks.

Chapter 5
Long Term Evolution (LTE): An IPv6 Perspective... 77
    Nayef Mendahawi, Research In Motion (RIM), Ltd., Canada
    Sasan Adibi, Research In Motion (RIM), Ltd., Canada

The main characteristic of 4th Generation (4G) Networks is being based on all IP architecture, operating mainly on IPv6. This includes services such as voice, video, and messaging. LTE is considered to be a 3rd Generation (3G) network and one of 4th Generation (4G) roadmap mobile access technologies. LTE-Advanced (LTE-A), on the other hand, is a 4G technology concept with evolving features. Therefore LTE is the key feature in the understanding of LTE-A evolution. The main focus of LTE is the enhancement of the packet-switched (PS) mechanisms on top of the UMTS enhancements, based on All IP Network (AIPN). IPv6 networking provides maximum service delivery flexibility, user decoupling, and scalability improvements, while leveraging the existing IETF standards. This requires major focus on network simplification, end-to-end delay reductions, optimal traffic routing, seamless mobility, and IP-based transport provisioning. This chapter aims to present a survey and highlight specific IPv6-based features presented mostly in the 3GPP standard literature, and to provide a high-level discussion on the LTE-IPv6 requirements.

Chapter 6
HWN* Framework Towards 4G Mobile Communication Networks 100
    Chong Shen, Tyndall National Institute, Ireland
    Dirk Pesch, Cork Institute of Technology, Ireland
    Robert Atkinson, University of Strathclyde, Scotland
    Wencai Du, Hainan University, China

The objective of the Hybrid Wireless Network with dedicated Relay Nodes (HWN*) proposal is to interface the Base Station (BS) Oriented Mobile Network (BSON) and the 802.11X assisted Mobile Ad hoc Wireless Network (MANET) so that one system can be utilised as an alternative radio access network for data transmissions, while the incorporation of the Relay Node (RN) is to extend the communication coverage, optimise medium resource sharing, increase spatial reuse opportunity, stabilise MANET link and create more micro-cells. The HWN* keeps the existing cellular infrastructure and a end-user Mobile Terminal (MT) can borrow radio resources from other cells through secured multi-hop RN relaying, where RNs are placed at pre-engineered locations. The main contribution of this work is the development of a HWN* system framework and related medium access and routing protocols/algorithms. The framework dedicatedly addresses the transparent multiple interface traffic handover management, cross layer routing, RN positioning and network topology issues to increase communication system capacity, improve Quality of Service (QoS), optimise transmission delay and reduce packet delivery delay.

Forthcoming 4G Challenges: Problems and Solutions

Chapter 7
User Experience in 4G Networks 125
Pablo Vidales, Deutsche Telekom Laboratories, Germany
Marcel Wältermann, Deutsche Telekom Laboratories, Germany
Blazej Lewcio, Deutsche Telekom Laboratories, Germany
Sebastian Möller, Deutsche Telekom Laboratories, Germany

Forthcoming 4G networks will enable users to freely roam across different communication systems. This implies that formerly independent wireless and wired technologies will be integrated to deliver transparent access to a plethora of mobile services and applications. This will also involve changes in the user's experience mainly derived from (1) mobility across heterogeneous technologies, (2) drastic changes in the underlying link conditions, and (3) continuous adaptation of applications, e.g. flexible coding schemes. This chapter presents a detailed study of these so far unknown phenomena arising in the context of 4G networks. Current instrumental models employed to estimate user perception, such as PESQ (ITU-T Rec. P.862, 2001) for predicting the quality of transmitted speech, were designed to measure conditions that are common in today's wireless and wired systems. However, it is expected that new conditions encountered in 4G networks are not going to be accurately handled by today's models. Thus, they need to be adjusted, or new models should be proposed in order to predict the perceptual influence of new phenomena such as the three aspects aforementioned. The authors undertook this task and designed a novel methodology and experimental setup to measure user perception in future 4G networks. Moreover, the authors carried out an extensive set of subjective tests to accurately quantify user perception and derive conclusions for optimal user experience in 4G networks. These processes and initial results are included in this chapter.

Chapter 8
Fourth Generation Networks: Adoption and Dangers 146
Jivesh Govil, Cisco Systems Inc., USA
Jivika Govil, Carnegie Mellon University, USA

Mobile researchers are witnessing burgeoning interest in 4G wireless networks that patronize global roaming across diverse wireless and mobile networks. The pith of 4G mobile systems lies in seamlessly integrating the existing wireless technologies including Wideband Code Division Multiple Access (WCDMA). High Speed Uplink Packet Access (HSUPA)/ High-Speed Downlink Packet Access (HSDPA). 1×Evolution Data Optimized, (1×EVDO). Wireless LAN, and Bluetooth. However, migrating current systems to 4G engenders enormous challenges. With ever-changing specification and standards, developing a prototype requires flexible process to provide 4G system capabilities. The 4G system has its own advantages and associated dangers. This chapter intends to deal with adoption issues of 4G, the fundamentals as well as issues pertaining to 4G networks, standards, terminals, services of 4G and the vision of network operators and service providers. Besides, to overcome the challenges of sophisticated personal, session and service mobility, advanced mobility management (MM) is needed to fulfill the need for seamless global roaming. The chapter endeavors to make an evaluation on development, transition, and roadmap for fourth generation mobile communication system with a perspective of wireless convergence domain in addition to mobility management. Lastly, open research issues in 4G are succinctly discussed.

Chapter 9
Potential Scenarios and Drivers of the 4G Evolution 181
    Elias Aravantinos, Stevens Institute of Technology, USA
    M. Hosein Fallah, Stevens Institute of Technology, USA

Nowadays, the mobile Internet communications can play a significant role in the Telecommunications Sector, resolving certain issues and bottlenecks of personal communications, with most European countries close to 100% penetration and a global projection of 4 billion mobile users by 2011. As we are moving to the next generation, we are still lacking the precise definition of the architecture and the successful deployment path of the 4G technology. Several theories have been developed looking at different standards and aiming to select and develop the most promising one. In this paper the authors are introducing and presenting a study that aims to explain a new concept of "4G readiness" revealing long run national strategies for 4G deployment and suggesting some critical metrics that could describe the future of the mobile broadband environment. They describe the methodology, assumptions and discuss the expected results based on similar studies such as the e-readiness study.

Chapter 10
Knowledge Sharing to Improve Routing and Future 4G Networks 193
    Djamel F. H. Sadok, Federal University of Pernambuco, Brazil
    Joseilson Albuquerque de França, Federal University of Pernambuco, Brazil
    Luciana Pereira Oliveira, Federal University of Pernambuco, Brazil
    Renato Ricardo de Abreu, Federal University of Pernambuco, Brazil

Current networking trends show a rapid convergence of several nestled networks as GSM/UMTS, WLAN, Bluetooth, wired networks among others. Nonetheless, such a complex environment raises new challenges including information routing, high dynamicity and possible disconnections. For such reasons, nontraditional routing paradigms have been put forward while adopting innovative ideas based on models from areas as diverse as biological, epidemic and social behavior. The reader will learn about new forms of routing being considered for integrating these future networks, based on the use of different metrics, such as shared network knowledge and willingness in taking and forwarding it. In this chapter an overview of traditional routing and the requirements for 4G systems are first made. New directions for routing based on policy and the identification of stimulus to control and improve the message forwarding in addition to their efficiency in the new context of 4G networks are presented.

Next Generation Technologies

Chapter 11
Personal Environments: Towards Cooperative 4G Services... 228
    Tinku Rasheed, Create-Net, Italy
    Usman Javaid, Vodafone Group, UK

The Fourth Generation of wireless networks promises to offer a vast range and diversity of converged services in order to revolutionize the way we communicate today. 4G can not only offer ultra-high data rates, but would also enable the ubiquitous computing paradigm, particularly interesting for the end-user with the help of various personalized and user-friendly services and devices. This increase in short-range communication among users and introduction of personalized services would form a "Personal Ubiquitous Environment" around the user. Since in such environments, multiple users will come closer (without any third-party barriers); their cooperation would be the key to success. Several technological and social barriers have prevented so far an effective cooperation between technologies, systems or users. This chapter focuses on the potential impacts of cooperative ubiquitous services in 4G networking systems. The authors explain the technological implications of cooperative systems considering the personal environment ubiquity. Furthermore, it attempts to characterize the socio-technical dimension of the potentials and limits of cooperation in 4G systems.

Chapter 12
Next Generation Broadband Services from High Altitude Platforms.. 249
    Abbas Mohammed, Blekinge Institute of Technology, Sweden
    Zhe Yang, Blekinge Institute of Technology, Sweden

In this chapter the authors investigate the possibility and performance of delivering broadband services from High Altitude Platforms (HAPs). In particular, the performance and coexistence techniques of providing worldwide interoperability for microwave access (WiMAX) from HAPs and terrestrial systems in the shard frequency band are investigated. The WiMAX standard is based on orthogonal frequency-division multiplexing (OFDM) and multiple-input and multiple-output (MIMO) technologies and has been regarded as one of the most promising 4G standards to lead 4G market and deliver broadband services globally. The authors show that it is possible to provide WiMAX services from an individual HAP system. The coexistence capability with the terrestrial WiMAX system is also examined. The simulation results show that it is effective to deliver WiMAX via HAPs and share the spectrum with terrestrial systems.

Chapter 13
Radio-over-Fibre Networks for 4G.. 268
    Roberto Llorente, Universidad Politécnica de Valencia, Spain
    Maria Morant, Universidad Politécnica de Valencia, Spain
    Javier Martí, Universidad Politécnica de Valencia, Spain

Radio-over-Fibre (RoF) is an optical communication technique based on the transmission of standard wireless radio signals though optical fibre in their native format. This technique is an enabling step in the deployment of dense fourth generation (4G) cellular and pico-cellular wireless networks. The optical fibre provides a huge bandwidth that can support a variety of wireless systems, regardless of their frequency bands, being protocol-transparent which is reflected in an great network flexibility. Radio-over-fibre techniques enables a high user capacity by frequency reuse, simplifies the network operation as the signals are distribute in their native format, and permits to transfer signal part of the processing power from the base station units to the central control station, thus reducing the overall deployment cost and complexity. The principles of radio-over-fibre are presented in this chapter, including the key transmission impairments and the expected performance. The main application scenarios are discussed. These include the backhaul of 4G or base-stations, addressing 4G and 3G compatibility issues, and distributed-antenna system (DAS). Finally, emerging applications like radio-over-fibre in beyond-3G scenarios and transmission of 60 GHz wireless are also described in this chapter.

Section 2
Radio Access Protocols

Scheduling and Quality of Service

Chapter 14
MAC Protocol of WiMAX Mesh Network.. 292
    Ming-Tuo Zhou, National Institute of Information and Communications Technology, Singapore
    Peng-Yong Kong, Institute for Infocomm Research, Singapore

WiMAX based on IEEE std 802.16 is believed to be one of the important technologies of 4G. It aims to provide high-speed access over distance of several to tens kilometers. In IEEE std 802.16-2004, WiMAX defines an optional mesh mode, with which multi-hop, multi-route, self-organizing and self-healing communications can be achieved in metropolitan-level areas. This chapter presents medium access control (MAC) protocol of WiMAX mesh mode, on frame structure, network configuration, network entry, and scheduling algorithms. It also summaries the most recent progress on data slots resource scheduling and allocation algorithms. Finally, an application example of using WiMAX mesh network for high-speed and low-cost maritime communications is also presented in this chapter.

Chapter 15
Advanced Scheduling Schemes in 4G Systems... 313
    Arijit Ukil, Tata Consultancy Services Ltd., India

The deterministic factor for 4G wireless technologies is to successfully deliver high value services such as voice, video, real-time data with well defined Quality of Service (QoS), which has strict prerequisite of throughput, delay, latency and jitter. This requirement should be achieved with minimum use of limited shared resources. This constraint leads to the development and implementation of scheduling policy which along with adaptive physical layer design completely exploit the frequency, temporal and spatial dimensions of the resource space of multi-user system to achieve the best system-level performance. The basic goal for scheduling is to allocate the users with the network resources in a channel aware way primarily as a function of time and frequency to satisfy individual user's service request delivery (QoS guarantee) and overall system performance optimization. Advanced scheduling schemes consider cross-layer optimization principle, where to fully optimize wireless broadband networks; both the challenges from the physical medium and the QoS-demands from the applications are to be taken into account. Cross-layer optimization needs to be accomplished by the design philosophy of jointly optimizing the physical, media access control, and link layer, while leveraging the standard IP network architecture. Cross-layer design approaches are critical for efficient utilization of the scarce radio resources with QoS provisioning in 4G wireless networks and beyond. The scheduler, in a sense, becomes the focal point for achieving any cross-layer optimization, given that the system design allows for this. The scheduler uses information from the physical layer up to the application layer to make decisions and perform optimization. This is a fundamental advantage over a system where the intelligence is distributed throughout the all entities of the network. In this chapter, the authors present an overview of the basic scheduling schemes as well as investigate advanced scheduling schemes particularly in OFDMA and packet scheduling schemes in all-IP based 4G systems. Game theoretic approach of distributed scheduling, which is of particular importance in wireless ad hoc networks, will also be discussed. 4G wireless networks are mostly MIMO based which introduces another degree of freedom for optimization, i.e. spatial dimension, for which scheduling in MIMO systems is very much complicated and computation intensive. MIMO resource allocation and scheduling is also covered in this chapter. The key research challenges in 4G wireless networks like LTE, WiMAX and the future research direction for scheduling problems in 4G networks are also presented in this chapter.

Chapter 16
End-to-End Quality of Service in Evolved Packet Systems 361
    Wei Wu, Research In Motion, Limited, USA
    Noun Choi, Research In Motion, Limited, USA

The recent emergence of new IP-based services that require high bandwidth and low service latency such as voice over IP (VoIP), video sharing, and music streaming have motivated the 3rd Generation Partnership Project (3GPP) to work on the all IP-based cellular networks called Evolved Packet System (EPS). It is challenging for EPS not only to meet the Quality of Service (QoS) requirements of new services but also to make sure the QoS of existing services not impacted. In this chapter, the authors will first present an overview of EPS, and then focus on the aspects of QoS principles and mechanisms in EPS. End-to-end QoS models have been developed to analyze the application performance in EPS. Simulation results have shown that VoIP service requires resource reservation to guarantee its QoS requirement, and e-mail service does not experience significant performance degradation even when assigned a low service priority and the system experiences short period congestion. However, web browsing performance may not be improved proportionally to the network bandwidth increase due to the inherent network probing procedure of the transport protocol.

Mobility and Handover

Chapter 17
An End-to-End QoS Framework for Vehicular Mobile Networks 377
    Hamada Alshaer, University of Leeds, UK
    Jaafar Elmirghani, University of Leeds, UK

In recent years we have witnessed a great demand for high speed Internet access in vehicular environment, e.g., trains, buses and medical transport. This chapter introduces an integrated architecture for 4G vehicular mobile networks, which aims to guarantee high quality in provisioned triple-play traffic services (video, voice, and data) to road users. Within this architecture which is based on a cross layer design approach, our contributions can be described in three folds. Firstly, the authors introduce simple and efficient probing mechanisms which are integrated with network resource reservation policies for multihomed vehicular networks. Secondly, packet, flow and user splitting mechanisms have been integrated with end admission traffic control and scheduling mechanisms to guarantee even traffic load distribution among available air interfaces. Finally, the whole architecture has been evaluated under OMNeT++, where results illustrate the impact of network mobility on quality in provisioned services offered to a multihomed NEMO.

Chapter 18
LTE Mobility Solutions at Network Level for Global Convergence 405
    Titus-Constantin Bălan, Siemens SIS PSE, Romania
    Florin Sandu, "Transilvania" University of Brasov, Romania

One of the research challenges for next generation all-IP-based wireless systems is the design of intelligent mobility management techniques that take advantage of IP-based technologies to achieve global seamless roaming among various access technologies. Since Mobile IPv6 is considered a mature protocol, mobility management at the network layer is the frequent approach for heterogeneous networks. The tendency of future convergent scalable architectures is splitting the mobility management in two domains, global mobility and localized mobility management. This chapter presents the advantages of MIPv6, a global mobility protocol, and its enhancements. A case study based on MIPv6 for UMTS and WiFi convergence is also presented. Proxy MIPv6, the newest protocol of the MIPv6 family, already included in the roadmap of future 4G networks, will be analyzed as a solution for localized mobility management. The main goal of the chapter is describing the way mobility protocols (MIPv6 and PMIPv6) will be implemented for the 3rd Generation Partnership Project (3GPP) Long Term Evolution architecture. The chapter ends with the presentation of the interoperation between different network technologies using global and localized mobility management protocols, which provide flexibility, scalability and independence between mobility domains.

Chapter 19
Handover Optimization for 4G Wireless Networks.. 424
    Dongwook Kim, Korea Advanced Institute of Science and Technology, South Korea
    Hanjin Lee, Korea Advanced Institute of Science and Technology, South Korea
    Hyunsoo Yoon, Korea Advanced Institute of Science and Technology, South Korea
    Namgi Kim, Kyonggi University, South Korea

The authors present a velocity-based bicasting handover scheme to optimize link layer handover performance for 4G wireless networks. Before presenting their scheme, as related works, they firstly describe general handover protocols which have been proposed in the previous research, in terms of the layers of network protocol stack. Then, they introduce state-of-the-art trends for handover protocols in three representative standardization groups of IEEE 802.16, 3GPP LTE, and 3GPP2. Finally, they present the proposed bicasting handover scheme. Original bicasting handover scheme enables all potential target base stations for a mobile station (MS) which prepares for handover to keep bicasted data, in advance before the MS actually performs handover. This scheme minimizes the packet transmission delay caused by handover, which achieves the seamless connectivity. However, it leads to an aggressive consumption of backhaul network resources. Moreover, if this scheme gets widely adopted for high data rate services and the demand for these services grows, it is expected that the amount of backhaul network resources consumed by the scheme will significantly increase. Therefore, the authors propose a novel bicasting handover scheme which not only minimizes link layer handover delay but also reduces the consumption of backhaul network resources in 4G wireless networks. For the proposed scheme, they exploit the velocity parameter of MS and a novel concept of bicasting threshold is specified for the proposed mobile speed groups. Simulations prove the efficiency of the proposed scheme over the original one in reducing the amount of consumed backhaul network resources without inducing any service quality degradation.

Cross-Layer Designs

Chapter 20
Survey of Cross-Layer Optimization Techniques for Wireless Networks... 453
    Han-Chieh Chao, National Ilan University, Taiwan
    Chi-Yuan Chang, National Dong Hwa University, Taiwan
    Chi-Yuan Chen, National Dong Hwa University, Taiwan
    Kai-Di Chang, National Dong Hwa University, Taiwan

The explosive development of Internet and wireless communication has made personal communication more convenient. People can use a handy wireless device to transfer different kinds of data such as voice data, text data, and multimedia data. Multimedia streaming, video conferencing, and on-line interactive 3D games are expected to attract an increasing number of users in the future. The bandwidth requirement would be high and the heterogeneous terminals would generally provide limited resource, such as low processing power, low battery life and limited data rate capabilities. These applications would be the major challenge for wireless networks. Although the traditional layered protocol stacks have been used for many years, they are not suitable for the next generation wireless networks and the mobile systems. Due to the time varying transmission of the wireless channel and the dynamic resource requirements of different application, the traditional layered approach to the mobile multimedia communication is full of challenges to meet the user requirement on performance and efficiency. Cross-layer design is an interesting research topic that actively exploits the dependence between different protocol layers to obtain performance gains. The authors performed a survey and introduced the cross-layer design principles and issues for different research topics, including QoS, mobility, security, application, and the next generation wireless communication.

Chapter 21
Cross-Layer Joint Optimization of Multimedia Transmissions over IP Based
Wireless Networks 469
    Catherine Lamy-Bergot, THALES Communications S.A., France
    Gianmarco Panza, CEFRIEL, Italy

The traditional approach consisting in separately optimizing each module of a transmission chain has shown limitations in the case of wireless communications where delay, power limitation and error-prone channels are experienced. This is why modern designers focus on a more integrated strategy to establish the heterogeneous 21st century networks, such as 3G (i.e. UMTS) system and its evolutions (i.e. Beyond 3G or 4G like LTE or future 5G systems). Indeed, it was shown in several studies that optimal allocation of user and system resources could be effectively achieved with the co-operative optimization of communication system components. In this chapter, an innovative Joint-Source Channel Coding and Decoding (JSCC/D) system is described and its performance over an IPv6-based Network infrastructure is assessed. A particular focus is put on the application controller, the key component to realize the adaptation strategies. Conclusions and considerations about the system implementation are also proposed, and the interest of a possible extension to a point-to-multipoint scenario is explained.

Chapter 22
Video Streaming Based Services over 4G Networks: Challenges and Solutions 494
    Elsa Mª Macías, Universidad de Las Palmas de Gran Canaria, Spain
    Alvaro Suarez, Universidad de Las Palmas de Gran Canaria, Spain

4G networks must not only show high bandwidth but also provide an excellent user experience, especially for video streaming, which is a key technique for multimedia services on 4G networks like Voice over Internet Protocol (VoIP), Television over IP (TvIP), broadcatching, interactive digital television, and Video on Demand (VoD). These services are challenging because of the well-known problems of the radio channel. Efficient solutions are designed by considering cross layer techniques. In this chapter the authors firstly review a number of video streaming based services, and then they present the basic operation of the video streaming and its problems in 4G networks, emphasizing Wireless Fidelity (WiFi) technology. In order to solve these problems they propose two cross layer strategies (one for access networks and another for ad hoc networks) and integrate the first one into two application level solutions. The authors test the user experience that accesses a Web portal including a VoD with a mobile telephone equipped with WiFi and High Speed Downlink Packet Access (HSDPA) Wireless Network Card Interfaces (WNIC). Results invite them to be optimistic.

Section 3
Physical Layer Advances

Advanced Multiple Access Transmission Schemes

Chapter 23
Aspects of OFDM-Based 3G LTE Terminal Implementation 526
  Wen Xu, Infineon Technologies AG, Germany
  Jens Berkmann, Infineon Technologies AG, Germany
  Cecilia Carbonelli, Infineon Technologies AG, Germany
  Christian Drewes, Infineon Technologies AG, Germany
  Axel Huebner, Infineon Technologies AG, Germany

3GPP standardized an evolved UTRAN (E-UTRAN) within the release 8 Long Term Evolution (LTE) project. Targets include higher spectral efficiency, lower latency, and higher peak data rate in comparison with previous 3GPP air interfaces. The E-UTRAN air interface is based on OFDMA and MIMO in downlink and on SCFDMA in uplink. Main challenges for a terminal implementation include an efficient realization of fast and precise synchronization, MIMO channel estimation and equalization, and a turbo decoder for data rates of up to 75 Mbps per spatial MIMO stream. In this study, the authors outline the current 3GPP LTE standard and highlight some implementation details of an LTE terminal. Efficient sample algorithms are presented for key components in the baseband signal processing including synchronization, cell search, channel estimation and equalization, and turbo channel decoder. Their performances, computational and memory requirements, and relevant implementation challenges are discussed.

Chapter 24
The Use of Orthogonal Frequency Code Division (OFCD) Multiplexing in Wireless
Mesh Network (WMN) 565
  Syed S. Rizvi, University of Bridgeport, USA
  Khaled M. Elleithy, University of Bridgeport, USA
  Aasia Riasat, Institute of Business Management, Pakistan

In the present scenario, improvement in the data rate, network capacity, scalability, and the network throughput are some of the most serious issues in wireless mesh networks (WMN). Specifically, a major obstacle that hinders the widespread adoption of WMN is the severe limits on throughput and the network capacity. This chapter presents a discussion on the potential use of a combined orthogonal-frequency code-division (OFCD) multiple access scheme in a WMN. The OFCD is the combination of orthogonal frequency division multiplexing (OFDM) and the code division multiple access (CDMA). Since ODFM is one of the popular multi-access schemes that provide high data rates, combining the OFDM with the CDMA may yield a significant improvement in a WMN in terms of a comparatively high network throughput with the least error ration. However, these benefits demand for more sophisticated design of transmitter and receiver for WMN that can use OFCD as an underlying multiple access scheme. In order to demonstrate the potential use of OFCD scheme with the WMN, this chapter presents a new transmitter and receiver model along with a comprehensive discussion on the performance of WMN under the new OFCD multiple access scheme. The purpose of this analysis and experimental verification is to observe the performance of new transceiver with the OFCD scheme in WMN with respect to the overall network throughput, bit error rate (BER) performance, and network capacity. Moreover, in this chapter, the authors provide an analysis and comparison of different multiple access schemes such as FDMA, TDMA, CDMA, OFDM, and the new OFCD.

Chapter 25
A New Approach to BSOFDM: Parallel Concatenated Spreading Matrices OFDM 582
    Ibrahim Raad, University of Wollongong, Australia
    Xiaojing Huang, University of Wollongong, Australia

This chapter discusses a new concept for Block Spread OFDM called Parallel Concatenated Spreading matrices OFDM (PCSM-OFDM) which was first presented in (Raad, I. and Huang, X. 2007). While BSOFDM improved the overall BER performance on OFDM in frequency selective channels, this new approach further improves the BER of BSOFDM by over 3dB gain. This uses coding gain to achieve this and is similar in concept to the well known error correction codes Turbo Codes. This is done by copying the data at the transmitter n times in parallel and multiplexing.

Chapter 26
The Next Generation CDMA Technology for Futuristic Wireless Communications:
Why Complementary Codes? 596
    Hsiao-Hwa Chen, National Cheng Kung University, Taiwan

This chapter addresses the issues on the architecture of next generation CDMA (NG-CDMA) systems, which should offer a much better performance in terms of its capacity and transmission rate, etc., than that possible in all current 2-3G systems based on CDMA technology. The ultimate goal is to engineer a CDMA system, whose performance will no longer be interference-limited, for its application in futuristic wireless communications. To achieve this, many challenging issues should be tackled, such as innovated design approaches for CDMA codes, multi-dimensional spreading techniques, suitable CDMA signaling format for high-speed bursty traffic, and so forth. This chapter will review the author's ongoing research activities on the NG-CDMA technology, which can offer a performance never inferior to that of orthogonal frequency division multiple access (OFDMA) technology. In particular, the author will briefly introduce a new CDMA code design method, called Real Environment Adapted Linearization (REAL) approach, which can be used to generate CDMA code sets with inherent immunity against multipath interference and multiple access interference for both uplink and downlink transmissions. The chapter will also illustrate that an interference-free CDMA can only be made possible with the application of orthogonal complementary codes (OCCs). The use of traditional CDMA codes, such as Gold, Kasami, Walsh-Hadamard and OVSF codes, all working on an one-code-per-channel basis, will never help in this sense. Several other topics related to the NG-CDMA technology will also be addressed, such as system performance issues, other properties of the NG-CDMA technology, and so on.

Enhanced Decoding Techniques

Chapter 27
Configurable and Scalable Turbo Decoder for 4G Wireless Receivers 622
    Yang Sun, Rice University, USA
    Joseph R. Cavallaro, Rice University, USA
    Yuming Zhu, Texas Instruments, USA
    Manish Goel, Texas Instruments, USA

The increasing requirements of high data rates and quality of service (QoS) in fourth-generation (4G) wireless communication require the implementation of practical capacity approaching codes. In this chapter, the application of Turbo coding schemes that have recently been adopted in the IEEE 802.16e WiMax standard and 3GPP Long Term Evolution (LTE) standard are reviewed. In order to process several 4G wireless standards with a common hardware module, a reconfigurable and scalable Turbo decoder architecture is presented. A parallel Turbo decoding scheme with scalable parallelism tailored to the target throughput is applied to support high data rates in 4G applications. High-level decoding parallelism is achieved by employing contention-free interleavers. A multi-banked memory structure and routing network among memories and MAP decoders are designed to operate at full speed with parallel interleavers. A new on-line address generation technique is introduced to support multiple Turbo interleaving patterns, which avoids the interleaver address memory that is typically necessary in the traditional designs. Design trade-offs in terms of area and power efficiency are analyzed for different parallelism and clock frequency goals.

Chapter 28
Parallel Soft Spherical Detection for Coded MIMO Systems 644
    Hosein Nikopour, Huawei Technologies Co., Ltd., Canada
    Amin Mobasher, Stanford University, USA
    Amir K. Khandani, University of Waterloo, Canada
    Aladdin Saleh, Bell Canada, Canada

This Chapter briefly evaluates different multiple-input multiple-output (MIMO) detection techniques in the literature as the candidates for the next generation wireless systems. The authors evaluate the associated problems and solutions with these methods. The focus of the chapter is on two categories of MIMO decoding: i) hard detection and ii) soft detection. These techniques significantly increase the capacity of wireless communications systems. Theoretically, a-posteriori probability (APP) MIMO decoder with soft information can achieve the capacity of a MIMO system. A sub-optimum APP detector is proposed for iterative joint detection/decoding in a MIMO wireless communication system employing an outer code. The proposed detector searches inside a given sphere in a parallel manner to simultaneously find a list of m-best points based on an additive metric. The metric is formed by combining the channel output and the a-priori information. The parallel structure of the proposed method is suitable for hardware parallelization. The radius of the sphere and the value of m are selected according to the channel condition to reduce the complexity. Numerical results are provided showing a significant reduction in the average complexity (for a similar performance and peak complexity) as compared to the best earlier known method. This positions the proposed algorithm as a candidate for the next generation wireless systems. The proposed scheme is applied for the decoding of the rate 2, 4×2 MIMO code employed in the IEEE 802.16e standard.

Collaboration and Capacity

Chapter 29
Capacity Estimation of OFDMA-Based Wireless Cellular Networks 666
    André Carlos Guedes de Carvalho Reis, Universidade de Brasília, Brazil
    Paulo Roberto de Lira Gondim, Universidade de Brasília, Brazil

The usage of wireless cellular network architecture increases the capacity of a wireless system, by combining cells into clusters in which channels are uniquely assigned per cell and reusing such clusters throughout the network. Unfortunately, a cellular network system may become interference limited regarding its capacity instead of noise/range limited due to intensive resources reuse like time, frequency and space. Using as input the physical layer parameters and deployment scenario, an analytical approach is proposed for capacity estimation of networks based on Orthogonal Frequency Division Multiple Access (OFDMA) technology whose sub channels are composed of distributed subcarriers. This innovative approach is based on a new analytical method for SINR calculation based on a proposed subcarrier collision probability model. The usage of such method is exemplified for a single-hop sectorized Mobile WiMAX cellular network and the results are validated against published works.

Chapter 30
Wireless Collaboration: Maximizing Diversity through Relaying 684
    Patrick Tooher, Concordia University, Canada
    M. Reza Soleymani, Concordia University, Canada

To achieve performance gains in the wireless channel, spatial diversity is employed. These higher order transmit diversity gains generally require multiple transmit antennas at the source. This requirement is not always possible in real world applications, where practical concerns limit the number of antennas a wireless device can have. Recently, a new method to achieve transmit diversity has been proposed: collaborative communications. In this framework, a node in a wireless network can use the resources of other idle nodes and form what can be viewed as a virtual transmitting antenna array. This chapter presents an overview of the development of collaborative communications. Two-phase protocols that can achieve collaboration are presented. A discussion on the improvement of collaborative communications protocols is given. A broader perspective of collaborative communications is given by discussing ideas such as power allocation and multiple relays.

Compilation of References 710

About the Contributors 759

Index 777

Foreword: Wireless Communication—History and Visions

HOW DID WE ARRIVE AT OUR CURRENT STATE OF THE ART?

Pre-Cellular Mobile Telephony

In order to understand where we're going, we need to understand how we arrived at today's state of the art. The notion of reliable mobile telephone service was first introduced in the 1950's. The first widely deployed system was developed by AT&T Bell Laboratories in the United States and referred to as Mobile Telephone Service (MTS). MTS was fairly simple, comprising 1) a mobile analogue FM transceiver capable of operating on either 8 or 16 radio frequency channels in the vehicle, 2) a wide coverage FM base station transceiver and 3) an operator-assisted switching centre, by which calls were manually connected and disconnected to an outside party (Sarkar, Mailloux, Oliner, Salazar-Palma, & Sengupta, 2005). The mobile transceiver was extremely large, weighing 20 kilos or more and therefore was usually mounted in the trunk of the vehicle. A cable was run from the transceiver to the vehicle interior, where the user could operate the device using a control head and a telephone handset.

To place a call from the mobile, the user would first observe the "busy" light on the mobile control unit. If the system were available, the user would then lift the handset and press the "talk" switch to call the mobile operator and request that a number be dialled. The operator would then connect the audio lines from the Public Switched Telephone System (PSTN) to the mobile audio interface, and the call could proceed. Between 1964 and 1969, AT&T introduced the Improved Mobile Telephone Service (IMTS). The primary improvement was that automated switching replaced the need for manual involvement by a telephone operator. IMTS enabled users to place calls themselves using a dial installed on the mobile control head. Likewise, incoming calls were automatically routed to the mobile station, where the user was alerted by an alert tone (Gascoigne, 1974; Harte, 2006).

MTS and IMTS were fairly crude in contrast to the smart phones of today, but these systems were revolutionary for their time. They were truly technical substitutes for fixed-line telephony, and this is the way the consumer mobile communication industry remained for many years. Throughout the 1970's, adoption of MTS and IMTS service in the United States was rapid, and in the mean time similar services were introduced in Europe. From a market segmentation viewpoint, the subscribers of mobile telephone services between the 1950's and early 1980's were primarily high-end business users. The cost of equipment and service were extremely high compared to the cost of fixed-line phone service, prohibiting adoption by the general population. Nevertheless, mobile subscribers were individuals for whom economic utility value was of primary importance – they could afford it, and didn't mind paying the price one bit for mobile communication.

The rapid adoption of mobile telephony during the 1970's was somewhat of a paradox to many fixed-line telephone company executives who saw the mobile telephone market as an extremely small niche market, deserving very little attention. After all, there was a huge network of coin phones across Europe

Figure 1. MTS/IMTS mobile telephone: the actual radio device was trunk mounted and control head in the car is equipped with handset and dial (www.Motorola.com)

and the U.S. – if someone wanted to make a call, they could simply stop at a coin phone, drop in a relatively small amount of money and avoid the relatively large fixed cost of having mobile service. Of course, this was not how consumers viewed the situation. By 1980, there was a waiting list more than 5 years long for a mobile telephone number in every major city in Europe and the U.S. This was because the early systems had insufficient capacity. Each major city typically had either 8 or 16 FM voice frequencies, each of which could handle only one voice call at a time. To further compound the problem, the base stations were specifically designed to cover large geographical areas, typically a radius of 30 km or more, which limited the effectiveness of spatial frequency reuse. Indeed, the idea of aggressive frequency reuse and mobility control such as handover techniques were not even deployed until the advent of cellular technologies (Tabane, 2000).

Throughout the 1970's communication companies were hard at work to develop a next generation to systems like MTS and IMTS. The concept being explored was that of cellular radio, as it was called. The idea behind cellular systems was that of small base station coverage, enabling aggressive frequency reuse, resulting in many more times the available system capacity of the existing services. In order to achieve this capacity increase, mobility control techniques were required so users could be handed over between base stations as they traversed the geography over which they used their devices.

1st Generation Cellular: Analogue Voice Service

In the MTS/IMTS world, if a user travelled outside the coverage area of a base station, any ongoing call dropped and would have to be re-established when the user re-entered system coverage area. In the cellular world, users had smooth and relatively seamless mobility over multiple cells. A major underlying success factor for cellular and its seamless mobility control technique was the availability of the microprocessor, which provided sophisticated, intelligent control at both the mobile and network. In 1983, the first commercial cellular system, the Advanced Mobile Phone Service (AMPS) was deployed in the Chicago area. AMPS is typically referred to as 1st Generation Cellular. In addition to aggressive spatial frequency reuse and instantaneous mobility management techniques, regulators in the United States provided AMPS with a substantial quantity of radio spectrum. Instead of 8 or 16 channels per metropolitan area, AMPS now had 666 channels available which provided a capacity increase of over a million times in large metropolitan areas. AMPS was still FM in the beginning, but now many more phone numbers were available and adoption was rapid throughout the 1980's and early 1990's. Similar technologies were developed and deployed

Figure 2. IMTS brief case phone from the 1970's: very expensive, not much range or battery life but adoption by high-end business users was rapid (www.att.com)

around the globe, e.g. the Nordic Mobile Telephone Service (NMT) in 1981, Total Access Communication System (TACS) and Extended TACS (ETACS) in Europe.

During the years 1983 through about 1986, cellular mobile equipment was still expensive. A typical automotive installation brought a fixed cost of $2,000 to $4000 US Dollars plus the monthly subscription fees to the mobile operator. Incremental costs of making and receiving calls was on top of the cost of equipment and service. Therefore, even after the introduction of the AMPS cellular system, the primary market segment for mobile telephony was still largely commercial users. But with the availability of equipment and phone numbers, there was an element of high-end personal users entering the cellular user community as well. Throughout the 1980's and 1990's, the learning curve brought down the cost of manufacturing equipment (Freeman, 1997). With lower costs came lower prices, and with lower prices came greater demand. By the early 1990's, most middle class adults owned mobile telephone equipment of some kind.

The 2nd Generation: Digital Voice for Cellular

The first large global standard for digital mobile telecommunication was the result of work coordinated by the European Telecommunication Standards Institute (ETSI) in the standards body originally referred to as Groupe Speciale Mobile (GSM) during the late 1980's and early 1990's. The standard was referred to simply as GSM, but the acronym was later changed to the English phrase, "Global System for Mobile Communication" and later "Global System for Mobile Telecommunication" (GSM). GSM systems were deployed first in Europe during 1992. These first systems, referred to as GSM Phase 1, supported circuit-switched voice interchange and functioned much like their analogue cellular counterparts. In addition to the basic voice services, GSM offered some new functionality referred to collectively as "teleservices" which included extensible messaging features such as the Short Message Service (SMS) and Cell Broadcast (CB).

Work to extend the GSM feature base began in the mid-1990, as Phase 2 which included frequency hopping, support for global frequency bands and other enhancements. During the same period, other standards were deployed in the United States, e.g. Intermediate Standard 136 (IS-136) was a Time Division Multiple Access (TDMA) technique and Intermediate Standard 95 (IS-95) was a technique based on Code Division Multiple Access (CDMA) technology. These technologies were commonly referred to as "2nd Generation" or "2G" services. Other digital systems were deployed around the world during the same period, including the North American Digital Cellular (NADC) service in 1991 in the U.S., which was based on AMPS, and Personal Digital Cellular (PDC), which was deployed in Japan. Because GSM solved a very important business problem, i.e. the billing methods for users who "roam" among different

Figure 3. 1st generation analogue cellular equipment: fairly large and still expensive

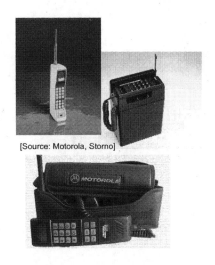

[Source: Motorola, Storno]

countries and network operators, these other systems eventually gave way to GSM's dominance (Mouly & Pautet, 1992; Mehrotra, 1997).

2.5 and 2.75G: Cellular Data is Added to GSM

The first digital cellular systems like GSM, IS-136 and IS-95 were developed for circuit-switched voice service. This meant that calls were made on a point-to-point basis through a mobile switching centre (MSC) that was much like a fixed-line telephone switch. These technologies did support some means of user data interchange, but they relied on a circuit-switched, connection-oriented approach which consumed a fair amount of wireless and network resources in contrast to the actual amount of data being sent and received. In 1992, the idea for a packet-based subsystem was introduced into ETSI standards called General Packet Radio Service (GPRS), but some time was needed for the industry to see commercial applicability before widespread standards support was achieved. Initially, GPRS was created for the transmission of mobile telematics information, e.g. truck and bus location data, for which the GSM circuit-switched Mobile Switching Centre (MSC) was a choke point. Often, telematics data comprised only small amounts of information that could be sent within a time period of under a tenth of a second, but the fact that the MSC was involved in setting up a circuit-switched connection resulted in latencies of up to 60 seconds or more on early systems because of the time required for the wire line modem on the remote end to synchronize with the mobile through the network. It was speculated that a true packet-based technique would allow the transfer of information without making a circuit-switched call at all, i.e. to set up a temporary packet data channel over which a small amount of data would flow followed by the rapid teardown of the channel. In this manner, data transfers would be completed rapidly while radio resources on the network would be conserved for other traffic (GSM-02.03, 1996; GSM-03.41, 1996).

The introduction of the World Wide Web (WWW) in the early 1990's created interest in extending the web to mobile users, and by the late 1990's, the concept of GPRS had gained general acceptance by the industry as a vehicle for a "mobile Internet". Because GPRS is packet-based, it creates the illusion of being "always on" by actually having both endpoints being "always off", i.e. the endpoints are only aware of each other as entries in each others' network-specific routing tables. Then, when there is a need to send data, i) a packet transfer is quickly set up, ii) data are transferred and iii) the transfer is torn down, returning resources to the network for subsequent use by other users or services.

Figure 4. GSM brought handset size down significantly and solved many of the infamous "roaming problems" in Europe with Home Location Register (HLR) architecture (www.Nokia.com; www.BellSouth.com)

Although telematics data interchange was the initial motivation for the creation of GPRS, the growth of the Internet and World-Wide Web in the mid 1990's generated far more interest and commercial intensity on the Standardisation Work Item. It was not until 1997 that Public Land Mobile Network (PLMN) operators began to take GPRS seriously as a means of generating additional revenues based on their excess capacity during non-peak usage period. The fact that GPRS had its first roots in telematics explains some of the technology decisions, e.g. the simplified method of mobility management during a packet transfer based on autonomous cell reselection by the mobile terminal. Some of these early technology decisions have introduced constraints on the system that remain today.

GPRS, and its superset, Enhanced Data for Global Evolution (EDGE), permit efficient use of radio and network resources when data transmission characteristics are i) packet based, ii) intermittent and non-periodic, iii) possibly frequent, with small transfers of data, e.g. less than 500 octets, or iv) possibly infrequent, with large transfers of data, e.g. more than several hundred kilobytes. User applications were originally envisioned to include Internet browsers, electronic mail, file transfers and other applications for which best efforts data transfer are appropriate. The first commercial release of GPRS specifications was Release 97, although the specifications were not complete until 1999 with a few corrections to Release 97 specifications noted as late as 2003 (GSM-04.06, 1996; GSM-02.60, 1996; GSM-03.60, 1996).

Enhanced Data for GSM Evolution (EDGE) was standardized as a parallel path to GSM evolution. EDGE is sometimes referred to in the 3GPP specifications as Enhanced General Packet Radio Service (EGPRS), and is a 3G superset of GPRS as defined by the International Telecommunications Union (ITU), although it is sometimes referred to as 2.75G because it was introduced early in the standardization cycle. EDGE enables higher data rates over the radio interface, but supports the same set of basic services as offered by GPRS. In order to support EDGE, the mobile terminal and network need to support GPRS. EDGE, extends the existing capability of GPRS, by the addition of three underlying technologies: i) high order modulation, ii) radio link adaptation and iii) incremental redundancy. These techniques result in higher user data rates and greater system capacity than was possible with basic GSM and GPRS. In addition, simultaneous voice and data operation was added to GSM with the introduction of Dual Transfer Mode (DTM) (3GPP Work Plan, 2009; Pecen & Howell, 2001).

In the days of MTS, IMTS, 1st and 2nd Generation Cellular, mobile devices were viewed largely as something to be used as telephones. Except for the introduction of SMS on the GSM system for short text

Figure 5. GPRS and EDGE added packet switched data service to GSM

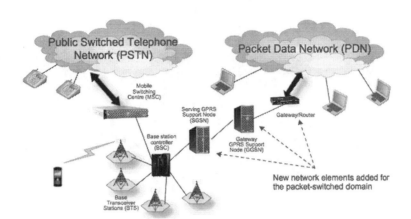

messages, the basic user perception was that mobile devices were something to talk on. The notion of major services other than voice had not even entered the scope of discussion within the industry until the early 1990's and the revenue produced by services like SMS was a tiny fragment of what revenues have become by the year 2009 for more sophisticated services such as email, browsing, enterprise applications, instant messaging, multimedia, social networking and others. So between the 1950's and the end of the 1990's the mobile device was primarily viewed, and utilized as a technical substitute for fixed-line voice telephone service.

3G: Evolution of 2G Architecture

Before GPRS standardization was complete, industry groups began work on the next generation of packet data systems, both for GSM and for a new standard referred to as Universal Mobile Telecommunication Service (UMTS) under the oversight of the 3rd Generation Partnership Project (3GPP), which was founded in 1997 by a number of mobile network operators and equipment manufacturers. The idea was to leverage the original 2G architecture of GSM and to extend the capabilities of the radio interface by moving to a Wideband Code Division Multiple Access (WCDMA). When 3GPP was created, the maintenance and evolution of the GSM standard was also placed under the management of 3GPP. UMTS was introduced as a completely new 3G standard with deployments beginning in 2001. Adoption of UMTS was slow, and by 2009, less than 6% of the total market had deployed UMTS (www.Informa.com, March 2009).

Simultaneously, it was recognized by the industry that some serious limitations to the user data interchange capabilities existed in UMTS, which prompted the development of an additional standard for High Speed Downlink Packet Access (HSDPA). HSDPA used similar techniques as did EDGE for increasing capacity and data rates, although within the framework of WCDMA. In 2005, participants in 3GPP began developing standards for the uplink equivalent to HSDPA, i.e. High Speed Uplink Packet Access (HSUPA) (3GPP FTP, 2009). By 2007, the combination of HSDPA/HSUPA was often referred to as High Speed Packet Access (HSPA) within the industry. By 2009, 90% of the global market for wide area wireless systems was occupied by the technologies standardized by 3GPP, 83% of which was GSM, GPRS and EDGE. In the first quarter of 2009, there were 3.9 billion GSM subscribers.

Other 3rd Generation systems were developed during the period between 1999 and 2007. 3rd Generation user data services were added to the IS-95 CDMA technology standards and introduced under the International Telecommunications Union (ITU) designator, IMT2000. This included such technologies

Figure 6. Backward compatibility of 3G to 2G GSM (3G Americas, 2009)

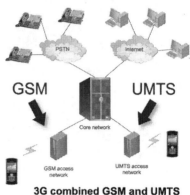

as Evolution for Data Only (EVDO), which technologically was fairly similar to HSDPA (3G Americas, 2009).

So how do we define 3G? It's essentially an evolution of 2G technology to the next level – same basic functionality, but with a focus on user data support and higher data rates. The support for mobile user data by dominant mobile technology such as GPRS/EDGE, UMTS/HSPA and CDMA2000/EVDO is all important, because it is the foundation of what we think of in the year 2009 as the "age of the Smartphone" – a mobile device that not only support voice service, but messaging, email, browsing, enterprise services, cameras, multimedia and music players.

In 2004 standardization work on an item referred to as Long-Term Evolution (LTE) was launched in 3GPP. This work item represented a drastic departure from earlier 3GPP technology both in terms of architecture and wireless interface. Beginning in 2002, the Institute for Electrical and Electronic Engineers (IEEE) standards bodies were developing standards for the 802.16 and 802.20 wireless standards, commonly referred to as Wireless Microwave Access or WiMAX. These technologies were intended to be wider-area economic substitutes for the IEEE 802.11 Wireless Local Area Network (WLAN) technologies commonly known as Wireless Fidelity or WiFi.

Even the original GSM-based technology standards continue evolving. Between 2005 and 2008, 3GPP industry participants developed what is referred to as Evolved EDGE, which is a faster and more efficient data service based on the original EDGE work but featuring broadband data rates. Why extend GSM and EDGE? With almost 4 billion GSM/EDGE subscribers in 2009, the technology switching costs of moving to new systems are extremely high. Even with the next generation of wireless systems, GSM is going to be around for some years yet (3GPP TR 45.912, 2007; Fuertes, 2009). At the end of 2008, 3GPP technologies (GSM and WCDMA) dominated the global subscriber market.

THE ROLE OF STANDARDS IN INTERNATIONAL TELECOMMUNICATION

Wireless communication comprises a wide range of technologies, services and applications that have come into existence to meet the particular needs of different deployments and user environments. Different systems can be broadly characterized by:

- content and services offered,
- frequency bands of operation,
- standards defining the systems,

Figure 7. The Age of the Smartphone: Mobiles offer more services then ever beforeEvolving beyond 3G in the 3rd Generation Partnership Project

- Email
- Browsing
- Navigation
- Music
- Cameras & multimedia
- Enhanced services
- Custom apps
- Instant messaging
- Calendar, contact and desktop sync

- data rates supported,
- bidirectional and unidirectional delivery mechanisms,
- degree of mobility,
- regulatory requirements, and
- cost.

In today's global society, it's essential for different countries and regions to agree on common frequency bands, technologies, regulatory requirements and expectation of services to enable user mobility and roaming. These commonalities create powerful economies of scope, which in turn create large economies of scale. The platform for such discussions and agreements is the International Telecommunications Union (ITU). The ITU is an agency of the United Nations and a general oversight body for global telecommunications standards. International Mobile Telecommunications-2000 (IMT-2000) is the global standard for 3rd Generation wireless as defined by a set of interdependent recommendations of the ITU. These recommendations include standards for frequency spectrum usage, wireless system technical specifications, tariffs and billing, technical assistance and studies on regulatory and policy aspects. .

Second generation systems were primarily designed to support voice service. IMT-2000 and enhanced IMT-2000 systems and systems beyond IMT-2000 were created to support multiple access technologies that compliment one another in an optimal way to provide a common, flexible platform for different services and applications.

A similarity of services and applications across the different systems is beneficial to users, and this has stimulated the current trend towards convergence. Furthermore, a broadly similar user experience across the different systems leads to a large-scale adoption of products and services, common applications and content. Access to a service or an application may be performed using one system or may be performed using multiple systems simultaneously. Such convergence should nevertheless impede any opportunities for competitive innovation. This is why it's appropriate to standardize some system aspects but not others.

Relationship of IMT-2000 and IMT-Advanced

IMT-2000 systems are intended to provide access to a wide range of telecommunication services, supported by the fixed telecommunication networks (e.g. PSTN/ISDN/IP), and to other services which are specific to

Figure 8. 3GPP technologies (GSM and WCDMA) dominated the global subscriber market at the end of 2008 (www.Informa.com, 2009)

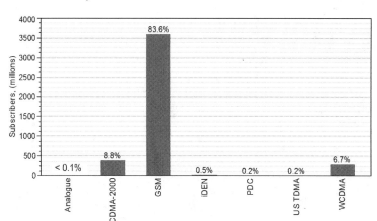

mobile users. To meet the ever increasing demand for wireless communication (e.g. increased no. of users, higher data rates, video or gaming services which require increased quality of service, etc.), IMT-2000 has been, and continues to be, enhanced.

The diagram in Figure 9 is taken directly from Recommendation ITU-R M.1645 and reflects the terminology in use at the time of its adoption. Resolution ITU-R 56 defines the relationship between 1) IMT-2000, 2)the future development of IMT-2000 and 3) "systems beyond IMT-2000" for which it also provides the new identifier, IMT-Advanced. Resolution ITU-R 56 resolves that the term IMT-2000 encompasses also its enhancements and future developments. The term IMT-Advanced should be applied to those systems, system components, and related aspects that include any new radio interface(s) that support additional capabilities of systems beyond IMT-2000. The term "IMT" is the root name that encompasses both IMT-2000 and IMT-Advanced collectively.

IMT-Advanced systems support low to high mobility applications and a wide range of data rates in accordance with user and service requirements in multiple user environments. IMT-Advanced also defines capabilities for high quality multimedia applications within a wide range of services and platforms to provide a significant improvement in performance and quality of service.

The primary features of IMT-Advanced are:

- a high degree of commonality of functionality worldwide while retaining the flexibility to support a wide range of services and applications in a cost efficient manner,
- compatibility of services within IMT and with fixed networks,
- capability of interworking with other radio access systems,
- high quality mobile services,
- user equipment suitable for worldwide use,
- user-friendly applications, services and equipment,
- worldwide roaming capability; and
- enhanced peak data rates to support advanced services and applications (100 Mbit/s for high and 1 Gbit/s for low mobility were established as targets for research)

Consumer demands will shape the future development of IMT-2000 and IMT-Advanced. Recommendation ITU-R M.1645 describes these trends in detail, some of which include the growing demand for

Figure 9. Illustration of capabilities of IMT-2000 and IMT-Advanced (ITU-R M.1645, 2008)

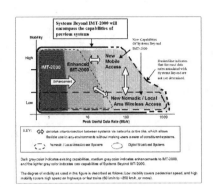

mobile services, increasing user expectations, and the evolving nature of the services and applications that may become available. Report ITU-R M.2072 details the market analysis and forecast of the evolution of mobile products and services for the future development of IMT-2000, IMT-Advanced and other systems. This Report provides forecasts for the year 2015, and 2020 timeframes.

IMT-Advanced Development Process and Timeline

Resolution ITU-R 57 on the "Principles for the process of development of IMT-Advanced" outlines the essential criteria and principles that are used in the process of developing the Recommendations and Reports for IMT-Advanced, including Recommendation(s) for the radio interface specification. The detailed procedure is illustrated in Figure 12. The timeline shown in Figure 13 is directly from ITU-R M.1645, which was originally drafted in 2003 and makes the assumption that most of the innovation on next generation technology has already been completed. This is far from the case. In spring 2008, the ITU issued an invitation for submission of technologies that will meet the requirements of IMT-2000 Advance (David, 2008; IMT2000, 2009; ITU, 2008; ITU-R M.1645, 2008). During each phase of technology development, there are problems that require innovative solutions that include additional research, product development and insertion into the value chain. It's extremely likely that we'll see further advances and deployment of next generation systems well past 2015.

WHAT'S NEXT?

An Aggressive Industry Vision

The various wireless standards development organizations have set forth an extremely aggressive vision for next generation wireless networks (3GPP Rel. 8, 2009; 3GPP Rel. 9, 2009; 3GPP TR 21.902, 2008; IEEE, 2009). The requirements were designed to approach the performance levels of broadband fixed line service with the addition of full mobility control, subscriber roaming and the other features that allow users to perform useful tasks while mobile. Following are the primary criteria as set forth by 3GPP and IEEE:

1. High data rates: > 100 Mbits/sec – This represents the per-sector throughput of an entire base station carrier. Each user would receive proportionally lower data rates depending on signal quality and system congestion. These high data rates are indeed achievable, given the industry's direction

Figure 10. Future network of systems beyond IMT-2000 including a variety of potential interworking access systems (ITU-R M.1645, 2008)

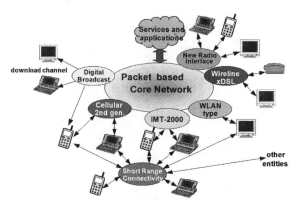

for greater RF spectrum occupation and the use of Orthogonal Frequency Division Multiplexing (OFDM). OFDM has other strengths as well, including the simplicity of spectrum sensing in the frequency domain, ability to limit co-channel interference instantaneously by selectively turning off sub-carriers as needed and ease of interfacing with smart antennas. There are nevertheless challenges to achieving 100+ Mbits/sec data rates in the practical world. Adequate radio link margin becomes more of a challenge as data rates increase, and we're seeing more technologies to address this issue such as advanced receivers and interference cancellation techniques.

2. Low latency: < 50 s – Depending on the application, latency is either totally unimportant or absolutely vital. An application for which low latency is vital is that of web browsing, and especially where web pages contain many pictures. Each picture downloaded by the browser must be acknowledged individually, and each individual acknowledgement is dependent on the latency of its turn-around time. Many systems such as GPRS have latencies of 600 ms or more, which severely cripples the transfer of web pages with several pictures. Imagine a web page having 10 pictures on a system with 600 ms of one way latency. Each picture requires 600 ms seconds to acknowledge and 600 ms begin sending the next one. If you have 10 pictures, latency would contribute 2 * 600 ms * 10 = 12 seconds, so the transfer could not be any quicker than 12 seconds. Contrast this to a system having 50 ms, where the same 10 pictures would require a minimum of 2 * 50 ms * 10 = 1 second to download – now latency becomes a less significant portion of the transmission time at 50 ms.

3. Mobility support: Mobility management controls and balances network resources incidental to the user's geographical movement. The next generation would support both macro-mobility and micro-mobility at relatively high user speeds, such as might be encountered in moving vehicles or trains. The notion of micro-mobility is that of managing and handing over calls and/or ongoing data transfers among clusters of related base stations, referred to as eNodeB's in 3GPP terminology. Macro-mobility is a higher-level abstraction whereby user mobility among un-related network components is achieved, e.g. from one Routing Area (RA) to another. Further to the mobility issue, the air-interface itself must be robust to Doppler shift and other fading channel effects. This is part of the basis for selecting Orthogonal Frequency Division Multiple Access (OFDMA) for the Long-Term Evolution (LTE) work item in 3GPP. OFDMA utilizes many relatively narrow band sub-carriers, which combined produce a broadband, high-speed channel, but may be individually treated as flat-fading channels thereby mitigating the impacts of broadband fading.

4. Scalable bandwidth: To produce the kinds of high data rates and trunking efficiency that are required for the next generation, network operators need a substantial amount of contiguous radio spectrum.

Figure 11. Illustration of complementary access systems (ITU-R M.1645, 2008)

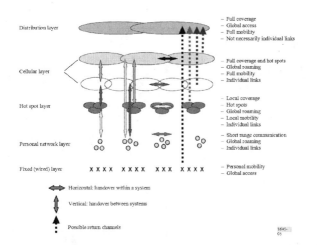

A potential issue is the immediate availability of spectrum for initial system rollout. An operator may not have e.g. 20 MHz of RF spectrum to immediately dedicate to the next generation system. Methods for incrementally adding RF bandwidth to a next generation system have been devised. These allow the operator to deploy their new systems in smaller segments of RF spectrum at first, and then gradually increase spectral occupation of the new system while reducing spectral occupation of their legacy systems. For example, a next-generation system may be first rolled out on 1.25 MHz of spectrum, and then gradually expanded to fill a full 20 MHz of RF bandwidth.

5. Optimized for data services: 1st, 2nd, and to some degree 3rd Generation wireless systems were optimized for voice traffic. This made complete sense at a time in history where mobile wireless was largely a technical and/or economic substitute for wired telephony. The use of data applications by both enterprise users and consumers is on a strong upward trend. Since approximately 2001, enterprise customers have led the consumption of user data capacity in cellular networks. This began to change in around 2005 as consumers brought non-voice mobile data applications into their life on a regular basis. In 2009, the top three user data applications were 1) email, 2) messaging and 3) browsing, with an exponential increase in browsing traffic between 2005 and 2009. Another strong trend since around 2007 is an ever increasing usage of streaming video, specifically YouTube videos over mobile terminals.

6. Backward compatibility with legacy systems, e.g. GSM and UMTS: With almost 4 billion subscribers in 2009, technology switching costs are a major force that the next generation wireless technology must mitigate in order to achieve substantial economic success. Compatibility between systems includes the ability to select among next generation and legacy systems as required based on coverage and available capacity. Looking back over the past 10 years, wireless systems have become more heterogeneous in general, with mobile devices now including such important auxiliary wireless interfaces as Bluetooth, WiFi and Near-Field Communication (NFC) techniques. As an industry, we've learned how these technologies can co-exist in a mobile terminal to a certain degree, and maybe how some of these vastly different wireless technologies can interwork, e.g. WiFi and cellular in the case of Unlicensed Mobile Access (UMA). An important issue for mobile terminal manufacturers as we move forward is how to cope with the number of new frequency bands and the required antennas, RF components and logic to manage their operation. We expect many more issues to be solved for

Figure 12. IMT-Advanced terrestrial component radio interface development process (IMT-ADV/1-E, 2008)

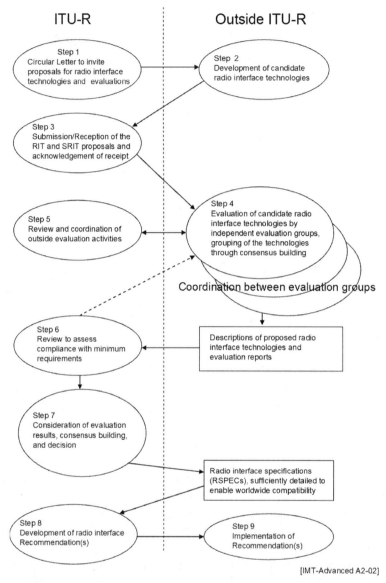

[IMT-Advanced A2-02]

the next generation on both the radio access side as well as the core network side, such as billing and Quality of Service (QoS) control.

7. High spectral efficiency, 5 bits/symbol/Hz: Radio spectrum is the wireless highway over which your information travels. There exists a finite amount of spectrum in the range of wavelengths appropriate for cellular communication. For the most part, this means from about 700 MHz to around 3500 MHz. Spectral efficiency may be increased in three fundamental ways: 1) information theoretic innovations, such as higher order modulation and advanced receivers, 2) statistical gains, such as fat pipe techniques that improve trunking efficiency and 3) reducing the cell sizes, which requires more equipment and real estate. The industry direction is to constantly search for ways that higher and higher order modulation can be used over the wireless channel, which often goes hand-in-hand with

Figure 13. Phases and expected timelines for future development of IMT-2000 (ITU-R M.1645, 2008)

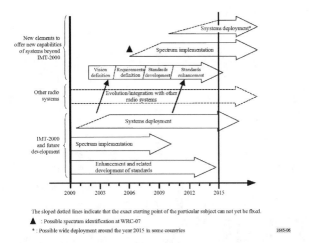

the use of an advanced receiver or interference cancellation technique on the receiving end. The industry trend toward use of OFDM also presents an opportunity to achieve statistical gains as well, as the bandwidth utilization increases with large amounts of contiguous spectrum. Such large spectral occupation means that there are no guard bands within the OFDM channel to consume spectrum. In addition, the network scheduler can quickly allocate capacity as-needed without incurring the overhead of a complex radio link setup procedure. Regulators around the globe also need to effectively coordinate the issuance and licensing of radio spectrum in such a way to help ensure that wide ranges of contiguous spectrum are in fact available. Reducing cell coverage increases spectral efficiency dramatically, but there are economic and basic physical limits. For more info regarding cell sizes, please see "About Wireless Capacity".

About Wireless Capacity

Between 1957 and 2008, wireless capacity has increased approximately 1,000,000 times (Chandrasekhar, Andrews, & Gatherer, 2008). This means that a given amount of radio spectrum over a given area of population will carry about 1,000,000 times more information than was possible in 1957. Let's examine the sources of this substantial increase in capacity:
    Since 1957, we've seen:

- A 25X information theoretic improvement (modulation and coding advances, equalizers, advanced receivers, interference cancellation, etc.)
- A 25X statistical improvement (wider contiguous spectrum, advanced scheduling techniques, etc.)
- A 1600X improvement due to reducing transmission distances, i.e. cell sizes

While reduction of the size of cell has provided most of the gains over these years, there are several factors that present practical limits to what can be achieved by this practice. For example, more base stations are needed, as is more real estate on which to deploy them. These factors become economic and logistic barriers at some point. Furthermore, there are practical limits to the granularity of mobility control, because a substantial amount of signalling capacity and mobile terminal battery life may be consumed by the mobile constantly informing the network where to reach it. A Work Item (WI) in 3GPP is related to the standardization of LTE Relays. These relays may provide vital low-profile links between primary cells and users who would receive a substantial increase in data rates in the presence of greater link margin, but only in places where they're needed.

What is 4G Anyway?

The authors of this book are helping to answer that question. Let's consider a cross-section of the research community's efforts to push wireless technology into the next generation as seen by the authors of the following chapters.

As of 2009, there was no precise formal definition of 4G, as was pointed out by authors Aravantinos and Fallah, who present an economic rank-based survey of market readiness for 4G later on in this book. The survey combines the dimensions of technology, business, consumer, legal/regulatory and R&D investment into a single rank-based index for what the market would require to support a next generation. Although not yet formally defined by the ITU, there existed many informal definitions for 4G in 2009, primarily based on the extension of the ITU definition for IMT-Advanced which defines the basic working assumptions for systems extending beyond IMT-2000.

New Business Designs and Opportunities

What 4G will eventually become depends on more than just technological forces. There are social, environmental, economic and political forces at work too, and these forces are vital in shaping the next generation of wireless networks and services. In the chapters to follow, our contributors explore technical innovations in the wireless domain, but they also go beyond in their exploration of the other strategic forces. For example, the next generation may also provide substantial economic opportunities for the industry to migrate from a Subscriber Business Model (SBM) to a Consumer Business Model (CBM). As applications become more creative and wireless becomes ubiquitous, an economic differentiation between the subscriber and the actual consumer of services has begun to appear. Today, the consumer and subscriber are largely treated as the same entity. This is to say that the person using the service is also the subscriber in the case of the SBM. But what if the subscriber is an individual, but that individual hosts multiple consumers for different services? This would be the case for a subscriber whose intelligent consumer devices or even other users may comprise the consumers of service, each consumer having differing service requirements and even commercial relationships that transcend the current generation of network capabilities. Authors Ganchev, O'Droma, Jakab, Ji and Tairov propose possibilities for business design based on the definition of a CBM for 4G networks.

Short-Range Wireless

In addition to ultra-high data rates and other features planned for 4G, a parallel evolution is occurring that is already enabling the ubiquity of short-range personal communication. Some personal area networks are already in use, e.g. Bluetooth, but as authors Rasheed and Javaid point out, these short-range technologies have not yet been used to their full potential. The implications of Personal Network Federation (PN-F) architectures and cooperative techniques are explored, which may have far-reaching effects in social, technological, economic and political domains. Why the impact in the political domain? Short-range technologies may operate on RF spectrum that is licence-free in some countries, or possibly spectrum that is licensed to a network operator. The proliferation of such miniature networks is already raising concerns regarding the coordination of spectrum regulation across the globe. Even today, users may carry devices that are legal to operate in one country and illegal in another, creating confusion for consumers, regulators and equipment manufacturers.

Cross-Layer Design

Before the year 2001 or so, the concept of state or control information crossing layers in a protocol stack was largely thought of as protocol violation. Today, there is a name for this: Cross-layer design, and it's expected to play an increasingly important role in next generation wireless networks. The cross-layer concept addresses a fundamental challenge that the industry has learned over time: that the needs of the wireless domain are vastly different from those of the wire-line. Transport and application layers behave differently than e.g. Radio Link Control (RLC) and Physical wireless layers. Authors Chao, C. Chang, Chen and K. Chang analyze the impacts of the design of new interfaces, merging multiple layers, design coupling without new layers and vertical collaboration across layers. These concepts may move 4G another step closer to having wireless-aware applications and application-aware wireless interfaces, providing the ability to better smooth out the effects of wireless channel variability and resulting race conditions over the span of protocol stacks.

Authors Alshaer and Elmirghani, who note that each network element may present large characteristic variability in itself which occurs across multiple dimensions, present further work on QoS control using cross-layer design. Highly variable characteristics include data rates, end-to-end latency, jitter and packet re-ordering. When variability cascades across multiple network elements, performance often degrades rapidly due to the adaptive nature of one element working against that of another. Authors Alshaer and Elmirghani examine the concept of cross-layer intelligent control in an end-to-end QoS architecture. They identify the possible extreme differences in behaviour among the wireless air-interfaces themselves, plus the impacts of cognitive radio, Medium Access Control (MAC) design and Link Layer active probing. They further propose some intelligent routing techniques to mitigate the effects of end-to-end variability, and present simulation results for their Network Mobility (NEMO) techniques. Routing is also the topic of work done by authors Sadok, J.A. de França, Oliveira and R.R. de Abreu, but within the context of knowledge sharing among the multiple, differing wireless interfaces. Data must be routed over interfaces having extremely different characteristics, and it's no longer the case that one method can be forced to fit a wide range of routing problems in the practical world. The authors explore the introduction of concepts learned from biological, epidemic and social behaviour into the development of a non-traditional routing technique.

Authors Lamy-Bergot and Panza note that the complexity of wireless user data traffic is increasing steadily over recent years, as users and application builders invent novel ways to utilize the available bandwidth of today's networks. As a result, better and better performance of wireless applications and services is expected by the time that next generation wireless networks are fully deployed. To address the trend of steadily increasing traffic, the authors explore the possibility of cooperative, cross-layer optimization with their proposal of a novel Joint-Source Channel Coding and Decoding (JSCC/D) technique that is specifically designed to address the impairments that are likely to be encountered by IP multimedia applications. Again, cross-layer design is central to their concept of end-to-end optimization based on a variety of traffic types and the variability of system loading. An entirely new architecture is proposed utilizing the concept of intelligent cross-layer controllers to address the problem of the lack of end-to-end supervisory control which is characteristic of current network architectures in which each layer operates autonomously, in a vacuum of system information of sorts. Their method and architecture show encouraging results, taking into account the limitations and variability of all system layers, from the wireless radio interface to the application.

Hybrid Wireless Environments

Cost of deployment and network maintenance is the issue addressed by Authors Dantu and Guturu. The propose a general network architecture using WiMAX as the air-interface, comparing cost of deployment to existing 2G/3G networks. They find that it may be economically advantageous to replace the traditionally expensive components of legacy networks with scalable IP-Radio Access Networks (RANs) configured to distribute mobility and QoS control across many network elements. The high degree of interest in integrating heterogeneous radio networks is further examined by authors Shen, Pesch, Atkinson and Du. These researchers examine some possibilities for developing a next generation Hybrid Wireless Network (HWN*) with a primary focus on mobility management and routing. A HWN* is an integrated architecture that combines multiple heterogeneous wireless air interfaces with seamless service delivery across the multiple networks. Building such a network is much easier said than done, due to the completely different characteristics of each wireless interface and their resource management demands. This challenge is especially relevant because a HWN* may be implemented from any off-the-shelf air-interface such as GSM, WiFi, CDMA-2000 and others, all under common control to some degree.

End-to-End Packet Architecture

The heterogeneous nature of networks goes far beyond the wireless air interface. Authors Wu and Choi present a comprehensive overview of the end-to-end Quality of Service (QoS) control in LTE's Enhanced Packet System (EPS) concept. The EPS is the first all-IP based mobile cellular system, which presents many challenges to managing QoS to both IP-based data services and the addition of packet-based voice services. Previously, voice over cellular networks was handled over a dedicated, circuit-switched environment. As we move toward all-IP systems, EPS must manage packet-based voice in a much more hostile network environment, where the quality of one voice call can depend on the characteristics and loading of non-voice data. Wu and Choi present a rigorous end-to-end QoS performance analysis of the basic services such as VoIP, e-mail and web browsing through simulation study of the EPS QoS model.

Consumers have exhibited a greater interest in mobile streaming services in recent years. Since around 2005, existing networks report an exponential increase in traffic created by users downloading YouTube videos, podcasts and short news clips. Other streaming services like VoIP and gaming have yet to be fully explored due to the tenuous ability of today's wireless networks to handle these services. The industry's vision of higher data rates and lower latencies is intended to address the need for better service quality, efficiency and availability of wireless streaming services. Primary impairments to streaming are the instantaneous variations in inter-packet arrival time, or jitter, and the periodic loss and/or re-ordering of data. Authors Macías and Suarez acknowledge and quantify a number of these challenges. Rather than proposing newer protocols to address the existence of a wireless component in a given service path, they propose the concept of a cross-layer optimization-theoretic framework. This framework may incorporate an objective function containing a set of parameters distributed across many unrelated network layers, many of which may not even physically reside in the same geographical area with one another. In this manner, end-to-end system characteristics may be optimised for a particular traffic type, provided that relevant aspects of system performance may be estimated in a manner timely enough to enable optimization. Macías and Suarez also foresee the introduction of traffic type adaption, so that networks not only optimize for the requirements of a particular traffic type such as streaming, but would also be able to sense subtle changes in traffic characteristics themselves.

Wireless Environments are Drastically Different from Wired Ones

Further related to the tremendous differences between the operational environment of the wired and wireless domains is the ability to quantify what the user may experience under certain conditions that may in fact be predictable. For example, the current state of the art for characterizing the voice quality experience makes use of Mean Opinion Scores (MOS), which are subjective evaluations on the part of the listener and therefore require a substantial number of observations in order to achieve reasonable confidence intervals. Common causes of user perceived degradation of application experience include changes in underlying network conditions. These include increases in Bit Error Rates (BER), handover artefacts, radio link CODEC rate adaptation to changing channel conditions and changes in system latency. Authors Vidales, Wältermann, Lewcio and Möller have developed techniques to extend the concept of MOS from the wireless voice domain to that of the user data domain. They introduce the Quality of Experience (QoE) abstraction for next generation services parametrically. This is done by constructing mappings between subjective a) MOS measurements and b) quantitative properties that can be acquired or objectively measured within the system. Using QoE is expected to streamline service evaluations, and therefore time to market of network equipment, devices, applications and services.

Scheduling and Routing

Next generation wireless networks will require highly effective scheduling. Author Ukil presents a comprehensive overview of 19 different scheduling techniques within a game-theoretic framework. Game theory has been used to predict behaviour of individuals and groups based on a set of preference functions, each party to the game having their own individual preference function and time horizon. As long as preference functions and time horizons are accurately identified, resulting behaviour of the parties may be predicted. Researchers have recently applied game theory to the analysis of network routing, radio link adaptation and scheduling. Author Ukil specifically concentrates on techniques to maintain overall system throughput while minimizing outage probability and maximizing the number of users obtaining service in heterogeneous QoS-bound traffic scenarios. Next generation scheduling is also the topic of authors Mohammed, Hashem, Gupta within the context of end-to-end coordination of system elements. They propose multi-level scheduling techniques, which would control both the radio access portion and core network portion of the systems. They present simulation results based on the existing HSPA system, and further propose a system architecture to incorporate two layer queuing and scheduling which is expected to provide lossless handover and QoS differentiation in future generation systems.

System Capacity and Data Rates

Radio system efficiency and information theory is at the heart of next generation wireless. Authors Raad and Huang discuss the characteristics of OFDM and present a comparison between OFDMA, Code Division Multiple Access (CDMA) and Time Division Multiple Access (TDMA) methods as a basis for introducing a new concept for information coding. The authors propose a Block Spread OFDM called Parallel Concatenated Spreading matrices OFDM (PCSM-OFDM). Their results show that this technique improves Bit Error Rate (BER) performance in frequency-selective channels by over 3dB by utilizing coding gain, and is similar to Turbo Coding. Coding efficiency is also the topic of research by author H. Chen, but within the context of CDMA systems. Chen introduces a new complementary code design called Real Environment Adapted Linearization (REAL), having inherent immunity against multipath interference and multiple access co-channel interference for both the uplink and downlink. Analysis demonstrates that REAL shows promise when contrasting its performance against the performance of traditional CDMA codes, such as Gold, Kasami, Walsh-Hadamard and Orthogonal Variable Spreading Factor (OVSF) codes. Authors Yang, Sun, Cavallaro, Zhu and Goel add scalability to the well-known Turbo Coding techniques. Their method uses scalable parallelism in order to enable coding/decoding at the high rates made possible by the requirements for the new generation air interfaces. They achieve scalable parallelism by using contention-free interleavers. They introduce a new address generation technique that supports multiple Turbo interleaving patterns and avoids requiring interleaver address memory typical of the traditional designs. The capacity of MIMO systems cannot be achieved unless by using an outer channel code concatenated with the space-time mapper acting like an inner code. Hence, authors Nikopour, Mobasher, Khandani, and Saleh proposed a list MIMO detector based on combining sphere decoder and m-algorithm approaches in conjunction with iterative turbo decoding.

Over the years, cellular systems have gone from being sensitivity bound to almost completely interference bound. Estimation of radio link and system capacity of a sensitivity bound system is fairly straightforward because it makes the assumption of a Gaussian-distributed noise floor. The case of an interference bound system is much more mathematically complex, as the underlying interference takes the form of a long-tailed distribution approximating a log-logistic curve at times. Authors A.C.G. de Carvalho Reis and P.R. de Lira Gondim present an analytical approach to capacity estimation of OFDMA networks using WiMAX as the example air-interface. The approach utilizes a new analytical method for computing Signal to Interference and Noise Ratio (SINR) based on the probability of the collision of subcarriers. The authors validate their results against published experimental data. As discussed previously, the greatest gains in wireless system capacity have been the result of aggressive frequency re-use by reducing the coverage of individual cells, (please see box, "About Wireless Capacity"). There exists a fabric of cellular tower installations throughout the developed world that is largely based on the delivery of voice services. While the number of cells and their coverage radii are far from completely static, it's questionable whether we'll see a large increase in cell density in the near future. Addressing data rates and system capacity by doubling or tripling cell density is just not economically feasible in most cases because of real estate availability, equipment and maintenance costs. Relays, in effect, may increase system capacity by adding small cells to a wireless network that not only extend coverage, but also increase SINR by their deliberate close proximity to the user. Authors Tooher and Soleymani explore ways to enhance radio system capacity by utilizing the concept of cooperative communications implemented as relays. They examine the differences between the amplify-and-forward approach and the decode-and-forward approach to relay design. The authors further extend the concept of the decode-and-forward method by introducing a cooperative approach to signal detection, fundamentally a distributed Multiple Input Multiple Output (MIMO) implementation where the reception and detection of information is split between multiple relay nodes. Traditional MIMO deployments utilize antenna spacing that is largely contained within some practical limits. Because these relays have substantial spatial separation when contrasted to traditional MIMO deployments, the channel transfer function may be dramatically different for the collaborative MIMO case implemented as relays, and therefore open many new possibilities for technical advancement such as collaborative coding. Because relays are a potential solution to the serious issue of system capacity, we're likely to see much more work in this area as we move toward the next generation.

Radio Over Fibre

In addition to the relay concept, Radio over Fibre (RoF) is another approach that may help solve the cell density issue for next generation networks. RoF is essentially the modulation and demodulation of a laser light source by a native RF signal, e.g. GSM, WCDMA, CDMA-2000. Rather than deploying base stations, the network would deploy remote antenna units that a) modulate a laser source with uplink signals within a specific RF passband and b) demodulate and amplify RF signals received from the fibre on the downlink. At a Central Office (CO), a bank of base stations would be connected to the other end of the fibre using a similar RF modulation/demodulation scheme. Authors Llorente, Morant and Martí explore this concept within the framework of next generation networks, and specifically some of the challenges and opportunities of RoF systems. For example, group delay can be problematic, but if managed properly it's possible to multiplex various frequency bands up to 150 GHz together on one fibre.

Challenges for Terminal Manufacturers

As of 2009, Long-Term Evolution (LTE) had greatest level of industry support as a next generation wireless technology. At mid-2009, some LTE standards were complete and a number of prototyping efforts and trials were underway among network equipment manufacturers, handset developers and network operators. LTE utilizes Orthogonal Frequency Division Multiple Access (OFDMA) technology for its air-interface, which provides both a number of advantages as well as challenges for the handset manufacturer. Authors Xu, Berkmann, Carbonelli, Drewes and Huebner provide a comprehensive overview to LTE air-interface technology and its continuing evolution LTE-Advanced (LTE-A) from the perspective of the handset. Mobile terminals must be small and portable, and a primary challenge for the terminal manufacturer is therefore battery life, which itself is challenged by the mathematical complexity of signal processing algorithms required by the LTE/LTE-A specifications. The authors provide an in-depth examination of system synchronization, timing estimation, equalization detection and other algorithmic requirements in terms of mathematical complexity and memory usage all within the context of mobile terminal implementation.

Management of User Mobility

Mobility management is fundamental to cellular technologies. The ability to select cells and handover calls and data sessions from one cell to another is the feature that made cellular frequency re-use a reality. Authors D. Kim, H. Lee, H. Yoon and N. Kim have identified some of the challenges to mobility control in the next generation, and they propose a bi-casting technique as an alternative or add-on to the hard handover method. Bi-casting is a technique that prepares multiple base stations for handover and requires each base station to keep bi-casted data until handover time, which reduces handover-related latency. Bi-casting techniques are well-known, but they're often cited as a challenge to signalling capacity and backhaul usage. The authors' proposed solution is one that conserves backhaul and signalling capacity. They supply evidence to support their proposal with simulation results.

Wireless Mesh Networks

Wireless Mesh Networks (WMN) have been a popular and promising research topic in recent years. These networks are based on the concept that multiple wireless devices such as user equipment and sensor nodes can act as dynamic routers. Authors Rizvi, Elleithy, and Riasat examine the possibility of applying Orthogonal Frequency Code Division (OFCD) to WMN systems. OFCD combine the multiple carrier aspect of OFDM with the multiple access mechanism of CDMA. Receiver design criteria and simulation results are presented that suggest a capacity gain in WMN systems by using OFCD, provided that the receiver is optimized correctly. The Wireless Mesh Network is also the topic of the work of authors Zhou and Kong, but from the perspective of MAC layer design for WiMAX. Their overview covers system architecture, frame structure, synchronization admission control, scheduling and scheduler performance and an analysis of Distributed Adaptive Timeslot Allocation (DATSA). Field trial results of a maritime mesh network system are also presented.

Additional Mobile IP Addresses

The main characteristic of 4th Generation (4G) Networks is being based on all IP architecture, operating mainly on IPv6. This includes services such as voice, video, and messaging. LTE is considered to be a 3rd Generation (3G) network and one of 4th Generation (4G) roadmap mobile access technologies. LTE-Advanced (LTE-A), on the other hand, is a 4G technology concept with evolving features. Therefore LTE is the key feature in the understanding of LTE-A evolution. The main focus of LTE is the enhancement of the packet-switched (PS) mechanisms on top of the UMTS enhancements, based on All IP Network (AIPN). IPv6 networking provides maximum service delivery flexibility, user decoupling, and scalability improvements, while leveraging the existing IETF standards. This requires major focus on network simplification, end-to-end delay reductions, optimal traffic routing, seamless mobility, and IP-based transport provisioning. Authors Mendahawi and Adibi provided an IPv6 perspective for LTE. The proliferation of Internet nodes has created a shortage of Internet addresses. Subscribers often have more than one Internet address, and we're moving into an era in which inanimate entities such as webcams and household appliances have Internet addresses as well. This issue is well-known, but within the context of mobile data systems, it's even more pronounced. Mobile Internet Protocol version 6 (MIPv6) is a fairly well-developed protocol with mobility control specifically designed in. Because it provides many more Internet addresses and is specifically suited to the mobility requirements of next generation networks, MPv6 is part of the roadmap for work on LTE/EPS in 3GPP. Authors Bălan and Sandu present a comprehensive survey of MIPv6 as it relates to the implementation in LTE.

High Altitude Platforms

High Altitude Platforms (HAPs) are planes or lighter-than-air craft operating in the stratosphere at altitudes of approximately of 20 km (about 75,000 feet). At these heights, the coverage areas provided by HAPs essentially split the difference between terrestrial systems and Low Earth Orbit (LEO) satellites, which operate at altitudes of around 600 km. The difference in altitudes provides a significant trade-off between LEOs and HAPs. The coverage of HAPs is smaller than LEOs, but the link margins for HAP systems are much higher for a given amount of uplink/downlink power. End-to-end delay for HAP systems is also substantially smaller for HAPs as well. Authors Mohammed, and Yang explore the potential use of HAPs for next generation systems, particularly for applications where large coverage areas are needed. This is important, because cellular systems are not well-suited to broadcast or multicast services, and it's often desirable to interface with a broadcast or multicast service from time to time, e.g. an emergency notification system.

Promises and Challenges

It's easy to look at the promises for the next generation. It's also easy to ignore many of the challenges that need to be addressed in order to commercially realize that next generation. Authors Govil and Govil present a comparison of the widely-deployed legacy wireless technologies to one another and to the 4G contenders, WiMAX and LTE. They make it clear that the next generation is not without some extremely serious challenges. They cite some major issues including mobility management – the ability to seamlessly control connectivity among heterogeneous radio interfaces is far from straightforward. 4G will require multi-technology vertical handover and system selection in addition to the fundamental mobility management functions. Even in 2009, we've experienced 2G/3G mobility issues just starting to manifest - as the adoption of 3G terminals continues, more users are able to notice 2G/3G handover and country boundary issues that didn't exist before. Other challenges include the practicalities of latency reduction, migration from conventional networks to the all-IP approach, system complexity, power consumption of the mobile terminal, spectrum issues, interference, intelligent billing and cost.

The next generation is all about possibilities and challenges – amazing possibilities, and some difficult challenges for sure. As researchers, we find the challenges to be an all important fuel for enthusiasm and innovation. Let's learn what we can from the experts who have contributed to this book and have helped to ease the industry forward into the next generation.

Mark Pecen
Research In Motion, Limited
June 24, 2009

REFERENCES

Sarkar, T. K.; Mailloux, R; Oliner, A. A.; Salazar-Palma, M.; Sengupta, D. L. (2005). *A History of Wireless Technologies*, Wiley-IEEE Press.

Gascoigne, G. (1974). *History of Radiotelephony and Telephony Telecommunications*, Ayer Co. Publishing.

Harte, L. (2006). *Introduction to Wireless Systems,* Althos Publishing.

Tabane, S. (2000). *Handbook of Mobile Radio Networks*, Artech House.

Freeman, R. L. (1997). *Radio System Design for Telecommunications*, Wiley Publishing.

Mouly, M.; Pautet, M.-B. (1992). *The GSM System for Mobile Communications*, Telecom Publishing.

Mehrotra, A. (1997). *GSM System Engineering*, Artech House, Inc.

GSM-02.03, Digital cellular telecommunications system (Phase 2+); Teleservices supported by a GSM Public Land Mobile Network (PLMN), In *European Telecommunications Standards Institute (ETSI), Global System for Mobile Communications (GSM) specifications.*

GSM-03.41, Digital cellular telecommunications system (Phase 2+); Technical realization of Short Message Service Cell Broadcast (SMSCB), In *European Telecommunications Standards Institute (ETSI), Global System for Mobile Communications (GSM) specifications.*

GSM-04.06, Digital cellular telecommunications system (Phase 2+); Mobile Station - Base Station System (MS - BSS) interface; Data Link (DL) layer specification, In *European Telecommunications Standards Institute (ETSI), Global System for Mobile Communications (GSM) specifications.*

GSM-02.60, Digital cellular telecommunications system (Phase 2+); General Packet Radio Service (GPRS); Service Description; Stage 1, In *European Telecommunications Standards Institute, (ETSI) Global System for Mobile Communications (GSM) specifications.*

GSM-03.60, Digital cellular telecommunications system (Phase 2+); General Packet Radio Service (GPRS); Service Description; Stage 2, In *European Telecommunications Standards Institute, (ETSI) Global System for Mobile Communications (GSM) specifications.*

3GPP Work Plan, (2009). 3rd Generation Partnership Project (3GPP) database files, Retrieved June 30, 2009, from http://www.3gpp.org/ftp/Information/.

Pecen, M.; Howell, A. (2001). Simultaneous voice and data operation for GPRS/EDGE: Class A Dual Transfer Mode (DTM), *IEEE Personal Communications*, 8(2), 14-29.

Rysavy, P. (2008). Mobile Broadband: EDGE, HSPA and LTE, *3G Americas whitepaper*; September.

3GPP FTP, (2009). 3rd Generation Partnership Project (3GPP) library site of all specifications and documents, Retrieved June 30, 2009, from http://www.3gpp.org/ftp/.

3G Americas, (2009). Retrieved June 30, 2009 from http://www.3gamericas.org/English/index.cfm.

3GPP TR 45.912 (2007). Technical Specification Group GSM/EDGE Radio Access Network; Feasibility study for Evolved GSM/EDGE Radio Access Network (GERAN), *3rd Generation Partnership Project (3GPP) Technical Report*, June.

David, K. (Ed.). (2008). *Technologies for the Wireless Future*: Wireless World Research Forum (WWRF) Book of Vision, John Wiley & Sons, Ltd.

IMT2000 (2009), *International Telecommunications Union (ITU)*, Retrieved June 30, 2009 from http://www.itu.int/home/imt.html.

ITU (2008). *Report of the Eighteenth Meeting of ITU-R Working Party 8F*, International Telecommunications Union, Retrieved June 30, 2009 from http://www.itu.int/home/imt.html.

ITU-R M.1645 (2008). Framework and overall objectives of the future development of IMT 2000 and systems beyond IMT 2000, *International Telecommunications Union*.

3GPP Rel. 8, (2009). 3rd Generation Partnership Project (3GPP), Release 8 Description, updated April.

3GPP Rel. 9 (2009). 3rd Generation Partnership Project (3GPP), Release 9 Description (draft), updated May.

3GPP TR 21.902 (2008). Evolution of 3GPP system, V8.0.0, *3rd Generation Partnership Project, Technical Specification Group Services and System Aspects,* Release 8, December.

IEEE (2009). IEEE Standards Association overviews, *Institute of Electrical and Electronic Engineers (IEEE)*, Retrieved June 30, 20009, from http://standards.ieee.org/.

Chandrasekhar, V.; Andrews, J. G.; & Gatherer, A. (2008). Femtocell Networks: A Survey, September.

Fuertes, A. (2009). EDGE Evolution: Competitive Air-Interface for the Next Decade, King's Park, New York: Visant Strategies Inc.

ITU-R 56 (2007). Naming for International Mobile Telecommunications.

ITU-R 57 (2007). Principles for the process of development of IMT-Advanced.

IMT-ADV/1-E (2008). Working Party 5D: Background on IMT-Advanced, Retrieved 7 March, 2008.

IMT-ADV/2-E (2008). Working Party 5D: Submission and Evaluation process and consensus building, Retrieved 7 August 2008.

Preface

This book presents a vision for the coming years, one that the editors of this book, active employees of Research In Motion (RIM) and AT&T Labs, and the collective chapter authors see ahead in terms of emerging fourth generation (4G) wireless technology trends and best practices. Each section explores the resulting challenges and technical opportunities that will arise in creating and delivering 4G networks for the emerging applications and services. The book also examines the fundamentals of advanced physical layer and radio resource management as the basis for cross layer and cross network optimization that will emerge for increased mobility and services in video, cloud computing virtualization, entertainment, education, health, and security.

The book editors have known each other for many years and they have worked together in various standards and university communities. They have organized this book especially for researchers, students, network engineers and designers and leaders of emerging companies, decision makers in standards, consumers, and product developers. Our goal is to provoke a discussion about the impact of orthogonal frequency division multiplexing (OFDM) technology trends in the coming years. This is essential in understanding a change, anticipating a change, even controlling a change, and ultimately making it work globally.

The fourth generation (4G) of cellular wireless is a successor to 3G and 2G standards. It is understood that 4G systems may upgrade existing communication networks and are expected to provide a comprehensive and secure IP based solutions where facilitate existing and emerging services such as voice, data, streamed multimedia, sensory, e-health, and social networking. It is expected that 4G will be provided to users on an "Anytime, Anywhere" basis and at much higher data rates compared to the previous generations. Moreover, the explosion of new applications over the mobile broadband networks (e.g., plethora of new BlackBerry and iPhone applications) has made it critical to have an efficient 4G network deployment to ensure widespread success of the emerging mobile broadband applications.

The springboard that we have organized for each section is a closer look at the dominant factors influencing the network topologies, their applications, and network practices in the next 2-5 years. These practices will transform the wireless evolution towards 4G world where seamless access to information will be the basic requirements in the emerging world.

The first part of this book is devoted to discuss future network architectures, deployments, and new technologies. This part deals with adoption issues and challenges of 4G systems. There are several advanced architectures that support efficient radio resource and mobility management for all-IP networking which is the ultimate goal of 4G wireless networks. These new designs provide possibilities for developing a next generation hybrid wireless network, which is discussed in detail in this part. This includes cross layer designs to provide quality of service, end-to-end delay reductions, seamless mobility, and IP-based transport provisioning. Several new technologies including cooperative services, services from high altitude platforms and radio over fiber are also considered as enabling techniques in hybrid networks. Moreover, some new business design models for 4G networks are introduced here, which they evaluate the market readiness for 4G and what requires supporting a next generation.

The second part of the book is devoted to radio access protocols, including; scheduling, quality of service, mobility, hand over issues, and cross layer designs. After a thorough review of radio access protocols, advanced multi-level scheduling techniques are introduced to control both radio access and the core network. An end-to-end quality of service control is considered for all-IP systems. Basic services such as VoIP, e-mail and web browsing are analyzed. Several challenges for streaming services are quantified, e.g. variations in inter-packet arrival time, or jitter, and the periodic loss and/or re-ordering of data. This part includes cross layer frameworks which incorporate several parameters distributed across network layers to optimize quality of service, mobility, and handover for different traffic types.

The third part of this book is devoted to advances in physical layer for 4G systems. Characteristics of OFDM along with a comparison between OFDM, code division multiple access (CDMA) and time division multiple access (TDMA) methods are the basis for this part. Several advanced multiple-access transmission schemes are introduced here, such as Block Spread OFDM or orthogonal frequency code division (OFCD). The OFCD is the combination of OFDM and CDMA which yields a significant improvement in 4G systems with the compensation of more sophisticated designs for transmitters and receivers. It is emphasized however that OFDM is the heart of wireless communication systems and CDMA can play a major role in 4G systems. The next generation of CDMA technology for 4G systems is also introduced based on a new complementary code design. Enhanced decoding techniques play a major role in the superior performance of an advanced physical layer. Reconfigurable and scalable turbo decoder architectures are presented here, which can be used in several standards with a common hardware module. In addition, new multiple antenna decoding techniques are introduced to enhance the joint iterative decoding in turbo decoding for 4G systems. Moreover, several new techniques are explored to enhance the radio system capacity such as interference measurement and cooperative communications by incorporating relays as part of physical layer enhances.

All these parts will lead to the emergence of what 4G will bring for us as the basis for a Virtual Society; a society in which everybody and everything will be virtually connected and linked to 4G mobile IP networks. Moreover, almost everyone and everything will be accessible from everywhere via the Internet using all sorts of devices.

Acknowledgment

The editors of this book would like to thank Mark Pecen, Jim Womack, Atul Asthana in Research In Motion (RIM) for their continuous support and encouragement during the preparation of this book. We thank Raj Jain (Washington University in St. Louis) and Shyam Parekh (Alcatel Lucent) for their helpful discussions and recommendations.

Our gratitude goes to IGI Global staff for the preparation of the book, especially Christine Bufton, David DeRicco, Erika Carter, Jan Travers, Jennifer Weston, and Neely Zanussi. We are also grateful of Vicki Livingston and Krissy Gochnour in 3G America, LLC for supporting this book.

This editorial book would not have been completed without the help of our reviewers: Shirook Ali (RIM), Hadi Baligh (Huawei), Alireza Bayesteh (RIM), Cagatay Buyukkoc, (AT&T), Masoud Ebrahimi (RIM), Hamid Farmanbar (Nortel), Nada Golmie (NIST), Farzaneh Kohandani (RIM), Heunchul Lee (Stanford), Xinhua Ling (RIM), Behrouz Maham (UNIK), Subhas Mondal (WIPRO), Biplab Sikdar (RPI), Pouya Taaghol (Intel), Wei Wu (RIM), Yi Yu (RIM). We appreciate their great help and support.

The second editor would also like to thank numerous friends and colleagues at the Stanford University for providing such a creative, intellectually stimulating and fun environment during his stay there. In particular, he would like to thank Prof. Arogyaswami Paulraj, Prof. John Cioffi, Heunchul Lee, Tae-Min Kim, and Bernd Bandemer.

The support and help of Prof. Amir K. Khandani in University of Waterloo is also greatly appreciated.

Last, and most importantly, we are indebted to our families. Their invaluable and relentless supports, encouragement, and love are without doubt the most important reasons behind our successes.

Sasan Adibi
Amin Mobasher
Tom Tofigh

Section 1
Network Architectures

Roadmaps and Architectural Models

Chapter 1
Evolution of Personal Wireless Broadband Services from 3G to 4G

Sudhir K. Routray
Krupajal Engineering College, India

ABSTRACT

The chapter covers the basic conceptual model of the 4G system and the operations of its physical systems. It starts from the very basics of the wireless communication services, and then the author goes though the different standards of the systems which provide wireless broadband services, like the 3G, and other wireless broadband systems, like WiMAX etc. The author then looks through the 3GP project and its visions, then goes through the 3GPP2. The vision and the achievements of the 3GPP LTE are then discussed including the 4G and its successor systems. After that the authors turn towards the technical ideas behind the wireless broadband services like the 3G and 4G. 4G system architecture and its features are looked into, the differences between the 3G and 4G are discussed, and then the whole chapter is concluded with the impact of the 4G system on the present mobile communication scenario.

1. INTRODUCTION

The mobile communication scenario throughout the world is growing in geometric progression. After the Internet, this is perhaps the second most important invention in globalizing the world. In almost every country, from the very advanced to the deepest corner of the world, we find mobile services everywhere. It has revolutionized the way people live and think. The 'connectivity everywhere' concept has given some extraordinary dimensions to the business community. Specifically, mobile computing related applications are like a treasure for business community and business related services.

Looking at the present demand, it is clear that the mobile communication systems needed to be evolved. In the first generation of mobile communications the services were mainly based on the analog communication techniques. However, there were many shortcomings in the analog version, so digital communication techniques were used and it was found that digital modes can handle different

DOI: 10.4018/978-1-61520-674-2.ch001

aspects properly. The second generation of the mobile systems was known as 2G. Of course the systems or technologies used in different countries were different. In Europe the GSM technology was quite popular and now it has stretched over the whole globe. GSM was nothing more than the TDMA system of the European countries with pan-European roaming facilities. It was a great revolution that a single phone could be used in different countries and thus was the main choice in many countries. In America CDMA emerged as the main competitor of the GMS. Though the CDMA technology was not as flexible as the GSM, it had some other advantages over the GSM technology, like easy dynamic channel allocation. The competition between these two technologies gave rise to many new standards and technologies. Some other versions like the Japanese PDC were also popular regionally, but later they parted with either of the two main technologies.

From the competitions between these technological systems it was expected that the evolution of the systems was going to happen sometime around the year 2000. That too happened. The GSM group or the European standards making body ETSI, was quite optimistic about that. So they proposed a project for the future of the GSM technologies. Then some other SDOs, or Standards Development Organizations, joined them from across the globe, and the project then went on to develop a new technology which we know as the UMTS or the Universal Mobile Telecommunications System. This technology that enabled GSM to leap forward and allowed high data rate services to become possible in the GSM framework. The new system that was developed through the UMTS is well known as the 3G system. Now almost all the developing countries are have 3G facilities. After 3G, people expected even more. There were some common bottlenecks in the 3G system and the rivals were trying to have a better system. They too have another project where the ITU or the International Telecommunications Union was taking the leader's role. They developed the IMT 2000. At the same time the air interface access technologies were evolving very fast. WCDMA was one of them. Similarly, CDMA2000, TD-SCDMA and their hybrid versions made it easy to provide the high broadband services. With all of these ingredients in place, business parties in the mobile communication systems started designing the next generation mobile systems. In the early years of this decade the 3G system emerged and people welcome it with enthusiasm. Although the UMTS-based 3G and the IMT2000-based 3G had some differences, they were very similar as far as the performances and the features are concerned. So it can be said the technologies are heading towards a common framework in which the better aspects of either will stay.

The ITU-supervised IMT2000, however, recommended the systems which were 3G technologies having some little differences. There are five such 3G radio interfaces.

The 3G systems were good enough in comparison to the 2G or the GPRS enabled 2.5G systems, but internet related services and anything that needed higher bandwidth were either out of reach, or the quality of service was not good enough. This led to demand for a system which could give some proper solution to these needs. 4G was first proposed by the team who were the members of the 3GPP LTE (Third Generation Partnership Project -Long Term Evolution). The front runner among the business group was the NTT DoCoMo of Japan, the first of many business partners in this project.

2. PERSONAL WIRELESS BROADBAND COMMUNICATION SYSTEMS

There are many personal communication systems evolving these days. Out of them the broadband services are the most popular ones. There are many kinds of broadband frameworks available today for public use, from traditional cable or wired broadband to more sophisticated systems. The most sophisticated ones are the wireless broadband services. There are many wireless broadband services like 3G, WiFi, WiMAX, etc. The main aim of these technologies is to provide broadband or high quality services to the customers at the best possible means. Of course, different standards are followed for different systems.

The initial 2G or digital mobile communication systems were using lower bandwidth and the data rate was only 16 kbps (though at first it was even lower). Then the initiative was taken to provide higher data rate services through the addition of the GPRS. GPRS was mainly used for non-voice services and the maximum data rate possible was around 160 kbps. But the GPRS was not the cure. So EDGE or the Enhanced Data Rates for GSM Evolution was adopted. It can be considered as the earliest broadband on the GSM framework. Some people consider EDGE to be 3G; however, this is not true, as EDGE is the 2.7G on the GSM infrastructure.

There was a need for mobility in the local area network as well, which led to the creation of the wireless LAN. There were many challenging issues in establishing this WLAN. In order to achieve the proper mobility and flexibility a new system came into existence, which is well known as WiFi or Wireless Fidelity. In WiFi, mobility and the range of coverage are limited; though it can provide broadband services wirelessly, the data rates are not very high. Nevertheless, it was very popular initially in the corporate sectors and universities. The main difficulties in the WiFi systems are the limited security and the difficulty of channel management.

The 3G services were a grand success for both the researchers and the service providers. The coverage areas were not limited and the mobility was not restricted. But the data rates were not enough for the video or other rich file formats. However it is still a great achievement and many countries are still trying to adopt this technology.

In the corporate sector, the need for mobility was immense, and WiFi based systems were not able to fulfill the demand. Because of this, there was a need for a technology which could take care of the mobility and provide a large range of coverage. Eventually that challenge was met and WiMAX, or Worldwide Interoperability for Microwave Access, was born. WiMAX boasts a high range of coverage, satisfactory mobility, and improved security compared to wireless communication. Sometimes WiMAX is confused the WiBro, but in reality there are some small differences, including the standards upon which they are based. WiMAX has two different versions: one is a fixed version, used mainly for static wireless LAN, and the other is a mobile version. WiMAX also incorporates OFDMA technology to combat the ISI and other co-channel and intra-channel interferences.

3. STANDARDS OF WIRELESS BROADBAND SYSTEMS

As we have seen in preceding sections, there are many wireless broadband technologies. In this section we shall have a look at the standards upon which they are based. Here it is worth discussing why we need the standards. There are many systems working in different countries, and in order to facilitate global communication or even local communications there must be proper compatibility between those systems. The MODEMS or other CODEC devices are designed according to those compatibilities and data rates. So there is need to follow a common standard throughout the world to avoid all these disparities and mismatches. IEEE makes standards for the communication technologies. For wireless communication and LANs also there are different standards, which utilize different protocols for proper communication.

WiFi or Wireless Fidelity used IEEE 802.11 as its backbone standard, the first version of the which was released in 1997 and made public in 1999. It specified two net bit rates of 1 or 2 megabits per second (MBPS) in addition to the forward error correction code. It also specified three alternative physical layer technologies: diffuse infrared operating at 1 MBPS; frequency-hopping spread spectrum operating at 1 MBPS or 2 MBPS; and direct-sequence spread spectrum operating at 1 MBPS or 2 MBPS. In the latter two, radio technologies used microwave transmission over the Industrial Scientific Medical frequency band at 2.4 GHz. Since the initial release IEEE 802.11 has undergone many revisions and amendments.

Unlike WiFi, WiMAX uses the IEEE 802.16 as its main standard. Of course these standards have been set forth by different research groups and the SDOs. The first 802.16 standard was approved in December 2001, and was set for wireless broadband transmission in the 10-66 GHz band, with only a LOS or line-of-sight operation. It also recommends a single carrier physical standard, which was updated into different versions over the course of the later meetings. These different versions account for the discrepancy between WiBro and WiMAX, as well as the different types of WiMAX (i.e., fixed or mobile). The amendments of these standards have given rise to better versions. The current WiMAX standard is the IEEE 802.16e, which enables mobility in WiMAX.

These IEEE standards mainly configure different physical and MAC layers for the communication, but they are also responsible for the collaboration between the upper and lower interfaces of these layers.

4. WHAT IS 3GPP?

The development of the GSM technology unified the Europe with a bond and realized a true European Union. Following the success of GSM, there was a need for a broadband system based on the GSM framework. This led to European countries looking forward to an advanced personal broadband service, which they called the 3G. So overall, the 3G standard is nothing more than personal broadband services on the GSM core network. Of course, the technology behind the 3G system is an advanced version known as the UMTS or the Universal Mobile Telecommunication System. Its main competitor is the IMT 2000, supervised by ITU.

4.1 Members of 3GPP

There are six organizational members of the 3GPP project. These members are the regional standard development organizations of different countries or continents. They are:

- European Telecommunications Standards Institute (ETSI, Europe)
- Association of Radio Industries and Businesses (ARIB, Japan)
- Telecommunication Technology Committee (TTC, Japan)
- China Communications Standards Association (CCSA, China)
- Alliance for Telecommunications Industry Solutions (ATIS, North America)
- Telecommunications Technology Association (TTA, South Korea)

The project was planned in 1997 and work started in December 1998. This 3GPP project started their work for the future of GSM. First the GPRS came in 2000, then the EDGE or the 2.7G (also the EGPRS) came in the year 2003. Later in the same year 3G came to the market commercially. Of course, the experimental versions of these technologies had come much before their public use. NTT DoCoMo first tried the experimental version of 3G in October 2001. Then in the US, the IMT 2000 based CDMA2000 1x EV-DO came into existence in 2003.

So overall, we can say that 3GPP is a European project in collaboration with the other leading members of the world, which seeks to create post-GSM networks on the GSM framework with the goal of meeting future mobile needs. The main contribution in the post-3G era which is coming to the public is 4G.

5. WHY 3GPP2?

3GPP2 stands for Third Generation Partnership Project 2. This project was started in the year 1998 when the then-new standard IMT2000 or the CDMA2000 was under development. The standards for this technology and its future were found to be really vast, resulting in the need for collaboration among the different standard making bodies around the world. The name 3GPP2 was derived from the 3GPP, and the supervising organization for the 3GPP2 was the ITU or the International Telecommunications Union.

The main motto of 3GPP2 was to form a collaborative third generation (3G) telecommunications specifications for global mobile communication. Because the Europeans were involved in 3GPP, they were not that interested in 3GPP2, so the 3GPP2 project is comprised of North American and Asian leaders of CDMA 2000. They are developing global specifications for ANSI/TIA/EIA-41 Cellular Radio telecommunication Intersystem Operations network evolution to 3G and beyond. The aim is also to develop and set the complete global specifications for the radio transmission technologies (RTTs) supported by ANSI/TIA/EIA-41.

3GPP2 was born out of the International Telecommunication Union's (ITU) International Mobile Telecommunications IMT-2000 initiative, covering high speed, broadband, and Internet Protocol (IP)-based mobile systems featuring network-to-network interconnection, feature/service transparency, global roaming and seamless services independent of location and mobility. The main idea behind this project is to bring high-quality mobile multimedia telecommunications to a worldwide mass market by achieving the goals of increasing the speed and ease of wireless communications. This is in response to the problems faced by the increased demand to pass data via telecommunications, and providing "anytime, anywhere" services.

As we have mentioned above, the concept of a "Partnership Project" was given by the European Telecommunications Standards Institute (ETSI) early in 1997 with the proposal to create a Third Generation Partnership Project (3GPP) using the Global System for Mobile (GSM) technology and its future versions. Although discussions took place between ETSI and the ANSI-41 community with a view to consolidating collaboration efforts for all ITU members, in the end it was deemed appropriate that a parallel Partnership Project be established - 3GPP2 which, like its sister project 3GPP, embodies the benefits of a collaborative effort (timely delivery of output, speedy working methods etc.). At the same time there was a need of recognition as a specifications-developing body, providing easier access of the outputs into the ITU after transposition of the specifications in a Standards Development Organization (SDO) into a standard.

5.1 Members of 3GPP2

The primary members of the 3GPP2 project are the Standard Development Organizations of both the east and west. In the west the leading players are the European and American standard development organizations, and in the east the front runners of mobile communication include countries like Japan, Korea and China. These countries' standard development organizations are participating in this project, whose five main members are:

- TIA - Telecommunications Industry Association (North America)
- TTA - Telecommunications Technology Association (Korea)

- TTC - Telecommunications Technology Committee (Japan)
- ARIB - Association of Radio Industries and Businesses (Japan)
- CCSA - China Communications Standards Association (China)

The project is a not only a collaboration of standard making organizations, but also an opportunity for experts of the technology creating organizations, like the IPv6 Forum, to share their advice. Besides the above SDO organizations, the technology building partners are also active players of this project. They are:

- The CDMA Development Group (CDG)
- IPv6 Forum
- MobileIgnite
- Femto Forum

The 3GPP2 recommended 3G systems are:

1. IMT-2000 CDMA Multi-carrier, also known as CDMA2000 (3X) developed by 3GPP2. (IMT-2000 CDMA2000 includes 1X components, like CDMA2000 1X EV-DO.)
2. IMT-2000 CDMA TDD, also known as UTRA TDD and TD-SCDMA. TD-SCDMA is developed in China and supported by TD-SCDMA Forum.
3. IMT-2000 TDMA Single Carrier, also known as UWC-136 (Edge) supported by UWCC.
4. IMT-2000 DECT supported by DECT Forum.

6. UMTS AND 3G

UMTS is well known as the technology of 3G; many people say it is the 3GSM. Basically, the UMTS technology is one of the advanced mobile communication technologies for wideband or broadband operations on the GSM infrastructure and its evolved versions. In its modern 3G version UMTS uses WCDMA as the air interface multiple access method.

UMTS system was mainly using the GSM framework for two reasons: first, to retain the existing customers in the same network, and second, to provide new dimensions which are able to give broadband services in personal communication systems.

At present all the major 3G Networks are using the FDD (frequency division duplex) mode of operation. As far as time division duplex is concerned, there are no commercial TDD networks at the moment, although some of the network service providers like T-Mobile have announced that they would install TDD Network in Europe and North America. The reason why TDD has not been used yet is the difficulty in its implementation; when both of them (TDD and FDD) stay together, the situation becomes even more complex.

4G became necessary when it was found that the 3G was lagging behind the competing technologies like the WiMAX. The 4G project and its road map was not very much related to the WiMAX development, but it was later found to be a basic need for the 3GPP members to come out with a better technology. Though there is not much difference between them, it is quite obvious that the 3GPP members do not want to lag behind.

The migration of the 2G to 3G was so smooth in the European countries that many people did not even realize that a transition had taken place. Just using the 3G handsets they were able to get the advanced services from their same service provider. Of course, UMTS subscribers differentiated from GSM subscribers based on a SIM card. For UMTS and GSM subscriber the SIM is different; UMTS subscribers use USIM while GSM subscribers use SIM. However, if the service provider wants, it can provide the 3G services on the same existing GSM SIM. Similarly, in case of the 4G it is expected that the UMTS's advanced version may make it quite flexible for the change from 3G to 4G.

7. 3GPP LTE/3GPP2 LTE AND THE ROAD AHEAD

In the above sections, we have seen different forms of the 3G technologies and their amendments. Recent revisions of both UMTS (3GPP LTE) and IMT-2000 (3GPP2 LTE) have recommended new dimensions to their technologies for the future growth. The seventh revision (M.145707) was approved by the ITU Radio communication Assembly in October 2007. The seventh revision was by far the most radical in the history of IMT-2000; for the first time, an entirely new radio interface was added. This sixth radio interface, entitled "IMT-2000 OFDMA TDD WMAN", introduced OFDMA technology into IMT-2000. At the same time, Revision 7 included 3GPP2's OFDMA-based UMB technology as well as an initial framework description of 3GPP's OFDMA-based "Long-Term Evolution" technology, in two different forms. As a result, Revision 7 of M.1457 suddenly contains six radio interfaces, of which four include OFDMA. These four radio interfaces share a number of other technological features as well, including support of packet-based (IP) networks. This radical change of technologies, broadly supported across a variety of independent standards organizations, marks the global recognition of the initiation of 4G mobile communications. IMT-2000 has expanded beyond its 3G origins.

7.1 IMT-2000 OFDMA TDD WMAN and Its Foundation: IEEE Standard 802.16

IMT-2000 OFDMA TDD WMAN is the version of IEEE Standard 802.16 that is specified in the WiMAX Forum Mobile System Profile. A summary description of the radio interface is provided in document IEEE L802.16-06/031r2, with which IEEE initially proposed the addition of a subset of 802.16 (designated as "IP-OFDMA") to ITU-R.

IEEE Standard 802.16, the Wireless MAN Standard for Wireless Metropolitan Area Networks, has been under development and evolution since 1999. The standard has always, as a fundamental design principle, supported differentiated QoS to allow a mix of simultaneous multimedia services on a single network. It originally supported only "fixed" (stationary) terminals, but it was enhanced for full mobility with the introduction of the IEEE 802.16e amendment, which was approved in 2005. The WiMAX Forum ("WiMAX Forum") has developed the WiMAX Forum Mobile System Profile to specify a particular version of the standard that could be tested for certification purposes. Certified products were announced in April 2008. IEEE Standard 802.16 is developed, maintained, and enhanced by the IEEE 802.16 Working Group on Broadband Wireless Access. The Working Group currently has 433 members and meets six times a year, with attendance recently running over 400. IEEE 802.16's expertise in the pioneering OFDMA technology is deep. The Working Group introduced OFDMA into its fixed-access standard with the amendment 802.16a in 2003. This was based on standardization work that began with contributions on OFDMA introduced into the Working Group in the year 2000.

The 802.16 Working Group is currently developing a revision of the base standard and all the subsequent amendments. Completion of this revision draft, unofficially and temporarily known as 802.16Rev2, is expected in late 2008. While the revision project is being completed, the Working Group is continuing its progress on the developments of three further amendments:

Recently, project 802.16h is developing improved coexistence mechanisms for license-exempt operations and for the further flexibilities in the system. Similarly, project 802.16j is developing a multi-hop relay specification and will specify a relay station that can communicate with mobile terminals. This will offer a valuable new tool to system operators for extending range and capacity.

Project 802.16m is developing an advanced air interface, as described in more detail below.

7.1.1 The Amendment in 802.16 and IMT-Advanced

After the IEEE 802.16e amendment was completed (mainly for the WiMAX), members of the IEEE 802.16 community began to consider how an enhanced version of IEEE 802.16 could satisfy the emerging requirements of IMT-Advanced. In late 2006, following a significant effort, the Working Group was authorized to develop the IEEE 802.16m Project, which has the stated scope of amending the IEEE 802.16 Wireless MAN-OFDMA specification to provide an advanced air interface to meet the cellular layer requirements of IMT-Advanced next generation mobile networks while providing "continuing support for legacy Wireless MAN-OFDMA equipment." This was an extraordinary initiative for better mobility.

While the ITU's view of the IMT-Advanced process timeline has varied over time, the Task Group's view of the 802.16m project schedule has remained mostly independent. The basic intent of the project, and the planned 2009 completion date for the 802.16m amendment, have remained constant.

The 802.16m task Group has generated a set of system requirements that reflects the evolving IMT-Advanced requirements but also adds unique demands. Primary among these additions is a requirement for support for legacy Wireless MAN-OFDMA systems. The 802.16m Task Group has also developed an extensive "Evaluation Methodology Document". The Task Group is currently developing a System Description Document (SDD) before generating the draft standard. The primary purpose of the SDD is to allow alternative technical approaches to be assessed and agreed before detailed specifications are added to the draft standard. More information on the 802.16m and IMT-Advanced is available elsewhere.

7.1.2 Flexibility of IEEE 802.16 Technology for the Evolution of 4G

As mentioned previously, one of the key requirements of the IEEE 802.16m project is strong legacy support of Wireless MAN-OFDMA mobile stations and base stations. Fortunately, the flexibility of Wireless MAN-OFDMA allows for the possibility to satisfy these requirements. This flexibility is a distinct benefit of OFDMA technology and is a key reason that industry has turned to OFDMA for 4G. User demands for higher-rate services can be met partially by greater spectral efficiency, which is a benefit of OFDMA and of MIMO antenna technology that can be easily supported by OFDMA. Another way to increase throughput is to apply greater spectral bandwidths. OFDMA, because it subdivides the channel into many narrow sub-channels, is extremely scalable to broad as well as narrow channels, with little effect on the spectral efficiency. The technology is readily adaptable to multiple frequencies and to both paired and unpaired bands, using FDD and TDD duplexing, respectively. The technology lends itself to adaptability at the ASIC, allowing the possibility of very adaptable devices. Some ASIC designers are taking advantage of these features to provide chipsets that can operate with a broad range of channel bandwidths, sub-carrier counts, frequency ranges, and duplex methods.

The flexibility of IEEE 802.16 also provides new opportunities to operators regarding the services they wish to provide. Novel differentiating features can be introduced using the same basic network technology. For example, NextWave Wireless has introduced the MXtvTM mobile multicast and broadcast technology that runs using a portion of the time and frequency resources available on a normal two-way WiMAX network. This is another illustration of why it is difficult to define 4G from a service perspective. 4G technology, such as IEEE 802.16, will support a wide range of innovative services and applications.

7.2 4G Prospects of IMT 2000

Given the vast success of 2G systems, the 3G market has developed relatively slowly. Even though the original 3G standards were developed in the 1990s, global 3G operators typically report that, as of 2007, fewer than 10% of the customers are using 3G equipment. These operators have invested heavily in 3G technologies that are only recently beginning to fulfill their potential. In many cases, they see 4G not as an immediate prospect but as a long-term evolution that will require another round of investment, not only in the radio access equipment but also in the core network.

On the other hand, a number of other companies are ready to move forward with 4G on an earlier time scale. In general, those that are unburdened by legacy requirements in all or part of their spectra are more likely to see 4G as the best investment for mobile broadband networks. A number of these companies worldwide are implementing IEEE 802.16 Mobile WiMAX networks in 2008 and 2009.

The 4G mobile communication systems are based on several fundamental technology differentiators, including OFDMA and packet transport. 4G mobile communications was pioneered by IEEE Standard 802.16. The international community has recognized the transition to the new 4G technologies by approving Recommendation ITU-R M.1457-7 in October 2007.

7.3 LTE: Moving Towards Mobile Broadband

The LTE (Long Term Evolution) project has tried to bring out novel and efficient mobile system for the public use since its inception. At first it focused on a broadband service built on the GSM framework. Though it did not succeed, the UMTS technology and the system that came out of it was 3G, which is quite good in comparison to its predecessors.

The intention and future steps of LTE are quite clear now; the main aim is to achieve personal broadband services in all its standards. The LTE has already achieved 3G technologies and it has been used abundantly in most parts of the world where mobile communication is already advanced, it. The next project they are now looking forward to achieve is 4G, the advanced personal mobile communication technology that can enable the high data rate broadband communication in the personal service provider's network. It is expected that sometime between 2012 and 2014 this new technology will be able to provide the services in the more technologically advanced countries. The aforementioned frequency allocation as well as the other features like the TDD and FDD will be implemented, which will result in optimized services at a rate 100MBPS or more.

While the intentions and the expectations of 4G projects are partially clear now, the projects that will follow 4G are highly speculative. The 5G, or successor of 4G, would be a very different kind of system from the intelligence and operational point of views. It is difficult to predict what the features and requirements of a 5G system would be; however, in course of time the scenarios will gradually become clear. 5G would not only be a fast broadband system, but it would also incorporate many advanced technologies that are beyond our current imagination.

In brief, we can say the Long-Term (Radio) Evolution or LTE is also part of 3G technology. It is a 3GPP research item for the Release 8, also known as 3.9G or "Super 3G" by some researchers. It is planned to be commercialized in 2009, with an aim at peak data rates of 20 Mbps (for Down Link) and around 10 Mbps (for UL).

8. FEATURES OF 4G WIRELESS SYSTEMS

There are numerous features of 3G which needed to be modified for the future applications of the mobile communication networks and their service sectors. These collections of advanced versions, along with some others new advanced features, have been proposed for the forth coming 4G system. Of course, the complete picture is not clear yet, though it is believed that by the end of 2010 a better view of the complete 4G features will be known. Here we have captured the main features of the existing experimental 4G system and some of the essential future versions.

As per the announcements of the 4G working groups, the infrastructure and the transreceiver terminals of 4G system will have almost similar structures like that of 2G and 3G except for some advanced features. The previous legacy systems will be in place to keep the existing users. The major change in the infrastructure for 4G will be "all packet-based system" and the technology on which it will be based is the IPv6. There are some other proposals for an open platform in which the new innovations and evolutions of the future can fit. One of the first technology really fulfilling the 4G requirements as set by the ITU-R will be LTE Advanced as currently standardized by 3GPP. LTE Advanced will be an evolution of the 3GPP Long Term Evolution. The higher data rates needed are for instance, achieved by the aggregation of multiple LTE carriers that are currently limited to 20MHz bandwidth and there are many such changes have been recommended. In the following sections we have listed some of the important features of the 4G systems.

8.1 OFDM Based Physical Layer

The main aim of 4G technology is to Provide high speed wireless broadband services. Airport lounges, cafés, railway stations, conference arenas, and other such locations are required to have high speed internet services; in those places, 4G can play an important role. 4G is equipped with the proper arrangements at the physical layer to meet all the demands of those various scenarios. There are many difficulties, however, in providing high speed wireless internet services in these environments, such as multipath fading and the inter-symbol interferences generated by the system itself. As a result, OFDM technology is used to handle this problem.

8.2 Inter-Symbol Interference Due to Time Delay

In a multipath environment, the signals and their delayed versions arrive at different times. When the time delay between the different delayed signals is a large enough fraction of the transmitted signal's symbol period (actual time allotted for one symbol transmission), a transmitted symbol may arrive at the receiver during the next symbol period. This is well known as inter-symbol interference (or ISI). At higher data rates, the symbol period or duration is shorter; hence, it takes only a small time delay to introduce ISI. In case of broadband wireless, ISI is a big problem and reduces the quality of service significantly. In conventional situations, statistical equalization is the method for dealing with ISI, but at high data rates it is quite complex and requires considerable amount of processing power. OFDM appears as a better solution for controlling ISI in broadband systems like 4G.

OFDM deals with this problem in a very intelligent way by introducing a guard interval before each OFDM symbol. This guard interval is the duration in which no information is transmitted. Digitally, it is nothing but a certain number of zeros transmitted between each couple of symbols. Whatever signal comes during that interval is discarded by the receiver, but when the guard interval is properly chosen then the OFDM signal can be kept undistorted.

8.3 Effective Use of Bandwidth through OFDM

OFDM has the ability to optimize the consumption of resources. Extraneous bandwidths in the form of guard bands can, with proper implementation,

Figure 1. OFDM and bandwidth use

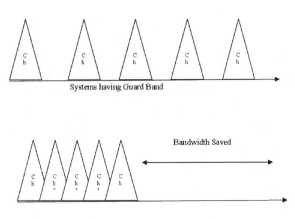

be reduced to zero. Due to the orthogonal nature of the carriers used for different channels, it is possible to overlap the bands on each other and still recover them in the receiver without losing any quality. Because of this, OFDM is very effective in saving bandwidth. In low bandwidth systems where the demand for spectrum is very high, OFDM comes naturally as the first choice. The bandwidth saving has been shown in Figure 1.

Besides the above advantages, OFDM based systems provide other facilities for digitalization and protocol supports. Processes like error correction and interleaving are easily supported by OFDM.

8.4 Software Defined Radio for the 4G System

Software defined radio (SDR) is an emerging radio technology that can be used in various digital networks and can be controlled and programmed through software.

According to the SDR Forum, SDR technology is "radio that provides software control of a variety of modulation techniques, wide-band or narrow-band operation, communication security functions (such as hopping), and waveform requirements of current and evolving standards over a broad frequency range." SDR has the ability to support wireless applications in various networks like Bluetooth, WLAN, GPS, radar, WCDMA and GPRS.

Currently, all of the major operations such as modulation, demodulation, coding, decoding, interference management, channel allocation and capacity management are done through the control software. One of the biggest advantages of the SDR is that it can ensure a secure communication network through implementation of encryption systems like the AES (Advanced Encryption Standard). This means that SDR is very reliable and useful for military and other high-level, secret communications. Due to these features, SRD is the most suitable method of data handling at the higher levels in 4G. With the link protocol standards now moving into 3G and 4G, networks differ dramatically in many ways. This is a big problem for both consumers and service vendors; while it can be handled through upgrading the handset, upgrading is usually not a good choice due to the high cost. Additionally, the wireless network operators face many interfacing problems during the migration of a network from one generation to another. Finally, the use of incompatible systems in different countries can hinder global communication. Through the use of SDR, all these scenarios can be handled smoothly.

The SDR system uses a generic hardware platform which has its own programmable units, microprocessors, digital signal processors, field programmable gate array and analog RF modules. The software modules of the SDR that implement link layer protocols and modulation/demodulation operations are called radio applications, and these applications provide link layer services to higher layer communication protocols such as WAP and TCP/IP. SDR has the ability to significantly reduce the life-cycle costs and can also support advanced capabilities in different portable networks. The SDR technology is also reconfigurable; it allows several software modules to co-exist, and also permits dynamic configuration on the handset as well as in the back-end equipment. As a result of this flexibility, the problem of discrepancies due to legacy handsets is solved, and the extra cost for a new handset is not required. SDR can also handle the implementation of multi-mode, multi-band and multi-standard terminals. All of these demonstrate that SDR is clearly the most desirable technology for 4G.

8.5 MIMO Antenna Systems for 4G

4G like its predecessor 3G would use the advanced versions of the MIMO Antennas. The antennas used for the 3G system were smart enough to take care of many advanced operations at the signal level. This system must continue for 4G as well, and may even be made more sophisticated for 4G, as the number of signal-level decisions would be far greater in the case of 4G compared to 3G.

8.6 IPv6 Based Packet Transmission

The all-packet infrastructure is quite popular in the wireless communication, and now it is also true for the 4G systems as well. The biggest difference between 3G and 4G is the all-IP network (AIPN) structure of 4G, which means that all communication will be controlled by TCP/IP protocols. As a result, the whole communication will be packet switched and the circuit switching part will be taken out of this advanced version. Not only can this make the system compatible with all digital devices, but internet access will be quite flexible and high data rates can be achieved. According to the 3GPP LTE team, this target will be achieved by the end of 2008. Similarly, the 3GPP2 LTE teams are also busy trying to keep pace with their competitors.

8.7 Presence of TDD and FDD

TDD (Time Division Duplex) and FDD (Frequency Division Duplex) are different modes of CDMA. In FDD transmission mode, both the transmitter and the receiver transmit simultaneously. This simultaneous transmission is possible because they are both on different frequencies. In TDD mode of operation either transmitter or receiver can transmit at one time. This is because they use the same frequency for the transmission.

At present all the major 3G Networks are using FDD mode of operation, but in the 4G system both the FDD and TDD will co-exist.

In the FDD mode of operation, the uplink and downlink use separate frequency bands. These carriers have a bandwidth of 5 MHz and are divided into 10-ms radio frames; each frame further id divided into 15 time slots. The frequency allocation consists of one frequency band at 1920-1980 MHz and one at 2110-2170 MHz. These frequency bands are used in FDD mode both by the UE (user equipment) and the Network. The lower frequency band is used for the Uplink (UL) transmission and the upper frequency band is used for the Downlink (DL) transmission. The frequency separation is specified with 190 MHz for the fixed frequency duplex mode and with 134.8MHz to 245.8MHz for the variable frequency duplex mode.

The TDD mode differs from the FDD mode in that both the uplink and the downlink use the same frequency carrier. There are 15 time slots in a radio frame that can be dynamically allocated between uplink and downlink directions. Thus the channel capacity of these links can be different which is very advantageous especially when people are downloading stuff on their mobiles. The chip rate of the normal TDD mode is also 3.84 Mbps, but there exists also a "narrowband" version of TDD known as TD-SCDMA. The carrier bandwidth of TD-SCDMA is 1.6 MHz and the chip rate 1.28 Mbps. TD-SCDMA has been proposed by China and potentially has a large market-share in China if implemented.

8.8 Self-Organizing Characteristics of 4G

The resource and duty management operations of 4G would be quite different from the present scenario. Extensive automation in the system and self-organizing characteristics can create an intelligent management. This is a quite strange feature unique to 4G that could lead the system to a complete new level.

8.9 Two-Tier Coverage

In case of 4G, the geographical coverage would consist of at least two tiers. The normal coverage would be through normal macro cells, but in order to handle the traffic and resources properly during the peak-hours, microcells would be kept in place. Depending on the traffic distribution, the transmission and control duties are switched to the appropriate cells. In some hot spots the coverage layering would be composed of multiple layers to improve the quality of service and resource management.

8.10 4G Uplink and Down Link Frequencies (Proposed)

Though the spectrum of 4G is still under planning, we have a rough idea about the uplink and downlink frequencies from the early developers. OFDM is used to divide the whole spectrum or bandwidth into thousands of small narrow bands, each having different frequencies. By doing this, the system becomes resistant to multipath fading and thus capable of providing better quality of service.

The 4G system also uses OFDMA for the downlink and single carrier FDMA (or SC-FDMA) for the uplink. It optimizes the data rate by using four MIMO antennas per station, which we have seen can provide tremendously high data rates. The channel coding schemes are chosen to be suitable for the OFDM signals. Turbo codes are preferred in this application.

8.10.1 Downlink

The OFDM system for the downlink uses maximum 2048 subcarriers. The subcarrier spacing in OFDM downlink is 15 kHz. The mobile device must have the ability to receive all the 2048 subcarriers but the base station needs only 72 subcarriers for transmission. The transmission is divided into subframes of 1.0 ms duration and each time slot is of 0.5 ms duration. The net length of a radio frame is 10ms. For downlink the popular modulation formats are QPSK, 16 QAM, 64 QAM and 256 QAM. The spectrum for the downlink has not been finalized; but it is expected to be wider than the mobile WiMAX and in the similar range of WiMAX.

8.10.2 Uplink

For uplink, the proposed multiplexing method is SC-FDMA, and proposed modulation methods are QPSK, 16 QAM and sometimes 64 QAM. SC-FDMA is used to suppress the high PAPR, as in the case of OFDMA. For high data rates the constellation size may go up to 256 QAM. Of course it is still under review and the current road map is considering 64 QAM as the proper choice. Uplink spectrum of 4G would be in the same range as the WiMAX but it would have more bandwidth for faster data rate.

9. ARCHITECTURE OF 4G

At this moment it is very difficult to predict the exact architecture of the 4G mobile communication system. Looking at the present scenario of the 3G and the likes of WiMAX etc. we can only predict the probable architecture of the fourth generation architecture. However in labs and on an experimental basis there are already some of the architectures available for the 4G. Of course, with the advances of the technology in both the UMTS and the CDMA 2000 and their evolved versions the architectures will be updated. Here some of the experimental architectures and some of the predicted models of the 4G architecture have been presented.

9.1 The OSI Model for 4G

The best way to represent any communication system architecture is the OSI model, and here the probable OSI model of 4G model has been presented (Figure 2) with the understanding that may be some differences in specific future 4G systems. The OSI model of 4G can properly explain the different operations of and the underlying technologies. It is similar to the various layers found in the OSI model of internet, but as a result of basic differences they are arranged in a different fashion and some of the layers are absent. Here the physical layer and the MAC (medium access control sublayer) are quite important.

The OSI model of 4G depicts the different layers and their functions in a proper sequence. The physical layer, or bottom one, deals with the signals in the OFDM format. Above it lies the transmission convergence sublayer (CS), which is between the physical layer and MAC layer. On top of that layer is the MAC layer, which has three sublayers. The uppermost layer of MAC, the convergence sublayer, supports both ATM services as well as IP based services. In 4G the MAC layer at the base station (BS) is responsible for the allocation of bandwidths to different users both in the uplink and downlink. MS only occasionally takes the control of bandwidth allocations when it has multiple sessions or it has connections with the BS. This is quite different from other services and ensures better quality of service. Most of the services of 4G would be IP based; as a result the optimization and QoS related improvements are done as per the IPv6 configuration and structure. ATM service facilities are also provided for the compatibility with other existing networks.

When we look at the first release of WiMAX standard in 2001, the IEEE 802.16 standard proposed applications for a fixed wireless scenario in licensed frequency bands in the range between 10 and 66 GHz, where the use of directional antennas were mandatory to obtain satisfactory performance figures. But difficulties were encountered in metropolitan sub-areas where line-of-sight operations cannot be ensured due to the presence of obstacles, buildings, towers etc. Thus, subsequent amendments to the standard (IEEE 802.16a and IEEE 802.16-2004) have extended the 802.16 air interface to non-line-of-sight applications in licensed and unlicensed bands in the 2–11 GHz frequency range. Additionally, after the revision of IEEE standard document 802.16e,

Figure 2. The OSI model of 4G

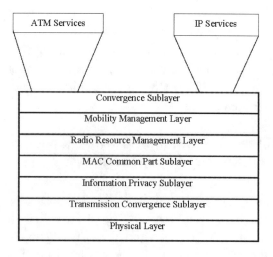

The OSI Model of 4G

some necessary mobility support will be provided. Revision of IEEE 802.16f is intended to improve multi-hop functionality, and 802.16g is supposed to deal with efficient handover and improved QoS. This revision also increased the range of WiMAX technology; according the WiMAX forum, it can reach up to a theoretical 50 Km coverage radius and achieve data rates up to 75 Mb/s. Of course, actual IEEE 802.16 equipment is still far from these performance figures, but it has been proved that with the use of MIMO antennas and OFDM based technologies the data rates can be made really high. For example, with 5 MHz bandwidth, a data rate of 18 MBPS is possible using this advanced MIMO technique. After looking at the success of these technologies in WiMAX, the 4G development research groups are ready to follow the same path. Most of the settings of 4G would be according to the IEEE 802.16m standards, and the new MAC layer bridging is waiting for some amendments of IEEE 802.16k.

Duplexing or bidirectional data transmission is provided by means of either Time Division Duplexing (TDD) or Frequency Division Duplexing (FDD). In TDD, the frame is divided into two sub-frames, one devoted to downlink and the other to the uplink. A Time-Division Multiple Access (TDMA) technique is used in the uplink subframe. The BS is in charge of assigning bandwidth to the SSs, while a Time Division Multiplexing (TDM) mechanism is employed in the downlink subframe. In case of FDD, the uplink and downlink sub-frames are concurrent in time, but are transmitted on separate carrier frequencies. There are supports for half-duplex FDD SSs, at the expense of some additional complexity. Each subframe is divided into physical slots. Each TDM/TDMA burst carries MAC Protocol Data Units (PDUs) containing data towards SSs or BS, respectively. The transmission convergence sublayer operates on top of the physical layer and provides the necessary interface with the MAC. This layer is specifically responsible for the transformation of variable length MAC PDUs into fixed length physical blocks. Here in 4G, the RR and the MM layers are different from the GSM RR and MM layers. Here the use of MIMO enabled antennas can manage the resources quite efficiently.

There is a big demand for secure data transmissions, which has led to the native inclusion of a privacy sub-layer in 4G (which is very similar to the WiMAX), at the MAC level. There are some well-organized protocols to take care of the security related processes. Those protocols are responsible for encryption/decryption of the packet payload, according to the rules defined in the standard. IEEE 802.16 uses a wireless medium for communications, and one of its main targets of the MAC layer is to manage the resources of the radio interface in an efficient way, while ensuring that the QoS levels negotiated in the connection setup phase are fulfilled. Of course the IEEE 802.16 MAC protocol is connection-oriented and is based on a centralized architecture. There is a need for segmentation and resemblance of frames for proper security monitoring of all the packets. The common part sublayer is responsible for this segmentation and the resemblance of MAC service data units (SDUs), the scheduling and the retransmission of MAC PDUs. The common part sublayer also provides the basic MAC rules and signaling mechanisms for different system access, bandwidth allocation and connection maintenance. The core function of the protocol is bandwidth requests/grants management. A SS may request more bandwidth, by means of a MAC message, to indicate to the BS that it needs (additional) up-stream bandwidth. The request of bandwidth is processed on a per-connection basis to allow the BS uplink scheduling algorithm, to consider QoS-related issues in the bandwidth assignment process. The bandwidth granting methods as per the original 2001 standard encompassed two operational modes: Grant per Connection (GPC) and Grant per Subscriber Station (GPSS). Later in the 2004 release, the term "grant" refers only to the GPSS mode. Whereas, in the GPC mode, the BS allocates different scalable bandwidths to individual destinations. With this revision, BS got the control of all the centralized mechanisms, with all the intelligence placed in the BS, while the SSs act as merely passive stations. On the other hand the bandwidth, in the GPSS mode, is granted to each individual SS, which is then in charge of allocating the available resources to the currently active flows.

Convergence sublayer (CS) is the uppermost sublayer of the MAC layer. The CS associates the traffic coming from the upper layer with an appropriate Service Flow (SF) and Connection Identifier (CID) which gives the idea about its destination. The CS also provides payload header suppression when some entity is sent and reconstruction at the receiving entity. CS delivers the resulting CS PDU to the MAC Common Part Sublayer to confirm the negotiated QoS levels.

The 4G standard defines two different Convergence Sublayers for mapping services to and from IEEE 802.16 MAC protocol like the WiMAX. The ATM convergence sublayer is there solely for ATM traffic, while the packet convergence sublayer is specific for mapping packet-oriented protocol suites, such as IPv4, IPv6, Ethernet and Virtual LAN etc. The IP Sublayer is as the name suggests is there to provide all IP enabled services. The classification of IP traffic and ATM traffic is done in the CS sublayer. However the system's IP architecture will be based completely on IPv6.

10. DIFFERENCES BETWEEN 3G AND 4G

Though the 4G is the evolved version of 3G there are many fundamental differences between the two systems, the greatest of these being speed enhancement of 3G network services.

The main differences between the 3G and 4G have been listed below.

10.1 The Speed or Data Rate

The maximum speed or the data rate of 3G system is 2MBPS. With that speed we can definitely get a better service than the 2G system or its advanced versions like the GPRS (where the maximum possible speed is 204 KBPS). Although 3G substantially enhanced the data rate, the data contents of the present wireless broadband services make it impossible to carry on far with the 3G system. 4G was the solution to overcome that bottleneck of the 3G technologies. In 4G the aim is to get a data rate of 100MBPS or even more in the indoor environment. In theory, 4G will be at least 50 times faster then the 3G system; this is the main difference between the 3G and 4G technologies. In other words, the bandwidth of 4G system would be much higher than the 3G system.

10.2 Packet Switched Infrastructure

In 4G, the whole network will be packet switched. The IP based infrastructure will be used for the 4G system exclusively. IPv6 is the version on which the whole protocol system will govern the different kinds of switching for the data transfer. 4G switching aspects will therefore be more sophisticated and complex than the 3G system.

10.3 Quality of Service or QoS

The quality of services in 4G networks is going to be much better than the 3G and its contemporary technologies. The improvement in main service factors will be due to the high broadband of the 4G systems, the improved quality of service of the IPv6 systems compared to previous IP versions, and better reception and transmission services from the smart antenna based MIMO system.

10.4 Network Security

Network security is another important aspect, one for which people are ready to pay. Network security in 4G is different from that of the 3G or 2G versions, in part because the security provided through the 4G networks is made up of two tiers. That means not only the MSC authentication is required but along with that there are some adder securities. We have seen the added security arrangement of the 4G systems in the OSI model. The Information Privacy Layer is there to take care of the security related aspects of the information that is exchanged in the 4G networks.

10.5 Management of Resources

The resource management in 4G is much better than 3G. Optimization is present in the 3G system, but most of the optimizations are not that adaptive and dynamic. In contrast to that, 4G would have very smart adaptations in the resource management sector. Adaptive algorithms are used to provide optimization everywhere, from the modulation and coding, to the individual scalable channel bandwidth allocation.

10.6 Differences Between the WiMAX and the 4G

Some people mistakenly believe that 4G and WiMAX are the same technology and only the application domains of the two are different. Here we would like to clarify that 4G and WiMAX are different types of technologies and the standards for the technologies are also different in many respects. Of course, there are some similarities between the two. The truth is that the WiMAX business group and the people who are working with the WiMAX technology want to popularize WiMAX as the 4G wireless technology, leading people to think that WiMAX and 4G are the same thing. In reality it is not accepted by the research community and the technology development group of either technology.

Like 4G, WiMAX can deliver high data rate (up to 70 Mbps over a 50Km radius). However, as mentioned in the earlier sections, with 4G wireless technology people would like to achieve up to 1Gbps (indoors). Additionally, WiMAX technology (802.16d) does not support mobility very well. To overcome this mobility problem 802.16e, or Mobile WiMAX, is being standardized in the same way as 4G. Ultimately though, when it comes to performance and speed 4G remains the winner.

The important thing to remember here is that all the research for 4G technology is based around the orthogonal technology OFDM. WiMAX is also based on OFDM, and this fact gives more credibility to those in the WiMAX lobby who would like to term WiMAX as a 4G technology. It is worth mentioning that 4G is going to be an ultramodern, high performance mobile system. Most of the advanced features of 4G are due to OFDM, MIMO and other advanced technologies. It is seen as a hybrid system which is based on both 3G and other advanced systems, including WiMAX.

11. THE IMPACT OF 4G ON OTHER SYSTEMS

The impact of 4G is obvious on contemporary wireless broadband services. Many researchers think that the 4G system would be nothing but a modified version of WiMAX due to the similarities in the architecture to a large extent, and because mobile WiMAX is able to handle wide range mobility. That is not true, however; in the previous sections we have seen that there are a lot of differences between a WiMAX system and a 4G system. Regardless, the impact of 4G is quite encouraging for all the mobile research community. Not only will the 3G GSM core network be evolved, but it would also be a tremendous system with a lot of advanced features for the future and capable of successfully handling the increased need.

The biggest impact will be on the business community, which is spending a huge amount for the 4G research in the hopes of receiving a good return. There is no doubt that 4G can generate a huge amount of revenue, but the timeline and other socio-political issues like spectrum distribution and health effects are making things complex. 4G would change the communication scenario to a large extent. Broadband services in the wireless mode would be appreciated by a large number of customers. Video transfer through wireless IP would be revolutionized.

Another interesting scenario for the 4G systems would be in the mobile computing arena. The 4G handsets would be nothing but some kind of advanced palmtops with broadband connection, so we can imagine what is going to happen and what kind of service can be expected. Business will go really mobile. That means that the true m-commerce would come to the market and true ubiquitous computing could be realized. Of course, the overall effects of 4G are now just predictions whose realities will only be witnessed in the future.

12. CONCLUSION

4G technologies would change many things in the technical world and every other sector. The migration of the mobile communication technologies from 3G to 4G would be more attractive than the switching from 2G to 3G. Both the 3GPP and 3GPP2 research groups are busy preparing their 4G road maps. The most important thing in the 4G would be the convergence of competitors' technologies. Though it has happened never before from 1G to 3G, 4G would bring the groups to a single table to have a unified platform of mobile communication along with some other contributing technologies like WiMAX. The unification of standards like the advanced LTEs and IEEE 802.16m can give rise to a global framework for wireless broadband and a unique, advanced composite radio environment for 4G.

4G would also be an intelligent system, and its performance can easily show what extra would be needed for its next version 5G. True mobile internet and mobile computing would hit the market once the 4G comes into existence. 4G would be more energy efficient than any other wireless networks which have come before due to its robust resource management. The evolution of personal wireless broadband from 3G to 4G is a great achievement and it is going to be a milestone in the history of communication.

REFERENCES

Armuelles, I., Robles, T., Madrid, S., & Chaouchi, H. (2004). On Ad Hoc Networks in the 4G Integration Process. In *Proc. Of The Third Annual Mediterranean Ad Hoc Networking Conference 2004*.

Ballon, P., Helmus, S., & Van de Pas, R. (2002). Business models for next-generation wireless services. [Mobile internet]. *Trends in Communications*, *9*, 2002.

Cianca, E., Sanctis, M., & Ruggieri, M. (2007). Convergence Towards 4G: A Novel View of Integration. S*pringer . Wireless Personal Communications*, *33*(3-4), 327–336. doi:10.1007/s11277-005-0577-y

Glisic, S. A. (2004). *Advanced Wireless Communications - 4g Technologies*. Chichester, UK: John Wiley and Sons.

3GPP. (n.d.). *UTRAN Iub Interface: General Aspects and Principles.* [Technical Specification 25.430]

3GPP. (n.d.). *UTRAN Iub Interface: Layer 1.* [Technical Specification 25.431]

3GPP. (n.d.). *UTRAN Iub Interface: Signaling* [Technical Specification 25.432].

Gunasekaran, V., & Harmantzis, F. (2005). Migration to 4G-Ubiquitous Broadband-Economic modeling of Wi-Fi with WiMAX. In *Proc. Of WWC, SFO,* USA.

Rubio, M. L., Garcia-Armada, A., Torres, R. P., & Garcia, J. L. (2002). Channel modeling and characterization at 17 GHz for indoor broadband WLAN. *IEEE Journal on Selected Areas in Communications, 20,* 593–601. doi:10.1109/49.995518

Takada, J., Fu, J., Zhu, H., & Kobayashi, T. (2002). Spatio-temporal channel characterization in a suburban non line-of-sight microcellular environment. *IEEE Journal on Selected Areas in Communications, 20*(3), 532–538. doi:10.1109/49.995512

WiMAX Forum. (n.d.). Retrieved from http://www.wimaxforum.org

Yano, S. M. (2002). Investigating the ultra-wideband indoor wireless channel. In *Proceedings of IEEE 55th Vehicular Technology Conference,* Vermont (Vol. 3, pp. 1200–1204).

Yu, W., & Lan, T. (2005 December). Transmitter optimization for the multi-antenna downlink with per-antenna power constraints. *IEEE Transactions on Signal Processing*.

Chapter 2
A New Global Ubiquitous Consumer Environment for 4G Wireless Communications

Ivan Ganchev
University of Limerick, Ireland

Máirtín S. O'Droma
University of Limerick, Ireland

Jený István Jakab
Tecnomen Ltd, Ireland

Zhanlin Ji
University of Limerick, Ireland

Dmitry Tairov
University of Limerick, Ireland

ABSTRACT

A changed wireless environment for 4G and future generations of wireless communications is addressed in this chapter. This change is primarily focussed on making the end user of wireless services more central and more a consumer in the global wireless environment than heretofore. In the 'Ubiquitous Consumer Wireless World (UCWW)'–the descriptive name for this new wireless environment paradigm– the global supply of wireless communications services is founded on the new Consumer-centric Business Model (CBM). This is a radical change and departure from present globally pervasive business model for service delivery based on the user being a subscriber, called appropriately the Subscriber-centric Business Model (SBM). The reasons and background for the drive to bring about this changed wireless environment are reviewed, with the main body of the chapter focusing on descriptions of the technological composition of two of the new core infrastructural enabling elements. These are the third-party authentication, au-

DOI: 10.4018/978-1-61520-674-2.ch002

thorization, and accounting (3P-AAA) service, and the service advertisement, discovery, and association (ADA) through newly defined wireless billboard channels (WBCs). The former, 3P-AAA, arises from the need to bring about a separation of the supply of the AAA service from the supply of the communications service, which is necessary to ensure the consumer character of the user and to promote and safeguard all the new benefits that will flow for users, new consumer-oriented wireless access network providers, and other stakeholders through this new wireless environment. As there will be restructuring implications for the operation and location of charging and billing functions, treatment of this aspect is also included. The latter, ADA & WBCs, arises in tandem with this separation of services, the consequent metric of business success changing from 'number of subscribers' to 'number of consumer transactions and service purchases' and the need, therefore, for a new direct 'push' advertisement means for service providers to attract consumers and for consumers to be continually up-to-date on new service offerings. The proposals for protocol interfaces and architectures for both these elements are explored and discussed, with those aspects needing to be addressed in global standardization activities highlighted.

INTRODUCTION

The migration towards next generation mobile communications, often referred to as 'Beyond 3G' (B3G), 4G, or wireless NGN, probably is one of the most challenging research topics of today's wireless world. For some 4G means higher-capacity new radio interfaces. For others it is an 'all-IP' integrated wireless environment. For others again it is being conceived as a new wireless environment encompassing existing, planned and future mobile and fixed wireless networks, both terrestrial and satellite, with harmonious internetworking. Most presume that this internetworking, which of its nature should be heterogeneous internetworking, will be network-driven, and will be seamless and largely transparent to the users. Some others more recently argue the opposite, that it will be user-driven, and seamless and transparent to the networks.

It is becoming difficult, however, to escape the dominant generation-defining category of an order of magnitude, or thereabouts, increase in data rates, with other attributes, e.g., all-IP, heterogeneous network integration, full multimedia service delivery, playing a less significant role in defining a new generation in wireless communications.

For 4G and for all further generations of wireless communications, one may also address the techno-business wireless environment for delivery of wireless services. This is the environment in which all new and existing generations of wireless communications will function, deliver services, make money for their owners, and also reflect, and be responsive to, socio-economic concerns of users and communication services' stakeholders. It is this underlying wireless environment – and its potential to evolve through architectural change in its core foundational infrastructural – which is the theme of this chapter.

The principal network provider goal in the evolution of the wireless world, as may discerned over its lifetime to date, is to sell services to a great common market of mobile users in overlapping local, regional, and global domains. Today, and typically for some this may be referred to as defining 4G, the connectivity and services should be available anytime, anywhere, and anyhow through the best connection to the user. Services should be rapidly deployed on-demand, customized to the user's needs, and adapted to the current wireless and connection environment. Accordingly, system and service integration is regarded as an important step towards the future 4G paradigm of 'Always Best Connected and best Served' (ABC&S)

mobile users (O'Droma, M. S., & Ganchev, I., 2004; O'Droma, M., et. al., 2006). This entails the interaction of different access technologies, each of which being optimized to operate in dedicated environments and conditions.

In 4G, mobile users' desires for greater choice and customization in accessing telecommunications services may be perceived as something greater than just a generation thing. Rather it easily be understood, and claimed, that this should apply to all present and future generations of wireless communications. Thus, it would better be categorized as a wireless environment attribute. Today users' freedom of choice, however, has yet to transcend the constraints of the legacy subscriber era and acquire the attributes of modern consumer freedoms. Their desires for user-controlled flexibility in service delivery, e.g., having it fully and automatically tailored to the user's needs, and for full portability freedom, e.g., the ability to move/migrate quickly to more competitive service providers offering better price/performance options or a wider selection of services, have yet to be addressed, not to mind be satisfied, through a formal re-appraisal and infrastructural overhaul of wireless communications from a techno-business perspective.

This chapter addresses some specific key elements of the new generic Consumer-centric Business Model (CBM), the construction and realization of which–with appropriate standardization–will be vital in creating a new wireless environment for upcoming 4G wireless communications systems evolution. 'Ubiquitous Consumer Wireless World (UCWW)' is the descriptive name for this new wireless environment paradigm. CBM may be considered as a natural evolution of the legacy Subscriber-centric Business Model (SBM). It may also be consider as a potential substitution or replacement for it, but this need not necessarily be the case as it is clear that both can co-exist. The characteristic claims of CBM-UCWW are its capacity to enable realization of truly consumer-oriented, user-friendly and user-driven ABC&S wireless communications environment, its full zero-cost anytime-anywhere-anyhow portability for consumers, its level playing field and relatively low entry-cost for new wireless access network providers (ANPs), its zero-cost international roaming, and its creation of new telecommunications services markets for such services as third-party authentication, authorization and accounting (3P-AAA, pronounced 'three-P-triple-A'), and wireless billboard channels (WBCs) used for advertisement and discovery of wireless services deployed in a particular area/location. The CBM-UCWW presentation here is complemented further by a description of the corresponding 4G reference communication model which facilitates the representation of cross-layer communications.

Supporting examples of the transition trend from SBM to CBM networking are presented on Figure 1. These include: cell phones with multiple (U)SIM slots (so the user can choose which card to use for each particular service), ticket-based Wi-Fi access (e.g., at hotels, airports, etc.); CBM-like products such as 'metakall' (www.metakall.com), and the opening up the wireless carriers' networks to third-party software, developers and mobile handsets (Perez, M., 2008).

The rest of the chapter is organized as follows. The next section presents the research background in the field of integrated networks, their architectures, and relevant business models. Then the new CBM and roles of business actors are considered. This is followed by detailed descriptions of some of the new core CBM-UCWW components. Future research directions are then discussed followed by a conclusion section summarizing the benefits of the proposed new 4G business plan.

BACKGROUND

A key solution in facilitating the fast growth and success of any new communications environment is building the foundation for business building blocks. Currently it seems that the many downsides

Figure 1. The 'SBM to CBM' transition trend

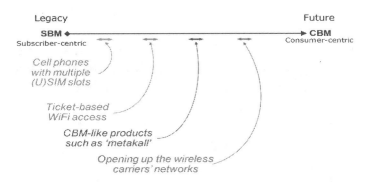

in wireless networking systems results from the rather constraining nature of SBM. For many access network providers (ANPs), it is the only really workable, underlying business model. However, when considered next to CBM, one could discern its weaknesses in providing the stimulus, the motivation and the glue for a real-life integrated 4G networking. For CBM to become a reality, it itself will have to have a business attraction, and not be something imposed. This will probably also have to come from outside the major cellular wireless ANPs as, initially at least, CBM may be perceived to be antagonistic to their business interests.

When addressing how to open the mobile communications environment to a wide range of new entrants, whether coming with new or old technology, whether filling niches in network accesses or provision of services, or challenging major ANPs with more modern, reconfigurable, adaptable systems, and so forth, it is clear that a new business foundation free from the constraints of the subscriber business culture is called for. In attempting to define such a new business foundation for the global wireless environment, appropriate standardization and regulatory support needs to be considered. Without such support, ensuring a global footprint for this new environment is not possible.

Since 2002 there has been significant research performed in the field of future 4G integrated networks, including architectures and business models. In 2003, a number of new projects were started within the frameworks of the European Union's Information Society Technology (EU-IST). Some examples are MOBY DICK, BRAIN/MIND, SCOUT, ETSI BRAN, MOBIVAS, and the Academic Network for Wireless Internet Research in Europe (ANWIRE). Further research was carried out in this area by different work groups (WG) of standardization bodies and forums such as the WG2 and WG3 of the Wireless World Research Forum (WWRF), and the ITU-T NGN Focus Group (http://www.itu.int/ITU-T/ngn/fgngn). Much of this research was focused on system and service integration in heterogeneous network environments creating new integration architectures founded on an all-IP infrastructure. This characterized the main trust of '4G' as perceived formally within the EU, differing in this to the definitions and perceptions of what 4G was in other economic regions. Within the ANWIRE project and the WWRF WG2 especially, some research work was supported on business models. An interesting aspect of this work was their investigations in to how to place the system users more at the centre of communication systems evolution.

In the ANWIRE project, researchers progressed this idea also by addressing the monopoly positions of the network providers –supporting the traditional subscriber-centric model–, and how this facilitated, or otherwise, the evolution and advanced development in cellular wireless networks and services. They found that regardless of the technological approach taken, in such systems, the subscribers are practically limited by the technology and service environment available with their subscriptions. For instance, the capability to use services from other service providers is inherently very limited. Such constraints have serious implications for future wireless access evolution. Much of the ANWIRE contributions on these matters have their source in the Telecommunications Research Centre (TRC) in the University of Limerick. The initial proposal for the new business model –Consumer-centric Business Model (CBM)– to remove such limitations by creating a new extra-network entity in which to locate the Authentication, Authorization and Accounting services originated from there, in 2004, (O'Droma & Ganchev, 2004). This new wireless environment entity, of which there could be many, was called the Third-Party Authentication, Authorization and Accounting Service Provider (3P-AAA-SP). The original model was later refined in (O'Droma & Ganchev, 2007). Also in the ANWIRE project, a Generic ANWIRE System and Service Integration Architecture (GAIA) was proposed and elaborated, (Ganchev et al., 2006); it was specifically referring to it as a new 4G architecture as then understood within the EU thinking. How this could form a basis for CBM was elaborated there. GAIA employs the principles of the "always best connected and best served" (ABC&S) paradigm (Ganchev et al., 2006). The ideas were further refined and re-positioned less as a specifically 4G technology and more as a new wireless communications environment and business foundation for 4G and for further wireless technological generations. Given, among other reasons, its radical placing of the user at the centre and the user-empowerment with normal consumer attributes unavailable to subscriber-users, this wireless environment was called a Ubiquitous Consumer Wireless World (UCWW), (O'Droma & Ganchev, 2007).

UCWW AND THE NEW CONSUMER-CENTRIC BUSINESS MODEL (CBM)

SBM and CBM: Underlying Differences and Technical Foundations

The new CBM model is probably best appreciated by contrasting it in with the legacy SBM model. SBM is founded on the user being a subscriber of a wireless access network provider (ANP), normally a cellular operator, usually designated the 'home ANP'. Of course, through multiple (U)SIM cards a user could be a subscriber to multiple home ANPs. All other ANPs are termed 'foreign ANPs' (Figure 2). Other forms of wireless access, e.g., Wi-Fi hotspots, when they offer communication services using the AAA of users' cellular home ANP for payment, become another form of foreign ANP.

The home ANP acts, through its AAA infrastructure, as both the effective manager and supplier of the mobile users' wireless access communications services. It may also supply some teleservices[1]. Teleservice providers (TSPs) can offer and sell their own services through ANP networks enabled by bilateral business agreements (full arrowed lines), e.g., for profit sharing. Mobile users normally have credit accounts with their home ANP(s) and receive from them itemized bills

Figure 2. The legacy network-centric, constrained SBM model

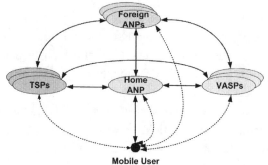

for all wireless access services, whether obtained locally or while roaming, through their home or foreign ANPs. Services obtained from other entities (dashed arrowed lines), besides foreign ANPs (for roaming and international call support), i.e., TSPs and perhaps value added service providers (VASPs), are more and more billed through these accounts.

The technical foundation of the 'subscriber' character of the wireless world today derives directly from the legacy fixed telecommunications world. At its core are the network ownership of the local loop and the legacy teleservice of telephony, especially the incoming call connection (ICC) service. The SBM is based simply on the user having a well-defined physical single point of attachment to the network. Whether mobile or fixed, it is mapped to a unique, globally significant, network-wide identity, i.e., a telephone number. The mobile user contracts with the home ANP for long-term connection to this point of attachment, through the unique identity, and thus receives services. Hence, the user is naturally identified as a "subscriber" and the user's unique identity is really jointly owned by the home ANP, at least for the duration of the contract.

The benefits and advantages of the SBM approach are apparent in today's everyday experience of wireless networking. Especially that of being able to have communications while on the move was really a phenomenal and radical paradigm shift from fixed network connectivity only. However, such benefits are taken for granted, and as they are the many downsides of SBM become ever more apparent. A subscriber of one ANP, for example, cannot easily access more attractive price/performance options for certain services of other ANPs. Changing ANPs –even with number-portability (whether legislated for or voluntary) – is problematic. Using a multiple (U)SIM card solution (e.g., to bypass roaming charges or get local charges applied for local calls) is cumbersome and problematic especially in respect of being contactable, i.e., receiving incoming calls.

Another example is roaming tariffs, which are perceived predominantly as being not cost-based and detailed information on them is usually not readily accessible, resulting in difficulties for economic management of wireless services when travelling.

Moreover, while the fixed point of attachment is a reality in the fixed communications infrastructure, it is not so in wireless communications. In wireless networks, the local loop only comes into existence for the duration of a service session. Thus, it is a *virtual* local loop, in that the apparent fixed local loop is using physical resources shared with others and which vary depending on the user location, but otherwise seems to have all the telecommunications' attributes of a fixed local loop. Furthermore, the used local loop medium is not owned by the wireless ANP, even if aspects of it, e.g., frequency bands, are leased to ANPs. It is in exploiting this rather simple distinction of the physical technical realities of the local loop that makes possible a new type of wireless communications business model. In effect, this is the underlying technical foundation for the new 'consumer' approach - the CBM, which in turn leads to a new wireless environment - the UCWW.

CBM Solutions and Recommendations

In CBM, a key infrastructural characteristic is the separation of the administration and management of users' authentication, authorization, and accounting (AAA) activity from the supply of a wireless access network service, and its re-location to trusted third-party AAA service providers (3P-AAA-SPs). The 3P-AAA-SP entities become the major players in CBM, Figure 3. With the assignment of this key role to 3P-AAA-SP entities the distinction between home ANP and foreign ANP disappears; all are simply Access Network Providers (ANPs). The various service-offering entities (ANPs, TSPs, and VASPs) will have arrangements only with the 3P-AAA-SPs (full arrowed lines)

Figure 3. The new user-centric CBM solution

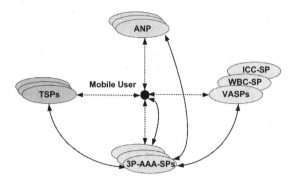

through which payments for services purchased by users (dashed arrowed lines) are transacted. This new structure creates the potential for the wireless access network market to be much more 'open' for both users and new ANPs and TSPs.

This approach and facility would be particularly attractive to new ANP entrants, and to existing ANPs and TSPs trying to extend their market share, streamline their business, fill niche ABC&S markets, and so forth. CBM has the potential to open the wireless communications market more, by providing easier and fairer access to ANP and TSP markets for both providers and users alike.

Also it has inherent business attributes to drive forward the evolution of ABC&S networking, as both competition, interoperation, and collaboration advantages will be underpinned by efforts to provide a wide range of quality of service (QoS) offerings with greater flexibility and with a wider range of price/performance ratio options in the efforts of providers to attract in greater numbers of service users, (O'Droma et. al., 2006). Enabling this will mean the growth of much new business seeking to serve and support this ABC&S capability.

CBM gives much greater freedom of movement to consumer-users. The legitimate modern consumer expectation of being able to move back and forth readily among ANPs at any time, for any and all services, is an inherent attribute under CBM. This is radical contrast with subscriber mobility in SBM. A further key enabler for this is *number ownership*. This inherently implies full anytime-anywhere-anyhow number-portability. Some kind of universal *Consumer Identity Module (CIM²) card* (or its software equivalent) is proposed in (Ganchev, I. & O'Droma M., 2007) through which consumers would own their personal globally significant, network-independent number (IPv6 address). Through this, with whatever terminal the users choose, they will behave as consumers, i.e., will obtain and securely pay for services from any ANP or TSP, anytime-anywhere-anyhow (whether local or roaming). It is not unlike entering a shop and making purchases using a credit card. Implicit in such business transactions is the availability of network independent, autonomous, trusted 3P-AAA service providers (3P-AAA-SPs) - new business entities in 4G.

A key support service for user-driven ABC&S is that provided by wireless billboard channel (WBC) advertisements. WBCs are used to push advertisements of ANP service offerings to mobile consumer-users so they could discover all available services of all consumer-oriented ANPs at any moment and location, and select the 'best' service. If the user moves to a new location, then his/her ABC&S-enabled mobile terminal –using the user profile[3] and information obtained from WBC– can select a new 'best' access network/provider or new 'best' teleservice/provider to be used, e.g., in the case of QoS change or when the signalling link (currently used) suffers from signal level degradation.

With CBM and through their CIM cards, consumers (wherever they are) would always appear as 'local users' to whatever networks they access. Implicit in being a 'local user' is that roaming charges will disappear.

Creation of a more open, fair, level playing field in the ANP market is another benefit of CBM. At present the market is dominated by a small number of large international-global ANPs. The present SBM technological infrastructure militates against 'start-up' ANPs (as distinct from virtual operators such as Mobile Virtual Network

Operators, MVNOs), e.g., directed at niche or specialized markets. In a CBM environment, the up-front cost for new ANPs is not so prohibitive. For instance, they will not need to have in place, before commencing business, a large network and associated infrastructure (including roaming agreements) over a wide geographic base. Rather they can start small and grow their networks and consumer-customer base, just like new businesses in other market sectors. Hence, instead of being tied to make-or-break numbers of signed-up subscribers, profitability and survival may be linked to the number of consumer transactions they attract.

Within CBM, a new type of an integrated heterogeneous networking (IHN) will be kick-started. This form of IHN is not driven by the networks. Rather it will be driven, managed and controlled by consumers and teleservice providers at the network edge and will mainly be transparent to access networks. The CBM will see a strong growth of wireless networking intelligence at the network edge - in the user terminals and teleservice providers' entities - leading to business opportunities for hardware and software manufacturers, and the likelihood that this form of IHN will become quite sophisticated in time.

Figure 4 illustrates the evolving CBM-based ABC&S networking environment.

Realization of the CBM wireless environment will require the underpinning of some strategic international standardization. This will be pivotal for the support of key new entities and functionalities, such as 3P-AAA-SPs, CIMs, wireless billboard channels (WBCs) and service advertisement, discovery and association (ADA), (O'Droma & Ganchev, 2007), and non-ANP incoming call connection service providers (ICC-SPs). Once these are in place the CBM-based 4G will begin to take shape and grow along an evolutionary path (as happened with, and in parallel with, the existing wireless world founded on the SBM) yielding social, economic and business benefits for users, access network providers, equipment (hardware and software) suppliers, the full range of teleservice providers and new business entities.

Main CBM Business Actors

The main actors in the CBM-based wireless environment are summarized in Table 1.

Figure 4. Typical CBM-based ABC&S networking environment

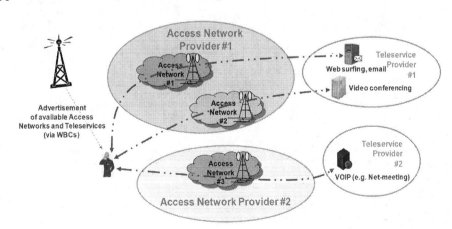

THIRD-PARTY AUTHENTICATION, AUTHORIZATION AND ACCOUNTING (3P-AAA)

The creation of trusted 3P-AAA services is essential for enabling a CBM-based 4G environment. This new secure payment infrastructure will be for all communication access services and teleservices. The 3P-AAA service is central in CBM and may be visualized in many of its features as a wireless equivalent of the credit-card service. As an essential infrastructural component, 3P-AAA will be integrated with a third-party charging and billing service (3P-C&B), described in the next section. The broad architectural and technological design criteria require it to be scalable, hierarchical and cognizant.

Three new application-layer wireless interfaces are envisaged: (a) terminal ↔ ANP/TSP, (b) terminal ↔ 3P-AAA-SP, and (c) ANP/TSP ↔ 3P-AAA-SP (Figure 5). Application autonomy (i.e., containing all necessary 3P-AAA functionality with defined messages for third-party authentication and authorization, purchaser transactions, mobility, user privacy, etc.) would also help the process of developing the standards across these interfaces. A possible approach we propose to meet this requirement is to implement a new 3P-AAA signaling application on top of IETF's Diameter base protocol (Calhoun, P., et. al. 2003). Since the latter defines only a general framework for implementation of AAA and it does not contain command codes for authentication and authorization (which are typically application-specific),

Table 1. The main CBM business actors

Consumer-user	The ABC&S consumer of available 'best' services offered by TSPs/VASPs by employing the networking services of 'best' ANPs.
Access Network Providers (ANPs)	The providers of access network infrastructure and transport medium for all consumer-users (currently in their coverage area / footprint) who were successfully authenticated, authorized and accounted by 3P-AAA-SPs for using their (wireless) communications services.
Teleservice Providers (TSPs)	The providers (directly or indirectly through VASPs) of teleservices such as VoIP, teleconferencing, eCommerce, eBusiness, eEducation, eGovernment, eLectures, eBooks, downloadable content, e.g., music files, video clips, MPEG movies, etc. A new part of the TSP service management support is envisaged that includes monitoring and measurement of service usage, the QoS delivered in general and to individual users, the quality of ANP connection currently used by the consumer, the ANP connection options available at anytime as well as the controlling parties of these options. A dynamic QoS management action could be to propose an alternate ANP connection to the consumer, either because quality degrades below a threshold or is likely to be so degraded. Depending on the teleservice being accessed, TSP policies, or consumer type, the additional service costs caused by the access change will be handled by the consumer and/or teleservice provider. In order to be able to offer better QoS to consumers, TSPs will probably need to form alliances with ANPs underpinned with service level agreements (SLAs), independent software vendors, platform vendors etc.
Value Added Service Providers (VASPs)	The providers of value-added services (i.e., provider of web-portal services) to one or more teleservices provided by a TSP; The WBC service providers (WBC-SPs) and the ICC service providers (ICC-SPs) are new VASP entities in CBM, c.f. Figure 3.
Third-Party AAA Service Providers (3P-AAA-SPs)	The providers of authentication, authorization and accounting services for consumer-users and all other providers. They handle the business agreements with TSPs, ANPs and VASPs, which apply charges to the consumer-user's account for service / network usage. The consumer-user's account (with a 3P-AAA-SP) acts like a credit-card account in eCommerce.
WBC Service Providers (WBC-SPs)	The providers of wireless billboard channels (WBCs) used by ANPs and TSPs for advertisement of their services (deployed in particular area/location), and by users - for discovery of these services followed by association with 'best' providers (for the duration of service usage).
ICC Service Providers (ICC-SPs)	The providers of incoming call connection (ICC) service for consumers (through their personal, geography- and network-independent IPv6 address) in a more dynamic and adaptive way (matching the users' roles and profiles) and with enriched consumer communications management with enhanced possibilities for customization (e.g. which ANP network to be used for which caller; etc.).

new command codes and attribute value pairs (AVPs) need to be defined. Unlike authorization and authentication, the messages for accounting are clearly defined within the Diameter base protocol. Open Diameter (www.opendiameter.org/) is envisaged as a primary candidate for the software implementation of the new 3P-AAA Diameter application. Open Diameter is based on a number of C++ libraries supported by both Linux and Windows platforms.

The new 3P-AAA signalling application will utilize a 3P-AAA protocol stack (Figure 6) which is built on the generic AAA server architecture as defined by de Laat et al. (2000).

The 3P-AAA protocol needs to be developed as a layered protocol in order to group service-specific functions together. Thus, the layered architecture flexibility is present, e.g., substitution of any particular layer with an equivalent functionally without disrupting the adjacent layers is possible. Figure 7 illustrates some possible 3P-AAA protocol stack implementations.

In general, the consumer, ANP, and TSP/VASP may belong to different 3P-AAA-SP administration domains. It would also be very likely that certain 3P-AAA-SPs want to be specialized in services for only one type of stakeholders in the 3P-AAA infrastructure (i.e., for consumers or for teleservice providers). Hence, three 3P-AAA-SP classes are proposed (Figure 8):

- Class A - specialized to provide 3P-AAA services to **a**ccess network providers and their access networks;
- Class B - specialized to provide 3P-AAA services to **b**road range of teleservice providers and their teleservices;
- Class C - specialized to provide 3P-AAA services to **c**onsumer-users and their mobile terminals.

A specific class of 3P-AAA-SP (i.e., Class A or Class B) stores and maintains an account, AAA policies, and business agreement information in relation to a specific type of stakeholder in the 3P-AAA infrastructure (i.e., consumer, ANP, or TSP). The charging and billing block (C&B) provides charging related functions (i.e., rating and pricing policy management and Online Charging Function (OCF) for Class A and Class B, or account balance maintenance together with top-up services for Class C.

The generic 3P-AAA architecture should use a policy-based accounting mechanism (Zseby, Zander, & Carle, 2002) to communicate accounting policies to TSP and ANP domains. These accounting policies are subject to business agreements among the various actors. The charging for access network and teleservice usage is based on the accounting record stream generated by the local accounting services of ANP and TSP, respectively.

Figure 5. The new application-layer interfaces for 3P-AAA

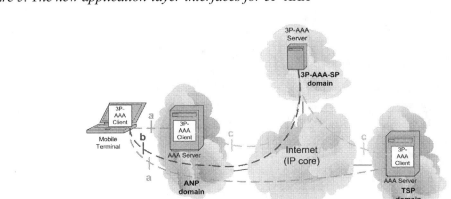

Figure 6. Generic AAA protocol stack

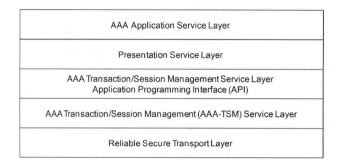

Figure 7. Possible 3P-AAA protocol stack implementations

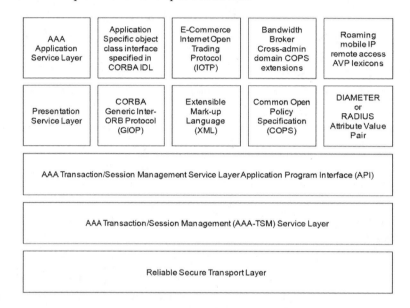

Figure 8. The three 3P-AAA-SP classes

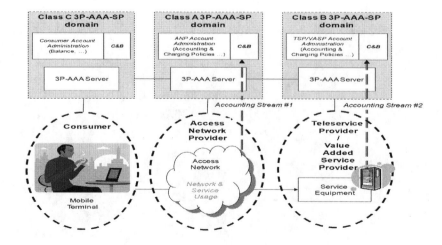

The generic 3P-AAA infrastructural design is shown in Figure 9 with an emphasis on the business perspective.

THIRD-PARTY CHARGING AND BILLING (3P-C&B)

An essential part of CBM is a new third-party charging and billing (3P-C&B) system. This should be based on the 3P-AAA architecture and utilize the AAA-protocol-based accounting and credit-controlling concept. The advantage of this approach is that the charging and accounting services can be built as a separate service and can be outsourced from the ANP and TSP/VASP into the 3P-AAA-SP domain as a 3P-AAA service.

The 3P-C&B architecture should contain two important parts:

- Credit control server located in the 3P-C&B domain (Figure 9);
- Credit control client (3P-CCC) commissioned near or co-located with the service equipment in TSP domain or with the access network resource in ANP domain. The 3P-CCC generates the credit control (CC) protocol messages based on the resource usage for which it has been commissioned. It covers some well-defined charging interfaces or reference points as input from the resource side (e.g., UMTS Ro).

Figure 9. The 3P-AAA service architecture and 'business tree'

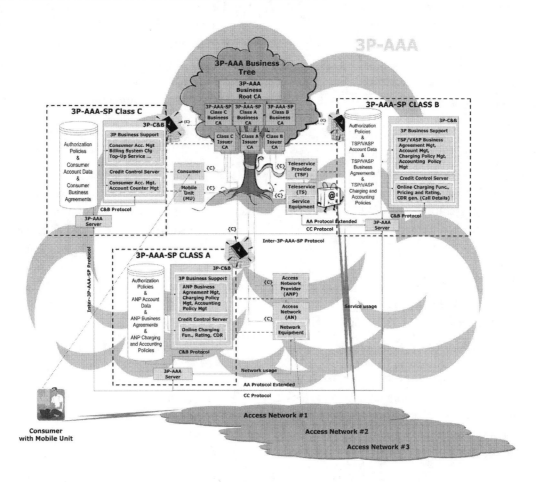

Charging interactions are strictly related to service and network usage or requested change in QoS (i.e., consumer- or TSP-initiated change of the access network). The following main charging scenarios are envisaged in the CBM wireless environment:

- Network usage: the consumer is charged for the use of (wireless) communications services of an access network owned by ANP;
- Service usage: the consumer is charged for the use of a teleservice provided by TSP;
- Network usage after an E2E executed 'Host Access network Change' (HAC[4]): when ANP seeks from TSP (or VASP) the extra-charge of changing the access network. This situation occurs when TSP (or VASP) switches the current user service session via a new ANP because the original ANP provided unsuitable or inadequate QoS.

In the new CBM model, the charging is based on: (i) rating and conversion between the service units and the monetary units (entirely on the 3P-AAA server side and using the 3P-AAA-SP's C&B subsystem) and (ii) the consumer being charged via its 3P-AAA-SP (holding its account). To support these charging interactions between the 3P-AAA-SPs, there is a need to introduce and develop a new signalling protocol – an Inter-3P-AAA-SP protocol (Figure 9).

SERVICE ADVERTISEMENT, DISCOVERY, AND ASSOCIATION (ADA) AND WIRELESS BILLBOARD CHANNELS (WBCS)

Advertisement, discovery and association (ADA) of access networks' communications services and teleservices are key aspects of the future 4G wireless communication world as mobile users/terminals need to discover all services available/deployed at a given area/location. Several different service discovery protocols have been proposed in the past with two basic modes of operation: *pull* mode and *push* mode (Benali, O., et. al, 2004). In the *pull* mode, clients query the environment about given services and dedicated servers express interest with a reply. In the *push* mode, servers periodically advertise their services while the clients listen for these advertisements.

With the potential for access network services in UCWW being sold to users on a consumer basis, transaction by transaction, rather than through a long-term subscriber contract as at present, wireless access networks competing for consumer business will seek new dynamic ways to advertise their services to potential customers. We believe that the future consumer-centric wireless communication environment should behave according to the *push* mode in which a network infrastructure (e.g., access network provider, ANP) or a source of data (teleservice provider, TSP) takes care of preparing connections and services, and proposes them to the mobile user. In this proactive mode of operation, the only user intervention is to choose the 'best' possible connection for each needed 'best' service.

Wireless Billboard Channels (WBC), (e.g., O'Droma, M. & Ganchev, I. 2007; Flynn, P., et. al. 2006), is a novel 'push' advertising concept, and new business opportunity, to cater for this need. For mobile users, as consumers, WBCs will be an effective means to discover who is offering what services; with what QoS options; at what price/tariff schedules; how to gain access to these services; etc. WBCs will have broad attraction for other Internet businesses and teleservice providers seeking to increase consumers' usage of their wireless networks and service offerings. WBCs will be:

- Broadcast, narrowband, and unidirectional channels;
- Local, regional, national, and international in their coverage regimes;

- Owned by WBC service providers (WBC-SPs), independent of wireless ANPs (for reasons of fairness); The existing digital TV and radio broadcasters are prime candidates for WBC-SPs;
- Typically hosted on suitable existing broadcast services' physical infrastructures, e.g., as an additional service. New potential broadcast platforms include but are not limited to Digital Radio Mondiale (DRM), Digital Audio Broadcasting (DAB), Digital Video Broadcasting - Handheld (DVB-H), Digital Multimedia Broadcasting (DMB), and Multimedia Broadcast Multicast Service (MBMS).

To broadcast datasets over narrowband and unidirectional WBC channels, the service advertisements represented by service descriptions (SDs) should be specified in a structured, efficient, and compact formal language. The typical SD structure consists of a set of attributes, such as ServiceType, ScopeList, Length, Composite Capabilities / Preferences Profile (CC/PP) (Klyne et. al., 2004), QoS, and AttrList. The ServiceType is used to group together all SDs performing the same function. Each ServiceType needs a corresponding ServiceTeplate to specify the SD attributes. The ScopeList, CC/PP, and QoS act as filters for SDs (together with the user profiles) and thus facilitating the blocking or receiving of SDs by the user mobile terminal. Since using as little bandwidth as possible is one of the WBC desired properties, in order to reduce the SD size the abstract syntax notation's packet encoding rules (ASN.1-PER) (Qiang et. al., 2006) are used to format SDs (Flynn et. al., 2006). A set of pre-defined basic and constructed types, such as INTEGER, IA5String, BIT STRING, LIST, SEQUENCE, CHOICE etc, are used to describe the complex SD data structure. In addition, with PER, the final data stream of SDs is smaller comparing with other formal languages, such as document type definition – extensible mark-up language (DTD-XML) and augmented Backus-Naur form (ABNF). Thus, the ASN.1-PER was selected to describe data structures in WBC. To integrate the ASN.1-PER scheme into a platform-independent environment, all SD templates were compiled to JAVA classes with an ASN.1 Java compiler. The ASN.1-PER encoder/decoder (Figure 10) depends on the Java classes used for the encoding a SD Java object to PER octets and decoding of PER octets to a SD Java object.

Considering the trend that the future mobile terminals will integrate IP-based transmission techniques, the WBCs should employ an IP datacasting (IPDC) such as that proposed by (Kornfeld M. & G. May 2007). To simplify the design and achieve compatibility with this technique, the WBCs are designed with a three-layer protocol architectural model (Figure 11) containing the following layers:

(i) *Service layer*, which describes the service discovery model, and data collection, clustering, scheduling, indexing, broadcasting, discovery, and association schemes;
(ii) *Link layer* concerned with the frame processing issues, such as addressing, forward error control, etc.;
(iii) *Physical layer*, which realizes the transmitter (in the WBC-SP node) and receiver (in the user mobile terminal).

Figure 10. The ASN.1-PER encoder / decoder

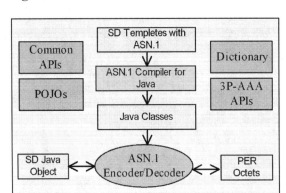

The WBC link layer and physical layer are hardware-dependent layers and thus have different structures depending on the carrier technology used. The service layer is the main WBC layer for research, design and development. It is a pure software layer, independent of the carrier technology. It must ensure smooth IPDC processing with different types of WBC link layer and physical layer, i.e., different carrier technologies.

WBC Service Layer

This is of course the most important layer from a service definition- and delivery point of view. It is realized in both the WBC-SP node and in the user mobile terminal. In the former, it is concerned with organizing the service advertisements in an efficient way in order to reduce the user access time and tuning time (Ji Zh., et. al., 2008). In the latter, the focus is on the MT discovering all services of interest to its user and associating with the 'best' wireless access network for each particular teleservice according to predefined user-policies or directly with live interactive user decisions – all of which are supportive of, and enabling of, full user-driven ABC&S paradigm realization. To achieve an extensible, flexible, and intelligent structure, the WBC service layer's architecture is built on three tiers, as shown in Figure 12 and explained below.

Service Discovery and Maintenance Tier

Taking into account that the Java 2 platform Enterprise Edition / Java 2 platform Micro Edition (J2EE/J2ME) is an effective, hardware-independent platform for building both enterprise applications and portable-devices applications, the WBC service layer was implemented in Java in order to provide system uniformity, distribution, and portability. There are three actors in this tier:

(a) Service providers (xSPs), who submit descriptions of their services to the WBC content database in a competitive way via the WBC-SP web portal;
(b) WBC service provider (WBC-SP), who maintains the roles and databases, and defines WBC broadcasting parameters;
(c) Mobile consumer-users, who discover and associate with 'best' services based on their own price/performance preferences specified in their user profiles.

The tier is sub-divided into three sub-tiers:

(i) *Presentation sub-tier* – which acts as an interface between the actors (WBC-SP, xSPs, users) and the system;
(ii) *Business sub-tier* – which follows the delegate design pattern to decouple the business logics and codes, and exposes simpler interfaces for other two tiers; and

Figure 11. The three-layer WBC architectural model

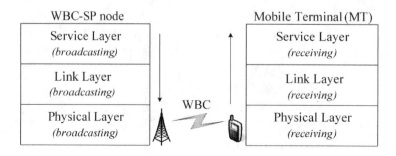

Figure 12. The WBC service layer's architecture

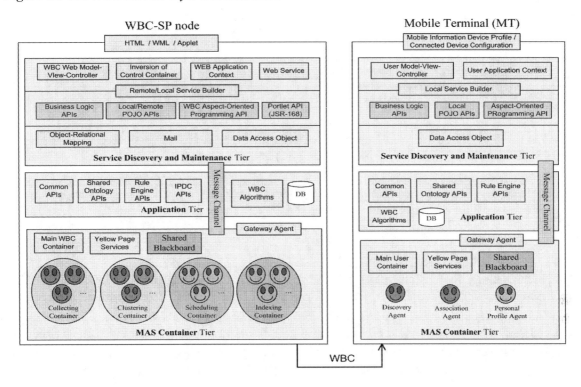

(iii) *Persistence sub-tier* – which is concerned with the storing/retrieving of plain old java objects (POJOs) in/from databases, i.e., service descriptions' database, xSP database, WBC system parameters' database, broadcasting database, WBC-SP/user intelligent rules' database, user/terminal profiles' database, etc.

Application Tier

To simply the WBC system design and enable decoupling, in this tier an expert system is implemented which has 'responsibility' for maintaining the business logic and WBC algorithms. Present also are a set of application programming interfaces (APIs), such as common APIs, ontology APIs, services' data processing APIs, intelligent rule-engine APIs etc, which are shared with the other two tiers.

The core WBC algorithms include WBC intelligent schemes for SDs' collection, clustering, scheduling, indexing, discovery, association, and WBC-IPDC. To uniform the expert knowledge base in the WBC system, an ontology technique is used to describe concepts and their relationships. The design pattern of ontologies follows the singleton design pattern to enable the sharing of the same ontology object among the three tiers. Two ontology types are used: data source ontology and task ontology. To achieve a loose-coupling system and enable WBC data organizing algorithms to run in an intelligent way, a rule-based expert system (knowledge-based system) was designed for facilitating the data broadcasting by the WBC-SP node and data reception by the user mobile terminal (Figure 13). The key advantage of using a rule-based expert system is that the business solutions could be found much more easily, i.e., the end user can change the rules (very close to

Figure 13. The WBC rule-based expert system

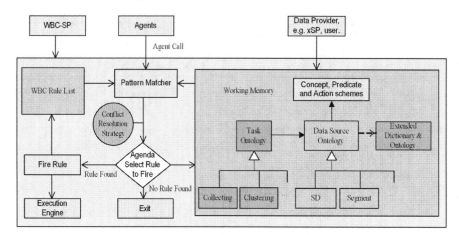

a natural language) without having to recompile the source code.

There are two basic elements in the WBC rule-based expert system: *facts* (all ontologies act as shadow-facts asserted into the working memory) and *rules* (different area and different time may cause running different rules). On the WBC-SP side, the collecting rule defines the broadcasting frequency of each service description (SD) based on the user access times and advertisements' fees paid by the service providers; the clustering rule clusters SDs with a category type, frequency, CC/PP, QoS, scope list etc, and defines the properties of segments (SDs are grouped into fixed-size segments for broadcasting over WBCs); the scheduling rule locates the position of each segment in the broadcasting sequence; the indexing rule generates indexed-segments; and the broadcasting rule determines the broadcasting time for each final sequence. On the mobile user's side, the user profile rule includes three sections: an *advertisement-filter rule* used for blocking of 'uninteresting' services, a *discovery rule* that discoveries "best" services within a received advertisement (group of services), and an *association rule* facilitating the association with the "best" (wireless) network/service.

A new advertisements delivery protocol (ADP) for wireless services' IP Datacasting was proposed in (Ji Zh., Ganchev, I., & O'Droma, M 2008), based on the modified asynchronous layered coding (ALC) protocol (Luby et. al., 2002). ADP uses a forward error correction with a Reed-Solomon coding scheme to guarantee packet-level reliable data delivery at the WBC service layer.

Multi-Agent System (MAS) Container Tier

This tier acts as an agent run-time environment for facilitating the (wireless) services advertisements' collection, clustering, scheduling, indexing, broadcasting, discovery, and association with application layer's APIs. A shared blackboard[5] and gateway agent are used for agents' communication and tiers' communication respectively.

Based on the foundation for intelligent physical agents (FIPA) framework (Bellifemine et. al., 2001), MAS job is to provide agent management system (AMS) and directory facilitator (DF) services, 'yellow pages' service, message transport service, etc. Each running instance of MAS is called a container with a set of agents. The main WBC container must be always active so that other containers can register with it. All WBC agents (i.e., the collecting agent, clustering agent, scheduling agent, indexing agent, and broadcasting agent - on the WBC-SP node, and the discovery agent, association agent, and personal profile agent - on the user mobile terminal) use the agent communication language (ACL) for interactions with each other. A content manager performs automatic conversion and check-up operations between the ACL byte stream and ACL ontology Java object. With this communication mechanism, the MAS tier runs in a peer-to-peer mode. A blackboard is used for sharing the information and common static objects. To communicate with other tiers, a gateway agent is used to receive/send messages via a shared message channel. All agents except the personal profile agent are logic-based agents that work with the knowledge-based system for ADA processing. The personal profile agent is a belief-desire-intention (BDI) agent, whose actions depend on the user history records, plans, beliefs/desires, and intentions.

FUTURE RESEARCH DIRECTIONS

The following directions are envisaged for future research. There is much research potential in the new fields of (i) third-party authentication, authorization, and accounting (3P-AAA) and third-party charging and billing (3P-C&B) and (ii) WBCs and ADA. Some ideas and directions are listed in the following. Included also is a suggestion to develop a cross-layer reference communications model to aid in design and analysis of open cross-layer functionalities.

3P-AAA and 3P-C&B

- *Extending 3P-AAA into the area of wireless Ad Hoc networks*. This can yield significant 3P-AAA use cases influencing the 3P-C&B requirements. Typical Ad Hoc domain scenarios involving hot-zone wireless heterogeneous architectures are envisaged where mobile terminals use multi-hop techniques to get to a hot zone using intermediate mobile terminals (the latter should benefit from their role in such scenarios, i.e., be paid properly).
- Research to date has identified and established the basic charging scenarios in CBM-based UCWW by employing inter-3P-AAA-SP signalling. However, when the inter-3P-AAA-SP signalling involves Internet usage, then charging interactions can experience *high network latency*. To eliminate this problem either further *optimization* is needed in the 3P-AAA-SP signalling (i.e., compressing the messages where possible) or a *new 'charging agent' concept* should be developed. This new concept would result in the following: (1) the charging occurs in the metering domain (TSP/ANP), (2) the charging agent is downloaded from the 3P-AAA-SP to provide the charging function in the TSP/ANP domain, (3) the charging agent imports the charging rule set from the 3P-AAA-SP, (4) the charging agent imports segments of the consumer account into the metering domain.
- *Elaboration of the C&B framework to support dynamic reconfiguration* of applicable metering and pricing policies for specific service, specific user or combination of both, and to support various pricing models according to the service profile, user profile and location, and one-stop billing schemes.
- *Implementation of a C&B system prototype* as a discrete service that can be provided by a trusted third-party authentication, authorization and accounting service providers (3P-AAA-SPs).
- *Running trial experiments* with the designed prototype in a 4G testbed environment showing good interfacing with the 3P-AAA service, WBC&ADA services, and other (new) types of 4G services (e.g., consumer-oriented ICC service).

WBC and ADA

- *At the WBC Service Layer* - Besides the intelligent software architecture already mentioned, other issues for future investigation include:
    - *Agent environment*: JADE has been used to date to act as an agent environment in the heterogeneous WBC software architecture. However, JADE is a heavy agent platform with a big footprint for executing both the SD collecting, clustering, scheduling, indexing, broadcasting on the server side, and the SD discovery and association on the mobile terminal side. In addition, it does not fully support the BDI agent. Therefore, investigation into lightweight BDI-based Java agent platforms (WBC-BDI) is recommended. Formatting the communication language's messages with WBC-ASN is also recommended, as well as ensuring that the agent platform functions correctly in the following environments: J2SE (Sun Java 2 platform, standard edition 2003), J2ME (Sun J2ME Specification 2009), Android (Google Android Software Development Kit 2008), WinCE (Windows Embedded CE Overview 2008), etc.
    - *SD formatting*: In order to encode SDs in a more compact way, an efficient abstract syntax notation language based on ASN.1 (WBC-ASN) is suggested. Any design should take into account the requirement for minimizing decoder's power consumption.
    - *Rule engine*: Resolution of the need to improve the flexibility and scalability could be approached by designing an intelligent SD self-organization lightweight Java rule engine. In suggesting this, we also recommend that the rules configuration file here could be defined with WBC-ASN.
    - *ADP*: Designing with system scalability in mind, the route of developing the ADP protocol in Java, together with a Java-based Reed-Solomon algorithm being fully implemented is suggested as worthy of investigation.
    - *Profile design*: To increase security and privacy for WBC-SPs, and mobility and personalization for mobile users, investigation of the benefits from this perspective of a well-structured rule-based profile developed and formatted with WBC-ASN is suggested.
- *At the WBC Link Layer and Physical Layer* – Potential broadcast platform solutions include WBC over DVB-H, over DRM, over DAB, etc. Investigations in the technical realization configurations for each have yet to be undertaken. There is little doubt about the potential consumer base into which WBC advertisement may be pushed. Today, for instance, there is an expectation of 300 million DVB-H capable handsets operational by 2009/10.

Cross-Layer Reference Communication Model

Some activities within UCWW, such as end-to-end (E2E) hot access network change (HAC) based on user-driven ABS&S policies, require cross-layer protocol functionality. Other examples include E2E reconfigurability (Z. Boufidis et. al., 2004, Sept), service adaptability (Houssos, N., et. al. 2003), E2E QoS support (Politis, C. al., 2004), ABC&S (O'Droma, M., Ganchev, I, et. al., 2006), user/network/service/terminal profile management, 3P-AAA and related 3P-C&B, and WBC & ADA operation. While this seems to contradict the layering architecture model for designing, planning, implementing and analysing communication protocols, nonetheless, it is the reality and it is worthwhile to structurally allow for it with suitable modifications of the reference models. Such a suitably modified reference communication model is presented in Figure 14. It has similarities with the B-ISDN/ATM reference communication model in that it is a 3D model consisting of three planes: user plane, control plane, and management plane. The new central element, which intersects all three planes, is added to allow for structured cross-layer functionality. This *cross-layer core cylinder* is a modification of that proposed in (Ganchev, I., O'Droma, M., et. al., 2006) and may be visualized as consisting of several parallel mini cylinders each with its own dedicated functionality, e.g., corresponding to the activities already listed above with cross-layer protocol functionality. Formal reflecting of these activities and their cross-layer functionalities into this model will assist their formal design and analysis, and facilitate development of formal and open primitives and APIs.

CONCLUSION

The technical foundations for the new consumer-centric business model (CBM), which will make the evolution to a truly user-driven always best connected and best served (ABC&S) wireless world possible, have been addressed. This chapter has set out some of the arguments for a directed, structurally planned, evolution of CBM. It has set out ideas on a selection of the network and protocol technological infrastructural changes and innovations needed in order to realize a new CBM-based wireless environment, called a ubiquitous consumer wireless world (UCWW). These include an infrastructural re-think on the way authentication, authorization and accounting (AAA) service is supplied, proposing the creation of a third-party AAA service provider entities (3P-AAA-SP), together with a number of new technological supports requiring global standardized including a service advertisement, discovery and association (ADA) support structure and a newly defined wireless billboard channels (WBCs). The former (3P-AAA) is an essential infrastructural component for the CBM wireless environment realization. The latter (WBC-ADA) technically is not so essential but from business perspective would be perceived to have an essential role. Both offer totally new telecommunication business opportunities and will be the foundation

Figure 14. The proposed cross-layer reference communication model

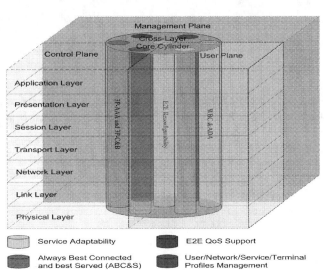

of a significant change to the business models of future wireless next generation networks (NGNs). The chapter includes detailed descriptions of the makeup of 3P-AAA and WBC, including the new wireless interfaces that will need to be globally standardized.

The new CBM-UCWW environment would bring benefits to the many wireless communications' stakeholders. It will be a great deal more open and facilitating than the present environment alone. This is so especially in relation to the improvement in ease of entry into the access network provider (ANP) and teleservice provider (TSP) markets for new entrants, to the realization of a more receptive innovation environment for 'disruptive' and creative technologies, and for stimulating competition driving forward the range and sophistication of ANPs and TSPs wireless services' product offerings. For the consumer, as already noted, benefits include: full consumer's freedom and independence in seeking value for money with the selection of 'best' access networks and 'best' teleservices to use, realization of truly consumer-oriented, user-friendly and user-driven ABC&S wireless communications environment, full zero-cost anytime-anywhere-anyhow portability, zero roaming charges, and new service offerings arising from the creation of new telecommunications services' markets.

REFERENCES

Bellifemine, F., & Poggi, A. A. & G. Rimassa, G. (2001, May). *JADE: A FIPA2000 Compliant Agent Development Environment.* Paper presented at the meeting of the AGENTS '01, Montreal, Canada.

Boufidis, Z., et al. (2004, September). *Actors, Management Plane, and Policy Provision Challenges for End-to-End Reconfiguration",* Paper presented at the ER Workshop on Reconfigurable Mobile Systems and Networks Beyond 3G, Barcelona, Spain. Broadband Radio Access Networks (ETSI BRAN). (n.d.). Retrieved December 30, 2008 from http://portal.etsi.org/portal_common/home.asp?tbkey1=BRAN

De Laat, C., Gross, G., Gommans, L., Vollbrecht, J., & Spence, D. (2000). *Generic AAA Architecture (RFC 2903).* Retrieved from http://www.ietf.org/rfc/rfc2903.txt

EU IST Project ANWIRE. (Academic Network on Wireless Internet Research in Europe) (n.d). Retrieved August 30, 2008, from http://www.ist-world.org/ProjectDetails.aspx?ProjectId=8f7d258e5ca542239e0b0b3d5ea16e33

EU IST Project BRAIN/MIND. (n.d). Retrieved December 30, 2008, from http://www.itu.int/osg/imt-project/docs/4.3_Wisely.pdf

EU IST Project MOBIVAS.)Downloadable Mobile Value-Added-Services through Software Radio and Switching Integrated Platforms) (n.d). Retrieved December 30, 2008 from http://www.ist-world.org/ProjectDetails.aspx?ProjectId=7a73337c09a3488d8947757b63db84c5

EU IST Project MOBY DICK. About the project (n.d). *About the project.* Retrieved December 30, 2008, from http://www.nw.neclab.eu/Projects/mobydick.htm

EU IST Project SCOUT (Smart user-centriC cOmmUnication environmenT) (n.d.). Retrieved December 30, 2008, from http://www.ist-world.org/ProjectDetails.aspx?ProjectId=3f4727fdc7ff444c87de64fe5c579084

Flynn, P., & Ganchev, I., I., & M. O'Droma, M. (2006). Wireless Billboard Channels: Vehicle and Infrastructural Support for Advertisement, Discovery, and Association of UCWW Services. *Annual Review of Communications, 59*, 493–504.

Ganchev, I., & O'Droma, M. M. (2007). " New personal IPv6 address scheme and universal CIM card for UCWW". In *Proc. of the 7th International Conference on Intelligent Transport Systems Telecommunications (ITST 2007)*, Pp. 381-386, 6-8 June, Sophia Antipolis, France (pp. 381-386). IEEE Catalog No: 07EX1765. Library of Congress: 2007923401. ISBN: 1-4244-1177-7.

Ganchev, I., O'Droma, M. S., Siebert, M., Bader, F., Chaouchi, H., & Armuelles, I. (2006). A 4G Generic ANWIRE System and Service Integration Architecture. *ACM SIGMOBILE Mobile Computing and Communications Review Journal, 10*(1), 13–30. doi:10.1145/1119759.1119761

Google Android Software Development Kit. (n.d.). Retrieved December 18, 2008, from http://code.google.com/android

Hakala, H., Mattila, L., Koskinen, J.-P., Stura, M., & Loughney, J. (2005). *Diameter Credit-Control Application (RFC 4006)*. Retrieved from http://www.ietf.org/rfc/rfc4006.txt

Ji, Zh. Ji, Z., I. GanchevGanchev, I., & O'Droma, M. (2008, June). *Efficient Collecting, Clustering, Scheduling, and Indexing Schemes for Advertisement of Services over Wireless Billboard Channels.* Paper presented at the meeting of the 15th International Conference on Telecommunications, St. Petersburg, Russia.

Ji, Z., Ganchev, I., & O'Droma, M. (2008, April). *Reliable and Efficient Advertisements Delivery Protocol for Use on Wireless Billboard Channels.* Paper presented at the meeting of the 12th IEEE International Symposium on Consumer Electronics, Algarve, Portugal.

Klyne, G., & Reynolds, F. F.C, Woodrow, CH., Ohto, HJ., Hjelm, J., & M. H. Butler, M. H. (2004). Composite Capability /Preference Profiles (CC/PP): Structure and Vocabularies, *W3C Recommendation*, from Retrieved from http://www.w3.org/Mobile/CCPP.

Kornfeld M., G. May. (2007) DVB-H and IP Datacast - Broadcast to Handheld Devices,. *IEEE Transactions on Broadcasting*, Vol. 53, Issue (2), (pp. 161-170).

Luby, M., & Gemmell, J. J., L. Vicisano, L., L. Rizzo, L., &and J. Crowcroft, J. (2002) Asynchronous Layered Coding (ALC) Protocol Instantiation. *IETF RFC 3450*, from Retrieved from http://www.ietf.org/rfc/rfc3450.txt.

Metakall - The Alternative to Cellular. (n.d.). Retrieved December 28, 2008, from www.metakall.com.

O'Droma, M., & Ganchev, I. (2004). Techno-Business Models for 4G. In *Int. Forum on 4G Mobile Communications*, King's College London, UK (pp. 1-30).

O'Droma, M., & Ganchev, I. (2007). Toward a ubiquitous consumer wireless world. *IEEE Wireless Communications Journal, 14*(1), 52–63. doi:10.1109/MWC.2007.314551

O'Droma, M., Ganchev, I., Chaouchi, H., Aghvami, H., & Friderikos, V. (2006). Always Best Connected and Served Vision for a Future Wireless World. *Journal of Information Technologies and Control, 4*(3-4), 25–37.

O'Droma, M. S., & Ganchev, I. (2004). Enabling an Always Best-Connected Defined 4G Wireless World. In []. Chicago, IL.: International Engineering Consortium.]. *Annual Review of Communications, 57*, 1157–1170.

Passas, N., Paskalis, S., Kaloxylos, A., Bader, F., Narcisi, R., & Tsontsis, E. (2006). Enabling technologies for the always best connected concept. *Wiley Wireless Communications and Mobile Computing Journal*, *6*(4), 523–540. doi:10.1002/wcm.392

Perez, M. (2008, September 11). Wireless Carriers Address An Open Future. *InformationWeek*. Retrieved from http://www.informationweek.com/news/mobility/business/showArticle.jhtml?articleID=210600964

Politis, C.; Oda, T.; Dixit, S.; Schieder, A.; Lach, H.-Y.; Smirnov, M.I.; Uskela, S.; & Tafazolli, R. (2004). Cooperative networks for the future wireless world. *Communications Magazine*, *42*(9), 70–79. doi:10.1109/MCOM.2004.1336723

Qiang, H. u XXue-cheng., & Zou Shi-min, Z. (2006). ASN.1 Application in Parsing ISUP PDUs., *Communications and Information Technologies,* (pp. 78-81).

Sun J2ME Specification. (n.d.). Retrieved January 19, 2009, from http://java.sun.com/javame

Sun Java 2 platform, standard edition (J2SE). (n.d.). Retrieved August 25, 2003, from http://java.sun.com/j2se

Windows Embedded, C. E. *Overview*. (n.d.). Retrieved December 30, 2008 from http://www.microsoft.com/windowsembedded

Wireless World Research Forum. (n.d). Retrieved December 30, 2008, from http://www.wireless-world-research.org

Zseby, T., Zander, S., & Carle, G. (2002). *Policy-Based Accounting (RFC 3334)*. Retrieved from http://www.ietf.org/rfc/rfc3334.txt

ADDITIONAL READING

Acharya, S., Alonso, R., Franklin, M., & Zdonik, S. (1995) Broadcast disks: data management for asymmetric communication environments. In *ACM SIGMOD Conference* (pp. 199-210).

Arbanowski, S., Ballon, P., David, K., Droegehorn, O., Eertink, H., & Kellerer, W. (2004, Sept.). I-centric communications: personalization, ambient awareness, and adaptability for future mobile services. *Communications Magazine*, *42*(9), 63–69. doi:10.1109/MCOM.2004.1336722

Arkko, J., & Haverinen, H. (2006). *Extensible Authentication Protocol Method for 3rd Generation Authentication and Key Agreement (EAP-AKA) (RFC 4187)*. Retrieved from http://www.ietf.org/rfc/4187.txt

Badia, L., Miozzo, M., Rossi, M., & Zorzi, M. (2007). Routing Schemes in Heterogeneous Wireless Network based on Access Advertisement and Backward Utilities for QoS Support. *IEEE Communications Magazine*, *45*(2), 67–73. doi:10.1109/MCOM.2007.313397

Benali, O., El-Khazen, K., Garrec, D., Guiraudou, M., & Martinez, G. (n.d.). A framework for an evolutionary path toward 4G by means of cooperation of networks. *Communications Magazine, IEEE*. 42(5), 82 – 89.

Calhoun, P., Loughney, J., Guttman, E., & Arkko, J. (2003). *Diameter Base Protocol (RFC 3588)*. Retrieved from http://www.ietf.org/rfc/rfc3588.txt

Calhoun, P., Zorn, G., Spence, D., & Mitton, D. (2005). *Diameter Network Access Server Application (RFC 4005)*. Retrieved from http://www.ietf.org/rfc/rfc4005.txt

Chandra, P. (2005). *Bullet-proof Wireless Security: GSM, UMTS, 802.11 and Ad Hoc Security*. Amsterdam: Elsevier.

Crisler, K., Turner, T., Aftelak, A., Visciola, M., Steinhage, A., & Anneroth, M. (2004). Considering the user in the wireless world. *Communications Magazine*, *42*(9), 56–62. doi:10.1109/MCOM.2004.1336721

Dong, X., & Beaulieu, N. C. (2002). Average level crossing rate and average fade duration of low-order maximal ratio diversity with unbalanced channels. *IEEE Communications Letters*, *6*, 135–137. doi:10.1109/4234.996033

Eronen, P., Hiller, T., & Zorn, G. (2005). *Diameter Extensible Authentication Protocol (EAP) Application (RFC 4072)*. Retrieved from http://www.ietf.org/rfc/rfc4072.txt

ETSI EN 300 744 v1.5.1. (2004. *Digital Video Broadcasting (DVB), Framing structure, channel coding and modulation for digital terrestrial television.*

Farrell, S., Vollbrecht, J., Calhoun, P., Gommans, L., Gross, G., Bruijn, B., et al. (2000). *AAA Authorization Requirements (RFC 2906)*. Retrieved from http://www.ietf.org/rfc/rfc2906.txt

Ferber, J. (1999). *Multi-Agent Systems*. Reading, MA: Addison-Wesley.

Friedman-Hill, E. (2003). *Jess in Action: Java Rule-based Systems*. Greenwich, CT: Manning Publications Company.

Gehrmann, C., Person, J., & Smeet, B. (2004). *Bluetooth Security*. Norwood, MA: Artech House.

Gennari, J. H. (2003). The Evolution of Protege: An Environment for Knowledge-Based Systems Development. *International Journal of Human-Computer Studies*, *58*, 89–123. doi:10.1016/S1071-5819(02)00127-1

Goldsmith, A. (2004). *Wireless Communications*. Cambridge, UK: Cambridge Univ. Press.

Housley, R., Polk, W., Ford, W., & Solo, D. (2002). *Internet X.509 Public Key Infrastructure Certificate and Certificate Revocation List (CRL) Profile (RFC 3280)*. Retrieved from http://www.ietf.org/rfc/rfc3280.txt

Houssos, N., Alonistioti, A., Merakos, L., Dillinger, M., Fahrmair, M., & Schoenmakers, M. (2003, August). Advanced adaptability and profile management framework for the support of flexible mobile service provision. *IEEE Wireless Communications*, *10*(4), 52–61. doi:10.1109/MWC.2003.1224979

Imielinski, T., Viswanathan, S., & Badrinath, B. R. (1994). Energy Efficient Indexing on Air. In *Proc of ACM-SIGMOD, Intl Conference on Management of Data*, Minnesota.

Imielinski, T., Viswanathan, S., & Badrinath, B. R. (1997). Data on Air: Organization and Access. *IEEE Transactions on Knowledge and Data Engineering*, 9.

ITU-T Draft Recommendation Q.3202.1. (2008). *Authentication Protocols based on EAP-AKA for Interworking among 3GPP, WiMax and WLAN in NGN*. Retrieved from www.itu.int/ITU-T/ngn/fgngn

Jayaraman, P., Lopez, R., Parthasarathy, M., & Yegin, A. (2008). *Protocol for Carrying Authentication for Network Access (PANA) Framework (RFC 5193)*. Retrieved from http://www.ietf.org/rfc/rfc5193.txt

Koutsopolou M., Kaloxylos A., Alonistioti A., Merakos L., & Philippopoulos, P. (n.d.). An integrated charging, accounting and billing management platform for the support of innovative business models in mobile networks. *Int. journal of Mobile Communications*, 2(4).

Linux, K. S. C. T. P. *(LKSCTP)*. Retrieved August 10, 2007 from http://lksctp.sourceforge.net/overview.html

Luby, M., & Vicisano, L. (2004). Compact Forward Error Correction (FEC) Schemes. *IETF RFC 3695*.

Paavola, J., Himmanen, H., Jokela, T., Poikonen, J., & Ipatov, V. (2007). The Performance Analysis of MPE-FEC Decoding Methods at the DVB-H Link Layer for Efficient IP Packet Retrieval. *IEEE Transactions on Broadcasting, 53*(1), 263–275. doi:10.1109/TBC.2007.891694

Paila, T., Luby, M., Lehtonen, R., Roca, V., & Walsh, R. (2004). FLUTE – File Delivery over Unidirectional Transport. *IETF (RFC 3926)*.

Proakis, J. G. (2001). *Digital Communications, 4th ed*. New York: McGraw Hill.

Rappaport, T. S. (2000). *Wireless Communications: Principles and Practice*. Reading, MA: Prentice Hall.

Reliable Multicast Transport (RMT). *Working Group Charter*. Retrieved from http://ietf.org/html.charters/rmt-charter.html

Rigney, C. (2000). *RADIUS Accounting (RFC 2866)*. Retrieved from http://www.ietf.org/rfc/rfc2866.txt

Rigney, C., Willats, W., & Calhoun, P. (2000). *RADIUS Extensions (RFC 2869)*. Retrieved from http://www.ietf.org/rfc/rfc2869.txt

Rigney, C., Willens, S., Rubens, A., & Simpson, W. (2000). *Remote Authentication Dial In User Service (RADIUS) (RFC2865)*. Retrieved from http://www.ietf.org/rfc/rfc2865.txt

Roca, V. (2007). FCAST: Scalable Object Delivery on Top of the ALC Protocol. *IETF RMT Working Group*. Uusitalo, M.A. (n.d.). Global Vision for the Future Wireless World from the WWRF. *IEEE Vehicular Technology Magazine, 1*(2), 4–8.

Vollbrecht, J., Calhoun, P., Farrell, S., Gommans, L., Gross, G., Bruijn, B., et al. (2000) AAA Authorization Framework (*RFC 2904*). Retrieved from http://www.ietf.org/rfc/rfc2904.txt.

Vollbrecht, J., Calhoun, P., Farrell, S., Gommans, L., Gross, G., Bruijn, B., et al. (2000). *AAA Authorization Application Examples* (*RFC 2905*). Retrieved from http://www.ietf.org/rfc/rfc2905.txt

Whetten, T., Vicisano, L., Kermode, R., Handley, M., Floyd, S., & Luby, M. (2001). Reliable Multicast Transport Building Blocks for One-to-Many Bulk-Data Transfer. *IETF (RFC 3048)*.

Wong, D. (2005). *Wireless Internet Telecommunication*. Norwood, MA: Artech House.

Yu, Y., & Miller, S. L. (2007). A Four-State Markov Frame Error Model for the Wireless Physical Layer. In IEEE Wireless Communications and Networking Conference, Hong Kong (pp. 2053-2057).

Zhang, Q., & Kassam, S. A. (1999). Finite-state Markov model for Rayleigh fading channels. *IEEE Transactions on Communications, 47*, 1688–1692. doi:10.1109/26.803503

ENDNOTES

[1] *Teleservice* – a generic term used here to encompass all non-access-network services, e.g., from eLearning, eGovernment to on-line Internet shopping, email, and web-browsing.

[2] *Consumer Identity Module (CIM)* - analogous to SIM card thinking but based on CBM.

[3] *User profile* stores user preferences, e.g., price range, preferred service provider, preferred QoS, etc.

4. *Hot Access Network Change (HAC)* is analogous to the network handover concept, but its structure, as a user-driven integrated heterogeneous networking, the reasons for it and consequences of it are quite different so a different term is needed. A typical ABC&S reason for HAC would be the availability of a better access option and offer for the same teleservice from another access network.

5. *Blackboard* - a basic element in expert systems and multi-agent systems, serving as sharing data structure during the execution cycle.

Chapter 3
4G Access Network Architecture

Young-June Choi
NEC Laboratories America, USA

ABSTRACT

Although all-IP networking is the ultimate goal of 4G wireless networks, 3G LTE and WiMAX systems have designed semi all-IP network architectures for efficient radio resource and mobility management. These semi all-IP networks separate layer 2 and layer 3 handoff operations by grouping many base stations (BSs) as a subnet, thus alleviating handoff, while the pure all-IP networks provide a simple network platform at the cost of high handoff overhead. The authors compare the semi all-IP networks to the pure all-IP networks, and provide an overview to WiMAX access service networks and 3G LTE backhaul networks. They then present advanced architectures that support efficient radio resource and mobility management. First, they present a semi hierarchical cellular system with a super BS that behaves like a normal BS as well as a supervisor over other BSs within the group. They further extend this model to a system that combines multiple access techniques of OFDMA and FH-OFDMA with microcells and macrocells. Also, to alleviate the handoff latency, a dual-linked BS model is presented in order to support seamless handoff. Finally, as an integrated approach to supporting diverse QoS requirements, the authors consider an IP-triggered resource allocation strategy (ITRAS) that exploits IntServ and DiffServ of the network layer to interwork with channel allocation and multiple access of MAC and PHY layers, respectively. These cross layer approaches shed light on designing a QoS support model in a 4G network that cannot be handled properly by a single layer based approach.

DOI: 10.4018/978-1-61520-674-2.ch003

1. INTRODUCTION

Fourth-generation (4G) networks are expected to deploy a simplified network architecture based on all-IP (ITU-R, 2003). The scenario of all-IP networking will alleviate the problem of third-generation (3G) access networks such as WCDMA and cdma2000, where there are many protocols to cover their complicated backhaul networks. While these 3G networks basically have evolved from a circuit-switched cellular network, 4G networks are expected to become an all-IP based packet-switched system where packets traverse across an access network and a backbone network without any protocol conversion. To set a goal for 4G networks, International Mobile Telecommunications (IMT) has defined *IMT-Advanced* of which requirements for supported data rates are 100 Mbps and 1 Gbps for high mobility and low mobility, respectively (ITU-R, 2007). Alongside this effort, proposals such as IEEE 802.16m and 3G LTE (Long-Term Evolution) Advanced are on the table to develop new systems towards 4G networks.

These proposals, despite the importance of all-IP networking, may not adopt pure all-IP due to some issues in terms of radio resource management (RRM) and mobility management. The all-IP scenario may enforce each base station (BS) to trigger the change of an IP address, when a mobile station (MS) switches its serving BS. In reality, it is known that changing an IP address incurs a too long delay to provide a seamless service for the MS (Yokota et al, 2002). Therefore, a semi all-IP network is considered, where changing an IP address is not executed within a subnet (i.e., a group of BSs).

In this chapter, we describe the difficulty in deploying all-IP networks for cellular systems by comparing pure all-IP and semi all-IP networks, and provide an overview of existing network architectures such as WiMAX and 3G networks. We then present advanced network architectures that solve the problems of existing network architectures. Finally, an IP-triggered resource allocation strategy is described.

2. OVERVIEW OF WIRELESS NETWORK ARCHITECTURES

A. All-IP Cellular Networks

In existing cellular networks, an access network consists of many entities for supporting radio resource management and mobility management. For example, in 2G GSM/GPRS networks, the base station subsystem (BSS) consists of the base transceiver system (BTS) that handles the physical layer and the base station controller (BSC) that

Figure 1. An example of protocol stacks in GSM systems

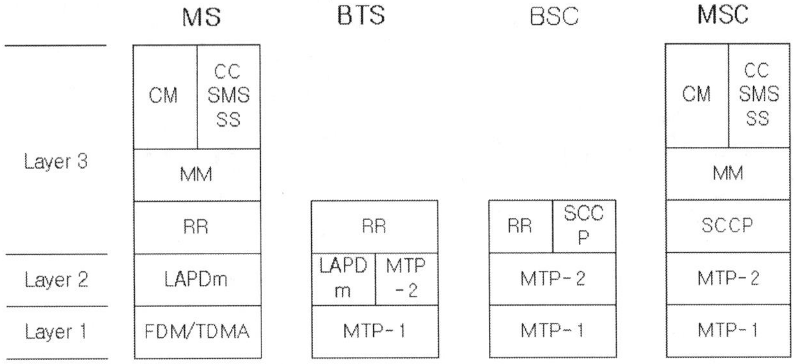

	MS	BTS	BSC	MSC
Layer 3	CM / CC SMS SS / MM / RR	RR	RR / SCCP	CM / CC SMS SS / MM / SCCP
Layer 2	LAPDm	LAPDm / MTP-2	MTP-2	MTP-2
Layer 1	FDM/TDMA	MTP-1	MTP-1	MTP-1

handles radio resource management and handoff. Also, the mobile service center (MSC) fulfills upper layer functionality and acts as the visitor location register (VLR) that is required to update the location of every MS for paging. Protocols defined in each layer in GSM systems are exhibited in Figure 1, where several protocols are defined for communication between any two entities.

4G networks, in contrast, will make such a complicated protocol stack much simpler, by enabling IP packets to traverse between a base station (BS) and a mobile station (MS). Each BS may need to perform all the functionalities required in BSS, BSC, and MSC. This makes the BS play a role of an access router (AR). This architecture is shown in Figure 2. It incurs high overhead, however, especially when an MS configures a mobile IP (MIP) address for handoff. As it is known that it takes several seconds to run the MIP handoff (Yokota et al, 2002), MIP hinders an MS from carrying out smooth handoff. In addition, the 4G network is expected to have a small cell radius due to use of high frequency band, which possibly results in short cell residence time. For this matter, reducing the latency in performing the MIP handoff is still a challenging issue. For instance, a *fast handoff scheme* (Koodli, 2004) proposes to decrease the address resolution delay by pre-configuration.

Another feature of such all-IP networks is their *flat* architecture. All the radio resource management and mobility management will be performed at each BS independently of the other BSs. Unlike traditional cellular networks of a hierarchical architecture, the flat all-IP network can be operated flexibly but at the cost of complexity in terms of intercell RRM (e.g., coordination among cells). There are increasing demands for intercell RRM for efficient network management; for example, fractional frequency planning for OFDMA wireless networks is needed to improve cell-edge performance. The upper entity such as the BSC in hierarchical cellular networks could be a good coordinator for such a scheme.

To alleviate the difficulty in radio resource and mobility management of all-IP cellular networks, a semi (i.e., subnet-based) all-IP cellular network can be considered as shown in Figure 3, an example of a simple network where an AR manages several BSs. The functionality of an AR is separated from that of a BS in order that each undertakes L3 and L2 protocols, respectively. This relation is similar to that between BSC and BTS in GSM networks. Then, an MS moving within the subnet (i.e., changing BSs) performs L2 handoff without changing MIP attachment. The MS only needs to trigger L3 handoff, when it moves into another AR area.

Figure 2. The pure all-IP 4G Network

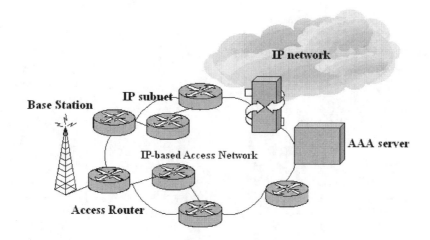

4G Access Network Architecture

Figure 3. The subnet-based 4G network

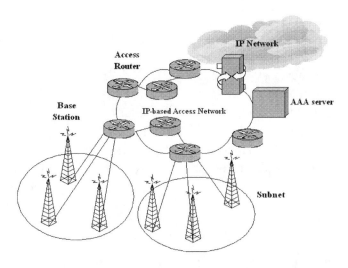

Table 1 compares the network architecture of pure and subnet-based all-IP cellular networks. A main difference is that the former is decentralized while the latter is centralized. Since the pure all-IP network incurs a L3 protocol in the end access link, it requires a long handoff latency and high signaling overhead. However, the architecture is simple and cost-efficient for implementation. On the other hand, the subnet-based all-IP network implements hierarchical architecture, so it is possible to fulfill efficient resource management in spite of its inflexibility. Both network architectures are being considered in WiMAX and 3G-LTE systems, which are described in the following.

B. WiMAX ASN Profiles

The WiMAX standard has defined three different profiles, Profile A, B, and C, for an Access Service Network (ASN) which consists of multiple BSs and an ASN gateway (WiMAX Forum 2008). The relation between a BS and an ASN gateway is also similar to that between a BTS and a BSC in GSM systems. A hierarchical ASN is defined in Profile A and C, whereas a flat ASN is defined in Profile B. Profile A is a hierarchical structure that is similar to traditional cellular networks. As shown in Figure 4, the radio resource controller (RRC) and the radio resource agent (RRA) are implemented at the ASN gateway and the BS, respectively, so most radio resources are managed by the ASN

Table 1. Comparison of two network types

	Pure all-IP network	Subnet-based all-IP network
Access network components	AR	BS + AR
Operation type	Flat	Hierarchical
Intercell RRM	Decentralized	Centralized (Efficient)
Handoff overhead	High	Low
Handoff protocol	L3	L2 + L3
Suitability	Fixed environments	Mobile environments

gateway. In Profile B, the functionalities of a BS and an ASN gateway are co-located on the same platform/solution, which makes the architecture flat. That is, R6 defined for the link between an ASN gateway and a BS does not exist. In Profile C (Figure 5), the RRC is implemented at each BS, so all the RRM functions are performed at each BS as in a flat architecture, although it is still based on a hierarchical structure. Thus, mobility can be managed by the ASN gateway or other upper entities.

C. 3G LTE

The 3G LTE standard (3GPP Release 8, 2009) has defined a simple network architecture of E-UTRAN (evolved universal terrestrial radio access network). The E-UTRAN consists of eNBs (evolved node Bs) which are interconnected with each other by the X2 interface. As presented in Figure 6, each eNB is connected with a S-GW (serving gateway). The S-GW terminates the S1 interface between an eNB and the MME (mobility management entity). The eNB hosts the functions of RRM and dynamic resource allocation as in other BSs, while the S-GW hosts the function of mobility anchoring by assigning an IP address to an end host. This architecture is similar to the Profile C of WiMAX ASN, since most RRM functions are fulfilled by the eNB in a flat manner while some mobility functions are fulfilled by the S-GW in a hierarchical manner.

D. WLAN

Similarly, it is also a controversial issue how to implement the access network in the IEEE 802.11 wireless LAN systems. A subnet is composed of an AR and APs, where the hierarchical structure is also similar to cellular networks. Three types of APs are considered---Fat AP, Thin AP, and Fit AP---according to the role assigned to the AP (Sridhar, 2006). The Fat AP provides router-like functions, so there is no backhauling of traffic. This scenario is very close to the all-IP networking. In contrast, the Thin AP is close to the BS in the WiMAX Profile A. The primary role of Thin APs is to receive and transmit wireless traffic, but in this case, a group of APs are managed by a centralized access controller which acts as an ASN gateway in the WiMAX ASN. In the Fit AP architecture, MAC functions are split between the AP and the access controller, so this architecture is compromised between the Fat AP and the Thin AP models.

Figure 5. WiMAX ASN Profile C

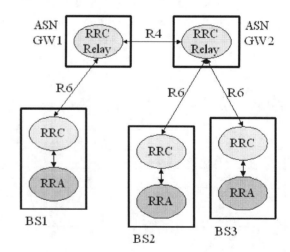

Figure 4. WiMAX ASN Profile A

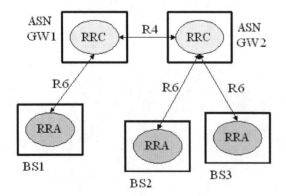

Figure 6. The overall architecture of 3G-LTE E-UTRAN

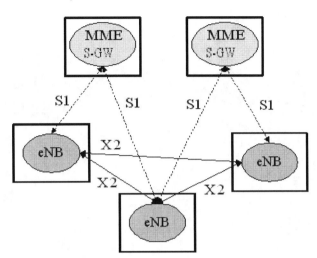

3. ADVANCED NETWORK ARCHITECTURES

Thus far, we have seen the ideal and the reality of 4G wireless access networks. Although ideal all-IP networks can simplify the complicated network architecture of 3G networks, they may be inefficient in terms of RRM and mobility management. Therefore, new wireless systems towards 4G such as WiMAX and 3G-LTE consider semi all-IP networks where BSs behave as in a flat architecture but mobility management is separately performed in a hierarchical manner. Considering this, we here present several advanced network architectures.

A. Semi Hierarchical Cellular Systems

As OFDM (orthogonal frequency division multiplexing) technology has been considered for most next-generation mobile wireless networks, such as IEEE 802.16-WiMAX and 3G LTE (Long-Term Evolution), and these systems are susceptible to the inter-cell interference problem (Li & Lu, 2006; Qualcomm, 2005; Elayoubi *et al*, 2008), there have been numerous approaches proposed to enhance the performance of *cell-edge* MSs. Recently, the 802.16m and 3G LTE Advanced standard groups proposed ways of supporting cell-edge MSs by deploying various network (multi-BS) MIMO antenna techniques (Karakayali *et al*, 2006; Foschini *et al*, 2006) as well as fractional frequency reuse schemes (Li & Lu, 2006; Qualcomm, 2005; Elayoubi *et al*, 2008). In case fractional frequency reuse is adopted, the designated neighboring BSs of a cell to which a certain cell-edge MS is attached, should avoid concurrent transmission over the same set of channels. If macro-diversity or network MIMO is used, one or more neighboring BSs should serve a certain cell-edge MS at the same time, thus requiring concurrent transmissions over the same set of channels from multiple BSs to the same MS.

Such network-level radio resource coordination is easier to achieve, when there is an upper-level management entity like the ASN gateway. When BSs fulfill RRM in a flat architecture, the intercell RRM should be coordinated at least by one BS, rather than in a distributed way. Therefore, we present a simple solution of semi hierarchical cellular systems in Figure 7 where a *super BS*

Figure 7. Semi hierarchical cellular systems

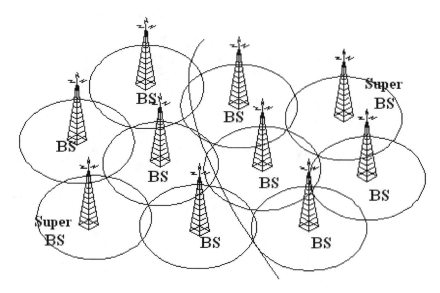

plays a role of a coordinator for intercell RRM. The super BS can be selected among a set of BSs in either a fixed or a dynamic way.

B. Macro/Micro Cellular Systems Combining Multiple Access

The semi hierarchical cellular systems can be extended for solving the mobility problem. Generally cells are categorized into macrocells, microcells and picocells depending on its size. Macrocells and microcells are usually deployed in rural and urban regions, respectively, while the picocells are in a building. In some region such as a hotspot zone, an MS can access both macrocells and microcells as in Figure 8. A service model is proposed (C.-L. I, *et al*, 1993) where macrocells and microcells cover high speed and low speed MSs, respectively. This structure is effective, because a high speed MS has to change cells frequently if covered by microcells.

We extend the hierarchical cell structure by integrating multiple access techniques. Some systems under development are based on OFDMA (e.g., WiMAX) that combines OFDM and frequency division multiple access (FDMA). Since OFDMA systems have lots of channels in a frequency domain, it has higher allocation granularity than OFDM system. It also has the ability of taking advantage of adaptive modulation and coding (AMC), but its application is limited to low mobility. If an MS using OFDMA has high mobility, it cannot perform coherent detection properly due to the long symbol.

Meanwhile, *a FH-OFDMA system*, which combines frequency-hopping and OFDMA, has the advantage of exploiting diversity (Y. Kim *et al*, 2003). Though it experiences difficulty in supporting high data rates and AMC, it can overcome channel fading and multiuser interference through a FH pattern. Accordingly, it is a viable combination that microcells for low mobility use OFDMA that has fine granularity, while macrocells for high mobility use FH-OFDMA that is robust to channel fading and interference.

Each cell plane can handle traffic classes differently as well. High rate data services are suitable for OFDMA that has high spectral efficiency and supports various data rates by AMC. In contrast, low rate services like voice are adequate for FH-OFDMA that is easy to use diversity. If an MS has the capability of supporting dual modes, it can

Figure 8. The model of a hierarchical cellular network

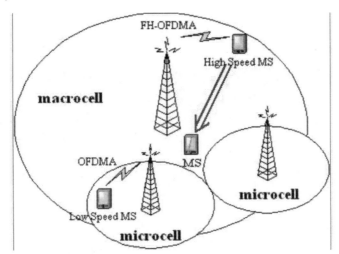

switch cells according to mobility and traffic type in a manner of using vertical handoff that offers an additional merit of load balancing.[1]

In summary, the hierarchical network that consists of OFDMA microcells and FH-OFDMA macrocells has the ability of supporting various users with different mobility and traffic types. Since the considered network is based on a common OFDMA platform, it is more manageable than other heterogeneous networks. Also, it provides the flexibility in network planning. 4G networks will be most probably overlaid with 2G or 3G cellular networks. As existing cellular networks are basically designed for circuit-switched voice service, in some regions, they will keep undertaking voice users and high mobility users like the FH-OFDMA macrocell system, while 4G networks focus on data traffic users by using the OFDMA microcell system.

C. Dual-Linked BS Model

WiMAX and 3G-LTE systems consider flat network architectures but mobility is still implemented in a hierarchical manner, as shown in Figure 3. An MS traveling within the subnet while changing BSs performs L2 handoff only without changing the MIP attachment. When it moves into another AR area, it triggers L3 handoff. On the other hand, the pure all-IP network suffers from a long handoff latency and high signaling overhead since it incurs L3 protocol at each handoff.

The subnet-based network reduces the frequency of L3 handoff that is accompanied by relatively long latency. Nevertheless, an MS still experiences a long latency when it performs L3 handoff. For seamless L3 handoff, we develop a *dual-linked BS model* where some BSs are connected to two neighboring ARs at the same time as shown in Figure 9. Obviously, this approach can be extended to support the case where a BS is linked with more than two ARs by adding more links as many as neighboring ARs to that BS. Here, we will use the terminology of "dual" as the general implementation term.

In the conventional subnet-based model, an MS performs L2 and L3 handoffs at the same time when it crosses the boundary of a subnet. This may cause a serious problem of communication blackout because L2 handoff typically exploits a conservative method in preventing the ping-pong effect. It happens like this. An MS starts an L2 handoff when the signal power of the corresponding BS is weak. As an L3 handoff is accompanied

Figure 9. The subnet-based access network that has a dual-linked BS

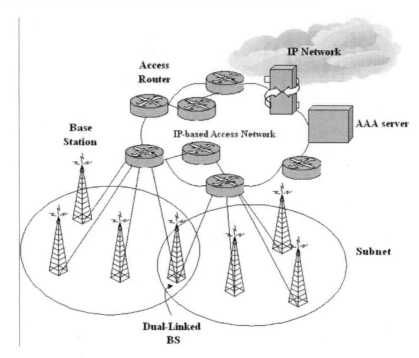

by a long latency, the signal may turn too weak during the L3 handoff, resulting in a blackout.

In contrast, the presented network model with some dual-linked BSs decouples L2 and L3 handoffs, thereby providing a flexible handoff mechanism. Since each dual-linked BS can access both ARs of new and old, it helps L3 handoff to be performed independently of L2 handoff when an MS stays in its coverage. An MS entering the area of the dual-linked BS will prepare the L3 handoff. More explanations are given next.

Future Movement Prediction

Generally an MS is able to sense the presence of neighboring BSs since each BS broadcasts its pilot signal. When the MS enters the service area of a dual-linked BS, it triggers L2 handoff. Completing the handoff, it predicts the movement by detecting the pilot strengths of neighboring BSs. When it is likely to move into some other subnet, it prepares L3 handoff. The handoff can be initiated by either the MS or the BS. If L3 handoff is triggered too early, there exists a possibility of too many L3 handoffs, resulting in the pingponging effect. On the other hand, if too late, L2 and L3 handoffs are incurred at the same time. In this case, the handoff delay may not be reduced, because L3 handoff dominates the overall delay.

This motivates to design an algorithm that initiates early L3 handoff following the concept of the existing L2 handoff algorithm. The graph in *Figure 10* shows an example of L2 and L3 handoff triggers. In this scenario, if the measured pilot signal strength at the MS from a new BS (BS3) exceeds that of the old BS (BS2) by Th_1 for the time interval I_1, the L2 handoff towards the new BS is triggered (Holma & Toskala, 2000). If the new BS belongs to a different subnet, the L3 handoff is initiated according to the thresholds Th_2 and I_2. In this case, the L3 handoff must start before the next L2 handoff. Deciding the threshold values is an implementation issue.

Figure 10. An example of handoff in the dual-linked BS model

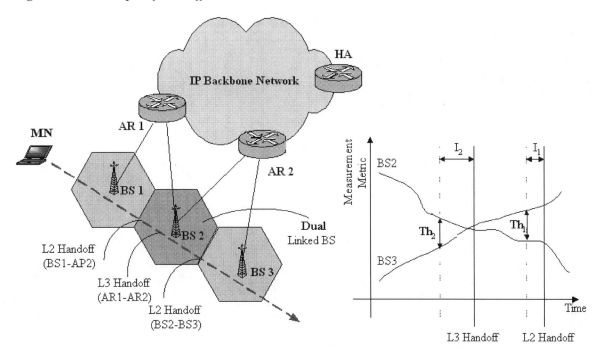

Mobile IP Handoff

Following a proper movement detection, an MS performs MIP handoff. The MS begins MIP handoff by sending a request message. After the MS obtains a CoA (care-of-address) for the new subnet, the AR forwards the request message to the MS's Home Agent (HA) to update the MS's location information. During this process, packets arriving at the BS via the old AR will be transferred to the MS by using the new CoA. This is possible because the BS can access the two ARs, which is a unique feature in this scheme. In contrast, in a conventional network, some packets arriving at the old BS or AR will be dropped or forwarded via old and new ARs, so the forwarded packets will experience some latency.

In the dual-linked BS model, each dual-linked BS maintains a table for mapping between old and new CoA during the handoff procedure. Indeed, the handoff latency in MIP is mainly incurred by HA registration. In the new architecture, however, there exists only a little handoff latency since every packet arrives at the same BS via either old or new AR during the handoff. *Figure 10* shows an example of the handoff scheme. An MS and its corresponding BS perform L2 handoff whenever it crosses a cell boundary, while performing L3 handoff separately from L2 handoff.

The advantage of the dual-linked BS model is exhibited in Figure 11. In the conventional model, the new BS or AR may receive packets for an MS to which it does not have a direct connection while the MS is performing the handoff between subnets. In this case, the new BS or AR may drop or buffer packets destined for the MS. This case also occurs when the whole handoff process has not completed yet even if the MS has established a new connection already. Therefore, as shown in Figure 11, some packets should be forwarded to the corresponding location during the handoff process. Unless the system supports

Figure 11. Packet forwarding comparison between the conventional and new models

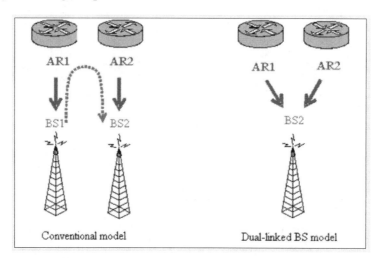

packet forwarding, packets will be dropped. In contrast, the dual-linked BS model has almost no packet loss during the handoff, because the BS has a connection to each of old and new ARs.

4. IP-TRIGGERED RESOURCE ALLOCATION STRATEGY (ITRAS)

To support IP QoS, the Internet Engineering Task Force (IETF) recommends *integrated services (IntServ)* (Braden *et al*, 1994) and *differentiated services (DiffServ)* (Blake *et al*, 1998). They are expected to be effective also in all-IP-based 4G networks. Since 4G networks will support multimedia traffic, we need to visit the issue of providing IP QoS in IP based wireless access networks, and propose ITRAS for QoS support in 4G networks, where the decision of radio resource allocation follows IntServ or DiffServ policy.

A. IP QoS

IntServ (Braden *et al*, 1994) uses Resource Reservation Protocol (RSVP) to reserve bandwidth during the session setup. As a first step of RSVP, the source sends a QoS request message of PATH to the receiver through intermediate routers which run an admission and a policy control. If the sender receives RESV returned from the receiver through the reverse route as an indication of QoS guarantee, it initiates the session. If each router along the path receives packets, it classifies and schedules them. IntServ ensures strict QoS, but each router has to implement RSVP and maintain per-flow state, which brings difficulty in a large-scale network.

DiffServ (Blake *et al*, 1998), on the other hand, does not need any signaling protocol and cooperation among nodes. As the QoS level of a packet is indicated by the DS field of IP header (TOS field in IPv4, Traffic Class field in IPv6), each domain can deal with it independently. Once the packet is classified, each router can mark, shape or drop it according to network status. Since DiffServ is not so rigorous as IntServ, it is scalable in supporting QoS statistically.

B. QoS of Wireless Access Networks

In general, a wireless access network has the capability of managing QoS independently of the IP network because it becomes a bottleneck for providing end-to-end QoS. QoS control can be made possible by using some access and scheduling methods. Recently the QoS of IEEE 802.11 WLAN system is supplemented by IEEE 802.11e standard (IEEE 802.11e, 2005). It defines Extended Distributed Contention Access (EDCA) that assigns a small backoff number to high priority traffic, and Hybrid Coordination Function (HCF) that improves the conventional polling scheme of Point Coordination Function (PCF). Also, cdma2000 1x EV-DO and WCDMA-HSDPA (High Speed Downlink Packet Access) adopted opportunistic scheduling to exploit channel fluctuation. The opportunistic scheduling has brought an implementation issue in designing various scheduling algorithms for QoS (Andrews et al, 2001).

The Third Generation Partnership Projects (3GPP and 3GPP2) define four traffic classes and their related parameters for QoS provisioning. There exist gateways between IP backbone and access networks that perform protocol conversion and QoS mapping between IP and access networks (Chen & Zhang, 2004). However, direct translation is difficult since access networks have their own QoS attributes that require strict QoS provisioning within them.

Meanwhile, the importance of unified QoS management grows in 4G networks as QoS management for both access network and IP network becomes cumbersome in all-IP networks. If each network has an individual QoS model, it needs a rule that integrates their QoS models to ensure end-to-end QoS. For the unified QoS management, we propose ITRAS that considers L1, L2 and L3 together. In ITRAS, L1 and L2 allocate radio resources and logical channels, respectively, according to the QoS indication of L3.

C. ITRAS

ITRAS concerns the information about IntServ and DiffServ for the resource management of L1 and L2. When IntServ sets up a real-time session, MAC reserves a dedicated channel. On the contrary, when DiffServ is used for low mobility users, MAC can exploit either a dedicated or shared channel. If the shared channel is allocated for DiffServ, the wireless scheduler runs a scheduling algorithm for QoS provisioning. In contrast, the dedicated channel allocation needs admission control that allows a limited number of users into the network for QoS support. Therefore IP QoS information helps MAC and PHY manage resources of the following in a flexible manner.

- Cell type - microcell or macrocell
- Multiple access - OFDMA or FH-OFDMA
- MAC channel - dedicated or shared
- PHY scheduling - priority or fairness

IntServ is easy to be involved in radio resource management because wireless access is usually accompanied by signaling. When an MS requests a real-time service in a 4G network, the corresponding AR can initiate IntServ and allocate a dedicated channel. For a downlink call, the AR can adjust the bandwidth of a dedicated channel with the aid of RSVP. As real-time traffic usually requires a constant data rate, the dedicated channel is recommended to use power control rather than AMC. In this aspect, FH-OFDMA and CDMA may have more suitability than OFDMA for real-time services.

Regarding DiffServ in 4G networks, it is enough for an MS to set the DS field properly for uplink packets because the BS controls radio resources before transferring them to the AR. For downlink traffic, the AR classifies packets according to the DS field and chooses a multiple access method, and accordingly the BS allocates a dedicated or shared channel. The dedicated channel has the advantage of simple management, while the shared channel goes well with the DiffServ because both require scheduling. Contrary to scheduling in routers which need to handle a lot of flows, wireless scheduling takes care of not many connections, which allows to use per-user buffer. So the wireless scheduler can exploit an

Figure 12. The coupled layering for resource management

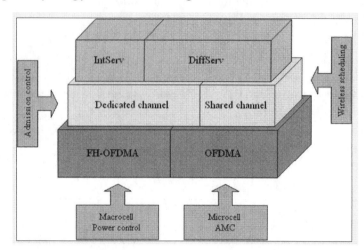

algorithm with high granularity of radio resources. Figure 12 and Table 2 summarize tightly coupled resource management among three layers through a unified QoS strategy.

D. Further Issues

Implementing ITRAS needs further study. Specifically, when the subnet-based all-IP network is deployed, an AR should cooperate with its subordinate BSs in performing ITRAS functions. While resource management functions are primarily handled by BSCs in the GSM networks, more functions will be imposed on BSs in 4G networks. Basically ARs will be responsible for IP QoS and BSs will play the primary role of resource management. Another challenge is the application of ITRAS to the macro/micro cellular network. In this case, a coordinator is needed in deciding whether an incoming session is served by a macrocell or a microcell. It will also have the capability of load balancing by triggering vertical handoff.

Along with the architectural evolution towards all-IP network, one of the most salient trends for future network design is emerging in the form of Fixed Mobile Convergence (FMC). The integration of wireline and wireless technologies and services realized by FMC is expected to offer benefits to both operators and consumers by delivering enhanced user experience over a unified framework. IP Multimedia Subsystem (IMS) (Poikselka, 2004) lies at the heart of this network convergence. It is a framework that provides a variety of IP based services. This framework enables wireline, wireless and cable operators to offer rich multimedia services across both

Table 2. An example of ITRAS

Traffic class	Mobility	IP QoS	Logical channel	Multiple access
Real-time	High	IntServ/DiffServ	Dedicated	FH-OFDMA
	Low	IntServ	Dedicated	FH-OFDMA / OFDMA
		DiffServ	Dedicated/shared	
Non-real-time	High	DiffServ	Shared	FH-OFDMA
	Low	DiffServ	Shared	OFDMA

legacy circuit switched and new packet switched network infrastructures. Also, together with QoS provisioning, security should be guaranteed in 4G mobile networks.

5. CONCLUSION

In this chapter, we discussed the ideal goal of all-IP 4G wireless networks and how they are being implemented in WiMAX and 3G LTE systems in reality. All-IP wireless networks, despite their ultimate usefulness, encounter constraints in providing seamless mobility and efficient RRM. Therefore, semi all-IP networks where IP connections are not directly managed through BSs have been considered. Further, we presented several advanced semi all-IP network models to deal with mobility and RRM efficiently. Finally, we designed a unified QoS strategy, named ITRAS that considers L1, L2 and L3 all together. In this approach, IP QoS such as IntServ and DiffServ can determine resource attributes of the wireless access network, i.e., MAC channel and PHY resource as well as multiple access type. Our approach of tight layer coupling provides a neat and flexible solution to accommodate various demands of 4G networks, which cannot be handled properly by a conventional simple layered approach.

REFERENCES

Andrews, M.(n.d.). Providing Quality of Service over a Shared Wireless Link. *IEEE Communications Magazine*, *39*(2), 150–154. doi:10.1109/35.900644

Blake, S., Black, D., Carlson, M., Davies, E., Wang Z., & Weiss, W. (1998). An Architecture for Differentiated Service. *RFC 2475*.

Braden, R., Clark, D., & Shenker, S. (1994). Integrated Services in the Internet Architecture: an Overview. *RFC, 1633*.

Chen, J.-C., & Zhang, (T. 2004). *IP-based Next Generation Wireless Networks*. Hoboken, NJ: Wiley.

Elayoubi, S.-E., Ben Haddada, O., & Fourestie, B. (2008). Performance Evaluation of Frequency Planning Schemes in OFDMA-based Networks. *IEEE Transactions on Wireless Communications*, *7*(5), 1623–1633. doi:10.1109/TWC.2008.060458.

Foschini, G. J., Karakayali, M. K., & Valenzuela, R. A. (2006). Coordinating Multiple Antenna Cellular Networks to Achieve Enormous Spectral Efficiency. *Proceedings of the IEEE*, *153*(4), 548–555. doi:10.1049/ip-com:20050423

Holma, H., & Toskala, A. (2000). *WCDMA for UMTS*. Hoboken, NJ: Wiley.

I, C.-L., Greenstein, L. J., & Gitlin, R. D. (1993). A Microcell/Macrocell Cellular Architecture for Low- and High-Mobility Wireless Users. *IEEE Journals on Selected Areas in Communications*, 11(6), 885-891.

IEEE 802.11e. (2005). Part 11: Wireless LAN Medium Access Control (MAC) and Physical Layer (PHY) specifications Amendment 8: Medium Access Control (MAC) Quality of Service Enhancements. *IEEE Standard for Information technology—Telecommunications and information exchange between systems—Local and metropolitan area networks—Specific requirements.*

ITU-R. (2003). *Framework and Overall Objectives of the Future Development of IMT-2000 and Systems Beyond IMT-2000* [Recommendation ITU-R M. 1645].

ITU-R. (2007). *Principles for the Process of Development of IMT-Advanced* [Resolution ITU-R 57].

Karakayali, M. K., Foschini, G. J., & Valenzuela, R. A. (2006). Network Coordination for Spectrally Efficient Communications in Cellular Systems. *IEEE Wireless Communications Magazine, 13*(4), 56–61. doi:10.1109/MWC.2006.1678166

Kim, Y. (2003). Beyond 3G: Vision, Requirements, & Enabling Technologies. *IEEE Communications Magazine, 41*(3), 120–124. doi:10.1109/MCOM.2003.1186555

Koodli, R. (2004). *Fast Handovers for Mobile IPv6.*

Li, G., & Lu, H. (2006). Downlink Radio Resource Allocation for Multi-cell OFDMA System. *IEEE Transactions on Wireless Communications, 5*(12), 3451–3459. doi:10.1109/TWC.2006.256968

Poikselka, M., et al. (2004). *The IMS: IP Multimedia Concepts and Services in the Mobile Domain.* Hoboken, NJ: Wiley.

Qualcomm R1-050896. (2005). *Description and Simulations of Interference Management Technique for OFDMA based E-UTRA Downlink Evaluation, 3GPP TSG-RAN WG1 #42.*

3GPP Release 8. (2009). *Overview of 3GPP Release 8 V0.0.4.*

Sridhar, T. (2006). Wireless LAN Switches -- Functions and Deployment. *The Internet Protocol Journal (Cisco), 9*(3), 2–15.

WiMAX Forum. (2008). *WiMAX Forum Network Architecture, Release 1, version 1.2.*

Yokota, H., Idoue, A., Hasegawa, T., & Kato, T. (2002). Link Layer Assisted Mobile IP Fast Handoff Method over Wireless LAN Networks. In *ACM MOBICOM '02.*

ADDITIONAL READING

Choi, Y.-J., Arora, A., & Bahk, S. (2007). Seamless Mobile IP handoff in Mobile WiMax Networks. In *Proceedings of World Wireless Congress 2007.*

Mani, M., & Crespi, N. (2007). Inter-Domain QoS Control Mechanism in IMS based Horizontally Converged Networks. In *Proceedings of IEEE International Conference on Networking and Services* (pp. 82-86).

Prasad, A., & Schoo, P. (2002). IP Security for Beyond 3G Towards 4G. In *Proceedings of WWRF 7.*

Sauter, M. (2009). *Beyond 3G: Bringing networks, terminals, & the web together.* Hoboken, NJ: Wiley.

ENDNOTE

[1] The interoperability between heterogeneous networks is being discussed in the IEEE 802.21 group.

Chapter 4
Architecture for IP-Based Next Generation Radio Access Network

Ram Dantu
University of North Texas, USA

Parthasarathy Guturu
University of North Texas, USA

ABSTRACT

High call volumes due to novel mobile data applications necessitate development of next generation wireless networks centered on high performing and highly available radio access networks (RANs). In this chapter, the authors present an innovative IP-based wireless routing architecture (for a RAN) with mechanisms for seamless handoff operations and high Quality of Service (QoS). Algorithms for dynamic configuration of the RAN, and efficacious network bandwidth management through traffic control are also presented. The authors establish the superiority of their system with real-life data indicating significant cost and availability improvements with our system over the traditional networks.

INTRODUCTION

Explosive growth in mobile data applications in recent times has motivated the mobile operators to explore efficacious and cost effective solutions for handling the backhaul traffic (from base-stations to their core networks). These solutions for the B3G (Beyond 3rd Generation) wireless networks include bandwidth optimizations in the current radio access networks (RANs) as well as IP (Internet Protocol)-based RANs over the T1/E1 and Ethernet backhaul for 1X-EVDO CDMA (1 time radio transmission technology Code Division Multiple Access) (TIA/EIA/IS-2000.1.A -2, 2002) and HSPA W-CDMA (High Speed Packet Access Wideband CDMA) networks (3GPP2 C-S0033 Rev 0 Ver 2.0, 2003). We propose here IP-based RANs which replace legacy mobile wireless systems with innovative wireless routers (WRs) equipped with traffic management and self-configuration capabilities. These attributes together with high system availability and high QOS makes our IP RAN an ideal choice for next generation mobile wireless networks.

The proposed IP-based RAN architecture can be adapted to new long term evolution (LTE) standard for mobile communications (Ekström *et al.*, 2006,

DOI: 10.4018/978-1-61520-674-2.ch004

Ergen, 2009, Fazel & Kaiser, 2009) based on orthogonal frequency division multiple access (OFDMA) technology developed by 3rd generation partnership project (3GPP). The evolved packet system (EPS) for the evolved universal mobile telephone system (UMTS) terrestrial radio network (E-UTRAN), which forms the access side of the LTE architecture, will not impact the IP-architecture proposed here. However, our proposal could be compatible with the IP-based flat core network architecture of LTE, variously known as extended packet core (EPC) and System Architecture Evolution (SAE). As a matter of fact, since LTE release 8's air interface is intended for use over any IP-network, compatibility of our IP architecture is no issue at all.

Now that LTE has been ratified in March 2009, adaptation of the current work to LTE can be carried out.

Worldwide Interoperability for Microwave Access (WiMAX) is another competitive next generation technology for mobile communications (Ergen, 2009, Fazel & Kaiser, 2009). It is based on the IEEE 802.16 standard for broadband wireless access. The WiMAX forum proposed an architecture for integration of a WiMAX network with an IP core network. This makes it possible to adapt our IP architecture to WiMAX networks also.

BACKGROUND

Since IP-based RANs is an emerging technology, the literature on flexible and reconfigurable next generation RAN architectures based on versatile wireless routers is rather sparse. An interesting paper in this direction, Ghosh, Basu & Das (2005) provides an insight into a flexible, reliable, and cost effective IP-based RAN architecture with BSs connected to form a mesh. Bu, Chan & Ramjee (2006) discuss connectivity, performance and resiliency of IP-based CDMA RANs focusing on a star topology between the BSs and RNC (Radio Network Controller). Vassiliou *et al* (2002) propose a RAN for next generation wireless networks based on multi-protocol label switching (MPLS) and hierarchical mobile IP. Kempf & Yegani (2002) propose a new architecture called open-RAN for mobile wireless IP RANs. Yasukawa, Nishikido, & Hisashi (2001) propose an IPV6-based wireless routing architecture with scalable mobility and quality of service (QOS) support. Chen & Hamalainen (2003) present a method for handovers in IP RANs.

The proposed wireless router (WR) architecture, its mechanisms for efficacious traffic and bandwidth management, and dynamic self-configuration are based on our earlier patented research work (Dantu, 2005; Dantu et al., 2006; Patel et al., 2006).

WIRELESS ROUTER ARCHITECTURE

1. Issues, Controversies, Problems

The legacy mobile wireless architecture shown on the right half of Figure 1 with a hierarchy of BS (Base Station)s, BSC (Base Station Controller)s, and MSC (Mobile Switching Center)s has a number of shortcomings with respect to handling of large volumes of data in B3G networks: i) wireless frame selection for handoff management is done at BSCs resulting in the duplicate traffic flow on the backhaul, ii) even during the ideal periods of a call, transmission resources are reserved resulting resources wastage in contrast to the IP-networks equipped with the statistical multiplexing scheme, iii) only 15% of the BS-BSC traffic is payload, and the rest is overhead, iv) BSs forward erroneous frames also BSC and this results in dead payload on the backhaul, v) uneven utilization of links makes the system inefficient, cost-ineffective, and unsuitable for deployment of new data intensive services, vi) transmission delays in long BS-BSC links could cause soft-handoff failures and call

Figure 1. A novel distributed wireless router based RAN communicating with a legacy network (each pink dot represents a wireless router)

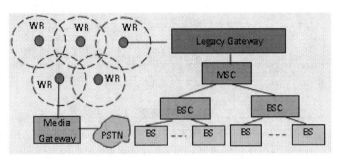

drops and thereby contribute to performance degradation, and vii) single-point failures of the legacy system entities or the links between them results in low system availability.

The demand for high quality services despite large volumes of call traffic necessitates a drastic reduction in the long backhaul (BS-MSC) control and data paths used in the legacy systems. An elegant approach to address the problem is to build the next generation RANs as distributed configurations of simple but functionally comprehensive wireless router (WR)s that could replace and at the same time inter-work, as shown in Figure 1, with the hierarchy of legacy wireless system entities *i.e.* the BSs, BSCs, and MSCs. Thus these new versatile WRs, need to have integrated into them several overlaying features of an all-IP 3G wireless network depicted in Figure 2. Even though the functional modules of the service layer are not detailed out in the figure in view of the ever growing number of wireless/wire-line services, typical entities that provide network services are call servers, bandwidth brokers, SLA (Service Level Agreement) managers, billing servers, HLR (Home Location Register)s, HSS (Home Subscriber Server)s, MGW (Media Gateway)s, SGW (Signaling Gateway)s, legacy servers, DNS (Domain Name Servers)s, and so on. The control layer supports the services with entities such as QOS (Quality of Service)/Mobility/Location/Power managers, Call Agents, and AAA (Authentication, Authorization, and Accounting) mangers. In the legacy systems, these two functional layers roughly correspond to the MSC and BSC functions, respectively. For execution of these functions, the legacy system entities (BSs, BSCs, and MSCs) need to communicate with one another. This is supported by backhaul IP networks with wire-line topologies. A part of the inter-BS communication is supported by wireless routing. Finally, the communication between the BSs at different cell sites and the mobile device constitutes the physical layer functionality of the wireless system.

Since the WR replaces the mobile network entities in the proposed RAN architecture, it needs to incorporate the control and routing functionalities of the legacy systems. In particular, it should have the following features: i) it should support all the data and signaling protocols for inter-router communication as well as communication with various service and control entities, ii) it should facilitate effective hand-off management to achieve nearly zero call drop rate, iii) it should provide high QOS by effective bandwidth management through traffic shaping, and iv) it should be capable of dynamically configuring its operational parameters in collaboration with its neighbors, and adapt itself to RF topology and other changes. Overall, it should render the RAN both high performing and highly available at a low cost.

Figure 2. Several overlaying features that need to be integrated into a WR

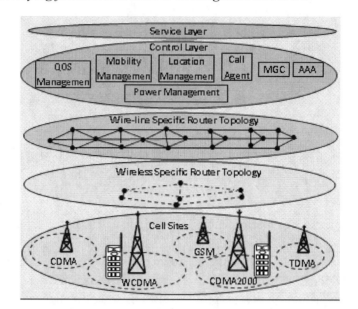

2. Solutions and Recommendations

In the proposed architecture, the wireless routers (WRs) situated in different cells perform message routing in addition to the traditional Base Station (BS) functions. In fact, the traditional functions of the base stations, BSC (Base Station Controller), and MSC (Mobile Switching Center) in the legacy systems are lumped together and distributed across various WRs in the new wireless architecture. Figure 3 is a very high level depiction of various inter-router links as well as the control links between the proposed WRs, and the traffic and control interfaces. The inter-router links include: i) wireless specific virtual tunnels *i.e.* Multi Protocol Label Switch (MPLS) paths based on IP packets, ii) RSVP (Resources Reservation Protocol)/LDP (label Distribution Protocol) signaling channel, iii) routing message channel, iv) wireless specific virtual channel to carry MPLS based radio frames, and v) one or more wireless-specific control channels for call setup and maintenance including a signaling channel for usage by any signaling protocol (SP) such as extended RSVP and a routing channel for usage by any radio routing protocol (RRP) such as extended OSPF (Open Shortest Path First), RIP (Routing Information Protocol), or BGP (Boarder Gateway Protocol). The bold and dashed lines in the figure indicate the data and control channels, respectively.

For supporting wireless traffic services, the WRs access traffic and control interfaces that include media gateway controllers, WAP (Wireless Application Protocol) servers, policy management servers, call agent controllers, mobility managers, and AAA (Authentication, Authorization, and Accounting) servers. WRs communicate with these interfaces through MGCP (Media Gateway Controller Protocol), COPS (Common Open Policy Service), and other suitable protocols.

The WR architecture proposed herein facilitates wireless-access technology (*e.g.* CDMA or TDMA) independent network routing with the help of wireless interfaces to disparate wireless peripherals as shown in Figure 4, and hence is pivotal to an all-IP (Internet Protocol) radio access network (RAN) that seamlessly inter-works with the backhaul IP network with interfaces to various network peripherals. For communication with the backhaul networks at the other end,

Figure 3. Proposed wireless router links to peer routers and various control interfaces

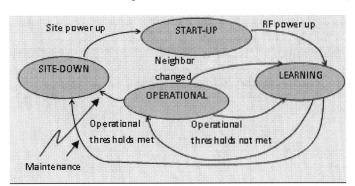

the WR includes wire-line interfaces to various network peripherals as well. At the heart of the WR lies the traffic control with various modules as follows: i) Quality of Service (QOS) engine for traffic conditioning and effective management of transmission resources, ii) Selection and Distribution Unit (SDU), iii) Central Processing Unit (CPU), iv) Call Processor, v) Timing Unit for synchronization purpose, vi) Communication Module with various traffic-controller interfaces to gateways, services, policy managers, IP routers, base-stations, call agents, and other remote nodes and resources, vii) Power and Interference Manager, viii) Radio resources Manager, ix) Mobility Manger, X) Packet Classification Unit, and xi) Security (IP SEC) Module.

In the traditional mobile networks, an SDU is placed centrally at the Base Station Controller (BSC) to manage the call processing at a number of base stations. It selects the best frame from a number of incoming radio frame instances from the mobile via different base-stations (BSs) for onward transmission to the intended destination through backhaul network, and distributes similarly the messages received from the backhaul to the target mobiles. Based on the quality of the frames received from different base-stations, an SDU also manages soft-handovers (that is, call redirections from one BS to the other). In our architecture, BS is replaced by the more versatile WR module. Additionally, it incorporates an innovative concept of a distributed SDU with every WR housing an SDU. These distributed SDUs in the present architecture facilitate distribution of intelligence, and switching and control functionality to individual cell sites.

Figure 4. Internal structure of the proposed wireless router

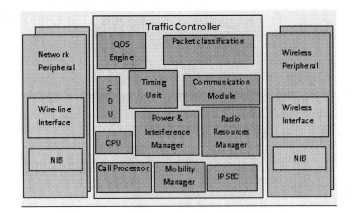

Figure 5. Data transfer from the mobile device to an edge router

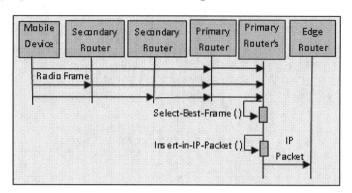

The main advantage here is that the radio frames need not be transmitted to the BSC over the backhaul links. A very efficacious use of the backhaul network bandwidth this way, in turn, results tremendous cost savings for the customer. Additionally, this approach helps in averting traffic congestion. Common switching points leading to delayed/dropped traffic are reduced if not altogether eliminated. However, these advantages could be realized only by effective inter-router communication methods for mobility management by call redirection (soft handoff) and MPLS path reconfiguration as the mobile device transitions between cells. The proposed innovation should also be implemented without compromising the quality of service (QOS). In the following subsections, we describe the mechanisms for soft handoff and QOS in the proposed architecture.

A. Call Processing and Soft Handoff Mechanism

Figure 5 depicts interactions among various network elements for ingress data (*i.e.* data flowing from the mobile device to the edge router. Figure 6 depicts the interactions for the egress data flowing in the reverse direction. These figures are OO (object-oriented) style sequence diagrams for call flow in an UML (Universal Modeling Language) like notation.

In both the scenarios described here, the first step is designation of one of the routers accessible to the mobile device as the primary router. In case of mobile originated calls, the mobile informs the WR from whom the strongest signal is received to take the responsibility as prime WR. On the other hand, in case of mobile terminated calls, the edge router determines the prime router based on the mobile location after locating and successfully paging the mobile. Once the primary router is determined, it initiates the process for setting MPLS tunnels between itself and the edge router as well as the secondary routers. The key innovation here is to employ these MPLS tunnels to emulate the BS-BSC and BSC-MSC (mobile switching center) A3/A7 interfaces in the legacy wireless systems. Distribution of call flow control this way among various routers this way results in enormous cost savings for the customers due to effective utilization network links by reduction of the call control and data paths. Figures 5 and 6 depict the call flows after creation of MPLS tunnels.

In case of ingress data, the radio frames transmitted by the device are received by all active routers including a primary router and a number of secondary routers as shown in Figure 5. The secondary routers simply forward the radio frames to the primary router. The SDU of the primary WR selects the best one among all such frames

Figure 6. Soft handoff and data transfer from an edge router to the mobile

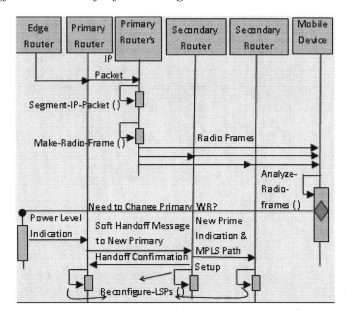

including the one directly received, inserts that into an IP packet, and transmits it to a back-haul network via an edge router for onward transmission to the other party.

The flow in case of egress data is naturally in the opposite direction as shown in Figure 6. The core network hands over the IP packets to an edge router for onward transmission to the primary router through a pre-established or dynamic MPLS path. The primary WR segments the packets into radio frames and multicasts them to all the secondary WRs in the active set via dynamically configured Label Switch Paths (LSPs). The primary and secondary WRs then transmit each one of these received radio frames after different amounts of delay offset to the mobile device so that the replicas of individual frames from different WRs arrive simultaneously at the destination. These synchronous radio frames received from different WRs are analyzed by the mobile device to not only obtain the best (correct) radio frame, but also assess the power levels of the frames. As the mobile device moves away from the primary WR, the power level of the radio frames from the mobile at the primary as well as that of the frames from the primary at the mobile drop. When the power level drops below a pre-configured threshold, the mobile device sends a control message to the primary WR indicating the power level of the prospective primary. The new primary WR could be one of the previous secondary wireless routers in an active set for the call. To achieve micro mobility, the current primary WR would signal the new primary an indication of the handover of its responsibility as primary WR. It would also supply to the latter the list of active WRs in the same control message. After receiving this message, the new primary WR receiving the strongest signals first confirms to the current primary that it is ready to take control of the traffic distribution, and then establishes multicast MPLS paths (LSPs) for the secondary routers in the active list. The LSPs provide synchronized framing for distribution and selection between neighbors of wireless traffic and fast rerouting for soft handoff using RSVP.

Figure 7. Feedback mechanism for traffic flow control

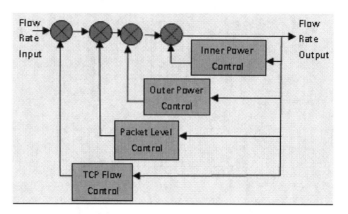

B. High QOS with Traffic Control for Effective Bandwidth Management

Overloading of a wireless network could occur because of the heavy data traffic. Traffic flow control for effective management of transmission resources, particularly the bandwidth, facilitates high quality of service (QOS). For example, by shaping (that is, spacing) the data traffic which usually comes in bursts unlike the voice traffic, the network bandwidth could be utilized more effectively. On the other hand, overloading of the wireless network even for short durations of time could result in degradation of QOS due to increased bit error rates.

Figure 7 depicts the different feedback controls used in our system for traffic flow adjustment for effective data transmission. Signal power of the mobile device or any other direct indicator of power for the RF link constitutes the inner power control just as in the traditional wireless systems. This feedback is provided every 1-2 milliseconds. Similarly, outer power comprises a link error rate and/or interference indicator for the wireless link and may account for soft handoff power. This feedback may be provided every 50-100 milliseconds. The inner and outer power control loops conjointly provide feedback based on the signal strength of the RF link. The packet level control is provided every several hundred milliseconds by the queuing system in the WR based on the congestion status of the queues. Finally, our repertoire of traffic control mechanisms included the well known and well studied TCP flow control, which is provided through the acknowledgment messages between the two end points of the TCP flow. Based on the acknowledgment messages received, the source (WR) adjusts its transmission rate. Thus, with this mechanism, traffic flow is controlled by the congestion and/or interference state of the wireless links.

In our WR, there is also a provision for the traditional end-to-end rate control with a queue mechanism to shape up bursty traffic from a source into a smooth traffic flow of radio frames into the sink (mobile). The ACKs from the sink are also similarly queued up, and used as feedback for the source so that it can control its egress traffic flow.

An innovative flow control mechanism in the present work is Gang (or Group) flow control which seeks to shape the TCP flows from various sectors of the wireless network simultaneously, shown as Figure 8 with N acknowledge shapers corresponding to N sectors of the WR. Each shaper accepts the packets from wireless network and stores the packets in the acknowledge queues inside the WR. The acknowledge shaper can transmit acknowledge message over time to change the traffic flow for the sector based on

Figure 8. Gang flow control for sectors of a wireless network

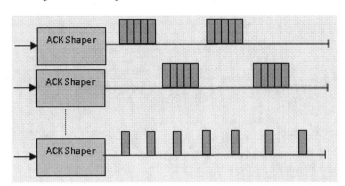

the flow's power indicator of the RF link from wireless network.

Figure 8 illustrates two kinds of shaped acknowledge messages. In the first type shown for flows 1 and 2, the acknowledge messages are arranged in small groups and the groups are dispatched periodically. In the second type depicted for Flow N, on the other hand, the ACKS are evenly distributed over time and transmitted. The difference of these two arrangements is that they have different transmit time. The first type of ACK shaping will affect the offsetting bursts for the traffic flow whereas the second type will affect the steady flow rate for the traffic flow.

The flows with unused bandwidth or lack of bandwidth will be identified for each TCP group for the interval related to retransmission time out for the TCP flow. The fair share of each TCP flow within the group will be calculated and used in adjusting the speed of acknowledge messages for the flows inside the gang. The fair share of each TCP flow may be used together with the RTT and arrival time for each traffic flow.

C. Mechanisms to Dynamically Configure the Router

The proposed WR has been designed for dynamic configuration of its operational parameters. Configuration parameters of a WR are typically related to the site and technology used. Site parameters may be classified as geo-location, network operation, service configuration, and antenna parameters. The technology specific parameters depend upon whether the CDMA or GSM is supported and include technology-specific site parameters, but may be broadly classified into coverage, spectrum, channel, interference, control, and threshold parameters. At a finer level of detail, site Id, number of sectors/beams, sector/beam ID, latitude and longitude, sector/beam location, maximum radius of influence are typical geo parameters. Similarly, network configuration parameters include network interfaces (*e.g.* T1, SONET, T3, *etc.*), site capacity, and network capacity. In the service configuration, we may have the list of various services supported, and the related directory agent (DA) addresses. The antenna parameters listed on a per sector/beam basis include the antenna type, digitized pattern, horizontal/vertical beam widths, max gain, and mechanical and electrical down tilts.

In the technology parameters class, maximum RTD (round trip delay), PER (packet error rate), FER (frame error rate), and percentages of blocked calls, access failures, dropped calls constitute the threshold parameters subclass. The coverage parameters subclass includes environment (*e.g.* rural or urban), path loss margin, technology specific hardware losses and gains, RF coverage prediction models, and traffic distribution maps. The spectrum parameters subclass consists of channel

Figure 9. A state machine for dynamic configuration of a wireless router

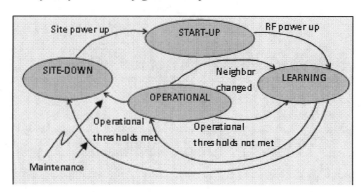

bandwidth, channel mask, channel number range, and maximum transmit power per channel. The channel parameters include the number of channels in the range, air capacity/bandwidth, minimum channel spacing, frequency use, frequency grouping, and hopping sequences. The interference parameters include interference thresholds, power control thresholds, channelization and sequencing, channel scheduling algorithms, RF interference prediction models, traffic distribution maps, and adjacent channel interference threshold. The control parameters subclass includes access parameters, intra-technology and inter-technology handoff parameters, and timing parameters.

In the following subsection, we present a very high level procedure by which a WR learns its parameters and configures itself in collaboration with its neighbors. The control logic and the operations performed in each high level state could be quite complex with several states for error paths and exceptions. For example, application of RF or IP discovery protocols in the startup state involves considerable information exchange between a WR and its neighbors. Hence, we present only an overview of our automatic WR configuration state machine here below, and refer to our patent [10] for finer details.

D. Automatic Router Configuration

Automatic configuration of the WR is performed using a 4-state machine including a start up state, a learning state, an operational state and a site down state as depicted in Figure 9.

The WR configures the RF/IP topology in the startup state, and refines the topology in the learning state. In the operational state, the wireless router handles a full traffic load, and continues to check if it meets the operational thresholds. Scheduled or unscheduled maintenance leads the WR into site-down state.

1) Start up state: In the starting state, connectivity is first established between the WR and the wire-line routers in the network. The wire-line connectivity is then used to establish connectivity between WR and its WR neighbors. Subsequently, MPLS paths or other suitable virtual circuits or IP tunnels are established among the WR neighbors to facilitate inter-router communication. The WR uses wire-line connectivity to learn from its neighbors the RF topology in the neighborhood, and establish wireless-specific connectivity with its neighbors.

After establishing the wireless connectivity with its WR neighbors, a WR exchanges RF impact information, including some or all of the parameters mentioned earlier. By exchanging this information, and negotiating the various operation parameters, the WR is able to determine or estimate a set of operating parameters that will help maximize radio coverage, minimize interference, and aid in providing a seamless coverage from cell to cell with smooth handovers. If no coordination could be achieved between the wireless routers in a neighborhood, Operation, Administration and Maintenance (OAM) server is contacted for resolving the differences. The OAM server then performs the RF impact analysis and responds with the operational parameters for the new site and the neighboring sites. The OAM server may also re-identify neighbor sites after parameters are agreed to, and store them in the configuration and parameter tables in the routers. The routers transition to the Learning state as shown the figure. Now, a RF system and network has been established by activating the wireless routers.

In the start up state, chores of the WR are: i) identification of the neighbors of the WR and preparation neighbor list, ii) interference impact, coverage, and other parameter analysis, iii) configuration of LSPs with neighbors, and iv) exchange and negotiation of power and handoff parameters with neighbors. In this way, the WR automatically configures itself for operation in the wireless network. Once these operations are completed, it transitions to the learning state.

2) *Learning State:* In the learning state, the WR continues to analyze, exchange and negotiate parameters, in order to minimize interference in the wireless network and to ensure that all the operational thresholds are met. The WR transitions into this state from the start-up state when RF power is up, and from operation state when either operational parameters change, or the WR neighbors change, or the operational thresholds are not met. In the learning state, parameters are re-negotiated and re-estimated based on the information given during transition from the operational state. Once operational thresholds are met for a specific period of time, the WR transitions to the operational state.

3) *Operational State:* In the operational state, the WR continues to monitor its operational thresholds periodically or otherwise exchanges information with its WR neighbor to ensure maximum efficiency and minimum RF interference within the wireless network. If the operational thresholds are not met, the wireless router transitions from the operational state back to the learning state for detailed analysis and evaluation of the configuration parameters and reconfiguration, as required, so that the operational thresholds can be met.

Also, if any of the neighboring routers change, affecting the topology of the network, such as a neighboring router failure, or a new router is added to the wireless topology, the router transitions from the operational state to the learning state, to reconfigure itself to suit to the new topology. In addition, if any parameters are changed due to any requests from its neighbors, the WR transitions to the learning state, for analysis and evaluation of operation using the new parameters.

4) *Site Down state:* The wireless router may enter the site down state from the learning state or the operational state if it requires either scheduled or unscheduled maintenance. Upon power up, the router will again transition back to the start up state for reloading and reconfiguration of the operational parameters. In this way, the wireless routers automatically adjust and account for changing conditions in the network to optimize operation of the network.

E. Results

Based on our analysis on a possible field deployment, we present in the following sections the results on cost and system availability improvements.

i) Improvement on Annual Costs

For this analysis, we use a typical configuration of traditional wireless network in the Denton area of Dallas metropolis, Texas, USA, with 16 base stations deployed at the "*" marks as shown in Figure 10. The big encircled "C" in the picture represents a central office (CO). We consider three wireless

Figure 10. Typical deployment of base-stations in the Denton area

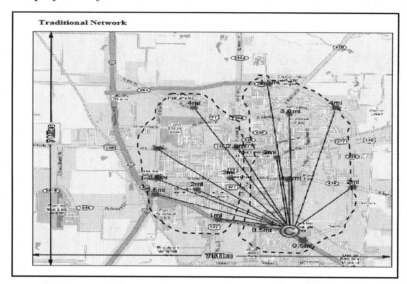

network architectures as shown in Figure 11 with identical deployment for our comparative analysis: i) a traditional network (Figure 11 (a)) with the CO in Denton connected to BSC in Dallas via two more Cos; MSC is assumed to be co-located with BSC, ii) a star network (Figure 11 (b)) in which the base-stations are partitioned into two groups as shown hatched line contours and connected to one CO cum WR each, and iii) a distributed WR mesh network (Figure 11(c)) with WRs replacing base stations and the same two groups of WRs as in the star network each connected to an edge router (ER). We assumed traditional T1 and T3 leased lines for communication between the COs, CO (CO cum WR) and BSC or BS, and WR and ER. In configuration (iii), WR-WR links are assumed to be fixed cost internet *e.g.* DSL (Digital Subscriber Loop) links. Table 1 depicts the cost structure used in our analysis. In order to make a fair comparison, we presumed that in configura-

Figure 11. Three network configurations with identical BS/WR deployment used in our analysis a) Traditional network, b) Star network, and c) Distributed WR mesh network

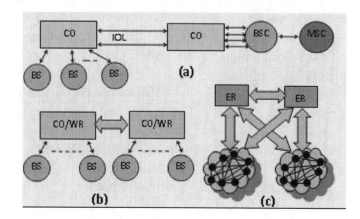

Table 1. Leased monthly class structure used in our analysis

Link	Channel Termination Cost	Inter-Office (Fixed Cost)	Cost/Mile
T1	150$	50$	15$
T3	1500$	500$	150$

tions (b) and (c), there is another CO/WR (ER, in (c)) at a distance 35 miles off (closer to BSC in (a)) for communication with users at that end, even though this is not depicted in the figure in order to keep it simple and clear.

Even though the inter-office connections depicted by IOL (Inter Office Links) in 11(a) and the links depicted by double arrows in Figure 11(b) and Figure 11(c) seem to have an identical functionality, there is a classic difference between them. The links represented by double arrows carry IP traffic and hence they take advantage of statistical multiplexing. Assuming this multiplexing factor to be 3, these links effectively carry only one-third of the traffic ensuing from the BSs. Further, since the handoff processing is performed at the CO/WR (or WR) level thanks to the distributed SDUs therein, the redundant traffic resulting from the need for simultaneous monitoring of the same call via multiple BSs for facilitating handoffs particularly at the cell boundaries, will not be present on the double arrow links. Assuming that the handoff factor is 1.75 (*i.e.* handoff processing results in 75% increase in the BS traffic), we will get a further reduction by a factor of 1.75 for the traffic on the double arrow links with a total reduction by a factor of 5.25. This assumption is consistent with the available statistical information that only 15% of the BTS-BSC traffic is pay load. Looking from a different angle, we need lesser number of T1/T3 (double arrow) links to handle a fixed BS traffic.

In our cost analysis, we used the maximum number of simultaneous subscribers at a BS at any time (shortly denoted by N_S as an independent variable and carried out for this variable varying in multiples 100, the cost estimates based on the T3/T1 links required to bear the traffic on different segments of the three network configurations under study. We performed our analysis for both 2G traffic with exclusively voice traffic, and 3G traffic with 70% voice traffic and 30% data traffic. Since bandwidth for a voice call is 64 Kbps, and that for a data call is on the average 700 kbps, it is straightforward to verify that the bandwidth requirements in mbps for N_S subscribers of a BS for these two cases is given by the two following two formula:

$$BW_{2G} = N_S * 0.064 \qquad (1)$$

$$BW_{3G} = N_S * (0.7*0.064 + 0.3*0.700) \qquad (2)$$

We estimated the number of T3s required by considering the multiples of 45 mbps in the estimated bandwidth. We tried to cover the residual bandwidth by T1s if it is less than 10 mbps. Otherwise, we used an additional T3 because, from Table 1, it is apparent that T3 is more cost effective if the number of T1s exceeds 9. As discussed before, the bandwidth estimates above have been scaled with handoff and statistical multiplexing factors as required in the relevant segments of the network. Now, the cost C_T for the traditional network, as per the ILEC (Incumbent Local Exchange Carrier) rate basis, is given by:

$$C_T = N_{BS} [\sum_{Link=T1,T3} N_{Link}(2.T_{link} + F_{Link} + M_{Link}D)] \qquad (3)$$

Here N_{BS} is the number of BSs, N_{Link} is the number of links of link type "link" = *T1* or *T3*, and T_{Link}, F_{Link} and M_{Link}, are the termination, fixed, and per mile costs, respectively, associated with the link. D is the average total distance from a BS to BSC (or farthest CO/WR or ER in the other two non-traditional configurations) and is assumed to be 35 miles. For computing C_S, the cost per star network, we need to consider the fact that, out of and N_{Link}s emanating from the BSs, only N^*_{Link} s are required to carry the traffic reduced by a factor 5.25 along the double arrow route as discussed above. Hence, C_S is given by:

$$C_S = N_{BS}[\sum_{Link=T1,T3}\{N_{Link}(T_{link} + F_{Link}) + N^*_{Link}M_{Link}D + (N_{Link} - N^*_{Link})M_{Link}d_{avg}\}] \quad (4)$$

Finally, C_M, the mesh network cost, may be estimated as the sum of the costs of the WR-WR links, WR to local ER links, WR to farther ER links (for better reliability), and ER-ER links. Upon aggregation, the total cost is given by:

$$C_M = 0.25N_{BS}(N_{BS} - 2)C_{Fixed-Link-Cost} + N_{BS}(\sum_{Link=T1,T3}N^*_{Link}(T_{link} + F_{Link} + M_{Link}D)) \quad (5)$$

In the above calculations, the N_{BS} WRs (in place of BSs) are assumed to be partitioned into two equal sized groups with total connectivity in each group. The first term on the right hand side represents the cost of the WR-WR links with a fixed cost $C_{Fixed-Link-Cost}$ (of 50$) for each link. The other three types of links across the double arrows can be treated as continuous pipes of total length D from WR to the farthest ER in downtown Dallas. The second term in the above equation represents the cost for these $N_{BS}N^*_{Link}$ IP-traffic bearer links.

Figure 12 contains the graphs showing variations of the total annual costs with the number of subscribers (N_S) in the three network configurations for the 2G and 3G scenarios, respectively. Obviously, the star and mesh configurations with the proposed router outperform the traditional

Figure 12. (a) Annual cost variations of the three networks with the size of the 2G subscribers that could access a base station simultaneously. (b) Annual cost variations of the three networks with the size of the 3G subscribers that could access a base station simultaneously

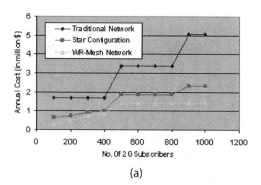

(a)

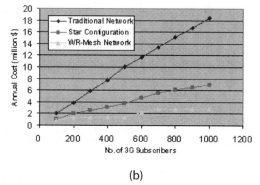

(b)

network configuration. The difference is much more pronounced in the 3G scenario compared the 2G one. In both the situations, the mesh is more cost effective compared to the star obviously because of handoff management at WR/BS level and inter-router IP communication.

B. Improvement on System Reliability

It is known that the availability of an individual system entity or link is given by *MTTF/(MTTF + MTBF)* where *MTTF* and *MTBF* are the mean time to failure and mean time between failures, respectively. In this analysis, we presume the individual availability values and compute the overall system availability by aggregating these

Table 2. Availability of different wireless network topologies

Traditional	Star	Mesh (N=2)	Mesh (N=3)
0.94	0.98	0.9999	0.999999

availabilities. It is obvious that the traditional network topology is the least reliable of all due to its susceptibility to single point failures along the long dedicated link paths from BS to MSC. Assuming there are M legacy entities (e.g. BSC, CO, MSC) on this path, each of availability A_E, interconnected by links each of availability A_L, the overall availability of the system can be estimated, based on the serial connectivity, to be $(A_E A_L)^M$. In the star topology, the dedicated path terminates at CO/WR. Hence, the availability of this topology is given by $A_E A_L$. For availability analysis of our mesh network, we need to consider three kinds of failures: (i) failures of individual WRs, and WR-WR links, (ii) WR-ER link failures, and (iii) ER failures. Multiple parallel paths in WR-mesh, and from WR-meshes to ERs make the effect failures of type (i) and (ii) on the overall availability negligible. Hence, we need to consider only the dominant ER failures. If A_{ER} is the availability of an ER, and each mesh has multiple parallel links to N ERs, the overall availability of the mesh configuration may be approximated as $(1-(1-A_{ER})^N)$. With $A_E = A_L = A_{ER} = 0.99$, $M = 3$, Table 2 gives overall availability of the topologies considered above. The 4 to 6 9s reliabilities (availabilities) of the WR-mesh networks as indicated in Table 2 satisfy the high availability requirement of next generation RANs.

FUTURE RESEARCH DIRECTIONS

In addition to the few network topologies explored in this section, a number of generic and application-specific topologies need to be explored in the future and evaluated based on the cost effectiveness, fault-tolerance, and performance. With the emergence of voice-over-IP (VOIP) technologies for multimedia communications, application of the IP wireless router technologies for development of next generation 911-service infrastructure becomes an important future direction for the research proposed herein. Analytical and simulation studies on quality of service and bandwidth efficiency tradeoffs with various wireless router configurations are an interesting direction for future research.

CONCLUSION

We present here a wireless IP-RAN architecture based on innovative wireless routers that provide message routing in addition to control functions of legacy system entities such as BSCs and MSCs. The tremendous savings these routers offer on backhaul network costs without sacrificing the QOS make the highly available and scalable IP-RANs based on a distributed configuration of these routers indispensable for next generation wireless networks.

REFERENCES

Bu, T., Chan, M. C., & Ramjee, R. (2006). Connectivity, performance, and resiliency of IP-based CDMA radio access networks. *IEEE Transactions on Mobile Computing*, 5(8), 1103–1118. doi:10.1109/TMC.2006.108

Chen, T., & Hamalainen, S. (2003). Handover in IP RAN. *In Proc. IEEE International Conf. Communication Technology, 2*, 812-815.

Dantu, R. (2005). *Method and System of Integrated Rate Flow Control for a Traffic Flow across Wireline and Wireless Networks.* US Patent No. 6904286 B1. Washington, DC: U.S. Patent and Trademark Office.

Dantu, R., Patel, P. R., Choksi, O. T., Patel, A. R., Ali, M. R., Miernik, J., & Holur, B. S. (2006). *Wireless Router and method for processing traffic in wireless communication network.* US Patent No. 7068624 B1. Washington, DC: U.S. Patent and Trademark Office.

Ekström, H., Furuskär, A., Karlsson, J., Meyer, M., Parkvall, S., Torsner, J., & Wahlqvist, M. (2006). Technical Solutions for the 3G Long-Term Evolution. *IEEE Communications Magazine, 44*(3), 38–45. doi:10.1109/MCOM.2006.1607864

Ergen, M. (2009). *Mobile Broadband - Including WiMAX and LTE.* New York: Springer.

Fazel, K., & Kaiser, S. (2008). *Multi-Carrier and Spread Spectrum Systems: From OFDM and MC-CDMA to LTE and WiMAX (2nd Ed.).* Hoboken, NJ: John Wiley & Sons.

Ghosh, S., Basu, K., & Das, S. K. (2005, September/October). What a Mesh! An Architecture for Next Generation Radio Access Networks. *IEEE Network*, 35–42. doi:10.1109/MNET.2005.1509950

3GPP2 C-S0033, Rev 0, Ver 2.0 (2003). *Recommended Minimum Performance Standards for cdma2000 High Rate Packet Data Access Terminal.* 3rd Generation Partnership Project [TIA-866-1 standard].

Kempf, J., & Yegani, P. (2002). OpenRAN: a new architecture for mobile wireless Internet radio access networks. *IEEE Communications Magazine, 40*(5), 118–123. doi:10.1109/35.1000222

Patel, P. R., Choksi, O. T., Davidson, K. W., & Dantu, R. (2006). *Method and System for Configuring Wireless Routers.* US Patent No. 7031266 B1. Washington, DC: U.S. Patent and Trademark Office.

TIA/EIA/IS-2000.1.A -2. (2002, February 11). The CDMA2000 Family of Standards for Spread Spectrum Systems.

Vassiliou, V., Owen, H. L., Barlow, D. A., Grimminger, J., Huth, H.-P., & Sokol, J. (2002). A radio access network for next generation wireless networks based on multi-protocol label switching and hierarchical Mobile IP. *Proc. 56th IEEE Vehicular Technology Conf., 2*, 782-786.

Yasukawa, S., Nishikido, J., & Hisashi, K. (2001). Scalable mobility and QoS support mechanism for IPv6-based real-time wireless Internet traffic. *Proc. IEEE GLOBECOM Conf., 6*, 3459–3462.

Chapter 5

Long Term Evolution (LTE):
An IPv6 Perspective

Nayef Mendahawi
Research In Motion (RIM), Ltd., Canada

Sasan Adibi
Research In Motion (RIM), Ltd., Canada

ABSTRACT

The main characteristic of 4th Generation (4G) Networks is being based on all IP architecture, operating mainly on IPv6. This includes services such as voice, video, and messaging. LTE is considered to be a 3rd Generation (3G) network and one of 4th Generation (4G) roadmap mobile access technologies. LTE-Advanced (LTE-A), on the other hand, is a 4G technology concept with evolving features. Therefore LTE is the key feature in the understanding of LTE-A evolution. The main focus of LTE is the enhancement of the packet-switched (PS) mechanisms on top of the UMTS enhancements, based on All IP Network (AIPN). IPv6 networking provides maximum service delivery flexibility, user decoupling, and scalability improvements, while leveraging the existing IETF standards. This requires major focus on network simplification, end-to-end delay reductions, optimal traffic routing, seamless mobility, and IP-based transport provisioning. This chapter aims to present a survey and highlight specific IPv6-based features presented mostly in the 3GPP standard literature, and to provide a high-level discussion on the LTE-IPv6 requirements.

1. INTRODUCTION TO IPV6

IPv6 is, by far, one of the most important and significant technology and network upgrades in the communication history. This upgrade is growing slowly and will eventually terminate IPv4's network deployment at the end of the transition phase. IPv6 is designed to work with high speed network and low bandwidth networks, particularly suitable for wireless networks. The IPv6 design accommodates new technology requirements, such as QoS, security, and of course extended addressing (Cisco Systems, 2008; Esposito, R., et al.).

Since the mid 90's many developers and researchers have been working on IPv6 and many RFCs are directly or indirectly discussing this technology.

DOI: 10.4018/978-1-61520-674-2.ch005

Figure 1. IPv6 penetration in vendors, ISPs, and users (adapted from Gallaher & Rowe, 2005)

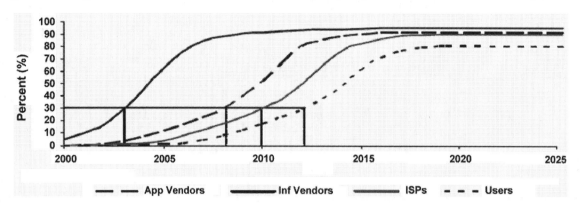

RFC 1883, released in 1995, is the first RFC in regards to IPv6. These efforts have been initiated and monitored by the Internet Engineering Task Force (IETF) (Deering & Hinden, 1995).

In the past several years IPv6 has deeply penetrated into the architecture of the Internet. Figure 1 (adapted from Gallaher & Rowe, 2005) presents a penetration estimates of IPv6 in the United States, which shows that by year 2010, almost 30% of ISPs and 18% of the users will be switching to IPv6.

The main issue with IPv6 integration is its interoperability with IPv4. IPv4 is going to be around for at least a decade before all the networks are purely running over IPv6. Therefore it is essential to ensure IPv6 traffic flows are not going to have any negative impacts on IPv4 and vice versa. Dual-stack systems are considered a solution for the IPv6/IPv4 interoperability issue, which are going to be utilized in the design and implementation of systems for sustaining both IPv4 and IPv6 parallel processing to guarantee interoperability amongst the two protocol suites (Punithavathani & Sankaranarayanan, 2009).

Mobility is another major features that is going to be impacted by this transition. 3G networks have been growing immensely in the past few years and IP connectivity has become an inseparable part of the 3G networks. The involvement of IP is going to increase more as 4G networks are becoming realizable. These requirements are being considered in the IPv6 applications (Ericsson Inc., 2009).

This chapter will explore IPv6 features, in particular, in conjunction with LTE architecture. At first, IPv6 is going to be discussed and compared to IPv4. Following this introduction, IPv4-to-IPv6 transition, IPv6 security, Mobile-IPv6, and QoS-IPv6 are covered. Various interactions of LTE and IPv6 will conclude this chapter.

1.1 IPv6 versus IPv4

Internet Protocol version 6 (IPv6), also known as the Internet Protocol next generation (IPng) is intended to sustain constant Internet growth with consideration of the number of users and functionality. The legacy IP version, IPv4, was implemented in the early 1980's based on stationary wired communication infrastructure (Postel, 1981). The IPv4 supports less than 2^{32} (over 4 billion) individual addresses, hence IPv4 suffers some limitations that may be inhibitors to growth of "tomorrow's" Internet, and use of the Internet as a global networking solution. Therefore, IPv6 is under development to take over IPv4's position by providing a greater expansion of IP address space; nonetheless, incorporating features of such include end-to-end security, mobile communications, Quality of Service (QoS), and system management burden reduction.

Without adequate global IP address space, applications have to work in such ways to afford local addressing. In a short-term, there have been various discretionary "workarounds" and extensions to IPv4 in and attempt to overcome its limitations. Network Address Translation (NAT) enables multiple devices to utilize local private addresses within an enterprise at the same time sharing one or more global IPv4 addresses for external communications. While NAT, to a certain degree, has postponed the exhaustion on IPv4 address space for the time being, it also complicates common application bi-directional communication. IPv6 simplifies the confusion of presenting an end-to-end security and eliminates the general incentive for using NAT since global addresses will be extensively accessible (Tsirtsis & Srisuresh, 2000).

IPv4 had numerous issues, one of which was not having sufficient geographical distribution; it currently has less than 50% coverage throughout the USA. One the other hand, routing was too complex for new technologies and features such as mobile computing had coverage areas issues. Most significantly, the number of IP addresses is reaching its limit and the time has come to adopt IPv6 to compensate for the technical and address space requirements. Figure 2 shows IPv4 and IPv6 header formats (IPv6, n.d.; IPv4, n.d.).

Table 1 shows the differences in the IPv4 and IPv6 headers fields' differences. As shown, IPv6 offers a few additional fields compared to IPv4.

1.2 The IPv6 Transition Mechanisms

IPv6 was designed with a long transition period in mind. Therefore there is a myriad of IPv4 to IPv6 transition mechanisms have been defined in the various RFCs. The transition mechanisms can be grouped into a few categories (3G Americas, 2008):

1. Dual Stack
2. IPv6 tunneling over IPv4
3. IPv4 – IPv6 translation

1.2.1 Dual Stack

In a dual stack mechanism, the device supports both IPv6 and IPv4, which means that the device is able to obtain both IPv6 and IPv4 addresses from the network and is able to choose either IP version to use to communicate depending on the IP version and the peer supports. The IP version and the peer supports can be discovered, for example, using a DNS service (3G Americas, 2008).

1.2.2 IPv6 Tunneling over IPv4

When two IPv6 domains are not directly connected over IPv6 but instead are connected through an IPv4 network, which is often the case during the initial transition of the Internet to IPv6, IPv6 traffic will be tunneled over IPv4. The following tunneling techniques that may be relevant to IPv6 migration are (Graveman et al., 2009; ISATAP, n.d.; 6to4, n.d.; Hagen, 2002):

- *Configured tunnels:* This is used for connecting two IPv6 domains that have native IPv6 connectivity. The tunnel (typically from router to router) is configured via administrative means. IPv6 packets are encapsulated in IPv4 packets.
- *Tunnel broker:* Tunnel broker is a technique that uses an IPv6 domain (a network or an individual host), which establishes an IPv6 connectivity using such a tunnel broker, serving as a virtual IPv6 ISP.
- *6to4:* This is used for deploying IPv6 in a network without waiting for the administrator of the network to provide native IPv6 connectivity. A 6to4 router advertises a global unicast IPv6 prefix to the network, constructed from a public IPv4 address assigned to the network. The 6to4 router also

Figure 2. IPv6 versus IPv4 header

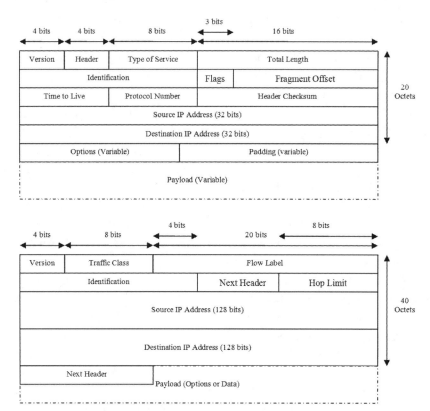

Table 1. IPv6/IPv4 header field differences

IPv4 Header field	IPv6 Header field
ToS	Traffic class filed (QoS parameter)
Total length	Payload length
TTL	The Next header
N/A	Flow label (QoS parameter)
N/A	Next Header
The source address and destination addresses are based on 32-bits address fields	The source address and destination addresses are based on 128-bits address fields

acts as a tunnel endpoint for the network. IPv6 packets are encapsulated in IPv4 packets. The IPv6-in-IPv4 tunnels between 6to4 sites are established automatically. The 6to4 relay routers allow 6to4 sites to communicate with native IPv6 sites.

- *Teredo:* Teredo is used for deploying IPv6 in a network without waiting for the administrator of the network to provide native IPv6 connectivity (similar to 6to4). This is particularly important when the network is behind a NAT and the NAT system is not upgradeable to provide 6to4 functionality, since the 6to4 router cannot sit behind an IPv4 NAT. In Teredo, a Teredo server deployed in front of the NAT, serves as the IPv6 router for the network, transmitting IPv6 router advertisements with a global unicast IPv6 prefix, constructed from a public IPv4 address assigned to the network. This helps the hosts behind the NAT (the Teredo clients) learn about their assigned public IP address and UDP port and enable hole punching through the NAT for Teredo clients between sites to establish direct connectivity. Teredo relays allow Teredo sites to communicate with native IPv6 sites.

- *ISATAP:* The ISATAP is an IPv6-in-IPv4 tunneling technique that allows dual-stack hosts and routers within a network segment to communicate over IPv6 when the IP infrastructure within the network segment does not support native IPv6. One example scenario is to allow a host to access the Internet via IPv6 when a site has native IPv6 access to the Internet, however the host happens to be within a network segment that has not been upgraded to IPv6.

1.2.3 IPv4 – IPv6 Translation

When an IPv6-only host needs to communicate with an IPv4-only, some form of protocol translation is needed. The following translation techniques have been developed:

- NAT-PT (Network Address Translation – Protocol Translation) and SIIT (Stateless IP/ICMP Translation Algorithm).
- IPv6-to-IPv4 Transport Relay Translator: A TRT is a TCP or UDP relay which additionally performs IPv4 – IPv6 translation.

2. LTE AND IPV6

The main network layer component of 4G networks is based on IPv6, which is the engine of all future higher level services, such as voice, video, and messaging. These services are carried on top of IPv6. Therefore the LTE's focus is to enhance and configure the packet-switched (PS) onto the future UMTS-based cellular network to form an All IP Network (AIPN). Using IPv6 networking, maximum flexibility is achieved via service delivery while decoupling the user and improving scalability to allow the leverage all the existing IETF standards. This requires network simplification, optimal traffic routing, seamless mobility, and efficient IP-based transport mechanism (*3GPP Long Term Evolution*, n.d.).

In considering the LTE-IPv6 transition, the Evolved Packet Core (EPC) is an important mechanism recommended by the vendor partners and the associated operators, which will be involved in a smooth transition implementation. As part of the 3GPP standard, since Circuit Switched Core is not supported in the EPC's structure, the IP Multimedia Subsystem (IMS) core will be used to support voice. It is therefore critical from operators' point of views to understand the IPv6's impact on the existing Voice Core and Signaling infrastructure due to the fact that slow and gradual transition to IMS-based Voice-over-IP (VoIP) will likely to take several years (*IPv6 Transition Considerations*, 2009).

2.1 Introduction to Evolved Packet Core (EPC)

A part of LTE contains EPC, which is a new system design based on the all-IP mobile core network architecture. EPC forms a converged framework based on real-time and non-real-time packet-based services. EPC is specified by 3GPP Release 8 standards (3G Americas, 2009).

The EPC is responsible for providing mobile core functionality, which used to be two separate sub-domains in previous mobile generations (2G, 3G). These two sub-domains are: Circuit-Switched (CS) (i.e., supporting voice) and Packet-Switched (PS) (i.e., supporting data). These two distinct sub-domains are used for individual mobile voice and data switching and processing under a unified single IP mobile domain. LTE offers an end-to-end IP-based architecture, which covers from the mobile handsets and other terminal devices offering embedded IP capabilities on top of LTE base stations. The LTE base station is an IP-based device often called Evolved NodeB.

LTE's EPC is an essential functional entity offering end-to-end IP service, as well as allowing the introduction and creation of new business models, including partnerships and revenue sharing with application and third-party content providers. EPC promotes the enablement of new applications and new innovative services.

EPC addresses those fundamental requirements of LTE that deal with media-rich and advanced real-time services offering enhanced Quality of Experience (QoE). Network performance can be improved using EPC. This is done by separating data and control planes by using a flat IP architecture that is able to reduce the hierarchy delay among mobile data elements. For instance, the data path traversing from eNodeB passes only through EPC gateways.

The introduction of the all-IP network architectural EPC in the design of mobile networks has caused various degrees of implications on the following mechanisms:

- All-IP mobile services, such as; IP-based voice, data and video communications.
- New mobile architecture interworking with previous mobile generations (2G/3G)
- Network scalability, which is required by all core elements to address any changes in the bandwidth and the number of user-terminal direct connections.
- Availability and reliability offered by each elements ensuring service continuity.

To address various service and network requirements, the EPC is designed in such a way to change the existing mobile network paradigms.

The following subcomponents are part of the EPC architecture ("System Architecture Evolution (SAE)," n.d.):

- *MME (Mobility Management Entity):* In LTE, the MME is a key control-node for the access network method. MME is responsible for paging procedure including retransmissions and the UE (User Equipment) tracking idle mode. It is also responsible for match an appropriate SGW for a UE at the time of intra-LTE handover that involves the Core Network (CN) node relocation and at the initial attach time. MME is further involved in the user (by interacting with the HSS) authentication and the bearer activation/deactivation procedures. The MME can terminate the non-Access Stratum (NAS) signaling can generate and allocate temporary identities for UEs. It also checks for the authorization of the UE for camping on the service provider's Public Land Mobile Network (PLMN) and enforcing UE roaming restrictions. The MME handles the security key management and is the termination point in the network for ciphering/integrity protection for NAS signaling. The MME also offers lawful signaling and LTE/2G/3G mobility control plane functions with the S3 interface that terminates at the MME from SGSN. The S6a interface is also terminated by MME towards the home HSS for UEs roaming purposes. The following interfaces were considered for MME:
  - S3: This interface enables bearer and user exchange information for inter 3GPP access network mobility between SGSNs.
  - S6a: This interface enables subscription and authentication data transfer for authenticating and authorizing user access.

- *PGW (PDN Gateway):* From the UE to the external packet data networks, PDN Gateway provides connectivity by providing the point of entry and exit for the UE's traffic. More than one PGWs may provide simultaneous connectivity for a UE providing multiple PDN access. The PGW performs packet filtering for each user, policy enforcement, lawful Interception, charging support, and packet screening. The PGW may act as the anchor for mobility between non-3GPP and 3GPP technologies, which is nother key role of PGW. These technologies include: 3GPP2 (CDMA 1X and EvDO), WiMAX and etc.
- *SGW (Serving Gateway):* User data packets are routed and forwarded by the SGW. During inter-eNodeB handovers the SGW acts as a mobility anchor for the user plane. It also acts as an anchor for mobility among LTE and other 3GPP technologies, where it can terminate the S4 interface and relay the traffic among PGW and 2G/3G systems. The SGW terminates the DL data path for the idle state of UEs and when DL data arrives for the UE, it triggers paging. The SGW also stores and manages UE contexts, such as parameters of network internal routing information and the IP bearer service. In case of lawful interception, it also performs replication of the user traffic.
- *PCRF (Policy and Charging Rules Function):* Though PGW, SGW, and MME were introduced in 3GPP Release 8, PCRF was part of 3GPP Release 7 introduction. Architectures using PCRF have not so far been widely adopted by standards, however the interoperability of PCRF's with the EPC gateways and the MME is essential for the operation of the LTE and mandated in Release 8.

2.2 EPC Challenges

The EPC changes key networking paradigms for previous mobile generations (2G/3G) core networks and the integration of EPC is expected to address a number of technological challenges successfully (Motorola Inc., 2007).

The radio side in LTE (eNodeB) has undergone significant technological advances to provide wider spectral bands and efficient use of the spectrum, which are reserved for LTE, which results greater performance and system capacity. At the same pace, the mobile core is required to change and to provide higher throughput while maintaining low latency; both due to the improved and simplified flat all-IP network architecture.

The important aspect of LTE is the introduction of new technologies and the delivery of the high performance LTE solution, which are both involved on the radio side.

The EPC needs to address the following key aspects of IP for the LTE deployment (Alcatel-Lucent, 2009):

1. Distributed versus centralized network architectures, including; SGW, PGW, and MME deployment.
2. Network addressing and IP routing, and real-time management for large IP domains.
3. The introduction, strategy, and coordination of IPv6 and its interoperability with IPv4.
4. End-to-end deployment for QoS and underlying transport coordination.
5. Data and control plane end-to-end security.
6. Layer 2 versus Layer 3 transport layer connectivity (eNodeB, PGW, SGW, MME).
7. External networks and VPNs interconnectivity.
8. Lawful Interception and Deep Packet Inspection (DPI).

There is a set of stringent requirements for scalability, reliability, and high-performance elements because of LTE's dynamic nature of user mobility, which are coupled with the large-scale deployment targets and short duration of multiple data sessions for each UE.

To satisfy these requirements, the EPC elements must have the best classification with high IP performance. In order to address all these fundamental aspects of EPC's and according to the network element and product level, a new generation of scalable mobile core equipment, purpose-built, and strong IP expertise are required.

It is important to integrate all these elements together to deliver the needed carrier-grade features for LTE. The EPC elements must fully interwork harmoniously while in both control and user planes, the fairly complex network procedures involve all EPC elements. The EPC is expected to address the demanding requirements for dynamic and multi-dimensional mobility management, policies and data bearers. This should be done in an orchestrated manner to enable the highest LTE performance, while offering interoperability and interworking with the legacy 3G/2G systems.

2.3 PDP Context

A PDP context is a data protocol structure that is present on both SGSN and GGSN. It contains information regarding the subscriber's session during subscriber's active session. The subscriber-related data within the PDP context information includes: IP address, IMSI information, and Tunnel ID (TEID). Therefore access to the external packet-switching network can be achieved through a PDP context specification. The data associated to the PDP context is consisted of information including the MS PDP address (IP address), the packet-switching network type, the GGSN reference, and the requested QoS. A PDP context is identified and handled by a mobile's PDP address within the following entities: SGSN and GGSN and the MS. Within a given MS several PDP contexts can be activated at the same time ("3GPP TS 23.060 V8.5.1," 2009).

A PDP context is created through a PDP context activation procedure, which may be initiated either by the network or by the MS. Before PDP context negotiation, the MS is always GPRS-attached.

To remove a PDP context a PDP context deactivation procedure is required. This procedure may be initiated either by the network (SGSN or GGSN) or through the MS itself. The PDP context deactivation procedure may be initiated either during the GPRS detach, delete subscriber data procedures, or during an application deactivation.

2.4 IPv6 Networks and LTE

In EPC architecture there are two gateways that may be combined into a single network element; the SGW and the PGW (Figure 3, adapted from 3G Americas, 2009). If these two gateways are separate network elements, there are two options for the protocol being used among them: GPRS Tunneling Protocol (GTP) and Proxy Mobile IP v6 (PMIPv6) (3G Americas, 2009).

The Release 8 specification identifies EPS bearers similar to the concept of PDP Context, which is defined in 3GPP's Release 7 specifications. An EPS bearer is a logical connection between a UE and a gateway, associated with a specific QoS class. As long as these Service Data Flows (SDFs) belong to the same QoS class, an EPS bearer can carry multiple SDFs ("3GPP TS 23.060 V8.5.1," 2009).

An EPS bearer can carry both native IPv6 and IPv4 traffic in contrast to the PDP Context defined in 3GPP Release 7. Therefore IPv4 and IPv6 stacks can be supported by a UE simultaneously as long as it's connected via a single EPS bearer.

Regarding the IPv4 scarce address resources, allocating both IPv6 and IPv4 addresses to a device does not solve the problem of IPv4 exhaustion. A service provider may therefore decide to assign only IPv6 addresses to certain devices, even when the device is able to support IPv4 and IPv6 simultaneously. In that case, NAT–PT or IPv6–

Figure 3. All-IP LTE structure (adapted from 3G Americas, 2009)

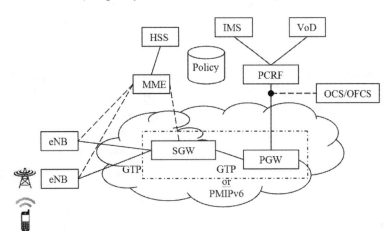

toIPv4 http-proxy functionality may be required to connect these IPv6-only devices with legacy IPv4 end–points. Such a decision needs careful consideration and the issues identified in RFC 4966 (Aoun & Davies, 2007) need to be taken into account.

A User Equipment device (UE) obtains an IP address in one of the following two methods:

1. As part of the attachment procedure
2. Via a separate assignment procedure, such as DHCP or IPv6 Stateless Address Autoconfiguration.

The attachment procedure consists of the following steps ("3GPP TS 23.060 V8.5.1," 2009):

1. The UE requests an attachment via sending a message to the MME, followed by an authentication procedure involving the HSS. The HSS sends to the MME, subscription data associated with the user as part of this procedure.
2. With a few exceptions, the MME is responsible for selecting the Serving and PDN Gateways that will be used for this UE. It sends a request for the establishment of the *default bearer* to the SGW, which forwards it to the PGW. This message exchanges results in the establishment of a GTP tunnel or a Mobile IP tunnel segment between SGW and PGW. As long as the user is attached, this segment remains up, even when the UE enters the idle state.
3. The MME orchestrates the establishment of the GTP tunnel segment between SGW and eNB and the (default) radio bearer between eNB and UE. The bearer between SGW and UE is torn down whenever the UE goes to idle state. If the SGW receives IP packets destined for the UE while it is in idle state, the SGW triggers the MME which starts a paging procedure.

When the IP address assignment is part of the attachment procedure, an IP address is allocated by the PGW to the UE according to step 2. This IP address is provided to the UE within the GTP control messages being used for establishing the EPS bearer according to step 3.

When a default bearer is established (disregarding the IP address assignment), DHCP or IPv6 Stateless Address Autoconfiguration (SLAAC) can be performed by the UE for obtaining an IP address. Therefore, a UE could receive an

IPv4 address during the attachment procedure and an additional IPv6 address through SLAAC procedure.

2.5 PDP Address Allocation

PDP addresses are handled differently for PDN interworking of type PPP and IP (IPv4 or IPv6).

2.5.1 Interworking with PDN Based on IP

During PDP context activation, the MS can configure an IPv4 address, or obtain an IPv6 interface identifier to be used during the IETF-based IP address allocation after PDP context establishment ("3GPP TS 23.060 V8.5.1," 2009).

The MS can obtain an IPv4 address or an IPv6 prefix via an IETF-based IP address allocation mechanism once the PDP context is established.

The EPS specifies the following IETF-based IP address/prefix allocation methods (3GPP TS 23.401 V8.1.0, 2008):

a) IPv4 parameter configuration and IPv4 address allocation through the DHCPv4.
b) IPv6 parameter configuration through a stateless DHCPv6
c) /64 IPv6 prefix allocation through IPv6 stateless address autoconfiguration

The MS releases any allocated IPv4 address or IPv6 prefix locally for the corresponding PDN connection upon deactivation of a default PDP context.

2.5.2 Interworking with PDN Based on PPP

No PDP address is configured during PDP context activation. Instead, such information is negotiated and configured during the NCP phase of PPP.

2.5.3 IP Address Allocation Via NAS Signalling

The MS sets the PDP type in the PDP address field using the ACTIVATE PDP CONTEXT REQUEST message. This is done when requesting default PDP context establishment..

If the MS requires using DHCPv4 for IPv4 address assignment, it indicates that to the network within the Protocol Configuration Options in the ACTIVATE PDP CONTEXT REQUEST message.

When the MS requests IPv6 address allocation, the network accommodates it in one of two parts: a /64 IPv6 prefix and an interface identifier of 64 bits length. The MSN doesn't use the IPv6 prefix part immediately by the MS; however, the network uses the same IPv6 prefix in IETF-based IP address allocation subsequent procedures. The interface identifier is only used to build a unique link-local IPv6 address.

2.6 Voice Core Network (VCN)

In a 3GPP network, voice telephony services are traditionally provided by through a Mobile–Services Switching Centre (MSSC) and supported infrastructure. This can also be provided through an IP Multimedia System (IMS), which is introduced here however further details are given in section 3 (*3GPP TS 23.221 V9.0.0*, 2009; *3GPP TS 23.228 V9.0.0*, 2009).

IMS relies heavily on IP protocol. According to 3GPPP standards, IMS should be based on IPv6, which avoids future IPv4-to-IPV6 migration requirements.

Though new IMS-compliant equipments will typically be deployed using IPv6, existing voice core equipments will still be utilizing IPv4.

The voice core IPv4 to IPv6 migration is a complex task, which requires the cooperation of various protocols and layers. Therefore alleviating the IP address consumption is one of the primary motivations for the IPv4 to IPv6 migration.

Due to the facts that the voice core migration from IPv4 to IPv6 is a complicated task and the end-user devices have almost consumed all IP addresses, it may not be feasible to migrate the entire 3GPP voice core of operators from IPv4 to IPv6. However instead, it would be feasible to perform such migration within the voice core at specific interfaces.

An operator, for instance, may wish to leave *intra*–call server MGW and all Iu–CS(IP) interfaces to the MGW traffic as IPv4 transfers all *inter*–call server MGW to MGW traffic using IPv6. This type of architecture requires dual stack capable MGWs only to be on the network bound interfaces and would doesn't require IPv6 capable RNCs.

For the LTE requirements, the carrier may dictate that the device to be assigned an IPv6 address whenever it attaches to the LTE network.

Through the LTE migration, IPv6's adoption will become increasingly important due to the fact that they will require more IP addresses for all the network connected devices. Therefore the number of unused IPv4 addresses will be reducing more and more, which requires a new standard to address new waves of data-capable devices.

2.7 Security Considerations

Privacy and security are among the most important and major concerns in LTE/SAE/IPv6 adoption strategy. The security mechanism involving IP networks slows down the further network technology adoption. Due to the fear of privacy breaches and putting the businesses at risk and potentially cause significant financial loss, operators and enterprises may be reluctant to adopt new network requirements though they are clearly aware of the cost saving benefits and productivity improvements of using convergent communication technologies on a single infrastructure, which enables universal connectivity for users. Therefore it is fundamentally essential to adopt an end-to-end system approach to security for the next-generation wireless networks. A number of mandatory security measures/requirements include (3G Americas, 2009):

- Secure protocols, infrastructure, communication, and data storage
- User authorization, authentication, and auditing
- End-to-end compliance
- Secure signaling, network management and control
- Unsolicited traffic protection
- Software integrity

2.9 Mobile-IPv6 Considerations

Mobile IPv6 (MIPv6) (Soliman, 2009) allow mobile nodes to roam between Internet domains while maintaining ongoing sessions and reachability. This is done by using an IPv6 home address or prefix. Since IPv6 is not widely deployed, however it is unlikely that mobile nodes will use IPv6 addresses just for their connections alone. It is assumed that mobile nodes be requiring an IPv4 home address for a long time, which can be used by upper layers. It is fair to assume that mobile nodes will be moving to networks that might not be IPv6-ready and would therefore require an IPv4 Care-of need capability support.

IPv6 compared to IPv4, offers a number of functional improvements, mostly due to its large address space and additional functional fields. The same goes for Mobile IPv6 compared to Mobile IPv4, which offers a number of improvements inheriting from IPv6 additional capabilities, including: Dynamic Home Agent Discovery (DHAD) and route optimization, which can only be achieved in Mobile IPv6 systems.

An advantage of having a large address space (as in IPv6) is to allows mobile nodes to obtain a globally unique care-of address disregarding their current location.. Hence, there is no need to use Network Address Translator (NAT) traversal techniques, which are required for Mobile IPv4.

This further simplifies Mobile IPv6 system architectures and increases the efficiency of mobility management protocol and bandwidth allocations. However for existing private IPv4 networks, NAT traversal needs to be considered during the transition towards IPv6.

In order to minimize the need to changing the mobility stack because of the IPv6introduction within a deployed network and to allow for a long lasting mobility solution Mobile IPv6 should be used with dual stack mobile nodes capability.

3. IMS-IPV6 MULTIMEDIA SUBSYSTEM

IMS, also known as IP Multimedia Subsystem is an IP-based multimedia and telephony core network technology, which was introduced by 3GPP and 3GPP2 standards and is based on IETF Internet protocols. IMS is based on a set of specifications describing the Next Generation Networking (NGN) architecture involving the implementation of IP based telephony and multimedia services. It contains the specification of a framework and a complete architecture enabling the integration of video, voice, data and mobile network technology on top of an IP-based infrastructure (Amirth, n.d.).

IMS is access technology independent since it supports IP-to-IP session over wired IP, wireless (i.e., 802.11, 802.15, CDMA, etc), and packet data along with GSM/EDGE/UMTS and other packet data applications.

3.1 IMS Architecture

In IMS, the networking infrastructure is subdivided into individual functions with standardized interfaces (reference point) between each of them. Every reference point defines both the operating functions and the protocol over the. Figures 1 and 2 show the IMS architecture overview.

As shown in Figure 4 (adapted from Amirth, n.d.) the IMS architecture is split into three main layers containing a number of individual entities.

Figure 4. IMS architecture (adapted from Amirth, n.d)

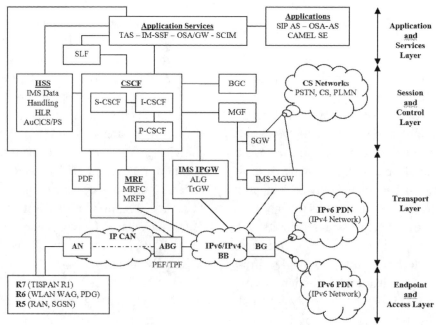

Figure 5 (adapted from Amirth, n.d.) presents a graphical overview of the IMS core entities. The combination of legacy mobile signaling networks, other IP multimedia networks, and the PSTN entities form the external interface functions. The combination of BGCF, CFCF, MGCF, and MRCF form the Sequencing and Control Functions. The combination of HSS and SLF form the Storage and Reference Functions. UE covers the User Interface Functions and the combination of MGW and MRFP form the Media Processing Functions. The interfaces between BGCF and MGCF and CDCF are of Ml type. The interface between MGCF and CSCF is Mg and the interface between WGW and MGCF is of M1 type.

The traffic types are of two major categories; Data/Bearer and signaling/control types. The PSTN link between PSTN and MGW entities contain CSD (TDM) traffic and the Data/bearer, such as Mb interface, contains IPv6 traffic. The rest of the interfaces contain signaling and control traffic, including: MEGACO (H.248), ISUP, SIP, QoS (COPS), MAP/TCAP, and other types of traffic.

3.2 Transport and Endpoint Layer

This layer is involved in setting up sessions and providing bearer services and initiating and terminating SIP signaling procedures. This layer is also responsible for providing the media gateways to convert the VoIP data to the PSTN TDM format.

3.3 Session Control Layer

This layer includes Call Session Control Function (CSCF), which provides the routing for the SIP signaling messages and the endpoints for the registration. The routing functionality enables the SIP signaling to be routed to the correct application servers. Through communicating with the transport and endpoint layer, the CSCF is able to guarantee QoS.

3.4 Types of CSCF

The first point of contact for the IMS terminal is a Proxy-CSCF (P-CSCF), which is a SIP proxy assigned to an IMS terminal during the registration, which does not change during the registration

Figure 5. IMS core network (adapted from Amirth, n.d.)

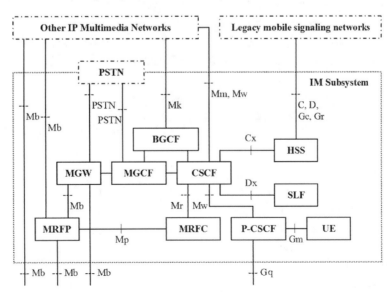

duration. It also involves authenticating the user and establishing IPsec security integration with the IMS terminal. The other nodes trust the P-CSCF, and do not have to re-authenticate the user.

The Policy Decision Function (PDF) is also involved, which is used for policy control, bandwidth management, and authorizes media plane resources, such as QoS over the media plane.

3.4.1 Proxy-Call Session Control Function (P-CSCF)

P-CSCF serves as a Back-to-Back User Agent (SIP B2BUA) and is the initial SIP signaling contact point for subscribers. The P-CSCF is responsible to forward SIP registration messages from the the User Element (UE) and subscriber's endpoint, to the Interrogating-CSCF (I-CSCF) and as a consequence; call set-up requests and responses to the Serving-CSCF (S-CSCF). The P-CSCF maintains the mapping between physical UE IP address and logical subscriber SIP URI address and a security association, for both confidentiality and authentication. It supports the E911 emergency call and local routing within the visited network, session timers, accounting, and admission control. The DIAMETER protocol (Rx interface) is used by the session admission control to query the external Policy and Charging (PC) Rule Function (PCRF) element for resource reservation and bandwidth-based admission control. The P-CSCF interacts with AGW (defined in subsection 3.4.5) for the boundary control at the media layers and signaling, including Port Translations (NAPT) lawful intercept and pinhole firewall, Network Address and etc.

3.4.2 Serving-CSCF (S-CSCF)

A S-CSCF serves as a central node in the signaling plane. It also serves as a SIP server, which performs session control. S-CSCF uses Diameter D_x and C_x interfaces in connection to the HSS to upload and download user profiles, without having a local storage for the user. All necessary data is loaded from the HSS.

The SIP registrations are handled at this signaling plane. This permits to bind the SIP address, the user location, and to decide the destination application servers for which the SIP message will be forwarded to. This is done in order to provide the routing services using Electronic Numbering lookups and enforcing the network operator's policy.

To offer load distribution and high availability, multiple S-CSCFs can coexist in the same network, however the S-CSCF is assigned to the user through the HSS when I-CSCF is queried.

3.4.3 Interrogating-CSCF (I-CSCF)

An I-CSCF is another SIP proxy that provides service locator functionality. The followings are within its major functions:

1. *Registration*: Registration is assigning a S-CSCF to a user via SIP registration.
2. *Session Flows (SFs)*: A routing message may include a session flow, which is part of a SIP request received at the S-CSCF from another network. It can also be a part of routing intra-domain SIP requests between users on different S-CSCFs.

3.5 Access Session Border Controller (ASBC)

Session border controllers (SBCs) offer required functionalities at the border to the applications and subscribers who access the IMS services. The SBC is responsible for connecting mobile devices to the SIP-based IMS services and related applications, including IPTV, interactive gaming, voice, video, and messaging. In SAE, the access SBCs are responsible to connect all access networks to the IMS network, including the 3G RAN, the LTE RAN, the trusted non-3GPP IP access networks (such as DSL, FTT_x and WiMAX), and un-trusted non-3GPP IP access network (such as the Internet and Wi-Fi networks).

The SBCs are located at the border point of IMS and SAE networks and integrate two functional elements from the IMS Release 8 architecture. A few architectural entities are introduced below:

- *Topology Hiding Internetwork Gateway (THIG)*: THIG can be used with I-CSCF, which hides the capacity, configuration, and topology of the network from the outside. The P-CSCF forwards the SIP messages received at the I-CSCF and/or the S-CSCF from the User Equipment, depending on the procedure and the type of message. The I-CSCF offers a contact point within an operator's network, which allows registration to the subscribers of that network operator and roaming subscribers. The S-CSCF will maintain session state for all IMS services, once registered. Other elements include: The Home Subscriber Server (HSS), or User Profile Server Function (UPSF), which is also called a Master User Database (MUD) that supports the IMS network entities, actually handling calls. The HSS contains the subscription-related information, which performs user-based authentication and authorization and provides user's physical location information.
- *BGCF (Border Gateway Control Function)*: The BGCF is used to select the network where PSTN connection is going to be made. It is responsible to forwards to another BGCF or to a MGCF controlling the access in regards to the PSTN.
- *MGCF (Media Gateway Control Function)*: The MGCF controls the media gateway call management, which is responsible of sending or receiving calls from or to PSTN or other circuit switched networks. MGCF uses SIP messages to or from the BGCF/CSCF and uses Media Gateway Control (MGWC) messages to or from the Media Gateway.
- *MGW (Media Gateway)*: The MGW is responsible for call media processing to or from PSTN/Circuit Switched Network (CSN).
- *MRF (Media Resource Function)*: The MRF provides media related functions such as media manipulation.
- *MRFC (Media Resource Function Controller)*: The MRFC is responsible for signaling plane node, which acts as a SIP User Agent to the S-CSCF. The MRFC also controls the MRFP.
- *MRFP (Media Resource Function Processor)*: The MRFP is a media plane node, which is responsible for the implementation of media-related functions.
- *MSG (Multiservice Security Gateways)*: The MSG delivers service provider's voice and data services securely over un-trusted Wi-Fi access networks and Internet links to dual mode handsets and femtocells.
- *SRP (Session Routing Proxy)*: The SRP selects the destination for incoming and outgoing SIP sessions and provides core session routing. It also deals with the traffic to or from media gateways and interconnects session border controllers.
- *AGWF (Access Gateway Function)*: The transport boundary at layers 3 and 4 between the service provider's network and the subscribers is being controlled by the AGWF. It acts as a NAT device and a pinhole firewall, which protects the service provider's IMS network. The AGWF also controls the access through IP address/port packet filtering and opening/closing gates into the network. It uses NAPT to hide the IP addresses/ports relate to the service elements in the IMS network. Other features of the AGWF include: bandwidth and signaling rate policing, QoS packet marking,

usage metering and media flows QoS measurements.

3.6 Interconnect Session Border Controller (ISBC)

This SBC addresses the boundary requirements where service provider networks interconnect and exchange inbound and outbound SIP sessions. It integrates the following three IMS functional elements related to the 3GPP Release 8 ("Exploring IMS security mechanisms"; Acme Packet):

- *Interconnect Border Control Function (IBCF):* The IBCF offers boundary control between various service provider networks, which provides IMS network security in terms of signaling information. This is done by implementing a Topology-Hiding Inter-network Gateway (THIG) sub-function, which performs signaling–based topology hiding, session screening and IPv4-IPv6 translations based on source and destination signaling addresses. When connecting non-SIP or non-IPv6 networks, the IBCF also invokes the Inter-Working Function and performs bandwidth allocation and admission control using local policies or through the interface to PCRF elements. The IBCF may also interact with TrGW for control of the boundary at the transport layer including NAPT, pinhole firewall, and other features.
- *Inter-Working Function (IWF):* The IWF provides signaling protocol mechanism between the SIP-based IMS network and other service provider networks, which uses other SIP profiles or H.323.
- *Transition Gateway (TrGW):* The transport boundary at layer 3 and layer 4 is controlled by the TrGW among various service provider networks having similar media functions as in the AGW.

3.7 Multiservice Security Gateway

In the SAE architecture, the Evolved Packet Data Gateway (ePDG) is the functional element, which delivers voice and data services over the un-trusted Wi-Fi networks and Internet to femtocells and dual-mode handsets. The ePDG is responsible to authenticate subscribers and use IPsec to tunnel voice and data securely to devices over the Wi-Fi networks and Internet. For conforming to the 3GPP standards, the ePDG system also supports the I-WLAN Tunnel Terminating Gateway (TTG) functional elements based on the 3GPP Release 7, and UMA/GAN Security Gateway (SeGW) based on the Release 6 and the Femtocell Security Gateway (SeGW) based on the Release 8 specifications.

3.8 Session Routing Proxy (SRP)

The SRP deals with interactive communication sessions between SIP network border points, including: SBCs, IMS subscriber call control elements, mobile switching centers (MSC), and controlling media gateways.

3.9 Application Server Layer (ASL)

The ASL undertakes the control of the end services required by the user. The functions and services supported at this layer include (Amirth, n.d.):

- *Telephony Application Server (TAS):* The TAS maintains the call state and is a back-to-back SIP user agent. The TAS supports routing, call setup, call forwarding, call waiting, conferencing and contains the service logic that provides the basic call processing services, which includes digit analysis.
- *IP Multimedia - Service Switching Function (IM-SSF):* The IM-SSF provides SIP message interworking, which corresponds to the Customized Applications for Mobile

Networks Enhanced Logic (CAMEL), ANSI-41, Transaction Capabilities Application Part (TCAP), or Intelligent Network Application Protocol (INAP) messages.
- *Supplemental Telephony Application Server (STAS)*: The STAS is a standalone independent server that provides supplemental telephony services at: the beginning, the end, or in the middle of a call, or by triggering.
- *Non-Telephony Application Server (NTAS)*: These application servers, such as NTAS, interwork with endpoint clients, which provide services such as PTT, IM, or presence-enabled services.
- *Open Service Access - Gateway (OSA-GW)*: The interworking between SIP and the Parlay API is provided in the OSA-GW, which is part of the 3GPP IMS architecture application server layer.

3.9.1 Benefits and Implementation of IMS

The benefits of IMS in regards to the existing cellular network infrastructure can be shown in the following forms (Amirth, n.d.):

- *Reduced time-to-market new multimedia services*: The IMS infrastructure provides reusable components and a standardized platform. The common features provided by IMS infrastructure and the related standardized interface help service provider to market new multimedia services in a relatively short period of time.
- *Quality of Service (QoS):* To improve and guarantee the transmission quality, IMS specifies Quality of Service within the IP network and therefore takes advantage of the QoS mechanism.
- *IMS Location Independence:* IMS offers service availability irrespective of the users' location. IMS uses specific protocols and Internet technologies, which allows users to roam across different countries and still be able to have access to all the services. Therefore all services are available to the users disregarding their location.

3.9.2 Push to Talk over Cellular (PoC)

The following PoC related functionalities are offered:

- *Simultaneous Ringing and Multiple / find-me, follow-me*: Which involves a predefined list of destinations (sequentially or in parallel) for routed calls.
- *Multimedia Push:* This service permits users to push some multimedia content.
- *Push Ring Tone:* Calling party selects a desirable ring tone on the destination number/address.
- *Real Time Video Sharing*: This deals with the real time peer-to-peer and multimedia streaming service.

3.9.3 Interactive Gaming

Interactive gaming deals with the followings (Amirth, n.d.):

- *Folder Sharing:* Content and folder sharing enables users to share files/folders among terminals.
- *Voice Messaging:* This involves sharing of audio files instant messaging.
- *Instant Messaging Services:* This refers to a general communication service that allows end-users to send and receive messages instantly.
- *Video-Conferencing:* This takes advantage of IP Multimedia Subsystem (IMS) Video-conferencing service, which extends the point-to-point video call to a multi-point service.

3.9.4 IMS Enabled Voice and Video Telephony

Multimedia enriched (i.e., voice and video calls) IMS traffic are carried over a packet core network (VoIP). In mobile networks, video telephony is counted as a critical end-user service. The Session Initiation Protocol (SIP) enables Voice and Video Telephony using Person-to-Person and Multiparty sessions over an IP network.

The main issue in VoIP and Video Telephony calls is the Quality of Service (QoS) in packet core networks, as well as on interoperability with the legacy phones and PSTN and inter-working with existing domains for Video Telephony such as H.324M and H.323.

User-based QoS schemes drive bandwidth requirements, thus, with similar QoS schemes, packet-switched video telephony may experience similar bandwidth allocations, similar to the circuit-switched video telephony scenario. However, packet switching provides more freedom in balancing the bandwidth and video quality requirements. The Quality of Service and bandwidth requirements for the connection are requested from the network by the terminal at connection set-up phase (PDP Context activation). The IMS infrastructure is responsible to provide several features to manage QoS.

3.9.5 Quality of Service (QoS) – Key to Quality Real-time Service Realization

IMS provides a standardized and effective solution for operators who require implementing real-time IP mobile services with guaranteed customer satisfaction in mind.

Maintaining real-time Mobile IP communication is difficult due to bandwidth fluctuations, which can severely affect the IP packet transmission through the network. In a QoS-disabled IP networks, IP transport is based on the 'best effort' setting, meaning that the network will do its best effort in a uniform sense to ensure the required bandwidths are provided without any guarantee. The result is that real-time mobile IP services may function poorly, such as the voice quality may sound poor or with garbled effects and video quality may include 'jitter' effects and so on, which depends on the network congestion and bandwidth availability.

The Quality of Service (QoS) mechanisms are used to lessen the degradation effects on the multimedia transmissions and to provide some type of guaranteed sense to the transmission compared to the case of 'best effort'. QoS ensures that critical elements of IP transmission such as gateway delay, transmission rate, and error rates can be guaranteed, measured, and improved in advance. Users should be able to specify the required quality level, depending on users' circumstances and the type of services used.

IMS is responsible to provide the 'intelligence' required to enable QoS within a mobile IP network, which is also known as the Policy Decision Function (PDF). Through the Go interface of the GGSN, the PDF interacts with and controls the underlying packet network.

3.9.6 Presence Server

A presence server is an element in the IMS network that offers presence aggregation that can provide network-based presence consolidation. The main components of the presence server include:

- *Interfaces:* Including Home Subscriber Server (HSS) with Diameter S_h interface to obtain subscription information and SIP interface with presentities and watchers.
- *Entities:* Including *HSS*, which is responsible to store service-specific information, *Presentities*, which can be either real-time services, user devices or in form of applications, which they send availability information to the server, and *Watchers*, which

are services interacting with the presence server and getting presence information from other services or users
- *Functional Modules:* Including *Presence core*, which handles all publish, subscription and notification events, *Presence database*, which is responsible for storing information given by presentities, *Presence aggregator* that consolidates presence information from multiple identities of the same user, *Transport layer*, which provides interfaces of SIP, Diameter S_h, XMPP (Extensible Messaging and Presence Protocol), if required, and *Presence subscription* and *watcher module*, which are storing a list of subscriptions (Valsala, 2007).

3.9.7 IP Centrex

Centrex provides PBX-type services, which provides switching functionality at the central office instead of the customer's premises. In a typical sense, the telephone company owns and manages all the communications hardware and software systems necessary for implementing the Centrex service.

Voice conversations can be digitized and packetized in IP telephony, for across network transmission. In such a context, an IP Centrex refers to a number of IP telephony solutions offering Centrex service for customers transmitting packetized voice streams across a broadband access facility. IP Centrex can be built on top of traditional networks combining them with the benefits of IP telephony. One of these IP telephony benefits is the access capacity increased utilization. Using IP Centrex, a single broadband access technology is used to carry the packetized voice streams for many simultaneous calls. During inactive call sessions (no active calls available), more bandwidth is available for high speed data sessions, such as Internet access. This is a much more efficient use of capacity compared to the traditional Centrex.

In analog Centrex, one pair of copper wires is required to serve each analog telephone station, disregarding if the phone has an active call; the bandwidth capacity of those wires is unused, or one of the phones is not engaged in a call.

Other IMS-related functionalities, which may be less IPv6 dependent include (Amirth, n.d.): Download parental control and media streaming, programmable control and incoming call screening, outgoing VoIP call control/barring, convergent instant messaging system, including: mixing SIP, SMS USSD, and text-to-speech), and enhanced policy controller, including: controls SIP message sequences and format.

3.10 Packet in 3GPP SAE

The Release 8 architecture of 3GPP incorporates both the previously introduced IMS functional elements and the new Service Architecture Evolution, which includes the Evolved Packet Core (EPC) and the evolved universal terrestrial RAN (EUTRAN).

CONCLUSION

Unlike 3rd Generation Mobile Networks (3G), which are based on two parallel infrastructures comprised of packet switched and circuit switched network and nodes, the 4th Generation Mobile Networks (4G, LTE roadmap "LTE-Advanced") will be based on packet switching only, which requires low-latency data transmission.

By the time that that 4G systems are readily deployed, the usage of IPv4 addresses is expected to be in its final stages. Therefore, the IPv6 support is essential in the context of 4G, in order to support a large number of wireless-enabled devices. IPv6 removes the need for Network Address Translation (NAT) by increasing the number of IP addresses, although NAT will still be required for communicating with devices that are using existing IPv4 networks.

There are numerous reasons for adopting IPv6 as the network layer mechanism, which have been discussed in this chapter. IPv4 addresses are likely to exhaust by 2012. As IPv4 addresses are being depleted, SIP-based applications (always-on services) are being deployed at an increasing rate, giving urgent need to move to IPv6, which continues to be a major issue for wireless-based vendors and operators.

This chapter's focus was on the LTE-IPv6 perspective, which studied various IP-based components used in the LTE system. This included an introduction to IPv6-based Evolved Packet Core (EPC) and IPv6 Multimedia Subsystem (IMS). It is recommended that LTE should consider making IPv6 a requirement from day one. This is due to the fact that EPC, which is a core LTE system, does not support a Circuit Switched Core as part of the 3GPP standard, which will however be supported by IMS core. Since the transition to IMS-based Voice-over-IP (VoIP) will likely take several years, it is critical for operators and vendors to understand the impact of IPv6 on the existing Voice Core and Signaling infrastructure.

GLOSSARY

2G: Second Generation Mobile Networks
3G: Third Generation Mobile Networks
3GPP: Third Generation Partnership Project
4G: Fourth Generation Mobile Networks
AGWF: Access Gateway Function
AIPN: All IP Network
ANSI: American National Standards Institute
ASBC: Access Session Border Controller
ASL: Application Server Layer
B2BUA: Back-to-Back User Agent
BGCF: Breakout Gateway Control Function
CAMEL: Customized Applications for Mobile Networks Enhanced Logic
CDCF: Content Delivery Control Function
CDMA: Code Division Multiple Access
CFCF: Central Flow Control Function
CN: Core Network
COPS: Common Open Policy Service
CS: Circuit Switched
CSD: Circuit Switch Data
DHAD: Dynamic Home Agent Discovery
DHCP: Dynamic Host Configuration Protocol
DNS: Domain Name System
DPI: Deep Packet Inspection
DSL: Digital Subscriber Line
EUTRAN: Evolved Universal Terrestrial Radio Access Network
eNB: eNodeB
eNodB; Evolved Node B
EPC: Evolved Packet Core
ePDG: Evolved Packet Data Gateway
EvDO: Evolution-Data Optimized
FTTx: Fiber To The x
GAN: Generic Access Network
GGSN: Gateway GPRS Support Node
GPRS: General Packet Radio Service
GTP: Tunneling Protocol
HSS: Home Subscriber Server
I-CSCF: Interrogating-CSCF
IBCF: Interconnect Border Control Function
IETF: Internet Engineering Task Force
IMS: IP Multimedia Subsystem
IM-SSF: IP Multimedia - Service Switching Function
INAP: Intelligent Network Application Protocol
IP: Internet Protocol
IPng: Internet Protocol - Next Generation, aka IPv6
IPTV: IP Television
IPv4: Internet Protocol version 4
IPv6: Internet Protocol version 6, aka Ipng
IPSec: IP Security
ISATAP: Intra-Site Automatic Tunnel Addressing Protocol
ISDN: Integrated Services Digital Network
ISP: Internet service provider
ISUP: ISDN User Application Part
IWF: Inter-Working Function
LTE: Long Term Evolution
LTE-A: LTE - Advanced

MAP: Mobile Application Part
MEGACO: Media Gateway Control Protocol
MGCF: Media Gateway Control Function
MGW: Media Gateway
MIPv6: Mobile IPv6
MME: Mobility Management Entity
MRCF: Media Resource Control Function
MRF: Media Resource Function
MRFC: Media Resource Function
MRFP: Media Resource Function Processor
MS: Mobile Station
MSC: Mobile Switching Centers
MSG: Multiservice Security Gateways
NAPT: Network Address Port Translation
NAT: Network Address Translation
NAT-PT: NAT - Protocol Translation
NCP: Network Control Protocol
NGN: Next Generation Networking
NTAS: Non-Telephony Application Server
OCS: Online Charging System
OFCS: Offline Charging System
OSA-GW: Open Service Access - Gateway
P-CSCF: Proxy-CSCF
PC: Policy Charging
PCRF: Policy and Charging Rules Function
PoC: Push to Talk over Cellular
PDN: Packet Data Network
PDF: Policy Decision Function
PDP: Packet Data Protocol
PGW: PDN Gateway
PLMN: Public Land Mobile Network
PMIPv6: Proxy Mobile IPv6
PPP: Point-to-Point Protocol
PS: Packet Switched
PSTN: Public Switched Telephone Network
QoE: Quality of Experience
QoS: Quality of Service
RFC: Request For Comment
RNC: Radio Network Controller
S-CSCF: Serving-CSCF
SAE: System Architecture Evolution
SBC: Session border controller
SeGW: Security Gateway
SDF: Service Data Flow

SGSN: Serving GPRS Support Node
SGW: Serving Gateway
SLAAC: Stateless Address Autoconfiguration
SRP: Session Routing Proxy
STAS: Supplemental Telephony Application Server
TAS: Telephony Application Server
TCAP: Transaction Capability Application Part
TDM: Time Domain Multiplexing
TEID: Tunnel ID
THIG: Topology Hiding Internetwork Gateway
ToS: Type of Service
TrGW: Translation Gateway
TTG: Terminating Gateway
TTL: Time To Live
UDP: User Datagram Protocol
UE: User Equipment
UMA: Unlicensed Mobile Access
UMTS: Universal Mobile Telecommunications System
URI: Uniform Resource Identifier
VoD: Video on Demand
VoIP: Voice-over-IP
VPN: Virtual Private Network
WGW: Wireless Gateway
Wi-Fi: Wireless Fidelity
WiMAX: Worldwide Interoperability for Microwave Access
WLAN: Wireless LAN
XMPP: Extensible Messaging and Presence Protocol

REFERENCES

Alcatel-Lucent. (2009). *Introduction to Evolved Packet Core, Strategic White Paper.* Retrieved from http://www.3g4g.co.uk/Lte/LTE_WP_0903_AlcatelLucent.pdf

3G Americas. (2008 March). *Introduction to IPv6*. Retrieved from http://www.3gamericas.org/documents/2008_IPv6_transition_3GA_Mar2008.pdf

3G Americas. (2009, February). *IPv6 Transition Considerations for LTE and Evolved Packet Core*. Retrieved from http://new.3gamericas.org/documents/3G%20Americas%20IPv6%20White%20Paper%20Feb%202009.pdf

Amirth. (n.d.). *IP IMS (IP Multimedia Subsystem)*. IPv6.com Tech Spotlight. Retrieved on September 8, 2009. Retrieved from http://www.ipv6.com/articles/general/IP_IMS.htm

Aoun, C., & Davies, E. (2007). *Reasons to Move the Network Address Translator - Protocol Translator (NAT-PT) to Historic Status*. RFC 4966. Retrieved from http://www.faqs.org/rfcs/rfc4966.html

Deering, S., & Hinden, R. (1995). *Internet Protocol, Version 6 (IPv6) Specification* [RFC 1883]. Retrieved from http://www.faqs.org/rfcs/rfc1883.html

Ericsson Inc. (2009, June). *LTE – an introduction* [White Paper]. Retrieved from http://www.ericsson.com/technology/whitepapers/lte_overview.pdf

Esposito, R., Hamilton, B. A., & Graveman, R. (2003). Security with IPv6 Explored. In *RFG Security, U.S. IPv6 Summit 2003*. Retrieved from http://www.usipv6.com/2003arlington/presents/Renee_Esposito_and_Rich_Graveman.pdf

Exploring IMS security mechanisms: will 3GPP and ETSI specs be enough to protect IP-based multimedia traffic from attack? (n.d.). *GOLIATH, Business Knowledge On Demand*. Retrieved on September 10, 2009, from http://goliath.ecnext.com/coms2/gi_0198-378937/Exploring-IMS-security-mechanisms-will.html

Gallaher, M., & Rowe, B. (2005). *The Costs and Benefits of Transferring Technology Infrastructures Underlying Complex Standards: The Case of IPv6*. Presented at Technology Transfer Society Annual Conference, Kansas City, MO. Retrieved from http://www.kauffman.org/uploadedFiles/Gallaher_Mike.pdf

3GPP Long Term Evolution. (n.d.). Techwritters Future, IPv6.com Tech Spotlight. Retrieved on September 8, 2009, from http://www.ipv6.com/articles/wireless/3GPP-Long-Term-Evolution.htm

3GPP TS 23.060 V8.5.1, Release 8. (2009 June). Technical Specification Group Services and System Aspects; General Packet Radio Service (GPRS); Service description; Stage 2, 3rd Generation Partnership Project. Retrieved from http://mailgroup.jp/3GPP/Specs/23060-851.pdf

3GPP TS 23.221 V9.0.0. (2009, June). Technical Specification Group Services and System Aspects; Architectural requirements, Release 9, 3rd Generation Partnership Project. Retrieved from http://www.3gpp.org/ftp/Specs/archive/23_series/23.221/23221-900.zip

3GPP TS 23.228 V9.0.0. (2009, June). Technical Specification Group Services and System Aspects; IP Multimedia Subsystem (IMS); Stage 2, Release 9, 3rd Generation Partnership Project. Retrieved from http://www.3gpp.org/ftp/Specs/archive/23_series/23.228/23228-900.zip

3GPP TS 23.401 V8.1.0, Release 8. (2008, March). 3rd Generation Partnership Project; Technical Specification Group Services and System Aspects; General Packet Radio Service (GPRS) enhancements for Evolved Universal Terrestrial Radio Access Network (E-UTRAN) access.

Graveman, R., Parthasarathy, M., Savola, P., & Tschofenig, H. (2009 May). Using IPsec to Secure IPv6-in-IPv4 Tunnels. *RFC 4891*. Retrieved from http://www.faqs.org/rfcs/rfc4891.html

Hagen, S. (2002 July). *IPv6 Essentials*. O'Reilly Publications. Retrieved from http://www.sunny.ch/downloadfiles/ipv6_sample.pdf

Hogg, S. (2007). IPv6: Dual stack where you can; tunnel where you must. *NetworkWorld*. Retrieved from http://www.networkworld.com/news/tech/2007/090507-tech-uodate.html

IPv6. (n.d.). Wikipedia. Retrieved on September 8, 2009, from http://en.wikipedia.org/wiki/IPv6

IPv6 Transition Considerations. (2009, February). 3G Americas. Retrieved from http://www.3gamericas.org/index.cfm?fuseaction=pressreleasedisplay&pressreleaseid=2150

ISATAP (Intra-Site Automatic Tunnel Addressing Protocol). (n.d.). Wikipedia. Retrieved on September 8, 2009, from http://en.wikipedia.org/wiki/ISATAP

Motorola Inc. (2007). *Long Term Evolution (LTE): A Technical Overview* [Technical White Paper]. Retrieved from http://www.motorola.com/staticfiles/Business/Solutions/Industry%20Solutions/Service%20Providers/Wireless%20Operators/LTE/_Document/Static%20Files/6834_MotDoc_New.pdf

Packet, A. (n.d.). *Acme Packet Defines Role of its Net-Net Product Family within IMS LTE Networks*. Retrieved on September 10, 2009, from http://www.ir.acmepacket.com/phoenix.zhtml?c=200804&p=irol-newsArticle_Print&ID=1254949&highlight=

Postel, J. (1981, September). *Internet Protocol*. RFC791. Retrieved from http://www.faqs.org/rfcs/rfc791.html

Punithavathani, D. S., & Sankaranarayanan, K. (2009). IPv4/IPv6 Transition Mechanisms. *European Journal of Scientific Research, 34(1), 110-124*. Retrieved from http://www.eurojournals.com/ejsr_34_1_12.pdf

Soliman, H. (2009). *Mobile IPv6 Support for Dual Stack Hosts and Routers*. RFC 5555. Retrieved from http://www.faqs.org/rfcs/rfc5555.html

System Architecture Evolution (SAE). (n.d.). Wikipedia. Retrieved on September 9, 2009. Retrieved from http://en.wikipedia.org/wiki/System_Architecture_Evolution

Systems, C. (2008, October). *Implementing QoS for IPv6*. Retrieved from http://www.cisco.com/en/US/docs/ios/ipv6/configuration/guide/ip6-qos.pdf

6to4. (n.d.). Wikipedia. Retrieved on September 8, 2009, from http://en.wikipedia.org/wiki/6to4

Tsirtsis, G., & Srisuresh, P. (2000). Network Address Translation - Protocol Translation. RFC 2766. Retrieved from http://www.faqs.org/rfcs/rfc2766.html

Valsala, K. (2007). Enabling Network-Based Presence Aggregation using IMS. *Infosys*. Retrieved from http://www.infosys.com/engineering-services/product-engineering/white-papers/enabling-networks-paper-using-IMS-final.pdf

Chapter 6
HWN* Framework Towards 4G Mobile Communication Networks

Chong Shen
Tyndall National Institute, Ireland

Dirk Pesch
Cork Institute of Technology, Ireland

Robert Atkinson
University of Strathclyde, Scotland

Wencai Du
Hainan University, China

ABSTRACT

The objective of the Hybrid Wireless Network with dedicated Relay Nodes (HWN) proposal is to interface the Base Station (BS) Oriented Mobile Network (BSON) and the 802.11X assisted Mobile Ad hoc Wireless Network (MANET) so that one system can be utilised as an alternative radio access network for data transmissions, while the incorporation of the Relay Node (RN) is to extend the communication coverage, optimise medium resource sharing, increase spatial reuse opportunity, stabilise MANET link and create more micro-cells. The HWN* keeps the existing cellular infrastructure and a end-user Mobile Terminal (MT) can borrow radio resources from other cells through secured multi-hop RN relaying, where RNs are placed at pre-engineered locations. The main contribution of this work is the development of a HWN* system framework and related medium access and routing protocols/algorithms. The framework dedicatedly addresses the transparent multiple interface traffic handover management, cross layer routing, RN positioning and network topology issues to increase communication system capacity, improve Quality of Service (QoS), optimise transmission delay and reduce packet delivery delay.*

DOI: 10.4018/978-1-61520-674-2.ch006

INTRODUCTION

The evolving cellular communications industry was originally focused on providing voice services. However, ever increasing demand for multimedia mobile services required new architectures to extend indoor and outdoor coverage, increase system capacity, improve service quality and reduce transmission delay. For example, a large amount of bandwidth is needed to support mobile broadband applications. Current solutions require the installation of expensive BSs to increase the system capacity with reduction of cellular cell sizes, which result in large costs in equipment, wiring and complex system management. The alternative approach is to install wirelessly linked multi-functional Relay Nodes (RN), which not only creates smaller cell sizes but also provides packet relay functions, introduces networking flexibility, saves end terminal power and largely mitigates the pass loss. In this work, we propose to realise high quality mobile communication services in a novel system infrastructure that includes a cellular network, a Mobile Ad hoc Network (MANET) and an integrated dual radio access interface on each RN and MT.

The proposal brings out many research challenges as the infrastructure lends itself to complex change in topology, medium access, resources sharing and routing path selection. The ad hoc interface has limited transmission range thus multi-hop communication and RN infrastructure support are required for data exchange. The cellular interface offers robust communication but the bottleneck is at the resource sharing between base stations and relay stations, which require the development of a traffic sharing approach between entities. In order to exploit the advantages of the infrastructure, algorithms and protocols have been proposed to tackle resource management problems e.g. inter-network traffic management and heterogeneous network route selection in a distributed manner.

Hybrid wireless networks with multiple active interfaces towards 4G wireless networks are still being developed and as yet no real world prototype has been deployed except the 3GPP Long Term Evolution (LTE) test operating by Motorola, Nortel and NTT DoCoMo, the IEEE 802.16j multimedia traffic relaying project and Alcatel-Lucent 3G Femto high speed mobile home access where a large number of Femto relay cells are required on per house basis. The HWN* represents another possible realisation of the hybrid wireless network concept. The HWN* and associated algorithms are proposed and evaluated by means of computer simulation (Rea, 2006) to analyse node mobility, scheduling algorithm, handovers, routing, resource sharing and topology design schemes. The complexity of the HWN* system is such that performance evaluation does not lend itself to pure mathematical treatment but more accurate evaluation is only possible by means of simulation, which also provides more practical options for parameter changes.

The following chapters will first briefly review the state of the art of the cellular network, MANET, relay concept, recent hybrid network system approaches, and hybrid wireless networks related algorithms to provide an understanding of the effectiveness of our HWN* infrastructure. The major achievement of our research is then discussed. Apart from the cost-effective HWN* architecture proposal with minimal change on existing cellular and MANET structures, other core contributions can be summarised as:

- **Network Design:** In order to maximise the spectrum usage and facilitate load balancing, RN positioning planning has been developed and investigated. Our approach focuses on heuristic relay placement to explore the node mobility pattern's impact, cellular system, MANET and relay characteristics.
- **Solving complexity:** For large scale hybrid networks, more components are added to the system. Therefore simple route management algorithms may produce larger

end-to-end transmission delays, which results in a slower application response time. The design of cascaded routing algorithm has contributed a novel approach to deal with complex systems over several layers. The scheme includes several route change triggers and considers cross layer design in overlayed networks.

- **Quality of Service**: The HWN* is expected to provide services with different QoS requirements, which results in complex user management. The research provides a user classification strategy with differentiated service class profiles for the evaluation of traffic admission, mobility management, handover management and cascaded routing. For the inter-network handovers, congestion control between service classes has been introduced, which discourages applications from using heavily loaded cellular resources and encourages the MANET usage.

The algorithms and strategies developed as part of this research aim to benefit resource management and routing for the HWN*. These methodologies, meanwhile, have potential to be applied on cooperative vehicle networks including vehicle-to-vehicle communication and vehicle-to-infrastructure communication, and wireless sensor network packet relay infrastructure and related medium access & route optimisation.

HYBRID WIRELESS NETWORK: CHALLENGES AND SOLUTIONS

Hybrid wireless networks provide combined medium access methods selected from cellular, MANET, IP and 802.1X. Effective resource management, route scheduling and relay strategies are key aspects in facilitating this heterogeneous environment. The hybrid relays have received significant research interest as a consequence of system capacity advantage and reasonable infrastructure cost (Zadeh, 2002; WINNER, 2005). The novel HWN* infrastructure proposed here is also expected to provide stable, high-speed and user satisfied communication services in most situations including urban city with both cellular and MANET coverage, indoor environment and remote areas without cellular coverage thus is a challenging design goal. The section starts with a state of the art review for hybrid wireless networks with different design objectives. It then provides the rationale which motivates us to develop HWN* with system concept and architecture. The HWN* infrastructure is further supported by the novel algorithms and protocols proposed at Medium Access layer (MAC), NETWORK transport layer, and cooperation between cellular network, MANET and relay structure. Therefore, algorithms and protocols related to hybrid wireless networks are also briefly discussed.

Hybrid Wireless Networks

Hybrid wireless networks are defined as an integrated infrastructure that provides seamless services over several networks. However, most of research focuses on the infrastructure research itself and few algorithms have been proposed to explore hybrid wireless network practical usage at medium access layer and routing layer. Current algorithm suites are normally proposed for system capacity, relay station placement plan, cooperative resource sharing and path discovery to improve only the cellular network performance (Aggelou, G. & Tafazolli, R. 2001). These solutions leverage the presence of persistent resources to support relay networks, but rely on the fact that ad hoc multi-hop services being underestimated and multiple interface technology is not comprehensively considered. The service-oriented algorithm for hybrid wireless networks was still left for further investigation as most previous proposals (De, S. 2002) only concentrate on the infrastructure design, which assume such services would be supported or extended based on existing cellular differentiated service methodologies.

For the design of effective resource and routing management algorithms in a complex system, centralised control approaches should be avoided as network scalability issues largely jeopardises the hybrid wireless network performance. The introduction of system flexibility by avoiding central control motivates this research to propose algorithms that consider cross layer communication issues and differentiated services in a decentralised manner that influence the global system performance through local conditional changes. The challenges for communication service provision have increased in many respects in recent years. Competition among providers demands a continuous reduction of production costs and improvement of service quality, which motives us to propose the cost-effective novel HWN* infrastructure and develop associated algorithms for the infrastructure realise with guaranteed QoS. The algorithm development does not require a brand new ubiquitous radio access technology or focusing on multi-hop based cellular networking technologies, but rather, the work integrally utilises existing cellular, ad hoc and relay technologies in an integrated fashion, and combines their advantages and overcomes their inefficiencies.

The research addresses several issues from handover management and routing perspectives, evaluates proposed algorithms in the HWN*, and finalises the HWN* management framework towards 4G mobile system. It is also concerned with the provision of fixed relay nodes to subscribers in both urban and remote areas and also implementing coordinated cellular and MANET radio access. It devises a framework for studying the trade-offs of interworking between the two active systems, identifies and answers the specific design space questions such as: How to share the resource between different service classes? How many relay nodes are sufficient in a reasonable cost? Where should we place RNs? What protocols are necessary to facilitate the load balancing between the two systems? How to handle network routing in a heterogeneous environment? Why the infrastructure uses relays and MANET to reduce the number of wired BSs? The study answers these questions in order to demonstrate that the proposed integrated infrastructure can be used for providing the enhanced data communication services.

Recent research attempts for MANET-cellular infrastructure integration include Multi-hop Cellular Network (MCN), Multi-Power Architecture for Cellular Networks (MuPAC), integrated Cellular and Ad hoc Relaying system (iCAR), Self-organising Packet Radio Networks with Overlay (SOPRANO) (Murthy, C. & Manoj, B 2004) and the IST-WINNER project. The basic rational of the MCN is a cellular network evolution that concentrates on cellular radio access technology. Traditionally, a MT and a BS have a direct link in 2G or 3G cellular networks, but in the MCN a MT may reach the serving base station by using multi-hop relaying. The relay is called soft MT based relay which refers to other MTs act as relay clients using ad hoc frequencies. The proposal also states that the relay node could be an infrastructure node if the condition allows. However, it only generally concludes that the relay has capabilities like those of the BS such like an ad hoc node access point. It does not exactly solve problems such as how to choose packet delivery route or when relay mode should be used other than original cellular communications. The analysis and simulation results present the throughput comparison between conventional Single hop Cellular Network (SCN) and MCN. The throughput of MCN is better than that of SCN and increases as the transmission range decreases. The node uses a transmission range r that is a fraction $1/k$ of the cell radius R where $r = R/k$. The parameter k is referred to as the reuse factor. The research explains these two observations by illustrating the different increasing orders, of mean number of channels such as simultaneous transmissions in a cell, and mean hop count, as the transmission range between source node and destination node decreases by *k* times. But unfortunately the actual gain will be lower. First of all, large control overheads are produced since every node may perform routing updates even when there is no topology change. If a MANET routing protocol e.g. ADOV is used for multi-hop routing, there is a high possibility of relay client MTs absence. The main disadvantage of the relay is the latency in route discovery and link failure. On the other hand, as the ad hoc frequency is assumed for packet relaying, a small topology change results in medium access failure on the fragile link.

MuPAC (Kumar. J, Manoj. K & Murthy 2002) is an extended MCN with focus on node flexible power management. The system architecture is exactly the same as MCN and each MT uses a separate 802.11X interface. The multiple transmission powers helps MuPAC achieve maximised spatial reuse without substantially increasing the number of hops. The work certainly improves the system capacity performance since the possibility of a path break in MuPAC is lower than MCN once the algorithm is used. Throughput Enhanced Wireless in Local loop (TWiLL) (Frank, C. Manoj, K & Murthy, S. 2002) is also an extended work based on MCN. It has been proposed by the same researchers and focuses on the power management of wireless multi-hop local loop to increase the hybrid system capacity. However, here it is highly possible that nodes increase the transmission power to reduce the path break probability and hop distance. It should also implement a distributed power management algorithm at each MT which is computationally expensive. Otherwise the power management has to be coordinated by BS central control because no dedicated relay node is used.

iCAR (Murthy, C. & Manoj, B 2004) is not difficult to be evolved from the cellular network and it can be also categorised as one type of MCNs. Other than increase the system capacity through node power management or a longer hop distance. The infrastructure works on adaptive traffic load balancing in this multi-hop cellular infrastructure. It is the first hybrid wireless network that looks at the horizontal handovers between cellular accesses through MANET access. Extra cellular bandwidth available in surrounding cells can be borrowed to the congested cell through dedicated ad hoc relay modes named primary, secondary and cascaded relaying. The channel borrowing evaluation results indicate the improvement on packet congestion delay over conventional cellular networks and MCN, and it verifies that with a limited number of relay nodes, the call blocking and dropping probability in a congested cell as well as the overall system can be reduced. However, the simulation evaluation of iCAR system suffers unfair packet contention problem as all packets are treated exactly the same. An extra bandwidth can be allowed to any packets without considering packet priority, packet transmission requirement and QoS. For example, an urgent communication request or instant request can be blocked or terminated due to the contention and unreasonable channel borrowing. And this is also why differentiated traffic input should be introduced for hybrid wireless network evaluations. The iCAR has introduced a novel concept called managed mobility for relay nodes based on its signalling protocol and node mobility model. The relay node movement provides assistance on channel borrowing in congested communication areas but at the expense of complex and two layered route management (MT layer and Relay Layer). Furthermore, each relay station in iCAR must be equipped with location tracking system with extra cost. Another drawback of iCAR is that the system does not explore or underestimate the dedicated relay node assisted MANET communication mode capability. It only intends to reuse the cellular resource via multi-hop ad hoc relaying.

SOPRANO (Zadeh, 2002) is a scalable architecture that assumes the use of multi-hop cellular and asynchronous CDMA with spreading codes to support high data rate Internet and multimedia traffic. The general idea of SOPRANO is not much different from iCAR other than IP network support and cross active network connections. It focuses on connection establishment and node power control based self-organisation, and investigates the formulation for an optimum transmission strategy. Again, a power management algorithm is proposed to adaptively selection transmission power to improve the system capacity, which is similar to MuPAC. The system presents high capacity bounds that illustrate how the technique helps in trading off conserved power for a multi-fold capacity advantage. However, both SOPRANO and iCAR rely heavily on soft MT based multi-hop traffic relay underestimate the usage of an alternative MANET with dedicated and location fixed RN infrastructure since fixed relay nodes are not considered.

European Commission Information Society Technologies (IST) WINNER project proposed the WINNER system. What WINNER focuses on is the Research & Development of a brand new beyond 3G radio interface technologies needed for a ubiquitous radio system. The RNs are deployed to incorporate with BSs realising an efficient and flexible spectrum usage and spectrum sharing environment, where the RNs only implement their new ubiquitous radio interface. The RNs are planned and share the same Radio Access Technology (RAT or refereed as interface) with BSs and MTs. Ad hoc communication is prohibited in this architecture and one can see the WINNER as an evolving node-oriented multi-hop cellular network with relay support. One disadvantage of WINNER can be the implementation cost. Significant hardware and software updates are required at base station radio network controller, relay node and end-user equipment to apply the novel radio interface. The telecom providers may not ready to replace a new radio access deployment without significant system performance improvement compared to current 3G cellular system.

The proposed HWN* has major differences from other hybrid wireless networks motioned previously. It has been summarised in a nutshell in (Shen, C. 2008) the important issues comparison between *HWN**, WINNER and SOPRANO architectures. Detailed comparison between iCar, MuPAC and MCN can be found in (Manoj, B. and Kumar, K 2006). Although advanced technologies such as location tracking make soft relay based infrastructure feasible, the route recalculating and reconfiguring in systems such as MCN, TWiLL and MuPAC, without dedicated relay nodes, are unstable. Meanwhile, there are other fundamental problems such as relay node absence and third party terminal relay security. Using dedicated and location fixed relay node provides straightforward method to enable reliable communication. The research will later compare system performance between HWN* with dedicated RNs, MCN with soft relay, SOPRANO and WINNER. SOPRANO is considered other than iCAR since CDMA based SOPRANO gives better results (Zadeh, 2002) in terms of system capacity and network delay.

With dedicated and location fixed RNs support, the adaptive and scalable HWN* has four basic communication modes, which is cellular communication (also named BSON), RN supported cellular communication (BSON RN), MANET communication and RN supported MANET. With the assumption of coded and modulated digital communications, **Node X** can transmit information to **Node Y** via one or more RNs. The dedicated RN can be part of a cellular network and a MANET. The **nodes X** and **Y** can be a BS, a MT or a dedicated RN. And, the term communications include uplink communications (link from MT to BS), downlink communications (link from BS to MT), MT to MT communications or BS to RN communications. Two MTs may communicate directly or through an intermediate node (The node can be a RN or a group of RNs). The MT can be also accommodated into cellular network with dedicated RNs assistance. Therefore, the relay structure is viewed as a means of extending the communication coverage of either a cellular network or MANET. MANET and cellular network are mutually supported through the use of RN structure. The performance of individual MANET or cellular network, respectively, is also enhanced by RN support. Figure 1 presents the topology of the HWN* used for handover mobility management and cross-layer routing. The RNs create a mesh structure to support node communication through RN infrastructure. The procedure is similar to 802.11X node-to-infrastructure communication but a virtual backbone is constructed between RNs. The BSs may connect to an IP network via fixed lines or switching nodes. The HWN* assumes that there exists wireless connections between RN and BS, and between RN and RN.

The HWN* is a generic system and works with any off-the-shelf air interfaces, as an example, for the cellular part of HWN*, Time Division Multiple Access (TDMA) based system is used. This means that every individual transmitting channel required as part of the chain between any two terminals is created by allocation of time slots such as multiple time slots in Enhanced Data rate for GSM Evolution (EDGE). Variability of data rate is achieved by allocating differing number of time slots. The TDMA cellular interface allows the HWN* to take advantage of multi hop connections formed through RNs with more flexible implementation compared to Frequency Division Multiple Access (FDMA) (3GPP). The HWN* network deployment scenarios against physical layer link duplex model, medium access method, spectrum usage, node movement speed, transmission rate and overall scenario capacity can be found in (Shen, C. 2008, pp 30-33). Traditional MANET scenario can not support high node mobility speed and data transmission rate without the presence of infrastructure node such as 802.11g access point or dedicated RN, thus high mobility and data rate can be realised in RN supported MANET. System capacity of cellular network or RN supported cellular are optimised as the introduction of RN optimises the resource sharing performance. We implement standard CSMA/CA for pure MANET and synchronised CSMA/CA for RN supported MANET mode considering IEEE 802.11e QoS

Figure 1. Hybrid wireless network with dedicated and fixed relay nodes (HWN)(© 2008, Chong Shen. Used with permission.)*

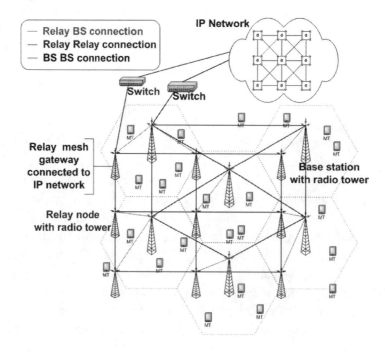

standard for delay-sensitive traffic and the RN is given priority in terms of medium access. The RN priority access will be detailed in cross-layer routing algorithm proposal introduced in the next section. Large scale deployment of dual-interfaced RNs is cost-effective as the equipment can be based on an integration of a modified 802.11 access point and cellular packet relaying function node. Each RN associates with one BS so that the radio resource usage in each RN can be coordinated. A decrease in BS density can be compensated by an increase in RN density, in order to maintain constant performance. The RN has properties and functionaries of:

- Two radio interfaces: The cellular interface and the MANET interface.
- The RN extends the cellular service range and optimises cell capacity.
- The RN minimises node transmitted power.
- The RN covers remote areas, supports inter network load balancing.
- The meshed RNs provide an alternative communication mode which is MANET with RN infrastructure for MANET based resource management and routing.

Theoretically, both the HWN* system capacity and the average packet delivery ratio per MT, compared to traditional 2G and 3G cellular networks, should be improved because the RNs provide relay capability as the substitution of a poor quality single-hop wireless link with a better-quality link encouraged whenever possible. The disadvantage of the RN integration is that whether in reality the infrastructure can be realised or not due to feasibility issues. Telecom providers should first agree, design and prototype such equipment. After identify air interfaces for MANET access and cellular access, both software and hardware are required to be upgraded on actual relay nodes.

HWN* MANAGEMENT FRAMEWORK AND ALGORITHMS

The previous section has examined state of the art hybrid wireless networks and has presented the HWN* infrastructure with system specification. In order to realise a HWN* deployment, algorithms and protocols to facilitate network management must be designed. The presented research here does not cover all aspects of HWH* in this regard but proposes selected key algorithms and protocols. Apart from the design of the HWN* infrastructure, the main research focus lies in resource management issues at the MAC layer and at the network layer.

HWN* Handover Management

In the HWN* context, each communication mode exhibits different capacity characteristics therefore selecting the most appropriate mode for a particular service request is critical to guarantee QoS to the end user. A balanced traffic load distribution between the available modes is also important to avoid one mode becoming excessively loaded, which leads to an unstable state. The HWN* requires an algorithm that includes the service transfer function between the cellular communication modes and MANET communication modes. Another novelty of the HWN* handover mobility management is that user segregation is provided to guarantee QoS to particular user classes by the prioritisation of user requests when the HWN* system approaches a congested state. As the RN provides stable infrastructure support for a MANET, it will be beneficial to switch more service from the cellular service to the MANET service. The HWN* concentrates on the mode selection considering differentiated user negotiation, multiple handover triggers and congestion control. Here the handover mobility management procedure (Shen, C. 2008) refers to a process of transferring a MT from its serving BS to another BS or from one medium access interface to another interface. The RN using cellular frequencies relays the handover traffic but does not have traffic admission functionality. While for the MANET mode and MANET with RN mode, the procedures refer to a new path finding process similar to re-routing, which is described as part of the cascaded ADCR routing algorithm. To mutually mitigate the traffic burden between communication modes, the HWN* dynamically allocates the radio resources without a fixed plan.

Three service classes are devised to describe subscriber behaviour: **High Profile Users** (HPUs), **Normal Profile Users** (NPUs) and **Low Profile Users** (LPUs). Principally, HPUs get the guaranteed QoS service and the class is assured with any amount of bandwidth, firm end-to-end delay bounds and limited queue loss for data flow. Next are NPUs with less QoS guaranteed medium access opportunities compared to HPUs, but the users will get a close service quality as the one received by HPUs. LPUs are a best-effort class with absence of QoS specification using currently available medium resources. HPUs have the highest access priority in any of the communication modes of the HWN*, and traffic admission of NPUs and LPUs has to consider the impact of ongoing HPUs sessions. NPUs are also configured to have a higher probability than LPUs in terms of resource acquisition and this probability is decided by an Association Level (AL) set, which will be described later. Inter-system and intra-system mobility management are addressed separately to reduce system complexity.

MANET RN and MANET modes can potentially become the primary communications methods if distributed management, link reliability and security problems of the MANET mode can be solved. Specifically for the MANET mode, the issue is that data relaying via a third party MT is not safe as discussed in MCN review. Data relaying via provider owned fixed RN provides a permanent solution. The RN integration largely moderates the problem by providing reliable relays between communication parities. Upon communication request, HPUs are configured to search for cellular based service, NPUs sfor a MANET RN service while LPUs for MANET service by default. The signalling, path discovery and route establishment for MANET and MANET RN modes are completed before the BSON and BSON RN modes, if all four modes are available. Inter-system and intra-system traffic handovers are only triggered when essential to avoid unnecessary network management overhead. (Shen, C. and Rea, S. & Pesch, D. 2007).

The control entity for inter system mobility cooperation is called the **HWN* Mobility Controller**, which is responsible for managing the modification of a route in an attempt to maintain or enhance the QoS level. The unit is located in each BS and a BS periodically monitors receives, specifies, filters and analyses frequencies in use by nearby MTs, RNs and BSs, geographically locates MTs so that the terrain blocks interfering signals, and may use directional antennas to reduce unwanted signals. The BS also gathers the information on the MANET with RN mode channel availability from its associated RN. A MT, either requiring differentiated service or not, makes a distributed decision on inter-network and intra-network handovers based on information gathered from its associated BS mobility controller as well as information gathered after probing one hop point to point direct communication. To obtain an effective handover process, while reducing the unnecessary handover rate, it is proposed to setup a dedicated **Status Check Point** embedded in the HWN* Mobility Controller where the necessary measurements are taken and then fed back to the handover algorithms of the nodes involved. The check result indicates the likelihood of a handover, which depends on interference level and physical layer information such as Bit Error Rates (BER), velocity, buffer size, etc. Since the HPUs applications have higher priorities over NPUs and LPUs, subscribers from this profile are more likely to get an accept ticket, which is issued by the **Negotiation Unit** in continuing with traffic handover (Shen, C. and Rea, S. & Pesch, D. 2007). If an **accept ticket** is not issued, the MT will not use the status check data to request continuing with the handover process unless the status check point data necessitates handover. Then the mobility controller will decide to **accept**, **queue** or **reject** the MT handover request. The negotiation process between HPUs, NPUs and LPUs is based on **Association Level** (AL), which makes decisions and feeds back to the **HWN* Mobility Controller**. The AL is a set of parameters monitoring channel availabilities, an AL that scores higher than the threshold means that the channels are already occupied by ongoing sessions. The AL set is further sub-classified as AL in the BSON, BSON RN, MANET and MANET (Shen, C. Rea, S. & Pesch, D. 2007).

The handover algorithms in the HWN* should allow subscribers to seamlessly move without dropping their communication session and considers differentiated QoS issues, for example, it must guarantee QoS for HPUs that agree to pay more than NPUs and LPUs. Two handover types are therefore used as described previously, **Intra Network Handover** occurs when a MT enters into another entity that belongs to the same network with a cellular TDMA MAC interface. An **Inter Network Handover** happens when a MT leaves the serving network and communicates with another entity that belongs to a different network. For all service classes, the intra network handover is selected before considering inter network handover as less traffic management overhead may be produced. In intra network handover, after obtaining pass tickets from the **Status Check Point** and the **Negotiation Unit**, the **Network Selector** entity that embedded in the mobility controller informs the MT if the RN should participate in handovers or not, then the MT makes a local decision. The inter network handover is seen as a switching process between ad hoc and cellular services which has different properties and procedures compared to the intra network handover. The **Status Check Point** is activated first to avoid extra expense and the **Negotiation Unit** keeps monitoring the channel availability status and updating the AL in a short interval to grant or reject handovers. The **Network Selector** always tries to divert the traffic back to intra network handover before it is informed that the intra network handover is not possible. If the MT is currently communicating in ad hoc modes, the selector will search for available BSs in neighbouring cells. If the MT is using cellular modes, it will look for either direct point-to-point communication or search for a fixed RNs involved multi-hop communications. The **Network Selector** also uses several short network search expire times for both intra-network search and inter-network search to make sure a MT is not long isolated during the network selection process (Shen. C, Rea. S, & Pesch. D, 2007).

HWN* Routing Management

Efforts must be invested on HWN* to avoid congestion, compute the next route, discover medium resources and gather data. This section proposes a novel cascaded Adaptive Distributed Cross-layer Routing (ADCR) scheme for the HWN* framework, using a minimal number of hops and considers dynamic routing models aimed at reducing latency, preserving communications and delivering good overall and per node throughput. A cross-layer network design that seeks to enhance the system performance by jointly designing MAC and network layers has been adopted.

The cascaded ADCR includes three sub packet transmission modes labeled as One-Hop Ad-Hoc Transmission (OHAHT) for point-to-point direct communication, Multi-Hop Combined Transmission (MHCT) for radio resource relaying using fixed RNs or MTs, and Cellular Transmission (CT) for a traditional cellular service. The RNs of the core network compose a mesh structure with fixed routing tables using ad hoc frequencies, while BSs are connected to each other via wires and the link between the RN and BS are established using cellular channels and directional antennae. In areas without infrastructure support, two MTs may communicate directly, or through intermediate MTs. When a MT transmits packets to a BS through RNs, the RNs extend the signalling coverage of the BSON thus enhanced resource sharing performance is expected. The QoS flows can consume all the bandwidth on certain links, thus creating congestion for, or even starvation of best effort sessions. Statically partitioning the link resources can result in low network throughput if the traffic mix changes over time. Thus, a mechanism that dynamically distributes link resources across traffic classes based on the current load conditions in each traffic class is critical for performance. By proposing the cascaded ADCR for HWN*, the framework discourages applications from using any route that is heavily loaded with low priority traffic. Traditional routing strategies that use global state information are not considered. Problems associated with maintaining global state information and the staleness of such information is avoided by having individual MTs infer the network state based on route discovery statistics collected locally, and perform traffic routing using this localised view of the network QoS state. Each application, categorised by the service class with the choice of three possible transmission modes, maintains a set of candidate paths to each possible destination and route flows along these paths. The selection of the candidate paths is a key issue in localised routing and has a considerable impact on how the ADCR performs. The high priority traffic is given high priority in accessing comparatively expensive cellular resources, while low priority traffic tries to access lower cost ad hoc resources. Per MT bandwidth is used as the metric for local route statistics collection since it is one of the most important metrics in QoS routing. Furthermore, important metrics such as end-to-end delay can be improved by an increased bandwidth as long as the traffic load is not largely increased. As in QoS based investigation for inter system handover management. Traffic sessions are divided by HPUs, NPUs and LPUs. In case of network congestion, CT mode may temporarily become unavailable to NPUs when HPUs are not fully accommodated, while LPUs sessions may be only granted MHCT and OHAHT mode access to mitigate network congestion, reduce transmission delay and improve per MT throughput (Shen. C, Rea. S, & Pesch. D, 2008).

To avoid having higher traffic classes being influenced by lower traffic classes in terms of queueing delay, a waiting time limitation is placed on each traffic class using a forced starvation packet switch model. A traffic flow maintains two queues: a slot queue and a packet queue. The slot queue is decoupled for traffic class identification from the packet queue for transmission. The RN is specially designed so that it reserves QoS guaranteed free channels for signalling information exchange in the slot queue. Each MT and RN maintains a routing table. A RN's routing table contains the information of other fixed RNs thus the routing delay and multi-hop signalling overhead in MANET RN mode are largely reduced. On the other hand, a RN can also analyse the current system traffic load condition through feed-back from other fixed RNs. The purpose of bandwidth reservation is to let RNs that receive the relaying discovery command in the slot queue check if they can provide the bandwidth required for the connection. Start and finish tags are associated with slots but not with packets. When a packet arrives for a flow, it gets added to the packet queue, and a new slot is added to the slot queue. Corresponding start and finish tags are assigned to the new slot. The way to raise priority in slot queue is that the packets related to a high profile have shorter back-off time to increase the probability of early medium access. As for the status table maintenance, information flooding is restricted to a limited scope. Once a positive acknowledgment message is confirmed by a requesting RN, the relay paths will not be changed unless resource contention happens. Given the fact that maintaining global RNs channel status in each RN slows down RN response time, each RN only updates neighbouring RNs' information, periodically.

In **OHAHT**, the requesting MT A first broadcasts SEARCH messages to every node in its transmission range including its associated RN and BS. If the destination MT B is within its transmission range and there is no ad hoc based media contention between MT A and MT B, MT B can respond to MT A with an ACK message. Once MT A confirms the acknowledgment, it starts a connection SETUP session immediately. The OHAHT transmission model can be extended to multi-hop ad hoc communication and it is only activated on demand.

The **MHCT** involves RNs acting as intermediate nodes for message relaying. BS provides assistance on cellular link establishment using cellular location registers. In the connection setup process for communication between MT A and MT B via the RN infrastructure, MT A first broadcasts SEARCH messages to every node to find MT B. After the SEARCH session, MT A may find that the cellular resources can be used through RNs by receiving three ACK messages from the serving BS of MT B, RNs and the MT B. It may also find that the ad hoc frequency based RN mesh is adequate for communications by receiving two ACK messages from the serving RN of MT B, and the MT B, respectively. These positive acknowledgments require MT B to send an ACK to its serving BS and serving RN, then the serving BS and the serving RN feedback the ACKs to MT A. Once the positive ACK is confirmed, MT A can either start a connection SETUP from MT A → RN, then RN → BS, and finally BS → MT B, or from MT A → RN then RN → M T B. The DATA transmission process follows the same packet delivery route, and further route discovery is prohibited to reduce the signalling overhead.

The label routing concept originated in Asynchronous Transfer Mode (ATM) network is introduced to MHCT mode since the position of RN is fixed and label based RN switching provides faster packet forwarding than routing because its operation is relatively simple.

The label is a fixed-length identifier. Multiple labels can identify a path or connection from source MT to destination MT. The structure of label has four parts (Acharya. A, Misra. A & Bansal. S 2002). The first part **CAST** of the label is 3 bits and relates to cast options. Only uni-cast is considered in the current research. As practically the RN mesh can be very large, the second part of the **Label** is 20 bits which includes a node identifier and is unique in the network ensuring conflicts do not occur. The third part is 3 bits **QoS**, which means class of service. The last part of the label is **TTL**. All label information has a time-related restriction. After time out for a label, all corresponding entries will be deleted from the label routing table. Label routing uses a label to directly index into a connection table entry to determine the next hop, lending itself to a simpler lookup implementation than the complex **IP** routing and hop-by-hop IP address lookup. All intermediate nodes in the virtual connection can forward packets more efficiently. The path from a source MT to a destination MT is identified by multiple labels. The protocol distributes labels and set up new route after the path is computed by the routing protocol. The path finding process dynamically initialised by the label request packet carrying a unique label and flow information, where low path setup delay is guaranteed (Shen. C, Rea. S & Pesch. D, 2008). The path between MT and destination MT is composed of multiple segments. The path is separated by segments and all data packets are relayed by these segments. Each segment is a real connection between two nodes and labeled by the sending-side node of label relay message in this segment. Figure 2 presents an example routing presentation and routing tables for three nodes MT A, RN B and MT C. The relay nodes only need to find the available entry indexed by a label in the packet, swap it with the respective Label out of this entry, and then send it out to

Figure 2. Presentation of routing presentation and routing tables for three nodes MT A, RN B and MT C (© 2008, Chong Shen. Used with permission.)

MT A Label routing table						
Labelin	Labelout	QoS	Sour.MT	Dest.MT	Dest. Sequence	TTL
N/A	B.8	HPUs	A	C	211	10
B.8	N/A					

RN B Label routing table						
Labelin	Labelout	QoS	Sour.MT	Dest.MT	Dest. Sequence	TTL
B.8	C.9	HPUs	A	C	211	10
C.9	B.8					

MT C Label routing table						
Labelin	Labelout	QoS	Sour.MT	Dest.MT	Dest. Sequence	TTL
C.9	N/A	HPUs	A	C	211	10
N/A	C.9					

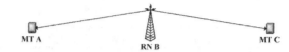

the next relay node. Furthermore, the fixed RN placement reduces the frequency of label entry changes with reliable service.

In the **CT** mode, the source node MT A first broadcasts SEARCH messages to every node to find destination MT B. After the SEARCH session, the MT A finds that it is able to communicate with MT B directly via BSs, while the connection can be setup through a virtual wireless backbone. The positive acknowledgment of a connection requires MT B to send an ACK to its serving BS, then the serving BS informs the serving BS of MT A or the BS feedbacks the ACK to MT B when both MT A and MT B share the same serving BS. Once the positive ACK is confirmed, MT A starts connection SETUP from MT A → BS, then BS → BS, and finally BS → MT B. The DATA transmission process follows the same packet switched delivery route. The cellular transmission also includes cellular transmission with RN support. For cellular based RN connection, the packet tries to establish a link with a BS first then the traffic will be connected to BS RN mode. Dynamic channel allocation can be realised in a distributed manner given that the channel usage does not break the two channel interference constraint which are cosite constraint where there are minimum channel separations within a cell and non-cosite constraint where minimum channel separation between two adjacent BSs is kept.

HWN* RN Site Planning

Plans have already been proposed to select RN sites for in other hybrid wireless networks. The solutions depend on the varied network performance objectives. Among these the packing based RN placement plan is simple, effective and straight-forward but is only suitable for ideal HWN* deployment scenarios where the BS sites are also packed with similar discipline plus the geographical site availability issues are not considered for both BS and RN. It is well known from planar geometry that to cover a two dimensional district with equal sized circles, the best possible packing solution can be obtained by surrounding each circle by six circles. But to have connections between the RNs to form a virtual RN backbone, an overlap is needed between relay cells. The framework therefore considers a situation where the locations of the RNs are centered with maximum coverage.

As the research addresses radio resource management, routing and node mobility problems, the RN site location planning should not be designed too idealistic such as in the packing based RN placement or not only include one or two parameters. We therefore propose a novel heuristic RN placement algorithm considers both physical distance through multi hop distance and Signal to Interference Ratio (SIR) through channel availability. The algorithm is devised in three steps described as:

1. Identify ideal RN locations based on radio resource management and MT mobility behaviour, then generate a set of RN position candidates.
2. Further formulate the RN site positioning as a constrained optimisation problem, of which the goal is to maximise the overall network throughput, the potential gain of MANET RN based services and minimise the hop distance and delay, so that more MTs can be served with guaranteed QoS.
3. Test RNs positioning sites combination recursively and update each RN's position based on performance result.

These three procedures are executed recursively until the algorithm converges. By imposing such a plan, in practice, one can expect the system performance at disadvantaged locations should be improved with enlarged network dimensioning. In order to provide fixed RN assistance for both cellular and MANET interfaces radio resource management and routing, initial RN placement site candidates and possible topologies are considered in the following scenarios as presented in Figure 3. The scenario I, II and III cover the mobility management problems, scenario IV discusses the routing problem and scenario V is concerned with the relay structure.

When RNs participate in cellular resource sharing and traffic handover along with the BSs and MTs (BSON RN mode), the situation is a little more complex than the BSON mode as more hops are involved during resource transfer, although RN does not have the traffic admission functionality. The heuristic algorithm first proposes to place RNs in positions within the coverage of several BSs, such as the shaded area presented in the Scenario I where the RN is located within the coverage of both BS1 and BS2. The RN can assist advanced communication mechanisms such as cell breathing traffic balancing and TDMA based soft handover. For example, suppose that in Scenario I the RN is associated with both BS1 and BS2. If BS1 reduces signal coverage radius to improve interference and capacity in considered area, the RN will lose its association to BS1, and it can transmit data to BS2. The BS2 may at this stage reduce or increase its coverage radius based on the information and traffic condition. The software handover support is similar to load balancing since the RN also acts as data relayer. Suppose a MT is moving from BS1 to BS2 and it currently receives data from the RN. If the received signal from BS2 becomes larger than it from BS1, the MT performs an intra cellular network handover from one cell to another cell without changing the serving RN. Scenario II places together two RNs at the coverage edge of BS1 and BS2 to facilitate the communication between two cells (It is assumed each cell has only one BS). From the SIR values combining with Shannon's formula, if the bandwidth allocation λB ratio to BS transmissions and the bandwidth allocation ratio to RN transmissions λR together is 1, Scenario II can have a slightly better system capacity performance compared to Scenario I because the average received signal power strength in Scenario I is lower than in Scenario II. However, the sites deployment of Scenario II can introduce much larger latencies, service interruption time and equipment cost, while most cellular coverage overlaps. To conclude, the shaded area in Scenario I is considered as a better site candidate plan other than Scenario I. Scenario III is also considered

Figure 3. The RN node test scenarios I, II and III for the cellular network, the test scenario IV for the MANET and virtual RN backbone scenario V for RN positioning evaluation (© 2008, Chong Shen. Used with permission.)

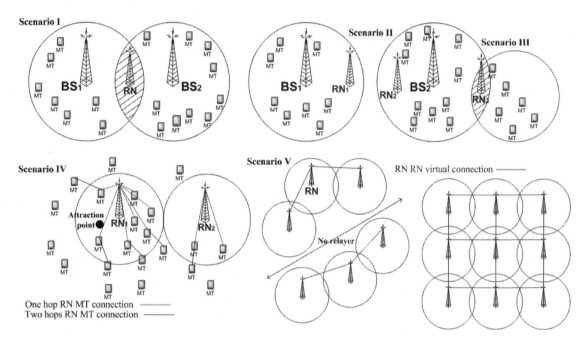

as an ideal location candidate. The RN at this position extends the cellular coverage to places without fixed infrastructure support.

As the attractor point mobility model is used to model MT movement, the MTs at some stage converge to the attraction points, dwell for a certain period of time and move within a short radius (more details can be found in next section). In order to mitigate resource contention in the "hot spot", reduce service interruption time and latencies for ad hoc communications, a RN should be positioned where more MTs can associate to the RN within one or two hop distance as illustrated in Scenario IV. It is also important to place RN in places based on MT traffic density prediction. Scenario V presents the RN candidate sites combination topologies, it is preferred that RNs can compose a mesh like service layer with virtual backbone connections using ad hoc frequencies thus Scenario V topology right is recommended other than Scenario V topology left. The next step of the heuristic algorithm is to decide the number of RNs needed to assist traffic relaying with guaranteed QoS. The strategy first updates system traffic load information and initialises the HWN* system. MTs then continue moving using attraction point mobility model. As BSs and attraction points sites are pre-defined by the system so only MTs trajectories are needed to be used for Scenario IV analysis. A system performance result for a RN site combination is recorded and compared with theoretical analysis. The RN's location will be changed at disadvantaged locations if the result is not satisfactory. The algorithm runs recursively until an optimal solution is found. Meanwhile, in the final sites selection, a hard distant limit δ is introduced to solve BS RN overlapping problem to regulate the distance between one RN candidate site and any BS.

HWN* SIMULATIONS

Accurate modelling of communication networks often results in models that are intractable to mathematical analysis and computer simulation becomes the option for modelling and analysis of complex systems. This section provides a description of the modelling concepts, simulation techniques and tools that have been used in the modelling and evaluation of the HWN* system. Algorithms proposed in conjunction with the HWN* infrastructure have been evaluated using OMNET++, which is a discrete event simulation environment with GUI support. OMNET++ is an object-oriented simulation tool, which consists of hierarchically nested modules written in C++. Nested modules are used that implement comprehensive and accurate modelling, which can reflect a good estimation of the actual system performance. In order to introduce events such as discontinuous transmission, session arrival & departure and etc, it is necessary to introduce the notion of time into the simulations. Events are triggered at some time instance in the simulation. For example, the event of session arrival at one node usually triggers the event of the arrival of that packet at a downstream node. Indeed the time delays are often random variables but events allow session arrival and departure processes to be modelled.

The HWN* physical layer needs to calculate the Bit Error Rate (BER) based on the SIR variation influenced by path loss, shadowing and fading models. The simulation of HWN* mainly considers packet level and session level modelling, but not bit level. Fast fading such as multipath fading is averaged out or captured in the BER and Packet Error Rate (PER). The propagation characteristics change from place to place when a MT moves. Thus, the transmission path between the transmitter and the receiver can be modelled from simple direct line of sight (LOS) to one that is severely obstructed by buildings, foliage and other terrain. Three mutually independent propagation phenomena considered are path loss, shadowing and multipath fading.

As complicated simulation models are evaluated, each simulation run may consume random numbers from several streams, which should be from several independent Random Number Generator (RNG) instances. Therefore different random streams or say simulation benches have been use for different tasks. For example, a random stream for generating packets and another random stream for simulating packet delivery rates in the transmission are different and not overlap. For all algorithm and system evaluation, we implement the Mersenne Twister RNG with 623-dimensional equal-distribution. The independent simulation processes run independently of one another and continuously send their observations to the central analyser and control process. This process combines the independent data streams and calculates from these observations an overall estimate of the mean value of each parameter. The Akaroa method provided by OMNET++ is also used which halts a simulation. It is decided by the 95% confidence intervals for all simulations to precise whether the result has enough observations.

Probabilistic models are applied to generate voice traffic, multimedia traffic and web traffic in this simulation environment for QoS oriented results evaluation. Voice traffic is modelled down to a packet activity with a different degree of granularity. Packet based voice source traffic models incorporate a voice activity aspect, which allows transmitting speech data when voice frames are detected, and facilitates discontinues data transmission. Hence, the voice source model has to include a higher degree of granularity and is represented as a sequence of consecutive talk spurts and silence periods within each call. The duration of the active and inactive periods reveals a negative exponential distribution for the duration of both active and inactive periods. The values applied for the investigations of this work are, a mean duration for the active periods of 1.35 seconds and a mean duration for the inactive periods of 1.70 seconds. The activity factor of the source is defined as the proportion of the time that the source is active. Several factors impact the nature of multimedia traffic and its transport requirements such as target quality, compression technique, coding time. The use of video traces is to facilitate algorithm evaluation with respect to QoS. As modeling the video sources always requires the original trace to be fully evaluated first, a simpler approach adopted in this research is to incorporate traces directly into OMNET++. Another benefit from incorporating the video traces directly into network simulators is the vast amount of video source models. Direct utilisation of video traces in OMNET++ facilitates the fastest method to incorporate video sources into existing network models. An interface has been provided for video trace is implemented, which is capable of detecting the different video trace file formats and to feed the data into the simulator accordingly. For particular environments and service types only specific traffic has to be instantiated. Whereas the evaluation of routing and inter-system traffic balancing of HWN* has instantiated all voice, multimedia and web services. Web traffic modelling is represented by a generic model with three levels session, activity and packet level. Session level consists of pages visited in a web session where a client starts an application, uses it for a time and then disconnects from the system. The moments when sessions arrive can be described by a Poisson model. The duration of sessions in real time applications depends on applications and in non real time is controlled by Transmission Control Protocol (TCP). Activity Level consists of a set of object applications such as images, sound and applet. The density of information in one application depends on the application itself. To describe this property, the activity level represents the application as a detailed succession of activity and inactivity periods in an ON/OFF model. ON represents page downloading time followed by an OFF period for the reading time. Packet Level decides the transmission of IP packets. If the User Datagram Protocol (UDP) is used, the packet inter arrivals can follow any distribution with packets following a truncated version of distributions to respect transmission limitation. In case of TCP, inter-arrivals of packets are determined by a TCP Pareto distribution.

To also include cellular network node movement characteristics, routing and handover algorithm evaluation of the HWN* system implements an Attraction (Attractor) Points oriented Mobility Model (APMM) based on the random waypoint model (Shen. C, Rea. S & Pesch. D, 2008). The algorithm first selects N attractors that are distributed at points where MTs will originate from or progress towards. Prior to heading for attraction points, nodes are grouped together using Cell Type Transition Probability, each subscriber selects a destination area with probabilities. In a typical MT movement, at the beginning, all MTs are scatted around in a metropolitan environment. After 100, 150 and 200 simulated minutes, the trend of MTs have been moved to several pre-defined attractor points which locate in the middle of northwest, northeast, southwest, southeast and the city centre. The northeast and southeast regions may have higher attractor probability than the rest of hot spots and therefore more MTs gather at the right hand side. It is flexible to change the geographical location of hot spots by revising the attractor points. Meanwhile, with a speed control mechanism, the current MT speed is configured correlated to the previous speed value and a smaller sampling time makes the speed change more smoothly.

HWN* configuration deals with message exchange, core network layer structure and individual algorithm implementation issues. The multiple access systems of the cellular component in HWN* deal with the inter-cell and intra-cell interference caused by common access to a shared band of frequencies. The mutual interference happens between BS and MT, RN cellular interface and BS, and RN cellular interface and MT. A TDMA based standard interface has been modelled. Each session is assigned a time slot within a frame which it keeps until it is handed off. No other sessions within the same cell are assigned the same slot, and thus users within a cell do not interfere. The MANET component of HWN* employs the contention based Carrier Sense Multiple Access/ Collision Avoidance (CSMA/CA) for the multiple access, between MT and MT, and MT and RN MANET interface, as it is the most adopted access protocol for MANETs and 802.1x networks. When regarding the RN MANET interface as a MANET node, if MANET node clusterhead selection is performed and along with RN provide the time beacon for synchronisation, a syncronised CSMA/CA can support fast data transfer, adopts ACK for successful transmissions and implements the handshaking mechanism between RN and MT to reduce collisions.

HWN* KEY EVALUATION RESULTS

Experiments are conducted to verify that the proposed handover management meets the goal of providing QoS differentiation among different users based on their class profile. To setup a comparison benchmark, a simple HWN* system is simulated without any dedicated handover management mechanisms and user classification.

In this system, each traffic class has the same privilege in terms of traffic admission and inter and intra system handover, arriving packets are accommodated on first-come-first-serve basis until all channels have been occupied. A MT either uses cellular mode or MANET mode. A MT terminates the handover process when it can not find an alternative route, or no free channels are available at the BS a MT roams to. A maximal packing based RN location plan is deployed where RN is placed at the centre point of intersection of the 3 hexagonal cells.

Figure 4 exhibits the average handover delay of three traffic profiles, HPUs, NPUs, LPUs and HWN* traffic without differentiation versus increasing traffic loads. Apparently, HWN* utilises existing medium resources efficiently and ensures the required level of service. Meanwhile, it indicates that the HWN* handover management mechanisms balance the traffic load between cellular and MANET components. This improvement becomes marginal after the system is heavily loaded; this is an interesting result since dynamic channel balancing usually presents even or worse performance compared to fixed channel balancing place under high traffic load. When traffic differentiation is not used, less control and negotiation overhead is introduced, the simple

Figure 4. Traffic class handover delay performance (© 2008, Chong Shen. Used with permission.)

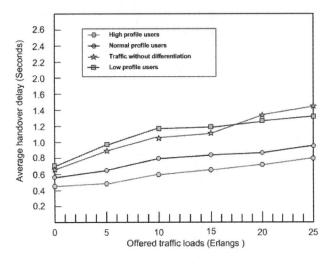

traffic class performs a similar result to LPUs at low and medium traffic loads between 5 to 15 Erlangs. However, the handover delay of this class is increased after 17 Erlangs since the handovers become largely influenced by the increasing traffic load. The medium access contention for both cellular interface and MANET interface is tense between ongoing sessions and handover sessions without coordination. Although with the control and negotiation overhead, the HPUs and NPUs still exhibit better results than traffic without differentiation and LPUs. Still, after the traffic differentiation, node per node handover delay and per node fairness in terms of handover delay can not be indicated through an averaged result.

The research is also interested in examining whether each traffic profile is satisfied with the QoS in terms of handover blocking probabilities. Therefore from a simulation for each traffic profile with increasing traffic loads. The performance of HPUs and NPUs are as expected, under any traffic load, the two premier traffic users are mostly guaranteed with QoS in terms of handover traffic blocking. Even under high traffic load, the blocking rates for HPUs or NPUs are less than 1 percent. Most traffic blocking attributes to LPUs with 2.7 percent blocking at high traffic load. The major reason contributing to this observation is that the excellent performance of HPUs and NPUs is attributed to the sacrifice of LPUs which use unstable ad hoc based services with a comparatively higher route change rate, when compared to cellular services.

The handover delay can be obtained in two ways and depends on whether service differentiation is considered or not. Without differentiation, handover delay includes the original communication mode detachment time and destination mode association time. The original mode and destination mode can use the same access interface for intra-system handover while they can be different for inter-system handover. However, when considering traffic differentiation, handover delay not only includes the mode detachment and association time, but also includes congestion control time and user class negotiation time. The average congestion control time is 0.1 seconds and the class negotiation time is 0.2 seconds.

Figure 5 presents the cumulative distribution function for various traffic profiles at target delay value 0.7 seconds. As the traffic load increases, for HPUs, one can see that over 90 percent of users

Figure 5. Traffic Cumulative distribution function of delay predominance at target 0.7 seconds (© 2008, Chong Shen. Used with permission.)

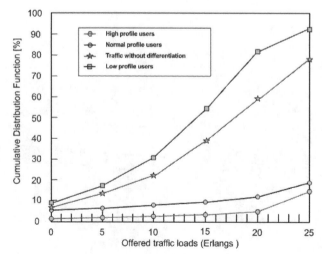

show a mapping of 20 Erlangs/cell or better and about 85 percent of users still reach a satisfactory handover delay performance within 0.7 seconds at high traffic load 25 Erlangs/cell. For NPUs, over 86 percent of users achieve target handover delay within 20 Erlangs/cell and still 80 percent of users present a mapping of high traffic load at 25 Erlangs/cell or better. However, the per node handover performance for LPUs or simple HWN* traffic profile is unsatisfactory. The simple HWN* traffic can only guarantee around 60 percent of users without QoS differentiation with this high delay performance target at medium 15 Erlangs/cell traffic load. more than 80 percent of LPUs can not delivery the target delay performance at 20 Erlangs/cell traffic load. Again, the result indicate the HPUs and NPUs exhibit largely better per node delay performance results when compared to traffic without differentiation and LPUs.

For the HWN* capacity evaluation, the service request portion is distributed and shared among these three user classes and the HWN* APMM has been implemented on each MT. A BS is located in the centre of each cell. The MT travels between 0 and 80 km/h. In order to increase HWN* system capacity, two simulation scenarios with different RN positioning plans have been evaluated, which are the packing the heuristic RN placement as proposed in previous section and maximal packing based RN placement plan where the packing is obtained by surrounding each BS serving circle by six other symmetric BS serving circles and a RN is place in the middle of each three BS serving circles. The simulation implements a modified AODV model for the multi-hop cellular network to be fairly compared with the proposed ADCR based HWN*. In the multi-hop cellular network, when paths are needed by MTs, which is demand-driven, the path query messages are unicast to the BS. If MTs do not obtain paths from the base, then path query messages will be flooded as in AODV. It is also configured that only the MANET clusterhead can act as traffic relay nodes in the evaluated multi-hop cellular network. Similar to traffic inter-network handover management evaluation, video streaming is generalised as a real-time service, while web transports are referred to as non real-time services. The default QoS traffic input profiles used consisting of streaming video, general voice calls and non real-time web services.

To present two pre-engineered RN positioning strategies' influence on the HWN* capacity under various traffic input, the MTs are added from 0 to 500 gradually as an input parameter to increase the offered load. When the heuristic and packing based RN positioning schemes are implemented in HWN*, per cell capacities are expected to be greater than the multi-hop cellular network, especially under high traffic volume. This is because MTs, not serviced in a cell, can use the optimised RN assistance of heuristic plan or packed relay path to access other media resources. As the traffic input is being increased, the relay nodes, after packing RN scheme implementations almost achieve complete connectivity considering both cellular service relay and MANET service relay. Figure 6 records per cell capacity performance of three scenarios as traffic load increases. The capacities of both heuristic and packing based HWN* increase until maximum throughput is reached at around 5.7 Mbps and 5.6 Mbps, respectively. As can be seen from the trend of the capacity lines, when the traffic load increases, the heuristic RN based HWN* outperforms the packing based RN HWN* largely at medium traffic load, and its maximum capacity is approaching the theoretical gain. The traffic is adaptively routed through both heuristic and packing plans while only the heuristic RN plan improves the system capacity at disadvantaged locations and reduces service interruptions. It also indicates that the heuristic plan, which considers realistic BSs arrangement, has been successfully adapted in HWN* system. In contrast, although the packing RN scheme's performance is fair, it is too idealistic and has difficulties in terms of implementation, which does not respect realistic BSs placement and installation. The RN BS as-

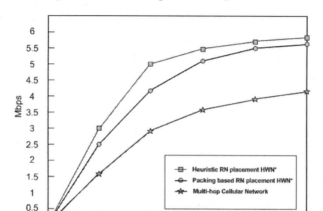

Figure 6. Average capacity comparison of the heuristic RN HWN*, packing RN HWN* and multi-hop cellular network (© 2008, Chong Shen. Used with permission.)

sociation in practice is not well-packed due to site availability problem.

Experiments are also conducted to verify that the proposed ADCR algorithm meets the goal of providing QoS differentiation among different users based on their class profile. To setup a comparison benchmark, a simple HWN* user profile without traffic differentiation is also simulated, which represents the average of all users and each packet has the same privileges when accessing the medium resource. A simple routing algorithm has been devised for the HWN* for the comparison. Packets first search for available cellular resource and the arriving packets are first accommodated on a first-come-first-serve basis until all available channels have been used. If packets can not get channels from the cellular interface, they then search for the MANET interface and the RN is treated as a normal ad hoc node. The routing decision is made by an individual MT and a MT terminates the routing process when it can not find an alternative route.

It is defined that all the medium access failures and routing failures contribute to the route acquisition success ratio. Figure 7 presents the route acquisition success ratio with increasing traffic load. The ADCR based HWN* scheme largely outperforms HWN* without ADCR in all traffic loads because the ADCR always returns a path on-demand and balances the traffic load between two networks with RN assistance. The ADCR continuously strives to provide an alternative route that will minimise the probability of route acquisition failure. The performance improvement is marginal when the system becomes heavily loaded. More traffic in the simple HWN* cannot find the route, the problem of this simple routing algorithm lies in its limitation on route selection, alternative paths provision and RN assistance. The result has also proved that the cascaded routing algorithm considering cross layer issues can be well-adapted to the HWN*, or may be deployed with modification for other hybrid wireless networks with multiple interfaces.

An interesting scenario is investigated where the multi-hop ad hoc communication (two hops maximum) is enabled along with ADCR routing, when considering the route acquisition ratio. It means once a route acquisition fails when using ADCR, the packets search for the destination node directly using ad hoc communication mode. Certainly, the incorporation of direct ad hoc communication has introduced routing, mode shifting and resource sharing overhead, but this evaluation

Figure 7. Comparisons of success route acquisition ratio between HWN* ADCR and HWN* without ADCR. (© 2008, Chong Shen. Used with permission.)

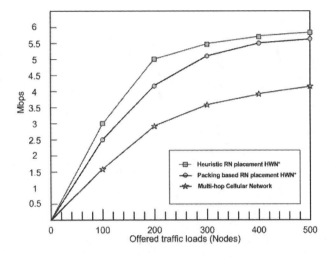

only tries to prove the capability of direct ad hoc communication in terms of MANET medium utilisation regardless overhead involved. For simulations, the system configuration is slightly different, 15% of the MTs are configured to reside in places outside infrastructure support, while the remaining MTs still migrate using the attractor point mobility model. Simulation parameters such as user profiles, traffic input etc. are kept the same as the parameters used in QoS based routing analysis. Figure 8 presents the results comparison between ADCR based HWN* and ADCR based HWN* + multi hop ad hoc. Different levels of route acquisition ratio degradation happen in both scenarios. However, under maximum traffic input, the HWN* + multi-hop ad hoc system outperforms the HWN*ADCR without multi-hop ad hoc communication by 11%. The performance improvement is marginal due to the reasons:

1. Because the system configuration has been changed and MT implements the attraction point mobility model, nodes converge in attractor points at some stage for long where more nodes are available to be used as relaying clients.
2. Under high traffic volume, more route acquisition failure takes place due to network congestion and routing failure without multi-hop ad hoc system.
3. HPUs and NPUs have the route acquisition priorities leaving a large number of LPUs unaccommodated in any communication modes.

To conclude, the ADCR may incorporate the multi-hop ad hoc communication as a backup or supplementary plan under high traffic volume.

FUTURE RESEARCH DIRECTIONS

Many research topics for hybrid wireless networks and relay structures are still open. Researchers are working on vertical handover between 802 networks and non-802 networks due to the benefit of higher bandwidth and lower delay with smaller cost. The next step towards this end is the media independent handover investigation, which is being developed by IEEE 802.21.

With primary focus on network resource management, routing and system optimisation,

Figure 8. Comparisons of success route acquire ratio between ADCR based HWN* and ADCR based HWN* + multi-hop ad hoc (© 2008, Chong Shen. Used with permission.)

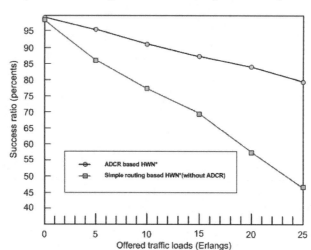

there are several specific topics that can be further investigated as follows:

1. The routing algorithm should be improved again considering more constrains posed by physical layer and medium access layer, especially the MT geographic location identification and resource allocation information to meet individual user QoS requirements.
2. As the multi-hop self-organised MANET has been integrated to HWN* as a communication mode, research should be conducted to address inter-connection between pure MANET and other HWN* communication modes since the traffic can be further balanced to the pure multi-hop MANET.

Another topic worthy of further investigation is the feasibility of using autonomic computing elements to monitor and manage overall HWN* system behaviour. Essentially, with this approach the relay node can act as an autonomic element where resource sharing and routing information is shared and managed in an on-demand and cost-oriented manner. MTs will be allocated appropriate services based on the decision made by the autonomic element. The promising part of this approach is that most computational work is accomplished in RNs instead of BSs or the MT itself. Hence the speed of decision making is expected to be largely improved with HWN* system complexity reduction.

CONCLUSION

To summarise, the chapter has investigated a novel hybrid infrastructure towards 4G mobile communication networks composed of a fully functional cellular network, a distributed MANET and relay structure using both active cellular and MANET interfaces. A management framework for the infrastructure has been proposed which highlights the importance of devising the radio resource management algorithms. The major contribution of this work has been the exploitation of the infrastructure and the development of related algorithms and protocols. The RELAY node element has been utilised to assist spectrum sharing, route configuration and active interface selection, while most modern mobile terminals have dual interfaces that can work under both MANET and cellular frequency bands.

The proposals of inter-network and intra-network mobility management, cascaded HWN* routing, and HWN* topology design are to fulfill the objectives of fair radio resource balancing, efficient traffic management and QoS guaranteed networking for the management framework. The dedicated and location fixed RN (that supports cellular-MANET hybrid wireless architecture co-development) and the proposed algorithms have not been investigated anywhere else and so are novel contributions as compared to other hybrid wireless networks. For example, the WINNER architecture deploys a ubiquitous interface for both BS and RN, the iCAR proposal focuses on multi-hop cellular networks using flexible RN, and the SOPRANO proposal, which is similar to iCAR, deploys CDMA with pre-configured RN to IP network connections. Another contribution of this chapter has been the use of constrained optimsation techniques to finalise the RN sites placement plan for the HWN* system.

REFERENCES

Acharya, A., Misra, A., & Bansal, S. (2002). *A Label-switching Packet Forwarding Architecture for Multi-hop Wireless Lans.* Paper presented in ACM WoWMoM, Atlanta, 2002.

Aggelou, G., & Tafazolli, R. (2001). On the Relaying Capability of Next Generation GSM Cellular Networks. *IEEE Communications Magazine*, 40–47.

Ananthapadmanabha, R., et al. (2001). *Multi-hop Cellular Networks: The Architecture and Routing Protocol.* Paper presented at IEEE PIMRC.

De, S., et al. (2002). Integrated Cellular and Ad Hoc Relay (iCAR) Systems: Pushing the Performance Limits of Conventional Wireless Networks. In *International Conference on System Sciences*, Hawaii.

Frank, C., Manoj, K., & Murthy, S. (2002). *Throughput Enhanced Wireless in Local Loop (TWiLL) - The Architecture, Protocols, and Pricing schemes.* Paper presented in IEEE LCN. 2002.

3GPP. (2001). *UTRAN overall description* [Technical Specification TS 25.401]. Retrieved January 3, 2008, from http://www.3gpp.org

Kumar, J., Manoj, K., & Murthy, S. *(2002). Multi-power Architecture for Cellular Networks.* Paper presented in IEEE PIMRC.

Manoj, B., & Kumar, K. (2006). On the Use of Multiple Hops in Next-Generation Wireless Systems. *Science Wireless Networks*.

Murthy, C., & Manoj, B. (2004). *Ad Hoc Wireless Networks: Architectures and Protocols.* Upper Saddle River, NJ: Prentice Hall.

Rea, S. (2006). *Dynamic Route Management Strategies for Mobile Ad Hoc Networks.* PhD Dissertation, Cork Institute of Technology, Ireland.

Shen, C. (2008). *Management Framework for Hybrid Wireless Networks.* PhD Dissertation, Cork Institute of Technology, Ireland.

Shen, C., Rea, S., & Pesch, D. (2008). Resource Sharing via Planed Relay for HWN*. *EURASIP Journal on Advances in Signal Processing, 2008*, 793126. doi:10.1155/2008/793126

Shen. C., Rea. S., & Pesch, D. (2007). HWN* Mobility Management Considering QoS, Optimisation and Cross Layer Issues. *IEEE Journal of Communication Software and Systems, 4*(3).

WINNER. (2005). *Identification, definition and assessment of cooperation schemes between RANs (D4.3).* IST-WINNER, Final deliverable.

Zadeh, A. (2002). Self-Organizing Packet Radio Ad Hoc Networks with Overlay (SOPRANO). *IEEE Communications Magazine*, 149–157. doi:10.1109/MCOM.2002.1007421

Forthcoming 4G Challenges:
Problems and Solutions

Chapter 7
User Experience in 4G Networks

Pablo Vidales
Deutsche Telekom Laboratories, Germany

Marcel Wältermann
Deutsche Telekom Laboratories, Germany

Blazej Lewcio
Deutsche Telekom Laboratories, Germany

Sebastian Möller
Deutsche Telekom Laboratories, Germany

ABSTRACT

Forthcoming 4G networks will enable users to freely roam across different communication systems. This implies that formerly independent wireless and wired technologies will be integrated to deliver transparent access to a plethora of mobile services and applications. This will also involve changes in the user's experience mainly derived from (1) mobility across heterogeneous technologies, (2) drastic changes in the underlying link conditions, and (3) continuous adaptation of applications, e.g. flexible coding schemes. This chapter presents a detailed study of these so far unknown phenomena arising in the context of 4G networks. Current instrumental models employed to estimate user perception, such as PESQ (ITU-T Rec. P.862, 2001) for predicting the quality of transmitted speech, were designed to measure conditions that are common in today's wireless and wired systems. However, it is expected that new conditions encountered in 4G networks are not going to be accurately handled by today's models. Thus, they need to be adjusted, or new models should be proposed in order to predict the perceptual influence of new phenomena such as the three aspects aforementioned. The authors undertook this task and designed a novel methodology and experimental setup to measure user perception in future 4G networks. Moreover, the authors carried out an extensive set of subjective tests to accurately quantify user perception and derive conclusions for optimal user experience in 4G networks. These processes and initial results are included in this chapter.

DOI: 10.4018/978-1-61520-674-2.ch007

INTRODUCTION

A basic characteristic in the landscape of wireless access technologies is and will continue to be the ever increasing heterogeneity in networking characteristics. This results from the fact that emerging mobile services and applications demand different Quality of Service (QoS) levels, forcing the deployment of new wireless access technologies to cope with the new demands. Therefore, it is not a surprise that Fourth Generation (4G) wireless networks or Next Generation Mobile Networks (NGMN) will be formed by multiple heterogeneous access technologies, placing seamless mobility within and across technologies as a fundamental property of these types of systems.

Users will demand more access, better quality, and less complexity when using mobile services and applications. This poses pressure on operators and service providers to design and deploy better solutions to support seamless mobility in future communications. During the design of these solutions two important facts should be considered: heterogeneity of the underlying systems and user perception (QoE, Quality of Experience). In order to solve this twofold equation strong collaboration between networking and quality research communities needs to be accomplished. This chapter is an example of joint work between these two fields of the Information Technology and Communication (ICT) industry, and it represents a seminal effort into solving the described challenges in 4G networks.

Clear cases of the need for this type of cooperation are the current mobility management solutions. It is evident that in order to advance these solutions we should consider information beyond network performance metrics, i.e. user perception, when managing connectivity resources. Part of the proposal in this chapter is to observe metrics derived from user perception when designing future mobile networks, services and, in particular, implementing mobility management solutions. In the following, a case study is briefly described to illustrate this.

In this chapter, it is explained how user perception may be included into the decisions about connectivity resources. Network quality is not enough anymore to base complex seamless mobility decisions in 4G networks on facts like overlaying heterogeneous networks, sudden and drastic changes in networking conditions, and independence between network operators (each network may offer different voice/video codecs, quality of service policies, etc.)

This work also discusses how user-driven evaluation of network phenomena can be used to improve existing speech quality prediction models such as PESQ or the E-model. Previous models were built to cope with the common conditions in existing (more homogeneous) communications systems, in which, for example, sudden drastic changes in lower layers are not so frequent. Therefore, current models do not cover the phenomena that occur in 4G networks. Moreover, users do not care where quality degradations stems from; they ask for at least the same quality they have been used to in the past – if the new NGMN technology does not provide this quality, the user may be dissatisfied, unless other advantages (mobility, costs) rule out the impact of degradations. As a consequence, the perceptual part of the mentioned quality models may still be valid, and the models just have to be adapted towards the new degradations found in NGMNs.

This chapter includes a detailed report of a case study and early results in the area of user-driven evaluation of 4G networks: the *Mobisense* project (Mobisense, 2007). It represents an initial hands-on evaluation of the concepts that are introduced in this chapter. For the purposes of this research, voice services, i.e. VoIP, were chosen as the main application. Thus, Mobisense is focused on the evaluation of the following scenario: *"In 4G networks, handovers between different wireless access technologies provide seamless roaming during voice over IP calls. The resulting speech quality depends on the audio bandwidth of the speech codecs used in the respective networks, as well as on degradations resulting from the handover, coding, and packet loss".* However, future research avenues include multimedia services and other phenomena that deserve separate studies.

The study includes listening tests to quantify speech quality as a function of network and codec characteristics, a comparison of the results of these tests versus estimations obtained from instrumental models, and the application of this analysis. Its main aim is to determine when and under which circumstances a network handover and/or codec changeover should occur in order to obtain better speech quality for Voice over IP services in 4G networks. This is fundamental for the development of high-quality roaming strategies and for the improvement of user experience in future networks.

The rest of this chapter is organized as follows. In Section *Background*, previous research on mobility management and speech quality is presented. In addition to the related work, prior contributions of the authors to this research field are summarized. It includes a subsection on phenomena in 4G networks that explains the key terms and definitions for the reader, in relation to the phenomena analyzed in the Mobisense project. Then, the core, Section *User Perception in 4G networks*, details the complete research process, designed, implemented and conducted by the authors, in order to unveil fundamental aspects that affect user perception and speech quality in 4G Networks. This section is divided into three parts corresponding to the main phases of the process. First, the experimental methodology designed for the purposes of this project is described. Second, the experimental setup build for the Mobisense project is presented. Finally, a summary of the test cases used to conduct the listening tests is included in the third part of this section. All the results and findings gathered from the experimental process are commented in Section *Experiments and Results*, paving the way to the next two sections: Section *Conclusions* and *Future Research*, in which the main observations are discussed and further research avenues for user perception of 4G Networks are proposed.

BACKGROUND

This section is organized into two main parts: mobility in 4G networks, and assessment of Quality of Experience (QoE) and user perception models. The first part includes basic concepts of mobility management, using Mobile IP version 4 as the enabling protocol. This section also includes a brief description of the Mobile IP protocol. The second part describes the state of the art of user perception assessment, including the most important methods and models in this area.

Mobility in 4G Networks

The Internet Protocol (IP) was developed in the early 1980s with the aim of supporting connectivity within research networks, as part of *Catenet* (Cerf, 1978). However, in the last decade IP has become the leading networking protocol. It is the basic tool for a plethora of client/server and peer-to-peer networks; it is predominant in both wired and wireless worlds, and now the current scale of deployment is straining many aspects of its more than 30 year old design. To overcome the limitations inherent in the original IP design, IPv6 has been proposed as the new protocol that will provide a firmer base for the continued growth of today's complex networks.

More people will access the Internet via wireless rather than via wired connections, and each user will have a set of wireless devices interconnected that will be accessing a great variety of IP-based services. There are over 3 billion mobile subscriptions in the world, and this number continues to grow, one for every two inhabitants. In the light of this growth, IPv4 faces many problems related to its limited address space and mobility capabilities applied to the mobile world.

Every mobile device is potentially capable of accessing IP services, Wi-Fi networks are becoming ubiquitous; the growth in the hotspot market is being accompanied by similar growth in other wireless technologies (e.g., Bluetooth, Ultra Wideband, and satellite), posing the urgent need for a new identification scheme and an adequate support for mobility.

Whereas the main thrust of IPv6 is to solve the addressing problem, it also provides important functions to enable mobility (e.g., scaling and ease-of-configuration). IPv4 has difficulties managing mobile terminals for several reasons such as address configuration and location management. However, in order to drive the evolution in the current mobile world and avoid the humongous effort to migrate all computers and equipments from IPv4 to IPv6, Charles Perkins, from Nokia Labs, originally proposed Mobile IPv4 (MIPv4). This protocol extension was projected to enable IPv4 devices to support micro and macro mobility (IETF RFC 3344, 2002). MIPv4 has disadvantages in comparison with its successor, Mobile IPv6 (built as a natural part of IPv6 protocol, and less of an extension, as in the case of IPv4), but neither Mobile IPv4 nor Mobile IPv6 were intended to support seamless roaming in heterogeneous environments such as 4G Networks.

The rapidly growing demand for "anywhere, anytime" high-speed access to IP-based services is becoming one of the major challenges for operators. As the demand for mobility increases, mobile terminals need to roam freely across heterogeneous systems forming the present wireless landscape. Currently, Mobile IP stands as the de facto solution for mobility management in 4G networks.

This work targets the user-driven evaluation of mobility in 4G networks, using MIPv4 as the underlying support protocol (Figure 1). In particular, it is of interest to evaluate the impact of *vertical handovers* (i.e. network handovers) and terminal mobility on user perception. A network handover occurs when a mobile device changes its point of attachment to the Internet and the former point belongs to a different wireless technology; for example, a mobile phone handing off from the cellular network to a public hotspot. A brief introduction to the basic concepts of Mobile IPv4 is included next to introduce readers to its basic concepts.

Mobile IP Version 4

Mobile IPv4 defines mechanisms that add support to a terminal (i.e. the Mobile Node) to change its point of attachment to the Internet whilst remain-

Figure 1. Mobile IPv4 protocol

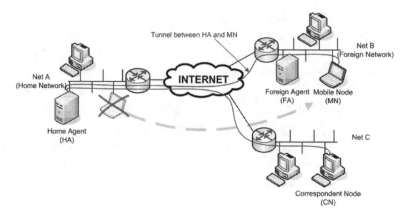

ing reachable through a permanent address (the Home Address, HoA) and preserving all the active connections it had before travelling to its new location. While a Mobile Node (MN) is connected to its Home Network (i.e. the network its Home Address belongs to), no special mode of operation is needed, and packets are forwarded (using normal IP routing) between the Mobile Node and any other node it is communicating with (the Correspondent Node, CN). When a MN is connected to a network other than its Home Network (i.e. it is visiting a Foreign Network), the MN acquires an IPv4 address belonging to the address space of the Foreign Network it is visiting (supported by the Foreign Agent, FA), called the Care-of Address (CoA).

The MN announces its CoA (by sending a Binding Update message, BU) to a special entity, called the Home Agent (HA) that is located in the MN's Home Network. The HA intercepts the packets addressed to the MN's Home Address while the MN is away from its Home Network and establishes a bidirectional tunnel with the MN's CoA, in order to redirect these packets to the MN's current point of attachment to the Internet. The MN also uses this tunnel to send its traffic to the Correspondent Node, avoiding in this way any ingress filtering.

Latency on MIPv4-enabled links can be very high, especially for interactive applications that require real-time response. Therefore, the research community is working on mechanisms that decrease this latency as much as possible, at least to levels that support real-time applications. Two of the most significant proposals are Fast Handovers for Mobile IPv6 (FMIPv6) (IETF RFC 4068, 2005) and Hierarchical Mobile IPv6 (HMIPv6) (IETF RFC 4140, 2005).

FMIPv6 aims to decrease the total latency to almost only the Layer 2 handover time. This approach has been shown to perform well in intra-technology (i.e. horizontal) handovers. The HMIPv6 approach is designed to reduce the degree of signaling required and to improve handover speed for mobile connections by managing local mobility in a more efficient way. Previous work (Costa et al., 2002) has analyzed which of these approaches (i.e. FMIPv6 and HMIPv6) performs better, the conclusion being that a combined approach would be optimal. However, given the implementation complexity that this would require, the FMIPv6 optimization by itself is good enough. This has been experimentally evaluated by (Bernardos et al., 2005).

Future 4G networks, where heterogeneity will be more the rule rather than the exception, have challenging characteristics when performing vertical handovers (also known as inter-technology). The present work focuses on the impact of this new type of mobility on the user perception. The enhancement (better performance, lower latencies, etc) of the underlying protocols (i.e. MIPv4) is not in the scope of this research. For further information about the latter issues on improving mobility management refer to (Vidales et al., 2005; Vidales et al., 2007).

Assessment of QoE and User Perception Models

Quality has been defined as the result of a perception and judgment process, during which the user compares the perceived characteristics of the speech sound (the so-called "auditory event") to the expected or desired characteristics (Jekosch, 2005). Because of the necessity of perception and judgment processes, subjective tests are still the only valid and reliable means for the purpose of quantifying the impact of different types of degradations on perceived overall quality (QoE). However, instrumental models such as PESQ and the E-model have shown to provide valid *estimations* of the results of perceptual tests, within the limits of applications they have been designed for. As a consequence, these models are widely used instead of auditory tests, e.g. for network planning and monitoring. Still, their range of validity needs to be respected, and this is why they cannot simply be applied without further validation to NGMN scenarios.

In the case of transmitted speech, which is the focus of the present chapter, the perceived quality of the given conditions is collected in auditory tests. In a listening-only situation, for instance, a selection of test participants is asked to judge the quality of a number of processed speech samples. The text material, read by different speakers, is chosen according to the aim of the experiment, and the recorded clean speech files are processed by the system of interest and are finally presented to the listeners. Their task is to judge upon the perceived quality of the processed speech sample, providing a quantitative measure of the QoE.

The Telecommunication Standardization Sector of the International Telecommunication Union (ITU-T) provides information about how such tests have to be performed in detail, e.g. in the Recommendation P.800 (1996). There, it is specified how to choose "balanced" speech material, "normal" speakers as well as "normal-hearing" test participants. In order to increase the reliability of the experiment, the recording and play-back situation is specified, as well as further test parameters which might have a significant impact on the measurement results.

The judgments of the listeners are usually limited to the identification or scaling of pre-defined properties of the percept. Therefore, a set of pre-defined scales are available. In the area of speech transmission, a 5-point category scale is usually employed (Absolute Category Rating, ACR). For the collection of overall quality ratings, this 5-point scale is labeled with the attributes "Excellent", "Good", "Fair", "Poor", and "Bad". Subsequently, the ratings are averaged over all participants, leading to an arithmetic mean (Mean Opinion Score, MOS) for each processing condition.

In contrast to the listening-only situation, a bi-directional communication system is employed in real-world conversation situations. Thus, the ecologically most valid method for quality assessment is the conduction of conversation experiments with two interacting participants. Here, the interlocutors are asked to have a conversation on an arbitrary or pre-defined topic. By means of, e.g., the abovementioned ACR scale, the opinion about the just finished connection is supposed to be judged. Such procedures are described in ITU-T Rec. P.805 (2007).

Another standardized paradigm, which can be regarded as a trade-off between the "artificial" listening-only experiments and the quite complex conversation tests, is defined in ETSI Technical Report 102 506 v.1.1.1 (2006) and it is also called the "CallQuality" method. In this process, the perceived quality of a "simulated" telephone conversation is assessed. The participants are asked to listen to short extracts of a normal conversation, and verbally answer questions regarding the content of the stimulus they just heard. After five of these stimuli, they rate the quality of the entire simulated conversation on the ACR scale. The answering part was introduced to come close to a real conversation with its turn-takings, and to distract subjects from concentrating on the quality until rating it (the test participants are instructed to try to put themselves into the position of an interlocutor).

Once the perceptual effects have been quantified, instrumental quality prediction models can be developed that are capable of estimating the subjective ratings. One type of recommended models by ITU-T, the so-called "Perceptual Evaluation of Speech Quality" (ITU-T Rec. P.862, 2001), is based on the application-layer speech signals. Here, the quality of transmitted speech in a listening-only situation is estimated by comparing the clean and degraded signals on a perceptual level, i.e. by taking advantage of psycho-acoustic knowledge, such as the Bark-scale transform, loudness functions, time/frequency masking, asymmetries of "positive" and "negative" error components, as well as insensitivities of certain variations in delay, the spectrum, and the amplitude. PESQ has been extended towards wideband speech by applying a flat input filter and a different mapping function —see ITU-T Rec. P.862.2 (2007).

Currently, the requirements for a successor of PESQ are discussed in Study Group 12 of ITU-T, which is supposed to improve known drawbacks of PESQ and is valid for an even wider range of distortions (e.g., audio bandwidth, time-warping). NGMN-specific degradations, however, will not explicitly be covered by this model.

For conversational speech quality, the ITU-T recommends the so-called E-Model (ITU-T Rec. G.107, 2005), which is a parametric model usually employed for offline quality estimation, e.g. for network planning. The MOS values are estimated on the basis of transmission channel parameters commonly known in classical telephonometry (e.g., loudness ratings, noise levels), but also in the context of packet-based networks (e.g., packet loss rates). These parameters are subsumed by so-called impairment factors, which are assumed to be additive on a psychological scale, the so-called transmission rating scale (R-scale). The eventual quality estimates are then obtained by a non-linear transformation of the R-values, i.e. the summations of the perceived impairments.

Until now, none of these models have been validated to correctly predict the effects of NGMN handovers and/or codec changeovers on user perception. However, such instrumental models are indispensable to rapidly design network handover and codec changeover strategies which provide an optimum quality to the user. The auditory investigations presented in this chapter can therefore be considered as a basis for the development of NGMN-capable quality models.

Phenomena in 4G Networks

After introducing the main concepts related to mobility in 4G networks and the relevant methods and models linked to the evaluation of QoE, the next section is focused on the potential effect that mobility across heterogeneous networks may have on the QoE of mobile users in 4G networks.

There are three main sources for potential degradations in overall call quality: (1) changes in the underlying network conditions, (2) network handover, and (3) codec adaptation to network changes. These phenomenona will be briefly described next, as they are the core conditions studied in this work.

Changes in Network Conditions

A main characteristic of 4G Networks is heterogeneity. This translates into potentially severe changes in network conditions when seamlessly roaming across wireless technologies. The variations in packet loss, delays and bandwidth to which mobile applications are confronted with in 4G networks may lead to changes in the user's quality experience. Variations in network conditions are common, but the magnitude of the changes will increase in 4G networks, and new methods to deal with this will be needed in order to maintain an acceptable QoE. A first step towards building these new methods is to understand the impact of these variations on user experience. This chapter discusses this impact for an ongoing VoIP call when a user moves between independent wireless access networks.

Network Handovers

In Next Generation Mobile Networks (NGMN), the convergence of a multitude of different network technologies will provide the user with transparent and ubiquitous access to data and media services. The independence of network and service layers also introduces a new level of mobility. Equipped with appropriate terminals, users may move through geographical areas covered by different wireless network technologies while service access is preserved. In order to guarantee the provision of seamless connectivity especially for time-critical services like media streaming or Voice-over-IP (VoIP), sophisticated mobility-enabling protocols like Mobile IP (IETF RFC

3344, 2002) are needed which ascertain a fast and robust roaming between heterogeneous wireless networks (Wältermann et al., 2008). This roaming between disparate wireless technologies (vertical handovers) while keeping ongoing sessions forces a new level of application level adaptation, causing an effect on user perception. This effect has not been thoroughly quantified as it is not present on existing communication systems.

Codec Adaptation

Current voice over IP codecs can adapt to variations in the underlying networks, but may not cope with the kind of changes caused by vertical handovers, as these could result in the need to change codec families (e.g., narrowband to wideband). In this case, for an ongoing VoIP call the network handover can occur in conjunction with the appropriate application layer adaptation in order to account for the newly encountered conditions. Most important, a codec re-negotiation might be enforced to meet sudden changes in network conditions. This should ideally be done in such a manner that neither the call nor the audio stream is interrupted during the handover. Thus, the continuity of the speech is maintained, resulting in an optimal speech quality.

Special attention has to be paid to these aspects as they produce new auditory artifacts that can invalidate speech quality models in the context of forthcoming communications systems, causing mismatches in the design of networks and mobile devices and applications. A deep dive into the methodology to study these phenomena is included in the next section.

USER PERCEPTION IN 4G NETWORKS

The present section is twofold. First, we will explain the methodology used all along this study, emphasizing the aspects that make it novel and that enable us to assess networking phenomena in 4G networks from the user's perspective. The second part details the experimental setup, which represents a core piece of the whole Mobisense project. The testbed emulates a 4G networking environment and makes it possible to experiment with multiple scenarios.

Methodology

The presented study follows a twofold evaluation approach in order to link NGMN characteristics to user perception (Lewcio et al., 2008). This approach is presented in Figure 2. It is based on speech samples processed under specified experimental conditions, using the Mobisense testbed described in the next section. According to the twofold evaluation approach, the process is divided into two orthogonal stages – perception and networking analysis. On the perception layer, the audio samples are recorded and judged in an auditory experiment to obtain average quality scores (MOSs). On the networking layer, network traces are collected on the involved network stations and used for a subsequent network trace analysis. Finally, the results of both layers are merged to develop a quality prediction model that provides speech quality estimates, including the impact of mobility events in NGMNs. Design of such models enables a successful service adaptation and mobility management in real-time.

It has to be emphasized that this experimental study is limited to the listening-only situation. The reason is that we would like to first understand what happens in the passive context, before stepping to a more interactive situation (i.e. conversational tests). As it was mentioned, interactive scenarios are more realistic, but they are also more difficult to evaluate because the results of conversational tests depend on a number of influencing factors, like the interactivity of the test scenario (free vs. more structured task-oriented conversation) and the instruction to the test subjects (i.e. whether they have been made aware of the fact that delay

Figure 2. Twofold speech sample evaluation approach for NGMNs

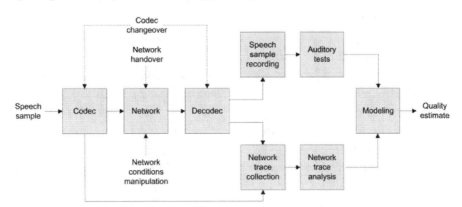

may occur). Thus, despite the online capability of our testbed, we left the evaluation of delay effects for further study.

Testbed

The Mobisense testbed was explicitly deployed to assess the user perceived quality of mobility in NGNs using VoIP as the case application. In order to do this, the Mobisense testbed should support experiments involving the collection of speech samples, the execution of network handovers and codec changeovers, gathering of network traces, and the modification of network conditions (Vidales et al., 2008). The following functional requirements can be derived from these processes:

- The mobile terminal should have connectivity to at least two different wireless access technologies like WiFi, Flash-OFDM, WiMAX, UMTS/HSDPA or GPRS/EDGE.
- The mobile terminal should be able to seamlessly hand off a connection between these networks, as it can be done with Mobile IPv4 or SIP (IETF RFC 3261, 2002).
- Network traces should be collected during the voice calls to allow further evaluation of the network conditions and codec changeovers.
- There should be the possibility to degrade the network conditions artificially, i.e., to increase delay, jitter, packet loss, and additional traffic to the network connection in use.
- It is desired to use a broadband sharing environment (using IEEE 802.11 technology).
- A VoIP call has to be established between the mobile terminal and a fixed host with a stable Internet connection to measure only the impact of the wireless path on audio quality.
- The audio stream of the VoIP call must be recorded for further analysis and for the listening and conversational tests.
- The VoIP client should support wideband and narrowband speech codecs for different transmission bandwidths.
- The VoIP client must be able to switch between different speech codecs during an ongoing call.
- A jitter buffer monitoring to track the codec frames in the VoIP application should be implemented.

Some of these requirements were realized using off-the-shelf hardware and software components. However, other parts were specially implemented for the Mobisense testbed. For example, the PJPROJECT client (PJSIP, n.a.) was extensively modified to cope with the demanded features such as codec changeover (Wältermann et al., 2008) and jitter buffer monitoring.

Hardware and Software Components

The Mobisense testbed uses Mobile IPv4 as a solution to enable seamless handovers between different radio access technologies. Mobile IPv4 requires a Home Agent, a Mobile Node, and a Correspondent Node to build a working system. The access networks emulate an integrated NGN conformed by the following technologies: LAN, WiFi, Flash-OFDM, and UMTS/HSDPA. The CN and MN were deployed on laptops with Linux 2.6.18.2, and the HA is a Cisco 2620XM router with Cisco IOS 12.2(8r), supporting MIPv4.

Mobile IPv4 was deployed as the mobility management protocol because all involved access networks support IPv4. As Mobile IP client, the SecGo implementation (SecGo, n.a.) was chosen because it provides a telnet interface for remote control (Telnet, n.a.) and supports NAT traversal (IETF RFC 3519, 2003). As the VoIP framework, the PJPROJECT was selected and extensive modifications have been made to fulfill the testbed requirements. The tcpdump (tcpdump, n.a.) and Wireshark (Wireshark, n.a.) tools are used for trace collection and evaluation. netem (netem, n.a.) is used to enable changing the network characteristics in terms of adding delays, packet loss, packet duplication, packet corruption, and packet re-ordering. Finally, a TCP-based client/server application has been implemented to centralize the overall control of the test settings. In addition, an UDP sender has been implemented to enable the remote control of the VoIP client.

Testbed Deployment

Figure 3 shows the Mobisense testbed network architecture and its hardware components. The Mobisense testbed supports connectivity to six networks as attachment points to the Internet, which are based on different (wireless) technologies: two LAN networks (one home and one foreign network), two foreign WLAN networks, one foreign UMTS/HSDPA network, and one foreign Flash-OFDM network. The home LAN network is directly connected to the Internet and constitutes the fixed attachment point for the HA and CN; the MN can roam between the five remaining networks. The MN could connect to the foreign LAN and can then roam to any of the two foreign WLAN networks, supplied by a DSL line as the backhaul. Moreover, the MN can log on to a public (foreign) UMTS/HSDPA network, as well as communicate with the Flash-OFDM RadioRouter that forwards packets to the Internet via the Flash-OFDM internal Home Agent. The Flash-OFDM access network is provided by the BIB3R testbed (Steuer et al., 2006) and it is based on Flarion RadioRouter version 1.1 (a testing license in the UMTS frequency band has been granted by the German regulator enabling operational measurements).

The Mobile Node may roam between LAN, WLAN, UMTS/HSDPA, and Flash-OFDM foreign networks. All of them apply IPv4 protocol stack. The mobility is served by the Mobile IPv4 protocol providing transparent mobility for overlying protocols. The deployed Mobile IPv4 infrastructure consists of a HA and a MN and does not involve additional Foreign Agents. Therefore, the MN individually obtains IP addresses and acts directly as one end of the Mobile IPv4 tunnel (reverse tunneling).

Arising from the complex setting described in this section, two further requirements were posed to the design: (1) the CN, acting as the counterpart of the VoIP call generated by the MN, has to be remotely controlled and (2) all software components should have an interface for remote management, connecting to the MN and CN. The central point of the architecture is a controlling script run on the MN that synchronizes the execution of all other software components. For example, the controlling script starts tcpdump to collect network traces, telnet to communicate with

Figure 3. Mobisense 4G network experimental setup

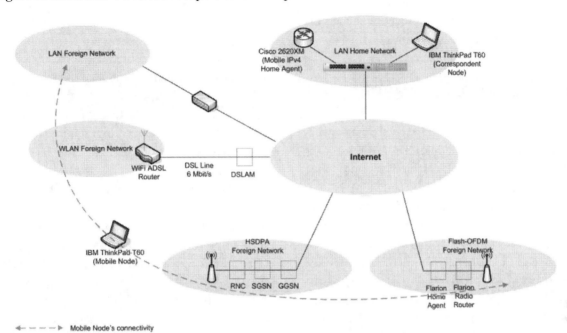

the Mobile IP client, a UDP sender to control the PJPROJECT client, and the TCP client/server. The TCP client/server is employed twofold. At first, it runs the PJPROJECT client as a background process. Secondly, it controls the software installed on the Correspondent Node. Thus, the TCP client/server is also integrated to the CN, where it starts tcpdump and the PJPROJECT client, and controls the UDP sender.

The Mobisense project aims to evaluate the effects of mobility on the perceived quality of real time services. Therefore, the testbed was designed to provide the following features:

- Access to different network technologies
- Ability to change the network connection (network handover)
- Possibility to artificially change network conditions
- Record network traces
- Share available bandwidth using DSL and WLAN technologies
- Perform VoIP calls over the networks
- Support for different voice codecs (e.g. narrowband and wideband)
- Possibility to change voice codecs (codec changeover)
- Ability to record voice samples
- Ability to record internal parameters of the VoIP application

With these features there is much potential to evaluate different effects and their impact on the quality of VoIP calls such as different network, different codecs, network handovers and codec changeovers. In order to evaluate the perceived quality, experiments with the above listed phenomena can be conducted. The resulting audio samples are recorded and then used to run auditory tests with human listeners. During these listening tests, people are asked to rate the quality of different samples. These results can be used as an input for quality prediction models. Beyond this, the network-centric effects of the network handovers and codec changeovers can be evaluated by means of the network trace recording ability of the testbed.

Through its open nature, the testbed can be easily extended towards different directions such as the evaluation of multimedia applications.

EXPERIMENTS AND RESULTS

This section, which is largely based on the investigations described by Möller et al. (2009) focuses on the effects of vertical handovers and codec changeovers on speech quality in NGMNs. Our aim is to quantify the effects of different network and codec characteristics (handover point within a call, networks used before and after the handover, audio bandwidth available in these networks, and packet loss) on the quality perceived by the user. More precisely, we would like to answer the following research questions: (1) which network and codec characteristics are the most relevant ones from a perceptual point-of-view? (2) Is it advantageous to switch from NB to WB whenever possible, or does audio bandwidth switching degrade perceived quality? If yes, under which circumstances? (3) Is it possible to predict the effects of network handovers and codec changeovers with existing quality prediction models?

In order to get analytic insights into the perceptually relevant effects, we limited our investigation to the listening-only situation. In order to control the network conditions, we employed the NGMN testbed described in the previous section. We additionally compared the subjective results to instrumental estimations using the cited models, in order to check whether we will be able to replace subjective listening-only tests in future studies. This way, we are able to provide initial answers to all three research questions.

Description of the Experiments

In NGMN scenarios, speech quality is expected to vary during a call, as a result of changing conditions in the connection, network handover and/or codec changeover. Thus, in order to quantify quality, the entire length of a call has to be considered. Standard listening-only tests which make use of speech samples of 4-8 s length (ITU-T Rec. P.800, 1996) are not suitable for this purpose. On the other hand, conversational tests – despite being comparable to normal telephone usage and thus being ecologically valid – place a content-related focus on the user's attention; in such a situation, users are generally less analytic in their judgments, and it might happen that subtle perceptual differences get blurred.

As a compromise, we opted for a twofold test protocol: (a) We simulated conversations of 60

Table 1. Test 1a conditions. H: HSDPA; W: WLAN; Ppl: packet loss in %; switching at the beginning (beg.), middle (mid.) or end of a simulated call.

No.	Network(s)	Codec(s)	*Ppl* per segment
1	H	G.711	0
2	W	G.722.2	0
3	H	G.711	0,10,20,10,10
4	W	G.722.2	0,10,20,10,10
5	H→W mid.	G.711→G.722.2	0,10,20,0,0
6	W→H mid.	G.722.2→G.711	0,10,20,0,0
7	H→W beg.	G.711→G.722.2	0
8	H→W end	G.711→G.722.2	0
9	W→H beg.	G.722.2→G.711	0
10	W→H end	G.722.2→G.711	0

s length by concatenating 5 meaningful speech segments alternating with pauses, playing them back to the test participants, asking them to answer content-related questions during the pauses, and asking for an overall quality judgment at the end of the simulated conversation; this approach has been developed by Berger et al. (2008) and is now recommended for call-quality measurement in ETSI Technical Report 102 506 v.1.1.1 (2006); (b) The composing segments of the simulated conversations and some additional segments of approx. 6 s length were presented to the participants in a standard listening-only context, asking for an overall quality rating after each sample. We carried out two tests of the first type (Tests 1a and 2a) and two corresponding tests of the second type (Tests 1b and 2b). The following subsections describe the test conditions, set-up and participant group in more detail.

Test Conditions

In the experimental work, Test 1a concentrated on WB/NB transitions and the effects of packet loss on perceived quality. It contained two conditions with pure NB and WB calls, 4 conditions where packet loss continuously increases until the middle (3rd segment) of the call, and then switching occurs to a loss-free network with a different codec (or not), and 4 conditions where NB→WB or WB→NB transitions occur at the beginning (2nd segment), or at the end (4th segment) of a call. We consider packet loss rates of 10-20% to be realistic constraints when a handover should be executed at latest. Table 1 summarizes the conditions. The corresponding Test 1b contains all segments of the simulated conversations, plus additional samples with similar degradations, addressing also Flash-OFDM networks. The resulting list of 26 segments for this test is presented in Table 3.

Test 2a was designed to put a magnifier on the most interesting findings of the first test. It focused on the switching position within a simulated conversation, as well as on additional packet loss rates (see Table 2). The corresponding Test 2b with short samples also included different network load and high packet-loss-rate scenarios for limited WLAN networks (overall 27 conditions).

Test Setup

Tests 1a/b and 2a/b were carried out at distinct points in time, with different participant groups. Test participants were invited to a sound-insulated laboratory, were instructed about the purpose of the test, and listened to the samples in three sessions of approx. 25 min. each (2 sessions for parts a, 1 session for parts b). Speech samples were

Table 2. Test 2a conditions. See Table 1 for explanations.

No.	Network(s)	Codec(s)	*Ppl* per segment
1	W	G.722.2	0
2	H	G.711	0
3	H→W beg.	G.711→G.722.2	0
4	H→W mid.	G.711→G.722.2	0
5	W→H mid.	G.722.2→G.711	0,3,3,0,0
6	W→H mid.	G.722.2→G.711	0,5,5,0,0
7	W→H mid.	G.722.2→G.711	0,10,10,0,0
8	W→H→W	G.722.2↔G.711	0
9	H→W→H→W	G.711↔G.722.2	0
10	W	G.722.2	0,0/5,5,5/0,0

Table 3. Test 1b conditions. H: HSDPA; W: WLAN; F: Flash-OFDM; Ppl: packet loss in %; codec switching directly before (b.) or after (a.) network handover.

No.	Network(s)	Codec(s)	Ppl per network
1	F->H	G.722.2	0
2	F->H	G.722.2	10, 0
3	F->H(a.)	G.722.2->G.711	0
4	F->H(b.)	G.722.2->G.711	0
5	H	G.711	0
6	H	G.711	10, 0
7	H	G.711	20, 0
8	H	G.711->G.722.2	0
9	H	G.722.2->G.711	0
10	H->F	G.711	0
11	H->F	G.711	10, 0
12	H->F(a.)	G.711->G.722.2	0
13	H->F(b.)	G.711->G.722.2	0
14	H->W	G.711	0
15	H->W	G.711	10, 0
16	H->W(a.)	G.711->G.722.2	0
17	H->W(b.)	G.711->G.722.2	0
18	H->W(b.)	G.711->G.722.2	20, 0
19	W	G.722.2	0
20	W	G.722.2	10, 0
21	W	G.722.2	20, 0
22	W->H	G.722.2	0
23	W->H	G.722.2	10, 0
24	W->H(a.)	G.722.2->G.711	0
25	W->H(b.)	G.722.2->G.711	0
26	W->H(b.)	G.722.2->G.711	20, 0

presented over a Sennheiser HMD 410 headset at a comfortable listening level, with a background level below 35 dB(A) (ITU-T Rec. P.800, 1996). At the end of each simulated conversation of part a, as well as after each sample of part b, participants had to rate the overall quality on a 5-point absolute category scale, with 5 corresponding to "excellent" and 1 to "bad" quality. The test set-up and scale followed mainly the requirements given in ITU-T Rec. P.800 (1996) and ETSI Technical Report 102 506 v.1.1.1 (2006). 13 participants took part in Test 1a, 24 in Test 1b, 14 in Test 2a, and 17 in Test 2b. They were recruited from the normal telephone-user population, did not report any hearing impairment, and received a voucher in return for their effort.

Analysis of Experimental Results

Figure 4 shows the auditory judgments of the simulated conversations of Test 1a, averaged over all test participants and samples used in

Figure 4. Results of Test 1a. See Table 1 for conditions

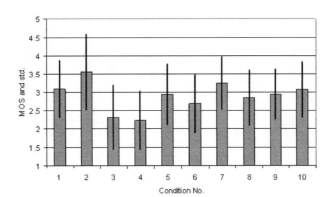

each condition (Mean Opinion Scores, MOS, and standard deviations, std. dev.) As expected, the pure WB condition (#2) is rated best, and better than the pure NB condition (#1). The packet loss of conditions #3 and #4 impacts quality significantly (analysis of variance with $F(9,27)=10.247$; Tukey HSD post-hoc test with $p<0.001$); these conditions have the lowest quality of the entire test, showing that packet loss seems to be the most dominant factor for quality degradation. Conditions #5 and #6 differ from #3 and #4 in that a network handover and a codec changeover occur, leading to a different audio bandwidth and no packet loss at the end of the call; in both cases the switching results in a significant improvement of perceived quality (Tukey HSD, $p<0.05$). Apparently, network handover can be an efficient tool for quality improvement in case that packet loss impairs quality significantly. This also holds when a transition from WB to NB is necessary, as the comparison between conditions #4 and #6 shows.

Even when no packets are lost, switching from NB to WB may be advantageous in case that a significant time period remains in order to take profit of the improved quality. A comparison between conditions #7, #8 and #1 shows that quality improves when switching from NB to WB at the beginning of a call; however, when switching occurs at the end of a call, the quality degrades compared to the pure NB case. For the opposite (WB→NB) direction, switching definitely degrades quality; the longer the WB connection remains established, the better the perceived quality.

Test 2a confirms most of the findings of the first test, cf. Figure 5. Once again, pure WB is rated best, and packet loss has the highest impact on speech quality. A comparison of conditions #2, #3 and #4 shows that switching from NB to WB is advantageous if it occurs early in the call. It was observed that switching at the middle of the call results in approximately the same quality as keeping NB for the entire call. Switching to loss-free NB because of packet loss in the WB network (#5, #6 and #7) does not improve quality: already with 3% packet loss in WB switching is worse than the pure NB case (#2). A two-segment-long period of 5% packet loss in the middle of a WB call (#10) is roughly equivalent in quality to the same period of NB in that call (#8). Multiple switching (#8 and #9) is always worse than single switching (#3 and #4).

The results of Test 1b help to quantify the trade-off between audio bandwidth and packet-loss degradations (see Figure 6.) As in Test 1a, packet loss is the most important quality degradation. In case of zero packet loss, a network handover without

Figure 5. Results of Test 2a

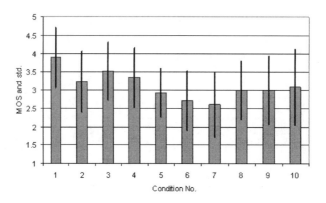

Figure 6. Results of Test 1b

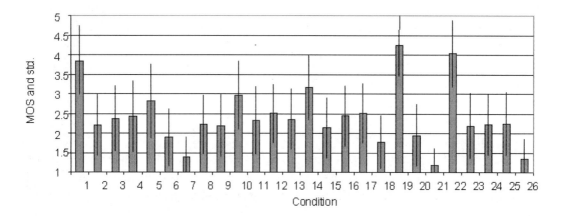

changing the codec degrades quality only slightly. However, if the codec has to be changed as well, this has a significant effect on perceived quality. For WB→NB transitions, the impact of bandwidth switching is roughly equivalent to 5-10% packet loss degradation. For the NB→WB transition, quality improves in all cases, and the improvement is again equivalent to 5-10% packet loss in the improved condition. These findings are valid both for the transitions between WLAN/HSDPA and Flash-OFDM/HSDPA. No significant difference was found between codec switching before and after the handover.

The results of Test 2b (not reproduced here) confirm the importance of packet loss, and show that the differences between WLAN and Flash-OFDM are negligible when the same codec is used. The impact of bandwidth switching depends in this case on the basic quality level provided by the WLAN: If the quality is high, then the impact of bandwidth switching is remarkably high as well; if the basic quality level is already low due to high packet loss in the network, then switching codecs does not have a strong effect. In other words: packet-loss degradations are able to mask the positive effect of switching to a higher audio bandwidth.

Estimation of Quality Judgments

Auditory tests like the ones described in the previous sections are expensive and time-consuming;

Table 4. Pearson Correlation and Root Mean Squared Error between auditory and estimated MOS

Test	WB E-model		WB-PESQ	
	r	σ	R	σ
1b	0.58	1.44	0.96	0.44
2b	n.a.	n.a.	0.89	0.70

thus, we checked the available quality prediction models as to whether they provide valid estimations also for NGMN handover situations. Two different types of models have been checked: A parametric model which estimates quality on the basis of the parameters describing the network conditions, and a signal-based model which estimates quality as a perceptually-weighted distance between the input and the output signals of the network under test.

We used an extended version of the E-model as a parametric model. It is based on the original algorithm described in ITU-T Rec. G.107 (2005) for NB networks and has been modified to take into account WB transmission (by linearly extending the underlying transmission rating scale from 100 to 129), WB speech codecs (by defining codec-specific equipment impairment factors), and packet loss. The necessary modifications are all described by Möller et al. (2006) and are recommended in the most recent update of ITU-T Rec. G.107, Amendment 1 (2006). As the E-model does not yet consider NB↔WB transitions, we decided to calculate two separate scores for the samples in which such NB↔WB transitions occur, then calculate an average of the underlying transmission ratings, and transform this back to the MOS scale. As a signal-based type of model, we used the WB extension of the PESQ model which is described in ITU-T Rec. P.862.2 (2007).

It has to be emphasized that both types of prediction models have not yet been validated for the scenarios investigated here. In addition, they estimate instantaneous speech quality of short samples, and not of entire conversations. Thus, we applied the predictions only to Tests 1b and 2b. The parametric E-model uses packet loss as an input parameter, which has not been manipulated in a controlled way in Test 2b; thus, for this test, only the signal-based model can be used. We compare the model estimations to the auditory test results in terms of MOS, and calculated the Pearson correlation r and the root mean squared error σ for each test. The results are given in Table 4.

The parametric E-model does only use general information on the network condition (average packet-loss percentage Ppl, codec type) as an input; consequently, the predictions are not accurate. In contrast to this, WB-PESQ is able to recognize speech signal degradations caused by the network handover, codec changeover and packet loss. For Test 1b, this leads to very good prediction accuracy, even better than the value of $r = 0.93$ which is obtained for in-scope data (Rix et al., 2006). The prediction accuracy is slightly lower for Test 2b which contains conditions with G.722.2 coding at 12.65 Kbit/s; WB-PESQ has been shown to have bigger problems in predicting the effects of this codec compared to the 23.05 Kbit/s bit-rate used in Test 1a/b, see e.g. by Côté et al. (2006).

Consequences for Handover Strategies

The auditory results help to answer the research questions raised at the beginning of this section:

1. The most important network characteristic is the packet loss rate. The second most important characteristic is the audio bandwidth. Switching the audio bandwidth is roughly

equivalent to the quality degradation of 5-10% packet loss, in both narrowband and wideband conditions. The degradations due to network handover (make-before-break) alone – without codec switching – are negligible in comparison to packet loss and bandwidth switching.

2. Switching codecs is advantageous if the packet loss rate is high, and if the changeover helps to reduce the packet loss impact. Switching codecs in order to take profit of a larger audio bandwidth is advantageous only if a sufficiently long period of WB speech transmission remains. From the limited results of our tests, it seems that this minimal length is around 30 seconds. Unfortunately, it is not possible to know the remaining length of a call. As a workaround, handover strategies could consider the perceived quality and use this value to estimate multiple scenarios (using different remaining lengths), and then select the most beneficial for the user, assuming past and current conditions. An interaction could be observed between packet-loss and audio-bandwidth degradations:
    ◦ If packet loss is high (low basic quality), the impact of audio bandwidth on perceived quality is low.
    ◦ If packet loss is low (high basic quality), the impact of audio bandwidth on overall call quality is high.
3. Parametric quality prediction models like the E-model are not yet able to estimate the speech quality resulting from codec changeovers. Signal-based models like WB-PESQ do a better job, but do not always perform as well on in-scope data.

The results may help to design efficient network handover and codec changeover strategies. In bad network conditions (high packet loss rate), a handover should be made if the packet loss rate can be reduced by this step. In this situation, it is not important whether the audio bandwidth can be maintained or not; the reduction of packet loss should be the ultimate goal. In contrast to this, a handover can also be fruitful in good network conditions (low packet loss rate). In this case, switching to a higher audio bandwidth can help to significantly improve quality. The improvement is most effective if it occurs early in the call; the remaining call duration should be more than 30 seconds. Changing from WB to NB is always linked to a loss in quality; the pure network handover without codec changeover, however, does not significantly impact the perceived quality.

FUTURE RESEARCH DIRECTIONS

We presented the first part of an in-depth study of the effects of network and codec characteristics in NGMNs on perceived speech quality. In order to obtain valid and analytic results, we decided to use an NGMN testbed to generate typical stimuli, and to judge them in a simulated conversation, as well as in a standard listening context. With the testbed working in real time, additional conversation tests will be carried out in order to validate the results in a more realistic setting. On the basis of these tests, instrumental quality prediction models like WB-PESQ (ITU-T Rec. P.862.2, 2007) and the E-model (ITU-T Rec. G.107, 2005) will be extended in order to better take into account handover effects and time-varying quality during a call, e.g. in the way described by Berger (2008). In this way, extended models can be used to improve handover strategies in real time, depending on measured network characteristics. In addition, the obtained results need to be mapped to actually observed network behavior.

The results included in this chapter can be explained further by using techniques for trace analysis. The interpretation included in this chapter may be extended or modified through the correlation of these results and the ones provided by network-level traces (that were also collected during the experiments, as explained in *Section Methodology*). This is ongoing work and subject of future studies. Furthermore, the experimental paradigms presented here can easily be extended for multimedia services in Next Generation Networks. In particular, the quality of video calls and video streaming as experienced by the mobile user is a current subject of research at Deutsche Telekom Laboratories. The experimental set-up is based on the testbed developed for the studies on speech quality presented in this chapter.

CONCLUSION

This chapter discussed the impact of different phenomena on the user's perception of 4G networks. We proposed a novel approach to QoE assessment and speech quality evaluation, based on the joint work between the networking and usability research groups. The idea behind is to combine subjective information gathered from auditory tests with objective performance metrics extracted from network trace analysis. We believe that out of this mishmash useful conclusions can be derived and used for the advancement of mobility management solutions, quality models, and the design of 4G networks and mobile services.

Moreover, we believe that the blending of these very different methodologies and sources of information is fundamental to keep up with demands on QoE and QoS in Next Generation Networks. The present work represents a seminal contribution to this inter-disciplinary field.

The proposed methodology proved useful in the evaluation of QoE, and together with a unique experimental setup, it allowed us to study and understand user perception in 4G networks in a much holistic way. This knowledge will be helpful when proposing new solutions for mobility management or speech quality models.

The results presented in this chapter will be applied in the design of speech quality models that may not be valid in 4G networks. There is an ongoing debate on the issue that current models may be extended or modified, or even completely replaced by new models that can cope with typical situations in future networks and applications.

REFERENCES

Berger, J., Hellenbart, A., Weiss, B., Möller, S., Gustafsson, J., & Heikkila, G. (2008). Estimation of Quality per Call in Modelled Telephone Conversations. In *Proceedings of IEEE International Conference on Acoustics, Speech and Signal Processing (ICASSP) 2008*, Las Vegas (pp. 4809-4812).

Bernardos, C. J., Soto, I., Moreno, J. I., Melia, T., Liebsch, M., & Schmitz, R. (2005). Mobile Networks Experimental Evaluation of a Handover Optimization Solution for Multimedia Applications in a Mobile IPv6 Network. *European Transactions on Telecommunications*, 16(4), 317–328. doi:10.1002/ett.1005

Cerf, V. (1978). *The Catenet Model for internetworking*. DARPA Information Processing Techniques Office, IEN 48.

Costa, X. P., Schmitz, R., Hartenstein, H., & Liebsch, M. (2002). A MIPv6, FMIPv6 and HMIPv6 Handover Latency Study: Analytical Approach. In *Proceedings of IST Mobile and Wireless Telecommunications Summit* (pp. 100-105).

Côté, N., Möller, S., Gautier-Turbin, V., & Raake, A. (2006). Analysis of a Quality Prediction Model for Wideband Speech Quality, the WB-PESQ. In *Proceedings of the 2nd ISCA/DEGA Tutorial and Research Workshop on Perceptual Quality of Systems*, Berlin (pp. 115-122).

ETSI Technical Report 102 506 v.1.1.1. (2006). *Speech Processing, Transmission and Quality Aspects (STQ); Estimating Speech Quality per Call*. Sophia Antipolis, France: European Telecommunications Standards Institute (ETSI).

IETF RFC 3261. (2002). *SIP: Session Initiation Protocol*. Internet Engineering Task Force (IETF).

IETF RFC 3344. (2002). *IP Mobility Support for IPv4*. Internet Engineering Task Force (IETF).

IETF RFC 3519. (2003). *Mobile IP Traversal of Network Address Translation (NAT) Devices*. Internet Engineering Task Force (IETF).

IETF RFC 4068. (2005). *Fast Handovers for Mobile IPv6*. Internet Engineering Task Force (IETF).

IETF RFC 4140. (2005). *Hierarchical Mobile IPv6 Mobility Management (HMIPv6)*. Internet Engineering Task Force (IETF).

Jekosch, U. (2005). *Voice and Speech Quality Perception. Assessment and Evaluation*. Berlin, Germany: Springer.

Lewcio, B., Möller, S., Vidales, P., Wältermann, M., & Kirschnick, N. (2008). Methods for Multimedia Service Adaptation in Next Generation Networks. In *Proceedings of the ETSI Workshop on Effects of Transmission Performance on Multimedia QoS*, Prague.

Mobisense. (2007). Retrieved January 20, 2009, from http://www.deutsche-telekom-laboratories.de/~vidales/mobisense

Möller, S., Raake, A., Kitawaki, N., Takahashi, A., & Waltermann, M. (2006). Impairment Factor Framework for Wideband Speech Codecs. *IEEE Transactions on Audio, Speech, and Language Processing*, *14*(6), 1969–1976. doi:10.1109/TASL.2006.883262

Möller, S., Wältermann, M., Lewcio, B., Kirschnick, N., & Vidales, P. (2009). Speech Quality while Roaming in Next Generation Networks. Accepted for *IEEE International Conference on Communications (ICC)*, Dresden. Netem. (n.d.). Retrieved January 20, 2009, from http://www.linux-foundation.org/en/Net:Netem

PJSIP. (n.d.). Retrieved January 20, 2009, from http://www.pjsip.org

Rec, I. T. U.-T. P.800. (1996*). Methods of Subjective Determination of Transmission Quality*. International Telecommunication Union (ITU-T), Geneva.

Rec, I. T. U.-T. P.862. (2001). *Perceptual Evaluation of Speech Quality (PESQ): An Objective Method for End-to-End Speech Quality Assessment of Narrow-Band Telephone Networks and Speech Codecs*. International Telecommunication Union (ITU-T), Geneva.

Rec, I. T. U.-T. G.107. (2005). *The E-model, a Computational Model for Use in Transmission Planning*. International Telecommunication Union (ITU-T), Geneva.

Rec, I. T. U.-T. G.107, Amendment 1. (2006). *New Appendix II - Provisional impairment factor framework for wideband speech transmission*. International Telecommunication Union (ITU-T), Geneva.

Rec, I. T. U.-T. P.805. (2007). *Subjective Evaluation of Conversational Quality*. International Telecommunication Union (ITU-T), Geneva.

Rec, I. T. U.-T. P.862.2. (2007). *Wideband Extension to Recommendation P.862 for the Assessment of Wideband Telephone Networks and Speech Codecs*. International Telecommunication Union (ITU-T), Geneva.

Rix, A. W., Beerends, J. G., Kim, D., Kroon, P., & Ghitza, O. (2006). Objective Assessment of Speech and Audio Quality - Technology and Applications. *IEEE Transactions on Audio . Speech & Language Processing*, *14*(6), 1890–1901. doi:10.1109/TASL.2006.883260

Secgo. (n.d.). Retrieved January 20, 2009, from http://www.secgo.com

Steuer, F., Elkotob, M., Albayrak, S., & Steinbach, A. (2006). Testbed for Mobile Network Operator Scenarios. In *Proceedings of 2nd International Conference on Testbeds & Research Infrastructures for the Development of NeTworks & COMmunities (TRIDENTCOM)* (pp. 256-266).

TCPDUMP. (n.d.). Retrieved January 20, 2009, from http://www.tcpdump.org

Telnet. (n.d.). Retrieved January 20, 2009, from http://www.telnet.org

Vidales, P., Baliosian, J., Serrat, J., Mapp, G., Stajano, F., & Hopper, J. (2005). Autonomic System for Mobility Support in 4G Networks. [JSAC]. *IEEE Journal on Selected Areas in Communications, 23*(12). doi:10.1109/JSAC.2005.857198

Vidales, P., Bernardos, C., Soto, I., Cottingham, D., Baliosian, J., & Crowcroft, J. (2007). MIPv6 Experimental Evaluation Using Overlay Networks. *Computer Networks, 51*(10), 2892–2915. doi:10.1016/j.comnet.2006.12.004

Vidales, P., Kirschnick, N., Steuer, F., Lewcio, B., Wältermann, M., & Möller, S. (2008). Mobisense Testbed: Merging User Perception and Network Performance. In *Proceedings of 4th International Conference on Testbeds & Research Infrastructures for the DEvelopment of NeTworks & COMmunities (TRIDENTCOM)* (pp. 1-9).

Wältermann, M., Lewcio, B., Vidales, P., & Möller, S. (2008). A Technique for Seamless VoIP Codec Switching in Next Generation Networks. In *Proceedings of International Conference on Communications (ICC)*, Beijing (pp. 1772-1776).

Wireshark. (n.d.). Retrieved January 20, 2009, from http://www.wireshark.org

Chapter 8
Fourth Generation Networks:
Adoption and Dangers

Jivesh Govil
Cisco Systems Inc., USA

Jivika Govil
Carnegie Mellon University, USA

ABSTRACT

Mobile researchers are witnessing burgeoning interest in 4G wireless networks that patronize global roaming across diverse wireless and mobile networks. The pith of 4G mobile systems lies in seamlessly integrating the existing wireless technologies including Wideband Code Division Multiple Access (WCDMA). High Speed Uplink Packet Access (HSUPA)/ High-Speed Downlink Packet Access (HSDPA). 1×Evolution Data Optimized, (1×EVDO). Wireless LAN, and Bluetooth. However, migrating current systems to 4G engenders enormous challenges. With ever-changing specification and standards, developing a prototype requires flexible process to provide 4G system capabilities. The 4G system has its own advantages and associated dangers. This chapter intends to deal with adoption issues of 4G, the fundamentals as well as issues pertaining to 4G networks, standards, terminals, services of 4G, and the vision of network operators and service providers. Besides, to overcome the challenges of sophisticated personal sessions and service mobility, advanced mobility management (MM) is needed to fulfill the need for seamless global roaming. The chapter endeavors to make an evaluation on development, transition, and roadmap for fourth generation mobile communication system with a perspective of wireless convergence domain in addition to mobility management. Lastly, open research issues in 4G are succinctly discussed.

INTRODUCTION

4G is short for fourth-generation cellular communication system. There is no set definition for the specifics of 4G (Young Kyun & Prasad, 2006). The 4G will be a fully IP-based integrated system of systems and network of networks, achieved after the convergence of wired and wireless networks as well as computer, consumer electronics, communication technology, and several other convergences. These will be capable of providing 100 Mbps and 1Gbps, respectively, in outdoor and indoor environments

DOI: 10.4018/978-1-61520-674-2.ch008

Fourth Generation Networks

with end-to-end QoS and high security, offering any kind of services anytime, anywhere, at affordable cost and one billing (Dixit, S. (2008)). The Wireless World Research Forum (WWRF) defines 4G as a network that operates on Internet technology, combines it with other applications and technologies such as Wi-Fi and WiMAX, and runs at speeds ranging from 100 Mbps (in cell-phone networks) to 1 Gbps (in local Wi-Fi networks). 4G is not just one defined technology or standard, but rather a collection of technologies and protocols to enable the highest throughput and lowest cost wireless network possible (Chang, 2007).

Fourth generation networks are likely to use a combination of WiMAX and WiFi. Technologies employed by 4G may include SDR (Software-Defined Radio) receivers, OFDM (Orthogonal Frequency Division Multiplexing). OFDMA (Orthogonal Frequency Division Multiple Access). MIMO (multiple input/multiple output) technologies, UMTS (Universal Mobile Telecommunications Service) and TD-SCDMA (Time Division Synchronous Code Division Multiple Access). All of these delivery methods are typified by high rates of data transmission and packet-switched transmission protocols. 3G technologies, by contrast, are a mix of packet and circuit-switched networks. When fully implemented, 4G is expected to enable pervasive computing in which simultaneous connections to multiple high-speed networks provide seamless handoffs throughout a geographical area. Network operators may employ technologies such as cognitive radio and wireless mesh networks to ensure connectivity and efficiently distribute both network traffic and spectrum

Moreover, the objective to 4G is to offer seamless multimedia services to users accessing an all IP-based infrastructure through heterogeneous access technologies. IP is assumed to act as an adhesive for providing global connectivity and mobility among networks. 4G will more resemble a conglomerate of the existing technologies rather than an entirely new standard. An all IP-based 4G wireless network has inherent advantages over its predecessors. It is compatible with, and independent of the actual radio access technology. With IP, one basically gets rid of the lock-in between the core networking protocol and the radio protocol. IP tolerates a variety of radio protocols. It lets one design a core network that gives complete flexibility in the access network type.

The goal of this chapter is to study development, transition, and challenges in 4G implementation and mobility management issues. Mobility management has been recognized as one of the most important and challenges problems for a seamless access to wireless network and mobile service. It is the fundamental technology used to automatically support mobile terminals and join their services while simultaneously roaming freely without the disruption of communication. Mobility management operations are introduced, along with the discussions of key research issues and possible solutions.

4G MOBILE COMMUNICATION SYSTEMS

The success of Second-generation (2G) mobile systems in the previous decade prompted the development of third-generation (3G) mobile systems. While 2G systems such as GSM, IS-95, and cdmaOne were designed to carry speech and low-bit-rate data, 3G systems were designed to provide higher data-rate services. During the evolution from 2G to 3G, a range of wireless systems, including General Packet Radio Services (GPRS). International Mobile Telecommunications-2000 (IMT-2000). Bluetooth, WLAN, and HiperLAN, have been developed. All these systems were designed independently, targeting different service types, data rates, and users. As these systems have their own merits and demerits, there is no single system that is good enough to replace all the other technologies. Researchers are making efforts to establish 4G systems that integrate existing and newly developed wireless systems as a more

Figure 1. Example of IP based networks

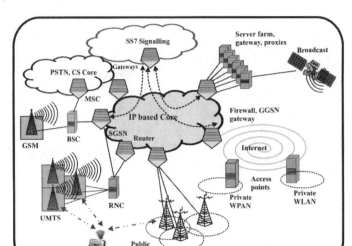

feasible option. Different research programs, such as Mobile Virtual Centre of Excellence (VCE), MIRAI, and DoCoMo, have their own visions for 4G features and implementations.

At present, plethora of wireless technologies with their own merits and demerits exist globally; the upcoming 4G mobile communications system is foreseeing potentially a smooth merger of these technologies with a goal to support cost effective seamless communication at high data rate supported with global roaming and user customized personal services.

As discussed 4G will be an IP based wireless network replacing the old Signaling System 7 (SS7) telecommunications protocol, which is considered massively redundant as shown in figure 1 This is because SS7 signal transmission consumes a larger part of network bandwidth even when there is no signalling traffic for the simple reason that it uses a call setup mechanism to reserve bandwidth, rather time/frequency slots in the radio waves. IP networks, on the other hand, are connectionless and use the slots only when they have data to send. Hence there is optimum usage of the available bandwidth.

The goal of 4G will be to replace the entire core of cellular networks with a single worldwide cellular network completely standardized based on the (Internet Protocol) IP for video, packet data utilizing Voice over IP (VoIP) and multimedia services. The newly standardized networks would provide uniform video, voice, and data services to the cellular handset or handheld Internet appliance, based entirely on IP (Internet Protocol).

The Goal of 4G

4G must be clearly more than 3G in terms of services, applications, and technology. As a comparison, 4G is not a combination of High Speed Uplink/Downlink Packet Access (HSUPA/HSDPA) or Wireless LAN (WLAN).

$3G + HSDPA = HSUPA(= 3.5G?) < 4G$

$3G + WLAN < 4G$

$3G + HSDPA + HSUPA + WLAN < 4G$

3G networks are inadequate to accommodate WLANs as access networks, which offer data rates of 11 Mbps. The goal of 4G will be to replace the entire core of cellular networks with a single worldwide cellular network completely

Table 1. Comparison of wireless communication technologies

	1G	2G	2.5G	3G	4G
Transmission	Analog	Digital	Digital	Digital	Digital
Architecture		Circuit Switch	Packed-Switch	Circuit and Packet Switch	Packet Switch
Speeds		9.6 to 14.4 Kbits/s	64 to 144 Kbits/s	384 Kbits/s to 2 Mbits/s	100 Mbits/s to 1000 Mbits/s
Standards	AMPS	TDMA, CDMA, GSM	GPRS, EDGE, 1×RTT	WCDMA, CDMA-2000	Single unified standard
Service	Mobile telephony (voice)	Digital voice, short messaging	Higher capacity, packetized data	Integrated high quality audio, video and data	Dynamic information access, wearable devices
Multiplexing	FDMA	TDMA, CDMA	TDMA, CDMA	CDMA	CDMA
Core Network	PSTN	PSTN	PSTN and Packet network	Packet network	Internet
Handoff	Horizontal	Horizontal	Horizontal	Horizontal	Horizontal and Vertical

standardized based on the IP for video, Voice over IP (VoIP) and multimedia services. The newly standardized networks would provide uniform video, voice, and data services to the cellular handset or handheld Internet appliance, based entirely on IP. As 4G standards are not defined so there is plenty of room for other applications to overlap into the 4G space.

The 4G providers of advanced cellular technology are adopting Concatenated Coding which has the capability of providing multiple Quality of Service (QoS) levels. forward error correction (FEC) coding adds redundancy to a transmitted coded signal through encoding prior to transmission. A major advantage of Concatenated Coding over the Convolution Coding method will be an improved network performance by the combining of two or more coding techniques, such as a Reed-Solomon and a Convolution Code into one Concatenated Code. The combination improves error correction combining with error detection. FEC using Concatenated Coding allows a wireless network to send much larger blocks of data while reducing bit-error rate, thereby increasing the overall through-put.

The primary goal of the planned 4G cellular services will include: Interactive Multimedia, Voice, Video Streaming, High Speed Global Internet Access. Virtual Private Network (VPN) Availability, Service Portability with Scalable Mobile Services, High Speed, High Capacity, Low Cost Services, Improved Information Security, QoS Enhancements, Multi-Hop Networking.

Table 1 compares here various Wireless Communication Technologies.

4G Objective

Before studying the objectives of 4G, let us understand some of the characteristics of 4G, which are summarized here in table 2.

The objective of 4G is to cater the quality of service and rate requirements set by the forthcoming applications like wireless broadband access, Multimedia Messaging Service, video chat, mobile TV, High definition TV content, DVB and minimal service like voice and data at anytime and anywhere 4G is being developed, the 4G working groups have defined the following as the objectives of the 4G wireless communication standard.

- Spectrally efficient system (in bits/s/Hz and bit/s/Hz/site)

Table 2. Characteristics of 4G

Achievable Data Rates	10 Mbps (wide coverage) to 1 Gbps (local area). These are design targets and represent cell overall throughput.
Networking	All-IP network (access and core networks). Plug & Access network architecture. An equal opportunity network of networks.
Ubiquitous	Mobile, seamless communications.
Cost Reduction	Cost per bit: 1/10-1/100 lower than 3G Infrastructure cost – 1/10 lower than 3G.
Connected Abilities	Person to person communication Person to Machine communication/Machine to machine communication.

- High network capacity, 10 times higher than 3G (Ibrahim, J. (2002))
- Nominal data rate of 100 Mbps at high speeds and 1 Gbps at stationary conditions as defined by the ITU-R
- Data rate of at least 100 Mbps between any two points in the world
- Smooth handoff across heterogeneous network (Hussain, S., Hamid, Z., Khagttak, N. (2006))
- Seamless connectivity and global roaming across multiple networks i.e. seamless services with fixed NW (Net Work) and private (NW Mohr, W. (2002))
- High quality of service for next generation multimedia support (real time audio, high speed data, High-Definition Television (HDTV) video content, mobile TV, etc)
- Higher frequencies: Microwave: 3–10 GHz
- Interoperable with the existing wireless standards (Schmitz, N. (March 2005)).
- All IP system, packet switched network
- Next-generation Internet support: IPV6, QoS, MoIP (Mobile over IP)
- Lower system costs: 1/10 of IMT-2000

In summary, the 4G system should dynamically share and utilize the network resource to meet the minimal requirements of all the 4G enabled users. Figure 2 illustrates here a prospective view of physical layer of 4G.

Key Parameters

The move to 4G is complicated by attempts to standardize on a single 3G protocol. Without a

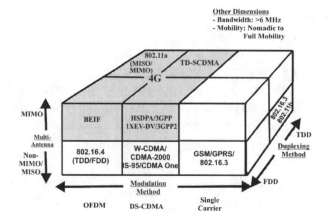

Figure 2. Prospective physical layer of 4G

Table 3. Comparison of key parameters of 3G and 4G

Key Parameters	3G	4G
Frequency Band	1.8 – 2.5 GHz	2-8 Ghz
Bandwidth	5-20 MHz	5-20 MHz
Data Rate	Upto 2 Mbps	Upto 20 Mbps
Access	W-CDMA	MC-CDMA or OFDM (TDMA)
Forward Error Correction	Convolutional Rate ½, ⅓	Concatenated coding scheme
Switching	Circuit/Packet	Packet
Mobile Top Speeds	200 km/h	200 km/h
Component Design	Optimized antenna design, multi-band adapters	Smarter Antennas, software multiband and wide-band radios
IP	A number of air link protocols, including IP 5.0	All IP (IP6.0)

single standard on which to built, designers face significant additional challenges.

Table 3 compares some of the key parameters of 3G and 4G. Though 4G does not have any solid specification as of yet, it is clear that some standardization is in order.

4G Network Requirement

From above it is clear that 4G is immensely complicated and hence there will be special requirement for future networks. Some of these tentative requirements are hereby summarized in table 4 (Toshio et al., 2005).

Development of 4G

There are many phases of developing 4G mobile communication systems. Let us study here two phases, i.e. development period and maturity period, which are described in table 5.

4G Transition Components

Some of the 4G transition components in brief are (Electrozoom, 2007):

- *Multi-Antenna Systems*: To foster the growing data rate needs of 4G, deploying

Table 4. Requirement for future networks (tentative)

Media	Transmission speed	Delay	Connection Latency	Terminal capabilities
Speech/ 3D Audio	< 1 Mbps	<50ms	<1 sec	3D sound field control High efficiency loud speakers
Video/ 3D video	10 Mbps (2D video) – 30 Gbps (3D video)	< 50 ms	< 1 sec	Real time hologram
Enhanced Reality	< 1 Mbps	<< 50 ms Should be predictable	N/A	Eyeglass display 3D and multimodal UI
Five senses communications	< 1 Mbps	< 50 ms	N/A	Five sense sensors
Tele-existence	< 10Mbps (Robotic I/F) < 1 Gbps (Virtual avatar) < 100 Mbps (Alter-ego existence)	< 10 ms < 30 ms < 5 ms (Small and known jitter)	1 Sec	Alter-ego robot

Table 5. Stepwise development of 4G mobile communications

4G Mobile Communications		
	Phase 1 (2009-2010): Developmental Period	**Phase 2 (2011): Maturity Period**
Core cellular systems	3.5G 3.5G mobile-communications system enhancing IMT-2000 (HSDPA/1xEV-DV)	4G 4G Mobile-communications systems
Transmission speed	30 Mbps	50 Mbps-100 Mbps
Service level	High-level application service	Service with higher-level authentication and security
Main users	Advanced users	General users
Functions	Basic functions	Fully-fledged system
Seam-lessness with other systems	Flexible realization of seamlessness with other systems	Seamlessness with no awareness thereof
Social impact	Positioning with social functions	Positioning as a factor inducing changes in social structure.

multiple antennas at the transmitter and receiver.

- *Software Defined Radio (SDR):* SDR is one form of Open Wireless Architecture (OWA). Since 4G is the collection of wireless standards, the final form of the 4G device will constitute all standards. SDR Technology offers one possible realization.
- *Smart antennas and beam forming:* These offer a significantly improved solution to reduce interference levels and improve the system capacity. With this technology, each user's signal is transmitted and received by the base station only in the direction of that particular user. This drastically reduces the overall interference in the system.
- *Adaptive Modulation and Coding Techniques:* The modulation and coding techniques change according to the network resource, user requirement and physical channel conditions.
- *Access Schemes:* The scarce resource frequency and network infrastructure is accessed using the channel accessing schemes. The existing wireless standards use TDMA (Time Division Multiple Access). FDMA (Frequency Division Multiple Access). CDMA (Code Division Multiple Access) and combinations of these. Recently, new access schemes like OFDMA (Orthogonal Frequency-Division Multiple Access) and MC-CDMA (Multi Carrier CDMA System) gained more importance in 802.16e and 802.20 standards.
- *IPv6:* It is generally believed that 4th generation wireless networks would support great number of wireless devices that are addressable and routable. Therefore in the context of 4G, IPv6 is an important network layer technology and standard that can support great number of wireless enabled devices. In addition to increasing the number of IP addresses, IPv6 also removes the need for Network Address Translation (NAT)—a technique used in 3G and other networks to make private IP addresses work with Internet applications. In the context of 4G, IPv6 also enables a number of applications with better multi-cast, security and route optimization capabilities. With the available address space and number of addressing bits in IPv6, many innovative coding schemes can be developed for 4G devices and applications that

could aid deployment of 4G networks and services.
- *Mesh Networks:* A mesh network is reliable and offers redundancy. If one node can no longer operate, all the rest can still communicate with each other, directly or through one or more intermediate nodes. Mesh networks work well when the nodes are located at scattered points that do not lie near a common line (Pinto, 2005).

Potential Applications of 4G

Some key potential applications of 4G are:

1) Virtual Presence: 4G system gives mobile users a "virtual presence"; for example, always-on connections that keep people involved in business activities regardless of whether they are on-site or off.
2) Virtual Navigation: A remote database contains the graphical representation of streets, buildings, and physical characteristics of a large metropolis. Blocks of this database are transmitted in rapid sequence to a vehicle, where a rendering program permits the occupants to visualize the environment ahead (Santhi et al., 2005).
3) Tele-medicine: 4G will support remote health monitoring of patients; for example, the paramedic assisting the victim of traffic accident in a remote location must access medical records and may need videoconference assistance from a surgeon for an emergency intervention. The paramedic may need to relay back to the hospital the victim's x-rays taken locally.
4) Tele-Geoprocessing Applications: The combination of Geographical Information Systems (GIS). Global Positioning Systems (GPS). and high-capacity wireless mobile systems will enable a new type of application referred to as tele-geoprocessing. Queries dependent on location information of several users, in addition to temporal aspects have many applications (Alexander, 2002).
5) Crisis-Management Applications: Natural disasters can affect the entire communications infrastructure is in disarray. Restoring communications quickly is essential. With wideband wireless mobile communications Internet and video services, could be set up in hours instead of days or even weeks required for restoration of wireline communications.
6) Education: Educational opportunities available on the internet for individuals interested in life-long education high speeds are unavailable to client in remote areas because of the economic unfeasibility of providing wideband wireline internet access. 4G wireless communications provides a cost-effective alternative in these situations

Advantages of 4G

Though 4G has many benefits like high usability, anytime, anywhere through a range of technologies, increased data transfer speed, and improved quality of service etc., other indirect advantages in alignment with ideal world are:

1) *Unregulated*: Presently 4G is unregulated, it requires no license.
2) *Wireless:* It can by-pass low capacity wired connection from the street to the home.
3) *Market*: Due to cheap bandwidth the market for PCs, consumer electronics, microprocessors and software will rise.
4) *Cheap*: Simple and cheap technology.
5) *Multi Media:* Users will experience multi media service at any place, at any time, at any acceptable cost.
6) *Reduce Network Load:* Conversion of public and 4G networks will reduce network cost by allowing capacity to be shared among Carriers and private users.

Figure 3. New business opportunities due to high data rates

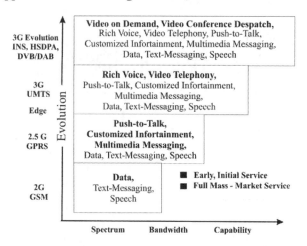

7) *Upgradability:* 4G is cheap and allow carriers to upgrade inexpensively and the success will depend on better packaging than DSL and better pricing.
8) *No Digging*: As 4G will be completely wireless no ditch digging requirements.
9) *Easy Buying:* 4G is so cheap that much of the network infrastructure will be bought by the consumers with their visa cards, saving carriers' balance sheet (Francis M. (2003)).
10) *Fast Data Exchange:* New type input/output devices for fast data exchange (glasses, displaying 3D virtual world, collapsible screens, e-paper, and voice and handwriting recognition).
11) *Easy Availability of Terminals:* New type semiconductor industry will rise (by means of plastic based chip technology, electronic tools will be common, 4G terminals will be available to everyone).
12) *Easy Access*: Access to the fourth generation mobile systems will be low-priced (advertisements will be displayed on the screen of 4G terminals).
13) *Good Number of Users:* Amount of users will reach a high level.
14) *Internet Access:* Quality of Internet access by wire or wireless will be equal or almost the same (quality of content providing will be excellent using a mobile terminal) (Shoewu, O.(2007)).
15) *Availability of Multimedia:* Multimedia will be required to the trivial work (means a kind of extra information).
16) *VPN (Virtual Private Network):* Some economic, social or state groups could maintain own part-networks (virtual private networks will be used well at administration, personal data-managing – for example mobile ID- and voting a president).
17) *More Opportunities:* New business opportunities will arise due to high data rates as shown in Figure 3. There will be a heavy competition between applications and service-providers for users.
18) *24-Hour Availability:* It follows that the mobile networks should be stable and dependable, should be available for 24 hour per day.
19) *Inter Connection*: Easy interconnection of different system (e.g. GPS, Internet, other communication networks).

MIGRATION TO 4G

Due to success of the second generation (2G) mobile system, the third generation (3G) system was developed. 3G systems were designed to provide higher data rate services. A range of wireless systems including GPRS, IMT–2000, Bluetooth, WLAN and Hyper-LAN have been developed, each with their own merits and short comings. No single system exists for integration of all these technologies; thus a 4G system that integrates existing and newly developed wireless systems is a more feasible option. 4G technology is still in the research & development stage and international standards do not exist. 3G technology has proper standards and many big companies have invested huge sums to acquire the needed spectrum space. The predicament is whether or not companies should bypass 3G and adopt 4G straight away. It is being argued that 3G and 4G technologies are not mutually exclusive but are complementary to each other (Sharma, 2002). Those countries that have made huge investments in 3G require a evolutionary path for migration to 4G, but developing countries which have not made investments for 3G need not follow the 3G-migration route, as they can easily by-pass 3G technology and adopt directly 4G networks. 4G networks are being designed to accommodate WLANs and PANs based on Bluetooth technologies. 3G has bandwidth limitations. 4G core networks are all IP networks which have been extended to radio access nodes, so the disadvantages of circuit switching are totally absent. These networks will be incorporating advanced IPv6; even the signaling will be done through IP. The setting up cost will be lower as they can be built on top of existing network and won't require operators to completely retool. Hence, 4G technology will be suitable to esp. those countries which have not yet adopted 3G.

Figure 4 gives an understanding of driving forces behind the adoption of 4G and impeding forces that forces the corporate house to use it commercially in future in adopting 4G Wireless networks.

Roadmap for Achieving 4G

Recently there have been major advances in wireless access technologies for planning roadmap to 4G. Among the new schemes of technology being proposed for 4G, 802.16e and 802.20 standards are OFDMA, Single Carrier FDMA, and MC-CDMA. The new technologies, while offering the efficiencies of the older technologies such as CDMA, also offer advantages in scalability. Current working assumptions for physical layer multiple access schemes is OFDMA for downlink and Single Carrier FDMA (SC-FDMA) for uplink.

One way to increase system capacity is to implement a Multiple-Input Multiple-Output (MIMO) antenna scheme. A wireless system with single antennas obeys Shannon's classical limit for capacity, which can be expressed as $C = \log_2(1+SNR)$. Ideal capacity therefore increases as the log of the signal-to-noise ratio. MIMO systems, on the other hand, are modeled to increase capacity linearly with respect to the number of transmit and receive pairs that are used.

Various Technologies

Though several technologies available today play a role in achieving roadmap for 4G as it material-

Figure 4. Driving and impeding forces for adoption of 4G wireless network

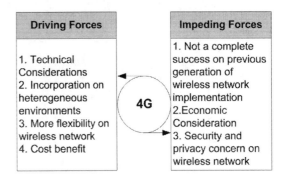

izes; figure 5 highlights few of them (Pappalardo, 2007):

1) *Orthogonal Frequency Division Multiplexing (OFDM)* and OFD Multiple Access (OFDMA) transmits data by splitting radio signals that are broadcast simultaneously over different frequencies. OFDMA, used in mobile WiMax, also provides signals that are immune to interference and can support high data rates. It utilizes power more efficiently than 3G systems while using smaller amplifiers and antennas. This all translates to expected lower equipment costs for wireless carriers. In OFDM and related modulations technology, multiple coherent sub-carriers are modulated and codes are used to insure that encoded bits can be decoded even if some of the sub-carriers arrive at very low SNR. OFDM is more resistant to inter-symbol interference. OFDM could be used both as multiple access technology and modulation scheme in 4G, but has many challenges. The OFDM has many uses for 4G:
   (a) Most efficient transmission technique for digital audio and video broadcasting system. Processing data at rates of 10 Mbps or higher results in a small computer inside the phone but requires higher power in the amplifier.
   (b) Provide robust, reliable broadband service to the most people at the greatest distance with the least amount of infrastructure,
   (c) OFDM is easier to implement then CDMA by small companies, as CDMA networks need more experienced engineers.
2) *Mobile WiMAX is an IEEE* specification also known as 802.16e and designed to support as high as 12Mbps data-transmission speeds. It uses OFDMA and is the next-generation technology of choice for many Service provider.
3) *Ultra Mobile Broadband* (UMB), also known as CDMA2000 EV-DO, is an expected path to 4G for legacy CDMA network providers. It's an IP-based technology that is said to support 100Mbps through Gbps data-transmission speeds.
4) *Multiple-input multiple-output (MIMO) wireless LAN* technology supports two or more radio signals in a single radio channel, increasing bandwidth. MIMO does this

Figure 5. Roadmap to 4G communication system

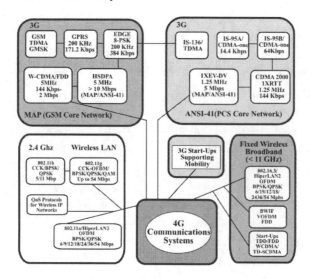

by using multiplexing. MIMO is expected to support data rates as high as 15Mbps in 36MHz of spectrum.

5) *Long Term Evolution* (LTE) is a modulation technique designed for GSM/UMTS-based technology that uses OFDM and MIMO. It's being developed by the Third Generation Partnership Project (3GPP) and is said to support 45M to 144Mbps in test networks today.

6) *Software Defined Radio* (SDR). sometimes shortened to software radio (SR). refers to wireless communication in which the transmitter modulation is generated or defined by a computer, and the receiver uses a computer to recover the signal intelligence. To select the desired modulation type, the proper programs must be run by microcomputers that control the transmitter and receiver (Bedell, 2005). Moreover, SDR is one form of Open Wireless Architecture (OWA). Since 4G is the collection of wireless standards, the final form of the 4G device will constitute all standards. This can be realized using SDR technology. Software-Defined Radio (SDR) is a radio communication technology that is based on software defined wireless communication protocols instead of hardwired implementations: frequency band, air interface protocol and functionality can be upgraded with software download instead of a complete hardware replacement. An SDR is capable of being re-programmed or reconfigured to operate with different waveform & protocols through dynamic loading of new waveforms and protocols. These waveforms and protocols can contain a number of different parts, including modulation techniques, security and performance characteristics defined in software as part of the waveform itself. Public safety radios, as well as commercial wireless applications, can use SDR for flexibility, and field upgradeability of their products. The ultimate goal of SDR technology is to provide a single radio transceiver capable of playing the roles of cordless telephone, cell phone, wireless fax, wireless e-mail system, pager, wireless videoconferencing unit, wireless Web browser, Global Positioning System (GPS) unit, and other functions still in the realm of science fiction, operable from any location on the surface of the earth, and perhaps in space as well. It is now possible that the future commercial viability of 3G and 4G wireless networks will depend upon capacity enhancing algorithms such as smart antennas and multi-user detection, and that these are prime candidates for implementation by SDR.

7) *All IP-based core networks:* 4G will resemble a convergence of existing technologies rather than an entirely new standard. An all IP based 4G wireless network has many advantages as it will be compatible as well as independent of actual radio access technology. IP radio protocols can be designed a core-network that give complete flexibility in the access network type, like 802.11, WCDMA, Bluetooth, hyper LAN and new CDMA protocols. Considering the cost of the equipments for 4G, IP based wireless networks is one-fourth to one-tenth of the equipments for 2G and 3G wireless infrastructure. There will be reduction in the cost by using interoperability in equipment for service provider.

Figure 6. Mesh network

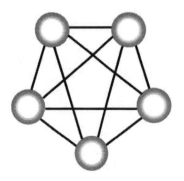

An IP Wireless network would completely replace the old SS7 signaling system because access signal transmission consumes a large part of network bandwidth even when there is no signaling traffic. IP networks use less bandwidth expensive mechanisms to achieve reliability (Sharma, 2002).

8) *Full adopted multi-layer protocol architecture:* The four major factors in achieving the high degree of integration, flexibility and efficiency envisioned in 4G are seamless integration, a high performance physical layer, flexible and adaptive multiple access, and high service and application adaptation.

9) *Mesh Network* is a local area network (LAN) that employs one of two connection arrangements, full mesh topology or partial mesh topology. In the full mesh topology, each node (workstation or other device) is connected directly to each of the others (Govil, 2008). In the partial mesh topology, some nodes are connected to all the others, but some of the nodes are connected only to those other nodes with which they exchange the most data. But above definition mention no dependency on any time parameter -- nothing is necessarily dynamic in a mesh. However, in recent years, and in connection with wireless networks, the term "mesh" is often used as a synonym for "ad hoc" or "mobile" network. Obviously, combining the two characteristics of a mesh topology and ad hoc capabilities is a very attractive proposition.

The figure 6 shows a full mesh network with five nodes. Each node is shown as a sphere, and connections are shown as straight lines. The connections can be wired or wireless.

STANDARDIZATION MOVES

Before elaborating on standardization, let us study some of the existing wireless standards which are summarized in table 6.

The two groups within the International Telecommunication Union (ITU) are specifically engaged to define the next generation of mobile wireless. These two groups are:

• Working Party 8F (WP8F) in section ITU-R.
• Special Study Group (SSG) "IMT 2000 and Beyond" in section ITU-T.

WP8F is focused on the overall radio-system aspects of 4G, such as radio interfaces, Radio-Access Networks (RANs) spectrum issues, service and traffic characteristics, and market estimations as shown in figure 7. The SSG "IMT – 2000 and Beyond" is primarily responsible for the network or wireless aspects of future wireless systems including wireless Internet, convergence of mobile and fixed networks, mobility management, internetworking and interoperability. (Toshio et al., 2005).

The main deliverable of WP8F is Recommendation ITU-R M 1645. This recommendation contains the overall goals for the future

Table 6. The existing wireless standards

1G	NMT, AMPS, Hicap, CDPD, Mobitex, DataTAC
2G	GSM, iDEN, D-AMPS, IS-95/cdmaOne. PDC, CSD, PHS, 2.5G - GPRS, HSCSD, WiDEN, 2.75G - CDMA2000, 1xRTT/IS-2000, EDGE (EGPRS)
3G	W-CDMA, UMTS (3GSM). FOMA, 1xEV-DO/IS-856, TD-SCDMA, GAN/UMA, 3.5G - HSDPA, 3.75G – HSUPA
4G	WiMax, WiBro

Figure 7. Structure of WP 8F

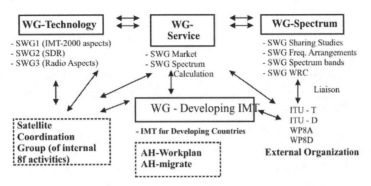

development of wireless communications. The list of the suggestions that are contained in the recommendation are:(i) the framework for 4G systems should fuse elements of current cellular systems with nomadic wireless-access systems and personal-area networks in a seamless layered architecture that is transparent to the user; (ii) data rates of 100 Mbps for mobile applications and 1 Gbps for nomadic applications should be achievable by the year 2010; (iii) Worldwide common spectrum and open, global standardization should be pursued.

Not only above, but the process of developing a standard is a long one, carried out by other several groups, which include Standards Development Organizations (SDOs), industry forums, and companies, such as OEMs, that have a stake in the end product. Some of the major SDOs are nonprofit regional or governmental bodies, such as ETSI (European Telecommunications Standards Institute) in Europe, CCSA (China Communications Standards Association) in China, and the TTA (Telecommunications Technology Association) in Korea. 3GPP and 3GPP2 are examples of industry SDOs (Service Data Objects) that develop and maintain standards for current 2G and 3G technologies. In April 2007, the ITU convened a global congress to set a course for the 4G standards development process. China has expressed interest to submit the standard in 2010. Presently it is doubtful that standard will emerge 2012. Nor are standards necessarily the final word on the subject. In the meantime, there is nothing to stop the various SDOs and wireless operators from deploying so-called 4G systems without waiting for the completion of the formal standards process.

The World Radio Communication Conference (WRC) in October/November 2007 at Switzerland discussed on the spectrum assignment for 4G. The road map of the ITU-R (International Telecommunication Union Radio communication Sector) targets the availability of 4G standard proposals for the year 2012. As soon as frequency bands for 4G are defined, 4G standardization activities are expected to start.

CHALLENGES TO 4G VISION

Simplicity and security are the main challenges for the consumers of wireless devices. All consumers have different needs but they all want simplicity i.e. use of their wireless device at any place any time. Corporate houses also want to ensure that their employees are accessing confidential data in a secure way. 4G technology will require massive innovation and substantial investment. It requires multiple changes in technology and network infrastructure, in handsets and software. A huge amount of innovation is necessary to deliver this kind of promise. As discussed the 4G network will be fully Internet protocol (IP) wireless infrastructure and thus this will give the ability to exploit the capabilities of Internet. The major challenges in realizing the 4G visions are:

1) *Power consumption:* Multiple processing and communication elements will increase the current drain. Managing power use in the infrastructure will play a vital role in achieving 4G technology. Additional hardware acceleration technology is required to manage power. OFDM based technology is crucial to manage some of the process streams and power challenges in such kinds of applications and devices. Some form of management of current drain and reduced battery drain is required for better battery life.

2) *Spectrum efficiency:* More spectrum is required to be made available. For this we have to refurbish existing spectrum. Use of cognitive radio can improve the spectral efficiency. Even with these steps, the 4G radio access network will need to provide significantly better spectral efficiency.

3) *Cost:* This is not only about infrastructure, operating or handset costs but also the cost of deployed services too. To deliver the spectral efficiency, the coverage is required. And for coverage there will be a dramatic growth in the number of base stations. Three times more base stations will be required to deliver a ten fold increase in data rate. Deployment of advanced antenna techniques such as MIMO and Space-Time Coding (STC) can improve spectrum efficiency, as well as reduction in growth rate of base stations; thus the capital cost will be reduced. On the handset side, there are significant challenges in continuing to drive down the cost of integrating greater and greater processing capability in multi mode RF technology.

4) *Miniaturization and Processing Challenges:* Miniaturization challenges include power reduction, cost, size and product development cycle. Multimode technology in 4G means availability of different types of radio access technologies in a seamless way. There are significant software, billing, carrier interoperability and enterprise carrier interoperability challenges. For bulk CMOS (Complementary Metal Oxide Semiconductor) process technologies, various companies are trying to provide higher performance at lower power as the semiconductor industry migrates from 90nm to 65nm to 45nm to 32nm technology. It is expected that in next five to ten years the evolving bulk CMOS technologies will completely resolved these complications. RF CMOS are continuing to scale for high-speed high-resolution A/D converters, and isolation techniques are being developed in the industry for multiple radio operation and single-chip die integration.

5) *Advanced Architectures for 4G:* In WCDMA, there are two variables –coding and time. With OFDM-based 4G, frequency, space and time are being used as variables to extract the data. With OFDM, information is separated into small sub-bands, and the information in each of these bands can be signal-processed independently in a parallel fashion. As consumers expect continuous service, so parallel processing is being increased so that the net result is that a high-speed, low-power drain OFDM engine architecture which can be developed to support 4G data throughput requirements. OFDM is typically discussed as being a WiMAX technology. WiMAX is an important technology as base business model behind broadband wireless access.

6) *Business Challenges:* 4G is not going to be driven by a single entity or organisation. It will require a tremendous number of partnerships and a robust ecosystem which can exploit the capabilities that are available in wireless technologies. Multiple standards bodies, corporations and government entities are required to come together

to drive standards-based interoperability and the opportunity to deliver 4G networks. Governments will have to manage the spectrum in different parts of the world, and this will have a dramatic impact on how we can exploit the capabilities available to us in wireless technologies. To enable triple-play merged services delivered over wired and wireless networks, the issue of equipment has also to be taken into account for affordability, simplicity, interoperability and reliability.

7) *Miscellaneous challenges:* As we move from circuit-switched networks to IP networks, some of the challenges include packet acceleration, traffic management, data integrity, security and quality of service, which all represent different challenges for us on the infrastructure side compared to the traditional network infrastructure. This notion of high performance in a very constrained power envelope continues to be a significant challenge open standards continue to be another issue.

8) *Challenges in IP Network Security and Traffic Control:* Wireline internet access is increasingly being challenged to improve security. Security has multiple elements, much more than just moving encrypted traffic at faster and faster rates across the network. Security is also about denial of service attacks and digital rights management. These are all becoming carrier problems. Improved security and quality of service require providers to be able to identify video packets and prioritize them so that viewers get an uninterrupted stream of video content, if they are watching a movie or TV show on demand. These security / quality of service capabilities are the key elements of managing the network in both wireless and wireline infrastructure. From the processor perspective, the network processor has been exceptionally efficient at driving performance at layers 2 and 3, but suffers a significant drop-off in performance in layer 4–7 protocols. A general-purpose microprocessor does not deliver the kind of performance networks require at layer 2, and does little better at layers 4–7. Today's communications processors use hardware acceleration techniques to achieve better performance in the lower levels and do slightly better in the higher levels. But there is still a significant content processing gap in terms of microprocessor, network processor and communications processor technologies. Clearly one of the requirements is the ability to provide a solution that does not just forward headers and IP packets. We need to inspect those packets; connect those packets and carry out stream processing instead of packet processing. In terms of base station size and cost constraints, there is a trend towards more base stations covering smaller areas while managing multiple power output limits, frequencies and standards.

9) *Financial Challenges:* There are various financial challenges in reducing the cost of power amplifier in the base station which manifests as a largest expense. Other financial challenges are cost of components, software or services providers or carriers; significant investment is required to enable the next generation 4G network (Perkins, 2005).

Key Challenges for 4G

4G must be dynamic and adaptable with built-in intelligence. Key challenges will be personalization, seamless access, and quality of service, intelligent billing. Table 7 summarizes the major key challenges in migration to 4G (Yang, 2007; Govil & Govil, 2007).

Disadvantages of 4G

Though 4G comes with many advantages but has many disadvantages too, some of which are summarized here:

Table 7. Key challenges for 4G

4G Definition	Consensus on the 4G definition is needed for the purpose of the standardization.
Seamless Connectivity	Inter-network and intra-network connectivity is fundamental to the provision of temporally and spatially seamless services. Vertical and horizontal handovers are critical for 4G. In the former case, the heterogeneity and variety of networks exacerbate the problem.
Latency	Many 4G services are delay sensitive. Guaranteeing short delays in networks with different access architecture and coverage is far from straightforward.
New Access Architectures	More study is required to replace the non conventional access architectures to conventional ones.
Concealing Complexity	The complexity of 4G network needs to be hidden from the user.
Spectrum Issues	As the spectrum for 4G has not yet been allotted hence it is difficult to design a wireless system without knowing the channel.
Complex Resource Allocation	Management of time, frequency and spatial resources in a multi-network, multi-user environment is far from trivial.
Interference	Multiple access interference control and mitigation in heterogeneous environments (coexisting air interfaces, varied terminals and services) is an issue.
Power Consumption	Power consumption in future multi-function multi-standard 4G terminals will sharply increase. Usability is seriously compromised; hence heat management becomes an issue.
Personalization	In short required provisioning for advanced signaling & session control, AAA (authentication, authorization, accounting); open third party access (e.g. web services); communication (protocols); reconfigurable terminals; new strategies for pervasive/ubiquitous computing; programmable open platforms.
Seamless Access	In short, requirements are seamless network integration based on IP, terminal mobility, personal mobility, service mobility, session mobility; dynamic resource allocation at all network/system levels, high adaptability/programmability of network components good security but simple service agreements .
Quality of Service	This encompasses the customer perception of QoS, the offered QoS and the QoS actually delivered. QoS modeling and signaling would be crucial factors for a system that integrates heterogeneous network types.
Intelligent Billing	(i) User requirement - QoS dependent charging; billing support to diverse access; support to real time billing information; support to interworking of prepaid systems; support to "per-call" service situations. (ii) Operator requirement - billing support to IP traffic; flexibility of costs calculations (time, volume, QoS dependent, access dependent); distribution of revenue by value chain operators; customer relationship management; reliability of billing operations; instant fraud detection and cut-off.
Security	A good security system needs to be designed, which needs co-operation amongst: Government regulator, Network infrastructure provider, Wireless service provider, Wireless equipment provider, Wireless user.
Cost	Cost of terminal and services are required to be kept low for practical mass implementation.

1) *Crowding of Bandwidth:* Shared free bandwidth may get too crowded.
2) *Security Issue:* Billing authentication and security will be an issue as seen today.
3) *Mobility Complexity*: 4G requires high mobility amongst various technologies: mobility requires a level of complexity that makes organic growth difficult.
4) *Capital Conservation:* 4G is seen as capital conservation tool with a part in global realignment of telecom assets. Carriers cannot afford to add capacity. They need easy to buy, easy to install, easy to maintain equipment that will rapidly expand their coverage (Francis, 2003).

Mobility Management Issues in 4G Networks

Mobility is a critical aspect of 4G. There are three main issues regarding mobility management in 4G networks:

1) *Choice of Access Technology i.e. how to be the best connected*: Given that a user may be offered connectivity from more than one technology at any one time, one has to consider how the terminal and an overlay network choose the radio access technology suitable for services the user is accessing as depicted in figure 8 (Frederic et al., 2002).

There are several network technologies available today, which can be viewed as complementary (Table 8). For example, WLAN is best suited for high data rate indoor coverage. GPRS or UMTS, on the other hand, are best suited for nation wide coverage and can be regarded as wide area networks, providing a higher degree of mobility. Thus a user of the mobile terminal or the network needs to make the optimal choice of radio access technology among those available. A handover algorithm should both determine which network to connect to as well as when to perform a handover between the different networks. Ideally, the handover algorithm would assure that the best overall wireless link is chosen. The network selection strategy should take into consideration the type of application being run by the user at the time of handover. This ensures stability as well as optimal bandwidth for interactive and background services (Mary, 2006).

2) *Mobility Between Access Technologies*: This is regarding the design of mobility enabled IP networking architecture. This includes fast, seamless vertical (between heterogeneous technologies) handovers (IP micro-mobility). quality of service, security and accounting.

Real-time applications in the future will require fast/seamless handovers for smooth operation.
Mobility in IPv6 is not optimized to take advantage of specific mechanisms that may be deployed in different administrative domains. Instead, IPv6 provides mobility in a manner that resembles only simple portability.
To enhance Mobility in IPv6, 'micro-mobility' protocols (such as Hawaii, Cellular IP and Hierarchical Mobile IPv6) have been developed for seamless handovers i.e. handovers that result in minimal handover delay, minimal packet loss, and minimal loss of communication state.

3) *Adaptation of Multimedia Transmission*: This issues concerns the adaptation of multimedia transmission the across 4G networks. Indeed multimedia will be a main service feature of 4G networks, and changing radio access networks may in particular result in drastic changes in the network condition. Thus the framework for multimedia transmission must be adaptive. In cellular networks such as UMTS, users compete for scarce and expensive bandwidth. Variable bit rate services provide a way to ensure service provisioning at lower costs. In addition the radio environment has dynamics that renders it difficult to provide a guaranteed network

Figure 8. Offered connectivity, choosing the access technology

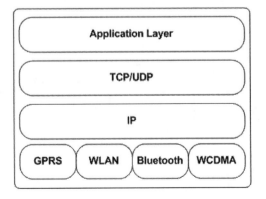

service. This requires that the services are adaptive and robust against varying radio conditions.

High variations in the network Quality of Service (QoS) leads to significant variations of the multimedia quality. The result could sometimes be unacceptable to the users. Avoiding this requires choosing an adaptive encoding framework for multimedia transmission. The network should signal QoS variations to allow the application to be aware in real time of the network conditions. User interactions will help to ensure personalized adaptation of the multimedia presentation.

TECHNOLOGICAL SOLUTIONS FOR 4G

4G Heterogeneous Networks General Architecture

The 4G Mobile communications will be based on the Open Wireless Architecture (OWA) to ensure that the single terminal can seamlessly and automatically connect to the local high-speed wireless access systems when the users are in the offices, homes, airports or shopping centers where the wireless access networks (i.e. Wireless LAN, Broadband Wireless Access, Wireless Local Loop, HomeRF, Wireless ATM, etc) are available. When the users move to the mobile zone, the same terminal can automatically switch to the wireless mobile networks (i.e. GPRS, W-CDMA, cdma2000, TD-SCDMA,etc.) (Willie, 2006).

The advantages of this converged wireless communications are:

1) *Spectrum efficiency* is greatly increased.
2) *Highest Data rate* to the wireless is mostly ensured.
3) *Best sharing* of the network resources and channel terminal utilization.
4) *Optimally manage* the service quality and multimedia applications.

The modules within the architectural framework should be able to incorporate the following high-level mobility issues (Figure 9):

- *Users*: This focuses on the movement of user, and allows user access to his/her home network while on the move, which involve the provision of personal communication.
- *Terminals*: This allows the provision of services at any time and anywhere. Terminal mobility allows mobile clients to roam across geographic boundaries of

Table 8. Current and emerging radio access technologies

Network	Coverage	Data Rates	Mobility	Cost
Satellite (B-GAN)	World	Max. 144 kb/s	High	High
GSM/GPRS	Aprox. 35 Km	9.6 kb/s up to 144 kb/s	High	High
IEEE 802.16a	Aprox. 30 Km	Max. 70 Mb/s	Low/Medium	Medium
IEEE 802.20	Aprox. 20 Km	1-9 Mb/s	Very high	High
UMTS	20 Km	up to 2 Mb/s	High	High
HIPERLAN 2	70 up to 300 m	25 Mb/s	Medium/high	Low
IEEE 802.11a	50 up to 300 m	54 Mb/s	Medium/high	Low
IEEE 802.11b	50 up to 300 m	11 Mb/s	Medium/high	Low
Bluetooth	10 m	Max. 700 kb/s	Very low	Low

Figure 9. Mobility dimensions

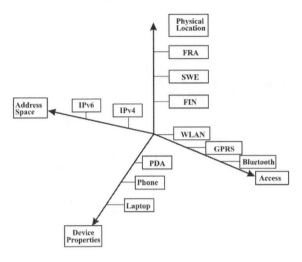

wireless networks. The greatest challenge in providing terminal mobility within a 4G Infrastructure is to locate and update the locations of the terminals in various systems.
- *Networks*: Network mobility is the ability of the network to support roaming of an entire subnet work, structured or ad hoc.
- *Applications*: Mobile application should refer to a user's profile so that it can be delivered in a way most preferred by the subscriber, such as context based personalized services.

The incorporation of new functions into existing mobility protocols and mechanisms does not appropriately solve the demands of future communication scenarios. Therefore a new *'Mobility Architecture'* needs to be defined, based on the following principles: Diversity, Harmonization among layers, Legacy Awareness, Concept of mobile entities and Naming and name management.

With the acceleration of technology, there is competition among commercial houses to create significant growth and they are looking for various technological solutions for Fourth Generation Mobile Communication Systems (Gazis et al., 2003). Some technical solutions are described here in brief in table 9.

Mobility Management in 4G

The 4G mobility management includes mobility related features, absent in previous generation networks, such as: Moving Networks, Seamless Roaming and Vertical Handover.

Mobility Management Operations

The operation of mobility management is divided into two related parts, location management and handoff management (Akyildiz et al., 1999).

Location Management

Location management involves two operations; location registration and call delivery as shown in figure 10. Location registration involves the mobile terminal periodically updating the network about its new location (access point). This allows the network to keep a track of the mobile terminal. In the second operation the network is queried for the user location profile and the current position of the mobile host is retrieved. Current techniques for location management involve database architecture design and the transmission of signaling messages between various components of a signaling network. Since location management deals with database and signaling issues, many of the issues are not protocol dependent and can be applied to various networks such as PLMN (Public Land Mobile Network) based networks, PSTN (Public Switched Telephone Network). ISDN (Integrated Services Digital Network). IP, Frame Relay, X.25, or ATM (Asynchronous Transfer Mode) Networks depending on the requirements.

Some key research issues for location management include:

- *Addressing*, i.e. how to represent and assign address information to mobile nodes. The problem is becoming more severe

Table 9. Technological solutions for 4G

Adaptable Capability – Aware Service Provision	Different wireless access networks differ significantly in terms of coverage area and supported bandwidth, mobile network, their capabilities should be considered so as to refine the list of applicable services.
Transparent Mobility and Universal Roaming Capability	Seamless user mobility across different wireless access technologies (e.g. WLAN, UMTS etc.) with minimal or zero user intervention must be supported by efficient inter-system mobility management and hand over procedures. Roaming should build on cross industries standard protocols and architecture, such as hierarchical Mobile IPv6. As different systems may entail different charges, it should include QoS and pricing information as part of mobility management signaling.
Automated Protocol Configuration Mechanisms	The multiple options capable of accommodated the same set of services will result in accruing different charges in 4G mobile environments, thus users to be informed regarding the pricing preferences
Policy Based Management and Information Models	Policy based management demarcates between enforcer entities and decision entities in the infrastructure which results in realization of flexible management architecture that spans across multiple administrative domains. Policy protocols also support both outsourcing and provisioning modes of operation, making policy based management an ideal approach for 4G mobile environments .
Flexible Pricing and Billing Mechanisms	Network related pricing models must be completely independent from service related ones, with regarding to formulation as well as application matters.
Application and Mobile Execution Environment Aspects	As million mobile terminals and different manufacturers with different characteristics and applications will use 4G environment so there is a need to develop universal hardware platform with hassle free application with interpreted languages. The independent service provider will be relieved from the burden of developing, supporting and maintaining multiple versions of their applications for each possible client.

since the 4G mobile communication systems will be based on the internetworking and interoperability of diverse and heterogeneous networks of different operators and/or technologies. A global addressing scheme is needed, e.g. IPv6 address, to locate the roaming nodes.

- *Database Structure*, i.e. how to organize the storage and distribution of the location information of mobile nodes. Database structure can be either centralized or distributed, or the hybrid of these two schemes. Tradeoff is needed between access speed, storage overhead, and traffic overhead due to the access to the related databases. Caching is also an important technique for the improvement of access performance.
- *Location Update Time*, i.e. when a mobile node should update its location information by renewing its entries in corresponding databases. Schemes for location update can be either static or dynamic. In a static scheme location update is triggered

Figure 10. Location management operations

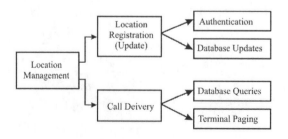

Fourth Generation Networks

by some fixed conditions like time period or network topology change. A dynamic scheme is more personalized and adaptive, and based on some situations such as counter, distance, timer, personal profile, or even predicted factors.

- *Paging Scheme*, i.e. how to determine the exact location of a mobile node within a limited time. Obviously an adequate tradeoff is needed between time overhead and bandwidth overhead. There are also both static and dynamic schemes for location paging. In static cases paging is simply done to the whole certain area where the mobile node must be in. For a dynamic method, the main problem is to firstly organize the paging areas into groups and then recognize the best sequence of the separated areas for paging, based on information like distance, probability, moving velocity, etc.

Handoff Management

Handoff management equals controlling the change of a mobile node's attachment point to a network in order to maintain connection with the moving node during active data transmission.

Operations of handoff management include (Figure 11):

- *Handoff Triggering*, i.e. to initiate handoff process according to some conditions. Possible conditions may include e.g. signal strength deterioration, workload overload, bandwidth decrease or insufficiency, new better connection available, cost and quality tradeoff, flow stream characteristic, network topology change, etc. Triggering may even happen according to a user's explicit control or heuristic advice from local monitor software.

- *Connection Re-establishing*, i.e. the process to generate new connection between the mobile node and the new attachment point and/or link channel. The main task of the operation relates to the discovery and assignment of new connection resource. This behaviour may be based on either network-active or mobile-active procedure, depending on which is needed to find the new resource essential to the new establishment of connection.

- *Packet Routing*, i.e. to change the delivering route of the succeeding data to the new connection path after the new connection has been successfully established.

Wireless networks vary widely in both service capabilities and technological aspects, so no single wireless network technology can fulfill the different requirements on latency, coverage, data rate, and cost. An efficient strategy is necessary for the management of such a wireless overlay architecture and mobility within the framework. In

Figure 11. Handoff of management operations

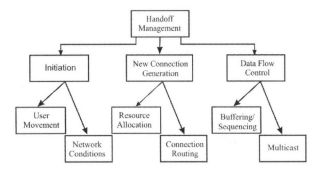

homogeneous environments, traditional horizontal handoff can be employed for intra-technology mobility. In heterogeneous environments, vertical handoff should be used for inter-technology mobility. Vertical handoff may be occur either upward (i.e. to a larger cell size and lower bandwidth) or downward (i.e. to a smaller cell size and higher bandwidth); and the mobile device does not necessarily move out of the coverage area of the original cell. Some packet-level QoS parameters become more important to real-time multimedia services, including packet latency, packet loss rate, throughput, signalling bandwidth overhead, and device power consumption.

Besides the basic functions that implement the goal of handoff management, there are many other requirements on performance and packet-level QoS that should be carefully taken into account when trying to design or select a handoff management scheme, including

- *Fast Handoff*, i.e. the handoff operations should be quick enough in order to ensure that the mobile node can receive data packets at its new location within a reasonable time interval and so reduce the packet delay as much as possible. This is extremely important to real-time services.
- *Seamless Handoff*, i.e. the handoff algorithm should minimize the packet loss rate into zero or near zero. Fast handoff and seamless handoff together are sometimes referred to as smooth handoff. While the former concerns mainly packet delay, the latter focuses more on packet loss.
- *Routing Efficiency*, i.e. the routing path between corresponding node and mobile node should be optimized in order to exclude possible redundant transfer or bypass path as triangle routing. Some distinct but complementary techniques exist for handoff management to achieve its performance and QoS requirements above, including:

- *Buffering and Forwarding*, i.e. the old attachment point can cache packets during the MN in handoff procedure, and then forward to the new attachment point after the operation of connection re-establishing of mobile node's handoff.
- *Movement Detection and Prediction*, i.e. mobile node's movement between different access points can be detected and predicted so that the next network that will soon be visited is able to prepare in advance and packets can even be delivered there before and/or during handoff simultaneously to the old attachment point.
- *Handoff Control*, i.e. to adopt different mechanisms for the handoff control. Typical examples include e.g. layer two or layer three triggered handoff, hard or soft handoff, mobile-controlled or network-controlled handoff, etc.
- *Domain-Based Mobility Management*, i.e. to divide the mobility into intra-domain mobility and inter-domain mobility according to whether the mobile host's movement happens within one domain or between different domains (Zhao & Sauvola, 2002).

Enhancements to IPv6 Mobility Management Protocols Required by 4G Networks

Although the features mentioned are suited for 4G networks, recently, there has been almost universal recognition that IPv6 needs to be enhanced to meet the need for future 4G cellular environments. In particular, the absence of a location management hierarchy (IPv6 uses only simple location updates for location management) leads to concerns about the signalling scalability and handoff latency. This is especially significant when we consider that 4G aims at providing mobility support to potentially billions of mobile devices, within the stringent performance bounds associated with real time multimedia traffic.

There are three main areas where IPv6 needs to be enhanced before being used as the core networking protocol in 4G networks:

1) *Paging Support:* The base IPv6 specification does not provide any form of paging support. Hence to maintain connectivity with the backbone infrastructure, the mobile node needs to generate location updates every time it changes its point of attachment, even if it is currently in dormant or standby mode. Excessive signaling caused by frequent motion leads to a significant wastage of the mobile node's battery power, especially in environments with smaller cell areas (such as 802.11 based cellular topologies). It is thus impractical to rely completely on location updates and is essential to define some sort of flexible paging support in the intra-domain mobility management scheme.

2) *Scalability:* IPv6 allows nodes to move within the Internet topology while maintaining reachability and on-going connections between mobile and correspondent nodes. To do this a mobile node (MN) sends Binding Updates (BUs) to its Home Agent (HA) and all Correspondent Nodes (CNs) it communicates with, every time it moves. Authenticating binding updates requires approximately 1.5 round trip times between the mobile node and each correspondent node (for the entire return routability procedure in a best case scenario, i.e. no packet losses). In addition one round trip time is needed to update the HA; this can be done simultaneously while updating CNs. These round trip delays will disrupt active connections every time a handoff to a new radio access technology is performed. Elimination of this additional delay element from the time-critical handover period will significantly improve the performance of IPv6. Moreover, in the case of wireless links, such a solution reduces the number of messages sent over the air interface to all CNs and the HA. A local anchor point will allow Mobile IPv6 to benefit from reduced mobility signaling with external networks. For these reasons a new Mobile IPv6 node, called the Mobility Anchor Point (MAP) is being suggested, that can be located at any level in a hierarchical network of routers. Unlike Foreign Agents in IPv4, a MAP is not required on each subnet. The MAP will limit the amount of Mobile IPv6 signaling outside the local domain. The introduction of the MAP provides a solution to the aforementioned problems in the following way:

a) The MN sends binding updates to the local MAP rather than the HA (which is typically further away) and CNs.

b) Only one binding update message needs to be transmitted by the MN before traffic from the HA and all CNs is re-routed to its new location. This is independent of the number of CNs that the MN is communicating with (Soliman et al., 2008). Thus by decreasing signaling traffic by having an intermediate level in the hierarchy helps accommodate a larger number of mobile nodes in the system.

3) *Technologies:* A mobile node switches from one network to another network in one of two cases: (a) when the signal from the network it is currently in starts to become weak or (b) when the mobile host detects another network which is better suited to its application compared to its current network.

The decision of the mobile device on the suitability of the network can be based on signal strength, network bandwidth or certain policies which the user might have stored in his profile based on which subsequent switching between networks of different access technologies may occur. For example, when a user is streaming a video, he/she may use WLAN and when he/she is listening to highly compressed audio, she might switch to GPRS (Ibrahim, 2002).

Further, another issue that needs to be resolved is that of informing the source (HA/CN) when the MN has moved. In such a situation, the MN does a location update to its HA, which then takes charge of sending IP datagrams to the MN's new location using standard Mobile-IP mechanisms.

In line with the 4G vision of bringing together wide-area networks and local-area packet-based technologies, mobile terminals are being designed with multiple physical or software-defined interfaces. This is expected to allow users to seamlessly switch between different access technologies, often, with overlapping areas of coverage and dramatically different cell sizes. Mobility management protocols should then be capable of handling vertical handoffs.

MOBILE/WIRELESS GRIDS FOR 4G MOBILE COMMUNICATION SYSTEMS

Mobile Grids

Mobile computing is an aspect that plays seminal role in the implementation of 4G Mobile Communication Systems since it primarily centers upon the requirement of providing access to various communications and services every where, any time and by any available means. Presently, the technical solutions for achieving mobile computing are hard to implement since they require the creation of communication infrastructures and the modification of operating systems, application programs and computer networks on account of limitations on the capability of a moving resource in contrast to a fixed one.

In the purview of Grid and Mobile Computing, Mobile Grid is a heir of Grid, that addresses mobility issues, with the added elements of supporting mobile users and resources in a seamless, transparent, secure and efficient way. It has the facility to organize underlying ad-hoc networks and offer a self-configuring Grid system of mobile resources (hosts and users) connected by wireless links and forming random and changeable topologies.

The mobile Grid needs to be upgraded from general Grid concept to make full use of all the capabilities that will be available; these functionalities will involve end-to-end solutions with emphasis on Quality of Service (QoS) and security, as well as interoperability issues between the diverse technologies involved. Further, enhanced security policies and approaches to address large scale and heterogeneous environments will be needed. Additionally, the volatile, mobile and poor networked environments have to be addressed with adaptable QoS aspects which have to be contextualized with respect to users and their profiles.

Wireless Grids

Grid computing lets devices connected to the Internet, overlay peer-to-peer networks, and the nascent wired computational grid dynamically share network connected resources in 4G kind of scenario. The wireless grid extends this sharing potential to mobile, nomadic, or fixed-location devices temporarily connected via ad hoc wireless networks. As fig. 12 shows, users and devices can come and go in a dynamic wireless grid, interacting with a changing landscape of information resources. Following Metcalfe's law, grid-based resources become more valuable as the number of devices and users increases. The wireless grid makes it easier to extend grid computing to large numbers of devices that would otherwise be unable to participate and share resources. While grid computing attracts much research, resource sharing across small, ad hoc, mobile, and nomadic grids draws much less (McKnight et al., 2004).

Wireless grids, a new type of resource-sharing network, connect sensors, mobile phones, and other edge devices with each other and with wired grids (Figure 12). Ad hoc distributed resource sharing allows these devices to offer new resources

Figure 12. Dynamic and fixed wireless grids

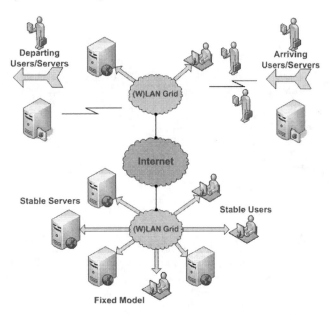

and locations of use for grid computing. In some ways, wireless grids resemble networks already found in connection with agricultural, military, transportation, air-quality, environmental, health, emergency, and security systems.

A range of institutions, from the largest governments to very small enterprises, will own and at least partially control wireless grids. To make things still more complex for researchers and business strategists, users and producers could sometimes be one and the same. Devices on the wireless grid will be not only mobile but nomadic — shifting across institutional boundaries. Just as real-world nomads cross institutional boundaries and frequently move from one location to another, so do wireless devices. The following classification offers one way to classify wireless grid applications.

- *Class 1*: Applications aggregating information from the range of input/output interfaces found in nomadic devices.
- *Class 2*: Applications leveraging the locations and contexts in which the devices exist.
- *Class 3*: Applications leveraging the mesh network capabilities of groups of nomadic devices.

The three classes of wireless grid applications conceptualized here are not mutually exclusive. Understanding more about the shareable resources, the places of use, and ownership and control patterns within which wireless grids will operate might assist us in visualizing these future patterns of wireless grid use.

The Grid, is a promising emerging technology that enables the simple "connect and share" approach analogously to the internet search engines that apply the "connect and acquire information" concept. Thus, mobile/wireless grids are an ideal solution for large scale applications which are the pith of 4G mobile communication systems. Besides, this grid-based-approach will potentially increase the performance of the involved applications and utilization rate of resources by employing efficient mechanisms for resource management in the majority of its resources, that is, by allowing the seamless integration of resources, data, services and ontologies. Figure 13 places wireless

Figure 13. Wireless grid issues and standard chart demonstrating their complex needs

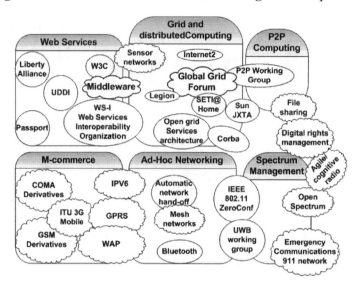

grids in context, illustrating how they span the technical approaches and issues of Web services, grid computing, P2P systems, mobile commerce, ad hoc networking, and spectrum management (McKnight, 2004). How sensor and mesh networks will ultimately interact with software radio and other technologies to solve wireless grid problems requires a great deal of further research.

CURRENT RESEARCH FOR MOBILITY MANAGEMENT IN 4G

There is considerable research being done on mobility management. Mobility solutions can be found by either developing improvements within the current architecture, or by revising the architecture to reflect the changing environment. The solutions propose different addressing and packet forwarding schemes. Almost all of them are IP based solutions, which allow interoperability and easy integration with the existing architectures. Within each of the solutions the relevance to mobility and their strengths and limitations, are discussed in brief:

1) *The Internet Indirection Infrastructure* is a scalable, self-organizing scheme which easily integrates with legacy systems. It proposes an architecture that offers a communication abstraction based on rendezvous points in an overlay network. When a host wants to send a packet, it forwards the packet to one of the servers it knows. A packet keeps traversing the network till the target server is reached; this leads to delay in route discovery and packet forwarding (Typpo et al., 2005).

2) *FARA* (Forwarding directive, Association, and Rendezvous Architecture) (Clark et al., 2003) is an ongoing project whose main purpose is to provide mobility by separating location from identity. One advantage is that neither an entirely new namespace nor a globally unique one is required for the entities. It allows several different forwarding mechanisms to co-exist in the network, resulting in variability in characteristics like mobility, identity, and anonymity. However, FARA model fails to take into consideration many packet forwarding issues like performance of network nodes, or the balance of

anonymity vs. identity for communicating endnodes. It does not accommodate for security either.

3) *Host Identity Payload* (HIP) (Moskowitz & Nikander 2003) provides another way of breaking the binding between identities and topological locations of network nodes. HIP introduces new cryptographic identities that can be dynamically mapped to IP addresses. However, HIP Host Identity (HI). being a public key, is not practical in all actions; it is somewhat long, it needs to be hashed before being used in IPv6 applications. While providing support for mobility and multi-homing with a major architectural change in the addressing concept, the solution requires only small changes in current host implementations.

4) *IST MIND* (Eardley et al., 2002) develops the concepts and protocols generated in *BRAIN* (Urban et al., 2001) by enabling hosts to cooperate with self-organizing wireless ad-hoc networks. It provides independent, interoperable solutions for local/micro-mobility from global mobility.

5) *DRiVE* (Paila et al. 2001) specifies a multi-access architecture allowing for seamless intersystem-handover. The concept of a host-controlled flow control was developed to enable parallel usage of different access systems. The architecture is based on Hierarchical Mobile IP, extended by an AAA (Authentication, Authorization and Accounting) component. *OverDRiVE* (Ronai et al., 2003) extends the scenario with moving networks (e.g. vehicles, trains, etc.) in a multiradio/multi-access environment, defines a Mobile IP-based solution, and focuses on multicast support. The project has strong influence on the ongoing work within the IETF *NEMO* (Ronai et al., 2003) (NEtwork MObility) group.

6) *The Architectural Principles of Ambient Networks* (Typpö et al., 2004; Selander, 2004; Kappler, 2004) require the integration of a multitude of different communication environments, rather than suffering from heterogeneity. The approach is to use *network composition* as the principle instead of terminals; networks as such can form the basic building block of the communication architecture. Network composition is a more powerful concept than the simple internetworking as enabled by the Internet Protocol. The current Internet assumes homogeneity in the environment in which to provide control. Ambient Networks have the potential to solve this issue of fragmented control.

7) *Developing Standards for Seamless & Secure Mobility* (Golmie, 2005): Several industry consortia and standard development organizations such as the IEEE 802 LAN/MAN Standards Committee and the Internet Engineering Task Force (IETF) are expending considerable efforts to develop a common framework and extend existing mobility protocols in order to facilitate and optimize handover performance. Various activities are currently under way, including extensions to Mobile IP at the IETF, and the formation of the *Media Independent Handover* (MIH) working group in IEEE 802, in addition to several task groups within IEEE 802.11 in order to deal with roaming (IEEE 802.11r) and interfacing to external networks (IEEE 802.11u) (Vassilis et al., 2005).

8) *Interference Alignment Techniques for Wireless Interference Channels:* The project is going at Samsung Advanced Institute of Technology, Korea from Nov. 2008 onwards.

9) *Transmission Techniques for Multiple MIMO Relay Channels:* This project is being developed at LG Electronics, Korea from Aug. 2008.

10) *Physical Layer Design for High-speed wireless Systems (9th):* The project is going

at Interstate Technology & Regulatory Council.

The best solution among the current and on-going projects will be the one that successfully addresses all the related challenges as well as allows scalability for future possibilities. A few open issues, however, need to be addressed in most of the existing projects; i.e. synchronization of the entire network and sound QoS.

SECURITY ISSUES IN 4G AND WIRELESS SECURITY

As discussed in earlier sections, 4G is the next generation after 3G. Some of the 4G services mentioned in this chapter are incorporation of quality of service (QoS) and Mobility. There is also a concept of always best connected which means that the terminal will always select the best possible access available. 4G will make use of the IPV6 address scheme. This might make it possible for each cell device to have its own IP address. Currently, the problem of security is solved by using multiple layers of encryption of the protocol stack. There are disadvantages in this scheme such as wasted power, wasted energy and a larger transmission delay. In 4G there will be a concept of interlayer security where only one layer will be configured to do encryption on data.

The infrastructure for 4G Cellular Networks is massive, complex with multiple entities coordinating together, such as the IP Internet coordinating with the core network; therefore it presents a challenge for the network to provide security at every possible communication path.

Before seeking to design and implement wireless security, however, one first need to understand the concept of security. In the present case, wireless security is a combination of wireless channel security (security of the radio transmission) and network security (security of the wired network through which the data flows). These collectively can be referred to as "wireless network security" (Russell 2001). All entities involved in creating the secure wireless system must cooperate which include government regulator, network infrastructure provider, wireless service provider, wireless equipment provider, and wireless user (Russell, 2001).

Important Security Issues in 4G

With the advent of advanced network of 4G, there are several security issues that are deemed important for consideration when deploying a wireless infrastructure like 4G:

1) *Authentication*: wireless networks have a large number of subscribers, and each has to be authenticated to ensure the right people are using the network. Since the purpose of 4G is to enable people to communicate from anywhere in the world, the issue of cross region and cross provider authentication becomes an issue.

2) *Integrity*: With services such as SMS, chat and file transfer it is important that the data arrives without any modifications.

3) *Confidentiality*: With the increased use of cellular phones in sensitive communication, there is a need for a secure channel in order to transmit information.

4) *Access Control*: The Cellular device may have files that need to have restricted access to them. The device might access a database where some sort of role based access control is necessary.

5) *Operating Systems* In Mobile Devices: Cellular Phones have evolved from low processing power, ad-hoc supervisors to high power processors and full fledged operating systems. Some phones may use a Java Based system, others use Microsoft Windows CE and have the same capabilities

as a desktop computer. Issues may arise in the OS which might open security holes that can be exploited.

6) *Web Services:* A Web Service is a component that provides functionality accessible through the web using the standard HTTP Protocol. This opens the cellular device to variety of security issues such as viruses, buffer overflows, denial of service attacks etc. (Fernandez, 2005)

7) *Location Detection*: The actual location of a cellular device needs to be kept hidden for reasons of privacy of the user. With the move to IP based networks, the issue arises that a user may be associated with an access point and therefore their location might be compromised.

8) *Viruses And Malware*: With increased functionality provided in cellular systems, problems prevalent in larger systems such as viruses and malware arise. The first virus that appeared on cellular devices was Liberty. An affected device can also be used to attack the cellular network infrastructure by becoming part of a large scale denial of service attack.

9) *Downloaded Contents*: Spyware or Adware might be downloaded causing security issues. Another problem is that of digital rights management. Users might download unauthorized copies of music, videos, wallpapers and games.

1) *Device Security*: If a device is lost or stolen, it needs to be protected from unauthorized use so that potential sensitive information such as emails, documents, phone numbers etc. cannot be accessed.

Security Requirements for 4G

Security is an important essential requirement for 4G (Zheng et al., 2005):

1) Security requirements on ME (Mobile Equipment)/USIM (Universal Subscriber Identity Module):
   a. Protection of integrity of the hardware, software and OS in mobile platform.
   b. Data control access in ME/USIM.
   c. Maintenance of confidentiality and integrity of data stored in the ME/USIM or transported on the interface between ME and USIM.
   d. User identity privacy retention to ME.

2) Security requirements on radio interface and network operator:
   a. Entity authentication: mutual authentication between user and network shall be implemented to ensure secure service access and provision.
   b. Ensure confidentiality of data including user traffic and signaling data on wired or wireless interface.
   c. Ensure integrity and origin authentication of user traffic, signaling data and control data.
   d. Security of user identity: It shall protect user identity confidentiality, user location confidentiality and user untraceability.
   e. Lawful interception: It shall be possible for law enforcement agencies to monitor and intercept every call in accordance with national laws.

3) Security visibility, configurability and scalability:
   a. Transparency of security features of the visited network to the user.
   b. Ability to negotiate acceptable security lever with the visited network when user roams outside HE (home environment).
   c. Scalability of the security mechanism to support increase of user and/or network elements.

Current Research in Security for 4G

Yoshihiro Ohba has divided security in 4G into Access Network security and Core Network security (Obha, 2007). For the access network security, a peer authentication mechanism across different link-layer technologies can be utilized for roaming. EAP (Extensible Authentication Protocol) is one such example of technology that can be recognized as unified PEA mechanism. For the core access network, security associations need be established between between a mobile and a middle box in the core network for different protocols such as Mobile IPv4/v6, SIP, Mobile IPv4/v6, SIP, IPsec IPsec, 802.21 MIH (Media, 802.21 MIH (Media-Independent Independent Handover) protocol. A single sign-on mechanism based on network access long term credentials may be needed to bootstrap security associations for different protocols.

Yu Zheng et al. have proposed trusted computing based security architecture for 4G networks (2005): The security framework based on Trusted Mobile Platform (TMP) and PKI is mentioned to provide a considerable robust platform for user's access to sensitive service and data in the scenario of 4G systems. Over this framework, with the combination of password and biometric identification (BI) as well as public key-based identification, an efficient hybrid authentication and key agreement (HAKA) scheme is presented to resist the possible attacks, particularly the attacks on/from ME. Compared with 3G architecture and other security schemes for 4G mobile networks, architecture and corresponding HAKA has been mentioned to be more secure, scalable and convenient to support globe mobility and capable of being employed to handle the complicated security issues in 4G mobile networks.

RESEARCH REQUIRED FOR 4G WIRELESS APPLICATIONS

The common research areas which are evolving for 4G of wireless communication systems are:

1) New decoding algorithms for turbo codes for wireless channels.
2) New coding/modulation techniques for reducing the peak-to-mean envelope ratio, maximizing the data rate and providing large coding gain.
3) New approaches to jointly designing modulation techniques, and power amplifiers to simultaneously obtain high power added efficiency along with bandwidth efficiency.
4) New demodulation / decoding techniques to simultaneously combat the near-far problem and do channel decoding in multi-rate DS-CDMA (Direct Sequence Code Division Multiple Access) systems.
5) Communication problems unique to high frequency systems (e.g., channel estimation).
6) Joint channel estimation and decoding/demodulation algorithms.
7) Multiple-access techniques for multi-rate systems with variable quality of service requirements.
8) Space-time coding for systems with multiple antennas.
9) Analog decoding techniques for high speed, low power systems.
10) Ultra wideband systems and hardware design. (xi) Research in methodologies for an integrated approach to wireless communications (device layer: e.g., power and low noise amplifiers, mixers, filters; physical layer: coding, modulation; medium access layer: CDMA/ FDMA/TDMA; data link layer: hybrid ARQ (Automatic Repeat Request); network layer: routing protocols).

CONCLUSION

As the history of mobile communications shows, attempts have been made to reduce a number of technologies to a single global standard. Projected 4G systems offer this promise of a standard that can be embraced worldwide through its key concept of integration. In spite of all the evolving technologies the final success of new 4G mobile communication systems will be dictated by the new services and contents made available to users. These new applications must meet user expectations, and give added value over existing offers. The Internet is the driving force for higher data rates and high speed access for mobile wireless users. This will be the motivation for an all mobile IP based core network evolution. OFDM and MIMO are the largest and the strongest candidates for 4G access technologies. The growth of the technology for 4G will be fueled by open standards. Because the future wireless networks will need to support diverse IP multimedia applications to allow sharing of resources among multiple users, there must be a low complexity of implementation and an efficient means of negotiation between the end users and the wireless infrastructure. The fourth generation promises to fulfill the goal of PCC (personal computing and communication)—a vision that affordably provides high data rates everywhere over a wireless network.

With this view this paper has highlighted the 4G system objectives, advantage, disadvantages, challenges and related standardization activities. The probable research issues have also been discussed. Moreover new challenges in the 4G mobile communication systems make the management mobility more complex. Many new problems need to be carefully taken into account. Besides above, this chapter focused on the prominent research issues and possible solutions in mobility management.

REFERENCES

Ahmad, J., Garrison, B., Gruen, J., Kelly, C., & Pankey, H. (2003). *4G Wireless Systems.* Next Generation Wireless Working Group. Retrieved April 9, 2009, from http://ckdake.com/files/4gwireless.pdf

Akhtar, S. (n.d.). *2G-4G Networks: Evolution of Technologies, Standards and Deployment.* College of Information Technology, UAE University.

Akyildiz, I. F., McNair, J., Ho, J. S. M., Uzunalioglu, H., & Wang, W. (1999). Mobility Management in Next-Generation Wireless Systems. *Proceedings of the IEEE, 87*(9), 1347-1384.

Alexander, J. H. (2002). *UMTS and Mobile Computing.* Norwood, MA: Artech House.

Bedell, P. (2005). *Wireless Crash Course, Edition 2.* New York: McGraw Hill Professional.

4G Beyond 2.5G and 3G Wireless Networks. (n.d.). MobileInfo.com. Retrieved February 13, 2009, from http://www.mobileinfo.com/3G/4GVision&Technologies.htm

Chang, W. Y. (2007). *Network Centric Service Oriented Enterprises.* Amsterdam, The Netherlands: SpringerLink.

Clark, D., Braden, R., Falk, A., & Pingali, V. (2003). Reorganizing the Addressing Architecture. In *Proceedings ACM SIGCOMM FDNA Workshop*, USA.

Dixit, S. (2008). *Should India go for 4G Wireless Networks Now?* Retrieved March 1, 2009, from http://www.shvoong.com/exact-sciences/1776578-india-4g-wireless-networks/

Eardley, P., Eisl, J., Hancock, R., Higgins, D., Manner, J., & Ruiz, P. (2002). Evolving Beyond UMTS - The MIND Research Project. In *Proceedings of IEE 3rd International Conference on Mobile Communication Technologies* (pp. 449-454).

Electrozoom. (2007). Association of Electronic and Communication Engineers.

Fernandez, E., et al. (2005). *Some security issues of wireless systems*. In *Advanced Distributed Systems: 5th International School and Symposium, ISSADS 2005*, Guadalajara, Mexico. Retrieved from http://www.cse.fau.edu/%7Eed/Fernandez_ISSADS2005Final.pdf

Francis, M. (2003). *Journal of Business Strategy*. American Sentinel University.

Frederic, P., Paal, E., Erik, V., Thomas, H., & Anne, M. N. Kjell, Mari. N. & Stein, S. (2002). *Mobility Aspects in 4G Networks* [White Paper]. Forskningsnotat/Scientific Document, Norway.

Gardezi, A. (n.d.). *Security In Wireless Cellular Networks*. Retrieved from http://www.cse.wustl.edu/~jain/cse574-06/ftp/cellular_security/#issues

Gazis, V., Housos, N., Alonistioti, A., & Merakos, L. (2003). Generic System Architecture For 4G Mobile Communications. In *Vehicular Technology Conference* (pp. 1512-1516).

Govil, J. (2008). An empirical feasibility study of 4G's key technologies. In *Electro/Information Technology, (EIT)-08* (pp. 267-270).

Govil, J., & Govil, J. (2007). 4G Mobile Communication Systems: Turns, Trends and Transition. In *International Conference on Convergence Information Technology*, Korea (pp. 13-18).

Haenggi, M. (n.d.). *4G Wireless Standard*. Retrieved March 1, 2009, from http://www.nd.edu/~mhaenggi/NET/wireless/4G/

Hussain, S., Hamid, Z., & Khagttak, N. (2006). Mobility Management Challenges and Issues in 4G Heterogeneous Networks. In *ACM Proceedings of the first international conference on Integrated internet ad hoc and sensor networks*. Retrieved from http://delivery.acm.org/10.1145/1150000/1142698/a14-hussain.pdf?key1=1142698&key2=8898704611&coll=GUIDE&dl=&CFID=15151515&CFTOKEN=6184618

Ibrahim, J. (2002). 4G Features. *Bechtel Telecommunications Technical Journal*.

Issues in Mobility Management in 4G Networks. (2001). *Computer, 34*(6). Retrieved from http://folk.uio.no/paalee/referencing_publications/ref-mob-4gissues.pdf

Kappler, C. (Ed.). (2004). *Scenarios, Requirements and Concepts*. IST–2002-507134-AN/WP3/D/3-1, Ambient Network WP3 deliverable.

Kappler, C. (Ed.). (2004). *Scenarios, Requirements and Concepts*. IST–2002-507134-AN/WP3/D/3-1, Ambient Network WP3 deliverable.

Katz1, M., & Fitzek, F. (2005). On the Definition of the Fourth Generation Wireless Communications Networks: The Challenges Ahead. In *International Workshop on Convergent Technologies*, Finland.

Magio Internet Security. (2009). Retrieved April 9, 2009, from http://t-com-eng.st.sk/Default.aspx?CatID=1248&section=home

Mary, N. (2006). *Implementation of Vertical Handoff Algorithm Between IEEE802.11 WLAN and CDMA Cellular Network*. Master Thesis, Georgia State University, USA.

McKnight, L.W., Howison, J., & Bradner, S. (2004). Wireless Grids – Distributed Resource Sharing by Mobile, Nomadic, and Fixed Devices. *Internet Computing*, 24-31.

Mobile, I. N. (n.d.). *What is 4G cellular?* Retrieved February 11, 2009, from http://www.mobilein.com/what_is_4GCellular.htm

Mobile Multimedia Laboratory. (n.d.). *Architectural, Economic, Security and Strategic Issues in 4G Wireless Networks.* Retrieved April 9, 2009, from http://mm.aueb.gr/research/4G.html

Mohr, W. (2002). *Mobile Communications Beyond 3G in the Global.* Siemens Mobile. Retrieved March 1, 2009, from http://www.cu.ipv6tf.org/pdf/werner_mohr.pdf

Moskowitz, R., & Nikander, P. (2003). *Host Identity Protocol Architecture.* Internet Draft draftmoskowitz- hip-arch-05.txt.

Ohba, Y. (2007). *MobiArch.* Retrieved April 9, 2009, from http://user.informatik.uni-goettingen.de/~mobiarch/2007/ slides/mobiarch07-panel-YoshiroOhba.pdf

Paila, T., Alladin, S., Frank, M., Goransson, T., Hansmann, W., Lohmar, T., et al. (2001). Flexible Network Architecture for Future Hybrid Wireless Systems. In *IST Mobile Summit*, Spain.

Pappalardo, D. (2007). What you need to know about 4G. *Network World.*

Perkins, D. (2005). Convergence Challenges and Solutions for Next-Generation Wireless Networking. In *Texas Wireless Symposium*, USA.

Ronai, M., Petrescu, A., Tönjes, R., & Wolf, M. (2003). Mobility Issues in OverDRiVE Mobile Networks. In *IST Mobile Summit*, Spain.

Russell, S.F. (2001). Wireless Network Security for Users. *Information Technology: Coding and Computing*, 171-177.

Santhi, K. R., Srivastava, V. K., Senthikumaran, G., & Butare, A. (2005). Goals of True Broad Band's Wireless Next Wave (4G-5G). In *Vehicular Technology Conference*, USA.

Sawal. (n.d.). *What is 4G technology?* Retrieved August 1, 2008, from http://sawaal.ibibo.com/computers-and-technology/what-4g-technology-464702.html

Schmitz, N. (March 2005). The Path to 4G Will Take Many Turns. *Wireless Systems Design.* Retrieved December 27, 2008, from http://www.wsdmag.com/Articles/ArticleID/10001/10001.html

Schreurs, W. (2006). How Dangerous is 4G. *FreNovation Online.*

Selander, G. (2004). *Ambient Networking: Concepts and Architecture.* IST–2002-507134-AN.

SFR SDR (Software Defined Ratio) Development Tools Product Bulletin. *(Rev. A).* (n.d.). Texas Instruments. Retrieved February 25, 2009, from http://focus.ti.com/lit/ml/sprt406a/sprt406a.pdf

Sharma, S. C. (2002). *4G Networks: A Case for Bypassing 3G.* Retrieved January 11, 2009, from http://voicendata.ciol.com/content/technology/102081902.asp

Shoewu, O. (2007). Evolution of Fourth Generation Mobile Networks: Trends and Tendencies. *The Pacific Journal of Science and Technology, 8*(2). Retrieved January 6, 2009, from http://www.akamaiuniversity.us/PJST.htm

Soliman, H., Elmalki, K., & Bellier, L. (2008). Network Working Group, USA.

Toshio, M., Tomoyuki, O., Hitoshi, Y., & Umeda, N. (2005). The Overview of the 3th Generation Mobile Communication System. In *Fifth International Conference on Information, Communications and Signal Processing (ICICS 2005)*, Bangkok.

Typpo, V., Fisl, J., Holler, J., Calvo, R. A., & Karl, H. (2004) Research Challenges in Mobility and Moving Networks: An Ambient Networks View. In *Workshop on Challenges of Mobility in conjunction with IFIP World Computing Congress*, France.

Typpo, V., Fisl, J., Holler, J., Calvo, R. A., & Karl, H. (2005). Research Challenges in Mobility and Moving Networks: An Ambient Networks View. In *Broadband Satellite Communication Systems and the Challenges of Mobility* (pp. 145-155). New York: Springer.

Urban, J., Wisely, D., Bolinth, E., Neureiter, G., Liljeberg, M., & Robles, T. (2001). BRAIN – An architecture for a broadband radio access network of next generation. *Journal of Wireless Communications & Mobile Computing*.

Willie, W. (2006). Open Wireless Architecture (OWA) – Defining China's Fourth Generation Mobile Communications. In *Fourth Generation Mobile Forum*, Hong Kong.

Yang, L. (2007). *Access Network Selection in a 4G Networking Environment*. Master Thesis, Department of Electrical and Computer Engineering Waterloo, Canada.

Young Kyun, K., & Prasad, R. (2006). *4G Roadmap and Emerging Communication Technologies*. Norwood, MA: Artech House.

Zafeiris, V. E., & Giakoumakis, E. A. (2005). An Agent-based Architecture for Handover Initiation & Decision in 4G Networks. In *Proceedings of the Sixth IEEE International Symposium on a World of Wireless Mobile and Multimedia Networks*.

Zhao, J., & Sauvola, J. (2002). Mobility and Mobility Management: A Conceptual Framework. In *IEEE International Conference on Networks*, Singapore (pp-205-210).

Zheng, Y., et al. (2005). Trusted Computing-Based Security Architecture For 4G Mobile Networks. In *Parallel and Distributed Computing, Applications and Technologies, PDACT* (pp. 251-255).

Zheng, Y., et al. (2005). Security scheme for 4G wireless systems. In *International Conference on Communications, Circuits and Systems, 2005* (pp. 397-401, Vol.1). Retrieved from http://ieeexplore.ieee.org/stamp/stamp.jsp?arnumber=1493433&isnumber=32104

Chapter 9
Potential Scenarios and Drivers of the 4G Evolution

Elias Aravantinos
Stevens Institute of Technology, USA

M. Hosein Fallah
Stevens Institute of Technology, USA

ABSTRACT

Nowadays, the mobile Internet communications can play a significant role in the Telecommunications Sector, resolving certain issues and bottlenecks of personal communications, with most European countries close to 100% penetration and a global projection of 4 billion mobile users by 2011. As we are moving to the next generation, we are still lacking the precise definition of the architecture and the successful deployment path of the 4G technology. Several theories have been developed looking at different standards and aiming to select and develop the most promising one. In this paper the authors are introducing and presenting a study that aims to explain a new concept of "4G readiness" revealing long run national strategies for 4G deployment and suggesting some critical metrics that could describe the future of the mobile broadband environment. They describe the methodology, assumptions and discuss the expected results based on similar studies such as the e-readiness study.

INTRODUCTION

In a world of increasing technological needs, the mobile Internet can play a significant role, meeting user's capacity and connectivity needs. There is a good deal of research around the 4G concept, where vendors and operators are trying to define it based on their preferred technology and strategic planning.

At the end of 2007, the global mobile subscribers reached 3 billion, with GSM based users accounting for over 2 billion. Several research reports have been predicting that WiMAX will be commercially deployed by 2009 and LTE (Long Term Evolution) by 2011. However, the debate on the standards for 4G continues and is a major concern. International Telecommunications Union (ITU), Institute of Electrical and Electronics Engineers (IEEE) and other similar associations and committees are

DOI: 10.4018/978-1-61520-674-2.ch009

working on securing a smooth transition to the new technology.

The 4G evolution as described in Figure 1, started in early 1990s transitioning into different stages, such 3.5G and 3.75G, ending to the 4G, meeting the market needs in most of the cases. The most recent transition that is expected is the migration from High Speed Packet Access (HSPA) to the 4G standard, which could be the WiMAX or the LTE or the combination of both.

In order to describe the market needs and behavior towards the 4G evolution, it deemed necessary to assess several countries' current readiness to deploy the 4G technology. Supporting the opinion that the LTE evolution will be the winning 4G, we have defined several metrics from different perspectives such as technology, business, and consumer spending to rank each market's 4G readiness in 16 countries. Our main objective is to use a ranking approach to shed light into the factors that are driving countries' progress in deployment of 4G, be able to estimate the deployment speed, and create future scenarios. We create three groups of countries 'established leaders', 'rapid adopters' and 'late entrants'. We also want to be able to compare 4G readiness results with existing similar studies for the same countries to provide observations and derive useful conclusions.

BACKGROUND

Currently, there is no formal definition for 4G. It is a term used to describe the next step in the evolution of wireless communication. Several terms are also describing the concept, such as "Super 3G" (Seizo & Nakamura 2007) or "Next Generation Wireless". ITU has been committed to announce a 4G definition. There is general agreement among experts that 4G is a new converged system that will provide at least 100Mbps connectivity to the broadband users. 4G is expected to offer data rates of 100 Mbps for mobile applications and 1 Gbps for nomadic applications and should be achievable between the years 2010 and 2011.

The current defined objectives for 4G include (Etoh 2005; Fratassi, et.al., 2006; Mohr 2008; Dursch, et.al., 2005):

- Fully integrated IP solution
- "Anytime, Anywhere"
- Seamless connectivity- wireless and wireline

Figure 1. 4G evolution into convergence (Adapted from Etoh, 2005)

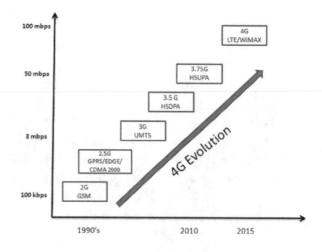

- Global access and interconnection
- Interoperability
- Data rates of at least 100Mbps
- Spectrally efficient system

There are several applications that could be supported and leveraged in the 4G due to the advanced environment. These include mobile commerce with an emphasis on mobile banking, peer-to-peer networking and full usage of the advanced Internet services in the converged cloud. This cloud be defined as a "Communications Technology Ecosystem" (Figure 2) with a plethora of different services that will give users a more convenient and easy lifestyle.

Since 4G is not well defined yet, there is no defined demand or market potential as yet. Therefore, we are lacking forecasts or predictions that could help the operators to strategically plan for the expected market with a time table. An interesting approach to this ill-defined problem is to evaluate each country's readiness to deploy 4G based on a set of criteria such as technological, business, legal, and policy considerations.

Investment in 3G in most cases and countries has not paid off yet and will not for the next 5 years. Nonetheless, operators need to decide on the best standard to invest in for the long run to provide for the future needs of their customers. Many are still debating on the WiMAX and the LTE choice.

In this chapter we describe an approach to study and evaluate the 4G readiness, using a quantitative analysis applied to a sample of two different groups of countries *developed* and *emerging*. We aim to assess the 4G predictions at a national level and answer the following research questions:

"Which countries are ahead in 4G adoption?" "Since the markets are still shaping up, how could one forecast demand from the operators perspective, and rank order the markets using the operators and vendors current trials and knowledge?"

To our knowledge, this is the first study that is trying to conduct an early assessment of the 4G status of some of the most important telecom markets. The "4G readiness" is defined as an index, building upon the literature and the e-readiness concept (Economist Intelligence Unit, 2008) as well as the non-market factors as described in Howells (2005).

Along the same lines, we expect that a country's 4G high ranking could be more an status and indicator of innovation, supported with an advanced digital environment rather than a natural path of technological evolution.

Also, there are two categories of factors that are considered to drive the 4G readiness of each market including the technology and the consumer/business spending. We proportionally weigh each factor to the highest value (percentage of each index). We expect to observe major differences in the spectrum law from country to country, that play a role into the index's estimation but not a significant one, since the migration from HSPA to the LTE does not require any major regulatory changes but rather investment planning. The scalable bandwidth will allow the operators to migrate their networks and users from HSPA to LTE over time.

Figure 2. A suggested heterogeneous digital ecosystem (Adapted from Mohr, 2008)

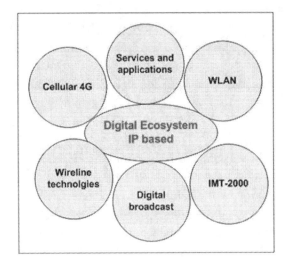

4G TECHNOLOGIES

Technological Feasibility

There are several technologies suggested to deploy in 4G and these may include (Fallah & Lechler, 2003)

- Software Defined Radio (SDR): is a radio communication system where components that have typically been implemented in hardware (i.e. mixers, filters, amplifiers, modulators/demodulators, detectors. etc.) are instead implemented in that allows changing the radio characteristics flexibly to meet specific requirements.
- Orthogonal frequency - division multiplexing (OFDM): is a frequency-division multiplexing (FDM) scheme utilized as a digital multi-carrier modulation method.
- Multiple-input and multiple-output, or MIMO), is the use of multiple antennas at both the transmitter and receiver to improve communication performance.
- Universal Mobile Telecommunications System (UMTS), standardized by 3GPP.
- Time Division-Synchronous Code Division Multiple Access, or TD-SCDMA, is a 3G mobile telecommunications standard, being pursued in the People's Republic of China by the Chinese Academy of Telecommunications Technology.

All these technologies are typified by high rates of data transmission and packet-switched transmission protocols. 3G technologies, by contrast, are a mix of packet and circuit-switched networks.

WIMAX VS. LTE

The LTE technology that Nokia and the Third Generation Partnership Project (3GPP) are pushing is an upgrade to existing GSM networks. The attraction to this technology had made even the CDMA operator, Verizon Wireless, to join the 3GPP trials. It is also a strategic decision, in order to be compatible with its European, GSM-based parent company, Vodafone. LTE looks like it can heal the GSM/CDMA rift that has divided the industry, as no major carrier has yet signed on with obvious CDMA 4G upgrade technology, Ultramobile Broadband (UMB).

LTE will have the following advantages:

- Fast, with peak data rates of 100 Mbps download and 50 Mbps upload.
- It makes CDMA and GSM debates moot.
- It offers both FDD and TDD duplexing, which means the upload and download speeds don't have to be synchronous, so operators can better optimize their networks to use more upload channels.
- LTE will have lower latency, which makes real-time interaction on high band-width applications using mobiles possible.

3GPP LTE, one of the most advanced mobile communication technologies to date, is currently undergoing 4G technology standardization by the 3GPP. This is the most likely technology to become the 4G standard, as many of the world's major operators and telecommunications companies are members of LTE/SAE (Long Term Evolution/ System Architecture Evolution) Trial Initiative (LSTI). These companies include operators, such as Vodafone, Orange, T-Mobile, NTT DoCoMo, China Mobile and Telecom Italia and vendors, Ericsson, Nortel, Alcatel-Lucent, Nokia Siemens and LG Electronics. These are also the companies that will be considered to have the advantage in deploying first the 4G services.

WiMAX has certain advantages mainly over the Fiber to the home (FTTH) technology. When bundled with broadband internet access and IPTV, a WiMAX triple play becomes very attractive to residential subscribers. Given the QoS, security and reliability mechanisms built into WiMAX, the users will find WiMAX VoIP as good as or even better than voice services from the telephone company. It also offers a cost effective infrastructure with efficient use of spectrum. Currently, the average cost of WiMAX 802.16-2004 baseband has decreased from $35 to almost $20 today per subscriber (Gartner, 2006).

4G proponents will serve as complements or upgrades to advance the 3G limitation to deliver video/TV and high speed Internet access. For WiMAX, there is a limitation of wireless bandwidth. For use in high density areas, it is possible that the bandwidth may not be sufficient to cater to the needs of a large clientele, driving potentially the costs high. But the main competitor for WiMAX today is the fiber and the wireline network that especially in the US is a real challenge for the residential users as the operators are deploying and growing really fast.

Trials and Status

In 2008, Nokia installed a prototype base station for ongoing LTE tests at the top of the Heinrich Hertz Institut building in the center of Berlin, where interference typically degrades bandwidth. The first-of-its-kind test featured multiple users connected to the new base station, giving the 173Mbps throughput number some credibility as a real-world peak. Nokia also tested LTE throughput by putting terminals into cars and driving them up to 1km away from the base station. Verizon and AT&T are also testing it with Motorola equipment (Strokes, 2007)

WiMAX on the other hand is ahead of LTE as a personal broadband option. CDMA-based operator Sprint-Nextel, for its part, is banking on WiMAX as a 4G solution. The Sprint-Nextel's WiMAX-based Xohm service in Chicago indicate that the bandwidth and pings are excellent (roughly 3Mbps/1.5Mbps and 70ms, respectively), but the numbers are nowhere near the +100Mbps /50Mbps that LTE promises in both directions. (Strokes, 2007)

4G Vendors and Players

A market restructuring with aggressive joint ventures and new players in broadband wireless could be a future highlight. The major 4G players in general are the current 3G players that invest in new Research and Development (R&D) projects and sources for the future of the mobile wireless (Fallah & Lechler, 2003). Recently, Alcatel-Lucent and Japan's NEC Corp formed a joint venture around LTE trials. Similarly, China Mobile joined Verizon Wireless and Vodafone in LTE standardization trials. China is a very interesting case, since their 3G networks is still in progress; soon the government will issue three licenses for high-speed third-generation mobile phone services and called for a merger of China Unicom and Netcom, two of its four biggest telecoms providers.

One of the WiMAX weaknesses is the lack of certification. The ITU recommendation adding WiMAX as an official 3G protocol is boosting the investment and the new spectrum auctions as in the US, for example, addressing the 700MHZ auctions. More than 100 WiMAX devices have been announced in 2008 and the fixed /portable wireless access equipment market has grown from $562 million to $1.2 billion in 2007. Additionally, Cisco is targeting WiMAX development as smart distributed wireless networking (Maravedis, 2008; Gartner, 2006).

4G Drivers and SCENARIOS

The rising mobile subscribers by 2011, estimating over 4 billion, in combination with converged systems and applications are the main contributors of the 4G evolution (GSM world 2009). Several services are expected to drive to the 4G converged ecosystem but the future operators revenues are data services and mainly entertainment services. Three services that exist in today's markets are expected to play a significant role in the future and into a more advanced mode. These are music, mobile games and mobile TV.

The new mobile user's lifestyle is increasing needs capacity, although the 'walled garden' might still be a limitation restricting the customer's experience. The users are changed from consumers to producers of content such as photos, videos etc. Several applications will drive the mobile broadband market globally, including:

- Web 2.0,
- Online blogs,
- Mobile music,
- Location Based Services (LBS),
- Multimedia messaging,
- Gambling and
- Mobile TV.

There are a few scenarios discussed including WiBro, which is expected to evolve during 2010 and 2015 and attempting to cover different markets through restructuring and transition into 4G (Onoe & Nakamura, 2007). For the next 5 years Verizon network will evolve into a 28Mbps download speed, leading to an early 4G LTE adoption compared to Vodafone (Business Week 2008).

These scenarios could be summarized as following:

1. Independent 4G system with one standard, the 3GPP LTE
2. Transition from 3G into 4G with existing (3GPP LTE) or new service providers WiMAX and WiBro
3. Co-existence of different standards
4. Spread of open transmission

To explain the above cases, we claim that history matters and the path dependent concept can really explain the long-term outcome based on initial conditions. The 4G development depends on the initial conditions as shaped from 3G in most of the cases. Based on the 'Increasing Returns' (Liebowitz &Margolis, 1994), and 'Path Dependency' (Arthur, 1994; Christensen, 2004; Howells, 2005), where alternatives are possible, and regarding the standards, "the one selected and heavily invested is good enough' or even optimal and remains in use because it becomes established in use". (Howells, 2005, p.71). This theory is matching the scenario of different standards co-existence that will interact in the ecosystem and complement each other referring to an advanced LTE or LTE+ and WiMAX that will be established and standardized as 802.16e that offers advanced mobility. This is what usually occurs in technological development scenarios.

4G READINESS ASSESSMENT

Methodology

The need for strategic planning and new services has urged the need for new studies that could give an idea of the current 4G status of the countries and their move towards the future 4G deployment.

The new 4G readiness concept could be defined as "state of play" of a country's mobile wireless 4G preparation status, and the ability of its potential and existing consumers, businesses and governments to use in the future the mobile wireless and broadband to their benefit. Based on the 4G readiness criteria we will rank the countries as a first step in estimating how soon they will close the gap to new 4G technological environment.

Also to describe our approach we are going to adopt Christensen (2004) theory, to measure each country's innovation using the Motivation/Ability framework (Figure 3). In this framework, we are describing the 4 different quadrants and how they are adjusted into our problem. The Motivation means that the 4G including the digital convergence should be the "pot of gold" and the new opportunity awaiting for the winners that most likely would be the first movers. The Ability describes the resources needed to develop 4G and craft them into business models for new products and services.

In the "Looking for a target" section, the operators are still undecided regarding the more beneficial choice or are lacking the spectrum to develop a new market. This hesitation also can derive from the "Looking for the Money" section, since the players are still expecting the 3G to pay off before they move into a new investment or wait for the LTE, in order to upgrade the GSM networks that might also include lower cost, much less than developing a WiMAX solution. "The Dilemma" is what we can quantify using our 4G readiness metrics and estimate per country, assuming basic innovation and ability. Finally "The Hotbed" is addressing all the innovative countries that feel confident and in the right path for the 4G adoption in the near future.

Other important non-market factors for 4G based on the framework development are:

- Industry standards
- Cultural norms
- State of technological development

Figure 3. Motivation/ability framework (Adapted from Christensen, 2004)

	Looking for a Target	The Hotbed
Motivation	Accessing resources (spectrum) or reaching potential consumers	Teeming for innovation
	The Dilemma	Looking for the Money
	No readily available avenues to create profitable businesses (unknown 4G readiness)	Firms struggle to find ways to monetize an opportunity (which standard choice will generate most revenues at low cost)
	Ability →	

- Government regulation
- Country's intellectual property infrastructure

We are applying non-market metrics and factors, because there are no markets structured shaped yet and the current 3G markets provide very little knowledge to support the new landscape for 10 years from today.

More specifically we will rank the following list of countries that are included in the top 20 of the referred study: Canada, Hong Kong, Netherlands, Switzerland UK, Denmark, Germany, USA, Japan, Sweden and Finland.

Also, some other countries will be added, such as China due to the China Mobile 4G activity and trials as mentioned before. Other strategic countries added in the sample: India, Russia, Brazil due to the following facts or figures:

- Almost represent half of the world's population
- Showing record in wireless adoption in the last few years

In our methodology we are using as input countries with the most efficient digital users, assuming they are more technologically innovative, demanding more advanced services than the rest of the world. The suggested categories and the data collected will be similar to the e-readiness study, adjusted more specifically when needed into the mobile wireless dataset, trying to include important non-market factors as already described. We assume in most cases that the 4G evolution can be considered as a digital subcategory developing similar categories and criteria.

Categories

In this section, we are describing the 6 different category and the weights as shown in Figure 4.

Using several resources and databases as listed in the Appendix, we build an index for each country

Figure 4. 4G readiness categories and weights

Category Weight	Connectivity and technology infrastructure	Business environment	Social and cultural environment	Legal environment	Consumer and business trend	Government policy and vision
Criteria	1. mobile phone penetration 2. current industry standards	Overall political environment; macroeconomic environment; market opportunities; policy toward private enterprise; foreign investment policy; foreign trade and exchange regimes; tax regime; financing; the labour market.	Level of education and literacy; level of Internet literacy; degree of entrepreneurship; technical skills of workforce; degree of innovation.	spectrum laws efficiency, Subscribers served per MHz of Spectrum	1. Consumer spending on communications services; 2. availability of mobile wireless services for citizens and businesses	1. Telecom investing per GDP 2. Total ICT national spending rate
Percentage	20%	20%	15%	10%	20%	15%

and based on its value, we rank it in our list into a certain position out of the 16. The index, when calculated is expected to be on a scale 1-10

The first of the categories is about connectivity and technology infrastructure, which is one of the most significant drivers, adding 20% to the overall score. The connectivity measures the extent to which businesses can access mobile networks, integrating the 4G path. The effective access uses the mobile-phone penetration that includes both 2G and 3G technology per country. Regarding the 4G path, most of the countries chose High Speed Download Packet Access (HSDPA) and High Speed Upload Packet Access (HSUPA) deployment before migrating to the LTE technology by. Most of the countries prefer not to deploy advanced HSPA or HSPA plus, but instead to spend the money on LTE testing as T-Mobile did in the USA. Thus in the figure if in a country both technologies are deployed, HSDPA and HSUPA the value '1' is assigned otherwise if only one of the two is in-service then value '0.5' is assigned. These two technologies are considered as 3.5G and 3.75G respectively.

For this category, the metrics are mobile phone penetration and industry standards in-service.

The second category is the business environment, weighing 15% in the overall score. As explained in the (Economist's Intelligence study, 2008, p.22) the criteria "cover such factors as the strength of the economy, political stability, taxation, competition policy, the labour market, and openness to trade and investment."

The third category refers to the social and cultural environment, weighing a 15% in the overall score. The environment again is expected to be similar to the one defined in the Economist's study, since the 4G evolution will not at least at this point and with the current knowledge to bring any major changes into that environment different from the digital study. Thus "e-literacy and basic education" could define this category's criteria. (Economist's Intelligence study, 2008, p.23).

The fourth category describes the legal environment, weighing a 15% in the overall score. It is very important to identify the legal and regulatory environment per country. The spectrum law policy with a combination of the spectrum's availability is considered to be the most important metric in that case, quantifying its effectiveness for each market. More specifically, the number of subscribers served per MHz is calculated. This shows the market's efficiency to adjust to the regulatory environment keeping up with growth and profits for the operators. Finally, all the values are normalized to the US one, which is the highest, proving that the US wireless carriers are able to get so much more use of the spectrum available.

The next category to be included is the Consumer's or the demand side, weighing an overall score of 20% into the index calculation.

Except for the supply side, we need to describe the demand side and the roadmap for the adoption of technology. Consumer spending on communications services is very important on each country's trends and the amount of money the subscribers are willing to spend. Additionally the market's efficiency as the consumers are demanding more services is described from the number of services that are activated in 3 general areas, mobile games, music and TV. If all three services are offered in the market then the value "1" is assigned, if one of the three services "0.33" and two of the three "0.66". A market is more efficient and is converging faster into new services and technologies, when it offers all of the three services, showing the consumer's demand and advanced needs. The category metrics are consumer spending on telecom services per household and active mobile services.

Finally, the last category is the R&D spending, accounting for 15% of the overall score.

The Telecom investing as a rate of the overall Gross Domestic Product (GDP) index is a significant metric that should be considered as part of the country's strategy especially in the long run. The investment could include trials, new networks and R&D projects etc. Additionally a country's profile could be completed with the Information and Communications Technology national spending rate. That metric could differentiate the countries and the special weight they put on the ICT sector overall. All this spending should be expected to pay back or to bring some results to the market after a certain amount of time that could range from 4-5 years. Thus the volume spent today, could pay back by 2011 and contribute into the 4G commercial deployment.

Data

The data for the different countries were collected from different, and to the extent possible, reliable resources. However in some cases the authors have used industry experts opinion, when the data was incomplete or missing.

In the first category, the mobile wireless penetration was collected from Campbell (2008) and the HSPA standards status from 3G Americas report (2008). The data for the business environment and also for the social and cultural categories were collected from the Economist's study (2008). The spectrum law effectiveness is calculated from the subscribers served per MHz of spectrum allocated. The available spectrum used for commercial use was collected from CTIA(2008) and the GSM world (2008) reports. In developed countries the available spectrum was usually over 300 MHz and in the emerging ones around 200MHz. To calculate the rate, the number of mobile wireless subscribers was divided to the available radio spectrum and then the data were normalized to the highest value. In the next category, the data on consumer spending on telecom services per household, was collected from Euromonitor database and then normalized. The active mobile wireless services were collected from Netsizeguide (2008). Finally in the last category both variables were collected from the OECD (2008) database.

Results

The results are presented in Figure 5. Most of the countries' score is in a scale between 2 and 5. The developed countries as expected are ranking in the top-10 list, with the rest of the countries, Brazil, Russia, China and India in the lower part.

The emerging markets such as China, and India, are expected to move up. Russia is surprisingly lagging behind the developing markets, revealing inefficiency to the LTE path.

Figure 5. 4G-readiness ranking, 2008

Category Weight	Country	Overall score	Connectivity and technology 20%	Business environment 15%	Social and cultural 15%	Legal 15%	Consumer 20%	ICT spending 15%
1	USA	3.619	1.86	8.53	9.00	1	1.49	1.13
2	Denmark	3.606	2.22	9.65	8.67	0.664	1.28	0.39
3	Australia	3.432	1.99	8.59	9.13	0.846	0.94	0.40
4	Finland	3.382	2.10	8.4	8.3	0.658	1.55	0.33
5	Germany	3.323	2.13	8.36	8.00	0.361	1.58	0.48
6	Netherlands	3.240	1.09	8.55	8.07	0.217	2.00	0.64
7	Sweden	3.227	1.12	8.52	8.6	0.658	1.65	0.28
8	Switzerland	3.357	1.06	8.57	8.27	0.292	1.80	1.44
9	UK	3.192	1.23	8.61	8.13	0.245	1.51	0.64
10	Japan	3.186	1.81	7.39	7.87	0.359	1.89	0.68
11	Canada	3.184	1.59	8.63	8.13	0.128	1.36	0.40
12	Hong Kong	3.176	2.22	8.64	7.47	0.188	1.18	0.34
13	Brazil	2.732	1.78	7.01	6.13	0.200	1.36	0.70
14	Russia	2.264	1.16	6.19	5.33	0.656	0.88	0.20
15	China	2.219	0.88	6.49	5.53	0.266	0.42	0.78
16	India	2.116	0.19	6.53	5.33	0.229	1.01	0.42

DISCUSSION

The 4G readiness scores rank the countries on the ability and current conditions to deploy 4G and who is in the running. We expect countries with higher mobile wireless penetration levels to score higher on this scale and in comparison to the e-readiness table (Figure 6). Similarly, countries that have more 4G operators and trials to be placed higher in the list compared to the digital study. The higher score is based on the fact that these operators in these countries have a higher readiness to deploy 4G as they are involved in the 4G trials. Furthermore higher score is also associated with higher R&D investment providing product and service innovations that could make it easier for the consumer adoption of 4G services. High R&D investments points to countries emphasize on innovation and perhaps strategic value of early adoption.

Another important factor is the effectiveness of the spectrum law and the market's reaction and adjustment over time. The closer the countries' legal efficiency to the ideal US spectrum law, the higher they are in the list with a few exceptions such as Russia. The business environment in those countries is rated low along with a low consumer's spending, making the evolution into the new technology more difficult and unclear.

As we see in the table of ranking, many of the developed countries have close scores represented by similar performance on many of the dimensions we have considered. With the same token, the emerging countries are not scoring as high, as we expected, since their capabilities and performance on some of the dimension are not as high.

Two important observations:

1. While one can argue whether our assignment of values to each dimension for each country is accurate, the purpose of the index is not the accuracy but relative positioning on the scale and we believe the baseline score is a reasonable ranking for these countries.
2. As countries take action to support their 4G deployment the ranking will shift and we should be able to see who is getting ahead and who is falling behind.

Finally, we close our discussion and analysis, comparing our results in Figure 6 with the e-readiness study. We identify several similarities in the top 10 of the last two columns. The common countries are bold highlighted and underlined. Thus 7 out of the 10 countries are in both top 10

Figure 6. Comparison with the e-readiness study

Ranking	Overall score	4G-readiness	e-readiness ranking (Economist Intelligence Unit)
1	3.619	USA	USA
2	3.606	Denmark	Hong Kong
3	3.432	Australia	Sweden
4	3.382	Finland	Australia
5	3.323	Germany	Denmark
6	3.240	Netherlands	Singapore
7	3.227	Sweden	Netherlands
8	3.357	Switzerland	UK
9	3.192	UK	Switzerland
10	3.186	Japan	Austria
11	3.184	Canada	Norway
12	3.176	Hong Kong	Canada
13	2.732	Brazil	Finland
14	2.264	Russia	Germany
15	2.219	China	S. Korea
16	2.116	India	New Zealand

lists, confirming our assumption that the digital ready countries would be the early adopters of a new technology such as 4G, since also connectivity is high and consumers can afford it.

FUTURE RESEARCH DIRECTIONS

The current study as presented in this chapter could be enriched with more countries and improved metrics. For example concentration measures such as the Herfindahl-Hirschman Index (HHI) could be applied in modeling the indices. Consumers' age groups, education levels and cultural factors can improve the measures of adoption for the 4G business environment.

Also, a more detailed market research using operators' data could derive more robust results. Further research can help develop predictive models using binary probit regression, in order to calculate the probability of each country to deploy 4G, and who the first adoptors. Similar conclusions could be derived from log linear models along the same research lines

CONCLUSION

This study provided an overview of the 4G evolution and technologies. It also described the 4G readiness ranking and the approach adopted in this study to develop the 4G readiness ranking.

The 4G readiness ranking combines dimensions of technology, business, consumer, legal/regulatory and R&D investment into a single index to rank order countries. The ranking at country level will help us identify if the biggest players in these countries play a significant role and having an impact as leaders in 4G. The study reveals which countries at this point have a competitive advantage for 4G. Some, countries score higher on 4G ranking than on e-readiness and vice versa because of greater emphasize on mobile technology.

Finally we need to pay special attention to the strong countries and perhaps identify the technology path, business and the long run strategies and planning horizon that will shape the 4G markets faster and accumulate more capital and investments.

REFERENCES

3GAmericas. (2008). *Report: Global UMTS HSPA Operator Status Update.* Arthur, W.B. (1994). *Increasing returns and path dependence in the economy.* Ann Arbor, MI: University of Michigan Press.

Campbell, G. (2008). *Global Wireless Matrix 2Q08.* Merill Lynch, Canada.

Christensen, C. (2004). *Seeing What's Next: Using Theories of Innovation to Predict Industry Change.* Cambridge, MA: Harvard Business School Press.

CTIA, 2008. http://files.ctia.org/pdf/080108_US-OECD_10_Comparison_Ex_Parte.pdf

Dursch, A., Yen, D. C., & Huang, S.-M. (2005). Fourth generation wireless communications: an analysis of future potential and implementation. *Computer Standards & Interfaces, 28,* 13–25. doi:10.1016/j.csi.2004.10.009

Etoh, M. (2005). *Next Generation Mobile Systems: 3G and Beyond.* Hoboken, NJ: Wiley.

Fallah, H., & Lechler, T. (2003). *Global innovation management in wireless communications industry.* IAMOT.

Frattasi, S., Fathi, H., Fitzek, F. H. P., Prasad, R., & Katz, M. D. (2006, January-February). Defining 4G technology from the users perspective. *Network, 20*(1), 35–41.

GSM World (2008). *Mobile Broadband, Competition and Spectrum Caps.* Retrieved from http://hspa.gsmworld.com/upload/resources/files/19012009094421.pdf

Howells, J. (2005). *The management of innovation and technology.* Thousand Oaks, CA: Sage.

Liebowitz, S., & Margolis, S. (1994). Path dependence, Lock-In and History. *Journal of Law Economics and Organization, 11*(1), 205–206.

Maravedis Research. (2008). *WiMAX, LTE and Broadband Wireless World wide Market Trends 2008-2014 – 5th edition.* Montreal, Canada: Maravedis.

Mohr, W. (2008, January). Vision for 2020? *Wireless Personal Communications, 44,* 27–49. doi:10.1007/s11277-007-9381-1

OECD Information Technology Outlook. (2008). OECD.

Redman, P., Dulaney, K., & Gutberlet, M. (2006). *Key Criteria for Wireless Carriers to Evaluate Mobile WiMAX.* Gartner Research.

Seizo, O., & Nakamura, T. (2007). 3G evolution scenario toward 4G: Super 3G concept. *Wireless Commununication and Mobile Computing, 7,* 1013–1019. doi:10.1002/wcm.511

Stokes, J. (2007). *Nokia 4G wireless tech hits 173Mbps in real-world test.* Retrieved December 28, 2007, from http://arstechnica.com/news.ars/post/20071228-nokia-4g-wireless-tech-hits-173mbps-in-real-world-test.html

The Economist Intelligence Unit. (2008). *Report: E-readiness rankings 2008.* Retrieved from http://a330.g.akamai.net/7/330/25828/20080331202303/graphics.eiu.com/upload/ibm_ereadiness_2008.pdf

UN e-Government Survey. (2008). New York: United Nations. Retrieved from http://unpan1.un.org/intradoc/groups/public/documents/UN/UNPAN028607.pdf

Vodafone's 'Long Term' Hesitance. (2008 April). *Business Week.*

Chapter 10
Knowledge Sharing to Improve Routing and Future 4G Networks

Djamel F. H. Sadok
Federal University of Pernambuco, Brazil

Joseilson Albuquerque de França
Federal University of Pernambuco, Brazil

Luciana Pereira Oliveira
Federal University of Pernambuco, Brazil

Renato Ricardo de Abreu
Federal University of Pernambuco, Brazil

ABSTRACT

Current networking trends show a rapid convergence of several nestled networks as GSM/UMTS, WLAN, Bluetooth, wired networks among others. Nonetheless, such a complex environment raises new challenges including information routing, high dynamicity and possible disconnections. For such reasons, nontraditional routing paradigms have been put forward while adopting innovative ideas based on models from areas as diverse as biological, epidemic and social behavior. The reader will learn about new forms of routing being considered for integrating these future networks, based on the use of different metrics, such as shared network knowledge and willingness in taking and forwarding it. In this chapter an overview of traditional routing and the requirements for 4G systems are first made. New directions for routing based on policy and the identification of stimulus to control and improve the message forwarding in addition to their efficiency in the new context of 4G networks are presented.

INTRODUCTION

Despite rapidly advancing research related to building 4G systems, there is still an apparent lack of general consensus of how to view their scope (Szczodrak, Kim & Baek, 2007). Some works mainly look at 4G as a mere improvement of 3G cellular networks or emerging mobile WiMax 802.16 in terms of performance, mobility, capacity and new

DOI: 10.4018/978-1-61520-674-2.ch010

advanced services. Others consider a bigger picture that includes the convergence of several nestled networks including GPRS – 2.5G, GSM/UMTS – 3G, Wifi - WLAN, WiMax – 802.16, Bluetooth – short range, and also fixed networks. 3GPP calls its work in such context as the Long Term Evolution (LTE). A third view is that of an all-IP user centric approach where services may be developed on end systems independently of the network infrastructure. This is similar to the end-to-end argument adopted by the Internet where TCP/IP and P2P applications may be developed to use the network while seeing it as a simple pipe. The Android initiative from Google is an example of such user centric view. Note that the different approaches for the design of future networks, not only have technological differences, but are also motivated and driven by new business models. The reader may see that the Telecom industry, the Internet community, the information technology (IT) companies all, understandably, would like to see a future where control over the network services and resources is part of their responsibility. Trying to figure out who is right or wrong is a religious discussion at best as all sides present their supporting their views and business models for future networks such as 4G and the Internet as a whole. Only the future may tell who is going to win such debate.

Instead of discussing which scenario may prevail, let us look at a common concern, one that explores how to make use of such convergent scenario and more importantly it tries to enhance knowledge sharing in future 4G networks. The reader is shown how user mobility and network dynamicity may be valuable allies for this endeavor rather than limiting factors as often seen by traditional cellular systems. More specifically, it is the intent of this chapter to view the impact and analyze the behavior of knowledge (seen here as an important new network metric) sharing in a complex 4G environment where traditional services such as routing and locating information raise new challenges.

Initially we review the current experience gained in the area of routing into fixed and dynamic networks as well as describe some challenges for next generation systems. Unlike traditional work in the area of communication system architectural design, a number of innovative and research ideas are based on biological models which have extensively been used to model knowledge sharing and node reachability, as well as to provide adaptability, scalability and robustness for systems (Babaoglu, Canright, Deutsch, Di Caro, Ducatelle, Gambardella, Ganguly, Jelasity, Montemanni, Montresor & Urnes, 2006). For instance, a Wasp model provides an advanced scheduling scheme for satisfying robustness when submitted to unexpected events as well as having considerably high performance tasks (Song, Hu, Tian & Xu, 2005; Cao, Yang & Wang, 2008). Ant models have been used for optimized routing support, i.e. leading to increasing throughput and reducing network delay (Gutjahr, 2008; Dorigo & Blum, 2005).

Considerable work has also gone into partially connected networks (Vahdat & Becker, 2000; Lindgren, Doria & Schelén, 2003) and epidemic routing (Kenah & Robins, 2007). Such works introduce Epidemic Routing, where random pair-wise exchanges of messages among mobile hosts ensure eventual message delivery (Jindal & Psounis, 2006). Under this type of routing, probabilistic delivery may be used to lead with an important aspect of future 4G nodes, that of disconnected operation. Given that messages are delivered probabilistically in epidemic routing, the application may require the use of acknowledgments to ensure some level of reliability.

Relevant ideas may also come from recent work on social phenomena such as SOLAR (Ghosh, Philip & Oiao, 2007 and Ghosh, J., Ngo, H.Q., Yoon, S. & Qiao, C, 2007), SimBet (Daly & Haahr, 2007) and Bubble (Hui, Crowcroft & Yoneki, 2008 and Hui, P. & Crowcroft, J., 2007) routing that will be discussed next. Bubble mimics the social communities way of life to create clusters and decreasing the number of routing messages.

Simbet is another protocol also based on the use of clusters. But it routes information according to node popularity and similarities between these Both Bubble and Simbet support message routing in Delay Tolerant Networks, an environment characterized by random or predictable disconnections, high latency and limited resources. DTNs are expected to be part of future network architectures and should therefore be considered as part of the 4G big picture. A different view is taken in (Cowan & Jonard, 2004) where a barter process is assumed between the members that can trade different types of knowledge. One may ultimately see shared knowledge as a mean to create a social group of interest or overlay, defining a trust relationship in order to increase the security of communication between nodes. In the context of security and information integrity, new semantics for information exchange may be adopted. For instance, one may need to lead with scenarios where a node may not accept information from other unknown nodes (Bernardos, Casar & Tarrio, 2006; Marti, Ganesan & Garcia-Molina, 2004; Marti, Ganesan & Garcia-Molina, 2005).

These untraditional algorithms commonly suggest the use of knowledge as a new network metric for routing and path selection. They are based on associating to paths information on their nodes, pheromone levels and location among others to guide the choice of the next best hop. Node knowledge influences routing by such algorithms. One can call such effect as stimulus for network behavior. A specific routing algorithm is evaluated in this chapter to shed light on the behavior of its nodes when submitted to stimulus. To summarize, knowledge sharing presents itself as a new paradigm for network route management in future 4G networks.

In this chapter, we build the case for the need for new routing metaphors in the context of 4G networks. We also show how future 4G nodes could cooperate and enhance their communications by learning from known epidemic, biological and social behavior. Such knowledge sharing reflects high level information exchange dynamics seen today across networks. We will show the dynamic changes to a community of devices constantly conducting knowledge diffusion due to stimulus activities. This is by far, a wider scope than traditional mere routing as it embraces service level advertising and selection, QoS information, node decision and discovery.

BACKGROUND

Routing algorithms are mechanisms to build paths and forwarding messages by selecting one or more intermediary nodes between source and destination of data messages. At the early days of the Internet, theses algorithms were designed to often lead with a single routing metric: the number of hops between routers. This was primarily due to the homogenous nature of the early IP backbone of the ARPANET. A general rule of thumb was adopted: the fewer hops a packet goes through, the less time it stays in the network and the fewer resources it uses. With time, the Internet evolved to a heterogeneous melting pot where such simple hop routing could no longer be sufficient. Let us consider that a network, expressed by N, is constituted by a set of heterogeneous routers (n nodes), then $N = \{n_1, n_2, ..., n_{n-1}, n_n\}$, where each node n_i can have a distinct total network capacity c_i and a set of known routes $\{r_1, r_2, ..., r_{n-1}, r_n\}$ that consumes the node's resources. Given that each router can have different capacities and routes known to it, its available capacity to forward messages can be represented by formula (1). Consequently, the network is constituted by a set of available capacities of nodes $X = \{x_1, x_2, ..., x_{n-1}, x_n\}$, where the network balance can be measured by the average and variance according to formulas (2) and (3) respectively.

$$x_i = \frac{c_i - \sum r_i}{c_i} \quad (1)$$

$$\overline{X} = \frac{X}{|N|} \quad (2)$$

$$\sigma^2 = \frac{1}{|N|} \sum (x_i - \overline{X}) \quad (3)$$

However, performance measurements itself generate overhead that should be known and controlled. Node and network measurement, as well as, the routing algorithms also consume resources. Overhead control may be subject to what is being evaluated, the operating capacities of each node, its dynamicity, mobility, constraints on the resources and the underlying network switching, transmission capabilities among other factors. As a result, several routing algorithms have emerged, where each one is better when executed in specific environment(s) according to some pre-established. On the other hand, there are some common characteristics among them, allowing their grouping and classification. Following (Tanenbaum, 2003) and (Royer & Toh, 1999), the traditional classification of routing algorithms looks for similarity in terms of three routing aspects: the process of route selection, whether it concerns itself or not with network balancing and process of route building, when routes are announced, including also what is announced and to whom.

Optimal or Shortest Path Routing

In terms of route choices, the traditional routing may be classified as Optimal or Shortest Path routing. An algorithm is said to present optimal routing, only when it has a convex function $f_{routing}(X): X \to D$, where D is always the best routing for all X, the convex set of parameters for this function. Although seemingly highly desirable, it suffers from some design limitations. It is difficult to build an optimized routing function that accurately captures all the performance measures and balances all data flows in a network. This task could lead to the use of routing parameters with incorrect or insufficient information, generating a probable unbalanced environment (Di Caro, 2004).

In general, many works define a function $f(x)$ which expresses how to measure the network performance using metrics such as delay, latency, packet loss, link throughput and so on. Next, $f(x)$ is minimized over variable x by applying one or more derivatives and defining the inequality of the form $g_i(x) \leq \alpha$ and/or equality of the form $h_i(x) = \omega$ as *constraints*, where g_i is a convex function and h_i is an affine function (i.e. one that consists of a linear transformation then a translation). As a result, an optimal routing algorithm controls the traffic and load balancing according to constraints of the minimum global value. The formula (4) could be a generic optimal function defined to choose the best route r_i (one with minimal load among neighbors of node k expressed by N_k) in terms of network load (expressed using constraints over the total mean and variance flow over the network).

$$C_{hoose}(d) = \{ r_i = \min(X) | r_i \in N_k \text{ and } \overline{X} \leq \alpha \text{ and } \sigma^2 \leq \omega \} \quad (4)$$

Although not strictly a routing protocol, acting between the link and network layers, the IETF multiprotocol label switching (MPLS) may be seen as one that falls into the optimal routing class. MPLS has emerged as a very promising traffic engineering tool used in the control of traffic within current broadband core networks.

The routing protocols, classified as Shortest Path Routing (SPF), are characterized as those where each network node k selects routes following a formula such as the one in (5): here node k chooses the neighbor that can route its message with the shortest metric between source (node k) and destination. The number of hops is the usually used metric, but others also may used in addition to the attribution of weight for each metric such delay, throughput and others. In this case, the shortest path is the route with lowest weight instead to be the one with smallest number of hops. Consequently, we could classify SPF as an optimal routing when the network traffic is always stable, where, for example, the user's network maintains $\overline{X} \leq \alpha$ and $\sigma^2 \leq \omega$.

$$C^k_{hoose}(d) = \{r_i = \min(X) | \; r_i \in N_k, \; \forall \overline{X}, \sigma^2 \} \quad (5)$$

However, this scenario is not a common one. Given that each node leads with estimated costs of link and does not worry whether its choices will unbalance the network traffic, SPF nodes tend to generate points of congestion on the network on a shortest path, along which most traffic is directed. Consequently, this class of algorithms is not an interesting one when dealing with highly dynamic network and disconnections. Dynamicity may turn the environment unstable with some nodes underutilized and others overloaded.

Distance Vector or Link-State Routing

Many routing strategies rely on mechanisms for sharing and announcing their routes among routing nodes. Traditional routing algorithms may be classified as Distance Vector and Link State under this angle.

Distance Vector algorithms are characterized by the fact that each network node announces the content of its routing table, known as a vector in this context, to "only" its neighbors. Although such limited scope announcement saves networking resources, it is also the root for some known route convergence latency related problems suffered by large networks. Each vector entry contains local information on some node k and how this may be reached in terms of number of hops and delay for example. Each neighbor for node k receives its advertisement (routing vector or information from node k) to build a routing database with size m.m$_{etric}$, where m is the number of routers known by node k and m$_{etric}$ is a number of metrics announced about each node. So, the traffic size generated by each node is given by formula (6) in order for a network node to get to know the network's topology. Once, a node acquires this knowledge through these advertisements, each node locally calculates the importance of each other node in the network according to its own vector and those received vector from the neighbors. Hence there is a combination of both local and external information to perform shortest paths selection. The result of such regular calculus updates the routing table based on algorithms such as Bellman-Ford and Ford-Fulkerson (Bellman, 1958; Ford & Fulkerson, 1962; Bertsekas & Gallager, 1992).

Given that the advertisement is relayed from neighbor to neighbor until all network nodes acquire network topology and path information, the protocols classified as Distance Vector suffer from slow convergence. The following protocols are example of Distance Vector: the Routing Internet Protocol (RIP) executed in intra-domain, Border Gateway Protocol (BGP) executed in inter-domain, DVMRP (Distance Vector Multicast Routing Protocol) for multicast applications, AODV (Ad-hoc On-demand Distance Vector) (Perkins, Belding-Royer & Das, 2003) for mobile ad hoc networks (MANETs) and DSDV (Destination-Sequence Distance-Vector) (Perkins & Bhagwat, 1994) also for MANETs.

$$A^k_{nounce} = |N_k| \times \sum_{i}^{n} |(n_i, m_{etric})| \quad (6)$$

link-state algorithms are different from distance vector ones in terms of the scope of their announcements. Unlike the former one where each node notifies the neighbors only about its vector, link-state routing floods such announcement to all the network nodes. Further, link-state updates are conveyed over small messages to all nodes. Each message contains the link-state of its neighbor's node such as occupied bandwidth, throughput, delay and so forth. A network node then builds a routing database with size m.n.m$_{etric}$, where m is the number of neighbor nodes, n is the number of routers in the network and m$_{etric}$ the number of metrics announced for each node. As a result, the traffic size generated by each node is given by formula (7) in order for network node to build its database and get to know the current topology. With this knowledge at hand, each node locally then calculates the importance of each neighbor to forward messages over the shortest paths determined for each destination. This may be achieved usually using Dijkstra's algorithm (Dijkstra, 1959), although there are others as described in (Cherkassky, Goldberg, & Radzik, 1994). The most common and widespread link state protocols used are the OSPF (Open Shortest Path Frist) (Moy, 1998) and IS-IS (Callon, 1990). Similarly, Multicast OSPF (MOSPF) (Moy, 1994) has been developed to lead with multicast transmission. Other link state protocols include SPF (McQuillan, Richer & Rosen, 1980).

$$A^k_{nounce} = n \times \sum_i^{|N_k|} |<n_i, m_{etric}>| \qquad (7)$$

Proactive or Reactive Routing

Proactive and Reactive are ways to classify a routing mechanism in terms of when an algorithm announces the routing route selection information. The routing mechanism may *anticipate* having this information or select a route on-demand due to the high dynamicity and unpredictability of a network environment.

The proactive routing announces periodically network information generating a predictability amount of messages. In a worst case, the time to select the correct path $t_s = t_{period}$ (period used to update paths). For example, formula (8) presents a possible generic function that may be adopted to calculate the number of messages generated by each node k when using proactive algorithms, where T is a set of announcement periods *t* independently of node or network faults. Hence, if there is large interval between path announcements, the number of messages will be small, nonetheless leading to a routing scenario practically lacking reactivity to possible network failures. Examples of pro-active routing include: DSDV (Destination-Sequenced Distance Vector) (Perkins & Bhagwat, 1994) and WRP (Wireless Routing Protocol) (Murthy & Garcia-Luna-Aceves, 1995).

$$M^k_{essage} = \sum_i^{|T|} A_{nnounce} \mid T = \{t_1 U \ldots U\, t_{n-1} U\, t_n\} = t \qquad (8)$$

On the other hand, the algorithms classified as reactive do not issue their route announcement messages following any periodicity. They rely on observed failures or user demand to update routes. Consequently the number of messages announced could be much lower than that of proactive algorithms. Given that they do not need to wait a whole period to update their routing information, they may have offer better performance in unpredicted environment. They may also however suffer from low performance when failures are predictable. In worst the case, the time t_s it takes to select the correct path may be given as the sum of $t_s = t_{fault}$ (time needed to identify a failure) + $t_{request}$ (to search for a correct path) + $t_{response}$ (someone response the node with the correct path). For example, formula (9) presents a possible generic function to calculate the number of messages generated by each node k using a reactive algorithm, where F is a set of independent failures. So, one cannot know the number of message without knowing the number of failures. When some fault occurs, node k is notified, next it announces a request to find the route and others nodes announce or send back their responses. Examples that fall into this routing category are AODV (Ad-Hoc On-Demand Distance Vector) (Perkins, Belding-Royer & Das, 2003), DSR (Dynamic Source Routing) (Johnson, Hu & Maltz, 2007) and LMR (Lightweight Mobile Routing) (Corson & Ephremides, 1995).

$$M^k_{essage} = \sum_i^{|F|} A_{nnounce} \mid F = \{f_1 U \ldots U\, f_{n-1} U\, f_n\} \qquad (9)$$

WHY DO WE NEED NEW ROUTING MECHANISMS?

If anything is certain, the fact that 4G system will be completely IP-based is one. Although homogeneous at the higher levels of the architecture, 4G most likely will integrate several nested networks with heterogeneous offered services, features, applications and service providers represented by mobile operators, companies or even individual users and content providers. The role of 4G systems in the wireless arena will be parallel to that of the Internet for fixed networks. It will be a common place to find the same service being offered by different sub-networks though maybe under varying security, bandwidth, delay, availability and pricing features.

While the prices for multi-radio and multi-interface devices are continuing to drop, new network topologies are also emerging including ad-hoc, mesh, hybrid and multi-dimensional ones. The latter are networks where a node may take part into different technology dependent networks at the same time, leading hence to a multi-layer view of the new topology. For this reason, new routing algorithms are designed to lead with new topologies and network challenges.

New Network Topologies

One expects that existing traditional network topologies such as those based on fixed infrastructure, ad hoc networks and others, to blend into 4G networks in a highly dynamic cooperating environment. 4G is also embracing highly dynamic networks such as delay tolerant or disruptive networks in addition to opportunistic communications which may be exploited in disaster recovery scenarios and military applications. The emergence of vehicular networks also presents new challenges to the design of 4G edge capacity and mobility support. VANs are expected to have bit rates close to those of fixed networks. New VAN topologies depict dynamic multi-hop short range vehicle-to-vehicle and vehicle-with-infrastructure communications.

Similarly to fixed networks, overlay topologies can run over 4G to offer virtual high level structures among individuals across 4G networks. Examples of such overlays are emerging new social networks, where topology information is primordial to improving data forwarding, privacy and security as seen in (Bernardos, Casar & Tarrio, 2006). Unlike current networks, a 4G one should open up further possibilities for individuals and groups to work, discuss and play together often by self-structuring driven by common interests, needs and incentives, instead of mere connectivity. The success of future 4G systems may depend to a great extent on their capacity to accommodate these new data services using social, biological and other nature inspired.

Challenges for Future 4G Networks

We have established that 4G convergence is expected to deal with several environments, including short range networks, which, in general, involve devices with capabilities subject to CPU, memory and disk limitations. Thus, with regard to routing, the next generation network should consider strategies that conserve battery power not only from a device survival point of view but more importantly in an attempt to use greener environment friendly solutions. These are also known as power-aware routing. For example, these devices should avoid the use of highly demanding table updates.

Routing mechanisms using location information have been proposed in order to improve energy consumption and data delivery. Note that the excessive use of location information from devices such as GPS for routing decisions can lead to a considerable increase in energy consumption when compared to location agnostic solutions. Moreover, vehicular networks are also expected to maintain location servers for vehicles to help in information routing.

In existing 2G and 3G networks, there is the horizontal cell connectivity handoff, one that involves cell changes by a terminal within the same type of network. 4G introduces also the vertical handoff. Here devices are expected to change networks they attach to. Hence, the challenge is to design routing protocols that are capable of handling vertical handoffs between pairs of *different types* of networks, while maintaining QoS requirements and optimizing common radio resources availability.

Under 4G networks, security is also seen as a paramount concern. Traditional attacks on existing IP networks will certainly migrate to 4G networks encouraged by both the heterogeneity and the open air interfaces of such networks. Routing information may be targeted in such attacks on system security, there is a need to design secure routing approaches to ensure the integrity of routing in 4G systems. This is not however the object of this chapter.

In summary, among the many routing-related challenges this work has identified, we mentioned QoS, security, energy and location information. The following section will present the details of new routing algorithms, with new network metrics and routing mechanisms, which motivate the extension of traditional classification to include epidemic, biological and social class of routing ideas, one expects to see in 4G networks.

NEW ROUTING AND CLASSIFICATION

There is an increasing interest in using non-traditional knowledge sharing and routing strategies in future 4G networks. This Chapter shows that information flow across 4G networks may benefit from analogies to many biological, physical and natural phenomena. Even if not all 4G services may rely on the concepts shown here, many should find in these a useful way for sharing knowledge and disseminating information in future 4G networks. The reader is first introduced to the main concepts and inner working of many of these new routing strategies.

Epidemic

Protocols classified as Epidemic behave similarly to viruses: infecting individuals within their access range. They may replicate a given piece of information as much as possible, ultimately leading to its spreading across a wide contamination area. Epidemic style information transfer does not require the maintenance of connections among the infected nodes.

Typically, each node n_i, executing an epidemic routing algorithm, sends messages according to its number of encounters with other node (devices within its range), consequently each announce of node k corresponds to the formula (11), where M_j and M_k are the index list of the messages from node respectively n_i and n_k. Both nodes receive and store the messages, assuming that each message has a unique identifier which is used as the access key in the list. When a node sends a message, each intermediary node updates its own list accordingly. In the route selection process represented by formula (10), each node requests $\neg(M_k \cap M_j)$ (copies of messages that it has not yet seen). The idea here is that each node should now obtain from the other one a copy of the messages it does not hold in its local storage space.

$$C^{k,e}_{choose}(d) = \{r_i \mid \neg(M^d_k \cap M^d_j) \neq \emptyset\} \quad (10)$$

$$A^{k,e}_{nounce} = \sum_i^{|E|} (|M_i| + |M_k|) \quad (11)$$

$$M^{k,e}_{essage} = \sum_i^{|E|} A_{k,e\ nnounce} \mid E = \{e_1 \cup \ldots \cup e_{n-1} \cup e_n\} \quad (12)$$

Epidemic protocols are executed according to the number of encounters expressed by formula (12). Consequently, the information is routed under *uncertainty*! There are no guarantees for message delivery and these protocols do not know of all possible paths from source to destination. As a result, some variations make use of acknowledgments to signal successful message delivery. The main feature of this class of routing protocols is a good delivery rate and robustness, they have been extensively used in the contexts of ad hoc, sensor, vehicular and Delay Tolerant networks (DTN) (Vahdat & Becker, 2000), where the delivery rate is more important than the delay. They also are part of solutions for disaster recovery and military deployment networks where no infrastructure is available or intermittent connections exist (Jindal & Psounis, 2006). The Fluid Nexus project[1] provides an Epidemic routing implementation for short-range wireless communication like Bluetooth. Adaptive storage and forwarding techniques are employed in order to support the disconnected transitive communications (Lindgren & Phanse, 2006), commonly present in wireless networks.

On the other hand, their flooding approach consumes a great deal of resources including energy, bandwidth and memory. As a result, variations of the Epidemic approach have been suggested in an attempt to minimize the latency and resource consumption, important to other scenarios such as sensor networks.

To avoid redundant connections, each host may additionally maintain a cache of nodes that it has contacted recently to within a configurable time period. This is useful to ensure that nodes that meet very frequently do not keep exchanging message lists wastefully. Moreover a given message can have a hop number or time to live field associated to it. This is seen as an important limiting threshold for the duration of message resource allocation. Once reached, no new copies are made of the existing message and this could be discarded even if not delivered yet. Other possible mechanisms may be considered for improving such scheme. For example, PRoPHET (Lindgren, Doria & Schelén, 2003) and HUM (Li & Shen, 2008) are hybrid algorithms that while using epidemic routing, include probabilistic techniques to improve their performance. Instead of broadcasting messages to all possible nodes as in the case of epidemic routing, these hybrid protocols use probabilistic analysis to make a selection, as explained in the following.

The PRoPHET algorithm is seen as an epidemic routing protocol that uses new artifacts to improve resource allocation in forwarding and announcing information (Lindgren, Doria & Schelén, 2003). It uses a probabilistic metric (p_{kj}) to determine whether a message is forwarded to a given device, changing the formulas (10) and (11) from epidemic routing to those enumerated as (15) and (16). The variable p_{kj} predicts the likelihood for a message arriving to n_j using n_k, for this reason, the node forwards data messages as long as $p_{ij} \geq T_{hreshold}$. It exploits buffered techniques since it does not send or store any additional messages for this purpose as well as the mobility pattern of nodes (number of nodes encounters) instead of the assumption of a random mobility. This last approach can be seen as a social approach that will further be presented when talking about social routing. In PRoPHET, when node n_j meets node n_k, the probabilistic metric p_{kj} involving n_j and n_k is updated according to the first formula bellow (12), where p_0 is an initial probability, a design parameter choice for a given network. When n_j does not meet n_k for some given time, p_{kj} is then decreased as shown in the next formula (13), where α^w represents the aging factor (< 1), and w is the elapsed time since the last update. Hence, each PRoPHET node exchanges its M_i message list and also P_i representing the set of all probabilistic metrics for each destination message from M_i. When the node receives M_i and P_i, it may compute the transitive delivery probability through n_j to node n_z through the use of formula (3), where β is a design parameter that highlights the impact of transitivity.

$$p'_{kj} = (1 - p_{kj}) p_0 + p_{kj} \qquad (12)$$

$$p'_{kj} = \alpha^w p_{kj} \qquad (13)$$

$$p'_{kz} = p_{kz} + (1 - p_{kz}) p_{kj} p_{kz} \beta \quad (14)$$

$$A^{k,p}_{nounce} = \sum_{i}^{|E|} (|M_i| + |P_i| + |M_k| + |P_k|) \quad (15)$$

$$C^{k,p}_{hoose}(d) = \{r_i \mid \neg(M^d_k \cap M^d_j) \neq \emptyset \text{ and } m^d_{etric} = p^{kj} \geq T_{hreshold}\} \quad (16)$$

The HUM algorithm is a second hybrid epidemic routing protocol that also uses a probabilistic forwarding mechanism of replicated messages to a chosen set of nodes. This protocol improves the delay performance and decreases the overhead due to replications and buffer management. It has three phases, namely, replication, forwarding and a clearing phase. In the first step (replication), N_c packet copies are sent to some of N_c's distinct neighbors chosen randomly. If the destination node $d \in N_k$, the transmission is obviously finalized with a successful delivery. Otherwise the second phase is executed, where each node n_k stores and forwards the utility ($U_{tility}(i,j)$) of another node n_j it has met. This utility provides reachability information to the destination node (correct path), where $T_{k,j}$ is the total meeting time between n_k and n_j in a time interval T_k. These steps change the formula (10) and (11) we have seen from epidemic routing to those shown in (18) and (19). During the last phase, when the destination node receives the information, the protocol sends a message to clear (free) the node's buffer.

$$U_{tility}(k,j) = T_{k,j} / T_k \quad (17)$$

$$A^{k,h}_{nounce} = \sum_{i}^{|E|} (|M_i| + |U_{tility}(i,k)| + |M_k| + |U_{tility}(k,i)|) \quad (18)$$

$$C^{k,h}_{hoose}(d) = \{r_i \mid \neg(M^d_k \cap M^d_j) \neq \emptyset \text{ and } m^d_{etric} = U_{tility}(k,j) \geq T_{hreshold}\} \quad (19)$$

These algorithms can yet be improved using policies to manage and decrease device resource consumption, including, bandwidth, buffer, memory and disk space. A section is dedicated in this chapter to the use of polices in the context of untraditional routing while also giving some insights on policy management for such limited resources.

Biological

Biological systems are seen as those composed of a large number of autonomous, distributed and dynamic units, generating effective adaptive behaviors as a result of self-organization, through local interactions. Even when submitted to constant environment changes, a biological system is able to effectively adapt. Colonies of insects are capable of self-organizing to solve a number of daily tasks such as looking for food and spreading knowledge about it, dividing work among their members efficiently, protecting their brood from external threats, building their nests, etc.

For this reason, social insects have revealed themselves as a useful inspiration for the design of routing algorithms by striking a parallel between the nodes in a network environment and the units of a biological system. The robustness of biological systems against internal perturbations or loss of units, and their capability to evolve and survive in a wide range of environments are some of the important and very attractive features that biological routing seeks to successfully mimic as seen next.

The ant model, for example, has been used as a good routing metaphor to find optimized paths. The ants, when searching for food use the services of a special first Aunt known as the Forward Ant. This Ant is responsible for finding a path by being the first to reach a place that has food. A Backward Ant is then sent back along the same path that the Forward one has used. Next, each Ant that follows the newly found path, deposits a chemical substance known as pheromone. Subsequent Ants will use the pheromone concentration level to ensure that they follow the same path to the food place. On other hand, the pheromone evaporates along the time, so the path may have no pheromone left after a while due to the dropping number of Ants that utilize it.

(Di Caro, 2004) is a work example that associates the Ant model and routing algorithm. To do this, their work allows each node to send three message types to improve the routing: m_{fa} (Forward Ant) used to find a route to a given destination, m_{ba} (Backward Ant) which updates routes and m_{data} (data message) to carry the data. The pheromone corresponds is used as a quality metric for each path and each network node k has a pheromone table $T^k_{|i| \times |j|}$, where p^k_{ij} is the probability of a neighbor's node k (say node i) forwards the message with a high quality metric to node j. It is obtained by combining the set of neighbors to node k and the set of network nodes different from node k. A constraint for $T^k_{|i| \times |j|}$ is stipulated that each column must obey (20).

$$\sum_{i \in Nk} p^k_{id} = 1, d \in [1, N], N_k = \{neighbors(k)\} \quad (20)$$

Given that each node forwards m_{data} to the node with the highest p^k_{ij}, this section describes only routing processes related to the messages m_{fa} and m_{ba}. In terms of the traffic level is generated by this protocol, it is similar to that shown earlier in (8) from proactive routing. Each node sends m_{fa} at regular intervals, to a random destination node d. For each m_{fa} forwarding, the node receives m_{ba}, consequently equation (23) expresses the traffic generated by each Ant node. When m_{fa} passes hops through several nodes until finding d, all visited nodes and their respective pheromone p_{id} are stored in m_{fa}. During the forwarding process, the route is chosen according to message type received, as in (24). If the message type is m_{data}, the node chooses the route that has the highest pheromone level from table $T^k_{|i| \times |j|}$. When the message is m_{ba}, the route used is the one carried into this message (backward Ant) and if messages is that of a Forward Ant m_{fa}, each intermediary node n_k forwards m_{fa}, where $l_i \in [0,1]$ and $\alpha \in [0,1]$. The variable α is the weight of importance with respect to the pheromone stored and l_i is the status of local link and the intermediary node (neighbor) must not be in V_{mfa} which is the list of already visited nodes by message m_{fa}. When there is already some duplicate node in V_{mfa}, m_{fa} is dropped When m_{ba} is forwarded from destination node d to the source node, the pheromone table of each node n_k in V_{mfa} and source node are updated with information from m_{ba}. For each node visited by m_{ba}, the table $T^k_{|i| \times |j|}$ is updated by incrementing the probability and decrementing this later using a constant r according to (21) and (22).

$$p^k_{id} = p^k_{id} - r(1 - p^k_{id}), \text{ if } \forall i \in N_k \wedge i \in V_{mfa} \quad (21)$$

$$p^k_{id} = p^k_{id} - r(p^k_{id}), \text{ if } \forall i \in N_k \wedge i \notin V_{mfa} \quad (22)$$

$$A^{k,a}_{nounce} = \sum^{|T|} |m_{fa}| + \sum^{|T|} |m_{ba}| \quad (23)$$

$\min(T^k_{|i| \times |j|})$, if $i \in N_k \wedge$ message $= m_{data}$

$$C^k_{hoose}(d) = \frac{p^k_{id} + \alpha l_i}{1 + \alpha(|N_k| - 1)} \text{ if }$$
$$\forall i \in N_k \wedge i \notin V_{mfa} \wedge \text{ message } = m_{fa} \quad (24)$$

Although the Ant model was defined in 1992, many new work initiatives have taken the task of improving this. For example, (Heegaard, Helvik & Wittner, 2008) uses the same AntNet principles to control the creation of m_{fa} messages, but it designed a mechanism to reduce the number of Backward Ants m_{ba} often seen as a potential for a high traffic cost. Their work also exchanged the decision function taken at intermediate nodes to detect an increase in m_{fa} to some destination, when signaling network changes by the replication of m_{fa}.

The routing algorithm based on a known Wasp model is another biological routing analogy that inspires 4G routing. The Wasp model provides a scheduling mechanism to improve the environment via local interactions. Wasps coordinate and distribute nest tasks depending of the amount of brood stimulus (response threshold) present in their zone and social hierarchies (the strength of wasp). Each wasp (θ_{wasp}) has a response threshold (t) for each zone (z) and may be or not allocated to perform a task according to stimulus (S_j - set of attributes relevant to a given request), where $\theta_{wasp} = \{ \theta_{t,z} \mid \theta$ is a response threshold t for nest zone z$\}$. If in an area or zone, its threshold is low the wasp engagement probability is increased for its activities on a stimulus.

Potentially, such simple reinforcement of response threshold may be easily used to describe any task allocation problem and the force variation may be applied to determine priority among the tasks. For example, in the context of a packet routing system, (Song, Hu, Tian & Xu, 2005) proposed a Dynamic Vehicular Routing (DVRP) algorithm based on the Wasp model to deal with the allocation and re-optimization of routes according real requests and the capacities of nodes (i.e. vehicular routers).

In a dynamic routing system highly frequent route updating is needed due to constant mobility and topology changes. More precisely, the Wasp model helps striking a balance by determining when to execute this task to avoid unnecessary overhead or inefficient routing due to outdated information. Hence a Wasp-like Agent Strategy (WAS) was used in (Song, Hu, Tian & Xu, 2005). This strategy is adapted to the routing system to collect some information about the nodes, or vehicles, including their geographic location and time window. The information (the zone of wasp) is then used to define when a node (or vehicle wasp in this scenario) should execute route re-optimization. Just like wasps, the vehicle wasp has $\{ \theta_{t,z} \mid \theta$ is a response threshold t for nest zone z$\}$. To distribute the requests of users for vehicular wasp and this re-optimizes the routes of vehicular wasps, the vehicular wasp routes the messages according to a stimulus, as presented in (25), that is based on the threshold of wasp ($\theta_{t,z}$) which is modified according the next rules.

$$P(\theta_{t,z}, S_j) = S_j^2 / (S_j^2 + \theta_{t,z}^2) \qquad (25)$$

- Rule 1 – ($\theta_{t,z} = \theta_0$) - If customer solicitations exceed the vehicle wasp capacity, then it needs to return to the depot and this threshold is set back to the initial state.
- Rule 2 – ($\theta_{t,z} = \theta_{t,z} + \partial_1$) - If after serving a customer's solicitations and still having capacity. If when this vehicle continues to serve another real-time customer it becomes overloaded, then its threshold is increased by the result of an undefined function (∂_1).
- Rule 3 – ($\theta_{t,z} = \theta_{t,z} - \partial_2$) - If after serving a customer's solicitations, and not running out of capacity, and this vehicle continues to serve another real-time customer while still without being overloaded, then its threshold is decreased in by an amount of undefined function (∂_2).

Therefore, $\theta_{t,z}$ is modified according the node's waiting time and its geographic location. In other words, if the node A has been waiting for a long period of time and is adjacent to node C, then node A has a better stimulus than node B that has been waiting for a smaller period of time and is located distant to node C.

However, if two or more vehicle wasps have the same stimulus at the same time, then the wasp hierarchy system is used to resolve what vehicular node must receive the user request. Just as in nature, each vehicle wasp force will be used to define the winner. Thus, this strength (F) is defined by the quantity of customers served by each vehicle Wasp and the one with the greatest number of served customers, consequently it will have a stronger or bigger force (F). The user request is routed for the vehicular Wasp with F_a with the probability (26) given as:

$$P(F_a, F_b) = F_a^2 / F_a^2 + F_b^2 \qquad (26)$$

During the process of route selection according to this wasp model, firstly a set of routes R_i is chosen according source node zone and highest stimulus of each neighbor node N_z. Secondly, if $|R_i| > 1$, then the force F_j of each node in R_i is evaluated. Hence, the messages are forwarded according to (27), where the chosen route r_i consists of nodes n_k and n_i while taking into account the fact that both are localized in same zone z, the highest stimulus and force value. In terms of when and to who announces routes, this algorithm may follow announcement mechanisms from any traditional routing mechanism given that that the stimulus $P(\theta_{t,z}, S_j)$ and force F_j at each node are part of the information announced.

$$C^{k,w}_{hoose}(d) = \{r_i \mid R_i \in N_z \text{ and } R_i = \max(P(\theta_{t,z}, S_j), R) \text{ and } r_i = \max(F_j, R_i)\} \quad (27)$$

Therefore, biological routing algorithms associate the routing challenges to animal tasks, such as Wasp task allocation coordination and Ant food search, providing solutions based on animal behavior. Typically, these algorithms make their announcements using the traditional approaches to routing but instead of announcing traditional metrics such as delay and hops, they advertise biological metrics (i.e. stimulus, pheromone and strength) and also improve the process of route selection.

Social Routing Models

Social routing mimics the way people exchange information within their friendship, work and other social circles. It is based on the observation that information reaches more people and destinations when going through popular nodes. This fact increases the probability of finding suitable information forwarding carriers to any given destination. For such reason, this section examines the concepts behind the design of such routing metaphor.

Centrality, similarity and graph theory are extracted and combined by social algorithms to provide message forwarding. Unlike the previous routing strategies, this one deals with several types of social knowledge in order to choose the next hop and establish whether a given message must be dropped or forwarded. This is a departure from using random or predictable routing decisions. The social routing is concerned with performance, offering ways to decreasing the forwarding cost, as well as the overhead generated by routing mechanisms such as flooding. Although social ideas can be applied at network or application levels (i.e. social overlay), initially this section presents ideas and algorithms adopted at the network level and the next section provides some useful metrics and information that could be extracted from a social overlay and applied to network routing.

Graph theory is widely used in the modeling of network and application level information exchange. The researches firstly identify the graph represented by $G = (V, E)$, where V represents a set of nodes (or vertexes) and E represents a set of links between the nodes in N. Moreover, a Boolean function ($c_{onnected}$) maps the vertices to show the relationship among the nodes. If v_i and v_j are two nodes in V and $c_{onnected}(v_i, v_j) = 1$, then they constitute a direct link in E connecting both nodes, otherwise $c_{onnected}(v_i, v_j) = 0$, but $c_{onnected}(v_i, v_j) = 1$ by default. The connectivity level or number of the v_i neighbors is seen in (28), the closeness value is calculated by the addition of all distances between a node and each neighbor and is expressed by (29) whereas the *betweenness* value is presented by the expression in (30) where g_{jk} is the number of shortest geodesic paths from j to k, and $g_{jk}(v_i)$ is the number of shortest geodesic paths from j to k that pass through a node v_i.

$$n_{eighbors}(v_i) = \sum_{k=1}^{N} C_{onnected}(v_i, v_k) \quad (28)$$

$$C_{loseness}(v_i) = \frac{N-1}{\sum_{k=1}^{N} \text{int}_{erval}(v_i, v_k)} \quad (29)$$

$$betweenness(v_i) = \sum_{j=1}^{N}\sum_{k=1}^{j-1} \frac{g_{jk}(v_i)}{g_{jk}} \qquad (30)$$

Random graphs, small world (or power law) and scale-free networks are among the most commonly used models of the graph theory. Under random networks the nodes end up having roughly the same connectivity level (Erdös & Rényi, 1959). Small world networks, also known as six degrees of separation in the case of the Web, preach the fact that a small number of nodes has most links and is therefore highly connected. Consequently, the average distance among any two nodes is lower than a given value, hence the name small world (Kochen, 1989). Under this connectivity model, the rich get richer (Barabasi & Albert, 1999). Similarly, Michalis, Petros and Christos show that Inter-domain Autonomous System (AS) connectivity follows a Power-Law (Willinger, Taqqu, Sherman & Wilson, 1995).

The centrality concept expresses the popularity and importance of a node in a social network. We can divide centrality into three metrics: degree, which is the node popularity obtained by formula (28), closeness ($c_{loseness}$), which represents the distance of some node to others expressed by (29), and betweenness, which reflects the node with a high communication level to other nodes from its environment being presented by (30). These metrics are important to identify the "centre" of a network in an attempt to increase the successful forwarding among sources and destinations.

Similarity expresses nodes proximity through their common neighbors see equation (31), where x and y are two vertices and N is a set of neighbors for some given node. This can be easily evaluated using historical encounters and due to high transitivity in social networks. This metric is important to creating clusters and to route messages between nodes with low degree separation. We can identify similarity among a source and a next hop in order to choose the next neighbor that will receive the message and forward it next.

For example, all nodes could be have some label associated to them to specify their community and each one only forwards those messages from nodes with similar label values. This way, a community that constrains routing to nodes with some specific similarity, such as support for security or traffic shaping for example, may be created and enforced. Since this algorithm mimics social behavior, it could be used to control performance and security in disruptive, opportunistic, dynamic and vehicular networks. This is in a way similar to the use of the IETF MPLS in wide area networks to offer similar traffic engineering benefits when set adequately.

$$S_{imilarity}(x,y) = |N(x) \cap N(y)| \qquad (31)$$

(Hui, Chaintreau, Scott, Gass, Crowcroft & Diot, 2005) studied the contact disruptive and human mobility pattern among the participants of some conferences. They identified a power-law model in terms of the duration of people's contacts at the Infocom 2005 conference and that there was always a group, called a bubble, which could be used to influence the forwarding of information. In general, the social algorithm calculates a social metric before taking a decision on whether to send or drop messages.

The Bubble algorithm is an example of social routing. It classifies the nodes in groups and communities before routing any messages. Firstly, its authors establish that each node has one or more groups (e.g. school, work, sports or accumulative contact duration). A set of local neighbors is able to constitute a community provided they contain similar groups. For example, each node in a network classifies its encounter nodes (n) according to proximity and by group $G = \{G_i \cup \check{G}_i|$ where n is a neighbor node and $n \in G_i$, or n is not a neighbor and $n \in \check{G}_i$ if the accumulated contact duration between the node and encounter node has value i. Consequently each node has its own community (C_0) constituted by its neighbors. In Bubble routing, when two nodes find each other, they exchange their communities and groups following the formula (32). Each node may then accept or decline the other nodes in its communities or/and perform a merger of the communities according to similarity. A node accepts another one n_i, if $|G_i \cap C_0|/|G_q| > \lambda$, where λ is an acceptation threshold in a community. The communities of two nodes are merged when $|C_i \cap C_0|/|C_q \cup C_0| > \gamma$, where γ is a threshold to merge two communities. Therefore, a community is a concept drawn from the real world, where people may play several roles (groups) and participate of several communities (e.g. school, work, sports or accumulative contact duration), while having a possibly different ranking in each of these. For instance, a given popular node, according to the centrality metric, is not necessarily popular in other groups. Bubble forwards messages according formula (33), it uses the popular nodes and members of destination communities as relays or the multi-homed nodes i.e. nodes belonging to more than a community. The current given node sends its message to another one if both have some group in common $s_{imilarity}(G_k, G_j)$ as shown by formula (31) and this one has more communities than the current node using formula (28), where $N = |C|$.

$$A^{k,b}_{nounce} = \sum_{i}^{|E|} (|M_i| + |C_i| + |G_i| + |M_k| + |C_k| + |G_k|) \quad (32)$$

$$C^{k,h}_{hoose}(d) = \{r_i \mid s_{imilarity}(G_k, G_j) \neq \emptyset \text{ and } n_{eighbor}(n_j) > n_{eighbor}(n_i)\} \quad (33)$$

The SimBet (Daly & Haahr, 2007) proposal is yet another social-based routing algorithm. It improves the standard epidemic algorithm by reducing resource usage overhead. Here, SimBet nodes meet each other through the traditional use of the Hello message. Upon receiving such message, a node verifies if it has some message to deliver to this new neighbor, it calculates the social metrics according to its set of encountered nodes before sending an encounter request. Nonetheless, it does not use the formulas (28, 29 and 30), because it would need the complete network knowledge. Each SimBet node stores a symmetric matrix {n x n}, where n is a number of encountered nodes and $A_{i,j}$ = { 1 if there is a encounter between i and j ; 0 otherwise}. The betweenness metric (Bet_n) of node n is calculated by $A^2[1-A]_{i,j}$; and the similarity ($Sim_n(d)$) between nodes n and d is given by the number of non-zero equivalent row entries of the product matrix $a \times b$, where a is the number of the common direct contacts of node n and b is the number of direct contacts of node n that reach the node d. Theses social metrics are used to calculate the utility of node n for delivering a message to destination node d compared to another node m, for example, according to formulas (34, 35 and 36), where α and β are tunable parameters and $\alpha + \beta = 1$.

The SimBet routing algorithm is similar to the standard epidemic algorithm in that both nodes exchange a summary vector with their current destination nodes. In addition, this SimBet algorithm updates the social metrics ($SimUtil_n(d)$ and $BetUtil_n(d)$) for each destination being identified in the exchanged messages. However, SimBet nodes do not exchange all messages from a summary vector. Luckily, they transfer only those messages that actually have the destination with high $SimUtil_n$ and $BetUtil_n(d)$ values. This is identified after evaluating the entire summary vector according to the established social metrics. For this protocol, the cost of route announcement is similar to the one for epidemic routing given by formulas (11) and (12). The difference lies in the metric used to choose a route. Unlike epidemic routing, SimBet selects the route according to (37). The objective here is to route messages from unknown destinations to a popular node (one with high $SimUtil_n$ and $BetUtil_n(d)$) in order to increase the probability of finding a suitable forward carrier.

$$SimUtil_n(d) = Sim_n(d) / (Sim_n(d) + Sim_m(d)) \quad (34)$$

$$BetUtil_n(d) = Bet_n / (Bet_n + Bet_m) \quad (35)$$

$$SimBetUtil_n(d) = \alpha SimUtil_n(d) + \beta BetUtil_n \quad (36)$$

$$C^{k,s}_{hoose}(d) = \{ri \mid \neg(M^d_k \cap M^d_j) \neq \emptyset \text{ and } SimBetUtil_j(d) > SimBetUtil_k(d)\} \quad (37)$$

The reader may note that although PRoPHET was previously classified as an Epidemic type of routing solution earlier in this chapter, it nonetheless portrays a social feature too. It evaluates previous node encounters to decide on routing. Similarly to PRoPHET, the algorithm in (Leguay, Friedman & Conan, 2006) takes its routing decision based on meeting frequencies for message forwarding and may be described with the same formula (12). However, (Leguay, Friedman & Conan, 2006) proposal counts the number of meetings between a node n_k and a location l_n, where $n_k = m_k(l^k_1, l^k_2, ..., l^k_{n-1}, l^k_n)$, because n_k is expressed as a n-dimensional space to all n possible node locations, creating a virtual space through a high-dimensional Euclidean space (called MobySpace). Hence, when the nodes meet themselves, they exchange theses probabilities (n_k) as in (40) and according to the number of meetings between n_k and l^k_i, each node k updates its probability of meeting again l^k_j, where $\sum_{k=1}^{n} l^k_i = 1$ and the distance between n_i and n_j is given by the Euclidian distance between them as shown in (38). Considering such issues, the routing decision ($C^{k,sm}_{hoose}$) is shown in (40), where $W_k(t)$ is a set of neighbors to node n_k at a given time and $W_k^+(t) = W_k(t) \cup \{k\}$. The messages are forwarded closer to the neighbors or routed to nodes that have similar mobility patterns. An important issue here is that location information is not necessarily in the form of geographic coordinates, it can be any information that characterizes some given location name, i.e. a University laboratory, shopping, cafeteria and so on. In this algorithm, the nodes can learn about their location in MobySpace by evaluating their contacts and the frequency of visits to several other locations. So, this work involves mobility pattern information, similarity of mobility and the ability of nodes to learn about their own mobility.

$$d(n_i, n_j) = \sqrt{\sum_{k=1}^{n}(l^k_i - l^k_j)^2} \quad (38)$$

$$A^{k,sm}_{nounce} = \sum_{i}^{|E|}(|m_i(l_1,...,l_{n-1},l_n)| + |m_k(l_1,...,l_{n-1},l_n)|) \quad (39)$$

$$C^{k,sm}_{hoose}(d) = \{r_i \mid \neg(M^d_k \cap M^d_j) \neq \emptyset \text{ and } n_j \in W_k^+(t): d(n_j, n_d) = \min_{nj \in W_{k+}(t)}(d(n_i, n_d))\} \quad (40)$$

The authors of Sociological Orbit aware Location Approximation and Routing (SOLAR) protocol observed that user movements often followed a pattern around a set of specific places (hubs or attraction points). Each SOLAR node has a profile that describes the mobility pattern of a given node - called "orbital" - a list of hubs and node probabilities to move or stay during some time at each hub. Differently from other works on Epidemic routing that rely on the use of the meeting encounter information to predict the mobility and occurrence of other meetings, SOLAR believes in partially deterministic movement of nodes (or memoryless), where the next node movement depends only its last state. For this reason, a Markov chain model is used to model the movement of a node k. It is represented by a matrix M^k_{ij}, containing all possible mobility states of node k, the probability p^k_{ij} of node's state in hub i moving to hub j according to equation (41), where $\beta^k_{hh'}$ is the hub transitional probability, $\beta^k_{hh'} > 0$, $\sum_{h' \neq h} \beta^k_{hh'} = 1, \forall h \in S$ and S is the set of hubs. When two nodes meet, each node sends its matrix M^k_{ij} in order for each one to calculate the contact probability ($A^{contact}_{i,j}(X,Y)$) between nodes x and y by establishing the Cartesian product of their matrices (42). Moreover, the nodes update the weight of contact probability according to (43) and calculate the delivery probability of other node delivery of its message according to

(44), where there are i nodes between node k and destination d. Consequently this proposal offers a new scheme to find such users if their individual orbital is known. Their approach is based on the use of a geographic algorithm for data forwarding. For instance, if some student spends a lot of time at home, the computing laboratory and cafeteria, then the student's profile will register these locations at S. In this protocol, each node sends a Hello message with its position and $M^k_{i,j}$ (mobility profile) periodically (same proactive formula (8)) to its neighbors, consequently the nodes can exchange their mobility profiles once within the same radio range. When a source node wants to send some data to a destination d that is not among its neighbors, it selects the next node with highest f^k_{id} value according to (45), where f^k_{id} is the message delivery probability (obtained by (44)) from outgoing link i from k to destination d. However, if there is no information about d in its hub list, the source node sends a query about the likely hubs of the destination to some acquaintances that can propagate the query until finding d or that there is a time out. The response message to such query contains the current hub of the destination, its hub list in addition to geographic forwarding information. Alternatively, the answer may simply contain an update message with a hub list instead of the exact geographic positions of each node. Clearly this protocol differs from the one in (Leguay, Friedman & Conan, 2006) in terms of metric and routing approach. SOLAR depends on a partially (limited to knowledge of the last node) deterministic mobility of nodes to calculate the delivery probability and consequently to select the next hop. One problem found with this approach is that it does not protect node privacy in terms of location as other nodes will know all possible geographic positions (list of hubs) of a given one. A workaround may be for users to decide on the type of location information they want the routing strategy to turn public about them. This however may limit the efficiency of this type of social routing and limit its applicability in real networks.

$p^k_{ij} = \beta^k_{hh'}$, i = hub h and j = other hub h'

$M^k_{i,j} = p^k_{ij} = 1$, i = hub h' and j = same hub h'

$p^k_{ij} = 0$, otherwise, (41)

$A^{contact}_{i,j}(X,Y) = M^X_{i,j} \times M^Y_{i,j}$ for only common set of hubs (42)

$\omega_{contact}(k,j) = -1 * \log(p^k_{k,j})$, where $k \neq j$ (43)

$f^k_{id} = \omega_{contact}(k,i) * (1 - \prod^n_1(1 - f^1_{zd}))$, where z is outgoing links from node i (44)

$C^{k,so}_{hoose}(d) = \{r_i \mid r_i \in N_k$ and $r_i = \max(f^k_{id})\}$ (45)

Therefore, the social algorithms look for human behavior and human habits to benefit from them (such as mobility around a set of places), searching to identify popularity, similarity, communities, preferences from people in order to apply these into the routing network. However such social metrics can be applied at both network or application levels (i.e. social overlay). We already presented the ideas and algorithms associated to routing at the network level. The next section in this Chapter examines metrics and information that could be extracted from a social overlay and applied to network routing.

SOCIAL OVERLAY NETWORKS

The overlay network approach allows the creation of several virtual networks over physical ones. Here one hop in a virtual network may correspond to several hops from one or more physical networks (underlays). In turn, a virtual network may even be built over other virtual networks. However the overlay routing does not have direct control over how the underlay forwarding of packets is actually performed. As a result one may not classify solutions such as Buble and SimBet as mechanisms to build overlay networks, since their nodes have routing control over the physical layer. Other routing algorithms found in widely spread Peer-to-Peer (P2P) social networks such as Gnutella, Chord and Tapestry are examples of peer-to-peer protocols that can be used to create overlay networks. In order to improve these overlay networks, (Motani, Srinivasan & Nuggehalli, 2005; Banks, Ye, Huang & Wu, 2007; Tomiyasu, Maekawa, Hara, & Nishio, 2006) and other works applied social mechanisms and consequently we named these proposals as social overlay networks. They deal with security and optimization using social ideas extracted from the underlays.

DSL (Davis Social Links) in (Banks, Ye, Huang & Wu, 2007) is an example of a proposal that applied social approaches to deal with security. It has a social overlay mechanism that creates trust relationships of social networks based on the small-world phenomenon in order to control, trace and separate address and identity information. The idea is to allow communication between nodes with a direct link or a social path linking them. Messages in the DSL social network contain a set of keywords describing node properties. The social path is created by the exchange of keywords between nodes connected by direct links. The nodes can accept or drop messages from some social path or modify these along another one. There is consequently a recipient controllability being exerted. For example, nodes A and B have "red" and "yellow" as part of their keyword sets respectively. These strings are cryptographically exchanged prior to the exchange of data. A Node can propagate the keywords in order to increase routability.

Work in (Tomiyasu, Maekawa, Hara, & Nishio, 2006) is a bit similar to social routing while not applied at the physical layer however. SOLAR, seen earlier, imports a user's profile information to forward data. In SOLAR, the profile is defined by a node itself and describes node mobility and its likely locations. Differently from SOLAR, (Tomiyasu, Maekawa, Hara, & Nishio, 2006) acts at the application layer where the profile manipulates the dissemination and routing of messages. An even greater difference is to do with the fact that the profile in (Tomiyasu, Maekawa, Hara, & Nishio, 2006) may be updated and created by friends and acquaintances of a given node instead of being the sole responsibility of the node as in SOLAR. In (Tomiyasu, Maekawa, Hara, & Nishio, 2006) when a node creates a link, the user must attribute one or more keywords to describe the new friend. This one, the friend, needs then to agree with such description to allow for the effective link creation. As described by an application in (Tomiyasu, Maekawa, Hara, & Nishio, 2006), it is important to find a best person to answer some questions about some given keywords. This idea for creating profiles by friends could be implemented at the network level in terms of performance and made available to view by other nodes. Consequently, a profile that describes the options and possibilities for other nodes could be an interesting mechanism to help the node decide if another encountered node is a good relay to forward its messages.

PeopleNet (Motani, Srinivasan & Nuggehalli, 2005) is another social mechanism that propagates information using overlay networks. It has only three types of messages: request propagation, request and response. The request and response messages are always forwarded over long range connectivity such as over a cellular infrastructure; the propagation messages are always broadcasted with shorter range connections until some node matches the request propagation with some response. Whenever a propagation message matching occurs, the user who placed the request message receives the response message via long range connectivity. Moreover, the users can pre-determine the type of queries to handle in some specific geographic context, called Bazaar. So, any person close or distant to a Bazaar can send requests through the cellular infrastructure to other users in a specific geographic (Bazaar). Information is spread around to users in a specific geographic location, but it does not benefit from the number of meetings among nodes in a specific geographic area as in SOLAR.

PeopleNet differs from the proposal defined in (Leguay, Friedman & Conan, 2006) as it does not concern itself with any mobility pattern, similarity of mobility and the ability of nodes to learn about their own mobility. PeopleNet relies on the innovative idea of using the widely deployed cellular infrastructure and Bluetooth devices to propagate information search. A second peculiar contribution involves the overlay routing according to the meta-information (i.e. Bazzar and message type) to choose what connection must be used to forward the messages. So, one could extract the importance from high level information that could be used in routing.

SPROUT (Marti, Ganesan, & Garcia-Molina, 2004) presents a social mechanism to route messages in overlay networks such as Distributed Hash Tables (DHT). It is based on the use of the knowledge of a trust relationship among social nodes to choose what node must receive a message and to associate message priority. SPROUT presents possible trust function according to the number of hops in a social overlay (the distance d_{ij} between social nodes n_i and n_j). The relation in (46) is one of its trust functions used to choose the next hop, where f is a static probability for two nodes to be trusted friends. The reader may note that the probability of two nodes being friends is limited by the value r. For example, if $f = 0.95$ and $r = 0.6$, then the trust function assumes that the friends with high proximity of node n_i are best friends (reflecting a high trust relationship) and consequently very likely to correctly route a message and when $d_{ij} > 8$ the trust will be maintained as 0.6. The objective of this work is to reduce the number of several network attack types that may drop the packets or forward the data to any different node other than the correct destination. For instance, in a DHT structure, a malicious node may exchange messages in order to disseminate unwanted information. SPROUT locates the trustiest friend of a given node that has a closer identifier, but not greater than, a key value until finding the destination node for that given key. Should this fail, then the source node executes the traditional DHT process. Although this work has been implemented in overlays, it could also be applied at the network layer. A node should evaluate the trust relationship of its neighbor instead of choosing the shortest path, which may be an unsafe path in terms of optimization and integrity. Further, trust identification remains a hard undertaking.

$$t_{rust}(d_{ij}) = Max(1 - (1-f)d_{ij}, r) \qquad (46)$$

On the other hand, the trust relationship is not immune to attacks completely. Malicious nodes may convince a small number of honest ones trough the creation of several and false identities to increase their influence and credibility in the network. Looking at this problem SybilGuard (Yu, Kaminsky, Gibbons & Flaxman, 2008) and SybilLimit (Yu, Gibbons, Kaminsky & Xiao, 2008) map users and nodes, separating the network in two groups: a honest region with nodes with only one identity and a Sybil region populated by malicious nodes with more than a single identity. They also established that the number of links between these regions (called attack edges) is independent of the number of malicious identities. Moreover, if a trust route contains only nodes into the honest region then all the routes that cross the same node or edge will converge. Therefore, one may observe that there are several works applying social approaches in order to improve information propagation in overlay networks, security and optimization.

Propagation in social overlay and underlay is very similar, but there is a little difference. Both the optimization and security are interlaced in the context of social networks. This may be the case when messages are dropped or wrongly forwarded.

A number of security enhancements have been suggested in (Marti, Ganesan, & Garcia-Molina, 2004; Yu, Kaminsky, Gibbons & Flaxman, 2008; Yu, Gibbons, Kaminsky & Xiao, 2008) to improve underlay security. User's devices are used to route data in DTN scenarios, increasing the likelihood to disseminate worms and viruses, as their users are often inexperienced with regard to such security threats. Moreover, networks with high-degree nodes tend to connect to other high-degree node networks (the famous often move in the same circles) and are therefore more likely to be subject to epidemics. Indeed a single infected high degree node will quickly infect other high-degree ones. On the other hand, networks where high-degree nodes tend to connect to low-degree nodes show the opposite behavior; a single infected high-degree node will not spread an epidemic very far.

POLICIES FOR UNTRADITIONAL ROUTING

The untraditional routing algorithms, as seen in the previous section, deal with different quality metrics associated with nodes, location or other information used to choose a suitable relay node. On the other hand, most traditional routing policies deal with only address and interface metrics, where the objective is the management of network traffic in order to decrease its latency, cost, congestion and other metrics. For example, policies for BGP, the current inter-domain routing protocol in the Internet, may be used to update routing tables and consequently to control the traffic forwarding, taking into account path lengths, local preference values, traffic origin among others factors. In terms of software, the IPFilter[2], iproute2[3] and netfilter[4] are utilities commonly used to create and apply network policies in order to control the routing in local area networks.

However, these approaches are based on the origin-destination pair of IP addresses and interfaces and are not suitable for 4G networks where routing involves often multi-homed end-hosts, intermittent connection, unpredicted mobility and other dynamic characteristics. We need new policies and mechanisms to deal with buffer, message replication and scheduling schemes.

In untraditional routing algorithms, information forwarding can be based on replication (usually broadcast) or on the basis of the pair: source and destination nodes. When using replication, one needs to evaluate and establish the necessary number of replicas according to the underlying environment being used (network size, mobility model, etc.). This is important to avoid network congestion and unnecessary overhead and resource usage. Note that a replication level or value must be carefully set as a small number of replicas may very well become insufficient for the dissemination of information and discovering all nodes and paths. Changes to the number of neighbors over a short period of time indicate a high mobility scenario and consequent frequent topology updates including the network size. This information is a very determinant factor in setting the replication policy and its parameters.

One may choose to apply a policy over the forwarding mechanism based on source and destination information, whereby a node is not required to send all messages. Forwarding may be subject to priority in order to give preferential treatment to some messages for example or offer differentiated traffic. Work in (Lindgren & Phanse, 2006) evaluated four such strategies: GRTR, GRTRSort, GRTRMax and COIN. GRTR defines a mechanism that allows forwarding to a neighbor if the encountered node has higher delivery predictability to the message destination. When a node A uses GRTRSort with its neighboring node B, node A subtracts both delivery predictability values for this message destination and then forwards the message for node B using the highest subtraction result it could find. The GRTRMax strategy is also a little similar to GRTR and the difference between these lies in the ordering executed by GRTRMax which is before comparing the delivery predictability. The last policy example is given by COIN which generates a random value for each message. Depending on whether this is higher than 0.5, then the node forwards the message which is similar to tossing a coin and taking a decision according to the outcome.

Independently of whether the message forwarding policy is based on broadcast, replication use, or according to source and destination, the buffer (where messages are stored before being forwarded) is obviously not an infinite resource. Therefore, one needs to have one or more policies in place to determine what to do with the messages. Common per hop behavior or queue management techniques such as FIFO (First In First Out), MOFO (Most Forwarded first) – messages, MOPR (Most favorably forwarded first), SHLI (Shortest Life Time First) or LEPR (Least Probable First) may apply.

Policies may also determine for how long a message could be stored. For example, the Data Mule project considers that a mobile node must store information until finding or coming into contact with a destination (in this case, a base station within a sensor network). Moreover, Data Mule chooses the next hop according to the mobile's behavior, all the way from the fixed nodes, by forwarding the data to intermediary mobile nodes that collect and store the information until the base station is reached.

(Motani, Srinivasan & Nuggehalli, 2005) is another routing work involving policies. In this social overlay shaped by groups, a node needs to decide which query could be dropped, when the buffer is full or simply choose randomly queries to swap. However the authors suggest that the nodes should firstly exchange meta-information, before deleting and adding queries (buffer management). Although it uses a simple decision mechanism, it follows the policy approach to choose what queries must be dropped from the buffer and exchanged with its neighbors. Instead of using such a simple policy model, it could adopt a social policy that gives higher priority to messages forwarded from friends or partners over strangers or unknown senders.

In summary, this section presented how important it is to work towards reducing costs and the requirement for policy changes according to behavior modification in order to provide more flexibility for routing mechanisms. Moreover, it has shown that policies may be applied to control several device features (i.e. buffer, message replication and message scheduling schemes) to determine, configure and improve 4G routing.

ANALYSES OF UNTRADITIONAL ROUTING

In previous sections, the restrictions imposed by traditional network technologies were presented and we showed how new ways for thinking about routing have emerged to overcome these. They include insights and parallels made from observing a number of biological, social and epidemic behaviors. A number of proposals, associated to these metaphors, make use of mobility patterns, pheromone levels, user habits and profiles, relationships and other types of stimulus to offer self-organization, load balancing, adaptability and advanced technology dependent routing. This section is going to perform some concrete evaluations to show and determine the impact of some of network and other important parameters and examine their configuration. To achieve this, the reader is invited to review some optimization and evaluation techniques that are very much relevant to the context of routing in future networks,

Percolation

The percolation theory (Grimmett, 1999) is inspired from the observation that there is a limit value for a physical material to make a transition between two states called by "critical phenomenon". For instance, water (a fluid) has two states: liquid and gas. A bottle of water may transition from the state liquid to gas when submitted to a higher temperature, namely, at 100°C at sea level. Another example is that of a filter where there is a given alpha number of porous in a stone. When the number of porous reaches a threshold, water, then, passes to the other side of this stone. These probabilistic changes of states are defined according to a percolation model that uses a threshold to determine such transitions. Hence, such strategy would help determining which routing parameter values would cause percolation, or successful knowledge sharing in the context of future 4G networks.

Some works set up a static percolation coefficient value in order to improve routing. The spatial gossip (Kempe, Kleinberg & Demers, 2001) is an example of a routing algorithm that used this to select the forwarding node. Other works chose to evaluate the environment to discover when such algorithm percolates. For instance, one could seek the relation between buffer size and the success delivery rate. Otherwise, one could check if there is a limit buffer size that determines success or no delivery of messages. The analogy in this example associates messages to a fluid in a percolation scenario and nodes to the surface. Consequently, when all the messages start going from the source and reach their destination, one says that routing has percolated.

Diffusion and Chemotaxy

Adolph Fick was among the pioneering researchers who studied extensively the diffusion process. He observed that salt movement occurs from high to low concentration in liquids and defined an equation to express the proportionality between the flow and the spatial gradient of diffusion. Other researchers also studied the diffusion observing a spontaneous particles movement from low to high concentration (Narasimhan, 2004). However, there is a common concept among these equations: they expressed the movement of cell or substance to obtain equilibrium, considering, in general, the position as a variable or both time and position.

Similarly, Chemotaxy is a movement behavior according to the gradient of concentration, but it is not a spontaneous event. Chemotaxy represents the attraction or retraction among cells due to some substance. It is commonly used in biology to analyze the behavior of human cell, virus or bacteria. However, such behavior has been analyzed and shown to also benefit the routing environment. Routing policies could be seen as the substance that modifies spontaneous movement.

Given that some message forwarding is based on a probabilistic mechanism set according to the encounter frequency of nodes (i.e. PRoPHET). We could evaluate the diffusion by modifying node movement in order to verify whether node mobility could be a stimulus to influence this behavior or not. In other words, we could check whether node mobility increases or not the message delivery rate.

Considering that PRoPHET could also be executed in sensor networks, policies are likely to move a node by several spaces in order to increase the encounter frequency and as a result may be used to improve the delivery ratio. Alternatively, one could set fake information altering encounter frequency, the message delivery decreases, because messages are removed from a buffer before actually finding their destination.

Stigmergy

Pierre-Paul Grassé (Grassé, 1959) introduced the Stigmergy concept after studying nest building. He observed that there is an indirect communication used by social individuals in order to coordinate their efforts towards some objective. For example, Ants lay down more pheromone when they find food to enable other ants to detect and react to this stimulus. In summary, they indirectly interact and cooperate to feed (or finding a path in the routing analogy). Although it is a comprehensive behavior, there is lack of mathematical models or equations to describe Stigmergy. Typically, the stimulus is not reached by some well established known equation, one may consider a given variable as stimulus to Stigmergy behavior and verify whether only a node with a fake variable can modify the Stigmergy of all individuals of a group and consequently the environment

Given that the decision mechanism of PRoPHET routing evaluates the number of encounters of neighbors, we could set up the encounter frequency for a single given node with fake information and next observe the success ratio. The encounter frequencies are used by such node as a Stigmergy where the nodes collaborate with this information to route the information.

Case Study

The 4G routing algorithms presented in this Chapter are evaluated in terms of their delivery success rate, level of message overhead and delay they incur. One or more scenarios are defined according to some parameters chosen mostly as fixed variables or a combination of factors (various values). The selected scenarios depict a number of situations such future networks may operate in. Not only are these metrics going to tell the reader more about the relationship between some existent and future routing approaches for 4G, but these are also expected to give some helpful and explicit insights into the understanding of routing issues and solutions in the new 4G context. As a result, the simulations are going to span a number of configurations in an attempt to identify and separate those states that offer better performance while reducing overhead and resource requirements. For instance, some specific factor value may be set to induce chemotaxy, stigmergy, diffusion or percolation, used in biological, social and physics modeling.

One objective is to show how percolation stimulus may be discovered for a given 4G scenario. The next step is then to determine the routing improvement a network engineer may obtain once this understands and knows how to handle the stimulus. Given that this is a case study, public domain software and solutions will be used whenever possible to allow the reader to repeat parts of the study. The algorithms and simulation tools may be openly obtained from the Web as well as information on the traffic and topology models used in the scenarios. With this in mind, the OMNet simulator, the PRoPHET algorithm and its set of data - mobility and traffic models have been selected to compose the network scenario. This scenario contains 50 nodes with random mobility and its simulation takes around 3995 seconds to run. The initial topology is shown in Figure 1.

To evaluate percolation in the PRoPHET scenario, both the buffer size and the number of message replication in routing information were chosen as the percolation variables that should establish a percolation threshold in the model. Next, it was necessary to check the stimulus selected and see if it was able to percolate in this scenario. In other works, this work checked if there is a buffer size limit and a message replication level that determine a stable number or a successful delivery rate of the network. In this case, messages were

Figure 1. Initial scenario

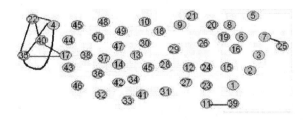

associated to the fluid, and both buffer and message replication to the surface when considering the analogy with the liquid percolation model used to establish when a fluid starts running through a given surface.

Firstly, the buffer size is a variable that was changed between 2 and 100, with increments that also varied between 2 to 10 with an increment of 2, while all the other variables were maintained fixed. Message replication was also set to 1. Figure 2 shows the results of this first evaluation with a stable state reached for a buffer size of 12. Hence this is the limit value of the percolation model, but the best configuration, the one with highest delivery rate, is obtained with an even smaller buffer size equal to 6.

Next, the number of the message replication was changed in the simulations from 2 to 100, with increments of 2. In turn these increments where also set to values from 2 until 10 with a step of 2. Similarly the other scenario variables were maintained constant including the buffer size set to 100. Figure 3 illustrates the results with stable state when the number of message copies is 24 or any other larger value. Hence, 24 is seen as the percolating number of copies, but the best delivery rate occurs when the number of copies is 2.

Both the variation of buffer size and message replication of only one node are also evaluated.

Firstly, the buffer size of only one node is a variable that was changed from 2 to 100, incrementing this by 2 to 10 increments, while all other variables were maintained. Message replication was also maintained with a single value and buffer size of 100 for all nodes. Figure 4 presents the results of this evaluation with a stable state reached when the buffer size is 10. So this value represents the limit value of the percolation model.

Next, the effect of message replication is studied. In these simulations, a single node was selected and subjected to changes in its level of message replication. In this case, node 24 had his message replication mechanism changed to use values between 2 and 100, with increments between 2 and 10. In the scenario presented here, the buffer size was set to 100 for this special node while all variables were left unchanged. For the other nodes, message replication was fixed at 1, i.e. a single replication was used, while their buffer size had the value of 100. One can see from the results that, for example, when node 24 (the differentiated node in this scenario) sends 2 replications, it already obtained almost a 50% delivery rate. Percolation was achieved when the level of copies reached 4 (Figure 5).

In a different evaluation, the number of message replication and nodes that may replicate messages were changed. The number of nodes

Figure 2. Percolation of one node according to set up buffer size for all node

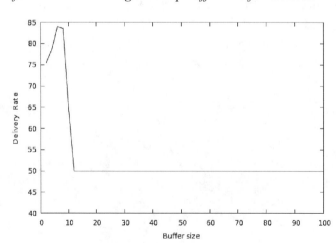

Figure 3. Percolation of one node according to set up message replication for all node

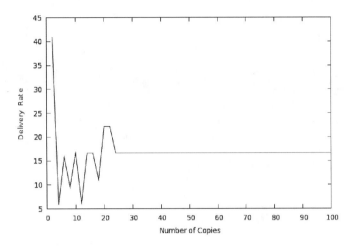

allowed to replicate messages was incremented until reaching 50 nodes representing the total nodes of the network scenario. The other nodes continued with their default behavior where only a single copy of any given message was injected into the network. So, initially, a unique node was making 20 copies of messages it was sending, in the next round of simulation two nodes could make 20 copies of a message and send this, then 3, 4,.., nodes until all the 50 nodes where doing this. Strangely, the results have shown that even when there were different numbers of nodes with the same number of replication (heterogeneous network), when the number of copies was 50, 100 and 150, the delivery rate of node 24 presented the same behavior as in Figure 6 (a special and identical function). There is a stable behavior when the number of copies is 50, 100 and 150. For example, when there are 3 nodes with 50, or 100, or 150 copies, the delivery rate is 27%. When the number of copies for 7 nodes is 20, 50, 100 or 150, the delivery rate is 43%. This shows that it is necessary to run extensive simulations, for any given scenario, in order to determine the minimum number of copies as well as the minimum number of nodes making message copying that

Figure 4. Percolation of one node according to set up buffer size of the only one node

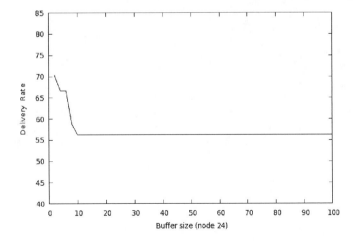

Figure 5. Single node percolation according to its own message replication

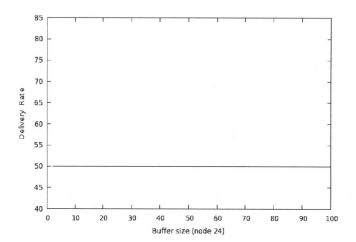

offer a good delivery level that ensures a required quality of service level to node 24 in this case. It is clear that one needs to establish the percolation pair (replication level, number of nodes that do message replication) in order to ensure that:

- All the nodes of a network are able to achieve their minimum QoS requirements in terms of message delivery in this scenario. Other QoS parameters such as bandwidth, delay and jitter may also be observed as reference levels to achieve;
- To avoid clogging the network with unnecessary message copies and possibly leading to the congestion of parts of the network, one needs to consider the traffic overhead and network occupation as new metrics for choosing the right percolation pairs that may achieve similar levels of message delivery or other QoS requirements.

Actually, by observing only node 24 delivery rate, the simulations have shown that a configuration where 8 nodes make 20 message replicas each gives the best delivery to this node.

New studies and simulations are needed to gain more insights on how to build some rules of thumb for the optimized routing configuration

Figure 6. Delivery rate for 20, 50, 100 and 150 replicas

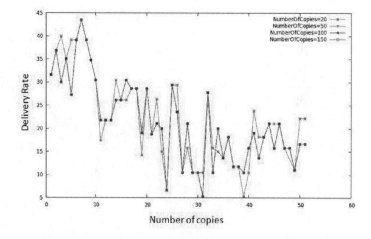

for this type of algorithms. Although each topology may need a different setup, but one expects to hopefully find some general rules that apply across the spectrum of 4G topologies and provide adequate routing performance.

COMPARING INNOVATION ROUTING AND 4G REQUIREMENTS

Advanced Scheduling Schemes and Resource Allocation

Although the fourth generation is homogeneous under the IP umbrella, the participating 4G devices may have multiple interfaces and radios. A 4G device could bind each one of its interfaces to a distinct network, or use multiple interfaces to access a single network. With these features it will be possible to increase traffic bandwidth using link aggregation or execute hand-off without latency and instability, and maximize processing time available for functions such as data-error correction. Hence future routing protocols cannot assume the presence of static links between device interfaces and networks. They must deal with these heterogeneities under the IP layer and dynamic binding. A new class of challenges emerges, mainly in terms of QoS guarantees. So, how could the user achieve good acceptable performance when using several interfaces submitted to varying working conditions seen at several networks?

Mobile wireless devices often need to maintain data or voice communication across different access points and radio base stations. This process is known as hand-off. Current cellular systems implement handoff over a single interface and only for phone calls. However, the next generation wireless system (4G) supports seamless handoff for data traffic and should be able to manage radio resources efficiently. VoIP continuity is another requirement in LTE especially when using the 3GPP IP Multimedia Service (IMS). In such a multi-radio environment, there is space to optimize bandwidth radio resources usage, signal quality and reduce information loss.

An important role for nontraditional routing approaches to play in improving future 4G communication systems is foreseen. The Wasp model has shown to be able to schedule tasks and reallocate resources, following a hierarchical and threshold based approach. Each wasp is stimulated to execute its task when its variable value becomes bellow a given threshold. We have seen this model being applied in the dynamic routing of vehicles, a prominent component of future 4G networks. Here, each vehicle is seen as a wasp with a threshold that waits before finding new optimized paths. When a node receives a request from two or more wasps with the same threshold, it then reserves the necessary resources and improves the routing path to the wasp with the highest hierarchy.

The wasp model may be used to resolve another important problem: that of interface selection. The individual force variation (F) is used for decision making, and to determine the best interface to use. This same wasp characteristic has also been associated to model signal strength, stability, efficiency and power consumption.

Since the binding between network and interface could be seen as a task, then wasp routing (Song, Hu, Tian & Xu, 2005) could also be a good approach to improve routing performance. Moreover, this scheme could also be used to manage and improve robustness by allocating messages to different networks when some paths may become unreachable.

Routing Optimization

At the early days of the Internet, routing took into account the number of hops an as important metric for path selection. This was a wise decision at the time as most of the Internet was still homogenous in terms of its links, router capacity and traffic. Soon later, weights were associated to links giving autonomous systems a new criterion to decide on the best routes and a mean to engineer their traffic and balance this. The creation of labels by MPLS provided a similar traffic engineering mechanism capable of controlling and improving the routing and service delivery through path selection.

In the new context of 4G networks, routers must deal with different dynamic link stability levels, security and QoS levels and network handoff. It is for such reasons that the 4G networks will certainly need to also consider new routing metrics and change these according to their environment or context. Under some scenarios, reachability could be more important than performance whereas QoS may become the metric of choice in other circumstances. This may also be service and application driven. Electronic mail transfer is a store and forward application that requires information integrity mainly whereas video conferencing considers low delay and bandwidth as primordial network resources.

Future 4G will certainly embrace disruptive connections and delay tolerant networks, high mobility users resulting in a challenging mix with different routing metaphors and techniques thriving within a single 4G unifying architectures. A simple, "*one hat fits all*" approach to routing cannot be the way forward. Therefore 4G needs to consider multi-metric optimization following different innovative routing approaches instead of merely reusing traditional strategies. It is believed that new routing and resource management insights borrowed from areas as diverse as biology, social phenomena, random and probabilistic diffusion models are expected to lead the way ahead.

But this is nonetheless not a complete breakaway from routing as we know it. In fact one expects to continue making use of useful traditional concepts such as clustering and hierarchical structures to simplify, organize and improve 4G routing. Following a dynamic approach, social algorithms exchange messages to find popular nodes and establish similarity among them in order to create clusters and hierarchical structures. Similarly to traditional routing algorithms from fixed networks, messages can be forwarded from any social node to a popular one judged to be in a better position to disseminate the information and capable of increasing the probability of a message reaching its destination. This offers ways to increase the delivery rate, but differently from the flooding algorithm, the social strategy reduces the number of message replication as these messages only are forwarding among a restricted number of nodes. Moreover, approaches such as SOLAR and (Leguay, Friedman & Conan, 2006) work by extracting location information to identify mobility patterns in order to improve their routing efficiency. They rely therefore on the understanding of user's behaviors in terms of mobility patterns

4G networks are expected to collaborate with each other independently of their underlying technologies. For example, a user with Bluetooth and GPRS devices can choose one or another technology to disseminate a given type of information according to application level criteria such as urgency and destination distance. One could use an epidemic algorithm to send a simple message to a friend through a Bluetooth interface while selecting a GPRS interface to transfer credit (possible future money) to a distant family member.

Self-Organization

Communication in 4G is expected to live with frequent disconnections due to wide ranging mobility patterns supported including vehicular, user planned disconnection, cellular, short-range and delay tolerant networks. Hence, 4G devices should always be on the lookout to find the best alternative for message delivery depending on availability and application requirements.

Both epidemic and Ant models are leading models for dealing with these challenges by continuously adapting and self organizing. The epidemic approach has proven efficiency for spreading information to all devices with high delivery rates. It nonetheless suffers from some QoS restrictions that could be reduced through buffer management and the use of adequate traffic engineering and routing policies. The probabilistic, random and social metrics can be used to decide whether and when a node must send some information.

On the other hand, the Ant model mimics the search for food, providing a mechanism for finding interesting routing paths according to QoS, security and others requirements. Moreover, the Ant model can increase or decrease the update process of the routing table according to traffic and network stability. When the scenario at hand is unstable, then the algorithm sends many "Forward Ants" to find new paths. Hence, the Ant model may then increase throughput while decreasing delay.

Given that a 4G device may be part of several networks, then it can execute a vertical handoff in order to obtain and exchange information from these different environments, improve its routing and also collaborate with other network users.

In the social routing approach, the nodes can exchange messages in order to identify popular ones and similarity behavior and consequently self-organize to improve their security and performance levels as seen in sections about social routing and social overlay networks.

Management

Policy management is a common tendency in the control and monitoring of a number of networking aspects including information routing. The Internet inter-domain Border Gateway Potocol (BGP) filters and controls path and reachability announcements made across domains according to administrative interests such as billing, performance and security.

4G networks are highly dynamic with users and networks composing networks and services at will. This puts a heavy management burden on 4G networks and should be subject to policy negotiation.

For example, users could define policies to control the usage of their radio and buffering resources in order to accommodate their prioritized services according to possible changing QoS and security requirements. Policy enforcement is a complex effort and there are currently a number of researches to implement this over varying devices and resources.

Security

The use of social and biological routing opens the way for attacks very much similar to those traditionally seen in fixed networks such as distributed denial of services DDoS. Social models may be used to provide a mechanism for creating and enforcing the use of trust relationships among nodes and networks the like in order to protect resources from the possible damages and losses triggered by malicious nodes.

Under social routing, a node can find similarities in order to create a confidence circle for message forwarding while ensuring some security guarantees or even improve routing robustness against malicious nodes. These are likely to drop messages or inject false or harmful ones in an attempt to break routing, increase latency or reduce available resources by sending large amounts of bogus messages.

A second security contribution made by social algorithms is possibly obtained from the association of a credibility level for each node through profile information generated by the node itself or its closest friend nodes. This metric could be an important one for path selection according to the associated nodes security level. For example, some application profile may enforce the requirement for a level of security with a set of acceptable cryptographic methods.

Although social algorithms contribute with new ideas to security, they are hardly sufficient enough on their own for ensuring path protection. Malicious nodes can create fake identities in social networks and consequently break this simple security approach. For this reason, there are a number of works looking for solutions that may deal with such attacks. They work by identifying and removing existing fake trust circles.

FUTURE RESEARCH DIRECTIONS

Untraditional routing algorithms remain an open field for future applications and networks. New research can seek to integrate different routing mechanisms, or improve the untraditional algorithms, or yet to develop new routing proposals based on such social, biological and epidemic concepts. For example, there is a limited routing experience based on the Wasp model which may also be applied in the allocation of radio resources when there are multiple interfaces, as well as the optimization of routing and selection of networks.

The social model was presented as a way to extract the relationship and mobility pattern in order to improve the security and optimization over routing. Another interesting issue, not presented here, is with regard to building negotiation models. What information should be used to negotiate the communication between social networks? How to negotiate and apply policy routing among theses environment?

Although this chapter presented and evaluated a number of non-traditional routing strategies that future 4G networks could dwell from, there is still a considerable need for more studies of routing stimulus. This could be especially important for the integration of different 4G environments that internally may work with varying routing algorithms. For this reason, future work should seek to make a better analysis of some of the proposed routing ideas such as diffusion, chemotaxy, stigmergy and percolation. Such evaluations should be more complete in looking for the behavior of each algorithm in different 4G environments, including as disruptive network, delay tolerant network, short-range network, disaster recovery and overlay network. Moreover, new mathematical and possibly better calibrated models should accurately describe the result of blending and breeding this new class of routing algorithms.

A deeper knowledge of stimulus could be a first step to define a new policy architecture to control and manage future 4G networks. Further, work is also needed to understand the impact of policies on stimulus in order to enable their dynamic adjusting and routing customization. One expects to see studies showing how these classes of routing protocols may be calibrated and taken advantage of to suit different operating environments and requirements.

CONCLUSION

This chapter explored new ideas for 4G routing and presented a possible classification for traditional protocols, theirs features and pinpointed some cumbersome restrictions for their extension into 4G networks. We have seen that these new requirements for future networks include dealing with intermittent connections, nestled networks, vertical handoff and other characteristics presented in background section.

On the other hand, untraditional routing mimics biological, social and epidemic behavior to offer solutions for some niche scenarios. Among these, many are related with such 4G's new requirements. They provide important and flexible behavior to route information in future environment and new ideas to undertake the allocation of resources, their optimization, security, self-organization and management.

Given that future 4G complex wireless environments embrace several technologies and infrastructures, the coexistence of different routing solutions working together is expected at different levels. The use of each of these untraditional routing methods shows its benefits in some given context. Wasp routing is a biological model that provides an advanced scheduling scheme for decision taking in terms of network and interface selection when there are several of these available to choose from. Epidemic models are interesting for environments with low connectivity such as DTN. Social models provide interesting mechanisms for creating and managing trust relationships and routing optimization. This is a result of taking routing decisions based on information, including mobility patterns, encounter history and profiles.

4G management was seen as another important work point. Untraditional algorithms require new information and parameters that needs to be continuously monitored and bound according to well established policies. Buffer configuration and mobility information are examples of routing relevant concerns that need to be handled by future a 4G management architecture.

Finally a set of behavior based strategies, namely, chemotaxy, percolation, diffusion and stigmergy, may also contribute to the search for new 4G routing algorithms. When some stimulus correspondents to these behaviors is found, policies can be put in network to set the right amount of stimulus values and thresholds and consequently control the environment. The PRoPHET algorithm was used to demonstrate how to analyze untraditional routing and identify the presence or lack of percolation stimulus.

REFERENCES

Babaoglu, O., Canright, G., Deutsch, A., Di Caro, G., Ducatelle, F., & Gambardella, L. (2006). Design patterns from biology for distributed computing. [TAAS]. *ACM Transactions on Autonomous and Adaptive Systems*, *1*(1), 26–66. doi:10.1145/1152934.1152937

Banks, L., Ye, S., Huang, Y., & Wu, S. F. (2007). Davis Social Links: Integrating Social Networks with Internet Routing. In *Applications, Technologies, Architectures, and Protocols for Computer Communication. 2007 workshop on Large scale attack defense* (pp. 121-128).

Barabasi, A., & Albert, R. (1999). Emergence of scaling in random networks. *Science*, *286*, 509–512. doi:10.1126/science.286.5439.509

Bellman, R. (1958). On a routing problem. *Quarterly of Applied Mathematics*, *16*(1), 87–90.

Bernardos, M., Casar, J., & Tarrio, P. (2006. July). *Efficient social routing in sensor fusion networks.* Paper presented at 9th International Conference on Information Fusion, Florence, Italy.

Bertsekas, D., & Gallager, R. (1992). *Data Networks*. Upper Saddle River, NJ: Prentice-Hall.

Callon, R. (1990). *Use of OSI IS-IS for Routing in TCP/IP and Dual Environments.* Network Working Group, RFC 1195.

Cao, Y., Yang, Y., & Wang, H. (2008). Integrated Routing Wasp Algorithm and Scheduling Wasp Algorithm for Job Shop Dynamic Scheduling. In *International Symposium on Electronic Commerce and Security* (pp. 674-678).

Cherkassky, B. V., Goldberg, A. V., & Radzik, T. (1994). Shortest Paths Algorithms: Theory and Experimental Evaluation. In D. D. Sleator (Ed.), *Proceedings of the 5th Annual ACM-SIAM Symposium on Discrete Algorithms (SODA 94)* (pp. 516-525). Arlington, VA: ACM Press.

Corson, M. S., & Ephremides, A. (1995). Lightweight Mobile Routing protocol (LMR), A distributed routing algorithm for mobile wireless networks. *Wireless Networks*, 1.

Cowan, R., & Jonard, N. (2004). Network structure and the diffusion of knowledge. *Journal of Economic Dynamics & Control*, *28*(8), 1557–1575. doi:10.1016/j.jedc.2003.04.002

Daly, E. M., & Haahr, M. (2007). Social network analysis for routing in disconnected delay-tolerant MANETs. In *8th ACM international symposium on Mobile ad hoc networking and computing* (pp.32-40). New York: ACM.

Di Caro, G. (2004). *Ant colony optimization and its application to adaptive routing in telecommunication networks*. Ph.D. thesis, Université Libre de Bruxelles, Belgium.

Dijkstra, E. W. (1959). A Note on Two Problems in Connection with Graphs. *Numer.Math, 1*, 269–271.

Dorigo, M., & Blum, C. (2005). Ant colony optimization theory: A survey. *Theoretical Computer Science, 344*(2-3), 243–278. doi:10.1016/j.tcs.2005.05.020

Erdös, P., & Rényi, A. (1959). On Random Graphs I. *Publicationes Mathematicae, 6*, 290–297.

Ford, L., & Fulkerson, D. (1962). Flows in Networks. Upper Saddle River, NJ: Prentice-Hall.

Ghosh, J., Ngo, H. Q., Yoon, S., & Qiao, C. (2007). On a Routing Problem Within Probabilistic Graphs and its Application to Intermittently Connected Networks. In *26th IEEE International Conference on Computer Communications (INFOCOM)* (pp. 1721-1729).

Ghosh, J., Philip, S. J., & Qiao, C. (2007). Sociological orbit aware location approximation and routing (SOLAR) in MANET. *Ad Hoc Networks, 5*(2), 189–209. doi:10.1016/j.adhoc.2005.10.003

Grassé, P.-P. (1959). La reconstruction du nid et les coordinations inter-individuelles chez Bellicositermes natalensis et Cubitermes sp., La théorie de la Stigmergie: Essai d'interprétation du comportement des Termites Constructeurs. *Insectes Sociaux, 6*, 41–80. doi:10.1007/BF02223791

Grimmett, G. (1999). *Percolation (2nd ed.)*. Grundlehren der Mathematischen Wissenschaften. Berlin: Springer-Verlag.

Gutjahr, W. (2008). First steps to the runtime complexity analysis of ant colony optimization. *Journal Computers & Operations Research, 35*(9), 2711–2727. doi:10.1016/j.cor.2006.12.017

Heegaard, P. E., Helvik, B. E., & Wittner, O. J. (2008). The Cross Entropy Ant System for Network Path Management. *Telektronikk, 1*, 19–40.

Hui, P., Chaintreau, A., Scott, J., & Gass, R. Crowcroft & Diot, C. (2005). Pocket switched networks and human mobility in conference environments. In *ACM SIGCOMM workshop on Delay-tolerant networking*, Philadelphia, Pennsylvania, USA (pp. 244 - 251).

Hui, P., & Crowcroft, J. (2007). Bubble rap: Forwarding in small world dtns in ever decreasing circles. Technical Report UCAM-CL-TR-684, Cambridge Univ., Comp. Lab., 2007.

Hui, P., Crowcroft, J., & Yoneki, E. (2008). BUBBLE Rap: Social Based Forwarding in Delay Tolerant Networks. In *9th ACM International Symposium on Mobile Ad Hoc Networking and Computing (MobiHoc)* (pp. 241-250). New York: ACM.

Jindal, A., & Psounis, K. (2006). Performance Analysis of Epidemic Routing under Contention. In *International Conference On Communications And Mobile Computing* (pp. 539-544). New York: ACM.

Johnson, D., Hu, Y., & Maltz, D. (2007). *The Dynamic Source Routing Protocol (DSR) for Mobile Ad Hoc Networks for IPv4*. Network Working Group, RFC 4728.

Kempe, D., Kleinberg, J. M., & Demers, A. J. (2001). Spatial gossip and resource location protocols. In *Thirty-third annual ACM symposium on Theory of computing* (pp. 163-172). New York: ACM.

Kenah, E., & Robins, J. (2007). Network-based analysis of stochastic SIR epidemic models with random and proportionate mixing. *Journal of Theoretical Biology, 249*(4), 706–722. doi:10.1016/j.jtbi.2007.09.011

Kochen, M. (Ed.). (1989). *The Small World*. Norwood, NJ: Ablex.

Leguay, J., Friedman, T., & Conan, V. (2006). Evaluating Mobility Pattern Space Routing for DTNs. In *25th IEEE International Conference on Computer Communications (INFOCOM)* (pp. 1-10).

Li, Z., & Shen, H. (2008). Probabilistic Routing with Multi-Copies in Delay Tolerant Networks. In *28th International Conference on Distributed Computing Systems Workshops* (pp. 471-476). Washington, DC: IEEE Computer Society.

Lindgren, A., Doria, A., & Schelén, O. (2003). Probabilistic Routing in Intermittently Connected Networks. [New York: ACM.]. *ACM SIGMOBILE Mobile Computing and Communications, 7*(3), 19–20. doi:10.1145/961268.961272

Lindgren, A., & Phanse, K. S. (2006). Evaluation of Queueing Policies and Forwarding Strategies for Routing in Intermittently Connected Networks. In *International Conference on Communication System Software and Middleware (Comsware)* (pp. 1-10).

Marti, S., Ganesan, P., & Garcia-Molina, H. (2004). Sprout: P2p routing with social networks. In *1st International Workshop on Peer-to-Peer Computing and Databases* (pp. 425-435).

Marti, S., Ganesan, P., & Garcia-Molina, H. (2005). DHT Routing Using Social Links. In *3rd International Workshop* (pp. 100-111). New York: Springer.

McQuillan, J. M., Richer, I., & Rosen, E. C. (1980). The New Routing Algorithm for the ARPANET. *IEEE Transactions on Communications, 28*, 711–719. doi:10.1109/TCOM.1980.1094721

Motani, M., Srinivasan, V., & Nuggehalli, P. S. (2005). PeopleNet: engineering a wireless virtual social network. In *11th annual international conference on Mobile computing and networking* (pp. 243-257). New York: ACM.

Moy, J. (1994). *Multicast Extensions to OSPF*. Network Working Group, RFC 1584.

Moy, J. (1998). *OSPF Version 2*. Network Working Group, RFC 2328.

Murthy, S., & Garcia-Luna-Aceves, J. J. (1995). A Routing Protocol for Packet Radio Networks. In *1st Annual International Conference on Mobile Computing and Networking* (pp. 86-95). New York: ACM.

Narasimhan, T. N. (2004). Fick's insights on liquid diffusion. *Transactions - American Geophysical Union, 85*(47), 499–501.

Perkins, C., Belding-Royer, E., & Das, S. (2003). *Ad hoc On-Demand Distance Vector (AODV) Routing*. Network Working Group, RFC 3561.

Perkins, C., & Bhagwat, P. (1994). Highly dynamic Destination-Sequenced Distance-Vector routing (DSDV) for mobile computers. In *ACM SIGCOMM Computer* []. New York: ACM.]. *Communication Review, 24*, 234–244. doi:10.1145/190809.190336

Royer, E. M., & Toh, C.-K. (1999). A review of current routing protocols for ad hoc mobile wireless networks. *IEEE Personal Communications*.

Song, J., Hu, J., Tian, Y., & Xu, Y. (2005). Re-optimization in dynamic vehicle routing problem based on Wasp-like agent strategy. *Intelligent Transportation Systems, 2005*, 231–236.

Szczodrak, M., Kim, J., & Baek, Y. (2007). 4GM@4GW: Implementing 4G in the Military Mobile Ad-Hoc Network Environment. *International Journal of Computer Science and Network Security, 7*(4), 70–79.

Tanenbaum, A. (2003). *Computer Networks (4th ed.)*. Upper Saddle River, NJ: Prentice Hall.

Tomiyasu, H., Maekawa, T., Hara, T., & Nishio, S. (2006). Profile-based Query Routing in a Mobile Social Network. In *7th International Conference on Mobile Data Management* (pp. 105-108). Washington, DC: IEEE Computer Society.

Vahdat, A., & Becker, D. (2000). *Epidemic routing for partially-connected ad hoc networks* [Tech. Rep. CS-2000-06]. Durham, NC: Duke University.

Willinger, W., Taqqu, M. S., Sherman, R., & Wilson, D. V. (1995). Self-Similarity Through High-Variability: Statistical Analysis of Ethernet LAN Traffic at the Source Level. In *ACM SIGCOMM* (pp. 100-113). Cambridge, MA: University of Massachusetts.

Yu, H., Gibbons, P. B., Kaminsky, M., & Xiao, F. (2008). SybilLimit: A near-optimal social network defense against sybil attacks. In *IEEE Symposium on Security and Privacy* (pp. 3-17).

Yu, H., Kaminsky, M., Gibbons P.B., & Flaxman, A. (2008). SybilGuard: Defending against sybil attacks via social networks. *IEEE/ACM Transaction on Networking, 16*(3), 576-589.

ENDNOTES

[1] *Fluid Nexus*. (n.d.). Retrieved from http://www.inclusiva-net.es/fluidnexus/

[2] Reed, D. (n.d.). *IPFilter*. Retrieved from http://coombs.anu.edu.au/~avalon/

[3] Kuznetsov, A. N. (1999). *IP command reference*. Technical report, Institute for Nuclear Research, Moscow, April 1999.

[4] Welte, H. & Ayuso, P. N. (n.d.). *netfilter*. Retrieved from http://www.netfilter.org/

Next Generation Technologies

Chapter 11
Personal Environments:
Towards Cooperative 4G Services

Tinku Rasheed
Create-Net, Italy

Usman Javaid
Vodafone Group, UK

ABSTRACT

The Fourth Generation of wireless networks promises to offer a vast range and diversity of converged services in order to revolutionize the way we communicate today. 4G can not only offer ultra-high data rates, but would also enable the ubiquitous computing paradigm, particularly interesting for the end-user with the help of various personalized and user-friendly services and devices. This increase in short-range communication among users and introduction of personalized services would form a "Personal Ubiquitous Environment" around the user. Since in such environments, multiple users will come closer (without any third-party barriers); their cooperation would be the key to success. Several technological and social barriers have prevented so far an effective cooperation between technologies, systems or users. This chapter focuses on the potential impacts of cooperative ubiquitous services in 4G networking systems. The authors explain the technological implications of cooperative systems considering the personal environment ubiquity. Furthermore, it attempts to characterize the socio-technical dimension of the potentials and limits of cooperation in 4G systems.

1. INTRODUCTION

The goal of the original Internet was to provide a unified communication platform for different kind of devices and networks as well as future technologies, where every single host would be an equal player. However, this fundamental design radically changed over time with the emergence of the client/server architecture, with relatively small number of privileged servers serving a huge mass of consumer hosts. This emerged architecture was totally opposite to the fundamental design of the Internet i.e. "a cooperative network of peers". However, in late 90s, with the appearance of the music-sharing application, Napster, the Internet experienced another drastic change, where the architectural design of the

DOI: 10.4018/978-1-61520-674-2.ch011

Internet reverted and pushed back to its original "peer to peer" notion. The millions of hosts connected to the Internet, inspired by the culture of cooperation and openness, started connecting to each other directly, forming collaborative groups, sharing their resources to become user-created powerful information clusters. Currently, the peer to peer applications are using the Internet much as it was originally dreamed for; a common platform for hosts to collaborate and to share information as equal computing peers.

Wireless communication has simply revolutionized the way we communicate today and is not less than a magic for someone who does not know how it works. It enables us to communicate anytime, anywhere in any form (data, voice). However, wireless technology is not only limited to communication, it can offer much more than just a phone call. The limits of wireless communication are still unpredictable and unimaginable. The father of radio communication Heinrich Hertz once said "I do not think that the wireless waves I have discovered will have any practical applications." The inventor of first wireless telegraph system Guglielmo Marconi said "Have I done the world good; or have I added a menace?" These early giants of wireless communications were not so sure about the usefulness of their work and were underestimating the power of wireless. They might have envisaged that without the essence of cooperation and sharing, no technology can be economically and socially viable.

The cooperation in wireless technologies is a key to discover a variety of unforeseen innovative applications (Nash, 1951; Borcea, 2002; Frattasi, 2004; Gupta & Kumar, 2002). This latter is the core reason, why the cooperation is gradually increasing with the progress in the generation of mobile systems. Cooperative and distributed wireless techniques have received significant attention in the past decade and a large body of research both highly useful and contradicting has emerged. Today we are at the doorstep of 4G systems, where collaborative services, technologies, environments and so on, are the major areas of research concern.

As it was originally expected, the future is not limited to cellular systems and 4G should not be exclusively understood as a liner extension of 3G. In concrete terms 4G is more about services than ultra-high speed broadband wireless connectivity. As predicted in Frattasi (2005), keeping the cellular core, the network architecture will be predominantly extended to short-range cooperative communication systems. Apart from the coverage extension, power and spectral efficiency, increased capacity and reliability, this enormous flexibility at the user end will help in the development of "personal ubiquitous environment" around the user. The 4G service and technology infrastructure will induce the user's devices to form cooperative groups and share information and resources in order to attain mutual socio-technical benefits. The whole bunch of unforeseen 4G cooperative services would enable the 4G technologies to recede into the background of our lives, making us a part of an intelligent and ubiquitous personal substrate.

Until recently, the cooperative services in 4G systems have received significant attention due to their high degree of technological and social flexibility, infinite freedom of choice and cooperation for the user and more importantly, a potential mega-revenue source for the industrial players. In this chapter, we focus on the services side of the cooperation in 4G systems and discuss how these personalized services would make use of the multitude of wireless systems and networks available under the auspices of 4G in a cooperative manner.

Cooperative Services in 4G

The widely agreed upon rule for success in 4G telecommunication markets is to visualize a cooperative service chain of multiple suppliers to satisfy the ever-growing requirements of end customers (Roussos, 2003). The evolution of 4G systems in

Figure 1. Cooperation in 4G, services perspective

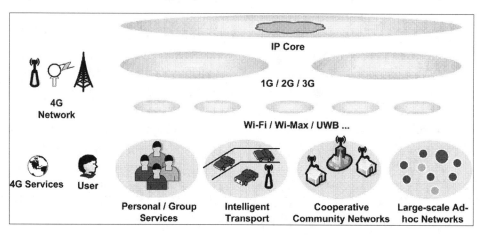

a multi-dimensional facet provided a scrupulous platform for deriving advanced and innovative user-oriented and cooperative services. Embossed to high level perspectives and equally leveraging on technical dimensions, we recognize several aspects of cooperative services; those related to personal (or group centric) services, intelligent transport network services, cooperative community networks and large scale ad hoc network services. As shown in Figure 1, these cooperative and heterogeneous services accounts for the efficient 4G convergence platforms that renders clear cut benefits in terms of bandwidth, coverage, power consumption and spectrum usage.

The personal and group-centric communication models put forth a multitude of interesting services, benefiting from the "cooperative clouds" formed as a result of multi-level social groups based on self-organizing common objectives (Sandvig, 2003). Within this context, various compelling services for smart-home networking, cooperative health care etc. are shaping up. One such service is the cooperative distribution of media content in stationary home networks, where the transparency enabled by the seamless and intelligent platform equips the home network to converge into an interdependent service ecosystem for the consumers (Borcea, 2002). Other services in group communication which exploits collaborative behavior include symbolic resource sharing among communication groups (for example, user-centric dynamic content sharing similar to popular web services like MySpace or YouTube), ubiquitous and collaborative healthcare monitoring at home or hospitals etc. The intelligent transport network is also a setting for providing collaborative 4G services from a user perspective. The most interesting among them is the development of evolutionary cooperative multi-player games as a massive collaborative constellation for vehicular networks (Hoebeke, 2006). These self-evolving games are targeted at intelligent transport networks which range from private vehicle owners to public transportation system users. Other envisaged services include varying location-based services in offer on a cooperative basis, where the consumers could either locate their intended footage leveraging on the collaborative platform or the customers could market their business availing on cooperative advertisement options. This creates an open service ecosystem beneficial for the entire service value chain in vehicular transportation networks.

Wireless community networks (commercial, public and non-profit), have matured enough through the continuing evolution of mesh networks (Lieu & Kaiser, 2005), which are now exploiting heterogeneity in a third generation mesh context with the use of multiple-radios (including different radios for downlink-uplink), dynamic interference detection and avoidance mechanisms, automatic location updating mechanisms etc. This, along with the introduction of inter-community networking aspects has given new dimensions to collaborative service distribution in community networks. This includes community-based IPTV services, cooperative web-radio, collective surveillance etc apart from common service attributes like resource sharing among users. In general, large-scale user cooperation is an important aspect to the success of community networks triggering the collaborative service-profit chain and introducing competitive differentiation. Mobile Ad Hoc networks applications have made appealing progress, particularly in the field of wireless sensor networks. Many distributed applications are envisaged in sensor networks where collaborative computing (Hoebeke, 2006) assumes the center stage; smart messaging services for sensors, collaborative objects tracking etc to name a few (Borcea, 2002).

In the search for niche markets and opportunity for 4G, large organizations and policy makers converge to accept that the 4G landscape will not just be about defining higher data rates or newer air interfaces, but rather will be shaped by the increasing integration and interconnection of heterogeneous systems, with different devices processing information for a variety of purposes, a mix of infrastructures supporting transmission and a multitude of applications working in parallel making the most efficient use of the spectrum. On the contrary, users are getting more vary about the services that they require and the modes with which they prefer to communicate and cooperate, which also hugely influences the future of 4G commercialization. These developments has led us to think in the lines of personal/group services as the most appealing and predominant platform for the development of 4G; where the users collaborate in a distributed and cooperative fashion. This user-centric cooperation and supporting issues which accounts for the development of cooperative, ubiquitous, personal communication models is the core motivation for further discussions in this chapter.

2. COOPERATIVE PERSONAL/ GROUP SERVICES IN 4G

Personal computing paradigm flourished faster than any other domain and with its marriage with the networking world, it gave birth to a new era of computing called ubiquitous computing. 4G is not the name of a single technology (Gupta & Kumar, 2002), rather it is a cooperative platform (Bohlin, 2004; Katz & Fitzek, 2005) where a large range of heterogeneous wireless networks and services coexist. The diverse devices, network and service elements find their way into the life of the end-user and this integration of 4G elements into the end-user environment should ideally go unnoticed to the user; so that the technology eventually focuses over the user and not the user focuses on the diversity of technology around him. Calm 4G technology integrated into user's world is only possible with the essence of cooperation, sharing, openness and trust, within the user's own devices and among the users. The notion of cooperation in personal/group services may take various dimensions ranging from technology and services to socio-physiological aspects.

There is a large array of actors in 4G service arena such as user, service/content provider, network operator, regulatory bodies, and so on, who bind their own proper stakes with 4G's success. However, economically speaking, user is a major player; a center of the entire 4G globe, whereas the other actors join hands to meet the expectations of the end-user. Taking the technological dimension, in the last few years, number of heterogeneous devices emerged and networked, ranging from mobile communication equipments to home electronics. This proliferation results into the availability of large range of choices to the user to communicate in highly diverse environments. As a result, in a 4G system, the user

Figure 2. User-centric cooperation

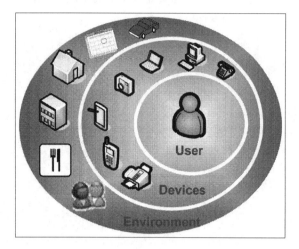

is surrounded by a variety of devices offering a multiplicity of different services (Javaid, 2006), as shown in Figure 2. Moreover, the utilization of these devices and services dramatically changes with the change in user's environment. Therefore, the devices and services in the 4G world should have a high deal of adaptation capabilities. "Personalization" (Borcea, 2002) is a key word in this context. Since every user is unique in his roles, taste and likings; the 4G systems should be intelligent enough to fully understand the user and adapt the network and service elements according to user's preferences.

In a user-centric model, the user is the focus of the whole system. The cooperation among his heterogeneous devices and his environment is vital for the seamless working of the entire 4G system. Here, we refer to the cooperation in two dimensions. At first, the devices themselves need to cooperate, for instance, while the user is busy working on his laptop and he receives an important voice message on his mobile phone, the mobile phone should track the activity of the user in order to notify him about the voice message. To this end, irrespective of their specifications, the user's devices should be able to cooperate in order to help the user in his daily life. And second, the devices should cooperate with the user's environment. Since the user preferences vary with the change in his environment therefore the devices should be capable to dynamically adjust themselves accordingly. For instance, if the user receives a video call while at home sitting in his TV lounge, the mobile phone should intelligently detect the activity/mood of the user and should propose to transfer the video flow on the higher resolution screen placed in front of the user. These both dimensions of cooperation are only possible when the 4G systems encircling the distinct end-user, fully understand the socio-physiological and the technological potentials and limitations of cooperation.

In 4G, towards personalization and user-centric cooperation, we generalize the concept of Personal Computers (PCs) and extend it towards Personal Networks (PN) (Hoebeke, 2006). It is a system/network owned and operated by one person i.e. the PN owner. The PN owner is the sole authority in his personal interconnected devices and can use the PN in a way he wants. The personal devices may be located, both in his close vicinity (forming a PAN) and at remote locations. Figure 3 presents the PN of Bob, which is composed of his home, office and car clusters. The owner of the PN can add new devices or personalized services in his personal network according to his will. The PN for its owner is a heaven of personalized services in the cyberspace and appears as a black box to the outside world.

Group-centric cooperation is also referred as cooperation among the end-users who are organized in groups. This is somehow fundamentally opposite to the user-centric cooperation, where only the user's devices and environments cooperate, and this cooperation appears as a dark cloud for the outside world (for other users). In fact, the 4G services which can be made available to a single user (with user-centric cooperation) are limited and the users need to cooperate with the each other to extend their global services repository. In addition, many service-oriented patterns need to extend the boundaries of "user-centric

Figure 3. Bob's personal network

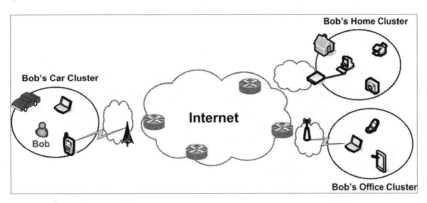

cooperation" and involve the secure interaction of multiple users having common interests for various professional and private services. Moreover, in this federated users environment towards group-centric cooperative model, the distinct users can offer services to each other promoting the concept of "give and take".

In order to promote the group-centric cooperation in 4G systems, the concept of Personal Network Federations (PN-F) has been recently introduced in the European MAGNET Beyond project (Hoebeke, 2006). PN-F addresses the interactions between multiple PN users with common interests for a range of diverse services. A PN federation can be defined as a secure impromptu, situation-aware or beforehand agreed cooperation between a subset of relevant devices belonging to different PNs for the purpose of achieving a common goal or service by forming an efficient collaboration. Consider the PN-F B in Figure 4, a simple example of PN-F is the federation of PNs belonging to a group of students in a classroom, sharing lecture notes.

Based on how cooperation between devices

Figure 4. Personal network federation architectures

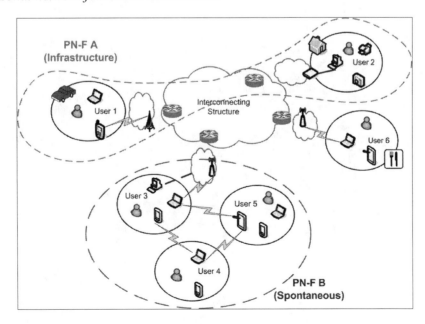

in different PNs is realized in order to establish the federation, we can differentiate between infrastructure and spontaneous PN federations. In an infrastructure based federation, PN-F is established between devices in PN clusters that are all connected to an infrastructure network. As shown in Figure 4, the infrastructure PN-F i.e. PN-F A is formed between the user 1 and user 2, who are located across the infrastructure network. On the other hand, in a spontaneous/ad-hoc PN-F, the federation is formed in the absence of a fixed infrastructure. This type of federation mostly occurs when nearby users collaborate within a federation.

The cooperation among the users, their devices and environments results into the development of a "Personal Ubiquitous Environment" around the user, which permits the "ubiquitous global access" to a vast number and variety of information resources. This uniform and comprehensive sense of cooperation results into a vast base of services for all the users who are part of this personal ubiquitous environment village. In the language of Personal Networking, we can collectively define PN and PN-F as a Personal Ubiquitous Environment. As shown in Figure 5, three users come closer to share devices, services and environments to form the cooperative group (PUE / PN-F). In PUE environment, the users believe in the essence of openness and sharing not only for their self-centric goals but also for the global benefits of the entire cooperative community. Those users, who are satisfied with their own proper resources and do not have any intention to cooperate; stays in their own user-centric environments i.e. PN, as shown in Figure 5.

3. COOPERATION IN THE PUE

In PUEs, we consider a scenario in which a group of users are located within each other's spatial proximity and are open to cooperate and share services and applications. However, some basic questions may arise here; why the user wants to extends his PUE in order to accommodate the other users, what he is interested in and more importantly, what he would be able to get after forming the PN-F with other users and what price the user may have to pay for these services. These questions would be answered and discussed throughout this section. We base our discussion around three fundamental stances outlined in the following:

Figure 5. Personal ubiquitous environment

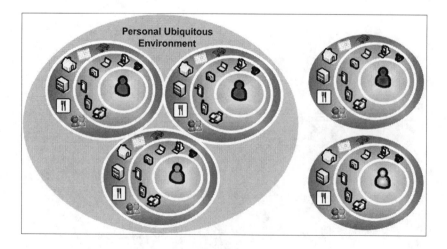

Before the Cooperation Begins

The PUE of a user first constitutes his own devices and services available in his PN. The user is the sole authority to extend his PUE (to form a PN-F) in order to accommodate the services and the devices available to other users in their own PNs. However, before really moving towards cooperating and forming groups, the user first looks at his motivation to cooperate. Adam Smith, the father of modern economics said, *"Every man, as long as he does not violate the laws of justice, is left perfectly free to pursue his own interest his own way"*. In terms of cooperative groups, if the user feels satisfied with the services he has in his own PN, no desire to cooperate and to from groups will come on his way. The user shall only devise ways into cooperation when he looks for some service which his own PN (current PUE) can not offer. The user's intent to cooperate can be classified in several ways: purpose-driven cooperation vs. opportunity-driven cooperation, short-lived cooperation vs. longer-term cooperation and proactive cooperation vs. reactive cooperation.

Purpose-driven cooperation means that the cooperative strategies are explicitly defined beforehand, whereas opportunity driven means that the users cooperate spontaneously when interesting circumstances to do so arise. In both cases, and especially in the second case, information about the user's context/environment/activities can play an important role. Next, depending on the lifetime of the cooperative groups, we can make the distinction between very *short-lived* cooperation and *longer term* cooperation. This distinction will have its implications on the complexity of the solutions to establish the cooperative groups. In the case of short-lived cooperative groups, solutions to setup and manage the cooperation need to be lightweight and simple. Longer term cooperation open up much more opportunities to introduce more complex and powerful management and definition mechanisms. Finally, based on the way the cooperation process is carried out, both *proactive* and *reactive* cooperative groups are possible. Proactive implies that the cooperative groups are established in anticipation of the use of the common goals or services provided by the cooperation strategies of each group user. Last but not the least, reactive cooperative groups are established upon request or when the opportunity arises.

Formation of Cooperative Groups

In precise terms, a cooperative group is a function of cooperation strategies defined by each participant of the group. First the group members define their local strategies and exchange them with the other members. The exchange of strategies is similar to negotiation between the end-users i.e. what each of the user wants to provide and consume as a part of the cooperative group. For instance, as shown in Figure 6, there are three distinct PNs who want to form a cooperative group (a PN Federation). Before forming the

Figure 6. Cooperation among PNs

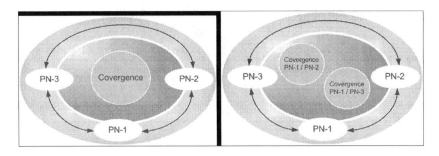

group, they negotiate on the terms and conditions of the PN-F. As an outcome of this negotiation, all of the potential cooperative group (PN-F) members converge at a certain point (a group of strategies), referred in Figure 8 as "convergence" point. Once the convergence point is attained i.e. the common strategies for the cooperative groups are defined, and further on the cooperative groups are actually formed.

Cooperative groups may vary on different scales such as age, profession, liking, needs, culture, so on and so forth. Therefore it is very less likely at times that they converge on a single point. The derivation of common strategies for the entire group gets more complicated and nontrivial with the increase in number of members of the cooperative group. Moreover, even if they finally converge to certain agreed upon strategies of the group, the time it would take to form a group would be considerably very high. Therefore, it would be quite efficient that the some group members converge on some strategies and does not converge on others. It is also possible that the cooperative group defines one single strategy as a "general" strategy for the group and other "specific" strategies for cooperation among group members. To this end, a cooperative group can have multiple convergence points. As shown in Figure 6, PN-1 defines two disjoint convergence points with each of the other PNs (i.e. PN-2 and PN-3) in the group. In concrete service terms, the cooperative group is formed by the PN-1 to consume/provide service to each of the other PNs, whereas other PNs i.e. PN-2 and PN-3 might not be interested in each others services. Therefore, in order for the group to achieve its goal, the convergence points of PN-1 with other PNs are essential. However, in this case, a much complex problem is to provide a secure and efficient interface between each of the convergence points defined within the scope of the cooperative groups. Moreover, during the lifetime of the cooperative group, due to the dynamism of the group and its members, the individual strategies can change. This dynamic nature of the group would certainly have its effect on the restructuring and reformation of the global group strategies. In this respect, coping up with the dynamism in the cooperative group environments is also a hard nut to crack.

Sharing Strategies in Cooperative Groups

In order to fully understand the sharing strategies in cooperative groups, it could be interesting to see how the economics of cooperation works in the society. Cooperation refers to the practice of people or greater entities working in common with commonly agreed upon goals and possibly methods, instead of working separately in competition. In the society, we cooperate when we want to accomplish something that we can not achieve working alone. In contrast, sometimes we cooperate not for obvious short-term benefits but for long-term gains. For instance, User A relays the traffic of User B so that in future, User B would be in a position to ask User A to relay his traffic. This type of cooperation involves the business, cultural and relationship development aspects. Even, sometimes in the society people cooperate just for social reasons and no obvious quantitative gains. Whatsoever the reason behind the cooperative behaviors is, the cooperation does not come for free and we always have to pay a certain price for it. The cost and the gains of cooperation can take many forms ranging from resources (man, money, machines) to moral and ethical support, referred as the potential of cooperation.

Even if all members of a group benefit from the cooperative group, individual self-interest may not favor cooperation. This theory of non-cooperative behaviors for self-interest in a cooperative group is referred as "prisoner's dilemma". There can be several reasons to be non-cooperative in a group. One of the major reasons is associated with the utility of being the part of the group. Everyone wants to have the best thing under the cost constraints he has. Therefore, the user would be cooperative to a certain limit where he sees that his total utility of being cooperative is greater or equal to the cost he is paying as a part of the cooperative group. Since the total utility and associated cost is associated with the satisfaction of the user, once the cost bypasses the total utility the user's satisfaction starts decreasing, and he becomes egoist or less cooperative member of the group.

4. TECHNICAL DIMENSION: ROUTING, A PRACTICAL ENABLER OF COOPERATION IN 4G

The key to successful realization of user-centric (Personal Network (PN)) and group-centric (PN Federation) cooperation is the general connectivity architecture which can seamlessly bridge heterogeneous personal devices, placed both in the close vicinity and at remote locations. The development, implementation and integration of global Personal Network architecture for connecting the devices geographically separated across the interconnection structures (intranet or Internet, for instance) has been explored recently in Hoebeke (2006). However, the ad-hoc seamless connectivity among the heterogeneous PN devices present in the close vicinity of each other, is still open to research. Moreover, the extension of PN concept to realize the group-centric Personal Network Federation (PN-F) i.e. connecting multiple PNs, towards PUE is a research theme that emerged recently. In this section, we present general connectivity architecture i.e. a routing protocol, which enables cooperation in PUE. At first, it facilitates the cooperation between the user's heterogeneous devices in order to form a Personal Network (PN). Moreover, at second, our cooperative routing protocol enables multiple PNs to join hands to form a PN Federation. In order to summarize, the routing solution enables the cooperation not only within the devices of a distinct user (Khadani, 2007; Maham, 2008), but also among the devices of different users, making the vision of PUE, a reality.

Preliminaries

4.1.1. Terminology

As described previously, we define a *"personal node (pn)"* to be the node that belongs to the owner of a PN. Each node is identified by its Personal Network Identification (PN-ID) and Node Identification (NID). All personal nodes of a PN owner share the same PN-ID. A *"Gateway Node (GN)"* is a personal node that enables the connectivity to the infrastructure network such as Internet or corporate LAN. A personal node is defined to be *"Federation Manager (FM)"*, if it enables connectivity to the personal nodes of the other PN(s). A *"PN Cluster"* is a network of personal nodes located within a limited geographical area (such as house, office or car). One or more than one "PN cluster" of a single owner contributes his PN.

4.1.2. Cooperative PN Federation Profiles

In order to interact among the PNs and to create trustable PN Federations (PN-Fs), rules and polices are needed to determine for instance, who is or can become member of the federation and how and which resources are made available to the PN-F members. Based on this, two different types of profiles have been identified to realise the concept of PN-F (PN to PN interaction), such as "PN-F profile" and "PN-F participation profile". As shown in Figure 7, the PN-F profile is common to the federation, created by the PN-F creator, which reflects the global information about the PN-F. Whereas, the "participation profile" is bound to the individual PN-F member and it reflects his local view regarding the PN-F. The PN-F is initiated by the PN-F profile, which is further updated

Figure 7. Personal network federation (PN-F) among Alice, Bob and Marc

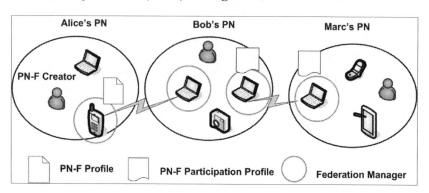

with the help of participation profiles during the course of PN-F's existence.

PN Cluster Formation and Personal Network Routing Protocol (PNRP)

The initial step towards the PN formation is the formation of PN clusters i.e. all the personal nodes that are in the close vicinity of each other discover the routes towards each other. Subsequently, the different geographically separated clusters of a PN are glued together to form a PN with the help of PN Agent. To this end, we propose the Personal Network Routing Protocol (PNRP) which helps in determining the routes among all the personal nodes in a PN cluster.

PNRP is a variant of link-state multi-hop routing protocol (Lieu, 2005) that adapts to the personal networking environments. It maintains the proactive topology of nodes which lie within the PN cluster boundary, at every personal node of the PN cluster. Moreover, the information on functionalities of the personal nodes such as Gateway Node and/or Federation Manager is also exchanged within the PN topology in order to facilitate PN's access to the outside world. In the following subsections, we discuss different steps that PNRP performs towards the formation of a PN cluster.

Integrated Topology Discovery

The role of Integrated Topology Discovery mechanism is to determine how the personal nodes are connected (using which interfaces in single/multi hop) in order to provide routes for any source/destination pair in the PN cluster. It first determines the direct connectivity among nodes and further exchanges this information to form a unified cluster topology.

Neighbor discovery is incorporated into PNRP by allowing every personal node to periodically transmit "Hello" packets on all of its interfaces. "Hello" packet contains the PN-ID and the Node-ID of the source node which is processed at the destination node to identify the source of the "Hello" packet. It is possible that the personal nodes may discover the non-personal nodes; therefore PN level authentication is indispensable. Every node maintains a 1-hop neighbour table and associated costs to each link with its direct neighbours and their PN identification. For the current implementation of PNRP, we have considered number-of-hops as a cost metric.

Once the 1-hop topology is formed, it is exchanged with other personal nodes in order to form a complete snapshot (table) of the PN cluster on its every single node. The topology information is only exchanged with the personal nodes i.e. among the nodes which belong to the same PN. To this end, every personal node periodically transmits

Personal Environments

the "Cluster topology (Ctopo)" message towards all its personal nodes (neighbours). On the reception of "Ctopo", the PN cluster routing tables are formed/updated and further exchanged with the other neighbouring nodes. Figure 8 shows the PN routing table at a personal node constructed after the exchange of "Hello" and "Ctopo" messages.

Gateway and Neighboring PN's Discovery

In PNRP, each Gateway Node (GN) advertises in the "Hello" message, whether it has connectivity with the infrastructure network or not. As can be seen in Figure 8, the exchange of integrated PN cluster topology with the help of "Ctopo" message permits each node to maintain routes to all the existing GNs and the cost to reach them, in the PN routing table.

Discovery of the neighbouring PNs is also intrinsic to Integrated Topology Discovery mechanism. During the exchange of "Hello" messages, if the destination node finds out that it's not the part of the source node's PN (with the help of PN-ID), the destination node sets itself as a Federation Manager (FM) to the source node's PN. The connectivity among the PNs is realised with the help of FMs. Once the integrated topology information is exchanged among all the nodes of the PN, every node knows the exit points (FMs) to communicate with other neighbouring PNs, which further helps in PN to PN (PN-F) routing. In case of multiple GNs or FMs, the minimum cost option is selected.

Route Discovery

PNRP differentiates the route discovery procedure when the destination is the part of same PN as the source and/or from the case when the destination is the part of different PN. In latter case, "PNRP for PN-F" mechanisms are triggered as discussed in the next section. In former case, since a proactive PN cluster-level topology is maintained at each PN node, the route to all the destinations in the PN will be known before time.

Figure 8. PNRP for PN routing

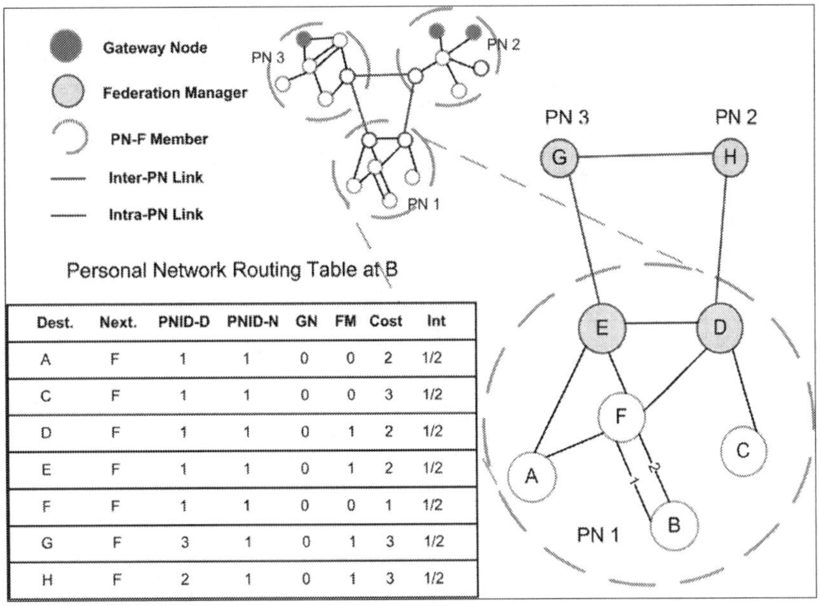

Dest.	Next.	PNID-D	PNID-N	GN	FM	Cost	Int
A	F	1	1	0	0	2	1/2
C	F	1	1	0	0	3	1/2
D	F	1	1	0	1	2	1/2
E	F	1	1	0	1	2	1/2
F	F	1	1	0	0	1	1/2
G	F	3	1	0	1	3	1/2
H	F	2	1	0	1	3	1/2

Initiation, Formation and Usage of Spontaneous PN-F

Since in the spontaneous PN-F, the federation is formed in the absence of a fixed infrastructure; therefore we extend our proposal for PN cluster formation i.e. PNRP, for the initiation, formation and usage of spontaneous PN-Fs. The Personal Network Routing Protocol (PNRP) extension for PN-F is a variant of on-demand multi-hop routing protocol such as AODV and DSR, which adapts to the personal networking environments for communication among PNs. The PN-F creator decides the PN-F members and sends them, the PN-F profile. On the reception of PN-F profile, if the initially proposed members decide to be the part of PN-F, they exchange their PN-F participation profile with the PN-F creator.

Therefore, PN-F initiation becomes an intrinsic capability of PN-F formation owed by PNRP. Moreover, PN-F routing mechanisms assist to determine the routes towards the desired PN-F members and provide means to form and to use PN Federations.

PN-F Topology Discovery, Initiation and Formation

PN-F routing builds routes using a join request/reply query cycle. When a PN desires to create a PN-F with certain defined PN-F members, it creates a PN-F profile and sends to its FMs, piggybacked with a Join-Request (JR) message, leveraging the inter-cluster routes; thanks to PNRP's PN-cluster formation capability. As show in Figure 9, in order to create a PN-F with PN2, PN3 and PN4, the PN1 (i.e. PN-F creator) sends the JR message to its FMs. The Join-Request (JR) message contains two lists of PN-IDs such as "destination-list", which stores the potential PN-F participants and "destination-attained-list", which represents the PNs already attained by the JR message. As FMs receive the JR, they investigate whether the PN-ID of their neighbouring PN is mentioned in JR message destination-list. In case of positive response, the JR is forwarded to the FM of the neighbouring PN. In contrast, if the FM does not find the adjacent PN in the destination-list and the potential participant PNs are not accessible through other FMs of the same PN, a "Connectivity PN-F" can be formed with the

Figure 9. PN-F topology discovery in PNRP

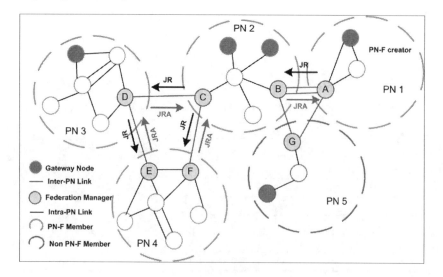

adjacent PN in order to relay the PN-F information towards the potential PN-F members.

On the reception of JR message, the neighbouring PN's FM first removes its PN-ID from JR's destination-list and then put it in the destination-attained-list. Moreover it also sets up backward pointers in PN-F routing table, towards the PN which sent the JR, as a next-hop to reach all the PNs mentioned in the destination-attained-list. PN1's FM (i.e. A) forwards the JR to PN2's FM (i.e. B), which sets the entries in its PN-F routing table that PN1 is reachable through its FM B.

The above presented mechanisms are repeated at each next PN until all the PNs mentioned in the JR destination-list are reached. Finally, on the reception of JR if the PN finds the JR's destination-list is empty, it will send a Joint-Request-Ack (JRA) message (by replacing the entries of destination-attained-list with destination-list) backwards to the PN which forwarded the JR message. As shown in Figure 9, the JRA is initiated by PN4, which receives the empty destination-list. The mechanisms of setting backward pointers to the PNs declared in destination-attained-list and moving the PN-IDs from destination-list to destination-attained-list at every next PN is also repeated for JRA message until it reaches the PN-F creator, who triggered the PN-F formation. The exchange of JR and JRA messages facilitate the establishment of PN-F routing tables at the Federation Managers of PNs, which are participating in the PN-F.

Data Forwarding and PN-F Use

During the integrated PN-F topology discovery process, the entire participant PNs learns the routes not only towards the PN-F creator but also towards each other. These routes are stored in the PN-F routing table and are leveraged to exchange, initially the PN-F profiles and then the PN-F participation profiles in order to form the PN-F. To this end, the data packets destined to any other PN are first forwarded to the FMs, which further routes the data with the help of PN-F routing table. The profiles are stored at the FMs (the entry points of PNs), which are used to enforce PN-F policies on the PN-F routing in order to ensure secure PN-F overlay concept.

Implementation and Performance Evaluation

We have implemented a Personal Ubiquitous Environment (PUE) with the proposed profile-based PN-F initiation, formation and usage schemes in Network Simulator, NS2. The main objective of our simulation study is to show the feasibility of our solution and to evaluate the performance of our spontaneous PN-F model.

The scenario for simulations (as shown in Figure 10) consists of three PNs connected with each other forming a spontaneous PN-F. Each PN is formed by five nodes connected in multihop fashion. Every PN has at least one FM, which connects it to at least one neighbouring PN. The simulation model parameters employed in our study are summarised in the Table 1.

For our simulation scenario, we adopted the Group Mobility Model. We investigate the performance of PRNP relative to classical proactive and reactive routing protocols, OLSR and AODV (Lieu, 2005) respectively. Simulations were carried out with the variable number of inter-PN flows. Figure 11 presents the simulation results.

The results show that our spontaneous PN-F initiation, formation and usage scheme i.e. PNRP, efficiently manages the user's personal environments (PN and PN-Fs). Moreover, it performs almost similar to traditional routing protocols such as OLSR and AODV, in terms of packet delivery ratio and delay under a scenario with small number of PNs/PN-Fs with low mobility. In contrast, PNRP consistently generates less routing load than any flat proactive routing protocol such as OLSR, because of its awareness of PN boundaries.

Figure 10. Simulation scenario

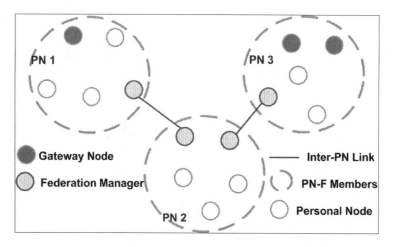

Table 1. Simulation model parameters

SIMULATION / SCENARIO		MAC / ROUTING	
Simulation Time	200s	MAC protocol	802.11 DCF
Simulation Area	600 x 600 m2	Channel Capacity	2 Mbps
Number of PNs	3	Trans. Range	100m
Nodes in each PN	5	Traffic Type	VoIP (G.729)
Mobility Model	Group Mobility	Hello Interval	2s
Node Speed	1-2 m/s	Ctopo Interval	5s
Inter-PN VoIP Flows: 1, 2, 3, 4, 5, 6, 7			

5. LIMITS AND POTENTIALS OF COOPERATION IN PUE: A SOCIO-TECHNICAL DIMENSION

PUE is particularly interesting in terms of social implications of cooperation since the user in a PUE is totally free to cooperate within his own network and with others, without undergoing through certain rigid set of obligations from the service provider or the network operator (which is the case in cellular/infrastructure-based networks today). We discuss the potentials and limits of cooperation by considering PUE as one of the services arena in 4G. We apply Nash Equilibrium (NE) theory (part of game theory) to Personal Ubiquitous Environment (PUE) concepts. John Nash introduced the concept of Nash equilibrium in his doctorate thesis and showed for the first time in his dissertation, *Non-cooperative games* (1950), that Nash equilibria must exist for all finite games with any number of players (Nash, 1951). In PUE, where different PNs join hands to form a cooperative group (PN Federation) in order to share certain services, let (S, f) be a cooperative group, where S is the set of strategy profiles and f is the set of payoff profiles. Let σ_{-i} be a strategy profile of all group members except for member i. When each member of the group $i \in \{1...n\}$ chooses strategy x_i resulting in strategy profile $x = (x_1,...,x_n)$ then member i obtains payoff $f_i(x)$. Note that the payoff depends on the strategy profile chosen, i.e. on the strategy chosen by member i as well as the strategies chosen by all the other members.

Figure 11. Simulation results, PDF, delay and load vs. inter-PN flows

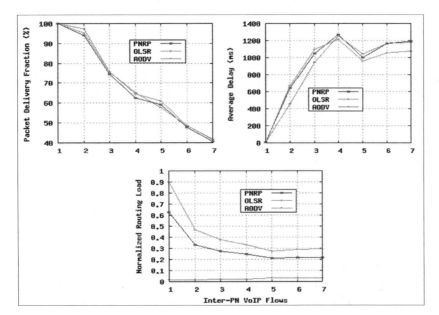

In descriptive terms, if there is a set of group strategies with the property that no group member can benefit by changing his strategy while the other members keep their strategies unchanged, then that set of agreed-upon group strategies and their corresponding payoffs constitute the Nash Equilibrium in the cooperative group. Therefore, in a Nash Equilibrium none of the group members can unilaterally change his strategy to increase his payoff.

We have analyzed the potentials and limits of cooperation with the help of NE theory under multiple group members (PNs) scenarios which form a Personal Network Federation (cooperative group). In this study the cooperative group strategies model referred as "consume/provide" is based on the basic supply/demand economics theory. Moreover, contradicting the basic NE concept, we have also studied the scenarios where multiple equilibrium points are possible.

Potentials of Cooperation in 4G's Personal Ubiquitous Environments

The potentials of cooperation in Personal Ubiquitous Environments (PUE) are associated with the strategic satisfaction of each cooperative group member. It implies that what percentages of his local strategies are reflected in the common group strategy. As discussed earlier, towards the formation of cooperative groups (PN Federation), each group member (PN) prepares his proper local strategy and then exchanges it with the other potential group members. A group member who first initiates the group formation process is referred as a group creator (PN-F creator). In concrete terms, a local strategy contains the information related to the participation of the member such as, which services he wants to consume/provide, what are his preferences (security, QoS, economic etc) for certain services, how much time he is willing to remain a member of the cooperative group and so on.

After the exhaustive exchange of local strategies among the cooperative group members, a final group strategy is prepared. This strategy is the convergence point in the entire cooperative group space. If all the group members agreed on certain group strategy upto an extent that none of the member none of the group members with to unilaterally change his strategy to increase his payoff, we can say that the cooperative group has

Figure 12. (a) (b) Potentials of cooperation in PUE, consumer/provider strategies

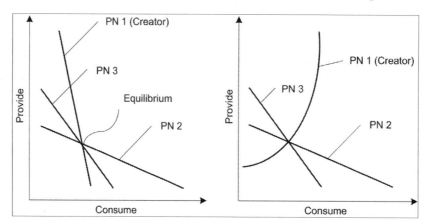

attained a Nash Equilibrium point as shown in Figure 12. The X-axis represents the "consume" strategy, whereas the "provide" strategy is on the y-axis. Here, three members (PNs) join hands to form a cooperative group (PN-F). PN-1 is the creator of the cooperative group. After some negotiations of their own local consume/provide willingness they all agree on a certain point, which is marked in Figure 12 (a) as "Equilibrium" point. Figure 12 (b) highlights a much different behavior of PN-1 (creator) in the cooperative group. Since PN-1 is the initiator of the group, it is quite possible that he might be more open to provide as much services as possible to the group with competitively very limited desired to consume services. This behavior is much justifiable in the society, as a manager or the front-liner is normally the center of focus of a group and his behavior has a strong impact on the strategies of the other group members. Therefore, for the success of a group, the initial strategy defined by the mentor of the group is highly important.

Its is important to note that the Nash Equilibrium point presents the minimum set of "provide" strategies owned by all the cooperative group members. It is of course possible that, at the later stage of cooperation, some of the member expresses a generous attitude and provide more services by keeping his "consume" strategy constant, as it can be the case with PN-1 in Figure 12 (b). To this end, the equilibrium point will shift keeping the entire group's "consume" strategies constant. This phenomenon of moving equilibrium point with variable "provide" strategy of one member and "constant" consume strategies of all the other members, can be clearly studied on a three-dimensional graph, where X-axis is "consume" strategy, Y-axis is "provide strategy and Z-axis is "equilibrium".

As we normally see in a society, some players in a group have their stakes associated with only certain members or certain goals of the group. They stay with the group for only such limited benefits as a part of the entire cooperative group's ecosystem. To this end, a group may have multiple equilibrium points satisfying all group members as a whole or some of them. As shown in Figure 13, a group consists of four members such as PN-1, PN-2, PN-3 and PN-4. Lets assume that PN-4's interest in the group is only associated with some services offered by PN-1 and he is not interested in any other service. In this respect, as in Figure 12, we have two equilibrium points such as Equilibrium-1 among all members except PN-4 and Equilibrim-2 between PN-1 and PN-4. In multiple equilibrium group cases, it is important that both the equilibrium strategies should have a certain level of interface among them. As in the example

Figure 13. Potentials of cooperation in PUE, multiple equilibrium strategies

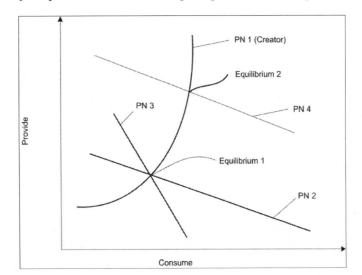

in Figure 13, a strong communication between both strategic equilibrium points would monitor and control the accurate working of the group. For example, here in this example, this interface would insure that the PN-4 only consumes the services of PN-1 as defined by Equilibrium-2 and does not interact with any other group services made available by other group members.

Limits of Cooperation in 4G's Personal Ubiquitous Environments

Sometimes certain group members either cooperate in a way that their cooperation is not useful for the group or they behave in a non-cooperative way (becomes egoist). Both of these situations limit the cooperation in the Personal Ubiquitous Environment. The former case is discussed in Figure 14 (a). A cooperative group has three potential members such as PN-1, PN-2 and PN-3. The strategies defined by PN-1 and PN-2 make them settled to a certain equilibrium point whereas PN-3 does not party to the common equilibrium. In Figure 14 (a), the strategy of PN-3 is represented by a straight line parallel to y-axis (provide). This implies that PN-3, is although very cooperative member of the group, but his cooperation is not interesting for the other group members. For example, PN-3 is providing such services which are not intended by the other members. In this case, an ideal equilibrium point among all the group members is blocked by the local strategies of PN-3.

In the latter case, as shown in Figure 14 (b), again three PNs are potential cooperative group members. But here we clearly see that PN-2 and PN-3 are extremely egoist in their cooperative behaviors i.e. they just are inclined towards consuming much more service than offering to other group members. This case is again a bottleneck in the formation of cooperative group with certain equilibrium point(s).

One way to overcome this bottleneck in cooperative groups is to reward more for cooperative attitude and to punish more for non-cooperative attitude. In the absence of any reward/punish mechanism; the non-cooperative behavior will have a tit-for-tat effect on the entire group. For instance, if a cooperative member's partner defects from cooperative behavior, it responds in a similar non-cooperative way towards other partners. This chronic behavior will rapidly spread within

Figure 14. (a) (b) Limits of cooperation in PUE, egoist/ineffective cooperation

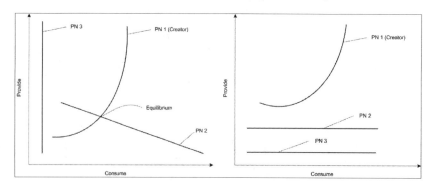

the group, and it might end-up with a total non-cooperative group, where no member is willing to cooperate.

6. SUMMARY

Cooperation is the buzz word in the communications industry today driving the notion of cooperative mechanism in future heterogeneous systems, including 4G. The 4G landscape is so diverse, and the industry leaders and policy leaders accept that 4G is not only about improved data rates or diverse air interfaces and unified standards, but rather is going to be shaped by increasing integration, collaboration and interconnection of heterogeneous systems. On the contrary, the widely agreed upon rule for success in 4G telecommunication markets is to visualize a cooperative service chain of multiple suppliers satisfying the ever-growing requirements of end customers. This intertwined and inspiring direction could facilitate the realization of large scale cooperative and ubiquitous wireless communities.

Personal or group communication environments, particularly, the PUE could eventually be the forerunner for exploiting the theoretical limits of cooperative systems, enabling the provision of niche cooperative systems and services. This needs to be explored in much detail, understanding the socio-technical aspects and potential limits of cooperation and developing efficient models to develop and nurture cooperative societies. Several socio-economic aspects need to carefully observed and studied, like human factors with respect to human nature, ego-centric human behaviors and social factors like the effects on society, economic competition etc. From a technical dimension, methods like the intelligent routing technique, PNRP act as enabler to cooperation not only within the devices of a distinct user but also among the devices of different users, making the vision of Personal Ubiquitous Environments, a reality. This stimulus for cooperation brings PUE more closely towards the 4G vision.

From a service perspective, we believe that the future of cooperative services in 4G largely depends on the result of cooperation of major players in the industry including the service providers, vendors etc on one hand and the policy makers, academia etc on the other. From a technology perspective, our opinion is that the large scale integration of coexisting applications and incorporation of emerging technologies, flexible models for spectrum allocation etc should be considered in depth. Finally, encouraging healthy inter-working between application research and technology research, supporting seamless cooperation should be the vision of future of 4G systems.

REFERENCES

Bohlin, E. (2004). *The Future of Mobile Communications in the EU: Assessing the Potential of 4G* [Tech, Rep. No. EUR21192 EN]. ESTO Publications.

Borcea, C. (2002). Cooperative Computing for Distributed Embedded Systems. *In Proceedings of 22nd International Conference on Distributed Computing Systems (ICDCS)*, Vienna, Austria.

Frattasi, S., Fathi, H., Fitzek, F., Chung, K., & Prasad, R. (2004). *4G: A User-Centric System*. Paper presented at the Mobile e-Conference (Me), Athens, Greece.

Frattasi, S., Fathi, H., Fitzek, F., Katz, M., & Prasad, R. (2005). A Pragmatic Methodology to Design 4G: From the User to the Technology. In *Springer LNCS: 5th IEEE International Conference on Networking (ICN)*, Reunion Island, France.

Frattasi, S., Fitzek, F., & Prasad, R. (2005). Cooperative Services for 4G. In *Proceedings of 14th IST Mobile and Wireless Communications Summit*, Dresden, Germany.

Gupta, P., & Kumar, P. R. (2002). The Capacity of Wireless Networks. *IEEE Transactions on Information Theory*, 388–404.

Hoebeke, J. (2006). Personal Networks Federations. In *Proceedings of Fifteenth IST Mobile and Wireless Summit*, Myconos, Greece.

Javaid, U., Meddour, D. E., & Ahmed, T. (2006). Towards Universal Convergence in Heterogeneous Wireless Networks using Ad Hoc Connectivity. In *Proceedings of of 9th International Conference on Wireless Personal Multimedia Communications (WPMC)*, San Diego.

Javaid, U., Rasheed, T., Meddour, D. E., & Ahmed, T. (2007). A Profile-based Personal Network Architecture for Personal Ubiquitous Environments. In *Proceedings of IEEE Vehicular Technology Conference (VTC)*, Dublin, Ireland.

Javaid, U., Rasheed, T., Meddour, D. E., & Ahmed, T. (2007). Personal Network Routing Protocol (PNRP) for Personal Ubiquitous Environments. In *Proceedings of IEEE International Conference on Communications (ICC)*, Glasgow, U.K.

Katz, M., & Fitzek, F. (2005) Cooperative Techniques and Principles Enabling Future 4G Wireless Networks. In *Proceedings of IEEE EUROCON 2005*. Serbia & Montenegro, Belgrade.

Khadani, E., Abounadi, J., Modiano, E., & Zheng, L. (2007). Cooperative Routing in Static Wireless Networks. *IEEE Transactions on Communications*, 2158–2192.

Knutsson, B. (2004). Peer-to-Peer Support for Massively Multiplayer Games. In *Proceedings of 23rd Annual IEEE Conference on Computer Communications (Infocom)*, Hong Kong.

Lieu, C., & Kaiser, J. (2005). A Survey of Mobile Ad-hoc Network Routing Protocols, University of Ulm Technical Report Series. Jacobsson, M. (2004). Network Layer Architecture for Personal Networks. In *Proceedings of MAGNET Workshop*, Shanghai, China.

Maham, B., Debbah, M., & Hjorungnes, A. (2008). Energy-efficient Cooperative Routing in BER Constrianed Multihop Networks. In *Proc. of Int. Conf. on Commun. And Networking in China (Chinacom)*, Hangzhou, China.

Moustafa, H., Javaid, U., Rasheed, T., & Meddour, D. E. (2006). A Panorama on Wireless Mesh Networks: Architectures, Applications and Technical Challenges. In *Proceedings of International Workshop on Wireless Mesh: Moving towards Applications (Wimeshnets)*, Waterloo, Canada.

Murota, K. (1999). *Mobile Communications Trends in Japan and DoCoMo's Activities Towards 21st Century*. Paper presented at the Fourth ACTS Mobile Communications Summit, Sorrento, Italy.

Nash, J. (1951). Non-Cooperative Games. *The Annals of Mathematics, 54*(2), 286–295. doi:10.2307/1969529

Pereira, J. M. (2000). *Fourth Generation: Now, it is Personal*. Paper presented at 11th IEEE International Symposium on Personal, Indoor, and Mobile Radio Communications (PIMRC), London, UK.

Rasheed, T., Javaid, U., Reynaud, L., & Al Agha, K. (2007). Cluster-Quality based Hybrid Routing in Large Scale Mobile Multi-hop Networks. In *Proceedings of IEEE Wireless Communications and Networking Conference (WCNC)*, Hong Kong.

Ray, S. K., & Mista, I. S. (2005). *Fourth Generation Networks: Roadmap – Migration to the Future*. Paper presented at IETE Technical Review, 23(4).

Roussos, G. (2003). *End-to-End Service architectures for 4G Mobile Systems, The Path to 4G Mobile*. Paper presented at IIR, London.

Sandvig, C. (2003). Assessing Cooperative action in 802.11 Networks. In *Proceedings of 31st International Conference on Communication, Information and Internet Policy*, Washington, D.C.

Weiser, M. (1992). *Does Ubiquitous Computing Need Interface Agents? No*. Invited talk at MIT Media Lab Symposium on User Interface Agents.

Chapter 12
Next Generation Broadband Services from High Altitude Platforms

Abbas Mohammed
Blekinge Institute of Technology, Sweden

Zhe Yang
Blekinge Institute of Technology, Sweden

ABSTRACT

In this chapter the authors investigate the possibility and performance of delivering broadband services from High Altitude Platforms (HAPs). In particular, the performance and coexistence techniques of providing worldwide interoperability for microwave access (WiMAX) from HAPs and terrestrial systems in the shard frequency band are investigated. The WiMAX standard is based on orthogonal frequency-division multiplexing (OFDM) and multiple-input and multiple-output (MIMO) technologies and has been regarded as one of the most promising 4G standards to lead 4G market and deliver broadband services globally. The authors show that it is possible to provide WiMAX services from an individual HAP system. The coexistence capability with the terrestrial WiMAX system is also examined. The simulation results show that it is effective to deliver WiMAX via HAPs and share the spectrum with terrestrial systems.

1 INTRODUCTION

The forthcoming fourth generation (4G) communication systems are expected to provide a variety of services based on IP solution to deliver high quality multimedia to users through high data rate wireless channels on an "Anytime, Anywhere" basis. It is thought that the demand for the capacity increases significantly when the next generation of multimedia applications are combined with future 4G wireless communication systems (Ohmori & Yamao et al., 2000). For the design of 4G communication systems, the possibility of employing High Altitude Platforms (HAPs) has been considered as a valid alternative to the traditional terrestrial or satellite based infrastructures (Dovis & Fantini et al., 2001; Mohammed & Arnon et al., 2008).

Wireless communication services are typically provided by terrestrial and satellite systems. The successful and rapid deployment of both wireless networks has illustrated the growing demand for wireless broadband communications. These

DOI: 10.4018/978-1-61520-674-2.ch012

networks are featured with high data rates, reconfigurable support, dynamic time and space coverage demand with considerable cost. Terrestrial links are widely used to provide services in areas with complex propagation conditions and in mobile applications. Satellite links are usually used to provide high speed connections where terrestrial links are not available. In parallel with these well established networks, a new alternative using HAPs has emerged recently and attracted international attention (Djuknic & Freidenfelds et al., 1997).

HAPs are airships or planes, operating in the stratosphere, at altitudes of typically 17-22 km (around 75,000 ft) (Collela & Martin et al., 2000; Grace & Thornton et al., 2005; Mohammed & Arnon et al., 2008; Hult & Mohammed et al., 2008). At this altitude (which is well above commercial aircraft height), they can support payloads to deliver a range of services: principally communications and remote sensing. A HAP can provide the best features of both terrestrial masts (which may be subject to planning restrictions and/or related environmental/health constraints) and satellite systems (which are usually highly expensive) (Mohammed & Arnon et al., 2008). This makes HAP a viable competitor/complement to conventional terrestrial infrastructures and satellite systems. Thus HAPs are regarded as a future candidates for next generation systems, either as a stand-alone system or integrated with other satellite or terrestrial systems. The integration and convergence of different technologies and services are key concepts in 4G systems. Current HAPs research and development activities include the COST 297 Action in Europe (Cost297, 2005; Mohammed & Arnon et al., 2008), along with government funded projects in Japan, Korea, and USA. Commercial projects are also underway in Switzerland, USA, China and the UK.

HAPs have been recently proposed as a novel approach for the delivery of wireless broadband services to fixed and mobile users. HAPs can act as base-stations or relay nodes, which may be effectively regarded as a very tall antenna mast or a very Low-Earth-Orbit (LEO) satellite (Karapantazis & Pavlidou, 2005). HAP systems have many useful characteristics including high receiver elevation angle, line of sight (LOS) transmission, large coverage area and mobile deployment. These characteristics make HAPs competitive when compared to conventional terrestrial and satellite systems, and furthermore they can contribute to a better overall system performance, greater capacity and cost-effective deployment (Yang & Mohammed, 2008). The major advantages for the 4G systems integrating HAPs will be the cost-effective coverage deployment, the system flexibility due to the platforms' mobility on demand (e.g., emergency situations) and the possibility of payload upgrading in order to reduce the risk of technology obsolescence experienced with traditional satellites.

Considering the above advantages and depending on the applications, HAPs may be an ideal complement or alternative solution when deploying next generation communication systems with high capacity demands. For example, supporting mobile TV as one of the 3G/4G multimedia applications based on multimedia broadcast and multicast services (MBMS) is subject to challenging issues due to high traffic load deriving from both signaling message exchange and data transmission between multicast sources and end users. When dealing with broadcast and multicast wireless services, key research issues are the effective exploitation of the limited radio spectrums, coordination of users accessing radio resources, delivering services with desired quality of service (QoS) and cost-effective delivery of services in different geographical coverage areas. The delivery of multicast services from HAPs have the following additional advantages compared to terrestrial 4G systems (Araniti & Iera et al., 2005):

- A wide coverage can be provided from the HAPs due to its unique position. A single

HAP is able to cover an area with a radius ranging between 30 km and 150 km (ITU-R, 2000). The coverage area is significantly important for multicast services, which need to include excessive number of terrestrial network infrastructures to deliver services to a large number of populations. Furthermore, spot beam coverage can also be achieved by phased array antenna from HAPs, which enables HAPs to enlighten only portions of a wide area when multicast users are not uniformly located.

- HAPs can provide a high elevation angle for most users. It implies a better quality of transmission when compared to terrestrial system in areas where buildings are intensively located (e.g., urban areas).
- HAPs can be deployed in a short time and they can extend the coverage to areas not covered by the terrestrial networks. These features are tremendously important for emergency situations requesting MBMS services delivery.

In this chapter we investigate the possibility and performance of delivering broadband services from High Altitude Platforms (HAPs). In particular, we investigate the performance and coexistence techniques of providing worldwide interoperability for microwave access (WiMAX) from HAPs and terrestrial systems in the shard frequency band. WiMAX is a standard-based wireless technology for providing high-speed, last-mile broadband wireless access up to 50 km for fixed stations, and 5-15 km for mobile stations in frequency bands ranging from 2 to 66 GHz (IEEE802.16, 2004). The standard is mainly based on OFDM and MIMO technologies, and has been regarded as one of the most promising 4G candidates to lead 4G market for the delivery of broadband wireless services.

Providing WiMAX from HAPs is a novel approach since it has the potential to deliver a significantly enhanced performance and coverage from HAPs in the frequency bands below 11 GHz, which leads to a more favorable propagation path compared with traditional base stations located on mountains or tall buildings due to the HAPs unique location. For this reason some preliminary research has been done to show the effectiveness of this novel approach. The coexistence performance of a single HAP and a single terrestrial base station in terms of modulation techniques has been examined in (Likitthanasate & Grace et al., 2005). The performance of an individual HAP system delivering WiMAX services with a seven-cell layout has been examined in examined in (Ahmed, 2006). The identification of suitable scenarios and a feasibility analysis of WiMAX delivered from HAPs has been addressed in (Giuliano & Luglio et al., 2008). The outcome from previous research shows that it is possible to deploy WiMAX from HAPs with the acceptable quality of downlink connection. In our scenarios, we focus on the application for delivering fixed services via HAPs. We assume fixed users with the antenna mounted on the roof with a directive antenna to receive signals from HAPs. It is anticipated that providing WiMAX from HAPs to be a competitive approach with a low complexity deployment of broadband services.

The chapter is organized as follows. In section 2 we give a description of the proposed coexistence system models, propagation and antenna models for the HAP and terrestrial deployment, and important system simulation parameters are listed. The criteria employed to measure the interference and system performance, e.g., downlink and uplink carrier to noise ratio (CNR) and carrier to interference plus noise ratio (CINR) are also defined. The system coexistence performance of HAP and terrestrial WiMAX is evaluated in section 3. In section 4, various techniques to improve the system coexistence performance and analysis for a single cell and multiple cell HAP system are investigated. Finally, conclusions and future research directions are given in section 5.

Figure 1. Coexistence model of a HAP with a single cell structure and a terrestrial base station

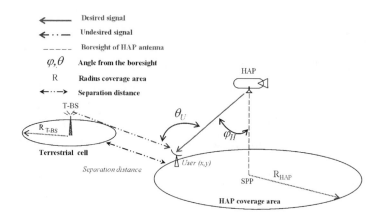

2 SCENARIOS OF DELIVERING BROADBAND SERVICES FROM HAPS AND COEXISTENCE PERFORMANCE

In this section the coexistence scenarios of HAP and terrestrial systems are discussed. The hypothetical coexistence communication scenarios are introduced. The system parameters which are considered to test the coexistence performance are discussed. The system coexistence performance is evaluated using CNR and CINR criterion defined in this section.

2.1 Coexistence Models of HAP and Terrestrial Systems

A general coexistence model consists of a HAP base station (H-BS), a terrestrial system with at least one base station and a receiver. The HAP is assumed to be deployed at an altitude of 17 km above the ground with a radius of coverage equal to 30 km. The terrestrial base station (T-BS) is assumed to be deployed on the ground with an appropriate separation distance away from the sub-platform point (SPP) of the HAP on the ground. The receiver, which we refer to as a 'user', is assumed to be located on the ground on a regular grid with 1 km distance. Two scenarios of HAP system with a single and multiple cell structure are considered and shown in Figure 1 and Figure 2, respectively.

2.1.1 A Single Cell Structure from HAP

The scenario in Figure 1 is composed of a HAP with a single antenna payload, a terrestrial base station (T-BS) and a user. The terrestrial base station is initially deployed on the ground with an appropriate separation distance 40 km away.

2.1.2 A Multiple Cell Structure from HAP

A scenario in Figure 2 is extended to include a cellular structure both for HAP and terrestrial systems. Terrestrial cellular architectures are based on division of the coverage area into a number of cells which are assigned different channels with respect to adjacent cells, in order to manage the cochannel interference and achieve frequency reuse. The scenario consists of a single HAP with multiple antenna payloads serving multiple cells. The number of antenna payloads is assumed to be equal to the number of cells, which consequently can be served independently. We assume that cells are hexagonally arranged and clustered in different frequency reuse patterns to cover the HAP service area.

Figure 2. Coexistence model of a HAP with a multiple cell structure and terrestrial WiMAX systems

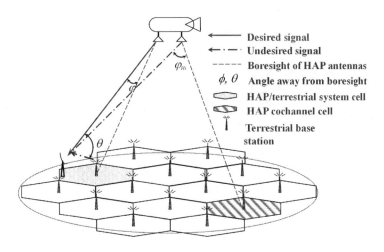

In a HAP cellular system with multiple antennas, interference is mainly caused by antennas serving cells on the same channel employing the terrestrial frequency reuse schedule (Thornton & Grace et al., 2003). The terrestrial WiMAX system is assumed to employ the same cellular configuration as the HAP system. Therefore, there are 19 terrestrial base stations considered in the scenario and all the base stations are located in the centre of HAP cells. A user communicating with the HAP in an arbitrary HAP cell is interfered by the HAP cochannel antennas and the terrestrial base station located in the centre of the same cell. In the examined scenario we assume that the user always receives interference from its nearest terrestrial base station.

2.1.3 Antenna Radiation Pattern of HAP and Receiver

Antenna gains of HAP AH (φ) at an angle φ with respect to its boresight and the ground receiver AU (θ) at an angle θ away from its boresight are approximated by a cosine function raised to a power roll off factor n with a flat side lobe level. They are represented in (1) and (2), respectively (Thornton & Grace et al., 2003):

$$A_H(\varphi) = G_H\left(\max\left[\cos(\varphi)^{n_H}, s_f\right]\right) \quad (1)$$

$$A_U(\theta) = G_U\left(\max\left[\cos(\theta)^{n_U}, s_f\right]\right) \quad (2)$$

where

- G_H and G_U are the boresight gain of the H-BS antenna and receive user antenna respectively.
- n_H and n_U control the rate of power roll off of the antenna main lobe individually.
- s_f in dB is a notional flat sidelobe floor.
- The beamwidth φ_{10dB} is initially set to be equal to the subtended angle ψ_{edge} at the SPP with a circular beamwidth as illustrated in Figure 3. This allows antenna directivities to be specified independently of the angle of the cell edge. Here φ_{10dB} is the 10 dB antenna beamwidth at which the directivity curve, controlled by a roll off factor n, is 10 dB lower than its maximum value.

In the single cell scenario of HAP system, the φ_{10dB} beamwidth of HAP antenna is specified to be equal to the subtended angle ψ_{edge} at the edge of the coverage area. In the multiple cell scenario of

Figure 3. HAP antenna beamwidth specifications

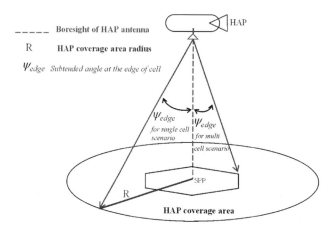

HAP base station with multiple antenna payloads, we assume the payload is composed of multiple antennas with the same pattern employed in the basic scenario, and the φ_{10dB} beamwidth is set to be equal to the subtended angle ψ_{edge} at the angle of the central cell edge. Therefore more power can be centrally radiated inside the HAP coverage area and produce less interference to the terrestrial systems. Antenna beamwidths for two scenarios are shown in Figure 3.

The antenna peak directivity, which is usually achieved in the direction of the boresight, is assumed to be achieved at the centre of each HAP cell corresponding to its serving antenna. The boresight gain is calculated as (Thornton & Grace et al., 2003):

$$G_{boresight} = \frac{32 \ln 2}{2\theta_{3dB}^2} \quad (3)$$

A cumulative distribution function (CDF) of CNR with idealized isotropic and directive antenna patterns adopted by a single cell HAP scenario is shown in Figure 4. It represents the CNR performance achieved from adopting isotropic and directive antenna patterns respectively by assuming that a user is situated at each point inside the HAP coverage area. It can be seen that by adopting a directive antenna on the HAP, an approximately a 3 dB increase is achieved on average over the entire coverage area. Furthermore, because the directional antenna points at the centre of the coverage area, the CNR is decreased at the edge of coverage (EOC) area.

Figure 5 shows the performance of the multiple HAP antenna payloads. It illustrates that the best performance is achieved at the centre, where the boresight of antenna is pointing. Since all the

Figure 4. CDF of CNR performance with isotropic and directive antenna patterns

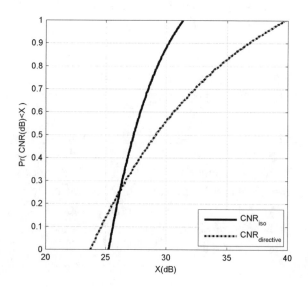

Figure 5. HAP cellular antenna payload performance

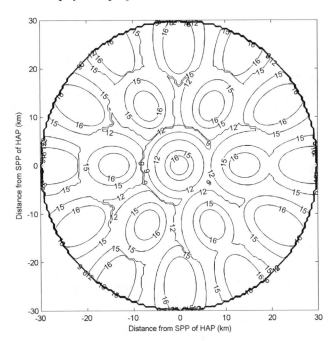

antennas employ the common beamwidth of the antenna serving the central cell, cells further away from the centre have a better performance due to a smaller subtended angle at the cell edge from its boresight.

2.1.4 Propagation Pathloss and Simulation Parameters

Currently most HAP research papers adopts the free space path loss (FSPL) in (4) as the propagation model used for HAP transmission, where d is distance from the transmitter and λ is the signal wavelength. Until now, no specific propagation model has been established for HAPs at 3.5 GHz, and therefore FSPL has been widely used in current research. Propagation models have been developed for HAPs in mm-wave band at 47/48 GHz, but they are not applicable in the 3.5 GHz frequency band. It should also be noted that directional user antennas are likely to be installed at a fixed location with this scenario. High elevation angles owing to the relatively small radius of HAP coverage also mean that the LOS propagation to the HAP is a reasonable assumption. Therefore FSPL is used, and diffraction and shadowing are not explicitly considered, without loss of general validity.

$$PL_H = \left(\frac{4\pi d}{\lambda}\right)^2 \qquad (4)$$

The propagation pathloss model PL_T is shown in (5) for the terrestrial signal propagation model (Erceg & Greenstein et al., 1999).

$$PL_T = PL_m + \Delta PL_f + \Delta PL_h \qquad (5)$$

PL_T is composed of a median path loss PL_m, the receiver antenna height correction term ΔPL_h and the frequency correction term ΔPL_f. The two-correction terms ΔPL_h and ΔPL_f are added in order to make PL_T more accurate by accounting for the antenna heights and frequencies (IEEE802.16, 2004).

Parameters in suburban environment are used for the simulations of a terrestrial deployment environment. We assume that the HAP carries the directional antenna payload and the terrestrial base stations use omni-directional antennas. Table 1 lists the most important system parameters for the downlink (DL) and uplink (UL) used in the simulations.

2.2 HAP Cellular Structure and Cochannel Interference Scenario

The WiMAX performance from a HAP cellular system is evaluated by assuming that the user inside the HAP coverage area communicates with the HAP and is interfered by the cochannel HAP antennas. In practice, a precise hexagonal pattern cannot be generated due to topographical limitations, local signal propagation conditions, and practical limitations on sitting antennas (Stallings, 2005). The frequency reuse scheme in a HAP cellular system allows the use of same frequency already employed in other cells nearby, thus allowing frequencies to be used for multiple simultaneous conversations.

2.2.1 HAP Cellular Structure

A plane view of a HAP ground cellular structure is shown in Figure 6. The 'x' markers in Figure 6 indicate the footprints of boresight of the HAP antennas.

We use a circular shape to approximately cover the ideally proposed hexagonal pattern in HAP coverage area. The frequency reuse pattern in HAP cellular system consists of N cells assigned the same number of frequencies, which are defined as cochannel cells. N is termed as a reuse factor and determines the number of cells in a repetitious pattern. Accordingly the minimum distance between the cochannel cells is given as (Stallings, 2005):

$$D = r_{cell} \cdot \sqrt{3N} \qquad (6)$$

Figure 6. Plane view of cell deployment of HAP coverage area

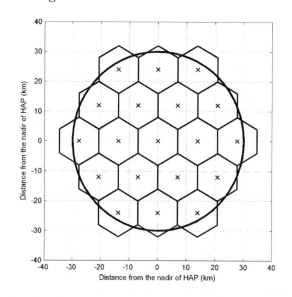

Table 1. Important system simulation parameters

Parameters	H-BS	T-BS
Coverage Radius Transmitter Height	30 km (R_H) 17 km (H_H)	7 km (R_T) 30 m (H_T)
Antenna Efficiency	80%	
User Roll Off Rate User Boresight Gain Sidelobe Level	58 (n_U) 18 dBi (G_U) -30 dB (s_l)	
Bandwidth Frequency	7 MHz /1.75 MHz (DL/UL) 3.5 GHz	

Figure 7 shows examples of frequency reuse patterns in HAP cellular system.

2.2.2 Cochannel Interference in a HAP System

In a HAP cellular system, the interference is due to HAP antennas serving cells in the same frequency band. A cochannel interference scenario is shown in Figure 8. The user is assumed to be located in HAP cell *i*. When it communicates with the HAP antenna *i*, the user is interfered by the HAP antenna *j* and *k*, which are assumed to operate in the same frequency band as the antenna serving the cell *i*. Since each HAP antenna has its boresight footprint in the centre of the corresponding cell, angles from the boresight can be calculated separately in order to evaluate the interference.

2.3 HAP System Performance Evaluation

Without considering interference from terrestrial system, we assume that a user in a location *(x, y)* is to be communicating with its serving HAP antenna and to be receiving interference signals from other antennas operating in the same frequency band. The performance of downlink and uplink communication can be evaluated as a function of CIR and CINR, respectively:

2.3.1 Downlink Performance Evaluation

The performance of downlink can be evaluated in (6) and (8), respectively:

$$CIR_H(x,y) = \frac{P_H A_H A_U PL_{HU}}{\sum_{i=1}^{N_H} P_{H_i} A_{H_i} A_{U_i} PL_{H_iU}} \quad (7)$$

$$CINR_H(x,y) = \frac{P_H A_H A_U PL_{HU}}{N_F + \sum_{i=1}^{N_H} P_{H_i} A_{H_i} A_{U_i} PL_{H_iU}} \quad (8)$$

where,

- P_H is the downlink transmission power of the HAP transmitter (40 dBm);
- P_{Hi} is the transmission power of the interfering HAP antennas (40 dBm);
- A_H and A_U are the antenna gains of the HAP and the user, and they depend on the angular deviation from the boresight;
- PL_{HU} is the propagation pathloss from HAP to user;
- N_H is the total number of cochannel cells in the HAP system. For the scenario with a single antenna payload, N_H is equal to 1;
- N_F is the noise power.

Figure 7. HAP cellular layouts with the reuse factor N at 3, 4 and 7

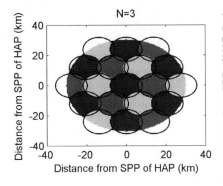
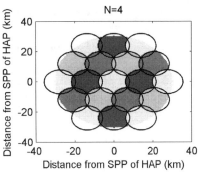

Figure 8. HAP cellular cochannel interference evaluation scenario

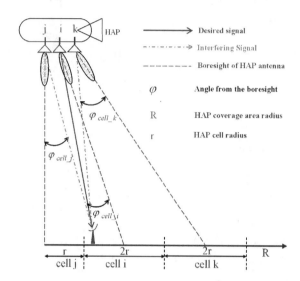

2.3.2 Uplink Performance Evaluation

The uplink performance of a HAP WiMAX system can be evaluated by CIR and CINR expressed in (9) and (10), respectively:

$$CIR_{UH}(x,y) = \frac{P_U A_U A_H PL_{UH}}{\sum_{i=1}^{N_H} P_U A_{U_i} A_{H_i} PL_{UH_i}} \quad (9)$$

$$CINR_{UH}(x,y) = \frac{P_U A_U A_H PL_{UH}}{N_F + \sum_{i=1}^{N_H} P_U A_{U_i} A_{H_i} PL_{UH_i}} \quad (10)$$

where,

- P_U is the transmission power of user in the target cell (30 dBm);
- PL_{UH} is the propagation pathloss from user to HAP;
- N_F is the noise power (-106.5 dBm for bandwidth at 1.75 MHz).

2.4 Coexistence Performance Evaluation of HAP and Terrestrial System

In the proposed coexistence scenario, we assume a user in location *(x, y)* communicates with HAP and always receives signals from its nearest terrestrial base station.

2.4.1 Downlink Coexistence Performance

The downlink coexistence performance of HAP and terrestrial system interfered by the other system can be expressed in (11) and (12), respectively:

$$CIR_{HT}(x,y) = \frac{P_H A_H A_U PL_{HU}}{\sum_{i=1}^{N_H} P_{H_i} A_{H_i} A_{U_i} PL_{H_i U} + P_T A_T A_{UT} PL_{TU}}$$

(11)

$$CIR_{TH}(x,y) = \frac{P_T A_T A_U PL_{TU}}{\sum_{i=1}^{N_H} P_{H_i} A_{H_i} A_{U_i} PL_{H_i U}} \quad (12)$$

where,

- P_T is transmission power of the interfering terrestrial base station;
- A_{UT} is the user antenna gain at an angle away from its boresight in terms of receiving interference from base stations.
- PL_{TU} is pathloss from terrestrial base station to user.
- N_H is equal to 1 for the scenario of a single cell structure from HAP.

2.4.2 Uplink Coexistence Performance

The uplink coexistence performance of HAP interfered by the terrestrial system can be expressed in (13) and (14), respectively:

$$CIR_{HT}(x,y) = \frac{P_U A_U A_H PL_{UH}}{\sum_{i=1}^{N_H} P_U A_{U_i} A_{H_i} PL_{UH_i} + P_U A_U A_T PL_{UT}} \quad (13)$$

$$CIR_{TH}(x,y) = \frac{P_U A_U A_T PL_{UT}}{\sum_{i=1}^{N_H+1} P_U A_{U_i} A_{H_i} PL_{UH_i}} \quad (14)$$

where,

- PL_{UT} is the pathloss from the user to the terrestrial base station.
- $N_H + 1$ is addition of the number of current serving HAP cell where the user is located and the number of corresponding cochannel cells in HAP system.

3 HAP AND TERRESTRIAL SYSTEM COEXISTENCE PERFORMANCE

3.1 Coexistence Performance of HAP with a Single Cell Structure

In the scenario of a single antenna payload, the terrestrial base station is deployed on the ground with an appropriate separation distance 40 km away from the SPP of the HAP on the ground. The CINR performance is shown in Figure 9 and Figure 10 to highlight the interference effects from T-BS.

The $CINR_H$ curve maintains a circular symmetry, since the signal from T-BS is heavily attenuated by the sidelobe of the user's antenna when it communicates with HAP. In contrast, the left half coverage area of T-BS the $CINR_T$ curve shrinks toward the base station under the interference from H-BS, because the signal from H-BS enters into the user's antenna main lobe which results in higher interference. However, on the other half of the coverage area the interference signal always enters into the user antenna's sidelobe which attenuates the interference, and so the contours are relatively circular. Consequently, in this case the HAP coverage area is less susceptible to interference.

The system performance is evaluated under different separation distance in order to justify deployment of WiMAX broadband from T-BS and H-BS at the same time in an appropriate service area. This case is modelled as shown in Figure 11. The separation distance is initially assumed to be 40 km, then we decrease the separation distance which brings the T-BS coverage area closer to the H-BS coverage area. When the separation distance becomes negative, the two coverage areas start to overlap.

The performance of this scenario is shown in Figure 12, and evaluated at the right and left edge of coverage (EOC) area of T-BS and the left EOC area of H-BS. When the terrestrial system coverage area starts to overlap the edge of H-BS coverage

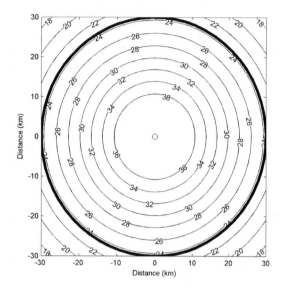

Figure 9. $CINR_H$ performance contour plot of HAP (marked as 'o') coverage area

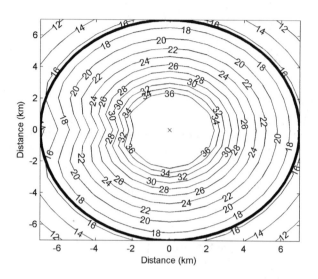

Figure 10. $CINR_T$ performance contour plot of T-BS (marked as 'x') coverage area

Figure 11. Edge of coverage area performance evaluation scenario with different separation distance

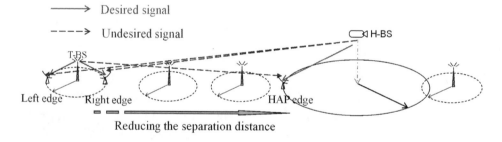

Figure 12. CINR at the edge of coverage of H-BS and T-BS with decreasing separation distance

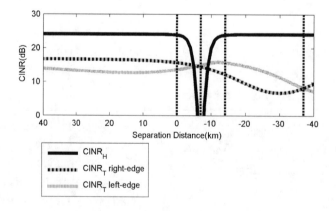

area (where separation distance is equal to 0 km), $CINR_H$ falls rapidly below 0 dB since the receive user on the EOC area of H-BS is much closer to the T-BS and receives much more interference power. When the coverage area of the terrestrial WiMAX system is totally contained inside the coverage area of H-BS, the $CINR_H$ (at the EOC area of H-BS) rapidly rises to the same level as before. For the EOC area of T-BS, $CINR_T$ on the right of the EOC always behaves better than the $CINR_T$ on the left of the EOC until the separation decreases to -7 km, which means the T-BS is just located in the left EOC area of H-BS. It is because the signal from H-BS enters into the test user's antenna main lobe on the left EOC which results in higher interference and lower CINR.

3.2 Performance Evaluation of HAP with a Multiple Cell Structure

3.2.1 Downlink Performance

The performance of HAP cellular system coexistence scenario is evaluated in Figure 13. The downlink contour plot clearly shows that in most of the cellular areas HAP system can provide stable services to the users, regardless of interference from the terrestrial system.

Figure 14 shows the CDF of the CIR value in the individual cellular HAP and its coexistence scenarios, respectively. On average the HAP system can be operated with a CIR larger or equal to 21 dB, but a slight decrease in performance in the coexistence scenario can be observed because of the interference from the terrestrial system.

3.2.2 Uplink HAP cellular system coexistence evaluation

Figure 15 shows the contour plot of the uplink coexistence scenario of the cellular HAP system. In most of cell areas the HAP can provide stable uplink services to the users, which are not susceptible to interference from the terrestrial system.

4 COEXISTENCE TECHNIQUES OF HAP AND TERRESTRIAL SYSTEMS

In this section, different coexistence and deployment techniques for reducing the interference from HAPs to terrestrial WiMAX system (e.g., adding new channels, cell splitting, cell sectoring, and deployment spacing distance) in order to improve the cellular system performance are investigated.

4.1 Increasing Cochannel Spacing Distance

Increasing D, the minimum distance between cochannel cells, is an effective approach since it sets the cochannel cells further away from each other and therefore reduces the interference without requiring more spectrum.

Figure 16 shows the downlink CIR_H performance of the HAP system for different values of D, obtained by increasing the reuse factor N. It shows that a CIR increase of approximately 2 dB can be achieved with each increment of N in the figure (from 3 to 4, or from 4 to 7). Usually the

Figure 13. Downlink system performance of user interfered by terrestrial deployments

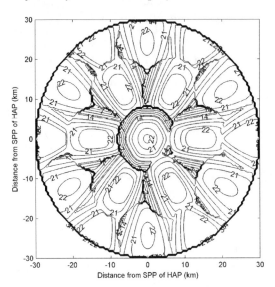

Figure 14. Comparison of the downlink CIR performance in HAP (CIR$_H$) and HAP/terrestrial (CIR$_{HT}$) coexistence scenarios

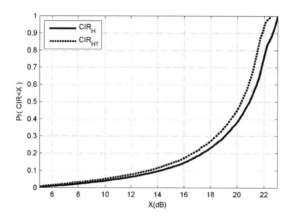

Figure 15. Uplink CIR performance of HAP system (CIR$_{HT}$) interfered by the terrestrial system

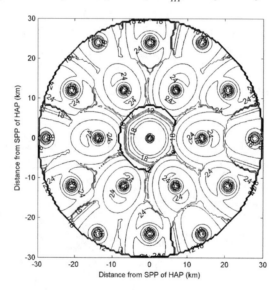

total number of frequencies allotted to the system is constant, so increasing D, on the other hand, decreases available channel resources in each cell of the system.

4.2 Varying HAP Spacing Radius

In the previous investigations, we assume that SPP of the HAP is in the center of the HAP coverage area and it has been shown to exhibit good system performance. Since a directional antenna is used on the HAP, it could allow HAPs to be deployed in the different parts of the sky while keeping the boresight of antenna pointing at the desired coverage area (Chen & Grace et al., 2005; Grace & Thornton et al., 2005). Furthermore, in practice it is difficult to keep HAPs absolutely stationary above the center of the coverage area. Therefore, we need to consider the system performance under the changeable HAP spacing distance, which means that the SPP of HAP doesn't overlap with the center of its service area. The location of the

Figure 16. Downlink CIR$_H$ performance of HAP system with reuse factor N (3, 4 and 7)

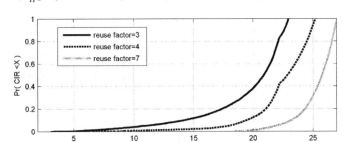

Figure 17. Illustration of changing of HAP spacing radius while setting the antenna pointing offset at the center of serving area

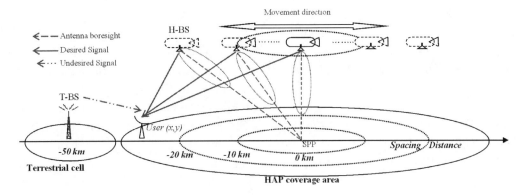

T-BS is fixed at 50 km away from the center of the HAP coverage area. This scenario is illustrated in Figure 17.

As the HAP antenna is not pointing at the SPP of HAP coverage area due to the variable HAP spacing distance, the antenna gain across the HAP coverage area changes accordingly. From Figure 18 we could see the antenna gain with different spacing distances. It shows that the curves fall more rapidly to the side lobe level with the wider spacing distance on the left side of the coverage area. For example, when the spacing radius is equal to -20 km, the signal from the left edge of the coverage area enters into its side lobe level. In this case, if the T-BS is deployed on the left side of the HAP coverage area as shown in Figure 18, users outside the HAP coverage area receives an interfering signal coming from the side lobe of the HAP antenna rather than the main lobe. Interference signals coming from terrestrial base stations are also suppressed by the HAP antenna sidelobe. On the right side of the HAP coverage area, the HAP antenna curve falls more slowly compared with the zero spacing distance case, which provides the higher gain with better performance to the users using HAP services. Interference signals coming from terrestrial base stations are decreased since they undergo a longer distance to the HAP antenna with a higher pathloss. Considering the efficient utilization of the antenna payload, this technique could be used in a multiple HAP deployment to serve multiple cells from HAPs by suppressing interfering signals into the side lobe of the HAP antenna.

4.3 Varying HAP Antenna Beamwidth

The antenna beamwidth is an important parameter affecting system performance. It determines the directivity of the antenna and hence controls the footprint on the ground. As shown in Figure 18, we can see a narrow beamwidth can bring a high peak gain and rapid roll off over the coverage area. At the edge of the HAP coverage area, the antenna gain is decreased to an appropriate level to create an acceptable coexistence environment with terrestrial systems.

Different antenna beamwidths are investigated in Figure 19 to show an improvement, which can be achieved by decreasing the HAP antenna beamwidth. When the beamwidth is narrowed to 43 degrees, less than 90% coverage area achieves a CINR of 35 dB and less than 10% area achieves a CINR of 10 dB at the EOC area. Compared with the 43 degree beamwidth performance, a 72 degree beamwidth antenna, which is adopted for simulation, gives 50% area inside the HAP coverage a higher CINR of 25 dB and a higher CINR at the edge of coverage area. The 72 degree beamwidth also provides a capability to extend the HAP coverage area by offering better link budgets at the edge of coverage.

4.4 Varying the User Antenna Beamwidth

Similar to changing the HAP antenna beamwidth, varying the user antenna beamwidth is also an effective means to improve the system performance as shown in Figure 20. We can see that with a narrower antenna beamwidth of the receiver the CINR performance is improved gradually. For example, the 17 degree beamwidth selected in the simulation achieves a mean CINR of 23 dB inside the HAP coverage area, when we specify that it is equal to its half power beamwidth (roll off at -3 dB). If we consider the movements of HAPs and receivers, a narrower beamwidth of the user antenna requires a higher antenna pointing accuracy.

5 CONCLUSION AND FUTURE RESEARCH DIRECTIONS

In this chapter, we have shown the performance of both downlink and uplink WiMAX broadband standard transmitted from a HAP cellular system in the 3.5 GHz band across a coverage area of 30 km radius, while operating in the same frequency band with terrestrial WiMAX deployments based on a proposed coexistence scenario. A single and cellular structure has been proposed for the HAP WiMAX system based on the typical WiMAX terrestrial system. The HAP coverage area was divided into 19 individual cells served

Figure 18. HAP antenna gain with different spacing distance (0 km, -10 km, -20 km) and different beamwidth roll off (-3 dB, -10 dB, -30 dB)

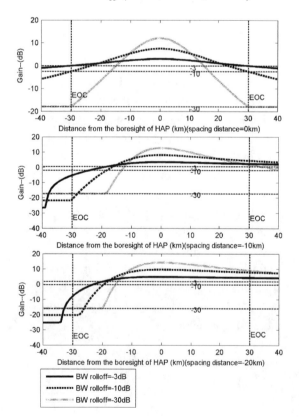

Figure 19. CINRH performance under different HAP antenna beamwidth

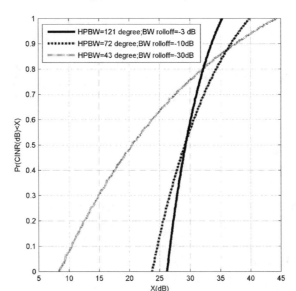

Figure 20. CINRH performance of different user antenna beamwidth

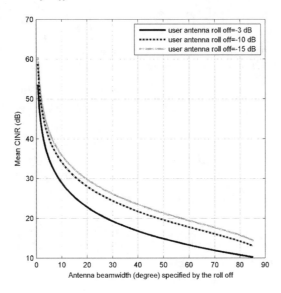

by multi-antenna payload. WiMAX broadband system performance of individual the HAP system was evaluated both separately and when taking into account of the cochannel interference from the antennas operating in the same frequency band. Coexistence capability was investigated based on a proposed coexistence scenario and examined by considering interference from the nearest terrestrial base station to HAP system. Simulation results clearly demonstrate that the internal cochannel interference was dominant when delivering WiMAX via HAPs, and HAP system can effectively share the spectrum with terrestrial WiMAX systems.

It is extremely beneficial to investigate other possibilities of providing mobile services from

HAPs since this would provide an important supplemental HAP application under the goal "Broadband for All". Previous HAP application investigations in the *CAPANINA* project mainly addressed the fixed-wireless application in the mm-wave band at 30/31 GHz or even higher. Delivery of mobile services from HAPs enables HAPs to exploit the highly profitable mobile market. The IEEE802.16e standard and beyond provide both stationary and mobile services. To extend the HAP capabilities to support full operations under the WiMAX standards brings a more competitive service especially in the mobile service field. Some 3G HAP mobile communication studies have also been carried out in the 2 GHz band. High Speed Downlink Packet Access (HSDPA), which is usually regarded as an enhanced version of Wideband Code Division Multiple Access (W-CDMA), and 3GPP Long Term Evolution (LTE) with MIMO and/or adaptive antenna systems capabilities for achieving higher data rates and improved system performance are also attractive directions for further investigations.

REFERENCES

Ahmed, B. T. (2006). *WiMAX in High Altitude Platforms (HAPs) Communications*. Paper presented at the 9th European Conference on Wireless Technology, Manchester, UK.

Araniti, G., Iera, A., & Molinaro, A. (2005). The Role of HAPS in Supporting Multimedia Broadcast and Multicast Services in Terrestrial-Satellite Integrated Systems. *Wireless Personal Communications*, *32*, 195–213. doi:10.1007/s11277-005-0742-3

Chen, G., Grace, D., & Tozer, T. C. (2005). *Performance of Multiple HAPs using Directive HAP and User Antennas*. Paper presented at the Wireless Personal Communications, Aalborg, Denmark.

Collela, N. J., Martin, J. N., & Akyildiz, I. F. (2000). The HALO Network. *IEEE Communications Magazine*, *38*(6), 142–148. doi:10.1109/35.846086

Cost297. (2005). *The European Union (EU) Cost 297 Action on High Altitude Platforms for Communications and Other Services*. Retrieved from http://www.hapcos.org/overview.php

Djuknic, G. M., Freidenfelds, J., & Okunev, Y. (1997). Establishing Wireless Communications Services via High-Altitude Aeronautical Platforms: A Concept Whose Time Has Come? *IEEE Communications Magazine*, *35*(9), 128–135. doi:10.1109/35.620534

Dovis, F., Fantini, R., Mondin, M., & Savi, P. (2001). 4G Communications Based on High Altitude Stratospheric Platforms: Channel Modeling and Performance Evaluation. In *Global Telecommunications Conference, GLOBECOM '01 IEEE* (Vol. 1, pp. 557-561).

Erceg, V., Greenstein, L. J., Tjandra, S. Y., Parkoff, S. R., Gupta, A., & Kulic, B. (1999). An empirically based path loss model for wireless channels in suburban environments. *IEEE Journal on Selected Areas in Communications*, *17*(7), 1205–1211. doi:10.1109/49.778178

Giuliano, R., Luglio, M., & Mazzenga, F. (2008, March). Interoperability between WiMAX and Broadband Mobile Space Networks. *IEEE Communications Magazine*, 50–57. doi:10.1109/MCOM.2008.4463771

Grace, D., Thornton, J., Chen, G., White, G. P., & Tozer, T. C. (2005). Improving the System Capacity of Broadband Services Using Multiple High Altitude Platforms. *IEEE Transactions on Wireless Communications*, *4*(2), 700–709. doi:10.1109/TWC.2004.842972

Hult, T., Mohammed, A., Yang, Z., & Grace, D. (2008). *Performance of a Multiple HAP System Employing Multiple Polarization.* Invited Paper, Special Issue on Wireless Personal Multimedia Communications (WMPC) 2006 Conference.

IEEE802.16. (2004). *IEEE802.16-2004 Part 16: Air Interface for Fixed Broadband Wireless Access System.* IEEE802.16 Broadband Wireless Access Working Group.

ITU-R. (2000). Feasibility of High Altitude Platform Station (HAPS) in the fixed and mobile services in the frequency bands above 3 GHz allocated exclusively for terrestrial radiocommunication. *Resolution 734 [COM5/14] Question ITU-R 212/9.*

Karapantazis, S., & Pavlidou, F. (2005). Broadband communications via high-altitude platforms: a survey. *Communications Surveys & Tutorials, IEEE, 7*(1), 2–31. doi:10.1109/COMST.2005.1423332

Likitthanasate, P., Grace, D., & Mitchell, P. D. (2005). Coexistence performance of high altitude platform and terrestrial systems sharing a common downlink WiMAX frequency band. *Electronics Letters, 41*(15), 858–860. doi:10.1049/el:20051930

Mohammed, A., Arnon, S., Grace, D., Mondin, M., & Miura, R. (2008). Advanced Communications Techniques and Applications for High-Altitude Platforms. *Editorial for a Special Issue, EURASIP Journal on Wireless Communications and Networking, 2008.*

Ohmori, S., Yamao, Y., & Nakajima, N. (2000, December). The Future Generationis of Mobile Communications Based on Broadband Access Technologies. *IEEE Communications Magazine,* 134–142.

Stallings, W. (2005). *Wireless Communications and Networking* (2nd ed). Upper Saddle River, NJ: Pearson Prentice Hall. Thornton, J., Grace, D., Capstick, M. H., & Tozer, T. C. (2003). Optimizing an Array of Antennas for Cellular Coverage from a High Altitude Platform. *IEEE Transactions on Wireless Communications, 2*(3), 484–492.

Yang, Z., & Mohammed, A. (2008). *On the Cost-Effective Wireless Broadband Service Delivery from High Altitude Platforms with an Economical Business Model Design.* Paper presented at the IEEE 68th Vehicular Technology Conference, 2008. VTC 2008-Fall, Calgary Marriott, Canada

Chapter 13
Radio-over-Fibre Networks for 4G

Roberto Llorente
Universidad Politécnica de Valencia, Spain

Maria Morant
Universidad Politécnica de Valencia, Spain

Javier Martí
Universidad Politécnica de Valencia, Spain

ABSTRACT

Radio-over-Fibre (RoF) is an optical communication technique based on the transmission of standard wireless radio signals though optical fibre in their native format. This technique is an enabling step in the deployment of dense fourth generation (4G) cellular and pico-cellular wireless networks. The optical fibre provides a huge bandwidth that can support a variety of wireless systems, regardless of their frequency bands, being protocol-transparent which is reflected in an great network flexibility. Radio-over-fibre techniques enables a high user capacity by frequency reuse, simplifies the network operation as the signals are distribute in their native format, and permits to transfer signal part of the processing power from the base station units to the central control station, thus reducing the overall deployment cost and complexity. The principles of radio-over-fibre are presented in this chapter, including the key transmission impairments and the expected performance. The main application scenarios are discussed. These include the backhaul of 4G or base-stations, addressing 4G and 3G compatibility issues, and distributed-antenna system (DAS). Finally, emerging applications like radio-over-fibre in beyond-3G scenarios and transmission of 60 GHz wireless are also described in this chapter.

INTRODUCTION

The huge number of subscribers to the latest media services with multimedia contents increased considerably the demand of broadband communication systems. These users must be able to access multimedia contents at any time and in any place. This bring out the paradigm to achieve high-capacity services always available for the users and pushed for the development of new communication systems using both wireline and wireless technologies (Stuckmann & Zimmermann, 2007). The main

DOI: 10.4018/978-1-61520-674-2.ch013

benefits of wireless services rely on its intrinsic mobility and fast deployment. For these reasons, wireless architectures have experienced a great development in the last years. But the radio spectrum scarcity has pushed the industry and operators to look for efficient ways of transporting and distributing the radio signals to remote locations, avoiding the use of high-frequency equipment which increases considerably the cost of the operator's architecture.

Radio-over-fibre (RoF) systems, eventually called microwave-photonics systems (Seeds & Williams, 2006; Capmany & Novak, 2007), offer a cost-effective solution when a dense distribution of the wireless signal is required, like in the cellular coverage of urban areas or when antenna clusters are employed. Using RoF technology, network architectures can be implemented in different scenarios. A description of different applications is shown in Figure 1. The first application scenario consists of distributing wireless services for the operator from the central office (CO) to several base stations (BS), as shown in Figure 1(a) called RoF transport. The second application scenario described in Figure 1(b) is used to interconnect several base stations and provide 4G services to the end costumer, i.e. last-mile access and femtocell networks. Radio-over-fibre provides the benefits of scalability and transparency to the radio service in these applications.

BACKGROUND

Fourth generation (4G) mobile broadband networks employing technologies like WiMAX, LTE (Long Term Evolution), and IMS (IP Multimedia Subsystem) transport high data rate signals to support high quality communication services. Pervasive presence of fourth-generation (4G) wireless requires an efficient, flexible and not bandwidth-limited backhaul. RoF optical technology addresses this need, being one of key enabling technologies for deployment of 4G networks.

In the last years, the migration to IP services has been increased considerable, i.e. WiMax and LTE are all IP, CDMA is migrating, and UMTS 3G is moving to IP via ATM and Ethernet. Ethernet is still the most successful and widely deployed local area network transport technology in the world due to its flexibility and cost-effective networks. On the other hand, wireless networks

Figure 1. Application scenarios of Radio-over-Fibre (RoF) systems for 4G distribution: (a) RoF for transport and (b) RoF for BS interconnection

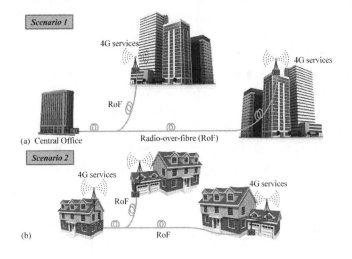

reduce significantly the backhaul costs. Optical Metro-Ethernet networks are an interesting approach to support 4G bandwidth needs. Metro networks transport digital data, which requires specific modulation/demodulation equipment to convert the digital data to a wireless signal and vice-versa. Nevertheless, RoF transports analog data, i.e. the wireless signal is transported in its final format. This approach is enabled by optics: An extremely large bandwidth is available in the fiber and electromagnetic compatibility is intrinsic to the transmission media. RoF transmission permits flexible, reconfigurable systems with centralized maintenance capable to serve a large number of users. By example, a RoF system was deployed in Sydney Olympics comprising more than 500 remote antenna units to provide multi-standard radio as GSM in 900/1800 MHz bands, 802.11 WLAN, CDMA2000 services demonstrating the RoF reliability and capacity, Henry, P.S. (2007). This system enabled dynamic allocation of the network capacity, providing service to more than 500,000 wireless calls on the opening day of the Olympic Games.

Comparing radio-over-fibre with microwave point-to-point link solutions, RoF solves the problem of high radio-wave attenuation when the frequency of operation of the wireless signal is higher than 3 GHz. RoF techniques applied on 4G systems based on pico-cells permits to decrease transmitting power and increase the efficiency of frequency reuse. It is also worth note that RoF links support a variety of wireless systems, regardless of their frequency bands and is signal-format and protocol transparent, i.e. 4G and third-generation (3G) wireless, wireless local-area networks (WLAN), digital multimedia broadcasting (DMB), and so-called beyond-third generation (B3G) can be seamless distributed on the same optical fibre, as described in Lee, K. (2005).

Radio-over-Fibre Principles

In the last twenty years, many techniques have been developed in the field of radio-over-fibre systems. These can be classified in optical signal processing techniques, including photonic analogue-to-digital converters, photonic-microwave filters, millimetre-wave and THz signal generation, beamforming and imaging. Nevertheless, from these applications the most successful relies on the use of optical fibre links to distribute wireless signal. This application leads to the so-called hybrid fibre-radio (HFR) transmission systems (Jager & Stohr, 2001). The HFR concept is similar to the distribution of cable-TV (CATV) signals over a hybrid-fibre coaxial (HFC) network, as described in Darcie and Bodeep (1990); and Wilson, Ghassemlooy and Darwazeh (1995).

The optical distribution of the wireless signal already in a format suitable for to be radiated at the end of the fibre allows centralizing the complex functions needed for RF signal processing in the central office (CO). In this way the base stations (BSs) architecture is simplified.

A simplified schematic of the proposed RoF application scenarios depicted in Figure 1 is shown in Figure 2. At the Central Office (CO) premises different wireless signals are generated in their native format. The wireless signals are distributed in both upstream and downstream directions. Different lights at different frequencies, i.e. wavelengths, can be employed to transmit the wireless signal intended for different users (downstream), or to provide the return channel (upstream). This arrangement is called wavelength-division multiplexing (WDM). Depending on the spectral separation of the optical channels, these are called coarse- or dense-wavelength division multiplexed systems, named CWDM and DWDM respectively. The specific case of DWDM system and the special requirements to minimise the crosstalk between the different lights is addressed, for example, in Kartalopoulos (2002). It is worth to note the availability of polarization-multiplexed (PM) systems

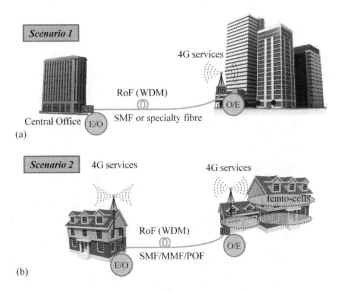

Figure 2. Main elements of a radio over fibre link for (a) scenario 1 and (b) scenario 2

which employ two orthogonal polarizations per wavelength, each transporting a given wireless signal. This approach maximizes the spectral efficiency and was first proposed in (Heidrich, Hoffmann, & Macdonald, 1986).

In the state of the art, analog transmission over fiber links has been demonstrated for frequencies up to 120 GHz (Hirata et al., 2003). The huge bandwidth of the optical fibre link and the availability of wireless commercial devices facilitate the standards-independent optical transmission and multiservice operation for existing cellular systems. Several transmissions of different wireless standards using radio-over-fibre techniques have been demonstrated so far, such as GSM (Ogawa et al., 1992), UMTS (Persson et al., 2006), wireless-LAN WiFi 802.11 a/b/g/n (Chia et al., 2003; Niiho et al., 2004; Nkansah et. al., 2006), WiMAX (Pfrommer et al., 2006), and ultra-wideband (UWB) radio (Llorente et al., 2008).

Technology Perspectives

Optimum network configurations for efficient signal transmission between the central office and the end user has been subject of continuous investigation. Mobile perators are investigating several radio access interfaces such as HSDPA, LTE, WIMAX, or UWB, not only from the technology point of view but also taking into account CAPEX and OPEX minimisation. It is not clear at this moment, which technology will dominate, as they are in a different stage of development and standardization. Moreover the coexistence and compatibility features of all these radio services are under the research scope of the principal operators. Taking into account this heterogeneous environment, huge efforts are made to develop convergent infrastructures able to cope with increasing bandwidth demand. This convergent infrastructure must be able to offer multi-standard transmission capabilities and to provide integration with both the transport network and wireless backhaul solutions in order to increase the network flexibility and the dynamic adaptation to changing traffic conditions.

This market is growing very fast as several market analysts have recently predicted. ABI Research estimates that the penetration of radio-over-fibre in-building systems will grow from

18% of the total in-building market in 2003 to 46% in 2009 (Anscombe, 2005). ABI states that distributed-antenna systems will make a significant impact in larger buildings because they offer multi-service broadband capabilities.

For these reasons, the use of RoF technologies is expected to be deeply used for indoor deployments due to the flexibility of the RoF system able to transmit multi-standard wireless signals over the same system with transparency and providing unlimited bandwidth thanks to the fibre-based architecture. In the case of UWB and WiMAX systems, recent field trial demonstrations have shown the potential of this technology to increase the coverage of the system. Currently, RoF systems adapted to UWB based in WiMedia MB-ODFM are under development in the frame of several European projects like UROOF (Ben-Ezra et al., 2008), EUWB (Giorgetti et al., 2008) and UCELLS (Llorente & Cartaxo et al., 2008).

RADIO-OVER-FIBRE TRANSMISSION

Optical Transmission

The main key elements of any radio-over-fibre system are the devices with electro-optical and opto-electric conversion functions. An electro-optic conversion is needed to up-convert the wireless signal to optical frequencies to be transmitted through fibre to the remote user, and afterwards an opto-electric conversion recovers the radio signal from the optical carrier to be transmitted wirelessly to the end user. These functions are illustrated in Figure 3.

The main systems for electro-optic conversion are broadband optical sources either based in direct or external modulation. Once the signal is adapted to the optical domain, it is transmitted over a suitable transmission media such as standard single-mode fibre (SSMF), specialty fibre (e.g. Teralight), multi-mode fibre (MMF), or plastic optical fibre (POF). Finally, the signal is converted to the electric domain using broadband photodetectors or photoreceivers (Capmany & Novak, 2007; Dagli, 1999), and filtering.

As it was commented previously, two main approaches can be used for electro-optic conversion functions (Dagli, 1999): direct modulation of a laser or a continuous wave laser with an external modulator. To point out the main advantages and disadvantages of both electro-optic conversions, it is worth to say that external modulation provides more bandwidth, with low drive voltage and good linearity. External modulation is usually performed using Mach-Zehnder electro-optic modulators (EOM) and electro-absorption modulators (Blumenthal, 1962). On the other hand, the cost of external modulators is nowadays more expensive that direct modulation devices, and for this reason this second method is usually is preferred for low-cost RoF systems deployments.

Concerning direct modulation techniques, the modulated signal changes the intensity of the laser output. In this case, the modulation bandwidth, the optical wavelength and the efficiency have to be considered. Since the first proposals of direct modulation (Ikegami & Suematsu, 1967) considerable work has been done devoted to the increase of the direct modulation bandwidth and semiconductor lasers at 1550 nm. For example, in Capmany & Novak, 2007, a modulation bandwidth greater than 30 GHz has been demonstrated.

When the wireless signal to be transported over the RoF system is limited to a frequency range between 800 and 2500 MHz, the detector/modulator can be achieve using waveguide electro-absorption modulators (Wake et al., 1997) or polarization independent asymmetric Fabry-Perot modulators (Liu et al., 2003; Liu & Polo et al., 2007). For the frequencies used in the major part of the current wireless standards as GSM, WiFi 802.11 a/b/g, UMTS and WiMAX, which can be up to 5-6 GHz, directly modulated semiconductor lasers are preferred (Qian et al., 2005) because these lasers has a lower production cost. At higher frequencies, only externally modulated transmitters can satisfy

Figure 3. Main elements in a Radio over Fibre link. Only one direction of communication (downlink/uplink) is shown

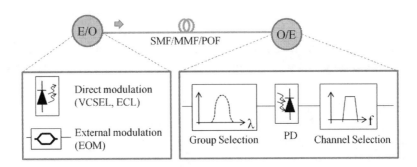

the required performance for adequate wireless transmission. Recently, important research efforts have been carried out targeting the development of low-cost/high-performance transmitters, for instance uncooled lasers (Ingham et al., 2003; Hartmann et al., 2003) or vertical-cavity surface emitting lasers (VCSEL) (Persson et al., 2006; Chia et al., 2003).

In the last years, vertical-cavity surface emitting lasers (VCSELs) have been proposed for low-cost direct modulation (see Persson et al., 2006; Chia et al., 2003). These VCSELs operate in a single longitudinal mode using of a very small cavity length (around 1 μm). These devices provide efficient coupling with optical fibre and have low cost production, which are very interesting features for the implementation of cheap radio-over-fibre systems. However, nowadays VCSEL is not a mature technology and usually present lower output power and bandwidth than the devices used for external modulation. Additionally, VCSEL are mainly available at wavelengths of 850 and 1310 nm, although devices at 1550 nm (the usual wavelength band for telecommunication applications) are beginning to be commercially available with increasing direct modulation bandwidth.

The second key element of the RoF system is the optical fibre. There are several types of fibres such as standard single-mode fibre (SSMF), multimode fibre (MMF) and plastic optical fibre (POF). The most suitable media for long-haul radio-over-fibre transmission links is SSMF while for indoor applications MMF or POF are preferred due to its easier installation and reduced cost. The main difference between the different media is the fibre core radius (typically around 9 μm in SSMF and 62.5 μm in MMF) which results in the propagation of a single or multiples modes through the fibre. POF is made of plastic instead of glass and shows larger core radius than MMF. The main benefit of POF is that the cost is lower that glass fibres (which is an attractive solution for indoor applications), although the insertion loss and dispersion is higher.

SSMF is usually selected by operators for long haul links of optical communication core networks. This kind of fibre provides huge capacity, with capacity per distance products up to 41 Petabit/s.km as was demonstrated in (Chang, Fu, & Su, 2008). On the other side, MMF and POF is intended for short links inside buildings, airports, shopping malls and corporate office premises. Due to its low cost and ease of handling this types of fibre can be installed in-building with typical lengths of up to 300 m. Recent advances in photonic crystal and microstructure fibre geometries suggest that tight light confinement (i.e. making it ultrabendable fibre) may be possible whilst maintaining a low modal order, i.e. exhibiting a relatively low v number, similar to that of SMF.

The latest advanced were presented in February 2008, ClearCurve fibre from Corning (Corning, 2008), in September 2009, EZ-Bend fibre, OFS (OFS Optics, 2008)) and BendBright-XS fibre from Draka Comteq (Draka, 2008) and in January 2009 the ClearCurve multimode fibre (Corning, 2009).

Finally, the opto-electronic conversion is performed by a photodetector (PD) (Agrawal, 1997), which key requirements are high quantum efficiency and bandwidth. High-speed photodetectors in the 1310-1550 nm bands have been reported in the literature with 3-dB bandwidths up to 300 GHz (Ito et al., 2000). Nowadays there are commercially available devices that operate at frequencies up to 100 GHz.

Transmission Impairments and Expected Performance

In the down-link path shown in Figure 3, the 4G signal directly modulates a laser diode. The optical modulated signal travels through an optical fibre span with a length equal to the distance between the base station (BS) and the wireless signal extractor (WSE). For the up-link path the procedure would be similar to the downlink, the received UWB signal from the antenna directly modulates a laser diode and the modulated optical signal is transmitted to the BS through the same optical fibre span.

According to the scheme shown in Figure 3, both optical links (down-link and up-link) present a similar configuration. The detected signal power at the output of the photodiode in the WSE from the down-link path is calculated in (Agrawal, 1997)

Moreover, 4G-over fibre transmission has to deal with the optical fibre attenuation (around 0.2 dB/km in SSMF at 1550 nm) and other transmission impairments which limits the transmission reach. These transmission impairments are summarized below:

Attenuation and Effective Bandwidth

The network capacity, known as the product of the aggregated bitrate x reach, is limited at first instance by the bandwidth of the optical fibre and the crosstalk produced in the network nodes.

The standard single-mode fibre (SSMF), also known as optical fibre type G.652.A, exhibits a high attenuation band (1380 to 1400 nm) due to OH^- absorption. The new version G.652.C suppressed this attenuation band, and for this reason it is recommended for networks deployments. The fibre characteristics of G.652.C optical fibre are shown in Table 1 compared with SSMF G.652.A.

The attenuation of the fiber is, apart from the hydrogen absorption peak, is also due to Rayleigh scattering. The fact that this physical limit has been reached, explains the extremely low loss of 0.36 dB/km at 1.3 μm and 0.2 dB/km at 1.5 μm (e.g. 0.36 dB correspond to the reflection loss of a clean window pane). Nevertheless to achieve transmission lengths of more than several hundred kilometers, optical amplifiers have to be implemented in the transmission lines. This solves the loss problem but one has to deal with the noise of optical amplifiers and the effective bandwidth. Obviously, the optical signal to noise ratio (OSNR) has to be kept above a certain value to guarantee the required transmission quality.

The optical link bandwidth is ultimately limited by the optical amplifier. Erbium-doped fibre based amplifiers (EDFA) (Giles & Desurvire, 1991) or Raman effect amplification (Mizuochi et al., 2002) are employed for amplification. EDFAs exhibit a gain spectrum bandwidth of 30 to 40 nm (typically in the range from 1530 to 1560 nm). Raman amplification exhibits an amplification bandwidth, i.e. high values of the Raman gain profile, of 20 to 30 nm.

Table 1. ITU G.652 fibre characteristics

	G.652.A	G.652.C
Cut-off wavelength	≤ 1260 nm	≤ 1260 nm
Attenuation	≤ 0.50 dB/km @ 1310 nm ≤ 0.40 dB/km @ 1550 nm	≤ 0.4 dB/km @ 1310 nm ≤ 0.35 dB/km @ 1550 nm ≤ 0.4 dB/km @ 1625 nm
Mode diameter	8.6-9.5±0.7 μm @ 1310 nm	8.6-9.5±0.7 μm @ 1310 nm
Cladding diameter	125.0±1 μm @ 1310 nm	125.0±1 μm @ 1310 nm
GVD	+ 17 ps/nm/km	+ 17 ps/nm/km
GVD slope	≤ 0.093 ps/nm²·km	≤ 0.093 ps/nm²·km
PMD	≤ 0.2 ps/√km	≤ 0.5 ps/√km
Macrobend loss	≤ 0.50 dB @ 1550 nm	≤ 0.50 dB @ 1550-1625 nm

Group Velocity Dispersion (GVD)

The group velocity dispersion is originated by the fibre refractive index dependence with the wavelength. The different spectral components of a given optical signal propagate at different velocity, broadening the transmitted signal and potentially causing inter-symbol interference (ISI) in digital transmission and severe frequency-selective attenuation in radio-over-fibre transmissions (Schmuck, 1995). SMF exhibits a GVD values around 16 ps/nm/km at 1550 nm wavelength. Fibers with lower dispersion than standard single mode fibers have been developed (non zero dispersion shifted fibers) to reduce the length of dispersion compensating fiber or to enhance the transmission span without dispersion compensation.

GVD can be compensated employing so-called dispersion compensation fibres (DCF), which exhibits negative GVD values. Combining transmission through SMF and DCF, the overall effect is mitigated. This approach requires expensive fibres, and it must be implemented at fibre link deployment. Another approach shifts the GVD compensation problem from the optical to the electrical domain by means of electronic compensators/equalizers (Jansen et al., 2007).

Polarization-Mode Dispersion (PMD)

PMD is a complicated process that cannot be solved by link design. PMD is due to the fiber core birefringence influenced by mechanical (fiber stress, vibration) or environmental (temperature) factors. Birefringence is a characteristic of silica optical fibre originated in the manufacturing process. State-of-the-art fibres are manufactured following enhanced processes achieving a minimal asymmetry in the core. PMD levels are typically lower that 0.1 ps/√km. PMD compensation in electrical domain has been recently reported suitable for OFDM-based wireless signals, as described in Djordjevic, 2007.

If an optical pulse is injected in a fiber with constant birefringence, at the end of the fiber the pulse is split up in two pulses which have orthogonal polarizations and a delay against each other. The polarization states and delay depend on the birefringence of the fiber. On a real fiber the birefringence varies locally and the local birefringence may fluctuate in time, due to the variety of effects inducing the birefringence. The output pulse of a long fiber will therefore be the result of a very complex superposition of all these local birefringence effects.

It has been shown, that for a small enough optical bandwidth the overall effect can be described as split up of the input pulse in two output pulses with distinct polarization states. This effect is called first order PMD. All more complex effects are called higher order PMD. The PMD increases with the square root of fiber length, which is due to random accumulation of the effect like in a random walk.

PMD exhibits a random nature different than, by example, group velocity dispersion (GVD) or the GVD slope, which both can be statically compensated. PMD requires dynamic compensation (Sunnerud et al., 2002) which in turn requires a fast monitoring signal from the optical transmission link (Bulow et al., 1999).

PMD become an important impairment in high-bitrate long-reach optical transmission systems. It is worth note that OFDM-based wireless like LTE (downstream), WiMAX and UWB radio is well suited for the compensation of the fibre transmission impairments, namely chromatic dispersion, intra-channel nonlinear effects, and nonlinear phase noise (Mayrock & Haunstein, 2007).

Modal Dispersion

Distortion mechanism present in multimode fibres (MMF). The optical signal transmitted is distorted because the different propagation velocity of the electromagnetic modes present in the optical media. This effect severely limits the available bandwidth of the MMF. Bye example, step-index fibres (50 µm core) exhibits a bandwidth of 20 MHz·km. Typical off-the-shelf graded-index MMF (50 µm) bandwidth is close to 1 GHz·km. Electronic compensation of modal dispersion has been demonstrated for OFDM communications (Lowery & Armstrong, 2005).

(4G, B3G AND 3G)-ON-FIBRE

4G communication systems could be defined as a combination of a higher-capacity (e.g., more than 100 Mb/s) new radio interface and an interworking of cellular and different wireless technologies like WiMAX, UMTS and UWB. There is no doubt that 4G systems will provide higher data rates. Traffic demand estimates suggest that, to accommodate the foreseen amount of traffic an economically viable way, 4G mobile systems must achieve a manifold capacity increase compared to their predecessors.

Radio-over-fibre transmission systems can provide service to 4G technologies like WiMAX, 3G technologies like UMTS, and converged B3G technologies like UWB. Nowadays, radio access and in particular radio services distribution is needed almost everywhere, even if small number of clients will use it. Especially in closed environments (e.g. tunnels), this becomes more difficult when high data rates are demanded (e.g. 4G communications). The RoF solution means that radio signals are directly sent through the fibre in a transparent manner. In this way, radio signals can be transported to places that could be difficult to reach with other method and at the same time, as we described before, the antennas become simpler due to the nature of the modulated signals.

Regarding 4G, WiMAX is a wireless access transmission technology which targets a medium to long range communications. WiMAX is expected to replace large wireless local-area network installations, like University campus, commercial areas, etc. WiMAX radio-over-fibre targets to improve the coverage of broadband services in distributed antenna systems (DAS) applications (Saleh, Rustako, & Roman, 1987), where several remote antenna units are distributed around the service area. IEEE 802.16d WiMAX distribution over low cost MMF radio-over-fibre systems and its further wireless transmission has been demonstrated (Alemany et al., 2008), obtaining EVM values keep almost constant (around 1.25% EVM) for 400 m multimode fibre (MMF) transmission and 20 m wireless radio.

Beyond-3G (B3G) applications extend the cellular network to the pico- and femto-cell scenarios. Ultra-wideband (UWB) radio has been indicated as one of the most promising technique for B3G implementation. This is due to the low self-interference, tolerance to multi-path fading, low probability of interception and capability of passing through walls while maintaining the communication (Kohno, 2004) inherent characteristics of UWB radio. Market-available UWB products target the replacement of high definition (HD) video/audio cabling (Duan, 2006).

UWB is defined as a radio modulation technique with more than 500 MHz bandwidth or with, at least, 20% fractional bandwidth. UWB is regulated in the United States in the band between 3.1 and 10.6 GHz for indoor communications, and in the band from 22 to 29 GHz for vehicular radar applications (FCC 02-48 (2002). The maximum radiated power spectral density is 41.3 dBm/MHz to minimize the potential interference on licensed or un-licensed wireless services. The low spectral power density of this wireless signal requires a careful system design o the radio-over-fibre system (Llorente et al., 2008).

The International Communication Union (ITU) is deeply studying the compatibility between devices using UWB technology and the mobile and satellite services, as UWB emissions spread over a very large frequency range and therefore may affect several radio services. The European Commission asked the European Telecommunications Standards Institute (ETSI) to produce of Harmonized Standards for UWB to. ETSI's standardization activity for short-range devices currently includes these UWB applications such as communications, ground- and wall probing radar and automotive radar applications. The generic Harmonized European Standard for UWB Communications (ETSI EN 302 065), published in February 2008, uses frequency ranges from 3.1 to 4.8 GHz and 6 to 8.5 GHz for implementation in road or rail vehicles. Detect-And-Avoid (DAA) specifications as a mechanism for the protection other radio services, are defined for UWB in 3.1 to 4.8 GHz and 8.5 to 9 GHz bands has been published in June 2008 (ETSI TS 102 754). The automotive radar group is developing the 77-81 GHz Harmonized European Standard for the anti-collision radar application (ETSI EN 302 264). The 24 GHz Harmonized European Standard (ETSI EN 302 288) was amended in Feb. 2008.

Two specific UWB implementations are mainstream nowadays: Impulse-radio (IR-UWB), which transmits data by short impulses (monopulses), and orthogonal frequency division multiplexing (OFDM-UWB), which divides the UWB spectrum into 14 channels of 528 MHz bandwidth. Impulse-radio UWB systems show more or less evenly distributed energy across a broad range of frequencies. With low levels of energy across a broad frequency range, UWB signals are extremely difficult to distinguish from noise, particularly for ordinary narrowband receivers. One significant additional advantage of short-duration pulses is that multipath distortion can be nearly eliminated. Multipath effects result from reflected signals that arrive at the receiver slightly out of phase with a direct signal, cancelling or otherwise interfering with clean reception. The extremely short pulses of UWB systems can be readily distinguished from unwanted multipath reflections. On the other hand OFDM-UWB benefits from the well-know implementation techniques for OFDM modulation and demodulation and the readily availability of silicon-radio, leading to cost-efficient, space-saving single-chip solutions like Elam and Yurdakul (2008).

Regarding 3G, the transmission of UMTS signals on RoF has been addressed in in-building coverage extension. In (Al-Raweshidy, 2002) the feasibility of a UMTS signal transmission on a bi-directional link using a unique optical source for both directions and an electroabsortion modulator is confirmed as a suitable solution for UMTS microcells and picocells using radio-over-fibre. In-building coverage extension has been confirmed using VCSEL reaching 500 m MMF (Persson et al., 2006). The distortion performance of radio over fibre links for UMTS W-CDMA signals can be analyzed considering the maximum allowed value for adjacent channel leakage ratio (ACLR) in UMTS of -45dBc for the downlink path (TS 25.104 v3.0.0 (1999) and -33dBc for the uplink path (TS 25.101 v3.0.0 (1999). In Wake and Schuh (2000), for 32 UMTS channels the maximum downlink fibre span is of 7 km SSMF for an input power of 12 dBm and 27 km SSMF for an input power of 9 dBm. The uplink path reaches 30 km SSMF for an input power of 12 dBm. For most of the applications of distributed antenna systems the fibre span will be below 10 km SSMF, which confirms 3G-over-fiber suitability.

Target Applications and Coexistence

4G-on-fibre technology has two main applications nowadays: first, serving the interconnection of 4G access points or base-stations via a backhaul of point-to-point fibre links and, second, increasing the user coverage deploying a distributed-antenna system (DAS) in a given area. These two applications are briefly described now.

Base-Station Backhaul

Access techniques include indoor and outdoor pico- and micro- cells to provide high user capacity density supporting bandwidth-intensive services. WiMAX base stations with integrated backhaul, i.e. WiMAX radio itself is employed as backhaul, can be deployed forming self-connected clusters. This technology is called "in-band backhaul". Nevertheless, when the number of users increases, it is necessary to reduce the WiMAX cell size and the spectral re-use in order to deliver an adequate bandwidth per subscriber. In this situation, out of band backhaul is required. Fibre is the best backhaul option in densely populated areas, like urban areas, when the maximum return on investment from WiMAX deployment can be obtained.

A WiMAX backhaul network can be used to transport high data rate information from the fiber node that provides connectivity to the network around the rest of the WiMAX access points using for example a ring configuration as depicted in Figure 4.

In most of the cases, a different frequency for backhaul is employed in order to simplify different issues as network design and interference planning. Moreover, this allows the service provider to use the entire WiMAX spectrum to the access portion of the network, and use a different frequency band for the network backhaul. This makes economic sense as well as engineering sense, as using a different frequency for backhaul will significantly simplify.

A backhaul alternative employing 70 or 80 GHz frequency was proposed in (Wells, 2006).

Figure 4. WiMAX backhaul network example

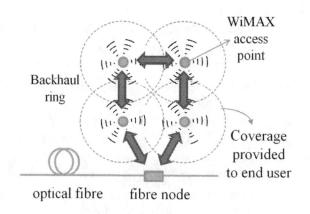

These 71 to 76 GHz and 81 to 86 GHz band allocations provides 5 GHz bandwidth of full duplex transmission that is enough to transmit a gigabit of data (e.g GigE) and even data rates of 10 Gbps employing more spectrally efficient modulations (e.g. OC-192, STM-64 or 10GigE). Low-cost radio architectures can be deployed using direct data conversion and low cost diplexers in this frequency bands. The inherent simplicity of a 70/80 GHz radio means that this approach could be offered at a market-competitive price to a conventional high data radio. Propagation characteristics at 70 and 80 GHz reaches up to 1.6 km transmission distances, as the atmospheric absorption at these frequencies is comparable to the popular microwave bands of 23 and 38 GHz.

Regarding commercial solutions, 3G and 4G solutions are provided with WiMax/LTE backhaul schemes that can be used by mobile providers or for last-mile connectivity. DragonWave's wireless Ethernet provides multi-point wireless systems as a viable option for last-mile access. Due to the critical backhaul requirements of service operators these solutions needs to ensure providing ultra low latency, native Ethernet, and up to 800 Mbps wire speed. For example, the AirPair native Ethernet solution avoids interference using licensed frequencies or the newly unlicensed 24 GHz band, eliminating costly downtime. Another available option is the backhaul module of Motorola Canopy system used by Xittel for fixed wireless access (Ménard, 2006), that with a long-range point-to-point link between base stations avoids the need to have optical fibres installed to each base station. This replaces the traditional primary line service and is able to provide VoIP with guaranteed Quality of Service using Ethernet traffic prioritization over this hybrid radio-over-fiber backhaul system.

Distributed Antenna Systems

In a distributed-antenna system (DAS) several point-source antennas are distributed along a fibre span to provide continuous coverage in heavy-usage areas, like tunnels, airports, buildings, and shopping areas. The antennas transmit the same frequency at different locations to provide a relatively continuous signal-level coverage employing reasonable power levels, thus keeping that signal low to avoid interference. This application has been demonstrated in a number of reported work, like in Xu et al. (2008), demonstrating a large increase of the overall system capacity.

DAS systems based on RoF technology are preferable over coaxial systems in the typical applications like covering an under-ground environment like tunnels, etc. due to its lower transmission losses. Fibre-based systems also exhibit greater flexibility. Moving the antenna location in a cable-based DAS means that the whole design must be changed because of noise contributions, impedance mismatching and power levels/amplifier placement restrictions.

Coexistence Issues

An important technology issue in radio-over-fibre is the coexistence between different wireless standards when transmitted through fibre. Coexistence of WiMAX with over wireless signals, e.g. ultra-wideband (UWB) or UMTS, when transmitted through the fibre usually does not pose a problem. The different licensed or un-licensed wireless services are allocated at different transmission bands and a careful design of the optical transmission system can minimize interference.

UWB wireless transmission technology is targeting short-range high bitrate communications. Pervasive presence of UWB transmitters are expected in the near future supporting a broad range of applications, from computer universal serial bus (USB) wireless links to home multimedia wireless communication systems. WiMAX and UWB are then fully complementary and are expected to coexist in a near future.

UWB radio can operate in the 3.1 to 10.6 GHz band as defined in current regulation (ECMA-368

Figure 5. 4G-on-fibre distributed antenna application

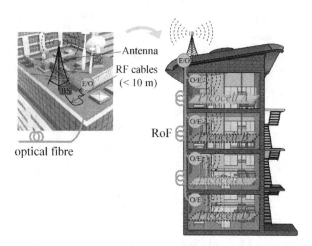

(2007). First-generation UWB systems already available in the market operate in the 3.1 to 4.9 GHz band. This band completely overlaps with IEEE 802.16d/e WiMAX systems operating in the band from 3 to 4.2 GHz (IEEE Std 802.16TM-2004 (2004).

The impact of UWB interference on WiMAX at 3.5 GHz band is a relevant research topic and raise regulatory concerns (Rahim, Zeisberg & Finger, 2007). For this reason, the coexistence problem of UWB radio with narrowband systems such as UMTS, GSM, GPS, etc. has been investigated in the literature (Cassioli 2005). WiMAX and UWB coexistence is a key issue in indoor scenarios. Indoor fibre installations are typically based on multi-mode fibres (MMF) and Vertical-Cavity Surface-Emitting Lasers (VCSELs) operating at 850 nm, and low-cost receivers (Das et al., 2006; Marozsak & Udvary, 2002).

The study of simultaneous RoF transmission of WiMAX and UWB on MMF has been carried out in (Alemany et al., 2008) to identify the transmission impairments in a spectral overlapping situation. Simultaneous UWB (employing standard multiband-orthogonal frequency division multiplex MB-OFDM, as defined in ECMA-368), and WiMAX (IEEE 802.16d) transmission over low cost MMF RoF systems has been demonstrated (Figure 6), including wireless transmission.

Stand-alone UWB transmission results show linear performance degradation due to the MMF modal dispersion at higher frequencies. This degradation translates to EVM degradation from 16.5% (no MMF) to 23% (100 m MMF) and a penalty of 1 m in the wireless reach.

Simultaneous transmission of WiMAX and UWB services in the same fibre using the same wavelength, increases the EVM at the WiMAX receiver from 1.25% on the stand-alone case to 2.74% when in coexistence for the analyzed 400 m MMF path and its 20 m wireless radio propagation. Therefore, WiMAX results satisfy standard limit, demonstrating successful transmission. On the other had, UWB EVM performance increases from 18% (stand-alone) to 31% (coexistence) situation for 100 m MMF transmission and 1 m wireless radio. In this case, UWB performance is also affected by the receiver transfer function and its limited bandwidth.

Figure 6. RF spectrum of WiMAX and UWB signals in coexistence (a) before VCSEL modulation, and (b) after radio-over-fibre transmission over 400 m of MMF (RBW: 181.068 kHz)

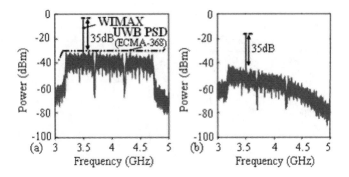

EMERGING TECHNOLOGIES

B3G in Fibre-to-the-Home Networks

Coexistence of 4G and B3G wireless can achieved in radio-over-fibre links by careful selections. Also, specific multiplexing techniques like, by example, polarisation division has been demonstrated suitable for simultaneous WiMAX and UWB transmission in SSMF at distances up to 25 km (Perez et al., 2009).

In Llorente, Thakur, Morant, et al. (2008), the radio-over-fibre distribution of UWB signals was extended to fibre-to-the-home (FTTH) access networks. This approach exhibits several advantages: FTTH networks provide bandwidth enough to distribute a large number of UWB signals, as each one of them can occupy up to 7 GHz in current UWB regulation (FCC 02-48, 2002). No trans-modulation is required at user premises. HD audio/video content is transmitted through the fibres in UWB native format. No frequency up-conversion is required at customer premises as the UWB signals are just photo-detected, filtered, amplified and radiated directly to establish the wireless connection. This approach can be extended to 60 GHz UWB technology which is under major attention for broadband communications (Cabric, D., et al., 2005).

The proposed UWB RoF technique is depicted in Figure 7. This figure shows an optical line terminal (OLT) located at the operator core network, typically in the central office (CO) that distributes the signal through a SSMF-based FTTH network to an optical network units (ONU) located at the customer premises. At the subscriber premises, the received UWB signals are photo-detected, filtered, amplified and radiated to broadcast the HD content to an UWB-enabled television set or computer. This technique benefits from the high bit-rate capabilities of UWB, supporting bit-rates up to 1 Gbit/s at a few meters range (Lunttila, Iraji, & Berg, 2006), which can be extended to 30 m by multiple-input multiple-output processing techniques covering a whole home (Zou et al., 2007). In Llorente, Alves, Morant, et al. (2008), the signal degradation IR-UWB and OFDM-UWB was analyzed with 1.25 Gbit/s bitrate adequate for uncompressed 1920×1080i×18bpp×60 Hz video (Shoji et al., 2006). The UWB signals were transmitted along different standard single-mode fibre (SSMF) links, ranging from 25 km to 60 km. The experimental results demonstrate the feasibility of the technique achieving a bit error rate (BER) lower than 10^{-9} operation at 50 km SSMF with both IR-UWB and OFDM-UWB implementations. Nevertheless, IR-UWB presents performance degradation, suggesting that UWB-over-fibre B3G should be accomplished with OFDM signals.

Figure 7. B3G UWB-on-fibre distribution in FTTH networks

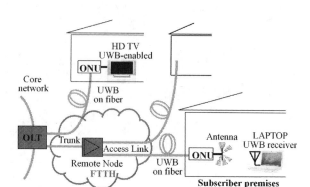

The spectral efficiency in these systems can be maximised by the distribution of polarization-multiplexed UWB (PM-UWB) signals is a suitable technique for the provision of wireless connectivity to a large number of users. This approach provides a higher spectral efficiency and the user capacity is doubled compared with UWB on a single wavelength. The maximum transmission reach of the proposed PM-UWB technique has been investigated in (Morant et al., 2009) demonstrating successful transmission distances up to 25 km.

Radio-over-Fibre of 24 GHz and 60 GHz Wireless

24 GHz wireless radio is an interesting solution for short-range radar (SRR), which aims target detection within a maximum range of 10 m (Meinnel et al., 2005). Nevertheless some research results indicate the potential of this frequency band for communication services. Reported work (Kuri et al., 2006; Guennec & Gary, 2007) exists proposing stand-alone UWB communications in the 24 GHz band. The simultaneous use of 24 GHz IR-UWB signals for SRR ranging and communication was proposed in (Beltran, Llorente, et al., 2009).

The experimental work reported in (Beltran, Llorente, et al., 2009) demonstrates a photonic generation and up-conversion technique for RoF systems capable of generating 24 GHz signals simultaneously bearing a 625 Mbit/s bitrate. This approach exhibits several advantages as radio-over-fibre is extended by optical generation and detection employing ultra-short pulses with low jitter. In addition, this system is frequency-flexible: The technique can inherently generate up-converted signals in the 60 GHz band (Beltran, Morant, et al., 2008) provided an adequate electro-optic modulator is employed. The up-conversion flexibility to upper frequency bands is of special interest as 60 GHz UWB is emerging as a viable approach for ultra-high rate wireless communications. From the application point of view, the optical detection approach is well suited for in-vehicle networks, as external interference from the car engine is strongly reduced.

The 60 GHz band has recently attracted a lot of interest for the huge available bandwidth allows reaching very high bit rates of several Gbits/s at ranges up to 10 meters. Researching studies seem to point out that the trend of this task could be the study of transmission techniques beyond 60 GHz, taking into account the technological constraints of generating and using this frequency band.

Probably, the most restrictive requirement for wireless services provision over RoF systems is the SFDR (Ng'oma, 2005). Nowadays SFDRs in excess of 100 dB/Hz$^{2/3}$ have been demonstrated experimentally, providing enough dynamic range to be employed in real applications (Seeds & Williams, 2006) and paving the way to the optical transport of wireless signal at 60 GHz.

CONCLUSION

In this Chapter, the basics of radio-over-fibre technology for the distribution of 4G wireless radio are presented. The technical principle, key optical and electro-optical devices, relevant transmission impairments and expected performance is described. RoF technology applied on 4G systems cells permits to decrease transmitting power and increase the efficiency of frequency reuse. RoF is employed in the interconnection of 4G access points or base-stations via a backhaul of point-to-point fibre links, and to increase the user coverage in distributed-antenna system (DAS). Emerging applications of RoF cover the provision of B3G services over FTTH networks, and the user of new frequencies, like the 24 GHz and 60 GHz bands.

REFERENCES

Agrawal, G. P. (1997). *Fibre-Optic Communication Systems*. Hoboken, NJ: John Wiley & Sons.

Al-Raweshidy, H., Glaubitt, K. M., & Faccin, P. (2002). In-building coverage for UMTS using radio over fibre technology. In *Proceedings of 5th International Symposium on Wireless Personal Multimedia Communications 2002* (Vol. 2, pp. 581- 585).

Alemany, R., Perez, J., Llorente, R., Polo, V., & Marti, J. (2008). Coexistence of WiMAX 802.16d and MB-OFDM UWB in Radio over Multi-mode Fibre Indoor Systems. In *Proceedings of International Topics Meeting on IEEE Microwave Photonics 2008*.

Anscombe, N. (2005). Demand for indoor coverage drives radio-over-fibre. *Wireless Europe*. Retrieved from http://www.nadya-anscombe.com/pdfs/radio%20over%20fibre.pdf

Bach, L., Kaiser, W., Reithmaier, J. P., Forchel, A., Berg, T. W., & Tromborg, B. (2003). Enhanced direct-modulated bandwidth of 37 GHz by a multi-section laser with a coupled-cavity-injection-grating design. *Electronics Letters*, *39*, 1592–1593. doi:10.1049/el:20031018

Beltran, M., Llorente, R., Sambaraju, R., & Marti, J. (2009). 60 GHz UWB-over-Fibre System for In-flight Communications. In . *Proceedings of*, *IMS2009*, TU1D–2.

Beltran, M., Morant, M., Perez, J., Llorente, R., & Marti, J. (2008). Photonic generation and frequency up-conversion of impulse-radio UWB signals. In *Proceedings of 21st Annual Meeting of The IEEE Lasers & Electro-Optics Society, 2008 LEOS Annual Conference,* Newport Beach, CA.

Ben-Ezra, Y., Ran, M., Borohovich, E., Leibovich, A., Thakur, M. P., Llorente, R., & Walker, S. D. (2008). Wimedia-Defined, Ultra-Wideband Radio Transmission over Optical Fibre. In *Conference on Optical Fiber communication/National Fiber Optic Engineers Conference OFC/NFOEC 2008*.

Blumenthal, R. H. (1962). Design of a microwave frequency light modulator. *In Proceedings of Institute of Radio Engineers IRE*, *50*, 452-456.

Bulow, H., Baumert, W., Schmuck, H., Mohr, F., Schulz, T., Kuppers, F., & Weiershausen, F. (1999). Measurement of the maximum speed of PMD fluctuation in installed field fiber. *Optical Fiber Communication Conference '99 Technical Digest*, *2*(2), 83-85.

Cabric, D., Chen, M. S. W., Sobel, D. A., Yang, J., & Brodersen, R. W. (2005). Future Wireless Systems: UWB, 60GHz, and Cognitive Radios. In *IEEE Custom Integrated Circuits Conference 2005*.

Capmany, J., & Novak, D. (2007). Microwave photonics combines two worlds. *Nature Photonics*, *1*, 319–330. doi:10.1038/nphoton.2007.89

Cassioli, D., Persia, S., Bernasconi, V., & Valent, A. (2005). Measurements of the performance degradation of UMTS receivers due to UWB emissions. *IEEE Communications Letters*, *9*(5), 441–443. doi:10.1109/LCOMM.2005.1431165

Chang, Q., Fu, H., & Su, Y. (2008). Simultaneous generation and transmission of downstream multiband signals and upstream data in a bidirectional radio-over-fibre system. *IEEE Photonics Technology Letters*, *20*, 181–183. doi:10.1109/LPT.2007.912992

Chia, M. (2003). Radio over multimode fibre transmission for wireless LAN using VCSELs. *Electronics Letters*, *39*, 1143–1144. doi:10.1049/el:20030724

Corning (2008). ClearCurve single-mode optical fibre. Retrieved from http://www.corning.com/opticalfibre/products/clearcurve_single_mode_fibre.aspx, February 2008.

Corning (2009). ClearCurve® OM3/OM4 multimode optical fibre. Retrieved from http://www.corning.com/opticalfibre/products/clearcurve_multimode_fibre.aspx, January 2009.

Cox, C. H., III. (2004). *Analog Optical Links - Theory and Practice*. Cambridge, MA: Cambridge University Press

Dagli, N. (1999). Wide-Bandwidth Lasers and Modulators for RF Photonics. *IEEE Transactions on Microwave Theory and Techniques*, *47*(7), 1151–1171. doi:10.1109/22.775453

Darcie, T. E., & Bodeep, G. E. (1990). Lightwave subcarrier CATV transmission systems. *IEEE Transactions on Microwave Theory and Techniques*, *38*(5), 524–533. doi:10.1109/22.54920

Das, A., Nkansah, A., Gomes, N. J., Garcia, I. J., Batchelor, J. C., & Wake, D. (2006). Design of low-cost multimode fibre-fed indoor wireless networks. *IEEE Transactions on Microwave Theory and Techniques*, *54*(8), 3426–3432. doi:10.1109/TMTT.2006.877835

Djordjevic, B. (2007). PMD compensation in fibre-optic communication systems with direct detection using LDPC-coded OFDM. *Optics Express*, *15*(7), 3692–3701. doi:10.1364/OE.15.003692

Draka (2008). Enhanced low macrobending sensitive Single Mode Fibre: BendBrightXS. Retrieved from http://www.drakafibre.com/draka/DrakaComteq/Drakafibre/Languages/English/Navigation/Markets%26Products/Single_mode/BendBright-XS/index.html, September 2008.

Duan, C., & Pekhteryev, G., & Fang, J., & Nakache, Y., & Zhang, J., & Tajima, k., & Nishioka, Y., & Hirai, H. (2006). Transmitting multiple HD video streams over UWB links. In *Proceedings of of CCNC'06*, Las Vegas (vol. 2, pp. 691-695).

ECMA-368. (2007). High rate ultra wideband PHY and MAC Standard. In *ECMA International Standard (2nd Ed.)*, December 2007.

Elam, E., & Yurdakul, S. (2008). Wisair's Wireless USB Single Chip Poised for Broader Global Distribution with Recent European Regulatory Approval and USB-IF Certification. *Wisair Ltd press release*. Retrieved from http://www.usb.org/press/WUSB_press/2008_03_04_wisair.pdf

Ellis, R. B., Weiss, F., & Anton, O. M. (2007). HFC and PON-FTTH networks using higher SBS threshold single mode optical fibre. *Electronics Letters*, *43*(7), 405–407. doi:10.1049/el:20070218

FCC 02-48. (2002). *Revision of Part 15 of the Commission's Rules regarding ultra-wideband transmission systems*. Federal Communications Commission.

Giles, C. R., & Desurvire, E. (1991). Propagation of signal and noise in concatenated erbium-doped fiber optical amplifiers. *IEEE Journal of Lightwave Technology, 9*, 147–154. doi:10.1109/50.65871

Giorgetti, A., Chiani, M., Dardari, D., Piesiewicz, R., & Bruck, G. H. (2008). The Cognitive Radio Paradigm for Ultra-Wideband Systems: the European Project EUWB. *IEEE International Conference on Ultra-Wideband ICUWB 2008* (vol. 2, pp. 169-172).

Guennec, Y. L., & Gary, R. (2007). Optical frequency conversion for millimeter-wave ultra-wideband-over-fibre systems. *IEEE Photonics Technology Letters, 19*, 996–998. doi:10.1109/LPT.2007.898745

Harjula, I., Ramirez, A., Martinez, F., Zorrilla, D., Katz, M., & Polo, V. (2008). Practical Issues in the Combining of MIMO Techniques and RoF in OFDM/A Systems. In *Proceedings of 7th WSEAS International Conference on Electronics, Hardware, Wireless and Optical Conference*.

Hartmann, P., et al. (2003). Low-cost multimode fibre-based wireless LAN distribution system using uncooled directly modulated DFB laser diodes. In *Proceedings of ECOC 2003*, Rimini, Italy (pp. 804–805).

Heidrich, H., & Hoffmann, D., & Macdonald, R. I. (1986). Polarization and wavelength multiplexed biderectional single fiber subscriber loop. *Journal of optical communications, 7*(4), 136-138.

Henry, P. S. (2007) Integrated Optical/Wireless Alternatives for the Metropolitan Environment. IEEE Communications Society Webminar. Retrieved from http://www.comsoc.org/webinar/1/PSH--Integ%20Opt%20and%20WlessFINAL.pdf

Hirata, A. (2003). 120-GHz wireless link using photonic techniques for generation, modulation, and emission of millimeter-wave signals. *IEEE Journal Lightwave Technology, 21*(10), 2145–2153. doi:10.1109/JLT.2003.814395

Hitachi, L. C. D.-T. V. (2008). Wooo UT-series Available in website www.hitachi.com and http://av.hitachi.co.jp/tv/l_lcd/ut/index.html.

Ho, L. T. W., & Claussen, H. (2007). Effects of User-Deployed, Co-Channel Femtocells on the Call Drop Probability in a Residential Scenario. In *Proceedings of IEEE Personal, Indoor and Mobile Radio Communications PIMRC 2007*.

Huang, M.-F., Yu, J., Qian, D., & Chang, G.-K. (2008). Lightwave Centralized WDM-OFDM-PON. In *Proceedings of ECOC 2008, Paper Th1F5*.

Ikegami, T., & Suematsu, Y. (1967). Resonance-like characteristics of the direct modulation of a junction laser. *Proceedings of the IEEE, 55*, 122–123. doi:10.1109/PROC.1967.5420

Ingham, J. D., et al. (2003). Wide-frequency-range operation of a high linearity uncooled DFB laser for next-generation radio-over-fibre. In *Proceedings of IEEE/OSA OFC 2003*, Atlanta, GA (pp. 754–756).

Ito, H., Furata, T., Kodama, S., & Ishibashi, T. (2000). InP/InGaAs uni-traveling-carrier photodiode with 310 GHz bandwidth. *Electronics Letters, 36*, 1809–1819. doi:10.1049/el:20001274

Jager, D., & Stohr, A. (2001), Microwave Photonics. In *Proceedings of European Microwave Conference 2001*, London (pp. 1-4).

Janjua, K. A., & Khan, S. A. (2007). A Comparative Economic Analysis of different FTTH Architectures. *Wireless Communications, Networking and Mobile Computing*, 4979-4982.

Jansen, S. L., Morita, I., Takeda, N., & Tanaka, H. (2007). *20-Gb/s OFDM transmission over 4,160-km SSMF enabled by RF-pilot tone phase noise compensation.* Paper presented at the Optical Fibre Communication Conference OFC2007, Anaheim, CA.

JDSU. (2007). *Test & Measurement Note, JDSU Triple-Play Service Deployment Guide.* JD Uniphase Corp.

Kartalopoulos, S. V. (2002). *DWDM: Networks, Devices, and Technology.* Hoboken, NJ: Wiley.

Kelley, B., & Rivas, E. (2008). OFDM location-based routing protocols in ad-hoc networks. *IEEE Wireless Hive Networks Conference, WHNC 2008.*

Kohno, R. (2004). State of arts in ultra wideband (UWB) wireless technology and global harmonization. In *Proceedings of 34th European Microwave Conf*, Netherlands (pp. 1093-1099).

Koonen, T., García Larrodé, M., Urban, P., Waardt, H., Tsekrekos, C., Yang, J., et al. (2007). Fibre-based Versatile Broadband Access and In-Building Networks. *IET Workshop From Access to Metro*, Europe.

Kuri, T., et al. (2006). Optical transmitter and receiver of 24-GHz ultra-wideband signal by direct photonic conversion techniques. *In Proceedings of International Topics Meeting on IEEE Microwave Photonics 2006.*

Lee, K. (2005). Radio over fibre for beyond 3G. *International Topical Microwave Photonics, MWP 2005.*

Liu, C., et al. (2003). Bi-directional transmission of broadband 5.2 GHz wireless signals over fibre using a multiple-quantum-well asymmetric Fabry–Pérot modulator/photodetector. In *Proceedings of IEEE/OSA OFC 2003*, Atlanta, GA (pp. 738–740).

Liu, C., & Polo, V. (2007). Full-duplex DOCSIS/wirelessDOCSIS fibre-radio network employing packaged AFPMs as optical/electrical transducers. *Journal of Lightwave Technology, 25*(3), 673–684. doi:10.1109/JLT.2006.889674

Llorente, R., Alves, T., Morant, M., Beltran, M., Perez, J., Cartaxo, A., & Marti, J. (2008). Ultra-Wideband Radio Signals Distribution in FTTH Networks. *IEEE Photonics Technology Letters, 20*(11), 945–947. doi:10.1109/LPT.2008.922329

Llorente, R., Cartaxo, A., Uguen, B., Duplicy, J., Romme, J., Puche, J. F., et al. (2008). Management of UWB picocell clusters: UCELLS project approach. In *IEEE International Conference on Ultra-Wideband ICUWB 2008* (vol. 3, pp. 139-142).

Llorente, R., Thakur, M. P., Morant, M., Walker, S. D., & Marti, J. (2008). Performance comparison of radio-over-fibre UWB distribution in SSMF and MMF optical media. In *Proceedings of ECOC 2008* (Vol. 2, pp. 119-120).

Lowery, A. J., & Armstrong, J. (2005). 10 Gbit/s multimode fibre link using power efficient orthogonal frequency division multiplexing. *Optics Express, 13*, 10003–10009. doi:10.1364/OPEX.13.010003

Lunttila, T., Iraji, S., & Berg, H. (2006). Advanced Coding Schemes for a Multi-Band OFDM Ultra-wideband System towards 1 Gbps. In *Proceedings of CCNC'06* (vol. 1, pp. 553-557).

Marozsak, T., & Udvary, E. (2002). Vertical cavity surface emitting lasers in radio over fibre applications. In *Proceedings of 14th International Conference on Microwaves, Radar and Wireless Communications 2002, MIKON-2002* (Vol.1, pp. 41-44).

Mayrock, M., & Haunstein, H. (2007). OFDM in optical long-haul transmission. In *Proceedings of of 12th Int. OFDM Workshop*, Hamburg, Germany.

McDonald, G. J., & Seeds, A. J. (2006). A novel pulse source for low-jitter optical sampling: a rugged alternative to mode-locked lasers. In *Proceedings of SPIE 6399,* October 2006.

Meinnel, H. H., et al. (2005). *Automotive radar: From long range collision warning to short range urban employment.* Paper presented in MINT-MIS2005/TSMMW2005.

Ménard, F. D. (2006). Xittel Combines Fiber and Motorola WiMAX to Serve as Few as 25 Customers. *Broadband properties*. Retrieved from http://www.broadbandproperties.com

Mi.Tel-Teleoptix. (n.d.). *43 Gbit/s DPSK PHOTO-RECEIVER with Limiting TIA (DualPIN-DTLIA Rx)*. Retrieved from http://www.teleoptix.com

Mizuochi, T., Kingo, K., Kajiva, S., Tokura, T., & Motoshima, K. (2002). Bidirectional unrepeatered 43 Gb/s WDM transmission with C/L band-separated Raman amplification. *IEEE Journal Ligthwave Technology*, 20, 2079–2085. doi:10.1109/JLT.2002.806767

Morant, M., Alves, T., Llorente, R., Cartaxo, A., & Marti, J. (2008). Experimental Comparison of Transmission Performance of Multi-channel OFDM-UWB Signals on FTTH Networks. *IEEE . Journal of Lightwave Technology*, 27(10), 1410–1416.

Morant, M., Pércz, J., Beltran, M., Llorente, R., & Marti, J. (2008). Integrated performance analysis of UWB wireless optical transmission in FTTH networks. In *Proceedings of 21st Annual Meeting of The IEEE Lasers & Electro-Optics Society, 2008 LEOS Annual Conference*, Newport Beach, CA.

Morant, M., Pérez, J., Llorente, R., & Marti, J. (2009). Transmission of 1.2 Gbit/s Polarization-Multiplexed UWB Signals in PON with 0.76 Bit/s/Hz Spectral Efficiency. In *Proceedings of IEEE/OSA OFC 2009.*

Ng'oma, A. (2005). *Radio-over-fibre technology for broadband wireless communication systems*. CIP-Data Library Technische Universiteit Eindhoven.

Niiho, T., et al. (2004). Multi-channel wireless LAN distributed antenna system based on radio-over-fibre techniques. In *Proceedings of IEEE LEOS Annual Meeting 2004*, Rio Grande, Puerto Rico (pp. 57–58).

Nkansah, A. (2006). Simultaneous Dual Band Transmission Over Multimode Fibre-Fed Indoor Wireless Network. *IEEE Microwave and Wireless Components Letters*, 16(11), 627–629. doi:10.1109/LMWC.2006.884899

Ogawa, H. (1992). Millimetre-wave fibre optic systems for personal radio communication. *IEEE Transactions on Microwave Theory and Techniques*, 40(12), 2285–2292. doi:10.1109/22.179892

Optics, O. F. S. (2009). Bending the rules: ofs to demonstrate ez-bend™ optical technology at the 2008 conference. Retrieved from http://www.ofsoptics.com/press_room/view_press_release.php?txtID=247. September 2009.

Paier, A., et al. (2007). First Results from Car-to-Car and Car-to-Infrastructure Radio Channel Measurements at 5.2GHZ. In *IEEE PIMRC 2007* (pp. 1-5).

Perez, J., Morant, M., Llorente, R., & Marti, J. (2009) Joint Distribution of Polarization-Multiplexed UWB and WiMAX Radio in PON. *IEEE Journal Lightwave Technology*.

Persson, K. A. (2006). WCDMA radio-over-fibre transmission experiment using singlemode VCSEL and multimode fibre. *Electronics Letters*, *42*, 372–374. doi:10.1049/el:20064130

Persson, K.-A., Carlsson, C., Alping, A., Haglund, A., Gustavsson, J. S., Modh, P., & Larsson, A. (2006). WCDMA radio-over-fibre transmission experiment using singlemode VCSEL and multimode fibre. *Electronics Letters*, *42*, 372–374. doi:10.1049/el:20064130

Pfrommer, H., Piqueras, M. A., Polo, V., Herrera, J., Martinez, A., & Marti, J. (2006). Radio-over-Fibre Architecture for Simultaneous Feeding of 5.5 and 41 GHz WiFi or WiMAX Access Networks. In *Microwave Symposium Digest 2006. IEEE MTT-S International* (pp. 301–303).

Qian, X., et al. (2005). Directly-modulated photonic devices for microwave applications. In *Proceedings of IEEE MTT-S Intl Microwave Symposium,* Long Beach, California, USA.

Rahim, A., Zeisberg, S., & Finger, A. (2007). Coexistence Study between UWB and WiMAX at 3.5 GHz Band. In *IEEE International Conference on Ultra-Wideband 2007, ICUWB 2007* (pp. 915-920).

Rahim, A., Zeisberg, S., & Finger, A. (2007). Coexistence Study between UWB and WiMAX at 3.5 GHz Band. In *Proceedings of ICUWB 2007 Conference* (pp. 915-920).

Saleh, A., Rustako, A., & Roman, R. (1987). Distributed antennas for indoor radio communications. *IEEE Transactions on Communications*, *35*(12), 1245–1251. doi:10.1109/TCOM.1987.1096716

Sambaraju, R., et al. (2008). Photonic Envelope Detector for Broadband Wireless Signals using a Single Mach-Zehnder Modulator and a Fibre Bragg Grating. In *Proceedings of ECOC 2008* (pp. 64).

Schmuck, H. (1995). Comparison of optically millimeter-wave system concepts with regard to chromatic dispersion. *IEEE Electron. Lett.*, *31*(21), 1848–1849. doi:10.1049/el:19951281

Schmuck, H. (1995). Comparison of optically millimeter-wave system concepts with regard to chromatic dispersion. *Electronics Letters*, *31*(21), 1848–1849. doi:10.1049/el:19951281

Seeds, A. J., & Williams, K. J. (2006). Microwave Photonics . *Journal of Lightwave Technology*, *24*(12), 4628–4641. doi:10.1109/JLT.2006.885787

Shoji, Y., Choi, C., Kato, S., Toyoda, I., Kawasaki, K., Oishi, Y., Takahashi, K., & Nakas, H. (2006). Re-summarization of merged usage model definitions parameters. *IEEE doc. 802.15-06-0379-02-003c.*

Siddiqui, A. S., & Zhou, J. (1991). Two-Channel Optical Fiber Transmission Using Polarization Division Multiplexing. *Journal of Optical Communications*, *12*(2), 47–49.

IEEE Std 802.16TM-2004. (2004). *802.16TM IEEE standard for local and metropolitan area networks Part 16: Air interface for fixed broadband wireless access systems*.

Strohm, K. M., Schneider, R., & Wenger, J. (2005). KOKON: A Joint Project for the Development of 79 GHz Automotive Radar Sensors. In *Proceedings of IRS 2005*.

Stuckmann, P., & Zimmermann, R. (2007). Towards ubiquitous and unlimited capacity communication networks: European research in framework programme 7. *IEEE Communications Magazine*, *45*(5), 148–147. doi:10.1109/MCOM.2007.358862

Sunnerud, H., Karlsson, M., Chongjin, X., & Andrekson, P. A. (2002). Polarization-mode dispersion in high-speed fiber-optic transmission systems. *IEEE Journal of Lightwave Technology, 20*(12), 2204–2219. doi:10.1109/JLT.2002.806765

TS 25.101 v3.0.0. (1999). UE radio transmission and reception (FDD). *3GPP TSG RAN WG4 document*.

TS 25.104 v3.0.0. (1999). UTRA (BS) FDD. radio transmission and reception. *3GPP TSG RAN WG4 document*.

Umbach, A., Waasen, S. V., Auer, U., Bach, H. G., Bertenburg, R. M., & Breur, V. (1996). Monolithic pin-HEMT 1.55 um photoreceiver on InP with 27 GHz bandwidth. *Electronics Letters, 32*(23), 2142–2143. doi:10.1049/el:19961421

Wake, D. (1997). Passive picocell-A new concept in wireless network infrastructure. *Electronics Letters, 33*(5), 404–406. doi:10.1049/el:19970277

Wake, D., & Schuh, R. E. (2000). Measurement and simulation of W-CDMA signal transmission over optical fibre. *Electronics Letters, 36*, 901–902. doi:10.1049/el:20000670

Wells, J. (2006). WiMAX Backhaul at 70/80 GHz [White paper]. *GigaBeam Wireless Fiber*.

Wilson, B., Ghassemlooy, Z., & Darwazeh, I. (1995). *Analogue Optical Fibre Communications*. Institution of Engineering and Technology.

WiMAX extension to isolated research data networks. (n.d.). Retrieved from http://www.ist-weird.eu/

WiMedia Alliance. (2008 January). UWB - best choice to enable WPANs. *WiMedia Alliance*.

Xu, Z., Zhou, C., & Wang, J. A (2008). Novel Cell Architecture Based on Distributed Antennas for Mobile WiMAX Systems. In *4th IEEE International Conference on Circuits and Systems for Communications, 2008. ICCSC 200* (pp. 172-176).

Zou, Q., Tarighat, A., & Sayed, A. H. (2007). Performance analysis of multiband OFDM UWB Communications with application to range improvement. *IEEE Transactions on Vehicular Technology, 56*, 3864–3878. doi:10.1109/TVT.2007.901957

Section 2
Radio Access Protocols

Scheduling and Quality of Service

Chapter 14
MAC Protocol of WiMAX Mesh Network

Ming-Tuo Zhou
National Institute of Information and Communications Technology, Singapore

Peng-Yong Kong
Institute for Infocomm Research, Singapore

ABSTRACT

WiMAX based on IEEE std 802.16 is believed one of the important technologies of 4G. It aims to provide high-speed access over distance of several to tens kilometers. In IEEE std 802.16-2004, WiMAX defines an optional mesh mode, with which multi-hop, multi-route, self-organizing and self-healing communications can be achieved in metropolitan-level areas. This chapter presents medium access control (MAC) protocol of WiMAX mesh mode, on frame structure, network configuration, network entry, and scheduling algorithms. It also summaries the most recent progress on data slots resource scheduling and allocation algorithms. Finally, an application example of using WiMAX mesh network for high-speed and low-cost maritime communications is also presented in this chapter.

1. INTRODUCTION

A number of technologies are under development toward next generation wireless networks beyond 3G, such as WiMAX (Worldwide Interoperability for Microwave Access), UMB (Ultra Mobile Broadband), and LTE (Long Term Evolution). Among of them, WiMAX is based on IEEE 802.16 standard, and it aims to provide high-speed data access using a variety of transmission modes, from point-to-multipoint links to portable and fully mobile Internet access.

IEEE std 802.16-2004 is an important member of WiMAX standard family IEEE (802.16-2004). It specifies layers of Physical (PHY) and Media Access Control (MAC), and it superseded earlier 802.16 documents. This version of standard supports line-of-sight (LOS) connections in 10-66 GHz and non-LOS (NLOS) communication in 2-11 GHz. Two multi-carrier modulation technologies are supported, i.e., OFDM with 256 carriers and OFDMA with 2048 carriers. In 2005, an amendment to 802.16-2004 was completed and named as 802.16-2005. This newer version of standard supports combined fixed and mobile operation in

DOI: 10.4018/978-1-61520-674-2.ch014

MAC Protocol of WiMAX Mesh Network

Figure 1. Typical topology of WiMAX mesh networks

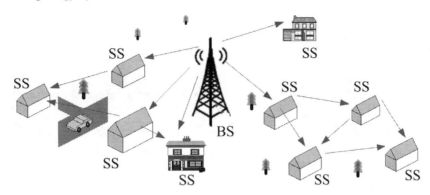

frequencies below 6 GHz, and it includes many new features such as scalable OFDMA (SOFDMA), Multiple Input Multiple Output (MIMO), Adaptive Antenna Systems (AAS), and hard and soft handoffs.

There are two operation modes defined in IEEE 802.16-2004, i.e., point-to-multipoint (PMP) and mesh. WiMAX PMP networks are based on cellular infrastructure, while WiMAX mesh network operates in a manner of multi-hop and multi-path communications. Some basic terms in WiMAX mesh network are different from that of cellular-like WiMAX PMP networks. In WiMAX mesh, a base station (BS) is a node that has direct connection to backhaul services outside the network and all other nodes are called mesh subscribers (SSs). Uplink and downlink in WiMAX mesh are defined as traffic in the direction to the mesh BS and traffic away from the mesh BS, respectively. In addition, a mesh mode node is different from a PMP mode node in frame structure, procedures of synchronization, network entry, data scheduling, ranging, and power control, etc. Although WiMAX mesh mode is not included in 802.16-2005, it attracts a lot of attention due to a number of advantages and potentials for applications in a variety of scenarios.

Figure 1 shows a typical topology of WiMAX mesh networks. A WiMAX mesh SS can be connected to backbone network through mesh BS. Peer-to-peer communications among mesh nodes is achievable through multi-hop relay. Each WiMAX mesh node is capable of playing a role of wireless router, and then the network capacity can be greatly increased. Since there is no a requirement of cellular infrastructure, it is relatively easy to deploy a WiMAX mesh network and the expensive cellular base stations can be saved, hence the cost can be relatively low. As there are multiple paths in a WiMAX mesh network, the network is robust and flexible. In cases of link failure, the network can recover by routing over redundant links. In addition to above, WiMAX mesh network has much longer connection distance (up to tens kilometers) than WiFi mesh network (up to several hundred meters) and ZigBee mesh network (up to several meters). Moreover, unlike WiFi mesh and ZigBee mesh, WiMAX mesh has a Time Division Multiplexing Access (TDMA) MAC protocol, by which easier quality-of-service (QoS) can be achieved

This chapter serves to introduce MAC protocol of WiMAX mesh mode and summarizes recent research progress on this technology (Figure 1). The frame structure, network configuration, and network entry of WiMAX mesh networks are presented in Section 2, 3, and 4, respectively. In Section 5, the three data scheduling schemes: coordinated centralized scheduling, coordinated distributed scheduling and uncoordinated distributed scheduling are introduced, as well as a number of new developments. In Section 6, a distributed adaptive time slots allocation algorithm is described. An example to use WiMAX mesh network for maritime communication is presented in Section 7. And the chapter is concluded by Section 8.

2. FRAME STRUCTURE

WiMAX mesh networks adopt OFDM physical layer and a TDMA-based MAC protocol to support multiple users. As illustrated in Figure 2, each periodic MAC frame is divided into two sub-frames, namely the control sub-frame and data sub-frame. Each data and control sub-frame is further divided into time slots. Each time slot consists of a number of OFDM symbols. Specifically, each time slot in the control sub-frame is made up of L_{cslot} = 7 OFDM symbols, part of which is used as guard time. Thus, the length of the control sub-frame is given by MSH-CTRL-LEN $\times L_{cslot}$ OFDM symbols, where MSH-CTRL-LEN indicates the number of time slots in the control sub-frame. Here, MSH-CTRL-LEN is a 4-bit variable and therefore, there can be at most 16 control time slots in each MAC frame. While the number of time slots in each control sub-frame is a controllable variable, the number of time slots in each data sub-frame is fixed at 256. Let L_{frame} be the number of OFDM symbols in each MAC frame. Then, the number of OFDM symbols in each data time slot, L_{dslot} is given as follows:

$$L_{dslot} = \left\lceil \frac{L_{frame} - MSH\text{-}CTRL\text{-}LEN \times L_{cslot}}{256} \right\rceil \quad (1)$$

There are two types of control sub-frame, namely network control sub-frame and scheduling control sub-frame. Network control sub-frame is used to create and maintain cohesion between different nodes in the network. Scheduling control sub-frame is used to facilitate time slot allocation and scheduling for transmissions in the data sub-frame. Each MAC frame only can has either the network control sub-frame or the scheduling control sub-frame.

As illustrated in Figure 2, network control sub-frame appears less frequently compared to scheduling control sub-frame. Specifically, there is only one network control sub-frame periodically in every SCH-FRM MAC frames, where SCH-FRM is a controllable variable takes a value in multiple of 4.

In a network control sub-frame, the first time slot is for network entry, followed by up to 15 network configuration time slots. In the network entry time slot, a new node may gain synchronization and initial entry to the network. In the network configuration time slots, each SS announces its basic information to its neighboring nodes.

In a scheduling control sub-frame, the first x (see Figure 2) time slots, where x = MSH-CTRL-LEN - MSH-DSCH-NUM are for centralized scheduling messages, while the remaining MSH-DSCH-NUM time slots are for distributed scheduling messages, since WiMAX mesh supports both centralized and distributed scheduling mechanisms.

Similar to scheduling control sub-frame, each data sub-frame is also divided into two sections as shown in Figure 2. The first section is reserved for data transmission allocated by centralized scheduling while the second section is for data transmission allocated by distributed scheduling. The variable MSH CSCH-DATA-FRACTION is 4-bit long and it specifies the number of data time slots in the first section. As such, there are only y (see Figure 2) time slots for distributed scheduling in the second section, where y = 256×(1- MSH-CSH-DATA-FRACTION×6.67). While each data sub-frame is divided into centralized scheduling section and distributed scheduling section, there is no clear separation between uplink and downlink. In WiMAX mesh, uplink and downlink generally refers to transmission directed to and from the BS, respectively.

Figure 2. WiMAX mesh MAC frame structure

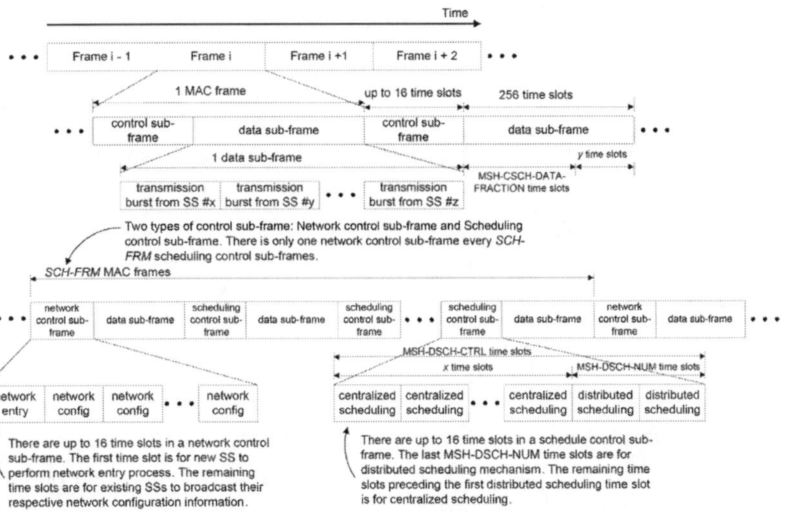

MAC layer control messages in WiMAX mesh are listed in Table 1.

3. NETWORK CONFIGURATION

WiMAX mesh networks configure the network by MSH-NCFG messages. A MSH-NCFG message contains physical layer information of the node, time synchronization information, scheduling information of MSH-NCFG messages, information of BS and neighbors, network parameters, and network entry information.

3.1 Network Description

The basic description parameters of a WiMAX mesh network are included in MSH-NCFG:NetworkDescriptor information element (IE). The parameters are information of frame duration, control sub-frame length, scheduling frames (SCH-FRM), number of burst profiles, operator ID, channels, and so on.

The frame duration is represented by a 4-bit entry FrameLengthCode and there are seven frame lengths, ranging from 2.5 to 20 ms as shown in Table 2. Control sub-frame length is the number of control slots in a control sub-frame and is indicated by a 4-bit MSH-CTRL-LEN. Among the MSH-CTRL-LEN control slots, the number of opportunities for MSH-DSCH messages is given by MSH-DSCH-NUM and the rest are for MSH-CSCF and MSH-CSCH messages, which are used for centralized scheduling.

The burst profiles are indicated by FEC-CodeType, each of which represents a type of

Table 1. WiMAX mesh control messages

MSH-NENT	Mesh Network Entry Message
MSH-NCFG	Mesh Network Configuration Message
MSH-DSCH	Mesh Distributed Scheduling Message
MSH-CSCF	Mesh Centralized Scheduling Configuration Message
MSH-CSCH	Mesh Centralized Scheduling Message

Table 2. Frame length codes and durations

FrameLengthCode	Frame Duration (ms)
0	2.5
1	4
2	5
3	8
4	10
5	12.5
6	20
7~255	reserved

modulation and coding format. For each burst profile, an exit and an entry threshold are specified. When the carrier-to-interference-and-noise ratio (CINR) is at or below an exit threshold, a more robust burst profile is required. The entry threshold is the minimum CINR that is required to start using this burst profile when a more robust profile is needed.

The number of logical channels of a WiMAX mesh network is given by a 4-bit entry Channels in MSH-NCFG:NetworkDescriptor IE. Each logical frequency channel is described by a MSH-NCFG:Channel IE. Both licensed and license-exempt frequency channels can be employed in WiMAX networks. For licensed frequency channels, the physical channel center frequency and channel width are indicated, as well as the channel reuse parameter – the minimum hop number that the channel can be reused. For license-exempt channel, a list is used to map physical channel to logical channel and the physical channel code, the corresponding reuse hops, the maximum transmission power and the maximum effective isotropic radiated power (EIRP) of the logical channel are included in the MSH-NCFG:Channel IE.

3.2 Neighbor and Base Station Information

A WiMAX mesh node broadcasts information of neighbors by MSH-NCFG:NbrPhysical IE and MSH-NCFG:NbrLogical IE. MSH-NCFG:NbrPhysical IE contains physical connection information of this neighbor, such as scheduling information of next MSH-NCFG message of the neighbor NbrNextNextMx and NbrXmtHoldoffExponent, hops to the neighbor, and estimated propagation delay to the neighbor, etc. MSH-NCFG:NbrLogical IE presents logical connection information such as burst profile, transmission power and antenna, and so on, of the reported neighbor. Existing of a physical connection means the neighbor node is able to exchange MSH-NCFG messages with other nodes. In order to exchange data a node requires a logical link with the desired neighbor. This node indicates whether it has a logical link with the reported neighbor, and if not and necessary, it can request to set up a logical link with this neighbor, or it can approve the request of setting up a logical link from the neighbor, all by entries contained in MSH-NCFG:NbrLogical IEs.

In WiMAX mesh network, base station is the node has access to backbone network. Base station information that a node is aware of is also broadcasted by MSH-NCFG messages. The information includes the node ID of the reported base station, the number of hops between this node and the base station, and the required energy per bit needed to reach the base station through this node. The number of hops is the lowest one reported by its one-hop neighbors plus one. The required energy per bit is calculated as the following:

$$E_{req} = E_{min} + \gamma, \qquad (2)$$

where E_{req} is the required energy per bit, E_{min} is the minimum value its one-hop neighbors reported, γ is the ratio between the transmission power and achievable physical data rate to that one-hop neighbor. This parameter (E_{req}) is helpful to balance the number of hops and the data rates to achieve better capacity.

3.3 MSH-NCFG Scheduling

WiMAX mesh nodes schedule MSH-NCFG messages among two-hop neighbors in a distributed manner by exchanging schedule information of next MSH-NCFG message of its own and its one-hop neighbors.

In order to make all WiMAX mesh nodes have chance to send MSH-NCFG messages, a WiMAX mesh node is not eligible to transmit in a number of MSH-NCFG opportunities after sending a MSH-NCFG packet. The number of MSH-NCFG opportunities XmtHoldoffTime is calculated by

$$\text{XmtHoldoffTime} = 2^{(\text{XmtHoldoffExponent} + 4)}, \qquad (3)$$

where XmtHoldoffExponent is a 5-bit parameter included in MSH-NCFG messages.

The eligibility interval of next MSH-NCFG message NextXmtTime is computed as the range

$$(2^{\text{XmtHoldoffExponent}} \cdot \text{NextXmtMx}, 2^{\text{XmtHoldoffExponent}} \cdot (\text{NextXmtMx}+1)] \qquad (4)$$

where NextXmtMx is a 3-bit parameter included in a MSH-NCFG message.

A WiMAX mesh node determines its next MSH-NCFG transmission time right before the current MSH-NCFG sending time using the following procedure:

1. For each neighbor, it adds the node's NextXmtTime to the node's XmtHoldoffTime to arrive at the node's EarliestSubsequentXmtTime.
2. It sets a TempXmtTime equal to its XmtHoldoffTime + its current XmtTime.
3. It determines the eligible competing nodes. Eligible competing nodes are all its neighbors that meet any of the following conditions:
   a. The neighbor's NextXmtTime interval includes TempXmtTime
   b. The neighbor's EarliestSubsequentXmtTime is equal to or smaller than TempXmtTime
   c. The neighbor's NextXmtTime is unknown
4. It holds Mesh Election among the eligible competing nodes and this node. Mesh Election is a procedure of comparison of values generated by a 32-bit reversible pseudo-random mixing function (http://home.comcast.net/ bretm/hash/, n.d.). The input of Mesh Election is TempXmtTime, this node's ID, and IDs of all competing nodes. If the values generated with this node's ID are biggest it wins Mesh Election and TempXmtTime is set as this node's NextXmtTime of its MSH-NCFG message. Otherwise, it sets TempXmtTime equal to the next MSH-NCFG opportunity and repeats steps of 3 and 4, until it wins Mesh Election or it has no competing nodes.

An example of determining competing nodes for MSH-NCFG message scheduling is illustrated in Figure 3. As illustrated, the node in consideration has three competing neighbors with temporary transmission time shown in the figure (TempXmtTime).

With above algorithm, in each MSH-NCFG opportunity only one node among two-hop neighborhood is allowed to transmit, i.e., MSH-NCFG sending is contention free within two-hop neighborhood.

Figure 3. Example of determining competing nodes for scheduling MSH-NCFG message

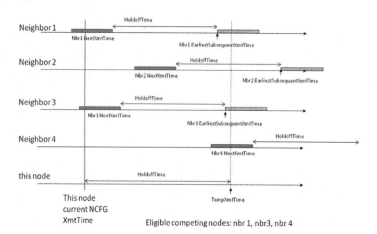

3.4 Coarse Synchronization

Time-synchronization information of a MSH-NCFG message is indicated by field TimeStamp, which consists of parameter FrameNumber, NetworkControlSlotNumberInFrame, and SynchronizationHopCount. As control slot number in one frame, control sub-frame duration, and frame duration are fixed, upon receiving a MSH-NCFG message, a new node can determine the start time of the frame. This is called coarse synchronization. Fine synchronization is achieved through PHY functions.

The parameter SynchronizationHopCount is used to determine superiority between nodes. A master time keeper node has a SynchronizationHopCount of zero. A WiMAX mesh nodes will keep synchronization to nodes with lower SynchronizationHopCount, or if hop counts are same, to the node with smaller node ID.

3.5 Multi-Channel Operation

A WiMAX mesh network may incorporate up to 16 logical frequency channels. The logical channels are a map of a subset of possible physical channels and are described in MSH-NCFG Network Descriptor.

The data schedule control information, i.e., MSH-DSCH, MSH-CSCF, and MSH-CSCH messages are broadcasted in a base channel that is indicated by a parameter NetworkBaseChannel in MSH-NCFG messages. All nodes use this common channel to exchange data schedule information.

To ensure that all nearby nodes are able to receive MSH-NCFG and MSH-NENT messages, the channel used for their transmission is cycled through the available frequency channels in the band. The channel selection is based on the FrameNumber. For FrameNumber i, the logical frequency channel is determined by the array lookup given by

$$\text{NetConfigChannel} = \text{Logical channel list}[(\text{FrameNumber} / (\text{SchedulingFrames} \times 4 + 1)) \% \text{Channels}] \quad (5)$$

where SchedulingFrames×4 (i.e., SCH-FRM in Fig. 2) defines how many frames have a schedule control sub-frame between two frames with network control sub-frames, and Channels is an entry indicating the number of logical channels in operation.

4. NETWORK ENTRY

Network entry in WiMAX mesh networks is accomplished by MSH-NENT and MSH-NCFG messages. When a new node wishes to join a network, it continuously scans all possible frequency channels to receive MSH-NCFG message, and then to establish coarse synchronization with the network and to acquire network parameters. When a new node receives MSH-NCFG messages from a node at least twice and at least one MSH-NCFG message contains NetworkDescriptor IE, it selects that node as potential sponsor node and synchronizes itself to the node. In the fist control slot of a network control sub-frame, the new node sends MSH-NENT:Request to the potential sponsor node. Upon receiving MSH-NENT:Request, the potential sponsor node can reject the request by replying a MSH-NCFG:NetEntryReject or accept the request with a MSH-NCFG:NetEntryOpen. The MSH-NCFG:NetEntryOpen IE contains a temporary schedule for transmission of higher layer authentication and configurations for the new node. This temporary schedule is in data sub-frame, and is called sponsor channel. Upon reception of MSH-NCFG:NetWorkEntryOpen, the potential sponsor node is confirmed as sponsor node, and the new node sends back a MSH-NENT:NetEntryAck to confirm the sponsor channel. And the new node uses the sponsor channel to perform basic capacity negotiation with the sponsor node, node authorization, registration, IP connectivity establishment, time-of-day establishment, and operational parameters transfer with the security sub-layer. When this procedure is finished, the new node sends a MSH-NENT:NetworkClose to the sponsor node. And the sponsor node ends the network entry by replying a MSH-NENT:NetworkAck. By now the new node is regular in the network.

5. SCHEDULING ALGORITHMS

Data transmission scheduling is the key to provide good quality of service and avoid congestion. WiMAX adopts a TDMA-based MAC protocol. In this context, data transmission scheduling equals to time slots allocation, where time slots are allocated to a link (link scheduling) or a packet (packet scheduling) to achieve a certain performance objective. In the literature, various data slots scheduling algorithms have been proposed for WiMAX PMP networks. However, these PMP scheduling algorithms are not directly applicable to WiMAX mesh networks because there are mainly packet scheduling algorithms limited to single hop scenarios where interference from concurrent transmissions over multiple hops is not considered.

In WiMAX mesh networks, three scheduling mechanisms have been standardized. The three mechanisms are called coordinated centralized scheduling, coordinated distributed scheduling and uncoordinated distributed scheduling (Ali et al., 2008).

5.1 Coordinated Centralized Scheduling

In coordinated centralized scheduling, a BS collects bandwidth requests from all mesh SSs within a certain hop count range and centrally performs the time slots allocation before sending out the allocation outcomes as grants. Thus, two scheduling components are required: (a) Scheduling to transmit requests and grants in control sub-frame, and (b) Scheduling to transmit data packets in data sub-frame.

In the control sub-frames, the requests and grants are transmitted as information elements within MSH-CSCH messages. These MSH-CSCH messages are transmitted in the first part of the scheduling control sub-frame (see Fig. 2) and it is required their transmission being collision free. To ensure collision free transmissions, a node needs to know exactly in which control time slots to transmit its messages. Given a routing tree that spans from the BS, a node transmits MSH-CSCH:Request to its parent toward BS in the order such that the node with the largest hop count transmits first. Similarly, a node transmits MSH-CSCH:Grant to its children away from BS in the order such that the node with the smallest hop count transmits first. In both cases, amongst the nodes with a same hop count, the node with the smallest node index transmits first. All nodes know the routing tree as determined and propagated from the BS in MSH-CSCF messages that are transmitted in the first part of the scheduling control sub-frame (see Fig. 2). Specifically, BS broadcasts the MSH-CSCF in control sub-frame to all its neighbors, and all the neighbors rebroadcast the message accordingly until all SSs have broadcasted the MSH-CSCF message once.

In transmitting MSH-CSCH:Request, each SS includes the requests from its children into its own request size. Upon collecting the requests, BS determines the fraction of data sub-frame allocated to each node such that all requests can be fulfilled within one or two MAC frames. If available data time slots are not sufficient to fulfil all requests, the BS scales down proportionally the sizes of all requests. Only the final allocated fraction of data sub-frame, but not the actual scheduled allocated to each SS is announced in MSH-CSCH:Grant. Each SS derives the actual allocated time slots based on the allocated fractions and the routing tree.

From the operation described above, there will be no transmission collision in control and data time slots within a routing tree. This collision free allocation is achieved at a high cost where each time slot is allocated to a single node within the routing tree such that no spatial reuse is allowed. In practice, two nodes can transmit concurrently if the interference at their respective receiver nodes is below a certain threshold. This type of interference which is caused by adjacent concurrent transmissions is called secondary interference. In the contrary, primary interference is due to multiple conflicting functions such as simultaneous transmit and receive, or transmit two different packets, at a node at the same time.

In the literature, there are algorithms proposed to deal with secondary interference to promote concurrent transmissions in the coordinated centralized scheduling although it is not sure if these algorithms are really compatible with WiMAX mesh standard. In (Du et al., 2007), Active Link Selection (ALS) algorithm has been proposed to eliminate secondary interference for concurrent transmission. In ALS, an interference graph is constructed at the BS. With the interference graph, a set of non-interfering and available links is determined for each time slot. For a given time slot, a link is considered available if the transmitter of the link has a token. Tokens are generated at the traffic source node according to its traffic rate, and are transferred to the receiver node of a link when the link is allocated time slot to transmit. From the set of non-interfering and available links, ALS allocates time slot to the link nearest in terms of hop count to its respective traffic destination node. Here, ALS allocates for as many links as possible from the set before moving on to the next time slot repeating the same process. The efficiency of ALS in improving concurrent transmission is measured in terms of transmission schedule length where a shorter schedule is produced by a more efficient algorithm.

Similar to ALS, in an earlier work, Han et al. (2006) has proposed the Transmission-Tree Scheduling (TTS) algorithms to reduce schedule length with concurrent transmissions. Compared to Du et al. (2007) that select links based on minimum hop count (nearest) to destination, Han et al. (2006) evaluated another three link selection criteria: (a) farthest (maximum hop count) to destination, (b) random selection, and (c) minimum interference. The maximum farthest link selection is simply the opposite of the nearest link selection criteria. In random selection, a link is selected randomly from the set of active links to find out if it can be scheduled concurrently. In the minimum interference selection, the link that interferes with least number of neighboring links is selected. The evaluation results show that these three selection criteria are not as efficient as the nearest link selection.

Concurrent transmissions (spatial reuse) can be improved by careful construction of the routing tree which is rooted at the BS and propagated through MSH-CSCF message as described earlier. In (Wei et al., 2005), an interference-aware routing tree construction algorithm and a time slot allocation algorithm are proposed. Here, a blocking value and a blocking metric are separately defined to quantify the interference level of a node and a route, respectively. The blocking value of a node is essentially the total number of neighboring nodes blocked by its transmission. This definition of blocking value is similar to the way of quantifying minimum interference in the concurrent link selection criteria in (Han et al., 2006). Then, the blocking metric of a route is simply the sum of blocking values of all its intermediate nodes. As such, routing tree should be constructed to minimize the blocking metric. This is achieved through a heuristic such that each new node will join the routing tree through a node with the minimum blocking metric to the BS. With the constructed routing tree, the centralized scheduling algorithm allocates a time slot to SS with the largest unfilled request size. For the same allocated time slot, the algorithm will search for other non-blocked SSs with the largest unfilled request size so that more than one transmission can happen concurrently. The scheduling algorithm will move to the next time slot only if no other unblocked SS is found. This process is repeated until all requests are fulfilled.

In (Ghosh et al., 2007), a centralized scheduling algorithm has been proposed to take into consideration flow (link) rate and delay requirements while providing concurrent transmissions. To capture the delay requirement, each link is assigned a start time, T_s which is the time slot within a MAC frame by which the link must be scheduled so that the flow can reach its destination by its deadline. Assuming each link has the same rate, T_s of a link is calculated as follows:

$$T_s = \frac{d - f_s - k \times t_1}{t_2} \quad (6)$$

where d is the deadline, f_s is the start of next MAC frame, k the number of hops to destination, t_1 is the propagation delay, and t_2 is the duration of each time slot.

For a route, its entire component links must be scheduled time slots in or before their respective T_s. This time slot allocation algorithm consists of two phases. In the first phase, each link is scheduled sequentially. For a link, it is allocated a sub-channel in all available time slots before T_s or until the link's maximum bandwidth requirement is fulfilled. Here, a time slot is considered available when this link does not cause secondary interference to other links already allocated the time slot. Here, we notice that Ghosh et al. (2007) allocates sub-channel in a time slot but this contradicts the WiMAX mesh standard. There is no notion of sub-channel in OFDM adopted by WiMAX mesh. At the end of first phase, all links of a route have been allocated time slots. If the allocated time slots meet or exceed the minimum bandwidth requirement of all links, there is no need to proceed to the second phase. In the second phase, all the time slots that have been allocated to a link but exceed the link's minimum bandwidth requirement, are taken away to be re-allocated to the links that have not had their minimum bandwidth requirement fulfilled.

5.2 Coordinated Distributed Scheduling

In coordinated distributed scheduling, a SS does not rely on the BS for time slot allocation. When a SS has a packet to send to another SS, it will transmit a bandwidth request to the SS as an information element within MSH-DSCH message. This MSH-DSCH:Request is transmitted in a time slot in the second part of the scheduling control sub-frame (see Fig. 2). The transmitted MSH-DSCH message, amongst others also announces which control time slot the current SS will use to transmit its next control message. No other nodes shall use the announced control time slot and thus, preventing collision. As such, the node must identify its next transmission time slot before transmitting the current MSH-DSCH message. The next transmission control time slot is identified using the Mesh Election Algorithm which ideally allows for only one node to transmit in each time slot. The algorithm is exactly same as the one used for next MSH-NCFG message scheduling.

A transmitter node sends it MSH-DSCH:Request after winning a time slot for its next transmission through the Mesh Election Algorithm. Upon receiving the request, the receiver node will allocate the first group of available time slots in data sub-frame for transmission. The outcome of this data time slot allocation will be announced by the receiver node as a grant transmitted as an information element in MSH-DSCH message. All the MSH-DSCH:Grants are transmitted in control sub-frame using the same mechanism described above. In response to the grant, the transmitter node will send a grant confirmation as an information element in MSH-DSCH message. This MSH-DSCH:Confirmation is transmitted in the control sub-frame following the same mechanism for a typical MSH-DSCH message. Usually, the data packet will be transmitted in the data sub-frame only after the successful transmission of MSH-DSCH:Confirmation. This process of exchanging MSH-DSCH:Request, MSH-DSCH:Grant and MSH-DSCH:Confirmation between a sender and its receiver is called three-way handshaking and is illustrated in Figure 4.

From the description above, coordinated distributed scheduling is similar to the coordinated centralized scheduling in the sense that it has to schedule for both control time slots and data time slots. Coordinated distributed scheduling uses the Mesh Election Algorithm to schedule control time slots to transmit request, grant and grant confirmation. For data time slots, a receiver oriented approach is adopted where the receiver node, in response to each received request, allocates the first available time slots.

5.2.1 Scheduling for Control Sub-Frame

In the literature, several algorithms have been proposed to improve control and data time slot allocation in coordinated distributed scheduling. We present the proposed control scheduling algorithms in this sub-section before introducing data scheduling algorithms in the next sub-section.

For control scheduling using the Mesh Election Algorithm, the interval between two consecutive successful control messages, τ_k of node k can be expressed as follows:

$$\tau_k = \text{XmtHoldoffTime}_k + S_k \qquad (7)$$

where XmtHoldffTime_k is the holdoff duration of node k and S_k is a random variable experienced by node k due to the Mesh Election Algorithm. Recall that $\text{XmtHoldoffTime} = 2^{\text{XmtHoldoffExponent}+4} \geq 16$. This means, after each successful transmission in the control sub-frame, a node has to backoff for at least 16 time slot before transmitting another control message. In (Bayer et al., 2006), it is argued that in a sparse network with few competing nodes, a node needs not to wait for at least 16 time slots to transmit another control message. Thus, Bayer et al. (2006) has proposed calculating the holdoff duration as follows:

$$\text{XmtHoldoffTime} = 2^{\text{XmtHoldoffExponent}+\alpha}, \qquad (8)$$

where α is an integer variable taking value between 0 and 4. It is shown that τ_k can be significantly reduced by choosing a small α.

The algorithm proposed in (Bayer et al., 2006) focuses on sparse networks. For a more general

Figure 4. Three-way handshake procedure

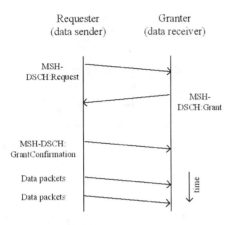

case with different node density, Cao et al. (2005) has analyzed the expect value of S_k, i.e., $E[S_k]$ assuming that XmtHoldoffExponent is fixed at 4. Assuming the transmission time sequence of the control messages of all nodes are independently and identically distributed renewal processes, Cao et al. (2005) shows that $E[\tau_k] = \text{XmtHoldoffTime}_k + E[S_k]$ and,

$$E[S_k] = \sum_{j=1, j\neq k, Exp_j \geq Exp_k}^{N_k^{known}} \frac{2^{Exp_k} + E[S_k]}{2^{Exp_j+4} + E[S_j]}$$
$$+ \sum_{j=1, j\neq k, Exp_j < Exp_k}^{N_k^{known}} 1 + N_k^{unknown} + 1 \quad (9)$$

where Exp_k is the XmtHoldoffExponent for node k, N_k^{known} is the number of competing neighbors of node k whose next transmission time is known, and $N_k^{unknown}$ is the number of competing neighbors of node k whose next transmission time is unknown. From Eq. (9), the expected interval between two consecutive successful control messages depends on the number of nodes in the network, XmtHoldoffExponent and network topology. Logically, when the number of nodes is increased or when using a larger XmtHoldoffExponent is used, the expected time interval between successful control messages increases.

From Cao et al. (2005), we understand that XmtHoldoffExponent configuration has a different impact in different network scenario. Thus, Bayer et al. (2007) has proposed an algorithm to set XmtHoldoffExponent of a node depending on its status and scenario. In (Bayer et al., 2007), a node can have one of the four status: Mesh Base Station (MBS), Active Node (AN), Sponsoring Node (SN) and Inactive Node (IN). Depending on the status, a node's XmtHoldoffExponent can be set as follows: $0 £ Exp_{MBS} £ Exp_{AN} £ Exp_{SN} £ Exp_{IN} £ 7$. For example, a node k will adapt its XmtHoldoffExponent, i.e., Exp_k such that $Exp_k £ £ Exp_{AN}$ where node k is an active node. In the adaptation, a node will increase its XmtHoldoffExponent if its NextXmtMx or its neighbor's NextXmtMx is larger than a threshold. Here, the value of NextXmtMx is used as an indicator to the congestion level. On the other hand, a node will decrease its XmtHoldofExponent if its status has change in one of the following ways: (a) An IN, SN or AN becomes a MBS, (b) An IN or SN becomes a AN, or (c) An IN becomes a SN. It is shown that such a dynamic adaptation to the XmtHoldoffExponent can reduce control message transmission interval as inactive nodes increases their XmtHoldoffExponent to reduce contention.

In (Loscri, 2007), the author has argued that Mesh Election Algorithm may result in under-utilization of control sub-frame. In order to utilize the control sub-frame more aggressively, Loscri (2007) has introduced an additional operation to redistribute unallocated control time slots. The proposed redistribution operation is carried out frame-by-frame where a node that has no allocated control slot in the current frame will randomly pick an unallocated time slot to transmit its control message. However, Loscri (2007) has not described how a node can find out the set of unallocated time slots. Also, we think that this additional redistribution operation may not be necessary because a properly performed Mesh

Election Algorithm should utilize all time slots except during the time interval where all nodes are backing off.

5.2.2 Scheduling for Data Sub-Frame

For uncoordinated distributed scheduling for data slots, it is important to ensure fairness in time slot allocation. Intuitively, fixed allocation is the simplest method in achieving fairness by allocating time slots in data sub-frame to the nodes. Let N_s^i denotes the number of time slots allocated to node i, N_{fr}^s denotes the total number of time slots in the data sub-frame and N denotes the total number of nodes in the mesh network. Then,

$$N_i^s = \left\lfloor \frac{N_{fr}^s}{N} \right\rfloor \qquad (10)$$

This algorithm is fair in the sense that each node gets equal throughput, but the disadvantage of this method is in the mesh network, nodes can be involved in different activities and hence has different demand for traffic. With fixed allocation, some inactive nodes may have nothing to send during its scheduled time slot while some active nodes may have lots of data waiting in the queue. Time utilization is very inefficient.

Therefore, Makarevitch (2006a) and Makarevitch (2006b) have proposed an improvement for fixed allocation algorithm. The proposed algorithm is called Proportional Scheduling (PS) algorithm, such that number of time slots allocated to a node is proportional to the traffic flow. Traffic flow information in a particular node can be propagated to other nodes in the control message. Let N_k^{fl} denotes the traffic flow on node k. Then,

$$N_i^s = \left\lfloor \frac{N_i^{fl} \times N_{fr}^s}{\sum_{k=1}^{N} N_k^{fl}} \right\rfloor \qquad (11)$$

This method addressed the problem of different traffic load among nodes. We can say that it is fair in the sense of normalized throughput. By using the traffic load information as a weighting factor, time slots in the data sub-frame are better utilized.

Although PS algorithm solves the problem of different traffic demands among nodes, it does not address the variation of traffic load in the mesh network. In practical situations, a node may frequently change its operation from active to inactive and vice versa. Therefore, an Adaptive Data Dependent Scheduling (ADDS) algorithm is further proposed in Makarevitch (2006a). The idea behind ADDS algorithm is to allocate more time slots in the data sub-frame to the nodes with the highest current traffic demand. This algorithm is distributed, because each node in the mesh network will run it. The ADDS algorithm is also coordinated in a sense that, it makes use of the traffic load information broadcasted by the local and neighboring nodes.

All the nodes will broadcast their request slot number in the MSH-DSCH messages. Eventually each node will get the request slots number of its 2 or 3 hop neighbors. In each node, the algorithm will sequentially compare the request slot numbers and assign slot to the node with the maximal request slot number. If one slot happens to be assigned to the local node itself, the local node will pick the receiver node that the local node has the maximum data queue to send to.

Consider two different nodes in the same mesh network, each running this algorithm. If these two nodes are not among the 3-hop neighbors of each other, they will have distinct set of competing nodes. Thus, it is highly possible that these two nodes assign the same time slot to different transmitters. Collision will occur. On the other hand, if both nodes are in the same neighborhood, they will have the same set of competing nodes. Two nodes will not assign the same slot to the same transmitter. The time slot assignment therefore is consistent for the two nodes. No collision will occur. Therefore, to achieve collision free scheduling, each node must include the nodes within a larger neighborhood. Naturally, to include a larger neighborhood with more hops there will be more signaling overhead as more MSH-DSCH messages need to be propagated.

All the algorithms described above are topology dependent, meaning changes in the network topology requires recalculation of a new schedule. To remove the topology dependency, a Finite Field Based Distributed Scheduling Algorithm is proposed in Makarevitch (2006a). This algorithm ensures each node will have one collision free time slot in the data sub-frame. An initial schedule is performed in advance and is robust to the topology changes in the mesh network. Subsequent schedule is performed whenever traffic load is changed, similar to traffic dependent scheduling.

A k degree polynomial $f(x) = \sum_{i=0}^{k} a_i x^i$ has at most k distinct roots. So, the difference of two polynomials will also have at most k distinct roots. In the context of WiMAX mesh network, Makarevitch (2006a) assigns each node a unique vector of coefficients, $a_0, a_1, \cdots, a_k$ over a Galois field GF(q), which is equivalent to assign a unique polynomial to each node. In the data sub-frame, a mapping of polynomials into sets of time slots can be done, which will produce sets having no more than k common time slots between them. So, as long as the number of time slot in one set is larger than k, we can always ensure each node will have at least one collision free time slot within the data sub-frame. To assign the unique coefficient vector to each node in the mesh network in a distributed manner, we can use a pseudo-random generator that takes in the MAC address of the nodes. By exchanging and comparing the coefficient vector, repeated coefficient vector can be regenerated.

5.3 Uncoordinated Distributed Scheduling

In distributed scheduling, each sender-receiver pair requests and allocates time slots locally. In WiMAX mesh networks, distributed scheduling can be further classified into coordinated distributed scheduling and uncoordinated distributed scheduling. For coordinated distributed scheduling, as presented in the previous section, the request-grant-confirmation three-way handshaking takes place in the scheduling control sub-frame. For uncoordinated distributed scheduling, the three-way handshaking is carried out opportunistically in the unused time slots in data sub-frame, and MSH-DSCH messages may collide. Thus, the performance of uncoordinated distributed scheduling is less predictable compared to the coordinated distributed scheduling.

Uncoordinated distributed scheduling can be used for fast, ad-hoc setup of schedules on a link-by-link basis. While transmitting MSH-DSCH:Request, MSH-DSCH:Grant and MSH-DSCH:Confirmation in data time slots, a SS must ensure that the MSH-DSCH and resulting data transmissions do not cause collisions with the data and control traffic scheduled by the coordinated centralized scheduling nor coordinated distributed scheduling. The collisions can be partly avoided by having a node to wait a sufficient number of time slots before responding to a MSH-DSCH:Request with its MSH-DSCH:Grant, such that nodes with earlier requests have an opportunity to respond. In the contrary, MSH-DSCH:Confirmation can be sent out immediately following the first successful reception of an associated MSH-DSCH:Grant.

6. DISTRIBUTED ADAPTIVE TIME SLOT ALLOCATION

In distributed scheduling of WiMAX mesh network, a three-way handshake procedure is defined for allocation of data slots. However, there is a lack of detailed algorithm to allocate the data slots. It also has no a solution to deal with conflicts of concurrent transmissions. Conflicts of transmissions may be caused by interference from nodes more than two hops away, and if nodes are mobile, from nodes moving close to the receiver node. To minimize these issues, a distributed adaptive time slots allocation (DATSA) algorithm is proposed in (Kong et al., 2009a).

With DATSA, each mesh node keeps track of the availability of its own and its one-hop neighbor's slots in data sub-frame. A granter marks all slots it granted "busy" and a requester marks all slots it accepted also "busy". One-hop neighbors of the granter and the requester mark their corresponding slots "busy" upon reception of MSH-DSCH:Grant from the granter and MSH-DSCH:Confirmation from the requester. Other slots in data sub-frame are marked as "free". No states "available for reception only" or "available for transmission only" is allowed. Further, each mesh node broadcasts their slots availability using MSH-DSCH:Availability that defined in the protocol, and each mesh node also keeps track of the slots availability of its one-hop neighbors. When grants or accepts slots, a granter and a requester checks record of its slots availability as well as the availabilities of its one-hop neighbors. If a mesh node is "busy" or any one of its one-hop neighbor is "busy" in a slots range, it cannot transmit or receive data packets in this slots range. By doing so, slots in use by a two-hop neighbor will not be used by this node, and then interference due to concurrent transmissions can be minimized.

A simple application example of DATSA for case with single frequency channel is shown in Fig. 5 (a). Node B is transmitting to its one-hop neighbor node C, and then both of them mark slots range s_1 as "busy". Node D and node A mark their slots range s_1 as "busy", after receiving MSH-DSCH:Grant from node C and MSH-DSCH:Confirmation from node B, respectively. Node D broadcasts DSH-DSCH:Availability, and then its one-hop neighbor node E knows node D is "busy" in slots range s_1, and it marks the corresponding slots range in its record of neighbor D's availability. As result, if node F requests slots to node E, node E will not grant s_1 to node F, but will grant other free slots range.

To extend the above method to case of multi-channel, following conditions should be met when grant data slots for a request (Zhou et al., 2009):

- In the slots range to be granted, there is no scheduled transmission or reception with the granter and requester in all frequency channels, as well as with their one-hop neighbors in the corresponding frequency channel.
- Note here single radio operation is assumed, i.e., a mesh node cannot transmit and receive simultaneously.

When a node scheduled transmission or reception in a slots range of a frequency channel, it marks the corresponding slots range in that frequency channel "busy". Then in one-hop neighbor's records of the requester and the granter, the slots range in the granted channel will be marked "busy" while in other frequency channels, the corresponding slots range is still "free". As results, the one-hop neighbors of both granter and requester can use the corresponding slots range in other frequency channels. Hence, concurrent transmission and reception can occur within two-hop neighborhood in different frequency channels, and the network capacity can be improved.

Figure 5 (b) shows a simple example to apply the above method in case of multi-channel. As shown in the figure, node B and node C are using slots range s_1 in frequency f_1 channel for data transmission. Marking and broadcasting availability of the granted channel f_1 at neighboring nodes A, D, and E are same as the case of single channel. For frequency channel f_2 and f_3, node B and C leave the corresponding slot range s_1 "free". However,

s_1 in channel of frequency f_2 and f_3 cannot be assigned for transmission or reception at node B and C due to single radio operation. Other nodes (A, D, and E) are "free" in s_1 of channel f_2 and f_3, and they can schedule transmission or reception of data in these data slots.

As illustrated in Figure 5, there is a delay between the time a receiver node receives a request and the time the receiver node sends its next scheduling control message with the respective grant. This delay is a random variable due to the Mesh Election algorithm. Instead of generating grant immediately upon receipt of a request, DATSA proposes to generate the grant right before transmitting the scheduling control message in the control slot. This helps to avoid spatial reuse conflicts when two neighboring nodes allocate a same time range due to unawareness of each other's allocation. However, conflict in slot allocation can still occur due to latency in sending confirmation. To handle this issue, DATSA proposes that a sender node checks for its slot availability right before sending a confirmation but not immediately after receiving a grant. Upon the checking, if an allocated slot is already marked unavailable, the node should send a cancellation instead of confirmation. In DATSA, the cancellation should piggyback the original request as re-request, and should inform the receiver node to release its respective allocated time slots. All neighboring nodes that overhear the cancellation should also release their respective time slots by marking them as "free".

In the example shown in Figure 4, if there is an absence of node D, then node E is not aware of the allocations of node C, and will possibly allocate the same slots range s_1 for transmission and reception with node F. In view of this issue, DATSA proposes to detect conflict in slot allocation by monitoring the number of successful transmissions within a time window T_d at the receiver node. The value T_d is determined based on probability of zero packet arrival at the sender node for transmission and the probability of transmission failure due to channel error. If there is no successful transmission within the time window, the sender node should send a cancellation. The cancellation message releases the allocated time slot. Different from the cancellation by sender node described earlier, this cancellation by receiver node does not act as a re-request. On the other hand, the sender node, upon receiving the cancellation, should send a re-request if there is still data packet to transmit.

When a cancellation transmitted from the receiver node to the sender node, DATSA interprets this as a detected allocation conflict and proposes to adapt to the detection by not allocating the same time slot to the same receiver node. To realize this, DATSA will only scan for available slot for this receiver node starting from the slot right after the slots range of the last allocation. If this operation reaches the end of the 256 slots data sub-frame, it starts from the first slot.

7. MARITIME WIMAX MESH NETWORKING

In this section, we illustrate an application example of using WiMAX mesh for high-speed and low-cost ship-to-ship/shore communications.

With the development of information technology and maritime industry, there is an increasing need of high-speed and low-cost wireless communications in maritime environment. Similar to the on-land vehicular networks that are under development for better vehicles safety and higher transportation efficiency, maritime wireless networks are required at sea to improve maritime safety and operation efficiency. Such a network can be used for better navigation, ship traffic management, sea condition surveillance, disaster rescue, location reporting and so on. In addition, ship crew and passengers need general communications like Internet, telephone, and FAX etc.; fleet and seaport managers need effective communication means for ship and seaport managements.

Figure 5. Example of DATSA slots and channel allocations, (a) in case of single channel; (b) in case of multi-channel

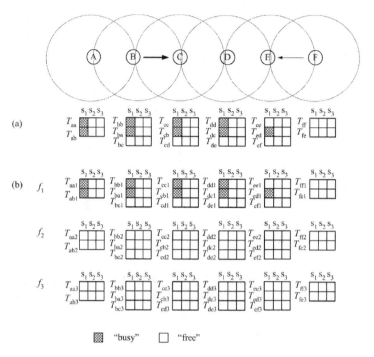

Unfortunately, there is a lack of such a broadband wireless network in current maritime environment. The present maritime communications are mainly based on analogue HF/VHF networks and satellite links. Marine HF/VHF networks have low-bandwidth and are mainly used for automatic identification system (AIS), weather radio broadcast, and distress rescue, etc. Clearly, they cannot fulfill requirements of modern maritime users. A typical satellite system is Inmarsat. The most recent system is Fleet Broadband that can provide services of voice, FAX, ISDN (64 kbps), standard IP (up to 432 kbps), steaming IP (up to 256 kbps), and text messages. However, all with higher price and lower speed compared to most of the current terrestrial networks.

On the other hand, broadband wireless technologies are actively studied for use on land. Developments include 3G networks, WiMAX networks, and Wireless LANs, etc. Outputs of these technologies can be used at sea. For this purpose, a project TRITON (TRI-media Telematic Oceanographic Network) is proposed (Pathmasuntharam et al., 2007). The network envisaged in TRITON for maritime communications is WiMAX mesh based on IEEE std 802.16-2004, as this technology allows long range access and multi-hop communications. It is a better choice in waters far from shores (like more than 100 km) as it is difficult to setup cellular infrastructure there. Moreover, a mesh network is capable of self-organizing, self-healing, and routing packets through multiple paths, thus it is robust and suitable for maritime environments in which radio links may experience frequent break due to sea wave movement and occlusion. In addition, WiMAX mesh networks employ TDMA based MAC protocol that can offer more efficient bandwidth utilization through time slot allocation or scheduling compared to contention based random access MAC like that of Wi-Fi.

7.1 High Level Architecture of TRITON

Figure 6 shows the high level architecture of TRITON. The coverage extension is achieved by forming a WiMAX mesh network amongst neighboring ships, marine beacons and buoys. The WiMAX mesh network will be connected to the terrestrial networks via land stations, which are placed at regular intervals along the shoreline. Each ship will carry a WiMAX mesh radio that has the capability of frequency agility where frequencies can be switched to suite the geographic location or sea conditions. In port waters or narrow water channels, the radio frequency usage will be based on radio frequencies that are limited by land based terrestrial communication systems. In locations far away from land, the frequencies could be in UHF, VHF or HF bands. While the mesh network is ideal when there are sufficient ships to relay the transmission, in locations where ships are sparse, the TRITON system can fall back to a satellite communication link.

7.2 Considerations of Maritime WiMAX Mesh Networking

WiMAX mesh network is initially designed for range extension in communities with sparse users. To apply WiMAX mesh network in maritime environment that is different from a terrestrial environment in radio propagation and nodes mobility, some considerations and improvements are necessary.

7.2.1 Distributed Scheduling vs. Centralized Scheduling

Coordinated distributed scheduling is adopted in TIRTON based on following analysis:

- A maritime WiMAX mesh network is highly dynamic due to sea wave movement and ship mobility. This will lead to frequent change of network topology. If centralized scheduling is uses, it will result in frequent change of scheduling tree and flush of MSH-CSCF and MSH-CSCH messages. There is no such an issue with coordinated distributed scheduling as it schedules data transmission in a distributed manner by exchanging MSH-DSCH messages.
- In areas like seaport, the number of nodes in coverage could be tens to hundreds. If centralized scheduling is used the size of MSH-CSCF and MSH-CSCH messages will be possibly larger than the capacity of a control slot.
- Concurrent transmission is not allowed with centralized scheduling with current standard. While, space reuse is possible in coordinated distributed scheduling and then higher capacity is achievable.

7.2.2 Supporting to Mobility

Mobility is not supported in the current WiMAX mesh standard while this is needed in maritime WiMAX mesh networks. This requires the maritime WiMAX mesh nodes can dynamically update the neighbor list and corresponding parameters such as hop counts, estimated propagation delay, transmission power, transmission antenna, and burst profile. Moreover, mechanism of detection of transmission conflicts is required. Conflictions may happen when ships far away moves into each other's interference range. This issue can be minimized by DATSA presented in Section 6.

7.2.3 Packets Retransmission

In maritime environment, the communication link quality is affected by sea wave movement, ship mobility, and propagation impairments. In a mild sea condition and with careful designs, the probability to have a good link is high; however, following a good link, a bad link may last several seconds due to the long period of sea waves. In

Figure 6. High level architecture of TRITON (Pathmasuntharam et al., 2007)

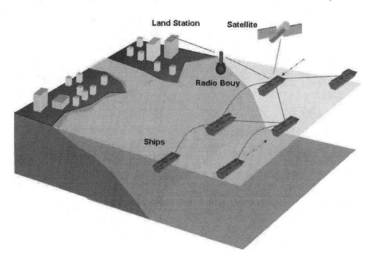

this case, retransmission for a failed packet over the same link may also fail and then it is not an effective solution for link failures. Instead, communication diversity can be used as that presented in (Kong et al., 2008a).

7.3 Routing Protocol

In a maritime multi-hop mesh network, a connection link may break often due to wave rocking and occlusion. As such routing protocol is very important in ensuring connectivity and reliable packet delivery. Three well known multi-hop routing protocols have been evaluated with a maritime simulator developed: Optimized Link State Routing Protocol (OLSR), Ad hoc On-demand Distance Vector (AODV) and Ad hoc On-demand Multipath Distance Vector (AOMDV). Simulations for packets transmission have been carried out using a maritime simulator. Simulated results show that OLSR has better performance in initial packet delay (i.e., time delay of the first packet arriving destination), because it is a proactive routing protocol and then routes are set up before traffic flow starts. However, in terms of average packet delay, OLSR is inferior to AODV and AOMDV because of its slow reaction to the link breakages that happen often in maritime environment. AODV has better performance in initial packet delay than AOMDV, but in average packet delay and packet delivery ratio, AOMDV has better performance. However, none of the three protocols is ideal for maritime multi-hop mesh networks. And TRITON developed and uses a new routing protocol called MAC-based Routing Protocol for TRITON (MRPT) (Kong et al., 2009b).

MRPT is an optimization of the classical routing algorithms tailored to the features of WiMAX mesh network and requirements of maritime communications. Designs and features of MRPT protocol include, 1) proactive; 2) deliver routing information by MSH-NCFG message; 3) maintain multiple routing paths. Routing information is piggybacked on MSH-NCFG messages, and because this message is periodically transmitted in dedicated control slots in a collision-free manner, the overhead and delay for routing information spreading can be greatly reduced. As in a maritime mesh network, nodes are mobile and the network performance is sensitive to routes, thus, MRPT considers both routing cost and stability in selecting route. Hop count is set as routing cost, and MRPT uses Received Signal Strength (RSS) as stability metric. Only if the link's RSS value is above an upper threshold it is eligible for being chosen for routing. A mesh node keeps several candidate links for next hops, and when a better route appears, it switches to it immediately. If a link breaks, a mesh node looks up the list of backup links and chooses the one with least hop count among links having sequence number larger than the broken one. Simulation results show MRPT is better than OLSR, AODV and AOMDV in initial packet delay, average packet delay and throughput.

8. CONCLUSION

MAC protocol of WiMAX mesh network defined in IEEE 802.16-2004 is presented in this chapter. The content includes frame structure, network entry and configuration, scheduling MAC control messages, as well as a bunch of data slot scheduling algorithms proposed recently. In addition, it also introduced a detailed algorithm DATSA for data slots allocations. Finally, as an application example, a maritime WiMAX mesh network is described, including some considerations and new developments to apply this technology in maritime environment.

REFERENCE

Ali, N. A. A., Taha, A.-E. M., Hassanein, H. S., & Mouftah, H. T. (2008). IEEE 802.16 Mesh Schedulers: Issues and Design Challenges. *IEEE Network*, *1*(22), 58–65. doi:10.1109/MNET.2008.4435904

Bayer, N., Sivchenko, D., Xu, B., Rakocevic, V., & Habermann, J. (2006). Transmission Timing of Signaling Messages in IEEE 802.16 based Mesh Networks. *European Wireless Conference*.

Bayer, N., Xu, B., Rakocevic, V., & Habemann, J. (2007). Improving the Performance of the Distributed Scheduler in IEEE 802.16 Mesh Networks. In *IEEE. IEEE Vehicular Technology Conference 2007 Spring (VTC-2007)* (pp. 1193-1197).

Cao, M., Ma, W., Zhang, Q., Wang, X., & Zhu, W. (2005). Modeling and Performance Analysis of the Distributed Scheduler for IEEE 802.16 Mesh Mode. In *The ACM International Symposium on Mobile Ad Hoc Networking and Computing* (pp. 78-89).

Du, P., Jia, W. J., Huang, L. S., & Lu, W. Y. (2007). Centralized Scheduling and Channel Assignment in Multi-Channel Single Transceiver WiMax Mesh Network. In *2007 IEEE Wireless Communications and Networks Conference* (pp. 1736-1741).

Ghosh, D., Gupta, A., & Mohapatra, P. (2007). Admission Control and Interference-Aware Scheduling in Multi-hop WiMAX Networks. In *IEEE Mobile Ad-hoc and Sensor Systems (MASS)* (pp. 1-9).

Han, B., Tso, F. P., Lin, L., & Jia, W. (2006). Performance Evaluation of Scheduling in IEEE 802.16 Based Wireless Mesh Networks. In *2006 IEEE Mobile Ad-hoc and Sensor Systems (MASS)* (pp. 789-794).

http://home.comcast.net/ bretm/hash/ (n.d.).

IEEE 802.16-2004. (2004, June). *IEEE Standard for Local and metropolitan area networks - Part 16: Air Interface for Fixed Broadband Wireless Access Systems*.

Kong, P.-Y., Pathmasuntharam, J. S., Wang, H.-G., Ge, Y., Ang, C.-W., Su, W., et al. (2009b). A Routing Protocol for WiMAX Based Maritime Wireless Mesh Networks. In *IEEE 69th Vehicular Technology Conference VTC 2009 Spring*.

Kong, P.-Y., Wang, H.-G., Ge, Y., Ang, C.-W., Su, W., Pathmasuntharam, J. S., et al. (2008b). A Performance Comparison of Routing Protocols for Maritime Wireless Mesh Networks. In *2008 IEEE Wireless Communications and Networking Conference*.

Kong, P.-Y., Wang Wang, H.-G., Ge, Y., Ang, C.-W., Pathmasuntharam, J. S., Su, W., et al. (2009a). Distributed Adaptive Time Slot Allocation for WiMAX Based Maritime Wireless Mesh Networks. In *IEEE Wireless Communications & Networking Conference (WCNC)*.

Kong, P.-Y., Zhou, M.-T., & Pathmasuntharam, J. S. (2008a). A Routing Approach for Inter-Ship Communications in Wireless Multi-Hop Networks. In *the 8th International Conference on Intelligent Transport System Telecommunications* (ITST 2008)

Loscri, V. (2007). A New Distributed Scheduling Scheme for Wireless Mesh Networks. In *IEEE International Symposium on Personal, Indoor and Mobile Radio Communications* (pp. 1-5).

Makarevitch, B. (2006a). Distributed Scheduling for WiMAC Mesh Networks. *IEEE International Symposium on Personal, Indoor and Mobile Radio Communications*

Makarevitch, B., (2006b). Jamming Resistant Architecture for WiMAX Mesh Networks. *IEEE MILCOM*, 1-6.

Pathmasuntharam, J. S., Kong, P.-Y., Joe, J., Ge, Y., Zhou, M.-T., & Miura, R. (2007). High speed maritime ship-to-ship/shore mesh networks. In *the 7th International Conference on ITS Telecommunications, (ITST2007)*, Sophia Antipolis, France (pp. 460-465).

Wei, H. Y., Ganguly, S., Izmailov, R., & Hass, Z. J. (2005). Interference-Aware IEEE 802.16 WiMax Mesh Networks. In *IEEE Vehicular Technology Conference 2005 Spring (VTC-2005)* (pp. 3102-3106).

Zhou, M.-T., Harada, H., Kong, P.-Y., Ang, C.-W., Ge, Y., & Pathmasuntharam, J. S. (2009). Multichannel Transmission with Efficient Delivery of Routing Information in Maritime WiMAX Mesh Networks. In *The 5th International Wireless Communications and Mobile Computing Conference*.

Chapter 15
Advanced Scheduling Schemes in 4G Systems

Arijit Ukil
Tata Consultancy Services Ltd., India

ABSTRACT

The deterministic factor for 4G wireless technologies is to successfully deliver high value services such as voice, video, real-time data with well defined Quality of Service (QoS), which has strict prerequisite of throughput, delay, latency and jitter. This requirement should be achieved with minimum use of limited shared resources. This constraint leads to the development and implementation of scheduling policy which along with adaptive physical layer design completely exploit the frequency, temporal and spatial dimensions of the resource space of multi-user system to achieve the best system-level performance. The basic goal for scheduling is to allocate the users with the network resources in a channel aware way primarily as a function of time and frequency to satisfy individual user's service request delivery (QoS guarantee) and overall system performance optimization. Advanced scheduling schemes consider cross-layer optimization principle, where to fully optimize wireless broadband networks; both the challenges from the physical medium and the QoS-demands from the applications are to be taken into account. Cross-layer optimization needs to be accomplished by the design philosophy of jointly optimizing the physical, media access control, and link layer, while leveraging the standard IP network architecture. Cross-layer design approaches are critical for efficient utilization of the scarce radio resources with QoS provisioning in 4G wireless networks and beyond. The scheduler, in a sense, becomes the focal point for achieving any cross-layer optimization, given that the system design allows for this. The scheduler uses information from the physical layer up to the application layer to make decisions and perform optimization. This is a fundamental advantage over a system where the intelligence is distributed throughout the all entities of the network. In this chapter, the authors present an overview of the basic scheduling schemes as well as investigate advanced scheduling schemes particularly in OFDMA and packet scheduling schemes

DOI: 10.4018/978-1-61520-674-2.ch015

in all-IP based 4G systems. Game theoretic approach of distributed scheduling, which is of particular importance in wireless ad hoc networks, will also be discussed. 4G wireless networks are mostly MIMO based which introduces another degree of freedom for optimization, i.e. spatial dimension, for which scheduling in MIMO systems is very much complicated and computation intensive. MIMO resource allocation and scheduling is also covered in this chapter. The key research challenges in 4G wireless networks like LTE, WiMAX and the future research direction for scheduling problems in 4G networks are also presented in this chapter.

INTRODUCTION

Next-generation wireless communication systems are expected to provide a wide range of services with high as well as time-varying bandwidth requirements, with various QoS constraints. Rapid growth of wireless technology, coupled with the explosive growth of the Internet, has increased the demand for wireless data services. Traffic on 4G networks like LTE, WiMAX is heterogeneous with random mix of real and non-real time traffic and applications requiring widely varying and diverse QoS guarantee. This enforces a robust and application specific optimization of limited system resources. The allocation and management of resources are crucial for 4G wireless networks, in which the scarce wireless spectral resources are shared by multiple users with the objective of satisfying demanding requirements. The requirement of providing end-to-end QoS with scarce resources calls for high spectral efficiency. To fulfill these two requirements of high spectral efficiency and QoS provision in the highly dynamic environment of mobile radio requires the collaboration of several layers in the system and effectively demands for an optimization scheme which is cross-layer adaptive. In wireless networks particular, the different layers interact in a nontrivial manner in order to support information transfer. In cross-layer design of wireless networks, a number of physical and access layer parameters are jointly controlled in synergy with higher layer functions like resource allocation, admission control and routing. In layered networking architecture, each layer is designed and operated independently to support transparency between layers. Among these layers, the physical layer is in charge of raw-bit transmission, and the medium access control (MAC) layer controls multiuser access to the shared resources. However, wireless channels suffer from time-varying multipath fading; moreover, the statistical channel characteristics of different users are different. The sub-optimality and inflexibility of this architecture result in inefficient resource use in wireless networks. We need an integrated adaptive design across different layers. Therefore, cross-layered design and optimization across the physical (PHY) and MAC layers are desired for wireless resource allocation and scheduling. In short, to achieve the prerequisite service guarantees like high minimum data rate, low latency, user fairness of next generation wireless networks, proper designing of cross-layer optimized system is very important. In a packet network, one important component to achieve the aforementioned efficiency goals is a properly designed scheduling and resource allocation algorithm. Scheduling plays an important role in providing QoS support to multimedia communications in various kinds of wireless networks, including cellular networks, mobile ad hoc networks, and wireless sensor networks. Scheduling is basically a kind of cross layer optimization method mainly involving PHY and MAC to manage the system resources adaptively to achieve the system goal. If we choose PHY and MAC layers to optimize the network resources, the best way to meet the objective is by exploiting the frequency and temporal dimension of the resource space. Scheduling optimization approaches attempt to dynamically match the requirements of data-link connections to the available physical layer resources to maximize some system metric.

Orthogonal Frequency Division Multiple Access (OFDMA) is widely considered one of the most promising multiple access solutions for today's high speed wireless networks. OFDMA is the de facto multiple access scheme for 4G wireless networks like LTE, WiMAX. OFDMA is based on OFDM (Bahai & Saltzberg, 2002), which is characterized by its ability to fight inter symbol interference (ISI) and frequency-selective fading. The frequency band is divided into a number of orthogonal subcarriers, which can be allocated to different users to provide flexible multiple access schemes. In OFDMA it is possible to efficiently exploit multiuser diversity: the subcarriers which are in deep fade for one user can be good for other users. Spectral efficiency of an OFDMA system can be greatly improved by an adaptive subcarrier allocation algorithm which takes channel quality indicator (CQI) into account. There is plenty of room to exploit the high degrees of flexibility of radio resource management in the context of OFDMA. Since channel frequency responses are different at different frequencies and for different users, data rate adaptation over each subcarrier, dynamic sub-carrier assignment, and adaptive power allocation can significantly improve the performance of OFDMA networks as the bit error probability of different OFDM subcarriers transmitted in time dispersive channels depends on the frequency domain channel transfer function. The occurrence of bit errors is normally concentrated in a set of several faded subcarriers, while in the other subcarriers less bit errors are observed. This opens the scope of maximizing multiuser diversity gain. In OFDMA, scheduler has to dynamically adjust the allocation of subcarriers to users based on their demands and link conditions to meet the intended performance and maintain user level QoS. In this chapter we will present the scheduling issue in OFDMA systems and discuss the different proposed scheduling algorithms.

Internet protocol (IP) is the traditional network-layer protocol for wireline packet networks, and is also considered the natural candidate in wireless systems. IP provides a globally successful open infrastructure for creating and providing services and applications. All-IP could make wireless networks more robust, scalable, and cost effective. In fact, next generation networks are invariably all-IP. IP is a connectionless datagram service. Its scheduling behaviour to support various QoS requirement will not provide the optimized performance, without which sustaining the 4G requirements of high data rate, low latency will not be possible. Scheduling will play a big role in order to provide alternative to conventional First Come First Serve (FCFS) or First in First out (FIFO) and round robin schemes for better and optimal control of network packets. Real-time traffic is characterized by low delay with reasonable packet drop, which demands for additional responsibility by the scheduler to minimize buffer overflow as well as providing minimum latency. This is achieved by incorporating another degree of freedom, i.e. queue management. Performance of all IP-based 4G wireless networks very much dependent on the packet scheduling algorithms implemented in the network layer. Here, we will discuss various packet scheduling algorithms and schemes outlined in the literature and make a comparative study to find out the packet scheduling algorithms and schemes suitable for wireless applications particularly with the objective of guaranteeing QoS by minimizing the packet error rate and probability of buffer overflow. The scenario assumed in the following is that of a packet scheduling algorithm operating on a computational node which is part of a sufficiently large network (e.g. WiMAX, LTE) and receives or generates packets which are forwarded to different destinations. These packets are buffered in one or more queues inside the network stack. We have to keep in mind that OFDMA scheduling and resource allocation is MAC initiated, where as packet scheduling is network layer initiated.

The next generation wireless communication systems will provide a wide range of multimedia services with real time applications while guaranteeing the required quality of service (QoS) of mobile users. Among all of the technical issues that need to be solved in a multimedia network, packet scheduling could be one of the most important issues, since packetized transmission over wireless links makes it possible to achieve a high statistical multiplexing gain and to provide strict delay guarantee. Scheduler design for delay sensitive applications like video conferencing, real time multimedia streaming is a challenging task, where apart from issues like resource constraints, hostile wireless channel, another degree of freedom is included, which is delay or latency. These parameters are very much difficult to control in unpredictable wireless channel. Even in famous Shannon-Hartley theorem from information theory, infinite delay is considered to reliably transmit error free data. In fact Shannon capacity of a channel provides the limit on the achievable rate. The channel capacity itself is not dependent on any delay considerations and is achievable in an asymptotic sense as delay tends to infinity. This notion enforces delay dependent scheduling to combine information-theoretic limits with QoS issues to maintain throughput and delay guarantee. Here, we have highlighted these issues and describe delay sensitive scheduling with the objective of maintaining the QoS related delay for the delay sensitive applications to retain Quality of Experience (QoE) and minimize the outage probability.

Ad hoc networks have occupied a paramount place in the wireless communications and are included in IEEE 802.16m. An ad hoc network is a self-configuring, multi-hop distributed network in which there is no central authority. Thus, every aspect of the configuration and operation of an ad hoc network must be completely distributed. Distributed adaptive behavior in wireless networks leads to recursive attribute, where the scheduling decisions of one user will subsequently influence the decisions of other users in a constructive or destructive way. Game-theoretic approach can provide effective framework development tool and deployment strategy to exploit the advantage of distributed implementation using only the local information, which provides more stability, fairness and even more overall throughput for the entire network by co-operative or non co-operative scheduling decisions. Game theory deals with multi-person decision making, in which each decision maker tries to maximize his utility. Game theory originates from economics, but it has been applied in various fields. In game theory, a solution of a game is a set of the possible outcomes. A game describes what actions the players can take and what the consequences of the actions are. The solution of a game is a description of outcomes that may emerge in the game if the players act rationally and intelligently. Generally, a solution is an outcome from which no player wants to deviate unilaterally. Game theory techniques are required to find solutions of many important design issues in wireless networks like scheduling, resource allocation, call admission control, power management, routing. In a multiuser network like 4G wireless (WiMAX, LTE), services are provided to the users with the assumption that each user is rational and has transparent idea of the overall objective function of the system. This requires game formulation to reach stable decision for finding equilibrium condition. Even in intelligent relay networks, distributed scheduling plays an important role in order to maximize the overall system capacity and enhance the network coverage area.

In order to achieve the high spectral efficiency and transmission rate, Multiple Input Multiple Output (MIMO) is an integral part of 4G wireless networks like LTE, WiMAX. For example, 2×2 and 4×2 MIMO configuration are selected in WiMAX standard. MIMO has the advantage of spatial multiplexing, array gain and spatial diversity gain. On the other hand, to effectively exploit the spatial multiplexing or spatial diversity of MIMO systems, the spacing of the antenna array must be larger than coherence distance of the channels. MIMO scheduler has the responsibility of maximize these advantages so that overall system performance can be enhanced as far as possible. MIMO scheduling has the additional degree of freedom of spatial dimension. In this chapter we will highlight the issues of MIMO scheduling mainly from the perspective of maximizing spatial diversity and multiplexing.

So, we observe that scheduling is responsible to provide mechanisms for bandwidth allocation, packet level multiplexing, queue management, distributed resource control and spatial allocation. In the next sections we discuss these issues in more details. As 4G wireless networks will mostly based on OFDMA, our discussion will refer OFDMA based system model. Section 2 describes the general system model, scenario and the broad level problem formulation with a description of cross-layer framework for advanced scheduler development. Section 3 describes the different basic scheduling algorithms from the perspective of different optimization schemes with system specific objectives like maximizing the instantaneous or long-term (time-average) overall system throughput, providing fairness among users etc in heterogeneous QoS constraint-bound conditions like minimum data rate of each user, maximum delay tolerance, elasticity or non-elasticity of the traffic etc. This section deals with the issue of dynamic scheduling which utilizes the time-varying channel conditions to the maximum extent to improve spectral efficiency by achieving multiuser diversity gain. The next section introduces packet scheduling concept, as the 4G networks are mainly all-IP. Packet scheduling is queuing analysis dependent. M/M/1 queue is to be taken into consideration for modeling and derivation purpose. A small introduction of traffic shaping is also presented. In section 5 game theoretic framework for wireless network scheduling in non co-operative and co-operative scenarios is studied. In non co-operative game, we will investigate the scenario where each user competes for the shared network resources to maximize individual throughput to achieve Nash Equilibria, where in co-operative scheduling, users bargain for network resources in a coalition mode to meet Pareto optimality and Nash Bargaining Solution Fairness will be discussed. Real-time traffic (e.g. voice, video streaming) is very much delay sensitive, can tolerate a certain level of packet loss, but computational time for scheduling and resource allocation algorithm implementation is very less. The services provided by 4G wireless networks will support a mixture of real-time streams (e.g., video/voice and multimedia) and best effort data transfers (like file downloads or web browsing). Section 6 discusses the issue of scheduling challenges for real-time traffic. This section comprises of system modeling, optimization and various algorithms and schemes to determine optimal scheduling decisions in a packet-centric real-time traffic (hard and soft QoS) oriented 4G wireless networks, which should ensure QoS to a real-time session as well as minimize transfer delays associated with best effort sessions with an overall system level optimization. In section 7, we will introduce MIMO scheduling and draw attention to the issues of MIMO scheduler optimization. A case study of practical importance regarding scheduling in LTE and WiMAX networks is given in section 8. We also highlight the future scope of research in 4G scheduling on which less works have been done but are highly important topics for further investigation.

ADVANCED SCHEDULING SCHEMES IN 4G SYSTEMS

General System Model and Cross-Layer Framework

Scheduling is concerned with the allocation of limited resources to tasks over time. It is a decision-making process that has as a goal to optimize one or more objectives (McCormick & Pinedo, 1995). For wireless transmissions, a scheduling policy is a rule or algorithm that specifies which user is to be allocated at each resource bin. Scheduling mostly corresponds to subcarrier and time-slot to user mapping in OFDMA, spreading code and rate adjustment in spread spectrum based CDMA and time-slot to user mapping in TDMA systems. Scheduling schemes for next generation wireless networks is to take the advantages of multi user diversity and channel diversity. To satisfy these requirements, scheduling scheme is to be dynamic and channel condition dependent. Apart from these, fairness and QoS constraint issues are to be considered in scheduler design. Here, we discuss generic system model for scheduler in OFDMA systems, packet-based IP networks and in distributed networks as well as cross-layer framework for scheduler design, which is of utmost important in 4G networks (Todini, 2006). As cross-layer design and optimization are involved in almost all the models we discuss, so we introduce cross-layer design framework first and other scheduler models like OFDMA, packet will follow.

Cross-Layer Based Scheduling Model

System capability is highly dependent on the proper utilization of its resources. In wireless systems, efficiency is highly dependent upon its physical resources like time, frequency and energy. This mandates the upper layer in network stack to incorporate the effect of physical layer and vice-versa, collapsing the traditional layered architecture. In fact, better system performance can be obtained from information exchanges across protocol layers through cross-layer design, which may not be available in the traditional layered architecture. This cross-layer viewpoint has potential to increase the system efficiency tremendously. Conceptual framework of MAC-centric cross-layer optimized resource allocation is shown in figure 1. Cross-layer design in general can be classified in four categories based on the order in which optimization is performed:

1. Top-down approach: Higher-layer protocols optimize their parameters and the strategies at the next lower layer. This cross-layer solution has been deployed in most existing systems, wherein the application dictates the MAC parameters and strategies, while the MAC selects the optimal PHY layer modulation and coding scheme.
2. Bottom-up approach: The lower layers try to insulate the higher layers from losses and bandwidth variations. This cross-layer solution is not optimal for multimedia transmission, due to the incurred delays and unnecessary throughput reductions.
3. MAC-centric approach: Here, the application layer passes its traffic information and QoS requirements to the MAC that decides which application layer packets, should be transmitted.
4. Integrated approach: In this approach, strategies are determined jointly by exhaustively trying all the possible strategies and their parameters in order to choose the composite strategy leading to the best quality performance.

Figure 1. Conceptual framework of cross-layer optimized resource allocation

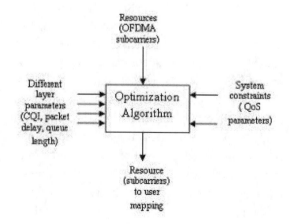

Advanced Scheduling Schemes in 4G Systems

We shall consider MAC-centric optimization, which is relatively simpler, easy to implement and most suitable for our problem of scheduling scheme development. The MAC-initiated model is based on the information like minimum data rate for the user, maximum packet-delay, CQI from higher and lower layers. Based on these parameters, MAC takes the scheduling decision with the help of an in-built algorithm. Generic cross-layer model of a conventional layered architecture based wireless system is shown in figure 2.

OFDMA Based Scheduling System Model

Scheduler in an OFDMA system is responsible for subcarrier to user mapping at every time-slot. The considered OFDMA system consists of single cell with one base station (BS) communicating simultaneously with K user terminals using N number of OFDMA subcarriers, which is shown in figure 3. The interference from adjacent cells is treated as background noise. Total noise power density including background noise and AWGN noise is taken as N_T. The mutually disjoint OFDMA subcarriers are denoted as: $\Omega_1, \Omega_2, ...\Omega_N$, where Ω_n = B/N and $\Omega_n \leq$ Bc, where Bc is the coher-

Figure 2. Generic cross-layer model in wireless systems

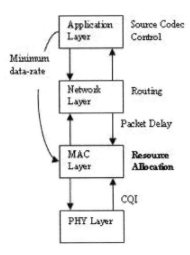

ence bandwidth of the channel and B is the total available bandwidth. P_T is the total available transmit power and P_{kn} is the transmit power for n^{th} subcarrier when transmitted to k^{th} user. If not otherwise equal power distribution among the sub-carriers is to be considered, i.e. $P_{kn} = P_T/N$, as performance can hardly be deteriorated by equal power distribution (Ergen, Coleri & Varaiya, 2003). Perfect channel characteristic is assumed in the form of CQI. Channel Gain for subcarrier n for user k is taken as h_{kn}, which is estimated from CQI information. Let ω_k be the achievable rate for k^{th} user

$$\omega_k = \sum_{n=1}^{N} \Omega_n \times \rho_{kn} \times \left(\log_2 \left(1 + \frac{h_{kn}^2 \times P_{kn} \times \Delta_{gap}}{N_T \times \Omega_n} \right) \right) \quad (1)$$

where ρ_{kn} is the sub-carrier assignment matrix, which is equal to 1 if n^{th} subcarrier assigned to k^{th} user, else equal to 0. Δ_{gap} is the imperfection of theoretical value of achievable data rate to the actual data rate, called as SNR gap, can be approximated as $\Delta_{gap} = \dfrac{-\ln(5BER)}{1.6}$ (Falahati & Ardestani, 2007).

The scheduler as shown in figure 4 performs the mapping subcarrier to user primarily based on QoS requirements and CQI information. OFDMA scheduler only considers subcarrier to user mapping and does not generally consider buffer state management, which is the part of packet scheduler.

Packet Scheduler System Model

Packet scheduling is to schedule the packets from the queues in such a way that the limited shared bandwidth and time-slots are fairly and optimally distributed among the competing classes of incoming traffic flows. Typically, packet scheduling is not standardized and is implementation dependent. Packet scheduling is an important part of

Figure 3. Generic downlink mobile wireless systems

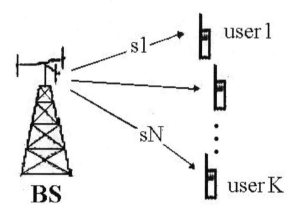

overall QoS mechanism and provides a measure of technology differentiation among manufacturers' products. In figure 5, typical packet scheduler system model is shown. In packet-switched computer networks the notion of scheduling algorithm and statistical multiplexing are used as an alternative to FCFS queuing of data packets. Packet scheduler consists of packet classification and queue generation engine as shown in figure 5. Both of them are governed by system specific optimization algorithm. If simple FCFS or FIFO scheduling is used, packet classifier is usually not required. In figure 6, a more detailed schematic of wireless packet scheduler along with other transmit-receive blocks like Forward Error Control (FEC), modulation, channel parameter estimator is shown. Here, packet scheduler is shown as a network layer-centric operation and simple FCFS algorithm is implemented. Adaptive modulation and coding (AMC) module has the responsibility of proper coding and modulation of the incoming packets according to the predicted channel condition estimated from the control channel. It can be observed from figure 6 that packet scheduler is an important part of higher MAC or network layer, particularly for output queue generation in the transmitter.

BASIC SCHEDULING ALGORITHMS IN 4G NETWORKS

Introduction

One of main challenges for 4G networks like IEEE 802.16, LTE is to simultaneously provide QoS for various services that have very different QoS characteristics due to the following reasons. First, different services have different QoS re-

Figure 4. OFDMA scheduler system model

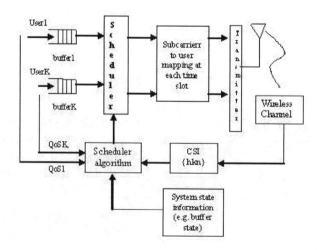

Advanced Scheduling Schemes in 4G Systems

Figure 5. Generic packet scheduler

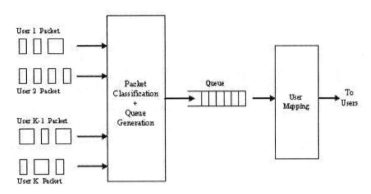

Figure 6. Detailed wireless packet scheduling system model

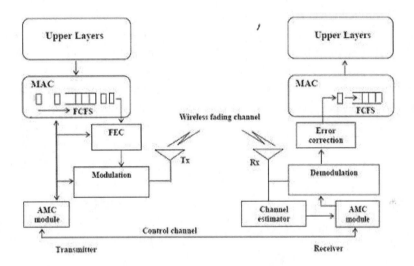

quirements, such as minimum guaranteed rate, latency, blocking probability, maximum tolerant bit error rate (BER), etc. Next, the traffic of different services exhibits a wide range of diverse characteristics, such as various packet sizes, different inter-packet arrive times, etc. Thereafter, enabling QoS in wireless networks is much more difficult than in wired networks because the characteristics of a wireless link are highly variable and unpredictable, on both a time-dependent basis and a spatial-dependent basis. To cope with these issues, the QoS in wireless networks is usually managed at the MAC layer and partly in network layer. In fact, radio resource allocation is a MAC-initiated process, where as packet scheduling is traditionally network-layer driven process. Here, our main focus is cross-layer design as well as cross-layer optimization, which enables us to integrate these two processes under one common framework. So, we will not distinguish between these two and concentrate on integrated resource allocation and scheduling system. So, by scheduling, we will mean both radio resource allocation and packet scheduling, unless otherwise stated.

Wireless scheduling techniques have emerged as versions of wireline scheduling schemes tailored to cope with wireless link variability (Bhagwat, Bhattacharya, Krishna & Tripathi, 1996; Desilva & Das, 1998). Apart from path loss, wireless channel varies with high dynamic range by shadowing and fading effects (Sklar, 1997). In multi-user scenario where the scarce wireless spectral resources are shared by multiple users, allocation and management of resources are crucial for wireless networks. In traditional wireless networks, system is designed to accommodate the worst case situation, which inherently wastes lots of system resources most of the time. The challenge for dynamic resource allocation and scheduling is to find the feasible and optimal scheduling point dynamically in simple and practical way. Basically we have to formulate the resource allocation and scheduling as a constrained optimization problem. We represent optimization goal as Σ. The most important thing is to define the goal function that can represent the real network performances with practical constraints. We define the constraints sets as $\emptyset$ and the parameter set as Θ. The typical constraints are power, data-rate, l distortion, delay, outage probability. So the overall fundamental problem can be formulated as:

max Σ

Θ

such that: $\emptyset$

Considering the dynamic resource allocation and scheduling in OFDMA, we have to find the available resources, system constraints, parameters and scheduling objectives or goals. The resources available are mainly transmitter power, bandwidth or sub-carriers, transmission time or timeslots. These resources are to be optimally distributed among the users to meet individual QoS requirement and also to maximize overall system throughput. Scheduling system works with the assumption that the transmitters (BS, in the case of downlink communications) have knowledge of the instantaneous channel conditions, i.e., perfect CQI, of all users in the system. The assumption of perfect CQI is not always very realistic, especially when the channel conditions change quickly with time, as in a mobile radio network. In some cases, partial channel knowledge may be used for scheduling decision using uncertainty modeling like ergodic model or quasi static model (Yao, Giannakis, 2005). So, scheduling can also be performed based on statistical knowledge of the channel conditions. But this assumption of perfect CQI makes the problem very much tractable and implementable, otherwise the problem becomes highly complex and computationally expensive. By dynamically adapting the scheduling to the current network state we can benefit from the following beneficial effects: multiuser gain, multiplexing gain, diversity gain. Diversity techniques using the time, space, or frequency dimensions are used extensively in wireless communication to increase the outage capacity (Biglieri, Proakis & Shamai, 1998) of a fading channel, by minimizing the probability of deep fades as well as maximizing the probability of mean or instantaneous best channel condition. These traditional diversity methods are essentially applicable to a single-user link. In a wireless network with multiple users sharing a time-varying channel, multiuser diversity increases the channel capacity by maximizing the total Shannon capacity by allowing at any time-slot only the user with the best channel to transmit (Grossglauser & Tse, 2001). This kind of scheduling incurs fairness issue (Rhee & Cioffi, 2000) as the users with poor channel condition may not get allocated with the resources at all and get starved, which is not very much acceptable from a network provider point of view. There are different objective functions and system goals which should be considered to properly design a scheduling system. The basic scheduling schemes to be discussed are:

1. Raw-rate maximization
2. Max-Min Fairness
3. Proportional rate constraint scheduling
4. Margin Adaptive Scheduling
5. Proportional Fair
6. Long Term Proportional Fair

1. Raw-Rate Maximization

In this type of scheduling, the resources are scheduled to the users who can achieve the best performance, by opportunistically selecting the user with the best instantaneous channel condition. The drawback of this algorithm is that it is likely that a few users close to the base station, and hence having excellent channels, will be allocated all the system resources. In a wireless mobile network, channel condition varies with large dynamic range for different users. This scheme exploits multiuser diversity to the maximum, but lacks fairness. As some of the users with continuous bad channel quality may never get a chance to use the resources. Objective of the raw rate maximization algorithm is to maximize the sum rate of all users, given a total transmit power constraint (Zhang & Letaief, 2004). This algorithm is optimal if the goal is to get as much data as possible through the system. Let **W** be the total resource space (frequency, time-slot or subcarrier), where **W** = $\{\omega_1, \omega_2, ..., \omega_K\}$ and transmitted power to each MS is **P** = $\{P_1, P_2, ..., P_K\}$. This scheme can be mathematically modeled as:

$$\max \sum_{k=1}^{K} \omega_k \qquad (2)$$

such that,

$$\sum_{k=1}^{K} P_k \leq P_T \qquad (3)$$

Zhang and Letaief (2004) has discussed the raw rate maximization in multiuser OFDM system, where they have developed subcarrier and bit loading scheme to maximize the spectral efficiency. It is unlikely that subcarriers which appear in deep fade to one user may not be in deep fade for other users. In fact, it is also improbable that a subcarrier will be in deep fade for all users, as the fading parameters for different users are mutually independent. This motivates (Zhang & Letaief, 2004) to consider an adaptive multiuser subcarrier allocation scheme where the subcarriers are assigned to the users based on instantaneous channel information by "greedy" optimization. Jang and Lee (2003) investigated the problem with the objective of maximizing the total data rate over all users subject to power and Bit Error Rate (BER) constraints. They have shown that in order to maximize the total capacity, each sub-carrier should be allocated to the user with the best gain on it, and the power should be allocated using the water-filling algorithm across the subcarriers. The objective function (2) is an NP-hard combinatorial optimization problem with non-linear constraints. Hence, it is highly improbable that polynomial time algorithms can be used to solve it optimally.

2. Max-Min Fairness

In raw rate maximization, total system throughput is maximized. In a cellular system such as WiMAX, LTE in which the channel condition varies with high dynamic range between users, some users will be extremely underserved by maximum sum rate based scheduling. This raises fairness issue. Another extreme of raw rate maximization is maximum fairness, which allocates the system resources to the user with most poor channel condition, most often the farthest user from the transmitter or base station. The maximum fairness algorithm is also known as Max-Min problem, as the aim is to maximize the minimum data rate and delivers total fairness among the users. A scheduling scheme is max-min fair if without hurting users with lower service rate; no other user gets benefited by increasing its allocation. This kind of optimization is typically very much inefficient, as multi-user diversity gain is not utilized at all. Mathematically, this can be modeled by Rhee and Cioffi (2000), where total transmitter power is taken as optimization constraint.

$$\min \sum_{k=1}^{K} \omega_k \quad (4)$$

$$\sum_{k=1}^{K} P_k \leq P_T$$

Max-min fairness is more difficult to obtain and computationally more expensive than raw rate maximization because the objective function (4) is not concave. Therefore, for real-time as well as practical systems, low-complexity suboptimal algorithms are necessary. A common approach is to assume initially that equal power is allocated to each sub-carrier and then to iteratively assign each available subcarrier to a low-rate user with the best channel on it (Wong, Shen, Evans & Andrews, 2004). Ermolova and Makarevitch (2007) have proposed a heuristic non-iterative method to improve the average BER performance of the overall system and it also provides fairness between the users by a low complexity algorithm. User satisfaction is typically a concave function, satisfaction increases almost exponentially with increase in data rate for low-bandwidth region, where as, for high bandwidth region, insignificantly user satisfaction increases with increase of data rate. The fairness policy of max-min scheduling is to distribute resources as evenly as possible with least over provisioning of resources. (Tassiulas & Sarkar, 2002) proposes max-min fair scheduling for ad hoc networks.

3. Proportional Rate Constraint Scheduling

In wireless broadband network different users applications require different data rate. Proportional rate constraint scheduling is a generalized fairness algorithm, where the objective of maximizing the sum rate and the fairness is based on user's long-term target data rate. The problem is formulated as:

$$\max \sum_{k=1}^{K} \omega_k$$

$$\sum_{k=1}^{K} P_k \leq P_T$$

$$\sum_{k=1}^{K} \rho_k = 1$$

Subject to:

$$\omega_1 : \omega_2 : \ldots : \omega_K = \phi_1 : \phi_2 : \ldots \phi_K$$

$$\sum_{k=1}^{K} \phi_k = 1$$

φ_k is a set of predetermined values which are used to ensure proportional data rate among various users. ρ_k is resource assignment matrix, $\rho_k = 1$, if resource ω_k is assigned to user k, else $\rho_k = 0$. This also ensures orthogonally of resource assignment. The optimization problem stated above is very difficult to solve as it involves binary variable ρ_k, continuous variable P_k and one inequality constraint. The nonlinear constraints make the feasible set non convex and also the mixed involvement of continuous and binary variables makes the optimization as mixed integer problem.

In OFDMA system, there are N sub-carriers to be allocated to K users. Proportional rate constraint optimization for OFDMA systems can be formulated as:

$$\max \sum_{n=1}^{N} \sum_{k=1}^{K} \omega_{kn}$$

$$\sum_{n=1}^{N} \sum_{k=1}^{K} P_{kn} \leq P_T$$

$$\sum_{n=1}^{N} \sum_{k=1}^{K} \rho_{nk} = N$$

The above problem is known as maximal bipartite matching problem where users and OFDMA sub-carriers form two disjoint sets of vertices of a graph with the objective to find a mapping between subcarriers and users such that only one edge connects two graph nodes and the total rate is maximized for all selected pair of user and sub-carrier. To cover K^N possible sub-carrier allocations for K users and N subcarriers in order to find global maximum is computationally very complex. More practical approach is to decouple the subcarrier and power allocation. By doing so, sub-optimal solution with low complexity can be formulated (Shen, Andrews, & Evans, 2005).

4. Margin Adaptive Scheduling

In the uplink of cellular networks, user terminals are limited by total available power, which is by far the scarcest resource in mobile handsets. So, the most of the scheduling optimization for uplink, devotes for providing solution of minimization of total transmit power. The objective of this kind of optimization approach is to achieve the minimum overall transmit power given the constraints on the users' data rate or BER. In downlink, power minimization is sometimes also an important optimization problem, particularly for relay stations. In (Kivanc, Li & Liu, 2003), problem of minimizing total power consumption with constraints on BER and transmission rate in OFDMA systems is presented. The proposed algorithm is shown to find the distribution of sub-carriers that minimizes the total power required when every user experiences a flat-fading channel. The objective is to find a subcarrier allocation which allows each user to satisfy its rate requirements while using minimum power. This can be modeled as:

$$\min \sum_{k=1}^{K} \sum_{n=1}^{N} P_{kn} \leq P_T$$

subject to:

$$\omega_k \geq \gamma_k, \ \forall k$$

The problem modeled above is computationally intractable, and a direct approach to solving this joint optimization problem does not yield a good algorithm. So, they (Kivanc, Li & Liu, 2003) proposed a computationally efficient 'sensible greedy' algorithm. Lagrangian multiplier based convex minimization solution of the same problem is proposed in (Wong, Cheng, Letaief & Murch, 1999). The above problem of sub-optimally, which results in relatively inefficient performance with some degree of complexity is dealt intelligently using linear programming method by considering subcarrier allocation and bit loading optimization as separate decoupled problem in (Kim, Park & Lee 2006).

5. Proportional Fair Scheduling

Designing a scheduler without a good fairness criterion, the system performance can be trivially optimized, but some users might be prevented from accessing the network resource as the advantages of multiuser diversity only contribute to a small portion of mobile users whose channel quality is good, which may not lead to a significant QoS performance improvement from entire network perspective. Proportional fair (PF) is considered to be a simple yet effective fairness notion. At the allocation instant t, suppose that user k has an average realized throughput $\overline{\omega_k(\Delta\tau)}$ over a window of length $\Delta\tau$. A scheduler is said to achieve proportional fairness if it selects for transmission the user satisfying the following criteria:

$$\max \left(\frac{\omega_k(t)}{\omega_k(\Delta\tau)} \right)$$

PF has the property that it cannot be replaced by any other arbitrary allocation that does not lead to a reduction in the aggregate fractional rate change. It should be noted that a PF scheduler heuristically tries to balance the fairness among the users in terms of outcome or throughput, while implicitly maximizing the system throughput in a greedy manner. PF optimization is a pure outcome fairness metric, which is simple to use, but does not guarantee fairness in a strict sense. For example, consider the situation where a session has experienced a prolonged period of poor channel condition and hence has a small value of $\omega_k(\Delta\tau)$. It may not get the desired allocation even though its channel condition improves (e.g., with a moderately large value of $\omega_k(t)$), if there is a "dominant" user that has a very good channel state. PF utility is defined as a logarithmic function of the rate allocated to a user. Because of the convex nature of the logarithmic function, diminishing return is modeled. According to (Kim & Han, 2005), PF in OFDMA system maximizes the sum of logarithmic mean user data rates:

$$\max \sum_{k=1}^{K} \ln \omega_k$$

Using Karush-Kuhn-Tucker conditions for optimality, (Nguyen, & Han, 2006) proposes an efficient subcarrier and power allocation algorithm based on the proportional fair concept with QoS provisioning. Qos is maintained by guaranteeing each user a minimum data rate, i.e:

$$\omega_k \geq \gamma_k, \forall k,$$

which leads to more generalized definition of PF as:

$$PF = \prod_{k \in K} \left(1 + \frac{\sum_{n \in N} \omega_{kn}}{(\Delta\tau - 1)\overline{\omega_k}} \right)$$

where PF is the proportional fairness index at t^{th} instant, $\Delta\tau$ is average window size or the period between successive allocations and $\overline{\omega_k}$ is the average data rate achieved of the user k at the preceding allocation instant. The generalized objective function is the optimal PF subcarrier allocation, but can not be implemented due to its high computational complexity. The suboptimal form (Wengerter, Ohlhorst & Elbwart, 2005) is where subcarrier n is allocated to k* user when the following condition is satisfied:

$$k^* = \arg\max_{k} \frac{\omega_{knt}}{\overline{\omega_{kt}}}$$

where, $\overline{\omega_{kt}} = (1 - \frac{1}{\Delta\tau})\overline{\omega_{k(t-1)}} + \frac{1}{\Delta\tau}\omega_{k(t-1)} \cdot \omega_{knt}$ denotes that nth OFDMA sub-carrier is assigned to kth user at tth time-instant. In more generalized form, above scheduling optimization can provide biasness between fairness and throughput maximization (Wengerter, Ohlhorst & Elbwart, 2005):

$$k^* = \arg\max_{k} \frac{(\omega_{knt})^\alpha}{(\overline{\omega_{knt}})^\beta}$$

Tuning parameters α,β are the biasing parameters, which decide the amount of trade-off between fairness and system throughput. Increasing α will increase the influence of the achievable instantaneous data-rate, which enhances the probability of a user in currently good condition to be scheduled. This results in higher overall system throughput, but fairness is reduced. Increasing β will increase the influence of the average data-rate, which increases the probability of a user with low average data-rate to be scheduled. This results in higher fairness, but lower spectral efficiency. With $\alpha = \beta = 1$, the optimization represents pure PF.

6. Long-Term Proportional Fairness

The channel dynamics of mobile wireless broadband system is very high. So instantaneous decision making may not be the best way for an optimization scheme. To utilize the channel dynamics in most useful way, particularly for non real-time applications, in (Ukil, 2008), PF optimization is modified as long-term maximization of user's mean achievable data rate subject to minimum data rate constraint. This optimization is termed as long-term proportional fair (LTPF).

Instantaneous decision making, i.e. satisfying optimization objective at every scheduling instant may not be the best way for an optimization scheme. Instead LTPF algorithm proposes long term maximization of user's mean achievable data rate subject to minimum data rate constraint. LTPF incurs the advantage of time diversity (TD) gain, when QoS parameter computation time (Δ_{ad}) is more than the channel coherence time. But TD technique is restricted to delay-tolerant applications in mobile wireless environment. LTPF optimization utilizes this performance gain for its attempt to converge to γ_k by computing PF metric over Δ_{ad} as well as relaxing the QoS constraint of minimum data rate as statistical mean value. Basically, instead of instantaneous optimization, LTPF allocates OFDMA sub-carriers to optimize the performance over few allocations and assures QoS guarantee in an average basis within that pre-defined allocation duration. The idea is formulated below:

$$\max \sum_{t=0}^{\Delta_{ad}} \sum_{n=1}^{N} \sum_{k=1}^{K} \frac{\omega_{knt}}{\overline{\omega_k}\big|_{t=0}^{\Delta_{ad}}}$$

$$\omega_k = \sum_{t=0}^{\Delta_{ad}} \sum_{n=1}^{N} \Omega_n \times \rho_{knt} \times f(h_{knt})$$

$$\sum_{t=0}^{\Delta_{ad}} \sum_{n=1}^{N} \sum_{k=1}^{K} \rho_{knt} = N \times M$$

subject to:

$$\overline{\omega_k} \geq \gamma_k, \forall k$$

where M= number of time-diversity branches.

The simulation results depict the performance of the algorithm and also show how user data rate follows the overall QoS profile at long term. Heterogeneous traffic model with variable QoS demand by the user is considered as in a broadband wireless scenario it is very practical to assume large differences in QoS requirement. Simulation scenario is given in Figure 7. The system parameters are roughly based on Mobile WiMAX Scalable OFDMA-PHY. The proposed LTPF algorithm is evaluated against the simulation scenario described as per Figure 7. Matlab simulation is made to support the claim. The results (plots) are shown below (Figure 8, Figure -9 and Figure 10). Figure 8 is the plot comparing QoS profile of users' achieved mean data-rate when Δ_{ad} is one unit. Here it is clear that the achieved mean data-rate profile deviates from the QoS profile considerably and also the individual data-rate achieved is substantially low, as in this case, LTPF algorithm becomes purely proportional fair. Figures 9-10 show the result when LTPF idea of sub-carrier allocation is incorporated. Figure 9 and Figure 10 clearly shows that the mean data-rate is now considerably improved and tries to follow the QoS profile. Figure 10 establishes the fact that achievable mean data rate is converging towards γ_k, when large value of Δ_{ad} taken.

Scheduling and resource allocation algorithms discussed so far, deals with the mechanism of exploiting channel variations across users to improve the spectral efficiency and QoS guarantee. These algorithms are mainly used to schedule frequency, time-slot or subcarriers and generally implemented in lower MAC (figure 1 and 2) in a cross-layer adaptive way. In the following section, we will present scheduling algorithms and schemes particularly for packet-data.

Figure 7. Simulation and system parameters

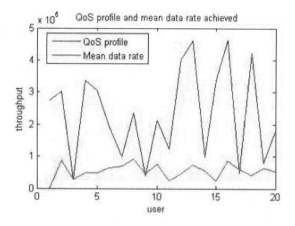

Figure 8. QoS profile and mean data rate achieved when M=1

Available Bandwidth	1.25 MHz
Total Transmitted Power	20 dBm
Number of users	20
Number of sub-carriers	72
BER	10^{-3}
Frame duration	5 msec
Allocation instant (T_f)	Frame duration
Δ_{ad}	$M \times T_f$
Channel model	Rayleigh
Modulation	16QAM
Channel sampling frequency	1.5 MHz
Maximum Doppler	100Hz

PACKET SCHEDULING ALGORITHMS FOR 4G NETWORKS

Introduction

Next generation wireless networks are all-IP. With the huge success of internet and Voice over IP, business and technology challenges drive to enable true all-IP network convergence. IP offers a connectionless datagram service without any guarantee of delivering the data packets and all data is treated with equal priority irrespective of the delay sensitivity of the traffic the datagram is carrying. The design of the current Internet is based on a paradigm that sufficed for simple data applications, the best-effort principle. It means that the network treats every packet equally and does not provide guarantees on the experienced packet losses or packet delays. When there are only data applications, there is indeed no need to differentiate between packets of different applications, or to provide strict loss or delay guarantees. However, various new applications have emerged with different, possibly more stringent requirements in terms of packet losses and delays. Clearly, the best-effort principle does not suffice when these new applications have to be supported over the same network as the existing data applications. 4G wireless networks has to carry heterogeneous traffic consisting of very low or new delay sensitive applications like FTP, email to highly stringent delay sensitive applications like video conferencing, VoIP. With strict QoS and reliable service requirement of 4G wireless systems, it is the job of the scheduler to allocate time-slots to the packet data in order to satisfy some objective function. Scheduling and resource allocation schemes discussed in the previous section are mainly concentrated on assigning frequency, time-slot or subcarriers to the users by using some algorithm. Here, the resource space is time-slots and users' data packets are allocated to these time-slots (fig. 5-6). Discussion of this section concentrates on packet level scheduling in time-slotted 4G wireless networks. In general, for each user separate buffers exist. Buffer size of individual user is mostly pre-assigned and fixed and varies for different users. The system can be characterized by one-dimensional birth and death models. The stochastic behavior of these systems at a particular point in time is completely described by a single number, which we shall think of as the

Figure 9. QoS profile and mean data rate achieved when M =4

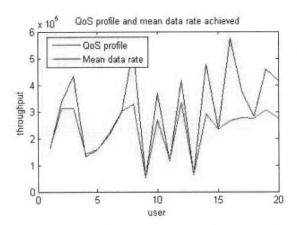

Figure 10. QoS profile and mean data rate achieved when M =10

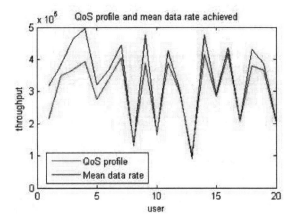

"occupancy" of the system. The dwell times for each state are drawn from exponential distributions independently, but, in general, the parameter of the exponential distribution depends upon the current state of the system. We examine the well known M/M/1 queueing system, which has Poisson arrivals and exponentially distributed service times, where M stands for Markovian. Due to the memory less property of the Poisson process and the exponential distribution, the dynamics of the process that counts the total number of arrivals to and departures from the system over very short periods of time are exactly the same as those of the Poisson process. Usually we assume that the inter-arrival times are independent and have a common distribution and service times are independent and identically distributed (i.i.d), and that they are independent of the inter-arrival times. In fact in many practical situations users' packets arrive according to a Poisson stream (i.e. exponential inter-arrival times). We know that for M/M/1 the probability of exactly n customers arriving during an interval of length t is given by the Poisson law, i.e.

$$P_n(t) = \frac{(\lambda t)^n}{n!} e^{-\lambda t}$$

where,

$P_n(t)$ = number of packets arrived at the input of the receiver in t time duration.

λ = mean arrival rate of packets.

If we define μ, ρ as mean service rate of packets and mean utilization factor by the scheduler, then:

$$\rho = \frac{\lambda}{\mu} \text{ for a stable system } \rho < 1.$$

M/M/1 queue can be modeled as per figure 11, where.

L = average number of packets in the system.
L_q = average number of packets in the queue.
W = average waiting time in the total system.
W_q = average waiting time in the queue.

In steady state:

$$L = \lambda \times W$$

$$L_q = \lambda \times W_q$$

Figure 11. M/M/1 queue model

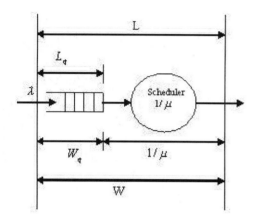

In general,

$$L = \sum_{n=0}^{\infty} n \times P_n$$

$$P_n = \rho^n(1-\rho)$$

This analysis is important to find out system parameter like buffer size in order to design optimum scheduler with queue management module.

For example: Packets arrive at a mean rate 125 packets/sec with 2 msec forwarding time. Find out:

1. Probability of buffer overflow if buffer size = 15.
2. How many buffers are needed to keep packet loss below one packet/million?

Solution:

$\lambda = 125$

$\mu = \dfrac{1}{.002} = 500$

$\rho = \dfrac{\lambda}{\mu} = .25$

Assuming M/M/1 queue,

$P(>15) = (.25)^{15} = 9.31 \times 10^{-10}$

Probability of buffer overflow = 9.31×10^{-10}.

For second problem:

$P_n \leq 10^{-6}$

$(.25)^n \leq 10^{-6}$

$n > \dfrac{\log(10^{-6})}{\log(.25)} = 9.96$

So, buffer size should be ≥ 10, optimum buffer size = 10.

The quantitative analysis of system performance and parameter evaluation leads to an optimum scheduling system design. In advanced packet radio systems like 4G, channel-dependent and constraint (queue length, QoS) optimized scheduling is used to take advantage of favorable channel conditions to make best use of available radio conditions to the deserving user. Below we will discuss some important packet-scheduling algorithms.

1. FIFO
2. Round Robin
3. Weighted Round Robin
4. Deficit Round-Robin
5. Stochastic Fair Queuing
6. Fair Queuing
7. Generalized Processor Sharing
8. Weighted Fair Queuing
9. Worst-case Weighted Fair Queuing
10. Random Early Detection
11. Channel-state Dependent Packet Scheduling
12. Proportional Fair Queuing
13. Capacity Queue Scheduling
14. Shortest Remaining Processing Time First (SRPT)
15. Wireless SRPT

16. Opportunistic Auto Rate (OAR) Scheduling
17. Wireless Credit-based Fair Queuing
18. Traffic Aided Opportunistic Scheduling (TAOS)
19. Scheduling by beam forming

Among these nineteen class of packet scheduling algorithms, 1-10 are channel state independent or non opportunistic scheduler, where as 11-19 are opportunistic scheduler. First we will discuss non opportunistic scheduling and then opportunistic scheduling.

Non Opportunistic Packet Scheduling

Non opportunistic packet scheduling does not consider the channel condition of individual user. So, PHY layer information like CQI is not used. These kinds of schedulers are also known as static schedulers and they are generally implemented in higher MAC.

1. FIFO Scheduling

FIFO is the traditional scheduling scheme mechanism in the absence of any specific packet scheduling algorithm. In FIFO, packets are queued in a single queue in the order they arrive and are sent out on the outgoing link in the same order they are queued. Since the first arriving packet is the first packet to be served, the FIFO queue is also referred to as the FCFS scheduling. The main concern of FIFO scheduler is queue management-coping with excessive burst of packets with finite buffer size. FIFO is in fact a work-conserving scheduling, i.e. it is never idle when packets are waiting for service. Consider K packets arriving at the scheduler with mean arriving rate λ_n and mean service rate μ_k for kth packet. So, mean utilization of kth packet is $\rho_k = \dfrac{\lambda_k}{\mu_k}$. Let W_{qk} be the average waiting time of kth packet in the queue. According to the law of work conservation law, the scheduler is work conserving if:

$$\sum_{k=1}^{K} \rho_k W_{qk} = C$$, where C=constant.

This work conservation in reality is the main drawback of FIFO scheduling. This property leads higher delay to one packet when another packet enjoys lower delay. Another drawback of FIFO as it does not distinguish the incoming traffic, which is very much unacceptable in 4G networks. For example WiMAX consists of five service or QoS classes: unsolicited grant service (UGS), real-time polling service (rtPS), extended real-time polling service (ertPS), non real-time polling service (nrtPS) and best effort (BE) and they are prioritized in the same order. FIFO treats all packets equally and, therefore, is suited only for the BE service. So, FIFO scheduling is not very useful in 4G networks.

2. Round Robin (RR) Scheduling

In RR Scheduling, users are cyclically scheduled irrespective of the channel condition and priority. Each link offers its packet transmission slots to its user sessions by polling them in round-robin order. In addition, window flow control is used to prevent excessive packet queues at the network nodes. As the window size increases, the session throughput rates are shown to approach limits that are perfectly fair in the max-min sense. That is, the smallest session rate in the network is as large as possible and, subject to that constraint; the second-smallest session rate is as large as possible, etc. If each session has periodic input (perhaps with jitter) or has such heavy demand that packets are always waiting to enter the network, then a finite window size suffices to produce perfectly fair throughput rates. RR scheduler is fair but very much inefficient in taking advantage of multi-user diversity and does not take into account user priority. Round-robin scheduling results in max-min fairness if the data packets are equally sized.

3. Weighted Round Robin (WRR)

Round-robin scheduling may not be desirable if the size of data packets are strongly varying and channel condition is highly user and location dependent. This results in modification in RR scheduling. WRR is a class-based queue scheduling. Here, queues are services round robin, in proportion to its assigned weight, so that higher priority user gets the privilege of better service. Let φ_k be the weight assigned to the data packet from kth user and τ be the RR time unit in second, and then data packet of kth user is served at every $\frac{\phi_k}{\tau}$ second, where,

$$\sum_{k=1}^{K}\phi_k = 1$$

In WRR, the weight assigned to each packet of same priority class is normalized by dividing the time-slot to be allocated with its average packet size. For many applications, it is not easy to evaluate average packet size apriory.

4. Deficit Round-Robin (DRR) Scheduling

DRR attempts to by-pass the requirement of advance knowledge of average packet size for each data packet. Each non-empty queue initially has a deficit counter (dc) which starts of at the value zero. As the scheduler visits each non-empty queue, it compares the packet at the head each of these queues and tries to serve one quantum of data. If dc is zero for a queue, then the packet is served if the size of the packet at the head of the queue is less than or equal to the quantum size. If a packet can not be served, then the value of the quantum is added to the dc for that queue.

If the scheduler visits a queue with a dc > 0, the packet from the queue is served if quantum + dc is greater than or equal to the size of the packet at the head of the queue. If the scheduler sees a queue with a non-zero dc, yet it is empty, then the dc is reset to zero so that it can not keep acquiring deficit. In brief, user's packet transmission deficit is kept track-off; queues that are short-changed in a round are compensated in the next. DRR as an ideal fair queuing approximation is shown by (Shreedhar & Vargese, 1996) with least computational complexity $O(1)$.

5. Stochastic Fair Queuing (SFQ) Scheduling

In order to avoid having a separate queue for each flow as in WRR, SFQ was developed (McKenney, 1991). In SFQ a hashing scheme is used to map incoming packets to their queues. Since the number of queues is considerably less than the number of possible flows, some flows will be assigned the same queue. Flows that collide this way will be treated unfairly - therefore the SFQ's fairness guarantees are probabilistic. The queues are served in round robin fashion (without taking the packet lengths into account). SFQ also uses a buffer stealing technique in order to share the available buffers among all queues: if all buffers are filled, the packet at the end of the longest queue is dropped.

6. Fair Queuing (FQ)

FQ scheduling algorithm is used in computer and telecommunications networks to allow multiple packet flows to fairly share the link capacity. The advantage over conventional FIFO scheduling is that a high-data-rate flow, consisting of large or many data packets, cannot take more than its fair share of the link capacity. FQ is used in routers, switches, and statistical multiplexers that forward packets from a buffer. The buffer space is divided into many queues, each of which is used to hold the packets of one flow, defined for instance by source and destination IP addresses. In the FQ, incoming packets are classified into K queues. Each queue is allocated 1/K of the output port bandwidth. The scheduler visits the queues according to the round robin order skipping empty queues. At each scheduled visit of a queue, one packet is transmitted out of the queue. The FQ is simple. It does not require a special bandwidth allocation mechanism. If a new queue is added to the existing K queues to create a new traffic class, the scheduler automatically adjusts the individual queue bandwidth to $1/(K+1)$ of the output port bandwidth. This simplicity is the FQ's main advantage. The FQ has two major drawbacks. First, since the output port bandwidth is divided to the K queues equally by $1/K$, if the input traffic classes have different bandwidth requirements, the FQ will not be able to distribute the output port bandwidth among the input flow classes according to their bandwidth requirements. Secondly, since one total packet is to be transmitted per each scheduled visit of a queue regardless of the packet size, the packet size will impact the actual bandwidth distribution among the queues even though each queue is visited equally by $1/K$. For example, if a particular queue tends to have bigger size packets than other queues, that queue would get more than the $1/K$ share of the total reserved bandwidth. With a link data-rate of R, at any given time the K active data flows (the ones with non-empty queues) are serviced each with an average data rate of R/K. In a short time interval the data rate may be fluctuating around this value since the packets are delivered sequentially. Fair queuing achieves max-min fairness, i.e., its first priority is to maximize the minimum data rate that any of the active data flows experiences, the second priority is to maximize the second minimum data rate, etc. This results in lower throughput (lower system spectrum efficiency in wireless networks) than maximum throughput scheduling, but scheduling starvation of expensive flows is avoided. In contrast to round-robin scheduling, fair queuing takes into account data packet sizes to ensure each flow is given equal opportunity to transmit an equal amount of data.

7. Generalized Processor Sharing (GPS) Scheduling

Fair queuing can be interpreted as a packet approximation of GPS, which was proposed by John Nagle (1985). GPS is an important paradigm for achieving service differentiation among heterogeneous applications in future communication networks. GPS scheduling is work conserving. In GPS, a flow is guaranteed a service share independent of the amount of traffic of other flows. With GPS, the service capacity is shared according to weights that are assigned to the various traffic classes. GPS is capable of offering protection among traffic classes, while achieving full statistical multiplexing. Let K classes are to be served and the total service rate is R. Each of the K classes has its own queue, and may consist of one or several flows. A class is backlogged if it has positive buffer content and hence, requires service. Class k is assigned a non-negative weight φ_k, $k = 1,\ldots,K$, and without loss of generality we take the weights such that they add up to 1, i.e., $\sum_{k=1}^{K}\phi_k = 1$. Hence, class k has a guaranteed minimum rate of $\varphi_k R$, i.e., if all classes are backlogged, then class k receives service at rate $\varphi_k R$. If some of the classes do not require service, then the surplus service rate is redistributed among the other classes in proportion to their respective weights. Denoting the set of backlogged classes with B, a backlogged class $k \in B$ receives service at rate in GPS such that:

$$\frac{\phi_k}{\sum_{j\in B}\phi_j} R \geq \phi_k R$$

During each infinitesimal interval of time, the GPS scheduler visits each backlogged flow once and schedules an equal and infinitesimal amount of data for transmission over the output link. If different weights are imposed on network flows, the amount of data transmitted from each flow by the GPS scheduler is proportional to its weight during each infinitesimal interval of time. By this means, the GPS scheduler achieves max-min fairness. The GPS scheduler, in the scheduling of bandwidth or resources over a link, is not implementable but ideally fair scheduler that exactly achieves the max-min fair distribution of the resources among the various flows.

8. Weighted Fair Queuing (WFQ) Scheduling

GPS is based on a fluid principle, and serves as a reference model for packet scheduling mechanisms like WFQ, which is a weighted version of FQ scheduling. The WFQ policy approximates the GPS scheduler in the sense that WFQ tends to serve packets in the order of their finishing time under GPS. WFQ is a data packet scheduling technique allowing different scheduling priorities to statistically multiplexed data flows. WFQ is a generalization of FQ. Both in WFQ and FQ, each data flow has a separate FIFO queue. In FQ, with a link data rate of R, at any given time the K active data flows are serviced simultaneously, each at an average data rate of R / K. Since each data flow has its own queue, an ill-behaved flow (who has sent larger packets or more packets per second than the others since it became active) will only punish itself and not other sessions. Contrary to FQ, WFQ allows different sessions to have different service shares. If K data flows currently are active, with weights $\varphi_1, \varphi_2,..., \varphi_K$, data flow number k will achieve an average data rate of $\dfrac{R\phi_k}{\phi_1 + \phi_2 + ...\phi_K}$

It can be proved (Stiliadis & Varma, 1998) that when using a network with WFQ switches and a data flow that is leaky bucket constrained, an end-to-end delay bound can be guaranteed. By regulating the WFQ weights dynamically, WFQ can be utilized for controlling the quality of service, for example to achieve guaranteed data rate. Proportional fairness can be achieved by setting the weights to $\varphi_k = 1 / c_k$, where c_k is the cost per data bit of data flow i. For example in CDMA spread spectrum cellular networks, the cost may be the required energy (the interference level), and in dynamic channel allocation systems, the cost may be the number of nearby base station sites that can not use the same frequency channel, in view to avoid co-channel interference.

9. Worst-Case WFQ (WF²Q) Scheduling

Alternative to WFQ, WF²Q is proposed by Bennett and Zhang (1996). In a WFQ system, when the server chooses the next packet for transmission at time t, it selects among all the packets that are backlogged at t, and selects the first packet that would complete service in the corresponding GPS. In a WF²Q system, when the server chooses the next packet at time t, it chooses only from the packets that have started receiving service in the corresponding GPS at t, and chooses the packet among them that would complete service first in the corresponding GPS. It can be noted that WFQ and WF²Q can achieve O(1) delay bound by keeping perfect track of the GPS clack and picking among the eligible (the one with smallest virtual GPS finish time) head-on packets. The worst case computational complexity for both WFQ and WF²Q are $O(\log_2 k)$. Basically, FQ scheduling is a special case of weighted fair queuing with equal weights for all queues.

10. Random Early Detection (RED) Scheduling

RED is another algorithm really implemented in commercial routers, which detects incipient congestion by computing the average queue size to drop packets before saturation. If the average queue size exceeds a fixed threshold, RED drops or marks each arriving packet with a certain probability, where the exact probability is a function of the average queue size. The scheme is to help transport layer protocols such as TCP, to avoid entering severe congestion. There are some modifications of the original scheme. Weighted Random Early Detection (WRED) drops packets selectively based on their priority (typically IP precedence) by following the rule that packets with a higher IP precedence are less likely to be dropped than packets with a lower precedence. Flow-based WRED is a particular feature of WRED, which forces it to afford greater fairness to all flows depending on how packets are dropped.

Opportunistic Packet Scheduling

Opportunistic scheduling algorithms maximize system throughput by making use of the channel variations and multi-user diversity. Opportunistic scheduling basically originates from a holistic view. The main idea is favoring users that are experiencing the most desirable channel conditions at each scheduling instant, i.e. riding the peaks. While maximizing capacity, such greedy algorithms may cause some users to experience unacceptable delays and unfairness, unless the users are highly mobile. Channel-aware constraint based opportunistic scheduling algorithms prove to be much practical in order to find the optimum performance and fairness balance. Possible constraints can be the basic QoS constraint, fairness constraint, power constraint and others. Let $U(n) = [U_1(n), U_2(n),...U_K(n)]'$ as the performance vector at time n and $U=\{U(n), n=1,2...\}$ be the time-sequence of the performance vectors. Let the system parameters be S and constraint at $C(U) \geq 0$. The system optimization goal ($O(.)$) is to maximize the system performance by exploiting the time-varying channel condition as well as maximizing multi user diversity gain such that

$$\arg\max_{S} O(U),$$
$$s.t.$$
$$C(U) \geq 0$$

Opportunistic scheduling by exploiting the time-varying wireless channel conditions with the objective of maximizing the system performance while satisfying various QoS requirements is discussed in (Liu, Chong & Shroff, 2003). They have proposed a framework to investigate different categories of scheduling problems involving two fairness requirements, viz. temporal fairness and utilitarian fairness. In particular, temporal fairness means that each user obtains a certain portion of the resource (e.g., timeslots) and utilitarian fairness means that each user obtains a certain portion of the overall system performance. The optimal scheduling solutions for these scheduling problems turn out to be index policies, and a stochastic-approximation-based algorithm is used to efficiently estimate the key parameters of the scheduling schemes in real-time.

11. Channel State Dependent Packet (CSDP) Scheduling

Bhagawat et al. (1996) proposed one of the earliest methods of channel state dependent packet (CSDP) scheduling to counter the unfairness of FIFO scheduling arising due to Head-of-Line (HoL) blocking when multiple sessions share the common wireless link. CSDP scheduling improves the system performance by taking into consideration the location and time-dependent channel conditions.

12. Proportional Fair Queuing (PFQ)

PFQ algorithm is one of the most popular channel-aware scheduling, where the user with the best channel condition (capacity) relative to its own average capacity is selected as based on proportional fair algorithm described earlier. In PFQ, the selected user k^* can be found as:

$$k^* = \arg\max_k \frac{\omega_k}{\overline{\omega_k}}$$

13. Capacity Queue (CQ) Scheduler

When the above opportunistic schemes are employed, users with high capacity links tend to have small queues, while users subject to poor channel conditions suffer from queue overflows and long delays. In (Neely et al., 2002), a scheduler termed as CQ scheduler is applied which maximizes the link rates weighted by queue backlog differentials for each link. In this downlink setting, the queue-weighted rate metric tries to select user k^* as

$$k^* = \arg\max_k \omega_k(t) Q_k(t)$$

where $Q_k(t)$ denotes the queue size of the user k at transmission time instant t. The inclusion of queue length in this scheme provides important insights for fairness. For instance, assume initially that the queue sizes are similar for all users, except for one user whose channel is superior to others. The user with the best channel will be selected and served so its queue size will be reduced; however, in the next scheduling instant, the advantage of better channel quality will be alleviated by the smaller queue size, yielding transmission to other users. The algorithm guarantees stability whenever the arrival rate vector lies within the stability region of the network.

14. Shortest Remaining Processing Time First (SRPT) Scheduling

One of the earliest queue length based channel-aware packet scheduling algorithm is proposed by Schrage and Miller (1966), called SRPT, which considers queue size together with capacity is method and the metric is defined as the amount of time it takes to serve all the packets from a given queue Here, the scheduler tries to choose the queue, which can be emptied in the shortest amount of time, i.e., the selected user k* at the time t is determined as:

$$k^* = \arg\max_k \frac{Q_k(t)}{\omega_k(t)}$$

15. Wireless SRPT (WSRPT) Scheduling

For wireless applications, SRPT can be modified as WSRPT, where the scheduler at the base station utilizes the knowledge of the average data rates and file size information of the users and particularly useful for connection control at Web servers. WSRPT aims for significant reduction in the average completion time by properly exploiting the file size information in scheduling by picking the user with the shortest file, so that the overall system performance can be improved. The WSRPT scheduler picks the user with the expected shortest remaining processing time, i.e,

$$k^* = \arg\min_k \left(\frac{Z_k(t)}{\overline{\omega_k}} \right)$$

where $Z_k(t)$ is the residual backlogged file size of user k at time instant t.

16. Opportunistic Auto Rate (OAR) Scheduling

OAR as an opportunistic scheduler to better exploit durations of high-quality channels conditions is introduced in (Sadeghi, Kanodia, Sabharwal & Knightly, 2002). OAR opportunistically sends multiple back-to-back data packets whenever the channel quality is good. When channel coherence times exceed multiple packet transmission times for both mobile and non-mobile users, OAR incurs significant time-diversity gain and thus increases the overall system performance. OAR takes into account the effect of aggregation, as the users are served in a round-robin fashion]. While serving each user, the number of packets transmitted for the user depends on the ratio of the user rate to basic rate, hence operating with larger aggregate sizes for users with better channel conditions. It is worthwhile to note that OAR provides temporal fairness since the packet transmission times for each user are equal.

17. Wireless Credit-Based Fair Queuing (WCFQ) Scheduling

WCFQ, a relatively new scheduler for wireless packet networks with provable statistical short- and long-term fairness guarantees is described in (Liu, Gruhl, Knightly, 2003). WCFQ provides a mechanism to exploit the inherent variations in wireless channel conditions and select low cost users in order to increase the system's overall performance (e.g., total throughput) to exploit temporal variations in the "cost" of scheduling different users to opportunistically select users with greater throughput potential, while also ensuring that the system's temporal fairness constraints are satisfied.

18. Traffic Aided Opportunistic Scheduling (TAOS)

TAOS scheme introduced by Hu, Zhang Sadowsky (2004) is a special kind of opportunistic scheduling in multiuser wireless networks to minimize the total completion time of the file to be downloaded, which consists of the processing time and the waiting time. It aims to improve the system throughput and thus reduce the total completion time by making use of file size information and channel variation in a unified manner. They have presented two TAOS algorithm, TAOS-1 and TAOS-2. They devise a cost function (or priority function), which increases with the file size and decreases with the instantaneous data rate, and scheduling is depended on the evaluation of this cost function at every scheduling instant. TAOS-1 is a generalization of the well-known PF scheduling, which is modified by taking into account the file size information. TAOS-1 picks the user k*, when

$$k^* = \arg\max_{k} \left(\frac{\omega_k(t)}{F_k \overline{\omega_k(t)}} \right)$$

where

F_k = initial backlogged file size of user k

$$\overline{\omega_k(t)} = \left(1 - \frac{1}{\Delta\tau}\right)\overline{\omega_k(t-1)} + \frac{\omega_k(t)}{\Delta\tau}$$

$\Delta\tau$ = sliding window length for average data rate computation

TAOS-2 finds the local optima to optimally characterize the completion time of the user's file by developing a cost function directly related to the completion time. TAOS-2 has two phases. In phase-1, sort all users in the ascending order of $\frac{F_k}{\overline{\omega_k(t)}}$ and sorting order is stored in $\mathbf{I}$, where $I_k(t)$ is the rank of the user k at time t. In phase-2, user k's cost function is computed as

$$D_k(t) = \left(I_k(t) - 1\right) - \left(M(t) - I_k(t) + 1\right)\left(\frac{\omega_k(t)}{\overline{\omega_k}} - 1\right)$$

where M(t)= number of remaining users in the system.

After computing the cost function $D_k(t)$ of all K users, schedule the transmission for user k which has the smallest cost function, i.e.,

$$k^* = \arg\min_{k}\left(D_k(t)\right)$$

19. Scheduling by Beam Forming

Exploiting multi-user diversity in a slow fading environment by inducing channel fluctuations through the use of multiple dumb antennas is presented in (Viswanath, Tse & Laroia, 2002), where the antennas are fed with randomly picked phase and amplitude. To inculcate multiuser diversity in a fairer and efficient way artificial diversity is introduced by beam-forming mechanisms, where each user feeds back the overall SINR of its "induced" channel to the BS. The BS selects the user with a large peak value of SINR. When there are a large number of users, the BS can always find a user with its peak SINR to transmit. Hence, the system performance is asymptotically as good as a solution with an optimal beam-forming configuration, while using only the overall SINR as feedback. Further, such a scheme can also be used opportunistically to null inter-cell interference.

Traffic Shaping

Traffic shaping, also known as "packet shaping," is the practice of regulating network data transfer to assure a certain level of performance or guaranteeing overall QoS. Traffic shaping policies limit flows to their committed rates, e.g. the flows need to be conforming to their traffic descriptors. They are very important to guarantee performance requirements. If flows or also single connections exceed their bandwidth consumption specifications, the network, which has dimensioned resources in strict dependence on the declarations, cannot guarantee any specified QoS requirement. An example of traffic shaping action is shown in Figure 12, where two, out of four, connections do not conform with their committed rates set to 16 Kbps. The two non-conformant flows that generate 64, 32 Kbps are cut down to the committed rate by the traffic shaper. Traffic shaping acts like a filter or regulator to change the rate of incoming traffic flow to regulate the rate in such a way that the outgoing traffic flow behaves more smoothly.

Figure 12. Simple traffic shaper

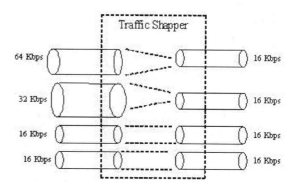

If the incoming traffic is highly bursty, it needs to be buffered so that the output of the buffer is less bursty and smoother. A traffic shaper smoothens out the bit-rate variations in a stream by delaying rather than by dropping or tagging cells. A user can avoid possible cell loss at the network access point by shaping its cell stream. All those actions will result in an effective traffic source. The practice involves delaying the flow of packets that have been designated as less important or less desired than those of prioritized traffic streams. Regulating the flow of packets into a network is known as "bandwidth throttling." Regulation of the flow of packets out of a network is known as "rate limiting." There are two types of traffic shaper: the pure traffic shaper and the leaky bucket traffic shaper. In pure traffic shaper, Incoming packets are put into a buffer, or a "bucket," with depth B, and are sent out on the outgoing link at a constant rate R. A leaky bucket is parameterized by a leak rate μ and a bucket size B. It is convenient to describe the operation of the leaky bucket in terms of a fictitious x/D/1/B queue for a deterministic input (D) served at constant rate μ with capacity B. In this model, each cell arrival to the leaky bucket generates a corresponding customer arrival to the fictitious queue. At the time of a cell arrival, if the number of customers in the fictitious queue is less than B, the cell is said to be conforming, otherwise the cell is nonconforming. Nonconforming cells are either dropped or tagged as low priority cells. A leaky bucket shaper parameterized by (μ, B) operates in a manner similar to the leaky bucket policer, except that nonconforming cells are delayed, rather than dropped or tagged. The operation of the leaky bucket shaper can be understood in terms of an x/D/1 queue. Upon arrival of a cell, if the number of customers in the fictitious queue is B or more, the cell is placed in an infinite capacity FIFO buffer. Whenever a customer leaves the fictitious queue, a cell (if any) is removed and transmitted from the FIFO buffer. Traffic shaping is very much essential in certain kind of packet scheduling, where statistical QoS guarantee is more important than deterministic QoS guarantee, like effective capacity based link-layer channel modeling to capture a generalized link-level capacity notion of the fading channel (Wu & Negi, 2003).

GAME THEORETIC FRAMEWORK FOR ADVANCED SCHEDULING

Introduction

The discussion on resource allocation and scheduling so far, considers BS as the central unit in deciding the strategy and optimization scheme based on the system objective and constraints. There are certain scenarios where centralized resource allocation is not possible, where distributed resource allocation scheme has to be adopted. Ad hoc networks have occupied a preeminent place in the wireless communications and networking literature for the last several years. An ad hoc network is a self-configuring, multi-hop network in which there is no central authority. Thus, every aspect of the configuration and operation of an ad hoc network must be completely distributed. Furthermore, nodes in an ad hoc network are often severely energy and power constrained. Since potentially there are many sources distributed in an ad hoc network which will be competing for

the use of the available resources, there are several issues like power minimization, fairness, QoS provisioning, overhead minimization, are to be dealt with. In the wireless scenario, an user might prefer outcomes that yield higher signal-to-noise ratios, lower bit error rates, more robust network connectivity, better QoS, lower power expenditure and least unfairness, although in many practical situations these goals will be in conflict. Appropriately modeling these preference relationships is one of the most challenging aspects of the application of game theory. Centralized scheduling has the advantage of implementation simplicity as well as overall system performance maximization, but lacks in reaching stabilization in the issues of individual user's fairness and changing utility function. Distributed control of network resources, on the other hand, has the tendency to eventually lead to a steady state condition, which is very much appreciable for long-term stability of the entire network. Distributed adaptive behavior in wireless networks leads to recursive attribute, where the decisions of one user will subsequently influence the decisions of other users in a constructive or destructive way. It is dependent on the deployment strategy to maximize the utilization of distributed decision-making in user and network-level optimized way. Game-theoretic approach can provide effective framework development tool and deployment strategy to exploit the advantage of distributed implementation using only the local information.

Non co-operative game theory assumes agents with distinct interests that interact through a predefined mechanism; it encompasses concepts such as normal and extensive form games, incomplete information, Nash Equilibrium. Co-operative game is another scenario where users constitute coalitions and negotiate via an unspecified mechanism to allocate resources. Cooperative game theory encompasses concept of Nash bargaining solution. In a non co-operative game framework, each user competes for a subset of network resources that maximizes its gain such that Nash equilibrium is reached. In a co-operative game approach, all users participate in a bargaining model for the use of network resources in order to achieve Pareto optimality. In most of the situation, co-operative game approach outperforms the non co-operative approach under similar network condition and user configuration. Non co-operative approach perform better only in extreme situation of maximum throughput allocation to greedy users, which lead to high degree of unfairness.

Game theory is a branch of applied mathematics, which deals with multi-person decision-making situations. The basic assumption is that the decision makers pursue some well-defined objectives and take into account their knowledge or expectations of the other decision makers' behavior. Game theory rationalizes the conflict and cooperation between intelligent and demanding users. A game involving two or more players, in which no player has anything to gain by changing only his or her own strategy unilaterally. If each player has chosen a strategy and no player can benefit by changing his or her strategy while the other players keep theirs unchanged, then the current set of strategy choices and the corresponding payoffs constitute Nash Equilibrium. Non co-operative games model the process of players making their own choices out of their own interest to achieve Nash Equilibrium, which is a kind of strategic equilibrium. Nash Equilibrium in non co-operative games is not optimal; the amount of efficiency loss is called Price of Anarchy, which is the ratio of aggregate utility to the maximum possible aggregate utility. Example of Nash equilibrium is Prisoner's Dilemma. Given a set of alternative allocations of, say, goods or income for a set of individuals, a movement from one allocation to another that can make at least one individual better off without making any other individual worse off is called a Pareto improvement. An allocation is Pareto efficient or Pareto optimal when no further Pareto improvements can be made. This is often called a Strong Pareto Optimum. The Nash bargaining solution is a standard tool in cooperative game theory, and is applied widely in network resource allocation. Proportional fairness is its special case. In the Nash Bargaining Game two players demand a portion of some good (usually some amount of money). If the two proposals sum to no more than the total good, then both players get their demand. Otherwise, both get nothing. Strategies are represented in the Nash bargaining game by a pair (x, y), x and y are selected from the interval [d, z], where z is the total good. If (x + y) is equal to or less than z, the first player receives x and the second y. Otherwise both get d. d here represents the disagreement point or the threat of the game; often d = 0. There are many Nash Equilibria in the Nash bargaining game. Any x and y such that x + y = z is a Nash equilibrium. If either player increases their demand, both players receive nothing. If either reduces their demand they will receive less than if had they demanded x or y. There is also Nash equilibrium where both players demand the entire good. Here both players receive nothing, but neither player can increase their return by unilaterally changing their strategy.

System Model

Mostly, game-theoretic framework is developed and implemented to find solutions to enhance the fairness of the achievable performance, like throughput among all users in a distributed way. Figure 13 shows a typical system model of game theoretic framework based distributed resource allocation and scheduling system. There K (in figure K =3 considered) number of users and N number of resources. By this model, user reaches a common agreement or equilibrium point for scheduling of resources in co-operative and non co-operative scenario respectively. Overall, a distributed algorithm is the key to develop game theory based solution.

Advanced Scheduling Schemes in 4G Systems

Figure 13. Distributed resource allocation and scheduling system model

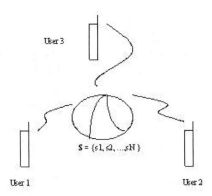

Scheduling in Non Co-Operative Game Framework

In a centralized method, the base stations know the channel usage of their neighbors. In a distributed method, each base station allocates resource to the users on its own and independent of the other base stations and hence the base stations are unaware of the resource usage of their neighbors. In another case, the users by themselves distribute the available resources in a co-operative or non co-operative way to collectively take control of the resource allocation and scheduling without or with least control of the base station. Game theoretic approach is the best-known tool to implement the distributed scheme of resource allocation, which as discussed earlier, provides more stability, fairness and even more overall throughput for the entire network in co-operative or non co-operative basis. Game theoretic framework for radio resource allocation has been introduced in many studies (Johari, Mannor& Tsitsiklis, 2005). 4G networks offering different levels of quality of service (QoS), both network performance and user satisfaction directly influences the user's choices of service-request. Since each user's choice of service may be influenced not only by the pricing policy but also by other users' behavior, the problem can naturally be treated under a game-theoretic framework, in which the operating point of the network is predicted by the Nash equilibrium. Pricing can be treated as a game between a network service provider through BS and a finite set of users or traffic flows. Different users or flows may have different QoS requirements, and each user must choose what level of service to request among all service classes supported by the network. For instance, in priority-based networks, a strategy may be the priority levels a user requests for his traffic; in networks that support delay or data rate guarantees, a strategy may be the minimum bandwidth to which a user requests guaranteed access. The tradeoff, of course, is that the higher the level of service requested the higher the price to be paid by the user. QoS-guaranteed flows are accepted as long as the network is capable of supporting users' requested service. The network service provider architects the Nash Equilibria by setting the rules of the game: the pricing structure and the dimensioning of network resources. A user's payoff is determined by the difference between how much a user values a given QoS level and how much he pays for it. The maximization of this payoff, given all other users' service choices, will determine the optimum strategy to be adopted by each user. Considering the scenarios where the users are selfish and act non co-operatively to maximize their own performance, which results in deteriorated and severely inefficient system performance. The greediness of the selfish users and distributed network structure challenge the feasibility of the traditional approach and require game theoretic techniques. In most non co-operative resource allocation problem, each user tries to maximize its reward, allocated resource units and utility function in order to achieve Nash Equilibria under total some system constraints. In (Johari, Mannor& Tsitsiklis, 2005), it is shown that Nash Equilibria suffer some efficiency loss compared to the total throughput optimization scheme. Distributed network resource allocation and scheduling as a non-cooperative Network Resource Allocation Game (NRAG) is described by Chen et al. (Chen. J, Yu. K, Ji. Y & Zhang, P. (2008). In NRAG each network unilaterally maximizes its individual utility. In this work by combining QoS satisfaction, communication link quality and price, the concept of arbitration probability is introduced, which reflects the level of willingness of an individual user to use particular network resources. With this concept, each user may select a network with the maximum arbitration probability whenever there is any outgoing traffic. They have proved that the resulting non-cooperative network resource allocation game has Nash Equilibria and this equilibrium is unique. By applying arbitration probability based network selection, the proposed network resource allocation algorithm successfully provides better revenue improvement for heterogeneous wireless data networks. Resource management through non co-operative game theory approach for optimal random channel access in IEEE 802.11 was discussed by Altman et al (Altman, Borkar & Kherani, 2004). The players or the users are the nodes of the network and the strategy is based on the probability of a transmission attempt if there is a packet in the queue. The payoff is defined as the utility gained due to successful packet transmission. They have proposed a distributed algorithm to achieve the Nash Equilibria with the constraint of total power.

Scheduling in Co-Operative Game Framework

Non co-operative resource allocation, where every user selfishly tries to maximize individual throughput, always tends to lower the total aggregate data rate as well as drastically reduce the fairness criteria. When, the solution does not come close to Pareto optimality, non co-operative scenario generates to severe unfairness and biasness to only few users. A set of solutions is considered Pareto optimal, if there are no wasted resources so that none of the users can be better off without making others worse off. Non-cooperative resource allocation to achieve Nash Equilibrium is discussed in (Niyato & Hossain, 2008). From the distributed resource allocation system model in Figure 10, co-operative game theoretic negotiation can be conceptualized. Let the total number of OFDMA subcarriers are N and each user occupies one at a time. According to channel condition, each user selects one subcarrier to maximize its throughput. But a conflict may arise as certain subcarriers are good for more than one user. So, the problem is to decide which user is to be allocated by negotiating with other users by a kind of bargaining, so that overall system performance and each user get satisfied in long-term. The Nash Equilibrium for this game gives the optimal allocation which maximizes the utilities of all the connections in the network by arbitration probability and a bargaining game formulated, which provides the capacity reservation thresholds so that the connection-level QoS requirements can be satisfied for the different types of connections on a long-term basis. In (Yaiche, Majumder & Rosenberg, 2000), a game theoretic framework for bandwidth allocation for elastic services in high-speed networks is presented, which is based on the idea of the Nash Bargaining Solution, to achieve Pareto optimality. Their bargaining framework can be used to characterize a rate allocation and a pricing policy which takes into account users' budget in a fair way and such that the total network revenue is maximized. Considering single-cell multiuser OFDMA systems, there is a Base Station (BS) serves as the central allocation unit to exchange and negotiate the user's demand and therefore corresponds to transactions of common market-place scenario. BS distributes the system resources (subcarrier) to the users such that each of them can operate at its optimum and joint agreements are made about their operating points. This notion demands the application of game theory, especially co-operative game theory, which can achieve the crucial idea of fairness and maximization of overall system rate. The concepts of the Nash bargaining solution (NBS) and coalitions provide a fair operation point such that most optimum allocation is reached in a co-operative manner. NBS is a natural framework that allows defining and designing fair assignment of bandwidth between applications with different utilities (Niyato & Hossain, 2008). Proportional fair has the property that it cannot be replaced by any other arbitrary allocation that does not lead to a reduction in the aggregate fractional rate change, which is basically the notion of Nash Bargaining Solution (NBS).

Let K users attempt to reach unanimous agreement on utilities: **u** = [u1, u2,…uK].

1. The utility region $U \subset \mathbf{R}_{++}^{K}$ is convex, comprehensive, closed and bounded.
2. Depending upon the chosen strategy, solution outcome ψ results.
3. If bargaining fails disagreement outcome d results.

Nash bargaining solution outcome is obtained when:

$$\max_{\{u \in U : u > d\}} \prod_{k=1}^{K} (u_k - d_k)$$

Often disagreement point is 0; d=0 ; so the outcome is:

$$\max_{\{u \in U : u > d\}} \prod_{k=1}^{K} u_k$$

For the case of OFDMA sub-carrier allocation, above **u** = **R**, i.e. $\mathbf{R} = [R_1, R_2, \ldots, R_K]$.

This solution is equivalent to PF:

$$\begin{aligned}\hat{R} &= \arg\max_{R} \prod_{k=1}^{K} R_k \\ &= \arg\max_{R} \ln \prod_{k=1}^{K} R_k \\ &= \arg\max_{R} \sum_{k=1}^{K} \ln R_k\end{aligned}$$

In a true dynamic wireless network, channel capacity varies with time and different users achievable data rate are often different at different time, i.e., total resources in the system also varies with time. So, it is important to consider the dynamics of resources also, which is considered in (Galstyan, Kolar, Lerman, 2003), where resource allocation with changing resource capacities is studied as El Farol Bar problem. El Farol Bar problem is inspired by real-life situations involve a set of independent entities competing for the same resource, in an uncoordinated fashion. Aram Galstyan, et al. have described a reinforcement learning model for adaptive resource allocation in a multi-use system, where the uses have a choice of several resources with time-dependent capacities with a learning scheme is based on minority games on networks and each user exercises a set of strategies to decide what resource to choose. One variant of the El Farol Bar problem is the minority game proposed by Challet et al. (Challet, Marsili & Zhang, 2004). The concept is very much suitable in the resource allocation problem we are discussing. A minority game is a repeated game in which K (odd) players have to choose between two alternatives (say A and B) at each time step. Those who happen to be in the minority win. Although it may seem rather simple at first glance, this game is subtle in the sense that if all players analyze the situation in the same way, they will all choose the same alternative and lose. Therefore, players have to be heterogeneous. Users end up on the minority side win, which is basically power minimization problem in up link. The user who decides to consume the least transmit power will get the opportunity to use the subcarrier. While the El Farol Bar problem was originally formulated to analyze a decision-making method other than deductive rationality, the minority game examines the characteristic of the game that no single strategy of any kind may be adopted by all participants in equilibrium. In (Wang, Han & Liu, 2006), Stackelberg game concept to jointly optimize the benefits of source nodes and relay nodes in a relay based cooperative communication is considered. Resource allocation in multi-hop relay stations or ad hoc networks can be modeled as a distributed and bargained decision based on the Stackelberg strategy. The game is divided into two levels: the source node plays the buyer-level game and the relay nodes play the seller-level game. Each player is selfish and wants to maximize own benefit. Specifically, the source aims to get most benefits at the least possible payment, where as each relay aims to earn the payment which can not only cover their forwarding cost but also gain as much extra profit as possible.

Delay-Sensitive Scheduling for Real-Time Applications

Communication services provided by future broadband wireless networks will range widely in their QoS requirements such as delay and BER. Applications such as file transfer and email services are relatively delay-tolerant but with stringent error rate requirement. On the other hand, applications such as voice transmission and real-time video streaming have a stringent delay constraint but less error rate requirement. Delay is a key parameter in QoS requirement particularly in real-time application and consequently proper scheduling mechanism is a key in guaranteeing delay-sensitive applications to run smoothly. The scheduling algorithms discussed so far are mainly delay-independent scheduling policies, which assigns system resources to the users (even if severely backlogged) based on a fairness notion or to maximize the overall system throughput or according to their queuing delays or based on individual traffic priorities in a channel-aware way. It is important to clarify that the term these algorithms does not completely ignore the delays experienced by traffic flows in the scheduling process, on the other hand the packets of this traffic family do not have strict deadlines. This means that packets of this family should be transmitted even if they experienced high queuing delay. However, the scheduler is required to reduce their queuing delay as much as possible. So these scheduling algorithms are particularly based on nrtps and BE traffic class of IEEE 802.16. The transmission scheme which maximizes long-term throughput transmits more power and information in good channel states, and less in poor conditions. Thus, it delays some parts of input traffic to wait for good channel states, buying more utility for the available power resources to achieve the maximal long-term throughput. Ukil (2008) has presented such kind of optimal scheduling algorithm to maximize the long-term throughput to cater delay-tolerant applications. Delay-sensitive scheduling needs some special attention in order to maintain the high volume of traffic with diverse and strict QoS requirements. The optimization problem of delay-sensitive scheduling is targeted to find a policy to minimize the average total queue length. By Little's law, minimizing the average queue length is equivalent to minimizing the average packet delay in steady state. in fact, introducing the delay parameter adds one more degree of freedom in scheduling algorithm, which in turn makes it more computationally complex and sometimes only sub-optimal solution will be available for implementation in real-time applications. In order to achieve the global optimality of delay-sensitive scheduler design, we need to consider design variables and the interactions among them as much as possible. However, more does not necessarily mean better. The more design variables we consider, the more difficult is orchestrating a large number of design variables to make them work harmonically and in proper synergy. As we know most of the scheduling problem is a kind of nonlinear optimization problem, where the number of design variables increases and the size of state space of the objective function will increase exponentially, making the optimization problem unmanageable. The most important possible constraint of delay-sensitive scheduling apart from rate constraint, fairness constraint and power constraint is the delay constraint. Let U(n) = [$U_1(n)$, $U_2(n)$,... $U_K(n)$]' as the performance vector at time n and U={U(n), n=1,2...} be the time-sequence of the performance vectors. Let the system parameters be S. We consider rate, power and delay constraints. The system optimization goal (O(.)) is to maximize the system performance in a fair way by exploiting the time-varying channel condition to maximize the multi user diversity gain as well as satisfying the system constraints such that

$$\arg\max_{S} O(U)$$

s.t.

$$\sum_{k=1}^{K} \omega_k \geq \gamma_k \tag{5}$$

$$\sum_{k=1}^{K} P_k \leq P_T \tag{6}$$

$$\Pr(\partial_k > \tau_k) \leq \Delta_k \tag{7}$$

where,

δ_k = delay encountered by a typical packet of user k,
τ_k = maximum delay that user k can tolerate,
Δ_k = largest probability with which the system is allowed to violate the delay requirement.

Delay-sensitive traffic is classified based on the values of τ_k and Δ_k, which are QoS class dependent. Below we will describe few important delay-sensitive scheduling algorithms.

Earliest Deadline First (EDF)

The QoS requirements of delay-sensitive traffic are usually defined in terms of a delay bound, generally termed deadline, before which the packets should be delivered to the receiver. Otherwise, the information contained in these packets will be of no use to the receiver, and thus should be dropped. Therefore, the cross-layer delay-sensitive scheduler should be able to minimize the number of packet deadline violations while maintaining fairness among users in channel-aware way. Dynamic priority algorithms are of important class of scheduling algorithms, where the priority of a task can change during its allocation. In EDF the priority of a user is inversely proportional to its absolute deadline, i.e., highest priority job is the one with the earliest deadline. The priority is dynamic since it changes according to the queue length of the user traffic. Given a task set of periodic or sporadic tasks, with relative deadlines equal to periods, the task set is schedulable by EDF if and only if

$$U = \sum_{k=1}^{K} \frac{\gamma_k}{Q_k} \leq 1,$$

where Q_k = instantaneous queue length of the kth user.

EDF is an optimal algorithm, in the sense that if a task set if schedulable, then it is schedulable by EDF. In fact, if U > 1 no algorithm can successfully schedule the task set and if U ≤ 1, then the task set is schedulable by EDF. EDF is relatively simple algorithm to minimize the delay-based outage probability, but lacks in providing critical or high priority traffic better privileges. When the system load is heavy and interference level is high, the EDF scheduler stalls on low throughput channels resulting further increase in load and interference. EDF algorithm schedules based on delay deadlines, it outperforms the WFQ algorithm with respect to delay under medium load conditions.

Largest Weighted Delay First (LWDF)

LWDF (Stolyar &. Ramanan, 2001) is a parameterized version of the first-input first-output (FIFO) that works as follows: at the beginning of the time slot starting at time t, serve at the maximal possible rate the queue of user k*, where

$$k^* = \arg\max_{k} \lambda_k \delta_k(t)$$

where, λ_k is an arbitrary positive constant (dependent on user class for example), $\delta_k(t)$ the delay of the head of line packet. If the delay QoS requirements for all users is as expressed above, it was proved that the choice of weights λ_k that makes LWDF discipline nearly throughput optimal at large values of the delay bound τ_k and very small values of $\delta_k(t)$.

Modified Largest Weighted Delay First (M-LWDF)

Due to time-varying properties of the wireless channel, EDF scheduling cannot always meet the deadlines of all packets. Consequently, a portion of these packets expires and is dropped. For providing minimum throughput guarantees in a channel-aware as well as delay-bound way, LWDF is modified to introduce multiuser diversity in order to increase the efficiency of channel utilization (and hence the system throughput) and also compensate delayed users. M-LWDF scheduling algorithm is described by Stolyar and Ramanan (2001) The M-LWDF algorithm picks a user k^* at time t, if:

$$k^* = \arg\max_{k} \lambda_k \delta_k(t) \frac{\omega_k(t)}{\omega_k}$$

It has been proven analytically that M-LWDF is throughput optimal. Suggested value of $\lambda_k = \frac{-\log(\Delta_k)}{\tau_k}$. Throughput optimal means that if a set of minimum rates are feasible, then M-LWDF achieves the minimum rate. In order to explicitly bring in the notion of minimum rates, the head of line packet delay is computed as:

$$\delta_k(t) = \frac{Number\_of\_token\_in\_Queue}{R_{k\min}}$$

where, a token is accumulated for each time slot that a user is not served, $R_{k\min}$ is the minimum rate for the user. M-LWDF scheduling was found to be highly dependent on the value of the parameters λ_k and its performance change significantly with the QoS requirements of users' flows.

Channel Dependent Earliest Deadline Due (CD-EDD)

A modified version of M-LWDF is suggested in (Khattab and Elsayed, 2004), termed as Channel Dependent Earliest Deadline Due (CD-EDD), where the performance is shown to be improved than traditional M-LWDF. CD-EDD is basically a channel-state dependent EDD policy where the scheduler chooses to schedule the queue whose HoL packet has the earliest time to expire and the best channel conditions, and consequently the highest transmission rate, among all queues. CD-EDD scheduler follows the following criteria to select the user to schedule:

$$k^* = \arg\max_{k} \left\{ \lambda_k \cdot \frac{\omega_k(t)}{\omega_k} \left(-\log(\Delta_k) \cdot \frac{\partial_k / \tau_k}{\tau_k - \partial_k} \right) \right\}$$

The behavior of the CD-EDD policy can be summarized as follows: when a certain queue has its HoL packet waiting in the system for a relatively long period (but have not expired yet), its time to expire will decrease significantly. An important feature of the CD-EDD algorithm is its weak dependency on the variance of the value of QoS weights λ_k, and thus it can be used for a wide variety of QoS requirements.

Priority Indexed Long Term Proportional Fair (PILTPF)

QoS class based scheduling algorithm to optimally handle delay differentiated traffic is presented by Ukil, Sen, Bera (2009), which is termed as Priority Indexed Long Term Proportional Fair (PILTPF). PILTPF has a priority index estimator, which based on user's QoS class maps the QoS metric of delay, buffer-size, minimum data rate requirement to a priority value, unique to each users. The weighting value (λ_k) of M-LWDF, CD-EDD is more realistically and traffic depended way evaluated in PILTPF algorithm. Priority estimation (λ_k) evaluates the dynamic priority of the user based on its current QoS class, available buffer–size, delay limit and minimum data rate requirement, which is highest for UGS class and for rest λ_k is estimated as below:

$$\lambda_k = \frac{Q_k \times \gamma_k}{\left(bs_{kMAX} - bs_k\big|_t\right) \times \left(\delta_{MAX} - \delta\big|_t\right)}$$

where $bs_{kMAX}, bs_k\big|_t, \delta_{MAX}, \delta\big|_t$ are the maximum buffer-size, used buffer-size, maximum delay-limit and elapsed delay for k^{th} user at current allocation instant and minimum individual rate requirement = $[\gamma_1, \gamma_2, \ldots\ldots, \gamma_K]$ in bits/sec. . For, non real-time traffic traffic, $\left|\delta_{MAX} - \delta\big|_t\right| = 1$. It can be clearly seen that:

$\lambda_k \rightarrow \infty$, when

$\left|bs_{kMAX} - bs_k\big|_t\right|$ or $\left|\delta_{MAX} - \delta\big|_t\right| \rightarrow 0$

As per PILTPF scheduling K* user is scheduled when:

$$k^* = \arg\max_k \left|\frac{\lambda_k \times \omega_k}{\overline{\omega_k}}\right|$$

The constraint of optimization utilizes the time diversity gain. Instead of instantaneous data rate guarantee, it assures long-term average data rate within the allocation duration, which consists of number of frames.

$$\overline{\omega_k} \geq \gamma_k$$

Actually allocation duration (τ_k) is upper-bounded by the delay constraint of user k (δ_{kMAX}).

$$\tau_k \leq \delta_{kMAX}$$

Considering OFDMA based IEEE 802.16 network, simulation results show that PILTPF performs better than PF. Figure 14 is showing the comparison of achievable data rate of the users by PILTPF and PF algorithm. It also shows the minimum data rate requirement of individual user as QoS profile. Figure 14 clearly shows that PILTPF algorithm follows QoS profile for all the users where as in PF algorithm some of the users get deprived and achieved very less data rate than their minimum requirement due to the inherent feature of its instantaneous PF metric computation. If the deprived user, say 10th user (figure 14) is a high priority customer, then the allocation is very much unacceptable from service provider as well as user perspective. In figure 15 a bar chart comparing the performance of PILTPF and PF algorithm is depicted. From figure 14 and 15, it can be interpreted that as PILTPF considers both the priority of the user and the time diversity gain, so it results in better performance both in terms of throughput and QoS guarantee. The unevenness in assigning subcarrier by conventional PF algorithm becomes more visible when more number of users exists. In that highly complex scenario of large number of users, it is depicted in figure 16 that PILTPF at least attempts to follow the QoS profile in order to preserve the importance of priority of the users. However, in a less number

Figure 14. User achievable throughput by PILTPPF and PF algorithm

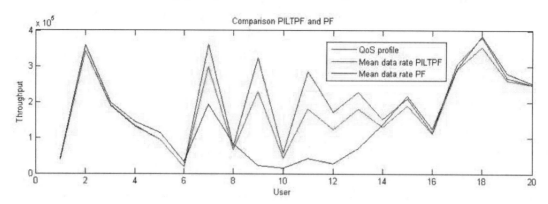

Figure 15. Bar-chart comparison between PILTPF and PF

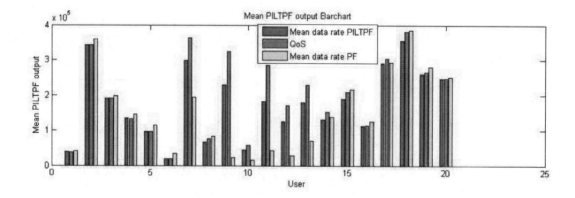

of users and good channel condition throughout the entire cell, the difference between the PF and PILTPF diminishes.

Apart from these there are few worth-mentioning contributions which need discussion. Tao, Liang and Zhang (2006) have formulated the resource allocation problem as a convex optimization problem and presented an iterative algorithm to compute the optimal solution, considering heterogeneous multiuser OFDM system with both delay constrained (DC) and no-delay-constrained (NDC) traffic. No-delay-constrained (NDC) traffic aims to maximize the long-term average transmission rate, which is basically an ergodic capacity problem where the instantaneous transmission rate and power level can be dynamically adjusted over time and frequency. The transmission of delay-constrained (DC) traffic is considered as the strictly delay limited capacity problem in which a target transmission rate should always be maintained irrespective of present channel condition and outage will take place if the instantaneous mutual information is lower than the target rate. Their proposed algorithm proposed simultaneously allocates power and rate for simultaneous transmission of delay differentiated traffic in a broadband wireless network and is intended to maximize the sum-rate of all the users with NDC traffic while maintaining guaranteed rates for the DC traffic under a total transmission power constraint over every frame irrespective of channel fading conditions. The wireless traffic can be broadly grouped into two categories as elastic (Best Effort Traffic, where

Figure 16. User achievable throughput by PILTPF and PF algorithm at large number of users

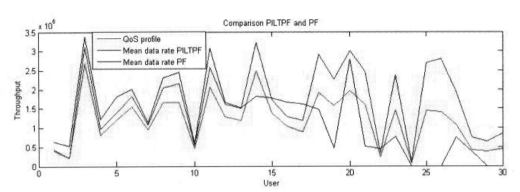

fairness and throughput are the performance objectives) and non-elastic traffic (Voice traffic with strict delay constraint) and proportional fairness provides a good balances between the two. Girici, Zhu, Agre and Ephremides (2008) have proposed delay and rate based scheduling algorithm for OFDMA based wireless systems. Buffer management at the base station plays a significant role to minimize the traffic outage of packet data. Buffer overflow, which is a function of individual users' traffic class and channel condition, if minimized, significant improvement of system throughput can be achieved as well as individual QoS can also be guaranteed as proposed by Guoqing Li and Hui Liu (2003). Maximizing instantaneous throughput (known as waterfilling in literature) is too myopic under realistic packet arrival. Queue backlog state information is the key factor for delay improvement in packet transmission. With this notion, the proposed heuristic algorithm by Somsak Kittipiyakul and Tara Javidi (2007) has considered optimal subcarrier allocation when the packet arrivals and channels are stochastic and time-varying in nature, respectively. The proposed algorithm under realistic packet arrivals and queue occupancies first maximizes the number of packets being served by instantaneous throughput maximization, while maximizing the expected number of non-empty queues (hence, served packets) in the future by load balancing. They have showed that in delay-optimal low-to moderate traffic condition, balancing the queues is more critical than opportunistically taking advantage of CSI. The opposite becomes true in the heavy traffic. Seungwan Ryu, Byung-Han Ryu, Hyunhwa Seo, Muyong Shin, and SeiKwon Park (2005) have proposed an Urgency and Efficiency based wireless Packet Scheduling (UEPS) algorithm in packet-based wireless OFDMA systems for simultaneous real-time (RT) and non-real-time (NRT) traffics, with the aim of maximizing the throughput of NRT traffics while satisfying Quality of Service (QoS) requirements of RT traffics. The UEPS algorithm has two distinct parts, urgency prioritization and efficiency maximization and uses the time-utility function as an urgency factor and the relative status of the current channel to the average one as an efficiency factor. The main objective of the UEPS algorithm is to maximize throughput of NRT traffics while satisfying the QoS requirements of RT traffics.

Resource Allocation and Scheduling in MIMO Systems

4G wireless networks are invariably Multiple Input Multiple Output (MIMO) based, where multiple transmit antennas (n_T) and multiple receive antennas (n_R) provide increased theoretical link capacity by m* = min (n_T, n_R) times relative to single-antenna wireless links. The $n_T \times n_R$ wireless MIMO channels are characterized by time varying fading $n_T \times n_R$ channel matrices, so that MIMO channel can be viewed as a vector channel with m* = $\min(n_T, n_R)$ spatial channels. In fact, the data rate of the MIMO link can be increased by spatial multiplexing of the m* spatial channels and therefore, the channel capacity increases with m* in a linear manner asymptotically. The basic advantage of having multiple antennas at the transmitter and the receiver is to transform the original single wireless fading channel into multiple wireless fading channels, as m spatial channels created as a result of the multiple antennas and the scattering and multipath environment surrounding the transmitter and the receiver. Hence, independent and uncorrelated information streams can be delivered on the m parallel spatial channels to realize the increased transmission capacity, which is called spatial multiplexing. On the other hand, one can deliver the same information bits over multiple spatial channels to exploit the spatial diversity so as to enhance the reliability of the transmission.

Scheduling for MIMO systems is more challenging than for Single Input Single Output (SISO) systems because of the spatial-interference between the orthogonal channels and the existence of spatial correlation matrix rather than the scalar channel response of Single Input Single Output (SISO) systems. The achievable signal-to-interference ratio (SIR) on a channel or subcarrier in OFDMA in MIMO systems is a function of the set of users that share the subcarrier. The general objective of MIMO scheduling and resource allocation is to maximize the spectrum efficiency while achieving a sufficient SIR and optimal sets of co-channel users should be identified, based on their spatial correlations and power distributions. In other words, the available degrees of freedom should all be configured to exploit spatial multiplexing in order to maximize the network capacity or network coverage, depending upon the objective function. They are known as capacity-optimized and the coverage optimized scheduler respectively. MIMO-OFDMA systems are able to multiplex the users in time, space and frequency domains, which makes scheduling more difficult, since we have to decide which dimension (space or frequency) should be occupied by which set of users. QoS requirements impose additional constraints on the optimization problem. MIMO-scheduling is in deed a challenging problem. With large number of degrees of freedom and complex system variables, efficient MIMO-scheduling for real-time systems in all-IP networks is one of the challenging future research topics.

System Model

The system model we considered a standard scenario involving a base station that simultaneously transmits data to K users, whose channels have been determined earlier either through the use of uplink training data (as in a time-division duplex system) or via a feedback channel (as in a frequency-division duplex system) as illustrated in Figure 17. The system consists of a single base station (BS) communicating with multiple mobile subscribers (MS) or users, shown as MIMO receiver. For each MS, the instantaneous matrix wireless channel conditions are assumed known perfectly at the BS. The scheduler at the BS may decide to schedule transmissions to one or more users based upon their current channel states. Figure 17 illustrates downlink data transfer for a cellular system with n_T transmit antennas per transmitter and n_R receive antennas per receiver. For each MS, the wireless channel conditions are assumed known (through feedback) at the BS.

MIMO Scheduling Design

Traditional wireless communications (SISO) exploit time or frequency domain pre-processing and decoding of the transmitted and received data respectively. Using additional antenna elements at either the BS or MS side at the downlink and/or uplink opens an extra spatial dimension for system level optimization. The so-called space-

Advanced Scheduling Schemes in 4G Systems

Figure 17. MIMO scheduling system model

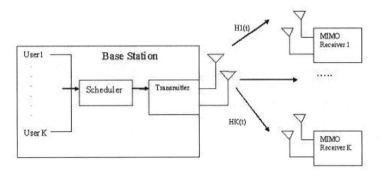

time processing methods exploit spatial dimension with the aim of improving the link's performance in terms of one or more possible metrics, such as the error rate, communication data rate, coverage area and spectral efficiency. In fact, the $n_T \times n_R$ wireless MIMO channels, however, are characterized by time varying fading $n_T \times n_R$ channels. As a result of the multiple antennas, the MIMO channel can be viewed as a vector channel with $m^* = \min\{n_T, n_R\}$ spatial channels.

The scheduling philosophy of MIMO based systems is mainly concerned with the maximization of one or more number of the following gains incurred:

1. Diversity gain
2. Array gain
3. Spatial multiplexing gain

Diversity gain corresponds to the mitigation of the effect of multipath fading, by means of transmitting or receiving over multiple antennas at which the fading is sufficiently uncorrelated. Array gain corresponds to a spatial version of the well-known matched-filter gain in time-domain receivers. Spatial multiplexing gain refers to the ability to send multiple data streams in parallel and to separate them on the basis of their spatial signature. The latter is much akin to the multiplexing of users separated by orthogonal spreading codes, timeslots or frequency assignments, with the great advantage that, unlike Code, Time or Frequency Division Multiple Access (CDMA, TDMA or FDMA), MIMO multiplexing does not come at the cost of bandwidth expansion; it does, however, suffer the expense of added antennas and signal processing complexity. The basic resource allocation and scheduling algorithms and schemes described previously are all applicable in MIMO systems with additional degree of freedom is added in terms of channel matrix instead of a single channel condition as in SISO. In SISO, we can consider received signal y(t) as:

$$y(t) = \alpha x(t) + n(t)$$

where x(t) is the transmitted signal, n(t) is the noise and α is the channel attenuation factor. So, channel capacity as per Shannon's theorem is:

$$C = \log_2\left(1 + \frac{P}{\sigma_n^2}|\alpha|^2\right)$$

where P = transmitter power, σ_n = noise voltage.

In MIMO case, channel capacity becomes:

$$C = \log_2\left[\det\left(I + \frac{\frac{P}{n_T}}{\sigma_n^2}.HQH^*\right)\right] \quad (8)$$

when

$y(t) = Hx(t) + n(t)$

where **H** is the channel matrix, **H***= transpose of **H** and **Q** = covariance matrix of the transmitted vector

We can apply traditional resource allocation and scheduling algorithms described earlier by considering channel capacity according to equation 8. In fact, The Gaussian MIMO channel capacity problem can be generally formulated into an optimization problem. Given a fixed channel or a random channel with certain probability distributions, the objective is to optimize the input covariance matrix under a power constraint so that the channel capacity is achieved. For simple illustration of the advantage of scheduling by exploiting the inherent MUD to enhance the spectral efficiency of a MIMO system, we can observe:

1. SISO spectral efficiency no diversity:
   $C = \log_2\left(1 + SNR\right)$ bps/Hz
2. MIMO spectral efficiency with n_T transmit and n_R receive antennas without diversity:
   $C = N \log_2\left(1 + SNR \times \frac{n_T}{n_R}\right)$ bps/HZ
3. MIMO spectral efficiency with n_T transmit and n_R receive antennas with diversity:
   $C = \log_2\left(1 + SNR \times n_T \times n_T\right)$ bps/Hz

There are also few specialized MIMO scheduling algorithms presented in the literature. Manish Airy et al. (2003) proposed spatially greedy scheduling in multi-user MIMO wireless systems which exploits opportunistic channel state dependency of spatially diverse MIMO wireless channel and integrates packet scheduling based on certain priority functions, where the priority functions capture the user QoS requirements quantified in terms of throughput and delay. The proposed MIMO scheduling schemes provide an optimal trade-off between

1. Higher system capacity by exploiting multiuser diversity.
2. Fairness among users based on their instantaneous channel conditions relative to average channel conditions.
3. Tolerable latency requirements specified by the user applications.

Recent progress in information theory suggests that so-called dirty paper coding (DPC) achieves the sum capacity of the multi-antenna Gaussian broadcast channel. DPC is a pre-coding technique that allows for pre-cancellation of interference at the transmitter. Further, when Gaussian inputs are optimal, the entire capacity region can be completely characterized by DPC. DPC achievable region is the largest known achievable region for the multiple-antenna broadcast channel (Jindal, et. al, 2005). However, implementation of DPC requires significant additional complexity at both transmitter and receiver, and the problem of finding practical dirty paper codes that approach the capacity limit is still unsolved. On the other hand, linear pre-coding is a low complexity but sub-optimal transmission technique with complexity roughly equivalent to point-to-point MIMO, and is able to transmit the same number of data streams as a DPC-based system. Linear pre-coding therefore achieves the same multiplexing gain which characterizes the slope of the capacity vs. SNR curve as DPC, but incurs an absolute rate or power offset relative to DPC. The contribution of this work is the quantification of this rate or power offset. The interaction between multi-user diversity and spatial diversity is assessed in (Gozali, et. al, 2003) by investigating the performance of STBC with scheduling algorithms. They consider a scenario in which K active data users, each equipped with multiple antenna elements, are served by a MIMO BS. Using different scheduling algorithms at the BS, time resources are allocated to the users in order to facilitate multiuser diversity which enhances the overall throughput of the system. MIMO scheduling can exploit the so-called random or dumb beam-forming approaches to provide a mechanism to induce fast and high fluctuations in a low-mobility environment via a time-varying common set of random weights. When the random weights are aligned with the channel vector for a given user, they provide a beam-forming gain, and the resulting SNR increases. If, instead, a user lies in a beam-former null, its SNR drops. Viswanath et al. (2002) show that for an increasing number of users, the performance of random beam-forming converges to that of coherent beam-forming, while the former only requires partial instead of full CQI. They have demonstrated the effect of introducing dumb antennas to maximize MUD. Dumb antennas in principle randomly sweep out a beam and opportunistically send data to the user closest to the beam which Improves performance in fast fading Rician environments by spreading the fading distribution.

Case Study: Scheduling in WiMAX and LTE

WiMAX and LTE are characterized by high data rate, low latency and stringent QoS. These demanding requirements call for highly optimized system design and scheduling plays an important role in order to maximize the network resource allocation and management to satisfy the intended throughput and QoS requirement.

WiMAX system uses OFDMA for multiple access at the MAC layer and is based on IEEE 802.16. WiMAX has gained much attention recently for its capability to support high transmission rates and QoS for different applications. In WiMAX, the BS (Base Station) centrally allocates the channels in different slots to different SSs (Subscriber Stations) for uplink and downlink which in turn allocates these resources to the various connections they are supporting at that time. The BS is aware of the channel state of all OFDMA subchannels for all the SSs. Thus the BS can exploit multi user diversity by allocating different subchannels to different SSs based on its objective function after satisfying some system constraints. WiMAX standard does not specify the type of scheduling algorithm to be used and instead leaves it to the discretion of the vendor. The algorithms we have discussed previously under OFDMA scheduling are all applicable here. These are not specified in WiMAX standard and developers are free to develop their own innovative procedures. WiMAX standard specifies the following scheduling types: Unsolicited Grant Service (UGS), real-time Polling Service (rtPS), extended real-time Polling Service (ertPS), non-real-time Polling Service (nrtPS) and Best Effort (BE). UGS has the highest priority and is designed for real-time service flows that will generate fixed size packets periodically at a constant bit rate (CBR). Examples of applications that may use this service are T1/E1 emulation and VoIP without silence suppression. The service offers fixed-size grants on a real-time periodic basis. The rtPS is designed to support real-time service flows that generate variable-size data packets periodically. It is suitable for VoIP, video streaming services. The service offers real-time, periodic, unicast request opportunities that meet the flows real-time needs and allow the SS to specify the size of the desired grant. This service requires more request overhead than UGS, but supports variable grant sizes for optimum data transport efficiency. The ertPS is an extension of rtPS to utilize the efficiency of both UGS and rtPS. It is suitable for voice with activity detection. The nrtPS is designed to support delay tolerant applications that consist of variable-sized data packets. It supports non-real-time flows such as file transfer protocol (FTP). The BE supports data streams, such as hypertext transfer protocol (HTTP), that do not require a minimum service-level guarantee and may be handled on a resource-available basis. In WiMAX, generally centralized scheduling is preferable. The central scheduling performed by the WiMAX BS and IP-level policing should be sufficient to enforce proper sharing of resources. The discussed OFDMA MAC scheduling algorithms like proportional fair, sum rate maximization, long term proportional fair along with link adaptation technique like Adaptive Modulation and Coding (AMC) can be implemented in WiMAX BS. The selection of a particular algorithm is depended up on the developer and system administrator as per the requirement. Network level IP-packet scheduler also needs to be implemented and round robin and FIFO should be avoided as far as practicable for system optimization. Many proposals and works are found in the literature particularly tailor-made for WiMAX, like in (Ukil, et al., 2009 & Belghith, et al., 2008), which mostly deal with QoS-aware scheduling in WiMAX.

Scheduling in LTE is similar to WiMAX except that LTE has different QoS architecture. QoS in LTE provides access network operators and service operators with a set of tools to enable service and subscriber differentiation. LTE core network is known as evolved packet system (EPS) for its support to all-IP configuration. QoS in LTE is primarily network-initiated and class-based, where a service is offered to a subscriber by the operator. The term service is used as the offering an operator makes to a subscriber. In order to distinguish the traffic types and treat them with respect to a predefined QoS, a QoS class identifier (QCI) is assigned for each traffic flow and associated with a Traffic Forwarding Policy (TFP). TFP denotes a set of pre-configured traffic handling attributes relevant within a particular user plane network element. In LTE bearer is the basic enabler for traffic separation to provide differential treatment for traffic with differing QoS requirements. A bearer in general is referred to as an edge-to-edge association between the User Equipment (UE) and the Gateway. As per functionality, two types of bearers exist; Guaranteed Bit Rate (GBR) and non-GBR and as per configuration, two another types of bearers exist; default and dedicated bear-

ers. The scheduling philosophy of LTE is to decide for establishing GBR bearer for a service request primarily based on expected traffic load versus predicted capacity. If sufficient capacity or low expected traffic load assumed, any service both real-time and non real-time, can be realized on non-GBR bearers. Like WiMAX, the scheduler in LTE is to be implemented in eNodeB and the eNodeB distributes the available radio resources in one cell among the UEs, and among the radio bearers of each UE. The details of the scheduling algorithm are left to the eNodeB implementation, but the signaling to support the scheduling is standardized.

FUTURE RESEARCH SCOPE

In this chapter, we have thoroughly discussed resource allocation and scheduling schemes for 4G wireless networks from the perspective of maximizing the overall system throughput, minimizing the outage probability, maximizing number of user in a fair way to guarantee users' satisfaction in a heterogeneous QoS-bound traffic scenario. There are few issue are still open and needs good amount of attention in future research work, which in effect help in meeting the target of 4G networks like LTE and WiMAX. Due to variable data rates and stochastic transmission inherent in channel-aware networks, the issue of admission control is becoming very challenging and interesting. Admission Control for channel-aware scheduling and its impact on MAC design is an open area. Admission control is responsible for accepting or rejecting the connection according to the available bandwidth that satisfies the connection and guarantees the required QoS without degrading the QoS for other existing connections. Admission control is not defined in the standard, although many propositions are made by different authors to establish admission control in the BS. Since the IEEE 802.16 MAC protocol is connection oriented, the application first establishes the connection with the BS as well as the associated service flow (UGS, rtPS, ertPS, nrtPS, BE). Connection request and response messages are exchanged between the SS and the BS in order to establish the connection. Integration of scheduling and admission control is in general tight and correlated optimization problem where the objective function is to maximize the network provider revenue while satisfying the subscriber. The objective functions of scheduler and admission control individually are almost mutually exclusive. While scheduler aims to assign the limited bandwidth and power to the user to maximize the system utility, admission control restricts the access of resources to the user. These two objective functions are very dissimilar as there exists contradiction between the expectation of the service provider and the subscriber. This is an open optimization problem which should guarantee the satisfaction of both the service provider and the subscriber.

The schemes and algorithms discussed so far assumed that CQI is perfectly known at the BS and the scheduler module through a separate feedback channel. The CSI is usually estimated at the receivers and, hence, prone to estimation errors. Moreover, feedback delays may cause outdated CQI used by the adaptive scheduling algorithm. However, when we have imperfect CSIT at the transmitter, the efficiency of multiuser scheduling is reduced because the wrong set of users may be selected for transmission, which results in packet transmission outage even if powerful error correction code is applied. The impact of imperfect and limited CQI to the system performance with adaptive scheduling and optimal algorithm development needs further amount of study.

Apart from WiMAX and LTE. IEEE 802.20 (Klerer, 2003), also known as Mobile Broadband Wireless Access (MBWA) is emerging as the standard which combines the benefits of mobility, standardization and multivendor support. IEEE 802.20 is an efficient packet based air interface to optimize the transport of IP-based services to enable world-wide deployment of affordable, ubiquitous, always-on and interoperable heterogeneous multivendor MBWA networks. Unlike, LTE and WiMAX, not much work has been done keeping in mind the issues of IEEE 802.20, particularly in scheduling and resource allocation. The key aspect in IEEE 802.20 scheduling is the integration of MAC and IP layer co-scheduling, which means cross-layer scheduling which should not be independently MAC or network layer centric.

Another issue which needs good attention is the design of a scheduler in presence of multiple networks, i.e. in heterogeneous networks, where vertical handoff takes place. This will be the common scenario when a user moves from LTE to WiMAX networks or vice versa but wants to seamlessly run the application with same level of QoS. The issues to be considered from heterogeneous scheduling are:

1. How does the scheduler deal with the changing resource environment and control end-to-end QoS for mobile users accessing different services over heterogeneous networks, given that each network possesses its own resource allocation schemes and QoS policies?
2. How to map the QoS of heterogeneous networks to facilitate the scheduler to work and optimize effectively?
3. Will the wireless environment in the context of heterogeneity be able to provide strict QoS guarantees to real-time applications and minimum QoS for the non real-time applications?
4. In case of service degradation, can it be graceful and be recovered very quickly to minimize the outage probability and QoS violation?

CONCLUSION

We have focus on introducing major aspects of advanced scheduling schemes and algorithms for 4G wireless networks. In essence, these scheduling algorithms are developed so that following objectives are achieved:

1. Maximizing overall system throughput, by higher wireless channel utilization, i.e. to increase spectral efficiency mostly by exploiting multi-user frequency and time diversity
2. Providing long term or short term fairness
3. Minimizing packet loss
4. Minimizing transmission power
5. Throughput Guarantee for individual users
6. Minimizing algorithm complexity for real-time implementation
7. Providing delay as well as jitter-bound for high priority traffic like UGS, rtps service classes in IEEE 802.16

In this chapter, we have introduced some general scheduling concepts and algorithms and discussed mostly through OFDMA system model. As the cross-layer design and optimization are very important in order to provide the best possible integrated solution to fulfill the above objectives in the most efficient manner, we have also presented cross-layer optimized framework. Section 3, 4 and 5 mainly deal with centralized scheduling method, where BS is responsible for taking the scheduling decision and allocate the resources to achieve local or global optimal point. From Shannon's theorem, rate-power relationship is convex in nature (shown in figure 18) which poses with the strategies of performance improvement by transmitting under better channel conditions require much less energy than under bad channel conditions for the same BER at the receiver and for delay constrained problem, transmit at rates just sufficient to meet the delay requirements. First strategy is to achieve MUD, which was the central theme of the scheduling algorithms in section 3. The second strategy is discussed in detail in section 5. In section 4, we have described the topic of packet scheduling, where we have assumed all-IP networks and network layer forwards packets to higher MAC or data link layer, where scheduling decision is taken. Different applications pose different requirements on the scheduling policy regarding tolerable complexity and distributed versus centralized implementations. In section 6, we have considered distributed scheduling, where scheduling decisions are taken in the absence of any centralized controller, like in ad hoc networks. For distributed scheduling we have considered game theoretic framework for both co-operative and non co-operative scenarios. In section 7, we have introduced the important topic of multi-layer resource allocation and scheduling, where we discussed the issues and complexities of MIMO scheduling. LTE and WiMAX are the two competing 4G wireless technology and scheduling is a significant part of their system design plan to meet up their target of high data rate, higher spectral efficiency, lesser latency and delay-jitter. Case study of LTE and WiMAX scheduler is described in section 8, which mainly covers the scheduling scheme and algorithm design, implementation issues primarily based on their QoS architecture. Finally, we have envisaged the future direction of research for developing advance scheduling scheme like, integrating admission control and

Figure 18. Data rate-power convex relationship

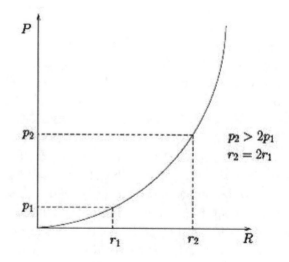

scheduling, scheduling under imperfect and limited CQI, scheduling challenges in emerging IEEE 802.20 and the important issue of the role of scheduler in heterogeneous environment.

REFERENCES

Airy, M., Shakkottai, S., & Heath Jr., R. W. (2003). Spatially Greedy Scheduling in Multi-user MIMO Wireless Systems. *IEEE PIMRC*, 982-986.

Altman, K., Borkar, V. S., & Kherani, A. A. (2004). Optimal random access in networks with two-way traffic. *PIMRC*, *1*, 609–613.

Bahai, A. R. S., & Burton, S. R. (2002). *Multi-Carrier Digital Communications Theory and Applications of OFDM*. New York: Kluwer Academic Publishers.

Belghith, A., & Nuaymi, L. (2008). Comparison of WiMAX scheduling algorithms and proposals for the rtPS QoS class. In *14th European Wireless Conference* (pp. 1-6).

Bhagawa, P., Bhattacharya, P., Krishna, A., & Tripathi, S. K. (1996). Enhancing throughput over wireless LANs using channel state dependent packet scheduling. In *INFOCOM*, San Francisco, USA, (pp. 1133-1140).

Biglieri, E., Proakis, J., & Shamai, S. (1998). Fading channel: information theoretic and communication aspects. *IEEE Transactions on Information Theory*, 2619–2692. doi:10.1109/18.720551

Challet, D., Marsili, M., & Zhang, T. (2004). *Minority Games Interacting Agents in Financial Markets*. Oxford, UK: Oxford Finance Series.

Chen, J., Yu, K., Ji, Y., & Zhang, P. (2008). Non-Cooperative Distributed Network Resource Allocation in Heterogeneous Wireless Data Networks. *IEEE Transactions on Mobile Computing*, *7*(3), 332–345. doi:10.1109/TMC.2007.70723

Desilva, S., & Das, S. R. (1998). Experimental evaluation of channel state dependent scheduling in an in-building wireless LAN. In *7th International Conference on Computer Communications and Networks* (pp.414-421).

Ergen, M., Coleri, S., & Varaiya, P. (2003). QoS Aware Adaptive Resource Allocation Techniques for Fair Scheduling in OFDMA Based Broadband Wireless Access Systems. *IEEE Transactions on Broadcasting*, *49*(4), 362–370. doi:10.1109/TBC.2003.819051

Ermolova, N. Y., & Makarevitch, B. (2007). Low Complexity Adaptive Power and Subcarrier Allocation for OFDMA. *IEEE Transactions on Wireless Communications*, *6*(2), 433–437. doi:10.1109/TWC.2007.05232

Falahati, A., & Ardestani, M. (2007). An Improved low-complexity resource allocation algorithm for OFDMA systems with proportional data rate constraint. In *International Conference on Advanced Communication Technology ICACT*, Gangwon-Do, South Korea (pp. 606-611).

Galstyan, A., Kolar, S., & Lerman, K. (2003). Resource allocation games with changing resource capacities. In *International Conference on Autonomous Agents, ACM* (pp. 145-152).

Girici, T., Zhu, C., Agre, J. R., & Ephremides, A. (2008). Practical Resource Allocation Algorithms for QoS in OFDMA-based Wireless Systems. In *IEEE International Broadband Wireless Access Workshop* (pp. 897-901).

Gozali, R., Buehrer, R. M., & Woerner, B. D. (2003). The impact of multiuser diversity on space-time block coding. *IEEE Communications Letters*, *7*(5), 213–215. doi:10.1109/LCOMM.2003.812182

Grossglauser, M., & Tse, D. (2001). Mobility increases the capacity of wireless adhoc networks. In *IEEE INFOCOM* (pp. 1360-1369).

Hu, M., Zhang, J., & Sadowasky, J. (2004). Traffic aided opportunistic scheduling for wireless networks: algorithms and performance bounds. *The International Journal of Computer and Telecommunications Networking, 46*(4), 505–518.

Jang, J., & Lee, K. B. (2003). Transmit Power Adaptation for Multiuser OFDM Systems. *IEEE Journal on Selected Areas in Communications, 21*(2), 171–178. doi:10.1109/JSAC.2002.807348

Jindal, N., & Goldsmith, A. (2005). Dirty-paper coding versus TDMA for MIMO Broadcast channels. In *ICC* (pp. 1783-1794).

Johari, R., Mannor, S., & Tsitsiklis, J. N. (2005). Efficiency loss in a network resource allocation game: the case of elastic supply. *IEEE Transactions on Automatic Control, 50*(11), 1712–1724. doi:10.1109/TAC.2005.858687

Khattab, A. K. F., & Elsayed, K. M. F. (2004). Channel-quality dependent earliest deadline due fair scheduling schemes for wireless multimedia networks. In *International symposium on modeling, analysis and simulation of wireless and mobile,* 2002, Venice, Italy, (pp. 31-38).

Kim, H., & Han, Y. (2005). A Proportional Fair Scheduling for Multi-carries Transmission Systems. *IEEE Communications Letters, 9*(3), 210–212. doi:10.1109/LCOMM.2005.03014

Kim, I., Park, I., & Lee, Y. H. (2006). Use of Linear Programming for Dynamic Subcarrier and Bit Allocation in Multiuser OFDM. *IEEE Transactions on Vehicular Technology, 55*(4), 1195–1207. doi:10.1109/TVT.2006.877490

Kittipiyakul, S., & Javidi, T. (2007). Resource Allocation in OFDMA with Time-Varying Channel and Bursty Arrivals. *IEEE Communications Letters, 11*(9), 708–710. doi:10.1109/LCOMM.2007.070672

Kivanc, D., Li, G., & Liu, H. (2003). Computationally efficient bandwidth allocation and power control for OFDMA. *IEEE Transactions on Wireless Communications, 2*(6), 1150–1158. doi:10.1109/TWC.2003.819016

Klerer, M. (2003). Introduction to IEEE 802.20: Technical and Procedural Orientation. *IEEE 802.20 PD-04.*

Li, G., & Liu, H. (2003). Dynamic resource allocation with finite buffer constraint in broadband OFDMA networks. IEEE *Wireless Communications and Networking,* Vol. 2, March 2003, 1037–1042.

Liu, X., Chong, E., & Shroff, N. (2003). A framework for opportunistic scheduling in wireless networks. *Computer Networks, 41*(4), 451–474. doi:10.1016/S1389-1286(02)00401-2

Liu, Y., Gruhl, S., & Knightly, E. W. (2003). WCFQ: an opportunistic wireless scheduler with statistical fairness bounds. *IEEE Transactions on Wireless Communications, 2*(5), 1017–1028. doi:10.1109/TWC.2003.816777

McCormick, S. T., & Pinedo, M. L. (1995). Scheduling independent jobs on uniform machines with both flowtime and makespan objectives: a parametric analysis. *ORSA Journal on Computing, 7*(1), 63–77.

McKenney, A. (1991). Stochastic fairness queuing. *Journal of Internetworking Research and Experience, 2,* 113–131.

Münz, G., Pfletschinger, S., & Speidel, J. (2002). An Efficient Waterfilling Algorithm for Multiple Access OFDM. In *Globecom* (pp. 681-685).

Nagle, J. (1987). On packet switches with infinite storage. *IEEE Transactions on Communications, 35*(4), 435–438. doi:10.1109/TCOM.1987.1096782

Neely, M., Modiano, E., & Rohrs, C. (2002). Power and server allocation in a multi-beam satellite with time varying channels. In *INFOCOM* (pp. 1451–1460).

Nguyen, T. D., & Han, Y. (2006). A Proportional Fairness Algorithm with QoS Provision in Downlink OFDMA Systems. *IEEE Communications Letters*, *10*(11), 760–762. doi:10.1109/LCOMM.2006.060750

Niyato, D., & Hossain, E. (2008). A Noncooperative Game-Theoretic Framework for Radio Resource Management in 4G Heterogeneous Wireless Access Networks. *IEEE Transaction on Mobile Communications*, *7*(3), 332–345. doi:10.1109/TMC.2007.70727

Rhee, W., & Cioffi, J. M. (2000). Increase in capacity of multiuser OFDM system using dynamic subchannel allocation. *VTC Spring*, Tokyo, Japan (pp. 1085–1089).

Ryu, S., Ryu, B. H., Seo, H., Shin, M., & Park, S. K. (2005, December). Wireless Packet Scheduling Algorithm for OFDMA System Based on Time-Utility and Channel State. *ETRI Journal*, *27*(6), 777–787. doi:10.4218/etrij.05.1005.0001

Sadeghi, B., Kanodia, V., Sabharwal, A., & Knightly, E. (2002, September). Opportunistic Media Access for Multirate Ad Hoc Networks. In *MOBICOM*, Atlanta, USA, (pp. 24-35).

Schrage, L. E., & Miller, L. W. (1966). The Queue M/G/1 with the Shortest Processing Remaining Time Discipline. *Operations Research*, *14*(4), 670–684. doi:10.1287/opre.14.4.670

Shreedhar, M., & Vargese, G. (1996). Efficient Fair Queuing Using Deficit Round-Robin. *IEEE/ACM Transactions on Networking*, *4*(3), 375-385.

Sklar, B. (1997). Rayleigh Fading Channels in Mobile Digital Communication Systems Part I: Characterization. *IEEE Communications Magazine*, *35*(7), 90–100. doi:10.1109/35.601747

Stiliadis, D., & Varma, A. (1998). Latency-rate servers: a general model for analysis of traffic scheduling algorithms. *IEEE/ACM Transactions on Networking*, *6*(5), 611–624.

Stolyar, A. L., & Ramanan, K. (2001). Largest Weighted Delay First Scheduling: Large Deviations and Optimality. *Annals of Applied Probability*, *11*(1), 1–48. doi:10.1214/aoap/998926986

Tao, M., Liang, Y. C., & Zhang, F. (2006, June). Adaptive Resource Allocation for Delay Differentiated Traffic in Multiuser OFDM Systems. In *ICC* (pp. 4403 – 4408).

Tassiulas, L., & Sarkar, L. (2002). Maxmin fair scheduling in wireless networks. In *IEEE INFOCOM*, (pp. 763-772).

Todini, A., et al. (2006). Cross-layer design of packet scheduling and resource allocation algorithms for 4G cellular systems, *WPMC*, San Diego, September 2006.

Ukil, A. (2008, August). Long-Term Proportional Fair QoS Profile Follower Sub-carrier Allocation Algorithm in Dynamic OFDMA Systems. In *13th International OFDM Workshop*, Hamburg, Germany (pp.1-5).

Ukil, A., Sen, J., & Bera, D. (2009, January). A New Optimization Scheme for Resource Allocation in OFDMA Based WiMAX Systems. In *International Conference on Computer Engineering and Technology (ICCET'09)*, Singapore, (pp. 145-149).

Viswanath, P., Tse, D., & Laroia, R. (2002). Opportunistic beamforming using dumb antennas. *IEEE Transactions on Information Theory*, *48*(6), 1277–1294. doi:10.1109/TIT.2002.1003822

Wang, B., Han, Z., & Liu, K. J. R. (2006). Stackelberg Game for Distributed Resource Allocation over Multiuser Cooperative Communication Networks, *Globecom*, (pp. 1-5).

Wengerter, C., Ohlhorst, J., & Elbwart, A. G. (2005). Fairness and Throughput Analysis for Generalized Proportional Fair Frequency Scheduling in OFDMA. In *VTC, Spring, 2005* (pp. 1903-1907).

Wong, C. Y., Cheng, R. S., Letaief, K. B., & Murch, R. D. (1999). Multiuser OFDM with adaptive subcarrier, bit and power allocation. *IEEE Journal on Selected Areas in Communications, 17*(10), 1747–1757. doi:10.1109/49.793310

Wong, I., Shen, Z., Evans, B., & Andrews, J. (2004). A low complexity algorithm for proportional resource allocation in OFDMA systems. In *IEEE Signal Processing Workshop*, TX, USA, (pp. 1–6).

Wu, D., & Negi, R. (2003). Effective capacity: a wireless link model for support of quality of service. *IEEE Transactions on Wireless Communications, 2*(4), 630–643.

Yaiche, H., Majumder, R., & Rosenberg, C. (2000). A game theoretic framework for bandwidth allocation and pricing for broadband networks. *IEEE/ACM Transactions on Networking,* 667-678.

Yao, Y., & Giannakis, G. B. (2005). Rate-Maximizing Power Allocation in OFDM Based on Partial Channel Knowledge. *IEEE Transactions on Wireless Communications, 4*(3), 1073–1083. doi:10.1109/TWC.2005.847022

Zhang, Y. J., & Letaief, K. B. (2004). Multiuser adaptive subcarrier-and-bit allocation with adaptive cell selection for OFDM systems. *IEEE Transactions on Wireless Communications, 3*(4), 1566–1575. doi:10.1109/TWC.2004.833501

Chapter 16
End-to-End Quality of Service in Evolved Packet Systems

Wei Wu
Research In Motion, Limited, USA

Noun Choi
Research In Motion, Limited, USA

ABSTRACT

The recent emergence of new IP-based services that require high bandwidth and low service latency such as voice over IP (VoIP), video sharing, and music streaming have motivated the 3rd Generation Partnership Project (3GPP) to work on the all IP-based cellular networks called Evolved Packet System (EPS). It is challenging for EPS not only to meet the Quality of Service (QoS) requirements of new services but also to make sure the QoS of existing services not impacted. In this chapter, the authors will first present an overview of EPS, and then focus on the aspects of QoS principles and mechanisms in EPS. End-to-end QoS models have been developed to analyze the application performance in EPS. Simulation results have shown that VoIP service requires resource reservation to guarantee its QoS requirement, and e-mail service does not experience significant performance degradation even when assigned a low service priority and the system experiences short period congestion. However, web browsing performance may not be improved proportionally to the network bandwidth increase due to the inherent network probing procedure of the transport protocol.

INTRODUCTION

This chapter gives an overview of the Evolved Packet Systems (EPS) including the system architecture and key network elements and their main functionalities. Some of the new essential features of EPS are highlighted. Among these features, this chapter will focus on the Quality of Service (QoS) in EPS. As EPS is the first all-IP based mobile cellular network, it is challenging to provide the QoS not only to traditional IP-based data services but also to services such as voice that has been usually provided through circuit-switched systems. The EPS QoS principles and mechanisms will be introduced in aspects such as service differentiation and admission control. We also present the end-to-end QoS performance analysis of the basic services such as

DOI: 10.4018/978-1-61520-674-2.ch017

VoIP, e-mail and web browsing through simulation study of EPS QoS model.

BACKGROUND

Current generation mobile cellular networks such as Global System for Mobile Communications (GSM), Enhanced Data Rates for GSM Evolution (EDGE) and Universal Mobile Telecommunications System (UMTS) have been designed for the circuit-switched voice service and low to medium bit rate data services such as e-mail and web browsing. The recent emergence of new IP-based services that require high bandwidth and low service latency such as voice over IP (VoIP), video sharing, and music streaming have motivated the 3rd Generation Partnership Project (3GPP) to work on the all IP-based cellular networks called Evolved Packet System (EPS). Compared to the existing systems, the EPS is targeting to provide higher user data rates, reduced latency, improved system capacity and coverage, reduced network complexity and lower operating costs. The following lists some of the major performance objectives of the EPS (3GPP TS 22.278, 2008; 3GPP TS 25.913, 2008):

- To support instantaneous peak data rates of 100 Mbps on the downlink and 50 Mbps on the uplink in the radio access network.
- To provide low user and control plane latency with a target of less than 10 ms user plane radio access network round-trip time (RTT) and less than 100 ms channel setup delay.
- To be capable of supporting large amount of mixed traffic including voice, data and multimedia.
- To optimize the level of system complexity and mobility management signaling in order to reduce infrastructure and operating costs. UE battery consumption shall also be minimized accordingly.
- The interruption time during handover shall not exceed the 300 ms for real-time services and 500 ms for non-real-time services.

EPS System Overview

Figure 1 shows the overall EPS system architecture, which consists of the evolved packet core (EPC) and long term evolution (LTE) radio access network (RAN), also known as, evolved

Figure 1. EPS system architecture

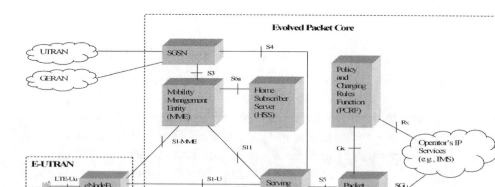

Universal Terrestrial Radio Access Network (E-UTRAN). As the EPS has to co-exist with the current 2G/3G mobile cellular networks, backward compatibility and inter-working are important aspects to consider while the EPS standards are being developed. The 2G and 3G 3GPP RANs including GSM/EDGE Radio Access Network (GERAN) and UTRAN are connected to the EPC via the SGSN. EPS also has the functions that support inter-working with non-3GPP radio access networks such as Worldwide Interoperability for Microwave Access (WiMAX).

Figure 2 shows the key EPS network elements including evolved Node B (eNB), Mobility Management Entity (MME), Serving Gateway (S-GW) and Packet Data Network Gateway (P-GW) and their main functionalities. The Serving General Packet Radio Service Support Node (SGSN), Home Subscriber Server (HSS) and Policy and Charging Rules Function (PCRF) are not included as they are inherited from the previous 3GPP standard releases with new network interfaces defined to communicate with EPS network elements. For a complete list and description of EPS network elements and functionalities, please refer to the EPS specifications (3GPP TS 23.401, 2008; 3GPP TS 23.402, 2008).

Evolved Node B (eNB)

The eNB is the only network node in E-UTRAN. An eNB is mainly responsible for the radio resource management (RRM) including radio bearer control, radio admission control, connection mobility control, and dynamic allocation of resources to UEs in both uplink and downlink. The eNB also has other responsibilities such as IP header compression and encryption, paging and broadcasting within the radio cell, UE measurement control and rate enforcement.

Mobility Management Entity (MME)

The MME is a signaling-only control plane entity. MME's main function is to manage each UE's mobility including UE location tracking and paging based on the tracking area list. An MME is in charge of the Non-Access Stratum (NAS) signaling, which refers to the signaling between the EPC and UE for procedures such as network attachment and service request. Other responsibilities of the MME include EPS bearer management including bearer establishment, modification and release, and UE authentication and authorization.

Figure 2. Key EPS network elements and main functionalities

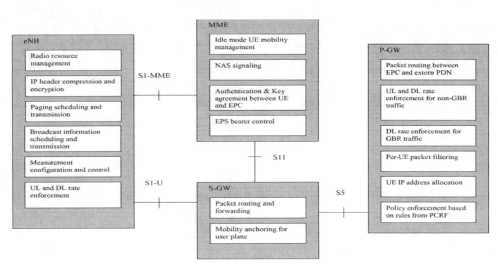

Serving Gateway (S-GW)

The S-GW is the user plane counterpart of the MME. An advantage of a separate network element for signaling is that signaling capacity can grow independently of the traffic capacity. The main function of the S-GW is to take care of the packet routing and forwarding between EPC and E-UTRAN.

Packet Data Network Gateway (P-GW)

The P-GW takes care of the packet routing between the EPC and external PDN such as the Internet and IMS. The P-GW has other functionalities including packet filtering, UE IP address allocation, and data rate and policy enforcement based on the rules provided by Policy Control and Charging Rules Function (PCRF).

EPS Essential New Features

EPS, as the name implies, is the system evolution based on the existing 2G/3G mobile cellular networks. The following is a list of the essential new features in EPS.

Higher Radio Bandwidth and Faster Radio Link Error Recovery

The target radio theoretical downlink peak rate is 100 Mbps, and the target theoretical radio uplink peak rate is 50 Mbps. By using very short radio subframes (1ms), a short round trip time (8 ms) between User Equipment (UE) and evolved Node B (eNB), and hybrid ARQ (HARQ), recovering lost packets over the radio link happens with low delay.

Flattened Network Architecture

An eNB is the only logical network entity connecting the UE and the EPC. The functions of Node B and RNC in 3G cellular networks are merged into the eNB, which directly reduces both user and control plane latency and overhead as there are few nodes involved in the packet processing and forwarding.

Packet-Switched Only Network

The interfaces between network elements in the EPS are all IP-based. The radio access network has been designed for scheduling and transmission of IP-based packet data including VoIP. EPS is a packet-switched (PS) only network.

Separate Core Network Control and User Plane

The MME is the control plane entity in EPC while the S-GW is the user plane counterpart of the MME. This allows the signaling capacity requirements to grow independently from that of traffic capacity.

Differentiated QoS

QoS Class Identifier (QCI) based differentiated services have been defined in EPS (3GPP TS 23.203, 2008). There are nine levels of QCI labels based on different traffic characteristics. The traffic priority associated with each QCI level will be used as one of the major parameters for packet scheduling.

Default Bearer for "Always-On IP Connectivity"

In EPS, each UE will be assigned and keep an IP address as long as the UE has registered with the network, i.e., has established a default bearer with a PDN. To realize the "Always-on IP Connectivity" in EPS, the default bearer (including the IP address) will be kept even when the UE is in the idle mode (the default bearer is explained

Figure 3. User plane bearer in EPS. ©2008, 3GPP™, Used with permission.

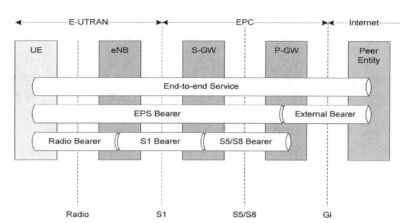

Service Differentiation

The QoS Class Identifier (QCI) label, shown in Table 1, is used to classify the packets and differentiate packet forwarding or dropping. An EPS bearer is associated with single QCI label, and an SDF aggregate gets a treatment based on the QCI label that the bearer is associated while they are queued and forwarded in the EPS. A QCI label has four characteristics, namely resource type (GBR, or Non-GBR), Priority, Packet Delay Budget (PDB), and Packet Error Loss Rate (PELR). The PDB defines the upper bound on the packet delay in the EPS, between UE and PDN-GW. The packet scheduling principle is that the PDB of an SDF with priority k should be met before meeting the PDB of an SDF with priority $k+1$ if congestion happens. The PELR defines the upper bound of non-congestion related link layer packet loss rate. The PELR is used when the link layer protocol configurations are determined.

In general, GBR bearers are given higher priorities than non-GBR bearers except for the bearers for real-time signalling like IMS signalling, which are given the highest priority. However, the assignment of priority to different types of packets is up to the operator. The last column in the above QCI table lists some example applications corresponding to each QCI level.

Admission Control and Resource Reservation

A service data flow on the GBR bearer has guaranteed bandwidth throughout the lifetime of the flow. The parameter GBR represents the guaranteed data rate for the SDF. Prioritized queue management based on QCI does not guarantee that GBR bearers get network resource to fulfill the bandwidth requirement. EPS utilizes admission control mechanism at eNB in addition to the QCI based differentiated queue management. While eNB performs admission control, the eNB only considers available radio bandwidth limitation. EPS assumes that wired links in the core network have enough capacity to carry all the traffic provided by the radio link in E-UTRAN.

An EPS bearer is associated with a QoS parameter called Admission and Retention Priority (ARP), and the ARP is used by eNB when it performs admission control. Suppose that a new bearer establishment request arrives at the eNB, and admitting the new bearer request will result in congestion on the radio link. The eNB may drop one or more bearers with the lowest ARP value among the competing bearers which include the new arrived bearer as well as the existing bearers. This makes sure that the GBR bearers have enough bandwidth to carry traffic at guaranteed bit rate.

in detail later). This is the fundamental change compared to the PDP context usage in the current cellular networks.

INTRODUCTION TO EPS QOS

As mentioned above, differentiated QoS principles and mechanisms have been used in EPS. The Differentiated Services (DiffServ) (Blake, Black, Carlson, Davies, Wang & Weiss, 1998) was originally defined by the Internet Engineering Task Force (IETF), and it has been applied in UMTS, with four defined traffic classes. However, as EPS is a PS only network, real-time services such as voice that are currently transported over the circuit-switched network, will be transported over the PS domain. One of the challenges is that the same level QoS for the real-time services (e.g., voice) is expected in the all IP-based EPS as in the circuit-switched cellular networks (e.g., GSM). On the other hand, the QoS for the traditional IP-based data services such as e-mail and web browsing should not be impacted under the QCI based EPS QoS mechanisms. In order to meet those requirements, EPS adopts the principles of Differentiated Services as well as Integrated Services (Braden, Clark & Shenker, 1994).

EPS Bearers

As shown in Figure 3, an EPS bearer between the UE and the P-GW consists of three concatenated bearers, namely Radio Bearer (between UE and eNB), S1 bearer (between eNB and S-GW), and S5/S8 bearer (between S-GW and P-GW). The communication peer in a PDN connects to the P-GW through the external bearer, in many cases in a form of virtual private network (VPN). This bearer architecture in EPS has a number of advantages. Because the next hop is indentified by the tunnel ID of the encapsulated packet, routing overhead is significantly reduced. The serving eNB might change over time due to the mobility of the UE. Routing path re-discovery process in that case is limited on the path between the S-GW and eNB by means of replacing the old S1 bearer with the new S1 bearer. Also, it is easy to enforce QoS. QoS mechanisms such as admission control and resource reservation can be performed on the control plane during the establishment of the EPS bearer.

An EPS bearer is called Guaranteed Bit Rate (GBR) bearer if the bearer requires certain amount of network resources throughout the lifetime of the bearer. Otherwise, the bearer is called Non-GBR bearer. When a UE connects to a PDN, it first establishes a Non-GBR bearer called default bearer. During or right after the default bearer establishment procedure, an IP address associated with the default bearer. All the other bearers established between the UE and this PDN are called dedicated bearers, and the IP address associated with the default bearer is used for all the traffic flows within the PDN connection in order to identify the UE.

One or more service data flows (SDF) can be associated to an EPS bearer if the QoS requirements of those SDFs are identical. An EPS bearer is associated with a pair of Traffic Flow Template (TFT), UL TFT and DL TFT (3GPP TS 23.203, 2008). A TFT can have zero or more packet filters. A packet filter has single value of a range of values for several parameters, such as source address, source port number, destination address, destination port number, and protocol type. When a packet arrives at P-GW, the packet is evaluated with packet filters of existing DL TFTs. The evaluation order of TFTs and packet filters in a TFT is predefined. If the arrived packet matches with a packet filter, the evaluation procedure is terminated, and then the packet is sent onto the EPS bearer that is associated with the TFT. For an uplink packet, similar procedure is performed at the UE. The S-GW and the eNB maintain mapping table between S5/S8 bearer and S1 bearer, and between S1 bearer and radio bearer respectively.

Table 1. QoS Class Identifier (QCI) Characteristics. © 2008, 3GPP™, Used with permission.

QCI	Resource Type	Priority	Packet Delay Budget	Packet Error Loss Rate	Example Services
1	GBR	2	100 ms	10^{-2}	Conversational Voice
2	GBR	4	150 ms	10^{-3}	Conversational Video (Live Streaming)
3	GBR	5	300 ms	10^{-6}	Non-Conversational Video (Buffered Streaming)
4	GBR	3	50 ms	10^{-3}	Real Time Gaming
5	Non-GBR	1	100 ms	10^{-6}	IMS Signalling
6	Non-GBR	7	100 ms	10^{-3}	Voice, Video (Live Streaming) Interactive Gaming
7	Non-GBR	6	300 ms	10^{-6}	Video (Buffered Streaming) TCP-based (e.g., www, e-mail, chat, ftp, p2p file sharing, progressive video, etc.)
8	Non-GBR	8	300 ms	10^{-6}	Video (Buffered Streaming) TCP-based (e.g., www, e-mail, chat, ftp, p2p file sharing, progressive video, etc.)
9	Non-GBR	9	300 ms	10^{-6}	Video (Buffered Streaming) TCP-based (e.g., www, e-mail, chat, ftp, p2p file sharing, progressive video, etc.)

The ARP can be used in another case where radio condition becomes poor and more robust modulation schemes are required. Not all the existing SDFs can be served properly in this case. The bearers with low ARP may be released in order to ensure the service quality of the other bearers. Although there is no guaranteed resource allocation for Non-GBR bearers, the admission control may also be applied to Non-GBR bearers.

Once a GBR bearer went through the admission control, the eNB reserves radio resource for the GBR bearer. Semi-Persistent Scheduling (SPS) is a way of reserving constant data bandwidth during the active span of the SDF. For example, for a VoIP call connection, radio resources will be reserved periodically as the VoIP packets will normally arrive in a constant interval (e.g., every 20 ms).

Aggregate Maximum Bit Rate (AMBR) defines the upper bound of aggregate bit rate for Non-GBR bearers of a UE. Access Point Name (APN)-AMBR limits the aggregate bit rate for Non-GBR bearers of a UE in a PDN connection while UE-AMBR limits the aggregate bit rate for all the Non-GBR bearers of a UE.

QoS Enforcement

Figure 4 depicts the network elements related to QoS in EPS. EPS bearer level QoS parameters include QCI, ARP, GBR and Maximum Bit Rate (MBR). GBR and MBR are applied only to GBR bearers. AMBR, including UE-AMBR and APN-AMBR, is defined for aggregate Non-GBR bearers. The values for those QoS parameters are decided as a part of Policy Control and Charging (PCC) decision, and PCC rules are, in general, made by a functional entity, Policy Control and Charging Rules Function (PCRF).

When a UE needs to establish an EPS bearer, the UE sends a bearer resource request message to the network. This message includes required QoS parameter values for the bearer. The message is routed through the control plane bearers to the P-GW. On receiving the bearer resource request message, Policy Control and Charging Enforcement Function (PCEF), which is a logical entity in the P-GW, relays the message to PCRF. The PCRF makes PCC decision for the request based on the available information. For the UE initiated bearer resource request, PCRF uses subscription related information such as subscriber's allowed services, priority for each allowed service, sub-

Figure 4. QoS related network elements in EPS

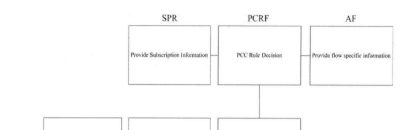

scriber's category, and charging related information, which is provided by Subscription Profile Repository (SPR). Since the UE has already provided required QoS parameters, the decision will be either accepting the request or rejecting it. Once the PCRF makes the PCC decision, PCEF initiates a bearer establishment procedure.

It is possible that the bearer is initiated by the network for some applications, such as incoming voice calls and data push services. When P-GW receives a downlink packet which does not belong to any existing EPS bearer, PCEF requests PCC rules for the new flow to the PCRF. The PCRF uses the flow related information provided by Application Function (AF) as well as subscription related information from SPR. The flow information from AF includes subscriber identity, source and destination IP address, media type, required bandwidth, and emergency indicator. After PCC rules are determined, PCEF initiates a bearer establishment procedure.

EPS utilizes several types of rate control. For a GBR bearer, data rate cannot exceed Maximum Bit Rate (MBR). MBR rate enforcement is performed by P-GW and eNB for both uplink and downlink. AMBR enforcement is applied to aggregate Non-GBR bearers. Uplink and downlink UE-AMBR is enforced by eNB while uplink APN-AMBR is enforced by UE and downlink APN-AMBR is enforced by P-GW. The P-GW also checks if uplink APN-AMBR is controlled properly by UE.

EPS QOS MODELING

To investigate the EPS QoS from end to end, QoS functions have been modeled at P-GW, eNB and UE. The EPS standards are not finalized at the time of the writing, and they will be evolving continuously. In addition, some system implementation details such as scheduler implementation at the eNB and bandwidth allocation policies at the P-GW are left up to the decisions of vendors and operators. Assumptions that reflect the updated standard status as well as realistic system configuration need to be made to model the EPS QoS functions.

Our approach for the simulation implementation is firstly to make the reasonable modeling assumptions. We reduced the network scale while allowing the simulation to represent the large scale network in deployment. Standard traffic models with the reference to the traffic statistics collected from currently deployed networks are used to construct the traffic models used in simulations. Packet scheduling principles, applied at P-GW and eNB, are assumed according to the latest

EPS specifications. Secondly, only the essential EPS features as listed below that interest us have been implemented.

- QCI based on differentiated services.
- Traffic policing based on Maximum Bit Rate (MBR) and Aggregate Maximum Bit Rate (AMBR).
- Medium Access Control (MAC) scheduling (Dahlman, Parkvall, Skold & Beming, 2008) at eNB including algorithms like Round Robin, Proportional Fair and Maximum Carrier-to-Interference Ratio (Max C/I).
- Faster radio link error recovery (i.e., Hybrid Automatic Request (HARQ) with 1ms sub-frame).
- Higher wireless bandwidth (i.e., DL 100Mbps/UL 50Mbps).

In addition, only the user plane functions have been modeled. The main purpose of the simulation is to investigate the end-to-end performance of applications when the system is loaded. It is assumed that before the simulation begins, the UE has attached to the network and created the EPS bearer, and the service session has been established.

We have used the network simulation tool OPNET (OPNET, 2008) to model the EPS QoS functions. As we are interested in the application performance from the viewpoint of network layer and above, some details of lower layer protocol operation could be hidden, and detailed modeling of every layer-2 (L2) protocol is not necessary. Instead, only the essential functions at L2 such as MAC scheduling and HARQ have been modeled. The physical layer models physical channel effects such as packet loss and radio channel conditions between the UE and the eNB. Because OPNET does not provide the wireless pipeline for EPS, we used the traces generated from the MATLAB physical channel modeling tool as the input to model the packet loss and radio channel condition.

For simplicity, the Ethernet connections have been used to simulate the transport pipelines between the network nodes.

Network Topology

Figure 5 shows the network topology used for the simulations. There are ten UEs, six of which generate GBR traffic whereas four UEs are used for non-GBR traffic. All the UEs are within the coverage of an eNB, and they are connected to two public data networks through the EPS.

Since the network system model is scaled down (e.g., with limited number of network entities), we assumed that the bandwidth of the links in the network is also scaled down. The bandwidth configure to each link is 10 Mbps. Note that GBR flows utilize up to 6 Mbps (i.e., 60%) of each link, and the aggregated maximum bit rate (AMBR) shared by non-GBR flows is set at 4 Mbps (i.e., 40%) of each link. As mentioned earlier, the wireless links between the UEs and the eNB have been simulated using wire line Ethernet links.

A traffic flow shall belong to one of the available QCI priority levels based on the type of the traffic. The following priority assignment has been used as default: Video conferences have priority 3, VoIP calls have priority 5, Web browsing is assigned at priority 7 and e-mail transferring flows have priority 9. The highest priority level is 1 while the lowest priority level is 9.

Simulation Configuration

A number of simulations have been performed. For microscopic observations, we generated one or two flows per traffic type and observed the round trip time, real time congestion window size, sequence number and acknowledge number at each end of connection. For macroscopic observations, we generated multiple connections per flow type and observed overall throughput of each traffic type, end-to-end delay of GBR traffic flows, and download / upload time of email and web pages.

Figure 5. Network topology used in simulations

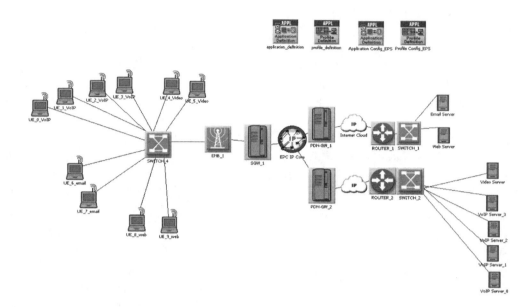

The number of active connections for a traffic type at any time is random and the size of an email or a web page follows statistical distribution as given below. Each simulation for macroscopic observations runs for 500 seconds.

Two types of applications are used for GBR traffic: Voice over IP (VoIP) and video conference. A video conference session generates 576 kbps of constant load in both directions. Two video conference sessions are active throughout the simulations. GSM codec was used as the coding scheme for VoIP and the talk spurt and silence lengths follow the default configuration of the simulation tool. The length of a VoIP session follows an exponential distribution with a mean of 45 seconds. The inter-repetition time also follows exponential distribution with a mean of 4.5 seconds. There could be up to 40 active VoIP sessions at any time, but the number of active sessions changes randomly over the time.

E-mail and web browsing are used as non-GBR traffic flows. The size of email can be categorized in to three classes: large, medium and small. The size of a small email is constant 500 bytes while the sizes of medium and large email follow uniform distributions with average 10,000 bytes and 150,000 bytes respectively. The maximum number of active email transmissions at any time is 60. The size of a web page can also be categorized into three classes: large, medium, and small. The maximum number of active web page download sessions is 120. The probability that an email is small, medium or large is 1/3 each. The size of a web page is also evenly distributed over the classes.

Performance Analysis

As mentioned earlier, one of the challenges in EPS is that the same level QoS for the real-time services (e.g., voice) is expected as in the circuit-switched cellular networks. On the other hand, the QoS for the traditional IP-based data services should not be impacted under the differentiated services-based EPS QoS mechanisms. In this section, we will present some of the simulation results with regarding to the application performance, where VoIP represents the real-time service and e-mail and web browsing represent the traditional IP-based data services.

Figure 6. VoIP E2E delay with and without QoS

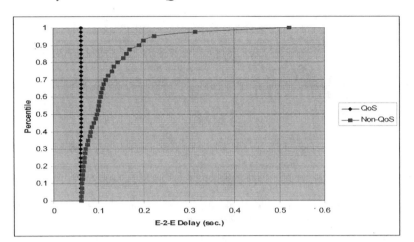

VoIP Performance

Figure 6 compares VoIP performance in terms of end-to-end delay with and without QoS cases. Resources are reserved for the VoIP traffic in the case of with QoS. VoIP traffic has to compete for the resources with other traffic in the case of without QoS. The results are as expected that in the case of with QoS, VoIP end-to-end delay is almost constant at about 70 ms while in the case of without QoS, the delay could be up to about 500ms. This justifies the needs of IntServ type of QoS mechanisms like resource reservation to guarantee the QoS for real-time applications like VoIP.

E-Mail Performance

One of the objectives of the simulation is to determine if there is any negative impact caused by the EPS service differentiation mechanisms. Figure 7 compares the medium-size e-mail download time in the case of with and without QoS. The system was heavily loaded during the simulation, i.e., applications servers generated traffic rate close to the maximum network bandwidth. In the case of with QoS, e-mail is given the lowest QCI priority compared to other non-GBR traffic, i.e., web traffic in the simulation. In the case of without QoS, email and web traffic are both treated as best-effort service with the same priority.

The results show the comparable e-mail download time in both cases. The e-mail performance is not significantly impacted with QCI based differentiated services, where the e-mail was assigned lowest priority, and the system was heavily loaded.

Web Performance

One of the objectives is to analyze the web browsing user experience in EPS. Figure 8 shows the web page transfer time as the function of available network bandwidth. The web page size ranges from 10 Kbytes to 1 Mbytes. The total network bandwidth is 10 Mbps, and the available bandwidth for web page transfer ranges from 5% (i.e., 500Kbps) to 80% (i.e., 8 Mbps).

It is intuitive that the web page transfer time decreases as the available network bandwidth increases. However, for the small size web pages (i.e., 10 Kbytes), the reduced web page transfer time may not be proportional to the increased available network bandwidth. In other words, the user may not notice the performance improvement as the available bandwidth is increased.

Figure 7. E-mail download time with and without QoS

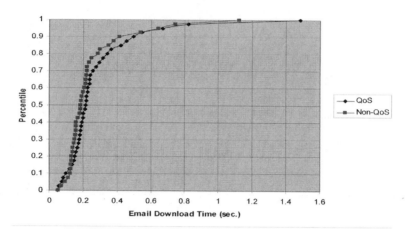

Also, suppose that the user has the maximum waiting tolerance, e.g., 7 seconds. Once a web page transfer exceeds this limit, from the user point of view, the service quality is not satisfactory. The user may give up the waiting or reload the web page after the waiting limit. For example, as shown in the Figure 8, for the 500 Kbytes web page transfer, even the available network bandwidth has been doubled from 5% to 10% (i.e., 500 Kbps to 1Mbps), the web page transfer time exceeds the 7 seconds (the maximum waiting tolerance) in both cases.

Figure 9 attempts to explain from the TCP perspective the fact that the web browsing performance improvement is not proportional to the increase of the available network bandwidth.

Let BW_Effective be the effective bandwidth used during the web browsing as defined as follows.

$$BW\_Effective = CW \times 8/RTT$$

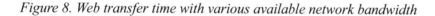

Figure 8. Web transfer time with various available network bandwidth

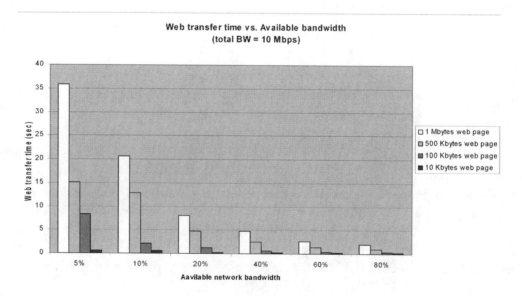

Figure 9. Effective bandwidth used for web page transfer

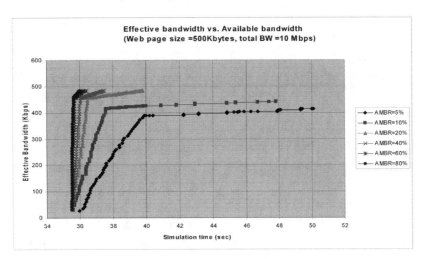

Where *CW* is the TCP congestion window size in bytes, and *RTT* is the round trip time in seconds for the TCP segment transfer.

As the web page or HTTP packets are transported by TCP, the web transfer time is controlled by the TCP sending rate. The TCP congestion windows determines how many TCP segments a sender can send without waiting for the acknowledgement form the receiver. So, the BW_Effective basically defines how much bandwidth the web server can utilize to deliver the web page to the client.

As shown in the Figure 9, during TCP slow start phase, the web server is probing the available network bandwidth by increasing the TCP congestion window size step by step until it reaches the slow-start threshold. As most of the web pages of mobile web browsing have small sizes, the web session may be already over before the web server figures out the maximum available network bandwidth. Therefore, web browsing performance improvement may not be proportional to the increase of the available network bandwidth, especially for the small-size web page transfer.

4G NETWORKS AND FURTHER RESEARCH

International Telecommunication Union – Radio Communications Sector (ITU-R) has set a list of requirements for the 4th generation cellular network systems (Report ITU-R M.2134, 2008). A number of key requirements are on the spectral efficiency, bandwidth, latency, mobility, handover, and VoIP capacity.

The 3GPP Technical Specification Group Radio Access Network (TSG RAN) is working toward improving radio network system beyond LTE, called LTE-Advanced in order to satisfy those requirements. A number of features are being investigated for LTE-Advanced. Those features include using Multiple-Input and Multiple-Output (MIMO) for uplink as well as for downlink, Carrier Aggregation, Relay, and Inter-Cell Interference Coordination. LTE already has the feature of downlink MIMO. Capturing the Uplink MIMO into LTE-Advanced is under study. Supporting bandwidths larger than 20 MHz is needed to meet the higher data rate requirement while backward compatibility with LTE should also be ensured. Carrier Aggregation (CA) has been introduced to satisfy both requirements. The concept of

Relay Node has been introduced to improve the throughput of the cell edge user and to expand the coverage of the high data rates. Control signals as well as user plane traffic between eNB and UE goes through the relay node. The relay nodes are connected to the eNB through wireless link. The neighboring eNBs may perform resource management procedure together to mitigate the inter-cell interference. The inter-cell interference coordination could improve the cell edge spectral efficiency.

The IP Multimedia Subsystem (IMS) is an important piece of the All-IP based wireless networks since network and service providers rely on IMS as the centralized controller for all the service creation and management. The 3GPP is currently studying to enhance the IMS in several aspects including improving the system efficiency, reliability, and scalability.

The EPS has been introduced in order to provide the users better IP based services. As the basic design work is about to finish, performance evaluation for a number of major applications such as VoIP, web browsing and multimedia streaming should be performed in the EPS. These studies may provide suggestions for the future system evolution.

CONCLUSION

This chapter has provided an overview of EPS including the system architecture and network entities and their main functionalities. The differentiated services-based EPS QoS principles and mechanisms have been introduced. End-to-end EPS QoS models have been developed to analyze the application performance in EPS. We have conducted simulation experiments to verify the VoIP and e-mail performance, and to analyze the performance of web browsing in the EPS.

The simulation results have shown that VoIP service requires the resource reservation to guarantee its QoS requirement. This is because that EPS is a packet-switched network, and only resource reservation can provide the QoS guarantee for the voice as in circuit-switched network.

Based on the simulation results, e-mail service does not experience significant performance degrading even being assigned low QCI priority when the system is heavily loaded. Given its delay-tolerant characteristics, the e-mail service should work fine in a well dimensioned network, where even congestion may happen, but should only last for a short period.

The web browsing performance may not be improved proportionally to the network bandwidth increase due to the inherent network probing procedure of the transport protocol. As most of the web pages of mobile web browsing have small sizes, the web session may be already over before the web server figures out the maximum available network bandwidth. One of the possible ways to speed up the network probing phase is to use large initial congestion window size; and, cross-layer information from the radio link layer may also help the transport protocol to speed up the network probing phase. For example, when the UE handovers from UTMS to EPS, which has much higher supported data rates, the transport protocol could use more aggressive ways to increase its congestion window size as long as this type of lower layer information is made available to the transport layer.

REFERENCES

3GPP TS 23.203 (2008). *Policy and charging control architecture*. Sophia Antipolis, France: the 3rd Generation Partnership Project.

3GPP TS 23.401 (2008). *General Packet Radio Service (GPRS) enhancements for Evolved Universal Terrestrial Radio Access Network*. Sophia Antipolis, France: the 3rd Generation Partnership Project.

3GPP TS 23.402 (2008). *Architecture enhancements for non-3GPP accesses*. Sophia Antipolis, France: the 3rd Generation Partnership Project.

3GPP TS 36.300 (2008). *Evolved Universal Terrestrial Radio Access (E-UTRA) and Evolved Universal Terrestrial Radio Access Network (E-UTRAN); Overall description; Stage 2*. Sophia Antipolis, France: the 3rd Generation Partnership Project.

Blake, S., Black, D., Carlson, M., Davies, E., Wang, Z., & Weiss, W. (1998). *An Architecture for Differentiated Services (RFC 2475)*. Fremont, CA: The Internet Engineering Task Force.

Braden, R., Clark, D., & Shenker, S. (1994). *Integrated Services in the Internet Architecture: an Overview (RFC1633)*. Fremont, CA: The Internet Engineering Task Force.

Dahlman, E., Parkvall, S., Skold, J., & Beming, P. (2008). *3G Evolution: HSPA and LTE for Mobile Broadband*. Burlington, MA: Academic Press.

OPNET. (2008). *OPNET Modeler Documentation*. Bethesda, MD: OPNET, Inc.

Report ITU-R M.2134 (2008). *Requirements related to technical performance for IMT-Advanced radio interface(s)*. Geneva, Switzerland: International Telecommunication Union.

Mobility and Handover

Chapter 17
An End-to-End QoS Framework for Vehicular Mobile Networks

Hamada Alshaer
University of Leeds, UK

Jaafar Elmirghani
University of Leeds, UK

ABSTRACT

In recent years we have witnessed a great demand for high speed Internet access in vehicular environment, e.g., trains, buses and medical transport. This chapter introduces an integrated architecture for 4G vehicular mobile networks, which aims to guarantee high quality in provisioned triple-play traffic services (video, voice, and data) to road users. Within this architecture which is based on a cross layer design approach, our contributions can be described in three folds. Firstly, the authors introduce simple and efficient probing mechanisms which are integrated with network resource reservation policies for multihomed vehicular networks. Secondly, packet, flow and user splitting mechanisms have been integrated with end admission traffic control and scheduling mechanisms to guarantee even traffic load distribution among available air interfaces. Finally, the whole architecture has been evaluated under OMNeT++, where results illustrate the impact of network mobility on quality in provisioned services offered to a multihomed NEMO.

1. INTRODUCTION

Fourth-Generation (4G) wireless networks technology is based on all-IP networks architecture which is envisioned to transparently and ubiquitously support triple traffic, voice, video and data for end users. This future technology is driven by the needs for more bandwidth, network capacity and new radio spectrum, calling for performance enhancement and quality-of-service (QoS) guarantees in wireless access networks. IETF has defined QoS as a service agreement to provide a set of measurable networking attributes, including e2e delay, jitter and bandwidth. Alternatively, the QoS can be defined as a set of specific requirements, e.g. delay, jitter, throughput, packet loss and de-sequencing (packets reordering), for a particular service, e.g. Expedited Forwarding (EF) service class (Davie & Charny & Bennett & et,

DOI: 10.4018/978-1-61520-674-2.ch018

2002) in Differentiated Services (DiffServ)(Blake & Black & Carlson & et,1998), Guaranteed Service (GS) class in Integrated Services (IntServ)(Braden & Zhang & Berson,1997)), Assured Forwarding (AF) service class (Heinanen & Baker, 1999) in DiffServ, control load service in IntServ and Best Effort service (Clark& Fang, 1998), which can be provisioned as point-to-point service (p2p) or e2e service by service providers (SPs)(Yokota & Idoue & Hasegawa & Kato, 2002) to subscribed customers (Jamalipour.& Lorenz, 2006).

One of the recognized goals in wide communication networking based on all-IP networks technology is to realize an efficient inter-operability between various types of networks: wired, wireless, radio, cellular and mobile networks to support the convergence of all communication services (data, voice, and video) onto a common IP platform(Bai & Atiquzzaman,2003). Such an IP-based unified communication platform will enable a mobile Internet network to get benefit of all the services that could be offered through Internet network (Agrawal & Zhang & Sreenan & Chen, 2004). Furthermore, this opens new arenas into the development of intelligent transport services for road users who can be drivers, passengers, manually driven vehicles or autonomous vehicles.

Network mobility has always been a determining factor in implementing a robust and efficient communication protocol for mobile networks. Most of the mobility protocols, which have been standardized so far, require QoS mechanisms to improve and add value to its performance. This chapter investigates some of those QoS mechanisms which are required to be incorporated into the network elements employed to run network mobility basic support protocol (NEMO-BSP) (Devarapalli & Wakikawa & Petrescu & Thubert, 2005; Ernst & Lach,2007) for vehicular communications, namely IP-based vehicle-to-infrastructure communications. Figure 1 depicts mobile networks (NEMOs) that communicate via infrastructure with external entities (correspondent nodes (CNs) or a server hosted by intelligent transport service providers (ITSPs)). NEMO-BSP (Devarapalli & Wakikawa & Petrescu & Thubert, 2005) has given a basic solution for a challenging problem with regard to network mobility, where a set of IPv6 enabled and non enabled mobile nodes moving together in one entity called NEMO, e.g. bus or ambulance shown in Figure 1, can continue to communicate with any external entities despite high and dynamic network mobility.

NEMO-BSP runs between two main network elements: home agent (HA) and mobile router (MR) to track and localize a set of nodes moving in one mobile network. One or more mobile routers can be configured on a mobile network, based on its size and amount of traffic, to manage IP addresses resolution, configuration and allocation, which are required for managing the whole network mobility. In this chapter, however, we will focus on NEMO managed by a single multihomed (multi-interfaced) MR and served by a single HA. Mobile networks can have a wide range of mobile network nodes (MNNs) which interact with one or more home agents based on IPv6 technology (Soliman, 2004; SOLOMON,1998) . MNNs can be sensors mounted on NEMO, which transmit critical data to another NEMO or a server somewhere to analyze it for safety purposes and fleet management. MNNs can be simple users surfing on Internet or sending emails and communicating through their PDAs with other external CNs. MNNs can be a group of doctors performing a surgical operation and getting assistance from others in a hospital while they are in an ambulance driving them to the hospital. MNNs can be travellers listening to music and watching video in a bus or train. MNNs might be fishermen, in their ship or boat, watching a film on TV or listening to on-line radio news. While the MR acts as a gateway for MNNs or an anchor point, the HA forwards every packet destined to MNNs to one of the different MR egress interfaces that enable it to access Internet through access routers.

The variety in MNNs and mobile network topologies require various communication wireless

Figure 1. Vehicular mobile networks interaction with all IP-network

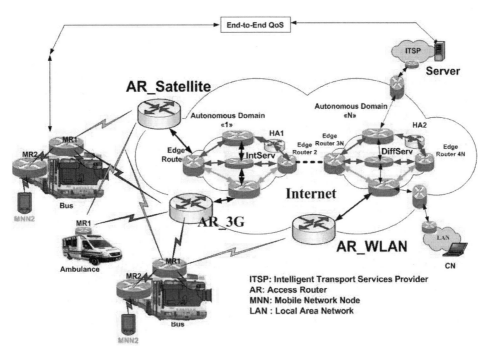

technologies to (i) cope with network mobility by maintaining a stable, continuous and transparent connection with external entities and (ii) enable wireless internet services providers (WISPs) and ITSPs to guarantee service level satisfaction for road users. The voice-oriented second generation cellular systems such as Carrier Division Multiple Access (CDMA) and Global System for Mobile Communication (GSM) are increasing their services and the General Packet Radio Service (GPRS) as well as Edge provide additional packet-switched data delivery on top of these enhanced features. Thus, there is a great interest in unifying the third generation (3G) wireless networks, wide CDMA (WCMA) and Universal Mobile Telecommunication System (UMTS) as well as its intermediate improvements, say 3G+, under the future all-IP networks (4G technology). Owing to the success of IP-based networks, it is therefore crucial to expand the IP features not only in the core network but also right through the access networks of the future 4G wireless networks. The future generation of mobile networks is application-driven, based dominantly on mobile Internet applications and services. Hence, an efficient support of these Internet-based services would not be possible without a core network totally based on IP, in which the wireless access networks constitute a main part (Eardley & Hancock, 2000).

The unification of all wireless access networks under the IPv6 layer will enable to utilize their advantages so that the shortcomings of individual wireless technologies could be complemented by other systems. Nevertheless, the all-IP networks technology will suffer from networks heterogeneity and interoperability which can deteriorate QoS provisioning (Akyildiz& Mohanty &Xie, 2004). To clarify this, let us take this scenario which is illustrated in Figure 1. Most of the public transport buses are equipped with a GPS which is used to localize the buses on the different routes and to calculate the waiting times at every stop. To this end, a satellite connection would be necessary to monitor and manage transport traffic congestion on the different routes. Although public buses run over determined routes, some of these across a city to reach rural place. To this end, wireless and radio networks with varied transmission ranges and capacities such as 3G and wireless local area network (WLAN) networks would be necessary to maintain uninterrupted access to Internet for these buses. However, the connections through 3G communication technologies, satellite and WLAN networks are very sensible to network mobility and its surrounding environment conditions. Furthermore, these networks suffer from scarce network resources and channel fading. Thus, a NEMO should be enabled to communicate simultaneously through all these communication networks so that the QoS required for upstream and down stream traffic can be guaranteed. This would enable reliable Internet services provisioning with variable service level agreements, reflecting network mobility on the scale of QoS provisioned for passengers.

When NEMO maintains multiple routing paths to reach Internet as shown in Figure 1, it is said a multihomed NEMO. NEMO can be multihomed through different configurations based on three parameters (Ng & Ernst & Paik & Bagnulo, 2007): the number of root MRs managing NEMO, the number of HAs serving NEMO and the number of mobile network prefixes (MNPs) advertised in the mobile network. In this chapter, we will focus on the configuration shown in Figure 1, where we consider there is only one root mobile router, one home agent and one MNP is advertised in the mobile network.

The remainder of this chapter is organized as follows. In Section 2, we explain NEMO-BSP operation and its interactions with NEMO and HA. In Section 3, we clarify the problem. In Section 4, we discuss the QoS required at the physical, link and IP layers, where we present probing techniques and evaluate a network resource reservation model. In Section 5, we introduce traffic engineering schemes for a multihomed NEMO. In Section 6, we introduce the simulation environment and discuss results, where further research directions are also emphasized. Finally Section 7 concludes the chapter and gives some perspectives for further research.

2. NEMO-BSP OPERATION AND ARCHITECTURE

When a mobile network like the public transport bus, shown in Figure 1, moves out of its station, home network, NEMO-BSP starts executing the tracking process based on IPv6, which is done through the configuration and registration processes. MR1 configures a temporary address called Care-of-Address (CoA) through each egress interface. Then, it sends a Binding Update (BU) to its home agent to bind its Home-of-address (HA) with the configured CoA, which enables the HA to reach NEMO anytime anywhere. The MNNs which are connected to the ingress interfaces of MR configure permanent addresses containing the Mobile Network Prefix (MNP) which is assigned previously to NEMO. In summary, NEMO-BSB (Devarapalli & Wakikawa & Petrescu & Thubert, 2005; Perera & Sivaraman & Seneviratne, 2004) procedure can be described in the following points:

- An IPv6 prefix, called Mobile Network Prefix (MNP), is delegated to a MR to advertise it in a mobile network. Consequently, MNNs configure IP addresses containing this prefix.
- If the MR is allocated multiple MNPs, then the MR home agent maintains a prefix table indexed by MR home address (HoA).
- When NEMO leaves the home link to a foreign network, the MR sends a Binding Update (BU) to the HA which creates a cache entry, binding the MR MNP to its CoA.
- MR can function as a mobile host (MH) or MR, according to the flag bit in the BU

message. This is set to R when the MR functions as a MR and it is set to zero when the MR functions as a MH.
- When the MR receives a BU ACK from the HA, a bi-directional tunnel is established between the HA address, tunnel entry point, to the MR CoA, tunnel exit point. All IPv6 traffic to/from NEMO is sent through this bi-directional tunnel.
- Packets destined to MNNs (NEMO) are encapsulated to the MR CoA by the HA, then they are de-capsulated by the MR and sent to MNNs in the mobile network.
- Packets destined to CNs are encapsulated to MR HA address by MR, then, they are de-capsulated by the HA and sent to CNs.

Although this protocol can guarantee transparent communication for MNNs when the bi-directional tunnel is established between the mobile router and home agent, different mechanisms are required to guarantee the quality in provisioned services for NEMO. This guarantees a minimum level of network performance for mobile applications running in mobile networks or on wireless devices carried by MNNs. In what follows we identify the different parts of the challenging problem which we consider throughout this chapter.

3. PROBLEM STATEMENT

Mobile networks can connect to Internet through multiple wireless and radio technologies, which have different communication characteristics in terms of traffic data rate, transmission range (communication coverage), QoS support, link reliability and communication stability. In addition, MNNs run different applications, generating heterogeneous traffic which requires different service classes and reliable QoS support. This chapter emphasizes the QoS mechanisms required to guarantee the e2e quality in provisioned services for NEMO, based on the multihoming configuration shown in Figure 1. Note we will focus more on the QoS mechanisms that can be developed in accessible network elements, namely MR and HA along the routing paths connecting MNNs with CNs or NEMO with servers hosted by Intelligent Transport Services Providers (ITSPs).

As shown in Figure 1, these routing paths are formed of different segments according to the employed communication technologies and link characteristics. MNNs and subnets in NEMO can connect to ingress interfaces of MR1 through Zigbee, Bluetooth, Wi-Fi IEEE 802.11a/b/g configured in access point mode or Ethernet cable technologies, meanwhile the MR1 egress interfaces can connect NEMO to Access Routers (ARs) through GPRS, satellite, 3G communication technologies or/and WLAN IEEE 802.11a/b/g configured in infrastructure mode. The Internet network is a federation of loosely connected autonomous network domains governed by different administrators and networking policies. Furthermore, various service architecture models and protocols are employed to manage traffic and communication services. This complex heterogeneity reveals the real challenges which all-IP networks technology faces in e2e QoS provisioning for NEMO.

When NEMO shown in Figure 1 performs IP handoff between heterogeneous networks, namely 3G and WLAN, NEMO-BSP enables the MR1 to switch its connection to Internet through another access router in order to maintain seamless and transparent communication for MNNs. This implies changing of the routing path segment between MR1 and AR as well as other segments along the routing path between the previous AR and HA. Network resource availability along the routing paths connecting NEMO with external entities influences the quality in provisioned services which can be evaluated in terms of some measurable parameters like throughput (bandwidth), delay, jitter, loss and de-sequencing (packet reordering) from which upstream and downstream traffic of NEMO suffers.

- **Throughput** (bandwidth) parameter refers to the maximum data rate which can be provided by MR to MNNs over a period of time. This is also called a bit rate and measured in order of bits per second (bps).
- **Delay** parameter refers to the time elapsed from where the first bit of a packet sent by a source (e.g. MR or external entities) to the last bit of the packet received by the destination (e.g. MR or external entities). Diverse delay sources contribute in the delay budget, so the delay components include: handover delay, queuing delay, processing delay, transmission delay and communication protocols delay. It is measured in order of milliseconds (ms).
- **Jitter** parameter refers to the variations in delay introduced by network elements (e.g. MR, AR or HA, etc) along the selected routing path from a source to a destination.
- **Packet loss** parameter refers to the number of lost packets along a selected routing path, which are caused by low communication quality between the MR and external entities (e.g., handover, bit error caused by a noisy channel, network congestion and network queuing overflow).
- **De-sequencing (packet reordering)** refers to the ordering process of packets arriving at the HA or MR in out of order so that voice traffic can be played or video frames can be reconstructed at the destination.

To this end, routing paths switching implies also alteration of above e2e QoS parameters unless the new and previous routing paths have enough available network resource to serve the already admitted and handing off traffic. Besides, the e2e QoS provisioning can be influenced by the fulfilment of QoS requirements at MNNs, CNs and MR side like transmission power, CPU processing/response time and wireless communication cards quality. In this chapter, we follow the bottom-up approach in the design of QoS architecture which we introduce to guarantee the e2e QoS required for real-time IP traffic destined to or generated by NEMO. This involves the following main tasks:

- Understanding the radio and wireless communication technologies characteristics which shape the 4G vehicular communication systems. We term this part as 4G applications for vehicular mobile networks.
- Determining a monitoring mechanism in MR to provide the network layer with information regarding link layers QoS capabilities such as bit error rate, loss and delay. We term this part of the design as link layer information-based probing.
- Determining which of the different IETF service models should be employed in wired, wireless, radio networks and NEMO. These service models differ based on their support of service classes, performance, scalability and robustness. The choice depends on the expected amount of traffic traversing these networks and services required for this traffic (Alshaer & Horlait, 2005). We term this part of the design as IP service model selection.
- Determining a network resource reservation policy which can cope with network mobility. We term this part of the design as network resource reservation support.
- Enabling MR to distribute optimally (or evenly) traffic among their input and output interfaces. We term this part of the design as intelligent route control.
- Configuring a Call Admission Control (CAC) mechanism and a Service Scheduling Discipline (SSD) in MR1 and HA so that they can decide about traffic flows admission based on their traffic class and network resources availability along the routing paths connecting them. We term this part of the design as traffic admission and Service scheduling Control.

These mechanisms are conceived to complement the operation of network mobility protocols, namely NEMO-BSP to guarantee seamless, transparent and high quality of communications between NEMO and CNs or servers hosted by ITSPs. The integration of 3G access networks, WLAN and satellite require such architecture to better use the overall radio resources and enjoy the diverse services supported in all–IP networks. In what follows we discuss the above parts which constitute the whole proposed architecture.

4. END-TO-END QOS ARCHITECTURE DESIGN

4.1 4G Applications for Vehicular Mobile Networks

There has been a serious debate on the air link technologies most suitable for 4G wireless communication systems. It has been proposed from the research that traditional CDMA technologies are suited only for slow-speed continuous transmission applications such as voice, but not for high-speed all-IP wireless applications. Therefore, multiple access/multiplexing technology, such as orthogonal frequency-division multiple access/multiplexing (OFDMA/OFDM), has been proposed to replace traditional CDMA as a prime air link design for 4G wireless communication systems (Huang & Zhuang,2002). Today, both CDMA and OFDMA have been adopted by various standards. CDMA is used in IS-95, CDMA2000, wideband CDMA (W-CDMA), time-division synchronous CDMA (TDSCDMA), and so on. Its spread spectrum multiplexing technique has found wide applications, including early IEEE 802.11 and 802.11b/g WLAN, Bluetooth, and cordless telephony. CDMA technology with its many unique features, such as universal frequency reuse, processing gain, and soft handoff, may have great potential for further technological evolution. Interleave-division multiple accesses (IDMA), which can be considered a special case of CDMA, have also gained attention, especially for wireless uplink applications. OFDMA/OFDM is used in IEEE 802.11a/g/n WLAN, HIPERLAN/2, WiMAX, DVBT, asymmetric digital subscriber line (ADSL), very high rate DSL (VDSL), and others. It has also been chosen as the physical layer architecture for 3GPP long-term evolution (LTE). The major advantage of OFDM is its ability to deal with multi-path fading and narrowband interference without using complicated channel equalization. Being independent of the powerful access technologies based on CDMA and OFDMA, multiple-input and multiple-output (MIMO) antenna systems have been identified as one of the key technologies to support higher data rates through spatial multiplexing and diversity in comparison to single-antenna systems. MIMO techniques can be elegantly combined with OFDMA as well as CDMA.

While 3G downlink data rates at the present time do not exceed 2 Mb/s and 384 kb/s at pedestrian and vehicular speeds, respectively, with a potential upgrade to 10 Mb/s, 4G systems are expected to attain data rates of 20 Mb/s or higher. Although cellular networks enable convenient voice communication and simple infotainment services to drivers and passengers, they are not well-suited for certain direct vehicle-to-vehicle (V2V) or vehicle-to-infrastructure (V2I) communications. Whereas, vehicular ad hoc networks (VANETs), which offer direct communication between vehicles and from roadside units (RSUs), can send and receive hazard warnings or information on the current traffic situation with minimal latency (Alshaer & Horlait, 2005). However, no central coordination or handshaking protocol can be assumed in V2V communication environment, where many applications will be broadcasting information of interest to many surrounding cars. Thus, the necessity of a single shared control channel can be derived, even when multiple channels are available using one or more transceivers, at least one shared control channel is required.

The ODFMA is used in WAVE IEEE 802.11p standard (Wireless Access for the Ability in Vehicular Environments) (Hartenstein & Laberteaux, 2008) which is a modified version of IEEE 802,11a. But, it uses 10MHz channels as opposed to the 20MHz channels for IEEE 802.11a. Subsequently, data rates ranges from 3 to 27 Mb/s for each channel, where lower rates are often preferred in order to obtain robust communication. It provides wireless communications over short distances between information sources and transaction stations on the roadside and mobile radio units, between mobile units, and between portable units and mobile units. The communications generally occur over line-of-sight distances of less than 1000m between roadside units and mostly high speed, but occasionally stopped and slow moving, vehicles or between high speed vehicles. It supports a full range of existing uses of IEEE 802.11, but also includes a number of new classes of applications related to V2V roadway safety (such as vehicle collision avoidance, hazard warning, lane change warning and highway merge assistant), to infrastructure-to-vehicle (I2V) safety (such as warning on road works, traffic signals violation, road condition and curve speed) and emergency services (such as those provided by police, fire departments, ambulances, and rescue vehicles).

Two-way satellite services can be considered as an alternative to provide wireless connectivity on vehicles due to their extended global coverage. However, the data services available and affordable so far have proved rather limited data capacity. Even if this limitation is not an obstacle to develop telematics applications using a short message data structure, such as probe vehicle monitoring, its operation mode as point-to-point communications channel and its large inherent system latency makes it incompatible with most of the vehicle safety applications identified.

Mobile WiMaX IEEE 802.16e standard uses OFDMA and supports several key features necessary for delivering mobile broadband services at vehicular speeds greater than 120 km/hour with QoS comparable to broadband wired access alternatives (She & Yu & Ho & Yang, (2009)). A typical on-board Bluetooth application is the Bluetooth hands free car kit, which enables one-button or even voice-activated use of the driver's mobile phone, therefore minimizing physical distractions. Bluetooth may serve as a V2I communications channel for stationary vehicles in close proximity to the desired communications point. Although Bluetooth performance in terms of range and latency preclude its ability to support vehicle safety applications, it could be used to perform safety-related tasks such as updating navigational databases when the vehicle is parked in the garage. Ultra-wideband (UWB) appears to be fairly immune to multipath interference, a significant benefit for moving vehicles makes UWB a reasonable candidate to monitor for further developments that may allow its use for vehicle safety applications. The largest limitation of Ultra-wideband (UWB) for vehicle communications is the limited range expected with the initial systems that are likely to become available (Pavn & Shankar & Gaddam & Chou, 2006).

4.1.1 Cognitive Radio

To support inter-vehicular communication (IVC), the dedicated short range communication (DSRC) in US, allocated by the Federal Communications Commission (FCC), spans over 75 MHz of spectrum in the 5.9 GHz band. In Japan, 5.8 GHz DSRC was used by DEMO 2000 and 60 GHz millimeter wave has been tested to evaluate its performance under the hidden terminals. The primary bands for WiMaX deployments in 2006 are allocated at 2.3, 2.5, 3.3 and 3.5 GHz (licensed bands) and at 5.8 GHz (unlicensed band) (She & Yu & Ho & Yang, 2009). The capacity of the frequency channels currently assigned or foreseen for VANET applications ranges from 10 to 20MHz. With a high vehicular traffic density, those channels easily could suffer from channel fading and congestion. Making use of more than one channel leads to multi-channel synchronization problems, in particular for the case of a single transceiver per vehicle and to co-channel interference problems.

Vehicles in urban places are likely to use WLAN hot spots for communication and those in rural places will likely use 3G technologies for communication. This may create large holes of unused spectrum depending on the places where vehicles are communicating, which leads to the tremendous interest in Cognitive Radio (CR), whose frequency agility gives it the ability to flexibly change channels and technologies. This allows in turn road users to adaptively utilize whatever wireless resources are available at their time and place, taking advantage of spectrum holes that would otherwise be wasted. Note that allocating a fixed frequency band to each wireless service based on the current frequency allocation policies, is an easy and natural approach to eliminate interference between different wireless services. However, extensive measurements reported indicate that the static frequency allocation results in a low utilization (only 6%) of the licensed. radio spectrum in most of the time (Mchenry, 2003). Even when a channel is actively used, the bursty nature of most data traffics are still implied that a great deal of opportunities exist in using the spare spectrum. In order to better utilize the licensed spectrum, the Federal Communication Committee (FCC) has recently suggested the cognitive radio concept/policy for dynamically allocating the spectrum (Karp & Kung, 2000) to alleviate the severe scarcity of spectrum bandwidth. Cognitive radio is typically built on the software-defined radio (SDR) technology, in which the transmitter's operating parameters, such as the frequency range, modulation type, and maximum transmission power can be dynamically adjusted by software to capture whatever available channel (frequency band) (Mitola, 2000). This would allow multi-homed vehicular NEMOs to dynamically access frequency bands assigned to different radio access networks. The biggest issue, however, is to ensure better management for bandwidth and spectrum allocation in order to support new emerging 4G applications and services offered in scalable manner for road users. More major obstacles facing 4G wireless system deployment involve (i) the mobile device's chipset for connecting to different frequencies and (ii) multi frequency base station units supporting various carrier frequencies communication. The underlying wireless networks and the mobile handset should be able to reconfigure themselves to adapt to the new communication scenarios requirements (Natarajan & Nassar & Shattil, 2001).

Numerous decentralized opportunistic MAC protocols (Su & Zhang, 2007) have been proposed to support cognitive radio concept in 4G wireless communication systems. In (Ma, & Han & Shen, 2005) the authors proposed a dynamic open spectrum sharing (DOSS) MAC protocol, which requires three separate sets of transceivers to operate on the control channel, data channel, and busy-tone channel, respectively. In (Hsu & Wei & Kuo, 2007) the authors proposed a cognitive MAC with statistical channel allocation, in which the secondary users select the channel that has the highest successful transmission probability to send packets based on the channel statistics. But, the computational complexity for determining the successful transmission probabilities increases quickly with the number of licensed channels. In (Mishra, 2006) the authors proposed a multi-channel opportunistic MAC protocol, which however targets only at the Global System for Mobile Communications (GSM) cellular networks. In (Corderio & Challapali, 2007) the authors developed a multi-channel cognitive MAC protocol, called C-MAC, which, however, does not differentiate the primary users and the secondary users.

4.1.2 MAC Layer Design for Vehicular Communications

The one-control channel paradigm, together with the requirement for distributed control, leads to some of the key challenges of VANET design, namely medium access control (MAC) design. Although time division multiple access (TDMA)- and spatial division multiple access (SDMA)-based mechanisms were proposed, the main focus today is on using the IEEE 802.11 carrier sense multiple access (CSMA)-based MAC for VANETs, because of its availability and low cost considerations. MAC mechanisms which have been developed for inter-vehicular communications can be broadly classified into two categories, depending on the adopted radio interface: One category is based on the existing wireless LAN physical layers, such as IEEE 802.11 or Bluetooth. The second category of MAC is based on extending 3G cellular technology, i.e., CDMA, for decentralized access. Although the first MAC category can support distributed coordination in ad hoc mode, the flexibility of radio resource assignment and of transmission rate control is low. On the contrary, 3G extensions have the potential of high granularity for data transmission and flexible assignment of radio resources due to the CDMA component, though it suffers from the complexity of designing coordination function in ad hoc mode.

To this end, the following problems have to be addressed in order to extend 3G technologies for IVC (i) Distributed radio resource management, (ii) Power control algorithms and (iii) Time synchronization. All these problems are due to the absence of centralized infrastructure, thus the solution should rely on distributed media access control mechanisms. It has been proposed the Reservation ALOHA (R-ALOHA) for distributed channel assignment (Menouar & Filali & Lenardi, 2006). R-ALOHA has higher throughput than slotted-ALOHA, since a node that catches a slots can use it in subsequent frames as long as it has packets to send. However, R-ALOHA has a potential risk of instability in the case of many participating nodes and frequent reservation attempts due to short packet trains and therefore requires some modification to adapt it to IVC. (Lott & Halfmann & Schulz & Radimirsch, 2001) have solved this problem by letting every node reserves a small part of transmit capacity permanently even if it has no packets to send. This results in a circuit-switched broadcast connection primarily used for signalling purposes. The time synchronization is built upon the information from Global Positioning System (GPS) and additional synchronization sequence in parallel to data transmission (Rudack & Meincke & Jobmann & Lott 2003). Traditional R-ALOHA needs a broadcast environment for all nodes to receive all the transmitted signals and, most important, to get the status information of slots. Since IVC suffers from the hidden terminal problem, destructive interference with already established channels can occur and accessing nodes have no idea about the outcome of their transmission. To overcome these problems, (Borgonovo & Capone & Cesana & Fratta 2003) introduced a new protocol, called Reliable R-ALOHA (or RR-ALOHA). This protocol transmits additional information to let all nodes be aware of the status of each slot, thus safely allows the same reservation procedure of R-ALOHA to happen in IVC. The two-hop relaying that propagates the status information is very similar to what is used in ad hoc routing to let a node know the neighbour information of its neighbours.

However, (Muqattash & Krunz, 2003; Menouar & Filali & Lenardi, 2006) pointed out that RA-CDMA (random access CDMA) suffers from multi-access interference (MAI), resulting in secondary collisions (also known as near-far problem in the literature) at a receiver. Thus, (Muqattash & Krunz, 2003; Menouar & Filali & Lenardi, 2006) improved CA-CDMA reservation mechanism by modifying the (request-to-send/clear-to-send) RTS/CTS signals. The channel is split into control and data channels. RTS/CTS are transferred over control channels to let all potentially interfering nodes be aware of the channel status. In contrast to IEEE 802.11, interfering nodes may be allowed to transmit concurrently depending on some criteria. The protocol also exploits knowledge of the power levels of the overheard RTS/CTS to perform power control that intends to alleviate near-far problem.

The IEEE 1609 framework which is built on top of the IEEE 802.11p standard provides two parallel stacks, one for UDP/Transmission Control Protocol (TCP) over IPv6 and one called Wave Short Message Protocol (WSMP). With WSMP, various low level parameters can be specified like data rate and transmit power level. The complete IEEE 1609 framework currently consists of four parts, which specify the networking services (1609.3), multi-channel operation (1609.4), security issues (1609.2), and a specific transponder-like application called resource manager (1609.1) (Hartenstein & Laberteaux, 2008).

In this Chapter we propose vertically coupled protocol architecture for 4G vehicular networks to provision QoS to road users who have subscribed to DiffServ. The proposed architecture combines the transport layer protocols and link layer resource allocation to both guarantee transport layer QoS requirements and achieve efficient resource utilization in the link layer. The transport layer QoS is guaranteed with minimal equivalent resources required at the link layer and therefore at the physical layer. Besides, the MAC scheduler in the link layer is based on per-flow information, thus reducing the computation complexity and system overhead compared to other MAC schemes (Huang & Zhuang, 2002), which use per-packet information to determine scheduling priority.

4.2 Link Layer Information-Based Probing

Flows generated by applications running in NEMO or on servers hosted by ITSPs traverse finally the link and physical layers of the communication technologies employed at ARs and MR which forward them to their destinations. We consider that the network segment which connects MR with ARs is the most critical part of the routing path with HA. Although the current wireless and radio technologies have been developed to support QoS, they cannot guarantee satisfactorily the QoS required for real-time IP traffic. Thus, the QoS mechanisms employed in the network layer should be able to determine the link layer QoS capabilities so that it can provide it with enough information regarding the coordination of network resources reservation on the air interface. The link layer should in return provide synchronous feedback to the network layer regarding the ongoing availability of network resource at the air interface, which should be taken into account when requesting resources from the network and providing feedback to applications. This feedback can be exploited by the service scheduling discipline employed at the network and link layers to transmit real-time IP packets with acceptable QoS bounds.

A probing mechanism which is introduced in (Vasudevan & Papagiannaki & Diot & Kurose & Towsley, 2005) can be employed in MR1 to get the status information of the different link technologies. Alternatively, Figure 2 illustrates that the MR1 can get this information by analyzing HELLO packets and IPv6 messages (RtSolPr and PrRtadv) received from the different Access Points(APs) or ARs (Soliman, 2004). In (Alshaer & Horlait, 2005), vehicle nodes could use HELLO packets to count the number of vehicle nodes in their neighbourhood to calculate the broadcast forwarding probability. Here IPv6 messages (Router Solicitation for Proxy Advertisement (RtSolPr) and Proxy Router Advertisement (PrRtadv)) are used for CoA resolution and reachability verification with ARs. Meanwhile, HELLO packets are used to enable NEMO through MR egress interfaces, namely WLAN egress interface to obtain information about the amount of downlink traffic forwarded by the available APs which piggyback

Figure 2. Active probing on link layer

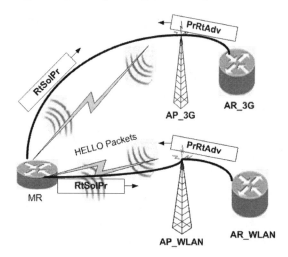

the load information in the data packets sent to NEMOs. These in turn record this information in their routing tables. Therefore, HELLO messages are sent on all channels at a time to enable NEMOs to update their routing tables based on the traffic load at APs. NEMO also receives HELLO messages from its neighbours; thereby, it becomes aware of how many NEMOs are connecting to its AP or a neighbouring AP. In addition, NEMO requires the estimation of traffic load at AP and MR so that the latter can decide to which AP among the available ones is better to connect. The MR and HA can estimate the upstream and downstream traffic along the different routing paths using one of the following techniques:

4.2.1 Locally Measured Load

NEMO measures and compares the amount of traffic it has received or forwarded during a recent time window through its egress interfaces. The HA can also measure and compare the upstream and downstream traffic destined to NEMO. For example, during the last 15 s, the average traffic load that MR has received or forwarded through the WLAN egress interface is 400 kbps, and through the 3G egress interface is 200 kbps. However, the average traffic load that HA has received is 520 kbps.

4.2.2 MR-Multihoming Measured Load

NEMO measures the average traffic it has received over a dynamic window time through its egress interfaces. Thereafter, traffic is stripped into egress interfaces to guarantee traffic load balancing among available egress links. The standard deviation of the amount of traffic accepted by egress interfaces should be minimized, while guaranteeing that the amount of traffic stripped into any egress link does not exceed the its capacity.

4.2.3 AP Measured Load

Since all traffics should pass through one of the available APs, the load of all NEMOs connected to this AP can be estimated as the load at the AP. For example, during the last 15s AP_WLAN observes that it has forwarded 5 Mbps of traffic, and AP_3G observes that it has forwarded 800 kbps of traffic. NEMO can use this information as a basis for egress interface selection. However, this is not enough because it does not indicate how many hops a packet needs to travel.

4.2.4 AP-Measured Weighted Load

The load is measured at an AP, but it is weighted according to the hop distance from the AP to the destination. For example, a NEMO is x distant from the AP_WLAN which observes that it has forwarded 5.5 Mbps and y distant from an AP_3G which observes that it has forwarded 600 kbps.

Therefore, the measured weighted load normalized by the hop distance can be used by NEMO to select the best AP. It can also be used by other NEMOs to choose better their relays to the available APs.

Note that our proposed probing mechanism is based on a specified policy. If NEMO is in a rural place or moves at a high speed, then probing should be done through the 3G egress interface. Meanwhile, if NEMO is in an urban environment, then probing should be done through the WLAN egress interface. By exploiting the status information collected by HELLO packets and IPv6 messages regarding the link technologies as well as estimated traffic load at AP, MR and HA, MR and HA can enable the network layer to select better (or switch to) a routing path among the available ones. In what follows we explain the support of QoS at the network layer.

4.3 IP Service Model Selection

On the network layer the IETF organization has standardized different IP-based service models such as DiffServ (Blake& Black & Carlson & et, 1998) or IntServ (Braden & Zhang& Berson, 1997). The consideration of an IP service model in a network domain depends on the expected amount of traffic entering in this network as well as traffic classes and QoS requirements which should be supported in this network. Although our NEMO scenario shown in Figure 1 indicates that we are focusing on non-transient NEMO (vehicle-to-infrastructure communication), our analysis can be developed to apply to a scenario where this NEMO functions as a relay to another NEMO intending to access the Internet using the VANET mode. In (Alshaer & Horlait, 2004), we explained the integration of these approaches in one platform called Client-Server Ad-hoc (CSAH). In such a scenario, although we can expect a relative increase in the amount of traffic in NEMO, the IntServ model would be the most appropriate to serve traffic. All wireless access networks should support IP traffic forwarding in 4G networks technology; however, traffic traversing these networks will be limited by their scarce capacity. Therefore, IntServ model would be appropriate to serve this traffic, where other works (Chiussi & Khotimsky& Krishnan, 2002; Langar & Thome & Grand, 2005) considered the MPLS technology to cope with mobility. Since a huge traffic traverses the Internet network, DiffServ (Blake & Black & Carlson, 1998) or MPLS can be employed. In this scenario, we assume that the CNs and ITSP belong to LANs, where IntServ can guarantee the e2e QoS required for traffic. Nevertheless, we only describe a scenario on which we will focus throughout this chapter; therefore, interested readers can select any other IETF service model based on the criteria mentioned so far.

4.4 Network Resource Reservation Support

Based on the scenario of ISP service model configuration shown in Figure 3, a policy for network resource reservation is required to guarantee the QoS required for mobile multimedia traffic. The Resource reSerVation Protocol (RSVP) (Braden & Zhang & Berson, 1997) is a signalling protocol for network resource reservation, which has been employed to estimate, guarantee, maintain and release network resource required for real-time traffic. For example, a passenger sitting in the bus, shown in Figure 1, should be able to check his emails and download some music from the Internet using his PDA. This could be the case for another passenger who uses his IPv6 camera to send live photos to his friends. The video installed in this bus should send images with a determined delay to a traffic or security centre for transport traffic control and security purposes. Consequently, traffic requiring various QoS requirements is generated by this NEMO, though we will focus on guaranteeing the e2e QoS required for real-time IP multimedia flows like video traffic in this scenario. The network resources required for this traffic should be reserved along a selected routing path between the MNN and CN or NEMO and ITSP so that the required e2e QoS can be guaranteed.

In the IntServ model, two aggregate reservations should be made which correspond to the Guaranteed Services (GS) and Control load Ser-

Figure 3. IETF service models selection scenario in e2e QoS architecture for NEMO

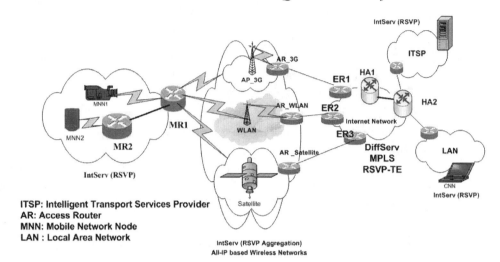

vices (CLS). The RSVP-TE (Baker & Iturralade & Faucheur & Davie, 2001) is employed to reserve network resources along the routing path connecting MNNs NEMO with external entities. In our proposed mechanism, the MR1 and HA represent the aggregator and de-aggregator points respectively, so they assign the same DiffServ Code Point (DSCP) in the packets header belonging to the same aggregate reservation. Figure 4 describes our mechanism, in which MNNs are passengers who would like to watch video-on-demand in NEMO (e.g. Eurostar train). The e2e PATH messages generated by the CN are intercepted, aggregated and assigned a DSCP and tunnelled by the HA to the MR_1. When this aggregated PATH message is received by the MR_1, it is decapsulated and sent to the MNNs which reply by sending RESV messages. These are intercepted, aggregated, assigned the same DSCP and tunnelled by the MR_1 to the HA which uses tunnel reservation (Terzis & Krawczyk & Wroclawski & Zhand, 2000) to confirm the network resources reservation required by the aggregated RESV message. Afterwards, the HA decapsulates the aggregated RESV message and sends to the CNs. In what follows we propose a policy to manage and control the bandwidth reservation for multimedia traffic aggregate along the route connecting MR_1 with HA.

4.4.1 Mobile RSVP Operation and Adaptation

To eliminate or minimize the delay required for establishing again RSVP over a new routing path when NEMO experiences a handover, various proactive mobile RSVPs (MRSVPs) have been proposed to tackle this problem. These mechanisms combine advanced reservation to cope with mobility and efficient reservation by minimizing network resource reserved for real-time traffic. In summary:

- The earliest RSVP extension to accommodate mobility was presented in (Awduche & Agu, 1997), which defines three reservation classes: committed, quiescent and transit reservation. In case of committed reservation, network resources are reserved and allocated to mobile nodes, whereas network resources are reserved but not allocated in case of quiescent reservation. In case of transit reservation, network resources are reserved but allocated to other users until MN arrival. A mobility management agent inter-operates with a virtual proxy in the MN domain to initiate and maintain these reservations on behalf of the MNs.

Figure 4. Signalling flow during the aggregated RSVP

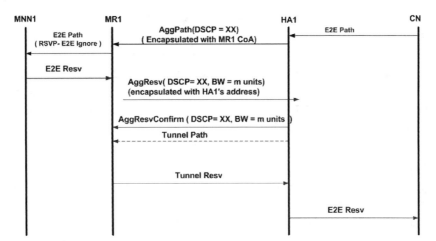

- An RSVP extension based on IP multicast to accommodate mobility is proposed in (Chen& Huang, 2000; Moon & Aghvami, 2002), where the node mobility is modelled as a transition in a multicast group formed of cells which may be visited by a MN during its active communication. A multicast tree is modified dynamically to cope with MN mobility in the multicast group, where transit reservation has been done in each member (cell) of this group. This approach suffers from high maintenance and cost due to signalling messages sent to maintain multicast tree and network resource reservation.

- A mobile RSVP (MRSVP) extension is introduced in (Talukdar & Badrinathm & Acharya,2001), where a MN obtains previously the list of addresses of all the network domains which may potentially visit. The different proxy agents in these domains make in advance transit network resource (passive) reservation for the MN. When it enters in a domain, it activates network resource already reserved and the local proxy discovers all proxy agents in the neighbouring subnets which will be visited by the MN to initiate transit reservation.

- Instead of having in advance the list of addresses of subnets which will be visited by the MN during its active communication, a modified MRSVP (Mahadevan & Sivalingam, 2001) is proposed to enable the MN to acquire progressively the addresses of its neighbouring cells. Note, however, this means that the MN will not be able to keep track of more than six addresses due to the cellular architecture. To ensure effective (minimum) network resource allocation in transit reservation, a rate reduction factor parameter is sent to inform neighbouring cell of the factor by which the resource request can be reduced if the demanded network resource reservation can not be completely satisfied.

- A Hierarchical MRSVP (HMRSVP) is presented in (Tseng& Lee & Liu & Wang, 2003), which combines both MRSVP and hierarchical MIP (HMIP) (Costa & Schmitz & Hartenstein& Liebsch, 2002) by using regional registration. The MN initiates transit reservation only when it is about to entering a neighbouring domain. This can reduce the time of establishing again the RSVP; however, it depends on the size of the overlapping area, mobile node speed and reservation latency, so the resource disruption may occur. A seamless handover can be retained only if the reservation process is completed while the MN is in the overlapped area, enabling the MN to make local reservation in the upcoming domain and limit the effect of IP handoff on those reserved in the network core. In addition, the MN does not need to have the list of all domains addresses. Therefore, this network resource reservation mechanism can be more efficient for NEMO than the previous mechanisms.

In what follows we discuss an intelligent network resource reservations policy to be integrated in the MR, which enables it to adapt to the vertical handover rate and maximize network resource utilization.

4.4.2 Traffic Class Based Stair Reservation Policy

This policy enables the MR_1 to adjust dynamically in a static increment manner the network resources reserved for each traffic class. For example, MR_1 reserves initially 200 kbps for a MNN requesting GS and 130 kbps for two MNNs requesting CLS which are already connected to NEMO. When a new MNN requesting GS joins NEMO, it reserves further 200 kbps. When NEMO serves as a relay for another NEMO, it reserves 4x200 kbps, which therefore depends on traffic class. But, the departure of only two MNNs receiving CLS enables MR_1 to release 130 kbps of the reserved bandwidth. This policy can provide high flexibility by altering the value of the static increment, and therefore, it can be employed for low and high varying NEMOs for example buses.

The RSVP-TE process based on this reservation policy which is integrated in MR_1 cane be modeled by M/G/1 queuing model. Based on this model, the inter-arrival of reservation requests which are generated by MNNs follows Poisson distribution, and the service time we can approximate by Hyper-exponential distribution as long as the standard deviation relative to the average reserved bandwidth value could be large (Nelson, 1995). Note we consider that the MR_1 can only make network resource reservation through a single egress interface. Thus, our problem can be described as follows. The RSVP-TE employed at MR_1 receives reservation requests at a varying average rate $\lambda = [0.1 \ldots 0.9]$ every 1s; the reservation response time follows a two-stage Hyper-exponential distribution with $\alpha_{GS} = 0.4$, $\alpha_{CLS} = 0.6, \mu_1 = 0.9s, \mu_2 = 0.75s$; calculate the total delay, T_D, and throughput, γ, of traffic flows that sent a number of reservation requests, $N_{Res_{req}}$, to the MR1, and therefore

$$\overline{S} = \frac{\alpha_{GS}}{\mu_{GS}} + \frac{\alpha_{CLS}}{\mu_{CLS}} \qquad (1)$$

$$\overline{S^2} = \frac{2*\alpha_{GS}}{\mu_{GS}^2} + 2*\frac{\alpha_{CLS}}{\mu_{CLS}^2} \qquad (2)$$

$$\rho = \lambda * \overline{S} \qquad (3)$$

$$T_D = \frac{\lambda * \overline{S^2}}{2(1-\rho)} + \overline{S} \qquad (4)$$

$$N_{Res_{req}} = \lambda * T_D \qquad (5)$$

$$\gamma = \frac{\rho}{N} \qquad (6)$$

where $\overline{S}$ denotes the average service time.

Figure 5 shows that the throughput distribution depends on many parameters among which ρ, μ_{GS}, and μ_{CLS}. In addition, this figure shows us that the bandwidth reservation is fairly distributed between GS and CLS reservation requests according to α_{GS}, α_{CLS}, μ_{GS} and μ_{CLS}. Figure 6 illustrates a promis-

An End-to-End QoS Framework for Vehicular Mobile Networks

Figure 5. Throughput vs. reservation requests intensity

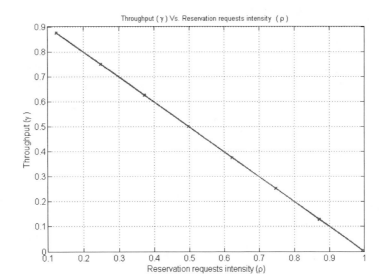

Figure 6. Distribution of successful requests reservation

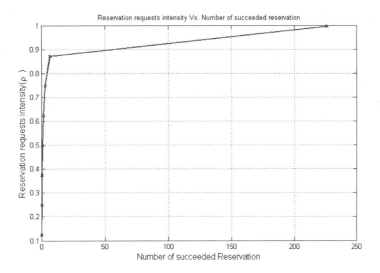

ing result where a high number of reservation requests have been admitted while being offered the required bandwidth. This proves to some extent the high utility of this proposed mobile network resource reservation model.

Note that the performance of the RSVP signalling protocol relies on the quality of the selected route. Besides, it requires that all the nodes along this route should support RSVP, unless they can guarantee the minimum network resources required for the demanding traffic flows aggregate. However, due to dynamic and high network mobility, the quality of wireless channels which can be allocated for NEMO is degraded and therefore bandwidth is reduced as well as varying on a very short time scale. To illustrate this impact on the number of MNNs which can potentially wait for successful network resource reservation, equation 5 can be rewritten as follows:

$$N_{MNNs} = \rho + \frac{\rho^2}{1-\rho} \frac{1+C_s^2}{2} \quad (7)$$

where C_s^2 denotes the parameter of service variations and ρ denotes the traffic load which represents the arrival of requests for network resource reservation at the MR. As a result, Figure 7 shows that when the fluctuation of bandwidth increases on the selected route, then, the number of successful requests for network resources decreases and subsequently number of MNNs waiting for network resource reservation increases in an exponential manner.

5. INTELLIGENT ROUTE CONTROL

Mobile network nodes experience various levels of communication quality, depending through which egress interface the MR routes their traffic to Internet. Each route mitigates the QoS enabled in the wireless technologies employed along the selected route. A mechanism is therefore required to be configured in the MR to intelligently control and manage traffic distribution over the MR egress interfaces so that the QoS required for traffic can be guaranteed as well as traffic is evenly distributed among these interfaces. Traffic splitting scheme and traffic scheduling constitute the main parts of this mechanism to realize efficient traffic engineering in a multi-homed NEMO.

5.1 Traffic Splitting Scheme

Service level specification (SLS) is associated with unidirectional flows where the flow is identified by a number of attributes including the priority value, source and destination IP address information, and application-related information such as protocol and port numbers. A micro-flow is a correlated set of packets that should be treated by the network in a prescribed way. A macro-flow (session) consists of a set of micro-flows with similar requirements, but not all of the flow attributes are specified. For example, for DiffServ behaviour aggregate classification, only the priority attribute is required. Similarly, for e2e QoS monitoring and analysis

Figure 7. Impact of service variations on requests for network resource reservation

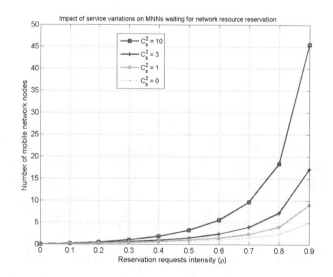

An End-to-End QoS Framework for Vehicular Mobile Networks

purposes, we identify an aggregate-flow as a correlated set of packets that are treated in the network in a similar manner (e.g., packets handled by the same Per-Hop Behaviour (PHB) in DiffServ or IntServ, or packets routed through the same Label Switched Path (LSP) in MPLS). In what follows we explain three schemes which can be employed in MR to distribute traffic among output wireless links: packet-based, flow-based and user-based traffic splitting schemes (Guo & Chen & Li & Chiuch, 2004;Yuen &Chung, 2007).

5.1.1 Packet-Based Traffic Splitting Scheme (PTSS)

This scheme consists of distributing traffic packets among multiple interfaces irrespective of their flows. Figure 8a illustrates the flows stripping process at packets level by MR which forwards traffic packets to the different physical interfaces: WLAN, UMTS or GPRS. Thus, packets of a single flow can travel through multiple routes; subsequently, they experience various network performances. For example, WLAN 802.11 has more than 7 Mbps available bandwidth, while the GPRS has normally less than 100 Kbps available bandwidth which is less than 70 times than the former. This causes a serious jitter problem and delay for flow packets, which can quickly bring down the quality of VoIP or video connections. On the other hand, the PTSS offers the finest level of traffic engineering control which enables the MR to make more efficient use of multiple wireless access links.

Because wireless links provide different delays and data rates as they change with network mobil-

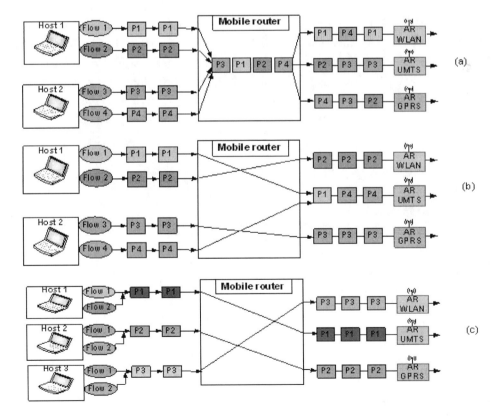

Figure 8. (a) Packet-based traffic splitting scheme (b) flow-based traffic splitting (c) user-based traffic splitting scheme

ity and environment conditions, the congestion window of the TCP keeps varying all the time which can drastically limit the e2e throughput required for real time applications. Furthermore, connections may suffer from different delays, resulting in unfair distribution of the bandwidth among competing connections. It is noteworthy that the packet traffic splitting scheme (PTSS) can only be used for uplink traffic. Since the MR is closer to the wireless links than the HA, then, it can easily distribute the flow packets among the available access links. The downlink traffic packets, however, may follow diverse routes to reach the MR (Guo & Chen & Li& Chiuch, 2004). Therefore, the performance of the PTSS also depends on the scheduling algorithm which is employed at the MR and HA to distribute traffic over multiple egress interfaces or multiple routes, respectively.

5.1.2 Flow-Based Traffic Splitting Scheme (FTSS)

This scheme consists of stripping traffic at the flow level, where flows aggregated can be striped across multiple connections (a connection refers to typically a UDP, TCP connection or ICMP packet). The whole connection is stripped on the physical layer for the whole duration on the link layer. Hence, the TCP connection semantics are preserved and no extra packet synchronisation needed, making this scheme simple to implement and transparent to the end users. The Flow Traffic Splitting Scheme (FTSS), however, has a drawback which is the connection stripping cannot make use of the aggregated bandwidth efficiently. For example, it is possible that an end-user would end waiting for a connection that was previously assigned to a congested channel (wireless link), although the other links are idle or less congested. The FTSS requires little support from the communication infrastructure and therefore it is a straight forward scheme to be implemented. Traffic flows are scheduled over the MR egress interfaces on per flow basis. Thus, traffic flows can be independently routed over different interfaces, where a single user may have multiple concurrent active flows sent over different interfaces, as it is illustrated in Figure 8b. Table 1 also clarifies the FTSS where three flows are generated and assigned to three physical interfaces (wireless links): WLAN, GPRS and UMTS.

The FTSS improves the reordering problem from which the PTTS suffers, because the packets of a flow are always forwarded to a single interface. Therefore, there is a very small probability that a packet will be received out of order. However, if there is only a single flow traversing a multi-

Table 1. Flow and user traffic splitting table

Flow-based traffic splitting scheme (FTSS)				
Source IP	Source Port	Destination IP	Destination Port	Access Link
2007:0ab8::1528:57db	1000	2007:1cb8::4000:17db	1007	*UMTS*
2007:0ab8::1528:57dc	1003	2007:1cb8::4000:17dd	1008	*WLAN*
2007:0ab8::1528:57dc	1006	2007:1cb8::4000:17ab	1009	*GPRS*
User-based traffic splitting scheme (UTSS)				
Source IP On-Board Host ISP				Access Link
2004:0ab8::1423:20ad				*UMTS*
2004:0ab8::1728:32dd				*WLAN*
2004:0ab8::1128:57ac				*GPRS*

interfaced MR, then, it may result in underutilizing the available bandwidth. Contrary to the PTSS, the FTSS can control as well as manage uplink and downlink traffic by enabling the MR and HA to lookup in the flow table and assign the flow to the appropriate link. Subsequently, traffic flows are routed on different routing paths based on their IP addresses.

5.1.3 User-Based Traffic Splitting Scheme (UTSS)

This is the simplest scheme of all traffic splitting schemes. Figure 8c illustrates the User-based Traffic Splitting Scheme (UTSS), where all flows generated by a single user are simply routed over the same link (egress interface). Note that MNNs are identified by the IP address allocated to their devices which communicate with the MR. Therefore all traffic generated by a MNN is always routed over a single link, making the probability that packets arriving out-of-order is very low. In this scheme, uplink and downlink traffic can be controlled based on user preferences recorded in the MR. This provides an opportunity to apply some cost and charging model to gain more profits out of the offered services. For example, users can be charged according to the usage of bandwidth resources at different time. It is noteworthy that the total volume charging model (Courcoubetis & Siris & Stamoulis, 1998; Yamai & Okayama & Shimamoto & Okamoto, 1999) is used in wireless access networks.

5.2 End Admission Traffic Control and Scheduling

Based on the collected link information and RSVP messages, a distributed coordinated Call Admission Control (CAC) mechanism can be employed in the MR_1 and HA to control upstream and downstream traffic, respectively. When NEMO shown in Figure 1 performs IP-handover or one of its MR_1 egress interfaces fails, traffic flows can be transmitted through the other egress interfaces based on the estimated available capacity. However, when these do not have enough capacity to serve already admitted flows and handing off traffic flows, then, the employed CAC decides which traffic flow should be admitted or rejected. This improves the network resource utilization and eliminates the influence of Best Effort (BE) traffic on real-time IP traffic (Alshaer & Elmirghani, 2008; Alshaer & Horlait, 2004). The round robin (RR) as well as weighted round robin (WRR) scheduling service disciplines (SSDs) (Alshaer & Elmirghani, 2008) have been integrated in the MR_1 to evaluate their performance with the FTSS and PTTS while serving admitted traffic flows according to their class and applications priority. While the RR SSD forwards packets/flows to egress interfaces based on solely available links in a round robin fashion, the WRR SSD forwards packets/flows to egress interfaces based on their measured available bandwidth and allocated weights to each of these interfaces.

6. SIMULATION AND RESULTS

We have conducted simulation experiments on NEMO-BSP integrated in the network topology shown in Figure 9 by using OMNeT++. Most of Internet applications use TCP/IP as an underlying transport protocol. To this end, our aims were to (i) measure and analyze the TCP congestion window (CWND) size when RR and WFQ SSDs are operating in the MR, (ii) evaluate the performance of traffic splitting schemes, namely packet and flow splitting schemes and (iii) evaluate the impact of network mobility on the TCP CWND performance. Note that the TCP CWND bounds the amount of data which can be sent per round-trip time (RRT) of any connection (Capone & Fratta & Martignon, 2004).

The slow start threshold (ssthresh) sets the value of the TCP CWND, which estimates the available bandwidth along the selected routing

Figure 9. Multi-homed NEMO simulation environment in OMNeT++

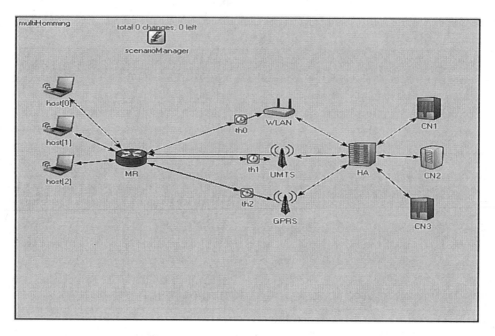

path and subsequently regulates the transmission rate, based on the measured bandwidth as follows:

$$\text{Bandwidth} = \frac{1.22 * \text{MTU}}{\text{RTT} * \sqrt{L}} \quad (8)$$

$$\text{ssthresh} = \text{Bandwidth} * \text{RTT}_{min} \quad (9)$$

where MTT denotes the maximum transmission unit, L denotes the loss probability along the routing path and RTT_{min} denotes the minimum propagation round trip time (delay) between the MNN and any CNs.

Figure 9 depicts our network topology in which NEMO is connected to Internet through WLAN, UMTS and GPRS communication technologies. A single HA serves NEMO and enables MNNs and CNs to communicate based on NEMO-BSP. Along the simulation time, 3MB to 6 MB files are constantly transferred from MNNs (host0, host1 and host2) to CNs. In our analysis, three performance metrics are measured: average end-to-end (e2e) delay, jitter and throughput of traffic flows, established between MNNs and CNs, which are served by the RR and WRR SSDs employed in the MR. NEMO is enabled to move at different speeds to measure the effect of network mobility on the communication performance metrics. In the first scenario, NEMO moves at speed 10 m/s, while in the second scenario it moves at speed 20 m/s, and in the third scenario it moves at 30 m/s.

As a result, table 2 summarizes the performance of RR SSD and WRR SSD with different TCP traffic intensities while FTTS and PTTS are employed at the MR. The results show that the WRR SSD outperforms the RR SSD. This can be attributed to the fact that the WRR SDD adapts to network mobility by allocating bandwidth to traffic based on the measured available bandwidth on egress interfaces. In addition, we notice that the FTSS associated with WRR SSD could remarkably improve the e2e QoS required for traffic flows in comparison to the FTSS associated with RR SSD. Based on these results, we can conclude that the PTSS outperforms the FTSS when traffic intensity is low, which can refer to the fact that in this network scenario traffic packets are better

Table 2. Summary of RR SSD and WRR SSD performance with FTSS and PTSS

	RR 3 TCP Flows PTSS FTSS	RR 6 TCP Flows PTSS FTSS	WRR 3 TCP Flows PTSS FTSS	WRR 6 TCP Flows PTSS FTSS
e2e delay (ms) Scenario 1	0.82626 0.83526	1.73382 1.65282	0.62633 0.31233	1.33215 0.67825
e2e delay (ms) Scenario 2	3.7476 8.07214	7.37914 16.10696	3.12476 1.51476	5.66624 2.99821
e2e delay (ms) Scenario 3	8.26556 8.123	17.214615 16.57812	13.33256 6.66123	14.98964 7.49482
e2e jitter (ms) Scenario 1	0.00122 0.22344	0.12542 0.41247	0.00133 0	0.35781 0.13278
e2e jitter (ms) Scenario 2	0.49245 1.01244	0.51235 3.12544	0.32456 0.12324	0.34578 0.24517
e2e jitter (ms) Scenario 3	1.24214 1.01244	2.21455 3.24578	0.56481 0.24545	1.25457 0.68774

distributed over egress interfaces. But, when the traffic intensity increases from 3 TCP flows to 6 TCP flows, then, the FTSS outperforms the PTSS, which can refer to the fact that in this network scenario more packets are likely to arrive out of order, requiring ordering and subsequently this increases their e2e delay and jitter.

Figure 10 shows a result based on the third scenario in which NEMO moves at 30 m/s. The result shows that the FTTS performs better than the PTSS in term of TCP CWND, offering MNNs higher throughput. In addition, the TCP CWND appears to be more stable in case of FTTS than that resulting from the PTSS. This can be attributed to the fact that traffic packets which are forwarded to different egress interfaces in case of PTSS, their TCP connections sat different ssthresh values based on equations 8 and 9 and therefore the CWND size varies on a short time scale, decreasing the e2e throughput. This exactly explains the problem of the TCP/IP CWND size, where competing connections might not be allocated equal network resources, because the TCP CWND size could converge to different CWND values.

Further Research Directions

- Extending QoS capabilities across multiple provider domains by introducing techniques which can propagate QoS-based agreements among a set of providers involved in the chain of inter-domain service delivery along the egress routing paths connecting NEMO with external entities.
- Design of a hierarchical resource reservation architecture which can be proactive and efficient at the same time.
- Determining a QoS negotiation mechanism in the MR to enable it, during the handoff time, to negotiate QoS (SLAs) which can be provisioned (guaranteed) to MNNs along the new selected routing path.
- Devising de-jitter algorithms to be coupled with the implemented SSDs in MR and HA to reorder packets in their corresponding traffic flows to eliminate jitter and therefore maximize network utilization.
- Devising joint optimal multi-hop communication and power transmission control algorithms for optimal vehicular communications.

Figure 10. Network mobility influences on TCP CWND and packet splitting schemes performance

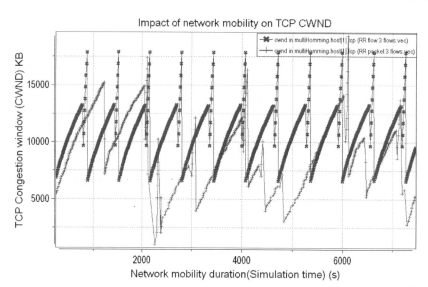

7. SUMMARY

This chapter introduced the principal components in a framework which provides a utility for intelligent service providers (ITSPs) and wireless services providers (WISPs) to provisioning voice, video and data at the required e2e QoS level to vehicular NEMOs. We mainly focused on the QoS required for vehicle-to-Infrastructure communication (V2I), so more components should be integrated in the framework to guarantee the e2e QoS required for inter-vehicle communications. For example, network policies, route optimization and vehicular transition mechanism from V2I to V2V and vice-versa. By following the bottom-up approach in the design of e2e QoS framework, PTSS, FTSS and UTSS, as well as, traffic admission and scheduling disciplines which are integrated in the IP layer become more effective in guaranteeing the e2e QoS for MNNs. This cross layer design approach enhances the TCP/IP stack layers by enabling applications and transport protocols to provide the required QoS for MNNs despite network mobility. Such a framework will enable all-IP networks to provision the e2e QoS for NEMOs, since all the different wireless and radio technologies will be able to transport IP traffic flows and therefore offer it adequate forwarding and QoS guarantee, making multimedia services available for MNNs anytime anywhere. Vehicular NEMOs still, however, require an integrated and complete architecture like the Continuous Air-interface for Long and Medium Range communication (CALM) (ISO,2006) to reduce the handover latency and guarantee seamless communication with CNs and servers hosted by ITSPs. Such architecture will enable vehicular NEMO to utilize 3G and WLAN technologies together and to roam between them while maintaining persistent, seamless and real-time communication with external entities. This view is supported by the simulation and mathematical modelling results obtained on studying the impact of network mobility on the TCP CWND performance and QoS performance metrics.

REFERENCES

Agrawal, P., Zhang, T., Sreenan, C. J., & Chen, J.-C. (2004). All-IP wireless systems. *IEEE Journal on Selected Areas in Communications*, *2*(4), 613–616. doi:10.1109/JSAC.2004.825950

Akyildiz, I. F., Xie, J., & Mohanty, S. (2004). A survey of mobility management in next-generation all-IP wireless systems. *IEEE Wireless Communication Magazine, 11*(4), 16–28. doi:10.1109/MWC.2004.1325888

Alshaer, H. (2005). *Priority Traffic Control and management in Multiservice IP Networks*. PhD thesis, Pierre et Marie Curie University, LIP6, Paris, France.

Alshaer, H., & Elmirghani, J. J. M. (2008). Expedited forwarding end-to-end delay and jitter in diffserv. *International Journal of Communication Systems, 21*, 815–841. doi:10.1002/dac.915

Alshaer, H., & Horlait, E. (2004). Emerging client-server and ad-hoc approach in inter-vehicle communication platform. In *Proc. of the IEEE VTC-Fall*, 6, (pp. 3955-3959), Los Angeles, CA.

Alshaer, H., & Horlait, E. (2004). Expedited forwarding delay budget through a novel call admission control. In *Proc. of the 3rd European Conference on Universal Multiservice Networks (ECUMN)*, (LNCS 3262, pp. 50-59), Porto, Portugal.

Alshaer, H., & Horlait, E. (2005). An optimized adaptive broadcast scheme for inter-vehicle communication. In *Proc. of IEEE VTC-Spring, 5*, (pp. 2840-2844), Stockholm, Sweden.

Awduche, D. O., & Agu, E. (1997). Mobile extensions to rsvp. In *Proc. 6th Int'l Conference computer Communication and Nets*, (Vol.7, pp. 132-136).

Bai, H., & Atiquzzaman, M. (2003). Error modeling schemes for fading channels in wireless communications. *IEEE Communications Surveys and Tutorials, 5*(2), 2–9. doi:10.1109/COMST.2003.5341334

Baker, F., Iturralade, C., Faucheur, F.L. & Davie, B. (2001). Aggregation of rsvp for ipv4 and ipv6 reservations. *IETF, RFC 3157*.

Blake, S., Black, D., Carlson, M., Davies, E., Wang, Z. & Weiss, W. (1998). An architecture for differentiated services. *IETF RFC 2475*.

Borgonovo, F., Capone, A., Cesana, M., & Fratta, L. (2003). Ad hoc MAC: A new flexible and reliable mac architecture for ad-hoc networks. In *Proc. of IEEE WCNC*.

Braden, R., Zhang, L., Berson, S., Herzog, S., & Jamin, S. (1997). Resource reservation protocol (rsvp)-version 1 functional specification. *IETF RFC 2205*.

Capone, A., Fratta, L. & Martignon, F.(2004). Banwidth estimation schemes for tcp over wireless networks. *Transactions on mobile computing, 3*(2), 1-15.

Chen, W. T., & Huang, L. C. (2000). Rsvp mobility support: a signaling protocol for integrated services interent with mobile hosts. In *Proc. of IEEE Infocom*, (Vol. 3, pp.1283-1292).

Chiussi, F., Khotimsky, D. A., & Krishnan, S. (2002). A network architecture for MPLS-based micro-mobility. In *Proc. of IEEE Wireless Communications and Networking (WCNC)*, (pp.549-555), Toulouse, France.

Clark D.D. & Fang, W.(1998). Explicit allocation of best-effort packet delivery service. *IEEE/ACM Transactions on Networking, 6*(4), 362-373.

Corderio, C., & Challapali, K. (2007). C-MAC: a cognitive MAC protocol for multichannel wireless networks. In *Proc. IEEE Symposium on New Frontiers in Dynamic Spectrum Access networks*, April 2007.

Costa, X. P., Schmitz, R., Hartenstein, H., & Liebsch, M. (2002). A MIPV6, fmipv6 and HMIPv6 handover latency study. In *Proc. of IST Mobile and Wireless Telcommunication*.

Courcoubetis, C., Siris, V. A., & Stamoulis, G. D. (1998). Network control and usage-based charging: is charging for volume adequate. In *Proc of the first international Conference on Information and Computation economics*, New York.

Davie, B., Charny, A., Bennett, J.C.R., Benson, K., Le Boudec, J.Y., Courtney, W., Davari, S., Firoiu, V., & Stiliadis, D. (2002). An expedited forwarding PHB (per-hop behavior). *IETF RFC 3246*.

Devarapalli, V., Wakikawa, R., Petrescu, A., & Thubert, P.(2005). Network mobility (NEMO) basic support protocol. *IETF RFC3963*.

Eardley, P., & Hancock, R. (2000). Modular IP architecture for wireless mobile access. In *1st International Workshop on Broadband Radio Access For IP Based Networks*.

Ernst, T., & Lach, H.(2007). Network mobility support terminology. *RFC 4885*.

Guo, F., Chen, J., Li, W., & Chiuch, T. (2004). Experiences in building a multihoming load balancing systems. In *Proc. of IEEE Infocom*, (Vol. 2).

Hartenstein, H., & Laberteaux, K. P. (2008). A tutorial survey on vehicular ad hoc networks. *IEEE Communications Magazine*, 164–171. doi:10.1109/MCOM.2008.4539481

Heinanen, J., & Baker, F. (1999). RFC 2597-assured forwarding PHB group. *IETF RFC 2597*.

Hsu, A., Wei, D., & Kuo, C. (2007). A cognitive mac protocol using statistical channel allocation for wireless ad-hoc networks. In *Proc. IEEE WCNC*, March.

Huang, V., & Zhuang, W. (2002). QoS-oriented access control for 4G mobile multimedia CDMA communications. *IEEE Communication Magazine*,118-125, March.

ISO. (2006). *The calm handbook: Continuous communication for vehicles*. Technical report . CALM Handbook ISO, TC204, WG16.

Jamalipour, A., & Lorenz, P. (2006). End-to-end qos support for ip and multimedia traffic in heterogeneous mobile networks. *Computer Communications*, 29, 671–682. doi:10.1016/j.comcom.2005.07.021

Karp, B., & Kung, H. T. (2000). GPSR: Greedy perimeter stateless routing for wireless networks. In *Proc. of ACM/IEEE MOBiCOM,* August.

Langar, R., Thome, S., & Grand, G. L. (2005). Micro mobile MPLS: A new scheme for micro-mobility manegement in 3G all-IP networks. In *Proc. IEEE Symposium Computers and Communications (ISCC),* (pp. 301-306).

Lott, M., Halfmann, R., Schulz, E., & Radimirsch, M. (2001). Medium access and radio resource management for ad hoc networks based on ULTRA and TDD. In *Proc. of the 2nd ACM/SIGMOBILE Symposium on Mobile Ad hoc Networking and Computing*.

Ma, L., Han, X., & Shen, C. (2005). Dynamic open spectrum sharing MAC protocol for wireless ad hoc networks. In *Proc. of IEEE Symposium on New Frontiers in Dynamic Spectrum Access networks*, November 2005.

Mahadevan, I., & Sivalingam, K. M. (2001). Architecture and experimental framework for supporting QoS in wireless networks using DiffServ. *Mobile Networks and Applications*, 6, 385–395. doi:10.1023/A:1011434813337

Mchenry, M.(2003). *Spectrum white space measurements*. New America Foundation Broadband Forum, June.

Menouar, H., Filali, F., & Lenardi, M. (2006). A survey and qualitative analysis of mac protocols vehicular ad hoc networks. In *Proc. IEEE Wireless Communications Magazine*, 30-35, October.

Mishra, A. (2006). A multi-channel mac for opportunistic spectrum sharing in cognitive networks. In *Proc. IEEE MILCOM,* October.

Mitola, J. (2000). *Cognitive radio: an integrated agent architecture for software defined radio.* PhD thesis, KTH Royal Inst. Technology, Stockholm, Sweden, 2000.

Moon, B., & Aghvami, H. (2002). Reliable RSVP path reservation for multimedia communications under an IP micromobility scenario. *IEEE Wireless Communications, 9*(5), 93–99. doi:10.1109/MWC.2002.1043859

Muqattash, A., & Krunz, M. (2003). CDMA-based mac protocol for wireless ad hoc networks. In *Proc. of the 4th ACM/SIGMOBILE Symposium on Mobile Ad hoc Networking and Computing.*

Natarajan, B., Nassar, C. R., & Shattil, S. (2001). High-performance MC-CDMA via carrier interferometry codes. *IEEE Transactions on Vehicular Technology, 50*(6). doi:10.1109/25.966567

Nelson, R. (1995). *Probability, Stochastic Processes, And Queuing Theory: The mathematics of computer performance Modeling.* Berlin: Springer-Verlag.

Ng, C., Ernst, T., Paik, E., & Bagnulo, M. (2007). Analysis of multihomming in network mobility support. *RFC 4980.*

Pavn, J.d.P., Shankar, N.S., Gaddam, V., & Chou, C-T.(2006). The MBOA-Wimedia specification for ULTRA wideband distributed networks. *IEEE Communication Magazine,* 128-134, June.

Perera, E., Sivaraman, V., & Seneviratne, A. (2004). Survey on network mobility support. *Mobilie Computer Communication Review, 8*(2), 7–19. doi:10.1145/997122.997127

Rudack, M., Meincke, M., Jobmann, K., & Lott, M. (2003). On traffic dynamical aspects inter-vehicle commuication(IVC. In *Proc. of the 57th IEEE Vehicular Technology Conference.*

She, J., Yu, X., Ho, P-H., & Yang, E-H.(2009). A cross-layer design framework for robust IPTV services over IEEE 802.16 networks. *IEEE Journal on Selected Areas in Communications, 27*(2).

Signh, J., Bambos, N., Srinivasan, B., & Clawin, D. (2002). Wireless LAN performance under varied stress conditions in vehicular traffic scenarios. In *Proc. of IEEE VTC Fall,* (vol. 2, pp.743-747).

Skordylis, A., & Trigoni, N. (2008). Delay-bounded routing in vehicular ad-hoc networks. In *Proc. of ACM/IEEE MobiHoc,* (pp. 341-350), Hong Kong, China.

Soliman, H. (2004). *Mobile IPv6: Mobility in Wireless Internet.* Reading MA: Addison-Wesley Professional.

Solomon, J. D. (1998). *Mobile IP: the Internet Unplugged.* Upper Saddle River, NJ: Prentice Hall Ptr.

Su, H., & Zhang, X. (2007). Opportunistic MAC protocols for cognitive radio based wireless networks. In *Proc. 41st Conference on Information Sciences and Systems,* Johns Hopkins University, March.

Talukdar, A., Badrinathm, B., & Acharya, A. (2001). MRSVP: A resource reservation protocol for an integrated services network with mobile hosts. *Wireless Networks,* (Vol. 7).

Terzis, A., Krawczyk, J., Wroclawski, J., & Zhand, L.(2000). RSVP: Operation over IP tunnels. *RFC 2746.*

Tseng, C. C., Lee, G. C., Liu, R. S., & Wang, T. P. (2003). HMRSVP: Hierarchial mobile RSVP protocol. *Wireless Networks, 9,* 467–472.

Vasudevan, S., Papagiannaki, K., Diot, C., Kurose, J., & Towsley, D. (2005). Facilitating access point selection in IEEE 802.11 wireless networks. *Internet Measurement Conference,* (pp. 293-298), Berkeley, CA.

Yamai, N., Okayama, K., Shimamoto, H., & Okamoto, T. (1999). TCP performance over GPRS. In *Proc. of the IEEE wireless communication and networking conference* (WCNC), (Vol.3, pp.1248-1252).

Yokota, H., Idoue, A., Hasegawa, T., & Kato, T. (2002). Link layer assisted mobile IP fast handover method over wireless LAN networks. In *Proc. ACM MobiCom,* (pp. 131 – 139), Atlanta, GA.

Yuen, A., & Chung, T. (2007). *Traffic Engineering for Mult-Homed Mobile Networks*. PhD thesis, The University of New South Wales Sydney, Australia.

Zhao, J., & Cao, G. (2006). VADD: Vehicle-assisted data delivery in vehicular ad hoc networks. In *Proc. of INFOCOM.*

Chapter 18
LTE Mobility Solutions at Network Level for Global Convergence

Titus-Constantin Bălan
Siemens SIS PSE, Romania

Florin Sandu
"Transilvania" University of Brasov, Romania

ABSTRACT

One of the research challenges for next generation all-IP-based wireless systems is the design of intelligent mobility management techniques that take advantage of IP-based technologies to achieve global seamless roaming among various access technologies. Since Mobile IPv6 is considered a mature protocol, mobility management at the network layer is the frequent approach for heterogeneous networks. The tendency of future convergent scalable architectures is splitting the mobility management in two domains, global mobility and localized mobility management. This chapter presents the advantages of MIPv6, a global mobility protocol, and its enhancements. A case study based on MIPv6 for UMTS and WiFi convergence is also presented. Proxy MIPv6, the newest protocol of the MIPv6 family, already included in the roadmap of future 4G networks, will be analyzed as a solution for localized mobility management. The main goal of the chapter is describing the way mobility protocols (MIPv6 and PMIPv6) will be implemented for the 3rd Generation Partnership Project (3GPP) Long Term Evolution architecture. The chapter ends with the presentation of the interoperation between different network technologies using global and localized mobility management protocols, which provide flexibility, scalability and independence between mobility domains.

INTRODUCTION

Even if Mobile IP popularity is growing, implementing mobility using these protocols for present mobile networks (GPRS/UMTS) still has some weaknesses.

Link layer mobility is wide used in 3G networks (GPRS Tunneling Protocol is a link layer protocol used for mobility management between the core network entities) and is also utilized in 802.11 wireless LANs (a device moving across 802.11 access points within the same distribution system continues to maintain its sessions uninterrupted). But these are mobility solutions that are technology-specific. Link-layer mobility solutions for seamless mobility across heterogeneous access media are extremely complex and, since the telecom world is heading to the concept of "all IP", it is necessary for 4'Th generation architectures to develop and deploy network-layer mobility solutions that are independent of the access technology.

Network layer mobility has been applied for a long period for mobility in 3G networks, but has some problems. Instead GTP is mainly used for these networks. The advantage of GTP is supporting micro-mobility through which the mobile station can access the network using one permanent IP address while moving in the area controlled by the same mobility management entity (SGSN for UMTS networks) and without necessity of re-registering to a home agent.

Keeping the same IP address in the mobility domain and not involving the mobile node in the mobility related signaling are two of the reasons that lead to the idea of splitting mobility management in two domains: Local Mobility Management (also called Network Mobility) who's exponent protocol is Proxy Mobile IPv6 and Global Mobility Management (or Host Mobility) represented by Mobile IPv6 and enhancements.

ALL-IP MOBILE NETWORKS: GLOBAL MOBILITY MANAGEMENT

MIPv6 Overview

Mobile IP is a layer 3 host based global mobility protocol that solves the routing problem for mobile users. Two versions of Mobile IP have been standardized for supporting host-based mobility on the Internet: MIPv4 and MIPv6. They support the mobility of IP hosts by allowing them to utilize two IP addresses: a home address (HoA) that represents the fixed address of a mobile node (MN) and a care-of address (CoA) that changes with the IP subnet to which an MN is currently attached. In terms of the fundamental architectural aspects, these two mobility support standards follow the same concept, but the IPv6 header is special designed to support mobility. This is one of the greatest advantages that IPv6 offers. Compared to Mobile IPv4, Mobile IPv6 offers the possibility for route optimizing, thus the tunneling through HA method (triangular routing) can be avoided. Also the MIPv4 protocol has a Foreign Agent, a server similar with the Home Agent, which transmits all the messages, once the MN is attached to that visited network. For IPv6 there is no need for the Foreign Agent, all the messages can be send directly by the MN, without the need of any intermediate router. Besides the advantages for the mobility process, IPv6 offers the possibility of using the Neighbor Discovery Algorithm, the mechanism for address reconfiguration that makes the handover transparent to the user. Mobile IPv6 offers support for multiple Home Agents. Thus a wide distributed network based on IPv6 can be produced, and the HA workload will be fragmented for easier implementation of the protocol for wide distributed networks.

MIPv6 has still revealed some problems: handover delays, excessive signaling and packet loss, involving the mobile node in mobility related signaling so special heavy processing at MN level is required. Signaling related communication with the MN using the air interface lowers the performance.

A typical scenario to reach MIPv6 limits is when the number of hops to reach the MN from the HA is big, so the MN location is far away from its home networks and the connections on the path are slow and unreliable. Furthermore, despite the reputation of this protocol, it has been slowly deployed in real implementations and does not appear to receive widespread acceptance in the market.

Figure 1. Proxy mobile IPv6 domain

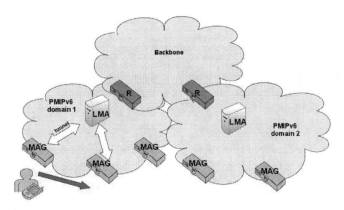

MIPv6 Enhancements

In addition to the base protocols, the IETF has standardized and is still standardizing several protocols that built on the same idea or provide enhancements to the base protocol. Most of the existing mobility support protocols have been developed for their own characteristic purposes and suitable environments so can be applied effective in some scenarios but can not be considered general efficient.

Fast Handover for Mobile IP (FMIP)

As the name states, this protocol is meant to reduce handover delays. The protocol enables a MN to quickly detect the new access point and the associated subnet prefix information of the next point of attachment when the MN is still connected to its current subnet. When a mobile node discovers the information about the next point of attachment to which it will attach, the mobile node sends a Router Solicitation for Proxy (RtSolPr) to the previous access router (PAR) with an identifier of the new point of attachment.

The new CoA (NCoA) generation is done by the PAR before movement based on the MN's interface ID and the New Access Router's (NAR) associated subnet prefix.

To reduce the Binding Update latency, the protocol specifies a tunnel between the Previous CoA (PCoA) and the NCoA. Send/receive binding updates are sent prior to movement: the MN sends a "Fast Binding Update" FBU to the PAR that responses with a FB Acknowledge and also initiates a Handover Initiate (HI) message to the NAR with a proposed NCoA. If the NCoA is accepted by the NAR, NAR responds with a Handover Acknowledgement (HAck). As a result, PAR begins tunneling packets arriving for PCoA to NCoA when the handover process is still ongoing. When the mobile node attaches to the NAR and its link layer connection is ready for network layer traffic, it sends a Fast Neighbor Advertisement (FNA) so the NAR knows that the MN is reachable and all the waiting packets will be forwarded from the NAR to the MN.

Hierarchical Mobile IPv6 (HMIPv6)

The goal of the hierarchical Mobile IP is to optimize routing in local mobility cases by introducing a hierarchy of mobility anchors and a new function, the MAP (Mobility Anchor Point) that performs the identical operations as the Home Agent in MIPv6 from the perspective of the MN, at a certain hierarchical level in the network. That means that MAP will receive all packets on behalf of the mobile node it is serving and will encapsulate and forward them directly to the mobile node's current address. The MN can bind its current location (on-Link CoA) with an address on the MAP's subnet (Regional CoA). This mapping is coordinated by the MAP through the use of Proxy Neighbor Advertisement messages.

When a mobile node performs a handover between two access routers within the same HMIPv6 domain, only the MAP has to be informed. If the MN moves within the same administrative domain, controlled by the same MAP, the RCoA is kept unchanged, only LCoA is updated. The real HA is only contacted if the MAP is changed.

To discover and configure different MAPs, HMIPv6 relies on a MAP option in the Router Advertisements. This option includes the distance vector from the mobile node, the preference for this particular MAP, the MAP's global address and subnet prefix. When the MN receives more than one MAP option, it needs to select an appropriate MAP to get registered to (the criteria is the higher Preference value and highest value in the Distance field).

MAP Discovery will be performed in the HMIPv6 domain only if the MN has HMIP capabilities, else the mobile node uses the Mobile IPv6 protocol for its mobility management.

HMIPv6, although it is an efficient localized mobility management protocol that can reduce handover latency significantly compared to MIPv6, it still requires movement detection and Duplicate Address Detection (DAD) because the MN's on-link CoA (LCoA) should be newly assigned whenever the MN moves to another subnet within a domain. It also has some other weaknesses because is not transparent for MN and all routing reconfiguration messages are sent over wireless link.

Dual Stack Mobile IP

DSMIP is another MIPv6 enhancement, based on the assumption that it is unlikely that mobile nodes will use IPv6 addresses. The advantage of DSMIP is that it supports MIPv4 and MIPv6 collocated/home addresses within one protocol. IPv4 will not soon be abandoned, support for this protocol is mandatory, so IPv4 home addresses could be used by upper layers.

It is also reasonable to assume that IPv6 mobile nodes will move to networks that might not support IPv6 and would therefore need an IPv4 Care-of Address.

DSMIP extends Mobile IPv6 capabilities and also modifies some of the most important messages of the mobility protocol like binding update and binding acknowledgement. Support for mobile nodes communicating with a MAP in HMIP is also included.

ALL-IP MOBILE NETWORKS: LOCAL MOBILITY MANAGEMENT

PMIPv6 Overview

A network-based mobility management protocol such as PMIPv6 offers the support for IP mobility of mobile nodes, but the MN itself is not required to participate in any mobility-related signaling. Proxy

Mobile IPv6 (PMIPv6) is being actively standardized by the IETF NETLMM (Network Localized Mobility Management) working group.

The serving network takes the responsibility of controlling IP mobility on behalf of the host: tracking the movements of the host and initiating the required mobility signaling. By this mechanism, mobility can effectively be hidden from the terminal. From the perspective of the MN, the entire PMIPv6 domain appears as its home network.

The main advantages of network mobility are:

- Efficient use of wireless resources - the air interface, which was main cause for

latency, is excluded from the mobility signaling path.
- Optimized handover delay – The home network prefix assigned by the serving network is unchanged in the PMIPv6 domain, so handover-related signaling overhead is reduced. Movement detection and Duplicate Address Detection (DAD) are not required in the PMIPv6 domain, except when the MN first enters the PMIP domain.
- Privacy – since the mobile node's Home Address is fixed over a PMIPv6 domain, the precise location of the mobile node is harder to be deduced by attackers
- Decreased MN complexity - No MIPv4/v6 stack should be implemented and executed on the mobile terminal. This provides flexibility (no software upgrade needed for nodes), reduces the processing requirements and as a consequence the power management for MN is more efficient.

Compared to host mobility protocols, no CoA address is needed to be configured on the MN, because the IP address is not changed in the localized mobility domain.

PMIPv6 extends MIPv6 signaling and reuses many concepts such as the HA functionality. Though, the original PMIPv6 protocol uses indirect routing via HA tunneling, so no route optimization is supported, as it was for MIPv6. Enhancements for PMIPv6 are being standardized so that optimization of routes is also used. While in the case of MIPv6 multiple MNs in the same subnet are configured with a common IPv6 network prefix in case of PMIPv6 a unique home network prefix is assigned to each MN.

The new principal functional entities of PMIPv6 are, as defined by RFC 5213:

- **Mobile Access Gateway** (MAG) is a function on an access router (AR) that manages the mobility-related signaling for a mobile node that is attached to its access link. It is responsible for tracking the mobile node's movements to and from the access link and for signaling the mobile node's local mobility anchor.
- **Local Mobility Anchor** (LMA) is the home agent for the mobile node in a Proxy Mobile IPv6 domain. It is the topological anchor point for the mobile node's home network prefix(es) and is the entity that manages the mobile node's binding state. The local mobility anchor has the functional capabilities of a home agent as defined in Mobile IPv6 base specification with the additional capabilities required for supporting Proxy Mobile IPv6 protocol. (Proxy Mobile IPv6, Request for Comments 5213, 2008, p.4)

PMIPv6 Operation

Steps of the attachment procedure, considering the MN is physically attached to an access link (Network discovery procedure is passed):

- MN sends a Access Initiation request (Attach request) to the MAG
- MAG authenticates the MN on the policy server (e.g., AAA-server), based on the MN subscriber identifier provided by the MN to the MAG. Proxy Mobile IPv6 domain must be able to identify a mobile node, using its MN-Identifier. This identifier must be stable and unique across the Proxy Mobile IPv6 domain. In some cases, the obtained identifier, as part of the access authentication, can be a temporary identifier and further that temporary identifier may be different at each re-authentication. With this identifier, the mobile node's profile can be retrieved by the MAG from the AAA server. The MAG will also determine if the MN is authorized for the network-based mobility management service.

Figure 2. Signaling call flow of mobile node attachment

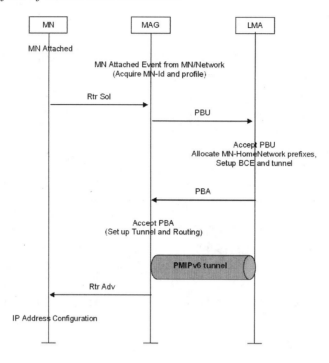

- The MN sends a Router Solicitation message
- The MAG sends a Proxy Binding Update (PBU) to the LMA in order to register the current point of attachment of the MN
- LMA creates a binding cache entry and responds with a Proxy Binding Acknowledgement (PBA) message including the mobile node's home network prefix(es) and the default-router address for the mobile node's home prefix. LMA also sets up its endpoint of the bi-directional tunnel to the MAG, so all necessary routes are set.
- After receiving the PBA, the MAG establishes the LMA as the endpoint for the tunnel. MAG adds forwarding rules, so all the traffic with the MN is forwarded to its home network. The MAG is emulating the mobile node's home interface on the access interface.
- MAG sends Router Advertisements (RA) with the home network prefix(es) learned from the PBA message.. It is also possible that the MN might have solicited an RA by sending a Router Solicitation message.
- The MN's interface will be configured either by stateful or stateless address configuration methods. Based on the policy profile information that indicates the type of address or prefixes, the MN address can be IPv6, IPv4 or dual IPv4/IPv6.
- After interface configuration the tunnel between the MAG and the LMA is used for packet routing (all the packages destined for the MN from outside the PMIPv6 domain are received by the LMA that uses the tunnel established before to forward the packages to the MN).

In the handover process, deregistration procedure should take place, so MN is detached from the previous MAG. The moment of detach, relative to the registration to the next MAG, depends on the implementation of the PMIP protocol. Fast PMIPv6 optimization solutions for PMIPv6, simi-

Figure 3. Architecture evolution from UMTS to LTE

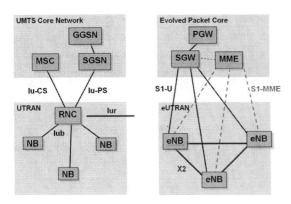

lar to FMIPv6 are being standardized so predictive logic can be used for soft handover.

If the MN changes its point of attachment, the MAG on the previous link will detect the mobil node's detachment from the link. It will signal the LMA with a DeRegistration PBU and the mobility anchor will remove or update the binding to the next MAG and the routing state for that MN. The MAG on the new access link, upon detecting the MN on its access link, will signal the LMA to update the binding state. The process is now the same to the normal attachment procedure. After completion of the signaling, the serving mobile access gateway will send the Router Advertisements containing the mobile node's home network prefix(es), and this will ensure the MN will not detect any change at network level relative to the attachment interface.

MOBILITY MANAGEMENT IN LTE NETWORKS

LTE Architecture with Respect to Mobility Protocols

Long Term Evolution (LTE) represents the next step in the development of the 3GPP (3rd Generation Partnership Project) global mobile broadband networks. LTE standardization started with the Release 8 of 3GPP.

The main principles of the LTE-SAE architecture include:

- High bitrates through efficient radio access technologies based on physical layer technologies like Orthogonal Frequency Division Multiplexing (OFDM) and Multiple-Input Multiple-Output (MIMO) systems (a 4x4 MIMO scheme will boost downlink rates up to 326 Mbps).
- delay optimizations and simplified architecture(basically there are two node types - base stations and gateways); a common anchor point and gateway (GW) node for all access technologies;
- IP-based protocols on all interfaces
- Harmonized architecture for 3GPP accesses and interworking with non-3GPP accesses: optimized interworking with 3GPP and CDMA accesses and the use of common subscriber and service control solutions

Compared to the 3G UMTS technology, LTE presents simplified and optimized architecture uses a minimum number of nodes in the user plane. The EUTRAN (Evolved UMTS Terrestrial Radio Access Network) is based on one single node, the eNodeB that has also routing capabilities through the X2 interface.

The new functional entities of the Evolved Packet Core (EPC) network are:

- **Mobility Management Entity (MME)** is responsible for NAS (Non-Access Stratum) signaling and NAS security and Idle mode Mobility Management. The NAS signaling terminates at the MME and it is also responsible for generation and allocation of temporary identities to UEs. MME is connected to the HSS so it manages authentication and authorization. The MME is directly connected to the SGSN through S3 interface and it takes care of signaling for Inter CN mobility in other 3GPP networks. MME is responsible for bearer management functions including dedicated bearer establishment and for selection of S-GW and PDN-GW.

- **Serving Gateway (SGW)** has the main role to route and forward user data packets. SGW acts as the mobility anchor for the user plane during inter-eNodeB handovers and as the anchor for mobility between LTE and other 3GPP technologies (terminating S4 interface and relaying the traffic between 2G/3G systems and PDN GW). It has connection to the policy and charging module PCRF, so it also offers Accounting & charging support. S-GW does Transport level packet marking in uplink and downlink.
- **Packet Data Network Gateway (PGW)** is of course the interface to the PDN. If a UE is accessing multiple PDNs, there may be more than one PDN GW for that UE.

Another key role of the PDN GW is to act as the anchor for mobility between 3GPP and non-3GPP technologies. PGW is responsible for the UE IP address allocation, Policy Enforcement Policy Enforcement Bearer binding and packet classification.

For handling mobility in LTE networks there are two versions defined. The basic Evolved Packet System architecture defines GTPv2 (GPRS Tunneling Protocol) as C-Plane protocol in the Core Network. 3GPP specifications were extended with support for PMIP. PMIP can be used as C-Plane protocol for the S5 interface and GRE tunneling is used for U-plane. Still, even in the PMIP based architecture, GTP is the C-Plane protocol from MME to SGW. Gxc interface is added since PMIP does not support QoS information.

S8b is the roaming variant of S5-PMIP, and used when P-GW is at Home PLMN (Figure 4b).

GTP is an IP based protocol used within GSM and UMTS networks. GTP can be considered

Figure 4. a) Non-roaming EPS architecture – S5 PMIP b) roaming for home routed traffic architecture – S5 PMIP

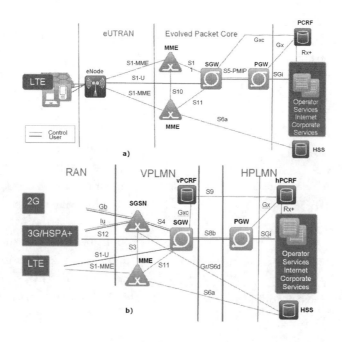

somehow similar with a network mobility protocol because the MSs are connected to a SGSN without being aware of GTP. One GTP tunnel endpoint is the GGSN and it is terminated on the SGSN not on the MS, so radio resources are saved.

GTP logic is similar to the one using foreign agent care-of address, as GTP tunnel is not terminated on mobile station but on SGSN. In GPRS, a tunnel is identified by a tunnel endpoint identifier and a SGSN/GGSN address.

Comparison between GTP and IETF PMIP implementations for EPC:

- While in the GTP version the bearer termination and bearer awareness takes place in PGW, in the PMIP case this is done in SGW.
- The GTP signaling is bearer based, while PMIP is binding (PDN connection) based.
- The policy and charging control element is connected with the PGW for the GTP based implementation, while for the PMIP implementation PCRF is reachable from both SGW and PGW. In roaming scenarios in the PMIP architecture both V-PCRF and H-PCRF are involved in charging related messaging

For roaming between a PMIP network and a GTP network there are two possibilities specified:

- Direct peering - one of the two roaming partners provide support for both variants of roaming
- Use of an Interworking Proxy (IWP) that is situated between the GTP-based PLMN and the PMIP-based PLMN to perform protocol conversion

LTE Support for Roaming Between Different Technologies

The great advantage of the LTE architecture is allowing for connections and handover to other fixed line and wireless access technologies, giving the service providers the ability to deliver seamless mobility.

Because of the correspondence between the functions of elements from different 3GPP technologies, some logical associations can be made. The common core SGSN/MME can be seen as only one element: uses shared subscriber data, MME and SGSN has same PDP/bearers for subscribers and the QoS model is similar. The network is optimized by usage of one GW. Same functional

Figure 5. LTE architecture and interconnection with other 3GPP/non-3GPP trusted/un-trusted networks

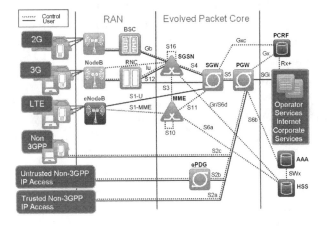

grouping can be done for the Serving and PDN Gateways as a common GW core.

3GPP Technical Specification 23.402 specifies interworking (mobility without IP address change) with networks that do not interface with 3GPP access networks on the access network level, for example WLAN and Wimax. Connection to other PDN networks is managed trough different types of interfaces (S2a/b/c) that also imply different logic for IP address preservation in case of handover.

The generic non-3GPP interworking provides connectivity to PDN and mobility between 3GPP access and any non-3GPP access that support one of the following scenario:

- Trusted non-3GPP access network with PMIPv6 (S2a interface)
- Un-trusted non-3GPP access networks with PMIPv6 (S2b interface)
- Trusted or un-trusted non-3GPP access networks with DSMIPv6 (S2c interface)
- Trusted non-3GPP access network with MIPv4 FA mode (S2a interface)
- Chained S8 with S2a/S2b

From the perspective of the PMIP mobility protocol:

- S-GW takes the role of a Mobile Access Gateway (MAG), if PMIP-based S5 or S8 is used. The MAG function shall be able to send UL packets before sending the PBU or before receiving the PBA. S-GW also decides if packets are to be forwarded (uplink towards PDN or downlink towards UE) or if they are locally destined to the S-GW (e.g. Router Solicitation). In the special case of PMIP-based S8-S2a/b chaining (concatenation of PMIP tunnels), the Serving GW includes functionality to link the user-plane of the PMIPv6 tunnel towards the PDN GW and the user-plane of the PMIPv6 tunnel towards the MAG function of the Trusted Non-3GPP IP Access or the ePDG. In this case, S-GW acts as a LMA towards the MAG function of the Trusted Non-3GPP IP Access or the ePDG and also act as a MAG towards the PDN GW.
- PGW represents the LMA if PMIP-based S5 or S8, or if S2a or S2b is used. The LMA function shall be able to accept UL packets from any trusted MAG without enforcing that the source IP address must match the CoA in the MN BCE. If S2cinterface is used PGW is a Dual stack MIPv6 (DSMIPv6) Home Agent.
- ePDG performs de-capsulation and en-capsulation, address translation and mapping. ePDG is also responsible for routing of packets from/to PDN GW (and from/to Serving GW if it is used as local anchor in VPLMN) to/from UE. ePDG can act as Mobile Access Gateway (MAG)if network based mobility (S2b) is used.

Depending on the type of network links that are involved in the handover process (trusted/non trusted, 3GPP/non 3GPP) and the protocols that are implemented in the networks involved in the handover process, the mobility schemes can be different and complex. This is why GTP and PMIP can be used in parallel, along with other MIP versions, like DSMIP and MIPv4 in FA mode. Implementing mechanisms for mobility protocols management are necessary.

During initial attach or handover attach a UE needs to discover the trust relationship (whether it is a Trusted or Untrusted Non-3GPP access network) of the non-3GPP access network. The trust relationship of a non-3GPP access network is made known to the UE with one of the following options:

- If the non-3GPP access supports 3GPP-based access authentication, the UE discovers the trust relationship during the 3GPP-based access authentication.

- The UE operates on the basis of pre-configured policy in the UE.

Mechanisms for Mobility Protocol Selection

IP session continuity between 3GPP and non-3GPP access types may not be provided in this case if there is a mismatch between what the UE expects and what the network supports in terms of mobility protocols. The Mobility Mechanisms supported between 3GPP and non-3GPP accesses within an operator and its roaming partner's network would depend upon operator choice.

In case of networks supporting multiple IP mobility mechanisms, IP Mobility Management Selection (IPMS) is used. 3GPP technical specification mentions that IPMS consist of two components:

- "IP MM protocol selection between Network Based Mobility (NBM) and Host based mobility (HBM - MIPv4 or DSMIPv6).
- Decision on IP address preservation if NBM is selected" (3rd Generation Partnership Project TS 23.402, 2009, p. 14)

IPMS is not responsible to the selection between PMIv6P and GTP over S5/S8.

In case of an initial attachment to a 3GPP access, no IPMS logic is necessary since connectivity to a PDN GW is always established with a network-based mobility mechanism. But on initial attachment to a trusted non-3GPP access or ePDG and upon handover from 3GPP to a trusted non-3GPP access, IPMS is performed before an IP address is allocated and provided to the UE.

When a NBM mechanism is used for establishing connectivity in the target access upon inter-access mobility, IP address preservation for session continuity based on NBM may take place additionally based on the knowledge in the network of UE's capability (if available) to support NBM.

IP address preservation for session continuity based on HBM may take place if the network is aware of the UE capability to support DSMIPv6 or MIPv4 (such knowledge may be based on an indication to the target trusted non-3GPP access or ePDG from the HSS/AAA):

- If IP mobility management protocol selected is DSMIPv6, in order to get IP address preservation for session continuity, the UE shall use DSMIPv6 over S2c reference point. This IP address shall be used as a CoA for DSMIPv6.
- If the IP mobility management protocol selected is MIPv4, the address provided to the UE by the non-3GPP access network is a FACoA and IP address preservation is performed over S2a using MIPv4 FACoA procedures.

A capability negotiation in terms of mobility protocols takes place: if the UE provided an explicit indication of the supported mobility mechanisms, the network shall also provide an indication to the UE of the selected mobility management mechanism.

Support of different IP mobility management protocols at local/home network is known by the AAA/HSS in one of the following ways:

- through static pre-configuration
- through the indication, received from the trusted non-3GPP access system or ePDG in the AAA exchange phase, about the supported IP mobility management protocols (PMIPv6 and/or MIPv4 FA CoA mode).

The final decision on the mobility management mechanism is made by the HSS/AAA upon UE authentication in the trusted non-3GPP access system or ePDG.

Figure 6. Handover from 3GPP Access to Trusted Non-3GPP IP Access with PMIPv6 on S2a and PMIPv6 or GTP on S5 interface

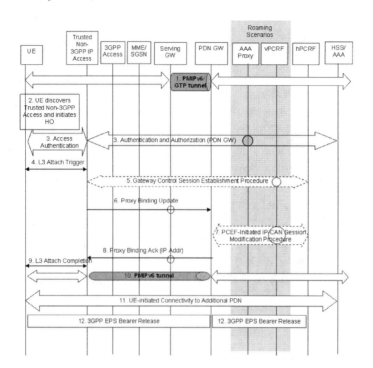

Upon selecting a mobility management mechanism, as part of the AAA exchange the HSS/AAA returns to the trusted non-3GPP access system or ePDG an indication on whether a local IP address shall be allocated to the UE, or if instead PMIPv6 shall be used to establish the connectivity, or an indication that the address of the MIPv4 Foreign Agent shall be provided to the UE.

Handover Process Between 3GPP and Non-3GPP Networks

For illustration of the handover process some particular cases will be presented, as described in 3GPP TS 23.402. First the case of the handover from 3GPP Access to Trusted Non-3GPP IP Access with PMIPv6 on S2a and PMIPv6 or GTP on S5 interface is illustrated.

It is assumed that while the UE is served by the 3GPP Access, a PMIPv6 or GTP tunnel is established between the S-GW and the PDN GW so this is the starting phase for the process description.

In the roaming case, the vPCRF acts as an intermediary, sending the QoS Policy Rules Provision from the hPCRF in the HPLMN to the Serving GW in the VPLMN. The vPCRF receives the Acknowledgment from the Serving GW and forwards it to the hPCRF. In the non-roaming case, the vPCRF is not involved at all.

1) The UE is connected in the 3GPP Access and has a PMIPv6 or GTP tunnel on the S5 interface.
2) The UE uses Network Discovery and Selection process to determine the handover from the 3GPP network currently used to the trusted non-3GPP IP access system
3) The UE performs access authentication and authorization in the non-3GPP access system. The 3GPP AAA server returns also the MN NAI (Network Address Identifier);

based on IMSI, to the trusted non-3GPP access system and this identifier will be used to identify the UE in the PBU message. The AAA server queries the HSS that has information about the PDN-GW identity or identities to the trusted non-3GPP access system.

4) The attach procedure is initiated and the UE and the Trusted non-3GPP start declaring there capabilities in terms of mobility protocols in order to preserve, if possible, the current IP address (method described in the previous paragraph). If the UE provides an APN, the Access verifies that it is allowed by subscription. If the UE does not provide an APN, and the subscription context from HSS contains a PDN GW identity and APN pair corresponding to the default APN, the trusted non-3GPP Access uses the default APN.

5) The Trusted Non-3GPP IP Access initiates a Gateway Control Session Establishment Procedure with the PCRF to obtain the rules required for the Trusted Non-3GPP IP Access to perform the bearer binding for all the active sessions the UE may establish as a result of the handover procedure.

6) The MAG in the Trusted non-3GPP IP Access sends a PBU message to the PDN GW, in order to establish the new registration. PBU specifies the MN-NAI, Access Technology Type, Handover Indicator, APN, GRE key for downlink traffic. Access Technology Type is set to a value matching the characteristics of the non-3GPP access. The APN may be necessary to differentiate the intended PDN from the other PDNs supported by the same PDN GW.

7) The PDN GW executes a PCEF-Initiated IP-CAN Session Modification Procedure with the PCRF. The Event Report indicates the change in Access Type. An IP-CAN session is the association between a UE represented by an IPv4 and/or an IPv6 address, and UE identity information, if available, and a PDN represented by a PDN ID (e.g. an APN). An IP-CAN session incorporates one or more IP-CAN bearers. The PCEF (Policy and Charging Enforcement Function) is responsible for service data flow detection, policy enforcement and flow based charging functionalities.

8) The PDN GW responds with a PMIP Binding Acknowledgement message to the Trusted Non-3GPP IP Access.

9) L3 attach procedure is completed at this point. The IP address(es) assigned to the UE by the PDN-GW is assigned to the UE.

10) The PMIPv6 tunnel is set up between the Trusted Non-3GPP IP Access and the PDN GW. The UE can send/receive IP packets at this point.

11) For connectivity to multiple PDNs, the UE establishes connectivity to all the other PDNs that the UE was also connected to before the handover.

12) For the previously used resources to be freed, the PDN GW shall initiate resource allocation deactivation procedure in 3GPP access

The second presented case is the reverse process of the first one, so a big part of the signalling is similar but the MME is the entity most involved in the handover process, while S-GW is an intermediary for communication between MME and PDN GW.

The optional interaction steps between the gateways and the PCRF in the procedures only occur if dynamic policy provisioning is deployed. Otherwise policy may be statically configured with the gateway.

Both the roaming and non-roaming scenarios are depicted in the figure.

The steps involved in the handover are discussed below:

Figure 7. Handover from trusted or untrusted non-3GPP IP access to E-UTRAN with PMIPv6 on S2a or S2b or MIPv4 on S2a and GTP on S5/S8 interfaces

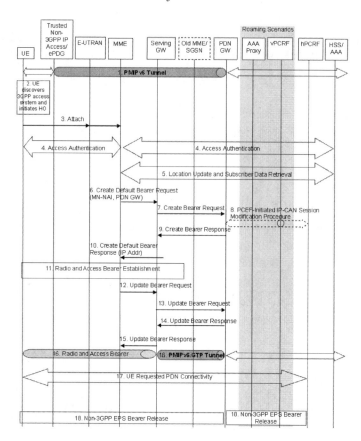

1) The UE uses a trusted or untrusted non-3GPP access system and is being served by PDN GW (as PMIPv6 LMA or MIPv4 HA).
2) UE discovers the E-UTRAN access and determines to handover to it.
3) The UE sends an Attach Request to the MME with Attach Type indicating "Handover" Attach. The message from the UE is routed by E-UTRAN to the MME.
4) The MME may contact the HSS and authenticate the UE.
5) MME performs location update procedure and subscriber data retrieval from the HSS. The MME receives information on the PDNs the UE is connected to over the non-3GPP access in the Subscriber Data obtained from the HSS.
6) The MME selects an APN, a serving GW and PDN and sends a Create Default Bearer Request (including IMSI, MME Context ID, PDN-GW address, Handover Indication, APN) message to the selected Serving GW. Since the Attach Type is "Handover" Attach, handover Indication information is included.
7) The Serving GW sends a Create Default Bearer Request message to the PDN-GW in the VPLMN or HPLMN. Since Handover Indication is included, the PDN GW should not switch the tunnel from non-3GPP IP access to 3GPP access system at this point.
8) The PDN GW executes a PCEF-Initiated IP CAN Session Modification Procedure with the PCRF to obtain the rules required

for the PDN GW to function as the PCEF for all the sessions with the new IP-CAN type as a result of the handover procedure. If the UE had disconnected from the default PDN before handover then the PDN GW executes a PCEF initiated IP CAN Session Establishment procedures.

9) The PDN GW responds with a Create Default Bearer Response message to the Serving that contains the IP address or the prefix that was assigned to the UE while it was connected to the non-3GPP IP access.

10) The Serving GW forwards the Create Default Bearer Response message to the MME. This message also includes the IP address of the UE. This response is an indication for the MME that the S5 bearer setup and update has been successful. The PMIPv6 or GTP tunnel(s) over S5 are established.

11) Radio and Access bearers are established at this step in the 3GPP access

12) The MME sends an Update Bearer Request (eNodeB address, eNodeB TEID, Handover Indication) message to the Serving GW.

13) The Serving GW sends an Update Bearer Request message to the PDN GW to prompt the PDN GW to tunnel packets from non 3GPP IP access to 3GPP access system and immediately start routing packets to the Serving GW for the default and any dedicated EPS bearers established.

14) The PDN GW acknowledges by sending Update Bearer Response to the Serving GW.

15) The Serving GW forwards the Update Bearer Response message to the MME.

16) The UE sends and receives data at this point via the E-UTRAN system.

17) For connectivity to multiple PDNs, the UE establishes connectivity also to all other PDNs used before the handover.

18) For the previously used resources to be freed, the PDN GW shall initiate resource allocation deactivation procedure in the trusted/untrusted non-3GPP IP access.

LTE Mobility Management States

As in UMTS 3G networks, from a mobility perspective, the UE can be in one of three states:

- LTE_DETACHED state is typically a transitory state in which the UE is powered-on but is in the process of searching and registering with the network.
- LTE_ACTIVE state, the UE is registered with the network and has an RRC connection with the eNB. In LTE_ACTIVE state, the network knows the cell to which the UE belongs with an accuracy of a serving eNodeB and can transmit/receive data from the UE.
- LTE_IDLE state is a power-conservation state for the UE, where typically the UE is not transmitting or receiving packets. In LTE_IDLE state, no context about the UE is stored in the eNB. In this state, the location of the UE is only known at the MME and only at the granularity of a tracking area (TA) that consists of multiple eNBs. Even in IDLE state, the UE performs periodic TA Update (concept similar to the Routing Area Update message used for 3G networks) and should be capable to send Service Request or respond to paging from the MME. When there is a UE-terminated call, the UE is paged in its last reported TA. One important detail is that during IDLE period or the UE, the IP address that was assigned to the terminal is not changed. The IDLE mode operation is facilitated by the possibility of communication directly between the UE and MME, without message processing in the eNodeB, so most part of the radio resources are released. This communication uses the NAS (Non

Figure 8. E-UTRAN protocol stack. NAS is transparent for the eNodeB

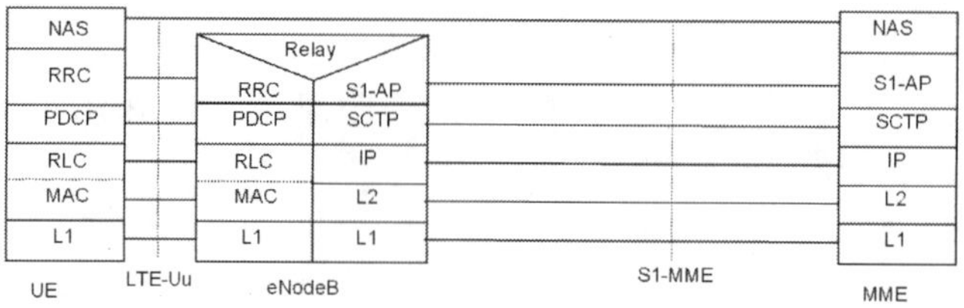

Access Stratum) communication, functionality also "inherited" from 3G networks. NAS operates between the UE and Core Network and the information transported is transparent for the eNodeB. Among the Layer 3 protocols, Radio Resource Control (RC) belongs to the Access Stratum, while Mobility Management and Call Control belong to the NAS.

As in 3GPP that TAs for LTE and for pre-LTE radio access technologies will be separate (an eNB and a UMTS Node-B will belong to separate TAs to simplify the network's handling of mobility of the UE when UE crosses from one 3GPP access technology to anther. The objective is to keep the UE camped in the idle state of the different technologies (LTE_IDLE in LTE and PMM_IDLE in UMTS/GPRS). When new traffic arrives for the UE, the UE is paged in both the technologies and depending on the technology in which the UE responds, data is forwarded through that RAT (but paging in networks based on two different technologies will be possible only for 3GPP networks).

It is very important to take into consideration that the states relate to terminal mobility management only. They are independent of the PDP contexts and the number of IP addresses allocated to the mobile terminal.

Case Study: GPRS/UMTS Core Networks Adaptation Towards All-IP Convergence Based on MIPv6

In this paragraph a MIPv6 based solution will be presented, a demonstrative implementation to prove that equipment that is still used and producing great income to the telecom operators (even if sometimes it's considered obsolete from the technology point of view), can be adapted for integration with next generation networks.

The Mobile IPv6 protocol implementation used is MIPL (Mobile IPv6 for Linux), an application developed by HUT (Helsinki University of Technology) as part of the GO-Core project. MIPL is one of the most appreciated software of this type, HUT also being responsible for the IPv4 mobility protocol implementation.

The experimental configuration tests the possibility of integration between GPRS and Wi-Fi networks. The home agent and mobile node are installed on Linux machines that are then integrated in the GPRS and Wi-Fi networks. The configuration was minimized as much as possible. Thus, the home agent (HA) it's considered the router of the home network and also as "access router" (the router that makes the connection to the foreign network).

For using typical GPRS or UMTS networks, and mainly for using public GSM networks that only generate IPv4 networks, a method for the "by-pass" of IPv4 networks is needed so the network will act as an IPv6 one. The solution is using OpenVPN tunneling that also makes the

Figure 9. Simplified experimental architecture for GPRS/UMTS and WLAN convergence using MIPv6

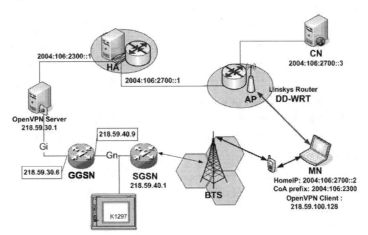

address conversion from IPv4 to IPv6 and from IPv4 to IPv6:

- The OpenVPN server can be installed directly on the home agent (HA) or on another terminal from the home subnet
- The OpenVPN client is installed on the mobile node (MN)

The Wi-Fi router/access point has to offer support for IPv6 communication. The wireless routers used is a modified Linksys WRT54GL, using the open source wireless router firmware DD-WRT. This router can be used as Access Router in the foreign network because it can send router advertisements to hosts for IPv6 address autoconfiguration and automatic setting of default routes. From the GUI of the Linksys router, the RADVD (router advertisement daemon) file can be edited, for setting the prefix that is advertised on a certain interface. RADVD listens to "router solicitation" messages and respond with "router advertisements" that are broadcast messages.

The GPRS network was used as the back-up network, because of the wide coverage area, so the connection to the GPRS network is initiated, but it is not used unless the wireless network is not available. Priorities are set in the mipv6.conf file, so Wi-Fi priority is set higher.

For handover tests, at the Application level there were used multimedia streaming programs (voice and video), file transfer (FTP) and control messages ICMP6, all of them based on IPv6. Also the QOS parameters of the GGSN were set to maximal values.

The conclusion is that handover is not seamless, but the packet handover time for multimedia streaming is short, under 30ms, which is acceptable. But if we consider that the test conditions were optimal, there was not any perturbation on the wireless link, since we were directly connected to the BTS through a cabled connection, the handover time should have been shorter. This also depends on application, some applications use buffers for data transmissions and thus the handover time is smaller.

As a conclusion, the demonstrator stated that, by using a low cost method for optimizing 2,5G and 3G networks, a scalable and pervasive MIPv6 network was developed and can be easily replicated by the communication network infrastructure of any public network operator.

FUTURE RESEARCH DIRECTIONS

The global trend towards scalable architectures that integrate, in some hierarchy level, different types of access technologies, regardless of the telecommunications associations involved in standardization. Splitting mobility management in two domains, the local and globalised mobility management, and making both solutions independent brings a lot of flexibility to operators.

Each different access network can manage the mobility of the terminals closer to them, and thus do so more efficiently and with less overhead. Moreover, they do not have to depend on functions of an external operator to provide their own mobility services. Local mobility domains can be more extended or quite limited. One of the advantages of combining global and local mobility is the easiness of integration of networks based on the concept of "mobile router": networks in transport vehicles, on ships or even networks on airplanes. The mobile router concept is described by RFC3963 - Network Mobility (NEMO).

The mobile router can have a similar functionality with the MAP in HMIPv6, having a double role:

- The mobile router acts a mobile node in a localized domain and receives an IP address from one LMA that is persistent while moving in the local mobility domain
- The mobile router behaves as a MAG to a VMN (visiting mobile node) and provides network prefixes for the mobile node and intermediate packets for the communication with LMA.

Integration of local and global mobility can enhance the functionality of other mobility concept, like the network mobility described by NEMO.

Advantages of Localized Mobility in Case of Vehicular Networks

MIPv6 was adopted by International Civil Aviation Organization (ICAO) as the global IP mobility management protocol for the future IPv6 based Aeronautical Telecommunication Network (ATN/IP). In case of the local mobility solutions, the service providers should have the implementation decision. A fundamental problem of the aeronautical environment is the utilization of the small bandwidth at the wireless link. PMIPv6 and HMIPv6 are the local mobility management approaches, essential for reducing the air-ground signaling overhead. A benefit of PMIPv6 is the delay performance since all the mobility signaling is done in the wired links which is faster and more reliable compared to wireless links. In PMIPv6 all mobility related signaling except MN identifier (MN-ID) provision to MAG, occurs on the ground, so there is a negligible overhead on the wireless channel. The backbone network that interconnects PMIP domains, presented in the classical case (Figure 1), is, in case of aeronautical networks, the ATN/IP backbone.

In the ATN network there are two types of correspondent nodes (CN):

- the Air Traffic Services Unit (ATSU), which is responsible for the safety of flight, changed during the flight, as the aircraft moves across airspace sector borders
- the Airline Operating Centre (AOC) for the exchange of airline operations data, that is mainly static during one flight.

Depending of the specific of the CNs, route optimization can have a different impact: in the case of AOC, this is not mandatory, but produces big improvements in case of multiple CNs, like multiple ATSUs.

Another interesting scenario is the one in which a user is connected via WLAN to a hotspot in a bus station, while waiting a bus. The bus provides connectivity, while it is moving, and users keep their connectivity when they hand off from the bus to the fixed hotspots available at the stations. This connectivity is maintained while the bus is travelling and moving (potentially moving from an administrative domain to another), in a transparent way to the users.

CONCLUSION

Network Localized Mobility Management protocols, like PMIPv6, represent a solution to outcome the weaknesses of the host mobility protocols and bring Mobile IP protocols closer to the mobile broadband architectures. A big advantage is the easiness of integrating non MIP-aware user equipment.

3GPP LTE architectures implement convergence mechanisms with any IP network, but due to the support offered to older 3GPP architectures, the algorithms for choosing the right mobility protocols are still complicated, despite of the simplification goal of LTE networks.

Even if now, because of the proxy mobility protocol, there is an alternative for mobility, LTE networks are still relying on the GTP protocol. If we take this aspect into consideration we can say that Long Term Evolution 3GPP networks are, as they are actually named, an evolution, and not a revolution, even if 4G networks will boost the mobile broadband usage.

REFERENCES

3rd Generation Partnership Project (2009). *Technical Specification Group Services and System Aspects. Architecture enhancements for non-3GPP accesses (Release 8)*, TS 23.402 v8.4.1. Retrieved January, 2009, from http://www.3gpp.org

Devarapalli, V., Gundavelli, S., Chowdhury, K., & Muhanna, A. (2007). *Proxy Mobile IPv6 and Mobile IPv6 interworking*. Retrieved June 2008, from http://tools.ietf.org/html/draft-devarapalli-netlmm-pmipv6-mipv6-00

Devarapalli, V., Wakikawa, R., Petrescu, A. & Thubert, P. (2005). *Network Mobility (NEMO) Basic Support Protocol, Request for Comments 3963*. Retrieved June 2008, from http://www.ietf.org/rfc/rfc3963.txt

Gundavelli, S., Leung, K., Devarapalli, V., Chowdhury, K., & Patil, B. (2008). *Proxy Mobile IPv6, Request for Comments 5213*. Retrieved September, 2008, from http://www.rfc-editor.org/rfc/rfc5213.txt

Johnson, D., Perkin, C., & Arkko, J. (2004). *Mobility Support in IPv6, Request for Comments 2775*. Retrieved September, 2005, from http://www.ietf.org/rfc/rfc3775.txt

Kong, K., Lee, W., Han, Y. & You, H. (2008). Mobility Management for All-IP Mobile Networks: Mobile IPv6 vs. Proxy Mobile IPv6. *IEEE Wireless Communications,* April.

Lei, J. & Fu, X. (2007). *Evaluating the Benefits of Introducing PMIPv6 for Localized Mobility Management*. Technische Berichte des Instituts für Informatik an der Georg-August-Universität Göttingen, 29.

Lin, Y., & Pang, A. (2005). *Wireless and Mobile All-IP Networks*, November. New York: Wiley Publishing Inc.

Nishida, K., & Yokota, H. (2006). *Mobility NETLMM Protocol Applicability Analysis for 3GPP SAE Network*. Retrieved January, 2008, http://tools.ietf.org/html/draft-nishida-netlmm-protocol-applicability-00

Patel, V., & McParland, T. (2008). *Proposed Guidance for IPS Mobility Management Aeronautical Communication Panel*. August, Montreal, Canada.

Chapter 19
Handover Optimization for 4G Wireless Networks

Dongwook Kim
Korea Advanced Institute of Science and Technology, South Korea

Hanjin Lee
Korea Advanced Institute of Science and Technology, South Korea

Hyunsoo Yoon
Korea Advanced Institute of Science and Technology, South Korea

Namgi Kim
Kyonggi University, South Korea

ABSTRACT

The authors present a velocity-based bicasting handover scheme to optimize link layer handover performance for 4G wireless networks. Before presenting their scheme, as related works, they firstly describe general handover protocols which have been proposed in the previous research, in terms of the layers of network protocol stack. Then, they introduce state-of-the-art trends for handover protocols in three representative standardization groups of IEEE 802.16, 3GPP LTE, and 3GPP2. Finally, they present the proposed bicasting handover scheme. Original bicasting handover scheme enables all potential target base stations for a mobile station (MS) which prepares for handover to keep bicasted data, in advance before the MS actually performs handover. This scheme minimizes the packet transmission delay caused by handover, which achieves the seamless connectivity. However, it leads to an aggressive consumption of backhaul network resources. Moreover, if this scheme gets widely adopted for high data rate services and the demand for these services grows, it is expected that the amount of backhaul network resources consumed by the scheme will significantly increase. Therefore, the authors propose a novel bicasting handover scheme which not only minimizes link layer handover delay but also reduces the consumption of backhaul network resources in 4G wireless networks. For the proposed scheme, they exploit the velocity parameter of MS and a novel concept of bicasting threshold is specified for the proposed mobile speed groups. Simulations prove the efficiency of the proposed scheme over the original one in reducing the amount of consumed backhaul network resources without inducing any service quality degradation.

DOI: 10.4018/978-1-61520-674-2.ch020

INTRODUCTION

Fourth-generation (4G) wireless networks aim at supporting further enhancement of mobile user experience through better wireless communication architecture in terms of quantity and quality. Various real-time services requiring high data rates and low delay constraints have been listed out in (Acx et al., 2003) as the main applications for the 4G wireless networks. For example, real-time gaming services and high-quality video conferences require data rates ranging from 1 to 20 Mbps and delay constraints of less than 20 ms. Geographic real-time data-casting services through real-time video streaming demand data rates ranging from 2 to 5 Mbps with 20 ms delay constraint. As well as the real-time services mentioned above, transport control protocol (TCP)-based services which require very high throughput performance will be also mainly supported in the 4G wireless networks. In wireless networks, handover is the mechanism by which an ongoing connection between a mobile station (MS) and a correspondent base station (BS) is transferred from one point of access to the fixed network to another (Pahlavan, Krishnamurthy, & Hatami, 2000). Hence, handover generally occurs due to the movement of MSs and it causes packet transmission delay. All the real-time and TCP-based services are significantly sensitive to this delay. This is because the packet transmission delay can directly impact on delivering real-time packets or indirectly impact on end-to-end TCP throughput requiring a low and stable round trip time. In (mITF, 2005), it is addressed that, in case of intra-system handover, packet transmission delay fluctuations for real-time streams are desired to be 30 ms or less. And in (Ericsson, 2005), it is noted that, for TCP-based services the interruption time by the handover should be below 50 ms in order to achieve robust TCP throughput performance.

By the way, 4G wireless networks will be envisaged as a heterogeneous network where various wireless access technologies are converged, and universal usage and broadband access to users are supported (Liu, Li, Guo, & Dutkiewicz, 2008; Mohanty & Akyildiz, 2006). In the heterogeneous network such as 4G, the type of handover is classified into two categories: horizontal handover and vertical handover. Horizontal handover occurs between different BSs which use the same radio access technology (RAT). On the other hands, vertical handover occurs when an MS performs handover from a system to another one which employs a different RAT. Thus the vertical handover is also referred to as inter-system handover. Even though there is the evident difference between the horizontal and vertical handovers, fundamental handover procedure in these handovers can be usually described in terms of four phases as follows (Verdone & Zanella, 2002). First, measurement of the link quality (e.g., received power, bit error rate (BER), and etc.) is carried out at MS or BS. Second, based on the link quality and/or other parameters such as network load and available resources, MS or BS decides whether handover is needed or not. Various handover decision algorithms such as received signal strength (RSS)-based (Marichamy, Chakrabarti, & Maskara, 2003; Zhang & Holtzman, 1996) and location-based (Itoh, Watanabe, Shih, & Sato, 2002; Juang, Lin, & Lin, 2005; Ozdural & Liu, 2007; Zaidi & Mark, 2004; Zhu & Kwak, 2007) have been proposed in the previous literature and we will introduce these algorithms in Section II. Third, if the handover is necessary, new wireless resources to support the on-going communication with required quality-of-service (QoS) may be selected in the target BS to which the MS desires to move. Finally, if these resources are selected, the handover is performed by means of break-before-make (BBM) procedure or make-before-break (MBB) procedure. The BBM procedure is referred to as hard handover and the MBB procedure is referred to as soft handover. Both the hard and soft handover methods will be introduced in Section II.

Previous research on the issue of minimizing handover delay has been classified into two categories: link layer approach and IP layer approach. The link layer approach is to minimize the interruption time due to the link layer handover. That is, the approach concentrates on the procedure for transferring on-going connection in the link layer. Thus it does not consider an interruption time due to the upper layer handover procedure. On the other hands, the IP layer approach is to minimize the interruption time due to the IP layer handover generated when an MS performs handover from a BS to another one which uses a different IP domain. To optimize the route establishment and registration procedures for the IP layer handover, a lot of mobility protocols have been proposed. We will also introduce these protocols in Section II. Since IP layer handover always occurs after link layer handover, it is worth noting that the interruption time caused by the link layer handover should be minimized to support high quality for both real-time and TCP-based service users. In addition, in order to minimize the total handover delay due to both link layer and IP layer handovers, cross-layer handover protocols have been also proposed. We will summarize various cross-layer handover protocols in Section II.

In this chapter, we propose a novel handover protocol referred to as velocity-based bicasting handover scheme (VBHS), to optimize link layer handover performance for 4G wireless networks. Before presenting the proposed scheme, we will describe a background for general handover protocols firstly. Then, we will introduce state-of-the-art handover protocol trends in three standardization groups of IEEE 802.16, third generation partnership project long-term evolution (3GPP LTE), and 3GPP2. After that, we will present the proposed VBHS which reduces the amount of consumed backhaul network resources without any degradation of QoS for both real-time and TCP-based services. Consequently, we will make a conclusion by summarizing and suggesting future research directions.

BACKGROUND

In this section, we describe general handover protocols which have been proposed in the previous literature. As mentioned in Section I, since the horizontal handover is performed when an MS moves between BSs which use the same RAT, the previous research on this type of handover has focused on improving the performance of link layer handover and managing the intra-domain IP layer handover. And the vertical handover is the inter-system handover which occurs when an MS performs a handover from the serving BS to the target BS whose RAT is different from that of the serving BS. Thus the research direction for the vertical handover has focused on managing the inter-domain IP layer handover and the mobility management for multi-modal devices. Especially, the mobility for multi-modal devices should be differently managed by a couple of modes, or RATs, installed in the device. Three representative standardization groups of IEEE 802.16, 3GPP-LTE, and 3GPP2 are actively researching on improving the performance of its own RAT and they are competing to broaden its own service coverage. Thus it is worth noting that the state-of-the-art trends for this mobility management in three groups of IEEE 802.16, 3GPP-LTE, and 3GPP2 should be investigated. We will introduce the brief summary for this in Section III. The remainder of this section is organized as follows. Firstly, we describe the link layer handover protocols in terms of hard and soft handover methods. Then we introduce various IP layer handover management schemes in terms of domain-based approach. Finally we summarize various cross-layer handover protocols.

1. Link Layer Handover

Link layer handover method is generally classified into hard handover and soft handover. The hard handover method is characterized by the BBM procedure, i.e., MS releases the connection with its serving BS and then establishes a new connection with the target BS. Since the MS experiences a temporary interruption of communication, it is important to reduce the interruption time not to degrade the required QoS. An advantage of hard handover method is that an MS maintains only one connection with one BS at any time, thus the hardware of the MS can be cheaply and simply implemented.

Since an MS is only served from one BS during hard handover, to serve the MS without any QoS degradation, the handover decision algorithm should determine the optimal time point of handover occurrence and the target BS which can support the best service quality. The handover decision algorithm generally exploits average signal strength and MS performs handover to the BS which gives the highest average signal strength to itself. As shown in Figure 1, if the instantaneous signal strength is used in the handover decision algorithm, an MS at the cell boundary may perform handover to the target BS and then back to its old serving BS repeatedly because of the signal fluctuation caused by fading effects. This phenomenon is called as ping-pong effect. Therefore, instead of the instantaneous signal strength, average signal strength is used in the handover decision algorithm to alleviate the ping-pong effect. The average signal strength is obtained by the moving average operation which averages a number of instantaneous signals measured for a predefined duration. The average signal strength is also referred to as the filtered signal strength since the moving average operation filters the fading effects in the strength. In (Brown, 1964), as an alternative of the moving average operation, the exponential average is used for improving handover performance. In (Holtzman & Sampath, 1995) and (Austin & Stüber, 1994), the moving average duration is adjusted according to the estimated velocity of MS. In (Mark & Leu, 2007), the local averaging scheme which samples the received signal strength at a faster rate than the handover decision rate is proposed. In addition, the timer-based handover decision scheme is proposed in (Leu & Mark, 2003).

However many unnecessary handovers may still occur at the cell boundary since the fading effects may not be completely eliminated by the moving average operation. To solve this problem, consequently, the concept of hysteresis is introduced. In the hysteresis-based handover decision algorithm, if the average signal strength from an adjacent BS is higher than that from the serving BS by a hysteresis value, an MS performs handover. However, even though this algorithm can reduce unnecessary handovers, it causes handover initiation delay as shown in Figure 2 (Ulukus & Pollini, 1998; Zonoozi & Dassanayake, 1997). In other words, instead of performing handovers for

Figure 1. Ping-pong effect caused by the fluctuation of the instantaneous signal strength

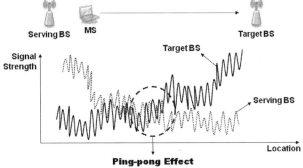

MSs when the MSs reach the midpoint location between the serving BS and target BS, the handovers are performed at the points further away from the serving BS. Since the handover initiation delay may lead to an increase in the call dropping probability, inter-cell interference, and BER, the hysteresis value should be carefully selected. In the previous literature, signal-to-noise-ratio (SNR), signal-to-interference-and-noise-ratio (SINR), BER, and other parameters have been used for handover decision criteria, instead of signal strength. Moreover, additional criteria considering mobility information such as location, velocity, and moving direction have been also used for handover decision. In (Itoh, Watanabe, Shih, & Sato, 2002), the distance from the serving BS as well as signal strength is used for the handover decision in order to reduce unnecessary handovers. In (Zaidi & Mark, 2004), a mobility aware handover decision scheme is proposed. The scheme predicts the future mobility state using location, velocity, and acceleration, and based on the mobility state, it also predicts the future signal strength. Then the scheme decides whether a handover is necessary with this signal strength. In (Juang, Lin, & Lin, 2005), a handover decision scheme which reduces unnecessary handovers caused by shadowing is proposed. The scheme considers the correlation among shadowing components which is obtained based on the estimated velocity. In (Zhu & Kwak, 2007), an adaptive handover scheme which adjusts a hysteresis value based on distance information is proposed. In (Ozdural & Liu, 2007), a predictive base station switching scheme using the moving speed and direction information is proposed.

On the other hands, in comparison with hard handover, soft handover method is characterized by the MBB procedure. With this method, MS establishes a connection with the target BS prior to releasing the connection with its serving BS, which enables the MS not to experience temporary interruption of communication during handover. The main advantage of soft handover method is that MS can receive data more reliably with diversity gain since it may receive the same copy of data from a couple of BSs. That is, since the probability that all downlink signals from BSs with which the MS maintains connections are degraded or faded simultaneously is low, the successful decoding probability in the MS increases by combining all received signals. However, this advantage incurs a victim of complex and expensive hardware of the MS. And, since several BSs should allocate resources for one MS, the efficiency of resource

Figure 2. Handover initiation delay caused by the hysteresis

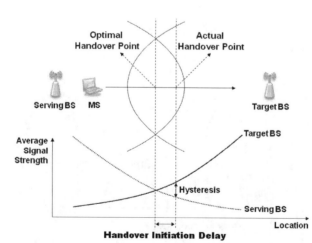

usage could decrease. The coordination between the BSs for the soft handover method also incurs high network load and lessens the flexibility.

In the soft handover method, both an MS and its serving BS maintain the active set including the BSs which can support the required QoS for the MS. The active set is determined based on the average signal strength parameter and, basically, two threshold values of T_ADD and T_DROP are used for managing the active set (Garg, 2000; Holma & Toskala, 2005). A BS is included in the active set if the average signal strength from the BS is greater than T_ADD and it is excluded from the active set if that from the BS is less than T_DROP. Various active set management schemes have been proposed and they use various criteria such as SNR, SINR, BER, and mobility information to determine the active set. Moreover, many researchers have proposed various BS-to-BS coordination schemes to efficiently utilize backhaul network resources (Lee, Mazumdar, & Shroff, 2006; Makaya & Aissa, 2003; Navaie & Yanikomeroglu, 2006).

2. IP Layer Handover

Emerging 4G wireless networks should coexist and harmonize with various existing wireless networks such as 3G systems, wireless local area network (WLAN), Bluetooth, and digital video broadcasting (DVB). And, in the networks, contemporary users will continually require not only voice-oriented service but also various multimedia services at anytime and anywhere. To achieve these goals, it is recommended that the 4G wireless networks should adopt the IP as a backbone network protocol. Considering this fact, handover schemes used in the networks should also be well matched and optimized with the IP.

The goal of IP layer handover management scheme is to provide the seamless connectivity by minimizing the service interruption time while an MS performs handover from a BS to another one that belongs to a different IP domain. A lot of schemes have been proposed to achieve this goal and they are generally classified into two categories depending upon the mobility scope: inter-domain handover and intra-domain handover. As shown in Figure 3, the inter-domain handover occurs when an MS moves from an administrative domain to another one, whereas the intra-domain handover occurs when the MS moves within an administrative domain.

2.1. Inter-Domain Handover Management

MSs perform inter-domain handovers when they move across different administrative domains in heterogeneous networks where different RAT

Figure 3. Handover management category

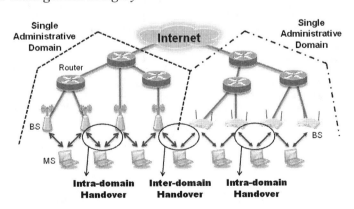

technologies coexist. Thus an inter-domain handover management scheme is implemented in the IP or upper layers. Since the scheme should transfer an on-going connection from an administrative domain to another one without any QoS degradation, previous research has mainly focused on maintaining the established connection rather than minimizing handover signaling overhead and service interruption time. Among various inter-domain handover management schemes, Mobile IP (Perkins, 1996) and session initiation protocol (SIP) (Rosenberg, Schulzrinne, Camarillo, Johnston, Peterson, Sparks, Handley, & Schooler, 2002) have been typically standardized by Internet engineering task force (IETF) (Das, Misra, & Agrawal, 2000). The Mobile IP supports the mobility of MS below the application layer and thus it can provide transparent mobility for the application layer protocols. On the other hands, the SIP is originally proposed as a signaling protocol for the session mobility. The protocol controls to establish, modify, and terminate a connected session in the application layer and thus it cannot support the transparent mobility for the TCP layer. Therefore the SIP is not suitable for 4G wireless networks which should be able to support the seamless mobility without any disruption to the application layer protocols. As a result, most proposed 4G wireless networks adopt the Mobile IP as an inter-domain handover management scheme.

Figure 4 gives the operation procedure of Mobile IP. In the figure, the MS originally has its home IP address allocated from the home agent (HA). When it performs handover from the home network to the foreign network, the foreign agent (FA) which manages the foreign network sends a registration message including a care-of-address (COA) to the home agent. The COA identifies the current location of the MS and it can be either the IP address of the foreign agent or a newly assigned IP address belonging to the foreign network. When the home agent gets the registration message from the foreign agent, the home agent updates the current location of the MS with the COA. After this operation, if a corresponding node sends packets to the MS with the home IP address, these packets are firstly delivered to the home agent. Then, the home agent redirects those packets to the current location of the MS with the COA. Based on the procedure, there are many extended and optimized schemes for efficient inter-domain handover management. For more information, refer to (Perkins, 1997) and (Solomon, 1998).

Even though the Mobile IP can cover diverse and heterogeneous network environment, it cannot support fast and seamless mobility due to the delay of address translation and update. Especially, when handover events occur frequently, the Mobile IP generates large signaling overhead. Therefore, it should be noted that supplementary techniques for fast and seamless IP layer handover management is necessary.

2.2. Intra-Domain Handover Management

The target of intra-domain handover management is for the movement of MS within a single administrative domain. Previous research on this management has focused on reducing handover signaling overhead and interruption time, and a lot of schemes such as Cellular IP (Campbell, Gomez, Kim, Valkó, Wan, & Turányi,

Figure 4. Operation procedure of Mobile IP

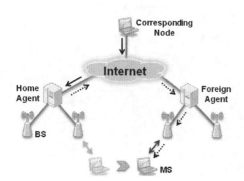

2000), handoff-aware wireless access Internet infrastructure (HAWAII) (Ramjee, Varadhan, Salgarelli, Thuel, Wang, & Porta, 1999), telecommunications enhanced Mobile IP (TeleMIP) (Das, Misra, & Agrawal, 2000), intra-domain mobility protocol (IDMP) (Das, McAuley, Dutta, Misra, Chakraborty, & Das, 2002), hierarchical Mobile IP (HMIP) (Fogelstroem, Jonsson, & Perkins, 2007; Soliman, Castelluccia, Malki, & Bellier, 2005), Pre-registration and Post-registration (Koodli, 2005; Malki, 2007) have been proposed in order to support efficient intra-domain handover. Although these protocols have different mechanisms in detail, they adopt similar fundamental techniques to achieve fast and seamless intra-domain handover management. These techniques are categorized into four approaches: hierarchical binding, tunneling, multicasting, and pre-registration.

Figure 5 shows the concept of hierarchical binding approach. The hierarchical binding approach is to isolate the scope of location update procedure according to the movement range of MS. In a domain-based hierarchical binding scheme, when an MS performs handover within a foreign domain, the location update message for the MS is not transferred to the home agent of the MS. The message only goes to the local mobility management agent within the domain.

In this approach, the packet sent from a node located in an outside domain is firstly delivered to the local mobility management agent (MMA). Then, the packet is relayed to the current location of the MS. Since the movement of the MS within a domain is transparent to the home agent and the signaling messages are localized in the domain, both the handover signaling overhead and interruption time are reduced. An enhanced hierarchical binding approach proposes the use of multiple hierarchical MMAs and, in the approach, the location update message is transferred up to the agent located in the crossover point between old and new paths. But, this approach has a trade-off between reducing location update signal overhead and increasing packet relaying overhead. Even though the hierarchical binding approach reduces the interruption time due to IP layer handover at some extent, the remained interruption time is still too large to provide the seamless mobility. Therefore other approaches such as tunneling, multicasting, and pre-registration have been proposed to further reduce the interruption time due to the IP layer handover.

Figure 6 shows the concept of tunneling approach. In the figure, since the MS performs handover from the serving BS to target BS which belongs to a different IP domain, the new IP layer path from the corresponding node to the MS via

Figure 5. Concept of hierarchical binding approach

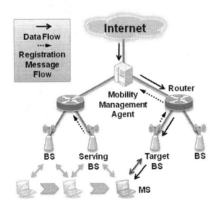

Figure 6. Concept of tunneling approach

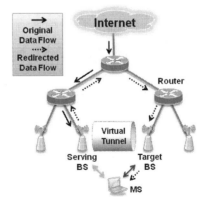

the target BS should be established. That is, the MS cannot receive packets from the corresponding node until the new path setup procedure is completed. This path setup delay is significantly large and increases handover interruption time. To solve this problem, the tunneling approach establishes a transient tunnel through the backhaul network between the serving BS and target BS. Through this tunnel, the serving BS forwards packets for the MS to the target BS during the new path setup time. After completing the path setup procedure, the transient tunnel is eliminated by serving or target BSs. This approach can effectively reduce the path setup delay. However, when the MS continuously performs handovers to other new target BSs due to its rapid movement, the tunnels will be chained and it causes the increase of packet transmission delay due to a roundabout path delivery.

Another approach to reduce the path setup delay during IP layer handover is multicasting. Figure 7 shows the concept of multicasting approach and the approach delivers the same copy of packets to multiple BSs to reduce the interruption time due to the path setup delay. In the figure, the MS is attached to the serving BS of BS 2 before handover and after some time, it decides to perform handover to the target BS of BS 3. In the multicasting approach, BS 2 sends the multicast starting message to the MMA of the MS before the MS switches the air-interface link to the BS 3. Upon reception of this message, the MMA delivers all packets to the neighbor BSs of BS 1 and BS 3 in duplicate. Thus, right after the MS switches the link to the BS 3, it receives the packets from the BS 3. After completing new path setup procedure, BS 3 may send the multicast stopping message to the MMA and then the multicasting operation ceases. The multicasting approach does not require any signaling overhead and delay which are needed to make a tunnel between the serving and target BSs, which can reduce the interruption time due to the IP layer handover further than the tunneling approach. However, the multicast approach wastes backhaul network resources to deliver duplicated packets to all neighbor BSs.

The pre-registration approach is a little different from the previous two approaches. The key concept of pre-registration approach is to simultaneously initiate the registration procedure for the IP layer handover and link layer handover procedure. Figure 8 shows the concept of pre-registration approach. When the pre-registration approach is adopted, the service interruption time due to the IP layer registration procedure can be reduced or eliminated. Moreover, if the IP layer registration procedure is completed before the link layer han-

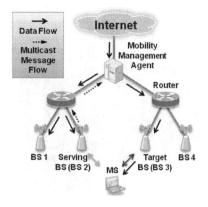

Figure 7. Concept of multicasting approach

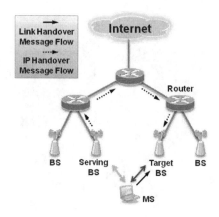

Figure 8. Concept of pre-registration approach

dover procedure, the MS can re-establish the connection with the target BS right after completing link layer handover. The pre-registration approach does not require additional signaling overhead to reduce the path setup delay. However, after initiating the IP layer registration procedure to a target BS, if an MS does not perform handover to the target BS due to the sudden change of link condition, it cannot receive packets until the path from a corresponding node to the current location of the MS is recovered.

Recently, the concepts of tunneling and multicasting (referred to as bicasting) approaches have been employed in the link layer handover schemes in order to reduce the interruption time. Moreover, since each intra-domain handover management scheme mentioned above has its strong and weak points, the standardization groups are trying to adopt multiple schemes in a combined manner for realizing 4G wireless networks.

3. Cross-Layer Handover

Cross-layer handover (CLH) protocols aim at performing fast handover by combining the information in more than one layer of network protocol stack. Figure 9 shows an example of integrated architecture which depicts a heterogeneous network scenario where two different wireless systems, System 1 and System 2, are connected through the IP backbone network. In the figure, we depict three types of handover: inter-BS handover, intra-domain handover, and inter-system handover. For the inter-BS handover, the update of IP address for the MS is not necessary, on the other hands, for both the intra-domain and inter-system handovers, the MS should update its IP address using a COA received from new Anchor BS (ABS) of ABS 2 (intra-domain handover) or ABS 3 (inter-system handover) which supports the FA functionalities.

According to the type of handover mentioned above, different layers are combined in the CLH protocols. In case of the inter-BS handover, link layer handover reduces the interruption time by exploiting the functionalities in lower or upper layers. And, in case of both the intra-domain and inter-system handovers, IP layer handover is performed fast with the help of link layer information. In addition, recently, a few commercial networks such as worldwide interoperability for microwave access (WiMAX), 3G networks, and WLAN networks have been deployed with the common service area. To perform fast inter-system handover in this environment, some researchers have proposed a few optimization techniques which change, reorder, and combine original link layer and IP layer procedures.

We briefly summarize the CLH protocols which have been proposed in the previous literature as follows: (1) Inter-BS handover: Chen, Cai, Sofia, & Huang (2007) present a cross-layer handover protocol for Mobile WiMAX. It performs fast link layer handover through a mechanism which incorporates information from several layers in network protocol stack. The mechanism uses an IP layer tunnel to redirect and relay link layer handover-related messages. (2) Intra-domain handover: Kim, Kim, Ra, Yoo, & Kim (2008) address the problem of the use of the conventional Mobile IP protocol for intra-domain handover in Mobile WiMAX networks. The authors also mention some disadvantages of the use of fast Mobile IPv6 (FMIPv6) and propose a novel IPv6 based cross-layer handover protocol which supports the seamless intra-domain mobility. The scheme reduces the likelihood of IP layer anticipation failure by simply modifying the conventional procedures of FMIPv6. (3) Inter-system handover: firstly, in (Zhu & McNair, 2005), a cross-layer movement detection technique is proposed to improve the inter-system handover. The authors propose a likelihood function which considers both the link and IP layers movement hints. They modify the original IP layer pre-registration protocol by applying this likelihood function. And Jo & Cho (2008) propose an inter-system handover scheme used in handover between Mobile WiMAX and 3G

Figure 9. Three types of handover in a heterogeneous network

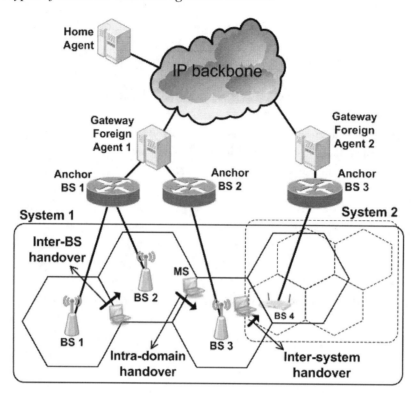

networks. The scheme reduces total handover delay and achieves high throughput by appropriately combining link layer and IP layer handover-related signaling messages which are conventionally used in each system. In addition, Luoto & Sutinen (2008) propose a framework which builds upon cross-layer architecture, in order to improve the performance of the Mobile IP. The framework exploits various cross-layer and cross-domain triggers. Through experiments with the prototype implementation, the authors show that combining of the information in each layer enhances the performance of handover decision-making of the Mobile IP protocol. Mohanty & Akyildiz (2006) also propose a cross-layer handover management protocol (CHMP) which provides the seamless connectivity during a course of intra-domain and inter-system handovers in heterogeneous networks. They analyze how the link layer information, network information such as MS's speed, and handover signaling delay parameters affect the performance of the original HMIP. Then, they propose the CHMP, as an extended HMIP protocol, which reflects the results of the analysis.

As a variation of cross-layer handover method, policy-based handover protocols have been proposed in the previous literature. The policy-based handover is proposed to perform the vertical handover in a heterogeneous network where various RATs such as universal mobile terrestrial service (UMTS), WLAN, DVB, and beyond 3G (B3G) co-exist. The policies for handover determine when the current serving network cannot satisfy the QoS requirements and which network holds the radio resources available (Mihovska, Luo, Mino, Tragos, Mensing, Vivier, & Fracchia, 2007). In (Nyberg, Ahlund, & Rojmyr, 2006), a cost function, the weighted sum of utilization cost, power consumption, and interface capacity is defined for the handover policies. And, in (Seo, Lee, & Song, 2007), four types of policies, maximum throughput, minimum energy consumption, maximum bit per energy, and adaptive configuration policies are used for the vertical handover. To obtain the information needed to define the policies, the cross-layer handover approach in which the serving network gathers the information of both the link layer and IP layer from candidate networks for handover has been proposed (Aust, Proetel, Fikouras, & Görg, 2003). And an integrated network architecture which efficiently gathers some information needed to perform a policy-based handover has been proposed (Mihovska, Luo, Mino, Tragos, Mensing, Vivier, & Fracchia, 2007; Pries, Mäder, & Staehle, 2006). Therefore, in 4G wireless networks, the policy-based handover can intelligently perform a vertical handover since it exploits not only the information of signals strength from candidate networks for handover but also other information such as the load status of network and user preference to the network selection.

STATE-OF-THE-ART TRENDS FOR HANDOVERS

We briefly introduce state-of-the-art handover protocols in three standardization groups of IEEE 802.16, 3GPP LTE, and 3GPP2.

1. Handovers in IEEE 802.16

The WLAN based on IEEE 802.11 (IEEE 802.11) networks provide very high data rate. However, they only support the limited user mobility and, in the network, the coverage per access point is narrow. The 3G networks such as UMTS fully support the user mobility and provide various data services with large coverage area. However, the data transmission rate is low and the service cost is considerably expensive in comparison with the WLAN networks. As a result, in order to achieve the higher data rate and lower service cost with full mobility support in a wide coverage area, the IEEE committee initiated a 802.16 project of wireless broadband access (WBA) (IEEE 802.16).

The IEEE 802.16e specification (IEEE 802.16e task group) which supports user mobility is an extended version of IEEE 802.16d (IEEE 802.16 task group d) in which fixed wireless access services are served with the carrier frequency ranging from 2 to 11 GHz in both line-of-sight and non-line-of-sight environments. To support the user mobility up to 60 km/h, the IEEE 802.16e specification presents a link layer handover scheme. In the specification, both MS- and BS-initiated handover procedures are given and in this section, we introduce the MS-initiate handover procedure as shown in Figure 10. In the figure, to perform the cell reselection procedure, an MS shall firstly obtain the information about neighbor BSs. This information is gathered by the serving BS and delivered to the MS with the MOB_NBR-ADV message. After recognizing the existence of neighbor BSs through the message, the MS measures the link condition from the each neighbor BS. Based on the link condition information, the MS determines whether handover is necessary or not. Since the handover decision algorithm is not only vendor-specific but also an implementation issue, the IEEE 802.16e specification does not present any algorithm. If the handover is necessary, the MS sends the MOB_MSHO-REQ message which includes the list of potential target BSs, to the serving BS. Then, the serving BS sends the HO-Request message to each potential target BS in order to know which BS can accept the connection with the MS without any QoS degradation. The BS receiving this message replies with the HO-Response message including the acceptance result. After gathering the results for acceptance, the serving BS informs the results to the MS with the MOB_BSHO-RSP message. After receiving the MOB_BSHO-RSP message, if the MS wants to perform handover to one of the positively acknowledged potential target BSs, the MS sends the MOB_HO-IND message to the serving BS

Figure 10. Handover procedure in the IEEE 802.16e system

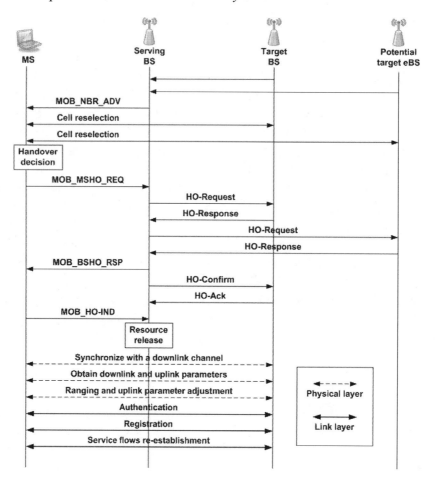

to identify the target BS. Upon reception of the MOB_HO-IND message, the serving BS sends the HO-Confirm message to this target BS. Then, the target BS sends the HO-Ack message if the handover is still acceptable. After sending the MOB_HO-IND message, the MS releases the connection with the serving BS and starts to establish a new connection with the selected target BS. To establish the connection with the target BS, the MS firstly synchronizes with a downlink channel of the target BS and obtains downlink and uplink parameters through the channel. Then, it performs the ranging and uplink parameter adjustment with the target BS. After that, the MS performs the authentication and registration processes. Lastly, the MS completes the handover operation by re-establishing the service flows with the target BS. In the 802.16e specification, there are no comments about the inter-system handover for a multi-mode MS. However the system requirement document (SRD) of the 802.16m which is proposed to an alternative system for the IMT-Advanced states that the 802.16m specification should support the inter-system handover (IEEE 802.16 task group m).

2. Handovers in 3GPP LTE

For the 3GPP LTE system standardization, various handover schemes have been proposed until now. To improve the system throughput and coverage, the standardization group will continually upgrade the original handover scheme. In this section, we briefly introduce the representative hard handover scheme proposed in (3GPP, 2008). Figure 11 shows the procedure of the scheme which adopts the mobile-assisted handover (MAHO) strategy, i.e., BS makes the handover decision based on the link condition measured and reported by MS.

Firstly, an MS sends the Measurement Report message which includes the link condition from neighbor BSs to the serving BS. Then the serving BS makes the handover decision based on the information of both Measurement Report and radio resource management. If handover is necessary, the serving BS determines the target BS with which the MS will make a new connection. As the IEEE 802.16e specification, handover decision algorithm is not specified in detail. Then, the serving BS starts the handover preparation phase by sending the Handover Request message to the target BS. The message gives all the information needed to prepare for handover to the target BS, e.g., radio resource control (RRC) context, evolved universal terrestrial radio access network (E-UTRAN) radio access bearer (E-RAB) context, access stratum (AS) configuration, and etc. Then, the target BS performs an admission control. If there are enough resources for the MS, the target BS configures the required resources and reserves a cell radio network temporary identifier (C-RNTI). After that, the target BS prepares for link layer handover and sends the Handover Request Ack message including the new C-RNTI and target BS security algorithm identifiers to the serving BS. After receiving the Handover Request Ack message, the serving BS sends the RRC Connection Reconfiguration message to the MS for making it perform handover. From now on, it is the start of handover execution phase. The serving BS no longer transmits and receives data to and from the MS since the MS detaches from the serving BS at this time. The serving BS sends the sequence number (SN) Status Transfer message which includes uplink and downlink packet data convergence protocol (PDCP) SN status for in-sequence delivery, and it starts to forward the downlink data to the target BS to avoid data loss during handover. Data forwarding continues as long as the serving BS receives data from evolved packet core (EPC) or until its buffer becomes to be empty. At the same time, the MS synchronizes

Figure 11. Handover procedure in the 3GPP LTE system

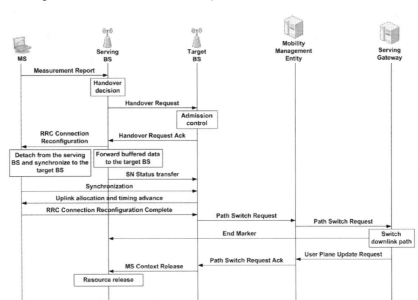

and accesses to the target BS through a random access channel (RACH). At the same time, the MS synchronizes and accesses to the target BS through a random access channel (RACH). Then, the target BS sends the uplink allocation and timing advance information to the MS. With this information, the MS sends the RRC Connection Reconfiguration Complete message to the target BS to inform that the handover procedure for the MS is completed. From this time, the target BS starts to send the buffered data forwarded from the serving BS to the MS. Now, the serving gateway (SG) should release the downlink data path to the serving BS and establish a new downlink path to the target BS. For this operation, the target BS sends the Path Switch message to the mobility management entity (MME) and the MME sends the User Plane Update Request message to the SG. After receiving the User Plane Update Request message from the MME, the SG sends the End Marker packets to the serving BS and releases any resources needed to forward packets to the serving BS. In addition, the SG establishes a new downlink data path toward the target BS and sends the User Plane Update Response message to the MME. Then, the MME sends the Path Switch Request Ack message to the target BS. After receiving this message, the target BS recognizes that the downlink data path is established. Then it should make the serving BS release the resources allocated to the MS. For this, the target BS sends the MS Context Release message to the serving BS and then the handover operation is completed after the serving BS releases the resources.

In addition, the 3GPP LTE system provides four approaches for supporting inter-system handover (MOTOROLA, 2008). (1) Mobile IP based approach for single transmit devices: Since single transmit devices cannot connect to different technologies at a time, only the BBM procedure is applicable. This approach does not require standard coordination, so it is suitable for handovers between 3GPP LTE and WiMAX, and between 3GPP LTE and evolution-data optimized (EV-DO). However, this approach causes long interruption time during handover thus it is not suitable for real-time services. (2) Mobile IP based approach for dual transmit devices: Dual transmit devices can communicate with two different technologies simultaneously. Thus, this approach prevents data loss and reduces interruption time during handover by adopting the MBB procedure. However, some problems such as cost, energy consumption, and interference need to be addressed. (3) Access network interconnect approach: This approach includes the exchange of information between two different technologies. Even though this approach employs the BBM procedure, the interruption time is very short thanks to information exchange between technologies. However, to do this, extensive works between standard activities are necessary. Examples are handovers between 3GPP LTE and its earlier system, and an optimized handover between LTE and EV-DO. (4) SIP based approach for dual transmit devices: This approach is suitable for SIP-based applications where Mobile IP is unavailable. However, it does not work in applications where IP address changes frequently. In addition, standardization for this approach has not been completed until now.

3. Handovers in 3GPP2

Recently, 3GPP2 has approved ultra mobile broadband (UMB) system as the successor of 1x EV-DO and the next generation of high speed wireless systems. The design objectives of UMB are to provide an improved mobile broadband experience using various state-of-the-art physical layer and link layer technologies to deliver industry-leading high spectral efficiency, low latency, and high service quality (3GPP2, 2008). One of the main principles of the UMB system is the support of seamless mobility. Novel concepts to facilitate the seamless mobility during handover are placed on not only the flat network architecture design but also the simplified core network and network interfaces. In addition, fast evolved

BS (eBS) switching scheme and new tunneling mechanisms enable fast connection establishment across BSs. Figure 12 shows an example of UMB system architecture. In the figure, each eBS performs all functionalities of BS, base station controller (BSC), packet data serving node (PDSN), and data accessing point (DAP) in the original 3G networks. Thus, the UMB system can be easily deployed due to the reduced number of entities and this advantage enables the system to be more reliable, flexible, and cost-effective.

We briefly introduce the handover scheme proposed in the UMB system. It is referred to as eBS switching scheme and Figure 13 shows the procedures of downlink and uplink eBS switching, respectively. If an MS detects higher signal strength from different eBS than the current serving eBS, it directly transmits the channel quality information (CQI) to the target eBS through the CQI channel. The target eBS requests the information about the state of radio link protocol (RLP) of the MS to the current serving eBS, through the link layer tunneling mechanism which is newly proposed in the UMB system. Then the target eBS sends the assignment message to the MS to indicate the completion of the downlink eBS switching. One of the key advances in UMB is to differently set the serving eBS in downlink and uplink directions, respectively. To select the best serving eBS in uplink direction, eBSs periodically sends the uplink pilot signal strength to MS. If the MS determines the target eBS which communicates with itself in uplink, it directly requests handover through the request channel (REQ). As the downlink eBS switching procedure, after the target eBS receives the RLP state of the MS, it sends the assignment message to the MS, and then the MS sends its data to the target eBS.

In (3GPP2, 2008), it is noted that the UMB system can support the seamless inter-system handover for a dual-mode MS. To perform a handover from an UMB system to EV-DO, a dual-mode MS firstly should negotiate an EV-DO session, create a point-to-point protocol (PPP) connection, and install a traffic flow template (TFT) through the UMB system by generating the link-layer tunnel between the MS and the EV-DO system. The dual-mode MS monitors the signals from both the UMB and EV-DO systems through a tune-away mechanism and it performs a physical layer handover based on these signals. Furthermore, the access gateway helps an IP layer

Figure 12. An example of UMB system architecture

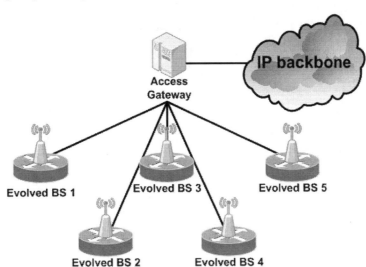

handover by establishing a connection with the HA for the MS using a proxy MIP (PMIP) or client MIP protocol.

ON OPTIMIZING HANDOVER PERFORMANCE FOR 4G WIRELESS NETWORKS

We present the proposed VBHS to optimize handover performance for 4G wireless networks. It should be noted that optimizing the handover performance is achieved by minimizing the handover delay. The proposed VBHS, especially, minimizes the link layer handover delay by adopting a cell switching scheme and a user-data-bicasting mechanism. The cell switching scheme enables all potential target BSs to keep the context information for an MS and to reserve control channel resources needed to handover in advance before the MS actually performs handover. The user-data-bicasting mechanism also makes all potential target BSs hold the data for the MS in advance before actual handover execution. We improve the original user-data-bicasting mechanism by adaptively controlling the bicasting time depending upon the velocity parameter of MS. Through the cell switching scheme and an enhanced version of user-data-bicasting mechanism, the proposed VBHS minimizes the link layer handover delay and it significantly reduces the amount of consumed backhaul network resources without any QoS degradation.

Figure 13. eBS switching procedures in both downlink (left) and uplink (right) directions in the UMB system

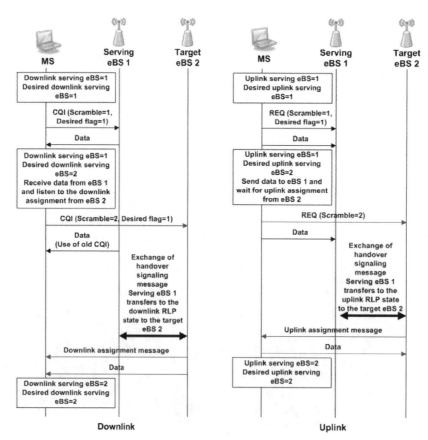

1. System Model and Problem Definition

The system architecture that we assume follows the one proposed in the E-UTRAN (3GPP, 2006). Figure 14 shows a hierarchical system model where anchor BS plays a role of the transmission of IP packets to its attached BSs and it exchanges various signaling messages to manage all the BSs. And they are fully connected through the backhaul network which makes a mesh topology. In the system, an MS receives its data from one BS at a time. Therefore, the MS should be able to change fast its serving BS which gives the best reliable and robust on-going communication, with minimal delay and signaling overhead. To satisfy this requirement, we adopt the cell switching scheme proposed in (Qualcomm Europe, 2006). In the scheme, both MS and BS configure and maintain a cell switching set (CSS) including the BSs which can be potential serving entities for the MS in the near future. And, each BS registered in the CSS has the context information of an MS which prepares for handover and it reserves control channel resources for the MS in advance before the MS actually performs handover. Then, handover operation is simply completed by changing the air-interface link from the serving BS to the target BS.

To determine whether a BS joins a CSS or not, we exploit the filtered SINR, instead of the instantaneous SINR reported by MS. The filtered SINR value at time t, $F_R(t)$ is obtained by the moving average operation which is expressed as $F_R(t) = \sum_{i=1}^{L} S_R(t-(i-1))/L$ where L is the number of instantaneous SINR values, $S_R(t)$, obtained for a predefined duration. Using the filtering SINR values, the cell switching is performed when the following condition is satisfied:

$$F_R^k(t) > F_R^S(t) + H, k \in \psi, \qquad (4.1.1)$$

where k is the identification (ID) of the target BS included in the CSS ψ and S is the ID of the serving BS. H is the hysteresis value and it reduces the ping-pong effect by generating the cell switching mechanism when the condition that the filtered SINR value of the target BS, $F_R^k(t)$ is higher than that of the serving BS, $F_R^S(t)$ by H is met.

We adopt the bicasting handover scheme proposed in the 3GPP LTE system to minimize the link layer handover delay. We call this scheme as the original bicasting handover scheme in this section. In Figure 14, the anchor BS sends the same copy of user data to the BSs of BS 2, BS 3, and BS 4, included in the CSS through the bicasting operation. The BSs keep the data in their respective buffers during a specified period. As soon as the switching from the serving BS of BS 2 to the target BS of BS 4 is completed, the MS indicates the packet that it is expecting to the BS 4. Then the MS receives the packet which is already held in the BS 4. All these procedures minimize the handover delay without any QoS degradation.

During the bicasting operation, BS 3 accumulates unnecessary data because the MS performs handover from the BS 2 to the BS 4 across the mobility path. And it is also inevitable that a certain amount of backhaul network resources are unnecessarily consumed. These two facts expose the inherent vulnerability of the bicasting handover scheme. Therefore, for the scheme, a novel mechanism that efficiently obtains necessary data, instead of just accumulating even unnecessary ones is needed to reduce the consumption of backhaul network resources. In this paper, we envision that the backhaul network resources can be efficiently utilized by reducing the bicasting time which directly impacts on the amount of resource usage. In other words, by controlling the execution time of bicasting performed until the cell switching occurs, we can efficiently utilize the resources in the backhaul network. We assume that the bicasting execution starts at time $t_B(m)$ when the filtered SINR value from a neighbor

Figure 14. An example of bicasting handover scheme

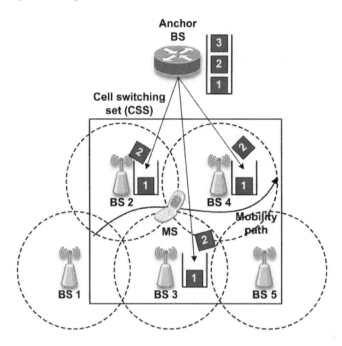

BS is higher than a specific threshold. Thus, $t_B(m)$ is expressed as $t_B(m) = T(F_R^k(t) > Th_m)$ where m is the index associated with each threshold value and $T()$ indicates the time when the condition within the parentheses is firstly satisfied. If the cell switching is performed at time t_{CS} when (4.1.1) is satisfied, our problem is mathematically defined as

$$m^* = \arg\min_{m \in M}\{t_{CS} - t_B(m) \mid t_{CS} - t_B > \tau\},$$
(4.1.2)

where M is a set which consists of the index for all threshold values and τ is the signaling delay needed to prepare for the bicasting execution.

The delay τ includes the time taken by the serving BS to recognize the BS which will receive bicasted data and to inform this to the anchor BS. The time taken by the anchor BS to prepare for bicasting to the newly recognized BS is also included in τ. That is, for the target BS which should hold user data, the anchor BS should bicast the data to the BS before τ time ahead of the cell switching time. Therefore, our aim is to find an optimal threshold value to initiate the bicasting execution as satisfying the τ requirement.

2. Velocity-Based Bicasting Handover Scheme (VBHS)

In the original bicasting handover scheme, if the filtered SINR value from a neighbor BS is higher than the set-update-and-add threshold, the BS gets included in a CSS, and it receives the bicasted data from the anchor BS. The filtered SINR changes with the mobility characteristics of MS thus it is impossible to exactly predict the starting time of bicasting with this SINR. Consequently, it is also impossible to derive the optimal bicasting threshold Th_m which minimizes the bicasting period while satisfying the τ requirement. The distance between BS and MS generally determines the filtered SINR value and it depends upon the velocity of MS. For example, if an MS moves to a neighbor BS with high velocity, it detects the rapidly increased signal strength from the BS.

But, for an MS moving with low velocity, the signal strength slowly increases. The execution of bicasting should be started earlier for an MS moving with high velocity in order to enable the target BS to get necessary data for the MS. If the bicasting for an MS moving with low velocity is started at the time when the bicasting is started for a fast moving MS, the BSs in a CSS for a slowly moving MS accumulate unnecessary bicasted data and lots of backhaul network resources are also wasted. Considering this fact, we exploit the velocity parameter in order to determine the appropriate bicasting threshold value. We assume that the velocity of MS is easily estimated by the various schemes proposed in the previous literature (Austin & Stüber, 1994; Azemi, Senadji, & Boashash, 2004).

Since it is not practical to determine bicasting threshold values for every velocity, we propose a criterion to group the velocities and we assign a bicasting threshold for the group. Given two velocities v_i and v_j, and two correspondent threshold values Th_i and Th_j, we define the following inequality:

$$|Th_i - Th_j| < \delta, \quad (4.2.1)$$

where δ is a criterion value for grouping. When (4.2.1) holds for an appropriate value of δ, we assume that $Th_i - Th_j$, and v_i and v_j form a group. Also, when $|Th_i - Th_j| > \delta$, v_j groups itself into a different group. Through (4.2.1), if some velocities are included in the same group, we use a representative bicasting threshold for the group. BS holds a look-up table which includes the specified mobile speed groups and correspondent bicasting threshold values. The information of the table can be obtained through simulations or empirical experiments while taking into account various usage scenarios. In this section, we introduce appropriate bicasting threshold values for the specified mobile speed groups in both urban and sub-urban environments where the values are obtained through extensive simulations.

We assume that MS moves with a relatively constant speed in the handover area which is the intersection part between two adjacent cells. And, during handover, the proposed VBHS with the MAHO strategy performs the following procedures. First, BS estimates the velocity of a served MS using a velocity estimation method. By relating the estimated velocity to a look-up table, the BS determines the appropriate bicasting threshold. For instance, if the estimated velocity of the MS is 40 km/h in urban scenario, the bicasting threshold of -6 dB is assigned to this MS using Table 4.3.2. The bicasting threshold for a mobile speed group is denoted as Th_{msg}. The filtered SINR value from BS k contained in a CSS at time t is denoted as $F_R^k(t)$. Then, BS checks whether the following inequality

$$F_R^k(t) > Th_{msg} \quad (4.2.2)$$

holds for all $k \in CSS$. If it does, the serving BS sends the bicasting execution message to the anchor BS where the message includes the ID of the BS which will receive bicasted data. Otherwise, the serving BS re-estimates the velocity and re-determines the bicasting threshold. Then it re-checks whether (4.2.2) is satisfied or not. The procedure ends when the anchor BS bicasts data for the MS to the BS indicated in the request message. When a BS is excluded from a CSS due to the decrease in the filtered SINR value, the anchor BS stops the bicasting for that excluded BS.

3. Performance Evaluation

We evaluate the performance of our scheme through system-level simulations and the simulations are performed based on the IEEE 802.20 mobile broadband time division duplex (MBTDD) system model (Tomcik, 2006). The simulation parameters are listed in Table 4.3.1. In the simulations, we assume that MS performs handover as it moves across the handover area which is the intersection part between the center BS and an adjacent neighbor BSs. Two different movement paths introduced in (IEEE 802.20, 2005) are exploited for the mobility model and we assume that these two paths are configured at the handover area. MS moves to the boundary area of the serving BS and performs handover to the target BS without any corner effect in the movement path 1 (MP1). In the movement path 2 (MP2), MS experiences the around-the-corner-effect as moving across the handover area. The around-the-corner-effect is augmented to model a Manhattan mobility scenario. It is defined as the effect that gets generated in urban area, usually at cell boundaries, due to the sudden change of direction. In the effect, that the signal strength from the serving BS abruptly decreases is called edge loss and the MS detects the abruptly increased signal strength from the target BS with edge gain. In the simulations, the edge loss or gain is set to 6 dB for 3 km/h ~ 15 km/h and 3 dB for 15 km/h ~ 90 km/h, respectively (IEEE 802.20, 2005). We use both the MP1 and MP2 for urban scenario while we only use the MP1 for sub-urban scenario. Since the probability that MS experiences the around-the-corner-effect in suburbs is very low, we only consider the MP1 for sub-urban scenario.

We carry out the simulations with the assumption of δ in (4.2.1) of 1 dB while increasing the velocity of MS by the unit of 5 km/h. Through the simulations, we establish three mobile speed groups for urban scenario: Pedestrian (~ 15 km/h), Vehicle (15 km/h ~ 90 km/h), and ExpressBus/Train (90 km/h ~). These groups are formed in reference to the mobility support mentioned by 3GPP E-UTRAN (3GPP™, 2006). Moreover, for sub-urban scenario, high speed mobile groups with the velocities ranging from 120 km/h to 200 km/h, from 200 km/h to 250 km/h, and from 250 km/h to above are formed. As we mentioned in (4.1.2), the execution of bicasting should be started in advance before the delay τ from the cell switching time. Therefore, we should consider this time requirement when we determine appropriate bicasting threshold values for the mobile speed groups. We perform extensive simulations while assuming the value of τ being 1 second and the results are given in (Kim, Lee, Kim, & Yoon, 2009) in detail. Based on the results, we make a look-up table of Table 1 which lists the bicasting thresholds suitable for both urban and sub-urban speed groups. Bicasting threshold values should be carefully determined since the total system performance significantly depends upon the values. Thus, in pragmatic mobile systems, various parameters which impact on the system performance should be considered to determine appropriate bicasting threshold values for mobile speed groups.

With the threshold values listed in Table 2, we obtain the results of the bicasting time performed until the cell switching occurs, for each mobile speed group in both urban and sub-urban scenarios. We compare our scheme with the original one which concurrently performs the bicasting and set-update-and-add procedures. The threshold for both procedures in the original scheme is set to -7 dB as shown in Table 1.

Figure 15 gives the results for the Pedestrian speed group with two mobility paths. We adopt two representative velocities of 3 km/h and 10 km/h for the group. In the MP1 scenario, our scheme reduces the bicasting time by 48.9% for 3 km/h and 41.3% for 10 km/h, respectively, when compared with the original scheme. Similarly, in the MP2 scenario, there are 84.8% and 76.9% reductions in the bicasting time with our scheme when the velocities are 3 km/h and 10 km/h, respectively. Consequently, in the Pedestrian speed group, our scheme outperforms the original one by an average of 63%, with the bicasting threshold of -5 dB. This gain directly leads to an improvement in the backhaul network resource utilization to that extent.

Figure 16 shows the results for the Vehicle speed group where three representative velocities, 15 km/h, 30 km/h, and 60 km/h, and both the MP1 and MP2 are considered, respectively. In the MP1 scenario, there are 18.3%, 18.1%,

Table 1. Simulation parameters

Items	Descriptions
Network topology	19 cells, hexagonal grid
BS-to-BS distance	Urban: 1 km, sub-urban: 2.5 km
Carrier frequency	1.9 GHz
Transmission bandwidth	10 MHz
BS transmit power	53 dBm (43 dBm/MHz)
Propagation model	31.5 + 35log(d), d in m
Log-normal shadowing	10 dB
De-correlation distance of the shadowing	3 km/h ~ 15 km/h: 5 m 15 km/h ~: 20 m
Multi-path fading	Jakes spectrum 5 km/h ~ 15 km/h: ITU Ped-B 15 km/h ~: ITU Veh-B
Filtering period	100 ms
Set-update-and-add threshold	-7 dB
Hysteresis	2 dB
Edge loss/gain	3 km/h ~ 15 km/h: 6 dB 15 km/h ~ 90 km/h: 3 dB

Table 2. Bicasting threshold values for both urban and sub-urban mobile speed groups

Mobile speed groups	Speed type	Bicasting threshold
Urban speed groups	Pedestrian (~ 15 km/h)	−5 dB
	Vehicle (15 km/h ~ 90 km/h)	−6 dB
	ExpressBus/Train (90 km/h ~)	−7 dB
Sub-urban speed groups	120 km/h ~ 200 km/h	−5 dB
	200 km/h ~ 250 km/h	−6 dB
	250km/h ~	−7 dB

and 24.7% reductions in the bicasting time with our scheme when the velocities are 15 km/h, 30 km/h, and 60 km/h, respectively. And, in the MP2 scenario, our scheme reduces the bicasting time by 50% for 15 km/h, 37.6% for 30 km/h, and 63.2% for 60 km/h, respectively. We conclude that our scheme outperforms the original one by an average of 35.3% with the bicasting threshold of -6 dB in Vehicle speed group.

Figure 17 illustrates the results for the sub-urban speed group with only the MP1 scenario. 49.8%, 54%, and 35% reductions in the bicasting time with our scheme are shown for the respective velocities of 120 km/h, 150 km/h, and 200 km/h. Therefore, our scheme reduces the bicasting time by an average of 46.3% in comparison to the original one with the bicasting thresholds determined for this group. Both the Express/Train speed group in urban scenario and the sub-urban speed group ranging 250 km/h to above use the bicasting threshold of -7 dB. Since this value is the same as the set-update-and-add threshold value, the bicasting is started with the set update procedure and thus there is no reduction of bicasting time with our scheme. This gives the reason why there are no results for both the Express/Train

Figure 15. Bicasting time vs. representative velocities for the Pedestrian speed group

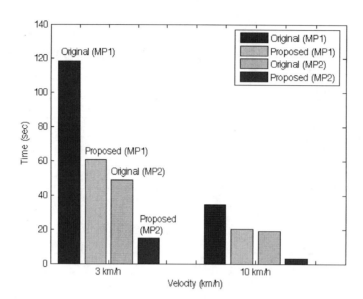

Figure 16. Bicasting time vs. representative velocities for the Vehicle speed group

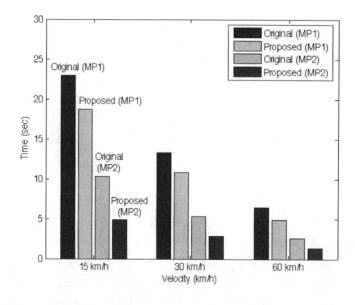

speed group in urban scenario and the sub-urban speed group ranging 250 km/h to above.

Through the simulation results, we conclude that the proposed VBHS efficiently reduces the bicasting time with the appropriate bicasting thresholds determined for the proposed mobile speed groups. The scheme also satisfies the time requirement of $t_{CS} - t_B > \tau$, i.e., it does not deteriorate the service quality by enabling the target BS to hold bicasted data before the occurrence of actual cell switching.

Figure 17. Bicasting time vs. representative velocities for the sub-urban speed group

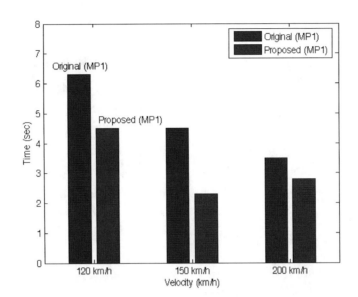

CONCLUSION

In this chapter, we have presented a novel bicasting handover scheme which optimizes link layer handover performance for 4G wireless networks. As related works, firstly, we have described various link layer, IP layer, and cross-layer handover protocols which have been proposed in the previous literature. Then, we have investigated the most advanced trends of handover protocols in IEEE 802.16, 3GPP LTE, and 3GPP2 standardization groups. Considering the fact that handover performance optimization is achieved by handover delay minimization, the proposed scheme minimizes link layer handover delay by adopting both a cell switching scheme and a user-data-bicasting mechanism. We have improved the performance of the original user-data-bicasting mechanism with the velocity parameter of MS. The proposed enhanced mechanism starts user data bicasting at the time which is determined by the appropriate velocity-based bicasting threshold, instead of performing the bicasting at CSS update-and-add time. For the performance evaluation of our scheme, we have established three mobile speed groups for urban and sub-urban scenarios, respectively. And for each speed group, we have determined the appropriate bicasting threshold, based on the empirical simulations assuming two different mobility paths. In the simulations, since we consider the signaling delay in determining bicasting threshold values, the target BS can receive bicasted data from the anchor BS at least a certain time ahead of cell switching initiation. As a result, our scheme does not incur any QoS degradation. Also, simulation results have shown that the proposed scheme significantly reduces the bicasting time in comparison to the original scheme. Therefore, we can directly conclude that the proposed scheme significantly enhances the resource utilization at the backhaul network.

For future works, we will investigate how the proposed scheme is affected by the diverse system parameters in 4G wireless networks. Then, we will find the optimal values of these system parameters and examine the optimal values for various criteria in the proposed scheme. Since the proposed scheme only considers the inter-

BS handover operation in the link layer, we will propose the extended schemes which minimize the handover delay while taking into account both the intra-domain handover and inter-system handover operations.

REFERENCES

3GPP2. (2008, December). Overview for ultra mobile broadband (UMB) air interface specification. *3GPP2 C.S0084-000-A v1.0*. Retrieved from http://www.3gpp2.org

3GPP. (2006, March). 3rd generation partnership project; Technical specification group radio access network; Requirements for evolved UTRA (E-UTRA) and evolved UTRAN (E-UTRAN) (Release 7). *3GPP TR 25.913 V7.3.0*. Retrieved from http://3gpp.org

3GPP. (2008, December). 3rd generation partnership project; Technical specification group radio access network; Evolved universal terrestrial radio access (E-UTRA) and evolved universal terrestrial radio access network (E-UTRAN); Overall description; Stage 2 (Release 8). *3GPP TS 36.300 V8.7.0*. Retrieved from http://www.3gpp.org

Acx, A., Henriksson, J., Hunt, B., Karetsos, G. T., Tragos, E., Mihovska, A., Kyriazakos, S., Moretti, L., & Lara, J. (2003). Final usage scenarios. *IST-2003-507581 WINNER D1.3 v1.0*.

Aust, S., Proetel, D., Fikouras, N. A., & Görg, C. (2003, October). Policy based Mobile IP handoff decision (POLIMAND) using generic link layer information. In *The 5th IEEE International Conference on Mobile and Wireless Communication Networks*, (pp. 201-204), Singapore.

Austin, M. D., & Stüber, G. L. (1994). Velocity adaptive handoff algorithms for microcellular systems. *IEEE Transactions on Vehicular Technology, 43*(3), 549–561. doi:10.1109/25.312791

Azemi, G., Senadji, B., & Boashash, B. (2004). Mobile unit velocity estimation based on the instantaneous frequency of the received signal. *IEEE Transactions on Vehicular Technology, 53*(3), 716–724. doi:10.1109/TVT.2004.827157

Brown, R. G. (1964). *Smoothing, forcasting and prediction of discrete time series*. Englewood Cliffs, NJ: Prentice Hall.

Campbell, A. T., Gomez, J., Kim, S., Valkó, A. G., Wan, C., & Turányi, Z. R. (2000). Design, implementation, and evaluation of cellular IP. *IEEE Personal Communications Magazine, 7*(4), 42–49. doi:10.1109/98.863995

Chen, L., Cai, X., Sofia, R., & Huang, Z. (2007, September). A cross-layer fast handover scheme for Mobile WiMAX. In *IEEE Vehicular Technology Conference 2007-Fall*, (pp. 1578-1582), Baltimore, MA.

Das, S., McAuley, A., Dutta, A., Misra, A., Chakraborty, K., & Das, S. K. (2002). IDMP: An intra-domain mobility management protocol for next generation wireless networks. *IEEE Wireless Communications Magazine, 9*(3), 38–45. doi:10.1109/MWC.2002.1016709

Das, S., Misra, A., & Agrawal, P. (2000). TeleMIP: Telecommunications enhanced Mobile IP architecture for fast intradomain mobility. *IEEE Personal Communications Magazine, 7*(4), 50–58. doi:10.1109/98.863996

Ericsson (2005, October), Handover interruption times. *3GPP TSG RAN WG2-WG3 Joint Meeting R3-051091*.

Fogelstroem, E., Jonsson, A., & Perkins, C. (2007, June). Mobile IPv4 regional registration. *IETF RFC 4857*.

Garg, V. K. (2000). *IS-95 CDMA and cdma2000, cellular/PCS systems implementation*. Upper Saddle River, NJ: Prentice Hall.

Holma, H., & Toskala, A. (2005). *WCDMA for UMTS, radio access for third generation mobile communications,* (3rd ed). New York: John Wiley & Sons.

Holtzman, J. M., & Sampath, A. (1995). Adaptive averaging methodology for handoffs in cellular systems. *IEEE Transactions on Vehicular Technology, 44*(1), 59–66. doi:10.1109/25.350270

IEEE. 802.20 (2005, September). 802.20 evaluation criteria – ver.1.0. *IEEE 802.20-PD-09.* Retrieved from http://www.ieee802.org/20/

IEEE. 802.11. (n.d.). *IEEE 802.11 wireless local area networks.* Retrieved from http://ieee802.org/11/

IEEE. 802.16. (n.d.). *The IEEE 802.16 working group on broadband wireless access standards.* Retrieved from http://ieee802.org/16/

IEEE. 802.16 task group d. (n.d.). Retrieved from http://ieee802.org/16/tgd/

IEEE. 802.16e task group. (n.d.).Retrieved from http://ieee802.org/16/tge/

IEEE. 802.16 task group m. (n.d.).Retrieved from http://www.ieee802.org/16/tgm/

Itoh, K., Watanabe, S., Shih, J., & Sato, T. (2002). Performance of handoff algorithm based on distance and RSSI measurements. *IEEE Transactions on Vehicular Technology, 51*(6), 1460–1468. doi:10.1109/TVT.2002.804866

Jo, J., & Cho, J. (2008, August). A cross-layer vertical handover between Mobile WiMAX and 3G networks. *International Wireless Communications and Mobile Computing Conference,* (pp. 644-649), Crete Island, Greece.

Juang, R., Lin, H., & Lin, D. (2005, March). An improved location-based handover algorithm for GSM systems. *IEEE Wireless Communications and Networking Conference,* (pp. 1371-1376), Los Angeles, CA.

Kim, D., Lee, H., Kim, N., & Yoon, H. (2009). A velocity-based bicasting handover scheme for 4G mobile systems. *IEICE Transactions on Communications . E (Norwalk, Conn.), 92-B*(1), 288–295.

Kim, M., Kim, H., Ra, I., Yoo, J., & Kim, D. (2008, January). Cross-layer based fast handover mechanism for seamless macro-mobility support in WiBro networks. *The International Conference on Information Networking,* 1-5, Busan, Korea.

Koodli, R. (2005, July). Fast handovers for Mobile IPv6. *IETF RFC 4068.*

Lee, J. W., Mazumdar, R. R., & Shroff, N. B. (2006). Joint resource allocation and base-station assignment for the downlink in CDMA networks. *IEEE/ACM Transactions on Networking, 14*(1), 1-14.

Leu, A. E., & Mark, B. L. (2003, March). An efficient timer-based hard handoff algorithm for cellular networks. *IEEE Wireless Communications and Networking Conference,* (pp. 1207-1212), New Orleans, LA.

Liu, M., Li, Z., Guo, X., & Dutkiewicz, E. (2008). Performance analysis and optimization of handoff algorithms in heterogeneous wireless networks. *IEEE Transactions on Mobile Computing, 7*(7), 846–857. doi:10.1109/TMC.2007.70768

Luoto, M., & Sutinen, T. (2008, May). Cross-layer enhanced mobility management in heterogeneous networks. *IEEE International Conference on Communications,* (pp. 2277-2281), Beijing, China.

Makaya, C., & Aissa, S. (2003, October). Joint scheduling and base station assignment for CDMA packet data networks. *IEEE Vehicular Technology Conference 2003-Fall,* (pp. 1693–1697), Orlando, FL.

Malki, K. E. (2007, June). Low-latency handoffs in Mobile IPv4. *IETF RFC 4881.*

Marichamy, P., Chakrabarti, S., & Maskara, S. L. (2003, October). Performance evaluation of handoff detection schemes. *IEEE Region 10 Conference on Convergent Technologies for the Asia-Pacific*, (pp. 643-646), Taj Residency, Bangalore.

Mark, B. L., & Leu, A. E. (2007). Local averaging for fast handoffs in cellular networks. *IEEE Transactions on Wireless Communications*, 6(3), 866–874. doi:10.1109/TWC.2007.04080

Mihovska, A., Luo, J., Mino, E., Tragos, E., Mensing, C., Vivier, G., & Fracchia, R. (2007, July). Policy-based mobility management for heterogeneous networks. *IEEE Mobile and Wireless Communications Summit*, (pp. 1-6), Budapest, Hungary. mITF (2005). 4G Mobile System Requirement Document Version 1.1.

Mohanty, S., & Akyildiz, F. (2006). A cross-layer (layer 2 + 3) handoff management protocol for next-generation wireless systems. *IEEE Transactions on Mobile Computing*, 5(10), 1347–1360. doi:10.1109/TMC.2006.142

MOTOROLA. (2008, November). *LTE inter-technology mobility, enabling mobility between LTE and other access technologies* (White Paper).

Navaie, K., & Yanikomeroglu, H. (2006, June). Downlink joint base-station assignment and packet scheduling algorithm for cellular CDMA/TDMA networks. In *IEEE International Conference on Communications*, (pp. 4339-4344), Istanbul, Turkey.

Nyberg, G., Ahlund, C., & Rojmyr, T. (2006, July). SEMO: A policy-based system for handovers in heterogeneous networks. In *International Wireless Communications and Mobile Computing Conference*, (pp. 62-67), Vancouver, Canada.

Ozdural, O. C., & Liu, H. (2007, June). Mobile direction assisted predictive base station switching for broadband wireless systems. In *IEEE International Conference on Communications*, (pp. 5570-5574), Glasgow, UK.

Pahlavan, K., Krishnamurthy, P., & Hatami, A. (2000). Handoff in hybrid mobile data networks. *IEEE Personal Communications*, 7(2), 34–47. doi:10.1109/98.839330

Perkins, C. (1997). *Mobile IP: Design principles and practice*. Reading, MA: Addison-Wesley.

Perkins, C. (1996, October). IP mobility support. *IETF RFC2002*.

Pries, R., Mäder, A., & Staehle, D. (2006, August). A network architecture for a policy-based handover across heterogeneous networks. *OPNETWORK 2006*, Washington, DC, (pp. 1-9).

Qualcomm Europe (2006, May). Cell switching in LTE active state. *3GPP-RAN WG2 meeting #53 R2-061196*.

Ramjee, R., Varadhan, K., Salgarelli, L., Thuel, S. R., Wang, S., & Porta, T. L. (1999, October). HWAWII: A domain-based approach for supporting mobility in wide-area wireless networks. In *7th International Conference on Network Protocols*, Toronto, Canada (pp. 283-292).

Rosenberg, J., Schulzrinne, H., Camarillo, G., Johnston, A., Peterson, J., Sparks, R., Handley, M., & Schooler, E. (2002, June). SIP: Session initiation protocol. *IETF RFC 2543*.

Seo, S., Lee, S., & Song, J. (2007, November). Policy based intelligent vertical handover algorithm in heterogeneous wireless networks. *International Conference on Convergence Information Technology*, Gyeongju, Korea, (pp. 1900-1905).

Soliman, H., Castelluccia, C., Malki, K. E., & Bellier, L. (2005, August). Hierarchical Mobile IPv6 mobility management. *IETF RFC 4140*.

Solomon, J. D. (1998). *Mobile IP – The Internet unplugged*. Upper Saddle River, NJ: Prentice Hall.

Tomcik, J. (2006, January). MBTDD wideband mode performance report II. *IEEE C802.20-05/88r1*, from http://www.ieee802.org/20/

Ulukus, S., & Pollini, G. P. (1998, June). Handover delay in cellular wireless systems. *IEEE International Conference on Communications*, Atlanta, GA, (pp. 1370-1374).

Verdone, R., & Zanella, A. (2002). Performance of received power and traffic-driven handover algorithms in urban cellular networks. *IEEE Communications Magazine, 9*(1), 60–71. doi:10.1109/MWC.2002.986461

Zaidi, Z. R., & Mark, B. L. (2004, September). A mobility-aware handoff trigger scheme for seamless connectivity in cellular networks. *IEEE Vehicular Technology Conference 2004-Fall*, Los Angeles, CA, (pp. 3471-3475).

Zhang, N., & Holtzman, J. M. (1996). Analysis of handoff algorithms using both absolute and relative measurements. *IEEE Transactions on Vehicular Technology, 45*(1), 174–179. doi:10.1109/25.481835

Zhu, F., & McNair, J. (2005, May). Cross layer design for Mobile IP handoff. *IEEE Vehicular Technology Conference 2005-Spring*, Stockholm, Sweden, (pp. 2255-2259).

Zhu, H., & Kwak, K. (2007). Performance analysis of an adaptive handoff algorithm based on distance information. *Elsevier Computer Communications, 30*(6), 1278–1288.

Zonoozi, M., & Dassanayake, P. (1997, September). Handover delay and hysteresis margin in microcells and macrocells. *The 8th IEEE International Symposium on Personal, Indoor and Mobile Radio Communications,* Helsinki, Finland, (pp. 396-400).

Cross-Layer Designs

Chapter 20
Survey of Cross-Layer Optimization Techniques for Wireless Networks

Han-Chieh Chao
National Ilan University, Taiwan

Chi-Yuan Chang
National Dong Hwa University, Taiwan

Chi-Yuan Chen
National Dong Hwa University, Taiwan

Kai-Di Chang
National Dong Hwa University, Taiwan

ABSTRACT

The explosive development of Internet and wireless communication has made personal communication more convenient. People can use a handy wireless device to transfer different kinds of data such as voice data, text data, and multimedia data. Multimedia streaming, video conferencing, and on-line interactive 3D games are expected to attract an increasing number of users in the future. The bandwidth requirement would be high and the heterogeneous terminals would generally provide limited resource, such as low processing power, low battery life and limited data rate capabilities. These applications would be the major challenge for wireless networks. Although the traditional layered protocol stacks have been used for many years, they are not suitable for the next generation wireless networks and the mobile systems. Due to the time varying transmission of the wireless channel and the dynamic resource requirements of different application, the traditional layered approach to the mobile multimedia communication is full of challenges to meet the user requirement on performance and efficiency. Cross-layer design is an interesting research topic that actively exploits the dependence between different protocol layers to obtain performance gains. The authors performed a survey and introduced the cross-layer design principles and issues for different research topics, including QoS, mobility, security, application, and the next generation wireless communication.

DOI: 10.4018/978-1-61520-674-2.ch021

CURRENT PRINCIPLES FOR CROSS-LAYER DESIGN

The traditional Open System Interconnection (OSI) seven layer protocol stacks have been used for many years. The function of each layer is defined clearly. All those protocol layers are coordinated to complete the network communication. The OSI model could reduce the complex of network implementation and increase the flexibility. However, for the next generation wireless networks and mobile systems, the traditional approach to network design can not satisfy the user requirement on performance and efficiency. Because of the time varying transmission of the wireless channel and the dynamic resource requirements of different application, the mobile multimedia communication is very challenging. Based on limited frequency allocations and channel considerations, Shakkottai et al (2003) depicted the special properties of wireless networks that distinguish them from conventional wire-line networks.

In the layered OSI architecture model, the protocol at each layer is designed independently for the different layers. The layered architecture doesn't allow direct communication between nonadjacent layers. Communication between adjacent layers must follow the pre-defined interfaces through procedure calls and responses.

Cross-layer design is a new research topic that actively exploits the dependence and interaction between different protocol layers to obtain performance gains. Srivastava and Motani made a detail survey on cross-layer design and depicted that there are three main reasons to motivate designers to violate the layered architectures, which are the special problems under the wireless environment, the possibility of opportunistic communication on wireless links, and the new communication models offered by the wireless medium (Srivastava & Motani, 2005; Qusay, 2007).

An important cross-layer design aspect is the management of cross-layer interaction that can guarantee the system operation. Such cross-layer entities may reside within the protocol stack, in which case it is considered as an internal cross-layer entity or an external network node (Foukalas, Gazis, & Alonistioti, 2008). Figure 1 shows that the internal entity may be either an interlayer entity that coordinates the operation of all protocol stack layers or an intralayer entity that is located within a protocol layer. The external cases are illustrated in Figure 2. The external entities may be centralized by a specific network node or distributed over several network nodes.

Figure 1. The internal cross-layer entities

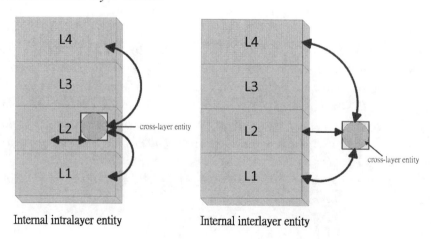

Internal intralayer entity   Internal interlayer entity

Kawadia and Kumar (2005) examined holistically the issues of cross-layer design and its architectural categories. They emphasized the importance of well constructed architecture. Modularity in system could provide the abstractions of system and makes designers easy to understand the overall system. It can also accelerate development of both design and implementation. Designers can focus their effort on a particular subsystem and assure that the entire system will operate correctly. A good architectural design can make system easy to maintain and lead to longer life cycle.

Based on the electronic transfer of information, Kawadia and Kumar (2005) depicted that operating a wireless network based on a layered stack is a natural way to implement the multi-hop decode and forward strategy. A layered architecture can achieve optimal performance, within a constant, with regard to network capacity. They also claimed that all cross-layer design can only improve throughput by at most a constant factor, but cannot result in any unbounded improvements.

In evaluating the cross-layer design proposals, the trade-off between performance and architecture needs to be fundamentally considered. Performance is considered to be more short-term and architecture is longer-term. Thus, a particular cross-layer suggestion may get an improvement in throughput or performance but the longer-term consideration is lack. However, it is very difficult to evaluate the merit of proposals on the overall architecture.

Kawadia and Kumar performed two simulation studies to show that unintended cross-layer interactions can lead to undesirable consequences on overall system performance. It is emphasized that unbridled cross-layer design can lead to sophisticated design, stifle further innovation and be difficult to maintain for longer-term operation.

Because wireless networks maybe attract an increasing number of users, the importance of good architecture needs to be kept in mind. The cautionary information proposed by Kawadia and Kumar (2005) should be examined while engaging in cross-layer design.

CATEGORIES FOR CROSS-LAYER DESIGN PROPOSALS AND IMPLEMENTATION METHODS

There are many literatures discussing about the cross-layer solution for different topics. After a survey of several internal cross-layer design proposals from the literature, Srivastava and Motani (2005) noted that the layered architecture could be violated in the following basic ways:

- Creation of new interfaces
- Merging of adjacent layers
- Design coupling without new interfaces
- Vertical calibration across layers

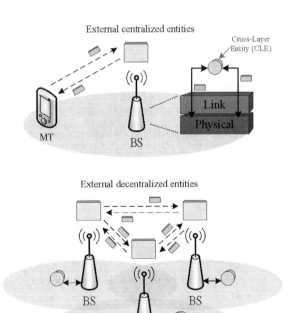

Figure 2. The external cross-layer entities

Figure 3. The models of creation of new interfaces

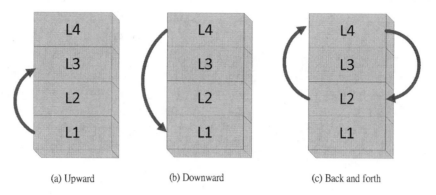

(a) Upward (b) Downward (c) Back and forth

Let us consider a system model with 4 layers, L1, L2, L3, and L4. L1 is the lowest layer and L4 indicate the highest layer.

- Creation of new interfaces

According to the interaction method between layers, there are three subtypes to create the new interfaces: upward, downward, and back and forth.

The upward model is a bottom-up approach, which is shown in Figure 3(a). Higher layer protocols (e.g. L3) require the parameters from the lower layer (e.g. L1) at runtime. It could also be said that the lower layer notifies the higher layer about the network condition information. Most of the existing cross-layer literatures are based on this approach.

The downward model is a top-down approach, which is shown in Figure 3(b). Higher layer protocols (e.g. L4) would set the parameters within the lower layer (e.g. L1) through the new interface at runtime. The lower layer could get the action hints of higher layer and take the more appropriate action.

Figure 3(c) shows the back and forth model which combines the bottom-up approach and top-down method. Higher layer (e.g. L4) and lower layer (e.g. L2) can cooperate more closely through the new interfaces.

- Merging of adjacent layers

In this approach, a super layer composed of two or more adjacent layers is used to achieve the cross-layer function. The super layer could provide the same interfaces to the rest protocol layers or it could also create new interfaces for use. Figure 4 shows the super layer cross-layer model.

- Design coupling without new interfaces

Figure 5 shows the model of design coupling without new interfaces. In this model, the design

Figure 4. The model of merging of adjacent layers

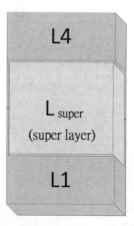

Merging of adjacent layers

Figure 5. The model of design coupling without new interfaces

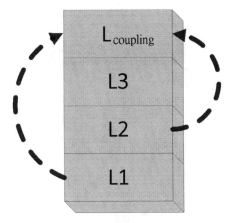

Design coupling without new interfaces

Figure 6. The model of vertical calibration across layers

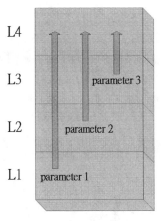

Vertical calibration

of some protocol layer (e.g. L4) would keep in mind about the functionality of the other layers (e.g. L1 and L2). The coupled layers would be highly correlated each other. Thus, the system architecture should be complex and those correlated layers need to be modified at the same time while some design changes need to be applied.

- Vertical calibration across layers

The vertical calibration model across layers is shown in Figure 6. The idea of this model is to find the optimal set of parameters used for different layers. Vertical calibration can be implemented using static or dynamic manners. For the static approach, the parameters across different layers are chosen at design time. In the dynamic manner, the parameters would be determined and adapted during run time. Unfortunately, the dynamic approach is very complex and incurs significant overhead and cost problems.

Most cross-layer design proposals in the current literatures could be fitted into one of the above basic categories. Moreover, considering the implementation methods for the cross-layer interactions, the techniques can be summarized into three categories:

- Direct communication between layers
- A shared database across the layers
- Completely new abstractions
- Direct communication between layers

The variables or parameters at one layer could be directly used by other layers through the new created interfaces or vertical calibration model. The architecture of direct communication is illustrated in Figure 7. Many methods could be used to make different layers directly communicate each other. For example, the protocol headers could be used to contain the information for cross-layer. The new packet format and information flow could be defined for cross-layer. Extra internal cross-layer entities could be designed to coordinate the cross-layer information.

- A shared database across the layers

Figure 8 demonstrates the architecture of shared database across the layers. A common shared database is used as the intermediate medium

Figure 7. The direct communication between layers

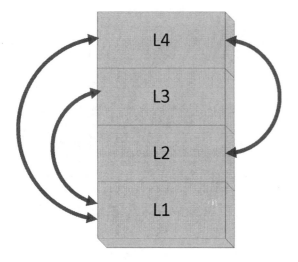

Direct communication

Figure 9. The completely new abstractions

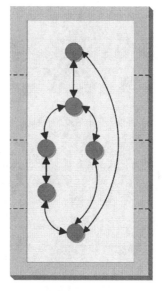

Completely new abstractions

for the communication of different layers. New interfaces or vertical calibration model could be realized through the shared database approach.

- Completely new abstractions

The idea of layers is broken thoroughly and the new operation mechanism is re-designed. Figure 9 shows the representative architecture of new abstractions without the boundary of layers.

The flexibility of this architecture is great and can provide rich interactive communications. However, the system complexity is very high. Almost a complete new system should be designed for this approach. The kernel system level modifications would be touched for implementation of this new architecture.

Some of the cross-layer issues are described in greater detail on the literature of Srivastava

Figure 8. A shared database across the layers

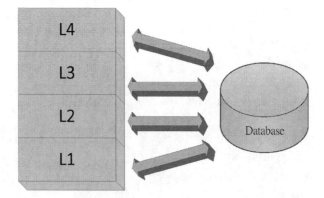

A shared database

and Motani (2005). What is the role the physical layer would play is suggested to be an important question in cross-layer designs. It is suggested that new researchers who would like to study cross-layer design proposals can start to resolve the open challenges such as cross-layer coexisting problem, the selection consideration between the environmental condition and different cross-layer models, the performance and architecture issues, and the possibility of standardized model for cross-layer design, etc.

ISSUES ASSOCIATED WITH CROSS-LAYER DESIGN IN TERMS OF QOS, MOBILITY, AND SECURITY

QoS

The big success of wireless network is attracting the attention of people and a lot of research projects are focused on this field. Basically wireless network is not one kind of reliable medium of communication due to the limited bandwidth, the mobility of user, high packet loss rate, and the lack of wireless security. That is the reason why QoS is so important in the wireless networks.

Basically the goal of cross-layer design is to optimize the usage of scarce wireless bandwidth. Many efforts have been done by researchers in the cross-layer design topic. There should be a QoS support architecture that could be fairly general and applicable to different wireless technologies.

The main idea of Chen et al (2004) is focused on "adaptation", which represents the ability of network protocols and applications to observe and respond to channel variation. The system design of the proposed method is focused on physical and MAC layer design. In the downlink physical layer, the idea of "shared pilot" is used, which eliminates the pilot overhead of data and control frame. The "address Residue Number System (RNS) code" is used to distinguish from different users. In the uplink physical layer, the pilot cannot be used because it is hard to establish phase reference for coherent modulation. About the MAC layer, the minimum transmission unit of data is set as a segment.

The cross-layer QoS design of Chen et al (2004) is focused on QoS-aware and power-adaptive MAC states. It is classified into three categories as follows.

- High-QoS state: It is not energy saving because users send and receive traffic actively.
- Medium-QoS state: Users have contention-free uplink request slots and shared downlink message slots. It is timing controlled but not power controlled.
- Low-QoS state: Users have shared downlink slots and they wake up periodically to listen for the incoming messages. It is the most power saving state.

Li and Cuthbert (2006) proposed a novel distributed cross-layer QoS (DCLQ) architecture based on node disjoint multipath routing. The proposed approach could provide QoS guarantees for real-time traffic flows and best-effort traffic flows in mobile ad hoc networks.

Multipath QoS routing protocol with low overhead for supporting DiffServ (MQRD) is an effective QoS mechanism to provide better service in the environments of limited wireless bandwidth. It is done by either raising the priority of a flow or limiting the priority of another flow. The scheduling and queue management are considered to be two important aspects of resource management. In MQRD, priority queuing is used to build a priority scheduler, which includes two queues: high-priority queue and low-priority queue. Although MQRD may provide the better service quality for real time flows than the best effort methods, it still cannot guarantee the desired service requirements of a real time flow. Hence, DCLQ is proposed to make real-time flows in mobile ad hoc networks achieve their desired service level.

In the DCLQ architecture, the data flows are classified into real-time flows or best effort flows in the network layer. Then the multipath routing information would be collected by the multipath routing component. Before entering the QoS-aware scheduler, the admission control component would check that if there are enough resources to support the flow. A real-time packet can enter the QoS-aware scheduler after it is admitted. The delay allowed for every hop can be obtained based on the maximum delay of a real time packet and the number of hops along a chosen routing path. The real-time packet and the hop waiting time are inserted into the high priority queue. The data in the high priority queue would be scheduled to the MAC layer. When the admission control component confirms that the low priority queue of QoS-aware scheduler is not full, the best effort traffic could be admitted to the low priority queue.

In order to achieve service differentiation, the contention such as window size and frame size, could be assigned to different classes. DCLQ can schedule packets of real-time flows according to the QoS requirements without any extra control overhead. Li and Cuthbert studied the performance evaluation and comparison between DCLQ and MQRD by simulations using OPNET.

Misic et al (2006) studied the activity of sensor nodes through the centralized and distributed algorithms. Sensor networks maintain the desired value of information throughput but are requested to maximize the lifetime of individual nodes. The lifetime of individual nodes can be maximized by adjusting the active and inactive periods of the radio. The characteristics of MAC and PHY layers should be taken into consideration during the design process.

Activity management can be achieved by sending the individual sensor nodes to sleep for variable time intervals. The desired packet rate at the network sink can be tuned by the activity scheduling policies that adjust the number of active sensors. The 1-limited scheduling, exhaustive scheduling, and Bernoulli scheduling disciplines are considered in the proposed policies.

Misic et al used an 802.15.4-compliant network to simulate and found that both the centralized and distributed activity management policies could achieve and maintain the desired network reliability while maximizing the lifetime of the entire network. The main problem of centralized approach lies in the excessive computational resources. Generally speaking, the sensor nodes are power limited and computation limited. The computational load of high complexity may exceed the capabilities of the sensor node. On the other hand, their proposed distributed algorithm could offer significant advantages in terms of computational complexity. It allows the network to function with the desired level of network reliability even in the presence of node failures. The lifetime gain could be improved and is declared to be close to the theoretical limit.

Mobility

Carneiro et al (2004) indicated that future wireless terminals are expected to be equipped with multiple wireless network interfaces, including 802.11, UMTS, and Bluetooth. People would use some handy wireless devices to transfer multimedia data while they are moving. The wireless terminals must have good mobility functions to meet the increasing requirements for wireless mobile communication. Mobility support at the IP level becomes very important.

The Internet was not designed with mobility in mind. Hence, the IP based protocols do not handle mobility well. Carneiro et al examined the TCP process with detail to understand the reason why TCP cannot operate efficiently under wireless networks. They indicated that cross-layer design is an approach to help solving the issues. A cross-layer mobility plane was proposed with the Mobile IP layer in the central to the protocol stack.

Mobile IP could shield upper layers from the operational details of mobility support. For example, the home address (HA) in the destination field of received packets would be automatically replaced as the care-of address (CoA), and the CoA in the source field of outgoing datagram would be replaced as the HA. The upper layers continue to work transparently in the presence of mobility. The mobility would still affect the performance of operations in the upper layers although the Mobile IP could shield them from the knowledge of handover. It is important that the Mobile IP layer could provide handover information to the upper layers such that each upper layer protocol can decide how to best cope with them. Given explicit notifications of handover using cross-layer information, the optimizations for horizontal handover and vertical handover processes could be employed.

IP-based technology is increasingly being adopted and wireless network such as 802.11 has been widely deployed as the infrastructure for mobile users. Voice over IP (VoIP) services on WLANs are still challenging because the handoff process of IEEE 802.11 MAC protocol is very long. The MAC layer handoff process is called a layer-2 handoff and involves AP probe, authentication, and association phases in 802.11 networks.

Mobile IP (MIP) could be adopted for network-layer mobility management and a layer-3 handoff process would be involved while mobile nodes move across different access points (AP). If the Session Initiation Protocol (SIP) is used as the application-layer mobility management method, a layer-7 handoff process would be involved to make the ongoing sessions continue without interruption.

Cross-layer designs are increasingly used to shorten the total handoff latency time in recent researches. Tseng et al (2005) proposed a topology-aided cross-layer fast handoff designs for Mobile IP over IEEE 802.11 networks. Layer-2 handoff delay includes the actions of AP probe, authentication, association, and 802.1x authentication. Layer-3 handoff delay contains the processes of movement detection, agent discovery / address configuration, and Mobile IP registration.

Many research projects have attempted to reduce the delay in different activity sections. Tseng et al (2005) proposed an approach to speed up the handoff process and shorten the total handoff latency time. Layer-2 triggers and cross-layer topology information are used to speedup Mobile IP handoffs. Tseng et al built their experimental environment and set up an independent location association server (LAS). The LAS maintains the information about handoff-to relationships, location information, the association of AP and mobility agent (MA), etc. The mobile node utilizes the layer-2 triggers and the cross-layer topology information, such as the association between 802.11 access points and Mobile IP mobility agents, to speedup handoff process. The layer-3 handoff process was started in parallel with or prior to layer-2 handoff operation.

Security

Cross-layer interactions are helpful to enhance network security. Based on the proposed researches, these cross-layer security issues could be classified into three categories: security policy, securing the communication for ad hoc networks and intrusion detection.

Security Policy

Multiple layers of encryption may be used to improve the security strength of communication. However, the additional security overhead will decrease system performance. Carneiro et al (2004) proposed the concept of a coordination plane using cross-layer design. This coordination plane is designed to eliminate the multiple encryptions in different layer. Several encryption protocols and security mechanisms based on different layers are currently in use. At the transport and the application layer, Security Shell (SSH), Secure Sockets

Layer (SSL) and Pretty Good Privacy (PGP) provide end-to-end encryption mechanism. At the network layer, Internet Protocol Security (IPSec) is an encryption mechanism for end-to-end hosts or a secured tunnel between two networks. Unlike upper layers, which security mechanism should be applied is according to the different network technologies. Wired Equivalent Privacy (WEP) is defined by the 802.11 standard to provide a wireless local area network (WLAN) with a minimal level of security and privacy. Bluetooth is another wireless technology for personal area networks (PANs) that provides encryption mechanism. The next generation mobile communication system, such as UMTS, also provides encryption mechanism. Under this mechanism, both data and signalling are encrypted between the user equipment and the radio network controller.

Enabling encryption mechanisms from multiple layers may enhance the security, but it would require more computing power and the delay would increase. Carneiro et al also suggested to disable encryption on all layers except one. Encryption by upper-layer protocols is preferable to lower-layer ones. Intermediate layer flows such as signalling should use encryption at the highest layer. The suggested solution is that each datagram or data unit keeps a flag to indicate if it has already been encrypted by an upper layer protocol or not.

Some results demonstrate that the stronger the security it is, the more signalling and delay overhead it suffers. Agarwal et al (2005) classified security protocols into individual and hybrid policies. They also introduced a new metric and relative security index to analyze security strength and overhead tradeoffs quantitatively. Finally, they concluded that there is a trade-off between security strength and performance. Their experimental results indicate that the IPsec polices provide the best trade-off for authentication time, and the 802.1x-EAP-TLS policy is suitable for those applications requiring low cryptographic overhead and better security strength. The implementation should be focused on the integration of different layers via security policies.

Securing the Communication for Ad Hoc Networks

For the problem of securing multicast under an energy constrained wireless environment, Lazos and Poovendran (2004) presented cross-layer algorithms that consider the node transmission power (physical layer) and the multicast routing tree (network layer) in order to construct an energy efficient key distribution scheme (application layer). Under the consideration of the physical and network layer, the authors formulated an optimization problem for minimizing the energy required for re-keying. They call the scheme as routing aware key distribution scheme (RAwKey).

Li and Ephremides (2006) presented an efficient system that provides anonymity by jointly considering the communication protocols and the security services it depends on. The new proposed scheme cuts across the traditional layer structure and is computationally efficient, requires simple key management, provides strong anonymity, and is most compatible with other secure routing schemes.

Intrusion Detection

Wireless ad hoc networks are vulnerable to various kinds of security threats. Most detection techniques of wired or wireless networks are not suitable for ad hoc networks. The major problem is due to the lack of a centralized infrastructure. Traditionally, most intrusion detection techniques are focused on extracting the misbehavior features from the network layer only. There are vulnerabilities of multiple layers in the wireless ad hoc network, and therefore we need an intrusion detection module that considers across each layer.

Zhang and Lee (2000) illustrated an IDS architecture to model the abnormal behavior in the ad hoc network. They described the concept of detecting abnormal activities in different layers and integrating multiple layers to detect and respond the intrusion. They also proposed two integration schemes at different layers to coordinate the intrusion detection and response. The detection on one layer can be initiated or aided by the evidence from the other layer.

Liu et al (2005) proposed a node-based anomaly IDS architecture for ad hoc networks, and validated it through NS-2 simulation. They adopted a rule based data mining technique and defined a feature set to record normal behavior by collecting information from the MAC and network layers. Considering the trade-off between effectiveness and efficiency, they chose the feature set for intrusion detection which is shown in Table 1. By using the single hop communication nature of MAC layer, the proposed IDS was able to localize the attack source within one-hop perimeter.

The cross-layer based detection of Denial of Service (DoS) attacks in the ad hoc networks have been proposed by G. Thamilarasu et al (2005). They indicated DoS attacks could be launched at multiple layers and Table 2 shows the cases. The goal of adopting cross-layer design approach is similar to Zhang and Lee (2000) and two schemes are proposed as following:

- Detecting intrusion at multiple levels of the protocol layers.
- Exploiting the information such as energy and congestion from one layer to make more accurate detection in another layer.

The authors provided a detection model to detect malicious behaviour. Utilizing the information from MAC and network layer, the proposed detection scheme could increase accuracy. The GloMoSim were used to simulate the proposed model.

CHALLENGES TO APPLICATIONS AND SERVICES

Multimedia streaming, video conferencing, and on-line interactive 3D games are expected to attract an increasing number of users in the future. Mobile multimedia applications require network environment that could optimally allocate resources and dynamically change their environments. The traditional layered protocols do not provide adequate support for multimedia applications in wireless networks, especially when the interference is high or the stations are mobile. Cross-layer design could be used to address the above challenges by making communication network architectures across traditional layer boundaries.

Schaar and Shankar (2005) indicated that the cross-layer optimization leads to an independent implementation, but results in suboptimal multimedia performance. In order to guarantee a predetermined quality at the receiver, the multimedia compression and transmission strategies for wireless mobile stations need to be carefully processed.

Schaar and Shankar proposed a new paradigm for wireless communications based on the interac-

Table 1. Feature set (Liu et al, 2005)

Feature	Value Space
Flow direction (Dir)	SEND, RECV, DROP
Send address (SA)	$sa_i, \forall i \in$ node set S
Destination address (DA)	$da_j, \forall j \in$ node set S
MACPktType	RTS, CTS, DATA, ACK
RoutingPktType*	routingDataPkt, routingCtrlPkt

Table 2. DoS attacks (G. Thamilarasu et al, 2005)

Protocol layer	DoS Attacks
Link Layer	Collision
Network Layer	Packet Drop Misdirection

tion of application layer, MAC layer and physical layer. Such a cross-layer design allows wireless mobile stations to obtain additional resources or free up resources optimally. Those stations could dynamically adapt their cross-layer transmission strategies to improve multimedia quality and reduce the power consumption.

Khan et al (2006) proposed a cross-layer optimization strategy that jointly optimizes the application layer, data link layer, and physical layer of the protocol stack. The proposed scheme could maximize user satisfaction through an application oriented objective function. An important idea was depicted that cross-layer design should not be viewed as an alternative to the layered approach, but rather as a complement. In order to get high performance in the wireless networks, the layering and cross-layer design should be used together.

Khan et al proposed a cross-layer architecture which was composed of N layers and a cross-layer optimizer (CLO). The CLO jointly optimizes multiple network protocol layers and makes predictions on their states to select optimal values for their parameters. There are different kinds of parameters involved, which are directly tunable parameters, indirectly tunable parameters, descriptive parameters, and abstracted parameters. Wireless video streaming application is evaluated on the cross-layer system. The system is optimized periodically at the beginning of each group of pictures. The expected PSNR at the receiver is used to measure the quality of user-perceived video. A parameter matrix was defined for the distortion and rate side information. These parameters would be sent along with the bit-stream and used at different layers. The parameter set that could maximize the objective function of the cross-layer approach would be finally chosen. Khan et al (2006) demonstrate the performance gain achievable with this approach. They also explore the trade-off between performance gain and additional computation.

CROSS-LAYER DESIGN FOR THE NEXT GENERATION WIRELESS COMMUNICATION

The Technologies for the Next Generation Wireless Communication

The official specification of 4th generation mobile communication has not been released yet. There are three major official organizations that are developing the prototype system and specifications for 4G. These three parties are (1) LTE (Long Term Evolution), which is based on the current 3G technology and is mainly supported by Erisson, (2) WiMAX (Worldwide Interoperability for Microwave Access) supported by Intel, and (3) UMB (Ultra Mobile Broadband) supported by Qualcomm.

In the trend of the development of broadband and wireless technologies, the 4G solutions mentioned above start to learn each other. Most of them add the same new radio frequency technologies in their architecture, e.g. MIMO (Multi-Input Multi-Output) and OFDM (Orthogonal Frequency Division Multiplexing). No matter what the broadband wireless technology is, e.g. Wi-Fi, WiMAX or cellular networks, there are some problems to those fundamental wireless technologies. For example, multi-path decay and frequency gain

Figure 10. The evolution from GSM/UMTS to LTE

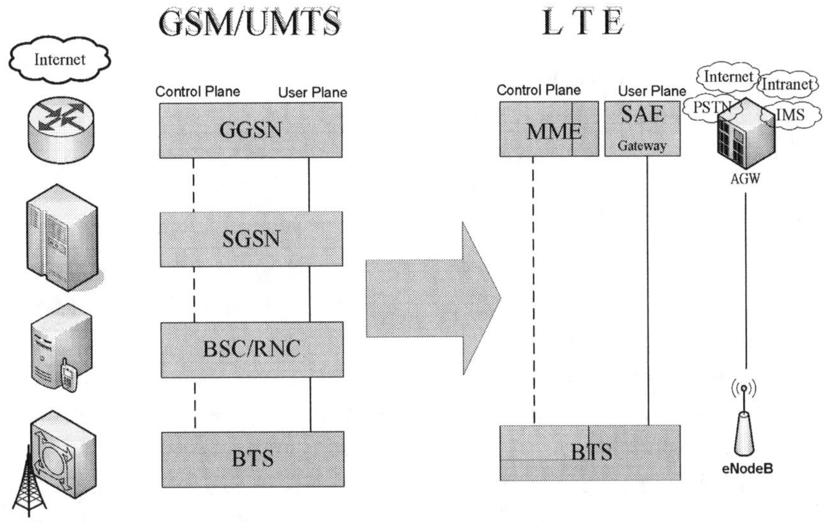

are the problems in wireless transmission. Under this situation, the working groups of broadband wireless technologies choose MIMO and OFDM as the solution to deal with the native limitations in wireless environments. Thus, the combination of MIMO and OFDM becomes a powerful key to broadband wireless communication. With the combination of MIMO and OFDM, people can get the maximal benefits in wireless communications.

In the three solutions to 4G communication, LTE is regarded as the most possible solution. LTE is the achievement of 3GPP Radio Access Network working group. The objectives of LTE are to develop the framework of 3GPP wireless access technologies, which can get higher transmission rate, lower delay and the packet optimization. The network capacity and performance in LTE is satisfied with the specification in 3GPP Release 3 to Release 5. LTE is a successful MIMO-OFDM application in wireless radio networks. For the MIMO-OFDM applications, the network architecture can be simplified to SAE (System Architecture Evolution). The operators can reduce the cost on node deployment and management by the SAE architecture. Figure 10 shows the evolution of traditional GSM/UMTS to LTE.

Cross-Layer Design in IP Multimedia Subsystem under Long Term Evolution

The MIMO-OFDM access technology plays an important role in LTE and many future wireless communications. MIMO-OFDM provides good technique to handle the problems in physical layer. However, the physical layer is still far away from users. The protocol layer which is close to users is application layer. The IMS (IP Multimedia Subsystem) is an important subsystem in LTE or future 4G communications system and is used to provide many multimedia services to users.

Keeping signaling flow in smooth is a challenge in IMS because of the text-based nature of the signaling in SIP (Session Initiation Protocol). The loss and constrained capacity caused by wireless links is also inevitable problems to IMS. In the next generation mobile communication system, the operator or the system developer must handle those problems well to keep the signaling correct and efficient. Melnyk et al (2007) proposed a cross-layer designed model to analyze the performance of IMS with EV-DO. Le & Li (2007) proposed a cross-layer approach to enhance the mobility management in SIP under the Mobile-IP environment in IMS.

The environment condition may change with time frequently in wireless transmission. Some attributes in IMS signaling may be modified to improve the transmission efficiency. What kind of attributes should be modified in IMS? How to modify the attributes in IMS signaling? What kind of information in the lower layer can be shared or exchanged with higher layer? These are still challenging issues for cross-layer design in IMS.

The implementation methods of direct communication between layers or a shared database across the layers would be the possible cross-layer design architectures for IMS in LTE. If the direct communication between layers method is used, the information of channel condition in physical layer can be passed to the SIP/SDP in the application layer. If the shared database architecture for cross-layer is used, each layer can access the information of the other layers through the shared database.

SUMMARY

According to the classifications in section II, we summarize the introduced literatures in Table 3. Most cross-layer design proposals can fit into one of the basic categories.

Wireless networks and mobile systems would still have explosive growth in the future. Although layered architectures have served well for wired networks in the past years, for the next generation wireless networks and the mobile systems, the traditional approach to network design can not satisfy the user requirement on performance and efficiency.

Table 3. Summary of the introduced literatures

Referenced works	Topic	Architecture category	Implementation category	Crossed layers
(Chen, Lv, & Zheng, 2004) (Misic, Shafi, & Misic, 2006)	QoS	back and forth	direct communication	layer 1, layer 2
(Li, & Cuthbert, 2006)	QoS	back and forth	direct communication	layer 2, layer 3
(Carneiro, Ruela, & Ricardo, 2004)	QoS Mobility Security	vertical calibration	a shared database	all layers
(Tseng, Yen, Chang, & Hsu, 2005)	Mobility	back and forth	a shared database	layer 2, layer 3
(Le, & Li, 2007)	Mobility	back and forth	direct communication	layer 3, layer 7
(Agarwal, Wang, & McNair, 2005)	Security	vertical calibration	none	all layers
(Lazos, & Poovendran, 2004)	Security	upward	direct communication	layer 1, layer 3
(Li, & Ephremides, 2006)	Security	back and forth	direct communication	layer 3, layer 7
Zhang, & Lee, 2000)	Security	back and forth	direct communication	all layers
(Liu, Li, & Man, 2005)	Security	back and forth	a shared database	layer 2, layer 3
(Thamilarasu, Balasubramanian, Mishra, & Sridhar, 2005)	Security	back and forth	direct communication	layer 2, layer 3
(Schaar, & Shankar, 2005)	Application	design coupling	direct communication	layer 1, layer 2, layer 7
(Khan, Peng, Steinbach, Sgroi, & Kellerer, 2006)	Application	vertical calibration	a shared database	all layers
(Melnyk, Jukan, & Polychronopoulos, 2007)	Application	back and forth	direct communication	layer 1, layer 7

Cross-layer design is a new research topic that actively exploits the dependence and interaction between different protocol layers to obtain performance gains.

In this article, some literatures proposed for the cross-layer design are surveyed. These topics include cross-layer design about QoS, mobility, security, application, and next generation wireless communication. There are a lot of other cross-layer related research literatures and could be referred for more information.

The open challenges and new opportunities discussed by Srivastava and Motani (2005) for cross-layer design could be the future research topics. It is suggested that researchers who would like to study the cross-layer design proposals can start to address those issues for the first step.

ACKNOWLEDGMENT

The authors would like to thank the National Science Council of the Republic of China, Taiwan for financially supporting this research under NSC 97-2219-E-197-001 and NSC 97-2219-E-197-002.

REFERENCES

Agarwal, A. K., Wang, W., & McNair, J. Y. (2005). An Experimental Study of Cross-Layer Security Protocols in Public Access Wireless Networks. *IEEE Global Telecommunications Conference (GLOBECOM'05),* (vol. 3, pp. 1747–1751).

Carneiro, G., Ruela, J., & Ricardo, M. (2004). Cross-Layer Design in 4G Wireless Terminals. *IEEE Wireless Communications, 11*(2), 7–13. doi:10.1109/MWC.2004.1295732

Chen, J., Lv, T., & Zheng, H. (2004). Cross-Layer Design for QoS Wireless Communication. *The 2004 International Symposium on Circuits and Systems, ISCAS'04,* (vol. 2, pp. II-217-220).

Foukalas, F., Gazis, V., & Alonistioti, N. (2008). Cross-Layer Design Proposals for Wireless Mobile Networks: A Survey and Taxonomy. *IEEE Communications Surveys & Tutorials, 10*(1), 70–85. doi:10.1109/COMST.2008.4483671

Kawadia, V., & Kumar, P. R. (2005). A Cautionary Perspective on Cross-Layer Design. *IEEE Wireless Communications, 12*(1), 3–11. doi:10.1109/MWC.2005.1404568

Khan, S., Peng, Y., Steinbach, E., Sgroi, M., & Kellerer, W. (2006). Application-Driven Cross-Layer Optimization for Video Streaming over Wireless Networks. *IEEE Communications Magazine, 44*(1), 122–130. doi:10.1109/MCOM.2006.1580942

Lazos, L., & Poovendran, R. (2004). Cross-Layer Design for Energy-Efficient Secure Multicast Communications in Ad Hoc Networks. *IEEE International Conference on Communications (ICC),* (vol. 6, pp. 3633-3639).

Le, L., & Li, G. (2007). Cross-layer Mobility Management based on Mobile IP and SIP in IMS. *International Conference on Wireless Communications, Networking and Mobile Computing, 2007. (WiCom 2007),* (pp. 803-806).

Li, S., & Ephremides, A. (2006). Anonymous Routing: A Cross-Layer Coupling between Application and Network Layer. *The Conference on Information Sciences and Systems 2006* (pp. 783-788).

Li, X., & Cuthbert, L. (2006). Node-Disjoint Multipath Routing and Distributed Cross-Layer QoS Guarantees in Mobile Ad hoc Networks. *The Seventh ACIS International Conference on Software Engineering, Artificial Intelligence, Networking, and Parallel/Distributed Computing (SNPD 2006),* (pp. 243-248).

Liu, Y., Li, Y., & Man, H. (2005). A Distributed Cross-layer Intrusion Detection System for Ad Hoc Networks. *1st International Conference on Security and Privacy for Emerging Areas in Communications Networks (SecureComm 2005)* (pp. 418-420).

Melnyk, M. A., Jukan, A., & Polychronopoulos, C. D. (2007). A Cross-Layer Analysis of Session Setup Delay in IP Multimedia Subsystem (IMS) with EV-DO Wireless Transmission. *IEEE Transactions on Multimedia*, 9(4), 869–881. doi:10.1109/TMM.2007.895680

Misic, J., Shafi, S., & Misic, V. B. (2006). Cross-Layer Activity Management in an 802.15.4 Sensor Network. *IEEE Communications Magazine*, 44(1), 131–136. doi:10.1109/MCOM.2006.1580943

Qusay, H. M. (2007). *Cognitive Networks: Towards Self-Aware Networks*. Chichester, UK: Wiley-Interscience.

Schaar, M. V., & Shankar, N. S. (2005). Cross-Layer Wireless Multimedia Transmission: Challenges, Principles, and New Paradigms. *IEEE Wireless Communications*, 12(4), 50–58. doi:10.1109/MWC.2005.1497858

Shakkottai, S., Rappaport, T. S., & Karlsson, P. C. (2003). Cross-layer Design for Wireless Networks. *IEEE Communications Magazine*, 41(10), 74–80. doi:10.1109/MCOM.2003.1235598

Srivastava, V., & Motani, M. (2005). Cross-Layer Design: A Survey and the Road Ahead. *IEEE Communications Magazine*, 43(12), 112–119. doi:10.1109/MCOM.2005.1561928

Thamilarasu, G., Balasubramanian, A., Mishra, S., & Sridhar, R. (2005). A Cross-Layer based Intrusion Detection Approach for Wireless Ad Hoc Networks. *IEEE International Conference on Mobile Adhoc and Sensor Systems Conference (MASS05)* (pp. -861).

Tseng, C. C., Yen, L. H., Chang, H. H., & Hsu, K. C. (2005). Topology-Aided Cross-Layer Fast Handoff Designs for IEEE 802.11/Mobile IP Environments. *IEEE Communications Magazine*, 43(12), 156–163. doi:10.1109/MCOM.2005.1561933

Zhang, Y., & Lee, W. (2000). Intrusion Detection in Wireless Ad Hoc Networks. In *6th Annual ACM/IEEE International Conference on Mobile Computing and Networking (MOBICOM)* (pp. 275-283).

Chapter 21
Cross-Layer Joint Optimization of Multimedia Transmissions over IP Based Wireless Networks

Catherine Lamy-Bergot
THALES Communications S.A., France

Gianmarco Panza
CEFRIEL, Italy

ABSTRACT

The traditional approach consisting in separately optimizing each module of a transmission chain has shown limitations in the case of wireless communications where delay, power limitation and error-prone channels are experienced. This is why modern designers focus on a more integrated strategy to establish the heterogeneous 21st century networks, such as 3G (i.e. UMTS) system and its evolutions (i.e. Beyond 3G or 4G like LTE or future 5G systems). Indeed, it was shown in several studies that optimal allocation of user and system resources could be effectively achieved with the co-operative optimization of communication system components. In this chapter, an innovative Joint-Source Channel Coding and Decoding (JSCC/D) system is described and its performance over an IPv6-based Network infrastructure is assessed. A particular focus is put on the application controller, the key component to realize the adaptation strategies. Conclusions and considerations about the system implementation are also proposed, and the interest of a possible extension to a point-to-multipoint scenario is explained.

INTRODUCTION

Following the path opened by GSM systems, the under-deployment 3G (*i.e.* UMTS) system and its evolutions are leading to more and more configurable, dependable, adaptable, intelligent, secure but also complex wireless solutions. Aiming at handling digital data of different nature (text, voice, image, video...) that will be used in various contexts (home, office, on the move...) these systems rely on inner software that make them more and more efficient and easy to use. However, the gap between what the actual systems propose and what the users envision and could use still remains. The lack is particularly noticeable in the domains of heterogeneous networks and systems interconnection (*i.e.* in beyond 3G, future 4G or even 5G networks), but also in the flexible management of resources and

DOI: 10.4018/978-1-61520-674-2.ch022

in the Quality of Service (QoS) and bandwidth optimization domains.

Noticeable progresses were made throughout the last decades of 20th century to individually optimize each module in modern communication systems. Still, although excellent results were obtained, the separate approach following Shannon's well known theorem has shown limitations in the case of wireless communications. Working on a more integrated strategy, for instance via optimal allocation of user and system resources is the key challenge for modern designers. Following the already known approach of JSCC or Joint-Source Channel Coding (Massey, 1978), strategies are and have been developed where the source coding, channel coding, modulation, ciphering, and, possibly, network parameters are jointly determined to yield the best end-to-end system performance.

For the user, such an approach should result in greatly enhanced perceived quality for multimedia communication, potentially allowing the development of currently too complex, expensive and/or time consuming video over wireless systems. The realization of a system operating under JSCC/D paradigm is however not simple: in particular, the delivering of the control and signaling information between the system components is a key point, as it affects the overall design and operation of the system. As a matter of fact, in the ISO model, only the modules at the sender and receiver sides that are at the same layer (*i.e.* peer entities) can communicate with each other. On the contrary cross-layer communication implies that the different layers can communicate in order to allow the system to adapt to the network changes and to increase the overall performance. When adding the condition of realizing an exchange of control and signaling messages in a backward compatible manner, the complexity of designing a JSCC/D system is obvious.

In this chapter, the case-study of the system proposed within the framework of the FP6 IST PHOENIX project[1] (Martini & Mazzotti, 2007; Lamy-Bergot & Panza, 2008) is presented and discussed. The pursued goal was to develop a scheme for point-to-point multimedia communications, offering the possibility to let the application world (source coding, ciphering) and the transmission world (channel coding, modulation) to talk to each other over an IPv6 protocol stack (network world), so that they can jointly develop an end-to-end optimized wireless communication link. To reach the goal, three main axes were identified, following the path of first to enhance each module of the chain while ensuring compatibility with overall optimization, of second to propose optimization strategies between several modules and finally to define the corresponding global network architecture necessary to realize the system. Said another way, the IST PHOENIX project goals were:

- To design innovative schemes to enable end-to-end joint optimization over wireless links. This includes the development of flexible channel coding and modulation schemes, the adaptation of existing source coding schemes with respect to their ability for the JSCC and the development of new ones specifically optimized for this purpose.
- To establish efficient and adaptive optimization strategies that jointly control the coding blocks and realistically take into account the system limitations and specifications, such as the presence of ciphering, the presence of one or several wireless hops, …
- To build a global network architecture based on joint optimization for future wireless systems. This also implies the support of transparent network communications, which allow applying the optimization strategies in any kind of fully IP-based network.

We will begin this chapter by providing a general overview of the Joint-Source Channel Coding and Decoding approach and related aspects. For that, we will explain how overall optimization can be envisaged and how it can be introduced into a complete and fully running system. From the authors' point of view, the usage of JSCC/D for multimedia communications is a small step in the direction of adaptive developments and improvements in wireless communication to be used over Next-Generation Networks (NGN).

We will in particular focus on the main component of the IST PHOENIX approach, which is called the joint controller. Its role is to select the best parameters to be used by the different modules at the current time. The controller decision is based on the knowledge of the current working point defined by the feedback information from the receiver side on the channel state and possibly network state. For this working point, the set of parameters jointly optimizing either the resource usage or the QoS for the user, expressed in terms of perceived visual quality, is decided by the controllers for the modules is manages. The main concepts and issues, together with the design trade-offs and overall performance of an instance of the interactive wireless video systems proposed within the FP6 IST PHOENIX project are outlined in the following sections, and simulation results are presented. It should be noticed that these simulation results were also backed up in the project by test results run made on PHOENIX real-time prototype.

Finally, main conclusions on the proposed JSCC/D approach will be reported, as well as some considerations about the implementation issues. Possible extensions to a point-to-multipoint scenario will also be presented, as it is appearing of particular interest for the current new value added applications such as video conferencing, video streaming to multiple recipients and e-learning.

BACKGROUND

The originality of the approach, when compared to a classical one, is that at each time-step or cycle, the joint controller performs its optimization based on the available feedback information. This implies that the module driven by the controllers are consequently able to update their parameters selection to follow the eventual evolutions of the transmission conditions. This will result in increasing the video quality when the channel and network are improving, and in reducing it smoothly when they are degrading, to ultimately offering always the best overall answer of the system to the instantaneous conditions. Actually, at the beginning of each cycle, the controller uses the expressions of video sensitivity (Bergeron & Lamy-Bergot, 2006) to determine for a given overall protection and compression bit rate as well as a transmission chain state (*e.g.* binary error rate over the channel, packet loss rate in the network...) the best combination of protection rates to be applied. This implies in practice that the adaptation is done systematically for the working point of a past observation time due to the system being causal. Still, for low varying transmission conditions, adaptation is done adequately most of the time.

Depending on the user preferences and terminal capabilities, the joint controller can apply equal or an unequal error protection (Lamy-Bergot, 2006). This last case is even more interesting when considering compression modes supporting data partitioning, where the standard has been specifically designed to isolate differently sensitive partitions. The application of the semi-analytical expressions established in (Bergeron & Lamy-Bergot, 2006) allows selecting the best trade-off between protection and compression for an instantaneous given working point, by comparing the overall compression and protection distortions resulting from the different configurations of source and channel coding for a global fixed bit rate over the channel, on the basis of optimization criterion.

Typically, the output of the application controller will consist at minimal of the settings for the source encoder, content protection module, and channel encoder.

Preliminary results obtained with first realizations implementing basic joint optimization techniques (Park, 2000; Perros-Meilhac, 2002) let indeed hope for peak video quality gains of about 3 dB in PSNR for equivalent bandwidth occupation, and even up to 8 dB in bad channel conditions (Wang, 2006; Chaddha, 1996).

Nevertheless, these results were in general obtained without actually considering a real system. A key issue to be considered to apply such joint optimization techniques in an implemented JSCC/D system is the cross-layer design techniques in order to support cross-layer communication from the application level (where source coding is performed) and lower or distant other layers (where protection, different QoS treatments, … can be applied), be it another application level, transport, data link or PHY layer, where protection or different QoS can be applied... In other words, one needs to enable a "network transparent" approach, which allows to realize communication exchanges between differently located entities into the network (including the end-terminals) without modifying the OSI paradigm. This transparent approach also aims at allowing JSCC/D devices to operate without degrading anyhow the performance of non-JSCC/D aware devices in the same network or transmission chain.

The first objective of this network transparency approach refers to the capability of transferring signaling and control information between both different network nodes and link layers as needed, in a transparent manner. "Transparent" means in spite of the strict rules of the ISO OSI model, which impose a modular and independent design of each link layer of a network node with well defined interfaces, from one side and the delivering through a telecommunication infrastructure that carries data only of a specific format (*i.e.* IP datagrams for the NGNs,) of the JSCC/D specific information without introducing relevant impairments, such as errors, loss, delay, delay-variations, on the other side. Furthermore, the design of a solution for the Network Transparency should take into account the deployed security options, *e.g.* providing authentication and encryption features, as well as compression mechanisms, even if working at different layers, as needed. The second objective of the network transparency approach is to ensure as much as possible backward compatibility of the system with standard implementations. Indeed, it is necessary to keep in mind that the nowadays telecommunication infrastructure that constitutes the basis for the NGNs cannot be swapped entirely to adapt for JSCC/D approach, and so that we need to allow for a smooth migration from current equipments to IPv6-enabled devices eventually supporting JSCC/D functionalities.

Examples of existing approaches to cross-layer design (Srivastava, 2005) are new interface creation, merging of layers, design coupling, and vertical calibration. Possible implementation methods are: direct communication between layers, a shared database across layers, a completely new abstraction, adaptation layers at the transmitter or receiver sides, as well as the exploitation of already existing and deployed ad-hoc signaling protocols (*e.g.* RTCP reports for the overall network state notification). Obviously, some mechanisms are more suitable for transferring certain kinds of information, depending on factors such as nature, frequency or synchronization constraints of the concerned information. Still, a first simple classification can be immediately introduced, distinguishing between internal (*e.g.* from the application to the network or data link layers and vice-versa of the same device) and external (*e.g.* from the source coder to the destination decoder and vice-versa, *i.e.* when the points of generation and reception are located in different network nodes) communications. Can then be considered to implement the cross-layer mechanisms the usage of different solutions such as IPv6 data packets and extension headers, ICMPv6 messages, direct socket-by-socket connections ..., or naturally an ad-hoc signaling protocol. It is worthwhile to point out that those different mechanisms are not anyhow equivalent, but based on different philosophies. They can require specific signaling generation (*e.g.* ICMP messages) or rather rely on an already active data or control flow (*e.g.* IP video data packets or ad-hoc signaling protocols), and will thus introduce diverse amount of overhead. This explains why it is interesting to consider different solutions, corresponding to needs as different as control/signaling information strictly coupled to the data (synchronous mode) or control/signaling information not directly coupled (asynchronous mode). A deeper analysis of such different approaches will be done later in this chapter, which provides details on the choices proposed by IST PHOENIX project.

CROSS-LAYER JOINT OPTIMISATION FOR MULTIMEDIA TRANSMISSION OVER AN IP WIRELESS LINK

Issues, Controversies, Problems

As stated before, the classical objection raised when dealing with optimization of a system from an overall point of view is the well known theory of separation established by Shannon in 1948 (Shannon, 1948). Since this date, the question of choosing to design separately or jointly compression (*i.e.* source coding) and protection (*i.e.* channel coding) operations has separated the digital communications community between people following Shannon's theorem and people pointing out the limitations of said theorem proofs. As a matter of fact, Shannon's theorem indicates that the separated approach performs as well as the joint one, but the proof is asymptotic, and no proof was ever given on whether the separated solution was the simplest one or the only one (McEliece, 1977). Naturally, the simplicity introduced by the possibility to design each layer separately, in particular in the modern context where different applications (video, visiophony, voice over IP, HTML downloading, images, ...) share the same medium was also a key factor leading many people to consider with interest a separated approach. Furthermore, it was often pointed out that realistic transmission chains in a networked environment include not only the source and channel coding operations, but also network layers. This may prevent from easily jointly design the two operations of compression and protection. For those different reasons, joint source and channel coding or more generally overall multimedia optimization approach has often been received with doubts.

Solutions and Recommendations

Despite the aforementioned doubts on JSCC/D approach, researchers have worked on approaches relying on cross-layer design to allow joint source channel coding and decoding (JSCC/D) (Zahir Azami, 1996). Such an approach is presented in the following sections. It allows developing strategies, where the source coding, channel coding, modulation, ciphering, and also network parameters are jointly determined to yield the best end-to-end system performance. This approach presents the advantage of keeping operations separated, while optimizing the parameters to be used by each module (source coding and channel coding ones in particular) in a joint manner according to an overall quality criterion. The analyzed system presents all the features and functionality that were announced as desired in the previous sections: optimization of the end user perceived quality or of the bandwidth usage, network transparent approach, ... Therefore, it constitutes a good case-study, by which to point out all the relevant issues and concepts, as well as provides effective solutions to each matter.

The developed system has been designed for end-to-end optimization of multimedia transmission over an IP wired/wireless channel and was made flexible enough to be utilized with different video coding standards (*e.g.* MPEG-4, H.264/AVC (ITU-T Rec. H.264,2003),...), for different network architectures (in particular different transport protocols, *e.g.* UDP (Postel, 1980),UDP-Lite (Larzon, 2004), DCCP (Kohler, 2006) and over existing radio access networks. This flexibility is extremely important, being the NGNs a seamless integration of several different architectures, platforms, protocols and standards in general, and even more importantly, the backward compatibility is a key point to be carefully considered. Indeed, one of the originalities of this work is to take into account very accurately the impact of networking protocol layers, which has often been neglected when considering JSCC systems in the literature. Only minimal effort has been made to find solutions for efficient inter-layer and network signaling mechanisms, except few counter-examples such as the work presented in (Van der Schaar, 2003), where a solution combining adaptively forward error correction (FEC) and medium access control (MAC) layer automatic repeat request (ARQ) is proposed. The discussed system has been developed within the framework of the FP6 IST PHOENIX project.

Figure 1 presents the end-to-end IP wireless multimedia communication chain used for transmitting video transmission over an interconnecting wired/wireless network. The data information (whether a file or live-captured video) is first encoded with respect to a multimedia standard (*e.g.* H.264/AVC). The data is further treated for insertion of new functionalities such as unequal error protection (UEP), content ciphering... and then encapsulated for transmission over the network that consists of routers and nodes using wired (*e.g.* an intranet or the "Internet") and wireless (*e.g.* WiFi, UMTS, WiMAX...) links. In our approach, the communication model also includes the insertion of specific JSCC/D control information that are used to provide the system components with information about user requirements and network and channel conditions, thus supervising their adaptation to changes. Next section introduces the modified transmission chain that can be obtained with JSCC/D optimization approach. In particular, the insertion of intelligent controllers that allow for joint optimization strategies to increase the useful bit-rate effectively transmitted is there detailed and the interest of the presented technique in the context of an IP wireless transmission is presented through simulation results done with and without optimization.

OPTIMISATION OF THE END-TO-END COMMUNICATION CHAIN BY INSERTION OF MODIFICATIONS IN THE DIFFERENT MODULES

Figure 2 presents the overall communication chain introduced in Figure 1, with highlighting of the modifications introduced by the cross-layer joint design. This system allows a video transmission over an interconnecting network that consists of routers and nodes using wired (*e.g.* Local Access Networks (LAN)) and wireless (*e.g.* 3G/4G system) links. It includes signals for transmitting JSCC/D control information used by specific functions such as unequal error protection (UEP) and JSCC adapted channel coding. Those controls are defined by the joint controlling module present at application level, and are used to provide the system components with information about user requirements and network and channel conditions, thus supervising their capability to adapt their functioning parameters to changes in the transmission chain. In particular, compression rate for video encoder and coding rates for the protection modules are set by this controller. Signaling mechanisms for both the transmitter and the receiver side are detailed in an up-coming sub-section.

For each block, a short summary on its respective role, current state-of-the-art capabilities and eventual introduced new feature are detailed in

Figure 1. End-to-end IP wireless multimedia communication chain

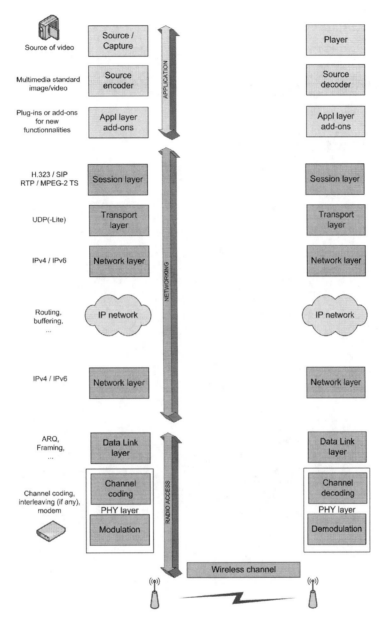

the following sections. Namely, source coding, protection of information (use of error correcting codes), networking issues for multimedia transmission, and the insertion of intelligent controllers that allow for joint optimization strategies to increase the useful bit-rate effectively transmitted are detailed hereafter.

It is to be noted that the modified blocs described in the following have been implemented into a simulation chain that has been used for producing the numerical results presented in the last section. This gives an idea of both the benefits of deploying a JSCC/D system and the cost in terms of introduced signaling overhead.

Figure 2. Optimized end-to-end IP wireless multimedia communication chain (location of proposed modifications: grayed modules)

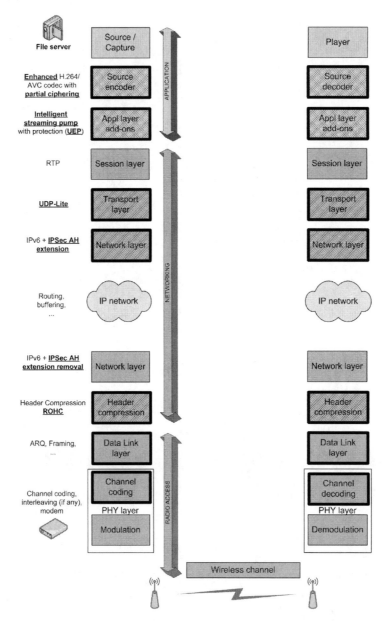

Source Coding

The most recent video coding schemes have improved the overall coding efficiency remarkably compared to previous video coding technology. Joining their efforts, ISO/IEC MPEG and ITU-T standardization bodies have developed the H.264/AVC standard (ITU-T Rec. H.264, ISOIEC 14496-10 AVC, 2003), which outperforms the previous technology nearly by a factor of two in terms of compression efficiency. This codec has consequently been selected for its performance efficiency, and the joint verification model developed by the Joint Video Team (JVT) of ISO/

IEC MPEG and ITU-T has been modified firstly to provide bit error resilience, secondly to offer new compression efficiency and/or temporal scalability functionality (Bergeron, Lamy-Bergot, Pau & Pesquet-Popescu, 2006). The joint model was also adapted to thirdly allow the establishment of sensitivity measurements enabling efficient unequal error protection, fourthly to permit the application controlling strategies driven by the joint controller at application and finally to be compatible with security requirements by offering in particular an embedded partial ciphering solution (Bergeron & Lamy-Bergot, 2005).

It should be mentioned that another source coding standard that can be of particular interest in the context of JSCC/D systems is the scalable extension of H.264/AVC—namely Scalable Video Coding (SVC). Currently under development by the JVT, SVC incorporates temporal, spatial and SNR scalability offering different spatio-temporal resolutions and bit-rates. It is expected that these enhancements could allow for adaptation of the video stream to usage requirements (*e.g.* network or terminal).

Application Layer Enhancement

Due to the relevance of security in a NGN, the application layer includes also a content ciphering module, which was inserted to make sure and prove that confidentiality was not incompatible with overall optimization and the corresponding needed cross-layer information exchanges. In other words, versatile communication applications will require the security services to be implemented in a versatile manner. In particular, these different security policies identify how different data types must be protected, possibly within a given format. For different channels and for different data types, the security functions (encryption, hashing...) must be optimized together with the joint coding approach or risk to be very costly. Considering an integrated framework will thus offer a versatile and adaptable secure infrastructure, which can satisfy the needs of rich multimedia based applications.

In a particular mode of operation, this ciphering can be done inside the H.264/AVC video source encoder thanks to the partial ciphering feature proposed in (Bergeron & Lamy-Bergot, 2005). However, other more classical confidentiality solutions for video streams can also be applied, and the system was also tested by applying ciphering as a plug-in of the video encoder, by application of a modified version of the Video Encryption Algorithm.

The application layer also includes the content level protection module that is directly driven by the joint controller to apply its decided protection rates over the different parts of the payload.

Transport and Packetisation

With JSCC, the underlying network architecture that is normally invisible to the separate source and channel worlds must in fact be adapted to transparently allow for cross-layer information exchange. This sub-section presents the most important system features and proposes some basic solutions on IPv6 networks (Deering, 1998).

Let us consider the layers along the media data route. First comes the transport layer, which provides a streaming video service for the application layer. In the system proposal, the transport protocols use partial checksum to maximize the number of delivered packets since most recent efficient audio/video decoders can use damaged packets to improve error concealment and decoding efficiency. Therefore only the important parts of the packets, such as the headers, are protected by the partial checksum mechanism, while general audio/video data in the payload remains unprotected. It must be noted that, contrary to IPv4, IPv6 implies a mandatory checksum mechanism at the transport layer, which can however be partial, as proposed by UDP-Lite (Larzon, 2004) and DCCP (Kohler, 2006). For media streaming, the RTP/RTCP (Casner, 1996) protocol can be used together with UDP-Lite. Furthermore, the RTCP control/signaling information has been identified as a potential network state information (NSI) source that, while meeting the overall JSCC/D approach, is compatible with existing networks. With DCCP, the use of RTP/RTCP is not needed, because DCCP itself can support all necessary features provided by RTP/RTCP.

IP Mobility and Network

Below the transport layer, one finds the network layer, with IPv6 protocol that has been considered in the communication chain with the possible addition of the IPSec Authentication Header (AH) extension, which in the realized implementation offers data origin authentication over the IP wired network. In practice, the IPSec AH is removed before the wireless link. This ensures that the packets with erroneous payloads are not dropped due to bad authentication.

A wired IPv6 network, modeled by an IP cloud composed of a configurable number of nodes crossed by the IPv6 packet which then introduce delay, loss and buffering is also introduced in the system simulation, to emulate the presence of a LAN or an autonomous system crossing. More specifically, in absence of an explicitly introduced bottleneck, the modeling of loss and delay is just based on statistical distributions (*e.g.* shifted Gamma for the latter) properly parameterized to fit well real world empirical data (Corlett, 2002).

Finally, it should also be noted that the network transparent cross-layer solutions proposed in a next sub-section could ultimately lead to modifying the IPv6 header, for instance by inserting a new hop-by-hop option (*e.g.* to carry source side information). Anyway, the described implementation of the transmission chain relies on the proprietary mechanism presented in Figure 3. As discussed in the previous section, the Network Transparency can be provided in several different ways. Typically, other than the already mentioned general explanations, each option has its own pros and cons, which means that the most suitable solution may depend on the specific system and application scenario to be considered.

After the IPv6 network, and in order to take into account the possible impact of user mobility, which allows the mobile terminal to change its access point while keeping a continuous connection, the emulation of IPv6 mobility impact has been implemented. The main purpose for the incorporation of an IP mobility module into the simulation chain was to be able to investigate the impact on the media flows of the handovers between wireless points of attachment in the IP mobility layer. This is also a key point, because the context of NGNs allows for user mobility in a seamless manner. Therefore, the robustness of a considerable JSCC/D system to such an issue must be proved. The effects of handovers on the IP mobility layer may vary from increased delay to packet loss that were simulated in a separate module since the rest of the chain is not meant to handle this mobility issue. As a matter of fact, the physical, data link and native IP modules prior the IP Mobility one cannot reveal the effects of the IP level handover that appear after a change of point of attachment of the mobile terminal in the IP layer, as the physical and the MAC layers are meant to take into account the effects of the establishment of a new physical access. On the other hand, the handover related processes in the IP layer (*e.g.* Address Auto configuration, Duplicate Address Detection (DAD), Neighbor Unreachability Detection, Mobile IP signaling) are taking place only after getting a new connection, so the modules below the IP layer cannot simulate these effects.

The logical function used to emulate the IP mobility function is a Finite State Machine (FSM) that contains all the states that can have any influence on the behavior of the IP Mobility protocol as a function. As a result the transitions between the different states reflect the possible events regarding the mobile terminal.

In order to be able to simulate a more realistic environment, the IP mobility simulator can be decomposed. The main goal with the decom-

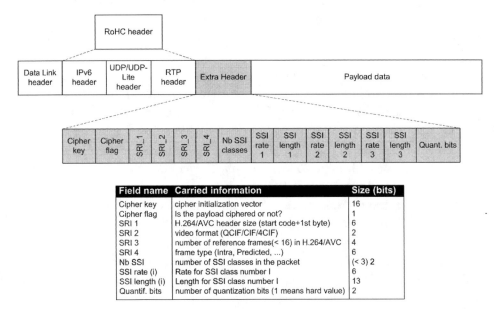

Figure 3. Insertion of cross-layer signaling information into the network packets

position of the module is to make it capable to incorporate the effects of different mobility scenarios and schemes into the simulation chain. In practice, to create a realistic simulation environment, the module relies on a Calculator sub module that allows generating input information retrieved from an already implemented and real-life testbed (MC2L IPv6 Mobility Testbed developed at Budapest University of Technology and Economics (Bokor, 2005)); this input information being then used to generate statistically probability variables.

Robust Header Compression

The overhead introduced by the protocol layers and extension headers (also to implement cross-layer design strategies) can be reduced by compression techniques. Consistently, after the IP network and below the IP layer, a header compression mechanism called Robust Header Compression (RoHC) (Bormann, 2001) has been deployed, that relies on fixed and known syntax of the various RTP/UDP-(Lite)/IP headers for their reconstruction from partial information.

RoHC's principle is to compress the transport and network headers by transmitting only non-redundant information. This allows reducing the bit-rate consumption over wireless links and consequently improves robustness to errors.

Radio Access

Compressed packets are then handed to the radio access, which includes data-link layer and PHY layer. In the optic of modifying this part as less as possible for compatibility issues, only critical modifications have been done. Similarly, the different PHY layers that have been considered were enhanced when compared to standard system as little as possible, the only critical improvement implemented being to apply the channel coding only on the PHY frame extended header, containing all network headers plus JSCC/D control information.

Firstly, in order to make the system robust, especially protocol headers should always be effectively protected, whereas erroneous unprotected audio or video payload go on undiscarded. Consequently, the link layer provides also unequal error detection with a partial CRC applied on headers only, while the payload is left unprotected and no fragmentation is applied.

To meet the hypotheses of 3G to 4G transmission the data link layer has been based on the IEEE 802.11 (IEEE 802.11, 1999) carrier sense multiple access (CSMA) MAC mechanism, with a new features of partial checksum for multimedia data, including MAC header and (possibly compressed) RTP/UDP(-Lite)/IP headers (Vehkaperä, 2006).

The designed solution enhances the IEEE 802.11 standard in multimedia delivery by proposing a new CRC length field into frame header and proposing a fast byte-by-byte cyclic redundancy check (CRC) instead of bit-by-bit hardware operations. In practice, this partial CRC and protection solutions are themselves a cross-layer mechanisms in the sense where they allow providing adapted reliability to the upper layers, by implicit use of their needs and mechanisms.

INSERTION OF INTELLIGENT CONTROLLERS FOR CROSS-LAYER DESIGN

Optimization solutions for increasing useful bit rate effectively transmitted are made feasible by introducing in the transmission chain a new module, called "joint controller", which intelligently controls jointly the different layer parameters. It leads to an increase in the efficiency of the communication link from the user point of view, and is obtained by taking jointly into account the different operations in the chain, *i.e.* the source coder, the network protocol stack and the applied protection to define an optimized set of functioning parameters.

Principle of JSCC/D Controlling

Joint Controller at Application Layer (APP)

The joint controller role is to select, for an instantaneous working point defined by the feedback information provided by the receiver side on the channel state and possibly network state, the best parameters to be used by the different modules for the current time-step. The peculiarity of the approach, when compared to a classical one, is that at each time step or cycle, the joint controller performs its optimization based on the available feedback information, and is consequently able to update its parameters selection and follow the eventual evolution of the transmission chain. This allows increasing the video quality when the channel and network are improving, and reducing it smoothly when they are degrading, offering always the best overall answer of the system to the instantaneous conditions.

In practice, the cycle time-step is chosen as a compromise between the availability and cost of regular updates of the feedback information and the efficiency of modules re-settings, in particular for the source codec, whose efficiency decreases a lot, if the source coding parameters (typically the GOP rate) change too often.

Actually, at the beginning of each cycle, the controller uses the expressions of video sensitivity (Bergeron & Lamy-Bergot, 2006) to determine for a given overall protection and compression bit rate as well as a transmission chain state (*e.g.* binary error rate over the channel, packet loss rate in the network...) the best combination of protection rates to be applied.

Depending on the user preferences and terminal capabilities, it can consider applying equal or an unequal error protection; this last case is in particular interesting when considering compression modes supporting data partitioning, where the standard has been specifically designed to isolate differently sensitivity partitions. Also techniques like Frame Shuffle (Bergeron, Lamy-Bergot, Pau & Pesquet-Popescu, 2006) support data partitioning, where hierarchical structure has been designed, which allow isolating different sensitivity frames.

The application of the semi-analytical expressions established in (Bergeron & Lamy-Bergot, 2006), following similar works done for other video codecs in (He, 2002; Martini & Chiani, 2004; Lamy-Bergot & Chautru, 2006), allows to select the best trade-off between protection and compression for an instantaneous given working point, by comparing the overall compression and protection distortions resulting from the different configurations of source and channel coding for a global fixed bit rate over the channel, based on the optimization criterion described in (Bergeron & Lamy-Bergot, 2006).

Typically, the output of the application controller consists at minimal of the settings for the source encoder, content protection module, and channel encoder. It can also consider a modification of the RoHC module parameters, for instance by adapting the timeout values in the compression process based on the feedback network state information.

The architecture contains also an alternative application controller for the SVC stream. To fully utilize the scalable video stream would require large amount of states to allow fine granular adjustments to the video stream, thus a different approach using fuzzy logic has been used. Proposed fuzzy logic controller sets data rate and frame rate at the video sender by truncating the pre-encoded scalable video stream during the transmission while taking into account several input parameters in decision making. Fuzzy logic-based controlling scheme is introduced in (Huusko, 2006), which concentrates more deeply in scalable video coding.

Physical Layer (PHY) Controller

In FP6 IST PHOENIX project, the authors propose to consider the deployment of a JSCC/D solution over an existing radio access with minimal modification, in order to better accomplish the general requirement of backward compatibility with the nowadays networks. Consequently, the possible cross-layer optimization of the data link and physical layers was not considered, and not implemented in the simulation that is presented later on in this section. Nevertheless, it is important to note that the proposed system can also be further enhanced when the radio access is also open for modification. In that case, as proposed by the same project team in (Martini & Mazzotti, 2005), the described architecture can be extended by the introduction and definition of a sub-controller at physical layer, which would adapt the channel coding and modulation parameters based on the instantaneous channel state information and the available bit rate decided by the application controller to set and update the PHY parameters during the application controller step-time.

As an example, the sub-controller can optimize the parameters for modulation (*e.g.* by using bit-loading in multi-carrier modulation), interleaver characteristics, adaptive antennas functions (*e.g.* by using MIMO solutions) and channel coding rates, to provide a smoother answer of the PHY layer from the upper layers point of view. Such a smoother answer will allow ensuring that the instantaneous bit error rates and channel answer are closer to the average value considered by the application controller for its optimization decision.

Cross-Layer Communication Solution

The reality of cross-layer communication for a JSCC/D system implies that the different signaling information is effectively transmitted over the communication chain.

More specifically, two different information types can be identified: first, one finds control signal highly synchronized with the bit stream (source side information (SSI), *i.e.* information related to video content and representing the sensitivity of the source bit stream to channel errors, source a priori information(SRI), *i.e.* information related to video content and representing statistical information on the source...), and second, one finds return information, with slight synchronization with the video data (CSI, *e.g.* information on the wireless channel state such as average signal-to-noise ratio (SNR), channel coherence time, and NSI, *i.e.* information on the transmission network state, represented in particular by packet loss rate and delay at network or session levels...).

For the direct channel, considering that the cross-layer information comes from the application (source encoder, application controller, content ciphering and content protection), it is consistent to transfer the extra data directly into the binary packets (see the considerations about the Network Transparency concept in general, and made in the previous sub-section),which are composed of the payload obtained after video encoding and application processing, with addition of an extra information field viewed as an additional header, as illustrated in Figure 3.

This forward direction information corresponds to the concatenation of security information (CIPHER Initialization vector(key), CIPHER flag), SSI, necessary for Content protection, which corresponds to puncturing rates and transitions length between the different sensitivity classes, SRI, *i.e.* specific video information that can be used at receiver side for better concealment, or soft-input decoding (the four SRI fields being typically used in the soft-input decoding process described in (Bergeron & Lamy-Bergot, 2004)), and quantization information fields (hard or quantified payload with number of quantization bits for performing this soft-input H.264/AVC decoding).

The actual sizes used in the simulations analysis are detailed in the figure. The maximum total size of the extra header, which is taken into consideration in the simulations to reflect the consumed bandwidth corresponding to this signaling information transmission, is consequently 96 bits i.e. 12 bytes. It must be noted that naturally, should this extra information transmission be standardized, a new RoHC compression profile for this header could be created, which would greatly reduce this cost.

One can also point out that the possibility to insert the SSI information in a new IPv6 hop-by-hop option. This would allow intelligent routers to implement DiffServ (Blake, 1994) services based on the importance of the carried payload data (*e.g.* providing loss guarantees in a relative manner by the Assured Forwarding Per Hop Behavior). The same concept could be also applied to Controlled Load services in an IntServ (Braden, 1994) architecture, by employing advanced queue management techniques, such as Random Early Detection (Floyd, 1993) with multiple thresholds.

In the return direction, one finds feedback information, whose purpose is to inform the transmitting side of the state of the wireless channel (Channel State Information: CSI), of the IPv6 network state (Network State Information: NSI) represented *e.g.* by packet loss rate, delay, jitter and possibly of a video quality measure, output of the source decoder and real system quality criterion. Nevertheless, the difficulty to obtain a good video quality measure without a reference (the standard PSNR measure needs reference to the original video frame), together with the possible delay introduced by the eventual non real-time decoders led the authors to focus on CSI and NSI information. Again, in designing new systems several trade-offs must be properly evaluated. A more precise measure of the perceived quality is really helpful, but it comes at a cost of further signaling overhead.

At the receiver side, it was also considered the possibility to perform soft-input source decoding

Table 1. Return direction and receiver side signal information, suitable mechanisms and overheads

Control signal	Suitable mechanism	Estimated overhead and comments
CSI	ICMPv6	Overhead of less than 1 kbyte/sec for CSI updating period up to 50 ms; slight synchronization with the video data.
NSI	RTCP	Low overhead with suitable frequency of 200 ms (less than 1 kbyte/sec).
DRI/SAI	IPv6 packets	Very high bandwidth consuming (even higher than the video data flow of a fixed multiplying factor). Those control signals should in practice be sent only when the wireless receiver is also the data traffic destination.

or even iterative soft-input soft-output source decoding with channel decoder, leading to the creation of two specific side information streams, namely decoder reliability information (DRI) for the soft values output of the channel decoder, to be transmitted to the input of the source decoder, and source a-posteriori information (SAI) for the soft values output of the source decoder, to be transmitted to the output of the channel encoder. However, the analysis of the resulting overhead suggests the implementation of soft-decoding techniques when the radio receiver and the source decoder are in the same device (*e.g.* a UMTS handset or WLAN PDA) only

The mechanisms identified to allow these information exchanges transparently for the network layers are summarized in Table 1. It must be noted that those selected solutions sometimes mirror already existing one. As examples of such control information carried over a return channel, one finds RTCP transporting NSI information or UMTS offering links estimation of Block Error Rate (BLER) to provide an estimation of the link quality (*i.e.* partial CSI information).

NUMERICAL RESULTS

To better illustrate the interest of the optimization approach, and provide examples of the gain achievable with the described system, some simulation results gathered with a demonstration platform, where all the involved system layers were realistically implemented, are reported.

Following architecture depicted in Figure 2, this simulation chain includes:

- at application level
  - the controller jointly optimizing all layers;
  - the video source encoder/decoder (H.264/AVC being considered in the following simulations);
  - content cipher/decipher unit;
  - content level protection block by means of rate-compatible puncturing codes (RCPC) (Hagenauer, 1988);
- in the network domain
  - real time transport protocol (RTP) header insertion/removal;
  - transport protocol header (UDP-Lite, UDP, or DCCP) insertion/removal;
  - IPv6 header insertion/removal;
  - IPv6 mobility modeling;
  - IPv6 network simulation;
- at the base station
  - Robust Header Compression (RoHC);
  - 802.11 like data link header insertion/removal (with CRC coverage modification);
  - 802.11 or UMTS like PHY layer including channel encoder/decoder by means of convolutional codes, interleaving, modulation (OFDM, TCM, TTCM, STTC), with soft and iterative demodulation allowed, and a simulated channel with various possible

configurations: AWGN, Rayleigh fading, shadowing, frequency selective channels.

Those different blocks, which constitute the integrated software simulation chain developed within the FP6 IST PHOENIX project, are driven by a set of parameters which are summarized in Table 2.

Consistently with the chosen approach to have a system with minimal modifications of an existing radio access, the data link layer is modified only by changing the MAC CRC coverage to protect only the frame header (MAC+RoHC), and the PHY layer is modified with regard to existing solutions only by changing the coverage of the channel coding by applying the code only to the frame extended header (MAC+RoHC+extra header), which is to the part of the transmitted PHY frame that could not be protected at content level, as it was not present there. The insertion of PHY protection is critical, because it ensures a minimum protection of the network headers (including the MAC header, which contains a CRC which could lead to packet discarding in presence of errors). No ARQ mechanism is here considered, which allows affirming that the bandwidth occupation estimation by the joint controller is accurate.

The optimization (*i.e.* the APP controller update rhythm, therefore its reaction time) and control signaling update is performed on a fixed time-step of 1 s. Consistently, H.264/AVC picture encoding is done in groups of pictures whose duration is equal to 1 s. The 1 s time step was chosen as a compromise between the need of regular updates of channel state information at transmitting side, and the compression efficiency. In particular, the time-step was chosen in adaptation with the channel coherence time of the shadowing, to ensure that the assumption of constant shadowing in one controller step holds.

In the following simulations, two different sets of parameters were considered, to validate the chosen approach over two types of radio access and for two different video reference sequences. The first settings, corresponding to video sequence 'Akiyo' in QCIF format at 15Hz, that is transmitted at a coded bit rate of 128 kbps at the channel level over an additive white Gaussian channel (AWGN) with variable signal to noise ratio, whose coherence time is set to 10 seconds (start joint controller conditions are then re-set to default values every 10 seconds).

The obtained behavior in terms of video quality (measured in terms of peak signal-to-noise ratio (PSNR)) versus time (expressed in seconds) is shown in Figure 4. The comparison is made over values obtained by averaging four simulation runs with different noise seeds and varying the average noise level on the AWGN channel from 1 dB to 6 dB (in random order).

From the curves in Figure 4, the gain achieved with adaptation is clear. It corresponds first to avoidance of loosing frames at lower layers thanks to a better header protection, and second to better final results obtained by a better tuning of the compression and protection rates, to ensure that the payload data at the output of the content protection module is as error free as possible. On average, PSNR gains of about 3.4 dB are observed in the conditions under analysis. When the channel conditions are improving, as in the right part of the figure, the two solutions become closer, with possibly sometimes better results achieved for "classical" scheme when protection level selected by the joint controller is too high, but at the recurrent cost of lost frames degrading unpleasantly the visual quality.

The second settings correspond to a WLAN scenario, with the WLAN link supporting a radio coded bit rate of 12 Mbit/s. In practice, the H.264/AVC encoded video stream 'Foreman' in QCIF format at 15Hz, is supposed to be multiplexed with other transmissions, and only uses a portion of this available bandwidth, corresponding to a coded bit rate of 256 kbps at the channel level. The considered channel follows here ETSI standard channel A, to which a log-normal flat fading

Table 2. Recapitulation of the considered simulation parameters

Test video sequence		
Video sequence:	ITU reference sequence "Akiyo"	ITU reference sequence "Foreman"
Video format:	QCIF (176x144)	QCIF (176x144)
Frame rate:	15 fps	15 fps
Duration:	10 seconds (then looped)	13 seconds (then looped)
Joint control		
Overall bit-rate:	128 kbps (at channel level)	256 kbps (at channel level)
controller mode:	streaming server with perfect knowledge of available streams	streaming server with perfect knowledge of available streams
Control update step:	1 second	1 second
Source coding		
Encoded sequence frame rate:	15 fps	15 fps
Intra frame refreshment period:	15 frames	15 frames
Source coding rate (classical):	fixed rate of ~61 kbps	fixed rate of ~124 kbps
Source coding rate (adapted)	variable rate from 40 to 100 kbps	variable rate from 80 to 200 kbps
H.264/AVC packet maximum size:	180 bytes	180 bytes
Encoding mode:	standard Data Partitioned mode	standard Data Partitioned mode
Extra header maximal size (for rate estimation):	96 bits (0 in classical mode with EEP)	96 bits (0 in classical mode with EEP)
Content ciphering		
Cipher mode:	H.264/AVC embedded partial ciphering (see (Bergeron & Lamy-Bergot, 2005))	
Content protection		
Channel encoder:	RCPC code (Hagenauer, 1998)	
Mother code rate:	1/3	
Constraint length:	6	
Code generators (in octal):	133;171;145	
Puncturing period:	8	
Code rates considered:	8/9, 4/5, 2/3, 4/7, 1/2, 4/9, 2/5, 4/11, 1/3	
Packetization and transport		
Real-time management:	RTP standard format	
Accepted delay for packet at receiver:	5000 (ms)	
Transport protocol:	UDP-Lite	
Transport checksum coverage:	UDP-Lite+RTP	
IPv6 wired network (emission side)		
IPv6 network nb of nodes:	10	
Mean node delay:	3 ms	
Mean node packet loss:	10 ppm	
Bottleneck rate:	10000 kbps	
Buffer size at bottleneck:	100000 bytes	

continued on the following page

Table 2. continued

Test video sequence		
IPv6 mobility		
additional delay mean value (ms):	10	
delay square standard deviation:	4	
handover latency mean (ms):	520	
handover square standard deviation;	100	
mean value between two handovers (ms):	820	
square standard deviation between two handovers:	34.5	
RoHC parameters		
Network headers considered:	RTP/UDP-Lite/IPv6	
Compression mode:	unidirectionnal (U)	
Compression rate:	average (~ 8 bytes, FO timeout=11, IR timeout=1000)	
Data Link layer parameters (802.11 based)		
Dynamic partial checksum	MAC + RoHC 36 to 96 bytes coverage, 4 bytes length	
Physical layer		
Channel coding:	Frame extended header (MAC+ RoHC+extra) protection only, with fixed rate. In classical mode, rate is 1/2, in adapted mode, rate is 1/3	
Channel encoder:	RCPC code of mother rate 1/3, constraint length 5	
Code generators (in octal):	23; 35; 27	
Code rates considered:	1/2, 1/3 with puncturing period 8	
Interleaving:	random (done packet by packet)	
Modulation:	BPSK	OFDM (48 carriers, frame duration 4 µs)
Number of RX/TX antennas:	1/1	1/1
Radio channel model:	Additive White Gaussian Channel	ETSI channel A
slow fading:	none	uncorrelated og-normal distribution, σ=4 dB
coherence time:	0s (duration of median Eb/N0 selected)	5s (slow-fading block duration)
Median Eb/N0:	1 to 6 dB AWGN noise	13.2 dB

component with coherence time of 5 seconds was added, to reflect fading effects due to large obstacles. The orthogonal frequency division multiplexing (OFDM) modulation is used with margin adaptive bit-loading techniques (Cioffi, 2004; Dardari, 2002).

The resulting behavior in terms of video quality versus time, obtained by averaging six simulation runs with different noise seeds, is shown in Figure 5 corresponding to an average coded bit signal-to-noise Eb/N0 ratio of 13.2 dB, with Eb the average energy per coded bit.

Again, the benefit of adaptation is clear, corresponding to an average value of 4.7 dB in PSNR. Some losses still can happen, due in particular to deep fading, but this time the non-adapted case also fails.

Sample visual results, in accordance with average visual impact, are depicted in Figure 5. Here again, the improvement with the adapted

Figure 4. Observed quality (PSNR) versus time for simulation with and without adaptation by application controller - Akiyo reference sequence

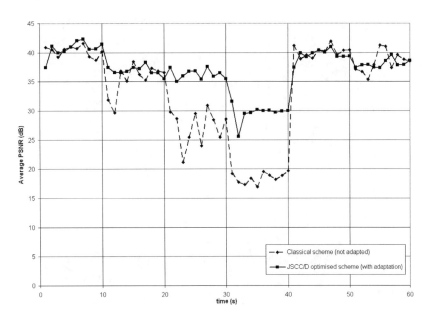

Figure 5. Observed quality (PSNR) versus time for simulation with and without adaptation by application controller - Foreman reference sequence

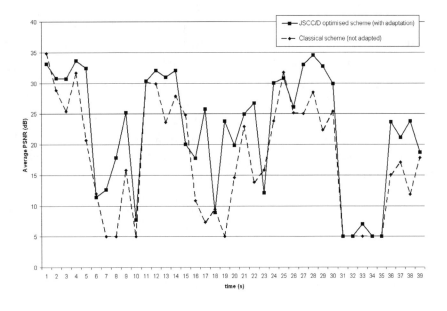

schemes (on the right) when compared to fixed classical approaches (on the left) is evident. The impact of the adaptation is particularly visible in Figure 6, where one sees that the adapted image is received with few errors, thanks to a better protection level obtained at the cost of a bigger compression rate, corresponding to a slightly blurred image.

Figure 6. Example of visual results obtained in both simulations: non-adapted (left) vs. adapted (right)

Furthermore, in (Lamy-Bergot & Panza, 2008) an analysis and optimization more at an architectural level, of the JSCC/D system developed within this IST PHOENIX framework has been provided. The assessment is based on several critical issues: feedback information overhead, reaction time of the application controller, impact of lost or delayed feedbacks and crossing of multiple wireless hops. The cost and benefits of the designed system, as well as the proposal of some trade-offs between the configuration parameters in order to optimize the QoS and resource utilization were reported.

These results justify the settings used for control update rate in our numerical results generation. As a matter of fact, it is shown in this last study that a 2 s Application Controller reaction time and refreshing periods of respectively 200 and 250 ms for CSI (feedback about the channel state) and NSI (feedback about the overall network state) messages lead to a good PSNR in both good and bad network conditions. Such settings also ensure robustness to delay and loss of feedback messages, thanks to the implicit filtering process of the received information within a time-step of the application controller. With the optimal configuration settings of reaction time and refreshing periods, the discussed system outperforms a traditional (i.e. non-adaptive) video streaming system of several dBs in PSNR, especially in bad network conditions.

An extension to the case of the crossing of multiple wireless hops is there also proposed, that shows that for this case a better choice is to average the CSI of the concerned radio channels, rather than taking the worst status as the reference. Indeed, a less conservative system can more effectively and fully exploit the available network resources for an improved PSNR.

FUTURE RESEARCH DIRECTIONS

Future work could be the design, implementation and assessments of a JSCC/D system for a point-to multipoint communication. In such a scenario, a novel critical issue to be considered is the way feedback information related to different users are generated, transmitted, possibly aggregated into the network and processed by the Application Controller. Therefore, some more challenging tasks should be efficiently performed, being the set of recipients of the multimedia streams potentially large.

Such a scenario is addressed by the FP7 IST OPTIMIX project[2]. In this project, feedbacks from all layers of the receiver terminal are collected with a "Mobile Unit Observer" and transmitted to the application controller. These feedbacks can be either objective measures such as the perceived quality from the application or subjective measures as the user experienced quality. The employment of the IEEE 802.21 framework (IEEE LAN MAN standards committee, 2008) is foreseen, in order to gather information from the lower layers (i.e. MAC and PHY). Access Points (APs) or Base-Stations (BSs) are possible points of feedback aggregation and processing, but other entities of the IP network could deal with such operations, as well, taking advantage of anycasting of IPv6 for routing purposes. A critical issue is also the policy to be applied to the received feedbacks by the application controller. The user heterogeneity, both in terms of requirements, terminal capabilities and bandwidth availability, must be carefully managed, with different target to consider and achieve (*e.g.* the maximization of the "worst-user", or "average-user" quality of service).

A key point in the realization of the discussed JSCC/D proposal, with adaptation at application level only, is that no modifications in the existing video, network or radio access standards are required. Indeed, the system relies on the addition of just JSCC/D controllers in the video delivery architecture, which will then drive the encoding process on the basis of the network status (i.e. on RTCP reports). Typical clients for such a solution are telecommunication companies and service providers, who could begin to integrate now this first realization for point to point adaptation in their own video delivery system.

For the more challenging problem of point to multi-point adaptation, new frameworks, such as that of FP7 ICT OPTIMIX project, are expected to develop first solutions in few year times, offering first implantable versions by the beginning of the next decade.

Active fields of research could be as follows.

- Innovation in the area of sophisticated multimedia source coding schemes devised to satisfy diverse design criteria and trade-offs in terms of source representation quality, bit rate, delay, encoding/decoding complexity etc..
- Design of attractive channel coding and modulation arrangements, which are capable of approaching the channel capacity at the lowest possible complexity and delay.
- Improvements to The Internet-Protocol based transport, which still have a number of deficiencies, such as the sensibility of transmission errors in the header that dramatically affect the bit stream.
- Identification of the most suitable feedback information and how to aggregate it, when related to different links, as well as the design of effective cross-layer communication techniques for a transparent delivery.
- Evaluation of joint optimization strategies for audio/video transmission to multiple receivers, based on the collection of side information on the system (in terms of channel state information, network conditions, source characteristics,..) and to consequently tune the parameters of the different system components.
- Design of simple audio/video assessment metrics, based on a partial- or no-reference model, to be used for the real-time user quality evaluation, helpful to achieve an overall system optimization (although, focusing on a subset of users might be reasonable or more efficient in given application scenarios).

CONCLUSION

Since the explosion of wireless telecommunication with the success of GSM in the nineties, the telecommunication world has known a very rapid and active evolution that has tried to both make an efficient use of any emerging technology while also going towards a global integrated system allowing for new services and applications, and trying to meet the requirements of both users and industry. Expectations of evolution are strong in terms of services and applications, which include, for end-users, good quality of service, easy access to applications and services, always enhanced security and reasonable cost. For the network deployment and service architecture, this translates into simple quality of service (QoS) and security management, flexibility of system configuration, and maximization of network capacity. Fulfilling these requirements is a challenging task for system designers aiming at producing flexible new wireless systems that must transparently interconnect a multitude of heterogeneous networks and systems and allow for a more integrated strategy.

In this chapter, the innovative system architecture designed by IST PHOENIX project for multimedia streaming over an IP-based wireless networks has been presented. This approach aims at jointly optimizing the different layers of the transmission chain to interconnect efficiently the application world (source coding, ciphering, and protection at content level), the network world (transport services, IP networking, and header compression) and the radio access (data link, channel coding, and modulation) thanks to a cross-layer design and joint source and channel coding approach. The proposed JSCC/D architecture considers the overall transmission chain from application level source coding to wireless and wired channels, the required JSCC/D signaling mechanisms and real network functionality. In particular, new solutions that effectively deliver control information through the network and the protocol stack are defined, in an as much as possible backward compatible manner, allowing for a cross-layer design strategy.

This global approach goes beyond the paradigm of classical telecommunications, where the settings used are finely tuned to given fixed transmission conditions, and proposes to adapt continuously the system over a large range of conditions, based on the feedback information received on channel or network state. To this end, several promising mechanisms for the adaptation of the layer settings and the delivery of JSCC/D controls, each best suited for easy adoption into different kinds of IP networks, and not requiring that system administrators and service providers change their whole infrastructure are proposed. The defined architecture presents the advantage of being adaptable with minor changes to already existing radio access solutions (such as WLAN), for which only modification on the coverage of the MAC CRC and channel coding application are required, but also to be applicable in future new systems, for which it could then be extended by controlling also the PHY channel coding and modulation as proposed in (Martini & Mazzotti, 2005). Moreover, this global approach can also be adapted for use with other video codec, as shown for H.264 scalable extension SVC, as presented in (Huusko, 2006).

Finally, the results obtained with a software prototype based on the proposed architecture, which encompasses not only the source coding and channel coding block, but also the whole network protocol stack including data link, header compression, network, and transport protocols are reported. Sample simulation results obtained using integrated simulation chains are provided, showing the improvements of video quality both visually and in terms of PSNR.

ACKNOWLEDGMENT

The authors would like to thank their colleagues, who have participated in IST PHOENIX project and given valuable work contribution for the development of PHOENIX system and simulation chain.

REFERENCES

Bergeron, C., & Lamy-Bergot, C. (2004). Soft-input decoding of variable-length codes applied to the H.264 standard. In *Proceedings of the IEEE MMSP'04 conference* (pp. 87-90), Siena, Italy. Washington, DC: IEEE.

Bergeron, C., & Lamy-Bergot, C. (2005). Compliant selective encryption for H.264/AVC video streams. In [Shanghai, China. Washington, DC: IEEE.]. *Proceedings of the IEEE MMSP, 05*, 477–480.

Bergeron, C., & Lamy-Bergot, C. (2006). Modelling H.264/AVC sensitivity for error protection in wireless transmission. In [Victoria, Canada. Washington, DC: IEEE.]. *Proceedings of the IEEE MMSP, 06*, 302–305.

Bergeron, C., Lamy-Bergot, C., Pau, G., & Pesquet-Popescu, B. (2006). Temporal scalability through adaptive M-Band filter banks for robust H.264/MPEG-4 AVC video coding. *EURASIP Journal on Applied Signal Processing, 5*, 21930–21940.

Blake, S., Black, D., Carlson, M., Davies, E., Wang, Z., & Weiss, W. (1998). An architecture for differentiated services. *IETF RFC 2475*. Retrieved on January 2009, http://www.ietf.org/

Bokor, L., Dudás, I., & Javier, R. O. (2005). *Streaming over mobile IPv6 networks*. Paper presented at the Ericsson High Speed Networks Laboratory Workshop, Mtrahza, Hungary.

Bormann, C. (2005). Robust header compression (RoHC): Framework and four profiles: RTP, UDP, EPS, and uncompressed. *IETF RFC 3095*. Retrieved on January 2009, http://www.ietf.org/

Braden, R., Clark, D., & Shenker, S. (1994). Integrated Services in the Internet Architecture: an Overview. *RFC 1633*. Retrieved on January 2009, http://www.ietf.org/

Chaddha, N., & Diggavi, S. A. (1996). Framework for joint source-channel coding of images over time-varying wireless channels. *IEEE Image Processing, 1*, 89–92.

Cioffi, J. M. (2004). *EE379C: advanced digital communications*. Lecture Notes, Stanford University.

Conta, A., & Deering, S. (1998). Internet control message protocol (ICMPv6) for the internet protocol version 6 (IPv6) specification. *IETF RFC 2463*. Retrieved on January 2009, http://www.ietf.org/

Corlett, A., Pullin, D. I., & Sargood, S. (2002). *Statistics of one-way internet packet delays*. Paper presented at the 53rd IETF meeting, Minneapolis, USA.

Dardari, D., Martini, M., Milantoni, M., & Chiani, M. (2002). MPEG-4 video transmission in the 5 GHz band through an adaptive OFDM wireless scheme. *Proceedings of the IEEE PIMRC'02*, Vol.4, (pp. 1680-1684), Lisboa, Portugal. Washington, DC: IEEE.

Deering, S., & Hinden, R. (1998). Internet protocol Version 6 (IPv6) specification. *IETF RFC 2460*. Retrieved on January 2009, http://www.ietf.org/

Floyd, S., & Jacobson, V. (1993). Random early detection gateways for congestion avoidance. *IEEE ACM Transaction on Networking, 1*, 397–413. doi:10.1109/90.251892

Hagenauer, J. (1998). Rate-compatible punctured convolutional codes (RCPC codes) and their application. *IEEE Transactions on Communications, 36*(4), 339–400.

He, Z., Cai, J., & Chen, C. W. (2002). Joint source channel rate-distortion analysis for adaptive mode selection and rate control in wireless video coding. *IEEE Transactions on Circuits and Systems for Video Technology, 12*, 511–523. doi:10.1109/TCSVT.2002.800313

Huusko, J., Vehkaperä, J., Amon, P., Lamy-Bergot, C., Panza, G., Peltola, J., & Martini, M.G. (2006). Cross-layer Architecture for Scalable Video Transmission in Wireless Network. *Signal processing: Image communication Special Issue on Mobile Video, 22*(3), 317-330.

IEEE 802.11 - 1999 edition, Part 11: Wireless LAN Medium Access Control (MAC) and Physical Layer (PHY) Specifications, *IEEE Standard 802.11*.

IEEE LAN MAN standards committee of the IEEE computer society (2008). Draft standard for local and metropolitan area networks: Media independent handover services. *IEEE P802.21/D10.00*. New York: IEEE.

Kohler, E., Handley, M., & Floyd, S. (2006). Datagram congestion control protocol. *IETF RFC 4340*. Retrieved on January 2009, http://www.ietf.org/

Lamy-Bergot, C., Chautru, N., & Bergeron, C. (2006). Unequal error protection for H.263+ bitstreams over a wireless IP network. In *Proceedings of the IEEE ICASSP'06* (pp. V-377/V-380), Toulouse, France. Washington, DC: IEEE.

Lamy-Bergot, C., Panza, G., Rotondi, A., & Fratta, L. (2008). Analysis and optimization of a JSCC/D system on 4G networks. In [Bologna, Italy. Washington, DC: IEEE.]. *Proceedings of the IEEE ISSSTA, 08*, 253–259.

Larzon, L. A., Degermark, M., Pink, S., Jonsson, L.-E., & Fairhust, G. (2004). The lightweight user datagram protocol (UDP-Lite). *IETF RFC 3828*. Retrieved on January 2009, http://www.ietf.org/

Martini, M. G., & Chiani, M. (2004). Rate-distortion models for unequal error protection for wireless video transmission. [Milan, Italy. Washington DC: IEEE.]. *Proceedings of the IEEE VTC, 04*, 1049–1053.

Martini, M. G., Mazzotti, M., Chiani, M., Panza, G., et al. (2005). *A demonstration platform for network aware joint optimisation of wireless video transmission: the PHOENIX basic demonstration platform*. Paper presented at the IST Mobile Summit 2005, Dresden, Germany.

Martini, M. G., Mazzotti, M., & Lamy-Bergot, C. (2007). Content adaptive network aware joint optimisation of wireless video transmission. *IEEE Communications Magazine, 45*(1), 84–90. doi:10.1109/MCOM.2007.284542

Massey, J. L. (1978). Joint source and channel coding. In J.K. Skwirzynski (Ed.), *Commun. systems and random process theory, NATO advanced studies institutes series E25*, (pp. 279–293). Alphen aan den Rijn, The Netherlands: Sijthoff & Noordhoff.

McEliece. (1977). *The Theory of Information and Coding*. London: Addison-Wesley.

Park, M., & Miller, D. J. (2000). Joint source-channel decoding for variable-length encoded data by exact and approximate MAP sequence estimation. *IEEE Transactions on Communications, 48*(1), 1–6. doi:10.1109/26.818865

Perros-Meilhac, L., & Lamy, C. (2002). Huffman tree based metric derivation for a low-complexity sequential soft VLC decoding. In *Proceedings of the IEEE ICC'02* (Vol. 2, pp. 783-787). New York: IEEE.

Postel, J. (1980). User datagram protocol. *IETF RFC 768*. Retrieved on January 2009, http://www.ietf.org/

Qiao, L., & Nahrstedt, K. (1997). A new algorithm for MPEG video encryption. In [Las Vegas, NV.]. *Proceedings of the CISST, 97*, 21–29.

Rec, I. T. U.-T. *H.264, ISOIEC 14496-10 AVC,* Doc JVT-G050r1. (2003). Final Draft International Standard of Joint Video Specification. Geneva, Switzerland: ITU.

Schulzrinne, H., Casner, S., & Frederick, R. (1996). RTP: A transport protocol for realtime applications. *IETF RFC 1889*. Retrieved on January 2009, http://www.ietf.org/

Shannon, C.E. (1948). A mathematical theory of communication. *Bell System Technical Journal, 27*, 379-423 and 623-656.

Srivastava, V., & Motani, M. (2005). Cross-layer design: A survey and the road ahead. *IEEE Communications Magazine, 43*(12), 112–119. doi:10.1109/MCOM.2005.1561928

Van der Schaar, M., Krishnamachari, S., & Choi, S. (2003). Adaptive cross-layer protection strategies for robust scalable video transmission over 802.11 WLANs. *IEEE Journal on Selected Areas in Communications, 21*, 1752–1763. doi:10.1109/JSAC.2003.815231

Vehkaperä, J., Peltola, J., & Huusko, J. Myllyniemi & Majanen, M. (2006). Evaluation of achieved video quality in wireless multimedia transmission using UDP-Lite. In *Proceedings of the IASTED IMSA 2006,* (pp.14-16), Honolulu, HI. New York: ActaPress.

Wang, H., Tsaftaris, S. A., & Katsaggelos, A. K. (2006). Joint source-channel coding for wireless object-based video communications utilizing data hiding. *IEEE Transactions on Image Processing, 15*, 2158–2169. doi:10.1109/TIP.2006.875194

Zahir Azami, S. B., Duhamel, P., & Rioul, O. (1996). Joint source-channel coding: Panorama of methods. In *Proceedings of CNES workshop on data compression*. Toulouse, France: CNES.

ENDNOTES

[1] IST PHOENIX project has been partially supported by the European Commission under the contract FP6-2002-IST-1-001812.

[2] The research leading to these results has received funding from the European Community's Seventh Framework Programme ([FP7/2007-2013]) under grant agreement n° [214625] ICT OPTIMIX.

Chapter 22
Video Streaming Based Services over 4G Networks:
Challenges and Solutions

Elsa Mª Macías
Universidad de Las Palmas de Gran Canaria, Spain

Alvaro Suarez
Universidad de Las Palmas de Gran Canaria, Spain

ABSTRACT

4G networks must not only show high bandwidth but also provide an excellent user experience, especially for video streaming, which is a key technique for multimedia services on 4G networks like Voice over Internet Protocol (VoIP), Television over IP (TvIP), broadcatching, interactive digital television, and Video on Demand (VoD). These services are challenging because of the well-known problems of the radio channel. Efficient solutions are designed by considering cross layer techniques. In this chapter the authors firstly review a number of video streaming based services, and then they present the basic operation of the video streaming and its problems in 4G networks, emphasizing Wireless Fidelity (WiFi) technology. In order to solve these problems they propose two cross layer strategies (one for access networks and another for ad hoc networks) and integrate the first one into two application level solutions. The authors test the user experience that accesses a Web portal including a VoD with a mobile telephone equipped with WiFi and High Speed Downlink Packet Access (HSDPA) Wireless Network Card Interfaces (WNIC). Results invite them to be optimistic.

1 INTRODUCTION

A real time multimedia system manages synchronized information with different semantic interpretations (usually video, audio and text). The processing and communication of this information can be naturally split in different units (images, a portion of audio between two silences, etc.). The processing results not only must be correct, but they also must be produced before a temporal deadline well established for each unit. Eventually the managed information must be communicated using any kind of network (internal to a machine or using the wide Internet). Working with multimedia communication deadlines is a very complex task.

DOI: 10.4018/978-1-61520-674-2.ch023

The convergence between telecommunication and computer communication has led to new set of multimedia services like videoconferencing, VoIP, instant messaging, and gaming. The massive use of these services speaks about their importance in the area of Communication and Information Technology and their impact on Information Society.

The communication of multimedia services results tends to be delay sensitive, bandwidth intense and loss tolerant. For this reason, it is very important to deliver the units of multimedia information such that the end user has a good experience. Whenever a real time communication or computation task (or a combination of both) can be split into others of less complexity, it will be possible to use the streaming technique to deliver (process) the units of information efficiently by overlapping communication and processing actions. Several examples are: Efficient Instruction Stream hiding memory latency compared with processor speed rate (Rodriguez & Campelo, 2003), Lot Streaming Technique for the Job-Shop Scheduling problem (Chan & Wong, 2005), and Media Streaming or Video Streaming techniques for multimedia information delivering in a traditional multimedia e-learning system (Ming, 2000). Media Streaming is used to reduce the delay requirements of multimedia services using efficiently the network bandwidth. Sometimes, this technique, when used for delivering multimedia information stored in a server is simply named streaming. This term is not used in cases where the server delivers real time video.

In the last ten years digital wireless networks have experienced a great evolution and social implantation (Pagani, 2005). Their evolution continues constantly due to operators and vendors that want to stay competitive, offer better provisioning, and provide more cost-efficient provisioning of old services as well as new services (Dahlman & Parkval, 2007). This means seeking a more efficient system for adapting and reconfiguring the wireless channel (while taking into account the number of active users), adapting the Medium Access Control (MAC), and considering the upper layers seeking a good routing algorithm, transport protocol, security procedures, mobility management, etc. (Savo, 2006). Moreover, they must satisfy the user's requirements presented in (Pagani & Schipani, 2005) specifically for multimedia services.

In the wireless network marketplace there is a true jungle of technologies: Zigbee (for multimedia wireless sensor networks), Bluetooth and WiFi (reviewed in (Ganz & Ganz, 2004)) Mobile Worldwide Interoperability for Microwave Access (WiMAX) (Cheng & Marca, 2008), wireless mesh networks (Aggelou, 2009), Radio Frequency Identification (RFID) (Ahson & Ilyas, 2008), and Satellite or similar (Zavala & Ruíz, 2008), among others. Evolution of other technologies like Long Term Evolution (LTE) (Holma & Toskala, 2007) is thought to be the core of the 4G cellular telephony network. In terms of transmission speed the objective of 4G wireless networks is to achieve 1 Gbps for Wireless Local Area Networks (WLAN) and 100 Mbps for mobile telephone networks.

The tremendous evolution of wireless networks has met the powerful hardware and software architecture of new mobile telephones. This architecture usually includes several on-chip WNIC like Bluetooth, WiFi, Global Positioning System (GPS), WiMAX, etc. Their computing powers allow them to receive multimedia streaming information at very low cost from a VoD server. This makes the efficient implementation of streaming techniques over wireless network a very important challenge. It is also an interesting challenge to achieve because there are a large number of mobile devices on the market.

The design of solutions to the challenge of streaming over 4G networks requires good knowledge about their characteristics, because it is well known that wireless channels present a set of problems: frequently packets are delayed, lost or even discarded. These problems lead to an unsatisfactory user experience. Moreover, all 4G wireless technologies will present this set of problems (or a variation of it) because all of them use a radio channel. For example, in Tse & Viswanath (2005), a presentation of these problems can be found for WiFi networks and in Ippolito (2008), the problem of wireless channel interference is studied for satellite networks. For this reason, we think it is not only important to provide a high speed transmission radio channel for 4G multimedia wireless networks, but also to explore the nature of wireless networks in order to efficiently solve their problems when delivering multimedia information using media streaming.

Moreover in 4G networks, mobile terminals will also experiment frequently sporadic disruptions when issuing a roaming process (Gupta & Williams, 2006), while stopped in a zone where the wireless channel is partially deteriorated, or while moving in a geographical area that limits radio coverage (Suárez & Menza, 2007). This would be a disaster in real time communications (sensors) or a user headache (firm real time multimedia communication). In the particular case where the terminal has agreed to be a part of a VoD session, a long disruption would provoke the VoD Server to terminate the session. This is an important problem because the Client must negotiate a new session with the corresponding loss of efficiency of multimedia streaming technique due to the overlapping of visualization, and communication will be broken, leading to user intervention to start the visualization again.

In order to satisfy the multimedia user experience, 4G networks will be able to take into account several requirements (Schaar, 2007) such as easy adaptability to wireless bandwidth fluctuations and robustness to partial loss of packets. This only can be achieved efficiently using a cross layer technique due to independent solutions that that do not provide adequate support for multimedia applications: network adaptive compression and bandwidth (Vandalore & Feng, 2001), channel condition bit stream adaptation, prioritization and layering mechanisms, error concealment strategies, rate-distortion modeling (Chakareski & Chou, 2006; Hassan & Vikram, 2008), joint channel coding, a single proxy based solution (Gao & Zhang, 2003), and simple buffering schemes (Bellavista & Corradi, 2005; Wu & Hsu, 2007), among others.

Cross Layer technique basically consists in obtaining a cooperative solution among different architecture levels globally optimizing certain function of n parameters (defined at different levels) (Kumwilaisak & Hou, 2003). This constitutes a complex problem that cannot always be optimally solved. Authors have paid attention to cross layer design using different visions. In Li and Li (2005) they try to predict the performance of the wireless communication system for streaming service by evaluating the rate of loss packets and Round Trip Time (RTT) at application level, the datagrams loss at network level, and the signal strength and channel capacity at physical level. They do not present any performance measure of this monitoring. In Tonev and Sunderam (2002) an application is presented that detects if the mobile terminal is in a geographical region where the performance of the wireless network decreases considerably. This software classifies parts of the above region combining parameters such as signal strength, loss of packets and latency. They show that the minor false alarms percentage is obtained when all of these parameters are combined to obtain a prediction of performance. They show the results of a few probes and do not show which the impact of the signaling and performance is. A three-level methodology is shown in Villalón & Cuenca (2007): physical, link and application levels. They use a two way methodology but only show simulation results for only one multicast session.

We have designed a cross layer strategy that has partially been implemented in practice in front of other authors that only simulate their proposals. We basically obtain the values of physical and application levels parameters in real time. With these values we build an strategy that consists of early optimization of the information process for the user, telling if the wireless device will be out of coverage in the recent future. We only issue this strategy if the user demands real time video or streaming from our server. That server is a collaborative Web portal dedicated to reduce the serious damages or even death provoked by nicotine poisoning (O'Connell, 2009). This collaborative server is intended for improving smokers' lives by allowing them and therapists to collaborate using video descriptions of clinical situations. These video descriptions can be downloaded to a mobile telephone or a laptop at any time and any place. In this way the performance of the clinical therapy is improved because a smoker can receive attention in critical situations. In this case it is very important to provide the best possible user experience in order for them not to abandon the therapy.

In Figure 1 we schematize our cross layer strategy for the nicotine poisoning problem. The different levels we consider are: Internet Application, Internet Transport and Physical levels.

As the interface between the Internet Transport and the Internet Application levels we consider on one hand the reliable sockets and in the other hand the Real Time Streaming Protocol (RTSP) (Schulzrinne & Rao, 1998) because it is the official standard protocol for streaming. We do not propose any change to the Internet network and transport protocols. We consider a set of parameters in each level and our objective is to propose an integral optimization for all these parameters. The parameters of the Application level are: Jump distance and Time off (Costa & Cunha, 2004) and the profile of users. At the Transport level we implement a middleware that is in charge to control different parameters: Bandwidth control, packets loss, control of RTT for each packet and battery consume in line with what is proposed in (Fei & Zhong, 2008). At the physical level there are several parameters that must be controlled in any current technology (WiFi, HSDPA, etc.) and future technology for 4G Wireless Networks (WiMAX, LTE, LTE-Advanced, etc.): Range

Figure 1. Cross layer strategy anticipating information about sporadic disruptions of media streaming

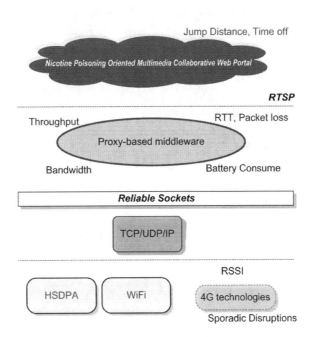

control in the mobile terminal, the Received Signal Strength Indication (RSSI) perceived by the terminal, and admission control and frequently sporadic disruptions of the streaming service. We integrate a global function optimization that predicts the performance of the different parameters by observing its values in the recent past and proposing that the user take corrective actions for the recent future in an anticipating way. For example, observing the RSSI, the RTT and the Time off that a typical user experiences during a set of streaming sessions in a concrete geographical place can indicate the alarms that will be sent to his terminal for improving his experience.

This cross layer strategy can be used with any current wireless technology and 4G technology because we only consider parameters that can be measured for all of them. The test of this cross layer strategy produces very good results in WiFi and HSDPA technologies.

The rest of this book chapter is organized as follows. In section 2 we outline the video streaming services in 4G networks. Section 3 is devoted to present our cross layer ideas and a mechanism to combine parameters of different levels of the network architecture to optimize the cross layer solution for video streaming. We present our further research directions in section 4. Finally, we summarize our conclusions and present additional readings for those interested in this research area.

2 VIDEO STREAMING SERVICES IN 4G NETWORKS

The evolution of wireless networking, multimedia and Internet in the past 10 years has been dramatic and very successful. Moreover, recent years also have produced a convergence of services and technologies in the field of telecommunications and computer communication. For example, telephony system can be implemented in a computer connected to Internet and computers are dedicated to control cellular telephony networks. The result of this convergence is a plethora of new telecommunications systems, more powerful multimedia services that meet the user experience and new techniques of communication and computing technologies such as Peer-to-Peer (P2P) and Grid Computing. These techniques can be used efficiently in the environment of multimedia services and 4G networks.

At the beginning of the twenty-first century basically three types of multimedia services were proposed, such as those that could have a high applicability (traditional niches of multimedia application): On line learning (Kurbel, 2003), Telemedicine (Murthy & Krishnamurthy, 2004) and e- commerce (Ghinea & Angelides, 2004) at any place and time. But nowadays, the evolution of wireless, multimedia and Internet has produced a big amount of firm real time multimedia services that are all based on media streaming. For example, Zhang & Mao (2008) presented wireless broadband multimedia communications such as mobile interactive Television, wireless video sensors, and multiparty audio conferencing as the current challenges to be implemented efficiently in the 4G wireless networks concept. As an example, we comment on the following types of services that are based on the media streaming technique and will impose some requirements to be achieved in 4G wireless networks:

- Video and music sharing applications. These applications are based on P2P technology. A lot of users can share their files in their local terminal memory (laptop or even a mobile telephone). The traffic generated by these applications represents a high percent of the Internet traffic nowadays. The challenge and new trends in these specific applications is to allow the creation of a Wireless Spontaneous Network (Silva & Salgado, 2006) that can allow to members of a Social Network to share instantaneously multimedia information when they

will meet in a common geographical area. To our knowledge there are types of these networks that only allow sharing profiles of users and short messages (see the Web site of Miraveo enterprise).
- Podcasting, vodcasting and broadcatching. The label <enclosure> of Really Simple Syndication (RSS) allows enclosing an audio file (podcast) or a video (vodcast) to be downloaded using traditional client access to Web servers. Broadcatching is the combination of RSS and P2P. The idea is to use an MP3 or MP4 gadget using a WNIC to access to this information; it would be downloaded in an area where there was a good level of coverage and then it would be used in a disconnected way or accessed using media streaming techniques. The challenge here is to design appropriate software in the gadget to allow it to use media streaming (needed for big music or video files, for example).
- Wireless multimedia streaming in sensor networks. The state of the art in wireless sensors allows them to deliver audio and video using the media streaming technique. This multihop network can be applied to surveillance. There are several challenges in these kinds of services, from low power coding, to routing algorithms, to adapting the existing ones to the specific characteristics of sensors. These kind of networks will explode in the next recent future taking into account that modern mobile telephones and new laptops normally include several sensors. They can naturally be deployed inside a Wireless Spontaneous Network for implementing a wireless game or a surveillance network.
- Internet radio and interactive television. Nowadays several ways to watch television or to listen radio in the mobile telephone are possible: a) including an analogical antenna, b) including a digital antenna and using one of the following technologies, among others: Digital Video Broadcasting-Handheld (DVB-H), Satellite-Digital Multimedia Broadcasting (S-DMB), Advanced Television Systems Committee Mobile/Handheld (ATSC M/H), One second (1seg) … c) Taking advantage of the HSDPA and WiFi antennas to access an Internet audio or video server. This option is the typical Internet multimedia access services using mobile telephones. It is still in its infancy the interactive digital television accessible from a mobile telephone. This service is also named Mobile Streaming Live Broadcasting Service.
- Multiparty audio and video conferencing. Cellular telephony networks offer multiparty audio conference and to a lesser extent for video. Both the WLAN as LTE will efficiently be able to handle these two types of services. In the WLAN would be very interesting to bear these services both in ad hoc mode and in infrastructure mode.
- Multimedia Content Management System (MCMS). Broadcasting systems with WiFi and Universal Mobile Telecommunications System (UMTS) (Ganz & Ganz, 2004) mobile telephones are revolutionizing the market for digital television and CMS. The idea is to distribute instantly multimedia content that is uploaded to a CMS by a mobile telephone directly. Real time live streaming sessions or VoD sessions can be accessed by other terminals that in this way can access multimedia information instantly. Examples of enterprises that define this new service are: Qik, Kyte, Upstream.tv and FlixWagon (see the Web site of these enterprises). This concept will be efficiently deployed in an ad hoc or a mesh network to define efficiently new MCMS local services (only accessible in the local area in where the network is defined). For example, these services could be implemented

for a local m-commerce within a shopping centre.
- Multimedia Collaborative Web Servers Applications. Typical groupwares allow a set of users to collaborate using the services provided by a cooperative server. Traditional services are: File and agenda sharing, community e-mail, messaging, forums, etc. Recently some authors think that also VoIP will be included in this kind of server (Yankelovich & Horan, 2009), and also video (VoD server) for augmenting the quality of the cooperation. The significance of this is that mobile telephones need to redefine the existing frameworks allowing the implementation of such applications on them. The nicotine poisoning Web portal we have implemented is in that line.

Therefore, it will be foreseeable in the next years to find new application niches of for 4G wireless networks in the area of media streaming. We classify them into two sets: a) The set in which the user must access Internet multimedia content when it is in a place where a fixed access network can not be implemented, and b) the set in which users can connect to local accessible gadgets, mobile telephones or computers using only a wireless network. For the first set, application areas of these services are wireless gaming, medical tele-care, on line learning, and so on. For the second set, application areas are: Wireless gaming using traditional commercial gaming consoles with WiFi antennas, home networking (Dixit &Prasad, 2008; Panton, 2008), and Wireless Spontaneous Networks.

From now on, we will suppose the following cases: a) a mobile terminal is accessing via Internet multimedia information; we only will observe the problems in the wireless last mile. b) The mobile terminal is in an ad hoc network and will observe the problems in the multihop communications.

2.1 The Video Streaming in Pictures

Media streaming techniques can be applied to a big part of the above multimedia services. That is, media streaming techniques can be applied to interactive audio and video, teleconferencing, and gaming services, all of them having an extremely low delay tolerance. Streaming media techniques can also be applied to non interactive applications like Internet radio with a large delay tolerance. Medium delay tolerance like Video on Demand (VoD) applications also can use this technique. Due to the massive use of the above services, the streaming technique is a key element on multimedia communications. In (Chou, 2007) a good technical introduction to the streaming technique for VoD and life broadcasting can be found. However, in this section we will briefly review the main ideas of the media streaming providing an alternative vision.

Media streaming is a technique used to visualize multimedia information (stored in a VoD server) while another part of the media object is being downloaded and no lasting information is stored in the Client device (only a buffering containing a few amount of information). As can be observed, three different parts can be defined: the Client terminal, normally issues commands similar to that of an old cassette tape, the Network (that can be anyone physical or a virtual one defined in terms of an overlay, or P2P, for example), and the Server that is supposed acts like an old cassette tape receiving the Client commands and returning the appropriate multimedia information. Firstly the Client must issue a connection command, and then the Server will answer with an OK signaling response. After this the Client issues a SETUP command for establishing one streaming session. When no new information is required by the Client, a TEARDOWN command is issued indicating the session must be finished. The Server and the Client must maintain a buffer for each stream and session and several active data structures while the information is communicated and new commands are issued from the Client, and they are freed when the session is closed. Due to the Client only needs a very small buffer, this technique can be applied to any kind of gadget and it is very appropriate for MP3, MP4, mobile telephones, gaming gadgets, etc. In Figure 2 (a) is shown an schematic example of the different parts in which some units of multimedia information (packets) have been named with numbers between 1 and n. In Figure 2 (b) is shown a temporary diagram which shows what packet is processed or communicated in each of the main resources: buffers in the Client and Server, the Network (considered as a single resource), the secondary Server's memory (disk), and the screen in the Client. Finally let us observe the overlapping of computation and communication packets.

There are well established commercial protocols: Microsoft© Media Services (MMS) with a reduced implementation for mobile telephones, Real Networks proprietary protocols with a great implementation on mobile telephones, Red5 (Allen & Arnold, 2008) that has not been fully implemented, Flash Media Server (Sanders, 2008), HiperText Transfer Protocol (HTTP), using the interleaved file format for lightweight playback over HTTP proposed as a technology named Hotmedia (Kumar & Lipscomb, 2001) that is still in use in the IBM Moving Picture Experts Group (MPEG-4) toolkit project.

2.2 Challenges of Video Streaming in 4G Networks

As noted in the introduction section, apart of the typical problems of wireless networks, there are other challenges associated to 4G networks and their terminals and software that we will briefly review in this section:

Figure 2. a) Definition of the main parts of a client-server media streaming system. b) a time-resource diagram of overlapped actions

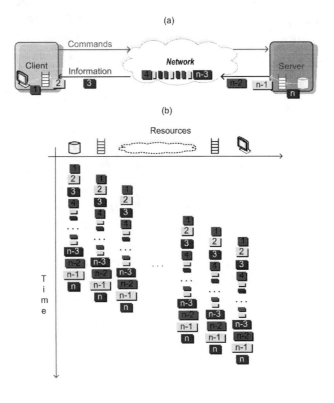

- The challenge of software deployment in mobile telephones and gadgets. The design and implementation of light applications on mobile telephones is tedious in some cases. In current market an application that runs in one mobile telephone does not run in another mobile telephone of the same firm but different model due to the heterogeneous hardware architecture and no efficient multiplatform language to program it.
- The jungle of media Servers. Not all the commercial media servers can stream media information for mobile telephones. While RealNetworks Multimedia Server is a standard for certain kind of mobiles, Darwin Streaming Server is the one for others, and Windows Media Server using new Silverligth technology and Flash Media Server are still in its infancy for mobiles telephones. The free software initiative named VideoLAN is also a serious alternative (especially for mobiles like iPhone). This supposes a problem with software design because it is impossible to obtain a real multiplatform solution for all mobile devices: each server defines its own programming framework in order to design the Client visualization part (player) to run in the mobile telephone.
- Portable players and frameworks. Frameworks like Mobile Media Application Programming Interface (MMAPI) for designing media streaming compliant with RealNetworks are intended to generate multiplatform players but also present problems (Goyal, 2006). An attempt done in (Belda, 2006) to adapt Java Media Framework (JMF) for mobile telephones has also some limitations. The roadmap of Freedom for Media in Java (FMJ) does not include mobile telephones architectures. A very promising scripting based framework over Java 2 Micro Edition (J2ME) named JavaFX 1.0 for mobile has been recently released and can work with Groovy, Grails and Android. Up to our knowledge it does not still work in commercial mobile telephones.
- The Internet streaming control protocol for mobile terminals. Each Server uses its own streaming control protocol and sometimes a proprietary data communication protocol. For this reason the design of a universal mobile player is a very complex task. Another problem is that typical protocols that run in fixed computers do not work in mobile telephones, for example, up to our knowledge the above IBM MPEG-4 framework can not be used in mobile telephones.
- The Internet transport level disruptions problem. In (Daniel & Macedo, 2008) can be found a recent survey of problems that arise in the context of Internet main protocols for gadgets and mobile telephones. Usually, the above servers use the User Datagram Protocol/Real Time Protocol (UDP/RTP) for data communication or a variation of it. In particular cases, they use a congestion control mechanism like Transmission Control Protocol Friendly Rate Control (TFRC) (Handley & Floyd, 2003). In the first case, there must be arranged an application level mechanism to detect the service disruptions. We have not detected the use of Datagram Congestion Control Protocol (DCCP) (Kholer & Handley, 2006) or Stream Control Transmission Protocol (SCTP) (Stewart & Xie, 2000) in the above servers. In the second case, reliable sockets can also be used to support long disruptions: The tests we have done using reliable sockets (Ansari & Sathyanath, 2007; Zandy & Miller, 2002) do not produce the expected results because there is no way to control the abort of the transport sockets in the Client and

Server when they are connected in ad hoc manner. Also when they are connected using a WiFi Access Point (AP) there are still some random failures with the implementation of reliable sockets (Moreno, 2009).
- The harmful wireless channel disruptions problem. Frequently a terminal can experiment sporadic disruptions due to the loss of coverage intermittently and due to unpredictably radio channel condition varies along time and space. Indeed, the same place can present different behaviors maintaining the same physical structure but varying the number of active terminals and weather conditions.

In Figure 3.a it is shown a terminal that sporadically experiments streaming disruptions while accessing to an Internet Server. Let us note that inside the coverage area also there can be zones with no radio coverage provoking the terminal suddenly could loss the coverage. The disruptions in Figure 3 occur if the mobile telephone will go out of the coverage area defined by its partner in the ad hoc network. But also this disruption occurs while a new network path is chosen due to a broken path between the Client and the Server. In any case, if the Server continues sending information while the Client is out of coverage, the wireless channel bandwidth will be inefficiently wasted. For PC and laptops (nor for mobile telephones), commercial media Server can detect the availability of the Client and try to reconnect it during 2 minutes, but other widely used free software media servers, like VideoLAN does not detect this availability.

We think it is important: The Server controls the availability of the mobile telephone using a clear mechanism (to the Client and the Server), the Client recovers the lost multimedia frames while it experiences disruptions providing an appropriate user experience.

2.3 The Software Architecture to Solve Problems

To meet the challenges of media streaming on 4G networks it is necessary to consider the following points:

Figure 3. a) Mobile telephone accessing Internet Streaming Server. B) Ad hoc Spontaneous Wireless Network with mobile telephones sharing a streaming session

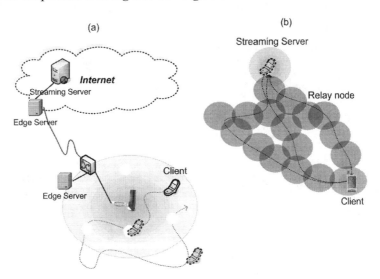

- It is unrealistic to propose the modifications to the Client and streaming media Server because they are commercial software and generally they don't allow third parties to modify them. In the case of free software it is very difficult to implement an efficient solution due to it is very time consuming task.
- Buffering policy on these clients and servers is usually a well-kept secret because it lies much of its performance. Therefore, it is not possible to propose changes in a realistic manner.
- The mechanism the Server uses to control the availability of the Client is often a closely guarded secret and is not likely to be changed.
- Sometimes servers use proprietary encapsulated data and multimedia data transport protocols which difficult the design of modifications due to it is very difficult to understand the protocol and encapsulation.

Proxy based architecture can be used to skip the above points. The proxies process only the signaling commands of the standard streaming protocols. In this way our architecture can be applied more generally to any type of Client and Server (whether commercial or not). Let us note that we can adapt our solution to any kind of data and commands protocols used by the streaming Client and Server and any programming platform they use.

In Figure 4 is showed a general diagram of this architecture. Let us note the components: Proxy Client and the Proxy Server. In addition to the channels of communication of commands and data we consider a new channel between the proxies, for signaling the Client availability (signaling channel). The Proxy Server is in charge to buffer the data while the Client is not available and to issue that data to the Client when it will be available again. In this way the Client can always recover the data it would loss.

As mentioned in the introductory section, a proxy-based solution is not efficient by itself. This architecture should be integrated as part of a cross layer strategy or it should take the value of parameters calculated with the cross layer strategy.

3 OUR CROSS LAYER STRATEGY FOR EFFICIENT VIDEO STREAMING

In this section we first present the state of the art in cross layer techniques. Then we will present our cross layer strategies. Finally a set of application level solutions that takes advantage of the cross layer strategies will be presented.

3.1 State of the Art in Cross Layer Techniques for Video Streaming

In Lin and Wong (2009), the IEEE802.11e and IEEE802.11n Physical and MAC level adaptation

Figure 4. General software architecture to solve the problems of wireless media streaming on 4G networks

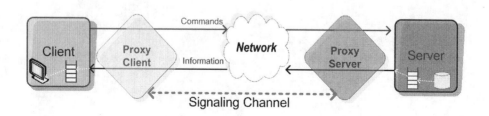

to support high throughput and cross layer design with higher layers are reviewed. Examples of adaptive schemes are: Adaptive carrier sensing range, dynamic tuning of the backoff process, frame aggregation adaptation, and cross layer with TCP.

In Schaar (2007), the cross layer problem for efficiently implementing multimedia applications over wireless networks is formalized presenting several existing solutions. For example they show the impact on throughput efficiency and delay for video streaming in an IEEE802.11e network as an example of performance of the cross layer technique. They defined formally the cross layer technique and classify the current different approaches: Top-down, bottom-up, application-centric, MAC-centric (adaptive retransmission using queuing theory) and integrated (using classification and machine learning techniques).

In Cho and Radha (2008), it is presented a method to obtain the media streaming optimal rate utilizing a rate-distortion model of the video source for the rates below the channel capacity, the distribution of channel prediction error process and the video rate quality/rate distortion model for rates exceeding the channel capacity. The problem with this work is that the model is static and can not be adapted to sporadic spatial-temporal variations of the channel, but in general is a good work. A similar work is presented in Choi and Yoo (2008). They propose a distortion measure of H.264 slice and a cross layer Unequal Error Protection scheme using the proposed distortion measure. Moreover they use a MAC-Level Automatic Retransmission request scheme in their cross layer scheme to improve transmission efficiency. Their simulation shows an improvement of the Equal Error Protection scheme. These two works present simulations of their proposals, because it is very difficult to modify these schemes in the IEEE802.11 for which they simulate.

The above methods are simulated providing good results, but it is not real to change some parameters of the link and physical level if they are not proposed by the manufacturer. We think it is better to propose practical solutions that can be experimentally tested, like our cross layer strategies.

A very interesting idea is presented in Djama and Ahmed (2008) that describes a new architecture for cross layer and experimental results. The idea is to observe, on one hand the network instantaneous conditions like link status, network status and transport level status, and on the other hand user and terminal triggered events that impose new quality adaptation events for the required Quality of Service (QoS) the user mark. They use MPEG-21 to provide a common support for implementing and managing the end-to-end QoS of audio/video streams. The authors show the benefits of their cross layer technique showing a good behaviour of the Peak Signal to Noise Ratio and Structural Similarity Index metrics. Our cross-layer technique is similar to this, except that we do not consider the specification of the user specific QoS required. Furthermore, our goal is to adapt to different control strategies at the application level for anticipating information about disconnections.

In Inés and Fujikawa, 2008) a preliminary work is presented in which the authors perceive the wireless channel parameters like: Link quality, signal strength at receiver (dBm), noise (no packet) level at the receiver (dBm), number of discarded packets due to different causes (invalid network id, unable to decrypt or fragment). They define quality states in order to the application can adapt their behaviour changing the quality of image, increasing or decreasing the frame rate, and activating or deactivating the colour information. They present their ideas and still have no results of their methodology. This is an interesting proposal which follows the trail of our cross layer strategies.

3.2 The Case of WiFi Infrastructure Access Network

We have developed a strategy to proactively estimate the quality of streaming video reception in a WiFi cell that a Client in motion receives from a VoD or a life broadcasting. The Client computer is connected to a WiFi AP to access the Internet Server. For implementing this strategy we deploy a RTSP/RTP filter based architecture that uses an ascendant cross layer technique to receive information from the physical and link levels. The main characteristic of our strategy is that our software is not intrusive to the network traffic because it generates reduced control traffic and requires little memory and low battery consumption.

The levels of congestion or poor radio coverage are detected analyzing a set of values of the recently observed RTT processing the UDP packets the mobile device has sent. These RTT values are mapped into three discrete values (0, 1 and 2). Depending on the average value of RTT a pre-alert state would be triggered. This state indicates the mobile device must be vigilant about the traffic it sniffs in case a degradation of the above state will be detected.

Detected a pre-alert, we try to distinguish a case of poor level of coverage or a high degree of congestion. To estimate the level of coverage we measure the following parameters of the physical level: RSSI, noise level and the PN Code Correlation Strength from the answer given by the driver that supports the Wireless Extensions API (WE) of Linux. To estimate the congestion at the network, we assess the global throughput (Tg) of the network, and the throughput (Tf), the rate of packets received (Tpr) and lost (Tpd) per each stream. To observe alert states thresholds are defined as follows: UTg, UTf, UTpr and UTpd whose values are dynamically updated every 3 minutes and estimated each second.

The following algorithm activates or deactivates an alert state and a possible situation of congestion is detected to be communicated to the disruptions control software at application level.

1. Calculate the average of the last RTT set. Let be RTTm this value.
2. Let be URTT = 0.5 a threshold for the values calculated for the RTT set using experimentation.
3. If RTTm >= URTT, Then CRTT = CRTT + 1 where CRTT is a counter that is updated each time RTT is calculated. If (CRTT reaches 10 after 1 second) Then the terminal switches to Pre-alert state. Considering the values of Tg, Tf, Tpr and Tpd:
    3.1. Distinguish the traffic used by the streaming Client (which should be alerted) from the rest of traffic.
    3.2. If (Tg, Tf, Tpr and Tpd) could be measured Then activate the alert state when the following restrictions are true:
    Tf< % Utf (95%)
    Tpd> %Utpd (3%)
    Tpr< % Utpr (95%)
    Else If (Tpd > %Tpr (20%)) Then activate alert state (packets are lost)
4. If (RTTm < URTT) Then CRTT = CRTT - 1. If (CRTT reaches 0) Then cancel the Pre-alert state.

On a practical level the following factors are taken into account:

- We do not use the Internet Control Message Protocol (ICMP), because if there is a firewall in the AP, it will not generate a response to the ICMP packets. Instead we use an architecture manager/agent, the manager installed in the AP and the agent in each WiFi device (Marrero & Macías, 2008).
- The values of Tg, Tf, Tpr and Tpd should not be obtained while in pre-alert state. If a Variable Bit Rate (VBR) compresor-decompresor (codec) is used then Tpd and

Tpr will indicate whether the stream would be received correctly (Tg and Tf values are unstable). To calculate Tf, Tpd and Tpr we consider the fragmentation of the IP datagrams.
- The various tests we conducted in a real environment show the reliability of this mechanism. We moved the mobile device along different zones: to an area of low coverage, no coverage and high level of coverage. We saturated the channel sending traffic establishing several streaming sessions simultaneously. In all cases we obtained a low overhead, low battery consumption and efficient use of the processor which determined the efficiency of this mechanism.

Figure 5 shows an example of different pre-alert and alert states to show the reliability of the mechanism. In the vertical axis is shown the mean normalized RTT values calculated from the discretized ones in the range 0 ... 2. For the horizontal axis we used the following notation: Pai is the i-th pre-alert, Pai is the cancelation of the i-th pre-alert, and Aliva is the i-th alert signalling that v (video) and/or a (audio) streams are affected.

3.3 The Case of Ad Hoc WiFi Network

Mobile ad hoc networks are prone by nature to path breaks and reconnections. Control of disrupting streaming sessions is challenging in ad hoc wireless networks. Moreover, it is very interesting to solve this problem due to the applicability of these types of networks to new scenarios for streaming in 4G networks: Spontaneous Wireless Networks and Wireless Sensor Video Networks.

We consider streaming sessions that use RTSP, stand-alone RTP or HTTP in presence of disconnections and reconnections and managing long disconnections and the TCP connection failures. We take into account UDP-based streaming video but also TCP-based in scenarios where both the Client and the Server are mobiles terminals interconnected by a physical path based on relay terminals (Fig. 3.b). In this scenario, path breaks take place frequently and quickly due to the movement of end nodes or intermediate nodes in the path. We use the Optimized Link State Routing Protocol (OLSR) (Clausen & Jacquet, 2003) for the quick reconfiguration of path breaks and to control the Client's availability and to announce the UDP services.

We consider the following software components in our strategy: an OLSR component and a Plugin for OLSR component installed in all kinds of nodes, a TCP control component, the Streaming command control protocol component and an Availability signalling component (installed in the Client and Server devices not affecting to the media Client and Server code).

The OLSR component lets use the OLSR optimized flooding mechanism to send information, routing related or not, from the application level using an OLSR plugin component. We just use this property to inject user defined packets (OLSR packets type 200) in the network. The OLSR plugin component on the Server, Client and intermediate nodes conveys to OLSR component the information to be sent into OLSR packets type 200.

The signalling channel for availability between the Client and the Server is implemented by the Availability signalling component. Since both the Client and the Server are mobile, this component is in charge of detecting if there is, or is not, a path between the Server and Client. In case the Server or the Client are not available some corrective actions that depend on the type of video streaming being served (VoD or live video) and the type of streaming protocol used to transport the data (RTSP, HTTP or RTP) are issued or signalled to the application level mechanism.

When establishing one TCP connection for a RTSP or HTTP session, three TCP connections are created: one between the Client and its Availability signalling component, another to communicate the Client's Availability signalling component

Figure 5. Examples of different pre-alert and alerts of our Internet access cross layer strategy

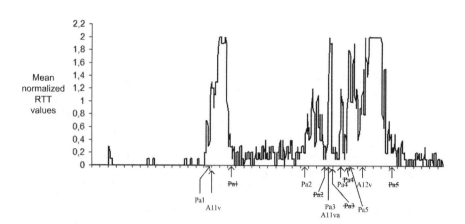

and the one in the Server, and the last one for the communication between the Availability signalling component of the Server and the Server. The former and the latter are local connections and they are not likely to fail during disconnections. The TCP Control component is the mechanism for transparently detecting TCP connection failures and to create a new TCP connection that avoids the streaming session release. Despite the fact there are enhancements to TCP for ad hoc wireless networks, we do not consider them for the reasons we explain in Macías and Suárez (2007) and moreover, using the official version of TCP we assure interoperability. The mechanism consists in four phases: a) Identify the TCP connection between Availability signalling components, b) session communications (the HTTP or RTSP session begins normal communication), c) disconnection detection. In this case, if the Client is out of coverage, then its Availability signalling component does the following: 1) Closes the TCP connection with the Server Availability signalling component, 2) detects the server's availability, 3) creates a new TCP connection with the Server Availability signalling component, 4) sends the new descriptor of the TCP connection and waits its acknowledge. If the server is out of coverage, it must order all Availability signalling components of the clients to close the TCP connections and create new TCP connections. For doing that, Server Availability signalling component broadcasts a message using the packet OLSR type 200 including in the message's content of the packet the command TCPFall to convey all the client agents that they must do the above steps 1 to 4, and d) ends sessions with long disconnections.

Basically we implement an indirect signalling channel with the help of OLSR. We implement a time out strategy that is specified by the user when it starts the OLSR plugin component in the Client. The specification of the best value for this time out is only a matter of knowing the physical environment in which the ad hoc will be deployed and the people that use them (mobile profiling). In this case, the application level scheme must receive information about the disconnections and reconnections in order it to infer possible behaviours of the network. Let us note that this will only take advantage in the case the ad hoc network repeats the connection pattern several times (in the same of consecutive days).

With our strategy we get the following:

- A system to control the Client's availability in order to avoid the sending of frames while it is temporarily disconnected.

- A system to announce the UDP services provided by the Server in order to let the clients request UDP streams on demand. Otherwise, the server can not control the Client's availability due to the connectionless nature of UDP.
- A mechanism to pause the VoD of ongoing RTSP sessions to save computing resources and battery on the server while the Client is disconnected. We are not concerned about the losing of frames during a disconnection for live video sessions or VoD over HTTP or RTP, i.e., streams that can not be paused with RTSP commands.
- A mechanism for transparently detecting TCP connection failures and to create a new TCP connection that avoids the streaming session release and to end sessions with clients disconnected for a long time that use TCP to transport data or commands (i.e. RTSP and HTTP).

We made a lot of tests and obtain the following general results: low overhead of the OLSR packets flooding including our packets type 200 in comparison with the data volume of the streaming, low usage of the processor and battery consumption due to the Availability signalling components, acceptable delay and jitter, and good benefits of signalling mechanism.

As an example, we show in Figure 6 the low overhead of the traffic due to OLSR messages compared with the streaming traffic in the network.

3.4 Application Level Adaptation Schemes Using Our Strategy

Once we explained the two strategies for cross layer information on the physical and link levels (for ad hoc networks using a mechanism based on time), now we present two schemes to take this information and make: a) an application that directly obtains benefits using the information generated by the previous strategies, b) the design of a middleware-based mechanism that enables rapid programming of applications.

Another possibility is based on a programming language that allows using information generated by our strategies directly through language primitives. The problem with this scheme is that it can not be directly implemented in mobile telephones due to the language and framework technologies are still not mature. We established the main ideas of this approximation in (Galván & Quintana, 2005). We used Aspect Oriented Programming (AOP) (Kiczales & Lamping, 1997) to rapidly and easily transform single user version video streaming Client and Server application to obtain the corresponding distributed one, in order to adapt its execution on wireless networks. This is done using refactorization technique. Refactorizing an existing middleware allow us to isolate some wireless concerns like channel failures, range control and spurious disconnections that can be connected using AOP in order to streaming video software support service disruptions. These concerns are represented by objects. At compilation time, the source code is woven using an aspect language with some objects that anticipate the wireless channel behaviour to avoid the consequences due to spurious disconnections. To save memory we avoid using Dynamic Link Libraries (DLLs) that generally must be adapted for these devices. In this way the functionality of the final application can be generated rapidly and efficiently for a lot of different heterogeneous terminal devices. Our cross layers strategies can be used for anticipating wireless channel behaviour.

The benefits of this refactorization of the code of software components of Fig. 4, and our cross layer components are: a) the application could choose an object (concern) of the middleware just as it would need in compilation time. For example, in one application the wireless channel must be controlled, but in other application this probably will not be used on demand. b) The resulting software system is very light because it only will

Figure 6. Overhead of our ad hoc strategy respecting the overall traffic

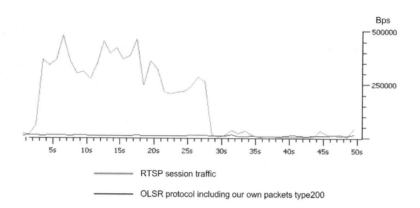

include code of the strictly main functionality. This is a clear advantage when programming wireless terminals for whose heavy DLL must be transformed into light ones, because heavy DLL can not be mapped in short memory. c) The reusability and flexibility of applications improve considerably. That is to say, we can change rapidly a concern in order to program the same application for a new kind of wireless device only changing the weaving of concerns and programming new objects (concern) in case it will be necessary.

In recent papers like Malte and Robert (2008) a similar approach is achieved but instead they refer the AOP as Context Awareness Programming.

3.4.1 A Simple Proxy Based Protocol for Mobile Telephones

This is the case we design applications that directly use information generated by our strategies. The main idea here is to program the Proxy Server and Proxy Client of the Fig. 4 such that they take advantage of the results of our Internet access cross layer mechanism. Both proxies cooperate in recovering automatically the streaming session in case the communication disruption lasts long time. The Proxy Server always proactively stores the multimedia information the Server sends to it. An important tip in this approximation is to provide buffer management technique implementation to manage the possible loss of data due to disruptions using the proxy based architecture taking into account the information provided by our Internet access cross layer mechanism.

We implemented the Client using the MMAPI which condition us to use the RealNetworks programming framework to program the proxies. Both proxies have a symmetric implementation based on similar modules which are: The RTSP Command Manager (RCM), Real Data Transport (RDT) Manager (RM), Buffer Manager (BM) and Disruptions Control Manager (DCM). The RCM in the Proxy Client manages the connection from the Player, initiates the clear connection negotiation (obtaining the numbers of ports used for RTSP) with the Proxy Server and echoes the RTSP commands to the Proxy Server (ports used for RTSP are also obtained during this process). The RM is controlled by the RCM and echoes the multimedia data information; in the Proxy Server, this component also controls the DCM. The BM in the Proxy Client simply receives RDT packets from the Proxy Server, stores them in its buffer and forwards them all at constant speed (intentionally) to the Player. In the Proxy Server the stored packets are sent to the Client at a higher speed than those received from the Server, in order to improve the control of disruptions. The Player Connection Control (PCC) in the Proxy Client receives indications of its DCM indicating if it must or not stop the Player. This command is received by the Proxy Client Connection Control (PCCC) in the Player. The Server Connection Control (SCC) in the Proxy Server does the same operation but with the Server. The visualizer in the Player is in charge of playing the multimedia information and presenting it to the user.

The buffering method consists of basically in the management of packets windows sent from the Proxy Server to the Proxy Client. The Proxy Server receives a continuous flow of RDT packets from the Server, it stores them all in its buffer and when it has received a set of RDT packets sequentially (packets window) it will send them to the Proxy Client waiting for an acknowledge. If it receives this acknowledge it will repeat the process with a new packets window (after a certain amount of time); else it will repeat the last packets windows sent. These actions are also overlapped with RDT packets reception from the Server and consequently stored in the buffer.

Both buffers in the proxies cooperate in reducing the effects of the sporadic and short disruptions. Depending on the computation power of the mobile telephone, Proxy Client will store these packets in its buffer and the Player will consume it at a certain speed (normally less than the one of the computer that executes the Proxy Server). Hence, the buffers of the proxies may be full but the Proxy Server buffer can accumulate more RDT packets than the Proxy Client. Moreover, the Proxy Client buffer must be limited as we stated before. This means that a sporadic or short disruption is solved rapidly because the Proxy Server will inject at a high speed the RDT packets in the WiFi link and the mobile telephone will have time to consume its buffer and receive the new packets. But this can not occur for disruptions that last much more than one second, because the mobile telephone can consume all the packets in its buffer and will not receive new packets for a long time. In this case the Proxy Server will send a PAUSE RTSP command to the Server (that will not teardown the streaming session), and the Proxy Client will also pause the Player. In parallel a message is shown to the user indicating that the wireless channel would experience problems and urging the user to move to another place where the channel quality is better. If this situation lasts for more than a minute the Proxies will restart the Server and Player. These last actions can be anticipated by the Internet access cross layer mechanism. It is important to anticipate this information because the warning to the user can avoid a bad experience.

In Figure 7 we show two example screenshots of the Graphic User Interface (GUI) of the Player, which is a standard Real Player with several additional messages indicating to the user the situation of the disruptions and forgetting him to use RTSP commands when the mobile telephone is not available.

Figure 7. Two images of a mobile telephone showing the corrective actions in the player

3.4.2 An Autonomic Solution

This is the case we consider a middleware based on software agents to use the information generated by our Internet access cross layer strategy in order to allow the rapid and efficient design and implementation of video streaming applications. The main idea is to use a software agent framework that could be deployed and executed in commercial mobile telephones. The proxies of Fig. 4 will be implemented in this framework. In this way, we ease the proxies programming getting a better time to market. This is because we only must specify their behaviour in term of framework primitives. The rest of controls are achieved by the framework itself, for example the availability of the Client. The behaviour will explain the different action to be done in term of the different kinds of disruptions we mention in the section 3.4.1. The buffering of packets windows is also done by the agents based framework.

At the moment there are several platforms enabling the deployment of software agents, both proprietary and free software. We impose the use of a free software consolidated agents platform implementable on mobile devices in a language that could be compiled on them. The platform that meets these requirements and has been used is the Java Agent DEvelopment framework-Lightweight Extensible Agent Platform (JADE-LEAP) (Bellifemine & Caire, 2003; Caire & Pieri, 2009). JADE is a completely open software platform implemented in Java. It is used for the development of agent-based distributed applications that meet the specification Foundation for Intelligent Physical Agents (FIPA). To enable the implementation of JADE on mobile telephones it was associated with a project called LEAP. It replaces some of the core JADE but does not change its API. It can be deployed in devices compatible with Java 2 Standard Edition (J2SE) from the Java Development Kit (JDK) 1.4 and J2ME from Mobile Information Device Profile (MIDP) 1.0 which opens the range of potential users.

To improve the deployment of JADE-LEAP on J2ME it is done in Split mode. This means that the container that displays the agent is divided into a FrontEnd in the mobile device, and a backend on Server, which are permanently connected. This achieves a better computational load balancing. Another benefit is an optimized use of the wireless connection between these two components.

JADE-LEAP implements a mechanism to handle disconnections of the FrontEnd and the BackEnd. Messages sent to mobile phone when it is out of coverage are stored in a buffer of the BackEnd, and they are sent when reconnecting. The FrontEnd attempts to reconnect automatically on a regular basis with the BackEnd. The concrete benefit of this mechanism is that the programmer should not make any extra effort to design a control mechanism and no disconnections deployment.

To deploy the disruptions control system the Proxy Client of Fig. 4 is split in two parts: the Agent Proxy Server (APS) and the Agent Proxy Client (APC). In addition, the APC will be divided into two parts, one to be executed in the mobile: The FrontEnd of the APC, and another running in a device housed in the fixed network (in the edge of Internet): The BackEnd of the APC. The server is connected by cable to the APS and this in turn with the APC through a WLAN. The links FrontEnd Server-APS and APC-Client messages are transmitted using RTSP, which uses the RTP for video communication. Between APS and the BackEnd of the APC, which are deployed within the same platform JADE, traffic FIPA Agent Communication Language (ACL) is communicated. RTSP requests and responses are encapsulated in messages FIPA ACL INFORM type and RTP packets PROPOSE message type. This is achieved by simplifying the communication between players avoiding to declare an ontology which would reduce the performance of the JADE platform.

We have shown that in practice this mechanism is efficient and enough to control the disruptions on mobile telephones. In Figure 8 it is shown a photograph of this system. At the lower side is shown the mobile phone and its display showing a video that the Client is playing. On the monitor of the laptop there is a window to the server in the upper right that shows the frame being sent. In the central part appears a succession of FIPA ACL messages exchanged between the APC and APS provided by the Sniffer utility of the platform JADE-LEAP. The first couple of INFORM encapsulates the request and the positive response of the order of the RTSP SETUP. The second pair of INFORM encapsulates the order of RTSP PLAY. The PROPOSE that comes after this encapsulates RTP packets each containing a frame, a Joint Picture Experts Group (JPEG) image.

3.5 Putting it All Together: The Nicotine Poisoning Web Portal in Pictures

The tele-medical care is a discipline to which can be applied ubiquitous wireless access obtaining spectacular results in terms of efficiency and effectiveness. Nicotine poisoning can be considered a technical in the tele-medical and clinical psychology care. To achieve high efficiency of the methods of nicotine poisoning is also important. In the literature, there are a large set of methods of nicotine poisoning that can be used via Web interfaces (all from fixed terminals and without itinerant wireless access). These servers only provide access to Client-Server following a disconnected operations model from fixed computers and very heavy Web interfaces. Although all of them provide subscribers, none takes this situation to provide information flow between patients who wish to share experiences among themselves and not permit them to share information between different therapists, thus limiting the effectiveness of the server greatly.

Smoking cessation methods that are applied through the clinical psychology usually consist of a series of activities planned for some time that the patient must perform a sequence followed by the day. Responsible for the implementation of these techniques is a therapist (psychologist specialized in these techniques). The therapist should check the activities are carried out every day because in that sequence is usually the success of the techniques applied. These techniques can be expensive to

Figure 8. An image of a mobile telephone and portable computer showing the corrective actions in the agent framework

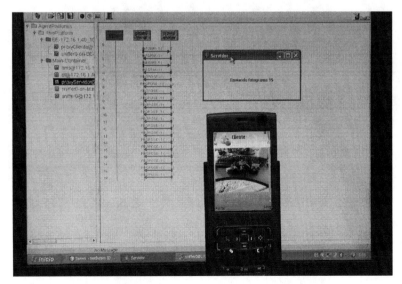

implement and therefore it is preferred to be apply to a group of people at once. This means having a management system for registered users, schedules and agendas both collective and individual. In addition, a therapist could publish a survey to be answered by the patients, to monitor whether the therapy is being implemented effectively. The results were reported after the trial of patients. This means that it must support asynchronous work through email, for example. Patients may also post through a forum or blog their achievements. They might even publish it in an instant by instant messaging system or comment on a common chat room or restricted to a group of them. This means you must also support the synchronous cooperative work. Finally it is very important to handle video streams stored therapists or patients could publish using a VoD. To support all of these characteristics has been chosen to deploy a Web application server called liferay, which had also installed the mercury e-mail server, the wildfire chat server and the helix video streaming server. All these components are free software.

The above software supports accesses by mobile telephones easily. Just one major problem we face is to access the Web using these devices; particularly from mobile telephones because not always available browsers provide efficient access to all types of Web pages.

To show the interactive mobile access to the Web site two photographs of a mobile telephone accessing the Web portal can be found in Figure 9. In Figure9 (a) is the personal agenda shown to the patient in order them to follow his particular therapy (a display has been used with a reduction factor of 50%). In Figure 9 (b) is shown the video player.

Figure 9. Two images of a mobile telephone showing the accessing to our Portal Web a) calendar, b) video player

(a)

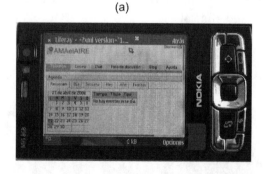

(b)

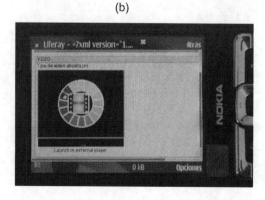

4 FUTURE RESEARCH DIRECTIONS

Media streaming is a key technique in modern multimedia applications, both in access to media servers from the Internet, as for the deployment of real time services. The implementation of these applications in 4G networks will be undoubtedly a very important area of research in the coming years.

The development of multimedia sensor networks is an area that will take ever more importance. New and exciting applications are foreseen in this field, as they solve the problems of location, efficient management of disconnections and deployment of the necessary software.

Wireless Spontaneous Networks will be of high interest among researchers and companies. The deployment of multimedia services on these networks in a given geographical location will be an interesting topic. For example, they can be deployed as an alternative to traditional cellular telephony and videoconferencing to avoid negative effects on failure. They will be used for medical tele-care located in a specific geographical location or they can be used for new types of wireless games.

High bandwidth wireless Internet access provides a wonderful opportunity to access the new Internet multimedia interactive services (radio, television, broadcatching, media content managers, etc.). This requires the development of new hardware based on a number of communication antennas, and defining standards of efficient communication. Some important efforts are currently done in this line.

It will also be necessary to develop new programming frameworks for wireless communication devices in order to support efficiently (at language level) physical aspects that surround them like battery consumption, wireless link and network behaviors, and so forth. In this area there is still a long way to go. For example, new technologies and multimedia frameworks such as multimedia design Silverligth, Flash and JavaFX should consider seriously the wireless devices efficient code generation. Besides the new devices and operating systems are revolutionizing the design of software for such devices (cell phones and gadgets for music, video and games). Some extra efforts will be needed in these topics. Especially taking into account the new terminals like iPhone and operating systems like Android that ease these tasks.

Finally, it will be necessary to provide a high quality service to the user mobility offered at an affordable price so that it can reach as many people as possible. It is necessary to control that service disruptions of mobile telephones will not affect (or at least have a small impact) in the user experience. In this field there will be much work to do to obtain an efficient cross layer solution.

Moreover, new marketing techniques should be investigated to achieve these services enter the market more than hitherto. One example is the marketing of Google telephone bearing Android installed.

In short, there is great scope in this area of research that is a multidisciplinary field in which professionals from various fields must bring their solutions.

5 CONCLUSION

In recent years there have been many developments in the context of wireless networks at different levels. At physical level the deployment of Multiple Input Multiple Output (MIMO) antennas, Ultra Wide Band (UWB) and Orthogonal Frequency-Division Multiple Access (OFDMA) have represented an improvement in the radio communications and control access to the wireless channel. The terminals have passed from primitive ones to a powerful gifted with powerful operating systems, great amount of internal memory and strong processors. However, the software development for these devices (especially for gadgets like MP3 and MP4) has not been deployed to be really multiplatform which represents a problem. Also the old protocols of Internet are not appropriate for these kinds of devices because they do not take into account the special characteristics of wireless networks and mobile terminals.

A problem that is important to be solved when the terminals use media streaming technique to communicate multimedia information is the sporadic disruptions in the service. Individual techniques are not efficient for this kind of network. The research community agrees that the efficient solutions for media streaming over 4G wireless networks will be given by cross layer techniques. These techniques combine different parameters at different layers to optimize the solution.

In this chapter we have presented two cross layer strategies for proactively inform the media streaming Client and Server about the state of the wireless channel and the link quality. For Internet wireless access networks (last mile) we presented an strategy consisting of to acquire in real time the values of physical and link level parameters to integrate an algorithm that inform application level about disruptions and the reason for this disruption. We differentiate physical disconnection from a link saturated events using a pre-alert and alert system. The application can take the corrective actions depending on application level criteria. For ad hoc networks we have presented a cross layer strategy that is able to recover in real time the disruptions when the path between the Client and the Server is broken using a time out programmed by the user and OLSR to inform the different node to discover another route from the Client to the Server.

To complete our cross layer strategy we design two solutions for Internet wireless access case at application level based in proxy architecture. The proxies can take advantage of the physical and link level parameters that our cross layer strategy optimizes to advance information about the physical channel and link status to the Client and Server. The first solution consists of designing a new protocol to discover the availability of the Client and a buffer strategy that supports different types of disruptions. The second one uses an open source software agents based framework to program the proxies. The behaviors of the agents are easily programmed; the buffering and Client availability is achieved by the agents' platform. We can obtain physical and link parameters values to proactively modify the behavior of the agents to adapt it to the wireless channel conditions. We used J2ME to program a mobile telephone for obtaining the performance in real experimentation. This is basically the difference with our method from others authors. To cope with the user experience we probed our cross layer method in a Web portal dedicated to interactive medical tele-care for the nicotine poisoning problem. The results obtained by our cross layer technique for the WiFi technology (we also has probed it on HSDPA technology with good results) and the different application level solution invite to optimism.

We think our strategies can be easily used in the future 4G wireless networks with minor or no changes because we consider the implementation of software only in the edge of the network (in the Client or final service provider side). It is assumed that the speed of 4G wireless networks is greater than or equal to that currently held by WiFi (54 Mpbs).

REFERENCES

Aggelou, G. (2009). Wireless Mesh Networking. New York: Mc Graw Hill.

Ahson, S., & Ilyas, M. (2008). RFID Handbook: Applications, Technology, Security, and Privacy. Boca Ratón, FL: CRC Press.

Allen, C., Arnold, W., Balkan, A., Cannasse, N., Grden, M., Gunesch, M., et al. (2008). The Essential Guide to Open Source Flash Development. Berkeley, CA: APRESS.

Ansari, F. & Sathyanath, A. (2007). STEM: seamless transport endpoint mobility. ACM Computing and Communications Review, 11(2), 1:13.

Austerberry, D. (2005). The Technology of Video and Audio Streaming. Oxford, UK: Focal Press.

Belda, A. (2006). Contribution to the multimedia communications in wireless networks for mobile devices (In Spanish). Unpublished doctoral dissertation, Technical University of Valencia, Spain.

Bellavista, P., Corradi, A., & Giannelli, C. (2005). Mobile Proxies for Proactive Buffering in Wireless Internet Multimedia Streaming. In IEEE International Conference on Distributed Computing Systems Workshop: (Vol. 3, pp. 297-304). Colummbus, OH: IEEE Press.

Bellifemine, F., Caire, G., Poggi, A., & Rimassa, G. (2003). JADE, A White Paper. *Journal of Telecom Italia Lab.*, *3*(3), 6–19.

Caire, G., & Pieri, F. (2009). LEAP USER GUI. Telecom Italia Lab. Retrieved Jannuary 30, 2009, from http://jade.tilab.com/doc/LEAPUserGuide.pdf.

Chakareski, J. & Chou, P. (2006). RaDiO edge: Rate-distortion Optimized Proxy- Driven Streaming from the Network Edge. IEEE/ACM Transactions on Networking, 14(6), 1302-1312.

Chan, F., Wong, T., & Chan, P. (2005). Lot streaming technique in job- shop environment, IEEE Mediterranean Conference on Control and Automation. (Vol. 1, pp. 364-369). Washington, DC: IEEE Press.

Cheng, K., & Marca, J. (Eds.). (2008). Mobile WiMAX. West Sussex, UK: John Wiley & Sons, Ltd.

Cho, Y., Radha, H., Yoo, J., & Hong, J. (2008). A rate-distortion empirical model for rate adaptive wireless scalable video. In International Conference in Information Sciences and Systems. (pp. 350- 355). Princeton, NJ: IEEE Press.

Choi, J. Y., Yoo, J. J., & Shin, J. (2008). Cross-Layer Transmission Scheme for Wireless H.264 Using Distortion Measure and MAC-Level Error-Control. In International Conference in Advanced Communication Technology. (pp. 1232-1237). Gangwon-Do, South Korea: IEEE Press.

Chou, P. A. (2007). Streaming Media on Demand and Live Broadcast. In M. Van der Schaar & P. A. Chou (Ed.), Multimedia over IP and Wireless Networks: Compression, Networking, and Systems (pp. 453-502). London: Academic Press.

Clausen, T. H., & Jacquet, P. (2003). Optimized Link State Routing Protocol. Request for Comments: 3626, Standard Track, The Internet Engineering Task Force.

Costa, C., Cunha, I., Borges, A., Ramos, C., Rocha, M., Almeida, J., & Ribeiro-Neto, B. (2004). Analyzing Client Interactivity in Streaming Media. In ACM International Conference on Wide World Web Conference. (pp. 534-543). Beijin, China: ACM Press.

Dahlman, E., Parkval, S., Skol, J., & Beming, P. (2007). 3G Evolution: HSPA and LTE for Mobile Broadband. London: Academic Press.

Dahr, S. (2005). MANET: Applications, Issues, and Challenges for the Future. *International Journal of Business Data Communications and Networking*, *1*(2), 66–92.

Daniel, F., Macedo, A. L., & Pujolle, G. (2008). From TCP/IP to Convergent Networks: Challenges and Taxonomy. *IEEE Communications Surveys & Tutorials*, *10*(4), 40–55.

Dixit, S., & Prasad, R. (Eds.). (2008). Technologies for Home Networking. Hoboken, NJ: John Wiley & Sons, Ltd.

Djama, I., Ahmed, T., Nafaa, A., & Boutaba, R. (2008). Meet In the Middle Cross-Layer Adaptation for Audiovisual Content Delivery. *IEEE Transactions on Multimedia, 10*(1), 105–120. doi:10.1109/TMM.2007.911243

Eichhorn, A. (2008). Middleware Abstractions for Cross-Layer controlled Media Streaming. In ACM Workshop on Middleware- Application Interaction. (pp. 13-18). Oslo, Norway: ACM Press.

Fei, Y., Zhong, L. & Jha, N. K. (2008). An Energy-Aware Framework for Dynamic Software Management in Mobile Computing Systems. ACM Transactions on Embedded Computing Systems, 7(3), 27:1-27:31.

Foukalas, F., Gazis, V., & Alonistioti, N. (2008). Cross-Layer Design Proposals for Wireless Mobile Networks: A Survey and Taxonomy. *IEEE Communications Surveys, 10*(1), 70–85. doi:10.1109/COMST.2008.4483671

Galván, S., Quintana, M. A., Macías, E. M., & Suárez, A. (2005). Multimedia Cooperative Applications on Wireless Networks using Aspect Oriented Programming. In IEEE International Conference on Wireless Networks. (Vol 1, pp. 268-274). Las Vegas, NV: CSREA Press.

Ganz, A., Ganz, Z., & Wongthavarawat, K. (2004). Multimedia Wireless Networks, Tecnologies, Standards and QoS. Upper Saddle River, NJ: Prentice Hall PTR.

Gao, L., Zhang, Z. & Towsley, D. (2003). Proxy-Assisted Techniques for Delivering Continuous Multimedia Streams. IEEE/ACM Transactions on Networking, 11(6), 884-894.

Ghinea, G., & Angelides, M. C. (2004). Quality of Perception in M- Commerce. In N. Shi (Ed.), Mobile Commerce Applications (pp. 284- 302). Hershey, PA: IRM Press.

Goyal, V. (2006). Pro Java ME MMAPI: Mobile Media API for Java Micro Edition. Berkeley, CA: APRESS.

Gupta, V., Williams, M., Johnston, D., McCann, S., Barber, Ph., & Ohba, Y. (2006). IEEE 802.21: Overview of standard for media independent handover services (Tech. Rep.). Washington, DC: IEEE Press.

Handley, M., Floyd, S., Padhye, J., & Widmer, J. (2003). TCP Friendly Rate Control (TFRC): Protocol Specification. Request for Comments: 3448, Standard Track, The Internet Engineering Task Force.

Hassan, M., Vikram, K., & Panos, N. (2008). Rate and distortion modeling of medium grain scalable video coding. In IEEE International Conference on Image Processing, (pp. 2564-2567). San Diego, CA: IEEE Press.

Holma, A., & Toskala, A. (Eds.). (2007). WCDMA for UMTS: HSPA Evolution and LTE. West Sussex, UK: John Wiley & Sons, Ltd.

Hossain, E., & Leung, K. K. (Eds.). (2008). Wireless Mesh Networks: Architectures and Protocols. New York: Springer Verlag.

Inés, D. E., Fujikawa, K., Kawai, E., & Sunahara, H. (2008). Video traffic optimization in mobile wireless environments using adaptive applications. In International Conference on Mobile Ubiquitous Computing, Systems, Services and Technologies. (pp. 376-379). Valencia, Spain: IEEE Press.

Ippolito, L. J. (2008). Satellite Communications Systems Engineering: Atmospheric Effects, Satellite Link Design and System Performance. West Sussex, England: John Wiley & Sons, Ltd.

Kholer, E., Handley, M., & Floyd, S. (2006). Datagram Congestion Control Protocol (DCCP) Request for Comments: 4340, Standard Track, The Internet Engineering Task Force.

Kiczales, G., Lamping, J., Mendhekar, A., Maeda, C., Lopes, C. V., Loingtier, J. M., & Irwin, J. (1997). Aspect- Oriented Programming. In European Conference on Object-Oriented Programming, (LNCS Vol. 1241, pp. 220-242). Jyväskylä, Finland: Springer Verlag.

Kumar, K.G., Lipscomb, J.S, Ramchandra, A., Chang, S.P., Gaddy, W.L., Leung, R.H., et al. (2001). The HotMedia Architecture: Progressive and Interactive Rich Media for the Internet. IEEE Transactions on Multimedia, 3(2), 253:267.

Kumwilaisak, W., Hou, Y., Zhang, Q., Zhu, W., Kuo, C., & Zhang, Y. (2003). A cross-layer quality-of-service mapping architecture for video delivery in wireless networks. *IEEE Journal on Selected Areas in Communications*, *21*(10), 1685–1698. doi:10.1109/JSAC.2003.816445

Kurbel, K. (2003). Video Streaming Solutions for Web-Based E-Learning Courses. In M. Khosrow-Pour, (Ed.), Information Technology and Organizations: Trends, Issues, Challenges, Solutions, Proceedings of the 2003 IRMA International Conference, Philadelphia, PA, (pp. 874- 876).

Li, M., Li, F., Claypool, M., & Kinicki, R. (2005). Weather forecasting: predicting performance for streaming video over wireless LANs. In ACM International Workshop on Network and Operating Systems Support For Digital Audio and Video. (pp. 33-38). Stevenson, WA: ACM Press.

Lin, Y., & Wong, V. W. (2009). Adaptive Techniques in Wireless Networks. In M. Ibnkahla (Ed.), Adaptation and Cross Layer Design in Wireless Networks (pp. 419-450). Boca Ratón, FL: CRC Press.

Macías, E. M., Suárez, A., Martin, J., & Sunderam, V. (2007). Using OLSR for Streaming Video in 802.11 Ad Hoc Networks to Save Bandwidth. *IAENG International Journal of Computer Science*, *33*(1), 101–110.

Malte, A., Robert, H., & Tobias, R. (2008). Dedicated Programming Support for Context-Aware Ubiquitous Applications. In International Conference on Mobile Ubiquitous Computing, Systems, Services and Technologies, (pp. 38 - 43). Valencia, Spain: IEEE Press.

Marrero, D., Macías, E. M., & Suárez, A. (2008). An Admission Control and Traffic Regulation Mechanism for Infrastructure WiFi networks. *IAENG International Journal of Computer Science*, *35*(1), 154–160.

Ming, L. X. (2000). Streaming technique and its application in distance learning system. IEEE International Conference on Signal Processing: Vol. 2, (pp. 1329-1332). Washington, DC: IEEE Press.

Moreno, E. (2009). Modification of the reliable sockets interconnection mechanism and its use in video streaming applications (In Spanish). Unpublished Master doctoral dissertation, University of Las Palmas de G.C., Spain.

Murthy, V. K., & Krishnamurthy, E. V. (2004). Multimedia Computing Environment for Telemedical Applications. In N. Shi (Ed.), Mobile Commerce Applications (pp. 95-115). Hershey, PA: IRM Press.

O'Connell, A. M. (2009). How Nicotine Works. How Stuff Works? Retrieved Jannuary 30, 2009, from http://health.howstuffworks.com/nicotine7.htm

Pagani, M. (Ed.). (2005). Mobile and Wireless Systems Beyond 3G: Managing New Business Opportunities. Hershey, PA: IRM Press.

Pagani, M. (Ed.). (2009). Encyclopedia of Multimedia Technology and Networking. Hershey, PA: IRM Press.

Pagani, M., Schipani, D., Vicari, V. & Al. (2005). Motivations and Barriers to the Adoption of 3G Mobile Multimedia Services: An End User Perspective in the Italian Market. In B. Montano (Ed.), Innovations of Knowledge Management (pp. 80-95). Hershey, PA: IRM Press.

Panton, M. (2008). DLNA for media streamers-- what does it all mean? Crave - CNET. Retrieved Jannuary 30, 2009, from http://news.cnet.com/8301-17938_105-10007069-1.html.

Perea, R. M. (2008). Internet Multimedia Communications Using SIP: A Modern Approach Including Java Practice. Burlington, MA: Morgan Kauffman Publishers.

Rao, K., Bojkovic, Z., & Milovanovic, D. (2006). Introduction to Multimedia Communications: Applications, Middleware, Networking. West Sussex, UK: John Wiley & Sons, Ltd.

Rodriguez, F., Campelo, J. C. & Serrano, Juan J. (2003). Improving the interleaved signature instruction stream technique, IEEE Canadian Conference on Electrical and Computer Engineering: Vol. 1, (pp. 93-96). Washington, DC: IEEE Press.

Sanders, W. B. (2008). Learning Flash Media Server 3. Sebastopol, CA: O' Reily Media, Inc.

Savo, G. G. (2006). Advanced Wireless Networks: 4G Technologies. West Sussex, UK: John Wiley & Sons, Ltd.

Schaar, M. (2007). Cross-Layer Wireless Multimedia. In M. Van der Schaar & P. A. Chou (Ed.), Multimedia over IP and Wireless Networks: Compression, Networking, and Systems (pp. 122-138). London: Academic Press.

Schulzrinne, H., Rao, A., & Lanphier, R. (1998). Real Time Streaming Protocol (RTSP) Request for Comments: 2326, Standard Track, The Internet Engineering Task Force.

Silva, V. H., Salgado, E. I., & Quintana, F. R. (2006). AWISPA: An Awareness Framework for Collaborative Spontaneous Networks. In ASEE/IEEE Frontiers in Education Conference. (pp. 1-6). San Diego, CA: IEEE Press.

Stauder, J., & Erbas, F. (2005). Mobile Multimedia over Wireless Networks. In S.M. Rahman (Ed.), Multimedia Networking: Technology, Management and Applications (pp. 422-471). Hershey, PA: IRM Press.

Stewart, R., Xie, Q., Morneault, K., Schwarzbauer, H., Taylor, T., Rytina, I., et al. (2000). Stream Control Transmission Protocol. Request for Comments: 2960, Standard Track, The Internet Engineering Task Force.

Suárez, A., Menza, M., Macías, E. M., & Sunderam, V. (2007). Automatic resumption of streaming sessions over WiFi using JADE. *IAENG International Journal of Computer Science*, *33*(1), 92–100.

Tonev, G., Sunderam, V., Loader, R., & Pascoe, J. (2002). Location and network quality issues in local area wireless networks. In International Workshop on Network and Operating Systems Support For Digital Audio and Video, (LNCS 2299, pp. 131-148). Kalsrue, Alemania: Springer Verlag.

Tse, D., & Viswanath, P. (2005). Fundamentals of Wireless Communication. Cambridge, UK: Cambridge University Press.

Vandalore, B., Feng, W., Jain, R., & Fahmy, S. (2001). A Survey of Application Layer Techniques for Adaptive Streaming of Multimedia. *International Journal of Real-Time Imaging*, *7*(3), 221–235. doi:10.1006/rtim.2001.0224

Villalón, J., Cuenca, P., Orozco-Barbosa, L., Seok, Y., & Turletti, T. (2007). Cross-layer architecture for adaptive video multicast streaming over multirate wireless LANs. *IEEE Journal on Selected Areas in Communications, 25*(4), 699–711. doi:10.1109/JSAC.2007.070507

Wang, W. (2005). Modeling and Management of Location and Mobility. In D. Katsaros, Nanopoulos & M. Yannis (Ed.), Wireless Information Highway (pp. 177-212). Hershey, PA: IRM Press.

Wu, S., Hsu, J., & Chen, C. (2007). Headlight Prefetching for Mobile Media Streaming. In ACM Workshop on Data Engineering for Wireless and Mobile Access. (pp. 67-74). Beijing, China: ACM Press.

Yankelovich, N., Horan, B., Kaplan, J., Simpson, N., & Slott, J. (2009). Collaboration tools for distributed work - projects include Project Wonderland, MPK20, jVoiceBridge, Porta-Person, Conference Manager, and the Sun Labs Meeting Suite. Collaborative Environments (Sun Labs). Retrieved Jannuary 30, 2009, from http://research.sun.com/projects/dashboard.php?id=85

Zandy, V., & Miller, B. P. (2002). Reliable Sockets (Computer Sciences Tech. Rep. 2002 LISA XVI November 3- 8). Philadelphia, PA: University of Wisconsin.

Zavala, A., Ruíz, J., & Penín, J. (2008). High-altitude Platforms for Wireless Communications. West Sussex, UK: John Wiley & Sons, Ltd.

Zhang, Y., Mao, S., Yang, L. T., & Chen, T. M. (Eds.). (2008). Broadband Mobile Multimedia: Techniques and Applications. Boca Ratón, USA: CRC Press.

ADDITIONAL READINGS

The implementation of streaming media in wireless networks has been studied and implemented in a set of cases but there are several problems for which no mature solution has been reached. For those readers interested in this area, we relate some recent and traditional works that can complete the formation of the future researchers and let other more experienced to seek for concrete solutions.

In the area of wireless networks there are several books that clarify the jungle of wireless technologies and evolution. In Hossain & Leung (2008), a high quality technical introduction to wireless mesh networks is introduced reviewing the challenges and issues, architectures and development strategies, implications in the MAC level, cross layer solutions for traffic forwarding and security among other topics. It is recommended for experienced researchers that want to introduce themselves into wireless mesh networking. An interesting book chapter presenting the evolution of multimedia ad hoc wireless networks and challenges is presented in Dahr (2005). They review the state of the art in 2005 presenting its applications (mobile conferencing, home networking, mobile commerce, emergency services, etc.), research in MAC protocols, routing and energy efficient mechanisms, location management and multimedia. Although not directly related to our work, it is interesting to advise to readers to study the book chapter presented in Wang, (2005), in which mobility models are presented for profiling the users in the wireless networks, especially when they stream multimedia information. This is interesting due to some cross layer methods are including this kind of information to inform the application how they can adapt their behavior.

It is also important to understand the key factors of cross layer techniques to design efficient solutions in wireless networks. Several papers and books are specially indicated to understand this. In (Foukalas & Gazis, 2008) is presented a very good introduction and taxonomy to cross layer design for multimedia mobile networks. They analyze the cross layer optimizations and signaling among adjacent or non-adjacent layers and classify a set of different models for doing that. For those interested in middleware for media streaming for 4G wireless networks, in Eichhorn (2008), one can find an interesting presentation of Noja middleware that presents a cross layer orientation to support failures in the delivery of the streaming. The developer can specify the grade of reliability using three classes: fully, semi and unreliable.

In the particular area of multimedia applications and streaming, there are some books. Rao and Bojkovic (2006) presentsd a complete vision of multimedia systems and network protocol implications. It is recommended for beginners to this research area. A set of interesting papers that review the wide multimedia topics are presented in Pagani (2009). It is very interesting for those that want to introduce themselves to this area of research. Let us note that they also present some papers in the field of multimedia wireless networks. Another interesting book is Perea (2008), in which a brief but interesting history of multimedia Internet is presented to arrive to Session Initiation Protocol (SIP). They also explain in some detail the implementation of some multimedia applications using the Java language. This book is very interesting because complex concepts are explained in a very clear way. Some basics ideas about streaming are presented in the second edition of the book by Austerberry (2005). It is very interesting for beginners to this area or researchers but also recommendable for experienced multimedia researchers that want to strengthen the knowledge of this research area.

Finally, a very good introduction to multimedia services on wireless networks can be found in Stauder & Erbas (2005). The importance of this book chapter is that it explains clearly the evolution from fixed to wireless networks that the authors planned at beginning of this century.

Section 3
Physical Layer Advances

Advanced Multiple Access Transmission Schemes

Chapter 23
Aspects of OFDM-Based 3G LTE Terminal Implementation

Wen Xu
Infineon Technologies AG, Germany

Jens Berkmann
Infineon Technologies AG, Germany

Cecilia Carbonelli
Infineon Technologies AG, Germany

Christian Drewes
Infineon Technologies AG, Germany

Axel Huebner
Infineon Technologies AG, Germany

ABSTRACT

3GPP standardized an evolved UTRAN (E-UTRAN) within the release 8 Long Term Evolution (LTE) project. Targets include higher spectral efficiency, lower latency, and higher peak data rate in comparison with previous 3GPP air interfaces. The E-UTRAN air interface is based on OFDMA and MIMO in downlink and on SCFDMA in uplink. Main challenges for a terminal implementation include an efficient realization of fast and precise synchronization, MIMO channel estimation and equalization, and a turbo decoder for data rates of up to 75 Mbps per spatial MIMO stream. In this study, the authors outline the current 3GPP LTE standard and highlight some implementation details of an LTE terminal. Efficient sample algorithms are presented for key components in the baseband signal processing including synchronization, cell search, channel estimation and equalization, and turbo channel decoder. Their performances, computational and memory requirements, and relevant implementation challenges are discussed.

DOI: 10.4018/978-1-61520-674-2.ch024

1. INTRODUCTION

The mobile radio network technology family of the 3GPP (3rd Generation Partnership Project) as well as its predecessor ETSI (European Telecommunications Standards Institute), including GSM/EDGE (Global System for Mobile communications/Enhanced Data rate for GSM Evolution) and UMTS/HSPA (Universal Mobile Telecommunication System/High Speed Packet Access) technologies, now accounts for nearly 90% of all mobile subscribers worldwide. The further increasing demand on high data rates in new applications such as mobile TV, online gaming, multimedia streaming, etc., has motivated the 3GPP to work on the long term evolution (LTE) project since late 2004. Overall target was to select and specify technology that would keep 3GPP's technologies at the forefront of mobile wireless well into the next decade.

Key objectives of the 3GPP LTE, whose radio access is called Evolved UMTS Terrestrial Radio Access Network (E-UTRAN), include substantially improved end-user throughputs, sector capacity, reduced user plane latency, significantly improved user experience with full mobility, simplified lower-cost network and reduced User Equipment (UE) complexity. Currently, first 3GPP LTE specification is being finalized within 3GPP release 8. Specifically, the physical layer has become quite stable recently for a first implementation.

The air interface of E-UTRAN is based on OFDMA (Orthogonal Frequency Division Multiple Access) and MIMO (Multiple-Input Multiple Output) in downlink (DL) and on SCFDMA (Single Carrier Frequency Division Multiple Access) in uplink (UL) direction. Main challenges for a terminal implementation include efficient realization of the synchronization, channel estimation and equalization, and the turbo decoding algorithms. We show that for quick and robust synchronization and cell search, algorithms based on auto- and cross-correlation provide the best performance-complexity trade-off. Although the inner receiver processing, mainly channel estimation and equalization, can nicely and straightforwardly be parallelized due to frequency domain processing, careful algorithm design is required to achieve low complexity and high performance. Due to the high data rate of up to 75 Mbps per spatial MIMO stream, the turbo decoder design demands a special consideration. In addition, flexibility for different MIMO modes, low power consumption and small silicon area need to be taken into account in the implementation of most of the core algorithms (Berkmann, Carbonelli, Dietrich, Drewes & Xu, 2008).

This chapter is structured as follows: In section 2, we first give an overview on the 3GPP LTE system and its evolution to LTE-advanced, especially the physical layer. Then in the following sections, according to the functional signal flow, efficient sample algorithms for key components in the baseband signal processing are presented. The computational and memory requirements for the example implementations are evaluated, and the challenges are highlighted. Specifically, section 3 deals with LTE-relevant synchronization and cell search, including symbol timing and identification, frame timing and cell identification based on the primary and secondary synchronization signals. In section 4, channel estimation based on the LTE reference signals is described. The algorithm used is the 2x 1-dimentional Wiener filtering In section 5, different equalization algorithms such as the linear MMSE, and the non-linear M-algorithm, the tree search based fixed sphere decoder (FSD) and parallel smart candidate adding (PSCA) algorithm are compared. The LTE-specific turbo coding and decoding algorithm including the newly specified turbo interleaver, and its implementation, are discussed in section 6. Finally, some concluding remarks are given in section 7.

Figure 1. Cellular radio systems

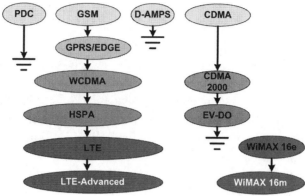

2. OVERVIEW OF LTE AND LTE-ADVANCED

Historical Roots

During the last twenty years the Global System for Mobile Communications (GSM) evolved into the most successful cellular radio system. It is followed by the Universal Mobile Telecommunications System (UMTS), sometimes also referred to as Wideband Code-Division Multiple-Access (WCDMA). Both systems have been specified by the 3rd Generation Partnership Project (3GPP). Until recently, most of the efforts in evolving cellular radio systems were spent on enhancing basic circuit switched voice telephony services. In the meantime, more focus is put on the respective packet-switched data extensions to GSM and WCDMA: General Packet Radio Service (GPRS), Enhanced Data Rates for GSM Evolution (EDGE), and High Speed Packet Access (HSPA).

Figure 1 shows the evolution of major digital cellular radio systems without considering first generation (analogue) cellular radio systems. Among them only a few were mainstream systems, including all 3GPP-based systems starting with GSM. By end of 2008, these 3GPP-based systems account to almost 90% of all worldwide cellular radio subscriptions. Operators have either started or planned to launch commercial services based on GSM/EDGE in 184 countries and based on WCDMA/HSPA in 114 countries (GSA, 2009).

The digital second generation systems, the Japanese Personal Digital Cellular (PDC) and American Digital Advanced Mobile Phone System (D-AMPS), were phased out a couple of years ago. They were replaced by systems based on WCDMA and CDMA2000. For third generation systems, 3GPP-based WCDMA including HSPA has the dominant market share. The only other notable third generation system is CDMA2000, which had evolved from cdmaOne. It further evolved into Evolution-Data Optimized (EV-DO) for high-speed data, comparable to HSPA. Work on a successor to CDMA2000, called Ultra Mobile Broadband (UMB), was not successful in getting any network design wins.

The so-called LTE or super 3G system is developed within the 3GPP long term evolution (LTE) project. It is solely based on the Internet Protocol (IP), thus moving all previously circuit-switched based applications into the packet-switched domain. The other relevant system competing with LTE is developed within IEEE, called Worldwide Interoperability for Microwave Access (WiMAX): IEEE 802.16e and IEEE 802.16m. However, LTE has currently achieved significantly more design wins than WiMAX. A commonality between LTE and WiMAX and the major part of the complete IEEE 802 family of standards is the use of or-

Figure 2. Evolution of 3GPP radio systems and implementation technologies

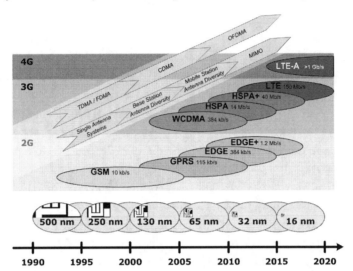

thogonal frequency division multiplex (OFDM). Within 3GPP, however, the use of OFDM is completely new, when compared to the time/frequency-division multiple-access (TDMA/FDMA) based GSM/EDGE and CDMA-based WCDMA/HSPA. Except for a few techniques like turbo coding most implementations of previous 3GPP physical layer implementations cannot be reused.

Figure 2 highlights the 3GPP-based systems together with some further evolutions. Data rates double roughly every 16 to 18 months, i.e. a factor of 10 every 4.5 to 5 years. Extrapolating the 7.2 Mb/s throughput achieved today by HSPA, we can expect 150 Mb/s devices in 2014 and devices supporting 1 Gb/s from 2018 on. The increase in data rate can be attributed to a few parameters that increase parallelism in a couple of dimensions: the system bandwidth increased from 200 kHz (GSM) via 5 MHz (WCDMA) and 20 MHz (LTE) to 100 MHz (LTE-Advanced, LTE-A); the fraction of time allocated to a user increased from 12.5% (GSM) to 100% (WCDMA); the number of simultaneous spatial data streams by using multiple-input multiple-output (MIMO) antenna systems increased from 1 via 2 (HSPA+) and 4 (LTE) to 8 (LTE-A).

It is noted that without the ongoing evolution in semiconductor technology the implementation of such advanced techniques would not be possible at all.

Introduction to LTE

Key objectives of LTE include amongst others, see also 3GPP TR 25.913 (2006-03), improved end-user experience by increased spectral efficiency and peak and cell-edge data rates. This results in an overall increased end-user throughput. Some specific properties of LTE are (3GPP TS 36.101, 2008-09; 3GPP TS 36.211, 2008-09; 3GPP TS 36.306, 2008-5)

- reduced latency for data transmission (<10 ms, user plane) and state transitions (50-100 ms, control plane),
- support of scalable bandwidths of 1.4, 3, 5, 10, 15, and 20 MHz,
- simplified networks and reduced user equipment (UE) complexity,
- support of one, two, or four antenna elements at the base station (eNodeB, eNB) and the UE, i.e. up to 4x4 DL MIMO,

- antenna switching in the UL, i.e., only a single transmit path is required, but with a selectable antenna, and
- peak data rate of 75 Mb/s in the UL (by using 64-QAM) and 300 Mb/s in the DL (by using 64-QAM and 4x4 MIMO) within 20 MHz channel bandwidth.

The air interface of E-UTRAN is based on orthogonal frequency division multiple access (OFDMA) and MIMO in the downlink (DL) and on single carrier frequency division multiple access (SCFDMA) in the uplink (UL) direction. All LTE terminals have to support at least 2 receive antennas and 64-QAM in the DL, and 16-QAM in the UL.

MIMO is supported with a couple of techniques (3GPP TS 36.211, 2008-09): transmit diversity (TD), spatial multiplexing (SM), and beamforming (BF). Both open-loop (OL) and closed-loop (CL) operation is possible for SM. TD is based on Alamouti coding (Alamouti, 1998) in the frequency domain on neighbouring OFDM subcarriers, i.e., space-frequency block coding (SFBC). TD maximizes diversity by transmitting dependent user bits over multiple antennas. When combining TD with the knowledge of the radio channel impulse response (e.g., via proper feedback signalling mechanisms), BF is achieved. BF maximizes a receiver metric, e.g., the signal-to-interference-plus-noise ratio (SINR), by individually and appropriately weighting the transmit signal at the individual transmit antennas. Other differences to TD are that no Alamouti coding is used with BF and that the number of transmit antennas needs not to be known to the UE. Neither TD nor BF increase the peak bit rate. SM, in contrast, exploits the spatial domain to increase the peak data rate by transmitting parallel independent data streams over multiple antennas.

SM can be combined with precoding to maximize the SINR at the receiver. This equals BF when only a single spatial data stream is transmitted. In the general case, a precoding matrix (PM) is multiplied with the vector of transmit signals for the multiple antennas. Highest gains are expected if the PM is carefully selected to match to the instantaneous radio channel conditions. Within the LTE specifications, the precoding operation is also used to derive TD schemes. Precoding in LTE is codebook based. The applicable PM is restricted to a predefined finite set, the codebook. The UE chooses the best PM to optimize its reception and signals it back to the eNB. Additionally, the rank of the channel is fed back to the eNB. The rank indicates the maximum number of independent spatial data streams that can be transmitted over that specific MIMO channel. In case of rank 1 MIMO collapses to BF or TD. The codebook for 4x4 MIMO is based on Householder reflections. Its elements have a nested property: Columns of a lower-rank PM are a subset of those of a higher-rank PM. This reduces complexity for proper PM selection at the UE and allows the eNB to easily choose a lower-rank PM, if it cannot grant all requested resources to a specific UE. The codebook for the 2-antenna case was designed more intuitively without any special structure.

OL-SM includes large-delay cyclic delay diversity (LD-CDD) (Dammann & Kaiser, 2001) resulting in a periodic permutation between PMs to increase the robustness to correlated channel. OL-SM is a fallback solution, if CL-SM is not applicable due to speed limitations and the inherent delay in feedback signalling. Finally, by scheduling multiple users onto different orthogonal layers of a codebook matrix multi-user MIMO can be deployed.

Baseband Implementation of an LTE Terminal

Here, we only consider baseband processing and omit all analogue components, higher layer protocols and application processing. From a baseband signal processing perspective, the main challenge for a UE implementation is in the downlink receiver. The receiver can be separated in a couple of more or less independent blocks regarding the respective implementation: an inner receiver and an outer receiver. The inner receiver includes functionality for synchronization, channel estimation, and equalization. Its basic goal is to reverse the impacts of the radio channel and to deliver estimates of the transmitted bits ("soft" bits) to the outer receiver. It typically delivers a data stream with a bit error rate (BER) of 1% to 10%. The outer receiver subsequently is responsible for reducing this BER to a block error rate (BLER) of typically 0.1% to 10%. For this purpose redundancy information is inserted into the data stream at the transmitter (channel encoding). The outer receiver uses this redundancy to correct the data stream delivered by the inner receiver (channel decoding).

The functional block diagram in Figure 3 shows an example of the internal data flows through an LTE UE with two receive (Rx) antennas and one transmit (Tx) antenna: The radio frequency (RF) signal is down converted, scaled by an automatic gain control, and digitized by an analogue to digital converter (ADC) within the analogue front-end (AFE). The digital front-end (DFE) may include a sample rate converter (SRC) and a numerically controlled oscillator (NCO). Afterwards, the cyclic prefix (CP) is removed and the signals are transformed into the frequency domain by a fast Fourier transform (FFT). Processing in the frequency domain can be parallelized per subcarrier. Since an OFDM assumption is that the subcarriers are mutually orthogonal, processing can be done independently on each subcarrier: This includes channel estimation (ChEst), equalization, MIMO combining, and log-likelihood-ratio (LLR) generation. The LLR values are generated at the interface between inner and outer receiver. The outer receiver consists of hybrid automatic repeat request (HARQ) combining and channel decoding for BLER minimization, and cyclic redundancy checking (CRC) to finally evaluate the correctness of a received data block. The different physical DL channels are: the physical broadcast channel (PBCH), the physical control format indicator channel (PCFICH), the physical HARQ indicator channel (PHICH), the physical downlink control channel (PDCCH), and the main data-bearing channel: the physical downlink shared channel (PDSCH). More details regarding all those dif-

Figure 3. Functional flow through an LTE terminal

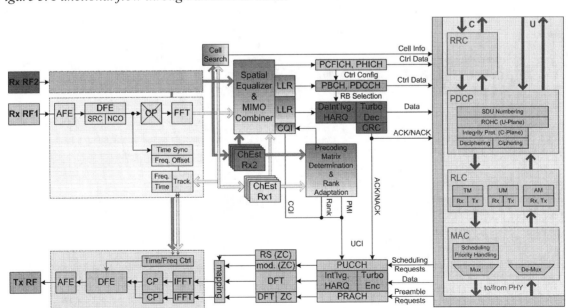

ferent channels can be found in 3GPP TS 36.211 (2008-09), 3GPP TS 36.212 (2008-09), 3GPP TS 36.213 (2008-09). Additionally, control information for DL CL-SM is generated for transmission on the UL: channel quality indication (CQI), PM information (PMI), and channel rank.

In UL direction, the main data-bearing channel, the physical uplink shared channel (PUSCH) data gets encoded, transformed into the frequency domain with a discrete Fourier transform (DFT), mapped onto neighbouring subcarriers for a subsequent inverse FFT (IFFT) together with a Zadoff-Chu (ZC) modulated physical uplink control channel (PUCCH) and ZC-modulated reference signals (RS). After IFFT and CP appendix, data is sent to the Tx front-end and the RF circuitry. For random access the corresponding physical random access channel (PRACH) is processed.

The protocol stack processing consists of medium access control (MAC) for scheduling different logical channels and multiplexing them onto transport channels, radio link control (RLC) with transparent, unacknowledged, and acknowledged modes (TM, UM, AM), packet data convergence protocol (PDCP) including ciphering, robust header compression (ROHC), and delivering service data units (SDU) from and to higher layers. Radio resource control (RRC) corresponds to layer 3 processing of control data, whereas MAC, RLC, and PDCP build layer 2 (3GPP TS 36.300, 2008-09).

Figure 4 highlights some timing relations and the LTE FDD frame structure. One LTE (radio) frame of length 10 ms includes 10 subframes. The subframe duration of 1 ms equals the duration of a transmission time interval (TTI), which is the basic time unit of the outer receiver. A subframe consists of two slots; each in turn consists of seven OFDM symbols in the case of a normal cyclic prefix and 15 kHz subcarrier spacing. The first up to three symbols in each subframe are reserved for control channels. In case of 1.4 MHz system bandwidth, four symbols are reserved for control channels. Synchronization channel (SCH) signals are only transmitted in the first and eleventh slot of every frame. The PBCH is transmitted only in the second slot once per frame (3GPP TS 36.211, 2008-09).

The Path Towards LTE-Advanced

Preliminary results indicate that the basic release 8 LTE can already fulfil most requirements for ITU IMT-Advanced in terms of the mobility

Figure 4. 3GPP LTE FDD frame structure

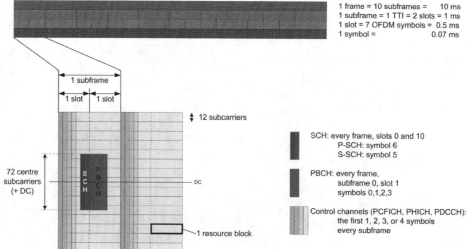

and the uplink/downlink spectrum efficiency. Challenges to meet ITU requirements remain in e.g. downlink urban macro and urban micro test environments (3GPP R1-091580, 2009). At the time of this writing, additional features are being standardized for LTE in the framework of 3GPP release 9, such as home (e)NodeB, multimedia broadcast multicast service (MBMS) support, network-based positioning support, etc. For future LTE-A, further enhancements including peak data rate of 1 Gb/s in the DL and 500 Mb/s in the UL (3GPP TR 36.913, 2009-03) are required. Some of the considered techniques for LTE-A are (3GPP TR 36.814, 2009-02)

- support of a wider bandwidth, e.g. up to 100MHz: aggregating multiple LTE carriers to support bandwidths larger than 20 MHz,
- UL SM with up to four spatial layers, i.e. extension up to 4x4 UL MIMO,
- DL SM with up to eight spatial layers, i.e. extension up to 8x8 DL MIMO,
- coordinated multipoint (CoMP) transmission (DL) and reception (UL) to increase mainly the cell-edge throughput.
- relaying to improve coverage or cell edge throughput.

LTE-A is planned to be backwards compatible to LTE. That means that any LTE terminal can operate in an LTE-A network. The current study item phase of LTE-A within 3GPP will take until the end of 2009 (3GPP RP-090067, 2009). Afterwards the 3GPP work item phase will start to produce a set of LTE-A specifications. First specifications are expected to be finalised by December 2010 within the scope of 3GPP release 10 (3GPP RP-080756, 2008).

3. SYNCHRONIZATION AND CELL IDENTIFICATION

In cellular systems, synchronization is the very first task when a UE tries to establish a radio connection with a network. It is well-known that OFDM-based systems are very sensitive to frequency and timing offsets/errors between the received signal and the local references used for signal demodulation. The frequency and timing offsets can be caused by for example oscillator mismatch at transmitter and receiver, Doppler effects, multipath propagation, etc. Frequency offset can destroy orthogonality among subcarriers and generate inter-channel interference (ICI) and multiple access interference. Timing offset can result in severe inter-block interference (IBI). To avoid a performance loss, frequency and timing offset must be precisely estimated and adequately compensated. Specifically in the OFDM-based LTE system, the whole synchronization is usually accomplished first in DL and then in UL.

1) *DL synchronization:* To facilitate the synchronization of a UE to the network, the eNBs periodically transmit a synchronization channel (SCH) and a physical broadcast channel (PBCH) signal[1] on pre-defined positions within a radio frame. The UE performs an initial timing and frequency estimation by scanning and detecting the strongest SCH signals. Afterwards, the UE can read basic system information such as the cell identity or the DL bandwidth from those signals. The master information block (MIB) is transmitted via the PBCH. It spans 4 consecutive radio frames. The MIB contains information required for connection establishment, like the DL cell bandwidth, the physical HARQ indicator channel (PHICH) configuration, and the system frame number (SFN).

2) *UL synchronization:* The UE transmits the physical random access channel (PRACH) signal at pre-defined time instants according

to the acquired frame structure during the initial DL synchronization. The eNB will then receive the PRACH and estimate the transmit timing of the UE. Based on this, the eNB informs the UE to adjust its transmit timing (timing advance command). Simultaneously, the UE identification is performed.

Here, we'll focus on DL synchronization in FDD mode. As shown in Figure 3, the coarse timing and frequency synchronization are initially performed to estimate the begin of the radio frame (BOF) and the carrier frequency offset (CFO). The begin of the OFDM symbol (BOS) needs also be estimated at the same time. Unless otherwise stated, a time-domain OFDM symbol includes the CP. The BOS therefore indicates the beginning of the CP. An overview on the OFDM synchronization can be found in (Morelli, Kuo & Pun, 2007).

System Model

We start with a perfect synchronized OFDM system with a DFT size N and CP length N_{CP}. Given the transmitted time domain signal $s(n)$ and a channel with the overall channel impulse response (CIR) $h(l)$ ($l = 0, 1, \ldots, N_{ch}-1$, N_{ch} = maximum channel delay spread), the received time domain baseband signal $r(n)$ in an OFDM symbol, after removing CP, can be written as

$$r(n) = \sum_{l=0}^{N_{ch}-1} h(l)s(n-l) + z(n), \qquad 0 \leq n \leq N-1 \tag{1}$$

where $z(n)$ is assumed to be a complex-valued zero-mean AWGN process. Furthermore, we assume the signal $s(n)$ is a zero-mean process independent of $z(n)$. This is normally true for the well-known systems where a random scrambling after coding is employed.

After N-point DFT, the corresponding received signal in frequency domain becomes

$$R(k) = H(k)S(k) + Z(k), \qquad 0 \leq k \leq N-1 \tag{2}$$

with $S(k)$, $R(k)$, $H(k)$ and $Z(k)$ being the N-point DFT of $s(n)$, $r(n)$, $h(n)$ and $z(n)$, respectively. The N-point DFT is defined here as

$$S(k) = DFT_N\{s(n)\} = \frac{1}{\sqrt{N}} \sum_{n=0}^{N-1} s(n) e^{-j\frac{2\pi nk}{N}} \tag{3}$$

Here we assume a multipath propagation channel with the channel coherence time much larger than the OFDM symbol duration, which indicates that the CIR remains constant at least for several consecutive symbols. To justify this assumption we take the worst case of the LTE as an example. The carrier frequency is $f_0 = 2.69$ GHz and the UE moves with $v = 500$ km/h (the highest vehicular speed considered in LTE), then the Doppler spread of the channel equals $f_D = v f_0 / c \approx 1.25$ kHz (c = speed of light). The channel coherence time becomes $t_\tau = 1/f_D \approx 0.8$ ms which is much larger than $T_{symb} = 0.07$ ms, the approximate OFDM symbol duration in LTE. For lower mobile speeds, the coherence time becomes proportionally larger.

Now consider the received signal has a timing offset θ in multiples of the sampling period T_s, and a normalized carrier frequency offset (CFO) $\varepsilon = N T_s f_d$ (f_d = CFO in Hz), compared with the local references. The received signal becomes

$$r(n) = e^{j\frac{2\pi n \varepsilon}{N}} \sum_{l=0}^{N_{ch}-1} h(l)s(n-\theta-l) + z(n), \tag{4}$$

Notice that we omit here the non-integer part of the timing offset and a possible initial carrier phase since they can be considered to be part of the CIR. The tasks of coarse frequency and timing estimation are, based on the received signal $r(n)$ as well as some known synchronization signals, to compute $\hat{\theta}$ and $\hat{\varepsilon}$, the estimates of θ and ε, respectively. Once θ and ε are estimated, the BOS (or BOF) and CFO can be determined, and the cell ID, etc. can be decoded. The tasks of fine frequency and timing estimation are to track the variation of θ and ε with time.

Theoretically, estimation of θ and ε belong to the class of standard parameter estimation problems. A common criterion used is the maximum likelihood (ML) or its variants, which usually lead to correlation based solutions. Sometimes timing and frequency offset can be estimated simultaneously. A critical point is that at the time of the initial synchronization, the CIR is usually unknown. Therefore, coarse synchronization must be robust against channel variations. Many research results have been obtained in last decades, see e.g. Morelli, Kuo & Pun (2007). In the following, the LTE-specific synchronization and cell search will be discussed.

Cyclic Prefix and Symbol Identification

In LTE, two types of cyclic prefix (CP) are specified, the normal CP defined for subcarrier spacing $\Delta f = 15$ kHz and the extended CP for $\Delta f = 15, 7.5$ kHz. The extended CP is typically used in time-dispersive environments to deal with long channel delay spread (e.g. in very large cells). And $\Delta f = 7.5$ kHz is used for multi-media broadcast over a single frequency network (MBSFN) Although a slot has a constant duration of 0.5 ms, the slot structure and CP length are different. We assume that for $\Delta f = 15$ kHz, the DFT size is N.

1) *Normal CP for $\Delta f = 15$ kHz:* Each slot has $K_{symb} = 7$ symbols (as shown in Figure 4). The first symbol has $N + N_{CP1}$ samples with N_{CP1} being the CP length, and the other 6 symbols have $N + N_{CP2}$ samples with N_{CP2} being the CP length.
2) *Extended CP for $\Delta f = 15$ kHz:* Each slot has $K_{symb} = 6$ symbols and each symbol has $N_{symb} = N + N_{eCP}$ samples where N_{eCP} is the CP length.
3) *Extended CP for $\Delta f = 7.5$ kHz:* Each slot has $K_{symb} = 3$ symbols and each symbol has an equal symbol length $N_{symb} = 2N + 2N_{eCP}$ where the CP length is $2N_{eCP}$.

Here, we have $N_{eCP} > N_{CP1} > N_{CP2}$. For example, for a 5 MHz bandwidth LTE with sampling rate $f_s = 1/T_s = 7.68$ MHz and $\Delta f = 15$ kHz, we have the DFT size $N = 512$, $N_{CP1} = 40$, $N_{CP2} = 36$, $N_{eCP} = 128$. All numbers are given in samples.

In order to find out the location of necessary information such as cell ID, PBCH within a slot, and to conduct the FFT for OFDM demodulation, the symbol timing, CP length and different slot structure need to be detected. For this purpose, a low-complexity blind method based on the lagged auto-correlation is proposed here.

Let $r(n)$ be the received data ($n = 0, 1, \ldots$), W the sliding window size, then the lag-P auto-correlation is in general defined as

$$C_{AC}(n) = \frac{1}{W} \sum_{m=0}^{W-1} r^*(n+m) r(n+P+m) \quad (5)$$

Correspondingly, the normalized lag-P auto-correlation is given as

$$\rho_{AC}(n) = \frac{C_{AC}(n)}{\sqrt{E_0(n) E_1(n)}} \quad (6)$$

with

$$E_0(n) = \frac{1}{W} \sum_{m=0}^{W-1} |r(n+m)|^2 \quad (7)$$

$$E_1(n) = \frac{1}{W} \sum_{m=0}^{W-1} |r(n+P+m)|^2 \quad (8)$$

When CP is used for detection, $P = N$ and $W = N_{CP}$ are usually chosen (N = DFT size, N_{CP} = CP length of a symbol). $C_{AC}(n)$ or $\rho_{AC}(n)$ can be employed to detect BOS and CFO (see e.g. van de Beek, Sandell, & Børjesson, 1997). In this case, the timing metric can e.g. be defined as

$$\Lambda(n) = |\rho_{AC}(n)|^2 \tag{9}$$

The BOS and CFO can then be estimated as

$$\hat{\theta} = \arg\max_n \{\Lambda(n)\} \tag{10}$$

$$\hat{\varepsilon} = \frac{1}{2\pi} \angle C_{AC}(\hat{\theta}) \tag{11}$$

The normalized correlation (coefficient) is preferred here since it is less dependent on signal energy and thus robust against time-varying fading. Notice that the obtained metric, say $|\rho_{AC}(n)|$, which is also called the correlation *profile*, is quasi-periodic. Within each symbol, a peak occurs at the position where CP is correlated with its duplicate. Specifically, when noise is low and no channel delay is present, then $\max\{\Lambda(n)\} \to 1$, and the peak position is exactly the desired BOS. For multipath channel, the peak will be delayed, depending on the channel spread. Due to the small window size $W = N_{CP}$, such a BOS estimate is very sensitive to the channel and noise. A robust estimate can be obtained by considering more symbols jointly. In general, when K symbols are available, we have an equivalent CP of KN_{CP} samples available as correlation window.

For simplicity, assume that K_{slot} length-N_{slot} slots plus at least one additional symbol of samples are available so that we can calculate the lag-P correlation for $n = 0, 1, ..., K_{slot} \cdot N_{slot} - 1$, and each slot has in turn K_{symb} length-N_{symb} symbols, then the lag-P auto-correlation for a window size of $K_{slot}K_{symb}N_{CP}$ samples can be re-written as

$$C'_{AC}(n) = \frac{1}{K_{symb}K_{slot}} \sum_{l=0}^{K_{symb}-1} \sum_{k=0}^{K_{slot}-1} C_{AC}(n + lN_{symb} + kN_{slot}) \tag{12}$$

$$E'_0(n) = \frac{1}{K_{symb}K_{slot}} \sum_{l=0}^{K_{symb}-1} \sum_{k=0}^{K_{slot}-1} E_0(n + lN_{symb} + kN_{slot}) \tag{13}$$

$$E'_1(n) = \frac{1}{K_{symb}K_{slot}} \sum_{l=0}^{K_{symb}-1} \sum_{k=0}^{K_{slot}-1} E_1(n + lN_{symb} + kN_{slot}) \tag{14}$$

$$\rho'_{AC}(n) = \frac{C'_{AC}(n)}{\sqrt{E'_0(n)E'_1(n)}} \tag{15}$$

for $n = 0, 1, ..., N_{symb} - 1$. The metrics for BOS and CFO can be calculated by replacing $C_{AC}(n)$ and $\rho_{AC}(n)$ in Eq. (9) – (11) with $C'_{AC}(n)$ and $\rho'_{AC}(n)$, respectively. Therefore, after computing $C_{AC}(n)$, $E_0(n)$, $E_1(n)$ for $n = 0, 1, ..., K_{slot}K_{symb}N_{symb} - 1$, $C'_{AC}(n)$, $E'_0(n)$, $E'_1(n)$ for $n = 0, 1, ..., N_{symb} - 1$ can be computed by averaging $C_{AC}(n)$, $E_0(n)$, $E_1(n)$ over all $K_{slot}K_{symb}$ symbols. Notice that in this case the correlation peaks of all symbols are added coherently.

The above method, however, can only be used for equal-length symbols as in the case of the extended CP. For unequal-length symbols such as in the case of the normal CP, it cannot be applied. To overcome this difficulty, a method is described below. First, set the window size $W = N_{CP2}$, the smaller CP length, and compute $C_{AC}(n)$, $E_0(n)$, $E_1(n)$, as defined in (5) – (8), where the resulting $C_{AC}(n), E_0(n), E_1(n)$ have K_{slot} length-N_{slot} slots. Then, omit or cut away any $(N_{CP1} - N_{CP2})$ consecutive samples at the same position within each of the K_{slot} slots[3]. In this way, $C_{AC}(n)$, $E_0(n)$ and $E_1(n)$ have K_{slot} slots, with each slot having $N_{slot} - (N_{CP1} - N_{CP2})$ symbols which correspond exactly to the total length of $K_{symb} = 7$ length-$(N_{CP2} + N)$ symbols. Now we simply consider that $C_{AC}(n)$,

$E_0(n)$, $E_1(n)$ are calculated from the equal-length symbols. They can then be averaged to result in $\rho'_{AC}(n)$ from which the BOS $\hat{\theta}$ and the CFO $\hat{\varepsilon}$ can be determined.

A careful analysis reveals that the BOS estimated in this way, $\hat{\theta}$, has a certain ambiguity even in a noise-free environment. In fact, for a given estimate $\hat{\theta}$, the *actual* BOS can be in the range of $\hat{\theta} \pm (N_{CP1} - N_{CP2})$, depending on which samples are cut. In case the $(N_{CP1} - N_{CP2})$ samples omitted coincide with the first or the last $(N_{CP1} - N_{CP2})$ samples of the *actual* first symbol, then no ambiguity in $\hat{\theta}$ exists. When the cut samples are not within the actual first symbol, the metric will have a plateau or a flat-peak of up to $(N_{CP1} - N_{CP2})$ samples in the first symbol. Since $(N_{CP1} - N_{CP2}) \ll N_{CP2}$, the ambiguity $\pm(N_{CP1} - N_{CP2})$ is tolerable in practice.

In what follows, we will detect the CP type as well as the slot structure. Detection of the slot structure corresponds to detection of different DFT and CP lengths. This can be done by a hypothesis test. To detect CPs with $\Delta f = 7.5$ kHz and $\Delta f = 15$ kHz, we can e.g. calculate the profile within a symbol, say $\{\Lambda(n); n = 0, 1, ..., N_{symb}-1\}$, for the extended CP with $W = 2N_{eCP}$ and $P = 2N$. If there is a significant peak, then it is $\Delta f = 7.5$ kHz. Otherwise, it is $\Delta f = 15$ kHz. The reason is that for a wrong lag P, no overlapping correlation between CP and its corresponding duplicate is possible. Therefore, a correlation peak for $P = 2N$ indicates no peak for $P = N$. On the other hand, a correlation peak for $P = N$ also indicates no peak for $P = 2N$.

To distinguish normal and extended CPs with $\Delta f = 15$ kHz, we can e.g. compute the metric, say $\{\Lambda(n); n = 0, 1, ..., N_{symb}-1\}$, for $W = N_{CP2}$ and $P = N$. We then check the width of the peak or plateau in $\{\Lambda(n)\}$ for $n = 0, 1, ..., N_{symb}-1$ to detect which CP it is. For the normal CP, the peak or plateau is narrow and has 1 to $(N_{CP1} - N_{CP2} + 1)$ samples, i.e. 1 to 5 samples for 7.68 MHz sampling rate.

But for the extended CP, the plateau will have a width of about $(N_{eCP} - N_{CP2} + 1)$ samples, i.e. 85 samples for 7.68 MHz sampling rate. Alternatively, we can use the parameters of the extended CP, i.e. $W = N_{eCP}$ and $P = N$ for detection. When a high peak is obtained in this case, it is the extended CP, otherwise the normal CP where a small peak usually exists. The position of the high peak is the desired BOS. We see that by properly employing the lagged auto-correlation method, we are able to simultaneously detect the BOS, CFO and the type of CP and slot.

Synchronization Signals and Cell ID

In the release 8 LTE, a total of 504 cell IDs are defined, where each cell ID can be expressed as

$$N_{ID}^{cell} = 3N_{ID}^{(1)} + N_{ID}^{(2)} \qquad (16)$$

with $N_{ID}^{(1)} = 0, 1, ..., 167$ being the cell *group ID* and $N_{ID}^{(2)} = 0, 1, 2$ being the physical layer ID (sometimes called *sector ID*) within a cell group.

The information on the physical layer ID $N_{ID}^{(2)}$ within a group and the group ID $N_{ID}^{(1)}$ are carried through two DL SCH signals, the primary synchronization channel (P-SCH) and secondary synchronization channel (S-SCH) signal, respectively. Both P-SCH and S-SCH occupy 72 center subcarriers, where in LTE no signal is transmitted on the DC subcarrier (see Figure 4). The P-SCH is imbedded in the last OFDM symbol and S-SCH in the second last OFDM symbol, of subframe 0 and 5 in each frame. For convenience, we divide the frame into two: the first half frame consisting of subframe 0 – 4 and the second half frame consisting of subframe 5 – 9. Each half frame has a duration of 5 ms and exactly one P-SCH in it. From the position of the P-SCH the begin of the half frame is known, but not the BOF. In LTE, N_{ID}^{cell} can be computed in two steps:

- Locate and detect P-SCH, and then decode $N_{ID}^{(2)}$.
- Based on the P-SCH position and $N_{ID}^{(2)}$, locate and detect S-SCH.

Then, the PBCH can be located and decoded, and the DL synchronization can be accomplished.

Detection of Physical Cell ID Within a Group

Primary Synchronization Signal

The P-SCH signal is chosen from a class of the odd-length Zadoff-Chu (ZC) sequences (Chu, 1972), which is defined as

$$d_u(k) = e^{-j\frac{\pi u k(k+1)}{N_{ZC}}} \qquad 0 \le k < N_{ZC} \qquad (17)$$

where N_{ZC} is the length of the ZC sequence, and u is the ZC root index relatively prime to N_{ZC}. In general, the ZC sequence has the constant-amplitude zero autocorrelation (CAZAC) except for a single maximum per period. Specifically, when $d_u(k)$ is an arbitrary length-N_{ZC} ZC sequence, then

- Its amplitude is constant $|d_u(k)| = 1$.
- The length-N_{ZC} ZC sequence $d_u(k)$ is orthogonal to its cyclic shift $d_u(k') = d_u(k' \bmod N_{ZC})$, $(k' \ne k)$.
- The length-N_{ZC} ZC sequences $d_u(k)$ and $d_v(k)$ ($u \ne v$) are orthogonal to each other.
- Its cyclic shift $d_u((k+l) \bmod N_{ZC})$ (l = integer) is a length-N_{ZC} ZC sequence.
- Its conjugation $d_u^*(k)$ is a length-N_{ZC} ZC sequence.
- Its linear phase shift $e^{j\frac{\pi l k}{N_{ZC}}} d_u(k)$ (l = integer) is a length-N_{ZC} ZC sequence.
- Its N_{ZC}-point DFT is a length-N_{ZC} ZC sequence.

Due to many attractive properties, the ZC sequence is used in LTE to generate the DL P-SCH, UL RACH preample as well as demodulation/sounding reference signals. For P-SCH signals in frequency domain, $N_{ZC} = 63$ is chosen. In order to avoid high interference from local oscillator leakage, $d_u(k)$ at DC subcarrier $K_1 = (N_{ZC} - 1)/2 = 31$ is set to zero. In this way, a P-SCH signal has actually 62 non-zero elements in frequency domain. In addition, 10 zero-valued subcarriers are inserted at the lower and higher subcarrier boundaries of the P-SCH, i.e. $S_u(k) = 0$ for $k = -5, \ldots, -1$ and $k = 63, \ldots, 67$, as guard regions. It can be shown that the ZC sequence is symmetric with respect to $k = K_1$, the index of the DC subcarrier, i.e.

$$d_u(k) = d_u(N_{ZC} - k) \qquad (18)$$

Let 0 be the index of the DC subcarrier, the P-SCH in frequency domain can be written as

$$S_u(k) = \begin{cases} d_u(k - K_1) & k = \pm 1, \pm 2, \ldots, \pm K_1 \\ 0 & otherwiese \end{cases} \qquad (19)$$

Then we have

$$S_u(k) = S_u(-k) = S_u(N - k) \qquad (20)$$

The latter equation results from the periodicity of $S_u(k)$ and $s_u(n)$ when an N-point DFT is applied. Consider

$$s_u(n) = IDFT_N\{S_u(k)\} = \frac{1}{\sqrt{N}} \sum_{k=-K_1}^{K_1} S_u(k) e^{j\frac{2\pi nk}{N}} \qquad (21)$$

It is easy to verify

$$s_u(n) = s_u(-n) = s_u(N - n) \qquad (22)$$

Aspects of OFDM-Based 3G LTE Terminal Implementation

Figure 5. Frequency and time domain P-SCH signals (N=128, u=34), with "ampl" = amplitude, "imag" = imaginary part, "real" = real part

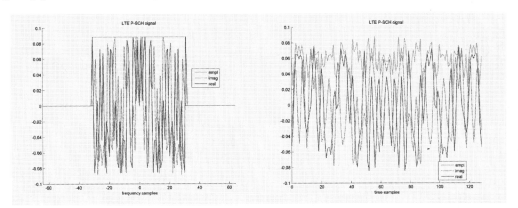

We see that $s_u(n)$ and $S_u(k)$ are symmetric with respect to $n, k = 0, N/2$. This property can be utilized, e.g. to realize the synchronization based on the reverse correlation (see below).

Figure 5, Figure 6 and Figure 7 show the frequency and time domain P-SCH signals, as well as the cross- and reverse correlation. Since the DC subcarrier $S_u(0) = 0$ and $\frac{1}{N}\sum_{n=0}^{N-1} s_u(n) = \frac{1}{\sqrt{N}} S_u(0) = 0$, $s_u(n)$ has a zero mean. It should be noticed that the P-SCH signal is not a ZC sequence any more. For different u, the P-SCH signals are in general no longer orthogonal to each other. However, they can still be considered as pseudo noise with many good properties, including

- small power variation in both time and frequency domains, an attractive property to achieve a low peak-to-average power ratio (PAR);
- very good auto-correlation property in both time and frequency domain, an ideal property for synchronization.

In the release 8 LTE, the three physical IDs within a group $N_{ID}^{(2)} = 0, 1, 2$ are represented by the P-SCH signals with three different ZC root indices $u = 25, 29, 34$, respectively, therefore decoding $N_{ID}^{(2)}$ means to determine u.

Detection of Physical Cell ID $N_{ID}^{(2)}$

To find out the P-SCH position, cross-correlation between the received signal and the P-SCH signal needs to be performed, either in time domain, or once the BOS is determined, in frequency domain. Here we take an implementation in time domain as an example. The three time domain P-SCH signals $s_u(n)$ for $u = 25, 29, 34$ are generated. The received signal $r(n)$ is then correlated with $s_u^*(n)$, the conjunction of $s_u(n)$, within a sliding window of size W, where $W = N$ is usually chosen. The normalized cross-correlation coefficient can be computed as

$$\rho_{XC}(n, u) = \frac{C_{XC}(n, u)}{\sqrt{E_0(u) E_1(n)}} \quad (23)$$

where

$$C_{XC}(n, u) = \frac{1}{W} \sum_{m=0}^{W-1} s_u^*(m) r(n + m) \quad (24)$$

Figure 6. Auto-correlation of P-SCH signal with its periodically extended duplicate in frequency and time domain (N=128, u=34)

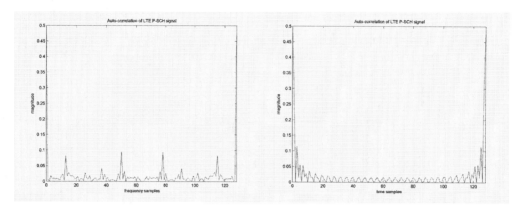

$$E_0(u) = \frac{1}{W}\sum_{m=0}^{W-1}\left|s_u(m)\right|^2 \quad (25)$$

$$E_1(n) = \frac{1}{W}\sum_{m=0}^{W-1}\left|r(n+m)\right|^2 \quad (26)$$

are the cross-correlation, energy of $s_u(n)$ and $r(n)$, respectively. The timing metric is then given as

$$\left|\rho_{XC}(n,u)\right|^2 = \frac{\left|C_{XC}(n,u)\right|^2}{E_0(u)E_1(n)} \quad (27)$$

The desired estimates for P-SCH position $\hat{n}$ and ZC root index $\hat{u}$ can be obtained by searching the maxima, namely

$$(\hat{n},\hat{u}) = \arg\max_{n,u}\left|\rho_{XC}(n,u)\right|^2 \quad (28)$$

Detection of Cell ID Group

Secondary Synchronization Signal

Similar to the P-SCH, S-SCH signal has 62 non-zero elements in frequency domain, with the DC subcarrier and 10 subcarriers at the lower and upper

Figure 7. Reverse correlation of the P-SCH signal in frequency and time domain (N=128, u=34)

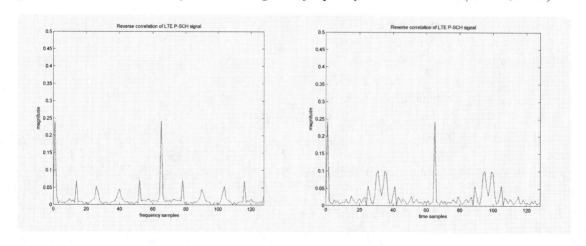

frequency boundaries set to zero. The 62 non-zero elements are an interleaved concatenation of two length-31 binary sequences, each taking 31 different values corresponding to 31 cyclic shifts of an m-sequence. The shifts are derived from the group ID $N_{ID}^{(1)}$. The two length-31 sequences are then scrambled with the scrambling sequences derived depending on $N_{ID}^{(2)}$, the physical layer ID within a group. Contrary to the P-SCH, the S-SCH signal in subframe 0 and 5 are different from each other. In fact, this property is used to distinguish the subframe 0 from the subframe 5, and thus to identify the BOF.

The frequency domain S-SCH sequence in subframe 0 can be expressed, for $0 \leq k \leq 30$, as

$$d(2k) = \tilde{s}(k+m_0)\tilde{c}(k+N_{ID}^{(2)})$$
$$d(2k+1) = \tilde{s}(k+m_1)\tilde{z}(k+(m_0 \bmod 8))\tilde{c}(k+N_{ID}^{(2)}+3)$$
(29)

where $\tilde{s}(k)$, $\tilde{z}(k)$, $\tilde{c}(k) \in \{\pm 1\}$ are the pre-defined length-31 binary m-sequences. For the index k outside of the range $0 \leq k \leq 30$, the modulo 31 applies, i.e. $\tilde{s}(k) = \tilde{s}(k \bmod 31)$, $\tilde{z}(k) = \tilde{z}(k \bmod 31)$ and $\tilde{c}(k) = \tilde{c}(k \bmod 31)$. The shift index pair m_0 and m_1 carry the information about the cell group $N_{ID}^{(1)} = 0, 1, ..., 167$ and are derived according to

$$m_0 = m' \bmod 31$$
$$m_1 = (m_0 + \lfloor m'/31 \rfloor + 1) \bmod 31$$
(30)

with

$$m' = N_{ID}^{(1)} + q(q+1)/2$$
(31)

$$q = \left\lfloor \frac{N_{ID}^{(1)} + q'(q'+1)/2}{30} \right\rfloor$$
(32)

$$q' = \lfloor N_{ID}^{(1)}/30 \rfloor$$
(33)

The relationship between $N_{ID}^{(1)}$ and the index pair m_0 and m_1 are also listed as a look-up table in the specification (3GPP TS 36.211, 2008-09). It can be verified that this ensures $0 \leq m_0 < m_1 \leq 30$ for $0 \leq N_{ID}^{(1)} \leq 167$. In cell identification procedure, we usually need to first determine the shifts m_0 and m_1, then obtain $N_{ID}^{(1)}$, say, by a simple table look-up.

The frequency domain S-SCH sequence in subframe 5 can in fact be obtained by simply exchanging the shift index m_0 with m_1 in the above equations, i.e.

$$d(2k) = \tilde{s}(k+m_1)\tilde{c}(k+N_{ID}^{(2)})$$
$$d(2k+1) = \tilde{s}(k+m_0)\tilde{z}(k+(m_1 \bmod 8))\tilde{c}(k+N_{ID}^{(2)}+3)$$
(34)

As a result, the same algorithm used to decode m_0 and m_1 of subframe 0 can be used to decode m_1 and m_0 of subframe 5, respectively. More specifically, we can employ the algorithm, say, for subframe 0, to compute the shift index m_0 and m_1. In case the result $m_0 < m_1$ is obtained, the subframe is indeed the subframe 0, otherwise it is the subframe 5 (In this case, we need to exchange the values of m_0 and m_1 to ensure $m_0 < m_1$). As such, we also solve the ambiguity between subframe 0 and 5. The structure of the radio frame including the BOF can be determined.

Detection of Cell ID Group $N_{ID}^{(1)}$

Determination of $N_{ID}^{(1)}$ means to detect which S-SCH is embedded in the received signal. Again, the cross-correlation can be used. In frequency domain, $\tilde{s}(k)$, $\tilde{z}(k)$, $\tilde{c}(k)$ and the resulting S-SCH signal are pre-defined binary pseudo noise taking the value ± 1 only. Therefore, the cross-correlation with the received signal can be computed without multiplication, which makes a low complexity implementation possible. We can therefore use the possible S-SCH sequences to directly correlate with the received signal. The S-SCH sequence having the highest correlation peak will be the S-SCH sequence embedded in the received signal. This can easily be done in frequency domain, e.g. with a procedure given in (Manolakis, Estevez, Jungnickel, Xu & Drewes, 2009).

Denote the received and detected length-62 S-SCH sequence in frequency domain (excluding the DC subcarrier) as $\{d(2k), d(2k+1); k = 0, 1, ..., 30\}$. $N_{ID}^{(1)}$ can be computed as follows

- Compute $\tilde{d}_0(k) = d(2k) \cdot \tilde{c}\left(k + N_{ID}^{(2)}\right)$ and the cyclic cross-correlation
$$\tilde{C}_0(m) = \sum_{k=0}^{30} \tilde{s}(k+m)\tilde{d}_0(k) \quad \text{for} \quad m = 0, 1, ..., 30,$$
then m_0 can be estimated as $\hat{m}_0 = \arg\max_m \left|\tilde{C}_0(m)\right|^2$.

- Compute:
$$\tilde{d}_1(k) = d(2k+1)\tilde{z}\left(k + (\hat{m}_0 \bmod 8)\right)\tilde{c}\left(k + N_{ID}^{(2)} + 3\right)$$

and the cyclic cross-correlation $\tilde{C}_1(m) = \sum_{k=0}^{30} \tilde{s}(k+m)\tilde{d}_1(k)$, then m_1 can be estimated as $\hat{m}_1 = \arg\max_m \left|\tilde{C}_1(m)\right|^2$.

- When $\hat{m}_0 < \hat{m}_1$, the S-SCH is in subframe 0. Otherwise, exchange the value of $\hat{m}_0$ and $\hat{m}_1$, and the S-SCH is in subframe 5.
- Determine the BOF and determine $N_{ID}^{(1)}$ by a table look-up.

Coarse Synchronization Algorithms

Timing Estimation

Since no timing information about the OFDM symbol boundary is available, coarse timing estimation is normally first performed in time domain prior to DFT. Here, we briefly overview the methods applicable to LTE, see also (Manolakis & Jungnickel, 2008).

- *CP based lagged auto-correlation:* This is one of the most widely used methods. As described before, this method has low complexity, but only detects the CP type, BOS and CFO. The BOF as well as cell ID $N_{ID}^{(1)}$ and $N_{ID}^{(2)}$, however, has to be estimated additionally
- *P-SCH based cross-correlation:* The P-SCH based cross-correlation can be used to locate the P-SCH, and thus the begin of the half frame and BOS, as well as to decode the physical ID $N_{ID}^{(2)}$ within a group. To resolve the ambiguity of whether the P-SCH is in subframe 0 or 5, S-SCH signal needs to be utilized. In addition, the frequency offset needs to be estimated additionally and cannot be directly obtained from the cross-correlation parameters.
- *P-SCH based reverse correlation:* As analyzed before, the P-SCH, say $s_u(n)$, is symmetric with respect to $n = N/2$. Based on this, the reverse correlation can be used to locate the P-SCH as well as estimate the CFO (Zhang, Long, Zhao & Liu, 2005). Specifically, the reverse correlation is defined as

$$\rho_{RC}(n) = \frac{C_{RC}(n)}{\sqrt{E_0(n)E_1(n)}} \qquad (35)$$

with

$$C_{RC}(n) = \frac{1}{W}\sum_{m=1}^{W-1} r^*(n+m)r(n+N-m) \qquad (36)$$

$$E_0(n) = \frac{1}{W}\sum_{m=1}^{W-1} \left|r(n+m)\right|^2 \qquad (37)$$

$$E_1(n) = \frac{1}{W} \sum_{m=1}^{W-1} |r(n+N-m)|^2 \qquad (38)$$

where the window size is $W - 1 = N/2 - 1$ since in an OFDM symbol no duplicates of the first and middle sample, $r(0)$ and $(N/2)$, are transmitted. The timing offset, $\hat{n}$, can be estimated by

$$(\hat{n}) = \arg\max_n |\rho_{RC}(n)|^2 \qquad (39)$$

With a few further steps the CFO can also be estimated. Once $\hat{n}$ is determined, the begin of the half frame and BOS are known. Again, we can not determine $N_{ID}^{(2)}$ since it is not known which P-SCH signal is embedded in the received signal.

- *P-SCH based lagged auto-correlation:* This is based on the fact that P-SCH signal is transmitted every 5 ms. Let P be the period in samples (= length of half frame), $W = N$ the size of the sliding window, then the lag-P auto-correlation defined in (5) – (8) can be used to estimate timing and frequency offset. Same as the method based on reverse correlation, the begin of the half frame and BOS can be obtained, but $N_{ID}^{(2)}$ needs to be determined additionally.

A disadvantage is the reduced performance for time-varying channel since 5 ms correlation lag can be greater than the channel coherence time. In this case, the two signals to be correlated may experience totally different channels. Due to the large correlation lag, the CFO estimated has also a relatively small range.

Frequency Offset Estimation

The carrier frequency offset (CFO) ε can be separated into an integer part ε_I which is a multiple of the subcarrier spacing, and a fractional part ε_F, i.e. $\varepsilon = \varepsilon_I + \varepsilon_F$ ($-1 < \varepsilon_F < 1$). ε_I causes a carrier frequency shift by ε_I subcarrier spacing, and ε_F leads to an inter-carrier interference. The fractional CFO can be estimated in time domain using auto-correlation based methods, e.g. by exploiting signal repetition such as CP or symmetry of the P-SCH. The integer CFO can be determined in frequency domain, say using known signal such as P-SCH, S-SCH, etc. This can be done as follows: First determine the fractional CFO and do the corresponding compensation in time domain, then perform FFT (assume symbol timing is estimated before). The integer CFO is reflected by cyclic shift of the signal in frequency domain, and can thus be determined, say using cross-correlation. The compensation of the integer CFO can be done accordingly.

Performance and Complexity

The cross-correlation based method has usually superior performance but it is typically computationally intensive, since the metric, say $|C_{XC}(n,u)|^2$ given in Eq. (24), needs to be computed for every sample n. On the other hand, $E_0(n)$ is constant and can be computed off-line.

Notice that the energy, e.g. $E_1(n)$ in the case of cross-correlation, can be computed recursively

$$WE_1(n+1) = WE_1(n) - |r(n)|^2 + |r(n+W)|^2 \qquad (40)$$

We take the LTE system with 20 MHz bandwidth (30.72 MHz sampling rate) and $W = N = 2408$ as an example. Since $E_1(n)$ and $|r(n)|^2$ have been calculated in previous stages, a low computational complexity of $3 \cdot 30.72 \cdot 10^6 \approx 93 \cdot 10^6$ MAC (multiply-accumulate) operations[4] per second are required, assuming for given n, 3 real MAC are needed to compute $E_1(n+1)$ in (40).

For the cross-correlation approach, $4W$ real MAC are required to compute $C_{XC}(n,u)$ for each n and each u. Then, a total of $4W \cdot 30.72 \cdot 10^6 \approx 252 \cdot 10^9$ MAC per second are required for each u. For three different u, the computational complexity amounts to $756 \cdot 10^9$ MAC per second. Such a computational complexity is usually too high for a software solution.

The reverse correlation based method is about half as complex as the cross-correlation based method for timing estimation, since $|C_{RC}(n)|^2$, similar to $|C_{XC}(n, u)|^2$, cannot be computed recursively, and the size of the sliding window is only about half as large as in the cross-correlation based method.

In contrast, the CP or P-SCH based lagged auto-correlation has low complexity since beside $E_0(n)$ and $E_1(n)$, $C_{AC}(n)|$ can also be computed recursively, namely

$$WC_{AC}(n+1) = WC_{AC}(n) - r^*(n)r(n+P) + r^*(n+W)r(n+P+W)$$
(41)

Notice that $C_{AC}(n)$ and $r^*(n)r(n+P)$ have been calculated before, therefore, computation of $C_{AC}(n+1)$ will need only a few MAC, say 6 real MAC operations. For the 30.72 MHz sampling rate, a total of $6 \cdot 30.72 \cdot 10^6 \approx 184 \cdot 10^6$ real MAC per second are required. Such a complexity requirement can easily be fulfilled nowadays.

Our simulations have shown that among the methods investigated, the P-SCH based cross-correlation method, mainly because of the use of noise-free replica in correlation, provides the highest accuracy in timing estimation. By properly averaging over more symbols, the CP based auto-correlation can not only estimate the CFO with high accuracy, but also achieve a robust BOS estimate. In practice, combined or mixed methods based on auto- and cross-correlation are employed. They usually provide a good compromise in terms of performance and computational requirements. In addition, over-sampling of the received data may significantly improve the performance of the synchronization and cell search, but demand more computational efforts.

Notice that initial acquisition of timing/frequency offset, cell ID, etc. is mainly performed during call setup, partly during handover, cell (re) selection, etc. It does not need to be performed continuously during data reception. Therefore, sharing computational resources with other receiver tasks is possible. In fact, after completion and switch-off of the coarse synchronization, the tracking algorithms can be switched on to deal with small timing and frequency variation. For LTE, this can for example be done, as in other OFDM based systems (see Speth, Fechtel, Fock & Meyr, 1999), by using the embedded reference signals which are also used for channel estimation.

4. CHANNEL ESTIMATION

Efficient data detection in OFDM systems operating in time and frequency selective channels requires accurate channel estimation. In LTE, like in many OFDM-based systems, in order to facilitate channel estimation, known symbols, called *pilots*, are inserted at specific locations in the time-frequency grid and channel estimation is performed by interpolation. The resulting two-dimensional pilot pattern is typically irregular, as shown in Figure 8, for the LTE case where the pilot spacing in the frequency direction equals six OFDM symbols, while in the time direction two OFDM symbols per slot (referred to as *reference symbols*) containing pilots are available, at a distance of 4 and 3 OFDM symbols from one another [5](3GPP TS 36.211, 2008-09).

Several pilot-aided channel-estimation schemes have been proposed for OFDM applications (Edfors, Sandell, van de Beek, Wilson & Borjesson, 1998; Le, 2000; Li, Cimini Jr. & Sollenberger, 1998; Negi & Cioffi, 1998; Rinne & Renfors, 1996). Some low complexity approaches (Rinne & Renfors, 1996) resort to a simple piecewise-constant and piecewise-linear interpolations between pilots. They require no a priori knowledge on the channel statistics but need a large number of pilots to achieve good performance. Other schemes are based on the frequency-domain linear minimum mean squared

Figure 8. Pilot grid for LTE SISO (top) and 2x2 MIMO (bottom) configuration

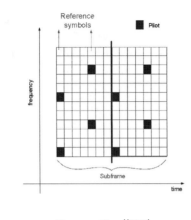

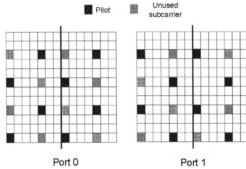

error estimator (MMSEE) (Edfors, Sandell, van de Beek, Wilson, & Borjesson, 1998) and achieve very good estimation accuracy at the expenses of complexity. In fact, we observe that the optimum pilot aided channel estimator in the MSE sense is represented by the Wiener interpolation filter (WIF). Unfortunately, exact computation of the WIF requires knowledge of the channel correlations in time and frequency domain and the operating signal-to-noise ratio (SNR). A frequency domain maximum likelihood estimator (MLE) has been discussed in (Negi & Cioffi, 1998) and compared to the MMSEE in (Morelli & Mengali, 2001). As shown in (Morelli & Mengali, 2001), the MLE can do without knowledge of the channel statistics and of the noise variance, and still achieve performance close to the MMSEE if the number of pilots is larger than the length of the channel impulse response. It is clear that a real time implementation of the MLE as described in (Morelli & Mengali, 2001) is not feasible due to the large memory requirements. On the other hand, the MMSEE is also not a viable solution since it requires knowledge of the channel statistics and operating SNR.

Robust MMSE channel estimators which can work in a mismatched mode have also been studied (see e.g. Hoeher, Kaiser & Robertson, 1997; Li, Cimini & Sollenberger, 1998). In contrast to the optimum frequency domain MMSEE discussed above, the 'robust' Wiener channel interpolator is matched to the worst case scenario. As a consequence, the Wiener interpolation filters can be computed off-line and stored. Mainly two types of robust Wiener estimators have been discussed in the literature (Hoeher, Kaiser & Robertson, 1997): 2-dimensional (2D) interpolators, where channel estimation is performed through a two-dimensional interpolation between pilots, and cascaded 1-dimensional interpolators (typically referred as 2x1D) where two one-dimensional interpolations are performed. Due to their straight-forward implementation and good performance, robust Wiener interpolation will be the focus of the remaining part of this section.

Wiener Filter Channel Interpolation

Consider now an OFDM system with N subcarriers and L OFDM symbols per slot operating in a time and frequency selective channel. Assuming perfect synchronization, the complex baseband representation of the received signal $y(n,l)$ for subcarrier n and OFDM symbol l reduces to:

$$y(n,l) = x(n,l)H(n,l) + z(n,l) \quad , \quad n = 0,1,...,N-1, \quad l = 0,1,...,L-1$$

(42)

where $x(n,l)$, $H(n,l)$ and $z(n,l)$ denote the transmitted symbol with energy (per symbol) E_s, the channel transfer function sample and the additive white Gaussian noise with zero mean and variance N_0, respectively.

Channel estimates are first obtained at the pilot positions using a simple least square (LS) demodulation, which for PSK pilot modulation reduces to

$$\hat{H}(n,l) = y(n,l)x^*(n,l) \quad , \quad \{n,l\} \in \mathsf{P} \tag{43}$$

where P is the set of all pilot locations and $()^*$ denotes the complex conjugate operation.

As for MIMO-OFDM, Figure 8 illustrates the LTE pilot grid for a 2x2 antenna configuration. When antenna port 0 is transmitting its pilot symbols, the other antenna is silent. This implies that pilot transmissions from the two antenna ports are completely orthogonal, i.e., MIMO channel estimation in LTE is a straightforward extension of SISO channel estimation techniques.

Once the channel coefficients at the pilot positions are known, the remaining channel coefficients are then calculated using interpolation techniques in both time and frequency directions (Hoeher, Kaiser & Robertson, 1997; Morelli & Mengali, 2001). Given the pilot grid in Figure 8, both 2D and 2x1D channel estimation are possible. It has been shown in (Hoeher, Kaiser & Robertson, 1997) that, for a given computational effort, the 2D approach has comparable performance to the 2x1D estimator where estimation is performed first in the frequency, and then in the time direction, or vice versa.

As mentioned above, both 2D and 2x1D methods rely on minimal *a priori* channel knowledge. Usually, uniform Doppler and delay power spectra are assumed, where the limit values (f_{max}, τ_{max}) are typically fixed to the maximum Doppler bandwidth $B_D = 2f_D$ (where f_D is the maximum channel Doppler frequency to be expected in the system, see section "Synchronization and Cell Search") and to the cyclic prefix length T_{CP}, respectively. Similarly, a large SNR is assumed and the noise variance is set accordingly. This allows to pre-compute the interpolation coefficients offline so that only multiplications by real-valued coefficients and summation operations are required in real time. Also, the minimum number of interpolating coefficient is typically $\ll N$ (with N being the size of the FFT) and can be chosen according to (Hoeher, Kaiser & Robertson, 1997). For the 2x1D approach, the Wiener (MMSE) coefficients are obtained as follows (N_f coefficients in frequency and N_t in time)

Frequency direction:

$$\mathbf{w}_f^{nT} = \left[w_f^n(0),\ldots,w_f^n(N_f-1)\right] = \mathbf{r}_f^{nT}\mathbf{R}_f^{-1} \quad , \quad n \in F \tag{44}$$

Time direction:

$$\mathbf{w}_t^{lT} = \left[w_t^l(0),\ldots,w_t^l(N_t-1)\right] = \mathbf{r}_t^{lT}\mathbf{R}_t^{-1} \quad , \quad l \in T \tag{45}$$

where $()^T$ and $()^{-1}$ denote the vector transpose and the matrix inverse operation, respectively. The elements of the cross- and auto-correlation matrices in (44)-(45) are given by (assuming uniform and symmetric Doppler and delay power spectra)

$$\mathbf{r}_f^n(i) = si(2\pi\tau_{max}\Delta F(n-i)) \quad , \quad i=0,1,\ldots,N_f-1$$
$$\mathbf{R}_f(i,j) = si(2\pi\tau_{max}\Delta F(i-j)) + \frac{N_0}{E_s}\delta(i-j) \quad , \quad i,j=0,1,\ldots,N_f-1 \tag{46}$$

$$\mathbf{r}_t^l(i) = si(2\pi f_{max}T_{symb}(l-i)) \quad , \quad i=0,1,\ldots,N_t-1$$
$$\mathbf{R}_t(i,j) = si(2\pi f_{max}T_{symb}(i-j)) + \frac{N_0}{E_s}\delta(i-j) \quad , \quad i,j=0,1,\ldots,N_t-1 \tag{47}$$

where si is the *sinc* function, while ΔF and T_{symb} denote the subcarrier spacing and the symbol duration, respectively.

Note that the indices n and l in equations (44)-(47) account for the fact that 1D Wiener filtering amounts to a window sliding operation along the frequency or time axis. Also, F and T denote the sets of frequency and time indices, respectively, at which interpolation is performed. The coefficients for the 2D method, although not provided in the following, are computed in a similar manner, see (Hoeher, Kaiser & Robertson, 1997) for a more detailed description.

Depending on the structure and size of the 2D pilot grid, the 2x1D estimator can noticeably reduce the book-keeping required by the 2D approach. In fact, the main drawback of the 2D approach lies in the fact that on the 'edges' of the two-dimensional grid, not all pilots are available and thus several sets of coefficients need to be pre-computed for each different subcarriers belonging to the 'edge' region, thus increasing the memory requirements of the receiver. The same is of course true also for the cascade estimator; however, the size of the 'edge' region is smaller in this type of implementation.

Given the LTE pilot grid in Figure 8, a possible 2x1D estimator could first interpolate in frequency direction at time 0 and $4T_{symb}$ and then, relying on these estimates, it would interpolate in time direction using, for instance, a sliding window of at least 4 reference symbols (see Figure 9). It is clear that in 2x1D approach, frequency interpolation can precede time interpolation (F-T) or, vice versa, time interpolation can precede frequency interpolation (T-F). The choice between F-T and T-F depends on the signal parameters (subcarrier spacing and symbol time), on the time-frequency pilot grid, and on the maximum allowed Doppler bandwidth/delay spread. For example, in the case of the LTE system, given the subcarrier spacing and the symbol length, the requirements on the channel delay spread appear to be more critical than the requirements on the Doppler bandwidth. This is illustrated in Table 1 where the maximum allowed delay spread and Doppler bandwidth (obtained according to the Nyquist theorem in time and frequency) are shown for the cases of frequency-time and time-frequency interpolation, respectively, based on the pilot grid in Figure 8 and a typical LTE parameter setting (3GPP TS 36.211,

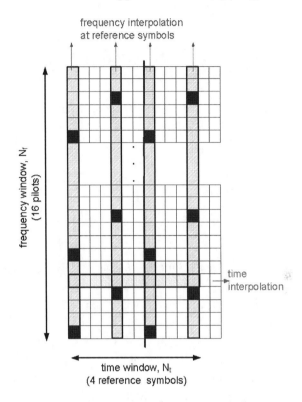

Figure 9. Possible implementation of a 2x1D Wiener estimator applied to the LTE pilot grid

2008-09) with symbol time T_{symb} = 0.07 ms and subcarrier spacing Δf = 15 kHz. It is seen that, for the LTE case, if the channel is interpolated first in time, the resulting pilot spacing in frequency direction will be halved and higher delay spreads could be tolerated. Table 2 provides Δf = 1.116 kHz similar results for a DVB-T/H 8K system with parameters and pilot grid given by and T_{symb} = 1.12 ms (ETSI EN 300 744, v.1.5.1, 2004-11).

At this point, a few comments on the performance of the robust Wiener channel estimator are in order. Because of its 'mismatched' nature, the performance of the 2x1D filter is not always completely satisfactory and BER floor is observed at high SNR values due to large interpolation errors (Auer, 2005; Carbonelli & Franz, 2008; Zhu & Murch, 2002). In (Zhu & Murch., 2002) it is shown that the BER floor caused by an irreducible estimation error (i.e., an error which does not

Table 1. Limit values for Doppler bandwidth and delay spread in a typical LTE system

LTE	1) Time 2) Frequency	1) Frequency 2) Time
$B_{d,lim} = 1/(2 \cdot DT)$	$1/(2 \cdot 6 \cdot T_{symb}) = 1167$ Hz	$1/(2 \cdot 4 \cdot T_{symb}) = 1745$ Hz
$T_{d,lim} = 1/(DF)$	$1/(3 \cdot \Delta f) = 22.2$ μs	$1/(6 \cdot \Delta f) = 11.1$ μs

Table 2. Limit values for Doppler bandwidth and delay spread in a typical DVB-T/H system (ETSI EN 300 744, v.1.5.1, 2004-11)

DVB-T/H	1) Time 2) Frequency	1) Frequency 2) Time
$B_{d,lim} = 1/(2 \cdot DT)$	$1/(2 \cdot 4 \cdot T_{symb}) = 112$ Hz	$1/(2 \cdot T_{symb}) = 446$ Hz
$T_{d,lim} = 1/(DF)$	$1/(3 \cdot \Delta f) = 298$ μs	$1/(12 \cdot \Delta f) = 74$ μs

decrease as the SNR increases) starts at lower SNR values as the number of transmitting antennas increases. As discussed in (Carbonelli & Franz, 2008) BER performance can be significantly improved (and not only in the high SNR region) if side information on the channel statistics and on the operating SNR is available. Specifically, an adaptive implementation of the channel estimator is envisaged where estimates of the maximum delay spread T_{mds}, the actual Doppler spread f_D and the noise variance are used to more judiciously select the Wiener coefficient sets. For example, a rough estimate of the channel length could be provided by the time tracking loop. Similarly, a Doppler estimator could be run in parallel to the channel estimator and feed a more correct estimate of the mobile speed. We recall that in a robust implementation of the Wiener filter, the time interpolation filters are pre-computed assuming a very large mobile velocity. If, however, the channel is static, significant performance improvement can be achieved by using this information and selecting the appropriate time filter, which, for the case of static channels, will be constant with coefficients equal to $1/N_t$. Finally, the noise variance estimates which the SNR estimator normally provides to the equalizer and LLR generation block could also be used at the channel estimator to better match the Wiener interpolation filters in the very low SNR region.

Figure 10 shows the uncoded BER performance of a 2x2 QPSK system employing ML detection. The channel is Extended Vehicular A (EVA) and the velocity is 100 km/h. The gain of the 'adaptive' approach over the 'robust' one is visible at all SNR values. Figure 11 shows some throughput results for a 1x2 Rx diversity (RxDiv) system with maximum ratio combining. The channel is Extended Typical Urban (ETU) and the 5 Hz Doppler spread corresponds to a speed of about 3 km/h. For such a configuration the maximum throughput is 4.584 Mbits/sec. The maximum throughput is achieved when *all* packets are correctly detected after HARQ processing. Again we observe almost a 1 dB gain when an adaptive channel estimator is used.

Computational Efforts and Memory Requirements

To give some insights on the complexity of the channel estimator introduced above we focus on a 1x2 receive diversity configuration for a transmission bandwidth of 10 MHz. Assuming a full bandwidth occupation, the number of used subcarriers is $N_{data} = 600$ and the number of pilots

Figure 10. Uncoded BER performance of QPSK 2x2 system: robust and adaptive 2x1D Wiener channel estimator

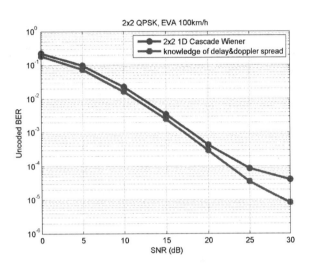

(reference symbols only) $N_{pilot} = 100$. As mentioned earlier, in a robust implementation of the channel estimator, the interpolation coefficients are real and can be computed offline. Specifically for an example analysis, the length of the frequency and time interpolation filter was fixed to $N_f = 12$ and $N_t = 4$, respectively. The overall number of complex additions and multiplications required per subframe (14 OFDM symbols) and per transmit-receive pair is then given by (including also the initial demodulation of the pilots)

$$N_{mlk} = N_{add} = 4\left(N_{pilot} + (N_{data} - N_{pilot})N_f\right) + 10 N_t N_{data} = 48400 \qquad (48)$$

We note that $N_t = 4$ implies an initial buffering (and delay) of at least 4 OFDM symbols. In other terms, at each reference symbol, we first perform frequency interpolation. Then, using the 4 most recent frequency estimates, we interpolate the previous 2 or 3 OFDM symbols, depending on whether we are in the first or second reference symbol of a slot.

Alternative choices are possible for the positioning and length of the Wiener filters. For instance, the time interpolation window could be symmetrically placed around the OFDM symbol to be interpolated so as to leave 2 reference symbols to the right and 2 reference symbols to the left. Such a solution allows to more judiciously exploiting the channel correlation properties. However, the delay requirements would be higher. Similarly, longer filters can be designed for improved noise averaging at the expenses of complexity.

Some considerations should be made also on the memory storage required for the different sets of Wiener coefficients. By exploiting the symmetries in the pilot grid, the number of coefficients to be stored for time and frequency interpolation can be drastically reduced. To this end, we remind that the arguments of the $si()$ in (46)-(47) are a function of the distance between pilots and between pilots and subcarrier to be interpolated. Since the pilot patterns repeats in frequency (and partly also in time) and the $si()$ function is a symmetric function, many coefficients can be reused for more than one subcarrier. At the edges of the bandwidth (or of the selected time interpolation window), symmetries are typically lost and a different set of coefficients for each subcarrier belonging to these regions need to computed.

Figure 11. Throughput performance of QPSK 1x2 system: robust and adaptive 2x1D Wiener channel estimator

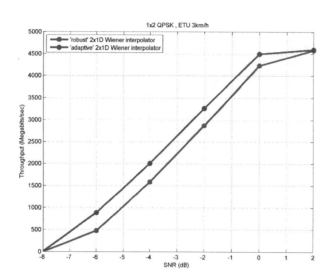

The longer the interpolation filter, the larger the number of 'edge' subcarriers will be.

Finally, it should be noted than when an adaptive channel estimator is employed, then, multiple coefficient sets need to be stored for each possible value of the delay spread, Doppler spread and SNR. Typically, 2 values are sufficient for the SNR (5 dB for smaller modulation orders, and 25 dB for higher modulation orders) and 3 values for the Doppler spread corresponding to the velocities 3 km/h, 100 km/h and 300 km/h. As far as the channel delay spread is concerned, it is sufficient to have at least two coefficient sets, one for the case of single path channel and one for the case of frequency selective fading, where the delay spread should be set greater or equal to the cyclic prefix.

5. EQUALIZATION AND DETECTION

Additive white Gaussian noise and channel distortions resulting in interferences cause high error rate if no equalization is employed compensating for it. This requires an accurate estimation of the channel distortions in the receiver (see last section "Channel Estimation"). Moreover, the separation of the spatially multiplexed data in MIMO systems, as specified in LTE (3GPP TS 36.211 2008-09), additionally makes high demands on the design of channel equalizer and data detector, respectively.

Suppressing interferences among layers results in a linear MIMO detector, such as, for instance, a linear minimum mean square error (MMSE) equalizer. The linear MIMO detector represents a low-complexity solution, but it suffers from the fact that it can only achieve a diversity order of one. The optimum detector for a MIMO system is the maximum-likelihood (ML) MIMO detector (van Nee, van Zelst & Awater, 2000). However, performance improvement[6] is achieved at cost of potentially high computational complexity. A brute force implementation of the soft-output ML MIMO detector, e.g., exhibits an exponential complexity with respect to the number of transmit antennas and the modulation order such that this approach is not yet suited for area-efficient hardware implementations (of course neither software implementations) – especially for a large number of transmit and receive antennas, respectively, and a high modulation order. Therefore, there is a need for a trade-off between performance and computational complexity so that the goal is to design a low-complexity detector with close-to-ML performance.

In (Murugan, El Gamal, Damen, & Caire, 2006), tree-search based schemes, such as sphere decoding or the M-algorithm, known from sequential decoding (Anderson & Mohan, 1984) have been applied to efficiently approach ML performance in MIMO detection with reduced complexity. Here, it is important to define the tree in such a way that the number of visited nodes is small. This can be achieved, with only small performance loss, by using the mean square error metric instead of the Euclidian metric, as shown in (Joham, Barbero, Lang, Utschick, Thompson & Ratnarajah, 2008). Layer sorting based on sorted QR decomposition (Wübben, Böhnke, Kühn & Kammeyer, 2003) or permuted Cholesky decomposition (Kusume, Joham, Utschick & Bauch, 2007) yields further complexity reduction. Furthermore, choosing the equivalent real-valued signal model representation instead of the complex baseband representation gives additional degree of freedom with respect to sorting, and hence potentially additional complexity reduction[7].

Since LTE (3GPP TS 36.211, 2008-09) also specifies SISO, SIMO, and SFBC, the corresponding equalizer realizations must be implemented in the receiver. For SIMO, e.g., a matched filter approach can be used to exploit the provided additional degree of freedom to enhance the SNR and improve Rx diversity. Moreover, a combining scheme based on the MMSE criterion can suppress strong inter-cell interference at the cell edges based on the correlation of the spatial noise-plus-interference. For the orthogonal Alamouti-type space-frequency block code (SFBC), a classical linear receiver is typically applied (Alamouti, 1998).

Beyond that, log-likelihood ratios (LLRs) must be generated in all equalizer modes as input for the channel decoder. This is particularly challenging for close-to-ML MIMO equalizers, since not only the ML solution or a solution close to that, respectively, must be found, but also the best counter-hypothesis (assuming max-log approximation) for each bit of a demodulated symbol. Indeed, this significantly increases complexity.

MIMO Detection

In the following, we consider an $N_T \times N_R$ MIMO system, i.e., a system with N_T transmit and N_R receive antennas, where $N_T \geq N_R$. The input signal to this system is given by the vector $\mathbf{s} = \begin{bmatrix} s_0, & \ldots, & s_{N_T-1} \end{bmatrix}^T$ of N_T symbols, each of which representing K independent identically distributed (i.i.d.) information bits and chosen from a complex Q-ary Quadrature Amplitude Modulation (QAM) alphabet with $Q = 2^K$. The symbol vector $\mathbf{s}$ is transmitted over the MIMO channel with transfer matrix $\mathbf{H}$ having complex gain factor $h_{i,j}$, $i = 0,1,\ldots,N_R - 1, j = 0,1,\ldots,N_T - 1$, and additionally perturbed by additive white Gaussian noise, with zero mean and variance σ_z^2, represented by $\mathbf{z} = \begin{bmatrix} z_0, \ldots, z_{N_R-1} \end{bmatrix}^T$. Hence, the received signal is given by

$$\mathbf{R} = \mathbf{Hs} + \mathbf{z} \tag{49}$$

The covariance matrix of $\mathbf{s}$ is given by $\Phi_{ss} = \mathrm{E}\begin{bmatrix} \mathbf{ss}^H \end{bmatrix} = \mathbf{I}_{N_T}$. The channel matrix is assumed to be known (see section "Channel Estimation"). The *a posteriori LLRs* are calculated from the received signal $\mathbf{r}$ as

$$L\left(c_{n,k} \mid \mathbf{r}\right) = \ln \frac{\Pr\left(c_{n,k} = +1 \mid \mathbf{r}\right)}{\Pr\left(c_{n,k} = -1 \mid \mathbf{r}\right)} \tag{50}$$

for the k-th bit of the n-th symbol.

After some intermediate steps, where we use the *max-log approximation*, and assuming statistically independent and equally likely bits $c_{n,k}$, we obtain the *soft-output ML LLR*

$$L(c_{n,k} \mid \mathbf{r}) \approx \min_{c \in C_{n,k}^{+1}} \left(\left\| \Phi_{zz}^{-1/2} (\mathbf{r} - \mathbf{Hs}) \right\|^2 \right) - \min_{c \in C_{n,k}^{-1}} \left(\left\| \Phi_{zz}^{-1/2} (\mathbf{r} - \mathbf{Hs}) \right\|^2 \right) \quad (51)$$

The LLRs in (51) can be approximately represented as the difference between the *metrics* $\Lambda(c_{n,k}^{+1})$ and $\Lambda(c_{n,k}^{-1})$:

$$L(c_{n,k} \mid \mathbf{r}) \approx \min_{c \in C_{n,k}^{+1}} \left(\Lambda(c_{n,k}^{+1}) \right) - \min_{c \in C_{n,k}^{-1}} \left(\Lambda(c_{n,k}^{-1}) \right) \quad (52)$$

which correspond to the *hypotheses* that $c_{n,k} = +1$ and $c_{n,k} = -1$, respectively, are transmitted. $C_{n,k}^{+1}$ and $C_{n,k}^{-1}$ are the sets of $2^{N_T K - 1}$ bit vectors c, having $c_{n,k} = +1$ and $c_{n,k} = -1$. Then, a so-called *ML hypothesis* is defined as

$$\mathbf{s}^{ML} = \arg\min_{\mathbf{s} \in S} \left\| \Phi_{zz}^{-1/2} (\mathbf{r} - \mathbf{Hs}) \right\|^2 \quad (53)$$

where S is the transmit symbol set. Therefore the ML hypothesis $\mathbf{s}^{ML}$ is the solution minimizing the metric $\Lambda(\mathbf{s})$ over all possible symbol vectors $\mathbf{s}$, i.e.,

$$\Lambda^{ML} = \min_{\mathbf{s} \in S} \left\| \Phi_{zz}^{-1/2} (\mathbf{r} - \mathbf{Hs}) \right\|^2 \quad (54)$$

with $\mathbf{c}^{ML}$ being the corresponding bit vector. With $S_{n,k}^{\overline{ML}}$ denoting the set S without the elements whose corresponding bit $c_{n,k} = c_{n,k}^{ML}$, i.e., the k-th bit of the n-th symbol in $\mathbf{s}^{ML}$, the second term in (51) is computed as

$$\Lambda_{n,k}^{\overline{ML}} = \min_{\mathbf{s} \in S_{n,k}^{\overline{ML}}} \left\| \Phi_{zz}^{-1/2} (\mathbf{r} - \mathbf{Hs}) \right\|^2 \quad (55)$$

where the element $c_{n,k}^{\overline{ML}}$ in $S_{n,k}^{\overline{ML}}$ is called *counter hypothesis* at symbol position n and bit position k (Studer, Burg & Bölcskei, 2006). Hence, the soft-output ML LLR is given by

$$L(c_{n,k}) \approx \begin{cases} \Lambda^{ML} - \Lambda_{n,k}^{\overline{ML}}, & c_{n,k}^{\overline{ML}} = +1 \\ \Lambda_{n,k}^{\overline{ML}} - \Lambda^{ML}, & c_{n,k}^{\overline{ML}} = -1 \end{cases} \quad (56)$$

From (56), it follows that solving the ML MIMO detection problem requires the calculation of the ML hypothesis as well as the minimum counter hypothesis. To improve the performance-complexity trade-off, the MMSE metric can be used instead of the ML metric, as shown in (Murugan, El Gamal, Damen & Caire, 2006). It can be derived from the ML metric. Using additionally a permuted channel transfer matrix and QR-decomposition the ML metric in (54) can be represented as a sum of partial distances and metric increments, respectively. For details, we refer to (Wübben, Böhnke, Kühn & Kammeyer, 2001).

Clearly, (56) cannot be solved directly in a brute-force search due to its high complexity – especially for large K and N_T as the cardinality of S grows exponentially with KN_T, i.e., $|S| = 2^{N_T K}$. Therefore, detection algorithms have been proposed that construct a subset $P \subseteq S$ to determine the LLRs. Again, the cardinality of P and the choice of its elements significantly impacts the performance and complexity, since P should be chosen to be sufficiently large to contain the ML hypothesis and appropriate counter hypothesis, but small enough to reduce complexity. These detection algorithms are based on a so-called *tree search*. In the following, we will consider the Schnorr-Euchner sphere decoder (SESD) (Studer, Burg & Bölcskei, 2006), the M-algorithm (Murugan, El Gamal, Damen & Caire, 2006, Anderson & Mohan, 1984), the fixed sphere decoder (FSD) (Barbero & Thompson, 2006), and the parallel smart candidate adding (PSCA) algorithm (Zimmermann, Fettweis, Milliner & Barry, 2008).

Tree Search Based MIMO Detection

The SESD with radius reduction (Studer, Burg, and Bölcskei, 2006) applies a depth-first strategy, i.e., it visits nodes branch-by-branch starting from the root node until it reaches a leaf node, and then applies backtracking. Child nodes to a specific parent node are visited and ordered in ascending distance increment order. Then, based on this order nodes are selected as new parent nodes and the procedure is repeated. The number of visited nodes is reduced by defining the search space as hyper-sphere of radius γ around the signal vector and considering only symbols within this space. With respect to the tree search this means that child nodes are only kept, if the partial distance for this node does not exceed γ; otherwise it is omitted. A possible strategy is to set the initial value of γ to ∞ and update it according to

$$\gamma_{new} = \min(\gamma_{old}, \Lambda(\mathbf{s})) \qquad (57)$$

when a leaf node s has been reached. If no more leaf nodes are found inside the sphere, the procedure stops and γ equals the current minimum distance.

Compared to the SESD, the M-algorithm (Murugan, El Gamal, Damen & Caire, 2006, Anderson & Mohan, 1984) represents a breadth first algorithm, i.e., nodes are visited on a layer-by-layer basis during the search procedure. Here, a set of M nodes is kept at each layer, i.e., starting from the parent nodes (and the root node in the first layer, respectively) the partial distances of all child nodes are calculated and the best M child nodes are chosen as parent nodes in the next layer; all other child nodes are omitted. Then, this procedure is repeated based on the newly chosen set of parent nodes. This is illustrated in Figure 12, where yellow nodes represent child nodes that are kept, i.e., that are parent nodes in the next layer, whereas blue nodes indicate omitted child nodes. Clearly, performance close to ML can be achieved, if M is chosen to be sufficiently large, while limiting M reduces the number of visited nodes, and hence the complexity on the other hand.

The FSD (Barbero & Thompson, 2006) is another type of breadth-first algorithm that additionally features a parallel structure, and hence offers the possibility to be implemented in a parallel architecture. Parent nodes are extended with child nodes in the next layer, similar to the M-algorithm. However, while the number of child nodes kept per layer may vary depending on the parent node they stem from in case of the M-algorithm, it is fixed for each layer in case of the FSD and given by b_i for layer $i = 1,...,I$, where I denotes the overall number of layers. Hence, the vector $\mathbf{b} = (b_1,...,b_I)$ is the essential design parameter that determines performance and complexity of the FSD at the

Figure 12. M-algorithm for M = 2 in a four layer system

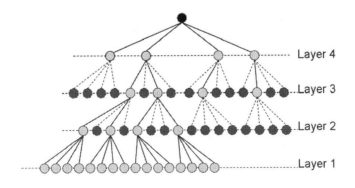

same time. An example of the FSD tree search procedure is shown in Figure 13.

The PSCA (Zimmermann, Fettweis, Milliner & Barry, 2008) algorithm also applies a breadth-first search as the M-algorithm and the FSD with the essential key feature that it concurrently finds the counter hypothesis with the ML search. In the PSCA algorithm, a number of child nodes b_i extended per layer i, $i = 1,...,I$, is fixed, too, comparable to the FSD. Here, the parameter N_{layer} gives the number of layers with $b_i = 2$ (counted from the root node); for all other layers $b_i = 1$. Starting from the root node, the best child nodes are determined; then the best of them is denoted as partial ML (PML) node (marked with a red circle in Figure 14). The additional child node has already provided a counter hypothesis and another counter hypothesis is added. With this set of nodes as parent nodes, the procedure is repeated in the next layer, as seen in Figure 14 for $N_{layer} = 2$.

Performance and Complexity

In the following, we consider performance and complexity of the algorithms previously presented. In Figure 15, the throughput curves of a 16QAM 2x2 LTE system are shown. As seen, all algorithms except for the linear MMSE detector perform comparably and very close to ML, which is due to the choice of the design parameters. However, the complexity of all methods significantly differs as seen from Figure 16.

Here, we especially focus on two detectors (Note that in case of the SD detector, only average complexity is shown – the worst case complexity, which is essential for hardware implementations,

*Figure 13. FSD for **b**= (1,1,2,4) in a four layer system*

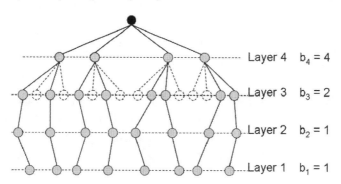

Figure 14. PSCA with $N_{layer} = 2$ in a four layer system

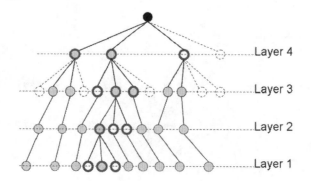

Figure 15. Performance comparison for different MIMO detection algorithms

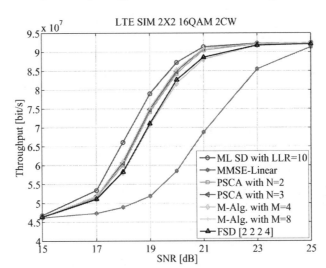

Figure 16. Complexity comparison for different MIMO detection algorithms

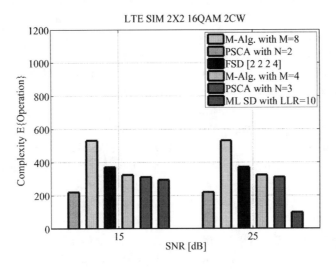

is much higher than depicted in Figure 16). While the M-algorithm with $M = 8$ exhibits higher complexity than most of the other detectors, the PSCA with $N = 2$ features lower complexity. Therefore, the PSCA represents a good choice with respect to the required performance-complexity trade-off of MIMO detectors.

6. CHANNEL CODING AND DECODING

While the control channel data is protected by a 64-state tail-biting convolutional code, LTE (3GPP TS 36.212, 2008-09) ultimately adopted the turbo channel coding scheme from 3GPP HSPA (3GPP TS 25.212, 2008-09) for the user data despite the numerous proposals that voted for low density parity check (LDPC) coding (Gallager, 1962) during standardization activities. Although its generator polynomials are the same as for the 3GPP HSPA turbo code, the turbo code internal interleaver was chosen differently. Whereas in HSPA the prunable prime interleaver (PIL) is used as the turbo internal interleaver, the so-called quadratic permutation polynomial (QPP) turbo code interleaver is standardized for LTE. With the QPP interleaver (Takeshita, 2006), the LTE turbo code has comparable or even better performances compared to the HSPA turbo code. However, the main reason for introducing a new turbo internal interleaver for LTE was to ease an implementation in hardware for achieving the high throughput (Tput) requirements.

From transmitter perspective the payload block arriving from the MAC layer is first extended by a length-24 CRC check. Then a subdivision into codeblocks (of equal or approximately equal length) is applied. Each codeblock independently undergoes turbo encoding where in case of multiple codeblocks to each codeblock an additional length-24 CRC is attached. The codeblock-specific CRC can serve as an early stopping criterion of the iterative turbo decoding process thereby helping to reduce receiver power consumption. The maximum codeblock size (and therefore also turbo interleaver size) was chosen to be 6144 bits being roughly 1000 bits more compared to 3GPP HSPA.

Like HSPA, LTE supports hybrid ARQ with incremental redundancy (IR) and soft combining at the receiver. The reasoning behind the usage of HARQ is to increases robustness on top of the running link adaptation based on channel quality information (CQI). Link adaptation based on CQI controls the modulation and coderate (and possibly transmit power) of a codeword while, loosely speaking, HARQ aims at squeezing out the best performance for the given modulation and coderate if the packet was not received successfully right away.

Despite the possibility of transmitting data over 4 spatial layers, for LTE no more than 2 codewords are transmitted in time/frequency and space per user and per subframe. Each codeword is associated with its own HARQ process and link adaptation. The presence of 2 codewords in case of multi-layer transmission allows to apply more advanced receiver structures like successive interference cancellation (SIC) with or without the turbo decoding iterations being involved in the SIC-loop.

Compared to HSPA the rate-matching algorithm (3GPP TS 36.212, 2008-09) to support arbitrary code rates and IR was simplified. Figure 17 shows the virtual buffer rate-matching used by LTE. Each encoded bit stream (systematic "sys", parity 1 "par1" and 2 "par2") of the rate-1/3 turbo encoder output of each code block is first permuted by a subblock interleaver and organized in a virtual buffer. The interleaving rule for the systematic and parity 1 streams are identical[8] while for the parity 2 part the rule is slightly different. Observe that the interleaved parity 1 and 2 bits are bit by bit alternately organized in the virtual buffer. Equidistant entry points to this buffer mark starting positions, denoted as idx(rv), rv = 0,1,2, in the figure, from which data is read out from the virtual buffer for the individual redundancy versions (RVs) in a wrap-around fashion. In case the HARQ buffer allocated for a particular process and codeblock is smaller than the encoded data stream, the wrap-around would occur earlier and the entry points are compressed as indicated in the left part of Figure 17. A two-bit field transmitted on the PDCCH indicates to the UE which redundancy version was used by the base station. On the receiver side softbits in the form of log likelihood ratios (LLRs) are combined with those already available from previous transmissions of the same packet. Since 8 parallel HARQ processes per (each of the up to 2) codeword(s) run in parallel and since the payload sizes can be roughly as large as 150000 bits for the highest LTE category, the HARQ buffer is a major memory contributor in the receiver. After HARQ-combining the packet undergoes turbo decoding and CRC checking.

The turbo code (Berrou, Glavieux & Thitimajshima, 1993) used for LTE (like HSPA) is composed of a parallel concatenation of 2 identical binary rate-1/2 systematic recursive 8-state convolutional encoders whose inputs are linked by an interleaver (the turbo-internal interleaver). The systematic part, being the raw data before encoding, is only transmitted once thereby creating a rate-1/3 turbo 'mother' code (before rate-matching). The decoding of turbo codes is done in an iterative fashion as indicated in Figure 18. Each of the two received data sequences $(L(X_k), L(Y_k^1))$ and $(L(I(X_k)), L(Y_k^2))$ is decoded with a probabilistic soft-in-soft-out (SISO) decoder operating on the respective decoding trellis where $L(X_k), L(Y_k^1), L(Y_k^2)$, denote the LLR of the k-th systematic, parity1, and parity2 bit, respectively, and where $I(.)$ denotes the turbo internal interleaver function. A particular part of the output of the SISO module, the so-called extrinsic information $L_{ext}(X_k)$, is passed on to the next SISO module as a priori information. After a certain number of iterations decoding is stopped and a hard decision is made on the final soft output.

Optimally, the SISO decoding module computes the maximum *a posteriori* (MAP) LLR $L_{app}(X_k) = \log P(X_k = 0|R)/P(X_k = 0|R)$ for each systematic bit given the entire received code sequence R. The trellis based implementation in the form of the well-known BCJR-algorithm (Bahl, Cocke, Jelink & Raviv 1974) operating in the log-domain (log-MAP or forward-backward algorithm) (Robertson, Hoeher & Villebrun, 1997) makes use of forward and backward recursions running on the decoding trellis and storage of partial path metrics associated with trellis states. One particularly important approximation thereof is known as the max-log-MAP algorithm (Rob-

Figure 17. Virtual buffer rate matching and incremental redundancy. Limited buffer rate matching left.

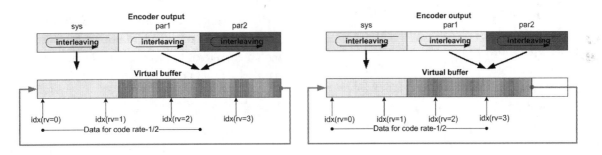

Figure 18. Iterative turbo decoding structure

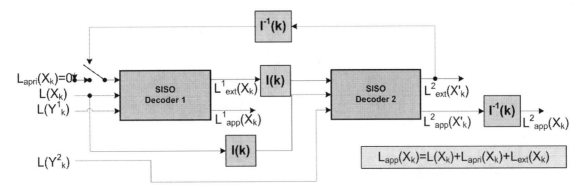

ertson, Hoeher & Villebrun, 1997) for which the output LLR is given by

$$L_{app}(X_k) = \max_{S^0}\left(\alpha_{k-1}(s_{k-1}) + \gamma_k(s_{k-1},s_k) + \beta_k(s_k)\right)$$
$$- \max_{S^1}\left(\alpha_{k-1}(s_{k-1}) + \gamma_k(s_{k-1},s_k) + \beta_k(s_k)\right) \quad (58)$$

where $\alpha_k(s_k)$, $\beta_k(s_k)$ denote the forward and backward state metrics and $\gamma_k(s_{k-1},s_k)$ the branch metrics associated with a state transition. S^0 and S^1 define the sets of state transitions associated with $X_k = 0$ and $X_k = 1$, respectively. The forward and backward recursions as well as the branch metrics are defined as follows

$$\alpha_k(s_k) = \max_{\forall s_{k-1}}\left(\alpha_{k-1}(s_{k-1}) + \gamma_k(s_{k-1},s_k)\right) \quad (59)$$

$$\beta_{k-1}(s_{k-1}) = \max_{\forall s_k}\left(\beta_k(s_k) + \gamma_k(s_{k-1},s_k)\right) \quad (60)$$

$$\gamma_k(s_{k-1},s_k) = \bar{X}_k\left(L(X_k) + L_{apri,k}(X_k)\right) + \bar{Y}_k L(Y_k) \quad (61)$$

with $\bar{X}_k$ meaning logical inversion of the bit. One nice property of the max-log-MAP algorithm is that all operations to be done are just additions and comparisons without the need for further table lookups as would be for the log-MAP. Moreover, the performance gap when using max-log-MAP instead of log-MAP can be closed up to a very few tenth of dBs if the extrinsic information is properly scaled down before passing it on to the subsequent SISO module (Claussen, Karimi & Mulgrew, 2005).

An exact implementation of (58) would be unsuitable for a practical implementation for two reasons. Firstly, the forward (or backward) recursion had to be stored over the entire decoding trellis and, secondly the decoding throughput would be limited by 1 decoded bit per 2 time units[9]. This is due to the fact that the exact implementation of (58) requires to execute a forward recursion from the very left corner of the trellis until the right end followed by a recursion and output LLR generation running backwardly as shown on the left part of Figure 19. With little loss in B(L)ER performance these limitations can be avoided by window-based methods for which many variants have been proposed in the literature (Benedetto, Dinoi, Montorsi & Tarable, 2006), (May, Neeb & Wehn, 2007). Here, the decoding trellis is subdivided into subblocks (windows) where multiple parallel sub-SISO-decoders (workers) work on different subblocks at the same time. One variant of a parallel SISO module (Kwak, Park, Yoon & Lee, 2003) is indicated in the right part of Figure 19. Here the forward and backward recursions are first "learned" over the acquisition length L_{acq}. Other variants use stored state metric information from previous iterations to initialize the recursion at the window borders (Bougard, Giulietti, Van der Perre & Catthoor, 2003). Observe from the figure that the decoding throughput has been roughly increased by a factor of M when going from (a) to (b) if the acquisition length is small compared to K/M. Of course more trivial measures for increased decoding throughput are thinkable, e.g. more parallel (but not-internally parallelized) SISO modules with pipelining of turbo half iterations. However, the advantage of the former approach is that the parallel workers can share the memories for the input and 'a-priori' LLRs.

Moreover, one problem is left when studying high throughput turbo decoders that try to increase the internal parallelism of a SISO module. It turns out that for a general turbo interleaver rule (and also for the turbo interleaver standardized for HSPA) the achievable throughput is limited by memory read/write access conflicts to the 'a-priori' memory which exchanges the feedback information between the two constituent decoders. However, for the chosen LTE interleaver these access conflicts can be avoided due to its contention freeness property (Takeshita, 2006).

Mathematically, a size interleaver $I(j)$ is said to be *maximum contention-free* if the following condition holds

Figure 19. (a) Exact implementation of (max-)log-MAP algorithm, (b) implementation with M subblock decoders

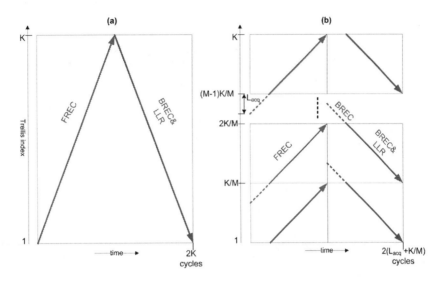

$$\lfloor I(i+jW)/W \rfloor \neq \lfloor I(i+kW)/W \rfloor, \quad (62)$$

for any window length W dividing the interleaver length K. It can be shown that the QPP interleavers chosen for LTE posseses this property. They are generated by the rule

$$I(k) = \left(ak + bk^2\right) \bmod K \quad (63)$$

for some integer values a and b depending on K.

Figure 20 illustrates the memory contention freeness property of the LTE interleaver. As an example consider a code block of length 16 which is subdivided into 4 windows which are processed simultaneously. Whenever the i-th index (the index having fixed offset $i=0,1,2,3$ within each window) within two or more windows try to fetch data from the memory banks in parallel, no memory access conflicts occur, no matter whether the first constituent decoder operates on the non-interleaved data or the second decoder operates on the permuted data given by the permutation rule $I(k)$. In Figure 20 the situation is shown in red and blue for offsets $i = 1, 2$, respectively.

Computational Efforts and Memory Requirements

The memory requirement of the outer receiver is dominated by the HARQ buffer size which is given by the LTE standard (3GPP TS 36.306, 2008-05) and the number of bits used to represent a softbit (LLR). As can be seen from Table 3 for e.g. a Cat-4 terminal, the HARQ buffer size is approximately 1.8M softbits which is used by up to 16 HARQ processes being active in parallel (8 per codeword).

The minimal buffer requirement for a high throughput MAP decoder with parallel processors working on the decoding trellis requires storage for 3·6144 soft input LLRs and 6144 extrinsic LLRs used for information exchange during the iterative decoding process. Typically, 4 to 8 bits are enough to represent an input or an extrinsic LLR. The number 6144 is given by the maximum code block size used for LTE.

Neglecting the computational effort for the address generation for the subblock deinterleaving required for HARQ processing, the soft combining amounts to N_{data} additions for a retransmission step. In case of a newly received packet, N_{data} softbits

Figure 20. Contention freeness property

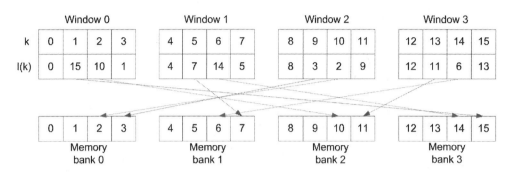

Table 3. Downlink physical layer parameter values

UE Category	Max payload per subframe	Max payload per codeword	HARQ buffer size [softbits]	Max number of layers for spatial Mux
1	10296	10276	250368	1
2	51024	51024	1237248	2
3	102048	75376	1237248	2
4	150752	75376	1827072	2
5	302752	151376	3667200	4

are copied to the HARQ buffer and additional $N_{ir} - N_{data}$ softbits are 'zero-flushed', where N_{data} denotes the number of bits transmitted in a particular subframe and N_{ir} is the number of softbits reserved for the current HARQ process.

The computational effort for turbo decoding shall now be given. For the branch metric calculation (61) 2 additions are required per trellis section. For the forward (or backward) recursion 16 additions and 8 max-operations are needed per trellis section. To compute the APP 32 additions, 16 max-operations and 1 subtraction is required. One further subtraction is needed to compute the apriori information for the next decoder run. For the ideal non-window-based implementation all in all 68 additions, 1 subtraction, and 32 max-operations sum up per bit to be decoded and per max-log-MAP decoder run (half iteration). For the highest LTE UE-category this leads to roughly $302752 \cdot 16 \cdot 101/1 \text{ms} \approx 490 \cdot 10^9$ operations per second for turbo decoding assuming 16 half iterations. Mind that assuming the ideal, non-window based method is still the best case with respect to computational complexity. Many window-based methods (not shown within this paper) that aim at minimizing the state metric storage requirements would require even more operations due to the multiple execution of backward recursions.

7. CONCLUSION

In this chapter, we overviewed the current 3GPP LTE standard (release 8) and its evolution to the LTE-A (release 9), and highlighted first implementation details in an LTE terminal. The sample core functional algorithms of the baseband signal processing, especially the synchronization, cell search, channel estimation and equalization, and the turbo channel decoder, were described and analysed. Their performances, computational efforts, memory requirements, and relevant implementation challenges were discussed. As a result, LTE terminals can be implemented with currently available semiconductor technologies. However, flexible architecture and low-complexity but high-performance algorithms are required to deal with quick and precise synchronization, spatial equalization and channel decoding in particular in different MIMO modes, and to ensure low power consumption and small silicon area.

ACKNOWLEDGMENT

The authors wish to gratefully acknowledge the support of their colleagues at Infineon Technologies. Parts of this work are carried out within the scope of the EUREKA MEDEA+ project MI-MOWA, which is partly funded by the German Federal Ministry of Education and Research.

REFERENCES

3GPP R1-091580 (2009). *LTE-performance and IMT-Advanced requirements*. Presented at 3GPP TSG RAN WG1 Meeting #56bis, Seoul, Korea.

3GPP RP-080756. (2008). *3GPP presentation on the LTE-Advanced as an IMT-Advanced technology solution*. Presented at the 41st 3GPP TSG RAN meeting, Kobe, Japan.

3GPP RP-090067. (2009). *Status report for WI to TSG RAN; work item name: further advancements for E-UTRA (LTE-Advanced)*. Presented at the 43rd 3GPP TSG RAN meeting, Biarritz, France.

3GPP TR 25.913, V7.3.0 (2006-03). *Requirements for evolved UTRA (E-UTRA) and evolved UTRAN (E-UTRAN) (release 7)*. Technical report, 3GPP, TSG RAN.

3GPP TR 36.814, V0.4.1 (2009-02). *Further advancements for E-UTRA; physical Layer aspects (release 9)*. Technical report, 3GPP TSG RAN.

3GPP TR 36.913, V8.0.1 (2009-03). *Requirements for Further Advancements for E-UTRA (LTE-Advanced) (release 8)*. Technical report, 3GPP TSG RAN.

3GPP TS 25.212, V8.4.0 (2008-12). *Multiplexing and channel coding (FDD) (release 8)*. Technical specification, 3GPP, TSG RAN.

3GPP TS 36.101, V8.3.0 (2008-09). *Evolved universal terrestrial radio access (E-UTRA); user equipment (UE) radio transmission and reception (release 8)*. Technical specification, 3GPP, TSG RAN.

3GPP TS 36.211, V8.4.0 (2008-09). *Evolved universal terrestrial radio access (E-UTRA); physical channels and modulation (release 8)*. Technical specification, 3GPP, TSG RAN.

3GPP TS 36.212, V8.4.0 (2008-09). *Evolved universal terrestrial radio access (E-UTRA); multiplexing and channel coding (release 8)*. Technical specification, 3GPP, TSG RAN.

3GPP TS 36.213, V8.4.0 (2008-09). *Evolved universal terrestrial radio access (E-UTRA); physical layer procedures (release 8)*. Technical specification, 3GPP, TSG RAN.

3GPP TS 36.300, V8.6.0 (2008-09). *Evolved universal terrestrial radio access (E-UTRA) and evolved universal terrestrial radio access network (E-UTRAN); overall description; stage 2 (release 8)*. Technical specification, 3GPP, TSG RAN.

3GPP TS 36.306, V8.2.0 (2008-05). *Evolved universal terrestrial radio access (E-UTRA); user equipment (UE) radio access capabilities (release 8)*. Technical specification, 3GPP, TSG RAN.

Alamouti, S. M. (1998). A simple transmit diversity technique for wireless communications. *IEEE Journal on Selected Areas in Communications*, *16*(8), 1451–1458. doi:10.1109/49.730453

Anderson, J. B., & Mohan, S. (1984). Sequential coding algorithms: A survey and cost analysis. *IEEE Transactions on Communications, 32*(2), 169–176. doi:10.1109/TCOM.1984.1096023

Auer, G. (2005), Modeling of OFDM Channel Estimation Errors. In *10th International OFDM Workshop*.

Bahl, L. R., Cocke, J., Jelink, F., & Raviv, J. (1974). Optimal decoding of linear codes for minimizing symbol error rate. *IEEE Transactions on Information Theory, 20*, 284–287. doi:10.1109/TIT.1974.1055186

Barbero, L. G., & Thompson, J. S. (2006). A fixed-complexity MIMO detector based on the complex sphere decoder, In *Proc. IEEE Intern. Workshop Signal Processing Advances Wireless Communications (SPAWC '06)*.

Benedetto, S., Dinoi, L., Montorsi, G., & Tarable, A. (2006). Design issues on the parallel implementation of versatile, high-speed iterative decoders. In *Proc. 4th Int. Symposium on turbo codes & related topics*, Munich, Germany, April, (pp. 341-351).

Berkmann, J., Carbonelli, C., Dietrich, F., Drewes, C., & Xu, W. (2008). On 3G LTE terminal implementation – Standard, algorithms, complexities and challenges. In *Proc. of IWCMC'08, IEEE International Wireless Communications and Mobile Computing Conference*, Crete, Greece, August.

Berrou, C., Glavieux, A., & Thitimajshima, P. (1993). Near Shannon limit error correcting coding and decoding: turbo codes. In *Proc. IEEE International Conference on Communications*, Geneva, May, (pp. 1064-1070).

Bougard, B., Giulietti, A., Van der Perre, L., & Catthoor, F. (2003). A low-power high-speed parallel concatenated turbo-decoding architecture. In *Proc. 3rd Intern. Symposium on Turbo codes and Related Topics*, Brest, France, (pp. 511-514).

Carbonelli, C., & Franz, S. (2008). Performance analysis of MIMO OFDM ML detection in the presence of channel estimation error. In *Proc. IEEE ISSSTA*.

Chu, D. C. (1972). Polyphase codes with good periodic correlation properties. *IEEE Transactions on Information Theory, 18*, 531–532. doi:10.1109/TIT.1972.1054840

Claussen, H., Karimi, H. R., & Mulgrew, B. (2005). Improved MAX-LOG-MAP turbo decoding by maximization of mutual information transfer. *EURASIP Journal on Applied Signal Processing, 6*, 820–827. doi:10.1155/ASP.2005.820

Dammann, A., & Kaiser, S. (2001). Standard conformable antenna diversity techniques for OFDM and its application to the DVB-T system. *IEEE Global Telecommunications Conference, San Antonio, Texas, USA, 25-29 November 2001*, (pp. 3100-3105).

Edfors, O., Sandell, M., van de Beek, J. J., Wilson, S. K., & Borjesson, P. O. (1998). OFDM channel estimation by singular value decomposition. *IEEE Transactions on Communications, 46*, 931–939. doi:10.1109/26.701321

ETSI EN 300 744, v.1.5.1 (2004-11), *Digital Video Broadcasting (DVB); Framing structure, channel coding and modulation for digital terrestrial television*.

Gallager, R. G. (1962). Low-density parity-check codes. *I.R.E. Transactions on Information Theory, 8*, 21–28. doi:10.1109/TIT.1962.1057683

GSA. (2009). *GSM/3G Market Update*. Retrieved from http://www.gsacom.com.

Hoeher, P., Kaiser, S., & Robertson, P. (1997). Pilot-symbol-aided channel estimation in time and frequency. In *Proc. IEEE Global Telecommunications Conference (GLOBECOM '97)*.

Joham, M., Barbero, L. G., Lang, T., Utschick, W., Thompson, J., & Ratnarajah, T. (2008), FPGA implementation of MMSE metric based efficient near-ML detection. In *Proc. Int. ITG Workshop on Smart Antennas* (pp. 139–146).

Kusume, K., Joham, M., Utschick, W., & Bauch, G. (2007). Cholesky factorization with symmetric permutation applied to detection and precoding spatially multiplexed data streams. *IEEE Transactions on Signal Processing, 55*(6), 3089–3103. doi:10.1109/TSP.2007.893978

Kwak, J., Park, S. M., Yoon, S., & Lee, K. (2003). Implementation of a parallel turbo decoder with dividable interleaver. In *Proc. Intern. Symposium on Circuits and Systems*, Bangkok, Thailand, (Vol. 2, pp. II65-II68).

Le, Y. (2000). Pilot-symbol-aided channel estimation for OFDM in wireless systems. *IEEE Transactions on Vehicular Technology, 48*, 1207–1215.

Li, Y., Cimini, L. J. Jr, & Sollenberger, N. R. (1998). Robust channel estimation for OFDM systems with rapid dispersive fading channels. *IEEE Transactions on Communications, 46*, 902–915. doi:10.1109/26.701317

Manolakis, K., Estevez, D. M. G., Jungnickel, V., Xu, W., & Drewes, C. (2009). A closed concept for synchronization and cell search in 3GPP LTE systems. In *Proc. WCNC'09, IEEE Wireless Communications and Networking Conference*, Budapest, Hungary.

Manolakis, K., & Jungnickel, V. (2008). Synchronization and cell search for 3GPP LTE. In *Proc. of 13th International OFDM Workshop (InOWo'08)*, Hamburg, Germany.

May, M., Neeb, C., & Wehn, N. (2007). Evaluation of high throughput turbo-decoder architectures. In *Proc. Intern. Symposium on Circuits and Systems*, New Orleans, USA, (pp. 2770-2773).

Morelli, M., Kuo, C. C. J., & Pun, M. O. (2007). Synchronization techniques for orthogonal frequency division multiple access (OFDMA): A tutorial review. *Proceedings of the IEEE, 95*(7), 1394–1427. doi:10.1109/JPROC.2007.897979

Morelli, M., & Mengali, U. (2001). A comparison of pilot-aided channel estimation methods for OFDM systems. *IEEE Transactions on Signal Processing, 49*(12), 3065–3073. doi:10.1109/78.969514

Murugan, A., El Gamal, H., Damen, M. O., & Caire, G. (2006). A unified framework for tree search decoding: Rediscovering the sequential decoder. *IEEE Transactions on Information Theory, 52*(3), 933–953. doi:10.1109/TIT.2005.864418

Negi, R., & Cioffi, J. (1998). Pilot tone selection for channel estimation in a mobile OFDM system. *IEEE Transactions on Consumer Electronics, 44*, 1122–1128. doi:10.1109/30.713244

Rinne, J., & Renfors, M. (1996). Pilot spacing in orthogonal frequency division multiplexing systems on practical channels. *IEEE Transactions on Consumer Electronics, 42*, 959–962. doi:10.1109/30.555792

Robertson, P., Hoeher, P., & Villebrun, E. (1997). Optimal and suboptimal maximum a posteriori algorithm suitable for turbo decoding. *Europ. Trans. Telecom., 8*, 119–125. doi:10.1002/ett.4460080202

Speth, M., Fechtel, S. A., Fock, G., & Meyr, H. (1999). Optimum receiver design for wireless broad-band systems using OFDM - part I. *IEEE Transactions on Communications, 47*(11), 1668–1677. doi:10.1109/26.803501

Studer, C., Burg, A., & Bölcskei, H. (2006). Soft-output sphere decoding: Algorithms and VLSI implementation. In *Proc. 40th Asilomar Conference on Signals, Systems, and Computers.*

Takeshita, O. Y. (2006). On maximum contention-free interleavers and permutation polynomials over integer rings. *IEEE Transactions on Information Theory, 52*(3), 1249–1253. doi:10.1109/TIT.2005.864450

van de Beek, J.-J., Sandell, M., & Börjesson, P. O. (1997). ML estimation of time and frequency offset in OFDM systems. *IEEE Transactions on Signal Processing, 145*, 1800–1805. doi:10.1109/78.599949

van Nee, R., van Zelst, A., & Awater, G. (2000), Maximum likelihood decoding in a space division multiplexing system. In *Proc. 51st IEEE Vehicular Technology Conference (VTC Spring)*.

Wübben, D., Böhnke, R., Kühn, V., & Kammeyer, K.-D. (2001). Efficient algorithm for decoding layered space-time codes. *IEEE Electronics Letters, 37*, 1348–1350. doi:10.1049/el:20010899

Wübben, D., Böhnke, R., Kühn, V., & Kammeyer, K. D. (2003). MMSE extension of V-BLAST based on sorted QR decomposition. In *Proc. IEEE Vehicular Technology Conference (VTC Fall)*.

Zhang, Z., Long, K., Zhao, M., & Liu, Y. (2005). Joint frame synchronization and frequency offset estimation OFDM systems. *IEEE Transactions on Broadcasting, 51*(3), 389–394. doi:10.1109/TBC.2005.851702

Zheng, L., & Tse, D. N. C. (2003). Diversity and multiplexing: A fundamental trade-off in multiple-antenna channels. *IEEE Transactions on Information Theory, 49*(5), 1073–1096. doi:10.1109/TIT.2003.810646

Zhu, X., & Murch, R. D. (2002). Performance analysis of maximum likelihood detection in a MIMO antenna system. *IEEE Transactions on Communications, 50*(2), 187–191. doi:10.1109/26.983313

Zimmermann, E., Fettweis, G., Milliner, D. L., & Barry, J. R. (2008). A parallel smart candidate adding algorithm for soft-output MIMO detection. In *Proc. 7th Intern. ITG Conference on Source and Channel Coding*.

ENDNOTES

[1] In practice, SCH, PBCH, etc. stand for "channel" as well as "signal".

[2] In LTE, a so-called basic time unit $T_S = 1/30720000$ sec is defined, which is adequate to be used as sampling period for the maximally supported 20 MHz bandwidth.

[3] For simplicity, we always omit the last ($N_{CP1} - N_{CP2}$) samples within each slot.

[4] A MAC operation is formally defined as $a \leftarrow a + b \cdot c$, where a, b, c are real numbers.

[5] We recall that an LTE resource block in the downlink and with normal cyclic prefix is defined as a box containing 12 consecutive subcarriers and 7 consecutive OFDM symbols (see Fig. 4), i.e., the resource block corresponds to one *slot*.

[6] Note that full diversity and spatial multiplexing cannot be achieved at the same time but are traded against each other, as shown in (Zheng & Tse, 2003).

[7] Serial interference cancellation represents an alternative approach for advanced MIMO detection with more than one codeword. Here, the re-encoding of the data stream detected first introduces additional latency.

[8] The subblock interleaver is a block interleaver with 32 columns, column permutation and possibly insertion of dummy bits to fill a rectangular matrix. The dummy bits are not transmitted.

[9] Time unit is meant as the time needed to process 1 trellis step in forward or backward direction.

Chapter 24
The Use of Orthogonal Frequency Code Division (OFCD) Multiplexing in Wireless Mesh Network (WMN)

Syed S. Rizvi
University of Bridgeport, USA

Khaled M. Elleithy
University of Bridgeport, USA

Aasia Riasat
Institute of Business Management, Pakistan

ABSTRACT

In the present scenario, improvement in the data rate, network capacity, scalability, and the network throughput are some of the most serious issues in wireless mesh networks (WMN). Specifically, a major obstacle that hinders the widespread adoption of WMN is the severe limits on throughput and the network capacity. This chapter presents a discussion on the potential use of a combined orthogonal-frequency code-division (OFCD) multiple access scheme in a WMN. The OFCD is the combination of orthogonal frequency division multiplexing (OFDM) and the code division multiple access (CDMA). Since ODFM is one of the popular multi-access schemes that provide high data rates, combining the OFDM with the CDMA may yield a significant improvement in a WMN in terms of a comparatively high network throughput with the least error ration. However, these benefits demand for more sophisticated design of transmitter and receiver for WMN that can use OFCD as an underlying multiple access scheme. In order to demonstrate the potential use of OFCD scheme with the WMN, this chapter presents a new transmitter and receiver model along with a comprehensive discussion on the performance of WMN under the new OFCD multiple access scheme. The purpose of this analysis and experimental verification is to observe the performance of new transceiver with the OFCD scheme in WMN with respect to the overall network throughput, bit error rate (BER) performance, and network capacity. Moreover, in this chapter, the authors provide an analysis and comparison of different multiple access schemes such as FDMA, TDMA, CDMA, OFDM, and the new OFCD.

DOI: 10.4018/978-1-61520-674-2.ch025

INTRODUCTION

Wireless Mesh Networks (WMNs) have become a major paradigm for construing a user access network that provides high speed network access to users in the context of enterprise and community networks (Akyildiz, Wang, & Wang, 2005; Tse & Grossglauser, 2002; Xiang, Peng-Jun, Wen-Zhan, & Yanwei, 2008). From the robust and stable connectivity perspective, WMN is considered as one of the most data efficient mesh topology which is constructed by, mesh routers, clients, and gateways (Kim & Bambos, 2002; Zhang, Honglin, & Chen, 2008). In WMN, mesh routers can play the roll of gateways whereas the clients provide connectivity with the Internet. One of the main advantages of WMN is the self-healing and self-configuring nature of mesh routers. For extending the geographical area of a network, mesh routers are equipped to provide connectivity between different networking technologies such as Wi-Fi, IEEE 802.11, mobile technology and wired Ethernets (Li, Qiu, Zhang, Mahajan, Zhong, Deshpande, & Rozner, 2007; Yu, Mohapatra, & Liu, 2008).

WMN itself brings many challenging issues from physical layer to application layer. Problems like network capacity, protocols used in different layers, network management and the network security are just some of the problems to point out. Recent theoretical studies and experimental verifications (Gupta & Kumar, 2001) have shown the current WMNs are severely limited in network capacity. This is due to the fact that when all nodes communicate using a single channel in a high speed wireless LAN (e.g., IEEE 802.11a), the number of simultaneous transmissions from multiple users is limited by interference. Since WMNs are multi-hop in their nature, interference causes a serious degradation in overall network capacity when adjacent hops on the same path start interfering with the neighboring paths (Kyasanur & Vaidya, 2005).

As of now, the scalability issue in WMN has not been fully solved yet (Xiang, Peng-Jun, Wen-Zhan, & Yanwei, 2008). Most of the existing multiple access schemes which are based on CSMA/CA solve only partial problems of the overall issue for WMN (Zhou & Lai, 2005). The implementation of such schemes raises other performance issues such as minimum network capacity, low end-to-end throughput, and scalability (Lin & Rasool, 2007). Thus, how to fundamentally improve the scalability and maximize the throughput performance in WMN is an interesting research issue now days. One of the efficient solutions that can be used for not only improving the network scalability but also maximizing the end-to-end network throughput is the use of a hybrid multiple access scheme for WMN (Sundaresan & Rangarajan, 2008). For networks based on techniques other than CSMA/CA, code division multiple access (CDMA) can be applied with orthogonal frequency division multiplexing (OFDM) in WMN as an efficient multiple access scheme to overcome some of the problems mentioned above. The combination of these two multiple access schemes allow us to take advantage of both CDMA and OFDM. OFDM has become widely adopted in many next generation cellular systems such as 3GPP Long Term Evolution (LTE) and IEEE 802.16m advanced WiMAX (Chang, Tao, Zhang, & Kuo, 2009; Zhang, Honglin, & Chen, 2008).

Recently, researchers have identified two fundamental problems that degrade the throughput of WMNs (Li, Qiu, Zhang, Mahajan, Zhong, Deshpande, & Rozner, 2007). First, they identified that if we do not consider the amount of data that the nodes can transmit, the network throughput may degrade when nodes start transmitting more than what the intermediate links can support. They highlighted that this is possible due to the presence of interference (i.e., multi-access interference (MAI)) which may cause additional traffic to reduce the capacity of bottleneck links. The second problem that they identified is that the current protocols are unable to accurately estimate link and path quality for the purposes of path selection. This inaccurate estimation is caused since the routing protocols do not consider the interference when they are determining the quality of links. However, due to interference, the quality can change arbitrarily with any change in the routing pattern. Based on this work (Li, Qiu, Zhang, Mahajan, Zhong, Deshpande, & Rozner, 2007), we can say that the use of OFDM as one part of the multi-access scheme (i.e., the OFCD) can solve the problem of interference since it provides strong resistance against both MAI and inter-symbol interference (ISI). The use of CDMA allows the fast and stable transmission of data. Combining the features of these schemes, we can significantly enhance the network throughput for the WMN.

The demand for high speed wireless applications and limited radio frequency (RF) signal bandwidth has spurred the development of power and bandwidth efficient air interference schemes. Therefore, the adoption of more sophisticated multiple access technologies such as OFDM coupled with the CDMA, provides two key benefits to WMN in the form of scalability and throughput gain. However, leveraging these benefits call for more sophisticated transmitter and receiver designs for WMNs that can provide greater system capacity with a reasonable bit error rate (BER) performance.

WMN have emerged to be a cost effective and performance adaptive paradigm for the next generation wireless Internet. The main reason why WMN is so attractive for several applications is its low cost of deployment and maintenance due to the absence of a wired infrastructure. However, the absence of a wired infrastructure causes communications among routers to suffer from noise and interference problem. The interference in WMN can be alleviated by using the OFCD as a multicarrier/multi-access scheme that provides strong resistance against noise and ISI which is especially occurred in wideband transmission over multipath fading channels (Gupta & Kumar, 2001; Yu, Mohapatra, & Liu, 2008).

OFDM is a multi-carrier multi-access scheme which was originally designed and developed in late 1960's (Ye & Gordon, 2006). In OFDM, the data is transmitted using a large number of sub-carriers that are completely orthogonal (Chuang & Sollenberger, 2000). Each sub carrier is modulated with a conventional modulation scheme at a low symbol rate. The transmitted data is typically divided into a large number of parallel data streams with respect to the number of sub carriers. Even though, the subcarriers used in OFDM are theoretically orthogonal, in practice, the orthogonality might not be ensured that results ISI that is the most important effect of multi-path delay spread (Xiang, Peng-Jun, Wen-Zhan, & Yanwei, 2008; Zhang, Honglin, & Chen, 2008). With the help of cyclic prefix, longer time duration symbols are transmitted in which the length of each transmitted bit is typically longer than the length of the impulse response of the channel (Ye & Gordon, 2006). This is one of the ways that can be used to mitigate the effects of ISI.

One of the objectives of this chapter is to analyze the potential use of a combined OFCD multiple access technique in a WMN. The OFCD is the hybrid of OFDM and CDMA. OFDM has become the popular choice for air interface technology in future local and wide area wireless networks. For instance, it has been applied to high speed wireless LAN (e.g., IEEE 802.11a) (Bing, Frank, James, Rainer, Hermann, & Adam, 2001), and high performance radio LAN type 2 HIPERLAN/2 (Johnsson, 1999).

Code division multiple access (CDMA) is originally designed as a multiplexing technique with strong spread spectrum characteristics. In CDMA, a pseudo random number generator (PRN) generates a spreading code that has comparatively large frequency components than the frequency components of an input narrow band signal. CDMA allows all users to transmit signals simultaneously by utilizing the entire available spectrum. CDMA is typically used with the direct sequence (DS) and the frequency hopping (FH) spread spectrums techniques. Most of the 3G cellular systems are proposing to use the combination of direct sequence (DS CDMA) as their modulation technique. In DS-CDMA, transmitter spreads the original data stream using a given spreading code in the time domain. The purpose of this spreading of the input signal using the spreading code is to increase the frequency range of the resultant signal. How effectively this technique can suppress MAI depends on the cross correlation between the spreading codes. Theoretically, spreading codes are perfectly orthogonal; however, in practice, the orthogonality can not be guaranteed.

The implementation of OFCD scheme in WMN requires the design of new sophisticated transmitter and receiver models that should be capable of using OFCD as an underlying multichannel subcarrier. This chapter provides a discussion on a new transmitter and a receiver model that uses OFCD as a multi-access multi-carrier modulation scheme for transmitting and receiving the signals at the transmitting and receiving ends, respectively. Moreover, the goal of this chapter is to provide an analysis of different multi-access techniques such as CDMA, OFDM, and OFCD with respect to their utilization with the WMNs. Our analysis could be used effectively to determine the performance differences between these multi-access schemes when implement with a WMN. For the sake of the performance analysis and the experimental verifications, this chapter adapted BER and the network throughput as performance measures.

COMBINING DIFFERENT MULTIPLE ACCESS SCHEMES

Based on OFDM, several multiple access schemes have been designed such as OFDM-frequency division multiple access (OFDM-FDMA), OFDM-time division multiple access (OFDM-TDMA), and OFDM-code division multiple access (OFDM-CDMA). The first two multiple access schemes (OFDM-FDMA and OFDM-TDMA) have been adopted by the IEEE 802.16 standard as two options for transmissions at the 2.11 GHz band (Akyildiz, Wang, & Wang, 2005). The last one is also referred as Multicarrier code division multiplexing (MC-CDMA) which has been drawing much attention as an alternative to conventional direct-sequence CDMA (DS-CDMA).

Both FDMA and TDMA schemes are used as resource management schemes in a multiaccess communication system where transmission resources are shared among multiple users (Wang & Xiang, 2006). The purpose of these techniques is to efficiently manage the resource sharing in a multiuser environment based on the principle of timesharing (TDMA) and frequency-sharing (FDMA). In other words, in a multicarrier system, the resources are shared among multiple users by allocating the subcarrier to them across the time and frequency domains.

In OFDM-FDMA scheme, OFDM can be implemented to allocate subcarriers to different users where each subcarrier can be multiplexed using the FDMA technique. In other words, each subcarrier can be considered as a narrowband communication channel that contains a portion of the entire bandwidth. Since each user has its own subcarrier, this allows users to transmit their data simultaneously over parallel channels. The principle advantage of OFDM-FDMA multiple access scheme is its ability to support simultaneous downlink data transmissions to different terminals. Based on this basic multiple access scheme, an OFDM-interleaved-FDMA scheme was proposed in which the subcarriers assign to users need not be consecutive in their order (i.e., the subcarriers allocated to users can be interleaved). Even though, the subcarriers are interleaved, they are fixed to users on time axis, and thus, making the recourse allocation non-flexible (Bing, Frank, James, Rainer, Hermann, & Adam, 2001).

In OFDM-TDMA scheme, we allocate predetermined time slots to users, where each allocated time slot contains ODFM symbols. In other words, the OFDM-TDMA scheme allocates the OFDM symbols while OFDMA allocates subcarriers to users. In addition, since the time slots are predetermined and pre-assigned, the subcarriers of one OFDM symbol can not be allocated to different user. In OFDM-TDMA scheme, the OFDM can be used in two modes for frame allocations: static and dynamic. In static modes, frames are allocated to users which are independent of their channel conditions whereas the dynamic mode allocates frames to users with the best channel gain. Based on the OFDM-TDMA, (Wang & Xiang, 2006) proposed a new multiple access scheme, which is called OFDM-TDMA with subcarrier allocation (OFDM-TDMA/SA). In this scheme, OFDM symbols are organized in a TDMA frame where each subcarrier for an OFDM symbols is assigned to a different user. The main aim of this scheme is to enhance the flexibility of recourse allocation for an OFDM based multiple access system. This work is slightly different from the work of (Cheong, Cheng, Lataief, & Murch, 1999), where different frame structures are used for large radio and small radios resources.

To improve the MAC performance of WMN, many multiaccess techniques are currently under development. For instance, Multichannel MAC (MMAC) (Bahl, Chandra, & Dunagan, 2004; Bahl, 2007), multiple radios (Adya, Bahl, Padhye, Wolman, & Zhou, 2004), directional and steerable antennas (Bahl, 2007) are example of some new multi-access techniques. Several efforts have been made by the researchers to combine the OFDM with the CDMA for wireless networks (Neishaboori & Kesidis, 2008; Xiang, Peng-Jun, Wen-Zhan, & Yanwei, 2008; Zhang & Tang, 2006). The principle reason for using the combined multiple access scheme (i.e., the OFDM-CDMA) is its ability to exploit frequency diversity in an explicit manner. This exploitation of frequency diversity is possible in this multiple access scheme due to the fact that the energy of the symbol is spread over several subcarriers. The combination of OFDM with the CDMA is refereed as multicarrier (MC) CDMA (MC-CDMA).

CDMA is a multiplexing technique with some strong spread spectrum characteristics. In this scheme, a number of users simultaneously access a channel by spreading their narrowband signals (i.e., input informational signal) with pre-assigned signature sequence. CDMA has become a prominent multiple access technique in mobile wireless systems (Prasanna & Ravichandran, 2007), because it has the capability to provide higher capacity over conventional techniques such as TDMA and FDMA, and to combat the hostile channel frequency selectivity (Neishaboori & Kesidis, 2008). Direct sequence (DS-) and frequency hopping (FH-) CDMA techniques have been subject to extensive research. When CDMA operates with DS, it suffers from the frequency selective fading problem, particularly in the downlink where orthogonal spreading codes are typically employed (Zhang, Tan, Chun, Laberteaux, & Bahai, 2008). The basic idea of CDMA is to maintain a sense of orthogonality among the signature waveforms in order to minimize the MAI. However, in practice, the orthogonality among the signature waveforms can not be guaranteed. In other words, in CDMA systems, the communication channels are defined by the pseudo-random codewords, which are carefully designed to cancel each other out as far as possible (i.e., maximizing the orthogonality between the codewords or signature waveforms) (Xiang, Peng-Jun, Wen-Zhan, & Yanwei, 2008). These codewords are then used to spread a narrowband signal (i.e., each bit of input signal) into a wideband signal by spreading it over a unique codeword. The bandwidth components of the resultant wideband signal are much wider than the minimum bandwidth required transmitting the original input narrowband signal.

On the other hand, OFDM is a multicarrier modulation scheme which has drawn a lot of attention in the field of radio communications. OFDM divides the available bandwidth into a large number of orthogonal bands or subcarriers. Each subcarrier has a bandwidth which is typically smaller than the coherence bandwidth and thus exposed only to frequency flat fading (Chang, Chien, & Kuo, 2007; Zhang, Tan, Chun, Laberteaux, & Bahai, 2008). OFDM addresses the ISI problem arising in channel where the signal bandwidth exceeds the coherence bandwidth of the fading process (Wang & Xiang, 2006; Ye & Gordon, 2006). The transmission of data using OFDM mitigates the problem of the frequency selectivity in multi-path fading channels while at the same time the use of CDMA provides good spectral properties of the transmitted data (Hottinen & Heikkinen, 2006; Zhang, Tan, Chun, Laberteaux, & Bahai, 2008). Another scheme for combining the OFDM with the CDMA is proposed in (Zhang, He, & Chong, 2005). In this scheme, different input symbols are transmitted using multiple subcarriers. This proposed technique offers all the advantages of multicarrier CDMA (MC-CDMA) scheme. However, this technique mainly depends in the derivations of subcarriers.

With the combination of these two multi-access schemes, the data can be transmitted over a large number of subcarriers where each subcarrier is assumed to be conventionally modulated. The combined multi-access scheme provides the spectral efficiencies, increases the network capacity, and minimizes the end-to-end delay to support 4G wireless systems. The principle advantage of using OFDM-CDMA as a multiple access scheme is its ability to satisfy some of the main requirements of 4G wireless systems such as the minimization of MAI and ISI and thus provides a better BER performance to the end user.

Broadband WMNs demand both high speed transmission rate and a more sophisticated multiple access technique that can be used to minimize MAI and ISI and maximize the BER performance. The high speed transmission rate requires higher frequency bands for transmission. However, due to sever frequency selectivity in broadband communications, the number of resolvable multiple paths fading degrades BER performance. The

MC-CDMA is a multiple access scheme that is typically used with an OFDM based system that allows system to support multiple users simultaneously. In MC-CDMA, each data symbol is transmitted at multiple narrowband subcarriers where each subcarrier is typically encoded with a phase offset of 0 or 180 degree (Hottinen & Heikkinen, 2006). It uses WALSH code which is an orthogonal spreading code sequence in a frequency domain (Zhang, Tan, Chun, Laberteaux, & Bahai, 2008). In MC-CDMA, all data symbols are not transmitted on each subcarrier. Instead, they can be transmitted on some of the selected channels. Those few channels on which data symbols can be transmitted are chosen after the channel assignment. This scheme not only transmits the data symbols but also resolves the problem of flat fading by minimizing the bit lost. Each subcarrier can be used by all users presented in the system whereas all the active users are differentiated by a WALSH code. In MC-CDMA, since each receiver uses a fast Fourier Transform (FFT) circuit with a variable gains diversity combiner, signal can be easily recovered (Lin & Rasool, 2007).

For more connectivity, mesh routers are equipped to provide connectivity between different networking technologies such as Wi-Fi, IEEE 802.11, mobile technology and wired Ethernet. In WMN, each client with same radio technology communicates via Ethernet links, for different clients communications are first made with their base station which has Ethernet connections to the mesh routers. Sometimes client nodes actually form network to perform routing and this kind of infrastructure is called client wireless mesh networks. Mesh clients can perform mesh functions with other mesh clients as well as accessing the network through routers yielding hybrid wireless mesh networks.

One of the main advantages of OFDM scheme is its robustness to frequency selective fading. However, this scheme also requires synchronization in each subcarrier and sensitivity to frequency offset and nonlinear amplification.

This problem is caused in OFDM due to the fact that it is composed of several subcarriers with their overlapping power spectra and exhibits a non constant nature in its envelop. However, the combinations of CDMA with the ODFM scheme can significantly minimize the symbol rate in each subcarrier to maximize the symbol duration which makes the synchronization process simple among the multiple transmissions of users. Using OFDM with CDMA, symbols are transmitted on many carriers. In addition, different spread input symbols are fed to the subcarriers, when OFDM is combined with the CDMA scheme (Zhang & Tang, 2006). With OFDM technique, frequency selectivity in multipath fading channel is resolved (Cheong, Cheng, Lataief, & Murch, 1999; Zhang, Tan, Chun, Laberteaux, & Bahai, 2008).

Features of OFCD Multiple Access Scheme for WMN

In this section of the chapter, we first give a logical reasoning of using OFCD in WMNs. Specifically, we discuss that what flexibilities this new multiaccess technique provides to WMN and what changes we may need to make in the framework of both WMN and OFCD in order to implement this new multiaccess scheme.

WMN's operation is similar to the way that packets are routed over the wired Ethernet (i.e., data hops from one device to another until it reaches its destination). This is possible only when each node shares its dynamic routing algorithm with every single node to which it is connected. The routing algorithm implemented in each node takes the fastest route to its destination. Since in WMNs there is no central server, each node (client) transmits data to the next node. As a result, each node behaves like a repeater that forms an externally big network which is analogous to the Internet. In today's scenario, hybrid WMN's have taken place of basic WMNs. The basic advantage of Hybrid WMN is that the network can be accessible either through mesh routers or through mesh clients. It supports all different kinds of network technologies like wired Ethernet, mobile communication, Wi-Fi, Wi-MAX, IEEE 802.11 etc.

The access points form a wireless backbone, providing connectivity in places otherwise it is difficult to access through traditional wired infrastructure. The wireless communication between the access points can use different technologies such as IEEE 802.11a/b/g or IEEE 802.16 and different hardware (directional or Omni-directional antennas). The use of multi channels in wireless network leads to throughput and reduced delay. One class of such protocols divides the available channels in two classes, control and data channels. Control channels are used to exchange network control information, while data channels are used for data transfer (Li, Qiu, Zhang, Mahajan, Zhong, Deshpande, & Rozner, 2007; Rappaport, 2002). The use of multi transceiver allows a node to scan all available channels concurrently, hence solves many complex problems.

Currently, in WMNs, most of the systems implement distributed multiple access schemes such as CSMA/CA as their multiaccess scheme. The principle advantage of using this scheme in WMNs is that it does not require the accurate timing synchronization within the global network. For WMN, as the size of the network grows, CSMA/CA systems suffer from scalability issues (i.e., when the size of the network increases, the network performance degrades significantly due to significant throughput reduction. For instance, current IEEE 802.11 MAC protocol and its variants cannot achieve a reasonable throughput as the number of hops increases to 4 or higher. This low scalability is due to the fact that the end-to-end reliability sharply drops as the scale of the network increases. This implies that in a large network, CSMA systems may suffer from high packet queuing delays.

Moreover, the current multiaccess scheme (CSMA/CA) has very low frequency spatial-reuse efficiency (Acharya, Misra, & Bansal, 2003), which significantly limits the scalability of

CSMA/CA-based multi-hop networks. In order to fundamentally resolve the issue of low end-to-end throughput in a WMN, innovative solutions are necessary. Determined by their poor scalability in WMN, random access protocols such as CSMA/CA are not an efficient solution (Kim & Bambos, 2002; Yu, Mohapatra, & Liu, 2008). Thus, revisiting the design of MAC protocols based on OFDM and CDMA is an important research topic (Acharya, Misra, & Bansal, 2003). As of now, only few TDMA or CDMA based MAC protocols have been proposed for WMNs. This is mainly because of two reasons. Firstly, it is relatively expensive from cost point of view to design and implement a distributed and cooperative MAC protocol with CDMA scheme. Secondly, a framework is needed to provide compatibility between CDMA and other existing MAC protocols.

CDMA offers several advantages to WMN when it works as a multiaccess scheme. First, nodes in CDMA networks can interfere each other but they do not damage each others' data as long as the degree of interference is relatively low. This implies that the CDMA does not impose any hard limit on the user capacity since we can continue adding users in a CDMA network as long as the level of interference can tolerate. This is one of the features that make CDMA unique when compared to the other conventional multiplexing schemes such as TDMA and FDMA. Another advantage that CDMA provides is that the data transmission rate for wireless networks can be increased by using the additional power control. This is especially true for 802.11 WLAN where the data rate is restricted due to the absence of a proper power control. The implementation of power control with CDMA allows us to increase the transmission rate of certain traffic flow which consequently increases the transmission range and decreases the number of hops between the transmitter and receiver. The end result of this additional power controller module in CDMA provides several advantages such as lower BER, less end-to-end delays, and higher throughput.

Based on the above discussion, one may conclude that the combination of CDMA and OFDM (we refer it as OFCD) may provide several advantageous to WMN as a multiaccess scheme such as minimizing the end-to-end delay, improving the BER performance, and increasing the network capacity. In order to increase the capacity of WMN, OFDM has significantly increased the speed of IEEE 802.11 from 11 Mbps to 54 Mbps (Chang, Chien, & Kuo, 2007; Neishaboori & Kesidis, 2008).

Implementation of OFCD Multiple access Scheme in WMN

In this section, we present a discussion on the implementation of OFCD multiaccess scheme in WMN. Specifically, we present the system model and the framework required to implement the OFCD with the WMN. To support the framework of OFCD, we also provide an analytical model to exhibit some of the strong characteristics of OFCD that it offers for WMN. All system variables, along with their definitions, are listed in Table 1. Before we present the implementation of OFCD in WMN, it is worth mentioning some of our key assumptions.

Assumptions and System Model

- There are n available channels where each channel has equal bandwidth.
- The access points or base station can receive data on multiple channels simultaneously. This is a reasonable assumption since the access points can be more specialized in higher end device as compared to a simple client that it serves.
- The channels are orthogonal and CDMA scheme is used (i.e., transmission on a channel does not interfere with transmission on any other channels). Here a channel may represent a code or a frequency band (OFDM-CDMA multi-access technique).

Table 1. Notation used in analysis

Notation	Related Quantities
n	Represents number of channels for signal transmission
N	Represents number of nodes in unit area (A)
T_f	Represents frame length (bits)
S_a	Number of contention slots
S_d	Number of data slots
T_a (interval length)	Contention duration
T_d (length of data interval)	Data duration

- Each network node including the base station is equipped with a multi radio multi transceiver, which is capable of performing in full duplex mode. Hence each node can either transmit or receive a signal on channel at any point in time. The nodes can however switch to different channels dynamically.
- The network is assumed to be fully synchronized (Prasanna & Ravichandran, 2007).

Basic Framework of OFCD in WMN

For multi hop mesh network, we consider the hybrid architecture as it is the most widely deployed architecture. This architecture is characterized by the fact that mesh clients do not need direct connection to a mesh route, but can connect multi hop over mesh clients to a route. The advantages are improved connectivity and coverage. And the disadvantages are that mesh clients need more resources because they also need to have routing capability.

Let's assume there are N sensor nodes distributed over an area A. Sensors are assumed to be independently and uniformly scattered over a region of interest (Yu, Mohapatra, & Liu, 2008). Number of nodes (n) is assumed to be distributed independently and uniformly over an area of Πr^2. Each node can communicate with every other node within the radius of r. Taking these factors into account, one can derive the following expression:

$$\pi r^2 = (\ln n + c(n))/n \qquad (1)$$

The networking is connected with unit probability if and only if the following expression exists: $\lim_{n \to \infty} c(n) \to \infty$. From this expression, one can choose the transmission range. After flooding the network traffic, network is organized into a tree with the observer at each possible route. In the first step of flooding, the observer first broadcasts a wakeup signal. All sensor nodes within the direct communication range receive this signal and reply to the observer. Once the observer receives one or more messages from sensor nodes, it registers them as first level nodes in the node tree and instructs them to repeat the process of broadcasting in a time shared manner to avoid collisions.

All sensor nodes that were not previously registered would register themselves as first level nodes when receives these broadcast messages from observer. Those sensor nodes that were at first level move to the second level. More over, the second level nodes continue to broadcast and repeat the same procedure until all nodes have been registered. Whenever a node broadcasts a wakeup signal, it also attaches its unique address and chain of nodes which leads to it from the observer. Nodes that are wakened up by the broadcast are designated as children of broadcasting nodes. These children nodes obtain the chain of nodes leading from observer to them by concatenating the last link with the chain, leading to there parent. Each node obtains a route from observer to reverse this chain.

The same reasoning suggests that a fewest-hop route should be optimal even in a mobile observer network. Since the observer does not stay at a fixed position, the fewest-hop route is time variant in nature, and so is the number of hops. Quite obviously, the best solution is to choose the route which consists of the fewest hops at any time. In other words, if $S = Ns_1, Ns_2, Ns_3, ...$ is the set of all the nodes that come within a direct communication range of the observer at any time, then the fewest-hop route to the observer is the shortest of the fewest-hop routes to any of the nodes in S joined with the link between the corresponding node in S and the observer.

Finding the shortest route in practice involves a procedure very similar to the flooding procedure described above except one difference. The 1st level nodes (those belonging to the set S just described) are discovered by moving the observer on its path while transmitting a wake-up signal to all nodes that are within the range. These nodes are registered as 1st level nodes. The remaining process of flooding proceeds exactly as described in this section. At the end of this registration procedure, each node in the network knows its route to one of the nodes in S that communicate directly with the observer.

OFCD System Model

As we briefly mentioned in Introduction Section, one of the objectives of this chapter is to describe a hybrid multiple access scheme for WMNs in which different users are transmitting signals simultaneously over multiple subcarriers where each subcarrier is the sum of orthogonal subcarriers. Such a hybrid multiple access technique is indeed very important in order to over come some of the deficiencies caused by the use of distributed multiaccess technique such as CSMA/CA. To show this, we present a discussion on the implementation of OFCD in WMN.

OFCD is the combination of OFDM and CDMA spread spectrum multiple access scheme. Next, we present an analytical model for both transmitter and receiver for a WMN. The design of transmitter and receiver help understanding the implementation of OFCD multiple access scheme for WMN. All system variables, along with their definition, are presented in Table 2.

Transmitter Design for OFCD in WMN

As shown in Fig. 1, $x(k)$ is the discrete digital data that transmitter receives from a digital data source where k represents discrete time. As mentioned above, in OFCD scheme, the transmitter spreads the original data stream (i.e., $x(k)$) over different subcarriers using the assigned spreading code. The assignment of spreading code to each user is done by CDMA technique.

The serial to parallel converter will convert $x(k)$ into n number of parallel data symbols with a symbol rate of $1/T$. The parallel symbols are $x_1(k)$ to $x_n(k)$. Quadrature phase-shift keying (QPSK) block is used which have the carrier frequency in the range of f_1 to f_n as its other inputs. The QPSK block will split the input bit stream into in-phase and quadrature phase components. The quadrature components and the in-phase components will be modulated with f_1 to f_n carrier frequencies. Based on these factors, the output of a QPSK block would be approximated as: $x_n(k) + f_n$.

This output block is then supplied to the mixer as shown in Figure 1. At mixer, PRN code C_b mixes with the incoming signal $x_n(k) + f_n$. Taking these factors into account, we can derive the following expression:

$$X_n = [x_n(k) + f_n] \cdot C_b \qquad (2)$$

The Inverse Discrete Fourier Transform (IDFT) would be used for modulation and it is described as:

Table 2. Notations used in proposed transmitter and receiver model

Notation	Related Quantities
x(k)	Input from digital data source
$x_1....x_N$	Parallel data symbols
$f_1....f_N$	Carrier frequency
Cb	PRN codes
X_n	Symbol mixed with PRN codes
χ_k	Output of the mixer
Y_k	Output of IDFT block
x(t)	Analog received signal
w(k)	Non negative weight function
X_k	Output of DFT
X(k)	Output of mixer at receivers end
n(t)	White additive Gaussian noise
G	Guard time matrix

$$Y_k = \sum_{n=0}^{N-1} X_n . e^{-\frac{2\pi.i}{N}kn} \quad (3)$$

$$<\chi_k, Y_k> = \int_a^b \chi_k . Y_k . w(k) dk \quad (4)$$

where $k = 0,..., N-1$.

The last symbol coming out of IDFT block is taken and added on the beginning of source code block to provide guard time which in fact provides orthogonality (this is analogous of an ideal case of OFDM). The proper choice of the number of subcarriers and the guard time is important in order to increase the robustness against the frequency selective fading.

If the last symbol coming out of IDFT block is X_n, then both (2) and (3) can be combined together.

where w(k) in (4) is the non-negative weight function of OFCD and χ_k and Y_k would be orthogonal if and only if the following expression exists:

$$\int_a^b \chi_k . Y_k . w(k) dk = 0 \quad (5)$$

As the input bit stream can be of infinite length, the infinite integral can be defined as follows:

$$\int_{-\infty}^{\infty} \chi_k . Y_k . w(k) dk = |X_k|^2 = |X_n|^2 \quad (6)$$

Figure 1. Transmitter model for WMNs with orthogonal frequency code division (OFCD) multiple access scheme

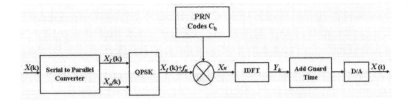

Figure 2. Transmitter model for WMNs with orthogonal frequency code division (OFCD) multiple access scheme

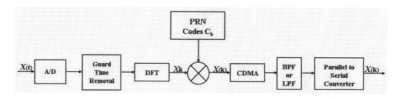

After digital to analog conversion, discrete digital signal would be converted into analog signal and then would be transmitted over the medium.

Receiver Design for OFCD in WMN

In Figure 2, $x(t)$ is the received signal which can be shown as:

$$x(t) = F^{-1}[X_k] + n(t) \tag{7}$$

where $n(t)$ in (7) represent noise which gets added during the transmission. The signal has been passed through the guard time removal block to remove the guard time G. the guard time can be expressed as:

$$G = (0_{Nxl}\ 1_{NxN}\ 0_{NxN}) \tag{8}$$

where G is a matrix of 0's and 1's.

Receiver performs Discrete Fourier Transform (DFT) to demodulate the signal. DFT can be represented as:

$$X_k = \sum_{n=0}^{N-1} x_n \cdot e^{-\frac{2\pi \cdot i}{N} kn} \tag{9}$$

where $k = 0,\ldots,N-1$. X_k is then fed into the mixer which will mix it with the PRN code C_b which was assumed to be deterministic to the receiver.

As we are deploying the CDMA's feature, the receiver should know the seeds of the PRN codes which were used at the transmitter. The output of mixer is represented by $X(k)$ and can be further expressed as:

$$X(k) = C_b X_k \tag{10}$$

The DS-CDMA is used to detect $X(k)$. DS-CDMA scheme simply performs the XOR operation between the PRN codes and the signal coming out of the mixer $X(k)$. The unique chip sequence is used by DS-CDMA depends on the number of seeds used by the transmitter. Band Pass Filter (BPF) will pass only a certain range of frequency by considering the maximum and minimum frequency components of the carrier frequency. If f_h is the highest frequency component and f_l is the smallest frequency component then $f_{BPF} = f_h - f_l$ and those detected signals will be passed which lie in the frequency range of f_h-f_l.

With the parallel to serial converter original transmitted signal $x(k)$ will be fully recovered. Given the property of a frequency selective fast multipath fading channel, there exists the optimal value to minimize the BER in the number of subcarriers and the length of the guard interval. This is one of the features that one can only achieve when a hybrid of CDMA and OFDM is implemented as a multiple access scheme in WMNs.

Performance Analysis

The OFCD scheme for WMN was implemented using NS2. We use frequency selective slow Rayleigh fading channels as channel model. The frame length of the OFCD frame is assumed to be 3 ms which is close to that of IEEE 802.11. A total of 8 time slots are used in which 5 time

The Use of Orthogonal Frequency Code Division (OFCD) Multiplexing in Wireless Mesh Network (WMN)

Figure 3. An illustration of end-to-end throughput without using OFCD for WMNs

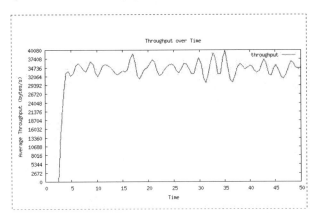

slots are dedicated for downlink frequency traffic where as rest of them are used for uplink traffic transmission. We assign exactly one OFDM symbol to each time slot whereas each ODFM symbol contains 128 subcarriers for the users/nodes. As mentioned in the previous section, QAM techniques was employed in each subcarrier Figure 3 shows the throughput of a WMN with the use of conventional distributed MAC protocol (i.e., CSMA/CS). This simulated computer network for WMN is congested with 10 nodes transmitting randomly to generate the throughput rate.

The peak end-to-end throughput that can be achieved without using the OFCD is upper-bounded by 40080 bytes per seconds as shown in Figure 3. Figure 4 shows the end-to-end throughput for the case where we implement OFCD as a multiple access scheme for WMN. As we can see in Figure 4 that the throughput increases significantly from 40080 bytes per seconds to 70440 bytes per second.

The BER performance comparison between the OFCD and CSMA/CA for WMN is presented in Figure 5. The *theoretical-exact 0* (i.e., it represents the BER performance of conventional distributed MAC protocol) is computed based on the implementation of CSMA/CA for WMN whereas the *theoretical-exact 1* is computed based

Figure 4. Implementation of OFCD for WMNs to maximize the end-to-end throughput

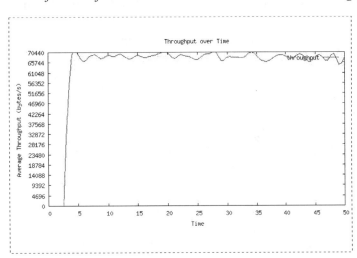

Figure 5. BER performance of OFCD for WMNs

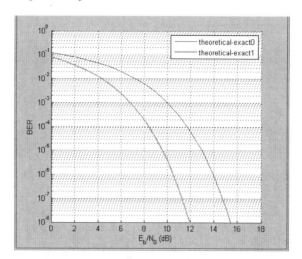

on the implementation of OFCD for WMN. We consider the same parameters as we discussed above for measuring the BER performance for implementing both CSMA/CA and OFCD for WMN. As one can see in Figure 5 that the OFCD outperforms the conventional MAC protocol for all values of E_b/N_o(dB). This significant gain in BER performance for OFCD is achieved due to the fact that both OFDM and CDMA exhibit strong characteristics that can effectively eliminate both MAI and ISI even in the presence of large subcarriers and users.

FUTURE RESEARCH IN WMN

Several efforts have been made to implement OFDM based technologies in WMN such as OFDM-FDMA (Cheong, Cheng, Lataief, & Murch, 1999), OFDM-TDMA (Bing, Frank, James, Rainer, Hermann, & Adam, 2001), and OFDM-CDMA (Prasanna & Ravichandran, 2007; Xiang, Peng-Jun, Wen-Zhan, & Yanwei, 2008; Zhang & Tang, 2006). The implementation of these multi-access techniques for the WMN brings many research challenges. For instance, one of the major issues in the design and engineering of OFDM based WMNs would be cross layer optimization involving OFDM based transmission, radio link level queue management, and network layer admission control management. Recently researchers have investigated the QoS performances in a solar-powered mesh network using OFDM-based radio transmission (Niyato, Hossain, & Fallahi, 2006). Specifically, they developed a queuing analytical model to analyze both the connection-level and the packet-level performances at a mesh node considering constrained power supply at that node. Their analysis demonstrates that their proposed analytical framework can be used to achieve the optimal connection admission control (CAC) threshold. The CAC threshold value can be further used to guarantee packet-level QoS.

As mentioned earlier, a WMN puts a hard limit on the number of users (i.e., the network capacity) that it can support at one time due to the presence of interference. The network capacity also degrades sharply when WMN is being used in multi-hop mode. One of the reasons of limited capacity in WMN is the presence of interference within a limited band of radio frequencies. Recently, researchers have identified that not only the channels but also the transmission rates of the links have to be properly selected to make a given set of flow rates schedulable (Kashyap, Ganguly, & Das, 2007). Specifically, they developed a forwarding paradigm to achieve a resulting set of flow rates while using a standard MAV protocol. Their work presented a measurement-based model that can capture the effect of interference in 802.11 based WMNs (Kashyap, Ganguly, & Das, 2007). In general, their proposed measurement-based model can characterize and model the impact of interference caused by active traffic from multiple surrounding nodes on the link capacity. Specifically, the main objective of their work is to model the network capacity of any given link in the presences of any given number of interferers in a deployed network, carrying any specified amount of traffic load.

Extensive research has done for improving the network capacity for WMN (Bahl, Chandra, & Dunagan, 2004; Raniwala & Chiueh, 2005). One scheme is to improve the network capacity is to utilize multiple channels that are available in the IEEE 802.11 a/b/g standards (Yu, Mohapatra, & Liu, 2008). In order to fully utilize the available multi-channels, the best solution is to equip each mesh node with multiple radios that are tuned to different frequencies. For instance, (Bahl, Chandra, & Dunagan, 2004) identified that the channel switching time could be decreased. This implies that it is possible for each node to use a different channel in each time slot. Recently, (Yu, Mohapatra, & Liu, 2008) proposed a new scheme in which they determine the highest gain by increasing the number of channels and radios with certain traffic demands under both dynamic and static channel assignments. However, the channel switching requires fine-grained synchronization among nodes in order to avoid the deafness problem (i.e., the transmitter and the receiver may be on different channels at one time) (Kashyap, Ganguly, & Das, 2007). Also, the time for channel switching which can be in the range of few milliseconds to a few hundred microseconds may be unacceptable for most real time multimedia applications (Kyasanur & Vaidya, 2005).

CONCLUSION

In this chapter, we have studied a multi radio multi channel WMN using OFCD which amalgamates the advantages of both OFDM and CDMA. A transmitter and a receiver model are described with the help of useful mathematical expressions. This is a simple and a realistic multi-access technique that has good performance properties. In very noisy multipath channel, the hybrid OFCD multiaccess scheme is expected to work efficiently and provides better BER performance and end-to-end throughput as compare to the conventional multiaccess schemes such as CSMA/CA. We have also highlighted some key advantages of multi-radio wireless mesh networks along with primary technical challenges that must be addressed for widespread deployment of such networks such as scalability, interference, and network throughput. WMNs present a promising solution by extending network coverage based on mixture of above wireless technologies through multi-hop communications. Finally, we also discussed several access schemes that can be developed and implemented based on the OFDM systems for WMN such as OFDM-FDMA, OFDM-TDMA, and OFDM-CDMA (also referred as MC-CDMA).

REFERENCES

Acharya, A., Misra, A., & Bansal, S. (2003). High-performance architectures for IP-based multi-hop 802.11 networks. *IEEE Wireless Communications*, *10*(5), 22–28. doi:10.1109/MWC.2003.1241091

Adya, A., Bahl, P., Padhye, J., Wolman, A., & Zhou, L. (2004). A multi-radio unification protocol for IEEE 802.11 wireless networks. In *Proc. of the IEEE BroadNets*, (pp. 344-354).

Akyildiz, F., Wang, X., & Wang, W. (2005, March). Wireless mesh networks: a survey. *Computer Networks*, *47*(4), 445–487. doi:10.1016/j.comnet.2005.01.002

Avallone, S., Akyildiz, I., & Ventre, G. (2009). A channel and rate assignment algorithm and a layer-2.5 forwarding paradigm for multi-radio wireless mesh networks. *IEEE/ACM Transactions on Networking, 17*(1), 267-280.

Bahl, P., Chandra, R., & Dunagan, J. (2004). SSCH: Slotted channel hopping for capacity improvement in IEEE 802.11 ad hoc wireless networks. In *Proc. of IEEE 802.11 Ad-Hoc Wireless Networks*, Philadelphia, USA.

Bahl, V. (2007). Wireless mesh networks. *IEEE INFOCOM Tutorial*.

Bing, C., Frank, F., James, G., Rainer, G., Hermann, R., & Adam, W. (2001). Framework for combined optimization of DLC and physical layer in mobile OFDM systems. In *Proc. of the 6th International OFDM workshop*, Hamburg, (pp. 32-36).

Chang, Y., Chien, F., & Kuo, J. (2007, May). Cross-layer QoS analysis of opportunistic OFDM-TDMA and OFDMA networks. *IEEE Journal on Selected Areas in Communications, 25*(4), 657–666. doi:10.1109/JSAC.2007.070503

Chang, Y., Tao, Z., Zhang, J., & Kuo, J. (2009, June). *A graph approach to dynamic fractional frequency reuse (FFR) in multi-cell OFDMA networks*. Accepted for publication in *IEEE ICC'09*, June 2009, from http://www-scf.usc.edu/~yujungc/.

Cheong, Y., Cheng, R., Lataief, K., & Murch, R. (1999). Multiuser OFDM with adaptive subcarrier, bit, and power allocation. *IEEE Journal on Selected Areas in Communications, 17*(10), 1747–1757. doi:10.1109/49.793310

Chuang, J., & Sollenberger, N. (2000). Beyond 3G: wideband wireless data access based on OFDM and dynamic packet assignment. *IEEE Communications Magazine, 38*(7), 78–87. doi:10.1109/35.852035

Gupta, P., & Kumar, R. (2001, January). The capacity of wireless networks. *IEEE Transactions on Information Theory, 46*(2), 388–404. doi:10.1109/18.825799

Hottinen, A., & Heikkinen, T. (2006, March). Subchannel assignment in OFDM relay nodes. In *Proc. of 40th Annual Conference on Information Sciences and Systems*, (pp. 1314-1317), Princeton, NJ.

Johnsson, M. (1999). HiperLAN/2-The broadband radio transmission technology operating in the 5GHz frequency band, technical specification. V 1.0. *HiperLAN/2 Global Forum*.

Kashyap, A., Ganguly, S., & Das, S. (2007). A measurement-based approach to modeling link capacity in 802.11-based wireless networks. In *Proc. of the 13th annual ACM international conference on Mobile computing and networking*, (pp. 242-253).

Kim, J., & Bambos, N. (2002). Power efficient MAC scheme using channel probing in multi-rate wireless ad hoc networks. In *IEEE Vehicular Technology Conference*, (pp. 2380–2384).

Kyasanur, P., & Vaidya, N. (2005). Routing and interface assignment in multi-channel multi-interface wireless networks. In *Proc. of IEEE Wireless Communications and Networking Conference*, (Vol. 4, pp. 2051- 2056).

Li, Y., Qiu, L., Zhang, Y., Mahajan, R., Zhong, Z., Deshpande, G., & Rozner, E. (2007). Effects of interference on throughput of wireless mesh networks: Pathologies and a Preliminary Solution. In *Proc. of HotNets-VI*, Atlanta, GA, USA.

Lin, X., & Rasool, S. (2007, May). Distributed joint channel assignment, scheduling and routing algorithm for multi-channel ad hoc wireless networks. In *Proc. of 26th IEEE International Conference on Computer Communications*, (pp. 1118-1126).

Neishaboori, A., & Kesidis, G. (2008). Wireless mesh networks based on CDMA. *Computer Communications*, *31*(8), 1513–1528. doi:10.1016/j.comcom.2008.01.020

Niyato, D., Hossain, E., & Fallahi, A. (2006). Solar-powered OFDM wireless mesh networks with sleep management and connection admission control. In *Proc. of the 2006 international conference on Wireless communications and mobile computing*, (pp. 653-658).

Prasanna, G., & Ravichandran, V. (2007). Performance analysis of MC-CDMA for wide band channels. *Information Technology Journal*, *6*(2), 267–270. doi:10.3923/itj.2007.267.270

Raniwala, A., & Chiueh, C. (2005, March). Architecture and algorithms for an IEEE 802.11 based multi-channel wireless mesh network. In *Proc. of the IEEE Infocom*, Miami, (pp. 13-17).

Rappaport, T. (2002). *Wireless communications: principles and practice*, (2nd ed.). New York: Prentice Hall.

Sundaresan, K., & Rangarajan, S. (2008). On exploiting diversity and spatial reuse in relay enabled wireless networks. In *Proc. of the 9th ACM international symposium on Mobile ad hoc networking and computing*, (pp. 13-22).

Tse, D., & Grossglauser, M. (2002, August). Mobility increases the capacity of ad hoc networks. *IEEE/ACM Trans. Net*, *10*(4), 477–486.

Wang, X., & Xiang, W. (2006, March). An OFDM-TDMA/SA MAC protocol with QoS constraints for broadband wireless LANs. *Wireless Networks*, *12*(2), 159–170. doi:10.1007/s11276-005-5263-1

Xiang, Y., Peng-Jun, W., & Wen-Zhan, S., & Yanwei Wu. (2008). Efficient throughput for wireless mesh networks by CDMA/OVSF code assignment. *Ad Hoc and Sensor Wireless Networks*, *5*(3-4), 265–291.

Ye, L., & Gordon, L. (2006, February). *Orthogonal frequency division multiplexing for wireless communications*. Berlin: Springer.

Yu, H., Mohapatra, P., & Liu, X. (2008, April). Channel assignment and link scheduling in multi-radio multi-channel wireless mesh networks. *Mobile Networks and Applications*, *13*, 169–185. doi:10.1007/s11036-008-0037-5

Zhang, X., & Tang, J. (2006, July). QoS-driven asynchronous uplink sub-channel allocation algorithms for space-time OFDM-CDMA systems in wireless networks. *Wireless Networks*, *12*(4), 411–425. doi:10.1007/s11276-006-6542-1

Zhang, Y., Honglin, H., & Chen, H. (2008, April). QoS differentiation for IEEE 802.16 WiMAX mesh networking. *Mobile Networks and Applications*, *13*(1-2), 19–37. doi:10.1007/s11036-008-0035-7

Zhang, Y., Tan, I., Chun, C., Laberteaux, K., & Bahai, A. (2008). A differential OFDM approach to coherence time mitigation in DSRC. In *Proc. of the fifth ACM international workshop on Vehicular Inter-Networking*, (pp. 1-6).

Zhang, Z., He, Y., & Chong, K. (2005, March). Opportunistic downlink scheduling for multiuser OFDM systems. *Wireless Communications and Networking*, *2*(13-17), 1206 – 1212.

Zhou, D., & Lai, H. (2005). A compatible and scalable clock synchronization protocol. *Synchronization scheme reference IEEE working group, a wireless LAN MAC and physical layer (PHY) specification*, (ICCP).

Chapter 25
A New Approach to BSOFDM:
Parallel Concatenated Spreading Matrices OFDM

Ibrahim Raad
University of Wollongong, Australia

Xiaojing Huang
University of Wollongong, Australia

1. ABSTRACT

This chapter discusses a new concept for Block Spread OFDM called Parallel Concatenated Spreading matrices OFDM (PCSM-OFDM) which was first presented in (Raad, I. and Huang, X. 2007). While BSOFDM improved the overall BER performance on OFDM in frequency selective channels, this new approach further improves the BER of BSOFDM by over 3dB gain. This uses coding gain to achieve this and is similar in concept to the well known error correction codes Turbo Codes. This is done by copying the data at the transmitter n times in parallel and multiplexing.

2. INTRODUCTION

In today's world, it has become extremely important to continue to develop wireless communications to maintain continued economic growth. This is only achievable by ensuring that businesses and their customers have the best possible communications available. It is very important to remember that many businesses have invested large amounts of capital into the existing communication systems and as such it is not possible to deploy new systems. Therefore, to achieve better use of existing solutions and make use of the existing bandwidth becomes the priority.

A number of wireless solutions for modulating symbols across frequency selective channels exist. One of these solutions is called Orthogonal Frequency Division Multiplexing (OFDM).

OFDM is a method used to implement mutually orthogonal signals and this is done by setting up multiple carriers at a suitable frequency separation and modulating each symbol stream separately (Kamilo, F. 1995). By increasing the number of carriers the data rate per carrier can be reduced for a given transmission. The symbol streams do not interfere with each other because of the carriers being mutually orthogonal. It is possible to mitigate fading through suitable interleaving and coding. One method of ensuring the signals are independent of

DOI: 10.4018/978-1-61520-674-2.ch026

each other is to select the frequency separation between each signal in a manner which will achieve orthogonality over a symbol interval.

While OFDM will combat the effect of multipath transmission, other methods need to be utilized to mitigate the effect of fading. One way of achieving this is called Diversity Transmission. Diversity transmission can be used to reduce or remove the effect of fading by the transmitted signal power being "split between two or more sub channels that fade independently of each other, then the degradation will most likely not be severe in all sub channels for a given binary digit" (Rappaport, T. S. 2002). Then when all the outputs of these sub channels are recombined in the proper way the performance achieved will be better than the single transmission. There are a number of ways to achieve this diversity and the main methods include "transmission over spatially different times (space diversity), at different paths (time diversity) or with different carrier frequencies (frequency diversity)" (Rappaport, T. S. 2002).

Block Spread OFDM (BSOFDM), also known as pre-coded OFDM, has been used to achieve frequency diversity and has shown significant improvement over conventional OFDM in frequency selective channels. This is done by dividing the N subcarriers into M sized blocks and spreading them by multiplying these blocks with spreading codes such as the Hadamard matrix.

This chapter introduces Parallel Concatenated Spreading Matrices which further improves the BSOFDM by employing coding gain.

This chapter is organized as follows. A detailed literature review is provided to discuss methods which are used in this kind of system. This begins with the discussion of OFDM which is followed by a comparison of OFDMA, CDMA and TDMA based on studies already available. In the same section a brief discussion about diversity is presented since Block Spread OFDM employs frequency diversity. The method and examples of how frequency diversity is used is presented next and gives examples of two different spreading matrices used widely in industry standards. These are the Hadamard and the Rotated Hadamard matrices.

Block Spread OFDM is discussed in detail in section 2.5. In Section 3 and 4, the Parallel Concatenated Spreading Matrices OFDM (PCSM-OFDM) is presented in detail and discussed. The comparison between OFDM, BSOFDM and PCSM-OFDM is presented and analyzed. The scalable version of PCSM-OFDM – higher order PCSM-OFDM, is presented and discussed in Section 5. The concluding remarks and future recommendations are given in Sections 6 and 7.

2.1 Orthogonal Frequency Division Multiplexing (OFDM)

OFDM is currently used in high speed DSL modems over copper based telephone access lines. Since this work is primarily a contribution to the wireless communications then it is important to discuss where this modulation scheme is used in this field. OFDM has been standardized as part of the IEEE802:11a and IEEE802:11g for a high bit rate 54Mbps data transmission over wireless LANs (WLANs) (Schwartz, M. 2005).

In today's world there is an increased need for higher bit rate and higher bandwidth data transmission over radio based communication systems. It has been established in (Schwartz, M. 2005) that as the transmission bandwidth increases, frequency selective fading and other signal distortions occur. One of the major signal distortions that do occur in digital transmission is inter-symbol interference.

OFDM, by dividing the signal transmission spectrum into narrow segments and transmitting signals in parallel over each of these segments, mitigates this effect. If the bandwidth of each of these frequency spectrum segments is narrow enough, flat or non-frequency-selective fading will be encountered and the signal transmitted over each segment will be received non-distorted.

For this reason OFDM has been widely used in wireless applications. Figure 1 depicts the simple procedure of OFDM.

The R is the rate in bps of the transmission of the binary digits. The B is the bandwidth required to transmit these bits defined to be R(1+r), where r is the Nyquist roll off factor, or the order of R Hz. Also, the parameters a_k, $1 \leq k \leq N$, represent the successive bits stored, while the frequencies f_k, $1 \leq k \leq N$, represent the N carrier frequencies transmitted in parallel. N is a sequence of these bits stored for an interval $T_s = N/R$. This interval is known as the OFDM symbol interval. Serial-to-parallel conversion is then carried out, with each of the N-bits stored used to separately modulate a carrier. All N modulated carrier signals are then transmitted simultaneously over the T_s long interval.

Figure 2 depicts a simplified block diagram of an OFDM system.

2.2 Comparison of OFDMA, TDMA and CDMA

Studies have been carried out to investigate which multiple access systems are more robust in the presence of Narrowband Interference and according to (Moeneclaey, et al 1998) OFDMA has more potential than CDMA and TDMA. The following are the conclusions derived from their work:

1. The BER is independent in TDMA and PN - CDMA. The BER in TDMA, for a given CJR_{tot} (carrier - to - jammer ratio for the nth user), does not depend on the jammer frequency (F_J) or on the (maximum) number N of users.
2. The BER in PN - CDMA is independent of Fj, but increases with the ratio $(M-1)/N$ of the number of interfering users and the spreading factor.

Figure 1. OFDM (a) serial to parallel conversion (b) OFDM spectrum

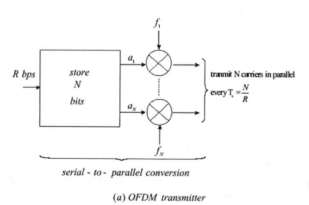

(a) OFDM transmitter

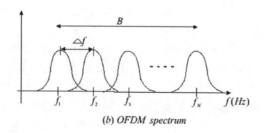

(b) OFDM spectrum

Figure 2. A simplified block diagram of an OFDM system

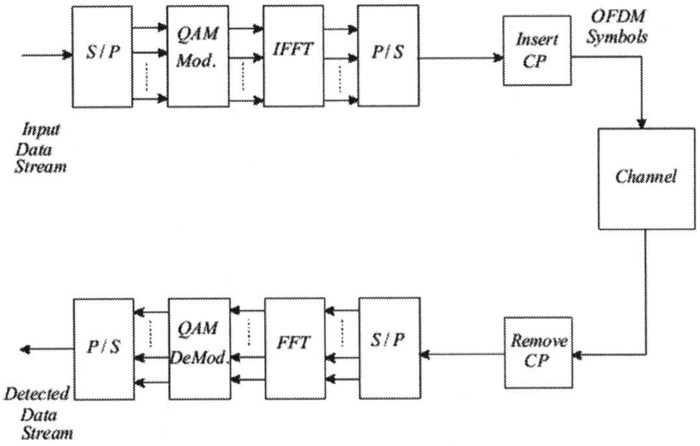

3. PN - CDMA is worse than TDMA even in the absence of interfering users due to the different statistics of the jammer term at the input of the decision device.
4. The BER for a given CJR_{tot} is a function of the user index, the maximum number of users N and F_J in OFDMA and OCDMA.
5. In TDMA and PN-CDMA, when CJR_{tot} is smaller than some threshold value, the fraction of users operating at a BER less than 10^{-3} is zero (resp. one).
6. In OFDMA and OCDMA, the fraction of users with a BER less than 10^{-3} is a gradually increasing function CJR_{tot}.
7. OFDMA and OCDMA systems have 50% users with BER $< 10^{-3}$ at values of CJR_{tot} far below the threshold values at which TDMA and PN-CDMA systems completely break down.
8. Finally, OFDMA performs significantly better than OCDMA.

2.3 Diversity

Although diversity is a form of redundancy, it is seen by many as a very useful solution to the problem of multipath fading in wireless communications. In basic terms, diversity is transmission of several replicas of the same information transmitted simultaneously over independent channels.

There are three types of diversity that have been studied, discussed and implemented in many different systems and include

1. Frequency diversity.
2. Time (signal -repetition) diversity.
3. Space diversity.

For frequency diversity, carriers are spaced sufficiently apart from each other so the system can provide independently fading versions of the channel which are used to transmit the signal. An example of this is frequency hopping; another is the use of spreading matrices. Since Block Spread OFDM uses spreading matrices to achieve frequency diversity, then only the frequency diversity will be discussed. This leads us to discuss some examples of spreading matrices.

2.4 Spreading Matrices

This section will discuss existing spreading matrices that are used, or can be used, to introduce frequency diversity in OFDM and specifically the system known as Block Spread OFDM (BSOFDM) or pre-coded OFDM. Some of the popular and known spreading matrices include the Hadamard matrix and the Rotated Hadamard.

Figure 3. After spreading using the Hadamard Matrix, the scatter plot of the data, QPSK modulation becomes a higher modulation scheme

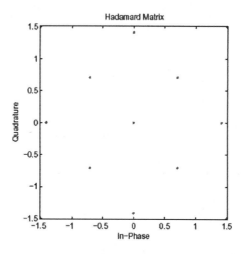

2.4.1 Hadamard Matrix

By selecting as code words the rows of a Hadamard matrix, it is possible to produce Hadamard codes. An N x N matrix of 1's and 0's (or -1's) is a Hadamard matrix "such that each row differs from any other row in exactly N/2 locations. One row contains all zeros with the remainder containing N/2 zeros and N/2 ones" (Rappaport, T. S. 2002). The minimum distance for these codes is N/2. An example for an N = 2 Hadamard matrix A is,

$$A = \begin{bmatrix} 1 & 1 \\ 1 & -1 \end{bmatrix} \quad (1)$$

In a BSOFDM system, after the modulated data is multiplied by the Hadamard matrix, a higher order modulation scheme is created which increases the correlation between the transmitted symbols, therefore achieving a better system performance. The new modulation scheme scatter plot is shown in Figure 3. The new scheme shows it has increased from the four QPSK symbols to nine, where the modulation at the transmission was carried using QPSK. This is an example of employing frequency diversity to achieve better performance for OFDM.

2.4.2 Rotated Hadamard

The rotated Hadamard matrix is a Hadamard matrix with the rotation described in the equation below,

$$U = \frac{1}{\sqrt{N}} H_{M \times M} diag\left(\exp^{\left(\frac{j\pi m}{c}\right)} \right) \quad (2)$$

Where C is the rotation value where the modulation is rotated back on to itself with. H is the Hadamard matrix and M is the size of the matrix.

The modulation data is multiplied by U and the rotation takes place producing a higher modulation scheme. This can be seen in Figure 4 depicting the modulated data after the rotated Hadamard matrix. The rotated Hadamard is capable of achieving 16QAM. So as can be seen this rotated Hadamard produces a higher order scheme than the traditional Hadamard. This is directly translated into a better BER performance in BSOFDM system of rotated Hadamard over Hadamard.

This leads into the description of Block Spread OFDM.

A New Approach to BSOFDM

Figure 4. After spreading using the Rotated Hadamard Matrix, the scatter plot of the data, QPSK modulation becomes 16QAM

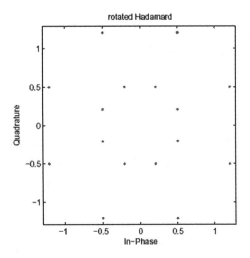

2.5 Block Spread – OFDM (pre-coded OFDM)

The main idea of BSOFDM (or pre-coded OFDM) is to "split the full set of subcarriers into smaller blocks and spread the data symbols across these blocks via unitary spreading matrices in order to gain multipath diversity across each block at the receiver" (McCloud, M. L 2004; Bury A., Engle J., and Linder J. 2003).

Block spreading matrices are used to introduce dependence among the subcarriers. As an example, N subcarriers are split into $N/2$ for blocks of size 2, then each of the blocks are multiplied by a 2×2 unitary matrix U_2. "The resulting length 2 output vectors are then interleaved to separate the entries in each block as far as possible across the frequency band so that they will encounter independent fading channels" (McCloud, M. L 2004). The transmitter's IFFT has the interleaved data passed through it and this data is sent across the frequency selective channel. The data is passed through an FFT processor at the receiver and de-interleaved before using block by block processing. A detailed block diagram of BSOFDM (also known as pre-coded OFDM) can be seen in Figure 5.

Referring to the transmitter model shown in Figure 5, let $x[i]$ where $i = 0, 1, ..., MN-1$, denote MN data symbols (M and N are integer powers of 2), which are modulated from the information data bits after binary phase shift keying (BPSK), quadrature phase shift keying (QPSK) or any other quadrature amplitude modulation (QAM) constellation mapping.

Before pre-coding, the MN data symbols are firstly divided into N groups of size M with the nth group denoted as a vector where $n = 0, 1, ..., N-1$ and $(.)^T$ denotes matrix transposition,

$$x = (x[nM], x[nM + 1], ..., x[nM + M-1])^T. \quad (3)$$

This is then expressed as a vector after serial-to-parallel conversion (S/P),

$$x = \begin{pmatrix} x_0 \\ x_1 \\ . \\ x_{N-1} \end{pmatrix} \quad (4)$$

Figure 5. A detailed block diagram representation of the BSOFDM/pre-coded OFDM system

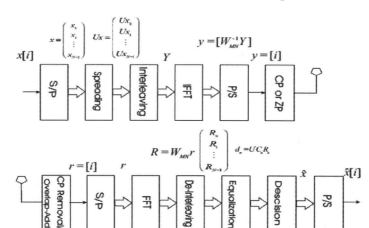

The pre-coding process is to apply an $M \times M$ unitary matrix U, which satisfies the property shown in the Equation 5, where (.)' denotes transposition and complex-conjugation operation. The 'I' is the identity matrix of order M.

$$UU' = U'U = I. \quad (5)$$

Each vector x_n produces a pre-coded vector where each element is a linear combination of the symbols in vector x_n.

The pre-coded symbols are mapped onto subcarriers equally spaced across the transmitted bandwidth to better exploit frequency diversity. This is equivalent to a block interleaving operation among N pre-coded vectors Ux_n, where $n = 0,1,...,N-1$, and then performing IFFT of length MN on the resulting pre-coded and interleaved vector Y.

A time domain sequence $y[i]$ where $i = 0,1,...,MN-1$ is produced after IFFT and parallel to serial conversion (P/S). Either a cyclic prefix (CP) or zero padding (ZP) of sufficient length (longer than the maximum channel multipath delays in samples) is added to $y[i]$ to form a pre-coded OFDM symbol. This is done to avoid interference between adjacent pre-coded OFDM symbols and turn the linear convolution of the transmitted signal with the channel impulse response into a circular one.

The pre-coded OFDM signal is then transmitted over a frequency selective multipath fading channel and received at the receiver baseband. An example of this type of channel is UWB.

By removing the CP or performing an overlap-add operation, MN- point received pre-coded OFDM samples $r[i]$ where $i = 0,1,...,MN-1$, will be produced. After FFT and de-interleaving, the discrete-time received signal can be expressed in the frequency domain as

$$R_n = H_n U x_n + V_n \quad (6)$$

where $n = 0,1, ..., N - 1$ and

$$R = (R[n], R[N + n], ..., R[(M\ 1)N + n])^T \quad (7)$$

is a vector of M elements which are decimated from $R[k]$, the MN - point discrete Fourier transform (DFT) or $r[i]$, by a down-sampling factor N.

$$H_n = \text{diag}(H[n], H[N + n], ..., H[(M - 1)N + n]) \quad (8)$$

A New Approach to BSOFDM

is an $M \times M$ diagonal matrix with diagonal elements decimated from $H[k]$, the MN - point DFT of the normalized discrete channel impulse response $h[i]$, and V_n is a zero-mean Gaussian noise vector with covariance matrix $E\{V_n V_n'\} = \sigma_v^2 I$ where $E\{\ \}$ denotes ensemble average.

To recover the transmitted data vector x_n, equalization and detection must be performed on the received signal R_n.

Due to the complexity of Maximum Likelihood Decoder (ML) which is not considered to be practical in many systems, Minimum Mean Square Error decoder (MMSE) is usually used since this simply applies a one tap equalizer for each subcarrier in the frequency domain.

This then leads to the Parallel Concatenated Spreading Matrices OFDM (PCSM-OFDM).

3. PARALLEL CONCATENATED SPREADING MATRICES OFDM

Coding gain has been previously used in error correction codes such as Turbo codes and Low Density Parity checks (LDPC) to achieve better performance in wireless communication systems. This type of gain is applied to BSOFDM in a new system described as Parallel Concatenated Spreading Matrices (PCSM) OFDM which is presented in (Raad, I. and Huang, X. 2007). While BSOFDM improved the overall BER performance on OFDM in frequency selective channels, this new approach further improves the BER of BSOFDM by over $3dB$ gain in the same frequency selective channels.

4. SYSTEM DESCRIPTION OF PCSM-OFDM

Figure 6 depicts the block diagram of the system PCSM-OFDM. The same data is copied into two streams of block size N/M, d_1 and d_2, where N is the number of subcarriers and M is the block size. Each stream is spread using a unitary spreading matrix U of size $M \times M$. The streams of M sized blocks can be described as $d_1 U_1$ and $d_2 U_2$. The streams are multiplexed and the same process which is applied to BSOFDM is applied to PCSM-OFDM after this point.

At the receiver, the same process applied to BSOFDM is applied to PCSM-OFDM except the de-multiplexing is used to separate the two streams apart.

To make full use of the coding gain the two data streams are combined before de-spreading (Figure 7) takes place and the combining is done

Figure 6. The new approach to BSOFDM, Parallel Concatenated Spreading Matrices

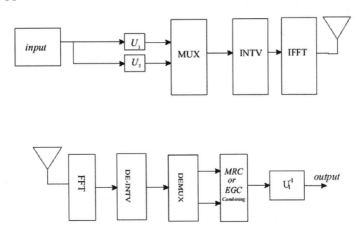

Figure 7. (a) PCSM-OFDM with the combining after the de-spreading and (b) an interleaver included before the second spreading matrix

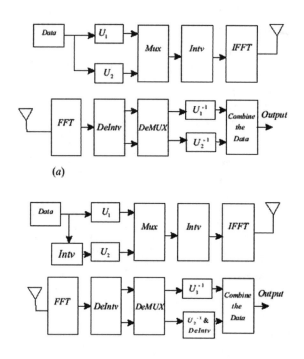

by using Maximum Ratio Combining (MRC) and can be represented by the following equation

$$R_1 = \frac{\left(\alpha_1^* r_1 + \alpha_2^* r_2\right)}{|\alpha_1|^2 + |\alpha_2|^2} \qquad (9)$$

where α_1 and α_2 are the estimated channel weights and r_1 and r_2 are the useful data from each stream. The system can also use the Equal Gain Combining (EGC) and is where the channel weights α_1 and α_2 are given equal priority of 1, then Equation 9 becomes

$$R_1 = \frac{\alpha_1^*}{|\alpha_1|} r_1 + \frac{\alpha_2^*}{|\alpha_2|} r_2. \qquad (10)$$

The decoder used for the simulation results is the Zero Forcing decoder and the channels used for these simulations are slow fading and frequency selective UWB channels. The combining unless otherwise specified is the EGC and the number of packets simulated is *10000* packets. The spreading matrices used in these studies are the Hadamard, Rotated Hadamard and the Rotation Spreading matrix which is first introduced in (Raad I., Huang X., and Raad R. 2006).

Figure 8 depicts the BER versus SNR in a slow fading channel when using the Hadamard matrix as the unitary spreading matrix. The number of subcarriers used is $N = 128$. The BER gain of PCSM-OFDM over BSOFDM is greater than *3dB*. This showed that employing this new structure greatly improved the performance of the OFDM system. The same performance is also seen when used with $N = 64$ subcarriers in the same environment. As expected the same performance is also seen when using different spreading matrices such as the Rotation Spread-

Figure 8. PCSM-OFDM compared with BSOFDM in slow fading channel using N=128

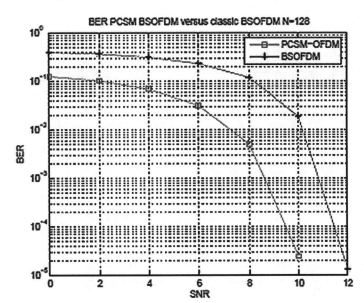

ing matrix and Rotated Hadamard in the same environment.

Figure 9 compares the PCSM-OFDM using *N = 16* subcarriers with classical BSOFDM *N = 32* subcarriers to ensure that there is a real improvement with this new approach and not just due to an increase in samples across the channel. The *N = 32* for BSOFDM is used to compare the same sample number across the channel, and as can be seen the improvement is still evident from the simulation results. It can be concluded that the improvement is not due to the increase in samples. In the slow fading channel there is a greater than *4dB* improvement in terms of BER.

Across frequency selective channels such as UWB, this new system also showed superior performance to the classical BSOFDM. UWB channels *CM3* and *CM4* (which represent medium to long distances of non-line of sight environments) are used for the purpose of studying the PCSM-OFDM system. Figure 10 shows the simulation results of BER versus SNR of the Hadamard matrix with *N = 64* subcarriers and again PCSM-OFDM outperforms the classical BSOFDM, with a gain in dB of over *2dB* this time shown in UWB channel *CM3*. It is important to note that these improvements are achieved at low SNR not at high SNR.

In Figure 7, two different configurations of structures for the PCSM-OFDM are shown. These include the use of a random interleaver at the transmitter seen in Figure 7 (b) and combining done at the receiver after de-spreading has as been carried out seen in Figure 7 (a). The simulation results do not show any difference in dB gain between the two structures. These results are depicted in Figure 11. The only real advantage of one over the other is when the combination of the two streams is done before the de-spreading, which will ensure that only one inverse unitary matrix is used at the receiver. This reduces the overall complexity of the system.

5. HIGHER ORDER PARALLEL CONCATENATED SPREADING MATRICES OFDM

This section studies higher order Parallel Concatenated Spreading Matrix OFDM (PCSM-OFDM)

Figure 9. PCSM-OFDM N = 16 compared with BSOFDM N = 32 to ensure that the correct comparison is made at the channel

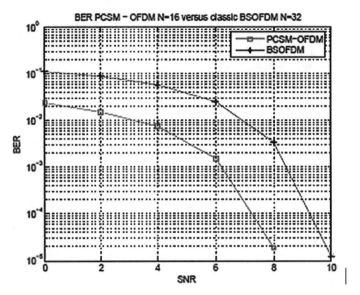

Figure 10. PCSM-OFDM N = 64 in UWB channel CM3

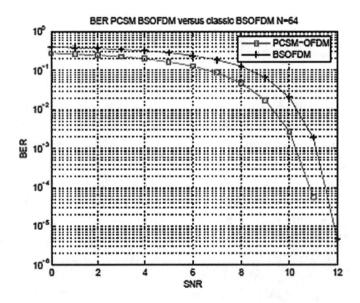

which is presented in (Raad I. and Huang X. 2008). PCSM-OFDM is shown to greatly improve on the classic Block Spread OFDM, this continued the work on PCSM-OFDM and studies higher order in terms of an increased number of streams. It is shown as the number of streams increase so does the BER gain.

Figure 11. PCSM-OFDM N = 128 in UWB channel CM4 compares interleaver and non-interleaver using first combination

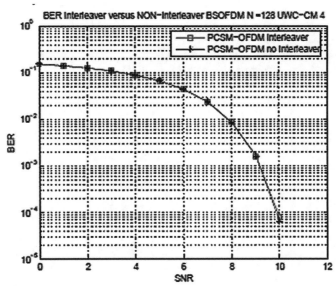

5.1 System Description of Higher Order PCSM-OFDM

A very important question in any system is scalability. This question is applied to PCSM-OFDM and is depicted in Figure 12, where n is the number of streams required. If the system is to use the higher order PCSM-OFDM, then at the receiver the combination using MRC Equation 9 is transformed into

$$R_1 = \frac{\left(\alpha_1^* r_1 + \alpha_2^* r_2 + \alpha_3^* r_3 + \ldots + \alpha_n^* r_n\right)}{|\alpha_1|^2 + |\alpha_2|^2 + |\alpha_3|^2 + \ldots + |\alpha_n|^2} \quad (11)$$

and if the system required the use of EGC, then Equation 11 would be transformed into

$$R_1 = \frac{\alpha_1^*}{|\alpha_1|} r_1 + \frac{\alpha_2^*}{|\alpha_2|} r_2 + \cdots + \frac{\alpha_n^*}{|\alpha_n|} r_n \quad (12)$$

where α_n are the estimated channel weights and r_n representing the useful channel information.

The PCSM-OFDM scalability improved the system performance with the increase of streams in terms of gain in dB. Figure 13 compares four, three and two streams for the PCSM-OFDM. The simulation result used $N = 64$ subcarriers with *10000* OFDM packets simulated.

As can be seen from the results, as the number of streams increased the BER performance and the dB gain improves. With each stream increase a gain of 1dB is achieved. All these experimental results through simulations are across frequency selective channels UWB and 2 ray slow fading channels. The decoder at the receiver used is the zero forcing with the unitary matrix used for spreading being the Hadamard matrix.

6. FUTURE RESEARCH DIRECTIONS

More research is required on combination techniques at the receiver. This research should focus on efficiency. This will be studied further down the line. Part of this research will include different types of decoders that are useful to such a system.

Figure 12. Higher Order Parallel Concatenated Spreading Matrices - OFDM

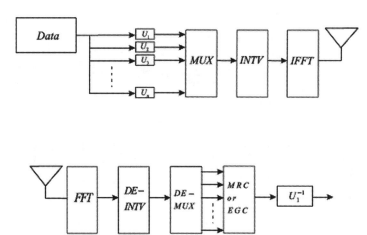

Figure 13. PCSM-OFDM higher order 4, 3 and 2 compared a slow fading channel using N = 64 sub-carriers

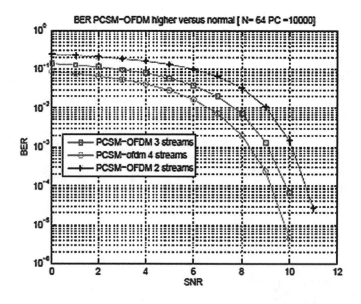

7. CONCLUSION

This chapter presented a new approach to Block spread OFDM called Parallel Concatenated Spreading Matrices OFDM (PCSM-OFDM) which is presented in (Raad, I. and Huang, X. 2007). The data is copied into two streams (which become parallel to each other) of M sized blocks and both these streams are spread using the same unitary spreading matrix. Then the two streams are multiplexed and the normal procedure is carried out in BSOFDM and OFDM systems. At the receiver, after channel equalization, the signal is de-multiplexed and combined using MRC or EGC before the de-spreading using the inverse of the unitary matrix. From the simulation results presented, this new approach to OFDM called PCSM-OFDM outperforms the BSOFDM (or pre-coded OFDM) by greater than 4dB in slow fading channels and over 3dB in frequency selective channels such as UWB channels. Other combinations of PCSM-OFDM included an interleaver between the two streams at the transmission side and carrying out the de-spreading before the two streams at the receiver are combined. The same performance is achieved using both combinations as the original configuration of PCSM-OFDM, but the de-spreading after the combining would reduce complexity as the system is only required to use one unitary matrix inverse. Higher order PCSM-OFDM is then presented in this chapter. This system showed that with the increase of the number of n streams the system gained an extra 1dB gain. Overall, this system is recommended for wireless communication systems were OFDM and BSOFDM are applied.

REFERENCES

Bury, A., Engle, J., & Linder, J. (2003). Diversity comparison of spreading transforms for multicarrier spread spectrum transmission. *IEEE Transactions on Communications*, *51*(5), 774–781. doi:10.1109/TCOMM.2003.811406

Kamilo, F. (1995). *Wireless Digital Communications - Modulation and Spread Spectrum Applications*. New York: Prentice Hall.

McCloud, M. L. (2004). Optimal binary spreading for block ofdm on multipath fading channels. *WCNC / IEEE Communicatio Socialis*, *2*, 965–970.

Moeneclaey, M., Van Bladel, M., & Sari, H. (1998). A Comparison of Multiple Access Techniques in the Presence of Narrowband Interference. In *International Symposium on Signals, Systems, and Electronics, 1998, ISSSE*, URSI, (pp. 223 – 228).

Raad, I., & Huang, X. (2007). A new approach to bsofdm-parallel concatenated spreading matrices OFDM. *The 7th IEEE International Symposium on Communications and Information Technologies, ISCIT07 2007*, October 16 – 19, 2007.

Raad, I., & Huang, X. (2008). Study of higher order parallel concatenated block spread OFDM. *The Third International Conference on Information and Communication Technologies: From Theory to Applications, IEEE ICTTA'08, Damascus, Syria*, April 7 – 11, 2008.

Raad, I., Huang, X., & Raad, R. (2006). A new spreading matrix for block spread OFDM. *The 10th IEEE International Conference on Communication Systems 2006, IEEE ICCS'06, Singapore*, October 30 - Nov 1.

Rappaport, T. S. (2002). *Wireless Communications - principles and practice* (2nd Ed.). New York: Prentice Hall.

Schwartz, M. (2005). *Mobile Wireless Communications*. Cambridge, UK: Cambridge University Press.

Chapter 26
The Next Generation CDMA Technology for Futuristic Wireless Communications:
Why Complementary Codes?

Hsiao-Hwa Chen
National Cheng Kung University, Taiwan

ABSTRACT

This chapter addresses the issues on the architecture of next generation CDMA (NG-CDMA) systems, which should offer a much better performance in terms of its capacity and transmission rate, etc., than that possible in all current 2-3G systems based on CDMA technology. The ultimate goal is to engineer a CDMA system, whose performance will no longer be interference-limited, for its application in futuristic wireless communications. To achieve this, many challenging issues should be tackled, such as innovated design approaches for CDMA codes, multi-dimensional spreading techniques, suitable CDMA signaling format for high-speed bursty traffic, and so forth. This chapter will review the author's ongoing research activities on the NG-CDMA technology, which can offer a performance never inferior to that of orthogonal frequency division multiple access (OFDMA) technology. In particular, the author will briefly introduce a new CDMA code design method, called Real Environment Adapted Linearization (REAL) approach, which can be used to generate CDMA code sets with inherent immunity against multipath interference and multiple access interference for both uplink and downlink transmissions. The chapter will also illustrate that an interference-free CDMA can only be made possible with the application of orthogonal complementary codes (OCCs). The use of traditional CDMA codes, such as Gold, Kasami, Walsh-Hadamard and OVSF codes, all working on an one-code-per-channel basis, will never help in this sense. Several other topics related to the NG-CDMA technology will also be addressed, such as system performance issues, other properties of the NG-CDMA technology, and so on.

DOI: 10.4018/978-1-61520-674-2.ch027

The Next Generation CDMA Technology for Futuristic Wireless Communications

INTRODUCTION

This chapter addresses the issues on the development of NG-CDMA technologies and contains the information on the subjects from both open literature and our own research activities in last fifteen years.

When we initially started to work on the research project on the NG-CDMA technology in 2003, the CDMA technology reached its climax of popularity and everybody was talking about CDMA, and its applications could be found then in various wireless and wired communication systems, virtually everywhere. It seemed to me at that time that CDMA technology will stay in its leading position for a long time. However, recently CDMA technology has faced a serious challenge from other multiple access technologies, in particular from orthogonal frequency division multiple access (OFDMA) technology, and many people have turned away from CDMA to OFDMA.

There are many reasons that CDMA technology has become less popular than it was a few years ago. One of the most plausible reasons is that, as quoted from some people's opinion, the concept of CDMA technology was developed more than ten years ago and it suits well only for slow-speed and continuous-time signal transmissions, which are relevant to voice-centric services, as carried by most 2G mobile cellular systems, such as IS-95, etc. Now, we are talking about high-speed burst-type traffic (such as 4G wireless applications) in the wireless channels, and thus CDMA technology is not suitable. For almost the same reason, the OFDMA technology came into the stage and seems to be a strong candidate to replace CDMA as the prime multiple access technology for futuristic wireless applications, such as exactly what 3GPP Long Term Evolution (LTE) and WiMax systems (Lu, Qian, Chen, & Fu, 2008) are doing.

However, behind the explanation on why CDMA technology can not continue taking the lead we have sensed some unrevealed truth, which might also be the cause that has made the CDMA technology lag behind. Let us take a look at the mobile cellular communication technologies, which have gone through 2G and 3G since the first commercial CDMA cellular systems was launched more than ten years ago. In Taiwan, as well as in many other regions or countries, we have actually entered 3.5G era with High Speed Downlink Packet Access (HSDPA) being put in place by several mobile service providers. On the other hand, the CDMA technology stays in the same place (with almost the same core technologies being used in both 2G and 3G systems) and we have not seen any substantial technological advancement related to CDMA so far. Therefore, it is natural and understandable that people have turned to some other better multiple access technologies to replace CDMA, if the current CDMA technology itself does not advance as fast as expected.

The technical requirements of future Gigabit wireless systems will be very much different from the current 2-3G mobile cellular systems in terms of their applications and working conditions. The current 2-3G systems were developed basically for slow speed transmission in continuous traffic, the major part of which is dedicated for the voice-centric services. The prime traffic in the current 2-3G systems is still in circuit-switching mode. On the other hand, the future Gigabit wireless systems are expected to run on an all-IP wireless platform with a transmission speed at Gigabit per second. Also, the packet switching data streams will be the major part of the traffic. Therefore, the technical requirements and thus the design methodologies should be necessarily innovated to address all the problems existing in the current 2-3G systems.

The appreciation of CDMA technologies in 2-3G systems was due partly to the fact that they could provide on the average a relatively high bandwidth efficiency than that possible by using other multiple access techniques, such as FDMA and TDMA. Unfortunately, the improvement on the bandwidth efficiency achieved in the 2-3G CDMA-based systems seems not to be enough to justify their applications in future Gigabit wireless systems. For instance, UMTS UTRA-FDD or W-CDMA system could offer only an 144 kb/s transmission rate at high mobility with a chunk of 10 MHz spectrum for full-duplex connection, ending up with a bandwidth efficiency of merely 0.0144 b/s/Hz, being far from satisfactory.

We have been fascinated by the research works on the NG-CDMA technologies as we have obtained a lot of interesting data and results. It will be shown in this chapter that the CDMA technology will have a great opportunity to stay as a leading multiple access technology for different communication systems (wireless and wired) if we can continue working hard to make it happen. We will show through the results given in this chapter that it is definitely possible to make CDMA systems interference-free (instead of being always interference-limited), which is one of the most important characteristic features for the NG-CDMA technology.

The rest of the chapter can be outlined as follows. Section II will summarize the important conclusions made from the code design method called REAL approach proposed by us. Section III will introduce different types of complementary codes used in the NG-CDMA technology. Section IV will provide necessary assumptions and system model that will be used throughout the chapter. Section V is dedicated to the discussions on interference-free properties of the NGCDMA systems. Section VI will study the performance of a CDMA system based on orthogonal complementary codes. Section VII will be focused to discuss the other useful properties of the NG-CDMA technology, followed by the conclusion of the chapter.

WHY COMPLEMENTARY CODES?

The innovation of current CDMA technology should start from the design approach of the signature codes or sequences. We have been working on this topic for many years and proposed a new method to generate the codes for the NG-CDMA systems. The method is called "real environment adaptation linearization" (REAL) approach. The salient feature of the REAL approach is that it takes into account many practical operational parameters, such as multipath propagation, asynchronous transmission, random bit signs, etc., in the design framework. In doing so, the obtained codes can address many problems existing in the unitary codes used in 2G and 3G CDMA based wireless systems. One of the most important conclusions obtained from the REAL approach is that an interference-free CDMA can be implemented only using orthogonal complementary codes. Due to their many unique desirable properties, orthogonal complementary codes surely will play an extremely important role in the development of NG-CDMA technology.

It should be emphasized here that the choice of orthogonal complementary codes for the NGCDMA technology is not an accident. As a matter of fact, we did not have any preemptive ideas on which type of signature codes might be suitable for the NG-CDMA technology before our study on the issues, which lasted for many years. As shown in the results obtained from our research (Chen, Chiu, & Guizani, 2006), we can show step by step the process which leads us to find the orthogonal complementary codes as the most desirable spreading codes in a natural way, which yields the conclusion that an interference-free CDMA system will be made possible if and only if the orthogonal complementary codes will be used as the spreading codes (or signature codes) for the system. We should not detail the REAL approach here, but just summarize the major contributions made in the REAL approach as follows.[1]

1) In the REAL approach for the first time we used the concept of joint code and system design in the code search process. All previous effort in this topic separated the codes design from the system design.
2) In the REAL approach, we have taken into account many real operational conditions when searching for the optimal signature codes for CDMA applications. Those real operational conditions include multipath propagation, synchronous and asynchronous transmissions, random bit signs in the data stream, bursty traffic, and so forth. Therefore, the solutions from this optimization problem can ensure an interference-free CDMA operation if a CDMA system operates under the same conditions.

3) There are two important conclusions obtained from the REAL approach. The first conclusion is that the REAL approach has shown that it is possible to implement an interference-free CDMA system, and only orthogonal complementary codes should be used for such an interference-free CDMA system. Second, in such an interference-free CDMA system the set size K of the orthogonal complementary codes must be made equal to the flock size M of the codes. It is noted that in a CDMA system using orthogonal complementary codes each user is assigned a particular flock of element codes for CDMA transmissions.

Therefore, we have to say that the selection of orthogonal complementary code as the core of the NG-CDMA technology is not an accident or an idea popping up from nowhere. Instead, the use of orthogonal complementary codes in the NG-CDMA technology is the results from solving an optimization problem, and no other type of codes should be used. The magic power of the orthogonal complementary codes comes from their unique correlation reconstruction properties, which were built up on much more degrees-of-freedoms, compared to the ways to reconstruct the correlation functions in all other traditional spreading codes, such as Gold code, Walsh-Hadamard code, Kasami code, m-sequence, OVSF code, etc.

DIFFERENT TYPES OF COMPLEMENTARY CODES

It is noted that complementary codes constitute a fairly large group of codes, whose autocorrelation functions and cross-correlation functions are built on their several element codes (whose number usually is even for obvious reason) jointly. Therefore, if they are used in a CDMA system, each user in the system has to be assigned a flock of element codes. On the contrary, all traditional spreading codes, such as Gold code, m-sequence, Walsh-Hadamard code, OVSF code, Kasami code, etc., are unitary codes, which are defined by us to characterize their properties when used in a CDMA system, and work on an one-code-per-user basis.

In this section, we will briefly introduce all complementary codes we have known so far. Those complementary codes (Chen, Chiu, & Guizani, 2006) include six different types listed as follows:

1) Primitive complementary codes;
2) Complete complementary codes;
3) Extended complementary codes;
4) Super complementary codes;
5) Pair-wise complementary codes; and
6) Column-wise complementary codes.

It should be noted that the first three types of complementary codes were introduced by others in the literature, and the last three were proposed by us. Some of the complementary codes are orthogonal complementary codes (OCCs) and some are not. In fact, only the first and fifth types, or the primitive complementary codes and pair-wise complementary codes, are not the orthogonal codes, and the rest are orthogonal complementary codes, which at least in theory can be used in CDMA systems. However, even among those orthogonal complementary codes, they may exhibit many different characteristic features in terms of their set sizes, their correlation properties and so forth. Therefore, we should say that the study on the complementary code based CDMA technology is only at its beginning and we have a very long way to go to work out the NG-CDMA technology suitable for futuristic wireless applications.

Here we would like to define the term named as ideally orthogonal set of the spreading codes as follows. If we say the spreading codes are ideally orthogonal, then both even and odd periodic cross-correlation functions of any two codes in the set are zero for any possible relative chip shifts. The even and odd periodic cross-correlation functions for any pair of codes is mathematically expressed as follows.

Assume that codes **A** and **B** are two codes in an ideally orthogonal set of the spreading codes. They can also be defined as $A = (A_1, A_2, ..., A_M)$ and $B = (B_1, B_2, ..., B_M)$, which states the fact that the codes **A** and **B** are complementary codes, each of which contains M element codes. Each element code consists of N chips, or $A_m = (a_{m1}, a_{m2}, ..., a_{mN})$ and $B_m = (b_{m1}, b_{m2}, ..., b_{mN})$, where m = 1, 2, ..., M. If **A** and **B** are two codes from an ideally orthogonal set, they must satisfy the following equations:

$$\rho(A,B;\tau) = \sum_{m=1}^{M}\left[\sum_{i=1}^{\tau} a_{mi}b_{m(N-\tau+i)} \pm \sum_{i=\tau+1}^{N} a_{mi}b_{m(i-\tau)}\right] = 0, \quad (\text{for any } \tau) \quad (1)$$

where τ is the relative delay between the two codes A and B. It is noted that the above equation gives the requirements on the even (if the positive sign is used in equation (1)) and odd (if the negative sign is used in equation (1)) periodic cross-correlation functions for the two ideally orthogonal codes. The equation (1)) is a general form for any code and letting $M=1$ implies that we are considering

the unitary spreading codes; otherwise it will refer to complementary codes.

We should note that not all complementary codes listed above are ideally orthogonal codes. For example, the primitive complementary codes and pair-wise complementary codes are not ideally orthogonal codes, as they do not satisfy the requirements on the perfect cross-correlation functions as specified in equation (1)). On the other hand, the rest complementary codes listed above, such as Complete Complementary Codes, Extended Complementary Codes, Super Complementary Codes, and Column-wise Complementary Codes, have their perfect cross-correlation functions and thus they are ideally orthogonal codes.

In addition to the cross-correlation function, which defines the ideal orthogonality of the set of spreading codes, we also need to define their auto-correlation function, which will affect the detection efficiency, especially in a multipath channel. The desirable auto-correlation function for an ideal spreading code set should be defined as follows.

$$\rho(A,A;\tau) = \begin{cases} \sum_{m=1}^{M}\left[\sum_{i=1}^{\tau} a_{mi}a_{m(N-\tau+i)} \pm \sum_{i=\tau+1}^{N} a_{mi}a_{m(i-\tau)}\right] = NM, & \tau=0 \\ \sum_{m=1}^{M}\left[\sum_{i=1}^{\tau} a_{mi}a_{m(N-\tau+i)} \pm \sum_{i=\tau+1}^{N} a_{mi}a_{m(i-\tau)}\right] = 0, & \text{elsewhere} \end{cases} \quad (2)$$

in which τ is the relative chip delay between the local correlator and the incoming code received at a receiver, and M can be any integer without losing generality. If M=1, the above autocorrelation requirement fits a unitary code set; otherwise it refers to a complementary code set. Equation (2) has included both even (for the positive sign in (2) and odd (for the negative sign in (2) periodic auto-correlation functions.

Obviously, if the code set size is K, we will have $C_2^K = \binom{K}{2} = \dfrac{K!}{2!(K-2)!}$ equations to specify the ideally orthogonal conditions as given in equation (1). We will also have K equations to specify the ideal auto-correlation functions for all K codes, as given in equation (2).

The definitions given here will form a basis for us to analyze the performance of the NGCDMA system based on complementary codes, as to be discussed in the section followed.

SYSTEM MODEL FOR PERFORMANCE ANALYSIS

Assume that an NG-CDMA system uses a set of orthogonal complementary codes, whose parameters are $N = 2^r$, $M = 2^R$ and $K = 2^R$, where r and R are any positive integers. Therefore, the parameter set for this orthogonal complementary codes can be written as $(N, M, K) = (2^r, 2^R, 2^R)$.

An important consideration for any complementary code based CDMA is that each user should be assigned a flock of M element codes and different element codes in the same flock should be sent via different channels to a receiver. Therefore, either time division multiplex (TDM) or frequency division multiplex (FDM) scheme can be used to send different element codes. For the discussions given in this chapter, we will only consider the use of FDM scheme to send different element codes in this chapter. Therefore, we need to use $M = 2^R$ sub-carriers to send 2^M different element codes in this NG-CDMA system. Due to the fact that output signal after DS spreading is still binary, we can use the simplest digital modulation scheme, such as BPSK, in the system. Of course, any other multi-level digital modulations can also be used here, as long as the link budget is enough for the application of a multi-level digital modulation scheme such as M-PSK and M-QAM, etc.

A conceptual block diagram for NG-CDMA system is shown in Figure 1 and Figure 2, where Figure 1 is to illustrate the multi-user signal

Figure 1. NG-CDMA multi-user signal formulation process, where 2^R users are present, each user is assigned a flock of 2^R element codes and 2^R sub-carriers are needed to send 2^R element codes

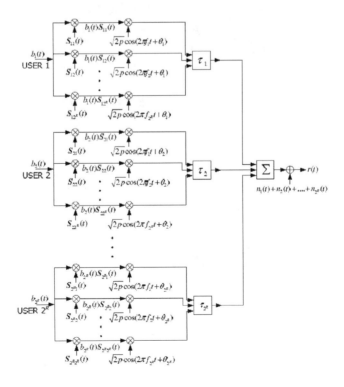

formulation process in the NG-CDMA system and Figure 2 is the receiver of the NG-CDMA system.

Assume that we will use a set of orthogonal complementary codes in this NG-CDMA system, and the parameters for this set of orthogonal complementary codes are N = 4, M = 4 and K = 4. Therefore the processing gain for any user will be 4 × 4 = 16. The detail information about the orthogonal complementary code set considered here is given as follows:

$$
\begin{aligned}
User\ 1 &: C_{11}:+++-\ ; C_{12}:++-+\ ; C_{13}:+++-; C_{14}:--+- \\
User\ 2 &: C_{21}:+-++\ ; C_{22}:+---\ ; C_{23}:+-++; C_{24}:-+++ \\
User\ 3 &: C_{31}:+++-\ ; C_{32}:++-+\ ; C_{33}:---+; C_{34}:++-+ \\
User\ 4 &: C_{41}:+-++\ ; C_{42}:+---\ ; C_{43}:-+--; C_{44}:+---
\end{aligned}
$$

(3)

INTERFERENCE-FREE OPERATION IN NG-CDMA SYSTEM

In order to illustrate how the NG-CDMA system using the orthogonal complementary code set works, we would like to use some figures to demonstrate several important characteristics of the system. To illustrate clearly, we should only consider two users (for instance, User 1 and User 2 with $C_1 = \{C_{11}, C_{12}, C_{13}, C_{14}\}$ and $C_2 = \{C_{21}, C_{22}, C_{23}, C_{24}\}$, respectively) in the system. In fact, the same results will apply if all four users are taken into account in the illustrations, with all figures becoming more complex. We further assume that the signal from user 1 is useful signal and that from user 2 will be considered as multiple access interference (MAI). Noise is omitted in the illustration for simplicity. In addition, we consider the case with signals being sent in very short packet, which consists of only three bits, [+1,−1,+1], for illustration conciseness.

In the following illustrations, we should first consider the situations where there is no multipath propagation effect, and then we extend the illustrations to the multipath channels. Both synchronous transmission and asynchronous transmissions will be considered here.

Isotropic MAI-Free Operation

In this subsection, we would like to use illustrations to show how a NG-CDMA system (based on a set of orthogonal complementary codes as given in (3)) could overcome the multiple access interference (MAI) to achieve an MAI-free operation. It should be noted that the MAI-free operation will be in place for both asynchronous transmissions and synchronous transmissions in the NG-CDMA system. Therefore, we would like to name this property as "isotropic MAI-free operation", which is in contrast with a conventional DS-CDMA system based on Walsh-Hadamard

Figure 2. NG-CDMA system receiver

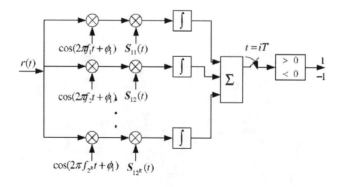

The Next Generation CDMA Technology for Futuristic Wireless Communications

codes, which can only offer the MAI-free operation in synchronous channels, but not in asynchronous channels, due to the fact that the cross-correlation functions among the Walsh-Hadamard codes become very bad in asynchronous transmission mode. We should also note that there is no real synchronous transmissions in any terrestrial wireless application scenarios. Even in downlink channels of a cellular system, the transmission in the downlink channels will not be synchronous if multipath propagation is present.

Figure 3 shows the process to despread the whole packet sent from users 1 and 2 at the receiver which is tuned to user 1. Therefore, the signals from the user 2 will contribute as interference in this detection process. In this illustration, we do not consider the noise and multipath and we only concern synchronous transmission, in which all bits from different users are aligned in time. It is seen from the figure that the detection to the three bits sent from the user 1 can be decoded successfully without being interfered by the signals sent

Figure 3. Despreading process for the whole packet sent from users 1 and 2 at the receiver which is tuned to user 1, where synchronous channel is considered

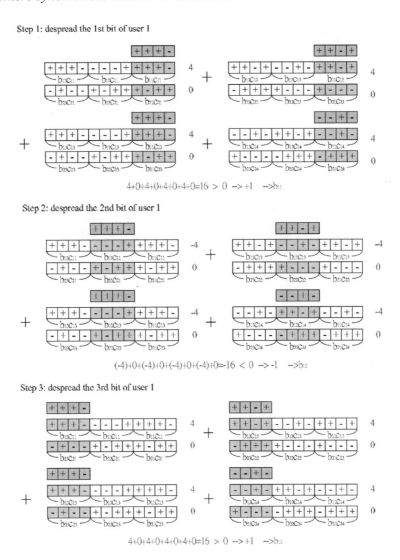

from user 2. Therefore, the MAI-free operation is guaranteed in this case.

Figure 4 shows the process to despread the whole packet sent from users 1 and 2 at the receiver which is tuned to user 1. Therefore, the signals from the user 2 will contribute as interference in this detection process. In this illustration, we do not consider the noise and multipath and we only concern asynchronous transmission, in which all bits from different users are not aligned in time. For illustration simplicity, we assume that the relative delay between users 1 and 2 is only one chip. However, it can be easily shown that the same result will apply if any other relative delays between the transmissions of users 1 and 2 are assumed. It is seen from the figure that the three bits sent from the user 1 can be decoded successfully without being interfered by the signals sent from user 2. Therefore, the MAI-free operation for a NG-CDMA system in asynchronous transmission mode is guaranteed in this case too.

Figure 4. Despreading process for the whole packet sent from users 1 and 2 at the receiver which is tuned to user 1, where asynchronous channel is considered

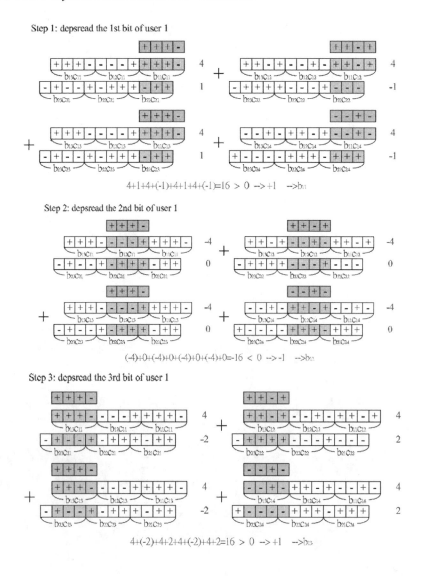

Isotropic MI-Free Operation

In the previous subsection, we have illustrated that a NG-CDMA system based on a set of orthogonal complementary codes can offer an MAI-free operation for both synchronous and asynchronous transmissions, namely isotropic MAI-free operation.

Next, we will illustrate another important operational feature for the NG-CDMA system. More specifically, a NG-CDMA system can also offer an isotropic MI-free operation, where the abbreviation "MI" stands for multipath interference. Let us consider a simple multipath channel, in which there are only two multipath returns and their relative delay is only one chip for illustration simplicity. It can be easily shown that the same results will be yielded if we use any other relative delays between the two multipath returns. Again, we assume that there are only two users in the system, and each will be assigned a flock of element codes, as given in (3), meaning that the same set of orthogonal complementary codes will be used here. In fact, the same results will apply if we use any other types of orthogonal complementary codes.

Our discussions will start with the synchronous transmission channel, such as a downlink channel of a mobile cellular system, and then the discussions will be extended to asynchronous transmis-

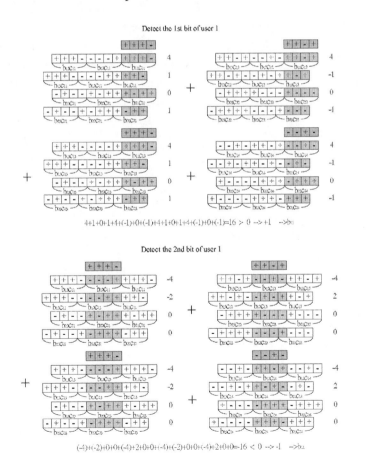

Figure 5. Detection for the first and second bits sent from users 1 at the receiver which is tuned to user 1, where synchronous two-return multipath channel is considered

sions, just as we did in the previous subsection. Figure 5 and Figure 6 show the detection of the first, second and third bits of the packet from the user 1, where there are two multipath returns and two users in a down link transmission channel. It is seen from Figure 5 and Figure 6 that the NG-CDMA system can offer an MI-free operation for this two-user system with synchronous transmissions.

Similarly, we have obtained Figure 7 and Figure 8 to show the detection of the first, second and third where there are two multipath returns and two users in an up link (asynchronous) transmission channel. It is seen from Figures 7 and 8 that the NG-CDMA system can offer an MI-free operation for this two-user system with asynchronous transmissions as well. Therefore, it is concluded that the NG-CDMA system is capable to provide isotropic MI-free operation.

The information revealed from Figures 7 and 8 is significant. It is seen from the figures that the detection efficiency at the edges of a short packet can be made exactly the same as that in the middle. In particular, we should note that the detection of the first bit shown in Figures 7 and 8 actually involves with partial correlation functions, instead of periodic correlation functions. This result reflects again the fact that the NG-CDMA can offer a unique operational feature (which all unitary codes do not have) such that the detection efficiency for a short packet can be made exactly the same as that for a long frame, making it in particular suitable for signal detection in high speed bursty traffic, an important characteristic feature for all futuristic wireless communication systems.

PERFORMANCE ANALYSIS

Having discussed the issues on isotropic MAI-free and MI-free operation for a NG-CDMA system using illustrations, we would like to go further to study its performance with the help of analysis in this subsection. The analysis can give us more generic study without being focused only on a few special cases, as we did in the illustrations. Therefore, the results obtained from the analysis

Figure 6. Detection for the third bit sent from users 1 at the receiver which is tuned to user 1, where synchronous two-return multipath channel is considered

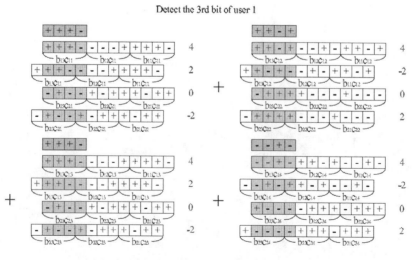

Figure 7. Detection for the first and second bits sent from users 1 at the receiver which is tuned to user 1, where asynchronous two-ray multipath channel is considered

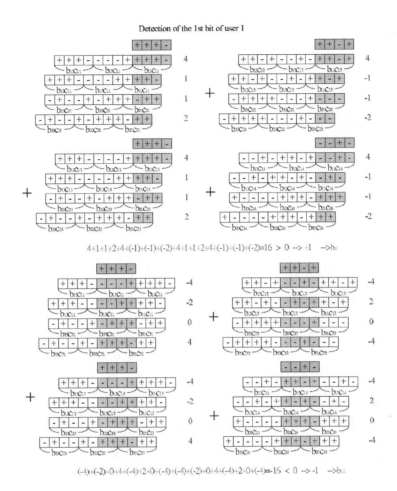

will be applicable to all NG-CDMA systems based on any orthogonal complementary codes.

Correlation with Fractional Chip Relative Delay

It is noted that in the illustrative study given in the earlier section we have assumed that the relative delays between users and multipath returns are always at multiple complete chips. However, it is possible that the relative delay may take any value, which may be a fractional chip, instead of a number of complete chips. Therefore, we should first look at the issue as the first step to extend our study to a more general case.

Figure 9 shows arbitrary relative delay between two sequences P and Q, with their relative delay being $1 + \tau$ chips, where τ is a positive real number less than one. It is noted that Figure 9 shows the case with two unitary sequences. However, it can be extended to the cases with complementary codes with multiple element codes. Now, let us look at the cross-correlation function between the two sequences P and Q with each consisting of four chips, as shown in Figure 9. The cross-correlation function between

Figure 8. Detection for the third bit sent from users 1 at the receiver which is tuned to user 1, where asynchronous two-ray multipath channel is considered

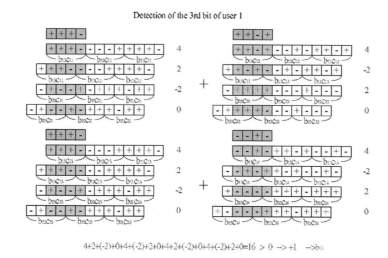

*Figure 9. Arbitrary relative delay between two sequences **P** and **Q**, each consisting of four chips*

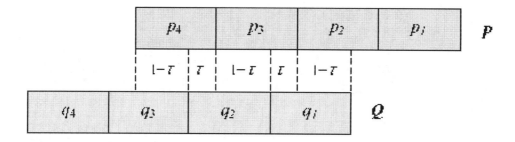

sequences P and Q with their relative delay being $1 + \tau$ can be written as

$$\rho(P,Q; 1+\tau)$$
$$= \frac{1}{4}\left[p_2 q_1^*(1-\tau) + p_3 q_1^* \tau + p_3 q_2^*(1-\tau) + p_4 q_2^* \tau + p_4 q_3^*(1-\tau)\right]$$
$$= \frac{1}{4}\left[\left(p_2 q_1^* + p_3 q_2^* + p_4 q_3^*\right)(1-\tau) + \left(p_3 q_1^* + p_4 q_2^*\right)\tau\right]$$
$$= \rho(P,Q;1)(1-\tau) + \rho(P,Q;2)\tau$$

(4)

from which it is seen that the cross-correlation function with fractional chip delay (which is equal to $1 + \tau$ chips) can be represented by the weighted sum of two decomposed terms, one being the cross-correlation function $\rho(P,Q;1)$ (which stands for the cross-correlation function with one chip relative delay) and the other $\rho(P,Q;2)$ (which stands for the cross-correlation function with two chips relative delay). The weights used in the two decomposed terms are just $(1-\tau)$ and τ. Therefore, if the cross-correlation functions with complete chips relative delays are zero, then the cross-correlation functions with any fractional chips relative delays are also zero, as shown in equation (4).

Based on (4), we can have the auto-correlation function for a flock of complementary codes $S_{nj} = \left\{S_{n1}, S_{n2}, \cdots, S_{n2^R}\right\}$ and the cross-correlation functions with fractional chips relative delay between two flocks of complementary codes $S_{nj} = \{S_{n1}, S_{n2}, \cdots, S_{n2^R}\}$ and $S_{mj} = \{S_{m1}, S_{m2}, \cdots, S_{m2^R}\}$ as follows:

$$\sum_{j=1}^{2^R} \rho(S_{nj}, S_{nj}; k+\tau)$$
$$= \sum_{j=1}^{2^R} \rho(S_{nj}, S_{nj}; k)(1-\tau) + \sum_{j=1}^{2^R} \rho(S_{nj}, S_{nj}; k+1)\tau = 0 + 0 = 0$$
(5)

where n = 1, 2, . . ., 2^R, k = 1, 2, . . ., $2^r - 1$ and $0 < \tau < 1$, and

$$\sum_{j=1}^{2^R} \rho(S_{nj}, S_{mj}; k+\tau)$$
$$= \sum_{j=1}^{2^R} \rho(S_{nj}, S_{mj}; k)(1-\tau) + \sum_{j=1}^{2^R} \rho(S_{nj}, S_{mj}; k+1)\tau = 0 + 0 = 0$$
(6)

where n, m = 1, 2, . . ., 2^R, n ≠ m, k = 0, 1, . . ., $2^r - 1$ and $0 < \tau < 1$.

From the discussions given above, we obtain an important conclusion that if the correlation properties for a sequence (any sequences, including unitary codes and complementary codes alike) are perfect with their relative delay being multiple chips, then its correlation properties with fractional relative delays will also be perfect. Therefore, the discussions on the correlation properties for a sequence with their relative delay being multiple chips are general enough. Therefore, also for this reason our discussions given in this chapter will be limited only to the correlation properties with their relative delay being multiple chips.

Correlation in Continuous Bit Streams

When discussing the correlation functions, we should also consider the situation where the input data streams can be continuous. Therefore, the correlation functions very often cover two consecutive bits, as shown in Figure 10, which shows that the correlation function across two consecutive bits can in fact be decomposed into two partial correlation functions.

It is seen from Figure 10 that both even (if two consecutive bits carry the same sign) and odd (if two consecutive bits carry different signs) periodic correlation functions can always be decomposed into two partial correlation functions, followed by a summation of them.

Therefore, if the partial correlation functions for a sequence (no matter what type of codes, either unitary codes or complementary codes) are ideal, then its even or odd periodic correlation functions should also be ideal. Here, we have used a general term, correlation functions to include both auto-correlation function and cross-correlation function. We say that a sequence has an ideal correlation function, if and only if its auto-correlation

Figure 10. Correlation across two consecutive bits, which can be decomposed into two partial correlation functions

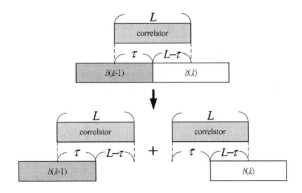

function is always zero except the zero relative shift and its cross-correlation function is also zero for any possible relative shifts.

BER in Synchronous Channel

Next, we would like to derive the bit error rate (BER) performance of a NG-CDMA system under the influence of both MAI and MI. We will use the results obtained from the discussions given in the previous subsections to simplify our analysis. The previous discussions have suggested that we need only to consider the partial correlation functions in performance analysis, as doing so will not loss generality. We need only consider the cases with multiple chip delays, as the results will be the same if we include the cases with fractional chip delays.

We still consider the NG-CDMA system model as shown in Figures 1 and 2. We assume that the receiver wants to receive the signal from user 1 and all other transmissions will be considered as interference. The NG-CDMA system should use a set of orthogonal complementary codes with both its flock size and set size being equal to K = M = 2^R. The element code length is 2^r. Therefore, we can write down the down link (synchronous) transmission signals via 2^R sub-carriers from a NG-CDMA transmitter as

$$\begin{cases} O_{f1} = b_1(t)S_{11} + b_2(t)S_{21} + \cdots + b_{2^R}(t)S_{2^R1} + n_1 \\ O_{f2} = b_1(t)S_{12} + b_2(t)S_{22} + \cdots + b_{2^R}(t)S_{2^R2} + n_2 \\ \vdots \\ O_{f2^R} = b_1(t)S_{12^R} + b_2(t)S_{22^R} + \cdots + b_{2^R}(t)S_{2^R2^R} + n_{2^R} \end{cases}$$

(7)

where $b_i(t)$ is the binary data signal from the ith user at the time t, and n_i is the noise term from the ith sub-carrier channel.

Assume that carrier demodulation process is lossless. Therefore, we can have the output signal from the bank of correlators, each of which is tuned to a particular element code in one subcarrier frequency as

$$\begin{cases} O_{c1} = b_1(t)\rho(S_{11},S_{11};0) + b_2(t)\rho(S_{21},S_{11};0) + \cdots + b_{2^R}(t)\rho(S_{2^R1},S_{11};0) + \rho(n_1,S_{11};0) \\ O_{c2} = b_1(t)\rho(S_{12},S_{12};0) + b_2(t)\rho(S_{22},S_{12};0) + \cdots + b_{2^R}(t)\rho(S_{2^R2},S_{12};0) + \rho(n_2,S_{12};0) \\ \vdots \\ O_{c2^R} = b_1(t)\rho(S_{12^R},S_{12^R};0) + b_2(t)\rho(S_{22^R},S_{12^R};0) + \cdots + b_{2^R}(t)\rho(S_{2^R2^R},S_{12^R};0) \\ + \rho(n_{2^R},S_{12^R};0) \end{cases}$$

(8)

where $\rho(S_{nj}, S_{nj}; 0)$ stands for the auto-correlation function of sequence S_{nj}, and $\rho(S_{nj}, S_{mj}; 0)$ stands for the cross-correlation function of sequences S_{nj} and S_{mj} (n ≠ m).

The decision variable is yielded after summing over all terms given in (8) to give

$$O_i = b_1(t)\sum_{j=1}^{2^R}\rho(S_{1j},S_{1j};0) + b_2(t)\sum_{j=1}^{2^R}\rho(S_{2j},S_{1j};0) + \cdots$$
$$+ b_{2^R}(t)\sum_{j=1}^{2^R}\rho(S_{2^Rj},S_{1j};0) + \sum_{j=1}^{2^R}\rho(n_j,S_{1j};0)$$
$$= 2^R b_1(t) + 0 + 0 + \cdots + 0 + \sum_{j=1}^{2^R}\rho(n_j,S_{1j};0)$$
$$= 2^R b_1(t) + \sum_{j=1}^{2^R}\rho(n_j,S_{1j};0)$$

(9)

which shows that the decision variable is corrupted only by noise, implying that all MAI terms do not affect the overall BER performance of the NG-CDMA system, where multipath propagation is not taken into account.

BER in Asynchronous Channel

Now, let us extend our discussion to an asynchronous transmission scenario. We still use the same orthogonal complementary code set as considered in the previous subsection, such that its element code length is 2^r, its set size and flock size are equal to K = M = 2^R, $b_i(t)$ is the ith user signal at the time t, and the relative delay between the first user and the jth user is $k_{1j}T_b + \tau_{1j}$. Here T_b is bit duration, and $0 < \tau_{1j} < 1$ stands for fractional bit duration. Also, n_i is the noise component carried in the ith sub-carrier frequency. Therefore, we will have the output signal from different correlators at the receiver as

$$O_{c1} = b_1(t)\rho(S_{11},S_{11};0) + b_2(t+k_{12}-1)\rho(S_{21},S_{11};-(2^r-\tau_{12}))$$
$$+ b_2(t+k_{12})\rho(S_{21},S_{11};\tau_{12}) + \cdots + b_{2^R}(t+k_{12^R}-1)\rho(S_{2^R1},S_{11};-(2^r-\tau_{12^R}))$$
$$+ b_{2^R}(t+k_{12^R})\rho(S_{2^R1},S_{11};\tau_{12^R}) + \rho(n_1,S_{11};0)$$
$$O_{c2} = b_1(t)\rho(S_{12},S_{12};0) + b_2(t+k_{12}-1)\rho(S_{22},S_{12};-(2^r-\tau_{12}))$$
$$+ b_2(t+k_{12})\rho(S_{22},S_{12};\tau_{12}) + \cdots + b_{2^R}(t+k_{12^R}-1)\rho(S_{2^R2},S_{12};-(2^r-\tau_{12^R}))$$
$$+ b_{2^R}(t+k_{12^R})\rho(S_{2^R2},S_{12};\tau_{12^R}) + \rho(n_2,S_{12};0)$$
$$\vdots$$
$$O_{c2^R} = b_1(t)\rho(S_{12^R},S_{12^R};0) + b_2(t+k_{12}-1)\rho(S_{22^R},S_{12^R};-(2^r-\tau_{12}))$$
$$+ b_2(t+k_{12})\rho(S_{22^R},S_{12^R};\tau_{12}) + \cdots + b_{2^R}(t+k_{12^R}-1)\rho(S_{2^R2^R},S_{12^R};-(2^r-\tau_{12^R}))$$
$$+ b_{2^R}(t+k_{12^R})\rho(S_{2^R2^R},S_{12^R};\tau_{12^R}) + \rho(n_{2^R},S_{12^R};0)$$

(10)

from which the decision variable can be generated as

$$O_i = b_1(t)\sum_{j=1}^{2^R}\rho(S_{1j},S_{1j};0) + b_2(t+k_{12}-1)\sum_{j=1}^{2^R}\rho(S_{2j},S_{1j};-(2^r-\tau_{12}))$$
$$+ b_2(t+k_{12})\sum_{j=1}^{2^R}\rho(S_{2j},S_{1j};\tau_{12}) + \cdots + b_{2^R}(t+k_{12^R}-1)\sum_{j=1}^{2^R}\rho(S_{2^Rj},S_{1j};-(2^r-\tau_{12^R}))$$
$$+ b_{2^R}(t+k_{12^R})\sum_{j=1}^{2^R}\rho(S_{2^Rj},S_{1j};\tau_{12^R}) + \sum_{j=1}^{2^R}\rho(n_j,S_{1j};0)$$
$$= 2^R b_1(t) + 0 + 0 + \cdots + 0 + \sum_{j=1}^{2^R}\rho(n_j,S_{1j};0)$$
$$= 2^R b_1(t) + \sum_{j=1}^{2^R}\rho(n_j,S_{1j};0)$$

(11)

Again, we can see from the above results that the decision variable contains only useful signal plus noise, meaning that performance of the NG-CDMA system is limited only by noise, but not by the interference.

BER in Multipath Channel

Finally, we want to look at the most general situation where an asynchronous multipath channel is considered. The multipath channel contains L different propagation paths and we will choose the first path as the useful signal and the rest will be treated as interference (i.e., MI). In the ith propagation path, the relative delay between the first user and the jth user is $k_{ij}T_b + \tau_{ij}$, where T_b is the bit duration and τ_{ij} (which is larger than zero and smaller than one) is the fractional chip delay. Therefore, the output signal from all correlators (corresponding to different carrier frequencies) at the receiver can be written as

$$O_{c1} = b_1(t)\rho(S_{11},S_{11};0) + b_2(t+k_{12}-1)\rho(S_{21},S_{11};-(2^r-\tau_{12})) + b_2(t+k_{12})\rho(S_{21},S_{11};\tau_{12})$$
$$+ \cdots + b_{2^R}(t+k_{12^R}-1)\rho(S_{2^R1},S_{11};-(2^r-\tau_{12^R})) + b_{2^R}(t+k_{12^R})\rho(S_{2^R1},S_{11};\tau_{12^R})$$
$$+ b_1(t+k_{21}-1)\rho(S_{11},S_{11};-(2^r-\tau_{21})) + b_1(t+k_{21})\rho(S_{11},S_{11};\tau_{21})$$
$$+ b_2(t+k_{22}-1)\rho(S_{21},S_{11};-(2^r-\tau_{22})) + b_2(t+k_{22})\rho(S_{21},S_{11};\tau_{22})$$
$$+ \cdots + b_{2^R}(t+k_{22^R}-1)\rho(S_{2^R1},S_{11};-(2^r-\tau_{22^R})) + b_{2^R}(t+k_{22^R})\rho(S_{2^R1},S_{11};\tau_{22^R})$$
$$+ \cdots + b_1(t+k_{L1}-1)\rho(S_{11},S_{11};-(2^r-\tau_{L1})) + b_1(t+k_{L1})\rho(S_{11},S_{11};\tau_{L1})$$
$$+ b_2(t+k_{L2}-1)\rho(S_{21},S_{11};-(2^r-\tau_{L2})) + b_2(t+k_{L2})\rho(S_{21},S_{11};\tau_{L2})$$
$$+ \cdots + b_{2^R}(t+k_{L2^R}-1)\rho(S_{2^R1},S_{11};-(2^r-\tau_{L2^R})) + b_{2^R}(t+k_{L2^R})\rho(S_{2^R1},S_{11};\tau_{L2^R})$$
$$+ \rho(n_1,S_{11};0)$$

(12)

$$O_{c2} = b_1(t)\rho(S_{12},S_{12};0) + b_2(t+k_{12}-1)\rho(S_{22},S_{12};-(2^r-\tau_{12})) + b_2(t+k_{12})\rho(S_{22},S_{12};\tau_{12})$$
$$+ \cdots + b_{2^R}(t+k_{12^R}-1)\rho(S_{2^R2},S_{12};-(2^r-\tau_{12^R})) + b_{2^R}(t+k_{12^R})\rho(S_{2^R2},S_{12};\tau_{12^R})$$
$$+ b_1(t+k_{21}-1)\rho(S_{12},S_{12};-(2^r-\tau_{21})) + b_1(t+k_{21})\rho(S_{12},S_{12};\tau_{21})$$
$$+ b_2(t+k_{22}-1)\rho(S_{22},S_{12};-(2^r-\tau_{22})) + b_2(t+k_{22})\rho(S_{22},S_{12};\tau_{22})$$
$$+ \cdots + b_{2^R}(t+k_{22^R}-1)\rho(S_{2^R2},S_{12};-(2^r-\tau_{22^R})) + b_{2^R}(t+k_{22^R})\rho(S_{2^R2},S_{12};\tau_{22^R})$$
$$+ \cdots + b_1(t+k_{L1}-1)\rho(S_{12},S_{12};-(2^r-\tau_{L1})) + b_1(t+k_{L1})\rho(S_{12},S_{12};\tau_{L1})$$
$$+ b_2(t+k_{L2}-1)\rho(S_{22},S_{12};-(2^r-\tau_{L2})) + b_2(t+k_{L2})\rho(S_{22},S_{12};\tau_{L2})$$
$$+ \cdots + b_{2^R}(t+k_{L2^R}-1)\rho(S_{2^R2},S_{12};-(2^r-\tau_{L2^R})) + b_{2^R}(t+k_{L2^R})\rho(S_{2^R2},S_{12};\tau_{L2^R})$$
$$+ \rho(n_2,S_{12};0)$$

(13)

$$\vdots$$

$$O_{c2^R} = b_1(t)\rho(S_{12^R},S_{12^R};0) + b_2(t+k_{12}-1)\rho(S_{22^R},S_{12^R};-(2^r-\tau_{12})) + b_2(t+k_{12})\rho(S_{22^R},S_{12^R};\tau_{12})$$
$$+ \cdots + b_{2^R}(t+k_{12^R}-1)\rho(S_{2^R2^R},S_{12^R};-(2^r-\tau_{12^R})) + b_{2^R}(t+k_{12^R})\rho(S_{2^R2^R},S_{12^R};\tau_{12^R})$$
$$+ b_1(t+k_{21}-1)\rho(S_{12^R},S_{12^R};-(2^r-\tau_{21})) + b_1(t+k_{21})\rho(S_{12^R},S_{12^R};\tau_{21})$$
$$+ b_2(t+k_{22}-1)\rho(S_{22^R},S_{12^R};-(2^r-\tau_{22})) + b_2(t+k_{22})\rho(S_{22^R},S_{12^R};\tau_{22})$$
$$+ \cdots + b_{2^R}(t+k_{22^R}-1)\rho(S_{2^R2^R},S_{12^R};-(2^r-\tau_{22^R})) + b_{2^R}(t+k_{22^R})\rho(S_{2^R2^R},S_{12^R};\tau_{22^R})$$
$$+ \cdots + b_1(t+k_{L1}-1)\rho(S_{12^R},S_{12^R};-(2^r-\tau_{L1})) + b_1(t+k_{L1})\rho(S_{12^R},S_{12^R};\tau_{L1})$$
$$+ b_2(t+k_{L2}-1)\rho(S_{22^R},S_{12^R};-(2^r-\tau_{L2})) + b_2(t+k_{L2})\rho(S_{22^R},S_{12^R};\tau_{L2})$$
$$+ \cdots + b_{2^R}(t+k_{L2^R}-1)\rho(S_{2^R2^R},S_{12^R};-(2^r-\tau_{L2^R})) + b_{2^R}(t+k_{L2^R})\rho(S_{2^R2^R},S_{12^R};\tau_{L2^R})$$
$$+ \rho(n_{2^R},S_{12^R};0)$$

(14)

from which we can have the decision variable as

$$O_i = 2^R b_1(t) + 0 + 0 + \cdots + 0 + \sum_{j=1}^{2^R}\rho(n_j,S_{1j};0) = 2^R b_1(t) + \sum_{j=1}^{2^R}\rho(n_j,S_{1j};0)$$

(15)

It is seen from the above results that the decision variable formed from the summation of the signals from correlator bank of the receiver depends only on the noise but not the multiple access interference and multipath interference, implying a perfect interference-free operation of the NG-CDMA system even under the influence of both MAI and MI. Therefore, the BER performance of the NG-CDMA system can be calculated simply by $P_b = Q\left(\sqrt{2SNR}\right)$, where a BPSK modulation is used.

Simulation Results

After having analyzed the BER performance of NG-CDMA system, we would like to countercheck the analytical results with the ones obtained from computer simulations.

Figure 11 shows the simulation results for a NG-CDMA system, which uses the orthogonal complementary codes with K = M = 8 and N = 8. The performance is compared with a conventional DS-CDMA system using Gold code and OVSF code, whose processing gains are 63 and 64, respectively. All systems carry the same number of users or eight users in the simulations. An asynchronous 3-ray multipath channel is considered in the simulations, with its inter-path relative delay being two chips and path gain coefficients vector being [1, 0.8564, 0.107]. The inter-user relative delay is four chips.

Figure 12 gives also a comparison among three different CDMA systems, one being NGCDMA with orthogonal complementary code set, and two DS-CDMA systems with Gold code and OVSF code. The same system set-up parameters are used, except for the multipath channel coefficients vector, which is a normalized channel gain of [0.7785, 0.6667, 0.0833] here, as the squared sum of three multipath return elements will be unit. On the other hand, Figure 11 uses non-normalized channel coefficients vector. Therefore, the simulation results are different if we compare the two figures. However, the general results are fairly consistent due to the fact that the NG-CDMA system offers a very robust BER performance against both MAI and MI, while the performance for the conventional DS-CDMA based on either Gold code or OVSF code fails to compete successfully with the NG-CDMA system in terms of their much worse BER, especially when the multipath propagation exists.

It is noted that the BERs for complementary coded CDMA with three paths in Figure 11 and Figure 12 are different due to the fact that the detection in the simulations took the first path as the signal of interest and treated all other path signals as interferences. In this manner, the BER

Figure 11. BER simulation for a NG-CDMA system with eight users and 3-ray multipath channel, where inter-user delay is four chips and inter-path delay is two chips. The performance is compared to that for a DS-CDMA system using Gold code and OVSF code, with their PG values being 63 and 64, respectively and the same channel model considered. The 3-ray multipath channel has its channel gain coefficients vector [1, 0.8564, 0.107].

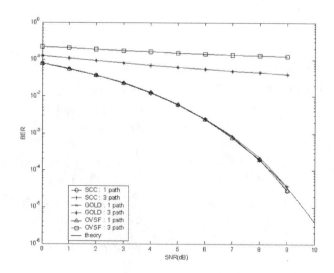

Figure 12. BER simulation for a NG-CDMA system with eight users and 3-ray multipath channel, where inter-user delay is four chips and inter-path delay is two chips. The performance is compared to that for a DS-CDMA system using Gold code and OVSF code, with their PG values being 63 and 64, respectively and the same channel model considered. The 3-ray multipath channel has its channel gain coefficients vector [0.7785, 0.6667, 0.0833].

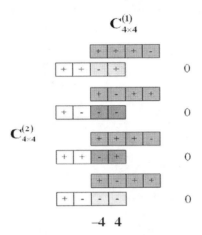

with a normalized delay spread profile (in Figure 12) is higher than that with a non-normalized delay profile (in Figure 11), because the first path powers are different.

TECHNICAL FEATURES OF NG-CDMA TECHNOLOGY

Having taken a look at the performance analysis using both illustrations and analytical derivations, we would like to discuss the major properties of the NG-CDMA system based on orthogonal complementary codes.

It is noted that there are many different types of complementary codes, and not all of them are orthogonal complementary codes. For instance, the generalized pair-wise complementary codes (Chen, 2006) are not orthogonal complementary codes due to their correlation functions are not perfect. However, many properties we will discuss here will also be applicable to those codes in general, subject to some necessary modifications. We would like also to stress that, although the results obtained in this section were based on the discussions on a particular orthogonal complementary code, the similar results will be obtained if any other orthogonal complementary codes will be used, such as complete complementary codes, column-wise complementary codes, etc. It should also be noted that the use of other types of orthogonal complementary codes may affect the properties of a NG-CDMA system in terms of its maximal number of users it can support due to the different set sizes. It may also affect the system implementation complexity due to the different flock sizes and element code lengths, etc. Nevertheless, the correlation properties of the NG-CDMA system based on different types of orthogonal complementary codes will be the same, and they will be discussed as follows.

Suitability for High-Speed Burst Traffic

As we can see from the results obtained in the previous subsections, a NG-CDMA system based on orthogonal complementary codes can offer many unique characteristic features if compared to a conventional DS-CDMA system based on unitary spreading codes, such as Gold codes, m-sequences, Walsh-Hadamard codes, OVSF code, Kasami codes, etc.

As an important requirement for a CDMA technology suitable for future Gigabit all-IP wireless communications, we should pay sufficient attention to detection efficiency at edges of a packet or burst. In this sense, a NG-CDMA scheme is in particular well suited for its applications in future Gigabit all-IP wireless communications, as shown in Figures 7 and 8, which illustrate the signal detection process for the first (or the rightmost) bit of a packet. In detecting those bits at the edges of a packet, partial auto-correlation functions (ACFs) and partial cross-correlation functions (CCFs) of the codes will become extremely important. It is seen from the figures that the NG-CDMA system yields zero partial CCFs, ensuring an ideal performance for signal detection even on the edges of a frame or packet. This observation is significant due to the dominating burst traffic in all future wireless systems.

We have to emphasize that the good detection efficiency for short packets in the NG-CDMA architecture is because of the use of orthogonal complementary codes, which offer a possibility for them to cancel all ACFs side lobes and CCFs in the process of summation of all correlation results generated from different element codes in the same flock. On the other hand, a conventional DS-CDMA system based on unitary codes will never have such an opportunity to cancel those ACFs side lobes and CCFs generated from a "single" correlator structure.

Resilience Against Time/frequency-Selective Fading

As demonstrated in the earlier discussions, the spreading process involved in a NG-CDMA system is actually a two-dimensional spreading process, which covers both the time and frequency domains. This gives us more degree of freedoms in choosing the ways to spread the same bit stream. This can never happen if a unitary spreading code is used in all conventional DS-CDMA system architecture, where only the time-domain spreading is possible.

Each user in a NG-CDMA system will be assigned a flock of M element codes, each of which has a length of N chips, thus resulting in a processing gain of M × N. Due to the use of two-dimensional spreading in the NG-CDMA systems, we find an interesting issue on how the orthogonality is established based on the two-dimensional spreading. It is well known that for any orthogonal complementary code set the correlation properties for an individual element code are never perfect. In fact, the perfect correlation functions in an orthogonal complementary code set are based on the summation of all correlation functions generated from individual element codes. In other words, the non-zero ACF side lobes and non-zero CCFs are all canceled in the process of the summation. Then, now comes an interesting question: how can those non-zero ACF side lobes and CCFs be canceled? To give an answer to this question, let us look at some simple examples from column-wise complementary codes. It should be noted that although we only take the column-wise complementary codes here as examples, the same results will apply if any other orthogonal complementary codes, such as complete complementary codes, orthogonal complementary codes, etc., will be considered.

Let us look at two column-wise complementary codes as

$$C^{(1)}_{4\times 4} = \begin{bmatrix} + & + & + & - \\ + & - & + & + \\ + & + & + & - \\ + & - & + & + \end{bmatrix}$$

$$C^{(2)}_{4\times 4} = \begin{bmatrix} + & + & - & + \\ + & - & - & - \\ + & + & - & + \\ + & - & - & - \end{bmatrix} \qquad (16)$$

whose parameters are M = K = 4 and N = 4. We only show two flocks from the code set (there are in total four flocks in the set) as examples. Figure 13 shows the time-domain CCFs cancelation process, based on which its orthogonality of the column-wise code set is established.

On the other hand, we can have another column-wise code set, whose two flocks are given as follows:

$$C_{4\times 4}^{(1)} = \begin{bmatrix} + & + & + & + \\ + & - & + & - \\ + & + & - & - \\ + & - & - & + \end{bmatrix}$$

$$C_{4\times 4}^{(2)} = \begin{bmatrix} + & + & - & - \\ + & - & - & + \\ + & + & + & + \\ + & - & + & - \end{bmatrix} \quad (17)$$

Figure 14 shows the CCFs cancelation process carried out in the frequency domain for the two column-wise codes given in (17). We can see that the cross-correlation functions for these two flocks of element codes are all zero due to the cancelation process carried out purely in the frequency domain.

Usually the cancelation can happen in both time and frequency domains at the same time, although with various proportions, whose percentage varies from code to code. It can be shown that the orthogonality of an orthogonal complementary code set can be established based solely on the frequency-domain cancelation (i.e., 100% frequency-domain cancelation), but can be based on at most 50% (which is a maximal value for the time-domain cancelation) on the time-domain cancelation.

The 100% frequency-domain cancelation can be an important property for an orthogonal complementary code set, as it can be used in many applications where mobility is an important fact to consider in the system design, such as vehicle-to-vehicle (V2V) communications (IEEE 802.11p standard) and high-speed railway systems. The property of the 100% frequency-domain cancelation in an orthogonal complementary code set can be translated into that the orthogonality of the orthogonal complementary codes will hardly be affected by time-selective fading because its orthogonality is established solely on the frequency-domain CCFs cancelation process, and thus the

Figure 13. Orthogonality of a column-wise code set is based on time-domain CCFs cancelation

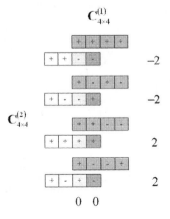

time domain changes in channel gains will not affect its detection process at the receiver.

Let us take a look at a particular example to show how much the time-varying fading will affect the detection process of a NG-CDMA receiver. Assume that time-varying channel coefficients vector at four different chips are $\{h(t_1), h(t_2), h(t_3), h(t_4)\}$, where t_1, t_2, t_3 and t_4 stand for the four chip sampling instances. We still use the two codes shown in (17) as the examples here.

Therefore, we will have the output from the local correlation process at a receiver as

$$\begin{aligned}-2h^{(1)}(t_3)h^{(2)}(t_1) + 2h^{(1)}(t_3)h^{(2)}(t_1) \\ -2h^{(1)}(t_4)h^{(2)}(t_2) + 2h^{(1)}(t_4)h^{(2)}(t_2) \\ = 0\end{aligned} \quad (18)$$

which tells us that the cross-correlation function is perfect after the frequency-domain cancelation process, which is illustrated in Figure 15.

On the other hand, we can also exploit the property of the time-domain CCFs cancelation process

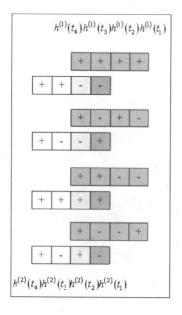

Figure 14. Orthogonality of a column-wise code set is based on frequency-domain CCFs cancelation

to design a NG-CDMA system for its applications in severe frequency-selective fading. Unfortunately, we can only achieve the time-domain CCFs cancelation to its maximal percentage 50%, which is different from the frequency-domain CCFs cancelation which can reach 100%. Therefore, a NG-CDMA system has a stronger time-selective fading resistance than its frequency-selective fading resistance. Nevertheless, both time-selective fading resistance and frequency-selective fading resistance can be exploited to greatly strengthen the capability of a NG-CDMA system against channel impairing factors.

Near-Far Resistance

Owing to the isotropic MAI-free and MI-free properties, near-far effect will virtually cause no harm to signal detection process at a correlator in an NG-CDMA system, as long as bit synchronization can be achieved prior to the data detection process. In other words, the NGCDMA is a system with an excellent near-far resistance. Therefore, complicated open-loop and closed-loop power control is no longer a necessity. More precisely, the power control in an NG-CDMA system is used merely to reduce unnecessary power emission at terminals, whose requirements on its response time and accuracy can be made much more relaxed than necessary in a conventional DS-CDMA system. In a NG-CDMA system, a similar conclusion can be drawn with respect to the power control requirement, due to its ideal MAI-free property.

RAKE vs. Matched-Filter

All current CDMA systems have to use RAKE receiver to mitigate otherwise formidable MI, which is caused due mainly to imperfect auto-correlation function of the CDMA codes used. Theoretically speaking, the orthogonal complementary codes virtually do not produce any autocorrelation side lobes if DS spreading modulation is considered,

Figure 15. Time-varying fading resistance for an orthogonal complementary code set with 100% frequency-domain CCFs cancelation property

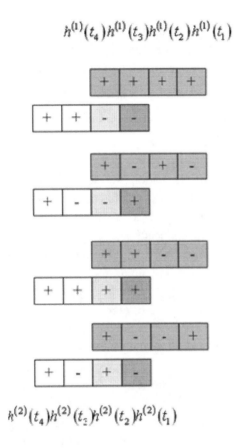

and thus an MI-free operation is guaranteed for either up-link and down-link transmissions.

To illustrate clearly how a NG-CDMA system with DS-spreading can overcome MI without the help of a RAKE receiver, we used Figures 7 and 8, where a three-bit data burst is sent into a two-user and two-path up-link asynchronous CDMA channel. The inter-path and inter-user delays are only one-chip for illustration simplicity (the same result will be given if any other delays are applied). A simple correlator receiver is used to detect incoming burst in the presence of both MAI and MI. It is seen from the figures that a simple correlator can successfully recover the original data information [+1,−1,+1] without any impairment caused by either MAI or MI. Similarly, we can show the same result for a down-link multipath channel, which usually causes much less problems than an up-link channel.

Thus, a simple correlator can solve MI and MAI problems in a NG-CDMA system. It should be noted that the save of a RAKE receiver is significant as it paves a way for a NG-CDMA receiver to work in a truly blind fashion without need of any prior channel information. On the other hand, a RAKE has to acquire virtually all channel state information, such as delays and amplitudes of all multipath returns, for its maximal-ratio-combining (MRC) operation, whose impact on implementation complexity in a mobile handset should never be underestimated. Obviously, for a NG-CDMA system we can also use a RAKE receiver to further boost up the signal-to-interference-ratio before decision. It is noted that the use of RAKE here can provide a much higher multipath-diversity gain than achievable in a conventional CDMA system due to the fact that output signal from each finger now contains only useful auto-correlation peak. On the other hand, the output from a finger of a conventional CDMA RAKE receiver contains both useful and unwanted signals caused by non-trivial autocorrelation side lobes.

Furthermore, if a RAKE has to be used in a NG-CDMA receiver, an equal gain combining (EGC) is already enough to yield a satisfactory detection efficiency. The advantage to use EGC rather than maximal ration combining (MRC) in a RAKE is to save a complicated multipath amplitude estimation unit.

MUD vs. Correlator Receiver

As shown in Figures 7 and 8, the isotropic MAI-free and MI-free properties in a NG-CDMA based on DS-spreading make it unnecessary to use multi-user detection (MUD) to de-correlate transmission signals from different users, due to the fact that the transmissions from different users in a NG-CDMA system have already been pre-decorrelated at transmitter side because of its unique MAI-free signaling structure.

FDM vs. TDM for Element Code Division

To implement a NG-CDMA system, an important requirement is to send a flock of M element codes, which are assigned to a specific user, via separate channels to an intended receiver, where each element code should be de-spread separately and their outputs should be combined to form a decision variable. The most straight-forward way to implement element codes division in a NG-CDMA system is to use frequency division multiplex (FDM) method, in which M element codes are sent through M different sub-carriers f_m, where $1 \leq m \leq M$. In this way, a NG-CDMA system looks just like a multi-carrier (MC) CDMA scheme. However, the major difference between a NG-CDMA/FDM system and a traditional MC-CDMA lies in the fact that the former uses different carriers to convey different information without providing any diversity in frequencies, while the latter usually does.

Obviously, another scheme to implement element code division in a NG-CDMA is to send element codes in different time slots or simply via time division multiplex (TDM) method. FDM and TDM implementations of element code division offer distinct system operational advantages. One of the benefits for using FDM is to allow a NG-CDMA system to work harmonically with frequency division duplex (FDD) operation mode used in most mobile cellular standards, such as W-CDMA, CDMA2000, etc. The another important benefit we get from the FDM method is to enable multi-dimensional spreading in a CDMA system based on the NGCDMA technology. Yet another salient feature for the FDM option is to reduce overall hardware complexity with the help of OFDM technology. Similar to any MC-CDMA system, a NGCDMA/ FDM system can also be implemented in an OFDM architecture, which can transform complicated multi-carrier RF circuitry into a base band signal processing unit.

On the other hand, the TDM implementation can fit TDD mode naturally, but the TDD operation suits only for covering a relatively small cell size due to its relatively low average transmission power.

CONCLUSION

This chapter has addressed several issues on the design of NG-CDMA architecture for future Gigabit wireless applications. To find better CDMA codes, we introduced "Real Environment Adapted Linearization" (REAL) approach, which takes into account many major impairing factors, such as multipath interference, asynchronous transmission, random symbol sign changes, etc., in the code generation procedure. It has been shown that the obtained orthogonal complementary code sets could offer MAI-free and MI-free operation for DS-spreading modulation under any multipath delay profile. A possible implementation scheme for both uplink and downlink transceiver in an NG-CDMA system is given. In addition, we have also discussed the technical features for the NG-CDMA technology. In particular, two-dimensional spreading used in the NG-CDMA technology offers us much more degree-of-freedoms to design a tailor-made CDMA system for some particular wireless applications to mitigate time-selective or frequency-selective fading. It is concluded that numerous desirable features of the NG-CDMA technology make it a very competitive candidate for future Gigabit wireless communications.

For those who are interested in the research activities on this topic, please refer to several special issues I edited and co-edited for several major technical journals as listed in the references (Chen, Chiu, & Guizani, 2006).

ACKNOWLEDGMENT

The author would like to acknowledge gratefully the research grant, NSC 97-2219-E-006-004, from National Science Council, Taiwan.

REFERENCES

ARIB/Japan. (1998, June). *Japan's Proposal for Candidate Radio Transmission Technology on IMT-2000: W-CDMA*.

CATT/China. (1998, June). *TD-SCDMA Radio Transmission Technology for IMT-2000*.

Chen, H-H. (2007, July). *The Next Generation CDMA Technologies*, (1st Ed.). New York: John Wiley & Sons.

Chen, H-H., Chiu H-W. & Guizani, M. (2006, February). Orthogonal complementary codes for interference-free CDMA technologies. *IEEE Wireless Communications*, 68-79.

Chen, H.-H., Chu, S. W., Kuroyanagi, N., & Han Vinck, A. J. (2007). An Algebraic Approach to Generate Super-Set of Perfect Complementary Codes for Interference-Free CDMA. [WCMC]. *Journal of Wireless Communications and Mobile Computing, 7*, 605–622. doi:10.1002/wcm.388

Chen, H.-H., Fan, C. X., & Lu, W. W. (2002). China's Perspectives on 3G Mobile Communications & Beyond: TD-SCDMA Technology. *IEEE Wireless Communications Magazine, 9*(2), 48–59. doi:10.1109/MWC.2002.998525

Chen, H.-H., Guizani, M., & Huber, J. F. (2006, February). Multiple Access Technologies for B3G Wireless Communications. *IEEE Communications Magazine, 43*(2), 65-67. Retrieved from http://www.comsoc.org/pubs/commag/cfpcommag205.htm

Chen, H.-H., Guizani, M., & Mohr, W. (2007). Evolution toward 4G Wireless Networking. *IEEE Network, 21*(1, January/February), 4-5. Retrieved from http://www.comsoc.org/pubs/net/

Chen, H.-H., Han Vinck, A. J., Bi, Q., & Adachi, F. (2006, January). The Next Generation of CDMA Technologies. [Retrieved from http://www.argreenhouse.com/society/J-SAC/Calls/cdma technologies.html]. *IEEE Journal on Selected Areas in Communications, 24*(1), 1–3. doi:10.1109/JSAC.2005.858872

Chen, H.-H., Hanky, D., Magana, M. E., & Guizani, M. (2006). Design of next generation CDMA using Orthogonal Complementary Codes and Offset Stacked Spreading. *IEEE Wireless Communications*.

Chen, H.-H., Lin, J.-X., Chu, S.-W., Wu, C.-F., & Chen, G.-F. (2003, September). Isotropic Air-Interface Technologies for Fourth Generation Wireless Communications. [WCMC]. *Journal of Wireless Communications & Mobile Computing, 3*(6), 687–704. doi:10.1002/wcm.150

Chen, H.-H., Wong, D., & Mueller, P. (2006, September). Evolution of Air-Interface Technologies for 4G Wireless Communications. *IEEE Vehicular Technology Magazine (Editorial), IEEE Vehicular Technology Magazine, 1*(3), 2-3. Retrieved from http://www.ieeevtc.org/vtmagazine/index.html.

Chen, H.-H., Xiao, Y., Li, J., & Fantacci, R. (2006, September). Challenges and futuristic perspective of CDMA technologies: the OCC-CDMA/OS for 4G wireless networking. *IEEE Vehicular Technology Magazine, IEEE Vehicular Technology Magazine, 1*(3), 12-21. Retrieved from http://www.ieeevtc.org/vtmagazine/index.html

Chen, H.-H., & Yeh, J.-F. (2003). A complementary codes based CDMA architecture for wideband mobile Internet with high spectral efficiency and exact rate-matching. *International Journal of Communication Systems, 16*, 497–512. doi:10.1002/dac.592

Chen, H.-H., Yeh, J. F., & Seuhiro, N. (2001). A Multi-Carrier CDMA Architecture Based on Orthogonal Complementary Codes for New Generations of Wideband Wireless Communications. *IEEE Communications Magazine, 39*(10), 126–135. doi:10.1109/35.956124

Chen, H-H. & Yeh, Y-C., (2005, April). Capacity of a Space-Time Block Coded CDMA System: Unitary Codes versus Complementary Codes. *IEEE Proceedings -Communications, 152*(2), 203-214.

Chen, H.-H., Yeh, Y.-C., Bi, Q., & Jamalipour, A. (2007, February). On a MIMO-based open wireless architecture: space-time complementary coding. *IEEE Communications Magazine, 45*(2), 104–112. doi:10.1109/MCOM.2007.313403

Chen, H-H., Yeh, Y-C., Zhang, X., Huang, A., Yang, Y., Li, J., & Xiao, Y., eta l. (2006, January). Generalized Pairwise Complementary Codes with Set-Wise Uniform Interference-Free Windows. *IEEE Journal on Selected Areas in Communications, 24*(1), 65–74. doi:10.1109/JSAC.2005.858878

Chen, H.-H., Zhang, X., & Xu, W. (2007). Next Generation CDMA vs. OFDMA for 4G Wireless Applications. *IEEE Wireless Communications Magazine, 14*(3), 6-7. Retrieved from http://www.comsoc.org/pubs/pcm/

ETSI/SMG2. (1998, June). *The ETSI UMTS Terrestrial Radio Access (UTRA) ITU-R RTT Candidate Submission.*

Li, J., Huang, A., Guizani, M., & Chen, H.-H. (2007). Inter-Group Complementary Codes for Interference- Resistant CDMA Wireless Communications. *IEEE Transactions on Wireless Communications.*

Li, X., Chen, H.-H., Qian, Y., Rong, B., & Soleymani, M. R. (2007). Welch bound analysis on generic code division multiple access codes with interference free windows. *IEEE Transactions on Wireless Communications.*

Lu, K., Qian, Y., Chen, H-H., & Fu, S. (2008). WIMAX NETWORKS: FROM ACCESS TO SERVICE PLATFORM. *IEEE Network, 22*(3, May/June), 38-45.

Magana, M. E., Rajatasereekul, T., Hank, D., & Chen, H.-H. (2007). Design of a MC-CDMA System that Uses Complete Complementary Orthogonal Spreading Codes. *IEEE Transactions on Vehicular Technology.*

TIA/US. (1998, June). *The cdma2000 ITU-R RTT Candidate Submission.*

ENDNOTE

[1] For those interested in the REAL approach, please refer to (Chen,2006), in particular the book I wrote under the title of "The next generation CDMA technology" (Chen,2007).

Enhanced Decoding Techniques

Chapter 27
Configurable and Scalable Turbo Decoder for 4G Wireless Receivers

Yang Sun
Rice University, USA

Joseph R. Cavallaro
Rice University, USA

Yuming Zhu
Texas Instruments, USA

Manish Goel
Texas Instruments, USA

ABSTRACT

The increasing requirements of high data rates and quality of service (QoS) in fourth-generation (4G) wireless communication require the implementation of practical capacity approaching codes. In this chapter, the application of Turbo coding schemes that have recently been adopted in the IEEE 802.16e WiMax standard and 3GPP Long Term Evolution (LTE) standard are reviewed. In order to process several 4G wireless standards with a common hardware module, a reconfigurable and scalable Turbo decoder architecture is presented. A parallel Turbo decoding scheme with scalable parallelism tailored to the target throughput is applied to support high data rates in 4G applications. High-level decoding parallelism is achieved by employing contention-free interleavers. A multi-banked memory structure and routing network among memories and MAP decoders are designed to operate at full speed with parallel interleavers. A new on-line address generation technique is introduced to support multiple Turbo interleaving patterns, which avoids the interleaver address memory that is typically necessary in the traditional designs. Design trade-offs in terms of area and power efficiency are analyzed for different parallelism and clock frequency goals.

DOI: 10.4018/978-1-61520-674-2.ch028

INTRODUCTION

The approaching fourth-generation (4G) wireless systems are promising to support very high data rates from 100 Mbps to 1 Gbps. This consequently leads to orders of complexity increases in a 4G wireless receiver. The high performance convolutional Turbo codes are employed in many 4G wireless standards such as IEEE 802.16e WiMax and 3GPP Long Term Evolution (LTE).

A Turbo decoder is typically one of the most computation-intensive parts in a 4G wireless receiver. Increased complexity and performance requirements and the need to reduce power and area are significant challenges for Turbo decoder hardware implementation. The push for multi-mode wireless physical layer (PHY) brings additional challenges for Turbo decoder design. While programmable DSP/SIMD/VLIW processors can offer great flexibility in supporting different types of Turbo codes, they have several drawbacks notably higher power consumption and lower throughput than the ASIC solutions, which make them unsuitable for handheld devices. The commonalities between these Turbo codes in 4G wireless standards allow resources to be shared thus reducing hardware area and making more efficient use of the data path. However, there are differences in the exact Turbo decoder implementations. In order to meet high-speed multiple 4G standards, a reconfigurable and scalable Turbo decoder (or coprocessor) is necessary. From an implementation point of view, there are many aspects of Turbo codes that make them still a very hot research topic. First, the original MAP algorithm is of great complexity, so it is impractical to implement it in hardware. So the Log-MAP and Max-Log-MAP algorithms were proposed later to reduce the arithmetic complexity while still maintaining good decoding performance. The long latency of MAP decoding has prevented it from being used in the real-time systems. One effective solution is to apply a sliding window algorithm to reduce the decoding latency. The scheduling of parallel sliding windows becomes the main challenge in parallel Turbo decoder design. Second, the non-binary Turbo codes are proven to have better performance than the binary Turbo codes. An area-efficient high-radix Turbo decoder architecture poses another design challenge. Finally, the new contention-free interleaver enables a very high level of parallelism in Turbo decoding, but on the other hand it creates an obstacle for the internal memory structure. The memories need to be partitioned and managed properly to avoid memory access conflicts introduced by the interleaver.

This chapter discusses several types of Turbo coding schemes that have recently been approved in IEEE 802.16e WiMax, 3GPP LTE, and some other 3G/4G standards. It describes a high-throughput, area- and power- efficient VLSI architecture for multi-mode Turbo decoders. A multi-banked memory structure and routing network between memories and MAP decoder cores are also introduced. Simulation and implementation results are presented which show that, with the aid of a unified trellis structure, a configurable and scalable Turbo decoder architecture provides a practical solution to the requirements of flexible and high data-rate reliable transmission for 4G wireless networks.

Table 1. Some applications of Turbo codes

Application	Code structure	Polynomials
CDMA, WCDMA, UMTS, LTE	8-state binary	13, 15, 17
WiMax, DVB-RCS	8-state double-binary	15, 13

BACKGROUND

The Turbo code (Berrou et al., 1993; Berrou et al., 1996) has become one of the most important research topics in coding theory since its discovery in 1993. The astounding performance of Turbo code has attracted a great deal of interest in the research activity in the area of iterative error correction codes. Due to its excellent error correction performance, many communication standards have chosen Turbo codes as the Forward Error Correction (FEC) codes, such as CDMA-2000, W-CDMA, DVB-RCS, HSDPA, UMTS, IEEE 802.16e WiMax, and 3GPP LTE. Turbo codes can be categorized into two classes: binary Turbo codes and non-binary Turbo codes. For example, Turbo codes in CDMA, HSDPA, UMTS and 3GPP LTE are binary types of Turbo codes, whereas Turbo codes in IEEE 802.16e and DVB-RCS are double-binary types of Turbo codes. Table 1 summarizes some of the Turbo codes in practice (Berrou, 2003). As we can see, there are many similarities between the Turbo codes employed in different standards. This motivates the design of a unified and flexible Turbo decoder which can support multiple standards. Without loss of generality, we will mainly focus on the Turbo codes defined in 3GPP LTE and WiMax in the following analysis. Note that these analyses can be applied to other systems directly because the encoder polynomials are same.

Binary Turbo Code in 3GPP LTE Standard

Turbo coding scheme in 3GPP LTE standard (3GPP TS 36.212, 2008) is a parallel concatenated convolutional code (PCCC) with two 8-state constituent encoders and one quadratic permutation polynomial (QPP) interleaver. The coding rate of the Turbo code is 1/3. The structure of the Turbo encoder is shown in Figure 1.

As seen in the figure, a Turbo encoder consists of two binary convolutional encoders connected by an interleaver. The basic coding rate is 1/3 which means N data bits will be coded into $3N$ data bits. The transfer function of the 8-state constituent code for PCCC is:

$$G(D) = \left[1, \frac{g_1(D)}{g_0(D)}\right]$$

where

$$g_0(D) = 1 + D_2 + D_3,$$

Figure 1. Structure of rate 1/3 Turbo encoder in 3GPP LTE

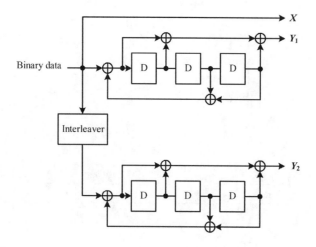

Figure 2. Structure of rate 1/3 double binary Turbo encoder in IEEE 802.16e

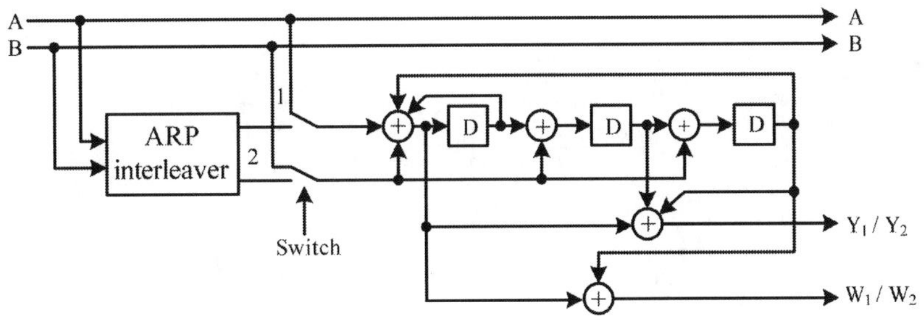

$g_1(D) = 1 + D + D_3$.

The initial value of the shift registers of the 8-state constituent encoders shall be all zeros when starting to encode the input bits. Trellis termination is performed by taking the tail bits from the shift register feedback after all information bits are encoded. Tail bits are padded after the encoding of information bits.

The function of the Interleaver is to take each incoming block of N data bits and shuffle them in a pseudo-random manner. One of the new features in the 3GPP LTE Turbo encoder is its quadratic permutation polynomial (QPP) internal interleaver. We will see later that this QPP interleaver is the key component enabling parallel decoding of Turbo codes.

Double Binary Turbo Code in IEEE 802.16e WiMax Standard

The convolutional Turbo encoder for the IEEE 802.16e standard (IEEE Std 802.16, 2004) is depicted in Figure 2. It uses a double binary circular recursive systematic convolutional code. Data couples (A, B), rather than a single bit sequence, are fed to the circular recursive systematic convolutional encoder twice, and four parity bits (Y1,W1) and (Y2, W2) are generated in the natural order and the interleaved order, respectively. The encoder polynomials are described in binary symbol notation as follows:

- For the feedback branch: $1 + D + D^3$,
- For the Y parity bit: $1 + D^2 + D^3$,
- For the W parity bit: $1 + D^3$.

The tail-biting Trellis termination scheme is used as opposed to inserting extra tail bits. In this termination scheme, the start state of the trellis equals to the end state of the trellis. Therefore, a pre-encoding operation has to be performed to determine the start state. This is not a complex problem because the encoding process can be performed at a much higher rate. A symbol-wise almost regular permutation (ARP) interleaver is used in the WiMax standard, which can enable parallel decoding of double binary Turbo codes.

Decoding Algorithm

The decoding algorithm employed in the Turbo decoders is the maximum *a posteriori* (MAP) algorithm proposed by Bahl *et al.* in 1974 and is also called the BCJR algorithm (Bahl *et al.*, 1974). The high complexity and long latency of the original MAP algorithm has made high-speed VLSI implementations extremely difficult to realize. Fortunately, many simplifications have been applied to the original MAP algorithm in order to reduce the implementation complexity.

The Turbo decoding concept is functionally illustrated in Figure 3. As discussed before, the decoding is based on the MAP algorithm and is usually calculated in the log domain (Robertson *et al.*, 1995) to avoid multiplications and divi-

Figure 3. Basic structure of Turbo encoder and decoder

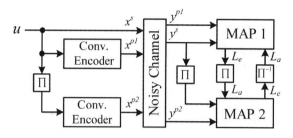

sions. During the decoding process, each soft-in soft-output (SISO) decoder receives the intrinsic log-likelihood ratios (LLRs) from the channel and the extrinsic LLRs from the other constituent SISO decoder through interleaving (Π) or deinterleaving (Π^{-1}). The main task of the Turbo internal interleaver is to generate a permutation of the input data sequence that is as uncorrelated as possible. The randomness of the interleaver not only affects the decoding performance, but also leads to decoding latency because one SISO decoder must wait for the other SISO decoder to finish decoding before it can start the next iteration.

An efficient representation of the Turbo decoding process is the trellis diagram which describes all the possible state transitions through a graph representation. Figure 4 shows a section of the trellis diagram for an 8-state binary Turbo code, where the dashed edges correspond to input bit $u_k=0$, and solid edges correspond to $u_k=1$.

The MAP algorithm is an optimal symbol decoding algorithm that minimizes the probability of a symbol error. It computes the *a posteriori* probabilities (APPs) of the information bits given the received sequence. The MAP algorithm can be summarized as follows (Soleymani, 2002, pp. 25-38; Moon, 2005, pp. 588-597):

$$L(\hat{u}_k) = \log \frac{P(u_k = +1 \mid y)}{P(u_k = -1 \mid y)} = \log \frac{\sum_{u_k=+1} P(s_{k-1}=s', s_k=s, y)}{\sum_{u_k=-1} P(s_{k-1}=s', s_k=s, y)}$$

For computing $P(s_{k-1}=s', s_k=s, y)$, BCJR algorithm (Bahl et al., 1974) can be applied:

$$P(s_{k-1}=s', s_k=s, y) = \alpha_{k-1}(s') \cdot \gamma_k(s', s) \cdot \beta_k(s)$$

where α_k and β_k are referred to forward and backward metrics and are computed as:

$$\alpha_k(s) = \sum_{s'} \gamma_k(s', s) \cdot \alpha_{k-1}(s')$$
$$\beta_k(s') = \sum_{s} \gamma_k(s', s) \cdot \beta_{k+1}(s)$$

In the above equations, γ is the state transition probability and is computed as:

$$\gamma_k(s', s) = P(s \mid s')P(y_k \mid s', s) = P(u_k)P(y_k \mid u_k)$$
$$= C_k \exp\left\{\frac{1}{2}u_k\left(L(u_k) + L_c y_k^s\right) + \frac{1}{2}L_c y_k^p x_k^p\right\}$$

where C_k is a constant and will not affect the calculation of $L(u_k)$. $L_c = 4E_s/N_0$. $L(u_k)$ is the log-likelihood ratio of u_k defined as:

$$L(u_k) = \log \frac{P(u_k = +1)}{P(u_k = -1)}$$

Now the *a posteriori* probability (APP) log-likelihood ratio (LLR) of the information bits can be expressed as:

$$L(\hat{u}_k) = \log \frac{P(u_k=+1 \mid y)}{P(u_k=-1 \mid y)} = \log \frac{\sum_{u_k=+1} \alpha_{k-1}(s') \cdot \gamma_k(s', s) \cdot \beta_k(s)}{\sum_{u_k=-1} \alpha_{k-1}(s') \cdot \gamma_k(s', s) \cdot \beta_k(s)}$$

Figure 4. Trellis diagram for an 8-state binary Turbo code

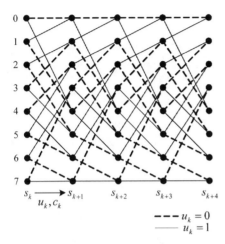

MULTI-STANDARD TURBO DECODER ARCHITECTURE

Related Research on Flexible Turbo Decoder Architectures

The ability to support multi-mode Turbo decoding is necessary for a multi-mode baseband physical layer (PHY) receiver. As some 3G/4G systems use different types of Turbo coding schemes (e.g. binary codes in CDMA, UMTS, HSDPA, and 3GPP LTE and double binary codes in WiMax), a general solution to supporting multiple code types is to use programmable processors. For example, a 2 Mbps Turbo decoder implemented on a DSP processor is proposed by Lin et al., (2007). Also, Shin & Park (2007) and Muller et al. (2006) develop a multi-mode Turbo decoder based on SIMD processors, where a 5.48 Mbps data rate is achieved by Shin & Park (2007) and a 100 Mbps data rate is achieved by Muller et al. (2006) at a cost of 16 processors. While these programmable SIMD/VLIW processors offer great flexibilities, they have several drawbacks, notably higher power consumption and lower throughput than ASIC solutions. A Turbo decoder is typically one of the most computation-intensive parts in a 4G receiver, therefore it is essential to design an area and power efficient flexible Turbo decoder in ASIC.

Due to the many commonalities between different Turbo coding schemes employed in 4G wireless standards, we will present a configurable VLSI architecture for multi-standard Turbo decoding. This architecture can be reconfigured to support both simple and double binary Turbo codes with up to eight states. The memory collision problem is addressed by applying contention-free parallel interleavers. The MAP decoder, memory structure and routing network are designed to operate at full speed with the parallel interleaver. The proposed architecture meets the challenge of multi-standard Turbo decoding at very high data rates.

Radix-2 Decoding of Binary Turbo Codes in the Log Domain

The original MAP algorithm is too complex for implementation in a practical system. To avoid the complicated multiplications and solve the numerical instability issues, one can calculate the MAP algorithm in the log domain. To explain the Log-MAP decoding algorithm, we first introduce the max* function which is defined as (Robertson et al., 1995):

$$\max{}^*(a,b) = \log\{e^a + e^b\} = \max(a,b) + \log(1 + e^{-|a-b|})$$

Consider the decoding process of a simple binary Turbo code, let s_k be the trellis state at time k, then the MAP decoder computes the LLR of the *a posteriori* probability (APP) of each information bit u_k by

$$\Lambda(\hat{u}_k) = \max_{u_k=1}{}^* \{\alpha_{k-1}(s_{k-1}) + \gamma_k(s_{k-1},s_k) + \beta_k(s_k)\} - \max_{u_k=0}{}^* \{\alpha_{k-1}(s_{k-1}) + \gamma_k(s_{k-1},s_k) + \beta_k(s_k)\},$$

where α_k and β_k are the forward and backward state metrics, respectively, and are computed as follows:

$$\alpha_k(s_k) = \max_{s_{k-1}}{}^{*}\{\alpha_{k-1}(s_{k-1}) + \gamma_k(s_{k-1}, s_k)\}$$

$$\beta_k(s_k) = \max_{s_{k+1}}{}^{*}\{\beta_{k+1}(s_{k+1}) + \gamma_k(s_k, s_{k+1})\}$$

where γ_k is the branch transition probability introduced earlier and is usually referred to as a branch metric (BM). To extract the extrinsic information, $\Lambda(\hat{u}_k)$ can be split into three terms: extrinsic LLR $L_e(u_k)$, a priori LLR $L_a(u_k)$ and systematic LLR $L_c(y_k^s)$ as:

$$\Lambda(\hat{u}_k) = L_e(u_k) + L_a(u_k) + L_c(y_k^s)$$

Radix-4 Decoding via One-Level Look-ahead Transform

For binary Turbo codes, the trellis cycles can be reduced 50% by applying a one-level look-ahead transform (Bickerstaff et al., 2003; Zhang & Parhi, 2006) as illustrated in Figure 5. Since two stages of the trellis can be processed at each time step, this process is referred to as the Radix-4 transform. For instance, the Radix-4 α recursion can be expressed as:

$$\alpha_k(s_k) = \max_{s_{k-1}}{}^{*}\left\{\max_{s_{k-2}}{}^{*}\{\alpha_{k-2}(s_{k-2}) + \gamma_{k-1}(s_{k-2}, s_{k-1})\} + \gamma_k(s_{k-1}, s_k)\right\}$$
$$= \max_{s_{k-2}, s_{k-1}}{}^{*}\{\alpha_{k-2}(s_{k-2}) + \gamma_k(s_{k-2}, s_k)\},$$

where $\gamma_k(s_{k-2}, s_k)$ is the merged branch metric for the two-bit vector $\{u_{k-1}, u_k\}$ connecting state s_{k-2} and s_k:

$$\gamma_k(s_{k-2}, s_k) = \gamma_{k-1}(s_{k-2}, s_{k-1}) + \gamma_k(s_{k-1}, s_k)$$

Similarly, the Radix-4 transform can be applied to the β recursion:

$$\beta_k(s_k) = \max_{s_{k+2}, s_{k+1}}{}^{*}\{\beta_{k+2}(s_{k+2}) + \gamma_k(s_k, s_{k+2})\}$$

Because this Radix-4 algorithm is based on the symbol level, we need to define the symbol reliability as:

$$L(\phi_{ij}) = \max_{s_{k-2}, s_k}{}^{*}\{\alpha_{k-2}(s_{k-2}) + \gamma_k^{ij} + \beta_k(s_k)\}$$

Figure 5. An example of one-level look ahead transform of a 4-state trellis (From Sun, Y. et al., IEEE International Conference on Application-Specific Systems, Architectures and Processors (ASAP), pp. 209-214, July 2008. © [2008] IEEE. Used with permission.)

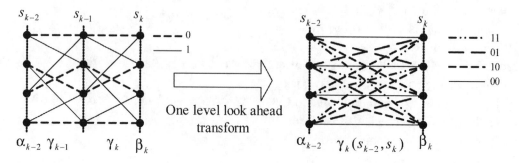

where γ_k^{ij} is the symbol branch transition probability with $u_{k-1} = i$ and $u_k = j$. After knowing the symbol probabilities, the bit LLRs can be computed as:

$$\Lambda(\hat{u}_{k-1}) = \max{}^*\{L(\phi_{10}), L(\phi_{11})\} - \max{}^*\{L(\phi_{00}), L(\phi_{01})\}$$
$$\Lambda(\hat{u}_k) = \max{}^*\{L(\phi_{01}), L(\phi_{11})\} - \max{}^*\{L(\phi_{00}), L(\phi_{10})\}.$$

Radix-4 Decoding of Double Binary Turbo Codes

The double binary Turbo codes were adopted in the IEEE 802.16e WiMax standard due to their better error correction performance than the binary codes. For double binary codes, the main difference with ordinary binary codes is the trellis termination scheme and the symbol-wise decoding scheme (Zhan, 2006). The double binary trellis is closed as a circle with the start state equal to the end state. This is also referred to as a tail-biting termination scheme, which is shown in Figure 6. The symbol wise MAP algorithm is applied with an anti-clockwise process for α state metrics update and clockwise process for β state metrics update. As shown in Figure 6, four branch transitions are associated with each α/β state update. The decoding algorithm for double binary codes is inherently based on the Radix-4 algorithm (Zhan, 2006), hence the same Radix-4 α, β and $L(\varphi)$ function units as used in the binary codes can be applied to the double binary codes in a straightforward manner. The only different parts are the branch metrics γ^{ij} calculations and the tail-biting trellis termination scheme. Three LLRs must be calculated for double binary codes:

$$\Lambda^1(\hat{u}_k) = L_k(\phi_{01}) - L_k(\phi_{00})$$
$$\Lambda^2(\hat{u}_k) = L_k(\phi_{10}) - L_k(\phi_{00})$$
$$\Lambda^3(\hat{u}_k) = L_k(\phi_{11}) - L_k(\phi_{00}).$$

Figure 7 compares the trellis structures (in their Radix-4 format) of IEEE 802.16e WiMax and 3GPP LTE standards. As can be seen, 25% of the trellis structures are identical. The similarities between these two trellis structure representations imply that a generic Turbo decoder can be efficiently designed to support multiple wireless standards with low hardware overhead.

Unified Log-MAP Decoder Architecture

Based on the observation that both binary and double binary codes can be decoded in a unified way, we introduce a flexible Radix-4 Log-MAP decoder architecture to support both types of decoding operations. To efficiently implement the Log-MAP algorithm in hardware, the sliding window technique (Masera, 1999) is adopted.

Figure 6. Circular trellis for double binary Turbo code (From Sun, Y. et al., IEEE International Conference on Application-Specific Systems, Architectures and Processors (ASAP), pp. 209-214, July 2008. © [2008] IEEE. Used with permission.)

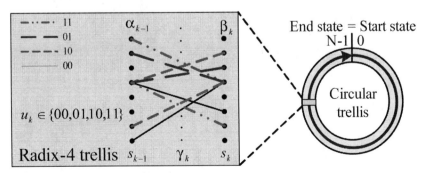

The two types of decoding operations can be generalized into one unified flow which is shown in Figure 8. Let us first use the binary Turbo codes as an example to explain the decoding process. In Figure 8(a), suppose a data sequence is divided into sliding blocks with a sliding window length of W. At the first time slot, the first sliding block I_0 is fed into the decoder and stored in scratch RAM 0. At the second time slot, sliding block I_1 is stored in scratch RAM 1; dummy β_0 recursion is executed on sliding block I_1, and simultaneously α recursion is executed on sliding block I_0 and the results are stored into the α-RAM (LIFO). Starting from the third time slot, α unit (working in the forward order), dummy β_0 unit (in the reverse order), effective β_1 unit (in the reverse order) and Λ unit (in the reverse order) are working in parallel to provide real-time decoding with a latency of $2W$. This decoding operation is based on three recursion units, two used for the backward recursions (dummy β_0 and effective β_1), and one for forward recursion (α). Each recursion unit contains full-parallel ACSA (Add-Compare-Select-Add) operators. To reduce decoding latency, data in a sliding window is fed into the decoder in the reverse order; α unit is working in the forward order; dummy β_0, effective β_1 and Λ units are working in the reverse order as shown in Figure 8. This leads to a decoding latency of $2W$ for binary codes and $3W$ for double binary codes. Double binary codes have an additional W delay because an additional acquisition is needed to obtain the initial α state metrics.

Figure 9 shows a multi-mode Radix-4 Log-MAP decoder ASIC architecture. Three scratch RAMs (with a depth of W) were required to buffer the input systematic, parity and *a priori* LLRs. And three branch metric calculation (BMC) units are used to compute the branch metrics for α, β_0 and β_1 function units. To support multiple Turbo codes, the decoder employs configurable BMCs and configurable α and β function units which can support multiple transfer functions by configuring the routing blocks. The routing block can be reconfigured to support different encoder polynomials. Each α and β unit consists of fully parallel ACSA units so the architecture can support up to 8-state Turbo decoding. The Radix-4 ACSA unit is implemented with four parallel adders followed by three max* units. In order to generate

Figure 7. Trellis structures of IEEE 802.16e WiMax and 3GPP LTE Turbo codes

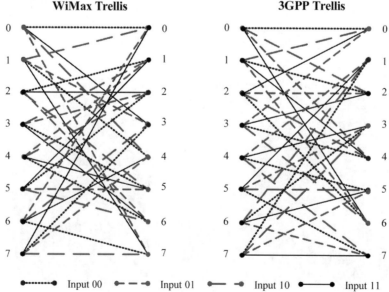

Figure 8. Sliding window tile chart for simple and double binary Turbo codes (From Sun, Y. et al., IEEE International Conference on Application-Specific Systems, Architectures and Processors (ASAP), pp. 209-214, July 2008. © [2008] IEEE. Used with permission.)

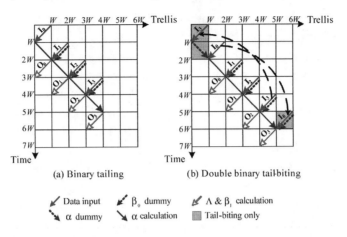

Figure 9. Unified Log-MAP decoder architecture (From Sun, Y. et al., IEEE International Conference on Application-Specific Systems, Architectures and Processors (ASAP), pp. 209-214, July 2008. © [2008] IEEE. Used with permission.)

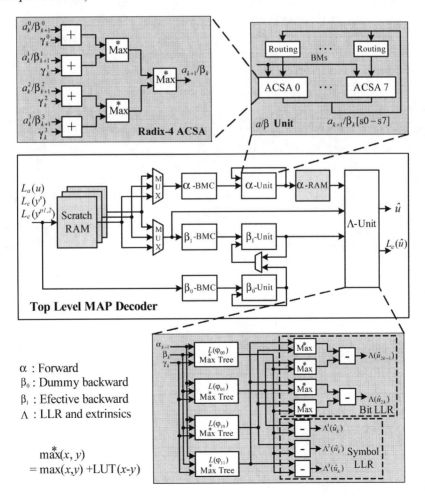

LLRs for both binary and double binary codes, the extrinsic Λ function unit implements both bit LLR and symbol LLR generation. In order to save logic, four max* trees were shared by both types of operations. The Λ unit can generate soft bit LLRs and symbol LLRs in real time with a fixed latency of $2W$ for binary codes or $3W$ for double binary codes.

In this architecture, many blocks can be shared between two decoding operations. For example, the α, β and $L(\varphi)$ units, the α RAMs, and scratch RAMs can be easily shared between two decoding operations. Table 2 compares the resource usage for a multi-mode decoder architecture with a single-mode decoder architecture (Sun, 2008). In Table 2, M is the number of trellis states, W is the sliding window length, B_m, B_b, B_c and B_e are the precisions of state metrics, branch metrics, channel LLRs and extrinsic LLRs, respectively. From table 2, we can see that the overhead for adding flexibility is very small, which is only about 7%. This overhead mainly comes from the multiplexers that were used in the routing networks in the recursion function units. Table 2 indicates that a configurable VLSI architecture is a promising solution to multi-standard Turbo decoder. As a proof of concept, the decoder has been synthesized for a 65nm CMOS technology. Table 3 summarizes the synthesis results at a 200MHz clock frequency which is a typical clock speed in an ASIC design (Sun, 2008).

Area Optimization

Although the multi-mode MAP decoder was designed with limited overhead, we investigated additional techniques to achieve further area-saving. When the MAP decoding kernel is designed, there are several implementation options for the $\log(e^a + e^b)$ function. We considered two options: Log-MAP where $\log(e^a + e^b) \approx \max(a,b) + C$, C is a correction factor, and max-Log-MAP where $\log(e^a$

Table 2. Complexity comparison

	Multi-mode	Single-mode
Storage (bits)	$(9B_e + 12B_c + MB_m)W$	$(9B_e + 12B_c + MB_m)W$
B_m-bit max*	$(25/2)M + 4$	$(25/2)M$
1-bit adder	$16MB_m + 10MB_b$	$16MB_m + 10MB_b$
1-bit flip-flop	$5MB_m + 2MB_b$	$5MB_m + 2MB_b$
1-bit mux	$16MB_m + 16MB_b$	$3MB_m$
Normalized area	1.0	0.93

1 four-input max4* is counted as 3 two-input max*
1 eight-input max8* is counted as 7 two-input max*

Table 3. Area distribution

Blocks	Gate count
α unit (including α BMU)	30.8K gates
β unit x 2 (including β BMUs)	66.2K gates
Λ unit	37.3K gates
α RAM	2560 bits
Scratch RAMs x 3	4224 bits
Control logic	13.4K gates

$+ e^b) \approx \max(a,b)$. If max-Log-MAP is used, the performance loss in comparison with Log-MAP is about 0.3 dB with about 15% logic area saving. To reduce the performance gap with the log-MAP decoder, we introduced a scaling factor applied to the extrinsic LLR values (Vogt & Finger, 2000) as shown in Figure 10 resulting in 0.1 dB loss with 15% logic area saving. With this optimization for the multi-mode MAP decoder using max-Log-MAP, the silicon area is comparable to the Log-MAP single-mode MAP decoder.

TURBO INTERNAL INTERLEAVER ARCHITECTURE

Interleaving has been frequently used in a variety of communication systems. Generally, an interleaver was used to randomize the error locations to combat with the fading or burst error channels. The Turbo internal interleaver is a device that takes its input bit sequence and produces an output sequence that is as uncorrelated as possible. Since this randomness directly affects the decoding performance, the best choice would be the random interleaver. However, the random interelaver is not only difficult to implement, but also is an obstacle to parallel Turbo decoding due to the memory access collision problem.

Figure 10. Extrinsic log-likelihood ratio (LLR) scaling method

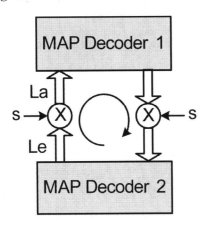

Therefore, the search for a structured interleaver, especially a contention-free interleaver, remains an active research topic in the coding community. Traditionally, memory collisions which occur due to interleaving are solved by having additional write buffers (Salmela, 2007). Recently, new contention-free interleavers have been adopted for the next-generation wireless standards, such as the quadratic polynomial permutation (QPP) interleaver (Sun & Takeshita, 2005) in the 3GPP LTE standard and the almost regular permutation (ARP) interleaver (Berrou, 2004) in the IEEE 802.16e WiMax standard.

Contention-Free Interleavers

An interleaver $\pi(i)$, $0 \leq i < K$, is said to be contention-free for a window size W if and only if it satisfies the following constraint for both $\psi = \pi$ (interleaver) and $\psi = \pi^{-1}$ (deinterleaver) (Nimbalker et al., 2008).

$$\left\lfloor \frac{\psi(j+tW)}{W} \right\rfloor \neq \left\lfloor \frac{\psi(j+vW)}{W} \right\rfloor$$

where $0 \leq j < W$, $0 \leq t$; $v < M (=K/W)$, and $t \neq v$. The terms in the above equation are essentially the memory bank indices that are concurrently accessed by the M processors. If these memory bank addresses are all unique during each read and write operations, then there are no contentions in memory access.

QPP Interleaver in the 3GPP LTE Standard

Given an information block length N, the x-th interleaved output position is given by (3GPP TS 36.212, 2008)

$$\Pi(x) = (f_2 x^2 + f_1 x) \bmod N,$$

where f_1 and f_2 are integers and depend on the block length N ($0 \leq x, f_1, f_2 < N$). The block length N defined in the 3GPP LTE standard ranges from 40 to 6144. Figure 11 depicts the bit error rate (BER) simulation result of the 3GPP LTE Turbo code for block lengths of 40, 240, 1024, and 6144. In Figure 11, both floating (corresponds to "opt" in Figure 11) and fixed point simulation results are shown. Parameters used in the fixed point simulation are as follows: channel input LLR bit precision = 6 (with 2 bit fractional bits), bit precision of the internal state metrics = 12, sliding window length = 64, parallel sliding window = 1, MAP decoding algorithm = 4-entry lookup table based Log-MAP, and maximum iteration = 6.

Hardware Implementation of QPP Interleaver

The direct computation of QPP interleaving is difficult due to the multiplication and modulo operations. A more efficient address generation method is to compute $\Pi(x)$ recursively:

$$\Pi(x+1) = ((f_2 x^2 + f_1 x) + (2f_2 x + f_1 + f_2)) \bmod N$$
$$= (\Pi(x) + \Gamma(x)) \bmod N,$$

where

$$\Gamma(x) = (2f_2 x + f_1 + f_2) \bmod N,$$

and $\Gamma(x)$ can also be computed recursively as:

$$\Gamma(x+1) = (\Gamma(x) + 2f_2) \bmod N.$$

Since $\Pi(x)$, $\Gamma(x)$ and $2f_2$ are all smaller than N, the modulo operation can be efficiently implemented with adders and multiplexers. To implement the QPP interleaver in hardware, we introduce an address generation circuit by cascading two Add-Compare-Choose (ACC) units as shown in Figure 12. As can be seen, no multipliers and dividers are required in this architecture. The critical path of this circuit only contains two adders and two multiplexers. After setting an initial value for $\Pi(x)$ and $\Gamma(x)$ at $x = x_0$, the circuit will continuously generate the interleaving address $\Pi(x)$ for $x = x_0+1, x_0+2, x_0+3, \ldots$ at each cycle. Although

Figure 11. Floating point and fixed point simulation result for 3GPP LTE Turbo codes

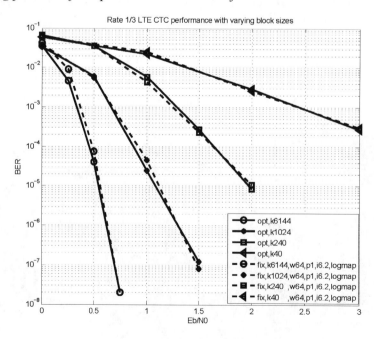

the simplest approach to implement an interleaver is to store all the interleaving patterns in ROMs, this approach becomes almost impractical for a Turbo decoder supporting multiple block sizes. For instance, the 3GPP LTE has defined 188 different Turbo code sizes, which makes the ROM based interleaver implementation very inefficient.

ARP Interleaver in IEEE 802.16e WiMax Standard

The interleaver adopted by the WiMax standard is called the almost regular permutation (ARP) interleaver which is also contention-free. The ARP interleaver employs a two-step interleaving process (IEEE Std 802.16, 2004). The first step switches the alternate couples as

$$[B_x, A_x] = [A_x, B_x], \text{ if } x \bmod 2 = 1$$

In the second step, the ARP interleaver computes

$$\Pi(x) = \begin{cases} P_0 x + 1, & \text{if } x \bmod 4 = 0 \\ P_0 x + 1 + N/2 + P_1, & \text{if } x \bmod 4 = 1 \\ P_0 x + 1 + P_2, & \text{if } x \bmod 4 = 2 \\ P_0 x + 1 + N/2 + P_3, & \text{if } x \bmod 4 = 3 \end{cases}$$

where parameters P_0, P_1, P_2 and P_3 are constants and depend on N.

Hardware Implementation of ARP Interleaver

The ARP interleaver can be also computed recursively. What is more interesting is that the ARP interleaver can be implemented in a similar manner as the QPP interleaver by reusing the same two ACC units, as shown in Figure 13. Let $\lambda(x) = P_0 x$, $Q_0=1$, $Q_1=1+N/2+P_1$, $Q_2=1+P_2$, $Q_3=1+N/2+P_3$. After setting an initial value for $\lambda(x = x_0)$, this circuit will then continuously generate the interleaving address $\Pi(x)$ for $x = x_0+1, x_0+2, x_0+3, \ldots$ at each clock cycle.

If we compare these two circuits shown in Figure 12 and Figure 13, both interleavers have the same logic structure. The differences between these two circuits are the initial values for the ACC units. As can be seen, this unified architecture only requires a few adders and multiplexers which leads to very low complexity and can support all QPP/ARP Turbo interleaving patterns. Compared to the traditional interleaver implementations, which need complex arithmetic units and/or RAMs/ROMs, the proposed QPP/ARP interleaver provides an efficient solution for supporting multi-standard Turbo interleaving.

Figure 12. QPP interleaver architecture (From Sun, Y. et al., IEEE International Conference on Application-Specific Systems, Architectures and Processors (ASAP), pp. 209-214, July 2008. © [2008] IEEE. Used with permission.)

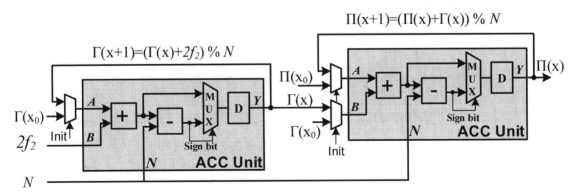

PARALLEL TURBO DECODING

In this section, we present the architecture challenges to design a high throughput parallel Turbo decoder for multi-mode functionality with small overhead. Due to the property of the convolutional trellis structure, one long trellis can be divided into P smaller trellises. Then, each smaller trellis is processed independently by a dedicated MAP decoder (Thul1 et al., 2005; Bougard et al., 2003; Prescher, 2005; Lee, 2005). Ignoring the small overhead introduced by the parallel MAP algorithm, the throughput can be increased almost by a factor of P. For example, Bougard et al. (2003) achieved a 75.6 Mbps data rate by employing 7 SISO decoders running at 160 MHz clock rate.

Scalable Turbo Decoder Architecture

The main issue in designing parallel Turbo decoder architecture is known to be the interleaver parallelism due to the memory access collision problem. To address the data channel decoding throughput issue, both 3GPP LTE and IEEE 802.16e WiMAX employ contention-free interleavers. One codeword can be divided into P sub-codewords and P maximum a posteriori (MAP) decoders can be employed to decode each sub-codeword concurrently which leads to P-level parallelism architecture. Figure 14(a) shows the proposed parallel decoder architecture based on contention-free 4G interleavers (Sun et al., 2008). This architecture is flexible in that it employs both intra-codeword and inter-codeword parallel decoding schemes to improve the overall efficiency and throughput. The intra-codeword mode is used for the decoding of large-size codewords. The parallelism is achieved by dividing the whole block N into P sub-blocks (SBs) and assigning P MAP decoders working in parallel to reduce the latency down to $O(N/P)$. The inter-codeword mode is used for small-size codewords by having P small codewords being decoded simultaneously and independently, so that the overall latency is reduced down to $O(N/P)$ as well.

The memory structure is designed to support concurrent access of LLRs by multiple MAP decoders in both linear addressing and interleaved addressing modes (Sun, 2008). This is achieved by partitioning the memory into P individual banks. Each bank has the same size and can be independently accessed. Because P MAP decoders always access data simultaneously at a particular offset x, it guarantees that no memory access conflicts occur due to the contention-free property of $\lfloor \Pi(x + jM) / M \rfloor \neq \lfloor \Pi(x + kM) / M \rfloor$, where x is the offset in the sub-block j and k ($0 \leq j < k < P$), and M is the sub-block length ($M=N/P$). A full crossbar is used for routing data between P MAP decoders and P memory banks. A parallel decoding example (in intra-codeword mode) for double

Figure 13. ARP interleaver architecture (From Sun, Y. et al., IEEE International Conference on Application-Specific Systems, Architectures and Processors (ASAP), pp. 209-214, July 2008. © [2008] IEEE. Used with permission.)

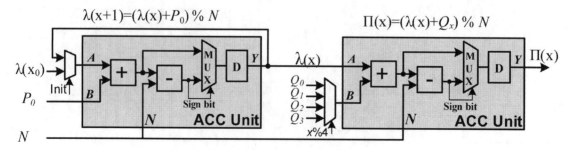

Figure 14. Parallel decoder architecture (From Sun, Y. et al., IEEE International Conference on Application-Specific Systems, Architectures and Processors (ASAP), pp. 209-214, July 2008. © [2008] IEEE. Used with permission.)

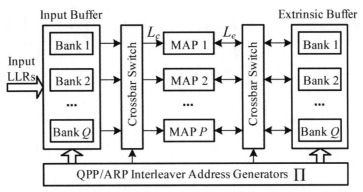

(a) Multi-MAP turbo decoder architecture

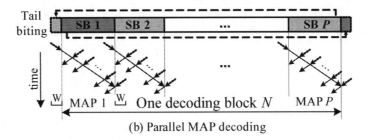

(b) Parallel MAP decoding

binary codes is shown in Figure 14(b). Note that this concept holds for binary codes as well.

Case Study: 100 Mbps Turbo Decoder for Category-3 3GPP LTE Device

The parallelism level P can be tailored to a given codeword size. For smaller codewords, it was found that $P=1$ decoding meets the throughput requirement with less logic complexity. There are several P values that can achieve contention-free memory access across all large codeword sizes. To support the category-3 3GPP LTE data rate of ~100 Mbps, we set the maximum parallelism to be $P=4$. As shown in Figure 15, we have $P=4$ memory banks and depending upon the codeword size, we configure $P=4$ MAP decoders accordingly to meet the throughput requirement for different codeword sizes.

Architecture Trade-Off Analysis

From the above descriptions, we know that high throughput can be achieved by using multiple MAP decoders and multiple memory banks. However, the throughput can not always increase linearly with the parallelism level. As SRAMs are getting smaller and smaller, the area efficiency will decrease. Also, there will be a fixed latency overhead for the sliding-window MAP decoding. In this section, we will analyze the impact of parallelism on throughput, area and power consumption. The maximum throughput is estimated as (Sun, 2008)

$$\text{Throughput} = \frac{N}{\text{Decoding Time}} \approx \frac{N \cdot f}{2 \cdot I \cdot (\frac{\tilde{N}}{P} + 3\tilde{W})}$$

Figure 15. Codeword-size scalable parallel Turbo decoder

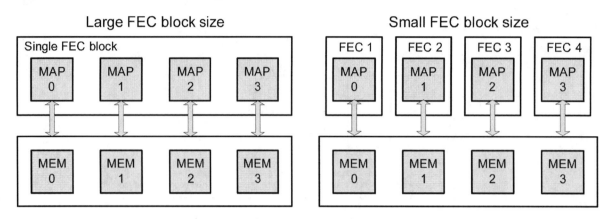

where $\tilde{N} = N/2$ is the code length (in Radix-4), I is the number of iterations (contains two half iterations), $3\tilde{W}$ is the decoding latency for each MAP decoder, and f is the clock frequency. The total area is estimated as (Sun, 2008)

$$\text{Area} \approx P \cdot A_{map}(f) + A_{mem}(P,N) + A_{route}(P,f),$$

where: A_{map} is one MAP decoder's area which will increase with f, A_{mem} is the memory area which will increase with both N and P, and A_{route} is the routing cost (crossbars plus interleavers) which will increase with both P and f. Note that the complexity of the full crossbar actually increases with P^2. To perform the area and power trade-off analysis, the decoder was described in Verilog RTL and synthesized for a 65 nm technology using Synopsys Design Compiler. The area tradeoff analysis is given in Figure 16 which plots the normalized area versus throughput for different parallelism levels and clock frequency goals (80-200 MHz, at a step of 20 MHz). Figure 17

Figure 16. Normalized area versus throughput (N=6144, I=6, W=32). (From Sun, Y. et al., IEEE International Conference on Application-Specific Systems, Architectures and Processors (ASAP), pp. 209-214, July 2008. © [2008] IEEE. Used with permission.)

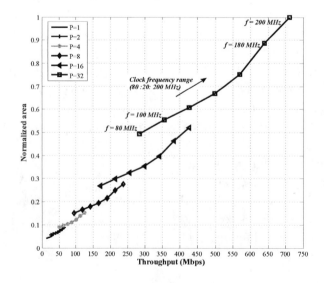

shows the dynamic power tradeoff analysis. As can be seen, for a given throughput there might be multiple configurations, not surprisingly higher parallelism (hence requiring lower frequency) is advantageous to the energy savings but unfavorable to the area utilization, and vice versa.

Table 4 compares the architecture flexibility and the decoding performance of the proposed decoder with existing state-of-the-art Turbo decoders. In Lin *et al.* (2007), Muller *et al.* (2006), and Ituero & Lopez-Vallejo, (2006), programmable VLIW/SIMD processors are designed to support the decoding of multiple Turbo codes.

In Thul *et al.*(2005) and Bougard *et al.*(2003), hardware ASIC architectures are proposed for simple binary Turbo codes based on the Log-MAP algorithm. In Prescher *et al.* (2005), a 757 Mbps decoding throughput is achieved by employing 64 sub-optimal Constant-Log-MAP SISO decoders. Though the decoder in Prescher *et al.* (2005) achieves high throughput at low cost, it has some limitations, i.e. the interleaver is non-standard compliant and it can not support double binary Turbo codes. The comparisons in Table 4 show that our proposed architecture not only has flexibility in supporting multiple Turbo codes (simple

Figure 17. Normalized power versus throughput (N=6144, I=6, W=32). (From Sun, Y. et al., IEEE International Conference on Application-Specific Systems, Architectures and Processors (ASAP), pp. 209-214, July 2008. © [2008] IEEE. Used with permission.)

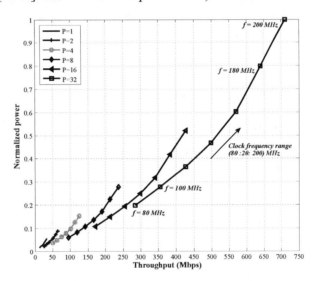

Table 4. Comparison with existing Turbo decoder architectures

Work	Architectures	Flexibility	Algorithm	Parallelism	Frequency	Throughput
Lin, 2006	32-wide SIMD	Multi-code	Max-LogMap	4 window	400 MHz	2.08 Mbps
Ituero, 2006	Cluster VLIW	Multi-code	LogMap	Dual cluster	80 MHz	10 Mbps
Muller, 2006	ASIP SIMD	Multi-code	Max-LogMap	16 ASIP	335 MHz	100 Mbps
Thul, 2006	ASIC	Single-code	LogMap	6 SISO	166 MHz	59.6 Mbps
Bougard, 2003	ASIC	Single-code	LogMap	7 SISO	160 MHz	75.6 Mbps
Prescher, 2005	ASIC	Single-code	Const-LogMap	64 SISO	256 MHz	758 Mbps
Our work	ASIC	Multi-code	LogMap	32 SISO	200 MHz	711 Mbps

binary + double binary) but also achieves a very high throughput (Sun, 2008). It is also scalable in terms of parallelism and data rates.

FUTURE RESEARCH DIRECTIONS

One of the key challenges in developing 4G mobile devices is low power design. The Turbo decoder is typically one of the most power consuming blocks in a 4G wireless PHY receiver. New techniques to dynamically configure the Turbo decoder hardware to achieve minimum energy consumption while guaranteeing quality of service (QoS) are extremely important for 4G wireless PHY design. Besides the semiconductor optimizations, algorithmic and architectural innovations are also required to reduce the power dissipation of the Turbo decoder. Thus, energy efficient Turbo decoder VLSI design is a very important future research topic.

Another key challenge in developing 4G mobile devices is to simultaneously support multiple wireless standards, which all employ forward error correction (FEC) coding schemes. Among these FEC code families, low-density parity-check (LDPC) codes and Turbo codes have received tremendous attention in the coding community. The success of LDPC and Turbo codes is mainly due to the efficient iterative decoding algorithm. Many efficient VLSI architectures for LDPC decoders have been investigated as well as for Turbo decoders. It is known that these two families of codes have similarities. For example, they can both be represented as codes on graphs which define the constraints satisfied by codewords. Both families of codes are decoded in an iterative way by using the sum-product algorithm or belief propagation algorithm. A few researchers have tried to connect these two codes by applying Turbo-like decoding algorithm for LDPC codes. Since both LDPC decoder and Turbo decoder will consume a significant amount of power and occupy a large portion of the silicon area in the wireless receiver PHY, a flexible VLSI decoder architecture supporting both LDPC and Turbo codes is very attractive. In our initial work, we have shown that a generic decoder that supports both LDPC and Turbo decoding is feasible and efficient (Sun & Cavallaro, 2008).

CONCLUSION

In this chapter we have introduced a flexible multi-mode Turbo decoder architecture together with a low-complexity contention-free interleaver. We have shown how to decode simple and double binary Turbo codes in a unified way by employing a Radix-4 Log-MAP decoding algorithm. Then based on the unified Radix-4 trellis, a multi-mode Log-MAP decoder is presented with a low area overhead. This multi-mode Log-MAP (constituent) decoder is the key component in the Turbo decoder architecture. We have shown that by employing certain area optimization techniques, the area of the multi-mode Log-Map decoder is comparable to that of the single-mode Log-Map decoder.

The interleaver parallelism is the key challenge in designing parallel Turbo decoder architectures due to the memory access collision issue. We have given a solution to generate the interleaving patterns in real time by designing low-complexity high-speed interleaver circuits. This circuit can be reconfigured to support both 3GPP LTE and IEEE 802.16e WiMax standards with negligible hardware overhead.

Based on the contention-free interleaver, we have shown a multi-MAP multi-memory parallel Turbo decoder architecture to support high data rates. The decoder parallelism is tailored to the given applications and is scalable for different code sizes. For large code sizes, multiple MAP decoders will be used to reduce the decoding latency and increase the decoding throughput, we called this scheme intra-codeword parallelism. For small code sizes, P small codewords are decoded simultaneously and independently, and we called this scheme inter-codeword parallelism. Given a target throughput, there are multiple choices of decoding parallelism levels. From the energy savings point of view, we tend to use a parallelism level that is as large as possible. However, large parallelism implies more MAP decoders which will therefore occupy more silicon area. We have shown a trade-off analysis of area and power versus decoder parallelism.

The major advantage of using configurable and scalable ASIC architectures is that the area and power are much lower than those of programmable DSP processors. Moreover, a configurable multi-mode decoder is more area and power efficient than multiple single-mode decoders.

Based on the fact that the Turbo decoder has a very regular datapath and there are many similarities among different Turbo code families, a flexible multi-mode Turbo decoder ASIC can be considered a key component toward of multi-mode 4G wireless modem design.

ACKNOWLEDGMENT

The first and second authors would like to thank National Science Foundation (under grants CCF-0541363, CNS-0551692, CNS-0619767, EECS-0925942 and CNS-0923479) for their support of the research.

REFERENCES

3GPP TS 36.212 (2008, September). *Evolved Universal Terrestrial Radio Access (E-UTRA): Mutiplexing and Channel Coding, version 8.4.0*.

Bahl, L., Cocke, J., Jelinek, F., & Raviv, J. (1974). Optimal Decoding of Linear Codes for Minimizing Symbol Error Rate. *IEEE Transactions on Information Theory*, *IT-20*, 284–287. doi:10.1109/TIT.1974.1055186

Berrou, C. (2003, Aug.). The Ten-Year-Old Turbo Codes are Entering into Service. *IEEE Communications Magazine*, *41*, 110–116. doi:10.1109/MCOM.2003.1222726

Berrou, C., & Glavieux, A. (1996, October). Near Optimum Error Correcting Coding and Decoding: Turbo-Codes. *IEEE Transactions on Communications*, 1261–1271. doi:10.1109/26.539767

Berrou, C., Glavieux, A., & Thitimajshima, P. (1993, May). Near Shannon Limit Error-Correcting Coding and Decoding: Turbo-Codes. *IEEE International Conference on Communications*, *2*, 1064-1070.

Berrou, C., Saouter, Y., Douillard, C., Kerouedan, S., & Jezequel, M. (2004, June). Designing Good Permutations for Turbo Codes: Towards A Single Model. *IEEE International Conference on Communications*, *1*, 341-345.

Bickerstaff, M., Davis, L., Thomas, C., Garrett, D., & Nicol, C. (2003). A 24Mb/s Radix-4 Log-MAP Turbo Decoder for 3GPP-HSDPA Mobile Wireless. *IEEE International Solid-State Circuits Conference (ISSCC)*.

Bougard, B., Giulietti, A., Derudder, V., Weijers, J.-W., Dupont, S., Hollevoet, L., et al. (2003). A Scalable 8.7-nJ/bit 75.6-Mb/s Parallel Concatenated Convolutional (Turbo-) Codec. *IEEE International Solid-State Circuits Conference*, *1*, 152-484.

Hagenauer, J., Offer, E., & Papke, L. (1996, March). Iterative Decoding of Binary Block and Convolutional Codes. *IEEE Transactions on Information Theory*, *42*, 429–445. doi:10.1109/18.485714

IEEE Standard for Local and Metropolitan Area Networks Part 16, (2004). *Air Interface for Fixed Broadband Wireless Access Systems, IEEE Std 802.16-2004*.

Ituero, P., & Lopez-Vallejo, M. (2006, September). New Schemes in Clustered VLIW Processors Applied to Turbo Decoding. *IEEE International Conference on Application-Specific Systems, Architectures and Processors* (pp. 291-296).

Lee, S.-J., Shanbhag, N. R., & Singer, A. C. (2005). Area-Efficient High-Throughput MAP Decoder Architectures. [VLSI]. *IEEE Transactions on Very Large Scale Integration, 13*, 921–933. doi:10.1109/TVLSI.2005.853604

Lin, Y., Mahlke, S., Mudge, T., Chakrabarti, C., Reid, A., & Flautner, K. (2007, October). Design and Implementation of Turbo Decoders for Software Defined Radio. *IEEE Workshop on Signal Processing Design and Implementation (SIPS)*, (pp. 22-27).

Masera, G., Piccinini, G., Roch, M., & Zamboni, M. (1999). VLSI architecture for turbo codes. [VLSI]. *IEEE Transactions on Very Large Scale Integration, 7*, 369–3797. doi:10.1109/92.784098

Moon, T. K. (2005). *Error Correction Coding*. Hoboken, NJ: John Wiley & Sons.

Muller, O., Baghdadi, A., & Jezequel, M. (2006, March). ASIP-Based Multiprocessor SoC Design for Simple and Double Binary Turbo Decoding. *IEEE Design . Automation and Test in Europe, 1*, 1–6. doi:10.1109/DATE.2006.244126

Nimbalker, A., Blankenship, Y., Classon, B., & Blankenship, T. K. (2008, March). ARP and QPP Interleavers for LTE Turbo Coding. In *IEEE Wireless Communications and Networking Conference* (pp. 1032-1037).

Prescher, G., Gemmeke, T., & Noll, T. G. (2005, March). A Parametrizable Low-Power High-Throughput Turbo-Decoder. *IEEE Internationa Conference on Acoustics, Speech, and Signal Processing (ICASSP, 5*, 25-28.

Robertson, P., Villebrun, E., & Hoeher, P. (1995, June). A Comparison of Optimal and Sub-optimal MAP Decoding Algorithm Operating in The Log Domain. *IEEE International Conference on Communications, 2*, 1009-1013.

Salmela, P., Gu, R., Bhattacharyya, S. S., & Takala, J. (2007, September). Efficient Parallel Memory Organization for Turbo Decoders. In *European Signal Processing Conference (EURASIP)* (pp. 209-214).

Shin, M. C., & Park, I. C. (2007, June). SIMD Processor-based Turbo Decoder Supporting Multiple Third-Generation Wireless Standards. [VLSI]. *IEEE Transactions on Very Large Scale Integration, 15*, 801–810. doi:10.1109/TVLSI.2007.899237

Soleymani, M. R., Gao, Y., & Vilaipornsawai, U. (2002). *Turbo Coding for Satellite and Wireless Communications*. Norwell, MA: Kluwer Academic.

Sun, J., & Takeshita, O. Y. (2005). Interleavers for Turbo Codes Using Permutation Polynomials Over Integer Rings. *IEEE Transactions on Information Theory, 51*, 101–119. doi:10.1109/TIT.2004.839478

Sun, Y., & Cavallaro, J. R. (2008, October). Unified Decoder Architecture for LDPC/TURBO Codes. *IEEE Workshop on Signal Processing Systems (SIPS)*, (pp. 13-18).

Sun, Y., Zhu, Y., Goel, M., & Cavallaro, J. R. (2008, July). Configurable and Scalable High Throughput Turbo Decoder Architecture for Multiple 4G Wireless Standards. *IEEE International Conference on Application-Specific Systems, Architectures and Processors (ASAP)*, (pp. 209-214).

Thul1, M. J., Gilbert1, F., Vogt1, T., Kreiselmaier1, G., & Wehn1, N (2005). A Scalable System Architecture for High-Throughput Turbo-Decoders. *The Journal of VLSI Signal Processing, 39*, 63-67.

Vogt, J., & Finger, A. (2000). Improving the Max-Log-MAP Turbo Decoder. *Electronics Letters*, 1937–1939. doi:10.1049/el:20001357

Zhan, C., Arslan, T., Erdogan, A. T., & MacDougall, S. (2006, May). An Efficient Decoder Scheme for Double Binary Circular Turbo Codes. *IEEE International Conference on Acoustics, Speech and Signal Processing (ICASSP), 4*, 14-19.

Zhang, Y., & Parhi, K. K. (2006, October). High-Throughput Radix-4 LogMAP Turbo Decoder Architecture. In *IEEE Asilomar Conference on Signals, Systems and Computers* (pp. 1711-1715).

Chapter 28
Parallel Soft Spherical Detection for Coded MIMO Systems

Hosein Nikopour
Huawei Technologies Co., Ltd., Canada

Amin Mobasher
Stanford University, USA

Amir K. Khandani
University of Waterloo, Canada

Aladdin Saleh
Bell Canada, Canada

ABSTRACT

This Chapter briefly evaluates different multiple-input multiple-output (MIMO) detection techniques in the literature as the candidates for the next generation wireless systems. The authors evaluate the associated problems and solutions with these methods. The focus of the chapter is on two categories of MIMO decoding: i) hard detection and ii) soft detection. These techniques significantly increase the capacity of wireless communications systems. Theoretically, a-posteriori probability (APP) MIMO decoder with soft information can achieve the capacity of a MIMO system. A sub-optimum APP detector is proposed for iterative joint detection/decoding in a MIMO wireless communication system employing an outer code. The proposed detector searches inside a given sphere in a parallel manner to simultaneously find a list of m-best points based on an additive metric. The metric is formed by combining the channel output and the a-priori information. The parallel structure of the proposed method is suitable for hardware parallelization. The radius of the sphere and the value of m are selected according to the channel condition to reduce the complexity. Numerical results are provided showing a significant reduction in the average complexity (for a similar performance and peak complexity) as compared to the best earlier known method. This positions the proposed algorithm as a candidate for the next generation wireless systems. The proposed scheme is applied for the decoding of the rate 2, 4×2 MIMO code employed in the IEEE 802.16e standard.

DOI: 10.4018/978-1-61520-674-2.ch029

1. INTRODUCTION

Recently, there has been considerable interest in multiple transmit and receive antennas because of capability to offer a high data rate over fading channels (Foschini & Gans 1998; Chuah, Tse, Kahn, & Valenzuel, 2002). By adopting many of the MIMO-OFDM wireless technologies, WiMAX or IEEE 802.16 (IEEE802.16, 2009; Ben-Shimol, Kitroser, & Dinitz, 2006; and references therein) and 3G LTE (3GPP TS22.146, 2009) standards are designed to achieve a peak downlink data rate of 100 Mbps. The next generations of wireless standards like LTE-Advanced are aiming to support 1 Gbps (3GPP TR36.913, 2009). Such capabilities have the potential to enable some significant new service opportunities including mobile TV and other important multicasting/broadcasting services (Hartung, Horn, Huschke, Kampmann, Lohmar, & Lundevall, 2007). Consequently, application of MIMO systems for higher-rate data transmission required by the next-generation broadcasting systems is studied in a number of articles (see Hartung, Horn, Huschke, Kampmann, Lohmar, & Lundevall, 2007; Zhang, Gui, Qiao, & Zhang, 2004; Baek, Kook, Kim, You, & Song, 2005; Baek, Kim, You, & Song, 2004; Qiao, Yu, Su, & Zhang 2005; and references therein).

In MIMO systems, a vector is transmitted by the transmit antennas. In the receiver, a corrupted version of this vector affected by the channel noise and fading is received. Decoding concerns the operation of recovering the transmitted vector from the received signal. MIMO detection techniques can be divided into two classes: i) hard detection and ii) soft detection. In the current standards, usually, MIMO detection and channel decoding are performed separately. Therefore, hard MIMO detection is mainly performed as the equalization. However, soft detection of MIMO systems in conjunction with iterative channel decoding results in a better performance. In order to achieve the capacity of the MIMO systems, the next generation of wireless systems should be able to implement APP MIMO detectors.

1.1. MIMO Decoding with Hard Information

In MIMO channels, as the received signal set has a regular structure, the maximum likelihood (ML) decoding problem is usually expressed in terms of "lattice decoding" which is known to be NP-hard. In other words, lattice decoding methods can be used for hard detection. However, the complexity of the optimum lattice decoding grows exponentially with the number of transmit antennas, and with the constellation size. Several sub-optimum MIMO detectors have been proposed. Zero-forcing (Schneider, 1979) and minimum mean-square-error (MMSE) (Xie, Short, & Rushforth, 1990) are the simplest MIMO detection methods that currently have been adopted for 3GPP LTE (3GPP TS22.146, 2009) or WiMAX (IEEE802.16, 2009) standards. These algorithms are considered the linear detection algorithms that offer low complexity with moderate performance.

More advanced algorithms include nulling and interference cancellation (IC) methods (Foschini 1996; Golden, Foschini, Valenzuela, & Wolnianky 1999), which are essentially based on ZF, MMSE, or QR decomposition (QRD) equalization. However, the performance of such detectors is significantly inferior to that of the ML detector.

Sphere decoding (SD) (Agrell, Vardy, & Zeger, 2002; Damen, El-Gamal, & Cairo 2003) is used as a detection method for MIMO systems with near ML performance. In SD, the lattice points inside a hyper-sphere are checked and the closest lattice point is determined. It is known that even the average complexity of the SD algorithm is exponential (Jalden, & Ottersten, 2005). Following this class of SD algorithms, several sub-optimal algorithms were proposed with a constraint on the complexity of SD algorithms. In these methods, by fixing the average/worst-case complexity of the SD method to a fixed and pre-defined limit, they avoided the exponential complexity. QRD with m-algorithm (QRDM) (Yue, Kim, Gibson, & Iltis, 2003) is an example of such techniques.

Recently, an alternative algorithm for hard detection of MIMO is developed with polynomial average complexity which is based on semi-definite programming (SDP) (Steingrimsson, Luo, & Wong 2003; Mobasher, Taherzadeh, Sotirov, & Khandani 2005; Mobasher & Khandani 2007). The exponential worst case complexity is still unsolvable by this approach.

1.2. MIMO Decoding with Soft Information

The capacity of MIMO systems cannot be achieved unless by using an outer channel code concatenated with the space-time mapper acting like an inner code. Iterative APP detection techniques (Benedetto, Montorsi, Divsalar, & Pollara 1996; Hagenauer, 1997), such as iterative joint detection and decoding with soft inputs and outputs (Damen 1999; Stefanov & Duman 2001; Sellathurai & Haykin 2002; Boutros & Caire 2002), can be used for decoding of such systems. In contrast to the ML detector which finds the closest valid point to the received noisy signal, the iterative soft MIMO detector provides probabilistic information about the transmitted bits. The soft information is passed to the decoder for the underlying error correction code (ECC), such as turbo (Berrou, Glavieux, & Thitimajshima, 1993) or low-density parity check (LDPC) (Gallager 1963) code. The soft outputs of the decoder can be used as the new soft inputs for the MIMO detector, and hence this method can work in an iterative fashion to improve the overall performance.

The optimum APP MIMO detector has a very large complexity, because it enumerates all the signal points of the lattice for the soft metric computation (Damen 1999). To reduce the complexity, several schemes are proposed based on finding a small set (list) of highly probable points for computing the soft values. List sphere decoder (LSD) (Hochwald & Brink 2003; Yin, Lee, Ahmed, Ryu, & Peterson 2004; Liu & Li 2005; Wong, Paulraj, & Murch 2005; Hong & Choi 2005) is a method in this category which uses a list of candidates inside a preset sphere for computing the soft information. The main drawback in the known LSD methods is the instability of the list size and the associated problem of the radius selection and reduction during the search. This significantly increases the complexity as compared to the original hard SD (HSD). In (Boutros, Gresset, Brunel, & Fossorier 2003), this problem is addressed by building a spherical list centered on the ML point.

Stack algorithm with limited stack size (Baro, Hagenauer, & Witzke 2003; Dong, Wang, & Doucet 2003; de Jong & Willink 2005; Reid, Grant, & Kind 2003) is also considered as a list detector method. The disadvantage of this class of detectors is that the complexity is only a function of the stack size and is independent of the received SNR and channel condition. The method in (Wang & Giannakis, 2004) is another class of soft detectors which is based on soft-to-hard transformation. This algorithm, however, imposes some limitations on the underlying modulation and coding schemes.

The main contribution of this work is a low-complexity list MIMO detector based on combining sphere and m-algorithm approaches. First, the traditional SD is modified to work in a parallel manner. The parallel sphere decoder (PSD) grows all the nodes at a given level of the tree simultaneously to find the best paths. During the parallel search, the m-algorithm helps the PSD to reduce the complexity by eliminating some branches. The radius of the sphere and the value of m are selected according to the channel condition to reduce the complexity. The proposed PSD has the benefits of both sphere and m-algorithm detectors, while it avoids their drawbacks. Numerical results are provided showing a significant reduction in the average complexity (for a similar performance and peak complexity) as compared to the best earlier

Figure 1. MIMO transmitter and iterative receiver

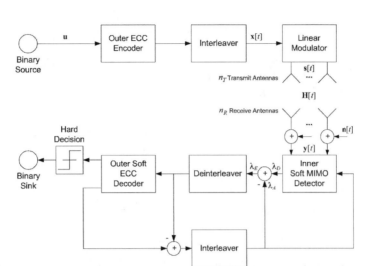

known methods reported in (Hochwald & Brink 2003), (Reid, Grant, & Kind 2003), and (Wang & Giannakis, 2004).

After introducing the MIMO system model under consideration and the iterative receiver in Section 2, we briefly review different hard and soft MIMO detection methods in the literature. Section 3 describes the details of the proposed parallel list sphere detector for the next generation wireless systems with some numerical results. Section 4 concludes this chapter.

2. BACKGROUND ON MIMO DECODING

2.1. System Model

We consider a typical bit interleaved coded modulation (BICM) MIMO system (Boutros, Boixadera, & Lamy 2000) with iterative APP receiver as shown in Fig. 1.

A block of information bits **u** is encoded by the outer code C with code rate R. Then, the encoded stream is permuted by an interleaver. At time t, the bits of the sequence $\mathbf{x}[t] = \left(x_1[t],\ldots,x_{M_c n_T}[t]\right)^T$ as a part of the permuted stream are mapped onto a complex vector $\mathbf{s}[t] = map\left(\mathbf{x}[t]\right) = \left(s_1[t],\ldots,s_{n_T}[t]\right)^T$ by n_T linear modulators. Each element $s_i[t]$ is taken from a complex constellation Q composed of 2^{M_c} distinct points. The outputs of the modulators are passed through a narrow-band multiple antenna channel with n_T transmit antennas and n_R receive antennas. Throughout this chapter, it is assumed that $n_T = n_R$.

Let $\mathbf{y}[t] = \left(y_1[t],\ldots,y_{n_R}[t]\right)^T$ be an n_R-dimensional vector of received signal given by,

$$y[t] = \mathbf{H}[t]s[t] = \mathbf{n}[t] \qquad (1)$$

where $\mathbf{H}[t] = (h_{ij}[t])$ is an $n_R \times n_T$ matrix with entries that are independent samples of a complex Gaussian random variable with zero mean and unit variance (Rayleigh fading). We assume that $\mathbf{H}[t]$ is known to the receiver. Entries of n[t] are the spatially and temporally complex white Gaussian noises with variance σ^2 per real component. Assuming $\mathbf{E}s_i[t]^2 = \mathbf{E}_s / n_T$ makes the total transmit power equal to $\mathbf{E}_s$ (Hochwald & Brink 2003). The overall bandwidth efficiency of such a system is

$RM_c n_T$ bits per channel use. In the following, for notation simplicity we ignore the time dependency of the samples.

In order to simplify the equations, we first extend the complex model of (1) to the real space (Damen, El-Gamal, & Cairo 2003). The channel matrix is changed to the real matrix $\mathbf{H} \leftarrow (H_{ij})$ using,

$$H_{ij} = \begin{bmatrix} \Re h_{ij} & -\Im h_{ij} \\ \Im h_{ij} & \Re h_{ij} \end{bmatrix} \quad (2)$$

where $\Re h_{ij}$ and $\Im h_{ij}$ denote the real and the imaginary parts of h_{ij}, respectively. Obviously, the matrix dimensions are extended to $n_T \leftarrow 2n_T$ and $n_R \leftarrow 2n_R$. The number of bits per real dimension is $M_c \leftarrow M_c/2$. The complex vector of the symbols is extended to the real vector $\mathbf{s} \leftarrow s_i$ as follows

$$s_i = \begin{bmatrix} \Re s_i \\ \Im s_i \end{bmatrix} \quad (3)$$

The other complex vectors **n** and **y** of the model (1) are extended in the same way. The derivations of the paper for the complex model hold for the extended real model, as well.

2.2. Hard Decoding Techniques

An important performance measure of the decoding algorithms is the probability of error, i.e. $\mathbb{P}\{\hat{\mathbf{s}} \neq \mathbf{s}\}$. An ML decoder finds the exact solution by minimizing the probability of error. In another word, the ML decoding rule is given by

$$\hat{\mathbf{s}} = \arg\min_{s_i \in Q} \|\mathbf{y} - \mathbf{H}\mathbf{s}\|^2 \quad (4)$$

The existing methods which solve the problem in (4) using hard information can be classified into four groups:

- Linear methods
- Successive cancellation methods
- Lattice reduction aided methods, and
- Sphere decoding methods

2.2.1. Linear Receivers

ZF (Schneider, 1979) and MMSE (Xie, Short, & Rushforth, 1990) are some of the well known linear receivers. These methods, which offer a low complexity with moderate performance, have been already adopted for 3GPP LTE (3GPP TS22.146, 2009) and WiMAX (IEEE802.16, 2009) standards. In ZF, the vector resulted by multiplying the received vector by pseudo inverse or Moore-Penrose inverse (Lancaster & Tismenetsky 1985) of matrix **H**, denoted by $\mathbf{H}^\dagger$, is rounded off to the closest integer (Schneider, 1979).

$$\hat{\mathbf{s}} = \lceil \mathbf{H}^\dagger \mathbf{y} \rfloor \quad (5)$$

MMSE has the same principal as ZF, but this detector considers the effect of the noise variance as well. By using inter-antenna interference in ZF, the $\mathbf{H}^\dagger$ in (5) is replaced by (Xie, Short, & Rushforth, 1990)

$$(\mathbf{H}^*\mathbf{H} + \sigma^2 \mathbf{I})^{-1}\mathbf{H}^* \quad (6)$$

where σ^2 is the channel noise variance.

2.2.2. Successive Cancellation Receivers

Decision feedback (DF) decoders are an improved version of linear receivers in which the components of the signal point are estimated recursively by one of the aforementioned answers (ZF or MMSE) and the effect of each detected component on the next components is cancelled. In this method, ZF or MMSE is used to estimate the first component of the signal, s_1, This component is assumed to be known and its effect is removed from the optimization problem (4) to obtain an equivalent problem with $n_T - 1$ unknown parameters. This procedure is repeated to find all the components. This procedure is also called nulling and cancellation method.

The DF decoder is highly suffering from error propagation. VBLAST detection algorithm (Foschini 1996; Golden, Foschini, Valenzuela, & Wolnianky 1999) is an example of successive cancellation methods which is based on DF algorithm with an appropriate ordering of components for detection purpose. The detection order is in accordance with the descending order of SNR of different elements in the received point in order to combat the error propagation problem. The VBLAST algorithm as described in (Golden, Foschini, Valenzuela, & Wolnianky 1999) is as follows:

Initialization: $\mathbf{G}_1 = \mathbf{H}^\dagger$ \hfill (7a)

$i = 1$ \hfill (7b)

Recursion: $k_i = \arg\min_{j \notin \{k_1,\ldots,k_{i-1}\}} \left(\mathbf{G}_i\right)_j^2$ \hfill (7c)

$\mathbf{w}_{k_i} = \left(\mathbf{G}_i\right)_{k_i}$ \hfill (7d)

$\mathbf{r}_{k_i} = \mathbf{w}_{k_i}^T \mathbf{y}_i$ \hfill (7e)

$\hat{s}_{k_i} = \mathbb{Q}\left(\mathbf{r}_{k_i}\right)$ \hfill (7f)

$\mathbf{y}_{i+1} = \mathbf{y}_i - \hat{s}_{k_i}(\mathbf{H})_{k_i}$ \hfill (7g)

$\mathbf{G}_{i+1} = H^\dagger_{\blacklozenge_{k_i}}$ \hfill (7h)

$i = i + 1$ \hfill (7i)

where $(\mathbf{G}_i)_j$ is the jth row of $\mathbf{G}_i$ and $\mathbb{Q}$ is the quantization operator to the constellation components. The equation (7c) determines the optimum order for detection of the components of s. Also, (7g) performs the cancellation of the decoded component.

Since VBLAST can handle high data rates with reasonable complexity, it can be easily implemented for the next generation receivers; however, the loss in performance as compared to ML decoding is usually significant. In addition, VBLAST transmits independent data streams on its antennas, so there is no built-in spatial or temporal coding, and the decoding scheme works only if the number of receive antennas is not less than the number of transmit antennas.

2.2.3. Lattice Reduction Aided Receivers

Lattice reduction techniques can be used in conjunction with any of the linear or successive cancellation methods to improve their performances. Lattice reduction technique transforms a given channel matrix into an equivalent channel matrix consisting short and fairly orthogonal vectors. In other words, the channel matrix $\mathbf{H}$ is decomposed to $\mathbf{H} = \mathbf{H}^T\mathbf{Q}$ such that $\mathbf{Q}$ is unimodular[1] and the matrix $\mathbf{H}^T$ consists of vectors which are shorter and more orthogonal compared to the vectors defined by $\mathbf{H}$.

General known procedures for lattice reduction methods find the solution in three steps:

- Reduce the channel matrix ($\mathbf{H} = \mathbf{H}^T\mathbf{Q}$).
- Perform finding the point in the reduced channel (finding the closest lattice point to y with the channel matrix $\mathbf{H}^T$) with any of the aforementioned methods.
- Transform the result to the original channel (multiplying $\mathbf{Q}^{-1}$ to the answer).

Several distinct notions of lattice reduction have been studied, including those associated to the names Minkowski, Korkin-Zolotarev (KZ) (Helfrich 1985), and LLL (Lenstra, Lenstra, & Lov´asz, 1982) reduced basis, which can be computed in polynomial time. Taherzadeh, Mobasher, & Khandani (2007) have shown that the decoding algorithm using LLL basis reduction achieves the same receive diversity as the ML decoding algorithm (which is equal to the number of receive antennas). Therefore, addition of the lattice reduction techniques to the current implementations of the standard receivers does not encounter any burden on the system in terms of complexity while improving the performance.

2.2.4. Sphere Decoding Receivers

ML decoder finds the optimal solution by searching over the entire constellation points. This decoder is practically infeasible due to its exponential complexity (Van Nee, Van Zelst, & Awater, 1999); however, its improvement in the bit error rate performance cannot be neglected.

A number of algorithms can be found in mathematical contexts for general lattice decoding. In Fincke and Pohst algorithm (Fincke & Pohst 1985), called sphere decoding (SD) algorithm, constellation points inside an N-dimensional hypersphere centered at **y** are searched. By finding a constellation point inside the sphere, the radius of the sphere is updated. Later, Schnorr and Euchner (1994) introduced an improved SD algorithm. They suggested enumerating the constellation points inside the N-D hypersphere in the order of increasing distance from the components of integer point corresponding to **y**.

Several applications in communications have used these algorithms. Damen et. al. used the Fincke and Pohst algorithm for decoding in MIMO channels (Damen, Abed-meriam, Belfiore 2000; Damen, Chkeif, Belfiore, May 2000; Damen, Chkeif, Belfiore, June 2000). They have explored the lattice representation of a multi antenna system and the algebraic space-time codes for any number of transmit and receive antennas.

Hassibi and Vikalo (Part I, 2005; Part II, 2005) in their two papers introduced a sphere decoding algorithm for MIMO systems based on Fincke and Pohst's algorithm. Instead of determining the lattice points in an N-dimensional space, this algorithm recursively determines components of the constellation in each dimension. The only problem here is selection of a proper radius. Radius is selected as a scaled version of the noise variance such that with a high probability a lattice point inside the sphere can be found (Hassibi & Vikalo 2005). In *Closest Lattice Point Search* Algorithm (Agrell, Vardy, & Zeger 2002), they have generalized Schnorr and Euchner's algorithm (Schnorr and Euchner 1994) for decoding of any MIMO system. In this method, the radius is ignored in the search algorithm due to the special ordering. There are several variants of these algorithms. In the following, the SD algorithm is presented as it is described in Hassibi & Vikalo 2005.

In order to compute the components of the vector s sequentially, the **QR** decomposition of the matrix $\mathbf{H} = \begin{bmatrix} \mathbf{Q} & \mathbf{Q}' \end{bmatrix} \begin{bmatrix} \mathbf{R} \\ 0 \end{bmatrix}$ is used in Hassibi & Vikalo 2005. The matrix $\mathbf{R} = (r_{ij})$ is an $n_T \times n_T$ upper triangular matrix with $r_{ij} \geq 0$ and $[\mathbf{Q} \quad \mathbf{Q}']$ is an $n_R \times n_R$ orthogonal matrix where the matrix $\mathbf{Q}$ is an $n_R \times n_T$ matrix. The SD algorithm is formalized as follows in Hassibi & Vikalo 2005:

Input. Inputs are $\mathbf{H} = \begin{bmatrix} \mathbf{Q} & \mathbf{Q}' \end{bmatrix} \begin{bmatrix} \mathbf{R} \\ 0 \end{bmatrix}$, **y**, z = **Q***y,

and C (radius of the SD algorithm)

Step 1. Set $k = n_T$, $C_k^{'2} = C^2 - \|\mathbf{Q}'^* \mathbf{y}\|^2$, $z_{k|k+1} = z_k$

Step 2. (*Bounds for* s_k) Set $UB(s_k) = \dfrac{C_k' + z_{k|k+1}}{r_{k-1,k-1}}$,

$s_k = \dfrac{-C_k' + z_{k|k+1}}{r_{k-1,k-1}} - 1$

Step 3. (*Increase* s_k) $s_k = s_k + 1$. If $s_k \leq UB(s_k)$ go to **Step 5**, else go to **Step 4**.

Step 4. (*Increase k*) $k = k + 1$; if $k = n_T + 1$ terminate algorithm, else go to **Step 3**.

Step 5. (*Decrease k*) If $k = 1$ go to **Step 6**, else $k = k - 1$, $z_{k|k-1} = z_k + \sum_{j=k+1}^{n_T} r_{kj} s_j$, $C_k'^2 = C_{k+1}'^2 - (z_{k+1} - r_{k+1,k+1} s_{k+1})^2$ and go to **Step 2**.

Step 6. Solution found. Save s and go to **Step 3**.

Note that this algorithm is developed for integer grid constellations; in other words, the constellation components belong to the set $\{...,-2,-1,0,1,2,...\}$. QAM constellations can be easily changed to the integer grid.

It has been shown that (Jalden, & Ottersten 2005), there is an exponential lower bound on the average complexity of sphere decoding, and its worst case complexity is exponential (Hassibi & Vikalo 2005; Agrell, Vardy, & Zeger 2002). However, it is experienced that over certain ranges of rate, SNR and dimension the average complexity is polynomial (Hassibi & Vikalo 2005).

Moreover, it has been shown that sphere decoding gains huge improvement over VBLAST decoding method, full diversity of coded multi antenna systems, high spectral efficiency, independency of the constellation size, and maximum likelihood performance. There have been several attempts to improve the SD algorithm especially by fixing a constraint on the complexity of the SD algorithm. These methods fix the average complexity or the worst-case complexity of the SD algorithm to a pre-defined limit; therefore, the exponential complexity of SD algorithm is avoided.

QRD with m-algorithm (QRDM) (Yue, Kim, Gibson, & Iltis, 2003) is an example of such techniques. First, QR decomposition (QRD) is performed on the estimated channel matrix. Then, components of s are found from the component with the weakest noise to the component with the highest noise in the m-algorithm (tree search algorithm) (Anderson & Mohan 1984; Schlegel 1997). The computational cost involved in the tree search is reduced by keeping only m minimum accumulated distance metrics at each level of the tree (Yue, Kim, Gibson, & Iltis, 2003). These new improvements place the SD algorithm and tree based search variants as a receiver candidate for the next generation of wireless technologies.

2.3. Soft Decoding Techniques

In order to achieve the capacity of the MIMO systems, soft detection of MIMO systems in conjunction with iterative channel decoding should be implemented instead of a simple hard decoding MIMO detection. Since the bits among the blocks of x are outputs of an ECC encoder that introduces redundancy, the block-by-block decision on the bits is no longer optimal. The APP MIMO detector should make decision jointly on all blocks using the knowledge of the correlation across blocks, and the channel code should decode using soft information on all the blocks obtained from the APP MIMO detector. Therefore, an iterative receiver that performs joint detection and decoding is needed.

An iterative receiver consists of two stages: the soft MIMO APP detector, followed by a soft ECC decoder, which are separated by a deinterleaver and an interleaver. Note that any of the algorithms in the previous sub-section can be used to extract the soft information required by the MIMO APP detector. However, here, the emphasis is on SD-based soft detection techniques. Fig. 1 illustrates how the soft information is iterated between the MIMO APP detector and the soft ECC decoder.

As the MIMO APP detector operates on each symbol vector separately, we omit the time index and work with the vector **x**. The optimal log-likelihood ratio (LLR) of the bit $x_k, k = 1,...,M_c n_T$, knowing the received vector **y**, is obtained by the APP detector as follows (Hagenauer 1997)

$$\lambda_D\left(x_k | \mathbf{y}\right) = \ln \frac{P\left(x_k = +1 | \mathbf{y}\right)}{P\left(x_k = -1 | \mathbf{y}\right)} \tag{8}$$

which constitutes the posterior information about x_k. Due to the existence of the interleaver, the bits within x are approximately statistically independent. Employing Bayes' theorem, exploiting the independence among the entries of x, and considering the channel model (1), the soft output values are given by

$$\lambda_D(x_k|\mathbf{y}) = \ln \frac{\sum_{\mathbf{x} \in \mathbb{X}_k^+} \exp(\Lambda(\mathbf{x}, \lambda_A; \mathbf{y}, \mathbf{H}))}{\sum_{\mathbf{x} \in \mathbb{X}_k^-} \exp(\Lambda(\mathbf{x}, \lambda_A; \mathbf{y}, \mathbf{H}))} \quad (9)$$

where

$$\Lambda(\mathbf{x}, \lambda_A; \mathbf{y}, \mathbf{H}) = -\frac{1}{2\sigma^2}\|\mathbf{y} - \mathbf{Hs}\|^2 + \frac{1}{2}\mathbf{x}^T \lambda_A \quad (10)$$

and $\mathbb{X}_k^{\pm} = \{\mathbf{x} \mid x_k = \pm 1\}$. With $\lambda_A(x_k) = \ln \frac{P(x_k=+1)}{P(x_k=-1)}$ denoting the a priori information of x_k, we define $\lambda_A = (\lambda_A(x_1),...,\lambda_A(x_{M_c n_T}))$. The a priori information is received from the previous iteration as the soft output of the ECC encoder. By definition, the extrinsic information of x_k is given by $\lambda_E(x_k|\mathbf{y}) = \lambda_D(x_k|\mathbf{y}) - \lambda_A(x_k)$

The ECC decoder uses the extrinsic information provided by the MIMO detector to update its knowledge about the transmitted bits. The new soft information of the bits serves as the prior information for the next iteration.

2.3.1. List Detector

The number of vectors x in the set $\mathbb{X}$ is equal to $2^{M_c n_T}$. Therefore, noting (9), the complexity of computing $\lambda_D(x_k|\mathbf{y})$ is exponential in terms of the length of the binary vector x. To reduce the complexity, several schemes are proposed based on finding a small set (list) of highly probable points to compute the soft values. List sphere decoder (LSD) (Hochwald & Brink 2003; Yin, Lee, Ahmed, Ryu, & Peterson 2004; Liu & Li 2005; Wong, Paulraj, & Murch 2005; Hong & Choi 2005) is a method in this category which uses a list of candidates inside a preset sphere for computing the soft information.

In (Hochwald & Brink 2003) or (Wong, Paulraj, & Murch 2005), the candidates are selected such that they have the smallest value for $\|\mathbf{y} - \mathbf{Hs}\|^2$. In other words, the candidates are the points inside a sphere centered at **y**, like Fincke and Pohst's (1985) algorithm. However, when the channel is not full rank, the estimator is not unique. Therefore, in (Yin, Lee, Ahmed, Ryu, & Peterson 2004), they have used a regularized version of the list square estimator. In other words, **H*H**, is replaced by **H*H** + σ^2**I**.

The list detector methods in (Hochwald & Brink 2003), (Yin, Lee, Ahmed, Ryu, & Peterson 2004), or (Wong, Paulraj, & Murch 2005) have ignored the second term in (10), which is the a priori information. In Liu & Li 2005, it has been shown that no matter how large the a priori information is, the candidates in the list are independent of the a priori information and the output of these MIMO detectors are bounded by $\frac{1}{2\sigma^2}\left[\max\{\|\mathbf{y} - \mathbf{Hs}\|^2\} - \min\{\|\mathbf{y} - \mathbf{Hs}\|^2\}\right]$. They have mitigated this problem by constraining the a priori information. In the LSD method in (Hong & Choi 2005), the a priori information is also utilized to improve the soft MIMO decoder. An approximation approach is proposed such that the function in (10) is approximated by a quadratic function. Thus, the LSD approach based on Fincke & Pohst (1984) is directly employed on the resulted function.

These LSD methods suffer from the instability of the list size and the associated problem of the radius selection and reduction during the search. Boutros, Gresset, Brunel, & Fossorier (2003) address this problem by building a spherical list centered on the ML point, instead of the received point. In their approach the effective list size is well controlled; however, the large list size is required to achieve a reasonable performance.

2.3.2. Stack Algorithm

Stack algorithm with limited stack size is another list detector proposed under different titles such as list-sequential (LISS) detector (Baro, Hagenauer, & Witzke 2003), deterministic sequential Monte Carlo (SMC) (Dong, Wang, & Doucet 2003), or iterative tree search (ITS) (de Jong & Willink 2005; Reid, Grant, & Kind 2003). This scheme evaluates only the 'good' candidate vectors with the aid of a sequential tree searching scheme, which is based on the m-algorithm (Anderson & Mohan 1984; Schlegel 1997). The disadvantage of this class of detectors is that the complexity is only a function of the stack size and is independent of the received SNR and channel condition.

2.3.3. Soft-to-Hard Transformation

Soft-to-hard transformation is another approach (Wang & Giannakis, 2004) to build a soft detector. The transformation converts a soft detection problem to a set of hard detection problems, which are less complex compared with the LSD. This approach, however, imposes some limitations on the underlying modulation and coding schemes. Therefore, this approach cannot be considered as a general solution for the soft MIMO detection.

3. PARALLEL LIST SPHERE DETECTOR

In this section, we propose a parallel SD (PSD) approach that can be used to find a 'good' list inside a sphere of a given radius. According to (10), in the absence of a priori information, the required candidate points of the list may be defined as follows

$$\forall \mathbf{x}' \notin \mathcal{L}, \forall \mathbf{x} \in \mathcal{L}, \|\mathbf{y} - \mathbf{H}\mathbf{s}\|^2 \ll \|\mathbf{y} - \mathbf{H}\mathbf{s}'\|^2 \quad (11)$$

The list $\mathcal{L}$ contains points inside a sphere centered at the received point $\mathbf{y}$. It means the sphere decoder can be used as a powerful tool to find a proper list of points around the received signal.

Here, we propose a soft-input soft-output list SD decoder combined with m-algorithm to achieve MIMO channel capacity. The traditional SD is implemented in a parallel manner. Then, the PSD grows all the nodes at a given level of the tree while the m-algorithm helps the PSD to reduce the complexity by eliminating some of the branches. The proposed APP PSD has the benefits of both sphere and m-algorithm detectors, while it avoids their drawbacks. A significant reduction in the average complexity is achieved as compared to the best earlier known methods reported in (Hochwald & Brink 2003), (Reid, Grant, & Kind 2003), and (Wang & Giannakis, 2004).

3.1. List Reduction

The empirical observation shows that the vast majority of the elements of $\mathbb{X}$ contribute a negligible amount to the summations in (9). We decrease the complexity of the calculation of $\lambda_D(x_k \mid \mathbf{y})$ by enumerating only the high-Λ-value subset denoted as $\mathcal{L}$. The list detection problem is formulated as follows. Find the set of points $\mathcal{L} \subset \mathbb{X}$ with cardinality $|\mathcal{L}| = m < 2^{M_c n_T}$ and the highest Λ-values (Kind & Grant, 2003):

$$\mathcal{L} = \arg\max_{\substack{\mathcal{L}' \subset \mathbb{X} \\ |\mathcal{L}'| = m}} \min_{x \in \mathcal{L}'} \Lambda(\mathbf{x}, \lambda_A; \mathbf{y}, \mathbf{H}) \quad (12)$$

In general, the problem of finding the optimum list as described in (12) is NP-hard (Ajtai 1998). Moreover, it should be mentioned that the SD algorithm must be modified in a manner that the search algorithm takes into account both the distance and the a priori information to find the highest Λ-value points.

A traditional HSD searches the closet point in a serial manner (Damen, El-Gamal, & Cairo, 2003) based on the distance metric. Starting with an initial radius, the SD finds a candidate point and updates the sphere radius which cannot exceed the initial radius. After that, the SD starts the search process over, using the newly computed radius to find any better candidate points. This process continues until finding the closet point. In the LSD (Hochwald & Brink 2003), a list of points is found within a preset sphere meaning that the radius cannot be decreased during the search process.

Note that when the radius is fixed, there will be no difference between the complexity of the serial and the parallel search schemes. The proposed parallel scheme, which is called PSD, grows all the branches inside the sphere simultaneously to find all the candidate points. The main advantage of the parallel scheme is that at each level of the sequential tree growing, like the m-algorithm approach (e.g. Baro, J. Hagenauer, & M. Witzke, 2003), we can limit our search to the signal paths with the highest Λ-values. The additive form of the Λ-value allows us to calculate the branch metrics sequentially in a tree structure. At each level of the tree growing, existing nodes are explored inside the sphere, the new extended branches are ordered, and finally, the m-best branches are selected and the rest are eliminated. Unlike (de Jong & Willink, 2005) and (Reid, Grant, & Kind, 2003) that use a fixed m, here, we select m adaptively according to the channel matrix as will be explained later.

3.2. Additive Metric

In order to compute Λ sequentially over the branches of the tree, we use the same QR decomposition of the matrix $\mathbf{H} = \begin{bmatrix} \mathbf{Q} & \mathbf{Q}' \end{bmatrix} \begin{bmatrix} \mathbf{R} \\ 0 \end{bmatrix}$ as in section 2.2.4. Therefore, the distance $\mathbf{y} - \mathbf{Hs}^2$ can be written as $\mathbf{z} - \mathbf{Rs}^2$ where $\mathbf{z} - \mathbf{Q}^*\mathbf{y}$. By expanding Λ in (10) to the summation form, we have an additive metric

$$\Lambda(\mathbf{x}, \lambda_A; \mathbf{z}, \mathbf{R}) = \sum_{i=1}^{n_T} \Lambda_i \qquad (13)$$

with the metric increments

$$\Lambda_i = -\frac{1}{2\sigma^2}\left(z_i - \sum_{j=i}^{n_T} r_{ij} s_j\right)^2 + \frac{1}{2}\sum_{j=1}^{M_c} x_{ij} \lambda_A(x_{ij}) \qquad (14)$$

in which $x_{ij} = x_{(i-1)M_c + j}$ for $i = 1, 2, \ldots, n_T$ and $j = 1, 2, \ldots, M_c$. Using this definition, $s_i = \mathrm{map}\left(\{x_{ij}\}_{j=1}^{M_c}\right)$. Referring to (14), Λ_i depends only on the transmit symbols s_j for $j \geq i$ and the a priori values corresponding to s_i.

3.3. The Proposed Algorithm

The proposed algorithm searches for the candidate points inside a sphere of radius C. This means $\mathbf{z} - \mathbf{Rs}^2 < C^2$ or

$$\sum_{i=1}^{n_T}\left(z_i - \sum_{j=i}^{n_T} r_{ij} s_j\right)^2 < C^2 \qquad (15)$$

Obviously, each element of the summation should satisfy the inequality. Starting from the n_T^{th} element and assuming $r_{n_T n_T} > 0$, we conclude that s_{n_T} belongs to the following interval (Damen, El-Gamal, & Cairo, 2003)

$$\left\lceil \frac{-C + z_{n_T}}{r_{n_T n_T}} \right\rceil \leq s_{n_T} \leq \left\lfloor \frac{C + z_{n_T}}{r_{n_T n_T}} \right\rfloor \qquad (16)$$

where $\lceil \cdot \rceil$ and $\lfloor \cdot \rfloor$ denote the ceiling and floor operations. For every s_{n_T} satisfying (16), we define $C^2_{n_T-1|n_T} = C^2 - (z_{n_T} - r_{n_T n_T} s_{n_T})^2$ and $z_{n_T-1|n_T} = z_{n_T-1} - r_{n_T-1,n_T} s_{n_T}$. Noting the $(n_T - 1)^{th}$ term in (15), we conclude that s_{n_T-1} belongs to the following interval

$$\frac{-C_{n_T-1|n_T} + z_{n_T-1|n_T}}{r_{n_T-1,n_T-1}} \le s_{n_T-1} \le \frac{C_{n_T-1|n_T} + z_{n_T-1|n_T}}{r_{n_T-1,n_T-1}} \quad (17)$$

One can continue in a similar manner for s_{n_T-2}, and so on until s_1, thereby going through all the points inside the sphere of radius C.

The radius C is chosen based on the statistics of the noise (Vikalo, Hassibi, & Kailath, 2004). Note that $\frac{1}{2\sigma^2}\mathbf{n}^2 = \frac{1}{2\sigma^2}\|\mathbf{y} - \mathbf{Hx}\|^2$ is a χ^2 random variable with $n_R/2$ degree of freedom. Thus, we can choose the radius proportional to the variance of the noise (Hochwald & Brink 2003; Vikalo, Hassibi, & Kailath, 2004),

$$C^2 = K_r n_R \sigma^2 \quad (18)$$

where the factor K_r controls the tradeoff between complexity and performance.

Noting that the number of points bounded within a sphere of a given radius is inversely proportional to $\det(\mathbf{HH}^*)$, we select m as follows

$$m = \min\left(K_m \frac{\prod_{i=1}^{n_T}\|\mathbf{h}_i\|^2}{\det(\mathbf{HH}^*)}, m_{max}\right) \quad (19)$$

where $\mathbf{h}_i$ is the ith column of $\mathbf{H}$. The projection of a lattice on different coordinates might have different number of points, because some points of the lattice might have the same projection on a specific coordinate. However, since $\mathbf{H}$ represents a random lattice, there is a low probability that two different points of the lattice have the same projection on a coordinate. Consequently, it is reasonable to use m as the number of nodes which are kept not only in the final level but also in the all previous levels of the tree search.

The values of K_r, K_m, and m_{max} are experimentally selected to minimize the complexity, while achieving the best possible performance. Note that imposing an upper bound on the allowed m limits the peak complexity. Noting above steps, the proposed algorithm can be summarized as follows:

Input. Inputs are $\mathbf{H}$, $\mathbf{y}$, λ_A, σ^2, n_T, n_R, and M_c, all in the real space. Maximum number of nodes at each level of the tree is m. The 2^M points of the real constellation are the elements of the real set Q_r. Note that $Q = \{Q_r + jQ_r\}$.

Step 1. (*Initialization*) Decompose $\mathbf{H}$ to $\mathbf{QR}$. $\mathbf{z} = \mathbf{Q}^*\mathbf{y}$. Each node has its own level, i, metric, α_i, label[2], $\mathbf{s}_i = (s_i,\ldots,s_{n_T})$ or $x_i = \left\{\{x_{kj}\}_{j=1}^{M_c} | k=i,\ldots,n_T\right\}$, sub-sphere radius for the next level, $C_{i-1|i}$, and received signal for the next level, $z_{i-1|i}$. Initialize root node (level $n_T + 1$) with $\alpha_{n_T+1} = 0$, $C_{n_T|n_T+1} = C$, $z_{n_T|n_T+1} = z_{n_T}$, and empty label. Set total number of nodes to 1. Set $i = n_T$.

Step 2. (*Node extension*) For each node of the level $i + 1$,

(*Lower bound*) Set $LB = \dfrac{-C_{i|i+1} + z_{i|i+1}}{r_{ii}}$.

(*Upper bound*) Set $UB = \dfrac{C_{i|i+1} + z_{i|i+1}}{r_{ii}}$.

(*Extended nodes*) Extend the current node to the nodes $s_i \in [LB, UB] \cap Q_r$.

(*Label and metric*) For each extended node s_i and its corresponding bit label $\{x_{ij}\}_{j=1}^{M_c}$:

(*Labeling*) Label the extended node as $\mathbf{s}_i = (s_i, \mathbf{s}_{i+1})$ or $x_i = \left\{\{x_{ij}\}_{j=1}^{M_c}, x_{i+1}\right\}$.

(*Metric increment*)
$$\Lambda_i = \frac{-1}{2\sigma^2}\left(z_{i|i+1} - r_{ii}s_i\right)^2 + \frac{1}{2}\sum_{j=1}^{M_c} x_{ij}\lambda_A(x_{ij}).$$
(*Metric Update*) $a_i = a_{i+1} + \Lambda_i$.

Step 3. (*Node selection*) If number of nodes of the level i is more than m, among all the nodes, select m nodes with the highest metrics, and eliminate the other ones. If number of nodes is less than m, keep all of them. Update number of nodes to the number of the selected nodes.

Step 4. (*Next Level*) Set $i \leftarrow i - 1$. If $i = 0$, go to **Step 7**.

Step 5. (*Update parameters*) For each node of the level $i + 1$:
$$z_{i|i+1} = z_i - \sum_{j=i+1}^{n_T} r_{ij}s_j$$
$$C_{i|i+1}^2 = C_{i+1|i+2}^2 - \left(z_{i+1|i+2} - r_{i+1,i+1}s_{i+1}\right)^2$$

Step 6. (*Loop*) Go to **Step 2**.

Step 7. (*List*) For each node of the level 1, save its label $(s_1,\ldots,s_{n_T})$ or $(x_1,\ldots,x_{M_c n_T})$ and corresponding metric $\Lambda = a_1$ in the set $\mathcal{L}$.

Output outputs are the set $\mathcal{L}$ and the corresponding Λ-values.

Note that the algorithm does not guarantee to deliver a list of m points, unless the initial radius is set sufficiently high. The output of the proposed PSD is a sub-optimum solution to the list detection problem of (12). Although the list that is found by the PSD contains the m points with the high Λ-values, these points are not necessarily the m best points. Some branches which are eliminated in the middle levels of the tree might be the paths which eventually tend to the optimum points. In other words, a path which tends to an optimum node of a middle tree level is not necessarily a portion of a path which tends to an optimum point of the last level of the tree search. Experimental results show that there is a large overlap among the sets of the best paths corresponding to the different levels of the tree. It means that with a high probability, the m nodes which are the best points of a middle level of the tree belong to the paths which represent the m best points of the optimum list.

3.4. Approximate APP Value

After finding the list $\mathcal{L}$ and its corresponding Λ-values, the list is applied to (9) to compute the APP values of the transmitted bits. To maintain a reasonable complexity, the exact exp() operation in (9) is implemented in the log-domain by deploying the max- or max*-approximation (Robertson, Villebrun, & Hoeher, 1995). With the max approximation, the posterior information of x_k becomes

$$\lambda_D(x_k|\mathbf{z}) \approx \max_{\mathbf{x}\in\mathbb{X}_k^+\cap\mathcal{L}}\left\{\Lambda(\mathbf{x}, \lambda_A; \mathbf{z}, \mathbf{R})\right\} - \max_{\mathbf{x}\in\mathbb{X}_k^-\cap\mathcal{L}}\left\{\Lambda(\mathbf{x}, \lambda_A; \mathbf{z}, \mathbf{R})\right\} \quad (20)$$

in which the maximizations are over the points of the list $\mathcal{L}$. If $\mathbb{X}_k^+ \cap \mathcal{L}$ (or $\mathbb{X}_k^- \cap \mathcal{L}$) is empty, then $\lambda_D(x_k|\mathbf{z})$ is set to the given minimum (or maximum) value of the LLR (Hochwald & Brink 2003).

3.5. Numerical Results

Numerical results are provided for the same scenario as described in (Hochwald & Brink 2003) and (Wang & Giannakis 2004). A parallel concatenated turbo code with rate $R = 1/2$ is used in the coded MIMO system. Each constituent convolutional code has memory 2, feedback polynomial $G_r(D) = 1 + D + D^2$, and feed forward polynomial $G(D) = 1 + D^2$. The interleaver size of the turbo code is *9214* information bits. The receiver performs 4 outer iterations between the turbo decoding and the MIMO channel detection blocks, and 8 inner iterations within the turbo decoder.

The signal power per transmitted information bits at the receiver side is defined as (Hochwald & Brink 2003, Wang & Giannakis 2004)

$$\frac{E_b}{N_0}\bigg|_{dB} = \frac{E_s}{N_0}\bigg|_{dB} + 10\log_{10}\frac{n_R}{RM_c n_T} \quad (17)$$

Figure 2 shows the performance results for a 4 × 4 MIMO system. The list size is set to $m = 512$ which is fixed regardless of the channel condition. According to Fig. 2, the performance of the proposed PSD ($K_r = 4$) in the same as the m-algorithm (or PSD with $K_r = \infty$). It means that the sphere detection does not degrade the performance of the PSD. By comparing the results of the PSD with the results presented in (Hochwald & Brink 2003) and (Wang & Giannakis 2004), there is no degradation between the PSD and the best results obtained by the LSD approach.

The advantage of the proposed method lies in its complexity of detection which is illustrated for a 16QAM 4 × 4 MIMO system in Fig. 3. As this figure shows, the complexity of the proposed PSD method for $m = 512$ and $K_r = 4$ and 8 is much less than the m-algorithm and the LSD with the same parameters. Computational complexity is measured in terms of the number of branches enumerated at each level of the tree. The higher is the number of constellation points and/or the number of antennas, the higher will be the complexity gain achieved due to using the proposed method.

According to Figure 3, the complexity of the m-algorithm is fixed independent of the channel condition and the SNR value. This causes the m-algorithm to have a very high complexity for the large SNR values, where the other two methods have lower complexities.

One advantage of the proposed PSD is that its complexity does not exceed a threshold which is determined by the m-algorithm, even at low SNR values where the sphere radius is large. In other words, in contrast to the LSD which has exponential worst case complexity, the worst case complexity of the PSD, which is equal to the complexity of the m-algorithm, is linear in terms of the number of antennas and the constellation size.

Figure 2. Performance results for a turbo coded MIMO system with 4 transmit and receive antennas

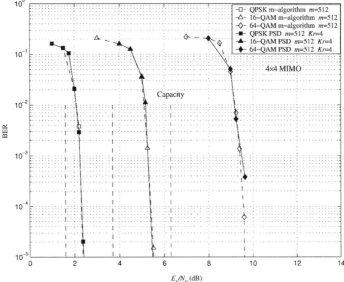

Figure 3 also compares sensitivity of the LSD and the PSD to the radius of the sphere. When K_r is increased from 4 to 8, there is an increase in the complexity of the PSD, but the complexity is limited to the threshold which is defined by the m-algorithm. However, the complexity of the LSD increases very fast and without any limitation by increasing the sphere radius.

The complexity of the PSD can be reduced further if m is selected adaptively according to (15). Figure 4 shows the complementary cumulative distribution function (ccdf) of m for $m_{max} = 512$

Figure 3. Complexity of different detection algorithms for a 16-QAM, 4 × 4 MIMO system

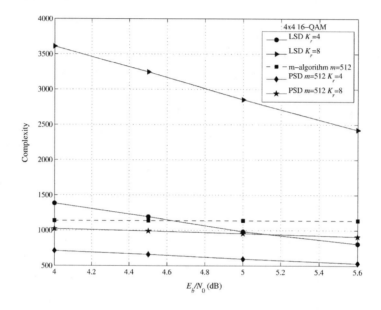

Figure 4. Effect of K_m on the probability distribution of m for $m_{max} = 512$

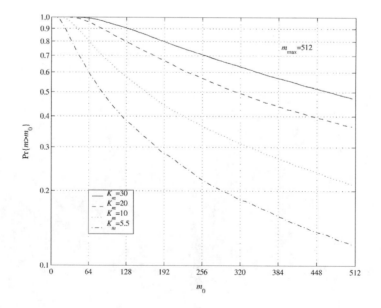

Figure 5. Effect of K_m on the complexity of a 16-QAM 4 × 4 MIMO system

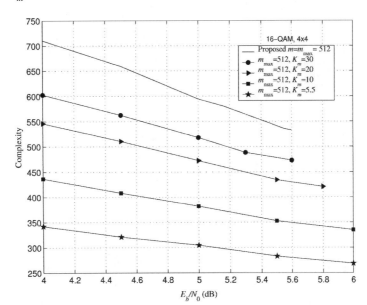

Figure 6. Effect of K_m on the performance of a 16-QAM 4 × 4 MIMO system

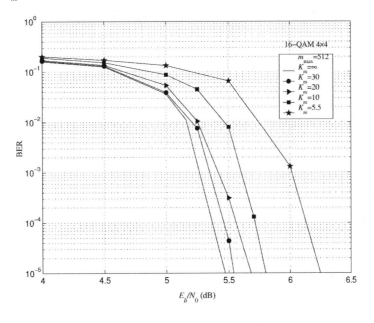

and different K_m values. With adaptive selection of m, there is no need to set $m = 512$ for all channel conditions. For example, when $K_m = 10$, only 36 percent of m's are greater than 256. It means that the average complexity of the PSD would be less than the case in which m is fixed at 512. This fact is shown in Figure 5, where the complexities of the PSD detectors are compared for different K_m values. As this figure shows, the complexity of the PSD for the 16-QAM 4 × 4 MIMO system

Figure 7. Performance result for IEEE 802.16e standard with a space-time code of rate 2, 16-QAM, 4 £ 2 MIMO. The result is compared with a 16-QAM 4 × 4 MIMO system.

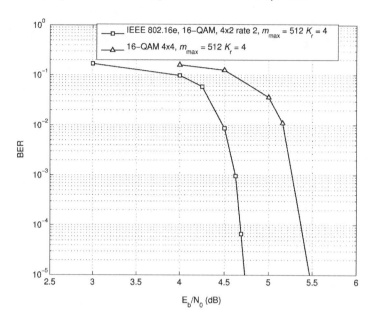

decreases to the half of the original PSD (with fixed $m = 512$) for $K_m = 5.5$.

The effect of the adaptive m on the performance of the PSD detector is illustrated in Figure 6 for 16-QAM 4 × 4 MIMO system. For $K_m = 5.5$, the degradation in performance is less than 0.8 dB. Consequently, with accepting only a small degradation, we can reduce the complexity of the original PSD by an adaptive selection of m.

As a practical example, Figure 7 shows the performance result for a 4 × 2 space time code of rate 2 adopted in the IEEE 802.16e standard (IEEE802.16, 2009). This space time code could be modeled as (1) with a 4 × 4 channel matrix. The performance result of the space time code is compared with a 16-QAM 4 × 4 MIMO system (rate 4). The complexity results are similar to those reported in Figure 3.

4. CONCLUSION

A soft MIMO detector is presented which achieves significant complexity reduction for a similar performance as the other relevant methods reported in the literature. Not only the average complexity, but also the worst case complexity of the proposed detector is linear. To further decrease the complexity of the detector, we propose to select the detector parameters adaptively based on the channel condition. In addition, the parallel structure of the proposed method is suitable for hardware parallelization (Guo & Nilsson, 2006).

REFERENCES

Ajtai, M. (1998). The shortest vector problem in L2 is NP-hard for randomized reductions. In *Proc. of 30th Annu. ACM Symp. Theory of Comput. (STOC)* (pp. 10–19).

Anderson, J., & Mohan, S. (1984). Sequential coding algorithms: A survey and cost analysis. *IEEE Transactions on Communications, 32*(2), 169–176. doi:10.1109/TCOM.1984.1096023

Baek, M., Kim, M., You, Y., & Song, H. (2004). Semi-blind channel estimation and PAR reduction for MIMO-OFDM system with multiple antennas. *IEEE Transactions on Broadcasting, 50*(12), 414–424. doi:10.1109/TBC.2004.837885

Baek, M., Kook, H., Kim, M., You, Y., & Song, H. (2005). Multi-antenna scheme for high capacity transmission in the digital audio broadcasting. *IEEE Transactions on Broadcasting, 51*(12), 551–559. doi:10.1109/TBC.2005.858426

Baro, S., Hagenauer, J., & Witzke, M. (2003, May). Iterative detection of MIMO transmission using a list-sequential (LISS) detector. In *Proc. Of IEEE Int. Conf. on Commun., ICC '03*, Anchorage, AK (pp. 2653–2657).

Ben-Shimol, Y., Kitroser, I., & Dinitz, Y. (2006). Two-dimensional mapping for wireless OFDMA systems. *IEEE Transactions on Broadcasting, 52*(9), 388–396. doi:10.1109/TBC.2006.879937

Benedetto, S., Montorsi, G., Divsalar, D., & Pollara, F. (1996, Nov.). A soft-input soft-output maximum a posterior (MAP) module to decode parallel and serial concatenated codes. In *NASA JPL TDA Progress Report* (pp. 42–127).

Berrou, C., Glavieux, A., & Thitimajshima, P. (1993, May). Near Shannon limit error-correcting coding and decoding: turbo-codes. In *Proc. Of IEEE Int. Conf. on Commun., ICC '93*, Geneva (pp. 1064–1070).

Boutros, J., Boixadera, F., & Lamy, C. (2000, Sept.). Bit-interleaved coded modulations for multiple-input multiple-output channels. In *Proc. Of IEEE 6th Int. Symp. on Spread Spectrum Techniques and Applications*, (Vol. 1, pp. 123–126).

Boutros, J., & Caire, G. (2002). Iterative multiuser joint decoding: unified framework and asymptotic analysis. *IEEE Transactions on Information Theory, 48*(7), 1772–1793. doi:10.1109/TIT.2002.1013125

Boutros, J., Gresset, N., Brunel, L., & Fossorier, M. (2003, December). Soft-input soft-output lattice sphere decoder for linear channels. In *Proc. of IEEE Global Telecommun. Conf., GLOBECOM '03* (Vol. 3, pp. 1583–1587).

Chuah, C. N., Tse, D. N. C., Kahn, J. M., & Valenzuela, R. A. (2002). Capacity scaling in MIMO wireless systems under correlated fading. *IEEE Transactions on Information Theory, 48*(3), 637–650. doi:10.1109/18.985982

Codes. In *ISIT2000,* Sorrento, Italy (p. 362).

Damen, M. O. (1999). *Joint coding/decoding in a multiple access system, application to mobile communications.* Ph.D. dissertation, Ecole Nationale Superieure des Telecommunications.

Damen, M. O., Abed-Meriam, K., & Belfiore, J.-C. (2000). A generalized lattice decoder for asymmetrical space-time communication architecture. In *Proceedings of the IEEE International Conference on Acoustics, Speech and Signal Processing 2000 (ICASSP2000)* (pp. 2581–2584).

Damen, M. O., Chkeif, A., & Belfiore, J.-C. (2000, May). Lattice Code Decoder for Space-Time Codes. *IEEE Communications Letters, 4*, 161–163. doi:10.1109/4234.846498

Damen, M. O., Chkeif, A., & Belfiore, J.-C. (2000 June). Sphere Decoding of Space-Time

Damen, M. O., El-Gamal, H., & Caire, G. (2003). On maximum-likelihood detection and the search for the closest lattice point. *IEEE Transactions on Information Theory, 49*(10), 2389–2402. doi:10.1109/TIT.2003.817444 de Jong, Y. L. C., & Willink, T. J. (2005). Iterative tree search detection for MIMO wireless systems. *IEEE Transactions on Communications, 53*(6), 930–935. doi:10.1109/TCOMM.2005.849638

Division Multiplexing System. In *IEEE VTC2000* (Vol. 45, pp. 6–10).

Dong, B., Wang, X., & Doucet, A. (2003). A new class of soft MIMO demodulation algorithms. *IEEE Transactions on Signal Processing, 51*(11), 2752–2763. doi:10.1109/TSP.2003.818155

Fincke, U., & Pohst, M. (1985, April). Improved methods for calculating vectors of short length in a lattice, including a complexity analysis. *Mathematics of Computation, 44*, 463–471. doi:10.2307/2007966

Foschini, G. J. (1996). Layered space-time architecture for wireless communication in a fading environment when using multi-element antennas. *Bell Labs Tech. J., 1*, 41–59. doi:10.1002/bltj.2015

Foschini, G. J., & Gans, M. J. (1998). On the limits of wireless communications in a fading when using multiple antennas. *Wireless Personal Communications, 6*, 311–335. doi:10.1023/A:1008889222784

Gallager, R. G. (1963). *Low-Density Parity-Check Codes*. Cambridge, MA: MIT Press

Golden, G. D., Foschini, G. J., Valenzuela, R. A., & Wolnianky, P. W. (1999). Detection algorithm and initial laboratory results using V-BALST space-time communication architecture. *Electronics Letters, 35*(1), 14–16. doi:10.1049/el:19990058

3GPP TS22.146. (2009). *Multimedia Broadcast/Multicast Service (MBMS); Stage 1*.

3GPP TR36.913. (2009). *Requirements for further advancements for Evolved Universal Terrestrial Radio Access (E-UTRA) (LTE-Advanced)*. Agrell, E., Vardy, A., & Zeger, K. (2002). Closest point search in lattices. *IEEE Transactions on Information Theory, 48*(8), 2201–2214. doi:10.1109/TIT.2002.800499

Guo, Z., & Nilsson, P. (2006). Algorithm and implementation of the K-best sphere decoding for MIMO detection. *IEEE Journal on Selected Areas in Communications, 24*(3), 491–503. doi:10.1109/JSAC.2005.862402

Hagenauer, J. (1997, September). The turbo principle: Tutorial introduction and state of the art. In *Proc. 1st Int. Symposium on Turbo Codes* (pp. 1–12).

Hartung, F., Horn, U., Huschke, J., Kampmann, M., Lohmar, T., & Lundevall, M. (2007). Delivery of broadcast services in 3G networks. *IEEE Transactions on Broadcasting, 53*(3), 188–199. doi:10.1109/TBC.2007.891711

Hassibi, B., & Vikalo, H. (2005). On Sphere Decoding Algorithm. Part I. Expected Complexity. *IEEE Transactions on Signal Processing, 53*(8), 2806–2818. doi:10.1109/TSP.2005.850352

Hassibi, B., & Vikalo, H. (2005). On Sphere Decoding Algorithm. Part II. generalization, second order statistics and applications to communications. *IEEE Transactions on Signal Processing, 53*(8), 2819–2834. doi:10.1109/TSP.2005.850352

Helfrich, B. (1985). Algorithms to construct Minkowski reduced and Hermit reduced lattice bases. *Theoretical Computer Science, 41*, 125–139. doi:10.1016/0304-3975(85)90067-2

Hochwald, B. M., & ten Brink, S. (2003). Achieving near-capacity on a multiple-antenna channel. *IEEE Transactions on Communications, 51*(3), 389–399. doi:10.1109/TCOMM.2003.809789

Hong, Y., & Choi, J. (May 2005). A new approach to iterative decoding on coded MIMO channels. In *Proc. of IEEE 61th Vehicular Tech. Conf., VTC '05-Spring* (Vol. 3, pp. 1676–1680).

IEEE802. *16 standards*. (2009). Retrieved from http://ieee802.org/16

Jalden, J., & Ottersten, B. (2005). On the complexity of sphere decoding in digital communications. *IEEE Transactions on Signal Processing, 53*(9), 1474–1484. doi:10.1109/TSP.2005.843746

Kind, A., & Grant, A. (2003, Feb.). Efficient list decoding for lattices. In *Proc. of 4th Australian Commun. Theory Workshop, AusCTW '03*, Melbourne, Australia (pp. 1–5).

Lancaster, P., & Tismenetsky, M. (1985). The theory of matrices. Academic Press.

Lenstra, A. K., Lenstra, H. W., & Lov'asz, L. (1982). Factoring polynomials with rational coefficients. *Mathematische Annalen, 261*, 515–534. doi:10.1007/BF01457454

Liu, J., & Li, J. (2005). Turbo processing for an OFDM-based MIMO system. *IEEE Transactions on Wireless Communications, 4*(9), 1988–1993.

Mobasher, A., & Khandani, A. K. (2007, June). Matrix-Lifting Semi-Definite Programming for Decoding in Multiple Antenna Systems. In *The 10th Canadian Workshop on Information Theory (CWIT'07)*, Edmonton, Alberta, Canada

Mobasher, A., Taherzadeh, M., Sotirov, R., & Khandani, A. K. (2005, September). A near maximum likelihood decoding algorithm for mimo systems based on graph partitioning. In *Proc. of IEEE Int. Symp. on Info. Theory, ISIT '05*, Adelaide, Australia.

Qiao, Y., Yu, S., Su, P., & Zhang, L. (2005). Research on an iterative algorithm of LS channel estimation in MIMO OFDM systems. *IEEE Transactions on Broadcasting, 51*(3), 149–153. doi:10.1109/TBC.2004.842524

Reid, A. B., Grant, A. J., & Kind, A. P. (2003, February). Low complexity list detection for high-rate MIMO channels. In *Proc. of 4th Australian Commun. Theory Workshop, AusCTW '03*, Melbourne, Australia (pp. 66–69).

Robertson, P., Villebrun, E., & Hoeher, P. (1995, June). A comparison of optimal and sub-optimal MAP decoding algorithms operating in the log domain. In *Proc. of IEEE Int. Conf. on Commun., ICC '95*, Seattle (Vol. 2, pp. 1009–1013).

Schlegel, C. (1997). *Trellis Coding*. New York: IEEE Press.

Schneider, K. S. (1979, January). Optimum detection of code division multiplexed signals. *IEEE Transactions on Aerospace and Electronic Systems, AES-15*, 181–185. doi:10.1109/TAES.1979.308816

Schnorr, C. P., & Euchner, M. (1994). Lattice basis reduction: Improved practical algorithms and solving subset sum problems. *Mathematical Programming, 66*, 181–191. doi:10.1007/BF01581144

Sellathurai, M., & Haykin, S. (2002). Turbo-BLAST for wireless communications: Theory and experiments. *IEEE Transactions on Signal Processing, 50*(10), 2538–2546. doi:10.1109/TSP.2002.803327

Stefanov, A., & Duman, T. (2001). Turbo-coded modulation for systems with transmit and receive antenna diversity over block fading channels: System model, decoding approaches, and practical considerations. *IEEE Journal on Selected Areas in Communications, 19*(5), 958–968. doi:10.1109/49.924879

Steingrimsson, B., Luo, T., & Wong, K. M. (2003). Soft quasi-maximum-likelihood detection for multiple-antenna wireless channels. *IEEE Transactions on Signal Processing, 51*(11), 2710–2719. doi:10.1109/TSP.2003.818203

Taherzadeh, M., Mobasher, A., & Khandani, A. K. (2007). LLL Reduction Achieves the Receive Diversity in MIMO Decoding. *IEEE Transactions on Information Theory*, *53*(12), 4801–4805. doi:10.1109/TIT.2007.909169

Van Nee, R., Van Zelst, A., & Awater, G. (July 1999). Maximum Likelihood Decoding in a Space

Vikalo, H., Hassibi, B., & Kailath, T. (2004). Iterative decoding for MIMO channels via modified sphere decoding. *IEEE Transactions on Wireless Communications*, *3*(11), 2299–2311. doi:10.1109/TWC.2004.837271

Wang, R., & Giannakis, G. B. (2004, March). Approaching MIMO channel capacity with reduced-complexity soft sphere decoding. In *Proc. of IEEE Wireless Commun. and Networking Conf., WCNC '04* (Vol. 3, pp. 1620–1625).

Wong, K. K., Paulraj, A., & Murch, R. D. (2005, May). List slab-sphere decoding: efficient high-performance decoding for asymmetric MIMO antenna systems. In *Proc. of IEEE 61th Vehicular Tech. Conf., VTC '05-Spring* (Vol. 1, pp. 697–701).

Woo, K. S., Lee, K. I., Paik, J. H., Park, K. W., Yang, W. Y., & Cho, Y. S. (2007). TA DSFBC-OFDM for a next generation broadcasting system with multiple antennas. *IEEE Transactions on Broadcasting*, *53*(6), 539–546.

Xie, Z., Short, R. T., & Rushforth, C. K. (1990, May). A family of suboptimum detectors for coherent multi-user communications. *IEEE Journal on Selected Areas in Communications*, *8*, 683–690. doi:10.1109/49.54464

Yin, J., Lee, H. N., Ahmed, M., Ryu, B., & Peterson, L. (2004, May). Iterative MMSE-sphere list detection and graph decoding MIMO OFDM transceiver. In *Proc. of IEEE 59th Vehicular Tech. Conf., VTC '04-Spring* (Vol. 2, pp. 1503–1507).

Yue, J., Kim, K. J., Gibson, J. D., & Iltis, R. A. (2003). Channel Estimation and Data Detection for MIMO-OFDM Systems. In *Proc. GLOBECOM 2003* (pp. 581–585).

Zhang, L., Gui, L., Qiao, Y., & Zhang, W. (2004). Obtaining diversity gain for DTV by using MIMO structure in SFN. *IEEE Transactions on Broadcasting*, *50*(3), 83–90. doi:10.1109/TBC.2003.822985

ENDNOTES

[1] A unimodular matrix is an integer square matrix whose determinant is one.

[2] Label of a node is defined as the path from the root node to that node.

Collaboration and Capacity

Chapter 29
Capacity Estimation of OFDMA-Based Wireless Cellular Networks

André Carlos Guedes de Carvalho Reis
Universidade de Brasília, Brazil

Paulo Roberto de Lira Gondim
Universidade de Brasília, Brazil

ABSTRACT

The usage of wireless cellular network architecture increases the capacity of a wireless system, by combining cells into clusters in which channels are uniquely assigned per cell and reusing such clusters throughout the network. Unfortunately, a cellular network system may become interference limited regarding its capacity instead of noise/range limited due to intensive resources reuse like time, frequency and space. Using as input the physical layer parameters and deployment scenario, an analytical approach is proposed for capacity estimation of networks based on Orthogonal Frequency Division Multiple Access (OFDMA) technology whose subchannels are composed of distributed subcarriers. This innovative approach is based on a new analytical method for SINR calculation based on a proposed subcarrier collision probability model. The usage of such method is exemplified for a single-hop sectorized Mobile WiMAX cellular network and the results are validated against published works.

INTRODUCTION

The Orthogonal Frequency Division Multiple Access (OFDMA) physical layer technology has been included in some wireless network standards like Digital Video Broadcasting - Return Channel Terrestrial (DVB-RCT) (ETSI EN 301 958, 2002), Worldwide Interoperability for Microwave Access (Mobile WiMAX) (IEEE 802.16e, 2005) and Evolved Universal Terrestrial Radio Access (E-UTRA) (ETSI LTE, 2008). OFDMA is also the candidate access method for the IEEE 802.22 (Wireless Regional Area Network) standard.

OFDMA provides some useful features like efficient usage of spectrum, robustness against narrow-band interference from co-channel cells (Einhaus, 2005), controlled inter-symbol interference, reduced intra-cell interference, non-line-of-sight (NLOS) operation capability and flexible allocation of radio resources.

DOI: 10.4018/978-1-61520-674-2.ch030

This chapter uses an analytical approach to determine the capacity of interference and noise limited OFDMA-based cellular networks. Given the subscribers geographical distribution, cellular architecture (cell size, sectoring and reuse strategy), base station (BS) and terminal characteristics (antenna system, MIMO support, scheduling and power allocation schemes), physical layer specifications (subchannel size, smallest allocable resource unit size, subcarrier permutation mode), scenario (urban, suburban, rural, LOS, NLOS), channel and path loss models, the quality of service (QoS) requirements like BER and delay must be matched providing an affordable quality of experience (QoE) for subscribers.

This work extends the previous work of (Hoymann & Göbbels, 2008) related to the dimensioning and capacity estimation of OFDM-based cellular systems by using an analytical interference method for SINR calculation based on a subcarrier collision probability model (Reis & Gondim, 2009). This method considers the effects of the OFDMA concentration and processing gains into the SINR calculation. It also considers the effect of subscriber station mobility by calculating an uplink correction factor using the same procedure as (Hoymann & Göbbels, 2008) but with slightly different results. Numerical results have already been obtained for DVB-T/RCT (Reis & Gondim, 2006). An example of the application of our model has been presented at the end of the Chapter for a single-hop sectorized Mobile WiMAX network.

BACKGROUND

This section provides basic concepts regarding wireless cellular systems.

Interference

In a cellular system, a large geographical area is divided into many smaller contiguous areas called cells which are served by its own radio base station. Each base station provides transmission resources to handle communication services to many mobile user terminals. Cells separation can occur in the time, frequency and space dimensions, allowing the reuse of transmission resources without producing interference.

Interference is defined as the effect of unwanted signal energy due to a combination of one or more emissions upon reception of the desired signal, manifested by any performance degradation which could be avoided by the absence of such combination. Interference is usually generated by the intensive reuse of transmission resources in the aforementioned dimensions within a cellular system.

Interference Classification

Interference can come from signals transmitted by the same system (intra-system interference) and/or signals transmitted by other systems (inter-system interference). Within a cellular system, interference among cells that use the same channel or adjacent channels is called co-channel interference or adjacent channel interference, respectively. Such interference can emanate from transmissions in the same cell (intracell interference) or in neighboring cells (intercell interference). Figure 1 summarizes this classification.

Usually intercell interference is dominated by co-channel interference and intracell interference between uplink and downlink transmissions can be neglected under the assumption of a proper separation either in the time (time division duplexing) or frequency (frequency division duplexing) dimensions.

Co-Channel Interference

In wireless cellular systems, co-channel interference is one of the main limiting factors for system capacity. System capacity can be expressed as the number of users the system can serve with quality of service (QoS) and quality of experience

Figure 1. Interference classification within a cellular system

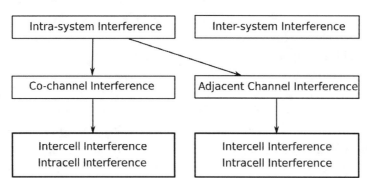

(QoE). It is directly related to the system spectral efficiency measured in Mbps/Hz/cell sector. The usual approach to improve system capacity is to mitigate co-channel interference by using frequency planning, transmission diversity, and/or sophisticated medium access methods like code division multiple access (CDMA) and orthogonal frequency division multiple access (OFDMA). Depending on the cell load, power control and user speed, intracell co-channel interference is nearly cancelled by the last two techniques.

Considering a set of subscriber stations (SS) served by a base station (BS), the uplink and downlink intercell co-channel interference can be calculated by Equations 1 and 2, respectively. Such equations show that the statistical properties of the CINR in both downlink and uplink channels have to be considered in the interference studies (Moiseev et al., 2006).

$$SINR_{UL}(i) = \frac{P_T^{SS(i)} \cdot G_A^{SS(i) \to BS(SS(i))} \cdot G_A^{BS(SS(i)) \to SS(i)} \cdot G_C^{SS(i)}}{\sigma^2 + \sum_{BS(SS(j)) \neq BS(SS(i))} I_j}$$

(1)

$$SINR_{DL}(i) = \frac{P_T^{BS(SS(i))} \cdot G_A^{BS(SS(i)) \to SS(i)} \cdot G_A^{SS(i) \to BS(SS(i))} \cdot G_C^{SS(i)}}{\sigma^2 + \sum_{BS(j) \neq BS(SS(i))} I_j}$$

(2)

where:

$P_T^{SS(i)}$ - transmit power per subcarrier of $SS(i)$ [dBm]

$G_A^{SS(i) \to BS(SS(i))}$ - antenna gain of $SS(i)$ in the direction of its serving BS [dB]

$G_A^{BS(SS(i)) \to SS(i)}$ - antenna gain of serving BS of $SS(i)$ in the direction of $SS(i)$ [dB]

$G_C^{SS(i)}$ - channel gain between $SS(i)$ and its serving BS [dB]

σ^2 - additive white Gaussian noise power relative to the signal bandwidth [dBm]

I_j - power of the received interference signal from $SS(j)$ (UL) or $BS(j)$ (DL) [dBm]

$P_T^{BS(SS(i))}$ - transmit power per subcarrier of serving BS of $SS(i)$ [dBm]

For a subscriber station located at the cell border and assuming that both subscriber and base stations transmit with equal power, use omnidirectional unit gain antennas, and that channel gain includes only path loss with exponentially decaying profile, Equations 1 and 2 can be further simplified to Equations 3 and 4, respectively.

$$SINR_{UL}(i) = \frac{P \cdot \beta \cdot R^{-\alpha}}{\sigma^2 + \sum_{BS(SS(j)) \neq BS(SS(i))} P \cdot \beta \cdot D_j^{-\alpha}}$$

(3)

Table 1. Uplink interference correction in [%] for LOS scenarios according to cluster size (k) and number of sectors per cell (s)

Sectors per cell (s)	Cluster order (k)					
	1	3	4	7	9	12
1	24.87	7.10	5.20	2.90	2.24	1.67
2	-21.50	-7.16	-13.58	-10.12	-9.84	-5.06
3	-41.05	-21.42	-24.30	-18.19	-17.25	-11.80
6	-48.19	-30.58	-30.15	-23.25	-21.88	-17.31

$$SINR_{DL}(i) = \frac{P \cdot \beta \cdot R^{-\alpha}}{\sigma^2 + \sum_{BS(j) \neq BS(SS(i))} P \cdot \beta \cdot (D_j - R)^{-\alpha}}$$

(4)

where:

P - transmitter power per subcarrier of subscriber station (UL) or base station (DL) [dBm]

R - cell radius [m]

D_j - distance from $SS(j)$ to serving BS of $SS(i)$ (UL) or from $BS(j)$ to $SS(i)$ (DL) [m]

α - path loss exponent

β - path loss coefficient

Based on Equations 3 and 4, a criterion can be defined to classify a cellular system as noise/range limited or interference limited based on the σ^2/P ratio:

- $\sigma^2/P \to 0$, system capacity is interference limited.
- $\sigma^2/P \gg \sum_{BS(SS(j)) \neq BS(SS(i))} \beta D_j^{-\gamma}$ and $\sigma^2/P \gg \sum_{BS(SS(j)) \neq BS(SS(i))} \beta (D_j - R)^{-\gamma}$, system coverage and capacity are noise limited.

Uplink Interference Correction

Regarding uplink SINR calculation, a simplification can be done placing all SSs in the center of the co-channel cell and using an uplink correction factor to account the non-linear behavior of the path loss. Considering such behavior, a SS close to the victim cell increases the co-channel interference more than a distant SS decreases it. Tables 1 and 2 summarize for LOS and NLOS scenarios, respectively, such correction factor in percentage of the calculated interference considering all SSs in the cell center. These values have been obtained using the formulation proposed in (Hoymann & Göbbels, 2008) and using the sectorization design of Figure 2.

Adjacent Channel Interference

Adjacent channel interference results from signals that are adjacent in frequency to the desired signal. Such interference is mainly due to imperfect transmission and receiver filters (interference mask) that allow nearby frequencies to leak into receiver pass band.

Considering the subscriber station located at the cell border and given the adjacent channel leakage ratio (ACLR) parameter, equations 5 and 6 allow the calculation of the uplink and downlink adjacent channel interference, respectively.

$$SINR_{UL}(i) = \frac{P \cdot \beta \cdot R^{-\alpha}}{\sigma^2 + \sum_{BS(j) \neq BS(i)} P \cdot ACLR(j) \cdot \beta \cdot D_j^{-\alpha}}$$

(5)

Table 2. Uplink interference correction in [%] for NLOS scenarios according to cluster size (k) and number of sectors per cell (s)

Sectors per cell (s)	Cluster order (k)					
	1	3	4	7	9	12
1	89.64	21.80	15.63	8.53	6.54	4.85
2	-21.99	-4.87	-18.73	-14.46	-14.59	-6.80
3	-57.47	-31.55	-36.74	-28.07	-26.93	-18.46
6	-65.99	-45.38	-44.93	-35.68	-33.81	-27.22

Figure 2. Uplink interfering co-channel cells architecture according to the cellular cluster size (k) and the number of sectors per cell (s)

Cluster size (k)	Number of sectors (s)				Victim cell coordinates		
	1	2	3	6	Id	x_0	y_0
1 (i=1, j=0)					1	0	$\sqrt{3}R$
					2	$\frac{3}{2}R$	$\frac{\sqrt{3}}{2}R$
					3	$\frac{3}{2}R$	$-\frac{\sqrt{3}}{2}R$
					4	0	$-\sqrt{3}R$
					5	$-\frac{3}{2}R$	$-\frac{\sqrt{3}}{2}R$
					6	$-\frac{3}{2}R$	$\frac{\sqrt{3}}{2}R$
3 (i=1, j=1)					1	$-\frac{3}{2}R$	$\frac{3\sqrt{3}}{2}R$
					2	$\frac{3}{2}R$	$\frac{3\sqrt{3}}{2}R$
					3	$3R$	0
					4	$\frac{3}{2}R$	$-\frac{3\sqrt{3}}{2}R$
					5	$-\frac{3}{2}R$	$-\frac{3\sqrt{3}}{2}R$
					6	$-3R$	0
4 (i=2, j=0)					1	0	$2\sqrt{3}R$
					2	$3R$	$\sqrt{3}R$
					3	$3R$	$-\sqrt{3}R$
					4	0	$-2\sqrt{3}R$
					5	$-3R$	$-\sqrt{3}R$
					6	$-3R$	$\sqrt{3}R$
7 (i=2, j=1)					1	$-\frac{3}{2}R$	$\frac{5\sqrt{3}}{2}R$
					2	$3R$	$2\sqrt{3}R$
					3	$\frac{9}{2}R$	$-\frac{\sqrt{3}}{2}R$
					4	$\frac{3}{2}R$	$-\frac{5\sqrt{3}}{2}R$
					5	$-3R$	$-2\sqrt{3}R$
					6	$-\frac{9}{2}R$	$\frac{\sqrt{3}}{2}R$
9 (i=3, j=0)					1	0	$3\sqrt{3}R$
					2	$\frac{9}{2}R$	$\frac{3\sqrt{3}}{2}R$
					3	$\frac{9}{2}R$	$-\frac{3\sqrt{3}}{2}R$
					4	0	$-3\sqrt{3}R$
					5	$-\frac{9}{2}R$	$-\frac{3\sqrt{3}}{2}R$
					6	$-\frac{9}{2}R$	$\frac{3\sqrt{3}}{2}R$
12 (i=2, j=2)					1	$-3R$	$3\sqrt{3}R$
					2	$3R$	$3\sqrt{3}R$
					3	$6R$	0
					4	$3R$	$-3\sqrt{3}R$
					5	$-3R$	$-3\sqrt{3}R$
					6	$-6R$	0

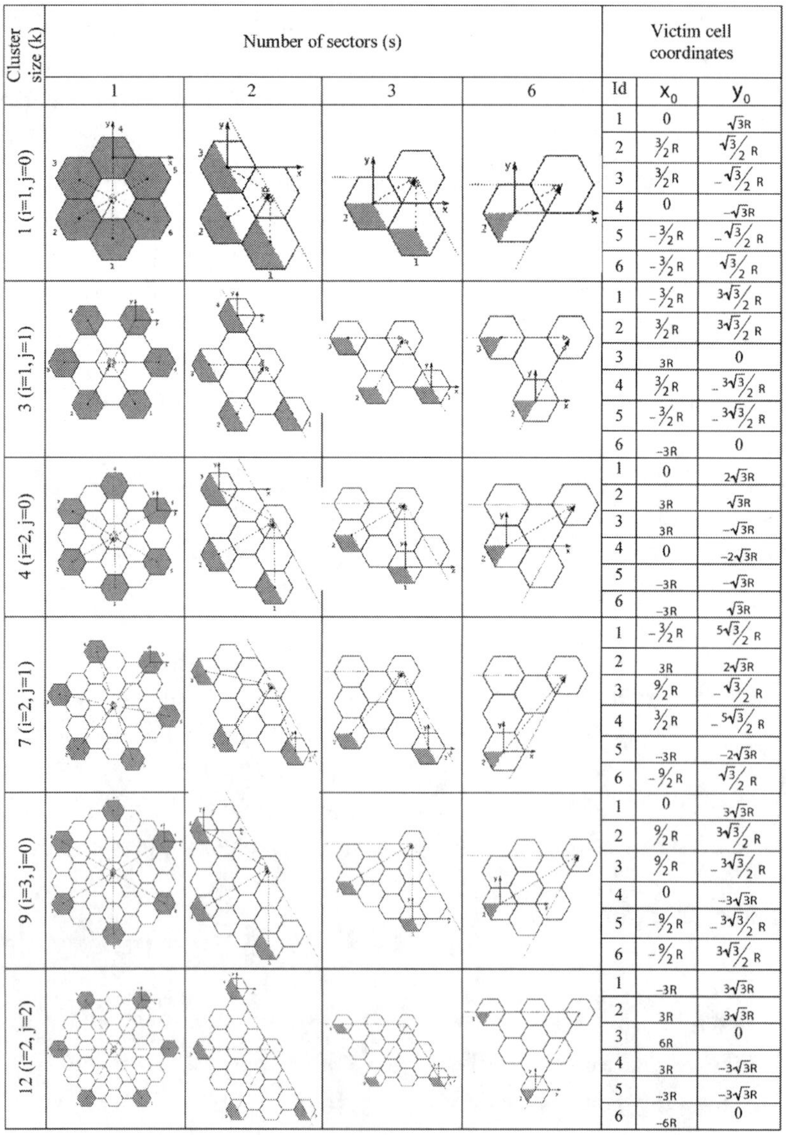

$$SINR_{DL}(i) = \frac{P \cdot \beta \cdot R^{-\alpha}}{\sigma^2 + \sum_{BS(j) \neq BS(SS(i))} P \cdot ACLR(j) \cdot \beta \cdot (D_j - R)^{-\alpha}}$$

(6)

Interference Mitigation Techniques in Cellular Systems

Interference is the major factor limiting capacity of a cellular system. Therefore, several mitigation techniques have been developed for capacity expansion like frequency planning and the use of multiple antennas.

Frequency Planning

Frequency planning techniques comprises cellular architectural change approaches like frequency reuse and cell sectoring using directional antennas. Other relevant frequency planning approaches are cell splitting and Lee's micro cell zone technique (Lee, 1991).

Frequency Reuse

Observing the huge demand for capacity and considering that the radio spectrum is one of scarcest resources available, reuse the spectrum so that several users separated by a distance can use the same frequency band is the basic approach for its efficient utilization. In this approach, radio channels are allocated to cells and frequency bands have to be divided into chunks and distributed among the cells of a cluster, that utilizes the entire available radio spectrum. The reason for clustering is that adjacent cells cannot use the same frequency spectrum because of interference.

Considering that the effects of interference are variable with distance, cells that use the same frequency band (co-channel cells) must be placed as far apart as possible for a given cluster size, being this distance named reuse distance (D). Thus, considering a cellular system modeled as a grid of hexagonal cells with a central BS, an interesting relationship holds between cluster size (k), the reuse distance (D) and the cell radius (R): $D = R\sqrt{3k}$. Values for k must obey the form $i^2 + j^2 + ij$, where i and j are integers.

Some other reuse-based techniques have been proposed in order to improve the capacity of a cellular system, such as reuse partitioning (Zander, 1993) and fractional reuse (Fujii, 2008).

Cell Sectoring Using Directional Antennas

Each cell is covered by one, two, three or six sectors but, regarding interference calculation for planning purposes, usually only the first tier of six, three, two or one interfering co-channel cells should be considered (Hoymann & Göbbels, 2008), respectively. In an interference limited cellular system, the reuse strategy should allow the increase of carrier to interference plus noise ratio (CINR) in the victim cell.

Multiple Antenna

The usage of multiple antenna techniques can further improve the performance of a wireless communications system by:

- Increasing its reliability or by decreasing its required transmit power due to spatial diversity gain (creation of multiple parallel channels for carrying dependent coded data streams);
- Reducing its interference to/from other nodes (beamforming);
- Increasing the transmission data rate through spatial multiplexing (creation of multiple parallel channels for carrying independent data streams)

OFDMA-Based Cellular Systems

The favorable characteristics of Orthogonal Frequency Division Multiplexing (OFDM) modulation technique for wireless communications systems motivated the development of an Orthogonal Frequency Division Multiple Access (OFDMA) scheme. Basically, in OFDMA, the OFDM symbol is divided into subsets of subcarriers, referred to as subchannels, and allocated during a specific time interval to a system node with an active MAC layer connection. The smallest allocable resource unit is denoted slot (DVB-RCT/WiMAX) or resource block (LTE) and expressed as a number of OFDMA symbols per number of subchannels. Within DVB-RCT and WiMAX standards, slot size varies according to medium access schemes (DVB-RCT, 2002) or operating modes (IEEE 802.16e, 2005), respectively.

In an OFDMA system, the subcarrier orthogonality condition requires that the subcarrier spacing C_s should be the inverse of the useful symbol time T_u and the effective channel bandwidth is given by $B = C_s \times N_{used}$. The OFDMA symbol duration is $T_s = T_u + T_g$, where $T_g = G \times T_u$ is the guard time that can completely eliminate Inter-Symbol Interference (ISI) as long as its duration is longer than the channel delay spread. The drawback of increasing the guard time is that it effectively reduces system spectrum efficiency and frequency offset immunity.

Subcarrier Permutation Modes

Regarding subcarrier permutation modes, both WiMAX and DVB-RCT standards have defined for mobile applications a distributed subcarrier permutation (DSP) mode to assign subcarriers to subchannels. In this mode, subchannels are equally adequate to all subscribers because are composed of subcarriers randomly distributed across the entire channel bandwidth (frequency diversity). For fixed, nomadic and low-mobility applications, WiMAX and E-UTRA also defines adjacent subcarrier permutation (ASP) mode where subchannels are constituted of contiguous subcarriers. Such subchannels allows multiuser diversity gains by means of adaptive modulation and coding (AMC) based on channel response estimation (Sternad, 2007) and inter-cell interference coordination policy (Fodor, 2008).

Resource Allocation

Concerning transmission resource allocation, an OFDMA-based cellular network allows that a BS node reserves resources, acting as a scheduler that allocates slots to subscriber stations, with the goal of maximizing throughput while satisfying demands of individual users with fairness. We note that scheduling techniques have not been exactly specified in some OFDMA systems (for example, WiMAX).

For IEEE 802.16 and 802.22 networks, the allocation of subcarriers can involve Partial Usage of Subchannel (PUSC) and Full Usage of Subchannels (FUSC). In FUSC all subcarriers have to be allocated in one cell or sector, while in PUSC, depending on the traffic conditions and to reduce interference, a set of the subcarriers can be allocated to one cell or sector. In order to form a subchannel and considering the use of DSP for addressing mobility, the diversity permutations include DL (downlink) FUSC, DL PUSC and UL (uplink) PUSC and additional optional permutations (IEEE 802.16e, 2005).

Concentration Gain

Usually, it is not a trivial task to design optimum subchannel and slot size that fully exploits channel frequency and time diversity (Sternad, 2007). Depending on design guidelines and goals, an affordable coverage with adequate net bit rate granularity for different applications (fixed, nomadic or mobile) and propagation scenarios (LOS or NLOS) can be obtained.

The concentration gain depends on the maximum number of simultaneous subchannels that can be allocated to a subscriber. Within an OFDMA-based cellular network, it is an important parameter for the dimensioning the cellular network because the total transmit power of the subscriber (UL) or allocated to subscriber (DL) must be divided among them. The effect of this operation is a subchannelization/concentration gain that reduces interference and increases signal power spectral density. The maximum concentration gain is $10\log_{10}(N_{sch})$, where N_{sch} is the number of subchannels.

The maximum number of simultaneous subchannels that can be allocated to a subscriber should also be defined based on QoS requirements because it determines the scheduling of transmission resources and the maximum data rate available to the applications.

Capacity

In order to guarantee a fair access to the transmission resources, a scheduler at the base station shall allocate slots to a SS based on its actual modulation and coding scheme (MCS). Therefore, usually more slots will be allocated to users located on the cell border due to the small capacity of its used MCS. As the MCS depends on the position of the SS relative to its serving BS, the capacity of a cell will depend on the subscriber's geographical distribution.

Several subscribers' geographical distribution model exists. Usually, for small cells covering small regions like streets, parks and towns, the uniform distribution is assumed. For large cells covering metropolitan regions, exponentially-based distributions may be used. Given the MCS distribution within a frame, the capacity can be calculated by a weighted mean of the transmission bit rate per MCS.

Channel Models

As the electromagnetic wave carrying the signal information travels from the transmitter antenna to the receiver antenna, several propagation medium characteristics and obstacles expose the wave to reflections, scattering and diffractions effects, causing the received wave to be a composition of several wave copies with different direction of arrival (multipath) and/or to have its energy reduced due to the distance between transmitter and receiver (path loss) as well as due to the relative receiver location (shadowing). Therefore, a channel model has three components: path loss, shadowing, and multipath. These components are random and only a statistical characterization is possible, being typically the first order statistics specified (mean and variance).

Path Loss Models

Several path loss models can be used to calculate the received signal level (Erceg, 1999) (IST-WINNER, 2006) and (ITU-R P.1546-3, 2007). The last model is valid only for distances ranging from 1km to 1000 km and, therefore, is useful only for large cells. Depending on the adopted model and cellular architecture, a different MCS distribution can be obtained.

Shadowing Model

If there are any objects (such buildings or trees) along the path of the signal, some part of the transmitted signal is lost through absorption, reflection, scattering, and diffraction. This effect is called shadowing. As a result of shadowing, power received at the points that are at the same distance d from the transmitter may be different and have a lognormal distribution. This phenomenon is referred to as lognormal shadowing.

Multipath Model

There are two commonly used multipath patterns models depending on the existence or not of the direct path: Ricean and Rayleigh. The Ricean model accounts for fixed Line-Of-Sight (LOS) reception, whereas the Rayleigh accounts for portable and mobile Non-Line-Of-Sight reception.

ANALYTICAL INTERFERENCE MODEL FOR OFDMA-BASED CELLULAR NETWORKS

Regarding inter-cell interference, an OFDMA-based cellular network behaves very much like a FHSS system affected by partial band jammer (Koffman, 2002). In this sense, subcarrier collisions in the victim cell due to the usage of the same subcarrier in an adjacent co-channel cell may increase the average number of subcarriers interfered per subchannel, which divided by the subchannel size, defines the processing gain of an OFDMA system. Due to temporal interleaving and error-correction coding, data from these interfered subcarriers (with low SINR) can be corrected. The impact of subcarriers collisions on the slot or data region capacity will depend not only on the average SINR per subcarrier but also on slot physical profile *i.e.*, modulation and coding scheme (Moiseev, 2006).

Collision Probability Model

An analytical subcarrier collision model for OFDMA-based cellular networks has been proposed in (Reis & Gondim, 2009). Using this model, the probability distribution function for the number of subcarriers collisions (interfered) per subchannel can be stated by Equation 7. Such distribution depends not only on the number of co-channel cells as well as their load, i.e., the expected number of simultaneously allocated slots, which can span one or more subchannels.

$$P_M^L(k) = \sum_{i=0}^{k} P_{M-1}^L(i) \times \Lambda_1^L(k'|i) \quad (7)$$

with:

$k' = k - i$

$$\Lambda_1^L(k'|i) = \sum_{\substack{A^L(k') \\ 0 < L < N_{sch}}} \left\{ \prod_{l=0}^{L-1} \lambda_1^l \left[k'(l) \left| \left(\sum_{i=0}^{l-1} k'(i) \right) u[l-1], i \right. \right] \right\}$$

$$\lambda_1^L(k'|c,i) = \binom{N_G - i - c}{k'} \left(\frac{1}{N_{sch} - l} \right)^{k'} \left(1 - \frac{1}{N_{sch} - l} \right)^{N_G - i - c - k'}$$

$$P_1^L(k) = \sum_{\substack{A^L(k') \\ 0 < L < N_{sch}}} \left\{ \prod_{l=0}^{L-1} p_1^l \left[k(l) \left| \left(\sum_{i=0}^{l-1} k(i) \right) u[l-1] \right. \right] \right\}$$

$$p_1^l(k|c) = \binom{N_G - c}{k} \left(\frac{1}{N_{sch} - l} \right)^k \left(1 - \frac{1}{N_{sch} - l} \right)^{N_G - c - k}$$

where:

$P_M^L(k)$ - probability of k subcarriers collisions per subchannel for load L and M co-channel cells

$A^L(k)$ - set of additive partitions of k with L non-negative integers terms

k' - number of new collisions generated by the M-th co-channel cell added to the system

N_G - number of subcarriers per subchannel

N_{sch} - number of subchannels

$u[l-1]$ - discrete unit step function with $u[l-1] = 1$ for $l \geq 1$, and zero otherwise.

Analytical Interference Model

The average SINR per subchannel can be calculated using Equation 8 and 9.

$$SINR_M^L = \sum_{k=0}^{N_G} SINR_M^{sch}(k) \cdot P_M^L(k) \quad (8)$$

$$SINR_M^{sch}(k) = \frac{(N_G - k) \cdot SNR^{sca} + k \cdot SINR_M^{sca}(k)}{N_G} \quad (9)$$

where:

$SINR_M^{sch}(k)$ - average subchannel signal to interference plus noise ratio for M co-channel cells and k collisions

SNR^{sca} - signal to noise ratio of a subcarrier not interfered (no collision)

$SINR_M^{sca}(k)$ - average subcarrier signal to interference plus noise ratio for M co-channel cells and k collisions

The formula to calculate the average subcarrier signal to interference plus noise ratio depends on the number of collisions (k) and number of interfering co-channel cells (M). For k=0, $SNR_M^{sca}(k) = SNR^{sca}$ and $SNR^{sca} = S/N$.

For k=1, the SINR can be expressed using the binomial expansion where each product factor represents a cell randomly choosing a subcarrier from a set with cardinality N_{sch} (Reis & Gondim, 2009). The probability of choosing exactly the same subcarrier is $1/N_{sch}$.

$$(x+1)^M = \binom{M}{0} + \binom{M}{1}x + \binom{M}{2}x^2 + \ldots + \binom{M}{i}x^i + \ldots + \binom{M}{M}x^M \quad (10)$$

The conditional probability of i collisions ($i \geq 1$) is given by Equation 11 which corresponds to the ratio between the binomial probability and the total probability for at least one collision. The expected value of $SINR_M^{sca}(1)$ is given Equation 12.

$$Q_i(M) = \frac{\binom{M}{i}p^i(1-p)^{M-i}}{\sum_{i=1}^{M}\binom{M}{i}p^i(1-p)^{M-i}} = \frac{\binom{M}{i}p^i(1-p)^{M-i}}{1-(1-p)^M} \quad (11)$$

$$SINR_M^{sca}(1) = \sum_{i=1}^{M} Q_i(M) \cdot \frac{S}{i \cdot I + \sigma^2} \quad (12)$$

For k>1, the binomial expansion has the form of Equation 13 where the product terms are independent.

$$\left[(x+1)^M\right]^k = \underbrace{(x+1)^M \cdot (x+1)^M \cdots (x+1)^M}_{k \text{ factors}} \quad (13)$$

Equation 14 gives the expected value of $SINR_M^{sca}(k)$. The double sum is done for all additive partitions $A^{M \cdot k - i - k}(k) | \forall k_i < M$, which represent the additive terms obtained after multiplication of k factors with at least one collision per factor. Because subcarrier choices are independent among co-channel interfering cells, the product of $Q_{k_j+1}(M)$ probability follows. Such probability must be multiplied by the average value of SINR over k subcarriers.

$$SINR_M^{sca}(k) = \sum_{i=0}^{M \cdot k - k} \sum_{A^{M \cdot k - i - k}(k) | \forall k_i < M} \left[\prod_{j=0}^{k-1} Q_{k_j+1}(M) \cdot \frac{\sum_{j=0}^{k-1}\frac{S}{(k_j+1)I+\sigma^2}}{k} \right] \quad (14)$$

MOBILE WIMAX EXAMPLE

An important feature of Mobile WiMAX OFDMA physical layer is that subcarrier spacing has been fixed at 10.94 kHz. Such design choice provides a good balance between range and mobility considering typical delay spread and Doppler spread of fixed and mobile environments. Using fixed subcarrier spacing implies that FFT size is scalable from 128 to 2,048 FFT when the available channel bandwidth increases from 1.25MHz to 20MHz, respectively. Furthermore, fixed subcarrier spacing keeps the OFDM symbol duration constant, thus minimizing the impact of the temporal granularity of resource allocation on higher layers applications. WiMAX also has a wide range of guard times allowing system designers to make appropriate choice between spectral efficiency and delay spread robustness.

The WiMAX link budget parameters are discussed below. Most parameters values have

Table 3. Base station downlink and uplink parameter values, according to its profile

Specification	Base Station Profile		
	Standard	MIMO 2x2	MIMO 2x2 AAS
Downlink tx power [dBm]	35	35	35
Downlink tx antenna gain [dBi]	16	16	16
Downlink tx diversity gain [dB]	0	9	9
Downlink tx adaptative antenna gain [dB]	0	0	6
Uplink rx antenna gain [dBi]	16	16	16
Uplink rx diversity gain [dB]	0	3	3
Uplink rx adaptative antenna gain [dB]	0	0	3
Uplink rx noise figure [dB]	5	5	5

been obtained from (WiMAX forum, 2007) considering today available technologies. Other parameters values have been borrowed from the WiMAX standard.

Base Station and Subscriber Station Specifications

Several base station and terminal characteristics related to cellular network dimensioning are discussed in this section. There are three BS and two SS profiles which have been summarized on Table 3 and Table 4. Some other parameters have not been taken into account like feeder, duplexer and cable losses. Only BS standard profile and SS mobile profile have been considered for simulation.

Receiver Sensitivity

The Equation 15 is used to define the required receiver sensitivity levels ($P_{R_{min}}$) considering only the power of data subcarriers. The channel bandwidth should be divided by the number of repetitions (R) if data repetition is used.

$$P_{R_{min}} = \underbrace{NF + 10\log_{10}(kT_0B) + 30}_{P_N} + L_{impl} + SINR$$

(15)

where:

P_N - receiver noise input power [dBm]
NF - receiver noise figure [dB]
k - Boltzmann constant ($1.3806504 \times 10^{-23}$ Ws/k)
T_0 - absolute temperature (290 K)
B - bandwidth [Hz]
L_{impl} - implementation loss (2 dB)
$SINR$ - signal to interference plus noise ratio [dB]

For dimensioning purposes, the minimum receiver SINR values from/at a SS located at the cell border must be known. The Table 5 summarizes the derived SINR values obtained in an additive white Gaussian noise (AWGN) channel with a bit error rate (BER) of 10^{-6} using two different forward error correction (FEC) codes. The simulation has adopted tail-biting convolution code values.

The MCS used by each user will depend on the relative position among its subscriber station, serving base station and interfering nodes. As SINR falls with distance between SS and BS, the most robust MCS will be used on the cell border when the BS is at the centre.

The implementation loss includes non-ideal receiver effects like channel estimation errors, adjacent channel leakage, phase noise, etc. The value suggested for IEEE 802.16e is 5 dB. Within WiMAX standard, four primitive parameters are

Table 4. Subscriber station downlink and uplink parameter values, according to its profile

Specification	Subscriber Station Profile	
	Portable/Nomadic	Mobile
Uplink tx power [dBm]	27	27
Uplink tx antenna gain [dBi]	6	0
Downlink rx antenna gain [dBi]	6	0
Downlink rx noise figure [dB]	7	7

used to define an OFDMA symbol: the nominal channel bandwidth B_{nom}, the number of used subcarriers N_{used}, the sampling factor n and the ratio G between the guard time T_g and useful symbol time T_u.

The sampling factor parameter depends on the nominal channel bandwidth. Its value is set to 28/25 for channel bandwidths that are a multiple of any of 1.25, 1.5, 2 or 2.75 MHz and 8/7 for channel bandwidths that are multiple of 1.75 MHz or other not otherwise specified. Table 6 shows the uplink and downlink parameters according to the nominal bandwidth. The simulation results refer to 1.25 MHz uplink and downlink parameters. The guard time 1/8 and a 5 ms frame has been used. The subcarrier spacing is given by F_s/N_{FFT} where F_s is the sampling frequency in MHz which is given by Equation 16.

$$F_s = \frac{\left\lfloor \frac{n \cdot B_{nom}}{8000} \right\rfloor}{10^6} \quad (16)$$

where:

B_{nom} - nominal bandwidth [MHz]
n - sampling factor

Margins

To calculate the link budget several margins can be considered like fade margin, interference margin and location correction factor margin. Interference margin for OFDMA-based cellular systems are load dependent (Ferneke β, 2008). The simulations have considered no margins.

Table 5. SINR values in dB according to the FEC code and MCS used

Modulation and Coding Scheme (MCS)	FEC code	
	Tail-biting Convolution Code	Convolution Turbo Code
QPSK 1/2	5	2.5
QPSK 3/4	8	6.3
16QAM 1/2	10.5	8.6
16QAM 3/4	14	12.7
64QAM 1/2	16	13.8
64QAM 2/3	18	16.9
QAM 3/4	20	18

Table 6. Uplink and downlink number of used subcarriers N_{used}, number of subcarriers per subchannel N_G and number of subchannels per channel N_{sch} within a PUSC frame zone

N_{FFT}	B_{nom} [MHz]	Uplink			Downlink		
		N_{used}	N_G	N_{sch}	N_{used}	N_G	N_{sch}
128	1.25	97		4	85		3
512	5	409	16	17	421	24	15
1024	10	841		35	841		30
2048	20	1681		70	1681		60

Maximum Cell Radius

Due to the fact that BS always has more transmission resources on the downlink to reach the SS within its coverage area, only the uplink direction has been considered for cell dimensioning. The same interference model used for the uplink could be used one by one to the downlink. The uplink interference correction has been taken into account.

The maximum cell radius has been obtained for the most distant SS located at the cell border using the link budget expressed by Equation 17.

$$\text{SINR} = P_t + G_C - (P_N + L) \quad (17)$$

where:

SINR - average signal to noise plus interference ratio per subchannel [dB]
P_t - transmit power [dBm]
G_C - concentration gain [dB]
P_N - receiver noise input power [dBm]
L - path loss [dB]

According to Equation 8, *SINR* depends on the load and the number of interfering co-channel cells. The parameters P_t, G_C and P_N should consider the power per subcarrier obtained after dividing the maximum transmit power by the maximum number of simultaneously allocated subchannels per subscriber.

Figure 3 and Figure 4 show the uplink SINR relative to a SS located on the cell border within a single sector cellular system according to propagation scenario, co-channel cell load, cluster order and cell radius without and with concentration gain, respectively. It can be observed that whenever the load increases the maximum cell radius decreases due to co-channel cell interference increase on the victim cell. Moreover, observing the influence of the concentration gain on the maximum cell radius, it can be noted that increasing the concentration gain, the maximum cell radius also increases. In the first case, the MCS distribution remains the same; in the second case, there is a percentual increase of the most robust scheme QPSK ½ (not shown).

Figure 3 and Figure 4 show that cluster order one do not lead to a sufficient SINR for 50% and 75% co-channel cell load. It can be observed that in both LOS and NLOS scenarios the level of interference relative to the noise level is so low that there is a small variation of the maximum cell radius for cluster orders greater than one. Therefore, the system is noise limited for cluster order greater than or equal to three and interference limited between one and three. It has also been observed that there is almost no significative difference between uplink SINR variation with cell radius and cluster sizes relative to 75% and 100% co-channels cells load (not shown). Note that the 100% load corresponds to a fully loaded OFDM system.

Figure 3. Uplink SINR relative to a SS located on the cell border within a single sector cellular system without concentration gain according to propagation scenario, co-channel cell load, cluster order and cell radius

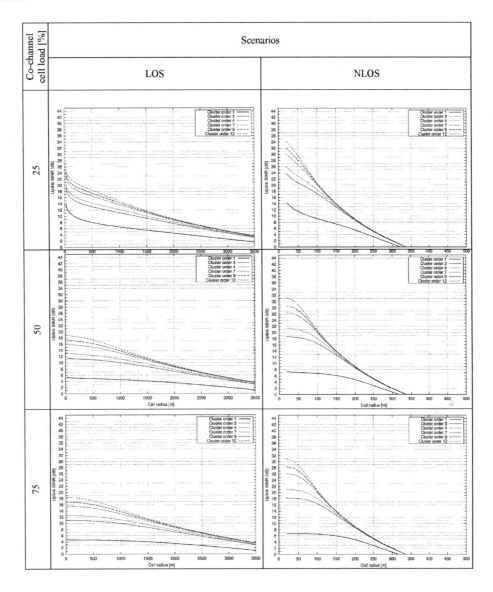

Cell Capacity

After calculating of the maximum cell radius, the distribution of SINR and, consequently, of MCS within the cell coverage area can be obtained varying the SS position. Figure 5 and Figure 6 illustrate the SINR distributions for LOS and NLOS scenarios, respectively.

Based on this distribution, the average cell capacity can be calculated using a weight mean of the average bit rate of each MCS. The weights will be the percentages summarized on Table 7.

Calculating the weighted mean, the average bit rates per cell sector of 1,350,155.52 bps and 1,911,755.52 bps have been obtained for LOS and NLOS scenarios, respectively. Dividing these

Figure 4. Uplink SINR relative to a SS located on the cell border within a single sector cellular system with concentration gain of 6 dB according to propagation scenario, co-channel cell load, cluster order and cell radius

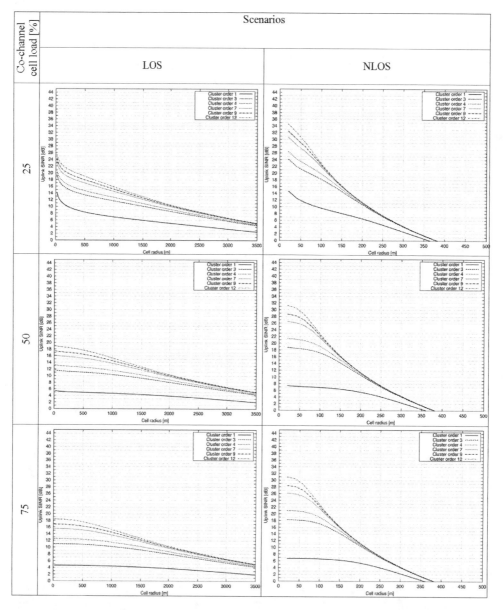

values by the cell nominal bandwidth, a spectral efficiency of 1,08 bit/s/Hz and 1,53 bit/s/Hz have been obtained for LOS and NLOS scenarios, respectively. Such results are very similar to those published on (Andrews, 2007).

CONCLUSION

This chapter has presented an analytical interference model for OFDMA-based cellular networks using DSP subcarrier assignment mode. Such model has been applied to determine the uplink capacity of an OFDMA-based wireless cellular

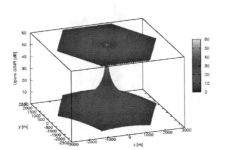

Figure 5. Uplink SINR distribution within the victim cell of a single sector cluster order three with a radius of 2280m and co-channel cell load of 25% for LOS scenario

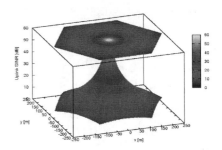

Figure 6. Uplink SINR distribution within the victim cell of a single sector cluster order three with a radius of 240m and co-channel cell load of 25%. NLOS

Table 7. Parameters values for capacity calculation of a 5 ms frame with 72 data subcarriers grouped into four subchannels with duration of 48 OFDM symbols

Parameter Specification	MCS profile						
	QPSK ½	QPSK ¾	16QAM ½	16QAM ¾	64QAM ½	64QAM ²/₃	64QAM ¾
LOS Percentage	29.70	25.25	21.75	7.47	5.08	3.44	7.30
NLOS Percentage	7.20	21.19	22.04	9.49	7.89	6.59	25.56
Bits per subcarrier	1	1.5	2	3	3	4	4.5
Data bit rate [bps]	691200	1036800	1382400	2073600	2073600	2764800	3110400
Number of slots LOS	58	49	42	15	10	7	11
Number of slots NLOS	14	41	43	19	16	13	46

network through the determination of its uplink SINR, considered as a bottleneck for the overall system performance. The methodology followed here can easily be applied one by one for the downlink SINR and capacity calculation.

An example of the use of the model for determining the capacity of a Mobile WiMAX network was presented for a single-hop sectorized case, including several cluster sizes and concentration gain values.

The results indicate that, depending on architectural parameters of the OFDMA-based cellular network, it is possible to devise situations in which the capacity and coverage of the system can be limited by noise or by interference.

Further research is necessary to extent this work for ASP mode and single carrier OFDMA.

REFERENCES

3GPP(2008). TS 36.211, Evolved Universal Terrestrial Radio Access: Physical Channels and Modulation, v8.4.0 ed.

Amzallag, D., Livschitz, M., Naor, J., & Raz, D. (2005). Cell Planning of 4G Cellular Networks: Algorithmic Techniques and Results. In *Proceedings of the 6th IEE International Conference on 3G and Beyond*, (pp. 1–5).

Andrews, J. G., Ghosh, A., & Rias, M. (2007). *Fundamentals of WiMAX: understanding broadband wireless networking.* Upper Saddle River, NJ: Pearson Education, Inc.

Einhaus, M., Klein, O., & Lott, M. (2005). Interference Averaging and Avoidance in the Downlink of an OFDMA System. In *Proceedings of 16th IEEE International Symposium on Personal, Indoor and Mobile Radio Communications,* (2), 905–910.

Erceg, V. (1999)... *IEEE Journal on Selected Areas in Communications, 17*(7), 1205–1211. doi:10.1109/49.778178

ETSI EN 301 958 (2002). *Digital Video Broadcasting (DVB): Interaction channel for digital terrestrial television (RCT) incorporating multiple access OFDM.*

Fernekeß, A., Klein, A., Wegmann, B., Dietrich, K., & Litzka, M. (2008). Load Dependent Inteference Margin for Link Budget Calculations of OFDMA Networks. *IEEE Communications Letters, 12*(5), 398–400. doi:10.1109/LCOMM.2008.080032

Fodor, G., Telek, M., & Koutsimanis, C. (2008). Performance analysis of scheduling and interference coordination policies for OFDMA networks. *Computer Networks, 52*(6), 1252–1271. doi:10.1016/j.comnet.2008.01.006

Fujii, H., & Yoshino, H. (2008). Theoretical Capacity and Outage Rate of OFDMA Cellular System with Fractional Frequency Reuse. In *Proceedings of IEEE Vehicular Technology Conference,* (pp. 1676–1680).

Hoymann, C., & Göebels, S. (2008). Dimensioning cellular multi-hop WiMAX networks. In K. Chen & J. R. B. de Marca (Ed.), *Mobile WiMAX,* (pp. 203-234). New York: John Wiley & Sons, Ltd.

IEEE 802.16e (2005). *IEEE Std 802.16e-2005, Part 16: air interface for fixed and mobile broadband wireless access systems,* Feb. 28, 2006.

IST-WINNER. (2006). *D5.4, Final Report on Link Level and System Level Channel Models. IST-4-027756 WINNER II.*

ITU-R. (2003). *P.1546-3. Method for point-to-area predictions for terrestrial services in the frequency range 30 MHz to 3000 MHz.*

Koffman, I., & Roman, V. (2002). Broadband Wireless Access Solutions Based on OFDM Access in IEEE 802.16. *IEEE Communications Magazine, 40*(4), 96–103. doi:10.1109/35.995857

Lee, W. C. Y. (1991). Smaller cells for greater performance. *IEEE Communications Magazine, 29*(11), 19–23. doi:10.1109/35.109660

Moiseev, S. N., Filin, S. A., Kondakov, M. S., Garmnonov, A. V., Yim, D. H., Lee, J., et al. (2006). Analysis of the statistical properties of the interference in the IEEE 802.16 OFDMA network. In *IEEE Wireless Communications & Networking Conference,* (pp. 1830-1835).

Reis, A. C. G. C., & Gondim, P. R. L. (2006). Performance Evaluation of DVB-RCT Standard. In *Proceedings of IEEE International Symposium on Broadband Multimedia Systems and Broadcasting,* Las Vegas, NV, April 2006.

Reis, A. C. G. C., & Gondim, P. R. L. (2009). *The OFDMA technology and its application in the in-band interaction channel of the terrestrial digital television systems.* PhD Thesis, Universidade de Brasília, February 2009.

Sternad, M., Svensson, T., Ottosson, A., Ahlen, A., Svensson, A., & Brunstrom, A. (2007). Towards systems beyond 3G based on adaptive OFDMA transmission. In *Proceedings of IEEE. Special Issue on Adaptive Transmission, 95*(12), 2432-2455.

WiMAX Forum. (2007). *WiMAX system evaluation methodology v.2.1.* Retrieved from http://www.wimaxforum.org/documents

Zander, J. (1993). Generalized reuse partitioning in cellular mobile radio. In *Proceedings of IEEE Vehicular Technology Conference*, (pp. 181–184).

Chapter 30
Wireless Collaboration:
Maximizing Diversity through Relaying

Patrick Tooher
Concordia University, Canada

M. Reza Soleymani
Concordia University, Canada

ABSTRACT

To achieve performance gains in the wireless channel, spatial diversity is employed. These higher order transmit diversity gains generally require multiple transmit antennas at the source. This requirement is not always possible in real world applications, where practical concerns limit the number of antennas a wireless device can have. Recently, a new method to achieve transmit diversity has been proposed: collaborative communications. In this framework, a node in a wireless network can use the resources of other idle nodes and form what can be viewed as a virtual transmitting antenna array. This chapter presents an overview of the development of collaborative communications. Two-phase protocols that can achieve collaboration are presented. A discussion on the improvement of collaborative communications protocols is given. A broader perspective of collaborative communications is given by discussing ideas such as power allocation and multiple relays.

INTRODUCTION

Multiple-input multiple-output (MIMO) communication (Foschini, Gans, 1998) is method by which several transmitting and receiving elements are used at the source and destination to achieve high spectral efficiency in wireless communications by using available spatial diversity. Spatial diversity is obtained by transmitting the same data through independent fading channels given by the multiple transmitting and receiving elements. The idea of MIMO led to the development of so-called space-time codes (Tarokh, Seshadri & Calderbank, 1998). MIMO technologies have been implemented in many existing wireless communication technologies (i.e. Wi-Fi), and is planned to be included in many future systems such as WiMAX.

Recent work has shown that traditional MIMO falls under a much bigger umbrella referred to as multi-user (MU) MIMO. This includes any tech-

DOI: 10.4018/978-1-61520-674-2.ch031

nology which provides spatial diversity in the wireless channel by allowing adjacent network nodes to collaborate. It is in this realm that we find collaborative communications – a system which allows transmitting nodes, each with a single antenna, to collaborate with nearby idle nodes to achieve similar performance as using multiple transmit antennas.

Collaborative communications is an attractive technology for the future deployment of 4G networks since it helps achieve several of the objectives set forth. Increasing spatial diversity allows one to improve the spectral efficiency (in b/s/Hz) of a system. For 4G to achieve its high data rate targets, spectrally efficient technologies, such as collaborative communications, are a must.

Fourth generation networks are expected to have seamless connectivity across multiple networks. Collaborative communication can be considered a component of ad hoc networking, where communication need not be point-to-point. As such, using collaboration can be a tool used to achieve connectivity between different networks.

Another objective of 4G networks is to achieve high data rate between any two points in the world. Having nearby nodes collaborating with each other significantly increases their useable range. Collaborative communications can also be deemed one step in a multi-hop system. Such a system can further ensure connectivity and thus achieve the "anytime, anywhere" credo of 4G networking.

It is because of all of the above that relaying concepts using some form of collaboration have been proposed to be included in future 4G technologies. Newer releases of the LTE standard (LTE Advanced) have discussed the use of various concepts for relay nodes. This is also true for future generations of WiMAX such as WiMAX-m.

The purpose of this chapter is to introduce the reader to the concept of collaborative communications and provide a mathematical framework to perform analysis of collaborative communications. First, some basic principles of collaboration are discussed: things such as decode-and-forward, coded collaboration, variable time-fraction, etc… This allows the reader to get a better understanding of the challenges presented in collaborative communications. Next we include more recent results on the performance of collaborative communications. From these the reader can better appreciate the value of using collaborative communications.

Using the tools provided in the discussion on performance, we provide some more advanced methods to further improve the performance of collaborative communications. These include using power allocation algorithms and multiple relaying nodes.

The topics discussed in this chapter provide a tutorial overview of collaborative communications. There are many complicated issues that must be addressed before collaboration can be implemented practically: the type of collaboration used, symbol timing, transmitting and receiving architecture of the wireless nodes, etc… Some of these are addressed in this chapter, and the interested reader is referred to the future work section and to the list of additional reading for more information.

The chapter is presented as follows: In section II, a review of the work which led to the development of collaboration is presented. Section II also provides the basic assumptions as well as basic results required to fully appreciate the benefits of collaboration. The use of two-phase protocols to achieve collaboration is presented in section III. Protocols such as Detect-and-Forward, Amplify-and-Forward, Decode-and-Forward as well as variable phase length and collaborative coding, are examined in this section.

The next two sections provide performance results of collaborative communications. Section IV provides results for both a coded collaborative communications scheme using variable time-fraction, as well as a higher order modulation collaborative communications scheme. In section

Figure 1. Collaborative communications

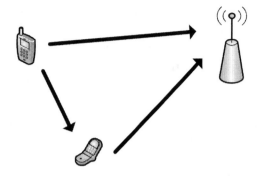

V, these two schemes are revisited and modified to use power allocation methods to further improve the performance. In section VI, the use of multiple relays in collaborative communications is discussed. Section VII provides some future work that deal with open problems in the field of collaborative communications. We conclude the chapter with section VIII, and then offer some references as well as some additional reading.

BACKGROUND

Even though transmit diversity is clearly advantageous in wireless communications, some practical issues arise when attempting to fully exploit it. Most wireless devices are built for some level of mobility. In the cellular sense this is obvious, since a mobile handset needs to be able to function while in motion. However, mobility is also implied in applications such as Wi-Fi; users expect to have terminals (i.e. laptops) that can be easily transported to different locations. The requirement for mobility and transportability is an obstacle to having multiple transmitting or receiving elements on wireless equipment. That is because mobility and transportability impose size limits on the wireless terminals. On the other hand, to ensure that the signals transmitted by different antenna elements are not correlated, restrictions on the closeness of the elements are required – thus increasing the required size. Another factor that hinders an increase in the number of transmitting elements is the required power. For example, nodes in wireless sensor networks are constrained by their available power. While using MIMO reduces the necessary transmission power, this can be offset by an increase in complexity of internal circuitry.

To alleviate these problems, collaborative communication, also referred to as cooperative communication, has been proposed. The basic premise of collaborative communication is that in a network filled with wireless nodes, a transmitting node with a single antenna may be aided by otherwise idle nodes to emulate a MIMO channel.

Collaborative communications has origins in the groundbreaking work of van der Meulen (1968; 1971) on three-terminal communication channels. Cover and El Gamal (1979) determined the relay channel's information theoretic properties by developing lower bounds on capacity, or achievable rates.

The work by Cover and El Gamal was not developed specifically for the wireless channel. For example, the capacity bounds are for the class of physically degraded relay channels. This implies the destination receives a corrupted version of what the relay receives. Secondly, work on the classical relay allows the terminals to transmit and receive simultaneously. In wireless communication, we deal with the half duplex constraint, explained in the next section. More importantly still, in wireless communication, the effect of fading is critical. The pioneering work on the relay channel assumed only additive white Gaussian noise (AWGN). It is because of fading that wireless communication can offer spatial diversity: the main reason behind the development of collaborative communications. All of these elements combine to show that while work on the relay was essential for the development of collaborative communications, recent work has taken a different emphasis.

The framework for collaboration in the wireless channel was first presented in (Sendonaris, Erkip & Aazhang, 2003a; Sendonaris, Erkip & Aazhang,

Wireless Collaboration

Figure 2. Network model for collaborative communications

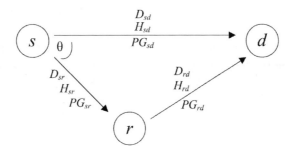

2003b). These two papers showed a basic method to achieve collaboration by taking advantage of CDMA architecture. Laneman, Tse and Wornell (2004) further refined the idea of collaboration by providing a mathematical background showing the potential gains. Through the derivation of outage probability Laneman shows that collaborative communication can achieve gains in spatial diversity. Laneman also proposes several collaborative protocols using methods such as Amplify-and-Forward and Decode-and-Forward. Hunter and Nosratinia (2006) and Stefanov and Erkip (2004) take the idea of Decode-and-Forward and propose using channel codes to achieve collaboration. They also verify the gains in spatial diversity by obtaining bounds on error performance.

To further improve the potential gains achieved with collaboration, Mitran, Ochiai and Tarokh (2005) and Ochiai, Mitran and Tarokh (2006) study the effect of modifying the amount of time the relay can collaborate. This leads to an improvement of coded collaboration found in (Tooher, Khoshneviss & Soleymani, 2007; Tooher & Soleymani 2008).

Furthermore, the Decode-and-Forward method is studied in (Su, Sadek & Liu, 2005), to show the performance of higher order modulation using collaborative communications.

Basic Assumptions

The most basic example of collaboration employs one source, one relay and one destination. The source has data it wishes to transmit to the destination, and since the relay is idle, it provides help to the source in order to maximize the transmit diversity. The network is set up as in Figure 2, where D_{ij}, H_{ij} and PG_{ij} denote the distance between nodes, the matrix of fading coefficients and path gain, respectively. The path gains are determined relative to the source-destination gain,

$$PG_{ij} = PG_{sd}\left(\frac{D_{sd}}{D_{ij}}\right)^{\alpha} \quad (1)$$

where α is the path loss exponent. In this chapter we use $PG_{sd}=1$, $D_{sd}=1$ and $\alpha = 2$ for all results.

Throughout this chapter the fading coefficients are assumed to be quasi-static block fading elements and, in the case of Rayleigh fading, are modeled as independent samples of a complex Gaussian random variable with zero mean and variance of 0.5 per dimension. In order to quantify the over-all attenuation of the signal suffered over any link, we have termed the channel coefficient to be the combination of the path gain and fading coefficient, namely

$$G_{ij} = \sqrt{PG_{ij}}H_{ij} \quad (2)$$

Direct Transmission

For comparative purposes, in this section we provide results pertaining to direct transmission of data from the source to the destination. We obtain the maximum average mutual information between the input and output. This is achieved by independent and identically distributed zero-mean circularly symmetric complex Gaussian inputs and is given by

$$I_D = \log(1 + \gamma |G_{sd}|^2), \qquad (3)$$

where $\gamma = \frac{E_s}{N_0}$. An outage event is defined as a channel realization where a specific rate of transmission cannot be supported by the link, $I_D < R$. At high signal-to-noise ratio (*SNR*) and for Rayleigh fading, the outage probability, averaged over all channel coefficients, is given by

$$P_D^{out}(\gamma, R) \triangleq \Pr[I_D < R] = 1 - \exp\left(-\frac{2^R - 1}{\gamma P G_{sd}}\right) \approx \frac{1}{P G_{sd}} \frac{2^R - 1}{\gamma} \qquad (4)$$

where the approximation is derived in (Laneman, Tse & Wornell, 2004). As expected, direct transmission achieves spatial diversity 1, which is obtained from the outage probability being proportional to γ^{-1}.

TWO-PHASE COMMUNICATION

Due to causality, the relay in collaborative communication is unaware of the source's data at the beginning of each frame. Therefore, the relay has two operations to perform: it must listen to the signal from the source and then it must transmit another signal to the destination. In practice, the relay cannot accomplish these two tasks on the same channel due to limitations in radio implementation. Severe attenuation over wireless channels, along with insufficient electrical isolation between the receiver and transmitter circuitry in the relay mean that any signal transmitted from a node will overwhelm any received signal. To deal with this, the so-called half-duplex constraint is assumed. Due to this constraint, two-phase protocols are proposed for collaboration.

In a two-phase protocol, the relay spends the first phase (also known as the exchange or listening phase) receiving the data from the source. In the second phase (also known as the collaborative phase) the relay transmits its own signal to the destination. Depending on the protocol under study, the destination may listen only to the collaborative phase or both phases to perform the decoding. Two-phase protocols eliminate the need for the relay to collaborate through a different wireless channel. However, there is still an increase in the bandwidth requirement. It is evident that any signal sent by the source will require twice the bandwidth (twice the time) for the relay to be able to collaborate.

When we compare collaboration to traditional direct transmission, as in Fig. 3, we see that for collaboration, the entire data transmitted from the source must be contained in a fraction of the channel allocation used in direct transmission. Note that this collaborative fraction, n_1/n, need not be ½. We also see that in the collaborative phase the source may or may not continue transmitting its signal (in the form of parity or repetition).

Detect-and-Forward

The idea behind two phases of communication to achieve collaboration was first proposed by Sendonaris et al. (2003a; 2003b). In this method, the relay attempts to detect the transmitted symbols from the source and then retransmits the detected symbols in the collaborative phase. In (Sendonaris et al. 2003a; Sendonaris et al. 2003b), wireless communication is separated into two parts: a non-collaborative part and a collaborative part. The trade-off between the amount of symbols transmitted with collaboration and those without, attempts to diminish the loss in throughput brought about by collaboration.

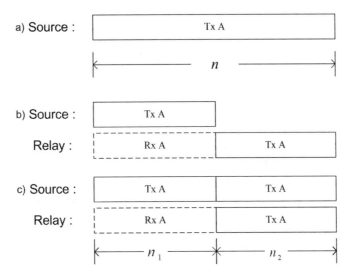

Figure 3. a) Direct transmission, b) collaboration with source silent in second phase, c) collaboration with source transmitting in second phase

As a simple example, in the first and second channel uses, a source transmits two symbols, while in the third, the source and relay retransmit the second symbol.

This method provides a simple way to achieve collaboration in the wireless channel. However it has some drawbacks. Firstly, if detection at the relay is incorrect, collaboration can be detrimental to the eventual detection at the destination. In addition, to achieve optimal decoding, the destination needs to be aware of the interuser channel characteristics.

The method was proposed in a framework to increase the over-all throughput of the system. This is why a balance is struck between symbols that are sent with collaboration and those without. This is the source of one of the main shortcomings of this method. The Symbol Error Rate (*SER*) performance of a wireless channel is dominated by the worst case scenario. In this case the symbols that are transmitted with collaboration can achieve full spatial diversity, whereas those transmitted without collaboration can only achieve unit diversity. Therefore the over-all performance of this method does not maximize diversity and hence cannot be assumed to maximize throughput.

Full Diversity Collaboration

To ensure full spatial diversity when using collaborative communications, Laneman et al. (2004) propose several collaborative communication protocols, termed cooperative diversity protocols. In this work, collaborative protocols require that only the relay transmits in the collaborative phase, while the source remains silent. Two main protocols are discussed in this section: Amplify-and-Forward (AF) and Decode-and-Forward (DF).

Amplify-and-Forward

The simplest method to achieve collaboration in the wireless channel is for the relay to listen to the source's signal, then simply amplify the signal and transmit it to the destination. This is basically a version of repetition encoding, where the source and relay transmit the same thing, except that here the relay's signal is corrupted by its own noise. For simplicity, in this section we assume the time-fraction, $\Delta = n_1/n = 1/2$. Therefore during the exchange phase of the AF protocol, the received signals are

$$y_r^e(k) = \sqrt{E_s} G_{sr} x_s^e(k) + z_r^e(k) \qquad (5)$$

$$y_d^e(k) = \sqrt{E_s} G_{sd} x_s^e(k) + z_d^e(k) \qquad (6)$$

where $y_i^e(k)$ denotes the kth received symbol at node i during the exchange phase, $x_i^e(k)$ denotes the kth transmitted symbol from node i during the exchange phase, $z_d^e(k)$ is a sample of a zero-mean complex Gaussian random variable with variance $N_o/2$ per dimension, E_s is the energy per transmitted symbol and G_{ij} is the channel coefficient between nodes i and j as defined in the previous section.

During the exchange phase, the relay processes the data and amplifies it. The received signal at the destination during the collaborative phase is

$$y_d^c(k) = \beta \sqrt{E_s} G_{rd} y_r^e(k - n/2) + z_d^c(k), \qquad (7)$$

where β is the amplifying gain used at the relay. To remain within its power constraints, the relay uses gain

$$\beta \leq \sqrt{\frac{E_s}{|G_{sr}|^2 E_s + N_o}} \qquad (8)$$

where we assume the relay can estimate with high accuracy G_{sr}.

In order to study the performance of the AF protocol, the outage probability is obtained and compared to that obtained for a direct transmission in (4). We provide the basic results without derivation, since this is beyond the scope of this chapter. For a full derivation the reader is encouraged to see (Laneman et al., 2004).

The maximum average mutual information between the source and the two outputs is given by

$$I_{AF} = \tfrac{1}{2} \log\left(1 + \gamma |G_{sd}|^2 + f\left(\gamma |G_{sr}|^2, \gamma |G_{rd}|^2\right)\right) \qquad (9)$$

where

$$f(x, y) \triangleq \frac{xy}{x + y + 1}$$

The outage event is the case where the maximum average mutual information is less than the desired rate of transmission, $I_{AF} < R$, or

$$|G_{sd}|^2 + \frac{1}{\gamma} f\left(\gamma |G_{sr}|^2, \gamma |G_{rd}|^2\right) < \frac{2^{2R} - 1}{\gamma} \qquad (10)$$

For Rayleigh fading the outage probability is approximated at high SNR to be

$$P_{AF}^{out}(\gamma, R) \triangleq \Pr[I_{AF} < R] \approx \left(\frac{1}{2PG_{sd}} \frac{PG_{sr} + PG_{rd}}{PG_{sr} PG_{rd}}\right) \left(\frac{2^{2R} - 1}{\gamma}\right)^2 \qquad (11)$$

which shows full spatial diversity of 2 is obtained since the outage probability proportional to γ^{-2}. Even though noise is amplified at the relay, the destination receives two copies of the signal through two independently faded channels, which maximizes the spatial diversity.

2) Decode-and-Forward

As it is implied in the name, the DF protocol requires the relay to decode the data transmitted by the source, re-encode it and transmit it to the destination in the collaborative phase. This extra step removes the effect of transmission errors in the source-relay channel. However, it comes with more complexity at the relay.

In the exchange phase, the signal received at the relay and destination remains the same as in the AF protocol (5) and (6). After the relay decodes

Wireless Collaboration

the codeword, it transmits its own signal, which is received at the destination as

$$y_d^c(k) = \sqrt{E_s} G_{rd} x_r^c(k) + z_d^c(k), \qquad (12)$$

where $x_r^c(k) = u\left(w\left(y_r^e(k)\right)\right)$, $u(.)$ is an encoding function and $w(.)$ is a decoding function. For the simple case of repetition-coded scheme, where the relay uses the same encoder as the source, we have $x_r^c(k) = \hat{x}_s^e(k)$.

In the derivation of the outage probability, a requirement is placed on the relay. It is required that the relay be able to fully decode the source's data without error in order to have a successful transmission. The maximum average mutual information for repetition-coded DF is given by

$$I_{DF} = \tfrac{1}{2} \min\left\{\log\left(1 + \gamma |G_{sr}|^2\right), \log\left(1 + \gamma |G_{sd}|^2 + \gamma |G_{rd}|^2\right)\right\} \qquad (13)$$

Since the mutual information depends on the minimum performance of either the source-relay channel or the source/relay-destination channel, it is simple to see that the outage probability is limited by the performance of the source-relay channel. Therefore this protocol does not achieve full spatial diversity when we require the relay to be able to decode the source's data without error.

In order to overcome the shortcomings of DF, Laneman et al. (2004) also propose selection relaying. In this case, when the relay is unable to decode the source's signal, the source falls back to direct transmission for the remainder of the frame. For the case of repetition coding at the relay, the maximum average mutual information for selection relaying using DF is given by

$$I_{SRDF} = \begin{cases} \tfrac{1}{2}\log\left(1 + 2\gamma |G_{sd}|^2\right), & |G_{sr}|^2 < g(\gamma) \\ \tfrac{1}{2}\log\left(1 + \gamma |G_{sd}|^2 + \gamma |G_{rd}|^2\right), & |G_{sr}|^2 \geq g(\gamma) \end{cases} \qquad (14)$$

where $g(\gamma) = [2^{2R} - 1]/\gamma$. The first row of (14) corresponds to the collaborative mode where the relay is unable to decode the source's signal and thus the source repeats its signal in the collaborative phase. The second row corresponds to the collaborative mode where the relay is able to decode the source's signal without error and then retransmits the same symbols. The outage probability derivation is involved, however it can be shown that the high *SNR* outage probability is identical to AF. Therefore selection relaying using DF achieves full spatial diversity since it does not rely only on the performance of the relay.

For the remainder of this chapter, all collaborative communication protocols presented and analyzed are assumed to be selection relaying DF protocols, unless stated otherwise.

Modifying the Time-Fraction

As mentioned previously, the fraction of time that a two-phase protocol resides in each time period need not be half. Suppose we wish to transmit data at a rate R bits per second in n channel uses, we may split the number of channel uses unevenly for each phase, such that $n = n_1 + n_2$. The time-fraction is defined as the ratio of the time spent in the exchange phase versus the over-all time. We have as time-fraction,

$$\Delta = \frac{n_1}{n} \qquad (15)$$

To maintain a constant over-all rate R, each phase is operated with rates R_1 and R_2.

In the simplest example, no channel coding is done in either phase. Therefore each phase must contain the same information (possibly with different coding rates) in order to ensure all data achieves full spatial diversity. We therefore select

$$R_1 n_1 = R_2 n_2 \qquad (16)$$

In (Ochiai et al. 2006), it is shown that different collaborative protocols using selection relaying and DF can achieve full spatial diversity at high *SNR* with any time-fraction.

1) Variable Time-Fraction

In the previous section we discuss modifying the time-fraction such that we may spend less time in the exchange phase. The performance at the destination can therefore improve due to the increase in collaboration. However, there is a cost associated with decreasing the time-fraction. Either we need to increase the energy from the source to ensure that a higher rate source-relay link can still be of high enough quality to allow for collaboration in the second phase, thus increasing the over-all energy cost. Or we suffer from a lack of collaboration from the relay, thus minimizing the potential gain promised by collaborative diversity.

In this section we present a bandwidth-efficient protocol without pre-determined fixed time-fractions. Here, the relay determines the time-fraction based on its receive channel for every instantaneous channel realization.

This method was first proposed with an information theoretic approach by Mitran et al. (2005). By allowing the relay to determine its own time-fraction, we can be sure to always maximize the amount of time the relay is collaborating. This is shown to improve Bit Error Rate (*BER*) performance to within a few dBs of the "genie" case where the relay can begin collaborating at the beginning of each frame.

We assume that none of the transmitting nodes have any prior information of the fading coefficients of the wireless channels. Therefore the source is unaware of the relay's time-fraction. In order to ensure a multitude of time-fraction possibilities, the source transmits a signal during the entire frame, never becoming silent. On the other hand, the relay listens to the source's signal until it can decode the data error-free and then begins collaborating. At the destination, to achieve optimal decoding, the destination is aware of all channel coefficients as well as the time-fraction.

A MIMO system with a Gaussian codebook and rate R can reliably communicate over a channel with channel gain matrix G as long as

$$R < \log_2 \det\left(I + \gamma G G^\dagger\right) \triangleq C(G) \qquad (17)$$

where I denotes the identity matrix, and $G^\dagger$ denotes the conjugate transpose of G.

The relay must be able to decode the entire transmitted codeword during the first phase of communication. The relay must decode nR bits in n_1 time slots. Since the relay is aware of its receive channel, it can select the time-fraction Δ such that $nR < n_1 C(G_{sr})$. At the destination, the data is received at a rate of $C(G_{sd})$ bits per time slot in the exchange phase and $C(G)$ bits per time slot in the collaborative phase; where we have used $G = [G_{sd}, G_{rd}]$. Since the destination uses the signals received in both phases to make its ultimate decision, it may reliably decode the data if $nR < n_1 C(G_{sd}) + (n - n_1) C(G)$.

There are two transmission possibilities, either the relay can or cannot collaborate during the time frame. The collaborative protocol is successful in transmitting the data, provided that either

$$R \leq \Delta C(G_{sr}) \qquad (18)$$

and

$$R \leq \Delta C(G_{sd}) + (1 - \Delta) C(G) \qquad (19)$$

where $0 < \Delta < 1$, or

$$R \leq C(G_{sd}) \tag{20}$$

Since the relay's collaborative status depends on the source-relay channel, the received energy at the destination is variable. Given that both the source and relay transmit at the same energy level, the received SNR at the destination, for a given channel realization, is given by

$$SNR_d = \frac{(PG_{sd} + (1-\Delta)PG_{rd})E_s}{N_o} \tag{21}$$

We see that the value of SNR is dependent on the relative location of the source and relay. As the relay moves, the same transmitted energy will not lead to the same received signal energy. Therefore this does not provide a proper comparative metric to the baseline model. We wish to have a measurement that will properly equate the energy used by two different transmission strategies so that they can be compared. In order to do this we introduce the concept of transmitted SNR, or TSNR. The transmitted energy per symbol for each channel realization is expressed as $(2 - \Delta)E_s$, hence the averaged TSNR at the destination is given by

$$TSNR_d = \frac{(2 - E[\Delta])E_s}{N_o} \tag{22}$$

where $E[\Delta]$ is the average time-fraction over all channel realizations.

To determine the outage probability of the variable time-fraction protocol, we first define an outage event as

$$E_o = [R > \Delta C(G_{sd}) + (1-\Delta)C(G)].$$

Therefore, the outage probability is defined as the probability, over all channel realizations, to experience an outage event,

$$P_{vtf}^{out} = \Pr[R > \Delta C(G_{sd}) + (1-\Delta)C(G)] \tag{23}$$

where *vtf* stands for variable time-fraction.

Figure 4 shows the outage probability of a $R=1/3$ collaborative communication system with $D_{sr}=0.3$ or $D_{sr}=0.7$. For comparison we provide the outage probability of direct transmission. It is evident from the steeper outage probability slope that collaboration provides greater diversity than direct transmission. We also provide "genie" results where we assume the relay is able to collaborate at the beginning of each frame since it has all the transmission data. The loss between the genie case and the realistic case is not very big when the relay is near the source ($D_{sr}=0.3$). On the other hand, when the relay is close to the destination, we see the loss - due to the relay being required to listen in the first phase - is much more evident. With a relay located close to the destination, if it were able to collaborate at the beginning of each frame, the system would require about 5 dB less TSNR to achieve the same outage. As expected, the genie case for a relay close to the destination has better performance than that of a relay near the source. That is because a relay with the same collaborative capabilities will provide more gain as it approaches the destination.

Collaborative Coding

The promise of spatial diversity gains from collaborative communications led to the development of practically achievable collaborative communication protocols. Collaborative (or cooperative) coding (Hunter & Nosratinia, 2006; Stefanov & Erkip, 2004) was developed by integrating channel coding together with collaboration. In coded collaboration, different portions of the codeword to be transmitted are sent via the source and relay's independently faded wireless channels.

At its simplest, the source encodes its data with a minimal channel code. The relay receives the

Figure 4. Outage probability for collaborative communications with R=1/3.

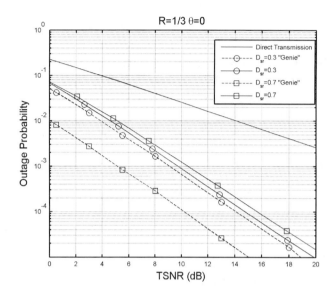

codeword and decodes the information bits. Either by the use of cyclic redundancy check (CRC) codes or by thresholds on its received *SNR*, the relay decides if decoding of the codeword will provide a reliable determination of the information bits. If so, the relay re-encodes the information bits obtaining a new set of parity bits to be transmitted to the destination. At the destination, the combination of parity bits obtained in the first phase from the source as well as parity bits obtained in the second phase from the source and/or relay, is used to obtain optimal decoding.

Different coding strategies may be used within this collaborative framework. Block and/or convolutional coding may be used at both the source and relay. In order to separate the codewords into two phases, puncturing, product codes or other forms of concatenation may be used.

We assume that the nodes have no access to channel state information at the transmitters (no CSIT) but have complete channel state information at the receivers (CSIR). This means that the destination is also aware of the collaboration capabilities of the relay.

In (Hunter & Nosratinia, 2006) an analysis of the pairwise error probability (*PEP*) is obtained to show the diversity advantage of collaborative coding. We assume that the relay is able to collaborate and show that in this collaborative mode, communication achieves full spatial diversity. The *PEP* is written as

$$P(\mathbf{c} \to \mathbf{e} \mid G) = Q\left(\sqrt{\frac{\gamma}{2} \|G(\mathbf{c}-\mathbf{e})\|^2}\right) \quad (24)$$

where $Q(x)$ denotes the Gaussian Q-function, $\|\cdot\|^2$ denotes the Frobenius norm, $\mathbf{c}$ denotes the transmitted codeword matrix and $\mathbf{e}$ denotes the erroneously decoded codeword matrix. Since we assume the relay is able to collaborate, the *PEP* can be rewritten as

$$P(\mathbf{d} \mid G) = Q\left(\sqrt{2\gamma\left(d_1 |G_{sd}|^2 + d_2 |G_{rd}|^2\right)}\right), \quad (25)$$

where d_i is the Hamming distance between codeword **c** and **e**, in phase i and $\mathbf{d}=(d_1,d_2)$ is a vector of Hamming distances.

To study the diversity advantage we need to obtain the average of the *PEP* over all channel realizations. Therefore we average (25) over all G_{sd} and G_{rd}. We use the alternate form of the *Q*-function and the MGF function method explained in (Simon & Alouini, 2000). For Rayleigh fading the average *PEP* is given by

$$P(\mathbf{d}) = E_G\left[P(\mathbf{d} \mid G)\right] = \frac{1}{\pi}\int_0^{\frac{\pi}{2}} \left(1 + \frac{d_1\gamma P G_{sd}}{\sin^2\theta}\right)^{-1} \left(1 + \frac{d_2\gamma P G_{rd}}{\sin^2\theta}\right)^{-1} d\theta \quad (26)$$

We can upper bound (26) by setting $\theta = \pi/2$, giving

$$P(\mathbf{d}) \leq \frac{1}{2}\left(\frac{1}{1 + d_1\gamma P G_{sd}}\right)\left(\frac{1}{1 + d_2\gamma P G_{rd}}\right) \quad (27)$$

At high *SNR*, the *PEP* becomes proportional to γ^{-2} as long as neither d_1 nor d_2 are zero. Therefore coded collaboration achieves full spatial diversity under specific coding criterion given that the relay is able to collaborate.

PERFORMANCE RESULTS OF COLLABORATIVE COMMUNICATIONS

In this section we obtain performance results for two collaborative communication methods. In the first method we study the combination of channel coding with collaboration and obtain Frame Error Rate (*FER*) results. In the second method we study higher order modulation in collaboration.

Variable Time-Fraction Collaborative Communications

We have seen that using channel codes to perform collaboration can provide a framework to obtain full spatial diversity in a practically implementable DF scenario. Collaborative communication as presented in (Hunter & Nosratinia, 2006; Stefanov & Erkip, 2004) contains two shortcomings that affect its performance. Firstly, the source is always silent in the collaborative phase of communication. This reduces some of the coding gain that could be obtained by having the source continue transmitting channel coded parity bits. Secondly, the time-fraction is held constant for any channel realization. We showed previously that having a variable time-fraction can lead to performances close to the "genie" bound.

It is with these two limitations in mind that variable time-fraction collaborative communication was proposed (Tooher et al. 2007; Tooher & Soleymani 2008). In this collaborative method, the source transmits a codeword to the destination, oblivious to the potential help a relay may provide. Therefore, the source encodes its data and transmits it to the destination for the duration of the frame. During this time, the relay listens to the transmitted signal and decides when it has received the data with a certain quality criterion (say when it can decode with $BER \leq 10^{-5}$). At this point, the relay encodes its own set of parity bits and transmits them to the destination. At the destination, decoding is done by using all bits received from both the source and relay during both phases. At the transmitting end, the nodes have no knowledge of channel state information (no CSIT); conversely, to achieve optimal decoding, we assume the nodes have complete CSIR.

For bandwidth considerations, we select a quantized set of time-fractions $\Delta = \{\Delta_1, \Delta_2, \ldots, 1\}$, where $\Delta = 1$ means the relay was unable to collaborate in that particular frame. To allow the possibility for the relay to collaborate with Δ_1, all of the *k* information bits must be represented in that fraction of the frame. The remaining time-fractions can be selected without any restrictions.

We can divide each frame of n bits into B blocks. At the end of each block, the relay determines its received *SNR* from the source and decides if it is strong enough to achieve the required decoding quality criterion. The length of these blocks therefore determines the set of time-fractions. Although not necessary, for simplicity of exposure we assume each block to be of equal length l, therefore the set of time-fractions is given as

$$\Delta = \left\{ \frac{1}{B}, \frac{2}{B}, \dots, \frac{(B-1)}{B}, 1 \right\} \quad (28)$$

The source and relay may use block or convolutional coding to perform collaboration, and coding criteria are outlined in (Tooher et al., 2007).

During the exchange phase, the received signals are

$$\mathbf{y}_r^e = \sqrt{E_s} G_{sr} \mathbf{x}_s^e + \mathbf{z}_r^e, \quad (29)$$

$$\mathbf{y}_d^e = \sqrt{E_s} G_{sd} \mathbf{x}_s^e + \mathbf{z}_d^e, \quad (30)$$

where $\mathbf{y}_i^e$, $\mathbf{x}_i^e$ and $\mathbf{z}_i^e$ denotes $1 \times n\Delta$ vectors of received signal, transmitted symbols and noise elements respectively, at node i. In the collaborative phase, the source and relay collaborate to transmit to the destination,

$$\mathbf{y}_d^c = \sqrt{E_s} G X^c + \mathbf{z}_d^c, \quad (31)$$

where $\mathbf{y}_i^c$ and $\mathbf{z}_i^c$ are $1 \times n(1-\Delta)$, and the transmitted matrix $X^c = \left[\mathbf{x}_s^c, \mathbf{x}_r^c \right]^T$ is the $2 \times n(1-\Delta)$ codeword matrix and $G = [G_{sd}, G_{rd}]$. We note that in the second phase, we do not require the source and relay to transmit the same symbols. In fact it is shown (Tooher et al., 2007) that doing so will not lead to the highest coding gain.

For comparative purposes we use the *TSNR* as explained previously. In this collaborative communication protocol, we have

$$TSNR_d = \frac{(2 - E[\Delta])nE_s}{kN_o}, \quad (32)$$

where we have added the cost of transmitting parity bits.

The relay decides on the quality of its received signal based on its received instantaneous SNR_r from the source. Achieving a certain SNR_r threshold means that it can transmit its own parity bits without having the negative effect of error propagation. The goal of the relay is to collaborate with the source the longest without adversely affecting the performance of the decoder at the destination. The relay is given a set of *SNR* thresholds, τ_{Δ_i}, each associated with a different time-fraction. With this set of thresholds, we can calculate $E[\Delta]$ in (32),

$$E[\Delta] = \sum_i \Delta_i p(\Delta = \Delta_i)$$

For the set of time-fractions given in (28) we have

$$E[\Delta] = \sum_{b=1}^{B} \frac{b}{B} p(\Delta = \tfrac{b}{B})$$

$$E[\Delta] = \sum_{b=1}^{B} \frac{b}{B} p(\tau_{b/B} \leq SNR_r \leq \tau_{b-1/B}) \quad (33)$$

where the instantaneous $SNR_r = \gamma |G_{sr}|^2$. With Rayleigh fading, the expected value of the time-fraction over all channel realization is given by

$$E[\Delta] = 1 - \sum_{b=1}^{B-1} \frac{1}{B} e^{\frac{-\tau_{b/B}}{PG_{sr}\gamma}} \quad (34)$$

Upper Bound on FER

An upper bound on the performance of variable time-fraction collaborative communication is determined in this section. The results provided here come from (Tooher & Soleymani, 2008) and can be considered a general solution to any interpretation of coded collaboration. By restricting the number of possible time-fractions to two and stopping the source from transmitting in the second phase, the FER results obtained in this section are equivalent to those presented in (Hunter & Nosratinia, 2006; Stefanov & Erkip, 2004).

The over-all FER of the scheme is obtained by solving for the FER of each collaborative mode (associated with a specific time-fraction) and then averaging over all the possible channel coefficients. The set of conditional FERs is given by

$$FER(G, \Delta) = \begin{cases} FER_{\Delta_1}(G) & \Delta = \Delta_1 \\ FER_{\Delta_2}(G) & \text{if } \Delta = \Delta_2 \\ \vdots & \vdots \\ FER_1(G) & \Delta = 1 \end{cases} \quad (35)$$

The over-all expected FER averaged over all channel realizations is

$$FER = E_{G_{sr},G}\left[FER(G,\Delta)\right] = \sum_{i=1}^{B} E_G\left[FER_{\Delta_i}(G)\right] p(\Delta = \Delta_i) \quad (36)$$

where the probability of each time-fraction depends on the thresholds and is provided in (33).

In order to solve for (36), we first obtain the conditional $FER_{\Delta_i}(G)$, then we average over all fading channel coefficients and then perform the weighted sum. The conditional FER are obtained by using an upper bound on terminated convolutional codes (Kellel & Leung, 1976),

$$FER_{\Delta_i}(G) \leq 1 - (1 - P_{E,\Delta_i}(G))^l \leq l \cdot P_{E,\Delta_i}(G) \quad (37)$$

where $P_{E,\Delta_i}(G)$ is the first error event probability defined as (Proakis, 1995)

$$P_{E,\Delta_i}(G) \leq \sum_{\mathbf{d}} a_{\Delta_i}(\mathbf{d}) P_{\Delta_i}(\mathbf{d}, G), \quad (38)$$

where $P_{\Delta_i}(\mathbf{d}, G)$ is the PEP between two codeword matrices, $a_{\Delta_i}(\mathbf{d})$ is the coefficient of the weight enumerating function and $\mathbf{d}=(d_1, d_2, d_3, f)$ is a distance vector between two codeword matrices. The values of d_1, d_2 and d_3 represent the Hamming distance between two codeword matrices for the part of the codewords transmitted by the source in the first phase, the source in the second phase and the relay in the second phase, respectively. The value of f is defined in detail in (Tooher & Soleymani, 2008) and is outside the scope of this chapter. Suffice it to say that f represents the effect of transmitting from both the source and relay in the second phase and is 0 for orthogonal transmission.

The PEP for collaborative communication is defined as

$$P_{\Delta_i}(\mathbf{d}, G) = Q\left(\sqrt{\frac{\gamma}{2} tr\left(GA(\mathbf{d})G^\dagger\right)}\right), \quad (39)$$

where

$$A(\mathbf{d}) = \begin{bmatrix} 4(d_1 + d_2) & 4f \\ 4f & 4d_3 \end{bmatrix}, \quad (40)$$

is the square of the codeword difference matrix between two codeword matrices.

To calculate the average, over all channel coefficients of the conditional FER, we first limit the first error event probability to make sure it doesn't exceed 1 even at low fading coefficients (Malkamaki & Leib, 1999). We get

$$E_G[FER_{\Delta_i}(G)] \leq \int \min\left(1, l\sum_{\mathbf{d}} a_{\Delta_i}(\mathbf{d})P_{\Delta_i}(\mathbf{d},G)\right) pdf_G(G) \cdot d(G) \quad (41)$$

where $pdf_G(G)$ denotes the probability density function of G. Solving for (41) for all time-fractions and then using (36) gives an upper bound on the performance of variable time-fraction collaborative communication.

Higher Order Modulation Collaborative Communications

The performance results obtained for the variable time-fraction collaborative communication framework assume BPSK and combine channel coding to guarantee the full spatial diversity of collaboration as well as increase the coding gain. This ensures that for a constant rate and energy one can improve the performance by decreasing the FER at the destination.

It is sometimes desirable to operate with a higher order modulation such as PSK or QAM. These allow the throughput to be increased and thus can provide an increase in the capabilities of the network. In this section we discuss the performance results of a collaborative communication scheme proposed by Su et al. (2005), which operates with M-PSK; the results are easily extended to M-QAM. For simplicity, we assume a static time-fraction of 0.5 and use a simplified version of the DF protocol. In the exchange phase of the protocol, the source transmits while the relay and destination receive signals as in (5) and (6). Using CRC codes, the relay decides if it can successfully decode the source's data. If decoding is successful, the relay retransmits the same symbols in the collaborative phase, while the source remains silent, as in (12). On the other hand, if decoding is not successful, both the source and relay remain silent in the collaborative phase.

At the end of the frame, the destination, having full knowledge of the channel coefficients, performs maximal ratio combining (MRC). The output of the MRC detector is given by

$$y = \frac{\sqrt{E_s}G_{sd}^*}{N_o} y_d^e + \frac{\sqrt{\bar{E}_s}G_{rd}^*}{N_o} y_d^c \quad (42)$$

where $\bar{E}_s = E_s$ if the relay can collaborate, otherwise $\bar{E}_s = 0$. The SNR of the output of the MRC is given by

$$\gamma_{MRC} = \frac{E_s|G_{sd}|^2 + \bar{E}_s|G_{rd}|^2}{N_o} \quad (43)$$

Using the MGF method (Simon & Alouini, 2000), the conditional SER for an M-PSK modulation is given by

$$SER(G_{sr}, G_{sd}, G_{rd}) = SER(\gamma_{MRC}) \triangleq \frac{1}{\pi}\int_0^{(M-1)\pi/M} \exp\left(-\frac{\sin^2(\pi/M)\gamma_{MRC}}{\sin^2(\theta)}\right)d\theta \quad (44)$$

The probability that the relay is able to decode the source's transmitted M-PSK symbol can also be obtained from (44) by modifying the SNR in the equation. The probability of incorrect decoding (no relay collaboration) is given by $SER(E_s|G_{sr}|^2/N_0)$, and the probability of correct decoding (relay collaborates) is thus $1 - SER(E_s|G_{sr}|^2/N_0)$.

The SER in (44) can be rewritten as

$$SER(G_{sr}, G_{sd}, G_{rd}) = SER(\gamma_{MRC})\big|_{\bar{E}_s=0} SER(\frac{E_s|G_{sr}|^2}{N_o}) \\ + SER(\gamma_{MRC})\big|_{\bar{E}_s=E_s}\left(1 - SER(\frac{E_s|G_{sr}|^2}{N_o})\right) \quad (45)$$

If we assume to have independently faded Rayleigh channels, the average of $SER(G_{sr}, G_{sd}, G_{rd})$

is given by

$$SER = F\left(1 + \frac{b\gamma PG_{sd}}{\sin^2\theta}\right)F\left(1 + \frac{b\gamma PG_{sr}}{\sin^2\theta}\right)$$
$$+ F\left[\left(1 + \frac{b\gamma PG_{sd}}{\sin^2\theta}\right)\left(1 + \frac{b\gamma PG_{rd}}{\sin^2\theta}\right)\right]\left[1 - F\left(1 + \frac{b\gamma PG_{sr}}{\sin^2\theta}\right)\right]$$
(46)

where $b = \sin^2(\pi/M)$ and

$$F(x(\theta)) = \frac{1}{\pi}\int_0^{(M-1)\pi/M} \frac{1}{x(\theta)}d\theta.$$

To find an upper bound on SER, we consider the expression in (46) and remove the negative term as well as all 1's in the denominator, this gives,

$$SER \leq \frac{A^2}{b^2\gamma^2 PG_{sd}PG_{sr}} + \frac{B}{b^2\gamma^2 PG_{sd}PG_{rd}}, \quad (47)$$

where

$$A = \frac{1}{\pi}\int_0^{(M-1)\pi/M} \sin^2\theta d\theta = \frac{M-1}{2M} + \frac{\sin\frac{2\pi}{M}}{4\pi},$$

$$B = \frac{1}{\pi}\int_0^{(M-1)\pi/M} \sin^4\theta d\theta = \frac{3(M-1)}{8M} + \frac{\sin\frac{2\pi}{M}}{4\pi} - \frac{\sin\frac{4\pi}{M}}{32\pi}$$

COLLABORATIVE COMMUNICATIONS WITH POWER ALLOCATION

The results presented in the previous section assume that the transmitting nodes have no knowledge of the channel characteristics. In such a case, the optimal power allocation is to keep the transmitting power constant and equal between the source and relay. This assumption is based on the fact that the relay can be located anywhere in the 2-dimensional plane where the source and destination are located, so on average its path gain to the destination is the same as the source's.

We now wish to see how the source and relay should behave if they are given some knowledge about CSIT. With either complete or statistical knowledge of the forward channels, the source and relay may use a power allocation algorithm to improve performance. As a basic example, if the transmitting nodes are aware that the link between the relay and destination is broken, all the available transmitting power may be used solely by the source and thus there is an improvement in performance versus equal power transmission.

Two power allocation methods are presented in this section. Power allocation depends on the knowledge of CSIT, however the type of CSIT available determines the type of power allocation that can be used. In the first part of this section we study a power allocation that is obtained by minimizing the overall SER that has been averaged over all channel realizations. This type of power allocation therefore depends only on knowledge of the variance of the channel coefficients G_{sr}, G_{sd} and G_{rd}. If we assume the nodes to be immobile, then the power allocation needs to be done only at the beginning of communication and is valid for the entire existence of the links.

In the second part of the section, we study a power allocation method that is obtained by having knowledge of every instantaneous channel realization. In such a case, with specific knowledge of CSIT, the source and relay optimize their power distribution to ensure the maximum achievable performance. This in turn minimizes the conditional FER. In this method, the power allocation needs to be done every time the fading coefficients change (i.e. at the beginning of each frame).

Power Allocation for Higher Order Modulation Collaborative Communications

We first determine the power allocation for a collaborative communication protocol using M-PSK as defined in the previous section. The power allocation method proposed by Su et al. (2005) is simple in that it depends only on the variance of the channel coefficients of the wireless links, namely PG_{sr}, PG_{sd} and PG_{rd}.

We rewrite the received signals during the first phase

$$\mathbf{y}_r^e = \sqrt{E_1} G_{sr} \mathbf{x}_s^e + \mathbf{z}_r^e \tag{48}$$

$$\mathbf{y}_d^e = \sqrt{E_1} G_{sd} \mathbf{x}_s^e + \mathbf{z}_d^e \tag{49}$$

and during the second phase,

$$\mathbf{y}_d^c = \sqrt{E_2} G_{rd} \mathbf{x}_r^c + \mathbf{z}_d^c \tag{50}$$

where we have $E_1 + E_2 = 2E_s$, to keep consistent with the previous results.

The power allocation is obtained by minimizing the averaged SER over all channel coefficients. The SER is obtained in the same manner as in (44)-(47) with the inclusion of the weighting terms. We get as an upper bound on SER,

$$SER \leq \frac{A^2 N_o^2}{b^2 E_1^2 PG_{sd} PG_{sr}} + \frac{BN_o^2}{b^2 E_1 E_2 PG_{sd} PG_{rd}} \tag{51}$$

The power allocation is obtained by taking the partial derivative of (51) over E_1 and setting the resulting derivation to 0, giving

$$B PG_{sr} \left(E_1^2 \frac{\partial E_2}{\partial E_1} + E_1 E_2 \right) + 2A^2 PG_{rd} E_2^2 = 0$$

Solving the above equation with the power constraint of $E_1 + E_2 = 2E_s$ gives the following optimal power allocation,

$$E_1 = \frac{2E_s \left(PG_{sr} + \sqrt{PG_{sr}^2 + 8(A^2/B)PG_{rd}^2} \right)}{3PG_{sr} + \sqrt{PG_{sr}^2 + 8(A^2/B)PG_{rd}^2}} \tag{52}$$

$$E_2 = \frac{4E_s PG_{sr}}{3PG_{sr} + \sqrt{PG_{sr}^2 + 8(A^2/B)PG_{rd}^2}} \tag{53}$$

We note in the above power allocation method that the results in (52) and (53) do not depend on the channel link between the source and destination, PG_{sd}. This result exposes the true purpose of collaborative communication, to provide an increase in spatial diversity. Since we are assured of obtaining the diversity from the source to the destination, the power allocation method is designed to maximize the performance of the source-relay-destination link, to achieve the second order diversity. In fact, if $PG_{sr} \gg PG_{rd}$, then $E_1 = E_2 = E_s$; on the other hand, if $PG_{sr} \ll PG_{rd}$, then $E_1 = 2E_s$ and $E_2 = 0$. This shows that as the decoding ability of the relay improves, we should provide the relay with more power since full spatial diversity is achievable. On the other hand, if the relay is unlikely to be able to decode the source's data since the source-relay link is weak, then most power should be used by the source.

We provide results on using a power allocation method versus equal power in Figure 5, where we use $TSNR = \frac{2E_s}{N_o}$. As the modulation level is decreased, we see that for a relay located near the source ($D_{sr} = 0.3$) there is not much gain to be obtained from using a power allocation method compared to the equal power method. This is because in this case we can say $PG_{sr} \gg PG_{rd}$. On the other hand, increasing the modulation level makes it more challenging for the relay to correctly decode the data in the first phase. Therefore the

Figure 5. SER performance of higher order modulation collaborative communications

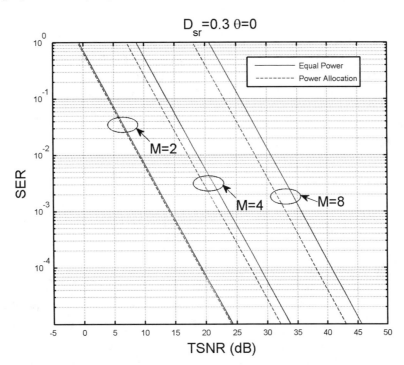

power allocation algorithm provides more power to the first phase to maximize the collaborative potential. In such a case equal power transmission is not optimal.

Power Allocation for Variable Time-Fraction Collaborative Communications

Having complete knowledge of the channel coefficients for all channel realizations allows us to develop a power allocation method that can minimize the conditional *FER*. In this section we modify the variable time-fraction collaborative communication protocol, presented in Section IV, to operate with a power allocation method (Tooher & Soleymani, 2009).

Using a variable time-fraction can in effect be considered a basic power allocation, the goal of which is to ensure the relay can collaborate. Therefore the power allocation method provided herein minimizes the conditional *FER* of all modes of collaboration by controlling the transmitted power from the source and relay during the collaborative phase only. In other words, at any given time during a frame, the transmitted power remains constant. This means that in the first phase, the source transmits with full power, while in the second phase the power allocation determines how much of the total power is transmitted from the source and relay.

During the first phase, the received signals at the relay and destination remain the same as in (29) and (30). In the collaborative phase, the received signal at the destination is given by

$$\mathbf{y}_d^c = \sqrt{E_s} W G X^c + \mathbf{z}_d^c \qquad (54)$$

where W is a matrix of complex channel weights determined by the power allocation algorithm and is given by

$$W = \begin{bmatrix} w_s & 0 \\ 0 & w_r \end{bmatrix}$$

where $|w_s|^2 + |w_r|^2 = 1$.

To obtain the power allocation method, we must minimize the *PEP* of (39) which in turn minimizes the conditional *FER* of (37). To solve for (39), we need the square of the codeword difference matrix $A(\mathbf{d})$, rewritten here with the channel weights,

$$A(\mathbf{d}) = A_e(\mathbf{d}) + A_c(\mathbf{d}) = 4d_1 + 4\begin{bmatrix} |w_s|^2 d_2 & w_s w_r^\dagger f \\ w_s^\dagger w_r f & |w_r|^2 d_3 \end{bmatrix} \quad (55)$$

where $A(\mathbf{d})$ is separated into two components, one for the exchange phase and one for the collaborative phase.

To make sure the power allocation algorithm isn't dependent on the codeword being transmitted, we force the source and relay to transmit the same parity bits in the collaborative phase, giving $d_2=d_3=f$.

To minimize the *PEP* in (39), we must maximize $tr\left(GA(\mathbf{d})G^\dagger\right)$. Since the power allocation only affects the collaborative phase, we can say that we wish to maximize $tr\left(GA_c(\mathbf{d})G^\dagger\right)$. This statement can be rewritten as solving for

$$\max_w tr(GA_c(\mathbf{d})G^\dagger), \text{ subject to } tr(R_{ww}) = 1 \quad (56)$$

where we have used $w=[w_1, w_2]^T$ and $R_{ww} = ww^\dagger$. The maximization of (56) is beyond the scope of this chapter and is provided in (Tooher & Soleymani, 2009). The final result states that the maximum is achieved when $w=v$, where v is the principal right singular vector of G. In such a case we have,

$$\max tr(GA_c(\mathbf{d})G^\dagger) = 4d_2 \lambda_{\max}(G^\dagger G), \quad (57)$$

where $\lambda_{\max}(G^\dagger G)$ is the maximum eigenvalue of $G^\dagger G$ and is given by

$$\lambda_{\max}(G^\dagger G) = |G_{sd}|^2 + |G_{rd}|^2.$$

Therefore the *PEP* in (39) can be rewritten as

$$P_{\Delta_i}(\mathbf{d}, G) = Q\left(\sqrt{\frac{\gamma}{2}\left(4d_1|G_{sd}|^2 + 4d_2\left(|G_{sd}|^2 + |G_{rd}|^2\right)\right)}\right) \quad (58)$$

The results of Figure 6 show the gain obtained from using a power allocation method. We first compare two codes (of memory order $m=5$) using the equal power method. The first is obtained when we don't constrain the source and relay to transmit the same data in the collaborative phase (max d_{min} code, given by $\{75,53,47,57,65\}_o$, where the first three blocks are transmitted by the source and the last two are possibly transmitted by the relay). The second code forces the source and relay to transmit the same data ($d_2=d_3$ code, given by $\{75,53,47,53,47\}_o$). These two codes assume a simple power allocation where the power is constant throughout the entire frame and is split evenly in the second phase. Forcing the source and relay to transmit the same parity bits in the collaborative phase is shown to give a coding loss of about 1.75 dB at $FER=10^{-3}$.

Next we show the improvement obtained when using the power allocation which requires full knowledge of CSIT. At $FER=10^{-3}$ the gain over the best equal power code is about 1.75 dB. This shows that the loss incurred by forcing the source and relay to transmit the same bits in the second phase of the power allocation method is offset by the gains provided by power allocation. We also compare the *FER* results to the outage

Figure 6. FER performance of variable time-fraction collaborative communications

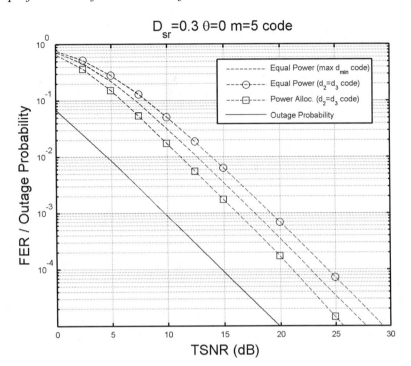

probability which can be seen as a benchmark value for optimal performance.

MULTIPLE RELAYS

We have shown how collaborative communication with one relay can increase the spatial diversity to order 2. The next step then is to add several more relays. In such a scenario one would hope that having L relays will lead to a diversity advantage of $L+1$. To do so, however, we need to be careful with code design.

The design of two-phase communication ensures that when one source and one relay are used we can achieve full diversity with basic criterion on the relay's signal. This is because full diversity is achieved with full rank of the codeword difference matrix (Tarokh et al., 1998) which is easily obtained since only the source transmits in the first phase; therefore any signal transmitted in the second phase will ensure full rank.

On the other hand, having multiple relays, each capable of transmitting a signal in the second phase, makes things more complicated. Whether we achieve full diversity when all the relays are transmitting in the second phase is not so clear.

In (Laneman & Wornell, 2003) two methods are proposed to achieve full spatial diversity with multiple relays. In the first method, we effectively use $L+1$ phases such that all relays have their specific phase and it is clear to see that the codeword difference matrix will achieve full rank since no two nodes are transmitting during the same channel use. This method provides an easy way to achieve full spatial diversity, but comes at a cost of increasing the bandwidth requirements.

The second method makes use of space-time coding and is thus termed space-time-coded cooperative diversity. In such a scheme, we use two phases. In the first phase the source transmits its data, while in the second phase, the relaying nodes re-encode the data using a space-time code that achieves full spatial diversity. This method achieves full spatial diversity in only two phases of communication. However, it requires complex space-time coding at the relays, for which some conditions may nevertheless lead to a bandwidth expansion.

Another method to use multiple relays is to select only the best one and use it for the remainder of the communication. In (Ibrahim, Sadek, Su & Liu, 2006), the *SER* for each relay (47) is used to determine which relay provides the best performance enhancement. Based on this metric, the source selects its optimal relay. In order for this scheme to function, it is assumed the source and relays have some means by which the selected relay can be notified. The analysis of the over-all *SER* is beyond the scope of this chapter. However, it is shown that full diversity (of order $L+1$) can be achieved even though only one relay is used.

The methods outline the impact that access to several relaying nodes can have on performance. However, the issue of fairness becomes much more vital when sources have access to multiple relaying nodes. An analysis of the performance of the entire network could lead to some intuition on methods that best make use of all the wireless nodes.

FUTURE WORK

Collaborative communication has been a main point of focus for wireless communication engineers for some time, and many results have been obtained. Nevertheless, many more issues remain to be addressed. For example, the effect of practical implementation of collaborative communication has not been fully researched. In practice, we can no longer assume to have perfect control over things such as CSI estimation or symbol timing. The effect of these can be two-fold, since they affect both the destination's ability to decode the transmitted signal properly, as well as the relay's ability to collaborate.

Another issue under study is partner allocation. In a typical wireless network, several sources and relays may be present. We may wish to determine the optimal partner allocation which will maximize performance given a specific cost constraint. One of the main challenges here is to develop a system which treats all nodes fairly and is in essence beneficial to all.

Having the ability to access multiple relays leads us to another potential topic of research. It is possible that in a two-phase protocol, the relays will have asymmetric data knowledge at the end of the first phase. There can still be value in having all the relays transmit their data, and it is important to determine methods which will not be detrimental to the decoding at the destination.

Instead of looking at collaboration as a transmission tool between the source and relay, we may think of it as a receiving tool between the destination and relay. This may seem like a redundant idea, however new methods can be developed that better take advantage of the receiver diversity. Collaborative methods such as quantize-and-forward may be used since the relay and destination have correlated data. Consequently, the relay uses Wyner-Ziv coding to achieve optimal compression.

In this chapter we have looked at collaboration as a physical layer method to increase spatial diversity. Collaboration, or cooperation, can also be applied to other layers or to achieve other goals. For example, network coding for the wireless channel (Yang & Koetter, 2007) provides a framework to integrate network coding as well as collaboration in a way to improve throughput of the over-all network. Moreover, some recent work on the relay-eavesdropper channel (Lei & El Gamal, 2008) has proposed methods to use collaboration to improve secrecy in wireless channels used for the transmission of sensitive information.

CONCLUSION

Practical limitations in wireless communication have slowed the eagerness to increase the number of antennas on wireless nodes. This chapter describes the burgeoning topic of collaborative communications: a technique allowing wireless nodes each equipped with a single antenna, the ability to obtain spatial diversity gains by sharing resources. Several two-phase protocols, necessary due to the half-duplex constraint, are presented in this chapter. Using outage probability, it is shown that collaboration can indeed maximize the spatial diversity to levels determined by the total number of available relays in the network.

Performance results are obtained for two types of collaborative communication protocols: a low-rate collaborative protocol combined with channel coding, as well as a higher order modulated collaborative communication protocol.

The chapter also outlines strategies to implement power allocation in collaborative communication, depending on the amount of knowledge of the channel state information available at the transmitting nodes. Some preliminary ideas on using multiple relays are provided.

All the results presented herein indicate the practicality of implementing collaborative communication in real world applications. They also validate the inclusion of collaborative communications in future wireless communications technologies, since it helps achieve several of the stated goals of 4G networking.

REFERENCES

Cover, T. M., & El Gamal, A. (1979). Capacity theorems for the relay channel. *IEEE Transactions on Information Theory*, *25*(5), 572–584. doi:10.1109/TIT.1979.1056084

Foschini, G., & Gans, M. (1998). On limits of wireless communication in a fading environment when using multiple antennas. *Wireless Personal Communications*, *6*(3), 311–335. doi:10.1023/A:1008889222784

Hunter, T. E., & Nosratinia, A. (2006). Diversity through coded cooperation. *IEEE Transactions on Wireless Communications*, *5*(2), 283–289. doi:10.1109/TWC.2006.1611050

Ibrahim, A. S., Sadek, A. K., Su, W., & Liu, K. J. R. (2006). *Relay selection in multi-node cooperative communications: When to cooperate and whom to cooperate with?* Paper presented at the IEEE Globecom conference, San Francisco, USA.

Kallel, S., & Leung, C. (1976). Efficient ARQ schemes with multiple copy decoding. *IEEE Transactions on Communications*, *COM-24*, 592–606.

Lai, L., & El Gamal, H. (2008). The relay-eavesdropper channel: Cooperation for secrecy. *IEEE Transactions on Information Theory*, *54*(9), 4005–4019. doi:10.1109/TIT.2008.928272

Laneman, J. N., Tse, D., & Wornell, G. W. (2004). Cooperative diversity in wireless networks: Efficient protocols and outage behavior. *IEEE Transactions on Information Theory*, *50*(12), 3062–3080. doi:10.1109/TIT.2004.838089

Laneman, J. N., & Wornell, G. W. (2003). Distributed space-time-coded protocols for exploiting cooperative diversity in wireless networks. *IEEE Transactions on Information Theory*, *49*(10), 2415–2425. doi:10.1109/TIT.2003.817829

Malkamaki, E., & Leib, H. (1999). Evaluating the performance of convolutional codes over block fading channels. *IEEE Transactions on Information Theory*, *45*(5), 1643–1646. doi:10.1109/18.771235

Mitran, P., Ochiai, H., & Tarokh, V. (2005). Space-time diversity enhancements using collaborative communications. *IEEE Transactions on Information Theory, 51*(6), 2041–2057. doi:10.1109/TIT.2005.847731

Ochiai, H., Mitran, P., & Tarokh, V. (2006). Variable-rate two-phase collaborative communication protocols for wireless networks. *IEEE Transactions on Information Theory, 52*(9), 4299–4313. doi:10.1109/TIT.2006.880055

Proakis, J. G. (1995). *Digital communication.* New York: McGraw-Hill Inc.

Sendonaris, A., Erkip, E., & Aazhang, B. (2003a). User cooperation diversity, part I: System description. *IEEE Transactions on Communications, 51*(11), 1927–1938. doi:10.1109/TCOMM.2003.818096

Sendonaris, A., Erkip, E., & Aazhang, B. (2003b). User cooperation diversity, part II: Implementation aspects and performance analysis. *IEEE Transactions on Communications, 51*(11), 1939–1948. doi:10.1109/TCOMM.2003.819238

Simon, M. K., & Alouini, M.-S. (2000). *Digital communication over fading channels: A unified approach to performance analysis.* New York: John Wiley and Sons.

Stefanov, A., & Erkip, E. (2004). Cooperative coding for wireless networks. *IEEE Transactions on Communications, 52*(9), 1470–1476. doi:10.1109/TCOMM.2004.833070

Su, W., Sadek, A. K., & Liu, K. J. R. (2005). *SER performance analysis and optimum power allocation for decode-and-forward cooperation protocol in wireless networks.* Paper presented at the IEEE Wireless Communications and Networking Conference (WCNC), New Orleans, USA.

Tarokh, V., Seshadri, N., & Calderbank, A. R. (1998). Space-time codes for high data rate wireless communication: Performance criterion and code construction. *IEEE Transactions on Information Theory, 44*(2), 744–765. doi:10.1109/18.661517

Tooher, P., Khoshneviss, H., & Soleymani, M. R. (2007). *Design of collaborative codes achieving space-time diversity.* Paper presented at the IEEE International Conference on Communications (ICC), Glasgow, UK.

Tooher, P., & Soleymani, M. R. (2008). *Performance of variable time-fraction collaborative communication.* Paper presented at the IEEE Personal, Indoor and Mobile Radio Communications (PIMRC) conference, Cannes, France.

Tooher, P., & Soleymani, M. R. (2009). *Power allocation for wireless communications using variable time-fraction collaboration.* Paper presented at the IEEE International Conference on Communications (ICC), Dresden, Germany.

van der Meulen, E. C. (1968). *Transmission of information in a T-terminal discrete memoryless channel.* Unpublished doctoral dissertation, University of California, Berkeley, CA.

van der Meulen, E. C. (1971). Three-terminal communication channels. *Advances in Applied Probability, 3*(1), 120–154. doi:10.2307/1426331

Yang, S., & Koetter, R. (2007). *Network coding over a noisy relay: A belief propagation approach.* A paper presented at the IEE International Symposium on Information Theory (ISIT), Nice, France.

ADDITIONAL READING

Alamouti, S. (1998). A simple transmit diversity technique for wireless communications. *IEEE Journal on Selected Areas in Communications, 16*(8), 1451–1458. doi:10.1109/49.730453

Alazem, F., Frigon, J.-F., & Haccoun, D. (2008). *Adaptive coded cooperation in wireless networks*. Paper presented at the IEEE International Wireless Communications and Mobile Computing Conference (IWCMC), Crete, Greece.

Anghel, P. A., & Kaveh, M. (2004). Exact symbol error probability of a cooperative network in a Rayleigh-fading environment. *IEEE Transactions on Wireless Communications, 3*(5), 1416–1421. doi:10.1109/TWC.2004.833431

Azarian, K., El Gamal, H., & Schniter, P. (2005). On the achievable diversity-multiplexing tradeoff in half-duplex cooperative channels. *IEEE Transactions on Information Theory, 51*(12), 4152–4172. doi:10.1109/TIT.2005.858920

Bletsas, A., Khisti, A., Reed, D. P., & Lippman, A. (2006). A simple cooperative diversity method based on network path selection. *IEEE Journal on Selected Areas in Communications, 24*(3), 659–672. doi:10.1109/JSAC.2005.862417

Cui, S., Goldsmith, A. J., & Bahai, A. (2004). Energy-efficiency of MIMO and cooperative MIMO techniques in sensor networks. *IEEE Journal on Selected Areas in Communications, 22*(6), 1089–1098. doi:10.1109/JSAC.2004.830916

Ding, Z., Ratnarajah, T., & Cowan, C. C. F. (2007). On the diversity-multiplexing tradeoff for wireless cooperative multiple access systems. *IEEE Transactions on Signal Processing, 55*(9), 4627–4638. doi:10.1109/TSP.2007.896276

Dohler, M., Rassool, B., & Aghvami, H. (2003). *Performance evaluation of STTCs for virtual antenna arrays*. Paper presented at the IEEE Spring Vehicular Technology Conference (VTC), Seogwipo, South Korea.

Gunduz, D., & Erkip, E. (2005). *Outage minimization by opportunistic cooperation*. Paper presented at the IEEE Wireless Communications Conference, Symposium on Information Theory, Maui, Hawaii.

Han, Z., Ji, Z., & Liu, R. J. K. (2008). A cartel maintenance framework to enforce cooperation in wireless networks with selfish users. *IEEE Transactions on Wireless Communications, 7*(5), 1889–1899. doi:10.1109/TWC.2008.061014

Host-Madsen, A. (2002). *On the capacity of wireless relaying*. Paper presented at the IEEE Fall Vehicular Technology Conference (VTC), Vancouver, Canada.

Hu, R., & Ti, J. (2006). *Practical compress-forward in user cooperation: Wyner-Ziv cooperation*. Paper presented at th IEEE International Symposium on Information Theory (ISIT), Seattle, USA.

Hunter, T. E., Sanayei, S., & Nosratinia, A. (2006). Outage analysis of coded cooperation. *IEEE Transactions on Information Theory, 52*(2), 375–391. doi:10.1109/TIT.2005.862084

Ibrahim, A. S., Sadek, A. K., Su, W., & Liu, K. J. R. (2005). *Cooperative communications with partial channel state information: When to cooperate?* Paper presented at the IEEE Globecom Conference, St-Louis, USA.

Janani, M., Hedayat, A., Hunter, T. E., & Nosratinia, A. (2004). Coded cooperation in wireless communications: Space-time transmission and iterative decoding. *IEEE Transactions on Signal Processing, 52*(2), 362–371. doi:10.1109/TSP.2003.821100

Jayaweera, S. K. (2006). Virtual MIMO-based cooperative communications for energy-constrained wireless sensor networks. *IEEE Transactions on Wireless Communications, 5*(5), 984–989. doi:10.1109/TWC.2006.1633350

Kramer, G., Gastpar, M., & Gupta, P. (2005). Cooperative strategies and capacity theorems for relay networks. *IEEE Transactions on Information Theory, 51*(9), 3037–3063. doi:10.1109/TIT.2005.853304

Kramer, G., Maric, I., & Yates, R. D. (2007). *Cooperative Communications*. Hanover, USA, now Publishers Inc.

Lai, L., & El Gamal, H. (2008). On cooperation in energy efficient wireless networks: The role of altruistic nodes. *IEEE Transactions on Wireless Communications, 7*(5), 1868–1878. doi:10.1109/TWC.2008.060568

Larsson, E., & Cao, Y. (2005). Collaborative transmit diversity with adaptive radio resource and power allocation. *IEEE Communications Letters, 9*(6), 511–513. doi:10.1109/LCOMM.2005.1437354

Larsson, E. G., & Vojcic, B. R. (2005). Cooperative transmit diversity based on superposition modulation. *IEEE Communications Letters, 9*(9), 778–780. doi:10.1109/LCOMM.2005.1506700

Lin, Z., Erkip, E., & Stefanov, A. (2006). Cooperative regions and partner choice in coded cooperative systems. *IEEE Transactions on Communications, 54*(7), 1323–1334. doi:10.1109/TCOMM.2006.877963

Liu, K. J. R., Sadek, A. K., Su, W., & Kwasinski, A. (2009). *Cooperative communications and networking*. Cambridge, UK: Cambridge University Press.

Liu, P., Tao, Z., Lin, Z., Erkip, E., & Panwar, S. (2006). Cooperative wireless communications: a cross-layer approach. *IEEE Wireless Communications Magazine, 13*(4), 84–92. doi:10.1109/MWC.2006.1678169

Luo, J., Blum, R. S., Cimini, L. J., Greenstein, L. J., & Haimovich, A. M. (2007). Decode-and-forward cooperative diversity with power allocation in wireless networks. *IEEE Transactions on Wireless Communications, 6*(3), 793–799. doi:10.1109/TWC.2007.05272

Nabar, R. U., Kneubuhler, F. W., & Bolcskei, H. (2004). *Performance limits of amplify-and-forward based fading relay channels*. Paper presented at the IEEE International Conference on Acoustics, Speech, and Signal Processing (ICASSP), Montreal, Canada.

Ng, C. T. K., & Goldsmith, A. J. (2006). *Capacity and power allocation for transmitter and receiver cooperation in fading channels*. Paper presented at the IEEE International Conference on Communications (ICC), Istanbul, Turkey.

Ng, C. T. K., Jindal, N., Goldsmith, A. J., & Mitra, U. (2007). Capacity gain from two-transmitter and two-receiver cooperation. *IEEE Transactions on Information Theory, 53*(10), 3822–3827. doi:10.1109/TIT.2007.904987

Nosratinia, A., & Hunter, T. E. (2007). Grouping and partner selection in cooperative wireless networks. *IEEE Journal on Selected Areas in Communications, 25*(2), 369–378. doi:10.1109/JSAC.2007.070212

Nosratinia, A., Hunter, T. E., & Hedayat, A. (2004). Cooperative communication in wireless networks. *IEEE Communications Magazine, 42*(10), 74–80. doi:10.1109/MCOM.2004.1341264

Ochiai, H., Mitran, P., Poor, H. V., & Tarokh, V. (2005). Collaborative beamforming for distributed wireless as hoc sensor networks. *IEEE Transactions on Signal Processing, 53*(11), 4110–4124. doi:10.1109/TSP.2005.857028

Ochiai, H., Mitran, P., & Tarokh, V. (2004). *Design and analysis of collaborative diversity protocols for wireless sensor networks.* Paper presented at the IEEE Fall Vehicular Technology Conference (VTC), Los Angeles, USA.

Sadek, A. K., Su, W., & Liu, K. J. R. (2007). Multi-node cooperative communications in wireless networks. *IEEE Transactions on Signal Processing, 55*(1), 341–355. doi:10.1109/TSP.2006.885773

Simic, L., Berber, S. M., & Sowerby, K. W. (2008). *Partner choice and power allocation for energy efficient cooperation in wireless sensor networks.* Paper presented at the IEEE Internal Conference on Communications (ICC), Beijing, China.

Stefanov, A., & Erkip, E. (2005). Cooperative space-time coding for wireless networks. *IEEE Transactions on Communications, 53*(11), 1804–1809. doi:10.1109/TCOMM.2005.858641

Steiner, A., Sanderovich, A., & Shamai, S. (2007). Broadcast cooperation strategies for two collocated users. *IEEE Transactions on Information Theory, 53*(10), 3394–3412. doi:10.1109/TIT.2007.904833

Tarokh, V., Jafarkhani, H., & Calderbank, A. R. (1999). Space-time block codes from orthogonal design. *IEEE Transactions on Information Theory, 45*(5), 1456–1467. doi:10.1109/18.771146

Telatar, E. (1999). Capacity of multi-antenna Gaussian channels. *European Transactions on Telecommunications, 10*(6), 585–595. doi:10.1002/ett.4460100604

Tooher, P., & Soleymani, M. R. (Manuscript submitted for publication). Analysis and code design of variable time-fraction collaborative communication. *Submitted to IEEE Transactions on Wireless Communications.*

Wolniansky, P. W., Foschini, G., Golden, G. D., & Valenzuela, R. A. (1998). *V-BLAST: An architecture for realizing very high data rates over the rich-scattering wireless channel.* Paper presented at the URSI International Symposium on Signals, Systems and Electronics, Pisa, Italy.

Xiao, L., Fuja, T. E., Kliewer, J., & Costello, D. J., Jr. (2007). *Signal superposition coded cooperative diversity: Analysis and optimization.* Paper presented at the IEEE Information Theory Workshop (ITW), Lake Tahoe, USA.

Compilation of References

3GAmericas. (2008 March). *Introduction to IPv6.* Retrieved from http://www.3gamericas.org/documents/2008_IPv6_transition_3GA_Mar2008.pdf

3GAmericas. (2008). *Report: Global UMTS HSPA Operator Status Update.* Arthur, W.B. (1994). *Increasing returns and path dependence in the economy.* Ann Arbor, MI: University of Michigan Press.

3GAmericas. (2009, February). *IPv6 Transition Considerations for LTE and Evolved Packet Core.* Retrieved from http://new.3gamericas.org/documents/3G%20Americas%20IPv6%20White%20Paper%20Feb%202009.pdf

3GPP R1-091580 (2009). *LTE-performance and IMT-Advanced requirements.* Presented at 3GPP TSG RAN WG1 Meeting #56bis, Seoul, Korea.

3GPP Release 8. (2009). *Overview of 3GPP Release 8 V0.0.4.*

3GPP RP-080756. (2008). *3GPP presentation on the LTE-Advanced as an IMT-Advanced technology solution.* Presented at the 41st 3GPP TSG RAN meeting, Kobe, Japan.

3GPP RP-090067. (2009). *Status report for WI to TSG RAN; work item name: further advancements for E-UTRA (LTE-Advanced).* Presented at the 43rd 3GPP TSG RAN meeting, Biarritz, France.

3GPP TR 25.913, V7.3.0 (2006-03). *Requirements for evolved UTRA (E-UTRA) and evolved UTRAN (E-UTRAN) (release 7).* Technical report, 3GPP, TSG RAN.

3GPP TR 36.814, V0.4.1 (2009-02). *Further advancements for E-UTRA; physical Layer aspects (release 9).* Technical report, 3GPP TSG RAN.

3GPP TR 36.913, V8.0.1 (2009-03). *Requirements for Further Advancements for E-UTRA (LTE-Advanced) (release 8).* Technical report, 3GPP TSG RAN.

3GPP TR 36.913. (2009). *Requirements for further advancements for Evolved Universal Terrestrial Radio Access (E-UTRA) (LTE-Advanced).* Agrell, E., Vardy, A., & Zeger, K. (2002). Closest point search in lattices. *IEEE Transactions on Information Theory, 48*(8), 2201–2214. doi:10.1109/TIT.2002.800499

3GPP TS 23.203 (2008). *Policy and charging control architecture.* Sophia Antipolis, France: the 3rd Generation Partnership Project.

3GPP TS 23.401 (2008). *General Packet Radio Service (GPRS) enhancements for Evolved Universal Terrestrial Radio Access Network.* Sophia Antipolis, France: the 3rd Generation Partnership Project.

3GPP TS 23.402 (2008). *Architecture enhancements for non-3GPP accesses.* Sophia Antipolis, France: the 3rd Generation Partnership Project.

3GPP TS 25.212, V8.4.0 (2008-12). *Multiplexing and channel coding (FDD) (release 8).* Technical specification, 3GPP, TSG RAN.

3GPP TS 36.101, V8.3.0 (2008-09). *Evolved universal terrestrial radio access (E-UTRA); user equipment (UE) radio transmission and reception (release 8).* Technical specification, 3GPP, TSG RAN.

3GPP TS 36.211, V8.4.0 (2008-09). *Evolved universal terrestrial radio access (E-UTRA); physical channels and modulation (release 8).* Technical specification, 3GPP, TSG RAN.

3GPP TS 36.212 (2008, September). *Evolved Universal Terrestrial Radio Access (E-UTRA): Mutiplexing and Channel Coding, version 8.4.0.*

3GPP TS 36.212, V8.4.0 (2008-09). *Evolved universal terrestrial radio access (E-UTRA); multiplexing and channel coding (release 8).* Technical specification, 3GPP, TSG RAN.

Compilation of References

3GPP TS 36.213, V8.4.0 (2008-09). *Evolved universal terrestrial radio access (E-UTRA); physical layer procedures (release 8)*. Technical specification, 3GPP, TSG RAN.

3GPP TS 36.300 (2008). *Evolved Universal Terrestrial Radio Access (E-UTRA) and Evolved Universal Terrestrial Radio Access Network (E-UTRAN); Overall description; Stage 2*. Sophia Antipolis, France: the 3rd Generation Partnership Project.

3GPP TS 36.300, V8.6.0 (2008-09). *Evolved universal terrestrial radio access (E-UTRA) and evolved universal terrestrial radio access network (E-UTRAN); overall description; stage 2 (release 8)*. Technical specification, 3GPP, TSG RAN.

3GPP TS 36.306, V8.2.0 (2008-05). *Evolved universal terrestrial radio access (E-UTRA); user equipment (UE) radio access capabilities (release 8)*. Technical specification, 3GPP, TSG RAN.

3GPP TS22.146. (2009). *Multimedia Broadcast/Multicast Service (MBMS); Stage 1*.

3*GPP(2008)*. TS 36.211, Evolved Universal Terrestrial Radio Access: Physical Channels and Modulation, v8.4.0 ed.

3GPP. (2001). *UTRAN overall description* [Technical Specification TS 25.401]. Retrieved January 3, 2008, from http://www.3gpp.org

3GPP. (2006, March). 3rd generation partnership project; Technical specification group radio access network; Requirements for evolved UTRA (E-UTRA) and evolved UTRAN (E-UTRAN) (Release 7). *3GPP TR 25.913 V7.3.0*. Retrieved from http://3gpp.org

3GPP. (2008, December). 3rd generation partnership project; Technical specification group radio access network; Evolved universal terrestrial radio access (E-UTRA) and evolved universal terrestrial radio access network (E-UTRAN); Overall description; Stage 2 (Release 8). *3GPP TS 36.300 V8.7.0*. Retrieved from http://www.3gpp.org

3GPP. (n.d.). *UTRAN Iub Interface: General Aspects and Principles*. [Technical Specification 25.430]

3GPP. (n.d.). *UTRAN Iub Interface: Layer 1*. [Technical Specification 25.431]

3GPP. (n.d.). *UTRAN Iub Interface: Signaling* [Technical Specification 25.432].

3GPP2 C-S0033, Rev 0, Ver 2.0 (2003). *Recommended Minimum Performance Standards for cdma2000 High Rate Packet Data Access Terminal*. 3rd Generation Partnership Project [TIA-866-1 standard].

3GPP2. (2008, December). Overview for ultra mobile broadband (UMB) air interface specification. *3GPP2 C.S0084-000-A v1.0*. Retrieved from http://www.3gpp2.org

3*GPPLong Term Evolution*. (n.d.). Techwritters Future, IPv6.com Tech Spotlight. Retrieved on September 8, 2009, from http://www.ipv6.com/articles/wireless/3GPP-Long-Term-Evolution.htm

3*GPPTS 23.060 V8.5.1, Release 8*. (2009 June). Technical Specification Group Services and System Aspects; General Packet Radio Service (GPRS); Service description; Stage 2, 3rd Generation Partnership Project. Retrieved from http://mailgroup.jp/3GPP/Specs/23060-851.pdf

3*GPPTS 23.221V9.0.0*. (2009, June). Technical Specification Group Services and System Aspects; Architectural requirements, Release 9, 3rd Generation Partnership Project. Retrieved from http://www.3gpp.org/ftp/Specs/archive/23_series/23.221/23221-900.zip

3*GPPTS 23.228 V9.0.0*. (2009, June). Technical Specification Group Services and System Aspects; IP Multimedia Subsystem (IMS); Stage 2, Release 9, 3rd Generation Partnership Project. Retrieved from http://www.3gpp.org/ftp/Specs/archive/23_series/23.228/23228-900.zip

3*GPPTS 23.401 V8.1.0, Release 8*. (2008, March). 3rd Generation Partnership Project; Technical Specification Group Services and System Aspects; General Packet Radio Service (GPRS) enhancements for Evolved Universal Terrestrial Radio Access Network (E-UTRAN) access.

3rd Generation Partnership Project (2009). Technical Specification Group Services and System Aspects. *Architecture enhancements for non-3GPP accesses (Release 8)*, TS 23.402 v8.4.1. Retrieved January, 2009, from http://www.3gpp.org

4*GBeyond 2.5G and 3G Wireless Networks*. (n.d.). MobileInfo.com. Retrieved February 13, 2009, from http://www.mobileinfo.com/3G/4GVision&Technologies.htm

6to4. (n.d.). Wikipedia. Retrieved on September 8, 2009, from http://en.wikipedia.org/wiki/6to4

Acharya, A., Misra, A., & Bansal, S. (2002). *A Label-switching Packet Forwarding Architecture for Multi-hop Wireless Lans*. Paper presented in ACM WoWMoM, Atlanta, 2002.

Acharya, A., Misra, A., & Bansal, S. (2003). High-performance architectures for IP-based multi-hop 802.11 networks. *IEEE Wireless Communications, 10*(5), 22–28. doi:10.1109/MWC.2003.1241091

Acharya, S., Alonso, R., Franklin, M., & Zdonik, S. (1995) Broadcast disks: data management for asymmetric communication environments. In *ACM SIGMOD Conference* (pp. 199-210).

Acx, A., Henriksson, J., Hunt, B., Karetsos, G. T., Tragos, E., Mihovska, A., Kyriazakos, S., Moretti, L., & Lara, J. (2003). Final usage scenarios. *IST-2003-507581 WINNER D1.3 v1.0*.

Adya, A., Bahl, P., Padhye, J., Wolman, A., & Zhou, L. (2004). A multi-radio unification protocol for IEEE 802.11 wireless networks. In *Proc. of the IEEE BroadNets*, (pp. 344-354).

Agarwal, A. K., Wang, W., & McNair, J. Y. (2005). An Experimental Study of Cross-Layer Security Protocols in Public Access Wireless Networks. *IEEE Global Telecommunications Conference (GLOBECOM'05)*, (vol. 3, pp. 1747–1751).

Aggelou, G. (2009). *Wireless Mesh Networking*. New York: Mc Graw Hill.

Aggelou, G., & Tafazolli, R. (2001). On the Relaying Capability of Next Generation GSM Cellular Networks. *IEEE Communications Magazine*, 40–47.

Agrawal, G. P. (1997). *Fibre-Optic Communication Systems*. Hoboken, NJ: John Wiley & Sons.

Agrawal, P., Zhang, T., Sreenan, C. J., & Chen, J.-C. (2004). All-IP wireless systems. *IEEE Journal on Selected Areas in Communications, 2*(4), 613–616. doi:10.1109/JSAC.2004.825950

Ahmad, J., Garrison, B., Gruen, J., Kelly, C., & Pankey, H. (2003). *4G Wireless Systems*. Next Generation Wireless Working Group. Retrieved April 9, 2009, from http://ckdake.com/files/4gwireless.pdf

Ahmed, B. T. (2006). *WiMAX in High Altitude Platforms (HAPs) Communications*. Paper presented at the 9th European Conference on Wireless Technology, Manchester, UK.

Ahson, S., & Ilyas, M. (2008). *RFID Handbook: Applications, Technology, Security, and Privacy*. Boca Ratón, FL: CRC Press.

Airy, M., Shakkottai, S., & Heath Jr., R. W. (2003). Spatially Greedy Scheduling in Multi-user MIMO Wireless Systems. *IEEE PIMRC*, 982-986.

Ajtai, M. (1998). The shortest vector problem in L2 is NP-hard for randomized reductions. In *Proc. of 30th Annu. ACM Symp. Theory of Comput. (STOC)* (pp. 10–19).

Akhtar, S. (n.d.). *2G-4G Networks: Evolution of Technologies, Standards and Deployment*. College of Information Technology, UAE University.

Akyildiz, F., Wang, X., & Wang, W. (2005, March). Wireless mesh networks: a survey. *Computer Networks, 47*(4), 445–487. doi:10.1016/j.comnet.2005.01.002

Akyildiz, I. F., McNair, J., Ho, J. S. M., Uzunalioglu, H., & Wang, W. (1999). Mobility Management in Next-Generation Wireless Systems. *Proceedings of the IEEE, 87*(9), 1347-1384.

Akyildiz, I. F., Xie, J., & Mohanty, S. (2004). A survey of mobility management in next-generation all-IP wireless systems. *IEEE Wireless Communication Magazine, 11*(4), 16–28. doi:10.1109/MWC.2004.1325888

Alamouti, S. M. (1998). A simple transmit diversity technique for wireless communications. *IEEE Journal on Selected Areas in Communications, 16*(8), 1451–1458. doi:10.1109/49.730453

Alazem, F., Frigon, J.-F., & Haccoun, D. (2008). *Adaptive coded cooperation in wireless networks*. Paper presented at the IEEE International Wireless Communications and Mobile Computing Conference (IWCMC), Crete, Greece.

Alcatel-Lucent. (2009). *Introduction to Evolved Packet Core, Strategic White Paper*. Retrieved from http://www.3g4g.co.uk/Lte/LTE_WP_0903_AlcatelLucent.pdf

Alemany, R., Perez, J., Llorente, R., Polo, V., & Marti, J. (2008). Coexistence of WiMAX 802.16d and MB-OFDM UWB in Radio over Multi-mode Fibre Indoor Systems. In *Proceedings of International Topics Meeting on IEEE Microwave Photonics 2008*.

Alexander, J. H. (2002). *UMTS and Mobile Computing*. Norwood, MA: Artech House.

Ali, N. A. A., Taha, A.-E. M., Hassanein, H. S., & Mouftah, H. T. (2008). IEEE 802.16 Mesh Schedulers: Issues and Design Challenges. *IEEE Network, 1*(22), 58–65. doi:10.1109/MNET.2008.4435904

Compilation of References

Allen, C., Arnold, W., Balkan, A., Cannasse, N., Grden, M., Gunesch, M., et al. (2008). *The Essential Guide to Open Source Flash Development*. Berkeley, CA: APRESS.

Al-Raweshidy, H., Glaubitt, K. M., & Faccin, P. (2002). In-building coverage for UMTS using radio over fibre technology. In *Proceedings of 5th International Symposium on Wireless Personal Multimedia Communications 2002* (Vol. 2, pp. 581- 585).

Alshaer, H. (2005). *Priority Traffic Control and management in Multiservice IP Networks*. PhD thesis, Pierre et Marie Curie University, LIP6, Paris, France.

Alshaer, H., & Elmirghani, J. J. M. (2008). Expedited forwarding end-to-end delay and jitter in diffserv. *International Journal of Communication Systems*, *21*, 815–841. doi:10.1002/dac.915

Alshaer, H., & Horlait, E. (2004). Emerging client-server and ad-hoc approach in inter-vehicle communication platform. In *Proc. of the IEEE VTC-Fall*, 6, (pp. 3955-3959), Los Angeles, CA.

Alshaer, H., & Horlait, E. (2004). Expedited forwarding delay budget through a novel call admission control. In *Proc. of the 3rd European Conference on Universal Multiservice Networks (ECUMN)*, (LNCS 3262, pp. 50-59), Porto, Portugal.

Alshaer, H., & Horlait, E. (2005). An optimized adaptive broadcast scheme for inter-vehicle communication. In *Proc. of IEEE VTC-Spring, 5*, (pp. 2840-2844), Stockholm, Sweden.

Altman, K., Borkar, V. S., & Kherani, A. A. (2004). Optimal random access in networks with two-way traffic. *PIMRC*, *1*, 609–613.

Amirth. (n.d.). *IP IMS (IP Multimedia Subsystem)*. IPv6.com Tech Spotlight. Retrieved on September 8, 2009. Retrieved from http://www.ipv6.com/articles/general/IP_IMS.htm

Amzallag, D., Livschitz, M., Naor, J., & Raz, D. (2005). Cell Planning of 4G Cellular Networks: Algorithmic Techniques and Results. In *Proceedings of the 6th IEE International Conference on 3G and Beyond*, (pp. 1–5).

Ananthapadmanabha, R., et al. (2001). *Multi-hop Cellular Networks: The Architecture and Routing Protocol*. Paper presented at IEEE PIMRC.

Anderson, J. B., & Mohan, S. (1984). Sequential coding algorithms: A survey and cost analysis. *IEEE Transactions on Communications*, *32*(2), 169–176. doi:10.1109/TCOM.1984.1096023

Andrews, J. G., Ghosh, A., & Rias, M. (2007). *Fundamentals of WiMAX: understanding broadband wireless networking*. Upper Saddle River, NJ: Pearson Education, Inc.

Andrews, M.(n.d.). Providing Quality of Service over a Shared Wireless Link. *IEEE Communications Magazine*, *39*(2), 150–154. doi:10.1109/35.900644

Anghel, P. A., & Kaveh, M. (2004). Exact symbol error probability of a cooperative network in a Rayleigh-fading environment. *IEEE Transactions on Wireless Communications*, *3*(5), 1416–1421. doi:10.1109/TWC.2004.833431

Ansari, F. & Sathyanath, A. (2007). STEM: seamless transport endpoint mobility. ACM Computing and Communications Review, 11(2), 1:13.

Anscombe, N. (2005). Demand for indoor coverage drives radio-over-fibre. *Wireless Europe*. Retrieved from http://www.nadya-anscombe.com/pdfs/radio%20over%20fibre.pdf

Aoun, C., & Davies, E. (2007). *Reasons to Move the Network Address Translator - Protocol Translator (NAT-PT) to Historic Status*. RFC 4966. Retrieved from http://www.faqs.org/rfcs/rfc4966.html

Araniti, G., Iera, A., & Molinaro, A. (2005). The Role of HAPS in Supporting Multimedia Broadcast and Multicast Services in Terrestrial-Satellite Integrated Systems. *Wireless Personal Communications*, *32*, 195–213. doi:10.1007/s11277-005-0742-3

Arbanowski, S., Ballon, P., David, K., Droegehorn, O., Eertink, H., & Kellerer, W. (2004, Sept.). I-centric communications: personalization, ambient awareness, and adaptability for future mobile services. *Communications Magazine*, *42*(9), 63–69. doi:10.1109/MCOM.2004.1336722

ARIB/Japan. (1998, June). *Japan's Proposal for Candidate Radio Transmission Technology on IMT-2000: W-CDMA*.

Arkko, J., & Haverinen, H. (2006). *Extensible Authentication Protocol Method for 3rd Generation Authentication and Key Agreement (EAP-AKA) (RFC 4187)*. Retrieved from http://www.ietf.org/rfc/4187.txt

Armuelles, I., Robles, T., Madrid, S., & Chaouchi, H. (2004). On Ad Hoc Networks in the 4G Integration Process. In *Proc. Of The Third Annual Mediterranean Ad Hoc Networking Conference 2004*.

Auer, G. (2005), Modeling of OFDM Channel Estimation Errors. In *10th International OFDM Workshop*.

Aust, S., Proetel, D., Fikouras, N. A., & Görg, C. (2003, October). Policy based Mobile IP handoff decision (POLIMAND) using generic link layer information. In *The 5th IEEE International Conference on Mobile and Wireless Communication Networks*, (pp. 201-204), Singapore.

Austerberry, D. (2005). The Technology of Video and Audio Streaming. Oxford, UK: Focal Press.

Austin, M. D., & Stüber, G. L. (1994). Velocity adaptive handoff algorithms for microcellular systems. *IEEE Transactions on Vehicular Technology, 43*(3), 549–561. doi:10.1109/25.312791

Avallone, S., Akyildiz, I., & Ventre, G. (2009). A channel and rate assignment algorithm and a layer-2.5 forwarding paradigm for multi-radio wireless mesh networks. *IEEE/ACM Transactions on Networking, 17*(1), 267-280.

Awduche, D. O., & Agu, E. (1997). Mobile extensions to rsvp. In *Proc. 6th Int'l Conference computer Communication and Nets*, (Vol.7, pp. 132-136).

Azarian, K., El Gamal, H., & Schniter, P. (2005). On the achievable diversity-multiplexing tradeoff in half-duplex cooperative channels. *IEEE Transactions on Information Theory, 51*(12), 4152–4172. doi:10.1109/TIT.2005.858920

Azemi, G., Senadji, B., & Boashash, B. (2004). Mobile unit velocity estimation based on the instantaneous frequency of the received signal. *IEEE Transactions on Vehicular Technology, 53*(3), 716–724. doi:10.1109/TVT.2004.827157

Babaoglu, O., Canright, G., Deutsch, A., Di Caro, G., Ducatelle, F., & Gambardella, L. (2006). Design patterns from biology for distributed computing. [TAAS]. *ACM Transactions on Autonomous and Adaptive Systems, 1*(1), 26–66. doi:10.1145/1152934.1152937

Bach, L., Kaiser, W., Reithmaier, J. P., Forchel, A., Berg, T. W., & Tromborg, B. (2003). Enhanced direct-modulated bandwidth of 37 GHz by a multi-section laser with a coupled-cavity-injection-grating design. *Electronics Letters, 39*, 1592–1593. doi:10.1049/el:20031018

Badia, L., Miozzo, M., Rossi, M., & Zorzi, M. (2007). Routing Schemes in Heterogeneous Wireless Network based on Access Advertisement and Backward Utilities for QoS Support. *IEEE Communications Magazine, 45*(2), 67–73. doi:10.1109/MCOM.2007.313397

Baek, M., Kim, M., You, Y., & Song, H. (2004). Semi-blind channel estimation and PAR reduction for MIMO-OFDM system with multiple antennas. *IEEE Transactions on Broadcasting, 50*(12), 414–424. doi:10.1109/TBC.2004.837885

Baek, M., Kook, H., Kim, M., You, Y., & Song, H. (2005). Multi-antenna scheme for high capacity transmission in the digital audio broadcasting. *IEEE Transactions on Broadcasting, 51*(12), 551–559. doi:10.1109/TBC.2005.858426

Bahai, A. R. S., & Burton, S. R. (2002). *Multi-Carrier Digital Communications Theory and Applications of OFDM*. New York: Kluwer Academic Publishers.

Bahl, L. R., Cocke, J., Jelink, F., & Raviv, J. (1974). Optimal decoding of linear codes for minimizing symbol error rate. *IEEE Transactions on Information Theory, 20*, 284–287. doi:10.1109/TIT.1974.1055186

Bahl, P., Chandra, R., & Dunagan, J. (2004). SSCH: Slotted channel hopping for capacity improvement in IEEE 802.11 ad hoc wireless networks. In *Proc. of IEEE 802.11 Ad-Hoc Wireless Networks*, Philadelphia, USA.

Bahl, V. (2007). Wireless mesh networks. *IEEE INFOCOM Tutorial*.

Bai, H., & Atiquzzaman, M. (2003). Error modeling schemes for fading channels in wireless communications. *IEEE Communications Surveys and Tutorials, 5*(2), 2–9. doi:10.1109/COMST.2003.5341334

Baker, F., Iturralade, C., Faucheur, F. L. & Davie, B. (2001). Aggregation of rsvp for ipv4 and ipv6 reservations. *IETF, RFC 3157*.

Ballon, P., Helmus, S., & Van de Pas, R. (2002). Business models for next-generation wireless services. [Mobile internet]. *Trends in Communications, 9*, 2002.

Banks, L., Ye, S., Huang, Y., & Wu, S. F. (2007). Davis Social Links: Integrating Social Networks with Internet Routing. In *Applications, Technologies, Architectures, and Protocols for Computer Communication. 2007 workshop on Large scale attack defense* (pp. 121-128).

Barabasi, A., & Albert, R. (1999). Emergence of scaling in random networks. *Science, 286*, 509–512. doi:10.1126/science.286.5439.509

Barbero, L. G., & Thompson, J. S. (2006). A fixed-complexity MIMO detector based on the complex sphere decoder, In *Proc. IEEE Intern. Workshop Signal Processing Advances Wireless Communications (SPAWC '06)*.

Baro, S., Hagenauer, J., & Witzke, M. (2003, May). Iterative detection of MIMO transmission using a list-sequential (LISS) detector. In *Proc. Of IEEE Int. Conf. on Commun., ICC '03,* Anchorage, AK (pp. 2653–2657).

Bayer, N., Sivchenko, D., Xu, B., Rakocevic, V., & Habermann, J. (2006). Transmission Timing of Signaling Messages in IEEE 802.16 based Mesh Networks. *European Wireless Conference.*

Bayer, N., Xu, B., Rakocevic, V., & Habemann, J. (2007). Improving the Performance of the Distributed Scheduler in IEEE 802.16 Mesh Networks. In *IEEE. IEEE Vehicular Technology Conference 2007 Spring (VTC-2007)* (pp. 1193-1197).

Bedell, P. (2005). *Wireless Crash Course, Edition 2*. New York: McGraw Hill Professional.

Belda, A. (2006). Contribution to the multimedia communications in wireless networks for mobile devices (In Spanish). Unpublished doctoral dissertation, Technical University of Valencia, Spain.

Belghith, A., & Nuaymi, L. (2008). Comparison of WiMAX scheduling algorithms and proposals for the rtPS QoS class. In *14th European Wireless Conference* (pp. 1-6).

Bellavista, P., Corradi, A., & Giannelli, C. (2005). Mobile Proxies for Proactive Buffering in Wireless Internet Multimedia Streaming. In IEEE International Conference on Distributed Computing Systems Workshop: (Vol. 3, pp. 297-304). Colummbus, OH: IEEE Press.

Bellifemine, F., & Poggi, A. A. & G. Rimassa, G. (2001, May). *JADE: A FIPA2000 Compliant Agent Development Environment.* Paper presented at the meeting of the AGENTS '01, Montreal, Canada.

Bellifemine, F., Caire, G., Poggi, A., & Rimassa, G. (2003). JADE, A White Paper. *Journal of Telecom Italia Lab., 3*(3), 6–19.

Bellman, R. (1958). On a routing problem. *Quarterly of Applied Mathematics, 16*(1), 87–90.

Beltran, M., Llorente, R., Sambaraju, R., & Marti, J. (2009). 60 GHz UWB-over-Fibre System for In-flight Communications. In . *Proceedings of, IMS2009,* TU1D–2.

Beltran, M., Morant, M., Perez, J., Llorente, R., & Marti, J. (2008). Photonic generation and frequency up-conversion of impulse-radio UWB signals. In *Proceedings of 21st Annual Meeting of The IEEE Lasers & Electro-Optics Society, 2008 LEOS Annual Conference,* Newport Beach, CA.

Benali, O., El-Khazen, K., Garrec, D., Guiraudou, M., & Martinez, G. (n.d.). A framework for an evolutionary path toward 4G by means of cooperation of networks. *Communications Magazine, IEEE. 42*(5), 82 – 89.

Benedetto, S., Dinoi, L., Montorsi, G., & Tarable, A. (2006). Design issues on the parallel implementation of versatile, high-speed iterative decoders. In *Proc. 4th Int. Symposium on turbo codes & related topics,* Munich, Germany, April, (pp. 341-351).

Benedetto, S., Montorsi, G., Divsalar, D., & Pollara, F. (1996, Nov.). A soft-input soft-output maximum a posterior (MAP) module to decode parallel and serial concatenated codes. In *NASA JPL TDA Progress Report* (pp. 42–127).

Ben-Ezra, Y., Ran, M., Borohovich, E., Leibovich, A., Thakur, M. P., Llorente, R., & Walker, S. D. (2008). Wimedia-Defined, Ultra-Wideband Radio Transmission over Optical Fibre. In *Conference on Optical Fiber communication/National Fiber Optic Engineers Conference OFC/NFOEC 2008.*

Ben-Shimol, Y., Kitroser, I., & Dinitz, Y. (2006). Two-dimensional mapping for wireless OFDMA systems. *IEEE Transactions on Broadcasting, 52*(9), 388–396. doi:10.1109/TBC.2006.879937

Berger, J., Hellenbart, A., Weiss, B., Möller, S., Gustafsson, J., & Heikkila, G. (2008). Estimation of Quality per Call in Modelled Telephone Conversations. In *Proceedings of IEEE International Conference on Acoustics, Speech and Signal Processing (ICASSP) 2008,* Las Vegas (pp. 4809-4812).

Bergeron, C., & Lamy-Bergot, C. (2004). Soft-input decoding of variable-length codes applied to the H.264 standard. In *Proceedings of the IEEE MMSP'04 conference* (pp. 87-90), Siena, Italy. Washington, DC: IEEE.

Bergeron, C., & Lamy-Bergot, C. (2005). Compliant selective encryption for H.264/AVC video streams. In [Shanghai, China. Washington, DC: IEEE.]. *Proceedings of the IEEE MMSP, 05,* 477–480.

Bergeron, C., & Lamy-Bergot, C. (2006). Modelling H.264/AVC sensitivity for error protection in wireless transmission. In [Victoria, Canada. Washington, DC: IEEE.]. *Proceedings of the IEEE MMSP, 06,* 302–305.

Bergeron, C., Lamy-Bergot, C., Pau, G., & Pesquet-Popescu, B. (2006). Temporal scalability through adaptive M-Band filter banks for robust H.264/MPEG-4 AVC video coding. *EURASIP Journal on Applied Signal Processing, 5,* 21930–21940.

Berkmann, J., Carbonelli, C., Dietrich, F., Drewes, C., & Xu, W. (2008). On 3G LTE terminal implementation – Standard, algorithms, complexities and challenges. In *Proc. of IWCMC'08, IEEE International Wireless Communications and Mobile Computing Conference*, Crete, Greece, August.

Bernardos, C. J., Soto, I., Moreno, J. I., Melia, T., Liebsch, M., & Schmitz, R. (2005). Mobile Networks Experimental Evaluation of a Handover Optimization Solution for Multimedia Applications in a Mobile IPv6 Network. *European Transactions on Telecommunications, 16*(4), 317–328. doi:10.1002/ett.1005

Bernardos, M., Casar, J., & Tarrio, P. (2006. July). *Efficient social routing in sensor fusion networks.* Paper presented at 9th International Conference on Information Fusion, Florence, Italy.

Berrou, C. (2003, Aug.). The Ten-Year-Old Turbo Codes are Entering into Service. *IEEE Communications Magazine, 41,* 110–116. doi:10.1109/MCOM.2003.1222726

Berrou, C., & Glavieux, A. (1996, October). Near Optimum Error Correcting Coding and Decoding: Turbo-Codes. *IEEE Transactions on Communications,* 1261–1271. doi:10.1109/26.539767

Berrou, C., Glavieux, A., & Thitimajshima, P. (1993). Near Shannon limit error correcting coding and decoding: turbo codes. In *Proc. IEEE International Conference on Communications,* Geneva, May, (pp. 1064-1070).

Berrou, C., Glavieux, A., & Thitimajshima, P. (1993, May). Near Shannon Limit Error-Correcting Coding and Decoding: Turbo-Codes. *IEEE International Conference on Communications, 2,* 1064-1070.

Berrou, C., Saouter, Y., Douillard, C., Kerouedan, S., & Jezequel, M. (2004, June). Designing Good Permutations for Turbo Codes: Towards A Single Model. *IEEE International Conference on Communications, 1,* 341-345.

Bertsekas, D., & Gallager, R. (1992). *Data Networks.* Upper Saddle River, NJ: Prentice-Hall.

Bhagawa, P., Bhattacharya, P., Krishna, A., & Tripathi, S. K. (1996). Enhancing throughput over wireless LANs using channel state dependent packet scheduling. In *INFOCOM,* San Francisco, USA, (pp. 1133-1140).

Bickerstaff, M., Davis, L., Thomas, C., Garrett, D., & Nicol, C. (2003). A 24Mb/s Radix-4 LogMAP Turbo Decoder for 3GPP-HSDPA Mobile Wireless. *IEEE International Solid-State Circuits Conference (ISSCC).*

Biglieri, E., Proakis, J., & Shamai, S. (1998). Fading channel: information theoretic and communication aspects. *IEEE Transactions on Information Theory,* 2619–2692. doi:10.1109/18.720551

Bing, C., Frank, F., James, G., Rainer, G., Hermann, R., & Adam, W. (2001). Framework for combined optimization of DLC and physical layer in mobile OFDM systems. In *Proc. of the 6th International OFDM workshop,* Hamburg, (pp. 32-36).

Blake, S., Black, D., Carlson, M., Davies, E., Wang, Z., & Weiss, W. (1998). An architecture for differentiated services. *IETF RFC 2475.* Retrieved on January 2009, http://www.ietf.org/

Bletsas, A., Khisti, A., Reed, D. P., & Lippman, A. (2006). A simple cooperative diversity method based on network path selection. *IEEE Journal on Selected Areas in Communications, 24*(3), 659–672. doi:10.1109/JSAC.2005.862417

Blumenthal, R. H. (1962). Design of a microwave frequency light modulator. *In Proceedings of Institute of Radio Engineers IRE, 50,* 452-456.

Bohlin, E. (2004). *The Future of Mobile Communications in the EU: Assessing the Potential of 4G* [Tech, Rep. No. EUR21192 EN]. ESTO Publications.

Bokor, L., Dudás, I., & Javier, R. O. (2005). *Streaming over mobile IPv6 networks.* Paper presented at the Ericsson High Speed Networks Laboratory Workshop, Mtrahza, Hungary.

Borcea, C. (2002). Cooperative Computing for Distributed Embedded Systems. *In Proceedings of 22nd International Conference on Distributed Computing Systems (ICDCS),* Vienna, Austria.

Borgonovo, F., Capone, A., Cesana, M., & Fratta, L. (2003). Ad hoc MAC: A new flexible and reliable mac architecture for ad-hoc networks. *In Proc. of IEEE WCNC.*

Bormann, C. (2005). Robust header compression (RoHC): Framework and four profiles: RTP, UDP, EPS, and uncompressed. *IETF RFC 3095.* Retrieved on January 2009, http://www.ietf.org/

Boufidis, Z., et al. (2004, September). *Actors, Management Plane, and Policy Provision Challenges for End-to-End Reconfiguration",* Paper presented at the ER Workshop on Reconfigurable Mobile Systems and Networks Beyond 3G, Barcelona, Spain. Broadband Radio Access Networks (ETSI BRAN). (n.d.). Retrieved December 30, 2008 from http://portal.etsi.org/portal_common/home.asp?tbkey1=BRAN

Bougard, B., Giulietti, A., Derudder, V., Weijers, J.-W., Dupont, S., Hollevoet, L., et al. (2003). A Scalable 8.7-nJ/bit 75.6-Mb/s Parallel Concatenated Convolutional (Turbo-) Codec. *IEEE International Solid-State Circuits Conference, 1*, 152-484.

Bougard, B., Giulietti, A., Van der Perre, L., & Catthoor, F. (2003). A low-power high-speed parallel concatenated turbo-decoding architecture. In *Proc. 3rd Intern. Symposium on Turbo codes and Related Topics*, Brest, France, (pp. 511-514).

Boutros, J., & Caire, G. (2002). Iterative multiuser joint decoding: unified framework and asymptotic analysis. *IEEE Transactions on Information Theory, 48*(7), 1772–1793. doi:10.1109/TIT.2002.1013125

Boutros, J., Boixadera, F., & Lamy, C. (2000, Sept.). Bit-interleaved coded modulations for multiple-input multiple-output channels. In *Proc. Of IEEE 6th Int. Symp. on Spread Spectrum Techniques and Applications*, (Vol. 1, pp. 123–126).

Boutros, J., Gresset, N., Brunel, L., & Fossorier, M. (2003, December). Soft-input soft-output lattice sphere decoder for linear channels. In *Proc. of IEEE Global Telecommun. Conf., GLOBECOM '03* (Vol. 3, pp. 1583–1587).

Braden, R., Clark, D., & Shenker, S. (1994). *Integrated Services in the Internet Architecture: an Overview (RFC1633)*. Fremont, CA: The Internet Engineering Task Force.

Braden, R., Zhang, L., Berson, S., Herzog, S., & Jamin, S. (1997). Resource reservation protocol (rsvp)-version 1 functional specification. *IETF RFC 2205*.

Brown, R. G. (1964). *Smoothing, forcasting and prediction of discrete time series*. Englewood Cliffs, NJ: Prentice Hall.

Bu, T., Chan, M. C., & Ramjee, R. (2006). Connectivity, performance, and resiliency of IP-based CDMA radio access networks. *IEEE Transactions on Mobile Computing, 5*(8), 1103–1118. doi:10.1109/TMC.2006.108

Bulow, H., Baumert, W., Schmuck, H., Mohr, F., Schulz, T., Kuppers, F., & Weiershausen, F. (1999). Measurement of the maximum speed of PMD fluctuation in installed field fiber. *Optical Fiber Communication Conference'99 Technical Digest, 2* (2), 83-85.

Bury, A., Engle, J., & Linder, J. (2003). Diversity comparison of spreading transforms for multicarrier spread spectrum transmission. *IEEE Transactions on Communications, 51*(5), 774–781. doi:10.1109/TCOMM.2003.811406

Cabric, D., Chen, M. S. W., Sobel, D. A., Yang, J., & Brodersen, R. W. (2005). Future Wireless Systems: UWB, 60GHz, and Cognitive Radios. In *IEEE Custom Integrated Circuits Conference 2005*.

Caire, G., & Pieri, F. (2009). LEAP USER GUI. Telecom Italia Lab. Retrieved Jannuary 30, 2009, from http://jade.tilab.com/doc/LEAPUserGuide.pdf.

Calhoun, P., Loughney, J., Guttman, E., & Arkko, J. (2003). *Diameter Base Protocol (RFC 3588)*. Retrieved from http://www.ietf.org/rfc/rfc3588.txt

Calhoun, P., Zorn, G., Spence, D., & Mitton, D. (2005). *Diameter Network Access Server Application (RFC 4005)*. Retrieved from http://www.ietf.org/rfc/rfc4005.txt

Callon, R. (1990). *Use of OSI IS-IS for Routing in TCP/IP and Dual Environments*. Network Working Group, RFC 1195.

Campbell, A. T., Gomez, J., Kim, S., Valkó, A. G., Wan, C., & Turányi, Z. R. (2000). Design, implementation, and evaluation of cellular IP. *IEEE Personal Communications Magazine, 7*(4), 42–49. doi:10.1109/98.863995

Campbell, G. (2008). *Global Wireless Matrix 2Q08*. Merill Lynch, Canada.

Cao, M., Ma, W., Zhang, Q., Wang, X., & Zhu, W. (2005). Modeling and Performance Analysis of the Distributed Scheduler for IEEE 802.16 Mesh Mode. In *The ACM International Symposium on Mobile Ad Hoc Networking and Computing* (pp. 78-89).

Cao, Y., Yang, Y., & Wang, H. (2008). Integrated Routing Wasp Algorithm and Scheduling Wasp Algorithm for Job Shop Dynamic Scheduling. In *International Symposium on Electronic Commerce and Security* (pp. 674-678).

Capmany, J., & Novak, D. (2007). Microwave photonics combines two worlds. *Nature Photonics, 1*, 319–330. doi:10.1038/nphoton.2007.89

Capone, A., Fratta, L. & Martignon, F.(2004). Banwidth estimation schemes for tcp over wireless networks. *Transactions on mobile computing, 3*(2), 1-15.

Carbonelli, C., & Franz, S. (2008). Performance analysis of MIMO OFDM ML detection in the presence of channel estimation error. In *Proc. IEEE ISSSTA*.

Carneiro, G., Ruela, J., & Ricardo, M. (2004). Cross-Layer Design in 4G Wireless Terminals. *IEEE Wireless Communications, 11*(2), 7–13. doi:10.1109/MWC.2004.1295732

Cassioli, D., Persia, S., Bernasconi, V., & Valent, A. (2005). Measurements of the performance degradation of UMTS receivers due to UWB emissions. *IEEE Communications Letters, 9*(5), 441–443. doi:10.1109/LCOMM.2005.1431165

CATT/China. (1998, June). *TD-SCDMA Radio Transmission Technology for IMT-2000.*

Cerf, V. (1978). *The Catenet Model for internetworking.* DARPA Information Processing Techniques Office, IEN 48.

Chaddha, N., & Diggavi, S. A. (1996). Frame-work for joint source-channel coding of images over time-varying wireless channels. *IEEE Image Processing, 1*, 89–92.

Chakareski, J. & Chou, P. (2006). RaDiO edge: Rate-distortion Optimized Proxy- Driven Streaming from the Network Edge. IEEE/ACM Transactions on Networking, 14(6), 1302-1312.

Challet, D., Marsili, M., & Zhang, T. (2004). *Minority Games Interacting Agents in Financial Markets*. Oxford, UK: Oxford Finance Series.

Chan, F., Wong, T., & Chan, P. (2005). Lot streaming technique in job- shop environment, IEEE Mediterranean Conference on Control and Automation. (Vol. 1, pp. 364-369). Washington, DC: IEEE Press.

Chandra, P. (2005). *Bullet-proof Wireless Security: GSM, UMTS, 802.11 and Ad Hoc Security.* Amsterdam: Elsevier.

Chang, Q., Fu, H., & Su, Y. (2008). Simultaneous generation and transmission of downstream multiband signals and upstream data in a bidirectional radio-over-fibre system. *IEEE Photonics Technology Letters, 20*, 181–183. doi:10.1109/LPT.2007.912992

Chang, W. Y. (2007). *Network Centric Service Oriented Enterprises.* Amsterdam, The Netherlands: SpringerLink.

Chang, Y., Chien, F., & Kuo, J. (2007, May). Cross-layer QoS analysis of opportunistic OFDM-TDMA and OFDMA networks. *IEEE Journal on Selected Areas in Communications, 25*(4), 657–666. doi:10.1109/JSAC.2007.070503

Chang, Y., Tao, Z., Zhang, J., & Kuo, J. (2009, June). *A graph approach to dynamic fractional frequency reuse (FFR) in multi-cell OFDMA networks.* Accepted for publication in *IEEE ICC'09*, June 2009, from http://www-scf.usc.edu/~yujungc/.

Chen, G., Grace, D., & Tozer, T. C. (2005). *Performance of Multiple HAPs using Directive HAP and User Antennas.* Paper presented at the Wireless Personal Communications, Aalborg, Denmark.

Chen, H.-H., & Yeh, J.-F. (2003). A complementary codes based CDMA architecture for wideband mobile Internet with high spectral efficiency and exact rate-matching. *International Journal of Communication Systems, 16*, 497–512. doi:10.1002/dac.592

Chen, H.-H., Chu, S. W., Kuroyanagi, N., & Han Vinck, A. J. (2007). An Algebraic Approach to Generate Super-Set of Perfect Complementary Codes for Interference-Free CDMA. [WCMC]. *Journal of Wireless Communications and Mobile Computing, 7*, 605–622. doi:10.1002/wcm.388

Chen, H.-H., Fan, C. X., & Lu, W. W. (2002). China's Perspectives on 3G Mobile Communications & Beyond: TD-SCDMA Technology. *IEEE Wireless Communications Magazine, 9*(2), 48–59. doi:10.1109/MWC.2002.998525

Chen, H.-H., Guizani, M., & Huber, J. F. (2006, February). Multiple Access Technologies for B3G Wireless Communications. *IEEE Communications Magazine, 43*(2), 65-67. Retrieved from http://www.comsoc.org/pubs/commag/cfpcommag205.htm

Chen, H.-H., Guizani, M., & Mohr, W. (2007). Evolution toward 4G Wireless Networking. *IEEE Network, 21*(1, January/February), 4-5. Retrieved from http://www.comsoc.org/pubs/net/

Chen, H.-H., Han Vinck, A. J., Bi, Q., & Adachi, F. (2006, January). The Next Generation of CDMA Technologies. [Retrieved from http://www.argreenhouse.com/society/JSAC/Calls/cdma technologies.html]. *IEEE Journal on Selected Areas in Communications, 24*(1), 1–3. doi:10.1109/JSAC.2005.858872

Chen, H.-H., Hanky, D., Magana, M. E., & Guizani, M. (2006). Design of next generation CDMA using Orthogonal

Compilation of References

Complementary Codes and Offset Stacked Spreading. *IEEE Wireless Communications.*

Chen, H.-H., Lin, J.-X., Chu, S.-W., Wu, C.-F., & Chen, G.-F. (2003, September). Isotropic Air-Interface Technologies for Fourth Generation Wireless Communications. [WCMC]. *Journal of Wireless Communications & Mobile Computing, 3*(6), 687–704. doi:10.1002/wcm.150

Chen, H.-H., Wong, D., & Mueller, P. (2006, September). Evolution of Air-Interface Technologies for 4G Wireless Communications. *IEEE Vehicular Technology Magazine (Editorial), IEEE Vehicular Technology Magazine, 1*(3), 2-3. Retrieved from http://www.ieeevtc.org/vtmagazine/index.html

Chen, H.-H., Xiao, Y., Li, J., & Fantacci, R. (2006, September). Challenges and futuristic perspective of CDMA technologies: the OCC-CDMA/OS for 4G wireless networking. *IEEE Vehicular Technology Magazine, IEEE Vehicular Technology Magazine, 1*(3), 12-21. Retrieved from http://www.ieeevtc.org/vtmagazine/index.html

Chen, H.-H., Yeh, J. F., & Seuhiro, N. (2001). A Multi-Carrier CDMA Architecture Based on Orthogonal Complementary Codes for New Generations of Wideband Wireless Communications. *IEEE Communications Magazine, 39*(10), 126–135. doi:10.1109/35.956124

Chen, H.-H., Yeh, Y.-C., Bi, Q., & Jamalipour, A. (2007, February). On a MIMO-based open wireless architecture: space-time complementary coding. *IEEE Communications Magazine, 45*(2), 104–112. doi:10.1109/MCOM.2007.313403

Chen, H.-H., Zhang, X., & Xu, W. (2007). Next Generation CDMA vs. OFDMA for 4G Wireless Applications. *IEEE Wireless Communications Magazine, 14*(3), 6-7. Retrieved from http://www.comsoc.org/pubs/pcm/

Chen, H-H. & Yeh, Y-C., (2005, April). Capacity of a Space-Time Block Coded CDMA System: Unitary Codes versus Complementary Codes. *IEEE Proceedings -Communications, 152*(2), 203-214.

Chen, H-H. (2007, July). *The Next Generation CDMA Technologies,* (1st Ed.). New York: John Wiley & Sons.

Chen, H.-H., Chiu H-W. & Guizani, M. (2006, February). Orthogonal complementary codes for interference-free CDMA technologies. *IEEE Wireless Communications,* 68-79.

Chen, H.-H., Yeh, Y-C., Zhang, X., Huang, A., Yang, Y., Li, J., & Xiao, Y., eta l. (2006, January). Generalized Pairwise Complementary Codes with Set-Wise Uniform Interference-Free Windows. *IEEE Journal on Selected Areas in Communications, 24*(1), 65–74. doi:10.1109/JSAC.2005.858878

Chen, J., Lv, T., & Zheng, H. (2004). Cross-Layer Design for QoS Wireless Communication. *The 2004 International Symposium on Circuits and Systems, ISCAS'04,* (vol. 2, pp. II-217-220).

Chen, J., Yu, K., Ji, Y., & Zhang, P. (2008). Non-Cooperative Distributed Network Resource Allocation in Heterogeneous Wireless Data Networks. *IEEE Transactions on Mobile Computing, 7*(3), 332–345. doi:10.1109/TMC.2007.70723

Chen, J.-C., & Zhang, (T. 2004). *IP-based Next Generation Wireless Networks.* Hoboken, NJ: Wiley.

Chen, L., Cai, X., Sofia, R., & Huang, Z. (2007, September). A cross-layer fast handover scheme for Mobile WiMAX. In *IEEE Vehicular Technology Conference 2007-Fall,* (pp. 1578-1582), Baltimore, MA.

Chen, T., & Hamalainen, S. (2003). Handover in IP RAN. *In Proc. IEEE International Conf. Communication Technology, 2,* 812-815.

Chen, W. T., & Huang, L. C. (2000). Rsvp mobility support: a signaling protocol for integrated services interent with mobile hosts. In *Proc. of IEEE Infocom,* (Vol. 3, pp.1283-1292).

Cheng, K., & Marca, J. (Eds.). (2008). Mobile WiMAX. West Sussex, UK: John Wiley & Sons, Ltd.

Cheong, Y., Cheng, R., Lataief, K., & Murch, R. (1999). Multiuser OFDM with adaptive subcarrier, bit, and power allocation. *IEEE Journal on Selected Areas in Communications, 17*(10), 1747–1757. doi:10.1109/49.793310

Cherkassky, B. V., Goldberg, A. V., & Radzik, T. (1994). Shortest Paths Algorithms: Theory and Experimental Evaluation. In D. D. Sleator (Ed.), *Proceedings of the 5th Annual ACM-SIAM Symposium on Discrete Algorithms (SODA 94)* (pp. 516-525). Arlington, VA: ACM Press.

Chia, M. (2003). Radio over multimode fibre transmission for wireless LAN using VCSELs. *Electronics Letters, 39,* 1143–1144. doi:10.1049/el:20030724

Chiussi, F., Khotimsky, D. A., & Krishnan, S. (2002). A network architecture for MPLS-based micro-mobility. In *Proc. of IEEE Wireless Communications and Networking (WCNC),* (pp.549-555), Toulouse, France.

Cho, Y., Radha, H., Yoo, J., & Hong, J. (2008). A rate-distortion empirical model for rate adaptive wireless scalable video. In *International Conference in Information Sciences and Systems*. (pp. 350- 355). Princeton, NJ: IEEE Press.

Choi, J. Y., Yoo, J. J., & Shin, J. (2008). Cross-Layer Transmission Scheme for Wireless H.264 Using Distortion Measure and MAC-Level Error-Control. In *International Conference in Advanced Communication Technology*. (pp. 1232-1237). Gangwon-Do, South Korea: IEEE Press.

Chou, P. A. (2007). Streaming Media on Demand and Live Broadcast. In M. Van der Schaar & P. A. Chou (Ed.), *Multimedia over IP and Wireless Networks: Compression, Networking, and Systems* (pp. 453-502). London: Academic Press.

Christensen, C. (2004). *Seeing What's Next: Using Theories of Innovation to Predict Industry Change*. Cambridge, MA: Harvard Business School Press.

Chu, D. C. (1972). Polyphase codes with good periodic correlation properties. *IEEE Transactions on Information Theory*, *18*, 531–532. doi:10.1109/TIT.1972.1054840

Chuah, C. N., Tse, D. N. C., Kahn, J. M., & Valenzuela, R. A. (2002). Capacity scaling in MIMO wireless systems under correlated fading. *IEEE Transactions on Information Theory*, *48*(3), 637–650. doi:10.1109/18.985982

Chuang, J., & Sollenberger, N. (2000). Beyond 3G: wideband wireless data access based on OFDM and dynamic packet assignment. *IEEE Communications Magazine*, *38*(7), 78–87. doi:10.1109/35.852035

Cianca, E., Sanctis, M., & Ruggieri, M. (2007). Convergence Towards 4G: A Novel View of Integration. S*pringer . Wireless Personal Communications*, *33*(3-4), 327–336. doi:10.1007/s11277-005-0577-y

Cioffi, J. M. (2004). *EE379C: advanced digital communications*. Lecture Notes, Stanford University.

Clark D.D. & Fang,W.(1998). Explicit allocation of best-effort packet delivery service. *IEEE/ACM Transactions on Networking*, *6*(4), 362-373.

Clark, D., Braden, R., Falk, A., & Pingali, V. (2003). Reorganizing the Addressing Architecture. In *Proceedings ACM SIGCOMM FDNA Workshop*, USA.

Clausen, T. H., & Jacquet, P. (2003). Optimized Link State Routing Protocol. Request for Comments: 3626, Standard Track, The Internet Engineering Task Force.

Claussen, H., Karimi, H. R., & Mulgrew, B. (2005). Improved MAX-LOG-MAP turbo decoding by maximization of mutual information transfer. *EURASIP Journal on Applied Signal Processing*, *6*, 820–827. doi:10.1155/ASP.2005.820

Codes. In *ISIT2000*, Sorrento, Italy (p. 362).

Collela, N. J., Martin, J. N., & Akyildiz, I. F. (2000). The HALO Network. *IEEE Communications Magazine*, *38*(6), 142–148. doi:10.1109/35.846086

Conta, A., & Deering, S. (1998). Internet control message protocol (ICMPv6) for the internet protocol version 6 (IPv6) specification. *IETF RFC 2463*. Retrieved on January 2009, http://www.ietf.org/

Corderio, C., & Challapali, K. (2007). C-MAC: a cognitive MAC protocol for multichannel wireless networks. In *Proc. IEEE Symposium on New Frontiers in Dynamic Spectrum Access networks*, April 2007.

Corlett, A., Pullin, D. I., & Sargood, S. (2002). *Statistics of one-way internet packet delays*. Paper presented at the 53rd IETF meeting, Minneapolis, USA.

Corning (2008). ClearCurve single-mode optical fibre. Retrieved from http://www.corning.com/opticalfibre/products/clearcurve_single_mode_fibre.aspx, February 2008.

Corning (2009). ClearCurve® OM3/OM4 multimode optical fibre. Retrieved from http://www.corning.com/opticalfibre/products/clearcurve_multimode_fibre.aspx, January 2009.

Corson, M. S., & Ephremides, A. (1995). Lightweight Mobile Routing protocol (LMR), A distributed routing algorithm for mobile wireless networks. *Wireless Networks*, 1.

Cost297. (2005). *The European Union (EU) Cost 297 Action on High Altitude Platforms for Communications and Other Services*. Retrieved from http://www.hapcos.org/overview.php

Costa, C., Cunha, I., Borges, A., Ramos, C., Rocha, M., Almeida, J., & Ribeiro-Neto, B. (2004). Analyzing Client Interactivity in Streaming Media. In *ACM International Conference on Wide World Web Conference*. (pp. 534-543). Beijin, China: ACM Press.

Costa, X. P., Schmitz, R., Hartenstein, H., & Liebsch, M. (2002). A MIPv6, FMIPv6 and HMIPv6 Handover Latency Study: Analytical Approach. In *Proceedings of IST Mobile and Wireless Telecommunications Summit* (pp. 100-105).

Costa, X. P., Schmitz, R., Hartenstein, H., & Liebsch, M. (2002). A MIPV6, fmipv6 and HMIPv6 handover latency study. In *Proc. of IST Mobile and Wireless Telcommunication*.

Côté, N., Möller, S., Gautier-Turbin, V., & Raake, A. (2006). Analysis of a Quality Prediction Model for Wideband Speech Quality, the WB-PESQ. In *Proceedings of the 2nd ISCA/DEGA Tutorial and Research Workshop on Perceptual Quality of Systems*, Berlin (pp. 115-122).

Courcoubetis, C., Siris, V. A., & Stamoulis, G. D. (1998). Network control and usage-based charging: is charging for volume adequate. In *Proc of the first international Conference on Information and Computation economics*, New York.

Cover, T. M., & El Gamal, A. (1979). Capacity theorems for the relay channel. *IEEE Transactions on Information Theory*, 25(5), 572–584. doi:10.1109/TIT.1979.1056084

Cowan, R., & Jonard, N. (2004). Network structure and the diffusion of knowledge. *Journal of Economic Dynamics & Control*, 28(8), 1557–1575. doi:10.1016/j.jedc.2003.04.002

Cox, C. H., III. (2004). *Analog Optical Links - Theory and Practice*. Cambridge, MA: Cambridge University Press

Crisler, K., Turner, T., Aftelak, A., Visciola, M., Steinhage, A., & Anneroth, M. (2004). Considering the user in the wireless world. *Communications Magazine*, 42(9), 56–62. doi:10.1109/MCOM.2004.1336721

CTIA, 2008. http://files.ctia.org/pdf/080108_US-OECD_10_Comparison_Ex_Parte.pdf

Cui, S., Goldsmith, A. J., & Bahai, A. (2004). Energy-efficiency of MIMO and cooperative MIMO techniques in sensor networks. *IEEE Journal on Selected Areas in Communications*, 22(6), 1089–1098. doi:10.1109/JSAC.2004.830916

Dagli, N. (1999). Wide-Bandwidth Lasers and Modulators for RF Photonics. *IEEE Transactions on Microwave Theory and Techniques*, 47(7), 1151–1171. doi:10.1109/22.775453

Dahlman, E., Parkval, S., Skol, J., & Beming, P. (2007). 3G Evolution: HSPA and LTE for Mobile Broadband. London: Academic Press.

Dahlman, E., Parkvall, S., Skold, J., & Beming, P. (2008). *3G Evolution: HSPA and LTE for Mobile Broadband*. Burlington, MA: Academic Press.

Dahr, S. (2005). MANET: Applications, Issues, and Challenges for the Future. *International Journal of Business Data Communications and Networking*, 1(2), 66–92.

Daly, E. M., & Haahr, M. (2007). Social network analysis for routing in disconnected delay-tolerant MANETs. In *8th ACM international symposium on Mobile ad hoc networking and computing* (pp.32-40). New York: ACM.

Damen, M. O. (1999). *Joint coding/decoding in a multiple access system, application to mobile communications*. Ph.D. dissertation, Ecole Nationale Superieure des Telecommunications.

Damen, M. O., Abed-Meriam, K., & Belfiore, J.-C. (2000). A generalized lattice decoder for asymmetrical space-time communication architecture. In *Proceedings of the IEEE International Conference on Acoustics, Speech and Signal Processing 2000 (ICASSP2000)* (pp. 2581–2584).

Damen, M. O., Chkeif, A., & Belfiore, J.-C. (2000 June). Sphere Decoding of Space-Time

Damen, M. O., Chkeif, A., & Belfiore, J.-C. (2000, May). Lattice Code Decoder for Space-Time Codes. *IEEE Communications Letters*, 4, 161–163. doi:10.1109/4234.846498

Damen, M. O., El-Gamal, H., & Cairo, G. (2003). On maximum-likelihood detection and the search for the closest lattice point. *IEEE Transactions on Information Theory*, 49(10), 2389–2402. doi:10.1109/TIT.2003.817444

Dammann, A., & Kaiser, S. (2001). Standard conformable antenna diversity techniques for OFDM and its application to the DVB-T system. *IEEE Global Telecommunications Conference, San Antonio, Texas, USA, 25-29 November 2001*, (pp. 3100-3105).

Daniel, F., Macedo, A. L., & Pujolle, G. (2008). From TCP/IP to Convergent Networks: Challenges and Taxonomy. *IEEE Communications Surveys & Tutorials*, 10(4), 40–55.

Dantu, R. (2005). *Method and System of Integrated Rate Flow Control for a Traffic Flow across Wireline and Wireless Networks*. US Patent No. 6904286 B1. Washington, DC: U.S. Patent and Trademark Office.

Dantu, R., Patel, P. R., Choksi, O. T., Patel, A. R., Ali, M. R., Miernik, J., & Holur, B. S. (2006). *Wireless Router and method for processing traffic in wireless communication network*. US Patent No. 7068624 B1. Washington, DC: U.S. Patent and Trademark Office.

Darcie, T. E., & Bodeep, G. E. (1990). Lightwave subcarrier CATV transmission systems. *IEEE Transactions on Microwave Theory and Techniques, 38*(5), 524–533. doi:10.1109/22.54920

Dardari, D., Martini, M., Milantoni, M., & Chiani, M. (2002). MPEG-4 video transmission in the 5 GHz band through an adaptive OFDM wireless scheme. *Proceedings of the IEEE PIMRC'02,* Vol.4, (pp. 1680-1684), Lisboa, Portugal. Washington, DC: IEEE.

Das, A., Nkansah, A., Gomes, N. J., Garcia, I. J., Batchelor, J. C., & Wake, D. (2006). Design of low-cost multimode fibre-fed indoor wireless networks. *IEEE Transactions on Microwave Theory and Techniques, 54*(8), 3426–3432. doi:10.1109/TMTT.2006.877835

Das, S., McAuley, A., Dutta, A., Misra, A., Chakraborty, K., & Das, S. K. (2002). IDMP: An intra-domain mobility management protocol for next generation wireless networks. *IEEE Wireless Communications Magazine, 9*(3), 38–45. doi:10.1109/MWC.2002.1016709

Das, S., Misra, A., & Agrawal, P. (2000). TeleMIP: Telecommunications enhanced Mobile IP architecture for fast intradomain mobility. *IEEE Personal Communications Magazine, 7*(4), 50–58. doi:10.1109/98.863996

Davie, B., Charny, A., Bennett, J.C.R., Benson, K., Le Boudec, J.Y., Courtney, W., Davari, S., Firoiu, V., & Stiliadis, D. (2002). An expedited forwarding PHB (per-hop behavior). *IETF RFC 3246.*

de Jong, Y. L. C., & Willink, T. J. (2005). Iterative tree search detection for MIMO wireless systems. *IEEE Transactions on Communications, 53*(6), 930–935. doi:10.1109/TCOMM.2005.849638

De Laat, C., Gross, G., Gommans, L., Vollbrecht, J., & Spence, D. (2000). *Generic AAA Architecture (RFC 2903).* Retrieved from http://www.ietf.org/rfc/rfc2903.txt

De, S., et al. (2002). Integrated Cellular and Ad Hoc Relay (iCAR) Systems: Pushing the Performance Limits of Conventional Wireless Networks. In *International Conference on System Sciences,* Hawaii.

Deering, S., & Hinden, R. (1995). *Internet Protocol, Version 6 (IPv6) Specification* [RFC 1883]. Retrieved from http://www.faqs.org/rfcs/rfc1883.html

Deering, S., & Hinden, R. (1998). Internet protocol Version 6 (IPv6) specification. *IETF RFC 2460.* Retrieved on January 2009, http://www.ietf.org/

Desilva, S., & Das, S. R. (1998). Experimental evaluation of channel state dependent scheduling in an in-building wireless LAN. In *7th International Conference on Computer Communications and Networks* (pp.414-421).

Devarapalli, V., Gundavelli, S., Chowdhury, K., & Muhanna, A. (2007). *Proxy Mobile IPv6 and Mobile IPv6 interworking.* Retrieved June 2008, from http://tools.ietf.org/html/draft-devarapalli-netlmm-pmipv6-mipv6-00

Devarapalli, V., Wakikawa, R., Petrescu, A. & Thubert, P. (2005). *Network Mobility (NEMO) Basic Support Protocol, Request for Comments 3963.* Retrieved June 2008, from http://www.ietf.org/rfc/rfc3963.txt

Devarapalli,V., Wakikawa, R., Petrescu, A., & Thubert, P.(2005). Network mobility (NEMO) basic support protocol. *IETF RFC3963.*

Di Caro, G. (2004). *Ant colony optimization and its application to adaptive routing in telecommunication networks.* Ph.D. thesis, Université Libre de Bruxelles, Belgium.

Dijkstra, E. W. (1959). A Note on Two Problems in Connection with Graphs. *Numer. Math, 1,* 269–271.

Ding, Z., Ratnarajah, T., & Cowan, C. C. F. (2007). On the diversity-multiplexing tradeoff for wireless cooperative multiple access systems. *IEEE Transactions on Signal Processing, 55*(9), 4627–4638. doi:10.1109/TSP.2007.896276

Division Multiplexing System. In *IEEE VTC2000* (Vol. 45, pp. 6–10).

Dixit, S. (2008). *Should India go for 4G Wireless Networks Now?* Retrieved March 1, 2009, from http://www.shvoong.com/exact-sciences/1776578-india-4g-wireless-networks/

Dixit, S., & Prasad, R. (Eds.). (2008). Technologies for Home Networking. Hoboken, NJ: John Wiley & Sons, Ltd.

Djama, I., Ahmed, T., Nafaa, A., & Boutaba, R. (2008). Meet In the Middle Cross-Layer Adaptation for Audiovisual Content Delivery. *IEEE Transactions on Multimedia, 10*(1), 105–120. doi:10.1109/TMM.2007.911243

Djordjevic, B. (2007). PMD compensation in fibre-optic communication systems with direct detection using LDPC-coded OFDM. *Optics Express, 15*(7), 3692–3701. doi:10.1364/OE.15.003692

Djuknic, G. M., Freidenfelds, J., & Okunev, Y. (1997). Establishing Wireless Communications Services via High-Altitude Aeronautical Platforms: A Concept Whose

Compilation of References

Time Has Come? *IEEE Communications Magazine, 35*(9), 128–135. doi:10.1109/35.620534

Dohler, M., Rassool, B., & Aghvami, H. (2003). *Performance evaluation of STTCs for virtual antenna arrays.* Paper presented at the IEEE Spring Vehicular Technology Conference (VTC), Seogwipo, South Korea.

Dong, B., Wang, X., & Doucet, A. (2003). A new class of soft MIMO demodulation algorithms. *IEEE Transactions on Signal Processing, 51*(11), 2752–2763. doi:10.1109/TSP.2003.818155

Dong, X., & Beaulieu, N. C. (2002). Average level crossing rate and average fade duration of low-order maximal ratio diversity with unbalanced channels. *IEEE Communications Letters, 6*, 135–137. doi:10.1109/4234.996033

Dorigo, M., & Blum, C. (2005). Ant colony optimization theory: A survey. *Theoretical Computer Science, 344*(2-3), 243–278. doi:10.1016/j.tcs.2005.05.020

Dovis, F., Fantini, R., Mondin, M., & Savi, P. (2001). 4G Communications Based on High Altitude Stratospheric Platforms: Channel Modeling and Performance Evaluation. In *Global Telecommunications Conference, GLOBECOM '01 IEEE* (Vol. 1, pp. 557-561).

Draka (2008). Enhanced low macrobending sensitive Single Mode Fibre: BendBrightXS. Retrieved from http://www.drakafibre.com/draka/DrakaComteq/Drakafibre/Languages/English/Navigation/Markets%26Products/Single_mode/BendBright-XS/index.html, September 2008.

Du, P., Jia, W. J., Huang, L. S., & Lu, W. Y. (2007). Centralized Scheduling and Channel Assignment in Multi-Channel Single Transceiver WiMax Mesh Network. In *2007 IEEE Wireless Communications and Networks Conference* (pp. 1736-1741).

Duan, C., & Pekhteryev, G., & Fang, J., & Nakache, Y., & Zhang, J., & Tajima, k., & Nishioka, Y., & Hirai, H. (2006). Transmitting multiple HD video streams over UWB links. In *Proceedings of of CCNC'06*, Las Vegas (vol. 2, pp. 691-695).

Dursch, A., Yen, D. C., & Huang, S.-M. (2005). Fourth generation wireless communications: an analysis of future potential and implementation. *Computer Standards & Interfaces, 28*, 13–25. doi:10.1016/j.csi.2004.10.009

Eardley, P., & Hancock, R. (2000). Modular IP architecture for wireless mobile access. In *1st International Workshop on Broadband Radio Access For IP Based Networks.*

Eardley, P., Eisl, J., Hancock, R., Higgins, D., Manner, J., & Ruiz, P. (2002). Evolving Beyond UMTS - The MIND Research Project. In *Proceedings of IEE 3rd International Conference on Mobile Communication Technologies* (pp. 449-454).

ECMA-368. (2007). High rate ultra wideband PHY and MAC Standard. In *ECMA International Standard (2nd Ed.)*, December 2007.

Edfors, O., Sandell, M., van de Beek, J. J., Wilson, S. K., & Borjesson, P. O. (1998). OFDM channel estimation by singular value decomposition. *IEEE Transactions on Communications, 46*, 931–939. doi:10.1109/26.701321

Eichhorn, A. (2008). Middleware Abstractions for Cross-Layer controlled Media Streaming. In *ACM Workshop on Middleware- Application Interaction.* (pp. 13-18). Oslo, Norway: ACM Press.

Einhaus, M., Klein, O., & Lott, M. (2005). Interference Averaging and Avoidance in the Downlink of an OFDMA System. In *Proceedings of 16th IEEE International Symposium on Personal, Indoor and Mobile Radio Communications*, (2), 905–910.

Ekström, H., Furuskär, A., Karlsson, J., Meyer, M., Parkvall, S., Torsner, J., & Wahlqvist, M. (2006). Technical Solutions for the 3G Long-Term Evolution. *IEEE Communications Magazine, 44*(3), 38–45. doi:10.1109/MCOM.2006.1607864

Elam, E., & Yurdakul, S. (2008). Wisair's Wireless USB Single Chip Poised for Broader Global Distribution with Recent European Regulatory Approval and USB-IF Certification. *Wisair Ltd press release.* Retrieved from http://www.usb.org/press/WUSB_press/2008_03_04_wisair.pdf

Elayoubi, S.-E., Ben Haddada, O., & Fourestie, B. (2008). Performance Evaluation of Frequency Planning Schemes in OFDMA-based Networks. *IEEE Transactions on Wireless Communications, 7*(5), 1623–1633. doi:10.1109/TWC.2008.060458.

Electrozoom. (2007). Association of Electronic and Communication Engineers.

Ellis, R. B., Weiss, F., & Anton, O. M. (2007). HFC and PON-FTTH networks using higher SBS threshold single mode optical fibre. *Electronics Letters, 43*(7), 405–407. doi:10.1049/el:20070218

Erceg, V. (1999)... *IEEE Journal on Selected Areas in Communications, 17*(7), 1205–1211. doi:10.1109/49.778178

Erceg, V., Greenstein, L. J., Tjandra, S. Y., Parkoff, S. R., Gupta, A., & Kulic, B. (1999). An empirically based path loss model for wireless channels in suburban environments. *IEEE Journal on Selected Areas in Communications, 17*(7), 1205–1211. doi:10.1109/49.778178

Erdös, P., & Rényi, A. (1959). On Random Graphs I. *Publicationes Mathematicae, 6*, 290–297.

Ergen, M. (2009). *Mobile Broadband - Including WiMAX and LTE*. New York: Springer.

Ergen, M., Coleri, S., & Varaiya, P. (2003). QoS Aware Adaptive Resource Allocation Techniques for Fair Scheduling in OFDMA Based Broadband Wireless Access Systems. *IEEE Transactions on Broadcasting, 49*(4), 362–370. doi:10.1109/TBC.2003.819051

Ericsson (2005, October), Handover interruption times. *3GPP TSG RAN WG2-WG3 Joint Meeting R3-051091*.

Ericsson Inc. (2009, June). *LTE – an introduction* [White Paper]. Retrieved from http://www.ericsson.com/technology/whitepapers/lte_overview.pdf

Ermolova, N. Y., & Makarevitch, B. (2007). Low Complexity Adaptive Power and Subcarrier Allocation for OFDMA. *IEEE Transactions on Wireless Communications, 6*(2), 433–437. doi:10.1109/TWC.2007.05232

Ernst, T., & Lach, H.(2007). Network mobility support terminology. *RFC 4885*.

Eronen, P., Hiller, T., & Zorn, G. (2005). *Diameter Extensible Authentication Protocol (EAP) Application (RFC 4072)*. Retrieved from http://www.ietf.org/rfc/rfc4072.txt

Esposito, R., Hamilton, B. A., & Graveman, R. (2003). Security with IPv6 Explored. In *RFG Security, U.S. IPv6 Summit 2003*. Retrieved from http://www.usipv6.com/2003arlington/presents/Renee_Esposito_and_Rich_Graveman.pdf

Etoh, M. (2005). *Next Generation Mobile Systems: 3G and Beyond*. Hoboken, NJ: Wiley.

ETSI EN 300 744, v.1.5.1 (2004-11), *Digital Video Broadcasting (DVB); Framing structure, channel coding and modulation for digital terrestrial television*.

ETSI EN 301 958 (2002). *Digital Video Broadcasting (DVB): Interaction channel for digital terrestrial television (RCT) incorporating multiple access OFDM*.

ETSI Technical Report 102 506 v.1.1.1. (2006). *Speech Processing, Transmission and Quality Aspects (STQ); Estimating Speech Quality per Call*. Sophia Antipolis, France: European Telecommunications Standards Institute (ETSI).

ETSI/SMG2. (1998, June). *The ETSI UMTS Terrestrial Radio Access (UTRA) ITU-R RTT Candidate Submission*.

EU IST Project ANWIRE. (Academic Network on Wireless Internet Research in Europe) (n.d). Retrieved August 30, 2008, from http://www.ist-world.org/ProjectDetails.aspx?ProjectId=8f7d258e5ca542239e0b0b3d5ea16e33

EU IST Project BRAIN/MIND. (n.d). Retrieved December 30, 2008, from http://www.itu.int/osg/imt-project/docs/4.3_Wisely.pdf

EU IST Project MOBIVAS. (Downloadable Mobile Value-Added-Services through Software Radio and Switching Integrated Platforms) (n.d). Retrieved December 30, 2008 from http://www.ist-world.org/ProjectDetails.aspx?ProjectId=7a73337c09a3488d8947757b63db84c5

EU IST Project MOBY DICK. About the project (n.d). *About the project*. Retrieved December 30, 2008, from http://www.nw.neclab.eu/Projects/mobydick.htm

EU IST Project SCOUT (Smart user-centriC cOmmUnication environmenT) (n.d). Retrieved December 30, 2008, from http://www.ist-world.org/ProjectDetails.aspx?ProjectId=3f4727fdc7ff444c87de64fe5c579084

Exploring IMS security mechanisms: will 3GPP and ETSI specs be enough to protect IP-based multimedia traffic from attack? (n.d). *GOLIATH, Business Knowledge On Demand*. Retrieved on September 10, 2009, from http://goliath.ecnext.com/coms2/gi_0198-378937/Exploring-IMS-security-mechanisms-will.html

Falahati, A., & Ardestani, M. (2007). An Improved low-complexity resource allocation algorithm for OFDMA systems with proportional data rate constraint. In *International Conference on Advanced Communication Technology ICACT*, Gangwon-Do, South Korea (pp. 606-611).

Fallah, H., & Lechler, T. (2003). *Global innovation management in wireless communications industry*. IAMOT.

Farrell, S., Vollbrecht, J., Calhoun, P., Gommans, L., Gross, G., Bruijn, B., et al. (2000). *AAA Authorization Requirements (RFC 2906)*. Retrieved from http://www.ietf.org/rfc/rfc2906.txt

Fazel, K., & Kaiser, S. (2008). *Multi-Carrier and Spread Spectrum Systems: From OFDM and MC-CDMA to LTE and WiMAX (2nd Ed.)*. Hoboken, NJ: John Wiley & Sons.

FCC 02-48. (2002). *Revision of Part 15 of the Commission's Rules regarding ultra-wideband transmission systems*. Federal Communications Commission.

Fei, Y., Zhong, L. & Jha, N. K. (2008). An Energy-Aware Framework for Dynamic Software Management in Mobile Computing Systems. ACM Transactions on Embedded Computing Systems, 7(3), 27:1-27:31.

Ferber, J. (1999). *Multi-Agent Systems*. Reading, MA: Addison-Wesley.

Fernandez, E., et al. (2005). *Some security issues of wireless systems*. In *Advanced Distributed Systems: 5th International School and Symposium, ISSADS 2005*, Guadalajara, Mexico. Retrieved from http://www.cse.fau.edu/%7Eed/Fernandez_ISSADS2005Final.pdf

Fernekeß, A., Klein, A., Wegmann, B., Dietrich, K., & Litzka, M. (2008). Load Dependent Inteference Margin for Link Budget Calculations of OFDMA Networks. *IEEE Communications Letters, 12*(5), 398–400. doi:10.1109/LCOMM.2008.080032

Fincke, U., & Pohst, M. (1985, April). Improved methods for calculating vectors of short length in a lattice, including a complexity analysis. *Mathematics of Computation, 44*, 463–471. doi:10.2307/2007966

Floyd, S., & Jacobson, V. (1993). Random early detection gateways for congestion avoidance. *IEEE ACM Transaction on Networking, 1*, 397–413. doi:10.1109/90.251892

Flynn, P., & Ganchev, I., I., & M. O'Droma, M. (2006). Wireless Billboard Channels: Vehicle and Infrastructural Support for Advertisement, Discovery, and Association of UCWW Services. *Annual Review of Communications, 59*, 493–504.

Fodor, G., Telek, M., & Koutsimanis, C. (2008). Performance analysis of scheduling and interference coordination policies for OFDMA networks. *Computer Networks, 52*(6), 1252–1271. doi:10.1016/j.comnet.2008.01.006

Fogelstroem, E., Jonsson, A., & Perkins, C. (2007, June). Mobile IPv4 regional registration. *IETF RFC 4857*.

Ford, L., & Fulkerson, D. (1962). Flows in Networks. Upper Saddle River, NJ: Prentice-Hall.

Foschini, G. J. (1996). Layered space-time architecture for wireless communication in a fading environment when using multi-element antennas. *Bell Labs Tech. J., 1*, 41–59. doi:10.1002/bltj.2015

Foschini, G. J., & Gans, M. J. (1998). On the limits of wireless communications in a fading when using multiple antennas. *Wireless Personal Communications, 6*, 311–335. doi:10.1023/A:1008889222784

Foschini, G. J., Karakayali, M. K., & Valenzuela, R. A. (2006). Coordinating Multiple Antenna Cellular Networks to Achieve Enormous Spectral Efficiency. *Proceedings of the IEEE, 153*(4), 548–555. doi:10.1049/ip-com:20050423

Foschini, G., & Gans, M. (1998). On limits of wireless communication in a fading environment when using multiple antennas. *Wireless Personal Communications, 6*(3), 311–335. doi:10.1023/A:1008889222784

Foukalas, F., Gazis, V., & Alonistioti, N. (2008). Cross-Layer Design Proposals for Wireless Mobile Networks: A Survey and Taxonomy. *IEEE Communications Surveys & Tutorials, 10*(1), 70–85. doi:10.1109/COMST.2008.4483671

Francis, M. (2003). *Journal of Business Strategy*. American Sentinel University.

Frank, C., Manoj, K., & Murthy, S. (2002). *Throughput Enhanced Wireless in Local Loop (TWiLL) - The Architecture, Protocols, and Pricing schemes*. Paper presented in IEEE LCN. 2002.

Frattasi, S., Fathi, H., Fitzek, F. H. P., Prasad, R., & Katz, M. D. (2006, January-February). Defining 4G technology from the users perspective. *Network, 20*(1), 35–41.

Frattasi, S., Fathi, H., Fitzek, F., Chung, K., & Prasad, R. (2004). *4G: A User-Centric System*. Paper presented at the Mobile e-Conference (Me), Athens, Greece.

Frattasi, S., Fathi, H., Fitzek, F., Katz, M., & Prasad, R. (2005). A Pragmatic Methodology to Design 4G: From the User to the Technology. In *Springer LNCS: 5th IEEE International Conference on Networking (ICN)*, Reunion Island, France.

Frattasi, S., Fitzek, F., & Prasad, R. (2005). Cooperative Services for 4G. In *Proceedings of 14th IST Mobile and Wireless Communications Summit*, Dresden, Germany.

Frederic, P., Paal, E., Erik, V., Thomas, H., & Anne, M. N. Kjell, Mari. N. & Stein, S. (2002). *Mobility Aspects in 4G Networks* [White Paper]. Forskningsnotat/Scientific Document, Norway.

Friedman-Hill, E. (2003). *Jess in Action: Java Rule-based Systems.* Greenwich, CT: Manning Publications Company.

Fujii, H., & Yoshino, H. (2008). Theoretical Capacity and Outage Rate of OFDMA Cellular System with Fractional Frequency Reuse. In *Proceedings of IEEE Vehicular Technology Conference,* (pp. 1676–1680).

Gallager, R. G. (1962). Low-density parity-check codes. *I.R.E. Transactions on Information Theory, 8,* 21–28. doi:10.1109/TIT.1962.1057683

Gallaher, M., & Rowe, B. (2005). *The Costs and Benefits of Transferring Technology Infrastructures Underlying Complex Standards: The Case of IPv6.* Presented at Technology Transfer Society Annual Conference, Kansas City, MO. Retrieved from http://www.kauffman.org/uploadedFiles/Gallaher_Mike.pdf

Galstyan, A., Kolar, S., & Lerman, K. (2003). Resource allocation games with changing resource capacities. In *International Conference on Autonomous Agents, ACM* (pp. 145-152).

Galván, S., Quintana, M. A., Macías, E. M., & Suárez, A. (2005). Multimedia Cooperative Applications on Wireless Networks using Aspect Oriented Programming. In IEEE International Conference on Wireless Networks. (Vol 1, pp. 268-274). Las Vegas, NV: CSREA Press.

Ganchev, I., & O'Droma, M. M. (2007). " New personal IPv6 address scheme and universal CIM card for UCWW". In *Proc. of the 7th International Conference on Intelligent Transport Systems Telecommunications (ITST 2007),* Pp. 381-386, 6-8 June, Sophia Antipolis, France (pp. 381-386). IEEE Catalog No: 07EX1765. Library of Congress: 2007923401. ISBN: 1-4244-1177-7.

Ganchev, I., O'Droma, M. S., Siebert, M., Bader, F., Chaouchi, H., & Armuelles, I. (2006). A 4G Generic ANWIRE System and Service Integration Architecture. *ACM SIGMOBILE Mobile Computing and Communications Review Journal, 10*(1), 13–30. doi:10.1145/1119759.1119761

Ganz, A., Ganz, Z., & Wongthavarawat, K. (2004). Multimedia Wireless Networks, Tecnologies, Standards and QoS. Upper Saddle River, NJ: Prentice Hall PTR.

Gao, L., Zhang, Z. & Towsley, D. (2003). Proxy-Assisted Techniques for Delivering Continuous Multimedia Streams. IEEE/ACM Transactions on Networking, 11(6), 884-894.

Gardezi, A. (n.d.). *Security In Wireless Cellular Networks.* Retrieved from http://www.cse.wustl.edu/~jain/cse574-06/ftp/cellular_security/#issues

Garg, V. K. (2000). *IS-95 CDMA and cdma2000, cellular/PCS systems implementation.* Upper Saddle River, NJ: Prentice Hall.

Gazis, V., Housos, N., Alonistioti, A., & Merakos, L. (2003). Generic System Architecture For 4G Mobile Communications. In *Vehicular Technology Conference* (pp. 1512-1516).

Gehrmann, C., Person, J., & Smeet, B. (2004). *Bluetooth Security.* Norwood, MA: Artech House.

Gennari, J. H. (2003). The Evolution of Protege: An Environment for Knowledge-Based Systems Development. *International Journal of Human-Computer Studies, 58,* 89–123. doi:10.1016/S1071-5819(02)00127-1

Ghinea, G., & Angelides, M. C. (2004). Quality of Perception in M- Commerce. In N. Shi (Ed.), Mobile Commerce Applications (pp. 284- 302). Hershey, PA: IRM Press.

Ghosh, D., Gupta, A., & Mohapatra, P. (2007). Admission Control and Interference-Aware Scheduling in Multi-hop WiMAX Networks. In *IEEE Mobile Ad-hoc and Sensor Systems (MASS)* (pp. 1-9).

Ghosh, J., Ngo, H. Q., Yoon, S., & Qiao, C. (2007). On a Routing Problem Within Probabilistic Graphs and its Application to Intermittently Connected Networks. In *26th IEEE International Conference on Computer Communications (INFOCOM)* (pp. 1721-1729).

Ghosh, J., Philip, S. J., & Qiao, C. (2007). Sociological orbit aware location approximation and routing (SOLAR) in MANET. *Ad Hoc Networks, 5*(2), 189–209. doi:10.1016/j.adhoc.2005.10.003

Ghosh, S., Basu, K., & Das, S. K. (2005, September/October). What a Mesh! An Architecture for Next Generation Radio Access Networks. *IEEE Network,* 35–42. doi:10.1109/MNET.2005.1509950

Giles, C. R., & Desurvire, E. (1991). Propagation of signal and noise in concatenated erbium-doped fiber optical amplifiers. *IEEE Journal of Lightwave Technology, 9,* 147–154. doi:10.1109/50.65871

Giorgetti, A., Chiani, M., Dardari, D., Piesiewicz, R., & Bruck, G. H. (2008). The Cognitive Radio Paradigm for Ultra-Wideband Systems: the European Project EUWB.

IEEE International Conference on Ultra-Wideband ICUWB 2008 (vol. 2, pp. 169-172).

Girici, T., Zhu, C., Agre, J. R., & Ephremides, A. (2008). Practical Resource Allocation Algorithms for QoS in OFDMA-based Wireless Systems. In *IEEE International Broadband Wireless Access Workshop* (pp. 897-901).

Giuliano, R., Luglio, M., & Mazzenga, F. (2008, March). Interoperability between WiMAX and Broadband Mobile Space Networks. *IEEE Communications Magazine*, 50–57. doi:10.1109/MCOM.2008.4463771

Glisic, S. A. (2004). *Advanced Wireless Communications - 4g Technologies*. Chichester, UK: John Wiley and Sons.

Golden, G. D., Foschini, G. J., Valenzuela, R. A., & Wolnianky, P. W. (1999). Detection algorithm and initial laboratory results using V-BALST space-time communication architecture. *Electronics Letters*, *35*(1), 14–16. doi:10.1049/el:19990058

Goldsmith, A. (2004). *Wireless Communications*. Cambridge, UK: Cambridge Univ. Press.

Google Android Software Development Kit. (n.d.). Retrieved December 18, 2008, from http://code.google.com/android

Govil, J. (2008). An empirical feasibility study of 4G's key technologies. In *Electro/Information Technology, (EIT)-08* (pp. 267-270).

Govil, J., & Govil, J. (2007). 4G Mobile Communication Systems: Turns, Trends and Transition. In *International Conference on Convergence Information Technology*, Korea (pp. 13-18).

Goyal, V. (2006). Pro Java ME MMAPI: Mobile Media API for Java Micro Edition. Berkeley, CA: APRESS.

Gozali, R., Buehrer, R. M., & Woerner, B. D. (2003). The impact of multiuser diversity on space-time block coding. *IEEE Communications Letters*, *7*(5), 213–215. doi:10.1109/LCOMM.2003.812182

Grace, D., Thornton, J., Chen, G., White, G. P., & Tozer, T. C. (2005). Improving the System Capacity of Broadband Services Using Multiple High Altitude Platforms. *IEEE Transactions on Wireless Communications*, *4*(2), 700–709. doi:10.1109/TWC.2004.842972

Grassé, P.-P. (1959). La reconstruction du nid et les coordinations inter-individuelles chez Bellicositermes natalensis et Cubitermes sp., La théorie de la Stigmergie: Essai d'interprétation du comportement des Termites Constructeurs. *Insectes Sociaux*, *6*, 41–80. doi:10.1007/BF02223791

Graveman, R., Parthasarathy, M., Savola, P., & Tschofenig, H. (2009 May). Using IPsec to Secure IPv6-in-IPv4 Tunnels. *RFC 4891*. Retrieved from http://www.faqs.org/rfcs/rfc4891.html

Grimmett, G. (1999). *Percolation (2nd ed.)*. Grundlehren der Mathematischen Wissenschaften. Berlin: Springer-Verlag.

Grossglauser, M., & Tse, D. (2001). Mobility increases the capacity of wireless adhoc networks. In *IEEE INFOCOM* (pp. 1360-1369).

GSA. (2009). *GSM/3G Market Update*. Retrieved from http://www.gsacom.com.

GSM World (2008). *Mobile Broadband, Competition and Spectrum Caps*. Retrieved from http://hspa.gsmworld.com/upload/resources/files/19012009094421.pdf

Guennec, Y. L., & Gary, R. (2007). Optical frequency conversion for millimeter-wave ultra-wideband-over-fibre systems. *IEEE Photonics Technology Letters*, *19*, 996–998. doi:10.1109/LPT.2007.898745

Gunasekaran, V., & Harmantzis, F. (2005). Migration to 4G-Ubiquitous Broadband-Economic modeling of Wi-Fi with WiMAX. In *Proc. Of WWC, SFO*, USA.

Gundavelli, S., Leung, K., Devarapalli, V., Chowdhury, K., & Patil, B. (2008). *Proxy Mobile IPv6, Request for Comments 5213*. Retrieved September, 2008, from http://www.rfc-editor.org/rfc/rfc5213.txt

Gunduz, D., & Erkip, E. (2005). *Outage minimization by opportunistic cooperation*. Paper presented at the IEEE Wireless Communications Conference, Symposium on Information Theory, Maui, Hawaii.

Guo, F., Chen, J., Li, W., & Chiuch, T. (2004). Experiences in building a multihomming load balancing systems. In *Proc. of IEEE Infocom*, (Vol. 2).

Guo, Z., & Nilsson, P. (2006). Algorithm and implementation of the K-best sphere decoding for MIMO detection. *IEEE Journal on Selected Areas in Communications*, *24*(3), 491–503. doi:10.1109/JSAC.2005.862402

Gupta, P., & Kumar, P. R. (2002). The Capacity of Wireless Networks. *IEEE Transactions on Information Theory*, 388–404.

Gupta, P., & Kumar, R. (2001, January). The capacity of wireless networks. *IEEE Transactions on Information Theory, 46*(2), 388–404. doi:10.1109/18.825799

Gupta, V., Williams, M., Johnston, D., McCann, S., Barber, Ph., & Ohba, Y. (2006). IEEE 802.21: Overview of standard for media independent handover services (Tech. Rep.). Washington, DC: IEEE Press.

Gutjahr, W. (2008). First steps to the runtime complexity analysis of ant colony optimization. *Journal Computers & Operations Research, 35*(9), 2711–2727. doi:10.1016/j.cor.2006.12.017

Haenggi, M. (n.d.). *4G Wireless Standard*. Retrieved March 1, 2009, from http://www.nd.edu/~mhaenggi/NET/wireless/4G/

Hagen, S. (2002 July). *IPv6 Essentials*. O'Reilly Publications. Retrieved from http://www.sunny.ch/downloadfiles/ipv6_sample.pdf

Hagenauer, J. (1997, September). The turbo principle: Tutorial introduction and state of the art. In *Proc. 1st Int. Symposium on Turbo Codes* (pp. 1–12).

Hagenauer, J. (1998). Rate-compatible punctured convolutional codes (RCPC codes) and their application. *IEEE Transactions on Communications, 36*(4), 339–400.

Hagenauer, J., Offer, E., & Papke, L. (1996, March). Iterative Decoding of Binary Block and Convolutional Codes. *IEEE Transactions on Information Theory, 42*, 429–445. doi:10.1109/18.485714

Hakala, H., Mattila, L., Koskinen, J.-P., Stura, M., & Loughney, J. (2005). *Diameter Credit-Control Application (RFC 4006)*. Retrieved from http://www.ietf.org/rfc/rfc4006.txt

Han, B., Tso, F. P., Lin, L., & Jia, W. (2006). Performance Evaluation of Scheduling in IEEE 802.16 Based Wireless Mesh Networks. In *2006 IEEE Mobile Ad-hoc and Sensor Systems (MASS)* (pp. 789-794).

Han, Z., Ji, Z., & Liu, R. J. K. (2008). A cartel maintenance framework to enforce cooperation in wireless networks with selfish users. *IEEE Transactions on Wireless Communications, 7*(5), 1889–1899. doi:10.1109/TWC.2008.061014

Handley, M., Floyd, S., Padhye, J., & Widmer, J. (2003). TCP Friendly Rate Control (TFRC): Protocol Specification. Request for Comments: 3448, Standard Track, The Internet Engineering Task Force.

Harjula, I., Ramirez, A., Martinez, F., Zorrilla, D., Katz, M., & Polo, V. (2008). Practical Issues in the Combining of MIMO Techniques and RoF in OFDM/A Systems. In *Proceedings of 7th WSEAS International Conference on Electronics, Hardware, Wireless and Optical Conference*.

Hartenstein, H., & Laberteaux, K. P. (2008). A tutorial survey on vehicular ad hoc networks. *IEEE Communications Magazine*, 164–171. doi:10.1109/MCOM.2008.4539481

Hartmann, P., et al. (2003). Low-cost multimode fibre-based wireless LAN distribution system using uncooled directly modulated DFB laser diodes. In *Proceedings of ECOC 2003*, Rimini, Italy (pp. 804–805).

Hartung, F., Horn, U., Huschke, J., Kampmann, M., Lohmar, T., & Lundevall, M. (2007). Delivery of broadcast services in 3G networks. *IEEE Transactions on Broadcasting, 53*(3), 188–199. doi:10.1109/TBC.2007.891711

Hassan, M., Vikram, K., & Panos, N. (2008). Rate and distortion modeling of medium grain scalable video coding. In IEEE International Conference on Image Processing, (pp. 2564-2567). San Diego, CA: IEEE Press.

Hassibi, B., & Vikalo, H. (2005). On Sphere Decoding Algorithm. Part I. Expected Complexity. *IEEE Transactions on Signal Processing, 53*(8), 2806–2818. doi:10.1109/TSP.2005.850352

Hassibi, B., & Vikalo, H. (2005). On Sphere Decoding Algorithm. Part II. generalization, second order statistics and applications to communications. *IEEE Transactions on Signal Processing, 53*(8), 2819–2834. doi:10.1109/TSP.2005.850352

He, Z., Cai, J., & Chen, C. W. (2002). Joint source channel rate-distortion analysis for adaptive mode selection and rate control in wireless video coding. *IEEE Transactions on Circuits and Systems for Video Technology, 12*, 511–523. doi:10.1109/TCSVT.2002.800313

Heegaard, P. E., Helvik, B. E., & Wittner, O. J. (2008). The Cross Entropy Ant System for Network Path Management. *Telektronikk, 1*, 19–40.

Heidrich, H., & Hoffmann, D., & Macdonald, R.I. (1986). Polarization and wavelength multiplexed biderectional single fiber subscriber loop. *Journal of optical communications, 7*(4), 136-138.

Heinanen, J., & Baker, F. (1999). RFC 2597-assured forwarding PHB group. *IETF RFC 2597*.

Compilation of References

Helfrich, B. (1985). Algorithms to construct Minkowski reduced and Hermit reduced lattice bases. *Theoretical Computer Science*, *41*, 125–139. doi:10.1016/0304-3975(85)90067-2

Henry, P. S. (2007) Integrated Optical/Wireless Alternatives for the Metropolitan Environment. IEEE Communications Society Webminar. Retrieved from http://www.comsoc.org/webinar/1/PSH--Integ%20Opt%20and%20 WlessFINAL.pdf

Hirata, A. (2003). 120-GHz wireless link using photonic techniques for generation, modulation, and emission of millimeter-wave signals. *IEEE Journal Lightwave Technology*, *21*(10), 2145–2153. doi:10.1109/JLT.2003.814395

Hitachi, L. C. D.-T. V. (2008). Wooo UT-series Available in website www.hitachi.com and http://av.hitachi.co.jp/tv/l_lcd/ut/index.html.

Ho, L. T. W., & Claussen, H. (2007). Effects of User-Deployed, Co-Channel Femtocells on the Call Drop Probability in a Residential Scenario. In *Proceedings of IEEE Personal, Indoor and Mobile Radio Communications PIMRC 2007*.

Hochwald, B. M., & ten Brink, S. (2003). Achieving near-capacity on a multiple-antenna channel. *IEEE Transactions on Communications*, *51*(3), 389–399. doi:10.1109/TCOMM.2003.809789

Hoebeke, J. (2006). Personal Networks Federations. In *Proceedings of Fifteenth IST Mobile and Wireless Summit*, Myconos, Greece.

Hoeher, P., Kaiser, S., & Robertson, P. (1997). Pilot-symbol-aided channel estimation in time and frequency. In *Proc. IEEE Global Telecommunications Conference (GLOBECOM '97)*.

Hogg, S. (2007). IPv6: Dual stack where you can; tunnel where you must. *NetworkWorld*. Retrieved from http://www.networkworld.com/news/tech/2007/090507-techuodate.html

Holma, A., & Toskala, A. (Eds.). (2007). *WCDMA for UMTS: HSPA Evolution and LTE*. West Sussex, UK: John Wiley & Sons, Ltd.

Holma, H., & Toskala, A. (2000). *WCDMA for UMTS*. Hoboken, NJ: Wiley.

Holma, H., & Toskala, A. (2005). *WCDMA for UMTS, radio access for third generation mobile communications*, (3rd ed). New York: John Wiley & Sons.

Holtzman, J. M., & Sampath, A. (1995). Adaptive averaging methodology for handoffs in cellular systems. *IEEE Transactions on Vehicular Technology*, *44*(1), 59–66. doi:10.1109/25.350270

Hong, Y., & Choi, J. (May 2005). A new approach to iterative decoding on coded MIMO channels. In *Proc. of IEEE 61th Vehicular Tech. Conf., VTC '05-Spring* (Vol. 3, pp. 1676–1680).

Hossain, E., & Leung, K. K. (Eds.). (2008). *Wireless Mesh Networks: Architectures and Protocols*. New York: Springer Verlag.

Host-Madsen, A. (2002). *On the capacity of wireless relaying*. Paper presented at the IEEE Fall Vehicular Technology Conference (VTC), Vancouver, Canada.

Hottinen, A., & Heikkinen, T. (2006, March). Subchannel assignment in OFDM relay nodes. In *Proc. of 40th Annual Conference on Information Sciences and Systems*, (pp. 1314-1317), Princeton, NJ.

Housley, R., Polk, W., Ford, W., & Solo, D. (2002). *Internet X.509 Public Key Infrastructure Certificate and Certificate Revocation List (CRL) Profile (RFC 3280)*. Retrieved from http://www.ietf.org/rfc/rfc3280.txt

Houssos, N., Alonistioti, A., Merakos, L., Dillinger, M., Fahrmair, M., & Schoenmakers, M. (2003, August). Advanced adaptability and profile management framework for the support of flexible mobile service provision. *IEEE Wireless Communications*, *10*(4), 52–61. doi:10.1109/MWC.2003.1224979

Howells, J. (2005). *The management of innovation and technology*. Thousand Oaks, CA: Sage.

Hoymann, C., & Göebels, S. (2008). Dimensioning cellular multi-hop WiMAX networks. In K. Chen & J. R. B. de Marca (Ed.), *Mobile WiMAX*, (pp. 203-234). New York: John Wiley & Sons, Ltd.

Hsu, A., Wei, D., & Kuo, C. (2007). A cognitive mac protocol using statistical channel allocation for wireless ad-hoc networks. In *Proc. IEEE WCNC*, March.

Hu, M., Zhang, J., & Sadowasky, J. (2004). Traffic aided opportunistic scheduling for wireless networks: algorithms and performance bounds. *The International Journal of Computer and Telecommunications Networking*, *46*(4), 505–518.

Hu, R., & Ti, J. (2006). *Practical compress-forward in user cooperation: Wyner-Ziv cooperation*. Paper presented at th IEEE International Symposium on Information Theory (ISIT), Seattle, USA.

Huang, M.-F., Yu, J., Qian, D., & Chang, G.-K. (2008). Lightwave Centralized WDM-OFDM-PON. In *Proceedings of ECOC 2008, Paper Th1F5*.

Huang, V., & Zhuang, W. (2002). QoS-oriented access control for 4G mobile multimedia CDMA communications. *IEEE Communication Magazine*, 118-125, March.

Hui, P., & Crowcroft, J. (2007). Bubble rap: Forwarding in small world dtns in ever decreasing circles. Technical Report UCAM-CL-TR-684, Cambridge Univ., Comp. Lab., 2007.

Hui, P., Chaintreau, A., Scott, J., & Gass, R. Crowcroft & Diot, C. (2005). Pocket switched networks and human mobility in conference environments. In *ACM SIGCOMM workshop on Delay-tolerant networking*, Philadelphia, Pennsylvania, USA (pp. 244 - 251).

Hui, P., Crowcroft, J., & Yoneki, E. (2008). BUBBLE Rap: Social Based Forwarding in Delay Tolerant Networks. In *9th ACM International Symposium on Mobile Ad Hoc Networking and Computing (MobiHoc)* (pp. 241-250). New York: ACM.

Hult, T., Mohammed, A., Yang, Z., & Grace, D. (2008). *Performance of a Multiple HAP System Employing Multiple Polarization*. Invited Paper, Special Issue on Wireless Personal Multimedia Communications (WMPC) 2006 Conference.

Hunter, T. E., & Nosratinia, A. (2006). Diversity through coded cooperation. *IEEE Transactions on Wireless Communications*, 5(2), 283–289. doi:10.1109/TWC.2006.1611050

Hunter, T. E., Sanayei, S., & Nosratinia, A. (2006). Outage analysis of coded cooperation. *IEEE Transactions on Information Theory*, 52(2), 375–391. doi:10.1109/TIT.2005.862084

Hussain, S., Hamid, Z., & Khagttak, N. (2006). Mobility Management Challenges and Issues in 4G Heterogeneous Networks. In *ACM Proceedings of the first international conference on Integrated internet ad hoc and sensor networks*. Retrieved from http://delivery.acm.org/10.1145/1150000/1142698/a14-hussain.pdf?key1=1142698&key2=8898704611&coll=GUIDE&dl=&CFID=1515 1515&CFTOKEN=6184618

Huusko, J., Vehkaper̈a, J., Amon, P., Lamy-Bergot, C., Panza, G., Peltola, J., & Martini, M.G. (2006). Cross-layer Architecture for Scalable Video Transmission in Wireless Network. *Signal processing: Image communication Special Issue on Mobile Video, 22*(3), 317-330.

I, C.-L., Greenstein, L. J., & Gitlin, R. D. (1993). A Microcell/Macrocell Cellular Architecture for Low- and High-Mobility Wireless Users. *IEEE Journals on Selected Areas in Communications*, 11(6), 885-891.

Ibrahim, A. S., Sadek, A. K., Su, W., & Liu, K. J. R. (2005). *Cooperative communications with partial channel state information: When to cooperate?* Paper presented at the IEEE Globecom Conference, St-Louis, USA.

Ibrahim, A. S., Sadek, A. K., Su, W., & Liu, K. J. R. (2006). *Relay selection in multi-node cooperative communications: When to cooperate and whom to cooperate with?* Paper presented at the IEEE Globecom conference, San Francisco, USA.

Ibrahim, J. (2002). 4G Features. *Bechtel Telecommunications Technical Journal*.

IEEE 802.11 - 1999 edition, Part 11: Wireless LAN Medium Access Control (MAC) and Physical Layer (PHY) Specifications, *IEEE Standard 802.11*.

IEEE 802.11e. (2005). Part 11: Wireless LAN Medium Access Control (MAC) and Physical Layer (PHY) specifications Amendment 8: Medium Access Control (MAC) Quality of Service Enhancements. *IEEE Standard for Information technology—Telecommunications and information exchange between systems—Local and metropolitan area networks—Specific requirements*.

IEEE 802.16-2004. (2004, June). *IEEE Standard for Local and metropolitan area networks - Part 16: Air Interface for Fixed Broadband Wireless Access Systems*.

IEEE 802.16e (2005). *IEEE Std 802.16e-2005, Part 16: air interface for fixed and mobile broadband wireless access systems*, Feb. 28, 2006.

IEEE LAN MAN standards committee of the IEEE computer society (2008). Draft standard for local and metropolitan area networks: Media independent handover services. *IEEE P802.21/D10.00*. New York: IEEE.

IEEE Standard for Local and Metropolitan Area Networks Part 16, (2004). *Air Interface for Fixed Broadband Wireless Access Systems, IEEE Std 802.16-2004*.

Compilation of References

IEEE Std 802.16TM-2004. (2004). *802.16TM IEEE standard for local and metropolitan area networks Part 16: Air interface for fixed broadband wireless access systems.*

IEEE. 802.11. (n.d.). *IEEE 802.11 wireless local area networks.* Retrieved from http://ieee802.org/11/

IEEE. 802.16 task group d. (n.d.). Retrieved from http://ieee802.org/16/tgd/

IEEE. 802.16 task group m. (n.d.). Retrieved from http://www.ieee802.org/16/tgm/

IEEE. 802.16. (n.d.). *The IEEE 802.16 working group on broadband wireless access standards.* Retrieved from http://ieee802.org/16/

IEEE. 802.16e task group. (n.d.). Retrieved from http://ieee802.org/16/tge/

IEEE. 802.20 (2005, September). 802.20 evaluation criteria – ver.1.0. *IEEE 802.20-PD-09.* Retrieved from http://www.ieee802.org/20/

IEEE802. *16 standards*. (2009). Retrieved from http://ieee802.org/16

IEEE802.16. (2004). *IEEE802.16-2004 Part 16: Air Interface for Fixed Broadband Wireless Access System.* IEEE802.16 Broadband Wireless Access Working Group.

IETF RFC 3261. (2002). *SIP: Session Initiation Protocol.* Internet Engineering Task Force (IETF).

IETF RFC 3344. (2002). *IP Mobility Support for IPv4.* Internet Engineering Task Force (IETF).

IETF RFC 3519. (2003). *Mobile IP Traversal of Network Address Translation (NAT) Devices.* Internet Engineering Task Force (IETF).

IETF RFC 4068. (2005). *Fast Handovers for Mobile IPv6.* Internet Engineering Task Force (IETF).

IETF RFC 4140. (2005). *Hierarchical Mobile IPv6 Mobility Management (HMIPv6).* Internet Engineering Task Force (IETF).

Ikegami, T., & Suematsu, Y. (1967). Resonance-like characteristics of the direct modulation of a junction laser. *Proceedings of the IEEE, 55*, 122–123. doi:10.1109/PROC.1967.5420

Imielinski, T., Viswanathan, S., & Badrinath, B. R. (1994). Energy Efficient Indexing on Air. In *Proc of ACM-SIGMOD, Intl Conference on Management of Data*, Minnesota.

Imielinski, T., Viswanathan, S., & Badrinath, B. R. (1997). Data on Air: Organization and Access. *IEEE Transactions on Knowledge and Data Engineering, 9*.

Inés, D. E., Fujikawa, K., Kawai, E., & Sunahara, H. (2008). Video traffic optimization in mobile wireless environments using adaptive applications. In International Conference on Mobile Ubiquitous Computing, Systems, Services and Technologies. (pp. 376-379). Valencia, Spain: IEEE Press.

Ingham, J. D., et al. (2003). Wide-frequency-range operation of a high linearity uncooled DFB laser for next-generation radio-over-fibre. In *Proceedings of IEEE/OSA OFC 2003*, Atlanta, GA (pp. 754–756).

Ippolito, L. J. (2008). Satellite Communications Systems Engineering: Atmospheric Effects, Satellite Link Design and System Performance. West Sussex, England: John Wiley & Sons, Ltd.

IPv6 Transition Considerations. (2009, February). 3G Americas. Retrieved from http://www.3gamericas.org/index.cfm?fuseaction=pressreleasedisplay&pressreleaseid=2150

IPv6. (n.d.). Wikipedia. Retrieved on September 8, 2009, from http://en.wikipedia.org/wiki/IPv6

ISATAP (Intra-Site Automatic Tunnel Addressing Protocol). (n.d.). Wikipedia. Retrieved on September 8, 2009, from http://en.wikipedia.org/wiki/ISATAP

ISO. (2006). *The calm handbook: Continuous communication for vehicles.* Technical report. *CALM Handbook ISO, TC204, WG16.*

Issues in Mobility Management in 4G Networks. (2001). *Computer, 34*(6). Retrieved from http://folk.uio.no/paalee/referencing_publications/ref-mob-4gissues.pdf

IST-WINNER. (2006). *D5.4, Final Report on Link Level and System Level Channel Models. IST-4-027756 WINNER II.*

Ito, H., Furata, T., Kodama, S., & Ishibashi, T. (2000). InP/InGaAs uni-traveling-carrier photodiode with 310 GHz bandwidth. *Electronics Letters, 36*, 1809–1819. doi:10.1049/el:20001274

Itoh, K., Watanabe, S., Shih, J., & Sato, T. (2002). Performance of handoff algorithm based on distance and RSSI measurements. *IEEE Transactions on Vehicular Technology, 51*(6), 1460–1468. doi:10.1109/TVT.2002.804866

Ituero, P., & Lopez-Vallejo, M. (2006, September). New Schemes in Clustered VLIW Processors Applied to Turbo Decoding. *IEEE International Conference on Application-Specific Systems, Architectures and Processors* (pp. 291-296).

ITU-R. (2000). Feasibility of High Altitude Platform Station (HAPS) in the fixed and mobile services in the frequency bands above 3 GHz allocated exclusively for terrestrial radiocommunication. *Resolution 734 [COM5/14] Question ITU-R 212/9.*

ITU-R. (2003). *Framework and Overall Objectives of the Future Development of IMT-2000 and Systems Beyond IMT-2000* [Recommendation ITU-R M. 1645].

ITU-R. (2003). *P.1546-3. Method for point-to-area predictions for terrestrial services in the frequency range 30 MHz to 3000 MHz.*

ITU-R. (2007). *Principles for the Process of Development of IMT-Advanced* [Resolution ITU-R 57].

ITU-T Draft Recommendation Q.3202.1. (2008). *Authentication Protocols based on EAP-AKA for Interworking among 3GPP, WiMax and WLAN in NGN.* Retrieved from www.itu.int/ITU-T/ngn/fgngn

Jager, D., & Stohr, A. (2001), Microwave Photonics. In *Proceedings of European Microwave Conference 2001*, London (pp. 1-4).

Jalden, J., & Ottersten, B. (2005). On the complexity of sphere decoding in digital communications. *IEEE Transactions on Signal Processing, 53*(9), 1474–1484. doi:10.1109/TSP.2005.843746

Jamalipour, A., & Lorenz, P. (2006). End-to-end qos support for ip and multimedia traffic in heterogeneous mobile networks. *Computer Communications, 29*, 671–682. doi:10.1016/j.comcom.2005.07.021

Janani, M., Hedayat, A., Hunter, T. E., & Nosratinia, A. (2004). Coded cooperation in wireless communications: Space-time transmission and iterative decoding. *IEEE Transactions on Signal Processing, 52*(2), 362–371. doi:10.1109/TSP.2003.821100

Jang, J., & Lee, K. B. (2003). Transmit Power Adaptation for Multiuser OFDM Systems. *IEEE Journal on Selected Areas in Communications, 21*(2), 171–178. doi:10.1109/JSAC.2002.807348

Janjua, K. A., & Khan, S. A. (2007). A Comparative Economic Analysis of different FTTH Architectures. *Wireless Communications, Networking and Mobile Computing*, 4979-4982.

Jansen, S. L., Morita, I., Takeda, N., & Tanaka, H. (2007). *20-Gb/s OFDM transmission over 4,160-km SSMF enabled by RF-pilot tone phase noise compensation.* Paper presented at the Optical Fibre Communication Conference OFC2007, Anaheim, CA.

Javaid, U., Meddour, D. E., & Ahmed, T. (2006). Towards Universal Convergence in Heterogeneous Wireless Networks using Ad Hoc Connectivity. In *Proceedings of of 9th International Conference on Wireless Personal Multimedia Communications (WPMC)*, San Diego.

Javaid, U., Rasheed, T., Meddour, D. E., & Ahmed, T. (2007). A Profile-based Personal Network Architecture for Personal Ubiquitous Environments. In *Proceedings of IEEE Vehicular Technology Conference (VTC)*, Dublin, Ireland.

Javaid, U., Rasheed, T., Meddour, D. E., & Ahmed, T. (2007). Personal Network Routing Protocol (PNRP) for Personal Ubiquitous Environments. In *Proceedings of IEEE International Conference on Communications (ICC)*, Glasgow, U.K.

Jayaraman, P., Lopez, R., Parthasarathy, M., & Yegin, A. (2008). *Protocol for Carrying Authentication for Network Access (PANA) Framework (RFC 5193).* Retrieved from http://www.ietf.org/rfc/rfc5193.txt

Jayaweera, S. K. (2006). Virtual MIMO-based cooperative communications for energy-constrained wireless sensor networks. *IEEE Transactions on Wireless Communications, 5*(5), 984–989. doi:10.1109/TWC.2006.1633350

JDSU. (2007). *Test & Measurement Note, JDSU Triple-Play Service Deployment Guide.* JD Uniphase Corp.

Jekosch, U. (2005). *Voice and Speech Quality Perception. Assessment and Evaluation.* Berlin, Germany: Springer.

Ji, Z., Ganchev, I., & O'Droma, M. (2008, April). *Reliable and Efficient Advertisements Delivery Protocol for Use on Wireless Billboard Channels.* Paper presented at the meeting of the 12th IEEE International Symposium on Consumer Electronics, Algarve, Portugal.

Ji, Zh. Ji, Z., I. GanchevGanchev, I., & O'Droma, M. (2008, June). *Efficient Collecting, Clustering, Scheduling, and Indexing Schemes for Advertisement of Services over Wire-*

less Billboard Channels*. Paper presented at the meeting of the 15th International Conference on Telecommunications, St. Petersburg, Russia.

Jindal, A., & Psounis, K. (2006). Performance Analysis of Epidemic Routing under Contention. In *International Conference On Communications And Mobile Computing* (pp. 539-544). New York: ACM.

Jindal, N., & Goldsmith, A. (2005). Dirty-paper coding versus TDMA for MIMO Broadcast channels. In *ICC* (pp. 1783-1794).

Jo, J., & Cho, J. (2008, August). A cross-layer vertical handover between Mobile WiMAX and 3G networks. *International Wireless Communications and Mobile Computing Conference*, (pp. 644-649), Crete Island, Greece.

Joham, M., Barbero, L. G., Lang, T., Utschick, W., Thompson, J., & Ratnarajah, T. (2008), FPGA implementation of MMSE metric based efficient near-ML detection. In *Proc. Int. ITG Workshop on Smart Antennas* (pp. 139–146).

Johari, R., Mannor, S., & Tsitsiklis, J. N. (2005). Efficiency loss in a network resource allocation game: the case of elastic supply. *IEEE Transactions on Automatic Control*, *50*(11), 1712–1724. doi:10.1109/TAC.2005.858687

Johnson, D., Hu, Y., & Maltz, D. (2007). *The Dynamic Source Routing Protocol (DSR) for Mobile Ad Hoc Networks for IPv4*. Network Working Group, RFC 4728.

Johnson, D., Perkin, C., & Arkko, J. (2004). *Mobility Support in IPv6, Request for Comments 2775*. Retrieved September, 2005, from http://www.ietf.org/rfc/rfc3775.txt

Johnsson, M. (1999). HiperLAN/2- The broadband radio transmission technology operating in the 5GHz frequency band, technical specification. V 1.0. *HiperLAN/2 Global Forum*.

Juang, R., Lin, H., & Lin, D. (2005, March). An improved location-based handover algorithm for GSM systems. *IEEE Wireless Communications and Networking Conference*, (pp. 1371-1376), Los Angeles, CA.

Kallel, S., & Leung, C. (1976). Efficient ARQ schemes with multiple copy decoding. *IEEE Transactions on Communications, COM-24*, 592–606.

Kamilo, F. (1995). *Wireless Digital Communications - Modulation and Spread Spectrum Applications*. New York: Prentice Hall.

Kappler, C. (Ed.). (2004). *Scenarios, Requirements and Concepts*. IST–2002-507134-AN/ WP3/D/3-1, Ambient Network WP3 deliverable.

Karakayali, M. K., Foschini, G. J., & Valenzuela, R. A. (2006). Network Coordination for Spectrally Efficient Communications in Cellular Systems. *IEEE Wireless Communications Magazine*, *13*(4), 56–61. doi:10.1109/MWC.2006.1678166

Karapantazis, S., & Pavlidou, F. (2005). Broadband communications via high-altitude platforms: a survey. *Communications Surveys & Tutorials, IEEE*, *7*(1), 2–31. doi:10.1109/COMST.2005.1423332

Karp, B., & Kung, H. T. (2000). GPSR: Greedy perimeter stateless routing for wireless networks. In *Proc. of ACM/IEEE MOBiCOM*, August.

Kartalopoulos, S. V. (2002). *DWDM: Networks, Devices, and Technology*. Hoboken, NJ: Wiley.

Kashyap, A., Ganguly, S., & Das, S. (2007). A measurement-based approach to modeling link capacity in 802.11-based wireless networks. In *Proc. of the 13th annual ACM international conference on Mobile computing and networking*, (pp. 242-253).

Katz, M., & Fitzek, F. (2005) Cooperative Techniques and Principles Enabling Future 4G Wireless Networks. In *Proceedings of IEEE EUROCON 2005*. Serbia & Montenegro, Belgrade.

Katz1, M., & Fitzek, F. (2005). On the Definition of the Fourth Generation Wireless Communications Networks: The Challenges Ahead. In *International Workshop on Convergent Technologies*, Finland.

Kawadia, V., & Kumar, P. R. (2005). A Cautionary Perspective on Cross-Layer Design. *IEEE Wireless Communications*, *12*(1), 3–11. doi:10.1109/MWC.2005.1404568

Kelley, B., & Rivas, E. (2008). OFDM location-based routing protocols in ad-hoc networks. *IEEE Wireless Hive Networks Conference, WHNC 2008*.

Kempe, D., Kleinberg, J. M., & Demers, A. J. (2001). Spatial gossip and resource location protocols. In *Thirty-third annual ACM symposium on Theory of computing* (pp. 163-172). New York: ACM.

Kempf, J., & Yegani, P. (2002). OpenRAN: a new architecture for mobile wireless Internet radio access networks. *IEEE Communications Magazine*, *40*(5), 118–123. doi:10.1109/35.1000222

Kenah, E., & Robins, J. (2007). Network-based analysis of stochastic SIR epidemic models with random and proportionate mixing. *Journal of Theoretical Biology, 249*(4), 706–722. doi:10.1016/j.jtbi.2007.09.011

Khadani, E., Abounadi, J., Modiano, E., & Zheng, L. (2007). Cooperative Routing in Static Wireless Networks. *IEEE Transactions on Communications*, 2158–2192.

Khan, S., Peng, Y., Steinbach, E., Sgroi, M., & Kellerer, W. (2006). Application-Driven Cross-Layer Optimization for Video Streaming over Wireless Networks. *IEEE Communications Magazine, 44*(1), 122–130. doi:10.1109/MCOM.2006.1580942

Khattab, A. K. F., & Elsayed, K. M. F. (2004). Channel-quality dependent earliest deadline due fair scheduling schemes for wireless multimedia networks. In *International symposium on modeling, analysis and simulation of wireless and mobile,* 2002, Venice, Italy, (pp. 31-38).

Kholer, E., Handley, M., & Floyd, S. (2006). Datagram Congestion Control Protocol (DCCP) Request for Comments: 4340, Standard Track, The Internet Engineering Task Force.

Kiczales, G., Lamping, J., Mendhekar, A., Maeda, C., Lopes, C. V., Loingtier, J. M., & Irwin, J. (1997). Aspect- Oriented Programming. In European Conference on Object-Oriented Programming, (LNCS Vol. 1241, pp. 220-242). Jyväskylä, Finland: Springer Verlag.

Kim, D., Lee, H., Kim, N., & Yoon, H. (2009). A velocity-based bicasting handover scheme for 4G mobile systems. *IEICE Transactions on Communications . E (Norwalk, Conn.), 92-B*(1), 288–295.

Kim, H., & Han, Y. (2005). A Proportional Fair Scheduling for Multi-carries Transmission Systems. *IEEE Communications Letters, 9*(3), 210–212. doi:10.1109/LCOMM.2005.03014

Kim, I., Park, I., & Lee, Y. H. (2006). Use of Linear Programming for Dynamic Subcarrier and Bit Allocation in Multiuser OFDM. *IEEE Transactions on Vehicular Technology, 55*(4), 1195–1207. doi:10.1109/TVT.2006.877490

Kim, J., & Bambos, N. (2002). Power efficient MAC scheme using channel probing in multi-rate wireless ad hoc networks. In *IEEE Vehicular Technology Conference*, (pp. 2380–2384).

Kim, M., Kim, H., Ra, I., Yoo, J., & Kim, D. (2008, January). Cross-layer based fast handover mechanism for seamless macro-mobility support in WiBro networks. *The International Conference on Information Networking*, 1-5, Busan, Korea.

Kim, Y. (2003). Beyond 3G: Vision, Requirements, & Enabling Technologies. *IEEE Communications Magazine, 41*(3), 120–124. doi:10.1109/MCOM.2003.1186555

Kind, A., & Grant, A. (2003, Feb.). Efficient list decoding for lattices. In *Proc. of 4th Australian Commun. Theory Workshop, AusCTW '03,* Melbourne, Australia (pp. 1–5).

Kittipiyakul, S., & Javidi, T. (2007). Resource Allocation in OFDMA with Time-Varying Channel and Bursty Arrivals. *IEEE Communications Letters, 11*(9), 708–710. doi:10.1109/LCOMM.2007.070672

Kivanc, D., Li, G., & Liu, H. (2003). Computationally efficient bandwidth allocation and power control for OFDMA. *IEEE Transactions on Wireless Communications, 2*(6), 1150–1158. doi:10.1109/TWC.2003.819016

Klerer, M. (2003). Introduction to IEEE 802.20: Technical and Procedural Orientation. IEEE *802.20 PD-04.*

Klyne, G., & Reynolds, F. F.C, Woodrow, CH., Ohto, HJ., Hjelm, J., & M. H. Butler, M. H. (2004). Composite Capability /Preference Profiles (CC/PP): Structure and Vocabularies, *W3C Recommendation,* from Retrieved from http://www.w3.org/Mobile/CCPP.

Knutsson, B. (2004). Peer-to-Peer Support for Massively Multiplayer Games. In *Proceedings of 23rd Annual IEEE Conference on Computer Communications (Infocom),* Hong Kong.

Kochen, M. (Ed.). (1989). *The Small World.* Norwood, NJ: Ablex.

Koffman, I., & Roman, V. (2002). Broadband Wireless Access Solutions Based on OFDM Access in IEEE 802.16. *IEEE Communications Magazine, 40*(4), 96–103. doi:10.1109/35.995857

Kohler, E., Handley, M., & Floyd, S. (2006). Datagram congestion control protocol. *IETF RFC 4340.* Retrieved on January 2009, http://www.ietf.org/

Kohno, R. (2004). State of arts in ultra wideband (UWB) wireless technology and global harmonization. In *Proceedings of 34th European Microwave Conf,* Netherlands (pp. 1093- 1099).

Compilation of References

Kong, K., Lee, W., Han, Y. & You, H. (2008). Mobility Management for All-IP Mobile Networks: Mobile IPv6 vs. Proxy Mobile IPv6. *IEEE Wireless Communications,* April.

Kong, P.-Y., Pathmasuntharam, J. S., Wang, H.-G., Ge, Y., Ang, C.-W., Su, W., et al. (2009). A Routing Protocol for WiMAX Based Maritime Wireless Mesh Networks. In *IEEE 69th Vehicular Technology Conference VTC 2009 Spring.*

Kong, P.-Y., Wang Wang, H.-G., Ge, Y., Ang, C.-W., Pathmasuntharam, J. S., Su, W., et al. (2009). Distributed Adaptive Time Slot Allocation for WiMAX Based Maritime Wireless Mesh Networks. In *IEEE Wireless Communications & Networking Conference (WCNC).*

Kong, P.-Y., Wang, H.-G., Ge, Y., Ang, C.-W., Su, W., Pathmasuntharam, J. S., et al. (2008). A Performance Comparison of Routing Protocols for Maritime Wireless Mesh Networks. In *2008 IEEE Wireless Communications and Networking Conference.*

Kong, P.-Y., Zhou, M.-T., & Pathmasuntharam, J. S. (2008). A Routing Approach for Inter-Ship Communications in Wireless Multi-Hop Networks. In *the 8th International Conference on Intelligent Transport System Telecommunications* (ITST 2008)

Koodli, R. (2005, July). Fast handovers for Mobile IPv6. *IETF RFC 4068.*

Koonen, T., García Larrodé, M., Urban, P., Waardt, H., Tsekrekos, C., Yang, J., et al. (2007). Fibre-based Versatile Broadband Access and In-Building Networks. *IET Workshop From Access to Metro,* Europe.

Kornfeld M., G. May. (2007) DVB-H and IP Datacast - Broadcast to Handheld Devices. *IEEE Transactions on Broadcasting,* Vol. 53, Issue (2), (pp. 161-170).

Koutsopolou M., Kaloxylos A., Alonistioti A., Merakos L., & Philippopoulos, P. (n.d.). An integrated charging, accounting and billing management platform for the support of innovative business models in mobile networks. *Int. journal of Mobile Communications,* 2(4).

Kramer, G., Gastpar, M., & Gupta, P. (2005). Cooperative strategies and capacity theorems for relay networks. *IEEE Transactions on Information Theory, 51*(9), 3037–3063. doi:10.1109/TIT.2005.853304

Kramer, G., Maric, I., & Yates, R. D. (2007). *Cooperative Communications.* Hanover, USA, now Publishers Inc.

Kumar, J., Manoj, K., & Murthy, S. *(2002). Multi-power Architecture for Cellular Networks.* Paper presented in IEEE PIMRC.

Kumar, K.G., Lipscomb, J.S, Ramchandra, A., Chang, S.P., Gaddy, W.L., Leung, R.H., et al. (2001). The HotMedia Architecture: Progressive and Interactive Rich Media for the Internet. IEEE Transactions on Multimedia, 3(2), 253:267.

Kumwilaisak, W., Hou, Y., Zhang, Q., Zhu, W., Kuo, C., & Zhang, Y. (2003). A cross-layer quality-of-service mapping architecture for video delivery in wireless networks. *IEEE Journal on Selected Areas in Communications, 21*(10), 1685–1698. doi:10.1109/JSAC.2003.816445

Kurbel, K. (2003). Video Streaming Solutions for Web-Based E-Learning Courses. In M. Khosrow-Pour, (Ed.), Information Technology and Organizations: Trends, Issues, Challenges, Solutions, Proceedings of the 2003 IRMA International Conference, Philadelphia, PA, (pp. 874- 876).

Kuri, T., et al. (2006). Optical transmitter and receiver of 24-GHz ultra-wideband signal by direct photonic conversion techniques. *In Proceedings of International Topics Meeting on IEEE Microwave Photonics 2006.*

Kusume, K., Joham, M., Utschick, W., & Bauch, G. (2007). Cholesky factorization with symmetric permutation applied to detection and precoding spatially multiplexed data streams. *IEEE Transactions on Signal Processing, 55*(6), 3089–3103. doi:10.1109/TSP.2007.893978

Kwak, J., Park, S. M., Yoon, S., & Lee, K. (2003). Implementation of a parallel turbo decoder with dividable interleaver. In *Proc. Intern. Symposium on Circuits and Systems,* Bangkok, Thailand, (Vol. 2, pp. II65-II68).

Kyasanur, P., & Vaidya, N. (2005). Routing and interface assignment in multi-channel multi-interface wireless networks. In *Proc. of IEEE Wireless Communications and Networking Conference,* (Vol. 4, pp. 2051- 2056).

Lai, L., & El Gamal, H. (2008). On cooperation in energy efficient wireless networks: The role of altruistic nodes. *IEEE Transactions on Wireless Communications, 7*(5), 1868–1878. doi:10.1109/TWC.2008.060568

Lai, L., & El Gamal, H. (2008). The relay-eavesdropper channel: Cooperation for secrecy. *IEEE Transactions on Information Theory, 54*(9), 4005–4019. doi:10.1109/TIT.2008.928272

Lamy-Bergot, C., Chautru, N., & Bergeron, C. (2006). Unequal error protection for H.263+ bitstreams over a wireless IP network. In *Proceedings of the IEEE ICASSP'06* (pp.V-377/V-380), Toulouse, France. Washington, DC: IEEE.

Lamy-Bergot, C., Panza, G., Rotondi, A., & Fratta, L. (2008). Analysis and optimization of a JSCC/D system on 4G networks. In [Bologna, Italy. Washington, DC: IEEE.]. *Proceedings of the IEEE ISSSTA, 08*, 253–259.

Lancaster, P., & Tismenetsky, M. (1985). The theory of matrices. Academic Press.

Laneman, J. N., & Wornell, G. W. (2003). Distributed space-time-coded protocols for exploiting cooperative diversity in wireless networks. *IEEE Transactions on Information Theory, 49*(10), 2415–2425. doi:10.1109/TIT.2003.817829

Laneman, J. N., Tse, D., & Wornell, G. W. (2004). Cooperative diversity in wireless networks: Efficient protocols and outage behavior. *IEEE Transactions on Information Theory, 50*(12), 3062–3080. doi:10.1109/TIT.2004.838089

Langar, R., Thome, S., & Grand, G. L. (2005). Micro mobile MPLS: A new scheme for micro-mobility manegement in 3G all-IP networks. In *Proc. IEEE Symposium Computers and Communications (ISCC)*, (pp. 301-306).

Larsson, E. G., & Vojcic, B. R. (2005). Cooperative transmit diversity based on superposition modulation. *IEEE Communications Letters, 9*(9), 778–780. doi:10.1109/LCOMM.2005.1506700

Larsson, E., & Cao, Y. (2005). Collaborative transmit diversity with adaptive radio resource and power allocation. *IEEE Communications Letters, 9*(6), 511–513. doi:10.1109/LCOMM.2005.1437354

Larzon, L. A., Degermark, M., Pink, S., Jonsson, L.-E., & Fairhust, G. (2004). The lightweight user datagram protocol (UDP-Lite). *IETF RFC 3828*. Retrieved on January 2009, http://www.ietf.org/

Lazos, L., & Poovendran, R. (2004). Cross-Layer Design for Energy-Efficient Secure Multicast Communications in Ad Hoc Networks. *IEEE International Conference on Communications (ICC)*, (vol. 6, pp. 3633-3639).

Le, L., & Li, G. (2007). Cross-layer Mobility Management based on Mobile IP and SIP in IMS. *International Conference on Wireless Communications, Networking and Mobile Computing, 2007. (WiCom 2007)*,(pp. 803-806).

Le, Y. (2000). Pilot-symbol-aided channel estimation for OFDM in wireless systems. *IEEE Transactions on Vehicular Technology, 48*, 1207–1215.

Lee, J. W., Mazumdar, R. R., & Shroff, N. B. (2006). Joint resource allocation and base-station assignment for the downlink in CDMA networks. *IEEE/ACM Transactions on Networking, 14*(1), 1-14.

Lee, K. (2005). Radio over fibre for beyond 3G. *International Topical Microwave Photonics, MWP 2005*.

Lee, S.-J., Shanbhag, N. R., & Singer, A. C. (2005). Area-Efficient High-Throughput MAP Decoder Architectures. [VLSI]. *IEEE Transactions on Very Large Scale Integration, 13*, 921–933. doi:10.1109/TVLSI.2005.853604

Lee, W. C. Y. (1991). Smaller cells for greater performance. *IEEE Communications Magazine, 29*(11), 19–23. doi:10.1109/35.109660

Leguay, J., Friedman, T., & Conan, V. (2006). Evaluating Mobility Pattern Space Routing for DTNs. In *25th IEEE International Conference on Computer Communications (INFOCOM)* (pp. 1-10).

Lei, J. & Fu, X. (2007). *Evaluating the Benefits of Introducing PMIPv6 for Localized Mobility Management*. Technische Berichte des Instituts für Informatik an der Georg-August-Universität Göttingen, 29.

Lenstra, A. K., Lenstra, H. W., & Lov'asz, L. (1982). Factoring polynomials with rational coefficients. *Mathematische Annalen, 261*, 515–534. doi:10.1007/BF01457454

Leu, A. E., & Mark, B. L. (2003, March). An efficient timer-based hard handoff algorithm for cellular networks. *IEEE Wireless Communications and Networking Conference*, (pp. 1207-1212), New Orleans, LA.

Lewcio, B., Möller, S., Vidales, P., Wältermann, M., & Kirschnick, N. (2008). Methods for Multimedia Service Adaptation in Next Generation Networks. In *Proceedings of the ETSI Workshop on Effects of Transmission Performance on Multimedia QoS*, Prague.

Li, G., & Liu, H. (2003). Dynamic resource allocation with finite buffer constraint in broadband OFDMA networks. IEEE *Wireless Communications and Networking*, Vol. 2, March 2003, 1037 – 1042.

Li, G., & Lu, H. (2006). Downlink Radio Resource Allocation for Multi-cell OFDMA System. *IEEE Transactions on Wireless Communications, 5*(12), 3451–3459. doi:10.1109/TWC.2006.256968

Li, J., Huang, A., Guizani, M., & Chen, H.-H. (2007). Inter-Group Complementary Codes for Interference- Resistant CDMA Wireless Communications. *IEEE Transactions on Wireless Communications*.

Li, M., Li, F., Claypool, M., & Kinicki, R. (2005). Weather forecasting: predicting performance for streaming video over wireless LANs. In ACM International Workshop on Network and Operating Systems Support For Digital Audio and Video. (pp. 33-38). Stevenson, WA: ACM Press.

Li, S., & Ephremides, A. (2006). Anonymous Routing: A Cross-Layer Coupling between Application and Network Layer. *The Conference on Information Sciences and Systems 2006* (pp.783-788).

Li, X., & Cuthbert, L. (2006). Node-Disjoint Multipath Routing and Distributed Cross-Layer QoS Guarantees in Mobile Ad hoc Networks. *The Seventh ACIS International Conference on Software Engineering, Artificial Intelligence, Networking, and Parallel/Distributed Computing (SNPD 2006)*, (pp. 243-248).

Li, X., Chen, H.-H., Qian, Y., Rong, B., & Soleymani, M. R. (2007). Welch bound analysis on generic code division multiple access codes with interference free windows. *IEEE Transactions on Wireless Communications*.

Li, Y., Cimini, L. J. Jr, & Sollenberger, N. R. (1998). Robust channel estimation for OFDM systems with rapid dispersive fading channels. *IEEE Transactions on Communications*, *46*, 902–915. doi:10.1109/26.701317

Li, Y., Qiu, L., Zhang, Y., Mahajan, R., Zhong, Z., Deshpande, G., & Rozner, E. (2007). Effects of interference on throughput of wireless mesh networks: Pathologies and a Preliminary Solution. In *Proc. of HotNets-VI*, Atlanta, GA, USA.

Li, Z., & Shen, H. (2008). Probabilistic Routing with Multi-Copies in Delay Tolerant Networks. In *28th International Conference on Distributed Computing Systems Workshops* (pp. 471-476). Washington, DC: IEEE Computer Society.

Liebowitz, S., & Margolis, S. (1994). Path dependence, Lock-In and History. *Journal of Law Economics and Organization*, *11*(1), 205–206.

Lieu, C., & Kaiser, J. (2005). A Survey of Mobile Ad-hoc Network Routing Protocols, University of Ulm Technical Report Series.Jacobsson, M. (2004). Network Layer Architecture for Personal Networks. In *Proceedings of MAGNET Workshop*, Shanghai, China.

Likitthanasate, P., Grace, D., & Mitchell, P. D. (2005). Coexistence performance of high altitude platform and terrestrial systems sharing a common downlink WiMAX frequency band. *Electronics Letters*, *41*(15), 858–860. doi:10.1049/el:20051930

Lin, X., & Rasool, S. (2007, May). Distributed joint channel assignment, scheduling and routing algorithm for multi-channel ad hoc wireless networks. In *Proc. of 26th IEEE International Conference on Computer Communications*, (pp. 1118-1126).

Lin, Y., & Pang, A. (2005). *Wireless and Mobile All-IP Networks*, November. New York: Wiley Publishing Inc.

Lin, Y., & Wong, V. W. (2009). Adaptive Techniques in Wireless Networks. In M. Ibnkahla (Ed.), Adaptation and Cross Layer Design in Wireless Networks (pp. 419-450). Boca Ratón, FL: CRC Press.

Lin, Y., Mahlke, S., Mudge, T., Chakrabarti, C., Reid, A., & Flautner, K. (2007, October). Design and Implementation of Turbo Decoders for Software Defined Radio. *IEEE Workshop on Signal Processing Design and Implementation (SIPS)*, (pp. 22-27).

Lin, Z., Erkip, E., & Stefanov, A. (2006). Cooperative regions and partner choice in coded cooperative systems. *IEEE Transactions on Communications*, *54*(7), 1323–1334. doi:10.1109/TCOMM.2006.877963

Lindgren, A., & Phanse, K. S. (2006). Evaluation of Queueing Policies and Forwarding Strategies for Routing in Intermittently Connected Networks. In *International Conference on Communication System Software and Middleware (Comsware)* (pp. 1-10).

Lindgren, A., Doria, A., & Schelén, O. (2003). Probabilistic Routing in Intermittently Connected Networks. [New York: ACM.]. *ACM SIGMOBILE Mobile Computing and Communications*, *7*(3), 19–20. doi:10.1145/961268.961272

Linux, K. S. C. T. P. *(LKSCTP)*. Retrieved August 10, 2007 from http://lksctp.sourceforge.net/overview.html

Liu, C., & Polo, V. (2007). Full-duplex DOCSIS/wireless-DOCSIS fibre-radio network employing packaged AFPMs as optical/electrical transducers. *Journal of Lightwave Technology*, *25*(3), 673–684. doi:10.1109/JLT.2006.889674

Liu, C., et al. (2003). Bi-directional transmission of broadband 5.2 GHz wireless signals over fibre using a multiple-quantum-well asymmetric Fabry–Pérot modulator/photodetector. In *Proceedings of IEEE/OSA OFC 2003*, Atlanta, GA (pp. 738–740).

Liu, J., & Li, J. (2005). Turbo processing for an OFDM-based MIMO system. *IEEE Transactions on Wireless Communications, 4*(9), 1988–1993.

Liu, K. J. R., Sadek, A. K., Su, W., & Kwasinski, A. (2009). *Cooperative communications and networking.* Cambridge, UK: Cambridge University Press.

Liu, M., Li, Z., Guo, X., & Dutkiewicz, E. (2008). Performance analysis and optimization of handoff algorithms in heterogeneous wireless networks. *IEEE Transactions on Mobile Computing, 7*(7), 846–857. doi:10.1109/TMC.2007.70768

Liu, P., Tao, Z., Lin, Z., Erkip, E., & Panwar, S. (2006). Cooperative wireless communications: a cross-layer approach. *IEEE Wireless Communications Magazine, 13*(4), 84–92. doi:10.1109/MWC.2006.1678169

Liu, X., Chong, E., & Shroff, N. (2003). A framework for opportunistic scheduling in wireless networks. *Computer Networks, 41*(4), 451–474. doi:10.1016/S1389-1286(02)00401-2

Liu, Y., Gruhl, S., & Knightly, E. W. (2003). WCFQ: an opportunistic wireless scheduler with statistical fairness bounds. *IEEE Transactions on Wireless Communications, 2*(5), 1017–1028. doi:10.1109/TWC.2003.816777

Liu, Y., Li, Y., & Man, H. (2005). A Distributed Cross-layer Intrusion Detection System for Ad Hoc Networks. *1st International Conference on Security and Privacy for Emerging Areas in Communications Networks (SecureComm 2005)* (pp. 418-420).

Llorente, R., Alves, T., Morant, M., Beltran, M., Perez, J., Cartaxo, A., & Marti, J. (2008). Ultra-Wideband Radio Signals Distribution in FTTH Networks. *IEEE Photonics Technology Letters, 20*(11), 945–947. doi:10.1109/LPT.2008.922329

Llorente, R., Cartaxo, A., Uguen, B., Duplicy, J., Romme, J., Puche, J. F., et al. (2008). Management of UWB picocell clusters: UCELLS project approach. In *IEEE International Conference on Ultra-Wideband ICUWB 2008* (vol. 3, pp. 139-142).

Llorente, R., Thakur, M. P., Morant, M., Walker, S. D., & Marti, J. (2008). Performance comparison of radio-over-fibre UWB distribution in SSMF and MMF optical media. In *Proceedings of ECOC 2008* (Vol. 2, pp. 119-120).

Loscri, V. (2007). A New Distributed Scheduling Scheme for Wireless Mesh Networks. In *IEEE International Symposium on Personal, Indoor and Mobile Radio Communications* (pp. 1-5).

Lott, M., Halfmann, R., Schulz, E., & Radimirsch, M. (2001). Medium access and radio resource management for ad hoc networks based on ULTRA and TDD. In *Proc. of the 2nd ACM/SIGMOBILE Symposium on Mobile Ad hoc Networking and Computing.*

Lowery, A. J., & Armstrong, J. (2005). 10 Gbit/s multimode fibre link using power efficient orthogonal frequency division multiplexing. *Optics Express, 13*, 10003–10009. doi:10.1364/OPEX.13.010003

Lu, K., Qian, Y., Chen, H-H., & Fu, S. (2008). WIMAX NETWORKS: FROM ACCESS TO SERVICE PLATFORM. *IEEE Network, 22*(3, May/June), 38-45.

Luby, M., & Gemmell, J. J., L. Vicisano, L., L. Rizzo, L., &and J. Crowcroft, J. (2002) Asynchronous Layered Coding (ALC) Protocol Instantiation. *IETF RFC 3450*, from Retrieved from http://www.ietf.org/rfc/rfc3450.txt.

Luby, M., & Vicisano, L. (2004). Compact Forward Error Correction (FEC) Schemes. *IETF RFC 3695.*

Lunttila, T., Iraji, S., & Berg, H. (2006). Advanced Coding Schemes for a Multi-Band OFDM Ultrawideband System towards 1 Gbps. In *Proceedings of CCNC'06* (vol. 1, pp. 553-557).

Luo, J., Blum, R. S., Cimini, L. J., Greenstein, L. J., & Haimovich, A. M. (2007). Decode-and-forward cooperative diversity with power allocation in wireless networks. *IEEE Transactions on Wireless Communications, 6*(3), 793–799. doi:10.1109/TWC.2007.05272

Luoto, M., & Sutinen, T. (2008, May). Cross-layer enhanced mobility management in heterogeneous networks. *IEEE International Conference on Communications*, (pp. 2277-2281), Beijing, China.

Ma, L., Han, X., & Shen, C. (2005). Dynamic open spectrum sharing MAC protocol for wireless ad hoc networks. In *Proc. of IEEE Symposium on New Frontiers in Dynamic Spectrum Access networks*, November 2005.

Macías, E. M., Suárez, A., Martin, J., & Sunderam, V. (2007). Using OLSR for Streaming Video in 802.11 Ad Hoc Networks to Save Bandwidth. *IAENG International Journal of Computer Science, 33*(1), 101–110.

Magana, M. E., Rajatasereekul, T., Hank, D., & Chen, H.-H. (2007). Design of a MC-CDMA System that Uses

Compilation of References

Complete Complementary Orthogonal Spreading Codes. *IEEE Transactions on Vehicular Technology*.

Magio Internet Security. (2009). Retrieved April 9, 2009, from http://t-com-eng.st.sk/Default.aspx?CatID=1248&section=home

Mahadevan, I., & Sivalingam, K. M. (2001). Architecture and experimental framework for supporting QoS in wireless networks using DiffServ. *Mobile Networks and Applications*, 6, 385–395. doi:10.1023/A:1011434813337

Maham, B., Debbah, M., & Hjorungnes, A. (2008). Energy-efficient Cooperative Routing in BER Constrianed Multihop Networks. In *Proc. of Int. Conf. on Commun. And Networking in China (Chinacom)*, Hangzhou, China.

Makarevitch, B. (2006). Distributed Scheduling for WiMAC Mesh Networks. *IEEE International Symposium on Personal, Indoor and Mobile Radio Communications*

Makarevitch, B., (2006). Jamming Resistant Architecture for WiMAX Mesh Networks. *IEEE MILCOM*, 1-6.

Makaya, C., & Aissa, S. (2003, October). Joint scheduling and base station assignment for CDMA packet data networks. *IEEE Vehicular Technology Conference 2003-Fall*, (pp. 1693–1697), Orlando, FL.

Malkamaki, E., & Leib, H. (1999). Evaluating the performance of convolutional codes over block fading channels. *IEEE Transactions on Information Theory*, 45(5), 1643–1646. doi:10.1109/18.771235

Malki, K. E. (2007, June). Low-latency handoffs in Mobile IPv4. *IETF RFC 4881*.

Malte, A., Robert, H., & Tobias, R. (2008). Dedicated Programming Support for Context-Aware Ubiquitous Applications. In International Conference on Mobile Ubiquitous Computing, Systems, Services and Technologies, (pp. 38 - 43). Valencia, Spain: IEEE Press.

Manoj, B., & Kumar, K. (2006). On the Use of Multiple Hops in Next-Generation Wireless Systems. *Science Wireless Networks*.

Manolakis, K., & Jungnickel, V. (2008). Synchronization and cell search for 3GPP LTE. In *Proc. of 13th International OFDM Workshop (InOWo'08)*, Hamburg, Germany.

Manolakis, K., Estevez, D. M. G., Jungnickel, V., Xu, W., & Drewes, C. (2009). A closed concept for synchronization and cell search in 3GPP LTE systems. In *Proc. WCNC'09, IEEE Wireless Communications and Networking Conference*, Budapest, Hungary.

Maravedis Research. (2008). *WiMAX, LTE and Broadband Wireless Worldwide Market Trends 2008-2014 – 5th edition*. Montreal, Canada: Maravedis.

Marichamy, P., Chakrabarti, S., & Maskara, S. L. (2003, October). Performance evaluation of handoff detection schemes. *IEEE Region 10 Conference on Convergent Technologies for the Asia-Pacific*, (pp. 643-646), Taj Residency, Bangalore.

Mark, B. L., & Leu, A. E. (2007). Local averaging for fast handoffs in cellular networks. *IEEE Transactions on Wireless Communications*, 6(3), 866–874. doi:10.1109/TWC.2007.04080

Marozsak, T., & Udvary, E. (2002). Vertical cavity surface emitting lasers in radio over fibre applications. In *Proceedings of 14th International Conference on Microwaves, Radar and Wireless Communications 2002, MIKON-2002* (Vol.1, pp. 41-44).

Marrero, D., Macías, E. M., & Suárez, A. (2008). An Admission Control and Traffic Regulation Mechanism for Infrastructure WiFi networks. *IAENG International Journal of Computer Science*, 35(1), 154–160.

Marti, S., Ganesan, P., & Garcia-Molina, H. (2004). Sprout: P2p routing with social networks. In *1st International Workshop on Peer-to-Peer Computing and Databases* (pp. 425-435).

Marti, S., Ganesan, P., & Garcia-Molina, H. (2005). DHT Routing Using Social Links. In *3rd International Workshop* (pp. 100-111). New York: Springer.

Martini, M. G., & Chiani, M. (2004). Rate-distortion models for unequal error protection for wireless video transmission. [Milan, Italy. Washington DC: IEEE.]. *Proceedings of the IEEE VTC*, 04, 1049–1053.

Martini, M. G., Mazzotti, M., & Lamy-Bergot, C. (2007). Content adaptive network aware joint optimisation of wireless video transmission. *IEEE Communications Magazine*, 45(1), 84–90. doi:10.1109/MCOM.2007.284542

Martini, M. G., Mazzotti, M., Chiani, M., Panza, G., et al. (2005). *A demonstration platform for network aware joint optimisation of wireless video transmission: the PHOENIX basic demonstration platform*. Paper presented at the IST Mobile Summit 2005, Dresden, Germany.

Mary, N. (2006). *Implementation of Vertical Handoff Algorithm Between IEEE802.11 WLAN and CDMA Cellular Network*. Master Thesis, Georgia State University, USA.

Masera, G., Piccinini, G., Roch, M., & Zamboni, M. (1999). VLSI architecture for turbo codes. [VLSI]. *IEEE Transactions on Very Large Scale Integration*, *7*, 369–3797. doi:10.1109/92.784098

Massey, J. L. (1978). Joint source and channel coding. In J.K. Skwirzynski (Ed.), *Commun. systems and random process theory, NATO advanced studies institutes series E25*, (pp. 279–293). Alphen aan den Rijn, The Netherlands: Sijthoff & Noordhoff.

May, M., Neeb, C., & Wehn, N. (2007). Evaluation of high throughput turbo-decoder architectures. In *Proc. Intern. Symposium on Circuits and Systems*, New Orleans, USA, (pp. 2770-2773).

Mayrock, M., & Haunstein, H. (2007). OFDM in optical long-haul transmission. In *Proceedings of of 12th Int. OFDM Workshop*, Hamburg, Germany.

McCloud, M. L. (2004). Optimal binary spreading for block ofdm on multipath fading channels. *WCNC / IEEE Communicatio Socialis*, *2*, 965–970.

McCormick, S. T., & Pinedo, M. L. (1995). Scheduling independent jobs on uniform machines with both flowtime and makespan objectives: a parametric analysis. *ORSA Journal on Computing*, *7*(1), 63–77.

McDonald, G. J., & Seeds, A. J. (2006). A novel pulse source for low-jitter optical sampling: a rugged alternative to mode-locked lasers. In *Proceedings of SPIE 6399*, October 2006.

McEliece. (1977). *The Theory of Information and Coding*. London: Addison-Wesley.

Mchenry, M.(2003). *Spectrum white space measurements*. New America Foundation Broadband Forum, June.

McKenney, A. (1991). Stochastic fairness queuing. *Journal of Internetworking Research and Experience*, *2*, 113–131.

McKnight, L.W., Howison, J., & Bradner, S. (2004). Wireless Grids – Distributed Resource Sharing by Mobile, Nomadic, and Fixed Devices. *Internet Computing*, 24-31.

McQuillan, J. M., Richer, I., & Rosen, E. C. (1980). The New Routing Algorithm for the ARPANET. *IEEE Transactions on Communications*, *28*, 711–719. doi:10.1109/TCOM.1980.1094721

Meinnel, H. H., et al. (2005). *Automotive radar: From long range collision warning to short range urban employment*. Paper presented in MINT-MIS2005/TSMMW2005.

Melnyk, M. A., Jukan, A., & Polychronopoulos, C. D. (2007). A Cross-Layer Analysis of Session Setup Delay in IP Multimedia Subsystem (IMS) with EV-DO Wireless Transmission. *IEEE Transactions on Multimedia*, *9*(4), 869–881. doi:10.1109/TMM.2007.895680

Ménard, F. D. (2006). Xittel Combines Fiber and Motorola WiMAX to Serve as Few as 25 Customers. *Broadband properties*. Retrieved from http://www.broadbandproperties.com

Menouar, H., Filali, F., & Lenardi, M. (2006). A survey and qualitative analysis of mac protocols vehicular ad hoc networks. In *Proc. IEEE Wireless Communications Magazine*, 30-35, October.

Metakall - The Alternative to Cellular. (n.d.). Retrieved December 28, 2008, from www.metakall.com.

Mi.Tel-Teleoptix. (n.d.). *43 Gbit/s DPSK PHOTORECEIVER with Limiting TIA (DualPIN-DTLIA Rx)*. Retrieved from http://www.teleoptix.com

Mihovska, A., Luo, J., Mino, E., Tragos, E., Mensing, C., Vivier, G., & Fracchia, R. (2007, July). Policy-based mobility management for heterogeneous networks. *IEEE Mobile and Wireless Communications Summit*, (pp. 1-6), Budapest, Hungary. mITF (2005). 4G Mobile System Requirement Document Version 1.1.

Ming, L. X. (2000). Streaming technique and its application in distance learning system. IEEE International Conference on Signal Processing: Vol. 2, (pp. 1329-1332). Washington, DC: IEEE Press.

Mishra, A. (2006). A multi-channel mac for opportunistic spectrum sharing in cognitive networks. In *Proc. IEEE MILCOM*, October.

Misic, J., Shafi, S., & Misic, V. B. (2006). Cross-Layer Activity Management in an 802.15.4 Sensor Network. *IEEE Communications Magazine*, *44*(1), 131–136. doi:10.1109/MCOM.2006.1580943

Mitola, J. (2000). *Cognitive radio: an integrated agent architecture for software defined radio*. PhD thesis, KTH Royal Inst. Technology, Stockholm, Sweden, 2000.

Mitran, P., Ochiai, H., & Tarokh, V. (2005). Space-time diversity enhancements using collaborative communications. *IEEE Transactions on Information Theory, 51*(6), 2041–2057. doi:10.1109/TIT.2005.847731

Mizuochi, T., Kingo, K., Kajiva, S., Tokura, T., & Motoshima, K. (2002). Bidirectional unrepeatered 43 Gb/s WDM transmission with C/L band-separated Raman amplification. *IEEE Journal Ligthwave Technology, 20*, 2079–2085. doi:10.1109/JLT.2002.806767

Mobasher, A., & Khandani, A. K. (2007, June). Matrix-Lifting Semi-Definite Programming for Decoding in Multiple Antenna Systems. In *The 10th Canadian Workshop on Information Theory (CWIT'07)*, Edmonton, Alberta, Canada

Mobasher, A., Taherzadeh, M., Sotirov, R., & Khandani, A. K. (2005, September). A near maximum likelihood decoding algorithm for mimo systems based on graph partitioning. In *Proc. of IEEE Int. Symp. on Info. Theory, ISIT '05*, Adelaide, Australia.

Mobile Multimedia Laboratory. (n.d.). *Architectural, Economic, Security and Strategic Issues in 4G Wireless Networks.* Retrieved April 9, 2009, from http://mm.aueb.gr/research/4G.html

Mobile, I. N. (n.d.). *What is 4G cellular?* Retrieved February 11, 2009, from http://www.mobilein.com/what_is_4GCellular.htm

Mobisense. (2007). Retrieved January 20, 2009, from http://www.deutsche-telekom-laboratories.de/~vidales/mobisense

Moeneclaey, M., Van Bladel, M., & Sari, H. (1998). A Comparison of Multiple Access Techniques in the Presence of Narrowband Interference. In *International Symposium on Signals, Systems, and Electronics, 1998, ISSSE*, URSI, (pp. 223 – 228).

Mohammed, A., Arnon, S., Grace, D., Mondin, M., & Miura, R. (2008). Advanced Communications Techniques and Applications for High-Altitude Platforms. *Editorial for a Special Issue, EURASIP Journal on Wireless Communications and Networking, 2008.*

Mohanty, S., & Akyildiz, F. (2006). A cross-layer (layer 2 + 3) handoff management protocol for next-generation wireless systems. *IEEE Transactions on Mobile Computing, 5*(10), 1347–1360. doi:10.1109/TMC.2006.142

Mohr, W. (2002). *Mobile Communications Beyond 3G in the Global.* Siemens Mobile. Retrieved March 1, 2009, from http://www.cu.ipv6tf.org/pdf/werner_mohr.pdf

Mohr, W. (2008, January). Vision for 2020? *Wireless Personal Communications, 44*, 27–49. doi:10.1007/s11277-007-9381-1

Moiseev, S. N., Filin, S. A., Kondakov, M. S., Garmnonov, A. V., Yim, D. H., Lee, J., et al. (2006). Analysis of the statistical properties of the interference in the IEEE 802.16 OFDMA network. In *IEEE Wireless Communications & Networking Conference*, (pp. 1830-1835).

Möller, S., Raake, A., Kitawaki, N., Takahashi, A., & Waltermann, M. (2006). Impairment Factor Framework for Wideband Speech Codecs. *IEEE Transactions on Audio, Speech, and Language Processing, 14*(6), 1969–1976. doi:10.1109/TASL.2006.883262

Möller, S., Wältermann, M., Lewcio, B., Kirschnick, N., & Vidales, P. (2009). Speech Quality while Roaming in Next Generation Networks. Accepted for *IEEE International Conference on Communications (ICC)*, Dresden. *Netem.* (n.d.). Retrieved January 20, 2009, from http://www.linux-foundation.org/en/Net:Netem

Moon, B., & Aghvami, H. (2002). Reliable RSVP path reservation for multimedia communications under an IP micromobility scenario. *IEEE Wireless Communications, 9*(5), 93–99. doi:10.1109/MWC.2002.1043859

Moon, T. K. (2005). *Error Correction Coding.* Hoboken, NJ: John Wiley & Sons.

Morant, M., Alves, T., Llorente, R., Cartaxo, A., & Marti, J. (2008). Experimental Comparison of Transmission Performance of Multi-channel OFDM-UWB Signals on FTTH Networks. *IEEE . Journal of Lightwave Technology, 27*(10), 1410–1416.

Morant, M., Pérez, J., Beltran, M., Llorente, R., & Marti, J. (2008). Integrated performance analysis of UWB wireless optical transmission in FTTH networks. In *Proceedings of 21st Annual Meeting of The IEEE Lasers & Electro-Optics Society, 2008 LEOS Annual Conference*, Newport Beach, CA.

Morant, M., Pérez, J., Llorente, R., & Marti, J. (2009). Transmission of 1.2 Gbit/s Polarization-Multiplexed UWB Signals in PON with 0.76 Bit/s/Hz Spectral Efficiency. In *Proceedings of IEEE/OSA OFC 2009.*

Morelli, M., & Mengali, U. (2001). A comparison of pilot-aided channel estimation methods for OFDM systems. *IEEE Transactions on Signal Processing, 49*(12), 3065–3073. doi:10.1109/78.969514

Morelli, M., Kuo, C. C. J., & Pun, M. O. (2007). Synchronization techniques for orthogonal frequency division multiple access (OFDMA): A tutorial review. *Proceedings of the IEEE, 95*(7), 1394–1427. doi:10.1109/JPROC.2007.897979

Moreno, E. (2009). Modification of the reliable sockets interconnection mechanism and its use in video streaming applications (In Spanish). Unpublished Master doctoral dissertation, University of Las Palmas de G.C., Spain.

Moskowitz, R., & Nikander, P. (2003). *Host Identity Protocol Architecture*. Internet Draft draftmoskowitz- hip-arch-05.txt.

Motani, M., Srinivasan, V., & Nuggehalli, P. S. (2005). PeopleNet: engineering a wireless virtual social network. In *11th annual international conference on Mobile computing and networking* (pp. 243-257). New York: ACM.

Motorola Inc. (2007). *Long Term Evolution (LTE): A Technical Overview* [Technical White Paper]. Retrieved from http://www.motorola.com/staticfiles/Business/Solutions/Industry%20Solutions/Service%20Providers/Wireless%20Operators/LTE/_Document/Static%20Files/6834_Mot-Doc_New.pdf

MOTOROLA. (2008, November). *LTE inter-technology mobility, enabling mobility between LTE and other access technologies* (White Paper).

Moustafa, H., Javaid, U., Rasheed, T., & Meddour, D. E. (2006). A Panorama on Wireless Mesh Networks: Architectures, Applications and Technical Challenges. In *Proceedings of International Workshop on Wireless Mesh: Moving towards Applications (Wimeshnets)*, Waterloo, Canada.

Moy, J. (1994). *Multicast Extensions to OSPF*. Network Working Group, RFC 1584.

Moy, J. (1998). *OSPF Version 2*. Network Working Group, RFC 2328.

Münz, G., Pfletschinger, S., & Speidel, J. (2002). An Efficient Waterfilling Algorithm for Multiple Access OFDM. In *Globecom* (pp. 681-685).

Muller, O., Baghdadi, A., & Jezequel, M. (2006, March). ASIP-Based Multiprocessor SoC Design for Simple and Double Binary Turbo Decoding. *IEEE Design. Automation and Test in Europe, 1*, 1–6. doi:10.1109/DATE.2006.244126

Muqattash, A., & Krunz, M. (2003). CDMA-based mac protocol for wireless ad hoc networks. In *Proc. of the 4th ACM/SIGMOBILE Symposium on Mobile Ad hoc Networking and Computing*.

Murota, K. (1999). *Mobile Communications Trends in Japan and DoCoMo's Activities Towards 21st Century*. Paper presented at the Fourth ACTS Mobile Communications Summit, Sorrento, Italy.

Murthy, C., & Manoj, B. (2004). *Ad Hoc Wireless Networks: Architectures and Protocols*. Upper Saddle River, NJ: Prentice Hall.

Murthy, S., & Garcia-Luna-Aceves, J. J. (1995). A Routing Protocol for Packet Radio Networks. In *1st Annual International Conference on Mobile Computing and Networking* (pp. 86-95). New York: ACM.

Murthy, V. K., & Krishnamurthy, E. V. (2004). Multimedia Computing Environment for Telemedical Applications. In N. Shi (Ed.), Mobile Commerce Applications (pp. 95-115). Hershey, PA: IRM Press.

Murugan, A., El Gamal, H., Damen, M. O., & Caire, G. (2006). A unified framework for tree search decoding: Rediscovering the sequential decoder. *IEEE Transactions on Information Theory, 52*(3), 933–953. doi:10.1109/TIT.2005.864418

Nabar, R. U., Kneubuhler, F. W., & Bolcskei, H. (2004). *Performance limits of amplify-and-forward based fading relay channels*. Paper presented at the IEEE International Conference on Acoustics, Speech, and Signal Processing (ICASSP), Montreal, Canada.

Nagle, J. (1987). On packet switches with infinite storage. *IEEE Transactions on Communications, 35*(4), 435–438. doi:10.1109/TCOM.1987.1096782

Narasimhan, T. N. (2004). Fick's insights on liquid diffusion. *Transactions - American Geophysical Union, 85*(47), 499–501.

Nash, J. (1951). Non-Cooperative Games. *The Annals of Mathematics, 54*(2), 286–295. doi:10.2307/1969529

Natarajan, B., Nassar, C. R., & Shattil, S. (2001). High-performance MC-CDMA via carrier inter-ferometry codes. *IEEE Transactions on Vehicular Technology, 50*(6). doi:10.1109/25.966567

Navaie, K., & Yanikomeroglu, H. (2006, June). Downlink joint base-station assignment and packet scheduling algorithm for cellular CDMA/TDMA networks. In *IEEE International Conference on Communications*, (pp. 4339-4344), Istanbul, Turkey.

Compilation of References

Neely, M., Modiano, E., & Rohrs, C. (2002). Power and server allocation in a multi-beam satellite with time varying channels. In *INFOCOM* (pp. 1451–1460).

Negi, R., & Cioffi, J. (1998). Pilot tone selection for channel estimation in a mobile OFDM system. *IEEE Transactions on Consumer Electronics, 44*, 1122–1128. doi:10.1109/30.713244

Neishaboori, A., & Kesidis, G. (2008). Wireless mesh networks based on CDMA. *Computer Communications, 31*(8), 1513–1528. doi:10.1016/j.comcom.2008.01.020

Nelson, R. (1995). *Probability, Stochastic Processes, And Queuing Theory: The mathematics of computer performance Modeling*. Berlin: Springer-Verlag.

Ng, C. T. K., & Goldsmith, A. J. (2006). *Capacity and power allocation for transmitter and receiver cooperation in fading channels*. Paper presented at the IEEE International Conference on Communications (ICC), Istanbul, Turkey.

Ng, C. T. K., Jindal, N., Goldsmith, A. J., & Mitra, U. (2007). Capacity gain from two-transmitter and two-receiver cooperation. *IEEE Transactions on Information Theory, 53*(10), 3822–3827. doi:10.1109/TIT.2007.904987

Ng, C., Ernst, T., Paik, E., & Bagnulo, M. (2007). Analysis of multihoming in network mobility support. *RFC 4980*.

Ng'oma, A. (2005). *Radio-over-fibre technology for broadband wireless communication systems*. CIP-Data Library Technische Universiteit Eindhoven.

Nguyen, T. D., & Han, Y. (2006). A Proportional Fairness Algorithm with QoS Provision in Downlink OFDMA Systems. *IEEE Communications Letters, 10*(11), 760–762. doi:10.1109/LCOMM.2006.060750

Niiho, T., et al. (2004). Multi-channel wireless LAN distributed antenna system based on radio-over-fibre techniques. In *Proceedings of IEEE LEOS Annual Meeting 2004*, Rio Grande, Puerto Rico (pp. 57–58).

Nimbalker, A., Blankenship, Y., Classon, B., & Blankenship, T. K. (2008, March). ARP and QPP Interleavers for LTE Turbo Coding. In *IEEE Wireless Communications and Networking Conference* (pp. 1032-1037).

Nishida, K., & Yokota, H. (2006). *Mobility NETLMM Protocol Applicability Analysis for 3GPP SAE Network*. Retrieved January, 2008, http://tools.ietf.org/html/draft-nishida-netlmm-protocol-applicability-00

Niyato, D., & Hossain, E. (2008). A Noncooperative Game-Theoretic Framework for Radio Resource Management in 4G Heterogeneous Wireless Access Networks. *IEEE Transaction on Mobile Communications, 7*(3), 332–345. doi:10.1109/TMC.2007.70727

Niyato, D., Hossain, E., & Fallahi, A. (2006). Solar-powered OFDM wireless mesh networks with sleep management and connection admission control. In *Proc. of the 2006 international conference on Wireless communications and mobile computing*, (pp. 653-658).

Nkansah, A. (2006). Simultaneous Dual Band Transmission Over Multimode Fibre-Fed Indoor Wireless Network. *IEEE Microwave and Wireless Components Letters, 16*(11), 627–629. doi:10.1109/LMWC.2006.884899

Nosratinia, A., & Hunter, T. E. (2007). Grouping and partner selection in cooperative wireless networks. *IEEE Journal on Selected Areas in Communications, 25*(2), 369–378. doi:10.1109/JSAC.2007.070212

Nosratinia, A., Hunter, T. E., & Hedayat, A. (2004). Cooperative communication in wireless networks. *IEEE Communications Magazine, 42*(10), 74–80. doi:10.1109/MCOM.2004.1341264

Nyberg, G., Ahlund, C., & Rojmyr, T. (2006, July). SEMO: A policy-based system for handovers in heterogeneous networks. In *International Wireless Communications and Mobile Computing Conference*, (pp. 62-67), Vancouver, Canada.

O'Connell, A. M. (2009). How Nicotine Works. How Stuff Works? Retrieved January 30, 2009, from http://health.howstuffworks.com/nicotine7.htm

O'Droma, M. S., & Ganchev, I. (2004). Enabling an Always Best-Connected Defined 4G Wireless World. In []. Chicago, IL.: International Engineering Consortium.]. *Annual Review of Communications, 57*, 1157–1170.

O'Droma, M., & Ganchev, I. (2004). Techno-Business Models for 4G. In *Int. Forum on 4G Mobile Communications*, King's College London, UK (pp. 1-30).

O'Droma, M., & Ganchev, I. (2007). Toward a ubiquitous consumer wireless world. *IEEE Wireless Communications Journal, 14*(1), 52–63. doi:10.1109/MWC.2007.314551

O'Droma, M., Ganchev, I., Chaouchi, H., Aghvami, H., & Friderikos, V. (2006). Always Best Connected and Served Vision for a Future Wireless World. *Journal of Information Technologies and Control, 4*(3-4), 25–37.

Ochiai, H., Mitran, P., & Tarokh, V. (2004). *Design and analysis of collaborative diversity protocols for wireless sensor networks*. Paper presented at the IEEE Fall Vehicular Technology Conference (VTC), Los Angeles, USA.

Ochiai, H., Mitran, P., & Tarokh, V. (2006). Variable-rate two-phase collaborative communication protocols for wireless networks. *IEEE Transactions on Information Theory, 52*(9), 4299–4313. doi:10.1109/TIT.2006.880055

Ochiai, H., Mitran, P., Poor, H. V., & Tarokh, V. (2005). Collaborative beamforming for distributed wireless as hoc sensor networks. *IEEE Transactions on Signal Processing, 53*(11), 4110–4124. doi:10.1109/TSP.2005.857028

OECD Information Technology Outlook. (2008). OECD.

Ogawa, H. (1992). Millimetre-wave fibre optic systems for personal radio communication. *IEEE Transactions on Microwave Theory and Techniques, 40*(12), 2285–2292. doi:10.1109/22.179892

Ohba, Y. (2007). *MobiArch*. Retrieved April 9, 2009, from http://user.informatik.uni-goettingen.de/~mobiarch/2007/slides/mobiarch07-panel-YoshiroOhba.pdf

Ohmori, S., Yamao, Y., & Nakajima, N. (2000, December). The Future Generationis of Mobile Communications Based on Broadband Access Technologies. *IEEE Communications Magazine*, 134–142.

OPNET. (2008). *OPNET Modeler Documentation*. Bethesda, MD: OPNET, Inc.

Optics, O. F. S. (2009). Bending the rules: ofs to demonstrate ez-bend™ optical technology at the 2008 conference. Retrieved from http://www.ofsoptics.com/press_room/view_press_release.php?txtID=247. September 2009.

Ozdural, O. C., & Liu, H. (2007, June). Mobile direction assisted predictive base station switching for broadband wireless systems. In *IEEE International Conference on Communications*, (pp. 5570-5574), Glasgow, UK.

Paavola, J., Himmanen, H., Jokela, T., Poikonen, J., & Ipatov, V. (2007). The Performance Analysis of MPE-FEC Decoding Methods at the DVB-H Link Layer for Efficient IP Packet Retrieval. *IEEE Transactions on Broadcasting, 53*(1), 263–275. doi:10.1109/TBC.2007.891694

Packet, A. (n.d.). *Acme Packet Defines Role of its Net-Net Product Family within IMS LTE Networks*. Retrieved on September 10, 2009, from http://www.ir.acmepacket.com/phoenix.zhtml?c=200804&p=irol-newsArticle_Print&ID=1254949&highlight=

Pagani, M. (Ed.). (2005). *Mobile and Wireless Systems Beyond 3G: Managing New Business Opportunities*. Hershey, PA: IRM Press.

Pagani, M. (Ed.). (2009). *Encyclopedia of Multimedia Technology and Networking*. Hershey, PA: IRM Press.

Pagani, M., Schipani, D., Vicari, V. & Al. (2005). Motivations and Barriers to the Adoption of 3G Mobile Multimedia Services: An End User Perspective in the Italian Market. In B. Montano (Ed.), Innovations of Knowledge Management (pp. 80-95). Hershey, PA: IRM Press.

Pahlavan, K., Krishnamurthy, P., & Hatami, A. (2000). Handoff in hybrid mobile data networks. *IEEE Personal Communications, 7*(2), 34–47. doi:10.1109/98.839330

Paier, A., et al. (2007). First Results from Car-to-Car and Car-to-Infrastructure Radio Channel Measurements at 5.2GHZ. In *IEEE PIMRC 2007* (pp. 1-5).

Paila, T., Alladin, S., Frank, M., Goransson, T., Hansmann, W., Lohmar, T., et al. (2001). Flexible Network Architecture for Future Hybrid Wireless Systems. In *IST Mobile Summit*, Spain.

Paila, T., Luby, M., Lehtonen, R., Roca, V., & Walsh, R. (2004). FLUTE – File Delivery over Unidirectional Transport. *IETF (RFC 3926)*.

Panton, M. (2008). DLNA for media streamers--what does it all mean? Crave - CNET. Retrieved Jannuary 30, 2009, from http://news.cnet.com/8301-17938_105-10007069-1.html.

Pappalardo, D. (2007). What you need to know about 4G. *Network World*.

Park, M., & Miller, D. J. (2000). Joint source-channel decoding for variable-length encoded data by exact and approximate MAP sequence estimation. *IEEE Transactions on Communications, 48*(1), 1–6. doi:10.1109/26.818865

Passas, N., Paskalis, S., Kaloxylos, A., Bader, F., Narcisi, R., & Tsontsis, E. (2006). Enabling technologies for the always best connected concept. *Wiley Wireless Communications and Mobile Computing Journal, 6*(4), 523–540. doi:10.1002/wcm.392

Patel, P. R., Choksi, O. T., Davidson, K. W., & Dantu, R. (2006). *Method and System for Configuring Wireless Routers*. US Patent No. 7031266 B1. Washington, DC: U.S. Patent and Trademark Office.

Patel, V., & McParland, T. (2008). *Proposed Guidance for IPS Mobility Management Aeronautical Communication Panel*. August, Montreal, Canada.

Pathmasuntharam, J. S., Kong, P.-Y., Joe, J., Ge, Y., Zhou, M.-T., & Miura, R. (2007). High speed maritime ship-to-ship/shore mesh networks. In *the 7th International Conference on ITS Telecommunications, (ITST2007)*, Sophia Antipolis, France (pp. 460-465).

Pavn, J.d.P., Shankar, N.S., Gaddam, V., & Chou, C-T. (2006). The MBOA-Wimedia specification for ULTRA wideband distributed networks. *IEEE Communication Magazine*, 128-134, June.

Perea, R. M. (2008). Internet Multimedia Communications Using SIP: A Modern Approach Including Java Practice. Burlington, MA: Morgan Kauffman Publishers.

Pereira, J. M. (2000). *Fourth Generation: Now, it is Personal*. Paper presented at 11th IEEE International Symposium on Personal, Indoor, and Mobile Radio Communications (PIMRC), London, UK.

Perera, E., Sivaraman, V., & Seneviratne, A. (2004). Survey on network mobility support. *Mobilie Computer Communication Review*, 8(2), 7–19. doi:10.1145/997122.997127

Perez, J., Morant, M., Llorente, R., & Marti, J. (2009) Joint Distribution of Polarization-Multiplexed UWB and WiMAX Radio in PON. *IEEE Journal Lightwave Technology*.

Perez, M. (2008, September 11). Wireless Carriers Address An Open Future. *InformationWeek*. Retrieved from http://www.informationweek.com/news/mobility/business/showArticle.jhtml?articleID=210600964

Perkins, C. (1996, October). IP mobility support. *IETF RFC2002*.

Perkins, C. (1997). *Mobile IP: Design principles and practice*. Reading, MA: Addison-Wesley.

Perkins, C., & Bhagwat, P. (1994). Highly dynamic Destination-Sequenced Distance-Vector routing (DSDV) for mobile computers. In *ACM SIGCOMM Computer* []. New York: ACM.]. *Communication Review, 24*, 234–244. doi:10.1145/190809.190336

Perkins, C., Belding-Royer, E., & Das, S. (2003). *Ad hoc On-Demand Distance Vector (AODV) Routing*. Network Working Group, RFC 3561.

Perkins, D. (2005). Convergence Challenges and Solutions for Next-Generation Wireless Networking. In *Texas Wireless Symposium*, USA.

Perros-Meilhac, L., & Lamy, C. (2002). Huffman tree based metric derivation for a low-complexity sequential soft VLC decoding. In *Proceedings of the IEEE ICC'02* (Vol. 2, pp. 783-787). New York: IEEE.

Persson, K.-A., Carlsson, C., Alping, A., Haglund, A., Gustavsson, J. S., Modh, P., & Larsson, A. (2006). WCDMA radio-over-fibre transmission experiment using singlemode VCSEL and multimode fibre. *Electronics Letters, 42*, 372–374. doi:10.1049/el:20064130

Pfrommer, H., Piqueras, M. A., Polo, V., Herrera, J., Martinez, A., & Marti, J. (2006). Radio-over-Fibre Architecture for Simultaneous Feeding of 5.5 and 41 GHz WiFi or WiMAX Access Networks. In *Microwave Symposium Digest 2006. IEEE MTT-S International* (pp. 301–303).

PJSIP. (n.d.). Retrieved January 20, 2009, from http://www.pjsip.org

Poikselka, M., et al. (2004). *The IMS: IP Multimedia Concepts and Services in the Mobile Domain*. Hoboken, NJ: Wiley.

Politis, C.; Oda, T.; Dixit, S.; Schieder, A.; Lach, H.-Y.; Smirnov, M.I.; Uskela, S.; & Tafazolli, R. (2004). Cooperative networks for the future wireless world. *Communications Magazine, 42*(9), 70–79. doi:10.1109/MCOM.2004.1336723

Postel, J. (1980). User datagram protocol. *IETF RFC 768*. Retrieved on January 2009, http://www.ietf.org/

Postel, J. (1981, September). *Internet Protocol*. RFC791. Retrieved from http://www.faqs.org/rfcs/rfc791.html

Prasanna, G., & Ravichandran, V. (2007). Performance analysis of MC-CDMA for wide band channels. *Information Technology Journal, 6*(2), 267–270. doi:10.3923/itj.2007.267.270

Prescher, G., Gemmeke, T., & Noll, T. G. (2005, March). A Parametrizable Low-Power High-Throughput Turbo-Decoder. *IEEE Internationa Conference on Acoustics, Speech, and Signal Processing (ICASSP, 5*, 25-28.

Pries, R., Mäder, A., & Staehle, D. (2006, August). A network architecture for a policy-based handover across heterogeneous networks. *OPNETWORK 2006*, Washington, DC, (pp. 1-9).

Proakis, J. G. (2001). *Digital Communications, 4th ed*. New York: McGraw Hill.

Punithavathani, D. S., & Sankaranarayanan, K. (2009). IPv4/IPv6 Transition Mechanisms. *European Journal of Scientific Research, 34(1), 110-124*. Retrieved from http://www.eurojournals.com/ejsr_34_1_12.pdf

Qian, X., et al. (2005). Directly-modulated photonic devices for microwave applications. In *Proceedings of IEEE MTT-S Intl Microwave Symposium,* Long Beach, California, USA.

Qiang, H, Xue-cheng, U., & Zou Shi-min, Z. (2006). ASN.1 Application in Parsing ISUP PDUs. *Communications and Information Technologies,* (pp. 78-81).

Qiao, L., & Nahrstedt, K. (1997). A new algorithm for MPEG video encryption. In [Las Vegas, NV.]. *Proceedings of the CISST, 97*, 21–29.

Qiao, Y., Yu, S., Su, P., & Zhang, L. (2005). Research on an iterative algorithm of LS channel estimation in MIMO OFDM systems. *IEEE Transactions on Broadcasting, 51*(3), 149–153. doi:10.1109/TBC.2004.842524

Qualcomm Europe (2006, May). Cell switching in LTE active state. *3GPP-RAN WG2 meeting #53 R2-061196*.

Qualcomm R1-050896. (2005). *Description and Simulations of Interference Management Technique for OFDMA based E-UTRA Downlink Evaluation, 3GPP TSG-RAN WG1 #42*.

Qusay, H. M. (2007). *Cognitive Networks: Towards Self-Aware Networks*. Chichester, UK: Wiley-Interscience.

Raad, I., & Huang, X. (2007). A new approach to bsofdm-parallel concatenated spreading matrices OFDM. *The 7th IEEE International Symposium on Communications and Information Technologies, ISCIT07 2007,* October 16 – 19, 2007.

Raad, I., & Huang, X. (2008). Study of higher order parallel concatenated block spread OFDM. *The Third International Conference on Information and Communication Technologies: From Theory to Applications, IEEE ICTTA'08,* Damascus, Syria, April 7 – 11, 2008.

Raad, I., Huang, X., & Raad, R. (2006). A new spreading matrix for block spread OFDM. *The 10th IEEE International Conference on Communication Systems 2006, IEEE ICCS'06*, Singapore, October 30 - Nov 1.

Rahim, A., Zeisberg, S., & Finger, A. (2007). Coexistence Study between UWB and WiMAX at 3.5 GHz Band. In *IEEE International Conference on Ultra-Wideband 2007, ICUWB 2007* (pp. 915-920).

Ramjee, R., Varadhan, K., Salgarelli, L., Thuel, S. R., Wang, S., & Porta, T. L. (1999, October). HWAWII: A domain-based approach for supporting mobility in wide-area wireless networks. In *7th International Conference on Network Protocols,* Toronto, Canada (pp. 283-292).

Raniwala, A., & Chiueh, C. (2005, March). Architecture and algorithms for an IEEE 802.11 based multi-channel wireless mesh network. In *Proc. of the IEEE Infocom*, Miami, (pp. 13-17).

Rao, K., Bojkovic, Z., & Milovanovic, D. (2006). *Introduction to Multimedia Communications: Applications, Middleware, Networking*. West Sussex, UK: John Wiley & Sons, Ltd.

Rappaport, T. (2002). *Wireless communications: principles and practice,* (2nd ed.). New York: Prentice Hall.

Rappaport, T. S. (2000). *Wireless Communications: Principles and Practice*. Reading, MA: Prentice Hall.

Rappaport, T. S. (2002). *Wireless Communications - principles and practice* (2nd Ed.). New York: Prentice Hall.

Rasheed, T., Javaid, U., Reynaud, L., & Al Agha, K. (2007). Cluster-Quality based Hybrid Routing in Large Scale Mobile Multi-hop Networks. In *Proceedings of IEEE Wireless Communications and Networking Conference (WCNC)*, Hong Kong.

Ray, S. K., & Mista, I. S. (2005). *Fourth Generation Networks: Roadmap – Migration to the Future*. Paper presented at IETE Technical Review, 23(4).

Rea, S. (2006). *Dynamic Route Management Strategies for Mobile Ad Hoc Networks*. PhD Dissertation, Cork Institute of Technology, Ireland.

Rec, I. T. U.-T. G.107, Amendment 1. (2006). *New Appendix II - Provisional impairment factor framework for wideband speech transmission*. International Telecommunication Union (ITU-T), Geneva.

Rec, I. T. U.-T. G.107. (2005). *The E-model, a Computational Model for Use in Transmission Planning*. International Telecommunication Union (ITU-T), Geneva.

Rec, I. T. U.-T. *H.264, ISOIEC 14496-10 AVC*, Doc JVT-G050r1. (2003). Final Draft International Standard of Joint Video Specification. Geneva, Switzerland: ITU.

Rec, I. T. U.-T. P.800. (1996). *Methods of Subjective Determination of Transmission Quality*. International Telecommunication Union (ITU-T), Geneva.

Compilation of References

Rec, I. T. U.-T. P.805. (2007). *Subjective Evaluation of Conversational Quality*. International Telecommunication Union (ITU-T), Geneva.

Rec, I. T. U.-T. P.862. (2001). *Perceptual Evaluation of Speech Quality (PESQ): An Objective Method for End-to-End Speech Quality Assessment of Narrow-Band Telephone Networks and Speech Codecs*. International Telecommunication Union (ITU-T), Geneva.

Rec, I. T. U.-T. P.862.2. (2007). *Wideband Extension to Recommendation P.862 for the Assessment of Wideband Telephone Networks and Speech Codecs*. International Telecommunication Union (ITU-T), Geneva.

Redman, P., Dulaney, K., & Gutberlet, M. (2006). *Key Criteria for Wireless Carriers to Evaluate Mobile WiMAX*. Gartner Research.

Reid, A. B., Grant, A. J., & Kind, A. P. (2003, February). Low complexity list detection for high-rate MIMO channels. *In Proc. of 4th Australian Commun. Theory Workshop, AusCTW '03,* Melbourne, Australia (pp. 66–69).

Reis, A. C. G. C., & Gondim, P. R. L. (2006). Performance Evaluation of DVB-RCT Standard. In *Proceedings of IEEE International Symposium on Broadband Multimedia Systems and Broadcasting*, Las Vegas, NV, April 2006.

Reis, A. C. G. C., & Gondim, P. R. L. (2009). *The OFDMA technology and its application in the in-band interaction channel of the terrestrial digital television systems*. PhD Thesis, Universidade de Brasília, February 2009.

Reliable Multicast Transport (RMT). *Working Group Charter*. Retrieved from http://ietf.org/html.charters/rmt-charter.html

Report ITU-R M.2134 (2008). *Requirements related to technical performance for IMT-Advanced radio interface(s)*. Geneva, Switzerland: International Telecommunication Union.

Rhee, W., & Cioffi, J. M. (2000). Increase in capacity of multiuser OFDM system using dynamic subchannel allocation. *VTC Spring*, Tokyo, Japan (pp. 1085–1089).

Rigney, C. (2000). *RADIUS Accounting (RFC 2866)*. Retrieved from http://www.ietf.org/rfc/rfc2866.txt

Rigney, C., Willats, W., & Calhoun, P. (2000). *RADIUS Extensions (RFC 2869)*. Retrieved from http://www.ietf.org/rfc/rfc2869.txt

Rigney, C., Willens, S., Rubens, A., & Simpson, W. (2000). *Remote Authentication Dial In User Service (RADIUS) (RFC2865)*. Retrieved from http://www.ietf.org/rfc/rfc2865.txt

Rinne, J., & Renfors, M. (1996). Pilot spacing in orthogonal frequency division multiplexing systems on practical channels. *IEEE Transactions on Consumer Electronics, 42*, 959–962. doi:10.1109/30.555792

Rix, A. W., Beerends, J. G., Kim, D., Kroon, P., & Ghitza, O. (2006). Objective Assessment of Speech and Audio Quality - Technology and Applications. *IEEE Transactions on Audio . Speech & Language Processing, 14*(6), 1890–1901. doi:10.1109/TASL.2006.883260

Robertson, P., Hoeher, P., & Villebrun, E. (1997). Optimal and suboptimal maximum a posteriori algorithm suitable for turbo decoding. *Europ. Trans. Telecom., 8*, 119–125. doi:10.1002/ett.4460080202

Robertson, P., Villebrun, E., & Hoeher, P. (1995, June). A comparison of optimal and sub-optimal MAP decoding algorithms operating in the log domain. In *Proc. of IEEE Int. Conf. on Commun., ICC '95*, Seattle (Vol. 2, pp. 1009–1013).

Roca, V. (2007). FCAST: Scalable Object Delivery on Top of the ALC Protocol. *IETF RMT Working Group*. Uusitalo, M.A. (n.d.). Global Vision for the Future Wireless World from the WWRF. *IEEE Vehicular Technology Magazine, 1*(2), 4–8.

Rodriguez, F., Campelo, J. C. & Serrano, Juan J. (2003). Improving the interleaved signature instruction stream technique, IEEE Canadian Conference on Electrical and Computer Engineering: Vol. 1, (pp. 93-96). Washington, DC: IEEE Press.

Ronai, M., Petrescu, A., Tönjes, R., & Wolf, M. (2003). Mobility Issues in OverDRiVE Mobile Networks. In *IST Mobile Summit*, Spain.

Rosenberg, J., Schulzrinne, H., Camarillo, G., Johnston, A., Peterson, J., Sparks, R., Handley, M., & Schooler, E. (2002, June). SIP: Session initiation protocol. *IETF RFC 2543*.

Roussos, G. (2003). *End-to-End Service architectures for 4G Mobile Systems, The Path to 4G Mobile*. Paper presented at IIR, London.

Royer, E. M., & Toh, C.-K. (1999). A review of current routing protocols for ad hoc mobile wireless networks. *IEEE Personal Communications*.

Rubio, M. L., Garcia-Armada, A., Torres, R. P., & Garcia, J. L. (2002). Channel modeling and characterization at 17 GHz for indoor broadband WLAN. *IEEE Journal on Selected Areas in Communications, 20*, 593–601. doi:10.1109/49.995518

Rudack, M., Meincke, M., Jobmann, K., & Lott, M. (2003). On traffic dynamical aspects inter-vehicle commuication(IVC. In *Proc. of the 57th IEEE Vehicular Technology Conference.*

Russell, S. F. (2001). Wireless Network Security for Users. *Information Technology: Coding and Computing,* 171-177.

Ryu, S., Ryu, B. H., Seo, H., Shin, M., & Park, S. K. (2005, December). Wireless Packet Scheduling Algorithm for OFDMA System Based on Time-Utility and Channel State. *ETRI Journal, 27*(6), 777–787. doi:10.4218/etrij.05.1005.0001

Sadeghi, B., Kanodia, V., Sabharwal, A., & Knightly, E. (2002, September). Opportunistic Media Access for Multirate Ad Hoc Networks. In *MOBICOM,* Atlanta, USA, (pp. 24-35).

Sadek, A. K., Su, W., & Liu, K. J. R. (2007). Multinode cooperative communications in wireless networks. *IEEE Transactions on Signal Processing, 55*(1), 341–355. doi:10.1109/TSP.2006.885773

Saleh, A., Rustako, A., & Roman, R. (1987). Distributed antennas for indoor radio communications. *IEEE Transactions on Communications, 35*(12), 1245–1251. doi:10.1109/TCOM.1987.1096716

Salmela, P., Gu, R., Bhattacharyya, S. S., & Takala, J. (2007, September). Efficient Parallel Memory Organization for Turbo Decoders. In *European Signal Processing Conference (EURASIP)* (pp. 209-214).

Sambaraju, R., et al. (2008). Photonic Envelope Detector for Broadband Wireless Signals using a Single Mach-Zehnder Modulator and a Fibre Bragg Grating. In *Proceedings of ECOC 2008* (pp. 64).

Sanders, W. B. (2008). Learning Flash Media Server 3. Sebastopol, CA: O' Reily Media, Inc.

Sandvig, C. (2003). Assessing Cooperative action in 802.11 Networks. In *Proceedings of 31st International Conference on Communication, Information and Internet Policy,* Washington, D.C.

Santhi, K. R., Srivastava, V. K., Senthikumaran, G., & Butare, A. (2005). Goals of True Broad Band's Wireless Next Wave (4G-5G). In *Vehicular Technology Conference,* USA.

Savo, G. G. (2006). Advanced Wireless Networks: 4G Technologies. West Sussex, UK: John Wiley & Sons, Ltd.

Sawal. (n.d.). *What is 4G technology?* Retrieved August 1, 2008, from http://sawaal.ibibo.com/computers-and-technology/what-4g-technology-464702.html

Schaar, M. (2007). Cross-Layer Wireless Multimedia. In M. Van der Schaar & P. A. Chou (Ed.), Multimedia over IP and Wireless Networks: Compression, Networking, and Systems (pp. 122-138). London: Academic Press.

Schaar, M. V., & Shankar, N. S. (2005). Cross-Layer Wireless Multimedia Transmission: Challenges, Principles, and New Paradigms. *IEEE Wireless Communications, 12*(4), 50–58. doi:10.1109/MWC.2005.1497858

Schlegel, C. (1997). *Trellis Coding.* New York: IEEE Press.

Schmitz, N. (March 2005). The Path to 4G Will Take Many Turns. *Wireless Systems Design.* Retrieved December 27, 2008, from http://www.wsdmag.com/Articles/ArticleID/10001/10001.html

Schmuck, H. (1995). Comparison of optically millimeter-wave system concepts with regard to chromatic dispersion. *IEEE Electron. Lett., 31*(21), 1848–1849. doi:10.1049/el:19951281

Schneider, K. S. (1979, January). Optimum detection of code division multiplexed signals. *IEEE Transactions on Aerospace and Electronic Systems, AES-15*, 181–185. doi:10.1109/TAES.1979.308816

Schnorr, C. P., & Euchner, M. (1994). Lattice basis reduction: Improved practical algorithms and solving subset sum problems. *Mathematical Programming, 66*, 181–191. doi:10.1007/BF01581144

Schrage, L. E., & Miller, L. W. (1966). The Queue M/G/1 with the Shortest Processing Remaining Time Discipline. *Operations Research, 14*(4), 670–684. doi:10.1287/opre.14.4.670

Schreurs, W. (2006). How Dangerous is 4G. *FreNovation Online.*

Schulzrinne, H., Casner, S., & Frederick, R. (1996). RTP: A transport protocol for realtime applications. *IETF RFC 1889.* Retrieved on January 2009, http://www.ietf.org/

Compilation of References

Schulzrinne, H., Rao, A., & Lanphier, R. (1998). Real Time Streaming Protocol (RTSP) Request for Comments: 2326, Standard Track, The Internet Engineering Task Force.

Schwartz, M. (2005). *Mobile Wireless Communications.* Cambridge, UK: Cambridge University Press.

Secgo. (n.d.). Retrieved January 20, 2009, from http://www.secgo.com

Seeds, A. J., & Williams, K. J. (2006). Microwave Photonics . *Journal of Lightwave Technology, 24*(12), 4628–4641. doi:10.1109/JLT.2006.885787

Seizo, O., & Nakamura, T. (2007). 3G evolution scenario toward 4G: Super 3G concept. *Wireless Commununication and Mobile Computing, 7,* 1013–1019. doi:10.1002/wcm.511

Selander, G. (2004). *Ambient Networking: Concepts and Architecture.* IST–2002-507134-AN.

Sellathurai, M., & Haykin, S. (2002). Turbo-BLAST for wireless communications: Theory and experiments. *IEEE Transactions on Signal Processing, 50*(10), 2538–2546. doi:10.1109/TSP.2002.803327

Sendonaris, A., Erkip, E., & Aazhang, B. (2003). User cooperation diversity, part I: System description. *IEEE Transactions on Communications, 51*(11), 1927–1938. doi:10.1109/TCOMM.2003.818096

Sendonaris, A., Erkip, E., & Aazhang, B. (2003). User cooperation diversity, part II: Implementation aspects and performance analysis. *IEEE Transactions on Communications, 51*(11), 1939–1948. doi:10.1109/TCOMM.2003.819238

Seo, S., Lee, S., & Song, J. (2007, November). Policy based intelligent vertical handover algorithm in heterogeneous wireless networks. *International Conference on Convergence Information Technology,* Gyeongju, Korea, (pp. 1900-1905).

SFR SDR (Software Defined Ratio) Development Tools Product Bulletin. *(Rev. A).* (n.d.). Texas Instruments. Retrieved February 25, 2009, from http://focus.ti.com/lit/ml/sprt406a/sprt406a.pdf

Shakkottai, S., Rappaport, T. S., & Karlsson, P. C. (2003). Cross-layer Design for Wireless Networks. *IEEE Communications Magazine, 41*(10), 74–80. doi:10.1109/MCOM.2003.1235598

Shannon, C.E. (1948). A mathematical theory of communication. *Bell System Technical Journal, 27,* 379-423 and 623-656.

Sharma, S. C. (2002). *4G Networks: A Case for Bypassing 3G.* Retrieved January 11, 2009, from http://voicendata.ciol.com/content/technology/102081902.asp

She, J., Yu, X., Ho, P-H., & Yang, E-H.(2009). A crosslayer design framework for robust IPTV services over IEEE 802.16 networks. *IEEE Journal on Selected Areas in Communications, 27*(2).

Shen, C. (2008). *Management Framework for Hybrid Wireless Networks.* PhD Dissertation, Cork Institute of Technology, Ireland.

Shen, C., Rea, S., & Pesch, D. (2008). Resource Sharing via Planed Relay for HWN*. *EURASIP Journal on Advances in Signal Processing, 2008,* 793126. doi:10.1155/2008/793126

Shen. C., Rea. S., & Pesch, D. (2007). HWN* Mobility Management Considering QoS, Optimisation and Cross Layer Issues. *IEEE Journal of Communication Software and Systems, 4*(3).

Shin, M. C., & Park, I. C. (2007, June). SIMD Processorbased Turbo Decoder Supporting Multiple Third-Generation Wireless Standards. [VLSI]. *IEEE Transactions on Very Large Scale Integration, 15,* 801–810. doi:10.1109/TVLSI.2007.899237

Shoewu, O. (2007). Evolution of Fourth Generation Mobile Networks: Trends and Tendencies. *The Pacific Journal of Science and Technology, 8*(2). Retrieved January 6, 2009, from http://www.akamaiuniversity.us/PJST.htm

Shoji, Y., Choi, C., Kato, S., Toyoda, I., Kawasaki, K., Oishi, Y., Takahashi, K., & Nakas, H. (2006). Re-summarization of merged usage model definitions parameters. *IEEE doc. 802.15-06-0379-02-003c.*

Shreedhar, M., & Vargese, G. (1996). Efficient Fair Queuing Using Deficit Round-Robin. *IEEE/ACM Transactions on Networking,* 4(3), 375-385.

Siddiqui, A. S., & Zhou, J. (1991). Two-Channel Optical Fiber Transmission Using Polarization Division Multiplexing. *Journal of Optical Communications, 12*(2), 47–49.

Signh, J., Bambos, N., Srinivasan, B., & Clawin, D. (2002). Wireless LAN performance under varied stress conditions in vehicular traffic scenarios. In *Proc. of IEEE VTC Fall,* (vol. 2, pp.743-747).

Silva, V. H., Salgado, E. I., & Quintana, F. R. (2006). AWISPA: An Awareness Framework for Collaborative Spontaneous Networks. In ASEE/IEEE Frontiers in Education Conference. (pp. 1-6). San Diego, CA: IEEE Press.

Simic, L., Berber, S. M., & Sowerby, K. W. (2008). *Partner choice and power allocation for energy efficient cooperation in wireless sensor networks*. Paper presented at the IEEE Internal Conference on Communications (ICC), Beijing, China.

Simon, M. K., & Alouini, M.-S. (2000). *Digital communication over fading channels: A unified approach to performance analysis*. New York: John Wiley and Sons.

Sklar, B. (1997). Rayleigh Fading Channels in Mobile Digital Communication Systems Part I: Characterization. *IEEE Communications Magazine*, *35*(7), 90–100. doi:10.1109/35.601747

Skordylis, A., & Trigoni, N. (2008). Delay-bounded routing in vehicular ad-hoc networks. In *Proc. of ACM/IEEE MobiHoc*, (pp. 341-350), Hong Kong, China.

Soleymani, M. R., Gao, Y., & Vilaipornsawai, U. (2002). *Turbo Coding for Satellite and Wireless Communications*. Norwell, MA: Kluwer Academic.

Soliman, H. (2004). *Mobile IPv6: Mobility in Wireless Internet*. Reading MA: Addison-Wesley Professional.

Soliman, H. (2009). *Mobile IPv6 Support for Dual Stack Hosts and Routers*. RFC 5555. Retrieved from http://www.faqs.org/rfcs/rfc5555.html

Soliman, H., Castelluccia, C., Malki, K. E., & Bellier, L. (2005, August). Hierarchical Mobile IPv6 mobility management. *IETF RFC 4140*.

Soliman, H., Elmalki, K., & Bellier, L. (2008). Network Working Group, USA.

Solomon, J. D. (1998). *Mobile IP – The Internet unplugged*. Upper Saddle River, NJ: Prentice Hall.

Song, J., Hu, J., Tian, Y., & Xu, Y. (2005). Re-optimization in dynamic vehicle routing problem based on Wasp-like agent strategy. *Intelligent Transportation Systems*, *2005*, 231–236.

Speth, M., Fechtel, S. A., Fock, G., & Meyr, H. (1999). Optimum receiver design for wireless broad-band systems using OFDM - part I. *IEEE Transactions on Communications*, *47*(11), 1668–1677. doi:10.1109/26.803501

Sridhar, T. (2006). Wireless LAN Switches -- Functions and Deployment. *The Internet Protocol Journal (Cisco)*, *9*(3), 2–15.

Srivastava, V., & Motani, M. (2005). Cross-layer design: A survey and the road ahead. *IEEE Communications Magazine*, *43*(12), 112–119. doi:10.1109/MCOM.2005.1561928

Stallings, W. (2005). *Wireless Communications and Networking* (2nd ed). Upper Saddle River, NJ: Pearson Prentice Hall. Thornton, J., Grace, D., Capstick, M. H., & Tozer, T. C. (2003). Optimizing an Array of Antennas for Cellular Coverage from a High Altitude Platform. *IEEE Transactions on Wireless Communications*, *2*(3), 484–492.

Stauder, J., & Erbas, F. (2005). Mobile Multimedia over Wireless Networks. In S.M. Rahman (Ed.), Multimedia Networking: Technology, Management and Applications (pp. 422-471). Hershey, PA: IRM Press.

Stefanov, A., & Duman, T. (2001). Turbo-coded modulation for systems with transmit and receive antenna diversity over block fading channels: System model, decoding approaches, and practical considerations. *IEEE Journal on Selected Areas in Communications*, *19*(5), 958–968. doi:10.1109/49.924879

Stefanov, A., & Erkip, E. (2004). Cooperative coding for wireless networks. *IEEE Transactions on Communications*, *52*(9), 1470–1476. doi:10.1109/TCOMM.2004.833070

Stefanov, A., & Erkip, E. (2005). Cooperative space-time coding for wireless networks. *IEEE Transactions on Communications*, *53*(11), 1804–1809. doi:10.1109/TCOMM.2005.858641

Steiner, A., Sanderovich, A., & Shamai, S. (2007). Broadcast cooperation strategies for two collocated users. *IEEE Transactions on Information Theory*, *53*(10), 3394–3412. doi:10.1109/TIT.2007.904833

Steingrimsson, B., Luo, T., & Wong, K. M. (2003). Soft quasi-maximum-likelihood detection for multiple-antenna wireless channels. *IEEE Transactions on Signal Processing*, *51*(11), 2710–2719. doi:10.1109/TSP.2003.818203

Sternad, M., Svensson, T., Ottosson, A., Ahlen, A., Svensson, A., & Brunstrom, A. (2007). Towards systems beyond 3G based on adaptive OFDMA transmission. In *Proceedings of IEEE. Special Issue on Adaptive Transmission*, *95*(12), 2432-2455.

Steuer, F., Elkotob, M., Albayrak, S., & Steinbach, A. (2006). Testbed for Mobile Network Operator Scenarios. In

Compilation of References

Proceedings of 2nd International Conference on Testbeds & Research Infrastructures for the Development of NeTworks & COMmunities (TRIDENTCOM) (pp. 256-266).

Stewart, R., Xie, Q., Morneault, K., Schwarzbauer, H., Taylor, T., Rytina, I., et al. (2000). Stream Control Transmission Protocol. Request for Comments: 2960, Standard Track, The Internet Engineering Task Force.

Stiliadis, D., & Varma, A. (1998). Latency-rate servers: a general model for analysis of traffic scheduling algorithms. *IEEE/ACM Transactions on Networking, 6*(5), 611–624.

Stokes, J. (2007). *Nokia 4G wireless tech hits 173Mbps in real-world test*. Retrieved December 28, 2007, from http://arstechnica.com/news.ars/post/20071228-nokia-4g-wireless-tech-hits-173mbps-in-real-world-test.html

Stolyar, A. L., & Ramanan, K. (2001). Largest Weighted Delay First Scheduling: Large Deviations and Optimality. *Annals of Applied Probability, 11*(1), 1–48. doi:10.1214/aoap/998926986

Strohm, K. M., Schneider, R., & Wenger, J. (2005). KOKON: A Joint Project for the Development of 79 GHz Automotive Radar Sensors. In *Proceedings of IRS 2005*.

Stuckmann, P., & Zimmermann, R. (2007). Towards ubiquitous and unlimited capacity communication networks: European research in framework programme 7. *IEEE Communications Magazine, 45*(5), 148–147. doi:10.1109/MCOM.2007.358862

Studer, C., Burg, A., & Bölcskei, H. (2006). Soft-output sphere decoding: Algorithms and VLSI implementation. In *Proc. 40th Asilomar Conference on Signals, Systems, and Computers*.

Su, H., & Zhang, X. (2007). Opportunistic MAC protocols for cognitive radio based wireless networks. In *Proc. 41st Conference on Information Sciences and Systems*, Johns Hopkins University, March.

Su, W., Sadek, A. K., & Liu, K. J. R. (2005). *SER performance analysis and optimum power allocation for decode-and-forward cooperation protocol in wireless networks*. Paper presented at the IEEE Wireless Communications and Networking Conference (WCNC), New Orleans, USA.

Suárez, A., Menza, M., Macías, E. M., & Sunderam, V. (2007). Automatic resumption of streaming sessions over WiFi using JADE. *IAENG International Journal of Computer Science, 33*(1), 92–100.

Sun J2ME Specification. (n.d.). Retrieved January 19, 2009, from http://java.sun.com/javame

Sun Java 2 platform, standard edition (J2SE). (n.d.). Retrieved August 25, 2003, from http://java.sun.com/j2se

Sun, J., & Takeshita, O. Y. (2005). Interleavers for Turbo Codes Using Permutation Polynomials Over Integer Rings. *IEEE Transactions on Information Theory, 51*, 101–119. doi:10.1109/TIT.2004.839478

Sun, Y., & Cavallaro, J. R. (2008, October). Unified Decoder Architecture for LDPC/TURBO Codes. *IEEE Workshop on Signal Processing Systems (SIPS)*, (pp. 13-18).

Sun, Y., Zhu, Y., Goel, M., & Cavallaro, J. R. (2008, July). Configurable and Scalable High Throughput Turbo Decoder Architecture for Multiple 4G Wireless Standards. *IEEE International Conference on Application-Specific Systems, Architectures and Processors (ASAP)*, (pp. 209-214).

Sundaresan, K., & Rangarajan, S. (2008). On exploiting diversity and spatial reuse in relay enabled wireless networks. In *Proc. of the 9th ACM international symposium on Mobile ad hoc networking and computing*, (pp. 13-22).

Sunnerud, H., Karlsson, M., Chongjin, X., & Andrekson, P. A. (2002). Polarization-mode dispersion in high-speed fiber-optic transmission systems. *IEEE Journal of Lightwave Technology, 20*(12), 2204–2219. doi:10.1109/JLT.2002.806765

System Architecture Evolution (SAE). (n.d.). Wikipedia. Retrieved on September 9, 2009. Retrieved from http://en.wikipedia.org/wiki/System_Architecture_Evolution

Systems, C. (2008, October). *Implementing QoS for IPv6*. Retrieved from http://www.cisco.com/en/US/docs/ios/ipv6/configuration/guide/ip6-qos.pdf

Szczodrak, M., Kim, J., & Baek, Y. (2007). 4GM@4GW: Implementing 4G in the Military Mobile Ad-Hoc Network Environment. *International Journal of Computer Science and Network Security, 7*(4), 70–79.

Taherzadeh, M., Mobasher, A., & Khandani, A. K. (2007). LLL Reduction Achieves the Receive Diversity in MIMO Decoding. *IEEE Transactions on Information Theory, 53*(12), 4801–4805. doi:10.1109/TIT.2007.909169

Takada, J., Fu, J., Zhu, H., & Kobayashi, T. (2002). Spatio-temporal channel characterization in a suburban non line-of-sight microcellular environment. *IEEE Journal on Selected Areas in Communications, 20*(3), 532–538. doi:10.1109/49.995512

Takeshita, O. Y. (2006). On maximum contention-free interleavers and permutation polynomials over integer rings. *IEEE Transactions on Information Theory, 52*(3), 1249–1253. doi:10.1109/TIT.2005.864450

Talukdar, A., Badrinathm, B., & Acharya, A. (2001). MRSVP: A resource reservation protocol for an integrated services network with mobile hosts. *Wireless Networks*, (Vol. 7).

Tanenbaum, A. (2003). *Computer Networks (4th ed.)*. Upper Saddle River, NJ: Prentice Hall.

Tao, M., Liang, Y. C., & Zhang, F. (2006, June). Adaptive Resource Allocation for Delay Differentiated Traffic in Multiuser OFDM Systems. In *ICC* (pp. 4403 – 4408).

Tarokh, V., Jafarkhani, H., & Calderbank, A. R. (1999). Space-time block codes from orthogonal design. *IEEE Transactions on Information Theory, 45*(5), 1456–1467. doi:10.1109/18.771146

Tarokh, V., Seshadri, N., & Calderbank, A. R. (1998). Space-time codes for high data rate wireless communication: Performance criterion and code construction. *IEEE Transactions on Information Theory, 44*(2), 744–765. doi:10.1109/18.661517

Tassiulas, L., & Sarkar, L. (2002). Maxmin fair scheduling in wireless networks. In *IEEE INFOCOM*, (pp. 763-772).

TCPDUMP. (n.d.). Retrieved January 20, 2009, from http://www.tcpdump.org

Telatar, E. (1999). Capacity of multi-antenna Gaussian channels. *European Transactions on Telecommunications, 10*(6), 585–595. doi:10.1002/ett.4460100604

Telnet. (n.d.). Retrieved January 20, 2009, from http://www.telnet.org

Terzis, A., Krawczyk, J., Wroclawski, J., & Zhand, L.(2000). RSVP: Operation over IP tunnels. *RFC 2746*.

Thamilarasu, G., Balasubramanian, A., Mishra, S., & Sridhar, R. (2005). A Cross-Layer based Intrusion Detection Approach for Wireless Ad Hoc Networks. *IEEE International Conference on Mobile Adhoc and Sensor Systems Conference (MASS05)* (pp. -861).

The Economist Intelligence Unit. (2008). *Report: E-readiness rankings 2008*. Retrieved from http://a330.g.akamai.net/7/330/25828/20080331202303/graphics.eiu.com/upload/ibm_ereadiness_2008.pdf

Thull, M. J., Gilbertl, F., Vogtl, T., Kreiselmaierl, G., & Wehn1, N (2005). A Scalable System Architecture for High-Throughput Turbo-Decoders. *The Journal of VLSI Signal Processing, 39*, 63-67.

TIA/EIA/IS-2000.1.A -2. (2002, February 11). The CDMA2000 Family of Standards for Spread Spectrum Systems.

TIA/US. (1998, June). *The cdma2000 ITU-R RTT Candidate Submission*.

Todini, A., et al. (2006). Cross-layer design of packet scheduling and resource allocation algorithms for 4G cellular systems, *WPMC*, San Diego, September 2006.

Tomcik, J. (2006, January). MBTDD wideband mode performance report II. *IEEE C802.20-05/88r1*, from http://www.ieee802.org/20/

Tomiyasu, H., Maekawa, T., Hara, T., & Nishio, S. (2006). Profile-based Query Routing in a Mobile Social Network. In *7th International Conference on Mobile Data Management* (pp. 105-108). Washington, DC: IEEE Computer Society.

Tonev, G., Sunderam, V., Loader, R., & Pascoe, J. (2002). Location and network quality issues in local area wireless networks. In International Workshop on Network and Operating Systems Support For Digital Audio and Video, (LNCS 2299, pp. 131- 148). Kalsrue, Alemania: Springer Verlag.

Tooher, P., & Soleymani, M. R. (2008). *Performance of variable time-fraction collaborative communication*. Paper presented at the IEEE Personal, Indoor and Mobile Radio Communications (PIMRC) conference, Cannes, France.

Tooher, P., & Soleymani, M. R. (2009). *Power allocation for wireless communications using variable time-fraction collaboration*. Paper presented at the IEEE International Conference on Communications (ICC), Dresden, Germany.

Tooher, P., & Soleymani, M. R. (Manuscript submitted for publication). Analysis and code design of variable time-fraction collaborative communication. *Submitted to IEEE Transactions on Wireless Communications*.

Tooher, P., Khoshneviss, H., & Soleymani, M. R. (2007). *Design of collaborative codes achieving space-time diversity*. Paper presented at the IEEE International Conference on Communications (ICC), Glasgow, UK.

Toshio, M., Tomoyuki, O., Hitoshi, Y., & Umeda, N. (2005). The Overview of the 3th Generation Mobile Communication

Compilation of References

System. In *Fifth International Conference on Information, Communications and Signal Processing (ICICS 2005)*, Bangkok.

TS 25.101 v3.0.0. (1999). UE radio transmission and reception (FDD). *3GPP TSG RAN WG4 document*.

TS 25.104 v3.0.0. (1999). UTRA (BS) FDD. radio transmission and reception. *3GPP TSG RAN WG4 document*.

Tse, D., & Grossglauser, M. (2002, August). Mobility increases the capacity of ad hoc networks. *IEEE/ACM Trans. Net, 10*(4), 477–486.

Tse, D., & Viswanath, P. (2005). Fundamentals of Wireless Communication. Cambridge, UK: Cambridge University Press.

Tseng, C. C., Lee, G. C., Liu, R. S., & Wang, T. P. (2003). HMRSVP: Hierarchial mobile RSVP protocol. *Wireless Networks, 9*, 467–472.

Tseng, C. C., Yen, L. H., Chang, H. H., & Hsu, K. C. (2005). Topology-Aided Cross-Layer Fast Handoff Designs for IEEE 802.11/Mobile IP Environments. *IEEE Communications Magazine, 43*(12), 156–163. doi:10.1109/MCOM.2005.1561933

Tsirtsis, G., & Srisuresh, P. (2000). Network Address Translation - Protocol Translation. RFC 2766. Retrieved from http://www.faqs.org/rfcs/rfc2766.html

Typpo, V., Fisl, J., Holler, J., Calvo, R. A., & Karl, H. (2005). Research Challenges in Mobility and Moving Networks: An Ambient Networks View. In *Broadband Satellite Communication Systems and the Challenges of Mobility* (pp. 145-155). New York: Springer.

Ukil, A. (2008, August). Long-Term Proportional Fair QoS Profile Follower Sub-carrier Allocation Algorithm in Dynamic OFDMA Systems. In *13th International OFDM Workshop*, Hamburg, Germany (pp.1-5).

Ukil, A., Sen, J., & Bera, D. (2009, January). A New Optimization Scheme for Resource Allocation in OFDMA Based WiMAX Systems. In *International Conference on Computer Engineering and Technology (ICCET'09)*, Singapore, (pp. 145-149).

Ulukus, S., & Pollini, G. P. (1998, June). Handover delay in cellular wireless systems. *IEEE International Conference on Communications*, Atlanta, GA, (pp. 1370-1374).

Umbach, A., Waasen, S. V., Auer, U., Bach, H. G., Bertenburg, R. M., & Breur, V. (1996). Monolithic pin-HEMT 1.55 um photoreceiver on InP with 27 GHz bandwidth. *Electronics Letters, 32*(23), 2142–2143. doi:10.1049/el:19961421

UN e-Government Survey. (2008). New York: United Nations. Retrieved from http://unpan1.un.org/intradoc/groups/public/documents/UN/UNPAN028607.pdf

Urban, J., Wisely, D., Bolinth, E., Neureiter, G., Liljeberg, M., & Robles, T. (2001). BRAIN – An architecture for a broadband radio access network of next generation. *Journal of Wireless Communications & Mobile Computing*.

Vahdat, A., & Becker, D. (2000). *Epidemic routing for partially-connected ad hoc networks* [Tech. Rep. CS-2000-06]. Durham, NC: Duke University.

Valsala, K. (2007). Enabling Network-Based Presence Aggregation using IMS. *Infosys*. Retrieved from http://www.infosys.com/engineering-services/product-engineering/white-papers/enabling-networks-paper-using-IMS-final.pdf

van de Beek, J.-J., Sandell, M., & Børjesson, P. O. (1997). ML estimation of time and frequency offset in OFDM systems. *IEEE Transactions on Signal Processing, 145*, 1800–1805. doi:10.1109/78.599949

van der Meulen, E. C. (1968). *Transmission of information in a T-terminal discrete memoryless channel*. Unpublished doctoral dissertation, University of California, Berkeley, CA.

van der Meulen, E. C. (1971). Three-terminal communication channels. *Advances in Applied Probability, 3*(1), 120–154. doi:10.2307/1426331

Van der Schaar, M., Krishnamachari, S., & Choi, S. (2003). Adaptive cross-layer protection strategies for robust scalable video transmission over 802.11 WLANs. *IEEE Journal on Selected Areas in Communications, 21*, 1752–1763. doi:10.1109/JSAC.2003.815231

van Nee, R., van Zelst, A., & Awater, G. (2000), Maximum likelihood decoding in a space division multiplexing system. In *Proc. 51st IEEE Vehicular Technology Conference (VTC Spring)*.

Van Nee, R., Van Zelst, A., & Awater, G. (July 1999). Maximum Likelihood Decoding in a Space

Vandalore, B., Feng, W., Jain, R., & Fahmy, S. (2001). A Survey of Application Layer Techniques for Adaptive Streaming of Multimedia. *International Journal of Real-Time Imaging, 7*(3), 221–235. doi:10.1006/rtim.2001.0224

Vassiliou, V., Owen, H. L., Barlow, D. A., Grimminger, J., Huth, H.-P., & Sokol, J. (2002). A radio access network for next generation wireless networks based on multi-protocol label switching and hierarchical Mobile IP. *Proc. 56th IEEE Vehicular Technology Conf., 2*, 782-786.

Vasudevan, S., Papagiannaki, K., Diot, C., Kurose, J., & Towsley, D. (2005). Facilitating access point selection in IEEE 802.11 wireless networks. *Internet Measurement Conference*, (pp. 293-298), Berkeley, CA.

Vehkapera, J., Peltola, J., & Huusko, J. Myllyniemi & Majanen, M. (2006). Evaluation of achieved video quality in wireless multimedia transmission using UDP-Lite. In *Proceedings of the IASTED IMSA 2006*, (pp.14-16), Honolulu, HI. New York: ActaPress.

Verdone, R., & Zanella, A. (2002). Performance of received power and traffic-driven handover algorithms in urban cellular networks. *IEEE Communications Magazine, 9*(1), 60–71. doi:10.1109/MWC.2002.986461

Vidales, P., Baliosian, J., Serrat, J., Mapp, G., Stajano, F., & Hopper, J. (2005). Autonomic System for Mobility Support in 4G Networks. [JSAC]. *IEEE Journal on Selected Areas in Communications, 23*(12). doi:10.1109/JSAC.2005.857198

Vidales, P., Bernardos, C., Soto, I., Cottingham, D., Baliosian, J., & Crowcroft, J. (2007). MIPv6 Experimental Evaluation Using Overlay Networks. *Computer Networks, 51*(10), 2892–2915. doi:10.1016/j.comnet.2006.12.004

Vidales, P., Kirschnick, N., Steuer, F., Lewcio, B., Wältermann, M., & Möller, S. (2008). Mobisense Testbed: Merging User Perception and Network Performance. In *Proceedings of 4th International Conference on Testbeds & Research Infrastructures for the DEvelopment of NeTworks & COMmunities (TRIDENTCOM)* (pp. 1-9).

Vikalo, H., Hassibi, B., & Kailath, T. (2004). Iterative decoding for MIMO channels via modified sphere decoding. *IEEE Transactions on Wireless Communications, 3*(11), 2299–2311. doi:10.1109/TWC.2004.837271

Villalón, J., Cuenca, P., Orozco-Barbosa, L., Seok, Y., & Turletti, T. (2007). Cross-layer architecture for adaptive video multicast streaming over multirate wireless LANs. *IEEE Journal on Selected Areas in Communications, 25*(4), 699–711. doi:10.1109/JSAC.2007.070507

Viswanath, P., Tse, D., & Laroia, R. (2002). Opportunistic beamforming using dumb antennas. *IEEE Transactions on Information Theory, 48*(6), 1277–1294. doi:10.1109/TIT.2002.1003822

Vodafone's 'Long Term' Hesitance. (2008 April). *Business Week*.

Vogt, J., & Finger, A. (2000). Improving the Max-Log-MAP Turbo Decoder. *Electronics Letters*, 1937–1939. doi:10.1049/el:20001357

Vollbrecht, J., Calhoun, P., Farrell, S., Gommans, L., Gross, G., Bruijn, B., et al. (2000) AAA Authorization Framework (*RFC 2904*). Retrieved from http://www.ietf.org/rfc/rfc2904.txt.

Vollbrecht, J., Calhoun, P., Farrell, S., Gommans, L., Gross, G., Bruijn, B., et al. (2000). *AAA Authorization Application Examples (RFC 2905)*. Retrieved from http://www.ietf.org/rfc/rfc2905.txt

Wake, D. (1997). Passive picocell-A new concept in wireless network infrastructure. *Electronics Letters, 33*(5), 404–406. doi:10.1049/el:19970277

Wake, D., & Schuh, R. E. (2000). Measurement and simulation of W-CDMA signal transmission over optical fibre. *Electronics Letters, 36*, 901–902. doi:10.1049/el:20000670

Wältermann, M., Lewcio, B., Vidales, P., & Möller, S. (2008). A Technique for Seamless VoIP Codec Switching in Next Generation Networks. In *Proceedings of International Conference on Communications (ICC)*, Beijing (pp. 1772-1776).

Wang, B., Han, Z., & Liu, K. J. R. (2006). Stackelberg Game for Distributed Resource Allocation over Multiuser Cooperative Communication Networks, *Globecom*, (pp. 1-5).

Wang, H., Tsaftaris, S. A., & Katsaggelos, A. K. (2006). Joint source-channel coding for wireless object-based video communications utilizing data hiding. *IEEE Transactions on Image Processing, 15*, 2158–2169. doi:10.1109/TIP.2006.875194

Wang, R., & Giannakis, G. B. (2004, March). Approaching MIMO channel capacity with reduced-complexity soft sphere decoding. In *Proc. of IEEE Wireless Commun. and Networking Conf., WCNC '04* (Vol. 3, pp. 1620–1625).

Wang, W. (2005). Modeling and Management of Location and Mobility. In D. Katsaros, Nanopoulos & M. Yannis (Ed.), Wireless Information Highway (pp. 177-212). Hershey, PA: IRM Press.

Wang, X., & Xiang, W. (2006, March). An OFDM-TDMA/SA MAC protocol with QoS constraints for broadband wireless LANs. *Wireless Networks*, *12*(2), 159–170. doi:10.1007/s11276-005-5263-1

Wei, H. Y., Ganguly, S., Izmailov, R., & Hass, Z. J. (2005). Interference-Aware IEEE 802.16 WiMax Mesh Networks. In *IEEE Vehicular Technology Conference 2005 Spring (VTC-2005)* (pp. 3102-3106).

Weiser, M. (1992). *Does Ubiquitous Computing Need Interface Agents? No.* Invited talk at MIT Media Lab Symposium on User Interface Agents.

Wells, J. (2006). WiMAX Backhaul at 70/80 GHz [White paper]. *GigaBeam Wireless Fiber*.

Wengerter, C., Ohlhorst, J., & Elbwart, A. G. (2005). Fairness and Throughput Analysis for Generalized Proportional Fair Frequency Scheduling in OFDMA. In *VTC, Spring, 2005* (pp. 1903-1907).

Whetten, T., Vicisano, L., Kermode, R., Handley, M., Floyd, S., & Luby, M. (2001). Reliable Multicast Transport Building Blocks for One-to-Many Bulk-Data Transfer. *IETF (RFC 3048)*.

Willie, W. (2006). Open Wireless Architecture (OWA) – Defining China's Fourth Generation Mobile Communications. In *Fourth Generation Mobile Forum*, Hong Kong.

Willinger, W., Taqqu, M. S., Sherman, R., & Wilson, D. V. (1995). Self-Similarity Through High-Variability: Statistical Analysis of Ethernet LAN Traffic at the Source Level. In *ACM SIGCOMM* (pp. 100-113). Cambridge, MA: University of Massachusetts.

Wilson, B., Ghassemlooy, Z., & Darwazeh, I. (1995). *Analogue Optical Fibre Communications*. Institution of Engineering and Technology.

WiMAX extension to isolated research data networks. (n.d.). Retrieved from http://www.ist-weird.eu/

WiMAX Forum. (2007). *WiMAX system evaluation methodology v.2.1*. Retrieved from http://www.wimaxforum.org/documents

WiMAX Forum. (2008). *WiMAX Forum Network Architecture, Release 1, version 1.2*.

WiMAX Forum. (n.d.). Retrieved from http://www.wimaxforum.org

WiMedia Alliance. (2008 January). UWB - best choice to enable WPANs. *WiMedia Alliance*.

Windows Embedded, C. E. *Overview*. (n.d.). Retrieved December 30, 2008 from http://www.microsoft.com/windowsembedded

WINNER. (2005). *Identification, definition and assessment of cooperation schemes between RANs (D4.3)*. IST-WINNER, Final deliverable.

Wireless World Research Forum. (n.d). Retrieved December 30, 2008, from http://www.wireless-world-research.org

Wireshark. (n.d.). Retrieved January 20, 2009, from http://www.wireshark.org

Wolniansky, P. W., Foschini, G., Golden, G. D., & Valenzuela, R. A. (1998). *V-BLAST: An architecture for realizing very high data rates over the rich-scattering wireless channel*. Paper presented at the URSI International Symposium on Signals, Systems and Electronics, Pisa, Italy.

Wong, C. Y., Cheng, R. S., Letaief, K. B., & Murch, R. D. (1999). Multiuser OFDM with adaptive subcarrier, bit and power allocation. *IEEE Journal on Selected Areas in Communications*, *17*(10), 1747–1757. doi:10.1109/49.793310

Wong, D. (2005). *Wireless Internet Telecommunication*. Norwood, MA: Artech House.

Wong, I., Shen, Z., Evans, B., & Andrews, J. (2004). A low complexity algorithm for proportional resource allocation in OFDMA systems. In *IEEE Signal Processing Workshop*, TX, USA, (pp. 1–6).

Wong, K. K., Paulraj, A., & Murch, R. D. (2005, May). List slab-sphere decoding: efficient high-performance decoding for asymmetric MIMO antenna systems. In *Proc. of IEEE 61th Vehicular Tech. Conf., VTC '05-Spring* (Vol. 1, pp. 697–701).

Woo, K. S., Lee, K. I., Paik, J. H., Park, K. W., Yang, W. Y., & Cho, Y. S. (2007). TA DSFBC-OFDM for a next generation broadcasting system with multiple antennas. *IEEE Transactions on Broadcasting*, *53*(6), 539–546.

Wu, D., & Negi, R. (2003). Effective capacity: a wireless link model for support of quality of service. *IEEE Transactions on Wireless Communications*, *2*(4), 630–643.

Wu, S., Hsu, J., & Chen, C. (2007). Headlight Prefetching for Mobile Media Streaming. In ACM Workshop on Data Engineering for Wireless and Mobile Access. (pp. 67-74). Beijing, China: ACM Press.

Wübben, D., Böhnke, R., Kühn, V., & Kammeyer, K. D. (2003). MMSE extension of V-BLAST based on sorted QR decomposition. In *Proc. IEEE Vehicular Technology Conference (VTC Fall)*.

Wübben, D., Böhnke, R., Kühn, V., & Kammeyer, K.-D. (2001). Efficient algorithm for decoding layered space-time codes. *IEEE Electronics Letters, 37*, 1348–1350. doi:10.1049/el:20010899

Xiang, Y., Peng-Jun, W., & Wen-Zhan, S., & Yanwei Wu. (2008). Efficient throughput for wireless mesh networks by CDMA/OVSF code assignment. *Ad Hoc and Sensor Wireless Networks, 5*(3-4), 265–291.

Xiao, L., Fuja, T. E., Kliewer, J., & Costello, D. J., Jr. (2007). *Signal superposition coded cooperative diversity: Analysis and optimization.* Paper presented at the IEEE Information Theory Workshop (ITW), Lake Tahoe, USA.

Xie, Z., Short, R. T., & Rushforth, C. K. (1990, May). A family of suboptimum detectors for coherent multi-user communications. *IEEE Journal on Selected Areas in Communications, 8*, 683–690. doi:10.1109/49.54464

Xu, Z., Zhou, C., & Wang, J. A (2008). Novel Cell Architecture Based on Distributed Antennas for Mobile WiMAX Systems. In *4th IEEE International Conference on Circuits and Systems for Communications, 2008. ICCSC 200* (pp. 172-176).

Yaiche, H., Majumder, R., & Rosenberg, C. (2000). A game theoretic framework for bandwidth allocation and pricing for broadband networks. *IEEE/ACM Transactions on Networking*, 667-678.

Yamai, N., Okayama, K., Shimamoto, H., & Okamoto, T. (1999). TCP performance over GPRS. In *Proc. of the IEEE wireless communication and networking conference (WCNC)*, (Vol.3, pp.1248-1252).

Yang, L. (2007). *Access Network Selection in a 4G Networking Environment.* Master Thesis, Department of Electrical and Computer Engineering Waterloo, Canada.

Yang, S., & Koetter, R. (2007). *Network coding over a noisy relay: A belief propagation approach.* A paper presented at the IEE International Symposium on Information Theory (ISIT), Nice, France.

Yang, Z., & Mohammed, A. (2008). *On the Cost-Effective Wireless Broadband Service Delivery from High Altitude Platforms with an Economical Business Model Design.* Paper presented at the IEEE 68th Vehicular Technology Conference, 2008. VTC 2008-Fall, Calgary Marriott, Canada

Yankelovich, N., Horan, B., Kaplan, J., Simpson, N., & Slott, J. (2009). Collaboration tools for distributed work - projects include Project Wonderland, MPK20, jVoiceBridge, Porta-Person, Conference Manager, and the Sun Labs Meeting Suite. Collaborative Environments (Sun Labs). Retrieved Jannuary 30, 2009, from http://research.sun.com/projects/dashboard.php?id=85

Yano, S. M. (2002). Investigating the ultra-wideband indoor wireless channel. In *Proceedings of IEEE 55th Vehicular Technology Conference,* Vermont (Vol. 3, pp. 1200–1204).

Yao, Y., & Giannakis, G. B. (2005). Rate-Maximizing Power Allocation in OFDM Based on Partial Channel Knowledge. *IEEE Transactions on Wireless Communications, 4*(3), 1073–1083. doi:10.1109/TWC.2005.847022

Yasukawa, S., Nishikido, J., & Hisashi, K. (2001). Scalable mobility and QoS support mechanism for IPv6-based real-time wireless Internet traffic. *Proc. IEEE GLOBECOM Conf., 6*, 3459 – 3462.

Ye, L., & Gordon, L. (2006, February). *Orthogonal frequency division multiplexing for wireless communications.* Berlin: Springer.

Yin, J., Lee, H. N., Ahmed, M., Ryu, B., & Peterson, L. (2004, May). Iterative MMSE-sphere list detection and graph decoding MIMO OFDM transceiver. In *Proc. of IEEE 59th Vehicular Tech. Conf., VTC '04-Spring* (Vol. 2, pp. 1503–1507).

Yokota, H., Idoue, A., Hasegawa, T., & Kato, T. (2002). Link layer assisted mobile IP fast handover method over wireless LAN networks. In *Proc. ACM MobiCom,* (pp. 131 – 139), Atlanta, GA.

Young Kyun, K., & Prasad, R. (2006). *4G Roadmap and Emerging Communication Technologies.* Norwood, MA: Artech House.

Yu, H., Gibbons, P. B., Kaminsky, M., & Xiao, F. (2008). SybilLimit: A near-optimal social network defense against sybil attacks. In *IEEE Symposium on Security and Privacy* (pp. 3-17).

Yu, H., Kaminsky, M., Gibbons P.B., & Flaxman, A. (2008). SybilGuard: Defending against sybil attacks via social networks. *IEEE/ACM Transaction on Networking, 16*(3), 576-589.

Compilation of References

Yu, H., Mohapatra, P., & Liu, X. (2008, April). Channel assignment and link scheduling in multi-radio multi-channel wireless mesh networks. *Mobile Networks and Applications, 13*, 169–185. doi:10.1007/s11036-008-0037-5

Yu, W., & Lan, T. (2005 December). Transmitter optimization for the multi-antenna downlink with per-antenna power constraints. *IEEE Transactions on Signal Processing*.

Yu, Y., & Miller, S. L. (2007). A Four-State Markov Frame Error Model for the Wireless Physical Layer. In IEEE Wireless Communications and Networking Conference, Hong Kong (pp. 2053-2057).

Yue, J., Kim, K. J., Gibson, J. D., & Iltis, R. A. (2003). Channel Estimation and Data Detection for MIMO-OFDM Systems. In *Proc. GLOBECOM 2003* (pp. 581–585).

Yuen, A., & Chung, T. (2007). *Traffic Engineering for Mult-Homed Mobile Networks*. PhD thesis, The University of New South Wales Sydney, Australia.

Zadeh, A. (2002). Self-Organizing Packet Radio Ad Hoc Networks with Overlay (SOPRANO). *IEEE Communications Magazine*, 149–157. doi:10.1109/MCOM.2002.1007421

Zafeiris, V. E., & Giakoumakis, E. A. (2005). An Agent-based Architecture for Handover Initiation & Decision in 4G Networks. In *Proceedings of the Sixth IEEE International Symposium on a World of Wireless Mobile and Multimedia Networks*.

Zahir Azami, S. B., Duhamel, P., & Rioul, O. (1996). Joint source-channel coding: Panorama of methods. In *Proceedings of CNES workshop on data compression*. Toulouse, France: CNES.

Zaidi, Z. R., & Mark, B. L. (2004, September). A mobility-aware handoff trigger scheme for seamless connectivity in cellular networks. *IEEE Vehicular Technology Conference 2004-Fall*, Los Angeles, CA, (pp. 3471-3475).

Zander, J. (1993). Generalized reuse partitioning in cellular mobile radio. In *Proceedings of IEEE Vehicular Technology Conference*, (pp. 181–184).

Zandy, V., & Miller, B. P. (2002). Reliable Sockets (Computer Sciences Tech. Rep. 2002 LISA XVI November 3- 8). Philadelphia, PA: University of Wisconsin.

Zavala, A., Ruíz, J., & Penín, J. (2008). High-altitude Platforms for Wireless Communications. West Sussex, UK: John Wiley & Sons, Ltd.

Zhan, C., Arslan, T., Erdogan, A. T., & MacDougall, S. (2006, May). An Efficient Decoder Scheme for Double Binary Circular Turbo Codes. *IEEE International Conference on Acoustics, Speech and Signal Processing (ICASSP), 4*, 14-19.

Zhang, L., Gui, L., Qiao, Y., & Zhang, W. (2004). Obtaining diversity gain for DTV by using MIMO structure in SFN. *IEEE Transactions on Broadcasting, 50*(3), 83–90. doi:10.1109/TBC.2003.822985

Zhang, N., & Holtzman, J. M. (1996). Analysis of handoff algorithms using both absolute and relative measurements. *IEEE Transactions on Vehicular Technology, 45*(1), 174–179. doi:10.1109/25.481835

Zhang, Q., & Kassam, S. A. (1999). Finite-state Markov model for Rayleigh fading channels. *IEEE Transactions on Communications, 47*, 1688–1692. doi:10.1109/26.803503

Zhang, X., & Tang, J. (2006, July). QoS-driven asynchronous uplink sub-channel allocation algorithms for space-time OFDM-CDMA systems in wireless networks. *Wireless Networks, 12*(4), 411–425. doi:10.1007/s11276-006-6542-1

Zhang, Y. J., & Letaief, K. B. (2004). Multiuser adaptive subcarrier-and-bit allocation with adaptive cell selection for OFDM systems. *IEEE Transactions on Wireless Communications, 3*(4), 1566–1575. doi:10.1109/TWC.2004.833501

Zhang, Y., & Lee, W. (2000). Intrusion Detection in Wireless Ad Hoc Networks. In *6th Annual ACM/IEEE International Conference on Mobile Computing and Networking (MOBICOM)* (pp. 275-283).

Zhang, Y., & Parhi, K. K. (2006, October). High-Throughput Radix-4 LogMAP Turbo Decoder Architecture. In *IEEE Asilomar Conference on Signals, Systems and Computers* (pp. 1711-1715).

Zhang, Y., Honglin, H., & Chen, H. (2008, April). QoS differentiation for IEEE 802.16 WiMAX mesh networking. *Mobile Networks and Applications, 13*(1-2), 19–37. doi:10.1007/s11036-008-0035-7

Zhang, Y., Mao, S., Yang, L. T., & Chen, T. M. (Eds.). (2008). Broadband Mobile Multimedia: Techniques and Applications. Boca Ratón, USA: CRC Press.

Zhang, Y., Tan, I., Chun, C., Laberteaux, K., & Bahai, A. (2008). A differential OFDM approach to coherence time mitigation in DSRC. In *Proc. of the fifth ACM international workshop on Vehicular Inter-Networking*, (pp. 1-6).

Zhang, Z., He, Y., & Chong, K. (2005, March). Opportunistic downlink scheduling for multiuser OFDM systems. *Wireless Communications and Networking, 2*(13-17), 1206 – 1212.

Zhang, Z., Long, K., Zhao, M., & Liu, Y. (2005). Joint frame synchronization and frequency offset estimation OFDM systems. *IEEE Transactions on Broadcasting, 51*(3), 389–394. doi:10.1109/TBC.2005.851702

Zhao, J., & Cao, G. (2006). VADD: Vehicle-assisted data delivery in vehicular ad hoc networks. In *Proc. of INFOCOM*.

Zhao, J., & Sauvola, J. (2002). Mobility and Mobility Management: A Conceptual Framework. In *IEEE International Conference on Networks*, Singapore (pp-205-210).

Zheng, L., & Tse, D. N. C. (2003). Diversity and multiplexing: A fundamental trade-off in multiple-antenna channels. *IEEE Transactions on Information Theory, 49*(5), 1073–1096. doi:10.1109/TIT.2003.810646

Zheng, Y., et al. (2005). Security scheme for 4G wireless systems. In *International Conference on Communications, Circuits and Systems, 2005* (pp. 397-401, Vol.1). Retrieved from http://ieeexplore.ieee.org/stamp/stamp.jsp?arnumber=1493433&isnumber=32104

Zheng, Y., et al. (2005). Trusted Computing-Based Security Architecture For 4G Mobile Networks. In *Parallel and Distributed Computing, Applications and Technologies, PDACT* (pp. 251-255).

Zhou, D., & Lai, H. (2005). A compatible and scalable clock synchronization protocol. *Synchronization scheme reference IEEE working group, a wireless LAN MAC and physical layer (PHY) specification,* (ICCP).

Zhou, M.-T., Harada, H., Kong, P.-Y., Ang, C.-W., Ge, Y., & Pathmasuntharam, J. S. (2009). Multi-channel Transmission with Efficient Delivery of Routing Information in Maritime WiMAX Mesh Networks. In *The 5th International Wireless Communications and Mobile Computing Conference*.

Zhu, F., & McNair, J. (2005, May). Cross layer design for Mobile IP handoff. *IEEE Vehicular Technology Conference 2005-Spring*, Stockholm, Sweden, (pp. 2255-2259).

Zhu, H., & Kwak, K. (2007). Performance analysis of an adaptive handoff algorithm based on distance information. *Elsevier Computer Communications, 30*(6), 1278–1288.

Zhu, X., & Murch, R. D. (2002). Performance analysis of maximum likelihood detection in a MIMO antenna system. *IEEE Transactions on Communications, 50*(2), 187–191. doi:10.1109/26.983313

Zimmermann, E., Fettweis, G., Milliner, D. L., & Barry, J. R. (2008). A parallel smart candidate adding algorithm for soft-output MIMO detection. In *Proc. 7th Intern. ITG Conference on Source and Channel Coding*.

Zonoozi, M., & Dassanayake, P. (1997, September). Handover delay and hysteresis margin in microcells and macrocells. *The 8th IEEE International Symposium on Personal, Indoor and Mobile Radio Communications*, Helsinki, Finland, (pp. 396-400).

Zou, Q., Tarighat, A., & Sayed, A. H. (2007). Performance analysis of multiband OFDM UWB Communications with application to range improvement. *IEEE Transactions on Vehicular Technology, 56*, 3864–3878. doi:10.1109/TVT.2007.901957

Zseby, T., Zander, S., & Carle, G. (2002). *Policy-Based Accounting (RFC 3334)*. Retrieved from http://www.ietf.org/rfc/rfc3334.txt

About the Contributors

Sasan Adibi is currently a Member of Technical Staff, Advanced Technology at Research In Motion (RIM). He is also expected to graduate from University of Waterloo in 2010 with a Ph.D. degree from Electrical and Computer Engineering Department. He has an extensive research background mostly in the areas of Quality of Service (QoS) and Security. He is the first author of +25 journal/conference/book chapter/white paper publications. He also +9 years of high-tech industry-based experience, having worked in numerous high-tech companies, including Nortel Networks and Siemens Canada.

Amin Mobasher has recently joined Research In Motion (RIM) as a Member, Technical Staff in Advanced Technology laboratory after his short visit at Stanford University. He earned his Ph.D. degree in electrical and computer engineering from University of Waterloo, Waterloo, ON, Canada in Dec. 2007. From Jan. 2008 until Jan. 2009, he was working as a member technical staff in the Advanced Technology Lab in RIM on LTE and LTE-A 3GPP standards. Between Jan. 2009 and Oct. 2009, he was with Smart Antenna Research Group (SARG) in Stanford University as a Visiting Scholar. His research interests are MIMO-OFDM systems, Optimization in Communication Systems, Network Coding, Relays, Interference Mitigation, and physical layer in 3GPP LTE and LTE-A standards. He is the recipient of several awards including Ontario Graduate Scholarship and University of Waterloo Presidential Award in 2006. He also received an NSERC Industrial R&D Fellowship and an NSERC post-doctoral Fellowship award in 2007 and 2008, respectively.

Tom Tofigh is a Principal Member of Technical Staff in the Radio Technology Architecture group Bell Labs at AT&T. He holds a JD (1995) and completed his Ph.D. requirements for electrical engineering & computer science at GWU(1990). He taught graduate courses from 1996- 2001 at various universities as an adjunct including GWU and Southeastern University. He chaired the WiMax Forum application working group forum 2004 – 2009. He has contributed broadly in major technical conferences and he is currently involved in application performance studies and cross layer optimization and radio layer APIs. He is a Senior Member of ACM, IEEE and a member of many industry standard forums.

* * *

Renato Ricardo de Abreu is a M.Sc. Student at Federal University of Pernambuco. His fields of interest are in Delay Tolerant Network; Social Routing; Next generation networks; Policy Based Management and Traffic Classification. He had published some papers on international Symposiums. He works at Information Technology.

Hamada Alshaer graduated from the Department of Electrical Engineering and Computer Science at Birzeit University, Palestine in 2001. He received the M.S. (DEA) degree in information technologies and Systems from Compiegne University of Technology, Compiegne, France, in 2002. He also received the Ph.D. degree in Computer Science and Telecommunications from Pierre et Marie Curie University in December 2005. Then, he jointed as a research fellow at the Electronic and Electrical Engineering department at Brunel

University, in west London. Later, he worked as a postdoctoral research fellow at the INRIA, France. Now, he is a research fellow in the school of Electronic and Electrical Engineering at the University of Leeds, in United Kingdom. His research interests include QoS, wireless sensor networks, inter-vehicular communications, multi-layer traffic engineering and resilience, optical communications and networking, networks security and social networks. Dr. Alshaer has served on the technical program committee of different IEEE conferences, including intelligent vehicles symposium, vehicular technology conference, Globecom, ICC and WCNC and chaired some of their sessions. He is the recipient of the 2009 Royal Academy of Engineering travel award, as well as, scholarships from UNRWA, French government and University Pierre et Marie Curie University for academic distinctions.

Elias Aravantinos is an active ICT consultant and IT Account Manager at the Secretariat General of Communication – Secretariat General of Information in Athens, Greece. His research interest is in the area of Technology Management with a focus on the telecommunications industry. Prior joining the Secretariat was an Adjunct Lecturer of Queens College, Economics department. He has about 10 years of experience in the areas of network management and engineering, systems analysis, IT new media strategies, project management and R&D innovation. He holds a Diploma and MS in Electrical Engineering from Patras Polytechnic School, an MBA from OCU and he is a PhD candidate in Technology Management of Stevens Institute of Technology.

Robert Atkinson is a Lecturer within the Mobile Communications Research Group within the EEE department. He specialises in Layer 3 mobility solutions and ad hoc networking. He has been an active researcher in various EPSRC, FP6 and industrially funded projects. He obtained his first degree in Electronic and Electrical Engineering in 1993 and a PhD in Mobile Communications Systems in 2003 from the University of Strathclyde. He is a member of the IET and a senior member of the IEEE.

Titus Constantin Bălan received the Dipl.-Ing. degree in Electronics and Telecommunications from the "Transilvania" University of Brasov and he is a PhD Candidate in Telecommunications at the "Transilvania" University, in the field of Mobility in Heterogeneous Networks. He works as telecommunications engineer for Siemens SIS PSE Romania since 2005, for the department of Telco Media Mobile and Fixed, involved in the development of a Mobile IPv6 demonstrator and virtual test-fields projects, also being member of international R&D teams in programs of the European Union. Starting with 2008 he is part of the Long Term Evolution Radio development team at the Nokia Siemens Networks R&D Centre from Düsseldorf, Germany.

Jens Berkmann received the Dipl.-Ing. and Dr.-Ing. (Ph.D.) degrees in electrical engineering from Technische Universität München (TUM), in 1991 and 2000, respectively. In 1992 he spent one year as a research fellow at the Computer Science Department of the University of Melbourne (Australia) sponsored by a scholarship of the Gottlieb-Daimler and Karl-Benz Foundation where he worked on computer vision and pattern recognition. From 1993 to 1999, he was a research assistant at the Institute of Communications of Munich University of Technology (TUM) working on channel decoding. In 2000, he joined Infineon Technologies in Munich, Germany, where he has been working on the simulation, algorithmic and architectural design, verification, and optimization of 3G modems. Dr. Berkmann is a recipient of the best paper award of the Verband der Elektrotechnik, Elektronik, Informationstechnik (VDE), Germany.

Cecilia Carbonelli received the Laurea degree (cum laude) in Telecommunications Engineering and the Ph. D. degree in Information Engineering from the University of Pisa, Italy (Department of Information Engineering) in 2001 and 2005, respectively. From 2005 to 2006, she was with the Department of Electrical Engineering, University of Southern California, Los Angeles, as a Post-doctoral Research Associate. During 2006 she held a visiting appointment at Qualcomm Inc., San Diego, USA. In December 2006, she joined Infineon Technologies AG, Munich, Germany. Her research interests include the area of digital communication theory, with special emphasis on MIMO OFDM systems, UWB communications, sensor networks, parameter estimation and synchronization techniques.

About the Contributors

Joseph R. Cavallaro received the B.S. degree from the University of Pennsylvania, Philadelphia, Pa, in 1981, the M.S. degree from Princeton University, Princeton, NJ, in 1982, and the Ph.D. degree from Cornell University, Ithaca, NY, in 1988, all in electrical engineering. From 1981 to 1983, he was with AT&T Bell Laboratories, Holmdel, NJ. In 1988, he joined the faculty of Rice University, Houston, TX, where he is currently a Professor of electrical and computer engineering. His research interests include computer arithmetic, VLSI design and microlithography, and DSP and VLSI architectures for applications in wireless communications. During the 1996–1997 academic year, he served at the USA National Science Foundation as Director of the Prototyping Tools and Methodology Program. He was a Nokia Foundation Fellow and a Visiting Professor at the University of Oulu, Finland in 2005 and continues his affiliation as an Adjunct Professor there. He is currently the Associate Director of the Center for Multimedia Communication at Rice University. He is a Senior Member of the IEEE. He was Co-chair of the 2004 Signal Processing for Communications Symposium at the IEEE Global Communications Conference and General Co-chair of the 2004 IEEE 15th International Conference on Application-Specific Systems, Architectures and Processors (ASAP).

Chi-Yuan Chang is a Ph.D. student of the Electrical Engineering, National Dong Hwa University, Hualien, Taiwan, R.O.C. His research interests include IPv6 based Networks, Wireless Networks and Network Processors. He received his M.S. degree from the Department of Computer Science and Information Engineering, National Chung Cheng University, Chia-Yi, Taiwan in 1994. He also works in the System Design Division, Computer and Network Center, National Dong Hwa University, Hualien, Taiwan, R.O.C.

Kai-Di Chang is a master student of Computer Science and Information Engineering, National ILan University, I-Lan, Taiwan, R.O.C. His research interest includes personal communications services and network security. He is going to join the Ph.D. program in Department of Electrical Engineering, National Taiwan University of Science and Technology at summer, 2009.

Han-Chieh Chao is a joint appointed Full Professor of the Department of Electronic Engineering and Institute of Computer Science & Information Engineering. He also serves as the Dean of the College of Electrical Engineering & Computer Science for National Ilan University, I-Lan, Taiwan, R.O.C. He has been appointed as the Director of the Computer Center for Ministry of Education on September 2009 as well. His research interests include High Speed Networks, Wireless Networks, IPv6 based Networks, Digital Creative Arts and Digital Divide. He received his MS and Ph.D. degrees in Electrical Engineering from Purdue University in 1989 and 1993 respectively. Dr. Chao is also serving as an IPv6 Steering Committee member and co-chair of R&D division of the NICI (National Information and Communication Initiative, a ministry level government agency which aims to integrate domestic IT and Telecom projects of Taiwan), Co-chair of the Technical Area for IPv6 Forum Taiwan, the Editor-in-Chief of the Journal of Internet Technology, Journal of Internet Protocol Technology and International Journal of Ad Hoc and Ubiquitous Computing. Dr. Chao has served as the guest editors for Mobile Networking and Applications (ACM MONET), IEEE JSAC, IEEE Communications Magazine, Computer Communications, IEE Proceedings Communications, the Computer Journal, Telecommunication Systems, Wireless Personal Communications, and Wireless Communications & Mobile Computing. Dr. Chao is an IEEE senior member and a Fellow of IET (IEE). He is a Chartered Fellow of British Computer Society.

Chi-Yuan Chen received his M.S. degree in electrical engineering from National Dong Hwa University, Taiwan, R.O.C. in 2007. He is currently pursuing his Ph.D. degree in electrical engineering at National Taiwan University, Taiwan, R.O.C. His research interests include wireless networking, personal communication services and network security.

Hsiao-Hwa Chen is currently a full-Professor in the Department of Engineering Science, National Cheng Kung University, Taiwan. He received BSc and MSc degrees with the highest honor from Zhejiang University, China, and a PhD degree from the University of Oulu, Finland, in 1982, 1985 and 1990, respectively, all in

Electrical Engineering. He worked with Academy of Finland as a Research Associate from 1991 to 1993, and the National University of Singapore as a Lecturer and then a Senior Lecturer from 1992 to 1997. He joined Department of Electrical Engineering, National Chung Hsing University, Taiwan, as an Associate Professor in 1997 and was promoted to a full-Professor in 2000. In 2001 he joined National Sun Yat-Sen University, Taiwan, as the founding Chair of the Institute of Communications Engineering of the University. Under his strong leadership the institute was ranked the second position of the country in terms of SCI journal publications and National Science Council funding per faculty member in 2004. In particular, National Sun Yat-Sen University was ranked the first place in the world in terms of the number of SCI journal publications in wireless LANs research papers during 2004 to mid-2005, according to a Research Report released by The Office of Navel Research, USA. He was a visiting Professor to Department of Electrical Engineering, University of Kaiserslautern, Germany, in 1999, the Institute of Applied Physics, Tsukuba University, Japan, in 2000, Institute of Experimental Mathematics, University of Essen, Germany in 2002 (under DFG Fellowship), the Chinese University of Hong Kong in 2004, and the City University of Hong Kong in 2007. His current research interests include wireless networking, MIMO systems, information security, and Beyond 3G wireless communications. He is the inventor of the next generation CDMA technology. He is a recipient of numerous Research and Teaching Awards from the National Science Council, the Ministry of Education and other professional groups in Taiwan. He has authored or co-authored over 200 technical papers in major international journals and conferences, five books and several book chapters in the areas of communications, including the book titled "Next Generation Wireless Systems and Networks" (512 pages) and "The Next Generation CDMA Technologies" (468 pages), both by John Wiley in 2005 and 2007, respectively. He has been an active volunteer for IEEE various technical activities for over 15 years. Currently, he is serving as the Chair of IEEE Communications Society Radio Communications Committee, and the Vice-Chair of IEEE Communications Society Communications & Information Security Technical Committee. He served or is serving as symposium chair/co-chair of many major IEEE conferences, including IEEE VTC 2003 Fall, IEEE ICC 2004, IEEE Globecom 2004, IEEE ICC 2005, IEEE Globecom 2005, IEEE ICC 2006, IEEE Globecom 2006, IEEE ICC 2007, IEEE WCNC 2007, etc. He served or is serving as Editorial Board Member or/and Guest Editor of IEEE Communications Letters, IEEE Communications Magazine, IEEE Wireless Communications Magazine, IEEE JSAC, IEEE Network Magazine, IEEE Transactions on Wireless Communications, and IEEE Vehicular Technology Magazine. He is the Editor-in-Chief of Wiley's "Security and Communication Networks" Journal (www.interscience.wiley.com/journal/security), and the Special Issue Editor-in-Chief of Hindawi Journal of Computer Systems, Networks, and Communications (http://www.hindawi.com/journals/jcsnc/). He is also serving as the Chief Editor (Asia and Pacific) for Wiley's Wireless Communications and Mobile Computing (WCMC) Journal and Wiley's International Journal of Communication Systems, etc. His original work in CDMA wireless networks, digital communications and radar systems has resulted in five US patents, two Finnish patents, three Taiwanese patents and two Chinese patents, some of which have been licensed to industry for commercial applications. He is an adjunct Professor of Zhejiang University, China, and Shanghai Jiao Tong University, China. Professor Chen is a recipient of the Best Paper Award in IEEE WCNC 2008.

Noun Choi currently is a Member, Technical Staff of the Advanced Technology group at Research in Motion, Inc. His research interest includes performance and efficiency of mobile IP networks from various aspects such as IP mobility, Quality of Service, and communication disruption during hand-over. He has obtained Ph.D. in Telecommunications Engineering and M.S. in Computer Science from the University of Texas at Dallas in 2008 and 2003 respectively. During his graduate study, he worked on link layer and network layer protocols of wireless networks including Wireless LANs, Mobile Ad Hoc Networks, and Wireless Mesh Networks. He also has a B.S. degree in Statistics from Soong Sil University in Korea. Noun Choi has worked as a software engineer in Korea before he went back to school for his graduate study. During that time, he mostly designed and implemented application layer protocols.

About the Contributors

Young-June Choi received B.S., M.S., and Ph.D. degrees from the Department of Electrical Engineering & Computer Science, Seoul National University, in 2000, 2002, and 2006, respectively. From 2006 through 2007, he was a postdoctoral researcher at the University of Michigan. In Aug. 2007, he joined NEC Laboratories America as a research staff member. He authored/coauthored over 30 technical papers and hold 10 US/Korean patents. He was awarded a Gold Prize at the Samsung Humantech Thesis Contest in 2006. His research interests include fourth generation wireless networks, radio resource management, and cognitive radio networks. He is a member of ACM and IEEE.

Ram (Ramanamurthy) Dantu has 20 years of experience in the networking industry. During his work at Cisco, Nortel, Alcatel, and Fujitsu, he was responsible for advanced technology products from concept to delivery. Later, as a technical director of IpMobile (acquired by Cisco), he was instrumental in the wireless IP product concept, architecture, design, and delivery. He was also a director at Netrake Technology, where architected the redundancy mechanism for VoIP firewalls. He is currently an assistant professor in the Department of Computer Science and Engineering at the University of North Texas (UNT). His research focus for the last 5 years has been on spam detection (in VOIP networks), network security, and next-generation networks. Recently, he has co-chaired 3 workshops in VoIP security. He is the founding director of the Network Security Laboratory (NSL) at UNT. The objective of NSL is to study the problems and issues related to next-generation networks. In addition to more than 50 research papers, he has authored several RFCs related to MPLS, SS7 over IP, and routing. His innovative work at Cisco and Alcatel culminated in 8 patents with 10 more pending.

Christian Drewes received Dipl.-Ing. and Dr.-Ing. degrees in electrical engineering and information technology from the Technische Universität München, Germany, in 1995 and 1999. His doctorate thesis was awarded by Texas Instruments and the VDE. He co-received the IEEE VTC'99-Fall best paper award. From 1996 to 2000, he was a research assistant at the Institute for Integrated Circuits at the Technische Universität München, where he headed a team of researchers working on xDSL and radio. In 2000, he joined Infineon Technologies in Munich, Germany, where he has been working on the simulation, algorithmic and architectural design, verification, and optimization of 3G modems. He is a member of VDE.

Wencai Du is the Dean of College of Information Science & Technology at Hainan University. He obtained his B.Eng from Peking University China and a PhD from the University of Adelaide, Australia, both in Electronic and Electrical Engineering.

Khaled Elleithy received the B.Sc. degree in computer science and automatic control from Alexandria University in 1983, the MS Degree in computer networks from the same university in 1986, and the MS and Ph.D. degrees in computer science from The Center for Advanced Computer Studies at the University of Louisiana at Lafayette in 1988 and 1990, respectively. From 1983 to 1986, he was with the Computer Science Department, Alexandria University, Egypt, as a lecturer. From September 1990 to May 1995 he worked as an assistant professor at the Department of Computer Engineering, King Fahd University of Petroleum and Minerals, Dhahran, Saudi Arabia. From May 1995 to December 2000, he has worked as an Associate Professor in the same department. In January 2000, Dr. Elleithy has joined the Department of Computer Science and Engineering in University of Bridgeport as an associate professor. Dr. Elleithy published more than seventy research papers in international journals and conferences. He has research interests are in the areas of computer networks, network security, mobile communications, and formal approaches for design and verification.

Jaafar Elmirghani is a Fellow of IEE and InstP, and was appointed to a chair in Communication Networks and Systems in Leeds in 2007. He was Professor of optical communications in the University of Wales Swansea between 2000 and 2007, head and founder of the Institute of Advanced Telecommunications (IAT) at the University of Wales Swansea, with about 40 full time staff. He is Chairman of the IEEE UK and RI Communications Chapter and was Chairman of IEEE Comsoc Transmission Access and Optical Systems Committee and Chairman of IEEE Comsoc Signal Processing and Communication Electronics (SPCE) Committee. He was a member of IEEE ComSoc Technical Activities Council (TAC) and is and has been on the technical program committee of thirteen IEEE ICC/GLOBECOM conferences between 1995 and 2007 including eight times as Symposium Chair. He received the IEEE Communications Society 2005 Hal Sobol award for exemplary service to meetings and conferences, the IEEE Communications Society 2005 Chapter Achievement award and the University of Wales Swansea inaugural Outstanding Research Achievement Award, 2006. He has published over 200 technical papers, co-edited Photonic Switching Technology- Systems and Networks, IEEE Press 1998, leads a number of research projects and has research interests in communication networks, wireless and optical communication systems.

Hosein Fallah is an Associate Professor of Technology Management at Stevens Institute of Technology in New Jersey. His research interest is in the area of Innovation Management with a focus on the telecommunications industry. Prior to joining Stevens, Dr. Fallah was Director of Network Planning and Systems Engineering at Bell Laboratories. He has over 30 years of experience in the areas of systems engineering, product/service realization, software engineering, project management, and R&D effectiveness. He holds a BS in Engineering from AIT, and MS and Ph.D. in Applied Science from the University of Delaware.

Joseilson França, System Analyst (UNICAP, 2000) has worked in the areas of DataCommunication Systems, Transmission Systems and Telecoms. Participated in the deployment of ISDN, Deterministic and Statistical Network, as well as network deployment of IP/Multiservice at TELPE/ Telemar/Oi. At Intelig Telecom worked in the areas of Communication and Data Transmission, participated of the closing of the northern ring DWDM network and the deployment of VoIP service. Now working at ATI (State Agency for Information Technology) in the implementation and analysis of network quality of PEMULTIDIGITAL, which provides data communication services and VoIP telephony for all administration of the state of Pernambuco

Ivan Ganchev, DipEng, PhD, is a Senior Member of the Institute of Electrical and Electronic Engineers (IEEE). He received his engineering and doctoral degrees from the Saint-Petersburg State University of Telecommunications in 1989 and 1994 respectively. He is an Associated Professor from the University of Plovdiv and currently a Lecturer and Deputy Director of the Telecommunications Research Centre (TRC), University of Limerick, Ireland. His previous activities include: founding partner and member, ANWIRE - Academic Network for Wireless Internet Research in Europe -, the EU FP5 Thematic Network of Excellence IST-38835, 2002-2004; member of two European 'COoperation in the field of Science and Technology research' Actions (COST 285 & 290). His research interests include: simulation and modeling of complex telecommunication systems, new communications paradigms for wireless NGN, 3P-AAA management, wireless billboard channels, and Internet tomography. Dr. Ganchev served on the Technical Program Committees of a number of international conferences, including IEEE Globecom (2006), IEEE VTC (2007–2009), and IEEE ISWCS (2006–2009).

Manish Goel joined Texas Instruments in 2000 and holds position of senior member of technical staff in Signal Processing systems R&D center at Texas Instruments. Manish's research interests are in low-power, signal processing, wireless communication and medical systems. Manish has led and contributed to several wireless R&D efforts at TI including mobile digital TV receiver, 4G WiMAX/LTE modem, multi-antenna wireless LAN and fixed-wireless access. He is currently managing an R&D team involved in low-power signal processing VLSI for medical and communication applications. Manish received his bachelor's degree from India Institute of technology in Delhi, and a master's degree and Ph.D. from the University of Illinois at Urbana-Champaign.

P. R. L. Gondim graduated in computer engineering (1987) and received his M.Sc. degree in Computing and Systems (1992) from Military Engineering Institute (IME), Rio de Janeiro RJ, Brazil, and got his Doctor-

About the Contributors

ate degree in Electrical Engineering (1998) from Pontifical Catholic University, Rio de Janeiro RJ, Brazil. His fields of study are Communication Networks and Digital Television. He is a Network-Engineering Professor at the Electrical Engineering Department, University of Brasília, Brasília DF 70910-900 Brazil. His current interests are wireless networking, quality of service/experience and convergence of wired/wireless networks.

Jivesh Govil is working with Cisco Systems, Inc. at California, USA. He completed Master of Science from University of Michigan, Ann Arbor, USA in 2007 with focus on communication networks. Prior to this he completed his Bachelor of Engineering from University of Delhi, India. He also did graduate certificate course in Digital Communication Networks from Massachusetts Institute of Technology, Cambridge, USA. He has been decorated with President Gold Medal for his outstanding academic performance and innovations. He is author of innumerous research papers and part of many Scientific Advisory Boards, Technical Program Committees of numerous IEEE/ACM conferences, symposiums, and journals worldwide. He is an editorial board member of International Journal of Computing and Information technology. In past, he also served as reviewer for ACM/Springer Wireless Networks: Journal of Mobile Communication, Computation, and Information. Mr. Govil has number of publications in IEEE/ACM related conference and scientific journals with special paper awards. He also devised a wireless solution for disabled while working with AT&T in the past. He has also been nominated for Who's Who in the World or 2010. Mr. Govil's research interest lies in Computer Networking, Convergence of technologies, Wireless and Satellite Network Architecture Planning, Cellular Mobile Radio Communication, Wireless Personal Communications, Broadband Wireless Access, 3G/4G Networks, MIMO Wireless Communications and Adaptive Signal Processing & Low Power-high Frequency Signal Processing for Wireless Communication. He is also IEEE Communication Society member.

Jivika Govil is pursuing Master of Information System Management (MISM) at Carnegie Mellon University at Pittsburgh, USA on partial scholarship. She has been invited as Key-Note speaker, technical program committee, reviewer, editor of many IEEE/other conferences and journals. She is also Editor-in-chief of the journal Applied Sciences Society for Education and Training (ASSET). She has been selected as editorial board member for International Journal of Information Technology, International Journal of Intelligent Technology, and International Journal of Computer, Informational, & Systems Science and Engineering. Due to her magnificent scientific knowledge, and publishing more than thirty five research papers at the age of 21 years, she has been nominated for Who's Who in the World, 2010. Jivika is an active member of promoting "Women in Engineering". She has received many accolades from International Conferences, University and from her college including Best Paper Awards, Distinguished Student and Best Student/Author during her UG studies at Maharshi Dayanand University, India. Her area of interests are IT, DBMS, MIS, AI, Information Architecture, Data Mining/ Authentication/Warehouse, Business Intelligence, Quantum Computing, Computer/Telecom Networking, IPv4 to IPv6, Security Issues in W/L Mobile Networks, 4G Communication Systems, W/L Access Technologies, Mobility Management, Swarm Intelligence, Transhumanism, Nanotechnology and Cyber Forensic etc. She is student member of IEEE, DAMA (Data Management Association) and ASEE (American Society of Engineering Education).

Parthasarathy (Partha) Guturu is currently a faculty member of the Electrical Engineering department of the University of North Texas (UNT), Denton. Prior to his recent return to academia, he has had experience of more than 7 years in corporate R&D and over ten years in teaching and academic research. His contributions to corporate R&D at the Nortel Corporation in the areas of Advanced Intelligent Networks and 3G Wireless Systems culminated into 3 US patents. In his earlier stint in academia at Indian Institute of Technology, Kharagpur, after Bachelor's though PhD (Eng.) degrees from the same place, he directed 4 PhD dissertations and a large number of graduate and undergraduate theses. Till date, he published over 40 papers in international journals and conferences, and contributed to disparate areas of Electrical and Computer Engineering including Networking, Pattern Recognition, Computer Vision, and Computational Intelligence. Dr. Guturu plans to integrate his past experience in Computational Intelligence and the latest experience in Wireless Networks and embark upon an ambitious research program in the area of Wireless Sensor Networks and Systems.

Xiaojing Huang was born in Hubei, China, on 3 October, 1963. He received his Bachelor of Engineering, Master of Engineering, and Ph.D. degrees from Shanghai Jiao Tong University, Shanghai, China, in 1983, 1986, and 1989, respectively, all in electronic engineering. From 1989 to 1994, he worked in the Electronic Engineering Department of Shanghai Jiao Tong University, where he had been a Lecturer since 1989 and an Associate Professor since 1991. From 1994 to 1997, he was the Chief Engineer with Shanghai Yang Tian Science and Technology Corporation Ltd., Shanghai, China. In 1998, he joined the Motorola Australian Research Centre, Sydney, Australia, as a Senior Research Engineer and had been a Principal Research Engineer since 2003. From 2004 to 2009, he was an Associate Professor in the School of Electrical, Computer and Telecommunications Engineering at the University of Wollongong, Wollongong, Australia. He is currently a Principal Research Scientist with the ICT Centre, Commonwealth Scientific and Industrial Research Organisation (CSIRO), Sydney, Australia. He is also an Adjunct Associate Professor at Macquarie University, Sydney, Australia. His research interests are in communications theory, digital signal processing, and wireless communications networks. He is a member of the Institute of Electrical and Electronics Engineers.

Axel Huebner received the Dipl.-Ing. degree in electrical engineering from the Technical University of Karlsruhe, in 1999 and the Ph.D. degree in electrical engineering from the University of Ulm, Ulm, Germany, in 2004. From 2005 to 2006, he was a Postdoctoral Research Associate in the Coding Research Group at the University of Notre Dame, Notre Dame, IN. During that period, he also was a Visiting Researcher at the Technical University of Munich, Munich, Germany (Summer 2005). In 2006, he joined the Connectivity Systems Group at Philips Research Europe-Aachen, Aachen, Germany, as a Research Scientist. Since 2007, he has been with the Wireless Communication Group at Infineon Technologies AG, Munich, Germany. His research interests include coding theory, with emphasis on concatenated convolutional coding schemes, and diversity techniques for wireless communication systems. He is a member of the IEEE Information Theory and Communications Societies.

Jenı István Jakab received his Masters Degree in Electrical Engineering from the Budapest University of Technology and Economics. He is a software engineer with Technotree Ltd., Ireland (formerly Tecnomen Ireland) while pursuing his PhD degree within the Telecommunications Research Centre (TRC), University of Limerick, Ireland. His company supplies messaging and charging solutions for telecom operators and service providers worldwide. He works in the Charging Business Unit and he has been making contributions to the company prepaid voice service (for voice call charging) and Data Charging service (for SMS, GPRS, MMS charging) solutions. He also worked for Nokia OY, Finland in a team to develop Network Management software called Node Manager to control and manage, at that time, newly developed network nodes with new radio FlexiHopper, MetroHopper outdoor units and Frogs indoor units. His research interests include wireless networks, mobile computing, AAA and charging mechanisms for ad-hoc wireless networks, charging and billing models for 4G, AAA management.

Usman Javaid is with New Technologies and Innovation department at Vodafone Group, Newbury, UK. His job responsibilities involve technical assessment of new technologies and innovative solutions which create new revenue opportunities for world's leading mobile operator and significantly reduce the cost of offering services to customers around the globe. Dr. Javaid earned Ph.D. degree in the field of Mobile Wireless Networks, from the University of Bordeaux, France in 2008 and Masters in Networks and Telecom from University of Paris in 2005. His research interests include 4G mobile networks testing and optimisation, Quality of Service, Mobility Management and Self-Organisable Networks.

Zhanlin Ji received his MEng degree in Telecommunications engineering from the Dublin City University, Ireland, and MSc degree in Software engineering from the Beijing University of Posts and Telecommunications, China, in 2005 and 2006 respectively. He is currently pursuing his PhD degree and researching within the Telecommunications Research Centre (TRC), University of Limerick, Ireland. His research is focused on the Wireless Billboard Channels development for the Ubiquitous Consumer Wireless World (UCWW).

About the Contributors

Amir K. Khandani is a Professor in Electrical and Computer Engineering, University of Waterloo, and holds an NSERC Industrial Research Chair (funded by NSERC and Nortel) on "Advanced Telecommunications Technologies" and a Canada Research Chair (Tier I) on "Wireless Systems". Dr. Khandani received his M.A.Sc. degree from University of Tehran in 1985, and his Ph.D. degree from McGill University in 1992. Dr. Khandani's current research interests are in the Physical Layer of Wireless Systems with emphasis on Source/Channel Coding, Multiple-Access, Multiple Antenna Systems, Co-operative Networking, and Information Theory. He is the author or co-author of more than 250 refereed articles and several patents. He has frequently served on technical program committees of major conferences in his area, and is currently serving as an Associate Editor for the IEEE Transactions on Communications in the area of Coding and Communication Theory.

Dongwook Kim received the B.S. degree in Information and Computer Engineering from AJOU University, South Korea, in 2002, and the M.S. and Ph.D. degrees in Electrical Engineering and Computer Science from Korea Advanced Institute of Science and Technology (KAIST), in 2004 and 2009, respectively. He is working for one year as a post-doctoral fellow in the School of Electrical and Computer Engineering at Georgia Institute of Technology, Atlanta, Georgia. His research interests include handover protocols, traffic management, cognitive radio networks, network coding, next-generation cellular networks, and sensor networks.

Namgi Kim received the B.S. degree in Computer Science from Sogang University, Korea, in 1997, and the M.S. degree and the Ph.D. degree in Computer Science from Korea Advanced Institute of Science and Technology (KAIST) in 2000 and 2005, respectively. From 2005 to 2007, he was a research member of the Samsung Electronics. Since 2007, he has been a faculty member of the Department of Computer Science, Kyonggi University. His research interests include distributed system, ad hoc network, wireless system, mobile communication, and network security.

Peng-Yong Kong received his PhD degree in Electrical and Computer Engineering from the National University of Singapore in 2002. Since 2001, he joined the Institute for Infocomm Research and he is now a Senior Research Fellow. Since 2003, he is also an adjunct faculty in the Department of Electrical and Computer Engineering, National University of Singapore. He lectures in wireless network protocols. His research interests are primarily in medium access control protocols, packet scheduling, network traffic control and quality of service provisioning for both wired and wireless networks.

Catherine Lamy-Bergot was born in Vernon, France, in 1972. She received the electrical engineering degree from the École Nationale Supérieure des Télécommunications (E.N.S.T.), Paris, France, in 1996, the Diplôme d'Études Approfondies (D.E.A.) in Telecommunications also in 1996 and the Ph.D. degree from the E.N.S.T. in 2000. She was with Philips Research France from 2000 to 2002 where she worked on joint source and channel coding techniques and participated to the European project JOCO. Since Sept. 2002, she joined THALES Communications as a Senior Scientist in digital communications. She participated to different French National Research projects and European projects both as technical manager and researcher. She is currently leading Celtic project BOSS and ICT project OPTIMIX. Her fields of interest include iterative decoding techniques, space time codes, high efficiency modulations, error correction codes, unequal error protection techniques, robust header compression, soft output decoding, joint source channel coding techniques and end-to-end wireless communication optimisation.

Hanjin Lee received the B.S. degree in Computer Science from Korea University, and the M.S. degree in Electrical Engineering and Computer Science from Korea Advanced Institute of Science and Technology (KAIST) in 2004 and 2006, respectively. He has been working toward the Ph.D. degree in the School of Electrical Engineering and Computer Science of KAIST. His research interests include single-hop and multi-hop cellular networks, ad hoc networks, and sensor networks.

Blazej Lewcio studied computer science and telecommunication at Deutsche Telekom Hochschule für Telekommunikation in Leipzig, Germany (University of Applied Science) with the focus on computer networks and distributed systems. He completed his study with a diploma thesis on "Assessment and adaptation of voice quality in All-IP heterogeneous networks". Since 2008, Blazej is a member of Deutsche Telekom Laboratories and works towards his PhD at Technical University Berlin. In his research on Quality of Service and Quality of Experience in Next Generation Networks, Blazej evaluates and merges networking QoS and perceptual QoE aspects of future multimedia services.

Roberto Llorente received the M.Sc. degree in telecommunication engineering from the Universidad Politécnica de Valencia, Valencia, Spain, in 1998. Since then, he has been in research positions within the Fiber-Radio Systems Group of the same university. In 2002 he joined the Valencia Nanophotonics Technology centre (NTC), where he has participated in several national and European research projects on areas such as bio-photonics, optical signal processing and OTDM/DWDM transmission systems working toward the Ph.D., received in 2006. Currently, he is Associated Professor of the Universidad Politécnica de Valencia in the Communications Department, teaching radio-communications related subjects. He has been the Technical Responsible of the European project FP6-IST-UROOF in the NTC, and, from January 2008, Coordinator of the European project FP7-ICT-UCELLS. His research interest includes optical and electro-optical processing techniques in the areas of transmission systems and hybrid wireless-optical access networks. Dr. Roberto Llorente has authored or co-authored more than 40 papers in leading international journals and conferences and has authored three patents.

Elsa María Macías López is an associate professor of Telecommunications at Las Palmas de Gran Canaria University, Department of Telematics Engineering, Spain. She received her Ph.D. in Telecommunications (2001) from Las Palmas of Gran Canaria University for her work on Parallel Computing on a LAN-WLAN Cluster Controlling at Runtime the Variation of the Number of Proccesses. She received her M.S. in Telecomunications (1997) from the same University for her work on Parallelization of Diffuse IR Radiation System Simulation for Indoor Applications. Her research interests are in parallel and distributed computing and infrastructure wireless networks for collaborative computing. Her current research efforts have focused on the management of wireless channel disconnections to prevent abrupt endings of applications. She has published about 8 papers in refereed journals, 40 papers in refereed conferences, 1 paper in Spanish magazine, one educational book and co-editor of one book. She is member of Program & Organizing Committees & Chair sessions for several international and Spanish conferences. She has collaborated in several research projects. Professor Macías teaches telecommunications at the beginning, advanced, and graduate levels, and advises graduate theses in the area of wireless communications and parallel and distributed computing.

Javier Marti received the Ingeniero de Telecomunicación degree from the Universidad Politécnica de Catalunya, Spain, in 1991, and the Ph.D. degree from the Universidad Politécnica de Valencia, Spain, in 1994. During 1989 and 1990, he was an Assistant Lecturer at the Universidad Politécnica de Catalunya. Since 1991 to 2000, he obtained the positions of Lecturer and Associate Professor at the Telecommunication Engineering Faculty, where he is currently a Full-Professor and leads the Fibre-Radio Group. Nowadays he is the Director of the Valencia Nanophotonics Technology Centre (NTC), a national research centre for photonic technologies in Spain. He has authored 7 patents and over 185 papers in refereed international technical journals in the fields of fibre-radio systems, technologies and access networks. He has led many national and international research projects (FP5-IST-TOPRATE) and has been the coordinator of the FP5-IST-OBANET and FP6-IST-GANDALF projects. He is currently participating IST-LASAGNE and coordination FP6-IST&NMP-PHOLOGIC. Prof. Marti is or has been a member of the Technical Program Committee of several conferences such as ECOC, LEOS, Microwave Photonics, and several other international workshops. He is currently involved in launching NTC spin-off companies addressing photonic-wireless technologies. He is also the recipient of several academic and industrial awards in Spain.

About the Contributors

Nayef Mendahawi holds a M. Sc. degree in Computer Engineer from King Fahd University of Petroleum and Minerals and B. Sc. degree in Electrical and Computer Engineer from Al-Balqa University. He has over 13 years of experience as a Senior Technical Consultant, including over 5 years of R&D experience in various wireless technologies. He is currently working for Research In Motion (RIM) as R&D in wireless technology.

Abbas Mohammed is a Professor of Telecommunications Theory at the School of Engineering, Blekinge Institute of Technology, Sweden. He was awarded the "PhD degree" from Liverpool University, UK, in 1992 and the Swedish "Docent degree" in Radio Communications and Navigation from Blekinge Institute of Technology in 2001. He was the recipient of the Blekinge Research Foundation Award "Researcher of the Year Award and Prize" for 2006. From 1993 to 1996, he was a Research Fellow with the Radio Navigation Group, University of Wales (Bangor), UK. From 1996 to 1998, he was with the University of Newcastle, UK, working on a European collaborative project within the ACTS (Advance Communications Technologies and Services) FP4 Research Programme that investigated 3G Satellite-UMTS systems. He was also employed by Ericsson AB, where he consulted on Power Control standardization issues for 3G. He has been a visiting lecturer to several Swedish universities. He is a Fellow of The Institution of Engineering and Technology. In 2006 he received Fellowship of the UK's Royal Institute of Navigation "in recognition of his significant contribution in navigation and in particular to advanced signal processing techniques that have enhanced the capability of Loran-C". He is an Associate Editor to the International Journal of Navigation and Observation, a Board Member of the IEEE Signal Processing Swedish Chapter, and an Editorial Board Member of the Radio Engineering Journal and the Mediterranean Journal of Electronics and Communications. He has also been a Guest Editor for several special issues of international journals. He is the author of over 170 publications in international journals, conference proceedings and book chapters, in the fields of signal processing, telecommunications and navigation systems. He has also developed techniques for measuring skywave delays in Loran-C receivers and received a Best Paper Award from the International Loran Association, USA, in connection to this work. His research interests are in space-time signal processing and MIMO systems, channel modelling, antennas and propagation, satellite and high altitude platform communications, and radio navigation systems. He is the Swedish representative and member of the management committee to the European Community COST 280, 296, 297 and IC0802 Actions.

Sebastian Möller was born in 1968 and studied electrical engineering at the universities of Bochum (Germany), Orléans (France) and Bologna (Italy). He received a Doctor-of-Engineering degree in 1999 for his work on the assessment and prediction of speech quality in telecommunications, and the Habilitation (venia legendi) with a book on the quality of telephone-based spoken dialogue systems in 2004, both at Ruhr-University Bochum. From 1994 to 2005, he held the position of a scientific researcher and Hochschuldozent at Ruhr-University Bochum. In 2005, he joined Deutsche Telekom Laboratories, Berlin University of Technology, and in 2007, he was appointed Professor for Quality and Usability at the same university. His primary interests are in the areas of speech signal processing, speech technology, communication acoustics, as well as in quality and usability evaluation. Since 1997, he has taken part in the standardization activities of ITU-T Study Group 12, where he is currently Co-Rapporteur of question Q.8/12.

Maria Morant received the M.Sc. degree in Telecommunication Engineering in 2008 from the Universidad Politécnica de Valencia, Spain. She is currently working towards the Ph.D degree in Telecommunications at wireless-photonics integration NTC area. Since 2006 she is collaborating with the Valencia Nanophotonics Technology Center (NTC) in Ultra-Wideband radio signals propagation over optical fibre and is working in European projects as FP6-IST-UROOF and FP7-ICT-UCELLS. Her current research areas of interest include ultra wideband communications on fiber optic networks.

Hossein Nikopour joined Nortel Networks, Canada, in June 2006 as a PhD. in communications systems. He worked on system design and product development of WiMAX IEEE 802.16e and 16m standards resulting in several standard contributions and patents. Hosein has worked for Huawei, Canada since April 2009, as a senior communications engineer. His current research interest is physical layer of MIMO-OFDM systems mainly 3GPP LTE standard. He has extensive experience in both link-level and system-level simulation of 4G communication systems.

Máirtín S. O'Droma, BE, PhD, C.Eng., is a Fellow of the U.K Institution of Engineers and Technology (IET), and Senior Member of the Institute of Electrical and Electronic Engineers (IEEE). He received his BE and PhD degrees from the National University of Ireland in 1973 and 1978 respectively. He is a Senior Lecturer and Director of the Telecommunications Research Centre, University of Limerick, Ireland. His previous activities include: founding partner and steering committee member, TARGET - Top Amplifier Research Groups in a European Team-, the EU FP6 Network of Excellence IST-507893, 2004-2008, and section head of the RF power linearization and amplifier modeling research strand; founding partner and steering committee member, ANWIRE - Academic Network for Wireless Internet Research in Europe -, the EU FP5 Thematic Network of Excellence IST-38835, 2002-2004; member of two European Cooperation in the field of Science and Technology research Actions (COST 285 & 290). Previous posts held by Dr. O'Droma include: lecturer in the University College Dublin and in the National University of Ireland, Galway, Director of Communications Software Ltd and ODR Patents Ltd. His research interests include: complex wireless telecommunication systems simulation and behavioral modeling, linearization & efficiency techniques in multimode, multicarrier broadband nonlinear microwave and mm-wave transmit power amplifiers; smart adaptive antenna arrays and MIMO channels; wireless network and protocol infrastructural innovations and new paradigms. Dr. O'Droma served on the Technical Program Committees of a number of international conferences and workshops, including IEEE VTC and IEEE ISWCS.

Luciana Oliveira is a PhD student at Federal University of Pernambuco. Her special fields of interest are in Delay Tolerant Networks; Social Routing; Next generation networks and Policy Based Management. She has a research experience in the communication field, with emphasis on dynamic networks, routing protocols and P2P. She received her M.Sc. degree and obtained honors as the best student in graduating class, both in Computer Science at the Federal University of Pernambuco. She had published some papers on International and Brazilian Symposiums, conferences and workshops.

Gianmarco Panza received the degree in Information Science Engineering at University of Padova, Italy in 1998 and a Master degree in ICT at CEFRIEL/Politecnico di Milan, Italy in 1998. In 1998, he was a researcher for ALCATEL ITALIA, Italy in the field of Quality of Service over IP networks. Since 1999, he has worked as a senior researcher and consultant on network systems at CEFRIEL Milan, Italy. Teacher in ICT master courses and company employee trainings for Quality of Service over IP networks, Voice over IP and MPLS (Multi-Protocol Label Switching). Participation to and coordination of several EU IST projects (e.g. BASS, MOICANE, JOCO, PHOENIX, Membrane, MULTINET, OPTIMIX). Participation and management of several national and international projects regarding all the aspects of network and service provisioning in LAN, MAN and WAN environments, from specification to simulation modeling, realization and testing. Several IEEE publications in Digital Signal Processing and Networking areas. Senior IEEE Member since 2008. Interests and expertise in Quality of Service in IP networks, Voice over IP, MPLS, multimedia systems, simulation modeling, and network planning, design, testing and assessment.

Mark Pecen serves as Vice President, Advanced Technology for Research in Motion Limited (RIM), makers of the BlackBerry wireless devices, systems and services. He reports to RIM CEO and founder Mike Lazaridis and is responsible for economic and strategic assessment of advanced wireless technology investments, commercialization of applied research, strategic technology partnerships and customer collaboration on future technology deployment. Pecen is the founder and General Manager of the RIM Wireless and Networking Advanced Research Centre. The Research Centre functions as a wireless technology incubator, and spans the

About the Contributors

value chain from applied research, analysis, simulation, proof-of-concept development, prototyping and the creation of initial products using internally developed technology. His labs are active in applied information theory, radio channel modeling, cross-layer wireless network design and protocols, mobility management, radio link control, statistical analysis and simulation and end-to-end modeling of wireless systems. Since 1988, Pecen has invented a number of technologies adopted in global standards for the Global System for Mobile telecommunication (GSM), General Packet Radio Service (GPRS), Enhanced Data for GSM Evolution (EDGE), Universal Mobile Telecommunication System (UMTS) and various Wireless Local Area Networks (WLAN) standards. He serves on the board of directors of Ecole Polytechnique Centre de Recherche En Electronique radiofrequence (CREER), School of Business and Economics of Wilfred Laurier University, Communitech Ontario, 3G Americas, LLC., QuantumWorks network for quantum information research and others. He also serves as advisory board member to several technology-focused industrial, academic and governmental associations in North America and Europe. Pecen holds more than 100 patents in the areas of mobile communication, networking and computing, and is a graduate of the University of Pennsylvania, Wharton School of Business and the School of Engineering and Applied Sciences

Dirk Pesch received a Dipl.-Ing. Degree from Aachen University of Technology, Germany in 1993 and a PhD from the University of Strathclyde, Glasgow, Scotland in 1999, both in Electrical and Electronic Engineering. From 1993 to 1995, he was with Nokia Mobile Phones in Bochum, Germany. From 1996 to 1999, he was a research fellow in the Mobile Communication Group at the University of Strathclyde in Glasgow, Scotland. In 1999, he moved to Cork Institute of Technology, Cork, taking up a position of lecturer, now a senior lecturer, in electronic engineering with a special focus on computer and communication engineering.

Ibrahim Raad was born in Beirut, Lebanon, on the 15 November 1979. He received his Bachelor of Engineering (Electrical), Master of Engineering - research (Telecommunications) and has submitted his PhD (waiting results) degrees from the University of Wollongong, Australia, in 2002, 2004 and 2008 respectively. From 2003 to 2007 he was an Associate Lecturer at the University of Wollongong teaching and researching in the field of Electrical, Telecommunications and Software Engineering. He joined Accenture in 2007 - 2008 as a software engineer. He is currently at Access Testing Australia as a senior test analyst. His research interests are in wireless communications networks and software engineering.

Tinku Rasheed is a Senior Research staff member of Create-Net's Pervasive Research Group at Trento, Italy. Before joining Create-Net, he was a research engineer with Orange Labs (previously France Telecom R&D) from May 2003 until December, 2006. During this time, Dr. Rasheed was also a graduate student at the Computer Science Labs (LRI) of University of Paris-Sud XI, Orsay, where I received my Ph.D. degree in 2007. Dr. Rasheed completed his Masters degree in 2004 from Aston University, U.K. specializing in Telecommunication engineering. His research interests include computer networks and protocols, large-scale distributed systems, heterogeneity, performance evaluation and implementation.

André C. G. C. Reis received B.S. and M.S. degrees in telecommunications engineering from Military Engineering Institute (IME), Rio de Janeiro-RJ, Brazil, in 1995, and 2002, respectively. He received D. Sc. degree in electrical engineering from Brasília University (UnB), Brasília-DF, Brasil, in 2009. His research interests include error correcting coding, digital transmission, datalink protocols and communications networks. He is a Brazilian Army's officer working at the Science and Technology Department, QGEx, Bl G, 3rd floor, Brasília-DF, 70630-901, Brazil.

Aasia Riasat is an Associate Professor of Computer Science at Collage of Business Management (CBM) since May 2006. She received an M.S.C. in Computer Science from the University of Sindh, and an M.S in Computer Science from Old Dominion University in 2005. For last one year, she is working as one of the active members of the wireless and mobile communications (WMC) lab research group of University of Bridgeport,

Bridgeport CT. In WMC research group, she is mainly responsible for simulation design for all the research work. Aasia Riasat is the author or co-author of more than 40 scholarly publications in various areas. Her research interests include modeling and simulation, web-based visualization, virtual reality, data compression, and algorithms optimization.

Syed S. Rizvi is a Ph.D. student of Computer Science and Engineering at University of Bridgeport. He received a B.S. in Computer Engineering from Sir Syed University of Engineering and Technology and an M.S. in Computer Engineering from Old Dominion University in 2001 and 2005, respectively. In the past, he has done research on bioinformatics projects where he investigated the use of Linux based cluster search engines for finding the desired proteins in input and outputs sequences from multiple databases. For last three year, his research focused primarily on the modeling and simulation of wide range parallel/distributed systems and the web based training applications. Syed Rizvi is the author of 68 scholarly publications in various areas. His current research focuses on the design, implementation and comparisons of algorithms in the areas of multiuser communications, multipath signals detection, multi-access interference estimation, computational complexity and combinatorial optimization of multiuser receivers, peer-to-peer networking, network security, and reconfigurable coprocessor and FPGA based architectures.

Sudhir K. Routray is now working as an Asst. Prof in Krupajal Engineering College, Bhubaneswar, Orissa, India. He got his degrees from the Utkal University and the University of Sheffield. He has 7 plus years experiences in teaching and research. He is a member of IEEE, Internet Society, IET and ISTE. He has several publications in journals and conference proceedings. He has been included in Marquis's Who is Who in the World and the Cambridge Biographies for his contributions to popular science. He has written two books in popular science.

Djamel Sadok received his PhD from the University of Kent at Canterbury in 1990. He has been working at the Federal University of Pernambuco since 1993. His research interests include traffic engineering, network management and architectures. Professor Sadok leads a very active research group in the area of computer networks and is currently involved with many research projects in cooperation with the Industry.

Aladdin Saleh earned his Ph.D. in Electrical Engineering at London University/ UK in 1984. He subsequently worked at several universities as an assistant professor then a professor. He served as a chairman of department and dean of engineering. In March 1998, Dr. Saleh moved to the industry where he started working with the wireless technology department of Bell Canada, the largest service provider in Canada. He started as a software engineer then as an application architect where he primed several key projects such as the wireless application protocol (WAP) and location-based services. He also led the work on several key projects in the broadband wireless access strategy group including the IEEE 802.16/ Wimax, the IEEE 802.11/ WiFi, and the integration of 802.11/802.16 with the 3G cellular network. More recently, he led several key projects related to mobile messaging and multimedia applications. Dr. Saleh also holds since January 2004 the position of Adjunct Professor at the ECE Department of Waterloo University/ Canada. He is a Fellow of IEE (IET) and a Senior Member of IEEE. His research interest is in the area of next generation wireless networks. Dr Saleh is also interested in the analysis and modeling of engineering systems.

Florin Sandu received the Dipl.-Ing. degree in Electronics and Telecommunications from the "Politechnica" University of Bucharest and the PhD degree in Electrical Engineering from the "Transilvania" University of Brasov, with a thesis on Data Acquisition for DSP. He works as professor in the academic field of electronic engineering, teaching Analysis & Synthesis of Electronic Circuits, Intelligent Networks and Industrial Telecommunications at "Transilvania" University. His research interests converge to mobile computing, mobile agents, service-oriented architectures, virtual instrumentation and remote test & measurement with DSP of acquired data.

About the Contributors

He published 3 books, 8 university hand-books & guides and 58 papers in scientific journals and proceedings of international conferences and visited many European universities in the frame of "Tempus", "Socrates" and "Leonardo da Vinci" programs of the European Union, as project manager or member of the international R&D teams. He is a member of IEEE - Communications Society. He joined Siemens SIS PSE Romania as consultant in 2001 and contributed to the development of mobile communication departments where he co-operates in innovation & technology management and in projects on tele-measurement and virtual test-fields.

Alvaro Suarez Sarmiento is Full Professor of Telecommunications, University of Las Palmas de Gran, Canaria, Spain. He is Member of the Experts Commission of research of the University of Las Palmas de Gran Canaria from 1999-2002. He in 1990 started working in systolic computing in the Technical University of Catalonia. Then he turned his attention to network computing and heterogeneous computing in 1994 when he returned to the University of Las Palmas de Gran Canaria where he founded the Concurrency and Architecture Group (GAC). His research interests are in parallel and heterogeneous distributed computing, infrastructure wireless networks for collaborative computing and collaborative frameworks. His current research efforts have focused on the management of wireless channel disconnections to prevent abrupt endings of applications. Professor Suarez teaches telecommunications at the beginning, advanced, and graduate levels, and advises graduate theses in the area of wireless communications and parallel and heterogeneous distributed computing. He also gives lectures at European universities. The Multimedia and Ubiquitous Wireless Access Networks and Services conference was lectured at University Ca' Foscari di Venezia (May 2005).

Chong Shen received his B.Eng. in Telecommunications from Wuhan University, China, an M.Phil. in Telecommunications from University of Strathclyde, Scotland, and a PhD in Electronics Engineering from Cork Institute of Technology, Ireland, in 2003, 2005 and 2008, respectively. He was postdoc at Tyndall national institute working on extreme low power wireless sensor network protocol and algorithm development. Now he is a research fellow at University of Strathclyde doing IPv6 enabled cross systems transparent handover testbed development for UK MobileVCE.

Mohammad Reza Soleymani received the B.S. degree from the University of Tehran, Tehran, Iran, in 1976, the M.S. degree from San Jose State University, San Jose, CA, in 1977, and the Ph.D. degree from Concordia University, Montreal, QC, Canada, in 1987, all in electrical engineering. From 1987 to 1990, he was an Assistant Professor with the Department of Electrical and Computer Engineering, McGill University, Montreal. From October 1990 to January 1998, he was with Spar Aerospace Ltd., Montreal, QC, Canada, where he had a leading role in the design and development of several satellite communication systems. In January 1998, he joined the Department of Electrical and Computer Engineering, Concordia University, where he is a Professor. His current research interests include wireless and satellite communications, information theory and coding.

Yang Sun received the B.S. degree in Testing Technology & Instrumentation from Zhejiang University, Hangzhou, Zhejiang, China, in 2000, the M.S. degree in Instrument Science & Technology from Zhejiang University, Hangzhou, Zhejiang, China, in 2003. From 2003 to 2004, he was with S3 Graphics Co. Ltd., Shanghai, China as a hardware design engineer. From 2004 to 2005, he was with Conexant Systems Inc., Shanghai, China as an ASIC design engineer. During the summer of 2007 and 2008, he worked at the Communication and Medical Systems Laboratory, DSP Solution R&D Center, Texas Instruments Inc., Dallas, TX, as an intern developing error correction codes (LDPC, Turbo) decoder for 3GPP LTE and IEEE 802.16e WiMax modem. Since August 2005, he has been studying toward his Ph.D. degree in the department of electrical and computer engineering at Rice University, Houston, TX. His research interests include parallel algorithms and VLSI architectures for applications in wireless communications. Yang Sun is the recipient of several best paper awards, including ACM/IEEE GLSVLSI Best Paper Award in 2009, IEEE SIPS "BOB OWENS MEMORIAL" Paper Award in 2008, and IEEE SOCC Best Paper Award in 2008.

Dmitry Tairov received his BEng degree in Computer Engineering from University of Limerick, Ireland in 2008. He is currently pursuing his MEng degree and researching within the Telecommunications Research Centre (TRC), University of Limerick, Ireland. His research is focused on the AAA services and their application in a Ubiquitous Consumer Wireless World (UCWW).

Patrick Tooher received the B.Eng. degree in 2000 with Great Distinction, and the M.A.Sc. degree in 2004, in electrical engineering and both from Concordia University, Montreal, Canada. He is currently pursuing his Ph.D. in electrical engineering at Concordia University, Montreal, Canada. He is the recipient of several scholarships and grants, including the NSERC and two FQRNT grants for graduate studies. In 2000 he worked at the Canadian Space Agency where he had a role designing antennas for micro-satellites. His current research is assisted by funding from InterDigital Canada Ltée, PROMPT Québec, FQRNT and NSERC. His research interests include wireless communications, collaborative communications, information theory and coding.

Arijit Ukil was born in India and has done his B.Tech in Electronics and Telecommunication Engineering from Haldia Institute of Technology, India in 2002. He is currently working as Scientist R&D in Wireless and Multimedia Innovation Lab, Tata Consultancy Services, Kolkata, India. His current research areas are wireless communication, resource allocation, scheduling, call admission control, OFDMA, cooperative communication and cross-layer optimization. Prior to joining Tata Consultancy Services, he has served as Scientist-C in Electronics and Radar Development Establishment (LRDE) of Defense Research and Development Organization (DRDO), India, where he was engaged in research and development activities in radar signal processing particularly for 3D Surveillance Naval Radar, where he was recognized for his outstanding contribution. He has number of international publications and is a member of IEEE. He has been reviewer of IEEE Transactions on Wireless Communications, IEEE VTC-Fall, 2009, IEEE WCNC, 2009, IEEE VTC-Spring, 2009. He has two patents filed in the area of OFDMA and QoS aware resource allocation and scheduling. He was invited for keynote speeches and invited lectures in number of conferences, including IEEE International Conference on Computer Engineering and Technology (ICCET), Jan 2009 and Emerging Trends in Computing and Communication (ETCC) at National Institute of Technology, Hamirpur, India in Dec, 2008. He is selected among world's who's who, to be featured in the 27th edition of Marquis Who's Who, 2010. He lives in Kolkata, India with his parents and wife.

Pablo Vidales is a Senior Research Scientist in the Networking & Distributed Systems Group at the Deutsche Telekom Laboratories (T-labs) in Berlin. His work focuses on Mobile Networking and Ubiquitous Computing. He joined T-labs in May 2005 just after completing a PhD at the University of Cambridge. His PhD dissertation was on "Seamless Mobility in 4G systems", working with Dr. Frank Stajano, Prof Jon Crowcroft and Prof. Andy Hopper. He holds a BSc in Computer Science and a BSc in Telecommunications from Instituto Tecnologico Autonomo de Mexico (ITAM). His present contribution to the research community includes more than 8 filed patents and over 40 scientific articles published in international journals and conferences, with over 300 citations. Recently, the National Researchers Institute in Mexico (SNI) has awarded Pablo Vidales the nomination to "National Researcher" in recognition of his ability to perform scientific research.

Marcel Waeltermann was born in 1978 and studied Electrical Engineering at Ruhr-University Bochum, Germany. In 2005, he graduated in the area of communication acoustics. After a two-year engagement at the Institute of Communication Acoustics, Ruhr-University Bochum, he now works as a scientific researcher at Deutsche Telekom Laboratories, Berlin University of Technology. There, he is currently continuing his work on quality models for transmitted speech on the basis of perceptual dimensions. Apart from that, his interests are in communication acoustics, quality evaluation and prediction, and speech signal processing. Since 2009, he is acting as a Co-Rapporteur for Question 8/12 of ITU-T Study Group 12.

About the Contributors

Wei Wu currently serves as a Member of Technical Staff, Advanced Technology for Research in Motion (RIM), Ltd. He has been involved in the standard research on 3GPP Evolved Packet Systems since he joined RIM in 2006. His research interests include quality of service (QoS) and mobility management for wireless Internet, mobile network and protocol simulation, and peer-to-peer networking. Before joining RIM, he has worked as a systems engineer at Alcatel USA, Inc. He was working on the system and architecture design of NGN wireless soft-switch for 2G/3G cellular networks including GSM, UMTS and GAN. Wei Wu holds a Ph.D. degree in computer science and engineering from the University of Texas at Arlington, and B.Eng. and M.Eng. degrees both in electrical engineering from Southeast University, Nanjing, China.

Wen Xu received a B.Sc. degree in 1982 and an M.Sc. degree in 1985 from Dalian University of Technology (DUT), China, and a Dr.-Ing. (Ph.D.) degree in 1996 from Technische Universität München (TUM), Germany, all in electrical engineering. From 1988 to 1994, he was a research assistant at Institute for Communications Engineering of TUM. From 1995 till 2006 he was with the Siemens Mobile (later BenQ Mobile), Munich, where he was head of the Algorithms and Standardization Laboratory. Since 2007, he is with the Infineon Technologies AG, Neubiberg, Germany. He has served as the Guest Editor for the IEEE Wireless Communications Magazine for the Special Issues on Next Generation of CDMA versus OFDMA for 4G Wireless Applications, and is an Associate Editor for the Security and Communication Networks. His research interests include source coding and processing, channel coding, equalization, cross-layer system design, wireless and wireline communications systems in general. Dr. Xu is a senior member of IEEE and a member of VDE.

Zhe Yang received the MSc. degree with distinction in communications from the Department of Electronics, University of York, U.K. in 2006. He is currently working toward the Ph.D. degree in the Department of Signal Processing, Blekinge Institute of Technology, Sweden. In 2007 and 2008, he was an invited guest researcher to Ben-Gurion University, Israel and University of York, UK. He is a member of IEEE and the author of over 40 publications including journal articles, conference papers and book chapters. He currently participates in the Cost 297 European project investigating broadband communications from high altitude platforms. His research interests include communications techniques from high altitude platforms, signal processing algorithms, mobile communication systems, and economical models of infrastructure deployment and analysis.

Hyunsoo Yoon received the B.S. degree in Electronics Engineering from the Seoul National University, Korea, in 1979, the M.S. degree in Computer Science from the KAIST, in 1981, and the Ph.D. degree in Computer and Information Science from the Ohio State University, Columbus, Ohio, in 1988. From 1978 to 1980, he was with the Tongyang Broadcasting Company, Korea, and from 1988 to 1989, with the AT\&T Bell Labs. as a Member of Technical Staff. Since 1989 he has been a Faculty Member of the Department of Computer Science, the KAIST. His research interests include parallel computer architecture, mobile communication, ad hoc networks, and information security.

Ming-Tuo Zhou received his bachelor degree in applied physics from Hunan University, China, in 1997, his master degree of engineering in communications from Chongqing University of Posts and Telecommunications, China, in 2000, and his doctor degree of engineering in telecommunications from Asian Institute of Technology, Bangkok, in 2003. During January to June 2004, he was with the Finland Government Program, Asian Institute of Technology, Bangkok, as a research specialist. He joined Wireless Communications Laboratory, National Institute of Information and Communications Technology, Singapore, in July 2004, and now is a research scientist. He has about 50 international publications as author or co-author and served as technical reviewers of more than 10 international journals and conferences, and as TPC chairs of several international workshops. He is editor of book Millimeter-Wave Technology in Wireless PAN, LAN, and WAN (CRC Press, USA, March 2007) and book Wireless Technologies for Intelligent Transportation Systems (Nova Science Publishers, 2009). His current research interests include wireless Mesh networks, wireless cognitive radio, and broadband maritime communications. He is a member of IEEE Communications Society, Vehicular Technology Society, and Lasers and Electro-Optics Society.

Yuming Zhu received the B.E. and M.S. degree in electronic engineering from Tsinghua University, Beijing, China, in 1999 and 2002 respectively, and the Ph.D. degree in electrical engineering from Arizona State University, in 2006. During the summer of 2005, he worked at the Communication and Medical Systems Laboratory, DSP Solution R&D Center, Texas Instruments Inc., Dallas, TX, where he was involved in WiMax modem development. Since September 2006, he has been working on WiMax and 3GPP long term evolution (LTE) modem development at the Communication and Medical Systems Laboratory, DSP Solution R&D Center, Texas Instruments Inc., Dallas, TX. His research interests include VLSI architectures for communication and signal processing systems.

Index

Symbols

3G networks 146, 148, 185
3GPP 1, 2, 4, 5, 6, 7, 9, 10, 12, 18, 43, 50, 57, 60, 62, 77, 81, 82, 83, 84, 85, 86, 87, 88, 90, 91, 92, 93, 95, 96, 98, 101, 106, 123, 157, 159, 184, 186, 361, 362, 363, 364, 365, 373, 374, 375, 383, 405, 411, 412, 413, 414, 415, 416, 417, 418, 419, 420, 423, 424, 426, 435, 436, 437, 438, 441, 444, 447, 448, 450
3GPP2 1, 5, 6, 7, 12, 18, 57, 61, 76, 83, 88, 424, 426, 435, 438, 439, 447, 448
3G system 2, 4, 12, 16, 17
4G network 1, 10, 12, 13, 16, 17, 46, 48, 49, 57, 125, 126, 127, 128, 129, 131, 132, 143, 193, 194, 195, 199, 200, 212, 214, 219, 220, 221, 222
4G nodes 194, 195
4G readiness 181, 182, 183, 186, 187, 188, 190, 191
4G receiver 627
4G standards 149, 159, 181

A

access gateway function (AGWF) 91, 96
adaptive antenna systems (AAS) 293
adaptive modulation 152
ad hoc network 316, 339, 459, 461, 462, 463, 494, 507, 509, 516
ad-hoc on-demand distance vector (AODV) 197, 198, 225, 240, 241
adjacent layers 454, 455, 456, 522
advertisement, discovery and association (ADA) 21, 27, 32, 37, 38, 39
all-IP network (AIPN) 12, 21, 23, 46, 47, 48, 49, 50, 51, 53, 56, 57, 58, 59, 77, 81, 96, 314, 315, 317, 328, 350, 354, 356, 361
amplify-and-forward 684, 708
antenna beamwidths 254, 264
ant model 202, 203, 221
application controller 469, 472, 481, 482, 487, 488, 489
application server layer (ASL) 92, 96
ARPANET 195, 225
assured forwarding (AF) 378

B

base-station backhaul 278
BCJR algorithm 625, 626
beam forming 152
betweenness 205, 206, 207
beyond-third generation (B3G) 270, 276, 277, 281, 282, 283
bicasting 424, 426, 433, 440, 441, 442, 443, 444, 445, 446, 447, 449
bicasting handover 424, 426, 441, 442, 447, 449
bicasting threshold 424, 442, 443, 444, 445, 447
birefringence 275
bit error rate (BER) 281, 425, 428, 429, 531, 547, 548, 549, 565, 567, 568, 570, 572, 576, 577, 578, 579, 582, 584, 585, 586, 589, 590, 591, 592, 593
Block Spread OFDM (BSOFDM) 582–595
boarder gateway protocol (BGP) 64
border gateway control function (BGCF) 89, 91, 96
border gateway protocol (BGP) 197, 212, 221

boresight 253, 254, 255, 256, 257, 259, 262
bottom-up approach 456
broadcatching 494, 499, 515
Bubble 194, 195, 206, 207, 224

C

call session control function (CSCF) 89, 90, 91, 96, 97
care-of address (CoA) 129
carrier to interference plus noise ratio (CINR) 251, 252, 257, 258, 259, 260, 261, 264
carrier to noise ratio (CNR) 251, 252, 254
Catenet 127, 143
cell search 526, 527, 535, 544, 560, 563
central processing unit (CPU) 65
channel allocation 2, 11, 46, 57
channel coding 684, 691, 693, 695, 698, 705
channel equalization 594
channel estimation 526, 527, 531, 544, 545, 546, 560, 562, 563
channel quality indicator (CQI) 315, 319, 322, 331, 353, 355, 357
circuit-switched (CS) 74, 81, 87, 96
ClearCurve 274, 284
cochannel cells 256, 257, 259, 261
CODEC 3
code division multiple access (CDMA) 2, 5, 6, 12, 14, 57, 61, 62, 64, 69, 75, 76, 83, 88, 96, 104, 105, 123, 149, 151, 152, 155, 156, 157, 158, 164, 176, 178, 184, 185, 529, 565, 566, 567, 568, 569, 570, 571, 572, 574, 576, 578, 579, 581, 583, 584, 585
coexistence 222, 249, 251, 252, 258, 259, 261, 262, 264, 265
common open policy service (COPS) 64, 89, 96
complementary metal oxide semiconductor (CMOS) 160
configured tunnels 79
consumer-centric business model (CBM) 20, 22, 23, 24, 25, 26, 27, 28, 31, 32, 37, 39, 40, 44
consumer identity module (CIM) 26, 41, 44
cooperative groups 229, 235, 236, 243, 245
cooperative network 228

cross layer 494, 496, 497, 498, 504, 505, 506, 508, 509, 510, 511, 512, 515, 516, 522
cross layer approaches 46
cross layer design 377, 400, 453, 454, 455, 457, 459, 460, 461, 463, 464, 466, 467, 469, 472, 473, 479, 490
cross layer optimization 313, 321
cumulative distribution function (CDF) 254, 261

D

decode-and-forward 684, 685, 706
deep packet inspection (DPI) 83, 96
default bearer 85
delay tolerant networks (DTN) 201, 212, 223
destination-sequence distance-vector (DSDV) 197, 198, 225
differentiated services (DiffServ) 364, 369, 370, 371, 374, 378, 387, 389, 390, 394, 395, 402
digital multimedia broadcasting (DMB) 270
disruptive networks 199
distance vector 197, 198, 225
distance vector multicast routing protocol (DVMRP) 197
distributed-antenna system (DAS) 268, 276, 278, 279, 283
downlink 12, 13, 14, 15, 19, 57, 231, 251, 256, 257, 258, 261, 262, 264, 267
duplex mode 12
dynamic home agent discovery (DHAD) 87, 96
dynamic networks 194, 199
dynamic source routing (DSR) 198, 224, 240

E

edge of coverage (EOC) 254, 259, 261, 264
electro-optic modulators (EOM) 272
end-to-end delay 77, 108, 110
enhanced data rates for GSM evolution (EDGE) 362, 363
equilibrium 214, 242, 243, 244, 245
Erbium-doped fibre based amplifiers (EDFA) 274
error correction codes 624

European Telecommunications Standards Institute (ETSI) 2, 4, 5, 23, 40, 43
evolved packet core 362
evolved packet system (EPS) 62, 84, 85, 86, 354, 361, 362, 363, 364, 365, 366, 367, 368, 369, 370, 371, 374
expedited forwarding (EF) 377
extended addressing 77
extended packet core (EPC) 62, 81, 82, 83, 84, 95, 96

F

fading channel 684, 705, 706, 708
fast binding update (FBU) 407
fast handover for mobile IP (FMIP) 407
fast neighbor advertisement (FNA) 407
fiber-to-the-home (FTTH) 184, 281, 282, 283, 284, 285, 286, 287
forward error correction (FEC) 149
frame error rate (FER) 69
free space path loss (FSPL) 255
frequency division duplex (FDD) 6, 8, 9, 12, 15
frequency division multiple access (FDMA) 149, 152, 155, 176
frequency estimation 533
frequency-selective fading 315

G

game theory 316, 339, 340, 342
geodesic paths 205
geographical information systems (GIS) 153
GGSN 84, 94, 96
globalizing 1
global mobility 405, 406, 422
global mobility management 406
global positioning systems (GPS) 153, 154, 157
global system for mobile communications (GSM) 2, 3, 4, 5, 6, 9, 15, 17, 42, 47, 48, 49, 58, 362, 363, 365, 370
GPRS 2, 3, 4, 11, 16, 47
GPRS tunneling protocol (GTP) 84, 85, 96
graph theory 205, 206
group communication 230, 246
group velocity dispersion (GVD) 275, 276

H

Hadamard matrix 583, 585, 586, 590, 591, 593
half duplex constraint 686
handoff latency 46, 49, 53, 55
handover 67, 82, 100, 102, 103, 106, 107, 108, 109, 110, 113, 116, 117, 118, 119, 121
handover delay 382, 409, 424, 425, 426, 434, 440, 441, 447, 448
handover initiate (HI) 407
hard detection 644, 645, 646, 653
heterogeneous networks 381, 405, 429, 434, 449, 450
hierarchical cellular network 46, 53
high altitude platforms (HAPs) 249, 250, 251, 252, 255, 261, 262, 263, 264, 265, 266, 267
high handoff 46
high speed downlink packet access (HSDPA) 133, 134, 136, 138, 140, 146, 148, 152, 158, 188, 494, 497, 498, 499, 516
high speed uplink packet access (HSUPA) 146, 148, 158, 188
home agent 378, 380, 381, 406, 409, 420, 421, 430, 431
home location register (HLR) 63
Home Subscriber Server (HSS) 63, 82, 85, 89, 90, 91, 94, 96
horizontal handover 425, 426
host mobility 406, 409, 423
hybrid-fibre coaxial (HFC) 270, 284
hybrid fibre-radio (HFR) 270
hybrid protocols 201
hybrid system 17
hybrid wireless network (HWN) 101, 102, 103, 104

I

IEEE 802.16 standard 4, 7, 8, 9, 14, 15, 16, 51, 62, 292, 293, 308, 311, 312, 320, 344, 347, 353, 354, 356
integrated services (IntServ) 378, 389, 395
inter-call server 87
interleaver 622, 623, 624, 625, 626, 627, 633, 634, 635, 636, 639, 640, 647, 651, 652, 656

intersymbol interference (ISI) 315, 567, 570, 578
intra-call server 87
IP multimedia subsystem (IMS) 81, 86, 88, 89, 90, 91, 92, 93, 94, 95, 96, 98, 99
IP protocol 86
IP-triggered resource allocation strategy (ITRAS) 46, 56, 57, 58, 59
IPv4 128
IPv6 76, 77, 78, 79, 80, 81, 83, 84, 85, 86, 87, 88, 89, 92, 95, 96, 97, 98, 99, 128
IPv6 versus IPv4 78, 80
iterative receiver 647, 651

L

label switch paths (LSPs) 67, 71
lattice decoding 644, 645, 650
lightweight mobile routing (LMR) 198, 223
line-of-sight (LOS) 250, 255, 292
link layer 382, 387, 388, 396, 405, 407, 424, 426, 432, 433, 434, 435, 437, 438, 439, 440, 441, 447, 448
link-state 197, 238
list detection 653, 656, 663, 664
localized mobility 405, 408, 409
location association server (LAS) 461
location based services (LBS) 186
long term evolution (LTE) 1, 2, 7, 9, 10, 12, 46, 47, 49, 50, 51, 53, 59, 61, 62, 76, 77, 78, 81, 82, 83, 84, 85, 87, 90, 95, 96, 97, 98, 99, 101, 157, 181, 182, 183, 184, 185, 186, 187, 188, 189, 192, 194, 219, 266, 292, 313, 314, 315, 316, 317, 320, 323, 353, 354, 355, 356, 362, 373, 375, 383, 405, 411, 412, 413, 419, 420, 423, 424, 426, 435, 436, 437, 438, 441, 447, 450, 453, 464, 465, 466, 469, 495, 497, 499, 517, 518
low-earth-orbit (LEO) satellite 250
LTE-A 77, 96

M

macrocells 46, 52, 53
m-algorithm 646, 651, 653, 654, 657, 658
MAP algorithm 623, 625, 626, 627, 629, 636, 639

maritime communications 292, 308, 310
maximum ratio combining 548
mean opinion score (MOS) 130, 131, 139, 141
media access control (MAC) 292, 293, 294, 295, 299, 300, 301, 305, 308, 310, 311, 314, 316, 318, 319, 320, 321, 327, 331, 353, 354, 355, 356, 369
media gateway control function (MGCF) 89, 91, 97
media gateway controller protocol (MGCP) 64
media gateway (MGW) 63, 87, 89, 91, 97
media resource function controller (MRFC) 91, 97
media resource function (MRF) 91, 97
media resource function processor (MRFP) 89, 91, 97
medium access 668, 672
medium access control 474
megabits per second (MBPS) 4, 15
mesh network 292, 293, 294, 295, 296, 298, 304, 305, 308, 309, 310, 311
microcells 13, 46, 52, 53
middleware 497, 509, 512, 522
mobile ad hoc wireless network (MANET) 100, 101, 102, 103, 104, 105, 106, 107, 108, 109, 110, 113, 114, 117, 118, 119, 120, 121, 122, 123
mobile IP 127, 128, 131, 134, 135, 144, 173
Mobile IPv6 (MIPv6) 403, 405, 406, 407, 408, 409, 420, 423, 433, 449, 450
mobile network nodes (MNNs) 378, 380, 381, 382, 390, 392, 394, 397, 398, 399, 400
mobile node 128, 129, 134
mobile router 378, 380, 381, 422
mobile-services switching centre (MSSC) 86
mobile switching center (MSC) 62, 63, 64, 66, 72, 75, 92, 97
mobile TV 644, 645
mobility anchor point (MAP) 407, 408, 422
mobility management entity (MME) 82, 83, 85, 97
mobility management (MM) 46, 47, 48, 50, 51, 146, 390, 401, 405, 406, 408, 409, 415, 416, 420, 422, 426, 431, 438, 448, 449, 450
Mobisense project 126, 127, 132, 133, 134, 135, 144, 145

modal dispersion 276, 280
motivation/ability framework 186
multi-antenna systems 151
multicast 644
multi-mode fibre (MMF) 272, 273, 276, 277, 280, 281, 286
multiple access interference (MAI) 566, 567, 568, 569, 570, 578
multiple input multiple output (MIMO) 8, 12, 13, 15, 16, 17, 51, 249, 251, 266, 293, 313, 314, 316, 317, 349, 350, 351, 352, 353, 356, 357, 358, 373
multiple-input multiple-output (MIMO) antenna 147, 155, 156, 157, 160, 173, 177, 184
multi-protocol label switching (MPLS) 62, 64, 66, 67, 70, 196, 206, 220
multiservice security gateways (MSG) 91, 97

N

NEMO 377, 378, 380, 381, 382, 383, 387, 388, 389, 390, 392, 393, 394, 397, 398, 399, 400, 402, 422, 423
network address translation (NAT) 79, 80, 81, 84, 87, 88, 91, 95, 97, 98, 134, 144, 152
network metric 194, 195
network throughput 565, 566, 567, 568, 579
network transparency 472
new radio spectrum 377
next generation mobile networks (NGMN) 126, 130, 131, 132, 136, 141, 142
next generation networking (NGN) 88, 97, 471, 477
noise levels 131
non-access stratum (NAS) 82, 86

O

object-oriented 66, 115
OFDMA 3, 7, 8, 9, 13, 46, 48, 52, 53, 57, 58, 59, 60, 315
OMNeT++ 377, 397, 398
open shortest path first (OSPF) 64
open system interconnection (OSI) 14, 17, 454, 472
open wireless architecture (OWA) 152, 157, 164, 180

optical communications 268, 285
optical line terminal (OLT) 281
optical links 274
optical signal to noise ratio (OSNR) 274
optic conversion 272
opto-electronic conversion 272, 274
orthogonal frequency code division (OFCD) 565, 567, 568, 571, 572, 573, 574, 575, 576, 577, 578, 579
orthogonal frequency division multiple access (OFDMA) 62, 147, 152, 155, 156
orthogonal frequency division multiplexing (OFDM) 249, 251
outage probability 687, 688, 690, 691, 693, 702, 705

P

packet-based networks 131
packet error rate (PER) 69, 115
packet loss 127, 131, 133, 134, 136, 137, 138, 139, 140, 141, 142, 163, 168
packet-switched (PS) 77, 81, 97
Pareto optimality 317, 340, 342
parity-check codes 562
partnership project 2, 5
PDC 2
PDP context 84, 86
performance evaluation 260
personalized services 228, 229, 232
personal ubiquitous environment 228, 234, 241, 242, 245
PESQ 125, 126, 129, 130, 131, 141, 142, 143, 144
photodetector (PD) 274
plastic optical fibre (POF) 272, 273
point-to-multipoint (PMP) 293, 299, 469, 471
polarization-mode dispersion (PMD) 275, 276, 283, 284
policy and charging rules function (PCRF) 83, 90, 92, 97
power allocation 667, 684, 685, 686, 699, 700, 701, 702, 705, 706, 708, 709
pre-coded OFDM 583, 585, 587, 588, 594
previous access router (PAR) 407
probabilistic analysis 201
probabilistic delivery 194

protocol stack 420, 424, 433
Proxy MIPv6 405
public land mobile network (PLMN) 82, 97

Q

quality of experience (QoE) 126, 127, 129, 130, 131, 143, 316
quality of service (QoS) 61, 76, 77, 78, 79, 80, 83, 84, 89, 90, 91, 92, 93, 94, 97, 99, 100, 102, 103, 104, 106, 107, 108, 109, 110, 111, 113, 114, 115, 116, 117, 118, 119, 120, 121, 122, 123, 126, 143, 144, 147, 149, 150, 162, 164, 166, 168, 170, 174, 184

R

radio access network (RAN) 61, 62, 63, 64, 75, 76, 90, 95, 362, 373
radio access technology (RAT) 420, 425, 426, 429
radio network controller (RNC) 62, 97
radio-over-fibre 268, 270, 271, 272, 273, 275, 276, 277, 279, 281, 282, 283, 284, 285, 286, 287, 288
radio resource management 46, 47, 48, 57
radio routing protocol (RRP) 64
radio transmission technologies (RTTs) 5
real-time response 129
receiver design 526, 563
relay node 103, 104, 105, 112, 122, 250
residue number system (RNS) 459
resource allocation 46, 47, 50, 56, 314, 315, 316, 317, 318, 321, 322, 327, 328, 339, 340, 341, 342, 343, 348, 350, 351, 352, 353, 354, 355, 356, 357, 358, 359, 360, 367
reverse tunneling 134
robust header compression (RoHC) 479, 481, 482, 483, 484, 486, 491
round trip delay (RTD) 69
round-trip time (RTT) 362, 372, 373
route optimizing 406
routing information protocol (RIP) 64, 197
routing mechanisms 199, 200, 205, 213, 222, 240
routing paradigms 193

R-scale 131
R-values 131

S

scalability 77, 81, 82, 83, 103
seamless mobility 77, 81
selection and distribution unit (SDU) 65, 66
sequential tree 653, 654
service data flows (SDFs) 84
service level agreement (SLA) 63
session flows 90
session initiation protocol (SIP) 461, 465, 466, 467, 520, 522
session routing proxy (SRP) 91, 92, 97
SGSN 82, 84, 97
shared database 457, 458, 466, 472
shared network knowledge 193
shortest path routing 196
signaling gateway (SGW) 63, 82, 83, 84, 85, 97
signaling protocol (SP) 64
SimBet 194, 207, 210
SIM card 6, 25, 44
smart antennas 152, 157
social overlay 205, 209, 210, 211, 213, 221
social routing 201, 205, 206, 209, 210, 221, 223
socio-technical limits 228
soft detection 644, 645, 651, 653
soft handoff 66, 67, 68, 293
software defined radio (SDR) 11, 12, 147, 152, 157, 179, 184
SOLAR 194, 208, 209, 210, 211, 220, 224
space-time codes 644, 650
spatial diversity 671, 684, 685, 686, 687, 688, 689, 690, 691, 692, 693, 694, 695, 698, 700, 703, 704, 705
spectrum efficiency 160
sphere decoding 551, 563, 644, 650, 651, 662, 663, 664
spreading matrices 583, 585, 587, 590, 595
standards development organization (SDO) 2, 4, 5, 6
standard single-mode fibre (SSMF) 272, 273, 274, 278, 281, 286

sub-platform point (SPP) 252, 253, 259, 262, 263
super layer 456
system architecture evolution (SAE) 62, 82, 87, 90, 91, 92, 95, 97, 99
system description document (SDD) 8

T

tele-medicine 153
telephonometry 131
telephony application server (TAS) 92, 97
Teralight 272
third-party authentication, authorization and accounting (3P-AAA) 21, 22, 24, 25, 26, 27, 28, 29, 30, 31, 32, 37, 38, 39, 40
third-party charging and billing (3P-C&B) 28, 31, 37, 38
time division duplex (TDD) 6, 7, 8, 9, 12, 13, 15
time division multiple access (TDMA) 2, 6, 15, 149, 151, 152, 176
top-down approach 456
topology hiding internetwork gateway (THIG) 91, 92, 97
traffic routing 77, 81, 110
traffic shaper 338
traffic splitting 395, 396, 397
transitivity 201, 206
transmitted speech 125, 130
tree search 527, 552, 553, 554, 563
triangular routing 406
tunnel broker 79
tunnel ID (TEID) 84, 97
tunneling protocol 405, 412
turbo codes 557, 562, 642
turbo coding 527, 529
turbo decoder 656
turbo decoding 527, 556, 557, 560, 562, 563

U

ubiquitous computing 228, 231
ubiquitous consumer wireless world (UCWW) 20, 22, 24, 25, 32, 37, 38, 39, 40, 41
ubiquitous environment 228, 234, 241, 242, 245

UDP-Lite 474, 477, 478, 483, 485, 486, 492, 493
ultra mobile broadband (UMB) 156, 184, 292
ultra-wideband (UWB) radio 271, 272, 274, 276, 277, 279, 280, 281, 282, 283, 284, 286, 287, 288, 289
universal mobile telecommunications system (UMTS) 2, 4, 6, 7, 9, 14, 31, 42, 59, 62, 77, 81, 88, 97, 133, 134, 147, 157, 158, 163, 164, 166, 177, 184, 192, 362, 365
universal modeling language (UML) 66
uplink 12, 13, 14, 15, 57, 231, 251, 256, 257, 258, 259, 261, 264
user decoupling 77
user equipment (UE) 82, 83, 84, 85, 89, 90, 97
user perception 125, 126, 127, 128, 129, 131, 132, 143
user profile server function (UPSF) 91
USIM card 6

V

variable time-fraction 685, 693, 695, 697, 698, 701, 703, 706, 709
vertical calibration 457, 458, 466, 472
vertical-cavity surface emitting lasers (VCSEL) 273, 277, 281, 288
vertical handover 128, 129, 132, 136, 392, 425, 426, 434, 435, 449, 450
video on demand (VoD) 494, 495, 496, 499, 500, 506, 507, 509, 514
video streaming 143, 149, 464, 471, 488, 494, 498, 505, 507, 509, 512, 514, 519
virtual navigation 153
virtual presence 153
virtual private network (VPN) 149, 154
voice core network (VCN) 86
voice over IP (VoIP) 81, 89, 94, 95, 96, 97, 126, 131, 132, 133, 134, 135, 145, 148, 149, 185, 328, 353, 361, 362, 364, 367, 369, 370, 371, 373, 374

W

Wasp model 194, 203, 204, 219, 222
WiBro 158, 186
wideband code division multiple access (WCDMA) 146, 149, 157, 160

Wiener filter 548
wireless application protocol (WAP) 64
wireless billboard channels (WBCs) 21, 22, 26, 27, 28, 32, 33, 36, 37, 39
wireless broadband 1, 3, 4, 10, 16, 17, 18
wireless grids 171
wireless mesh networks 565, 570, 579, 580, 581
wireless metropolitan area networks (WMAN) 7
wireless network card interfaces (WNIC) 494, 495, 499
wireless routing protocol (WRP) 198
wireless signal extractor (WSE) 274
workarounds 79
Worldwide Interoperability for Microwave Access (WiMAX) 1, 3, 4, 6, 7, 8, 9, 13, 14, 15, 16, 17, 18, 19, 46, 47, 49, 50, 51, 52, 53, 59, 60, 62, 76, 83, 90, 97, 133, 147, 156, 160, 181, 182, 183, 184, 185, 186, 187, 192, 249, 251, 253, 256, 258, 259, 261, 264, 265, 266, 267, 269, 271, 272, 276, 278, 279, 280, 281, 283, 287, 288, 289, 292, 293, 294, 295, 296, 297, 298, 299, 300, 301, 305, 307, 308, 309, 310, 311, 312, 313, 314, 315, 316, 317, 323, 327, 331, 353, 354, 355, 356, 357, 359, 363